S0-AIJ-380

STUDENT'S SOLUTIONS MANUAL

JEFFERY A. COLE
Anoka-Ramsey Community College

ALGEBRA FOR COLLEGE STUDENTS
SEVENTH EDITION

Margaret L. Lial
American River College

John Hornsby
University of New Orleans

Terry McGinnis

Addison-Wesley
is an imprint of

The author and publisher of this book have used their best efforts in preparing this book. These efforts include the development, research, and testing of the theories and programs to determine their effectiveness. The author and publisher make no warranty of any kind, expressed or implied, with regard to these programs or the documentation contained in this book. The author and publisher shall not be liable in any event for incidental or consequential damages in connection with, or arising out of, the furnishing, performance, or use of these programs.

Reproduced by Pearson Addison-Wesley from electronic files supplied by the author.

Copyright © 2012, 2008, 2004 Pearson Education, Inc.
Publishing as Addison-Wesley, 75 Arlington Street, Boston, MA 02116.

All rights reserved. No part of this publication may be reproduced, stored in a retrieval system, or transmitted, in any form or by any means, electronic, mechanical, photocopying, recording, or otherwise, without the prior written permission of the publisher. Printed in the United States of America.

ISBN-13: 978-0-321-71549-4
ISBN-10: 0-321-71549-7

1 2 3 4 5 6 BB 15 14 13 12 11

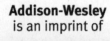

Addison-Wesley
is an imprint of

www.pearsonhighered.com

Preface

This *Student's Solutions Manual* contains solutions to selected exercises in the text *Algebra for College Students, Seventh Edition* by Margaret L. Lial, John Hornsby, and Terry McGinnis. It contains solutions to the Now Try Exercises, the odd-numbered exercises in each section, all Relating Concepts exercises, as well as solutions to all the exercises in the review sections, the chapter tests, and the cumulative review sections.

This manual is a text supplement and should be read along *with* the text. You should read all exercise solutions in this manual because many concept explanations are given and then used in subsequent solutions. All concepts necessary to solve a particular problem are not reviewed for every exercise. If you are having difficulty with a previously covered concept, refer back to the section where it was covered for more complete help.

A significant number of today's students are involved in various outside activities, and find it difficult, if not impossible, to attend all class sessions; this manual should help meet the needs of these students. In addition, it is my hope that this manual's solutions will enhance the understanding of all readers of the material and provide insights to solving other exercises.

I appreciate feedback concerning errors, solution correctness or style, and manual style. Any comments may be sent directly to me at the address below, at jeff.cole@anokaramsey.edu, or in care of the publisher, Pearson Addison-Wesley.

I would like to thank Mary Johnson and Marv Riedesel, formerly of Inver Hills Community College, for their careful accuracy checking and valuable suggestions; Karen Hartpence, for creating the new art pieces; and the authors and Maureen O'Connor and Mary St. Thomas, of Pearson Addison-Wesley, for entrusting me with this project.

Jeffery A. Cole
Anoka-Ramsey Community College
11200 Mississippi Blvd. NW
Coon Rapids, MN 55433

Copyright © 2012 Pearson Education, Inc. Publishing as Addison-Wesley.

Table of Contents

Copyright © 2012 Pearson Education, Inc. Publishing as Addison-Wesley.

Copyright © 2012 Pearson Education, Inc. Publishing as Addison-Wesley.

Copyright © 2012 Pearson Education, Inc. Publishing as Addison-Wesley.

CHAPTER 1   REVIEW OF THE REAL NUMBER SYSTEM

1.1   Basic Concepts

1.1   Now Try Exercises

N1. $\{p \mid p$ is a natural number less than $6\}$

$\{1, 2, 3, 4, 5\}$ is the set of natural numbers less than 6.

N2. $\{9, 10, 11, 12\}$

In set-builder notation, one answer is $\{x \mid x$ is a natural number between 8 and $13\}$.

N3. $\{-2.4, -\sqrt{1}, -\frac{1}{2}, 0, 0.\overline{3}, \sqrt{5}, \pi, 5\}$

(a) The whole numbers in the given set are in the set $\{0, 5\}$.

(b) The rational numbers in the given set are in the set $\{-2.4, -\sqrt{1}, -\frac{1}{2}, 0, 0.\overline{3}, 5\}$.

N4. **(a)** The statement "All integers are irrational numbers" is *false*. In fact, all integers are rational.

(b) The statement "Every whole number is an integer" is *true* since the integers include the whole numbers, the negatives of the whole numbers, and 0.

N5. **(a)** $|-7| = -(-7) = 7$

(b) $-|-15|$

Evaluate the absolute value. Then find the additive inverse.

$$-|-15| = -(15) = -15$$

(c) $|4| - |-4|$

Evaluate each absolute value, and then subtract.

$$|4| - |-4| = 4 - 4 = 0$$

N6. Security guards: $|16.9| = 16.9$

File clerks: $|-41.3| = 41.3$

Customer service representatives: $|24.8| = 24.8$

Since 16.9 is less than 24.8 and 41.3, security guards is expected to see the least change.

N7. **(a)** $-5 > -1$ is *false*.

Since -5 is to the left of -1 on a number line, -5 is less than, not greater than, -1.

(b) $-7 < -6$ is *true*

since -7 is to the left of -6 on a number line.

N8. $-|-12| \geq 2 \cdot 5$

Since $-|-12| = -12$ and $2 \cdot 5 = 10$, the statement becomes $-12 \geq 10$, which is *false*.

N9. $\{x \mid x < -2\}$ is written in interval notation as $(-\infty, -2)$. The parenthesis next to -2 means that -2 is not in the interval and is not part of the graph.

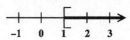

N10. $\{x \mid x \geq 1\}$ is written in interval notation as $[1, \infty)$. The bracket next to 1 means that 1 is in the interval and is part of the graph.

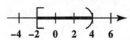

N11. $\{x \mid -2 \leq x < 4\}$ is written in interval notation as $[-2, 4)$. Here, -2 is in the interval and 4 is not.

1.1   Section Exercises

1. $\{x \mid x$ is a natural number less than $6\}$
The set of natural numbers is $\{1, 2, 3, \ldots\}$, so the set of natural numbers less than 6 is $\{1, 2, 3, 4, 5\}$.

3. $\{z \mid z$ is an integer greater than $4\}$
The set of integers is $\{\ldots, -3, -2, -1, 0, 1, 2, 3, \ldots\}$, so the set of integers greater than 4 is $\{5, 6, 7, 8, \ldots\}$.

5. $\{z \mid z$ is an integer less than or equal to $4\}$
The set of integers is $\{\ldots, -3, -2, -1, 0, 1, 2, 3, \ldots\}$, so the set of integers less than or equal to 4 is $\{\ldots, -1, 0, 1, 2, 3, 4\}$.

7. $\{a \mid a$ is an even integer greater than $8\}$
The set of even integers is $\{\ldots, -2, 0, 2, 4, \ldots\}$, so the set of even integers greater than 8 is $\{10, 12, 14, 16, \ldots\}$.

9. $\{x \mid x$ is an irrational number that is also rational$\}$
Irrational numbers cannot also be rational numbers. The set of irrational numbers that are also rational is $\emptyset$.

11. $\{p \mid p$ is a number whose absolute value is $4\}$
This is the set of numbers that lie a distance of 4 units from 0 on the number line. Thus, the set of numbers whose absolute value is 4 is $\{-4, 4\}$.

In Exercises 13 and 15, we give one possible answer.

13. $\{2, 4, 6, 8\}$ can be described by $\{x \mid x$ is an even natural number less than or equal to $8\}$.

15. $\{4, 8, 12, 16, \ldots\}$ can be described by $\{x \mid x$ is a multiple of 4 greater than $0\}$.

Copyright © 2012 Pearson Education, Inc. Publishing as Addison-Wesley.

17. Graph $\{-4, -2, 0, 3, 5\}$.
Place dots for -4, -2, 0, 3, and 5 on a number line.

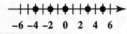

19. Graph $\left\{-\frac{6}{5}, -\frac{1}{4}, 0, \frac{5}{6}, \frac{13}{4}, 5.2, \frac{11}{2}\right\}$.
Place dots for $-\frac{6}{5} = -1.2$, $-\frac{1}{4} = -0.25$, 0, $\frac{5}{6} = 0.8\overline{3}$, $\frac{13}{4} = 3.25$, 5.2, and $\frac{11}{2} = 5.5$, on a number line.

21. $\left\{-9, -\sqrt{6}, -0.7, 0, \frac{6}{7}, \sqrt{7}, 4.\overline{6}, 8, \frac{21}{2}, 13, \frac{75}{5}\right\}$

(a) The elements 8, 13, and $\frac{75}{5}$ (or 15) are natural numbers.

(b) The elements 0, 8, 13, and $\frac{75}{5}$ are whole numbers.

(c) The elements -9, 0, 8, 13, and $\frac{75}{5}$ are integers.

(d) The elements -9, -0.7, 0, $\frac{6}{7}$, $4.\overline{6}$, 8, $\frac{21}{2}$, 13, and $\frac{75}{5}$ are rational numbers.

(e) The elements $-\sqrt{6}$ and $\sqrt{7}$ are irrational numbers.

(f) All the elements are real numbers.

23. Yes, the choice of variables is not important. The sets have the same description so they represent the same set.

25. The statement "Every integer is a whole number" is *false*. Some integers are whole numbers, but the negative integers are not whole numbers.

27. The statement "Every irrational number is an integer" is *false*. Irrational numbers have decimal representations that neither terminate nor repeat, so no irrational numbers are integers.

29. The statement "Every natural number is a whole number" is *true*. The whole numbers consist of the natural numbers and zero.

31. The statement "Some rational numbers are whole numbers" is *true*. Every whole number is rational.

33. The statement "The absolute value of any number is the same as the absolute value of its additive inverse" is *true*. The distance on a number line from 0 to a number is the same as the distance from 0 to its additive inverse.

35. (a) $-(-4) = 4$ (Choice **A**)

(b) $|-4| = 4$ (Choice **A**)

(c) $-|-4| = -(4) = -4$ (Choice **B**)

(d) $-|-(-4)| = -|4| = -4$ (Choice **B**)

37. (a) The additive inverse of 6 is -6.

(b) $6 > 0$, so $|6| = 6$.

39. (a) The additive inverse of -12 is $-(-12) = 12$.

(b) $-12 < 0$, so $|-12| = -(-12) = 12$.

41. (a) The additive inverse of $\frac{6}{5}$ is $-\frac{6}{5}$.

(b) $\frac{6}{5} > 0$, so $\left|\frac{6}{5}\right| = \frac{6}{5}$.

43. $|-8|$
Use the definition of absolute value.
$$-8 < 0, \text{ so } |-8| = -(-8) = 8$$

45. $\frac{3}{2} > 0$, so $\left|\frac{3}{2}\right| = \frac{3}{2}$.

47. $-|5| = -(5) = -5$

49. $-|-2| = -[-(-2)] = -(2) = -2$

51. $-|4.5| = -(4.5) = -4.5$

53. $|-2| + |3| = 2 + 3 = 5$

55. $|-9| - |-3| = 9 - 3 = 6$

57. $|-1| + |-2| - |-3| = 1 + 2 - 3$
$$= 3 - 3$$
$$= 0$$

59. (a) The greatest absolute value is $|35.6| = 35.6$. Therefore, Las Vegas had the greatest change in population. The population increased 35.6%.

(b) The least absolute value is $|-0.6| = 0.6$. Therefore, Detroit had the least change in population. The population decreased 0.6%.

61. Compare the depths of the bodies of water. The deepest, that is, the body of water whose depth has the greatest absolute value, is the Pacific Ocean ($|-12{,}925| = 12{,}925$) followed by the Indian Ocean, the Caribbean Sea, the South China Sea, and the Gulf of California.

63. True; the absolute value of the depth of the Pacific Ocean is
$$|-12{,}925| = 12{,}925$$
which is greater than the absolute value of the depth of the Indian Ocean,
$$|-12{,}598| = 12{,}598.$$

65. True; since -6 is to the left of -1 on a number line, -6 is less than -1.

67. False; since -4 is to the left of -3 on a number line, -4 is *less* than -3, not greater.

Copyright © 2012 Pearson Education, Inc. Publishing as Addison-Wesley.

69. True; since 3 is to the right of -2 on a number line, 3 is greater than -2.

71. True; since $-3 = -3$, -3 is greater than *or equal to* -3.

73. The inequality $6 > 2$ can also be written $2 < 6$. In each case, the inequality symbol points toward the smaller number so both inequalities are true.

75. $-9 < 4$ is equivalent to $4 > -9$.

77. $-5 > -10$ is equivalent to $-10 < -5$.

79. $0 < x$ is equivalent to $x > 0$.

81. "7 is greater than y" can be written as $7 > y$.

83. "5 is greater than or equal to 5" can be written as $5 \geq 5$.

85. "$3t - 4$ is less than or equal to 10" can be written as $3t - 4 \leq 10$.

87. "$5x + 3$ is not equal to 0" can be written as $5x + 3 \neq 0$.

89. "t is between -3 and 5" can be written as $-3 < t < 5$.

91. "$3x$ is between -3 and 4, including -3 and excluding 4" can be written as $-3 \leq 3x < 4$.

93. $-6 \overset{?}{<} 7 + 3$
$-6 < 10$ *True*

The last statement is true since -6 is to the left of 10 on a number line.

95. $2 \cdot 5 \overset{?}{\geq} 4 + 6$
$10 \geq 10$ *True*

97. $-|-3| \overset{?}{\geq} -3$
$-3 \geq -3$ *True*

99. $-8 \overset{?}{>} -|-6|$
$-8 > -6$ *False*

101. $\{x \mid x > -1\}$ includes all numbers greater than -1, written $(-1, \infty)$. Place a parenthesis at -1 since -1 is not an element of the set. The graph extends from -1 to the right.

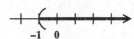

103. $\{x \mid x \leq 6\}$ includes all numbers less than or equal to 6, written $(-\infty, 6]$. Place a bracket at 6 since 6 is an element of the set. The graph extends from 6 to the left.

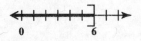

105. $\{x \mid 0 < x < 3.5\}$ includes all numbers between 0 and 3.5, written $(0, 3.5)$. Place parentheses at 0 and 3.5 since these numbers are not elements of the set. The graph goes from 0 to 3.5.

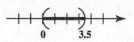

107. $\{x \mid 2 \leq x \leq 7\}$ includes all numbers from 2 to 7, written $[2, 7]$. Place brackets at 2 and 7 since these numbers are elements of the set. The graph goes from 2 to 7.

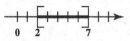

109. $\{x \mid -4 < x \leq 3\}$ includes all numbers between -4 and 3, excluding -4, but including 3, written $(-4, 3]$. Place a parenthesis at -4 and a bracket at 3 to show that -4 is not included, but 3 is. The graph goes from -4 to 3.

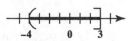

111. $\{x \mid 0 < x \leq 3\}$ includes all numbers between 0 and 3, excluding 0, but including 3, written $(0, 3]$. Place a parenthesis at 0 and a bracket at 3 to show that 0 is not included, but 3 is. The graph goes from 0 to 3.

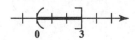

113. 2006: IA $= 13{,}811$, OH $= 7507$, PA $= 6687$

In 2006, Iowa (IA), Ohio (OH), and Pennsylvania (PA) had production greater than 6000 million eggs.

115. 2006: TX $= x = 5039$, OH $= y = 7507$
Since $5039 < 7507$, $x < y$ is true.

1.2 Operations on Real Numbers

1.2 Now Try Exercises

N1. (a) $-4 + (-9) = -(4 + 9) = -13$

(b) $-7.25 + (-3.57) = -(7.25 + 3.57)$
$= -10.82$

(c) $-\frac{2}{5} + (-\frac{3}{10}) = -(\frac{2}{5} + \frac{3}{10})$
$= -(\frac{4}{10} + \frac{3}{10})$
$= -\frac{7}{10}$

N2. (a) $-15 + 7$

Subtract the lesser absolute value from the greater absolute value $(15 - 7)$, and take the sign of the larger.

$$-15 + 7 = -(15 - 7) = -8$$

Copyright © 2012 Pearson Education, Inc. Publishing as Addison-Wesley.

(b) $4.6 + (-2.8) = 4.6 - 2.8 = 1.8$

(c) $-\dfrac{5}{9} + \dfrac{2}{7} = -\dfrac{5 \cdot 7}{9 \cdot 7} + \dfrac{2 \cdot 9}{7 \cdot 9} = -\dfrac{35}{63} + \dfrac{18}{63}$

$ = -\left(\dfrac{35}{63} - \dfrac{18}{63}\right) = -\dfrac{17}{63}$

N3. **(a)** $-4 - 11 = -4 + (-11) = -15$

(b) $-5.67 - (-2.34) = -5.67 + 2.34$
$ = -(5.67 - 2.34)$
$ = -3.33$

(c) $\dfrac{4}{9} - \dfrac{3}{5} = \dfrac{4}{9} - \dfrac{3}{5} = \dfrac{4 \cdot 5}{9 \cdot 5} - \dfrac{3 \cdot 9}{5 \cdot 9}$

$ = \dfrac{20}{45} - \dfrac{27}{45} = -\dfrac{7}{45}$

N4. $-4 - (-2 - 7) - 12 = -4 - (-9) - 12$
$ = -4 + 9 - 12$
$ = 5 - 12$
$ = -7$

N5. $|12 - (-7)| = |12 + 7| = |19| = 19$

N6. **(a)** $-3(-10) = 30$

The numbers have the same sign, so the product is positive.

(b) $0.7(-1.2) = -0.84$

The numbers have different signs, so the product is negative.

(c) $-\dfrac{8}{11}(33) = -\dfrac{8 \cdot 3 \cdot 11}{11} = -24$

N7. **(a)** $\dfrac{-10}{-5} = -10\left(-\dfrac{1}{5}\right) = 2$

(b) $\dfrac{-\frac{10}{3}}{\frac{3}{8}} = -\dfrac{10}{3} \div \dfrac{3}{8} = -\dfrac{10}{3} \cdot \dfrac{8}{3}$

$ = -\dfrac{10 \cdot 8}{3 \cdot 3} = -\dfrac{80}{9}$

(c) $-\dfrac{6}{5} \div \left(-\dfrac{3}{7}\right) = -\dfrac{6}{5} \cdot \left(-\dfrac{7}{3}\right) = \dfrac{2 \cdot 3 \cdot 7}{5 \cdot 3}$

$ = \dfrac{2 \cdot 7}{5} = \dfrac{14}{5}$

1.2   Section Exercises

1. The sum of a positive number and a negative number is 0 if the numbers are <u>additive inverses</u>. For example, $4 + (-4) = 0$.

3. The sum of two negative numbers is a <u>negative</u> number. For example, $-7 + (-21) = -28$.

5. The sum of a positive number and a negative number is positive if the positive number has the <u>greater</u> absolute value. For example, $15 + (-2) = 13$.

7. The difference between two negative numbers is negative if <u>the number with lesser absolute value is subtracted from the one with greater absolute value</u>. For example, $-15 - (-3) = -12$.

9. The product of two numbers with different signs is <u>negative</u>. For example, $-5(15) = -75$.

11. $-6 + (-13) = -(6 + 13) = -19$

13. $13 + (-4) = 13 - 4 = 9$

15. $-\dfrac{7}{3} + \dfrac{3}{4} = -\dfrac{28}{12} + \dfrac{9}{12} = -\dfrac{19}{12}$

17. The difference between 2.3 and 0.45 is 1.85. The number with the greater absolute value, -2.3, is negative, so the answer is negative. Thus, $-2.3 + 0.45 = -1.85$.

19. $-6 - 5 = -6 + (-5) = -(6 + 5) = -11$

21. $8 - (-13) = 8 + 13 = 21$

23. $-16 - (-3) = -16 + 3 = -13$

25. $-12.31 - (-2.13) = -12.31 + 2.13$
$ = -(12.31 - 2.13)$
$ = -10.18$

27. $\dfrac{9}{10} - \left(-\dfrac{4}{3}\right) = \dfrac{9}{10} + \dfrac{4}{3} = \dfrac{27}{30} + \dfrac{40}{30} = \dfrac{67}{30}$

29. $|-8 - 6| = |-14| = -(-14) = 14$

31. $-|-4 + 9| = -|5| = -5$

33. $-2 - |-4| = -2 - 4 = -2 + (-4) = -6$

35. $-7 + 5 - 9 = (-7 + 5) - 9 = -2 - 9 = -11$

37. $6 - (-2) + 8 = 6 + 2 + 8 = 8 + 8 = 16$

39. $-9 - 4 - (-3) + 6 = (-9 - 4) + 3 + 6$
$ = -13 + 3 + 6$
$ = -10 + 6$
$ = -4$

41. $-8 - (-12) - (2 - 6) = -8 + 12 - (-4)$
$ = -8 + 12 + 4$
$ = 4 + 4$
$ = 8$

43. $-0.382 + 4 - 0.6 = 3.618 - 0.6$
$ = 3.018$

45. $\left(-\dfrac{5}{4} - \dfrac{2}{3}\right) + \dfrac{1}{6} = \left(-\dfrac{15}{12} - \dfrac{8}{12}\right) + \left(\dfrac{2}{12}\right)$

$ = -\dfrac{23}{12} + \dfrac{2}{12}$

$ = -\dfrac{21}{12}, \text{ or } -\dfrac{7}{4}$

47. $-\dfrac{3}{4} - \left(\dfrac{1}{2} - \dfrac{3}{8}\right) = -\dfrac{6}{8} - \left(\dfrac{4}{8} - \dfrac{3}{8}\right)$

$ = -\dfrac{6}{8} - \dfrac{1}{8}$

$ = -\dfrac{7}{8}$

Copyright © 2012 Pearson Education, Inc. Publishing as Addison-Wesley.

49. $|-11| - |-5| - |7| + |-2|$
$= 11 - 5 - 7 + 2$
$= 6 - 7 + 2$
$= -1 + 2 = 1$

51. $A = -4$ and $B = 2$. The distance between A and B is the absolute value of their difference.

$$|-4 - 2| = |-6| = 6$$

53. $D = -6$ and $F = \frac{1}{2}$. The distance between D and F is the absolute value of their difference.

$$\left|-6 - \frac{1}{2}\right| = \left|-\frac{12}{2} - \frac{1}{2}\right|$$
$$= \left|-\frac{13}{2}\right|$$
$$= \frac{13}{2}, \text{ or } 6\frac{1}{2}$$

55. It is true for multiplication (and division). It is false for addition and subtraction when the number to be subtracted has the lesser absolute value. A more precise statement is, "The product or quotient of two negative numbers is positive."

57. The product of two numbers with *different* signs is *negative*, so

$$5(-7) = -35.$$

59. The product of two numbers with the *same* sign is *positive*, so

$$-8(-5) = 40.$$

61. The product of two numbers with the *same* sign is *positive*, so

$$-10(-\tfrac{1}{5}) = 2 \cdot 5 \cdot (\tfrac{1}{5}) = 2.$$

63. The product of two numbers with *different* signs is *negative*, so

$$\tfrac{3}{4}(-16) = -\tfrac{3}{4} \cdot 4 \cdot 4 = -12.$$

65. The product of two numbers with the *same* sign is *positive*, so

$$-\frac{5}{2}\left(-\frac{12}{25}\right) = \frac{5 \cdot 2 \cdot 6}{2 \cdot 5 \cdot 5} = \frac{6}{5}.$$

67. The product of two numbers with the *same* sign is *positive*, so

$$-\frac{3}{8}\left(-\frac{24}{9}\right) = \frac{3 \cdot 3 \cdot 8}{8 \cdot 9} = 1.$$

69. $-2.4(-2.45) = 5.88$

71. $3.4(-3.14) = -10.676$

73. The quotient of two nonzero real numbers with *different* signs is *negative*, so

$$\frac{-14}{2} = -14 \cdot \frac{1}{2} = -2 \cdot 7 \cdot \frac{1}{2} = -7.$$

75. The quotient of two nonzero real numbers with the *same* sign is *positive*, so

$$\frac{-24}{-4} = 24 \cdot \frac{1}{4} = 4 \cdot 6 \cdot \frac{1}{4} = 6.$$

77. The quotient of two nonzero real numbers with *different* signs is *negative*, so

$$\frac{100}{-25} = -100 \cdot \frac{1}{25} = -4 \cdot 25 \cdot \frac{1}{25} = -4.$$

79. $\frac{0}{-8} = 0 \cdot \left(-\frac{1}{8}\right) = 0$

81. Division by 0 is undefined, so $\frac{5}{0}$ is undefined.

83. The quotient of two nonzero real numbers with the *same* sign is *positive*, so

$$-\frac{10}{17} \div \left(-\frac{12}{5}\right) = \frac{10}{17} \cdot \frac{5}{12} = \frac{2 \cdot 5 \cdot 5}{17 \cdot 2 \cdot 6} = \frac{25}{102}.$$

85. $\dfrac{\frac{12}{13}}{-\frac{4}{3}} = \frac{12}{13} \div \left(-\frac{4}{3}\right) = \frac{12}{13}\left(-\frac{3}{4}\right)$
$$= -\frac{3 \cdot 4 \cdot 3}{13 \cdot 4} = -\frac{9}{13}$$

87. $\frac{-27.72}{13.2} = -2.1$

89. $\dfrac{-100}{-0.01} = \dfrac{100}{\frac{1}{100}} = 100 \div \frac{1}{100} = 100 \cdot 100$
$$= 10,000$$

91. $\frac{1}{6} - \left(-\frac{7}{9}\right) = \frac{1}{6} + \frac{7}{9} = \frac{3}{18} + \frac{14}{18} = \frac{17}{18}$

93. $-\frac{1}{9} + \frac{7}{12} = -\frac{4}{36} + \frac{21}{36} = \frac{17}{36}$

95. $-\frac{3}{8} - \frac{5}{12} = -\frac{9}{24} - \frac{10}{24} = -\frac{19}{24}$

97. $-\frac{7}{30} + \frac{2}{45} - \frac{3}{10} = -\frac{21}{90} + \frac{4}{90} - \frac{27}{90} = -\frac{44}{90} = -\frac{22}{45}$

99. $\frac{8}{25}\left(-\frac{5}{12}\right) = -\frac{2 \cdot 4 \cdot 5}{5 \cdot 5 \cdot 3 \cdot 4} = -\frac{2}{5 \cdot 3} = -\frac{2}{15}$

101. $\frac{5}{6}\left(-\frac{9}{10}\right)\left(-\frac{4}{5}\right) = \frac{5 \cdot 3 \cdot 3 \cdot 2 \cdot 2}{2 \cdot 3 \cdot 2 \cdot 5 \cdot 5} = \frac{3}{5}$

103. $\frac{7}{6} \div \left(-\frac{9}{10}\right) = \frac{7}{6} \cdot \left(-\frac{10}{9}\right) = -\frac{7 \cdot 2 \cdot 5}{2 \cdot 3 \cdot 9}$
$$= -\frac{7 \cdot 5}{3 \cdot 9} = -\frac{35}{27}, \text{ or } -1\frac{8}{27}$$

105. $\dfrac{-\frac{8}{9}}{2} = -\frac{8}{9} \div \frac{2}{1} = -\frac{8}{9} \cdot \frac{1}{2} = -\frac{2 \cdot 4}{9 \cdot 2} = -\frac{4}{9}$

107. $-8.6 - 3.751 = -(8.6 + 3.751) = -12.351$

109. $(-4.2)(1.4)(2.7) = (-5.88)(2.7) = -15.876$

111. $-24.84 \div 6 = -4.14$

113. $-2496 \div (-0.52) = 4800$

115. $-14.23 + 9.81 + 74.63 - 18.715$
$= -4.42 + 74.63 - 18.715$
$= 70.21 - 18.715$
$= 51.495$

Copyright © 2012 Pearson Education, Inc. Publishing as Addison-Wesley.

117. To find the difference between these two temperatures, subtract the lowest temperature from the highest temperature.

$$90° - (-22°) = 90° + 22°$$
$$= 112°$$

The difference is 112°F.

119. $48.35 - 35.99 - 20.00 - 28.50 + 66.27$
$= 12.36 - 20.00 - 28.50 + 66.27$
$= -7.64 - 28.50 + 66.27$
$= -36.14 + 66.27$
$= 30.13$

His balance is $30.13.

121. $-382.45 + 25.10 + 34.50 - 45.00 - 98.17$
$= -466.02$

His balance is $-$466.02.

(a) To pay off the balance, his payment should be $466.02.

(b) $-466.02 + 300 - 24.66 = -190.68$

His balance is $-$190.68.

123. (a) $-142 - 225 - 185 + 77 = -475$

The total loss for 2007–2010 was $475 thousand.

(b) $77 - (-185) = 77 + 185 = 262$

The difference between the profit or loss from 2009 to 2010 was $262 thousand.

(c) $-225 - (-142) = -225 + 142 = -83$

The difference between the profit or loss from 2007 to 2008 was $-$83 thousand.

125. (a)

Year	Difference (in billions)
2000	$538 - $409 = $129
2010	$916 - $710 = $206
2020	$1479 - $1405 = $74
2030	$2041 - $2542 = -$501

(b) The cost of Social Security will exceed revenue in 2030 by $501 billion.

1.3   Exponents, Roots, and Order of Operations

1.3 Now Try Exercises

N1. (a) $(-3)(-3)(-3) = (-3)^3$

(b) $t \cdot t \cdot t \cdot t \cdot t = t^5$

N2. (a) $7^2 = 7 \cdot 7 = 49$

(b) $(-7)^2 = (-7)(-7) = 49$

(c) $-7^2 = -(7 \cdot 7) = -49$

N3. (a) $-\sqrt{144} = -12$ since the negative sign is outside the radical symbol.

(b) $\sqrt{\frac{100}{9}} = \frac{10}{3}$ since $\frac{10}{3}$ is positive and $\left(\frac{10}{3}\right)^2 = \frac{100}{9}$.

(c) $\sqrt{-144}$ is not a real number.

N4. $15 - 3 \cdot 4 + 2 = 15 - 12 + 2$ *Multiply.*
$= 3 + 2$ *Subtract.*
$= 5$ *Add.*

N5. $-5^2 + 10 \div 5 - |3 - 7|$ *Work inside the absolute value bars.*

$= -5^2 + 10 \div 5 - |-4|$ *Subtract inside the absolute value bars.*

$= -5^2 + 10 \div 5 - 4$ *Take absolute value.*

$= -25 + 10 \div 5 - 4$ *Evaluate the power.*

$= -25 + 2 - 4$ *Divide.*

$= -23 - 4$ *Add.*

$= -27$ *Subtract.*

N6. $\dfrac{\sqrt{36} - 4 \cdot 3^2}{-2^2 - 8 \cdot 3 + 13}$

$= \dfrac{6 - 4 \cdot 9}{-4 - 8 \cdot 3 + 13}$ *Evaluate powers and roots.*

$= \dfrac{6 - 36}{-4 - 24 + 13}$ *Multiply.*

$= \dfrac{-30}{-15}$ *Add and subtract.*

$= 2$ *Reduce.*

N7. Let $x = -4$, $y = 7$, and $z = 36$.

$$\frac{x^2 - \sqrt{z}}{-3xy} = \frac{(-4)^2 - \sqrt{36}}{-3(-4)(7)}$$

$$= \frac{16 - 6}{-3(-4)(7)}$$

$$= \frac{16 - 6}{12(7)}$$

$$= \frac{16 - 6}{84}$$

$$= \frac{10}{84} = \frac{5}{42}$$

Copyright © 2012 Pearson Education, Inc. Publishing as Addison-Wesley.

1.3 Section Exercises

1. $-7^6 = -(7 \cdot 7 \cdot 7 \cdot 7 \cdot 7 \cdot 7) = -117{,}649$, whereas
$$(-7)^6 = (-7)(-7)(-7)(-7)(-7)(-7)$$
$$= 117{,}649.$$

Thus, the original statement, $-7^6 = (-7)^6$, is *false*. The following statement is true:
$-7^6 = -(7^6)$

3. The statement "$\sqrt{25}$ is a positive number" is *true*. The symbol $\sqrt{}$ always gives a positive square root provided the radicand is positive.

5. The statement "$(-6)^7$ is a negative number" is *true*. $(-6)^7$ gives an odd number of negative factors, so the product is negative.

7. The statement "The product of 10 positive factors and 10 negative factors is positive" is *true*. The product of an even number of negative factors is positive.

9. The statement "In the exponential -8^5, -8 is the base" is *false*. The base is 8, not -8. If the problem were written $(-8)^5$, then -8 would be the base.

11. (a) $8^2 = 64$

    (b) $-8^2 = -(8 \cdot 8) = -64$

    (c) $(-8)^2 = (-8)(-8) = 64$

    (d) $-(-8)^2 = -(64) = -64$

13. $10 \cdot 10 \cdot 10 \cdot 10 = 10^4$

15. $\frac{3}{4} \cdot \frac{3}{4} \cdot \frac{3}{4} \cdot \frac{3}{4} \cdot \frac{3}{4} = \left(\frac{3}{4}\right)^5$

17. $(-9)(-9)(-9) = (-9)^3$

19. $z \cdot z \cdot z \cdot z \cdot z \cdot z \cdot z = z^7$

21. $4^2 = 4 \cdot 4 = 16$

23. $0.28^3 = (0.28)(0.28)(0.28) = 0.021952$

25. $\left(\frac{1}{5}\right)^3 = \frac{1}{5} \cdot \frac{1}{5} \cdot \frac{1}{5} = \frac{1}{125}$

27. $\left(\frac{4}{5}\right)^4 = \left(\frac{4}{5}\right)\left(\frac{4}{5}\right)\left(\frac{4}{5}\right)\left(\frac{4}{5}\right)$
$$= \frac{4 \cdot 4 \cdot 4 \cdot 4}{5 \cdot 5 \cdot 5 \cdot 5} = \frac{256}{625}$$

29. $(-5)^3 = (-5)(-5)(-5) = -125$

31. $(-2)^8 = (-2)(-2)(-2)(-2)(-2)(-2)(-2)(-2)$
$$= 256$$

33. $-3^6 = -(3 \cdot 3 \cdot 3 \cdot 3 \cdot 3 \cdot 3) = -729$

35. $-8^4 = -(8 \cdot 8 \cdot 8 \cdot 8) = -4096$

37. $\sqrt{81} = 9$ since 9 is positive and $9^2 = 81$.

39. $\sqrt{169} = 13$ since 13 is positive and $13^2 = 169$.

41. $-\sqrt{400} = -(\sqrt{400}) = -(20) = -20$

43. $\sqrt{\frac{100}{121}} = \frac{10}{11}$ since $\frac{10}{11}$ is positive and $\left(\frac{10}{11}\right)^2 = \frac{100}{121}$.

45. $-\sqrt{0.49} = -\sqrt{(0.7)^2} = -(0.7) = -0.7$

47. There is no real number whose square is negative, so $\sqrt{-36}$ is not a real number.

49. (a) $\sqrt{144} = 12$; choice **B**

    (b) $\sqrt{-144}$ is not a real number; choice **C**

    (c) $-\sqrt{144} = -12$; choice **A**

51. If x is a positive number, then $\sqrt{x}$ is also a positive number, and $-\sqrt{x}$ is a *negative* number.

53. $12 + 3 \cdot 4 = 12 + 12$ *Multiply.*
$$\qquad\qquad = 24 \qquad\quad\; \textit{Add.}$$

55. $6 \cdot 3 - 12 \div 4 = 18 - 12 \div 4$ *Multiply.*
$$\qquad\qquad\qquad = 18 - 3 \qquad\;\; \textit{Divide.}$$
$$\qquad\qquad\qquad = 15 \qquad\qquad \textit{Subtract.}$$

57. $10 + 30 \div 2 \cdot 3 = 10 + 15 \cdot 3$ *Divide.*
$$\qquad\qquad\qquad = 10 + 45 \qquad\; \textit{Multiply.}$$
$$\qquad\qquad\qquad = 55 \qquad\qquad\;\; \textit{Add.}$$

59. $-3(5)^2 - (-2)(-8)$
$$= -3(25) - (-2)(-8) \quad \textit{Evaluate power.}$$
$$= -75 - 16 \qquad\qquad\quad \textit{Multiply.}$$
$$= -91 \qquad\qquad\qquad\;\; \textit{Subtract.}$$

61. $5 - 7 \cdot 3 - (-2)^3$
$$= 5 - 7 \cdot 3 - (-8) \quad \textit{Evaluate power.}$$
$$= 5 - 21 + 8 \qquad\quad\; \textit{Multiply; change sign}$$
$$= -16 + 8 \qquad\qquad\; \textit{Subtract.}$$
$$= -8 \qquad\qquad\qquad\; \textit{Add.}$$

63. $-7\left(\sqrt{36}\right) - (-2)(-3)$
$$= -7(6) - (-2)(-3) \quad \textit{Evaluate root.}$$
$$= -42 - 6 \qquad\qquad\;\; \textit{Multiply.}$$
$$= -48 \qquad\qquad\qquad \textit{Subtract.}$$

65. $6|4 - 5| - 24 \div 3$
$$= 6|-1| - 24 \div 3 \quad \textit{Simplify within absolute}$$
$$\qquad\qquad\qquad\qquad \textit{value bars.}$$
$$= 6(1) - 24 \div 3 \quad\; \textit{Take absolute value.}$$
$$= 6 - 8 \qquad\qquad\;\; \textit{Multiply and divide.}$$
$$= -2 \qquad\qquad\qquad \textit{Subtract.}$$

Copyright © 2012 Pearson Education, Inc. Publishing as Addison-Wesley.

67. $|-6 - 5|(-8) + 3^2$

$= |-11|(-8) + 3^2$ *Simplify within absolute value bars.*

$= 11(-8) + 3^2$ *Take absolute value.*

$= 11(-8) + 9$ *Evaluate power.*

$= -88 + 9$ *Multiply.*

$= -79$ *Add.*

69. $6 + \frac{2}{3}(-9) - \frac{5}{8} \cdot 16 = 6 + (-6) - 10$ *Multiply.*

$= 0 - 10$ *Add.*

$= -10$ *Subtract.*

71. $-14(-\frac{2}{7}) \div (2 \cdot 6 - 10)$

$= 4 \div (12 - 10)$ *Multiply.*

$= 4 \div 2$ *Work inside parentheses.*

$= 2$ *Divide.*

73. $\dfrac{(-5 + \sqrt{4})(-2^2)}{-5 - 1}$

$= \dfrac{(-5 + 2)(-4)}{-6}$ *Evaluate root and power; subtract.*

$= \dfrac{(-3)(-4)}{-6}$ *Work inside parentheses.*

$= \dfrac{12}{-6}$ *Multiply.*

$= -2$ *Divide.*

75. $\dfrac{2(-5) + (-3)(-2)}{-8 + 3^2 - 1}$

$= \dfrac{-10 + 6}{-8 + 9 - 1}$ *Evaluate power; multiply.*

$= \dfrac{-4}{1 - 1}$ *Add.*

$= \dfrac{-4}{0}$ *Subtract.*

Since division by 0 is undefined, the given expression is *undefined*.

77. $\dfrac{5 - 3\left(\dfrac{-5 - 9}{-7}\right) - 6}{-9 - 11 + 3 \cdot 7}$

$= \dfrac{5 - 3(\frac{-14}{-7}) - 6}{-9 - 11 + 21}$ *Work in numerator and denominator separately.*

$= \dfrac{5 - 3(2) - 6}{-20 + 21}$

$= \dfrac{5 - 6 - 6}{1}$

$= \dfrac{-7}{1}$

$= -7$ *Write in lowest terms.*

In Exercises 79–86, $a = -3$, $b = 64$, and $c = 6$.

79. $3a + \sqrt{b} = 3(-3) + \sqrt{64}$

$= 3(-3) + 8$

$= -9 + 8$

$= -1$

81. $\sqrt{b} + c - a = \sqrt{64} + 6 - (-3)$

$= 8 + 6 + 3$

$= 14 + 3 = 17$

83. $4a^3 + 2c = 4(-3)^3 + 2(6)$

$= 4(-27) + 12$

$= -108 + 12 = -96$

85. $\dfrac{2c + a^3}{4b + 6a} = \dfrac{2(6) + (-3)^3}{4(64) + 6(-3)}$

$= \dfrac{12 + (-27)}{256 + (-18)}$

$= \dfrac{-15}{238} = -\dfrac{15}{238}$

In Exercises 87–94, $w = 4$, $x = -\frac{3}{4}$, $y = \frac{1}{2}$, and $z = 1.25$.

87. $wy - 8x = 4(\frac{1}{2}) - 8(-\frac{3}{4})$

$= 2 + 6$

$= 8$

89. $xy + y^4 = -\frac{3}{4}(\frac{1}{2}) + (\frac{1}{2})^4$

$= -\frac{3}{8} + \frac{1}{16}$

$= -\frac{6}{16} + \frac{1}{16} = -\frac{5}{16}$

91. $-w + 2x + 3y + z$

$= -4 + 2(-\frac{3}{4}) + 3(\frac{1}{2}) + 1.25$

$= -4 - \frac{3}{2} + \frac{3}{2} + 1.25$

$= -4 + 1.25$

$= -2.75$

93. $\dfrac{7x + 9y}{w} = \dfrac{7(-\frac{3}{4}) + 9(\frac{1}{2})}{4}$

$= \dfrac{-\frac{21}{4} + \frac{9}{2}}{4} = \dfrac{-\frac{21}{4} + \frac{18}{4}}{4}$

$= \dfrac{-\frac{3}{4}}{4} = -\frac{3}{4} \div \frac{4}{1} = -\frac{3}{4} \cdot \frac{1}{4} = -\frac{3}{16}$

95. $(v \times 0.5485 - 4850) \div 1000 \times 31.44$

$= (100{,}000 \times 0.5485 - 4850) \div 1000 \times 31.44$

$= (54{,}850 - 4850) \div 1000 \times 31.44$

$= 50{,}000 \div 1000 \times 31.44$

$= 50 \times 31.44$

$= 1572$

The owner would pay $1572 in property taxes.

Copyright © 2012 Pearson Education, Inc. Publishing as Addison-Wesley.

97. $(v \times 0.5485 - 4850) \div 1000 \times 31.44$
$= (200{,}000 \times 0.5485 - 4850) \div 1000 \times 31.44$
$= (109{,}700 - 4850) \div 1000 \times 31.44$
$= 104{,}850 \div 1000 \times 31.44$
$= 104.85 \times 31.44$
$= 3296.484 \approx 3296$

The owner would pay \$3296 in property taxes.

99. number of oz $\times$ % alcohol $\times 0.075 \div$ body weight in lb $-$ hr of drinking $\times 0.015$

$= 36 \times 4.0 \times 0.075 \div 135 - 3 \times 0.015$
$= 144 \times 0.075 \div 135 - 3 \times 0.015$
$= 10.8 \div 135 - 3 \times 0.015$
$= 0.08 - 3 \times 0.015$
$= 0.08 - 0.045$
$= 0.035$

101. $1.909x - 3791$

(a) $1.909(1996) - 3791 \approx \$19.4$ billion

(b) $1.909(2002) - 3791 \approx \$30.8$ billion

(c) $1.909(2008) - 3791 \approx \$42.3$ billion

(d) \$42.3 billion is more than twice \$19.4 billion, so the amount spent on pets more than doubled from 1996 to 2008.

1.4 Properties of Real Numbers

1.4 Now Try Exercises

N1. (a) $-2(3x - y) = -2[3x + (-y)]$
$= -2(3x) + (-2)(-y)$
$= -6x + 2y$

(b) Use the second form of the distributive property.

$4k - 12k = [4 + (-12)]k$
$= -8k$

N2. (a) $7x + x = 7x + 1x$ *Identity property*
$= (7 + 1)x$ *Distributive property*
$= 8x$

(b) $-(5p - 3q)$
$= -1(5p - 3q)$ *Identity prop.*
$= -1(5p) + (-1)(-3q)$ *Distributive property*
$= -5p + 3q$

N3. $-7x + 10 - 3x - 4 + x$
(See Example 3 for more detailed steps.)
$= -7x - 3x + 1x + 10 - 4$
$= (-7 - 3 + 1)x + 10 - 4$
$= -9x + 6$

N4. (a) $-3(t - 4) - t + 15$
$= -3t + 12 - t + 15$ *Distributive prop.*
$= -3t - t + 12 + 15$ *Commutative prop.*
$= -4t + 27$ *Combine terms.*

(b) $5x(6y)$
$= [5x(6)]y$ *Associative property*
$= [5(x \cdot 6)]y$ *Associative property*
$= [5(6x)]y$ *Commutative property*
$= [(5 \cdot 6)x]y$ *Associative property*
$= (30x)y$ *Multiply.*
$= 30(xy)$ *Associative property*
$= 30xy$

1.4 Section Exercises

1. The identity element for addition is 0 since, for any real number a, $a + 0 = 0 + a = a$. Choice **B** is correct.

3. The additive inverse of a is $-a$ since, for any real number a, $a + (-a) = 0$ and $-a + a = 0$. Choice **A** is correct.

5. The multiplication property of 0 states that the product of 0 and any real number is $\underline{0}$.

7. The associative property is used to change the grouping of three terms or factors.

9. When simplifying an expression, only like terms can be combined.

11. Using the distributive property,
$$2(m + p) = 2m + 2p.$$

13. Using the distributive property,
$$-12(x - y) = -12[x + (-y)]$$
$$= -12(x) + (-12)(-y)$$
$$= -12x + 12y.$$

15. Using the second form of the distributive property,
$$5k + 3k = (5 + 3)k$$
$$= 8k.$$

17. $7r - 9r = 7r + (-9r)$
$= [7 + (-9)]r$
$= -2r$

19. $-8z + 4w$
Since there is no common variable factor here, we cannot use the distributive property to simplify the expression.

21. Using the identity property, then the distributive property,
$$a + 7a = 1a + 7a$$
$$= (1 + 7)a$$
$$= 8a.$$

Copyright © 2012 Pearson Education, Inc. Publishing as Addison-Wesley.

23. $-(2d - f) = -1(2d - f)$
$$= -1(2d) + (-1)(-f)$$
$$= -2d + f$$

25. $-(-x - y) = -1(-x - y)$
$$= -1(-x) + (-1)(-y)$$
$$= x + y$$

27. $-12y + 4y + 3 + 2y$
$$= -12y + 4y + 2y + 3$$
$$= (-12 + 4 + 2)y + 3$$
$$= -6y + 3$$

29. $-6p + 5 - 4p + 6 + 11p$
$$= -6p - 4p + 11p + 5 + 6$$
$$= (-6 - 4 + 11)p + 11$$
$$= 1p + 11 \quad \text{or} \quad p + 11$$

31. $3(k + 2) - 5k + 6 + 3$
$$= 3k + 6 - 5k + 6 + 3$$
$$= 3k - 5k + 6 + 6 + 3$$
$$= (3 - 5)k + 6 + 6 + 3$$
$$= -2k + 15$$

33. $-2(m + 1) - (m - 4)$
$$= -2m - 2 - m + 4$$
$$= -2m - m - 2 + 4$$
$$= (-2 - 1)m + 2$$
$$= -3m + 2$$

35. $0.25(8 + 4p) - 0.5(6 + 2p)$
$$= 0.25(8) + 0.25(4p) + (-0.5)(6) + (-0.5)(2p)$$
$$= 2 + p - 3 - p$$
$$= p - p + 2 - 3$$
$$= (1 - 1)p + 2 - 3$$
$$= 0p - 1$$
$$= -1$$

37. $-(2p + 5) + 3(2p + 4) - 2p$
$$= -2p - 5 + 6p + 12 - 2p$$
$$= (-2 + 6 - 2)p + (-5 + 12)$$
$$= 2p + 7$$

39. $2 + 3(2z - 5) - 3(4z + 6) - 8$
$$= 2 + 6z - 15 - 12z - 18 - 8$$
$$= 6z - 12z + 2 - 15 - 18 - 8$$
$$= (6 - 12)z - 13 - 18 - 8$$
$$= -6z - 31 - 8$$
$$= -6z - 39$$

41. $5x + 8x = (5 + 8)x = 13x$ *Distributive property*

43. $5(9r) = (5 \cdot 9)r = 45r$ *Associative property*

45. $5x + 9y = 9y + 5x$ *Commutative property*

47. $1 \cdot 7 = 7$ *Identity property*

49. $-\dfrac{1}{4}ty + \dfrac{1}{4}ty = 0$ *Inverse property*
A number plus its opposite equals 0.

51. $8(-4 + x) = 8(-4) + 8x$ *Distributive property*
$$= -32 + 8x$$

53. $0(0.875x + 9y - 88z) = 0$ *Multiplication property of 0*
Zero times any quantity equals 0.

55. Answers will vary. One example of commutativity is washing your face and brushing your teeth. The activities can be carried out in either order.
An example of non-commutativity is putting on your socks and putting on your shoes.

57. $96 \cdot 19 + 4 \cdot 19 = (96 + 4)19$
$$= (100)19$$
$$= 1900$$

59. $58 \cdot \dfrac{3}{2} - 8 \cdot \dfrac{3}{2} = (58 - 8)\dfrac{3}{2}$
$$= (50)\dfrac{3}{2} = \dfrac{50}{1} \cdot \dfrac{3}{2}$$
$$= \dfrac{150}{2} = 75$$

61. $4.31(69) + 4.31(31) = 4.31(69 + 31)$
$$= 4.31(100)$$
$$= 431$$

♦♦♦ Relating Concepts 63–68 ♦♦♦

63. The terms have been grouped using the associative property of addition.

64. The terms have been regrouped using the associative property of addition.

65. The order of the terms inside the parentheses has been changed using the commutative property of addition.

66. The terms have been regrouped using the associative property of addition.

67. The common factor, x, has been factored out using the distributive property.

68. The numbers in parentheses have been added to simplify the expression.

69. Is $a + (b \cdot c) = (a + b)(a + c)$?
No. One example is
$$7 + (5 \cdot 3) = (7 + 5)(7 + 3),$$
which is false since
$$7 + (5 \cdot 3) = 7 + 15 = 22$$
and
$$(7 + 5)(7 + 3) = 12(10) = 120.$$

Copyright © 2012 Pearson Education, Inc. Publishing as Addison-Wesley.

Chapter 1 Review Exercises

1. $\left\{-4, -1, 2, \frac{9}{4}, 4\right\}$

Place dots for -4, -1, 2, $\frac{9}{4} = 2.25$, and 4 on a number line.

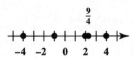

2. $\left\{-5, -\frac{11}{4}, -0.5, 0, 3, \frac{13}{3}\right\}$

Place dots for -5, $-\frac{11}{4} = -2.75$, -0.5, 0, 3, and $\frac{13}{3} = 4.\overline{3}$ on a number line.

3. $|-16| = -(-16) = 16$

4. $-|-8| = -[-(-8)] = -(8) = -8$

5. $|-8| - |-3| = -(-8) - [-(-3)]$
$$= 8 - [3]$$
$$= 5$$

In Exercises 6–9,

$$S = \left\{-9, -\frac{4}{3}, -\sqrt{4}, -0.25, 0, 0.\overline{35}, \frac{5}{3}, \sqrt{7}, \sqrt{-9}, \frac{12}{3}\right\}$$

6. The elements 0 and $\frac{12}{3}$ (or 4) are whole numbers.

7. The elements -9, $-\sqrt{4}$ (or -2), 0, and $\frac{12}{3}$ (or 4) are integers.

8. The elements -9, $-\frac{4}{3}$, $-\sqrt{4}$ (or -2), -0.25, 0, $0.\overline{35}$, $\frac{5}{3}$, and $\frac{12}{3}$ (or 4) are rational numbers. (Remember that terminating and repeating decimals are rational numbers.)

9. All the elements in the set are real numbers except $\sqrt{-9}$.

10. $\{x \mid x$ is a natural number between 3 and 9$\}$
The natural numbers between 3 and 9 are 4, 5, 6, 7, and 8. Therefore, the set is $\{4, 5, 6, 7, 8\}$.

11. $\{y \mid y$ is a whole number less than 4$\}$
The whole numbers less than 4 are 0, 1, 2, and 3. Therefore, the set is $\{0, 1, 2, 3\}$.

12. $4 \cdot 2 \overset{?}{\le} |12 - 4|$
$$8 \overset{?}{\le} |8|$$
$$8 \le 8 \qquad True$$

13. $2 + |-2| \overset{?}{>} 4$
$$2 + 2 \overset{?}{>} 4$$
$$4 > 4 \quad False$$

14. $4(3 + 7) \overset{?}{>} -|40|$
$$4(10) \overset{?}{>} -40$$
$$40 > -40 \qquad True$$

15. The longest bar represents the greatest change. Subaru had the greatest change of 45.9%.

16. The shortest bar represents the least change. Toyota had the least change of 2.14%.

17. Chrysler: $|21.3| = 21.3$

Hyundai: $|-23.8| = 23.8$

The absolute value of the percent change for Chrysler was *less* than the absolute value of the percent change for Hyundai, so the statement is *false*.

18. Subaru $\overset{?}{>}$ 2(Chrysler)
$$45.9 \overset{?}{>} 2(21.3)$$
$$45.9 > 42.6 \qquad True$$

19. $\{x \mid x < -5\}$
In interval notation, $x < -5$ is written as $(-\infty, -5)$. The parenthesis at -5 indicates that -5 is not included. The graph extends from -5 to the left.

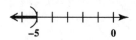

20. $\{x \mid -2 < x \le 3\}$
In interval notation, $-2 < x \le 3$ is written as $(-2, 3]$. The parenthesis indicates that -2 is not included, while the bracket indicates that 3 is included. The graph goes from -2 to 3.

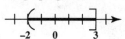

21. $-\frac{5}{8} - \left(-\frac{7}{3}\right) = -\frac{5}{8} + \frac{7}{3}$
$$= -\frac{15}{24} + \frac{56}{24} = \frac{41}{24}$$

22. $-\frac{4}{5} - \left(-\frac{3}{10}\right) = -\frac{4}{5} + \frac{3}{10}$
$$= -\frac{8}{10} + \frac{3}{10}$$
$$= -\frac{5}{10} = -\frac{1}{2}$$

23. $-5 + (-11) + 20 - 7$
$$= -16 + 20 - 7$$
$$= 4 - 7$$
$$= -3$$

24. $-9.42 + 1.83 - 7.6 - 1.8$
$$= -7.59 - 7.6 - 1.8$$
$$= -15.19 - 1.8$$
$$= -16.99$$

Copyright © 2012 Pearson Education, Inc. Publishing as Addison-Wesley.

25. $-15 + (-13) + (-11)$
$= -28 + (-11)$
$= -39$

26. $-1 - 3 - (-10) + (-6)$
$= -4 + 10 + (-6)$
$= 6 + (-6)$
$= 0$

27. $\frac{3}{4} - \left(\frac{1}{2} - \frac{9}{10}\right) = \frac{3}{4} - \left(\frac{5}{10} - \frac{9}{10}\right)$
$= \frac{3}{4} - \left(-\frac{4}{10}\right)$
$= \frac{3}{4} + \frac{4}{10}$
$= \frac{15}{20} + \frac{8}{20} = \frac{23}{20}$

28. $-|-12| - |-9| + (-4) - |10|$
$= -(12) - 9 + (-4) - 10$
$= -21 + (-4) - 10$
$= -25 - 10$
$= -35$

29. $11{,}049 - (-282) = 11{,}049 + 282$
$= 11{,}331$

The difference is 11,331 feet.

30. $2(-5)(-3)(-3) = (-10)(-3)(-3)$
$= (30)(-3)$
$= -90$

31. $-\frac{3}{7}\left(-\frac{14}{9}\right) = \frac{3}{7} \cdot \frac{2 \cdot 7}{3 \cdot 3} = \frac{2}{3}$

32. $\frac{75}{-5} = 15 \cdot 5\left(\frac{1}{-5}\right) = -15$

33. $\frac{-2.3754}{-0.74} = 3.21$

34. $\frac{5}{7-7} = \frac{5}{0}$, which is undefined.
$\frac{7-7}{5} = \frac{0}{5} = 0$

35. $10^4 = 10 \cdot 10 \cdot 10 \cdot 10 = 10{,}000$

36. $\left(\frac{3}{7}\right)^3 = \frac{3}{7} \cdot \frac{3}{7} \cdot \frac{3}{7} = \frac{27}{343}$

37. $(-5)^3 = (-5)(-5)(-5) = -125$

38. $-5^3 = -(5 \cdot 5 \cdot 5) = -125$

39. $\sqrt{400} = 20$, because 20 is positive and $20^2 = 400$.

40. $\sqrt{\frac{64}{121}} = \frac{8}{11}$ since $\frac{8}{11}$ is positive and $\left(\frac{8}{11}\right)^2 = \frac{64}{121}$.

41. $-\sqrt{0.81} = -0.9$ since $-0.9^2 = -0.81$.

42. $\sqrt{-49}$ is not a real number.

43. $-14\left(\frac{3}{7}\right) + 6 \div 3 = -2(3) + (6 \div 3)$
$= -6 + 2 = -4$

44. $-\frac{2}{3}[5(-2) + 8 - 4^3]$
$= -\frac{2}{3}[5(-2) + 8 - 64]$ *Evaluate the power.*
$= -\frac{2}{3}[-10 + 8 - 64]$
$= -\frac{2}{3}[-66] = \frac{2}{3}(22 \cdot 3) = 44$

45. $\frac{-5(3^2) + 9(\sqrt{4}) - 5}{6 - 5(-2)}$
$= \frac{-5(9) + 9(2) - 5}{6 + 10}$
$= \frac{-45 + 18 - 5}{16}$
$= \frac{-32}{16} = -2$

In Exercises 46–48, let $k = -4$, $m = 2$, and $n = 16$.

46. $4k - 7m = 4(-4) - 7(2)$
$= -16 - 14$
$= -30$

47. $-3\sqrt{n} + m + 5k$
$= -3(\sqrt{16}) + 2 + 5(-4)$
$= -3(4) + 2 - 20$
$= -12 + 2 - 20$
$= -10 - 20$
$= -30$

48. $\frac{4m^3 - 3n}{7k^2 - 10} = \frac{4(2)^3 - 3(16)}{7(-4)^2 - 10}$
$= \frac{4(8) - 3(16)}{7(16) - 10}$
$= \frac{32 - 48}{112 - 10}$
$= \frac{-16}{102} = -\frac{8}{51}$

49. **(a)** 6 ft 2 in. $= (6 \cdot 12 + 2)$ in. $= 74$ in.

704 × (weight in pounds) ÷ (height in inches)^2
$= 704 \times 200 \div 74^2$
$= 704 \times 200 \div 5476$
$= 140{,}800 \div 5476$
≈ 26

Grady Sizemore's BMI is 26.

(b) Answers will vary.

50. $2q + 18q$
$= (2 + 18)q$ *Distributive property*
$= 20q$

51. $13z - 17z$
$= (13 - 17)z$ *Distributive property*
$= -4z$

Copyright © 2012 Pearson Education, Inc. Publishing as Addison-Wesley.

52. $-m + 4m$
$$= -1m + 4m \quad \textit{Identity property}$$
$$= (-1 + 4)m \quad \textit{Distributive property}$$
$$= 3m$$

53. $5p - p$
$$= 5p + (-1)p \quad \textit{Identity property}$$
$$= [5 + (-1)]p \quad \textit{Distributive property}$$
$$= 4p$$

54. $-2(k + 3)$
$$= -2(k) + (-2)(3) \quad \textit{Distributive property}$$
$$= -2k - 6$$

55. $6(r + 3)$
$$= 6(r) + 6(3) \quad \textit{Distributive property}$$
$$= 6r + 18$$

56. $9(2m + 3n)$
$$= 9(2m) + 9(3n) \quad \textit{Distributive property}$$
$$= 18m + 27n$$

57. $-(-p + 6q) - (2p - 3q)$
$$= -1(-p + 6q) + (-1)(2p - 3q)$$
$$= -1(-p) + (-1)(6q) + (-1)(2p) + (-1)(-3q)$$
$$= p - 6q - 2p + 3q$$
$$= p - 2p - 6q + 3q$$
$$= -p - 3q$$

58. $-3y + 6 - 5 + 4y$
$$= -3y + 4y + 6 - 5$$
$$= y + 1$$

59. $2a + 3 - a - 1 - a - 2$
$$= 2a - a - a + 3 - 1 - 2$$
$$= 0$$

60. $-3(4m - 2) + 2(3m - 1) - 4(3m + 1)$
$$= -12m + 6 + 6m - 2 - 12m - 4$$
$$= -12m + 6m - 12m + 6 - 2 - 4$$
$$= -18m$$

61. $2x + 3x = (2 + 3)x = 5x \quad \textit{Distributive prop.}$

62. $-5 \cdot 1 = -5 \quad \textit{Identity property}$

63. $2(4x) = (2 \cdot 4)x = 8x \quad \textit{Associative property}$

64. $-3 + 13 = 13 + (-3) = 10 \quad \textit{Commutative prop.}$

65. $-3 + 3 = 0 \quad \textit{Inverse property}$

66. $6(x + z) = 6x + 6z \quad \textit{Distributive property}$

67. $0 + 7 = 7 \quad \textit{Identity property}$

68. $4 \cdot \frac{1}{4} = 1 \quad \textit{Inverse property}$

69.   **[1.2]**   For 2007 (in millions of dollars):
$$|9766 - 10{,}498| = |-732| = 732$$
The balance of trade (exports minus imports) is *negative* since the amount of money spent on imports is greater than the amount of money received for exports.

70.   **[1.2]**   For 2008 (in millions of dollars):
$$|12{,}190 - 11{,}094| = |1096| = 1096$$
The balance of trade (exports minus imports) is *positive* since the amount of money received for exports is greater than the amount of money spent on imports.

71.   **[1.2]**   For 2008 (in millions of dollars):
$$|7294 - 6495| = |799| = 799$$
The balance of trade (exports minus imports) is *positive* since the amount of money received for exports is greater than the amount of money spent on imports.

72.   **[1.3]** $\left(-\frac{4}{5}\right)^4 = \left(-\frac{4}{5}\right)\left(-\frac{4}{5}\right)\left(-\frac{4}{5}\right)\left(-\frac{4}{5}\right)$
$$= \frac{256}{625}$$

73.   **[1.2]** $-\frac{5}{8}(-40) = -\frac{5}{8} \cdot \frac{-40}{1}$
$$= 5 \cdot 5 = 25$$

74.   **[1.3]** $-25\left(-\frac{4}{5}\right) + 3^3 - 32 \div \sqrt{4}$
$$= -25\left(-\frac{4}{5}\right) + 27 - 32 \div 2$$
$$= 20 + 27 - 16$$
$$= 31$$

75.   **[1.2]** $-8 + |-14| + |-3| = -8 + 14 + 3$
$$= 9$$

76.   **[1.3]** $\dfrac{6 \cdot \sqrt{4} - 3 \cdot \sqrt{16}}{-2 \cdot 5 + 7(-3) - 10}$
$$= \dfrac{6 \cdot 2 - 3 \cdot 4}{-2 \cdot 5 + 7(-3) - 10}$$
$$= \dfrac{12 - 12}{-10 - 21 - 10}$$
$$= \dfrac{0}{-41} = 0$$

77.   **[1.3]** $-\sqrt{25} = -(5) = -5$

78.   **[1.2]** $-\dfrac{10}{21} \div -\dfrac{5}{14} = -\dfrac{10}{21} \cdot -\dfrac{14}{5}$
$$= \dfrac{2 \cdot 5}{3 \cdot 7} \cdot \dfrac{2 \cdot 7}{5}$$
$$= \dfrac{2 \cdot 2}{3} = \dfrac{4}{3}$$

Copyright © 2012 Pearson Education, Inc. Publishing as Addison-Wesley.

79. **[1.2]** $0.8 - 4.9 - 3.2 + 1.14$
$= -4.1 - 3.2 + 1.14$
$= -7.3 + 1.14$
$= -6.16$

80. **[1.3]** $-3^2 = -(3 \cdot 3) = -9$

81. **[1.2]** $\frac{-38}{-19} = 38(\frac{1}{19}) = 2 \cdot 19(\frac{1}{19}) = 2$

82. **[1.4]** $-2(k - 1) + 3k - k$
$= -2k + 2 + 3k - k$
$= -2k + 3k - k + 2$
$= (-2 + 3 - 1)k + 2$
$= 0k + 2 = 2$

83. **[1.3]** Since there is no real number whose square is -100, $-\sqrt{-100}$ is *not a real number*.

84. **[1.4]** $-(3k - 6h)$
$= -1(3k - 6h)$ *Identity prop.*
$= -1(3k) + (-1)(-6h)$
$= -3k + 6h$

85. **[1.2]** $-4.6(2.48) = -11.408$

86. **[1.3]** $-\frac{2}{3}(-15) + (2^4 - 8 \div 4)$
$= 10 + [16 - 8 \div 4]$
$= 10 + (16 - 2)$
$= 10 + 14 = 24$

87. **[1.4]** $-2x + 5 - 4x - 1$
$= -2x - 4x + 5 - 1$
$= -6x + 4$

88. **[1.2]** $-\frac{2}{3} - (\frac{1}{6} - \frac{5}{9}) = -\frac{2}{3} - (\frac{3}{18} - \frac{10}{18})$
$= -\frac{2}{3} - (-\frac{7}{18})$
$= -\frac{2}{3} + \frac{7}{18}$
$= -\frac{12}{18} + \frac{7}{18} = -\frac{5}{18}$

89. **[1.3]** **(a)** $-m(3k^2 + 5m)$
$= -2[3(-4)^2 + 5(2)]$ *Let $k = -4$, $m = 2$.*
$= -2[3(16) + 5(2)]$
$= -2[48 + 10]$
$= -2[58]$
$= -116$

(b) $-m(3k^2 + 5m)$
$= -(-\frac{3}{4})[3(\frac{1}{2})^2 + 5(-\frac{3}{4})]$ *Let $k = \frac{1}{2}$, $m = -\frac{3}{4}$.*
$= \frac{3}{4}[3(\frac{1}{4}) + 5(-\frac{3}{4})]$
$= \frac{3}{4}[\frac{3}{4} - \frac{15}{4}]$
$= \frac{3}{4}(-\frac{12}{4})$
$= \frac{3}{4}(-3) = -\frac{9}{4}$

90. **[1.3]** In order to evaluate $(3 + 5)^2$, you should work within the parentheses first.

Chapter 1 Test

1. $\{-3, 0.75, \frac{5}{3}, 5, 6.3\}$
Place dots at $-3, 0.75, \frac{5}{3} = 1.\overline{6}, 5,$ and 6.3.

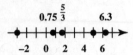

In Exercises 2–5,

$$A = \left\{-\sqrt{6}, -1, -0.5, 0, 3, \sqrt{25}, 7.5, \frac{24}{2}, \sqrt{-4}\right\}.$$

2. The elements $0, 3, \sqrt{25}$ (or 5), and $\frac{24}{2}$ (or 12) are whole numbers.

3. The elements $-1, 0, 3, \sqrt{25}$ (or 5), and $\frac{24}{2}$ (or 12) are integers.

4. The elements $-1, -0.5, 0, 3, \sqrt{25}$ (or 5), 7.5, and $\frac{24}{2}$ (or 12) are rational numbers.

5. All the elements in the set are real numbers except $\sqrt{-4}$.

6. $\{x \mid x < -3\}$
In interval notation, $x < -3$ is written as $(-\infty, -3)$. The parenthesis at -3 indicates that -3 is not included. The graph extends from -3 to the left.

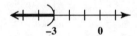

7. $\{x \mid -4 < x \le 2\}$
In interval notation, $-4 < x \le 2$ is written as $(-4, 2]$. The parenthesis indicates that -4 is not included, while the bracket indicates that 2 is included. The graph goes from -4 to 2.

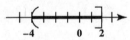

8. $-6 + 14 + (-11) - (-3)$
$= 8 + (-11) + 3$
$= -3 + 3 = 0$

9. $10 - 4 \cdot 3 + 6(-4)$
$= 10 - 12 + (-24)$
$= -2 + (-24) = -26$

10. $7 - 4^2 + 2(6) + (-4)^2$
$= 7 - 16 + 12 + 16$
$= 19$

11. $\dfrac{10 - 24 + (-6)}{\sqrt{16}(-5)}$
$= \dfrac{-14 + (-6)}{4(-5)}$
$= \dfrac{-20}{-20} = 1$

Copyright © 2012 Pearson Education, Inc. Publishing as Addison-Wesley.

12. $\dfrac{-2[3-(-1-2)+2]}{\sqrt{9}(-3)-(-2)}$

$= \dfrac{-2[3-(-3)+2]}{3(-3)-(-2)}$

$= \dfrac{-2[8]}{-9-(-2)}$

$= \dfrac{-16}{-7} = \dfrac{16}{7}$

13. $\dfrac{8\cdot4-3^2\cdot5-2(-1)}{-3\cdot2^3+1}$

$= \dfrac{8\cdot4-9\cdot5-2(-1)}{-3\cdot8+1}$

$= \dfrac{32-45+2}{-24+1}$

$= \dfrac{-11}{-23} = \dfrac{11}{23}$

14. $17,400-(-32,995) = 17,400+32,995$
$\qquad\qquad\qquad\qquad = 50,395$

The difference between the height of Mt. Foraker and the depth of the Philippine Trench is 50,395 feet.

15. $14,110-(-23,376) = 14,110+23,376$
$\qquad\qquad\qquad\qquad = 37,486$

The difference between the height of Pikes Peak and the depth of the Java Trench is 37,486 feet.

16. $-24,721-(-23,376) = -24,721+23,376$
$\qquad\qquad\qquad\qquad\quad = -1345$

The Cayman Trench is 1345 feet deeper than the Java Trench.

17. $\sqrt{196}=14$, because 14 is positive and $14^2=196$.

18. $-\sqrt{225}=-(\sqrt{225})=-(15)=-15$

19. Since there is no real number whose square is -16, $\sqrt{-16}$ is not a real number.

20. **(a)** If a is positive, then $\sqrt{a}$ will represent a positive number.

(b) If a is negative, then $\sqrt{a}$ will not represent a real number.

(c) If a is 0, then $\sqrt{a}$ will be 0.

21. $\dfrac{8k+2m^2}{r-2} = \dfrac{8(-3)+2(-3)^2}{25-2}$

Let $k=-3$, $m=-3$, and $r=25$.

$= \dfrac{8(-3)+2(9)}{23}$

$= \dfrac{-24+18}{23}$

$= \dfrac{-6}{23}$ or $-\dfrac{6}{23}$

22. $-3(2k-4)+4(3k-5)-2+4k$
$= -3(2k)+(-3)(-4)+4(3k)$
$\quad +4(-5)-2+4k$
$= -6k+12+12k-20-2+4k$
$= -6k+12k+4k+12-20-2$
$= 10k-10$

23. When simplifying

$$(3r+8)-(-4r+6),$$

the subtraction sign in front of $(-4r+6)$ changes the sign of the terms $-4r$ and 6.

$(3r+8)-(-4r+6)$
$= 3r+8-(-4r)-6$
$= 3r+8+4r-6$
$= 3r+4r+8-6$
$= 7r+2$

24. $6+(-6)=0$
The answer is **B**, *Inverse property*.
The sum of 6 and its inverse, -6, equals zero.

25. $-2+(3+6)=(-2+3)+6$
The answer is **D**, *Associative property*.
The order of the terms is the same, but the grouping has changed.

26. $5x+15x=(5+15)x$
The answer is **A**, *Distributive property*.
This is the second form of the distributive property.

27. $13\cdot0=0$
The answer is **F**, *Multiplication property of* 0.
Multiplication by 0 always equal 0.

28. $-9+0=-9$
The answer is **C**, *Identity property*.
The addition of 0 to any number does not change the number.

Copyright © 2012 Pearson Education, Inc. Publishing as Addison-Wesley.

29. $4 \cdot 1 = 4$
The answer is **C**, *Identity property*.
Multiplication of any number by 1 does not
change the number.

30. $(a + b) + c = (b + a) + c$
The answer is **E**, *Commutative property*.
The order of the terms a and b is reversed.

Copyright © 2012 Pearson Education, Inc. Publishing as Addison-Wesley.

CHAPTER 2   LINEAR EQUATIONS, INEQUALITIES, AND APPLICATIONS

2.1   Linear Equations in One Variable

2.1   Now Try Exercises

N1. **(a)** $2x + 17 - 3x$ is an *expression* because it does not contain an equals symbol.

(b) $2x + 17 = 3x$ is an *equation* because it contains an equals symbol.

N2.

$$5x + 11 = 2x - 13 - 3x \qquad \textit{Original equation}$$

$$5x + 11 = -x - 13 \qquad \textit{Combine terms.}$$

$$5x + 11 + x = -x - 13 + x \qquad \textit{Add x.}$$

$$6x + 11 = -13 \qquad \textit{Combine terms.}$$

$$6x + 11 - 11 = -13 - 11 \qquad \textit{Subtract 11.}$$

$$6x = -24 \qquad \textit{Combine terms.}$$

$$\frac{6x}{6} = \frac{-24}{6} \qquad \textit{Divide by 6.}$$

$$x = -4 \qquad \textit{Proposed solution}$$

Check by substituting -4 for x in the *original* equation.

$$5x + 11 = 2x - 13 - 3x \qquad \textit{Original equation}$$

$$5(-4) + 11 \overset{?}{=} 2(-4) - 13 - 3(-4) \qquad \textit{Let x = -4.}$$

$$-20 + 11 \overset{?}{=} -8 - 13 + 12$$

$$-9 \overset{?}{=} -21 + 12$$

$$-9 = -9 \qquad \textit{True}$$

The solution set is $\{-4\}$.

N3.

$$5(x - 4) - 9 = 3 - 2(x + 16)$$

$$5x - 20 - 9 = 3 - 2x - 32$$

$$\textit{Distributive property}$$

$$5x - 29 = -2x - 29$$

$$\textit{Combine terms.}$$

$$5x - 29 + 2x + 29 = -2x - 29 + 2x + 29$$

$$\textit{Add 2x; add 29.}$$

$$7x = 0$$

$$\textit{Combine terms.}$$

$$\frac{7x}{7} = \frac{0}{7}$$

$$\textit{Divide by 7.}$$

$$x = 0$$

$$\textit{Proposed solution}$$

We will use the following notation to indicate the value of each side of the original equation after we have substituted the proposed solution and simplified.

Check $x = 0$: $-20 - 9 = 3 - 32$ *True*

The solution set is $\{0\}$.

N4. $\dfrac{x - 4}{4} + \dfrac{2x + 4}{8} = 5$

Multiply each side by the LCD, 8, and use the distributive property.

$$8\left(\frac{x - 4}{4}\right) + 8\left(\frac{2x + 4}{8}\right) = 8(5)$$

$$2(x - 4) + 1(2x + 4) = 40$$

$$2x - 8 + 2x + 4 = 40$$

$$4x - 4 = 40$$

$$4x = 44 \qquad \textit{Add 4.}$$

$$x = 11 \qquad \textit{Divide by 4.}$$

Check $x = 11$: $\frac{7}{4} + \frac{13}{4} = 5$ *True*

The solution set is $\{11\}$.

N5. $0.08x - 0.12(x - 4) = 0.03(x - 5)$

$$\textit{Multiply each term by 100.}$$

$$8x - 12(x - 4) = 3(x - 5)$$

$$8x - 12x + 48 = 3x - 15$$

$$-4x + 48 = 3x - 15$$

$$-4x + 63 = 3x$$

$$63 = 7x$$

$$9 = x$$

Check $x = 9$: $0.72 - 0.60 = 0.12$ *True*

The solution set is $\{9\}$.

N6. **(a)** $9x - 3(x + 4) = 6(x - 2)$

$$9x - 3x - 12 = 6x - 12$$

$$6x - 12 = 6x - 12$$

This is an *identity*. Any real number will make the equation true.

The solution set is {all real numbers}.

(b) $-3(2x - 1) - 2x = 3 + x$

$$-6x + 3 - 2x = 3 + x$$

$$-8x + 3 = 3 + x$$

$$-9x + 3 = 3 \qquad \textit{Subtract x.}$$

$$-9x = 0 \qquad \textit{Subtract 3.}$$

$$x = 0 \qquad \textit{Divide by -9.}$$

This is a *conditional equation*.

Check $x = 0$: $-3(-1) = 3$ *True*

The solution set is $\{0\}$.

Copyright © 2012 Pearson Education, Inc. Publishing as Addison-Wesley.

(c)
$$10x - 21 = 2(x - 5) + 8x$$
$$10x - 21 = 2x - 10 + 8x$$
$$10x - 21 = 10x - 10$$
$$10x - 21 - 10x = 10x - 10 - 10x$$
$$\qquad\qquad\qquad\text{Subtract } 10x.$$
$$-21 = -10 \qquad \textit{False}$$

Since the result, $-21 = -10$, is *false*, the equation has no solution and is called a *contradiction*.

The solution set is $\emptyset$.

2.1 Section Exercises

1. **A.** $3x + x - 1 = 0$ can be written as $4x = 1$, so it is linear.

   C. $6x + 2 = 9$ is in linear form.

3. $3(x + 4) = 5x$ *Original equation*
$$3(6 + 4) \overset{?}{=} 5 \cdot 6 \quad \textit{Let } x = 6.$$
$$3(10) \overset{?}{=} 30 \quad \textit{Add.}$$
$$30 = 30 \quad \textit{True}$$

Since a true statement is obtained, 6 is a solution.

5. $-3x + 2 - 4 = x$ is an *equation* because it contains an equals symbol.

7. $4(x + 3) - 2(x + 1) - 10$ is an *expression* because it does not contain an equals symbol.

9. $-10x + 12 - 4x = -3$ is an *equation* because it contains an equals symbol.

In the following exercises, we show brief checks of the solutions. To be sure that your solution is correct, check it by substituting into the original equation.

11.
$$7x + 8 = 1$$
$$7x + 8 - 8 = 1 - 8 \quad \textit{Subtract 8.}$$
$$7x = -7$$
$$\frac{7x}{7} = \frac{-7}{7} \quad \textit{Divide by 7.}$$
$$x = -1$$

We will use the following notation to indicate the value of each side of the original equation after we have substituted the proposed solution and simplified.

Check $x = -1$: $-7 + 8 = 1$ *True*

The solution set is $\{-1\}$.

13.
$$5x + 2 = 3x - 6$$
$$5x + 2 - 3x = 3x - 6 - 3x \quad \textit{Subtract 3x.}$$
$$2x + 2 = -6$$
$$2x + 2 - 2 = -6 - 2 \quad \textit{Subtract 2.}$$
$$2x = -8$$
$$\frac{2x}{2} = \frac{-8}{2} \quad \textit{Divide by 2.}$$
$$x = -4$$

Check $x = -4$: $-20 + 2 = -12 - 6$ *True*

The solution set is $\{-4\}$.

15. $7x - 5x + 15 = x + 8$
$$2x + 15 = x + 8 \quad \textit{Combine terms.}$$
$$2x = x - 7 \quad \textit{Subtract 15.}$$
$$x = -7 \quad \textit{Subtract x.}$$

Check $x = -7$: $-49 + 35 + 15 = -7 + 8$ *True*

The solution set is $\{-7\}$.

17. $12w + 15w - 9 + 5 = -3w + 5 - 9$
$$27w - 4 = -3w - 4 \quad \textit{Combine terms.}$$
$$30w - 4 = -4 \quad \textit{Add 3w.}$$
$$30w = 0 \quad \textit{Add 4.}$$
$$w = 0 \quad \textit{Divide by 30.}$$

Check $w = 0$: $-9 + 5 = 5 - 9$ *True*

The solution set is $\{0\}$.

19. $3(2t - 4) = 20 - 2t$
$$6t - 12 = 20 - 2t \quad \textit{Distributive property}$$
$$8t - 12 = 20 \quad \textit{Add 2t.}$$
$$8t = 32 \quad \textit{Add 12.}$$
$$t = 4 \quad \textit{Divide by 8.}$$

Check $t = 4$: $3(4) = 20 - 8$ *True*

The solution set is $\{4\}$.

21. $-5(x + 1) + 3x + 2 = 6x + 4$
$$-5x - 5 + 3x + 2 = 6x + 4 \quad \textit{Distributive property}$$
$$-2x - 3 = 6x + 4 \quad \textit{Combine terms.}$$
$$-3 = 8x + 4 \quad \textit{Add 2x.}$$
$$-7 = 8x \quad \textit{Subtract 4.}$$
$$-\frac{7}{8} = x \quad \textit{Divide by 8.}$$

Check: Substitute $-\frac{7}{8}$ for x and show that both sides equal -1.25. The screen shows a typical check on a calculator.

```
-7/8→X
             -.875
-5(X+1)+3X+2
             -1.25
6X+4
             -1.25
```

The solution set is $\left\{-\frac{7}{8}\right\}$.

23. $-2x + 5x - 9 = 3(x - 4) - 5$
$$3x - 9 = 3x - 12 - 5$$
$$3x - 9 = 3x - 17$$
$$-9 = -17 \quad \textit{False}$$

The equation is a *contradiction*.

The solution set is $\emptyset$.

Copyright © 2012 Pearson Education, Inc. Publishing as Addison-Wesley.

25. $2(x+3) = -4(x+1)$

$2x + 6 = -4x - 4$ *Remove parentheses.*

$6x + 6 = -4$ *Add 4x.*

$6x = -10$ *Subtract 6.*

$x = \frac{-10}{6} = -\frac{5}{3}$ *Divide by 6.*

Check $x = -\frac{5}{3}$: $2(\frac{4}{3}) = -4(-\frac{2}{3})$ *True*

The solution set is $\left\{-\frac{5}{3}\right\}$.

27. $3(2x+1) - 2(x-2) = 5$

$6x + 3 - 2x + 4 = 5$ *Remove parentheses. Combine terms.*

$4x + 7 = 5$

$4x = -2$ *Subtract 7.*

$x = \frac{-2}{4} = -\frac{1}{2}$ *Divide by 4.*

Check $x = -\frac{1}{2}$: $3(0) - 2(-\frac{5}{2}) = 5$ *True*

The solution set is $\left\{-\frac{1}{2}\right\}$.

29. $2x + 3(x-4) = 2(x-3)$

$2x + 3x - 12 = 2x - 6$

$5x - 12 = 2x - 6$

$3x = 6$

$x = \frac{6}{3} = 2$

Check $x = 2$: $4 + 3(-2) = 2(-1)$ *True*

The solution set is $\{2\}$.

31. $6x - 4(3 - 2x) = 5(x-4) - 10$

$6x - 12 + 8x = 5x - 20 - 10$

$14x - 12 = 5x - 30$

$9x = -18$

$x = -2$

Check $x = -2$: $-12 - 4(7) = 5(-6) - 10$ *True*

The solution set is $\{-2\}$.

33. $-2(x+3) - x - 4 = -3(x+4) + 2$

$-2x - 6 - x - 4 = -3x - 12 + 2$

$-3x - 10 = -3x - 10$

The equation is an *identity*.

The solution set is {all real numbers}.

35. $2[x - (2x+4) + 3] = 2(x+1)$

$2[x - 2x - 4 + 3] = 2(x+1)$

$2[-x - 1] = 2(x+1)$

$-x - 1 = x + 1$ *Divide by 2.*

$-1 = 2x + 1$ *Add x.*

$-2 = 2x$ *Subtract 1.*

$-1 = x$ *Divide by 2.*

Check $x = -1$: $2[-1 - 2 + 3] = 0$ *True*

The solution set is $\{-1\}$.

37. $-[2x - (5x+2)] = 2 + (2x+7)$

$-[2x - 5x - 2] = 2 + 2x + 7$

$-[-3x - 2] = 2 + 2x + 7$

$3x + 2 = 2x + 9$

$x = 7$

Check $x = 7$: $-[14 - 37] = 2 + 21$ *True*

The solution set is $\{7\}$.

39. $-3x + 6 - 5(x-1) = -5x - (2x-4) + 5$

$-3x + 6 - 5x + 5 = -5x - 2x + 4 + 5$

$-8x + 11 = -7x + 9$

$-x = -2$

$x = 2$

Check $x = 2$:

$-6 + 6 - 5 = -10 - 0 + 5$ *True*

The solution set is $\{2\}$.

41. $7[2 - (3 + 4x)] - 2x = -9 + 2(1 - 15x)$

$7[2 - 3 - 4x] - 2x = -9 + 2 - 30x$

$7[-1 - 4x] - 2x = -7 - 30x$

$-7 - 28x - 2x = -7 - 30x$

$-7 - 30x = -7 - 30x$

The equation is an *identity*.

The solution set is {all real numbers}.

43. $-[3x - (2x+5)] = -4 - [3(2x-4) - 3x]$

$-[3x - 2x - 5] = -4 - [6x - 12 - 3x]$

$-[x - 5] = -4 - [3x - 12]$

$-x + 5 = -4 - 3x + 12$

$-x + 5 = -3x + 8$

$2x = 3$

$x = \frac{3}{2}$

Check $x = \frac{3}{2}$:

$-[\frac{9}{2} - 8] = -4 - [-3 - \frac{9}{2}]$ *True*

The solution set is $\left\{\frac{3}{2}\right\}$.

45. The denominators of the fractions are 3, 4, and 1. The least common denominator is $(3)(4)(1) = 12$, since it is the smallest number into which each denominator can divide without a remainder.

47. (a) We need to make the coefficient of the first term on the left an integer. Since $0.05 = \frac{5}{100}$, we multiply by 10^2 or 100. This will also take care of the second term.

(b) We need to make 0.006, 0.007, and 0.009 integers. These numbers can be written as $\frac{6}{1000}$, $\frac{7}{1000}$, and $\frac{9}{1000}$. Multiplying by 10^3 or 1000 will eliminate the decimal points (the denominators) so that all the coefficients are integers.

Copyright © 2012 Pearson Education, Inc. Publishing as Addison-Wesley.

49. $-\frac{5}{9}x = 2$

$\qquad -5x = 18 \qquad$ *Multiply by 9.*

$\qquad x = \frac{18}{-5} = -\frac{18}{5} \qquad$ *Divide by -5.*

Check $x = -\frac{18}{5}$: $\quad \left(-\frac{5}{9}\right)\left(-\frac{18}{5}\right) = 2 \quad$ *True*

The solution set is $\left\{-\frac{18}{5}\right\}$.

51. $\frac{6}{5}x = -1$

$\qquad 6x = -5 \qquad$ *Multiply by 5.*

$\qquad x = \frac{-5}{6} = -\frac{5}{6} \qquad$ *Divide by 6.*

Check $x = -\frac{5}{6}$: $\quad \left(\frac{6}{5}\right)\left(-\frac{5}{6}\right) = -1 \quad$ *True*

The solution set is $\left\{-\frac{5}{6}\right\}$.

53. $\qquad \frac{x}{2} + \frac{x}{3} = 5$

$\qquad$ *Multiply both sides by the LCD, 6.*

$\qquad 6\left(\frac{x}{2} + \frac{x}{3}\right) = 6(5)$

$\qquad 6\left(\frac{x}{2}\right) + 6\left(\frac{x}{3}\right) = 30 \qquad$ *Distributive property*

$\qquad 3x + 2x = 30$

$\qquad 5x = 30 \qquad$ *Add.*

$\qquad x = 6 \qquad$ *Divide by 5.*

Check $x = 6$: $\quad 3 + 2 = 5 \quad$ *True*

The solution set is $\{6\}$.

55. $\qquad \frac{3x}{4} + \frac{5x}{2} = 13$

$\qquad$ *Multiply both sides by the LCD, 4.*

$\qquad 4\left(\frac{3x}{4} + \frac{5x}{2}\right) = 4(13)$

$\qquad 4\left(\frac{3x}{4}\right) + 4\left(\frac{5x}{2}\right) = 4(13) \qquad$ *Distributive property*

$\qquad 3x + 10x = 52$

$\qquad 13x = 52 \qquad$ *Combine terms.*

$\qquad x = 4 \qquad$ *Divide by 13.*

Check $x = 4$: $\quad 3 + 10 = 13 \quad$ *True*

The solution set is $\{4\}$.

57. $\qquad \frac{x-10}{5} + \frac{2}{5} = -\frac{x}{3}$

$\qquad$ *Multiply both sides by the LCD, 15.*

$\qquad 15\left(\frac{x-10}{5} + \frac{2}{5}\right) = 15\left(-\frac{x}{3}\right)$

$\qquad 3(x - 10) + 3(2) = -5x$

$\qquad 3x - 30 + 6 = -5x$

$\qquad 8x = 24$

$\qquad x = \frac{24}{8} = 3$

Check $x = 3$: $\quad -\frac{7}{5} + \frac{2}{5} = -1 \quad$ *True*

The solution set is $\{3\}$.

59. $\qquad \frac{3x-1}{4} + \frac{x+3}{6} = 3$

$\qquad$ *Multiply both sides by the LCD, 12.*

$\qquad 12\left(\frac{3x-1}{4} + \frac{x+3}{6}\right) = 12(3)$

$\qquad 3(3x - 1) + 2(x + 3) = 36$

$\qquad 9x - 3 + 2x + 6 = 36$

$\qquad 11x + 3 = 36$

$\qquad 11x = 33$

$\qquad x = 3$

Check $x = 3$: $\quad 2 + 1 = 3 \quad$ *True*

The solution set is $\{3\}$.

61. $\qquad \frac{4x+1}{3} = \frac{x+5}{6} + \frac{x-3}{6}$

$\qquad$ *Multiply both sides by the LCD, 6.*

$\qquad 6\left(\frac{4x+1}{3}\right) = 6\left(\frac{x+5}{6} + \frac{x-3}{6}\right)$

$\qquad 2(4x + 1) = (x + 5) + (x - 3)$

$\qquad 8x + 2 = 2x + 2$

$\qquad 6x = 0$

$\qquad x = 0$

Check $x = 0$: $\quad \frac{1}{3} = \frac{5}{6} - \frac{3}{6} \quad$ *True*

The solution set is $\{0\}$.

63. $\quad 0.05x + 0.12(x + 5000) = 940$

$\qquad$ *Multiply each term by 100.*

$\qquad 5x + 12(x + 5000) = 100(940)$

$\qquad 5x + 12x + 60,000 = 94,000$

$\qquad 17x = 34,000$

$\qquad x = 2000$

Check $x = 2000$: $\quad 100 + 840 = 940 \quad$ *True*

The solution set is $\{2000\}$.

65. $\quad 0.02(50) + 0.08x = 0.04(50 + x)$

$\qquad$ *Multiply each term by 100.*

$\qquad 2(50) + 8x = 4(50 + x)$

$\qquad 100 + 8x = 200 + 4x$

$\qquad 4x = 100$

$\qquad x = 25$

Check $x = 25$: $\quad 1 + 2 = 3 \quad$ *True*

The solution set is $\{25\}$.

67. $\quad 0.05x + 0.10(200 - x) = 0.45x$

$\qquad$ *Multiply each term by 100.*

$\qquad 5x + 10(200 - x) = 45x$

$\qquad 5x + 2000 - 10x = 45x$

$\qquad 2000 - 5x = 45x$

$\qquad 2000 = 50x$

$\qquad 40 = x$

Check $x = 40$: $\quad 2 + 16 = 18 \quad$ *True*

The solution set is $\{40\}$.

Copyright © 2012 Pearson Education, Inc. Publishing as Addison-Wesley.

69. $0.006(x + 2) = 0.007x + 0.009$

Multiply each term by 1000.

$$6(x + 2) = 7x + 9$$
$$6x + 12 = 7x + 9$$
$$3 = x$$

Check $x = 3$: $0.03 = 0.021 + 0.009$ *True*

The solution set is $\{3\}$.

71. $2L + 2W$; $L = 10, W = 8$

$$2L + 2W = 2(10) + 2(8)$$
$$= 20 + 16 = 36$$

73. $\frac{1}{3}Bh$; $B = 27, h = 8$

$$\frac{1}{3}Bh = \frac{1}{3}(27)(8)$$
$$= 9(8) = 72$$

75. $\frac{5}{9}(F - 32)$; $F = 122$

$$\frac{5}{9}(F - 32) = \frac{5}{9}(122 - 32)$$
$$= \frac{5}{9}(90) = 50$$

2.2   Formulas and Percent

2.2   Now Try Exercises

N1. To solve $I = prt$ for p, treat p as the only variable.

$$I = prt$$
$$I = p(rt) \qquad \text{\textit{Associative property}}$$
$$\frac{I}{rt} = \frac{p(rt)}{rt} \qquad \text{\textit{Divide by rt.}}$$
$$\frac{I}{rt} = p, \ \text{ or } \ p = \frac{I}{rt}$$

N2. Solve $P = a + 2b + c$ for b.

$$P - a = 2b + c \qquad \text{\textit{Subtract a.}}$$
$$P - a - c = 2b \qquad \text{\textit{Subtract c.}}$$
$$\frac{P - a - c}{2} = \frac{2b}{2} \qquad \text{\textit{Divide by 2.}}$$
$$\frac{P - a - c}{2} = b, \ \text{ or } \ b = \frac{P - a - c}{2}$$

N3. Solve $P = 2(L + W)$ for L.

$$\frac{P}{2} = \frac{2(L + W)}{2} \qquad \text{\textit{Divide by 2.}}$$
$$\frac{P}{2} = L + W$$
$$\frac{P}{2} - W = L \qquad \text{\textit{Subtract W.}}$$
$$L = \frac{P}{2} - W, \ \text{ or } \ L = \frac{P - 2W}{2}$$

N4. Solve $5x - 6y = 12$ for y.

$$-6y = 12 - 5x \qquad \text{\textit{Subtract 5x.}}$$
$$y = \frac{12 - 5x}{-6}, \quad \text{\textit{Divide by −6.}}$$
$$\text{or } \ y = \frac{5x - 12}{6}$$

N5. Use $d = rt$. Solve for r.

$$\frac{d}{t} = \frac{rt}{t} \qquad \text{\textit{Divide by t.}}$$
$$\frac{d}{t} = r, \ \text{ or } \ r = \frac{d}{t}$$

Now substitute $d = 21$ and $t = \frac{1}{2}$.

$$r = \frac{21}{\frac{1}{2}} = 21 \times \frac{2}{1} = 42$$

Her average rate is 42 miles per hour.

N6. **(a)** The given amount of mixture is 5 L. The part that is antifreeze is 2 L. Thus, the percent of antifreeze is

$$\frac{\text{partial amount}}{\text{whole amount}} = \frac{2}{5} = 0.40 = 40\%.$$

(b) Let x represent the amount of interest earned.

$$\frac{x}{7500} = 0.025 \qquad \frac{partial}{whole} = percent$$
$$x = 0.025(7500) \quad \text{\textit{Multiply by 7500.}}$$
$$x = 187.50$$

The interest earned is \$187.50.

N7. Let x represent the amount spent on vet care.

$$\frac{x}{41.2} = 0.245 \qquad 24.5\% = 0.245$$
$$x = 0.245(41.2) \quad \text{\textit{Multiply by 41.2.}}$$
$$x = 10.094$$

Therefore, about \$10.1 billion was spent on vet care.

N8. **(a)** Let $x =$ the percent decrease (as a decimal).

$$\text{percent decrease} = \frac{\text{amount of decrease}}{\text{base}}$$
$$x = \frac{80 - 56}{80}$$
$$x = \frac{24}{80}$$
$$x = 0.3$$

The percent markdown was 30%.

(b) Let $x =$ the percent increase (as a decimal).

$$\text{percent increase} = \frac{\text{amount of increase}}{\text{base}}$$
$$x = \frac{689 - 650}{650}$$
$$x = \frac{39}{650}$$
$$x = 0.06$$

The percent increase was 6%.

Copyright © 2012 Pearson Education, Inc. Publishing as Addison-Wesley.

2.2 Section Exercises

1. Solve $I = prt$ for r.

$$I = prt$$
$$\frac{I}{pt} = \frac{prt}{pt}$$
$$\frac{I}{pt} = r, \text{ or } r = \frac{I}{pt}$$

3. Solve $P = 2L + 2W$ for L.

$$P = 2L + 2W$$
$$P - 2W = 2L$$
$$\frac{P - 2W}{2} = \frac{2L}{2}$$
$$\frac{P - 2W}{2} = L, \text{ or } L = \frac{P}{2} - W$$

5. **(a)** Solve for $V = LWH$ for W.

$$V = LWH$$
$$\frac{V}{LH} = \frac{LWH}{LH}$$
$$\frac{V}{LH} = W, \text{ or } W = \frac{V}{LH}$$

(b) Solve for $V = LWH$ for H.

$$V = LWH$$
$$\frac{V}{LW} = \frac{LWH}{LW}$$
$$\frac{V}{LW} = H, \text{ or } H = \frac{V}{LW}$$

7. Solve $C = 2\pi r$ for r.

$$C = 2\pi r$$
$$\frac{C}{2\pi} = \frac{2\pi r}{2\pi} \quad \text{Divide by } 2\pi.$$
$$\frac{C}{2\pi} = r$$

9. **(a)** Solve $A = \frac{1}{2}h(b + B)$ for h.

$$2A = h(b + B) \quad \text{Multiply by 2.}$$
$$\frac{2A}{b + B} = h \qquad \text{Divide by } b + B.$$

(b) Solve $A = \frac{1}{2}h(b + B)$ for B.

$$2A = h(b + B) \quad \text{Multiply by 2.}$$
$$\frac{2A}{h} = b + B \qquad \text{Divide by } h.$$
$$\frac{2A}{h} - b = B \qquad \text{Subtract } b.$$

Another method:

$$2A = h(b + B) \quad \text{Multiply by 2.}$$
$$2A = hb + hB \quad \text{Distributive prop.}$$
$$2A - hb = hB \qquad \text{Subtract } hb.$$
$$\frac{2A - hb}{h} = B \qquad \text{Divide by } h.$$

11. Solve $F = \frac{9}{5}C + 32$ for C.

$$F - 32 = \frac{9}{5}C$$
$$\frac{5}{9}(F - 32) = \frac{5}{9}\left(\frac{9}{5}C\right)$$
$$\frac{5}{9}(F - 32) = C$$

13.
$$A = \frac{1}{2}bh$$
$$2A = 2\left(\frac{1}{2}bh\right) \quad \text{Multiply by 2.}$$
$$2A = bh$$
$$\frac{2A}{b} = \frac{bh}{b} \qquad \text{Divide by } b.$$
$$\frac{2A}{b} = h$$
$$\frac{2A}{b} = \frac{2}{1} \cdot \frac{A}{b}$$
$$= 2\left(\frac{A}{b}\right) \qquad \text{This choice is } \mathbf{A}.$$
$$= 2A\left(\frac{1}{b}\right) \qquad \text{This is choice } \mathbf{B}.$$

To get choice $\mathbf{C}$, divide $A = \frac{1}{2}bh$ by $\frac{1}{2}b$.

$$\frac{A}{\frac{1}{2}b} = \frac{\frac{1}{2}bh}{\frac{1}{2}b} \quad \text{gives us} \quad h = \frac{A}{\frac{1}{2}b}.$$

Choice $\mathbf{D}$, $h = \dfrac{\frac{1}{2}A}{b}$, can be multiplied by $\dfrac{2}{2}$ on the right side to get $h = \dfrac{A}{2b}$, so it is *not* equivalent to $h = \dfrac{2A}{b}$. Therefore, the correct answer is $\mathbf{D}$.

15.
$$4x + 9y = 11$$
$$4x + 9y - 4x = 11 - 4x \quad \text{Subtract } 4x.$$
$$9y = 11 - 4x$$
$$\frac{9y}{9} = \frac{11 - 4x}{9} \quad \text{Divide by 9.}$$
$$y = \frac{11 - 4x}{9}$$

17.
$$-3x + 2y = 5$$
$$2y = 5 + 3x \quad \text{Add } 3x.$$
$$y = \frac{5 + 3x}{2} \quad \text{Divide by 2.}$$

19.
$$6x - 5y = 7$$
$$-5y = 7 - 6x \quad \text{Subtract } 6x.$$
$$y = \frac{7 - 6x}{-5}, \quad \text{Divide by } -5.$$
$$\text{or } y = \frac{6x - 7}{5}$$

21. Solve $d = rt$ for t.

$$t = \frac{d}{r}$$

To find t, substitute $d = 500$ and $r = 152.672$.

$$t = \frac{500}{152.672} \approx 3.275$$

His time was about 3.275 hours.

Copyright © 2012 Pearson Education, Inc. Publishing as Addison-Wesley.

23. Solve $d = rt$ for r.

$$r = \frac{d}{t}$$
$$r = \frac{520}{10} = 52 \quad \textit{Let d = 520, t = 10.}$$

Her rate was 52 mph.

25. Use the formula $F = \frac{9}{5}C + 32$.

$$F = \frac{9}{5}(45) + 32 \quad \textit{Let C = 45.}$$
$$= 81 + 32$$
$$= 113$$

The corresponding temperature is 113°F.

27. Solve $P = 4s$ for s.

$$s = \frac{P}{4}$$

To find s, substitute 920 for P.

$$s = \frac{920}{4} = 230$$

The length of each side is 230 m.

29. Use the formula $C = 2\pi r$.

$$480\pi = 2\pi r \quad \textit{Let C = 480}\pi.$$
$$\frac{480\pi}{2\pi} = r \quad \textit{Divide by 2}\pi.$$

So the radius of the circle is 240 inches and the diameter is twice that length, that is, 480 inches.

31. Use $V = LWH$.
Let $V = 187$, $L = 11$, and $W = 8.5$.

$$187 = 11(8.5)H$$
$$187 = 93.5H$$
$$2 = H \quad \textit{Divide by 93.5.}$$

The ream is 2 inches thick.

33. The mixture is 36 oz and that part which is alcohol is 9 oz. Thus, the percent of alcohol is

$$\frac{9}{36} = \frac{1}{4} = \frac{25}{100} = 25\%.$$

The percent of water is

$$100\% - 25\% = 75\%.$$

35. Find what percent $6300 is of $210,000.

$$\frac{6300}{210,000} = 0.03 = 3\%$$

The agent received a 3% rate of commission.

In Exercises 37–40, use the rule of 78.

$$u = f \cdot \frac{k(k+1)}{n(n+1)}$$

37. Substitute 700 for f, 4 for k, and 36 for n.

$$u = 700 \cdot \frac{4(4+1)}{36(36+1)}$$

$$= 700 \cdot \frac{4(5)}{36(37)} \approx 10.51$$

The unearned interest is $10.51.

39. Substitute 380.50 for f, 8 for k, and 24 for n.

$$u = (380.50) \cdot \frac{8(8+1)}{24(24+1)}$$
$$= (380.50) \cdot \frac{8(9)}{24(25)} \approx 45.66$$

The unearned interest is $45.66.

41. (a) Boston:

$$\text{Pct.} = \frac{W}{W+L} = \frac{95}{95+67} = \frac{95}{162} \approx .586$$

(b) Tampa Bay:

$$\text{Pct.} = \frac{W}{W+L} = \frac{84}{84+78} = \frac{84}{162} \approx .519$$

(c) Toronto:

$$\text{Pct.} = \frac{W}{W+L} = \frac{75}{75+87} = \frac{75}{162} \approx .463$$

(d) Baltimore:

$$\text{Pct.} = \frac{W}{W+L} = \frac{64}{64+98} = \frac{64}{162} = .395$$

43. $\frac{62.0 \text{ million}}{114.9 \text{ million}} \approx 0.54$

In 2009, about 54% of the U.S. households that owned at least one TV set owned at least 3 TV sets.

45. $0.88(114.9) = 101.112$

In 2009, about 101.1% of the U.S. households that owned at least one TV set received basic cable.

47. $0.32(221,190) = 70,780.80$

To the nearest dollar, $70,781 will be spent to provide housing.

49. Since 16% is twice as much as 8%, the cost for child care and education will be
$2(\$17,695) = \$35,390$.

51. Let x = the percent increase (as a decimal).

$$\text{percent increase} = \frac{\text{amount of increase}}{\text{base}}$$
$$x = \frac{11.34 - 10.50}{10.50}$$
$$x = \frac{0.84}{10.50}$$
$$x = 0.08$$

The percent increase was 8%.

Copyright © 2012 Pearson Education, Inc. Publishing as Addison-Wesley.

53. Let x = the percent decrease (as a decimal).

$$\text{percent decrease} = \frac{\text{amount of decrease}}{\text{base}}$$

$$x = \frac{134{,}953 - 129{,}798}{134{,}953}$$

$$x = \frac{5155}{134{,}953}$$

$$x \approx 0.038$$

The percent decrease was 3.8%.

55.

$$\text{percent decrease} = \frac{\text{amount of decrease}}{\text{base}}$$

$$= \frac{18.98 - 9.97}{18.98}$$

$$= \frac{9.01}{18.98} \approx 0.475$$

The percent discount was 47.5%.

57.
$$4x + 4(x + 7) = 124$$
$$4x + 4x + 28 = 124$$
$$8x + 28 = 124$$
$$8x = 96$$
$$x = \frac{96}{8} = 12$$

The solution set is $\{12\}$.

59.
$$2.4 + 0.4x = 0.25(6 + x)$$
$$240 + 40x = 25(6 + x)$$
Multiply by 100.
$$240 + 40x = 150 + 25x$$
$$15x = -90$$
$$x = -6$$

The solution set is $\{-6\}$.

61. "The product of -3 and 5, divided by 1 less than 6" is translated and evaluated as follows:

$$\frac{-3(5)}{6 - 1} = \frac{-15}{5} = -3$$

63. "The sum of 6 and -9, multiplied by the additive inverse of 2" is translated and evaluated as follows:

$$[6 + (-9)](-2) = -3(-2) = 6$$

2.3 Applications of Linear Equations

2.3 Now Try Exercises

N1. (a)

The quotient of a number and 10	is	twice the number.
$\downarrow$	$\downarrow$	$\downarrow$
$\frac{x}{10}$	$=$	$2x$

An equation is $\frac{x}{10} = 2x$.

The product

(b) of a number decreased and 5, by 7, is zero.

$\downarrow$		$\downarrow$	$\downarrow\ \downarrow$	$\downarrow$
$5x$	$-$	7	$=$	0

An equation is $5x - 7 = 0$.

N2. (a) $3(x - 5) + 2x - 1$ is an *expression* because there is no equals symbol.

$$3(x - 5) + 2x - 1$$
$$= 3x - 15 + 2x - 1 \quad \textit{Distributive property}$$
$$= 5x - 16 \quad\quad\quad\ \textit{Combine like terms.}$$

(b) $3(x - 5) + 2x = 1$ is an *equation* because it has an equals symbol.

$$3(x - 5) + 2x = 1$$
$$3x - 15 + 2x = 1 \quad \textit{Distributive property}$$
$$5x - 15 = 1 \quad\quad \textit{Combine like terms.}$$
$$5x = 16 \quad\quad \textit{Add 15.}$$
$$x = \tfrac{16}{5} \quad\quad \textit{Divide by 5.}$$

The solution set is $\{\frac{16}{5}\}$.

N3. *Step 2*

The length and perimeter are given in terms of the width W. The length L is 2 ft more than twice the width, so

$$L = 2W + 2.$$

The perimeter P is 34, so

$$P = 34.$$

Step 3

Use the formula for perimeter of a rectangle.

$$P = 2L + 2W$$
$$34 = 2(2W + 2) + 2W \quad \textit{P = 34; L = 2W + 2}$$

Step 4

Solve the equation.

$$34 = 4W + 4 + 2W \quad \textit{Distributive property}$$
$$34 = 6W + 4 \quad\quad\quad \textit{Combine terms.}$$
$$30 = 6W \quad\quad\quad\quad \textit{Subtract 4.}$$
$$5 = W \quad\quad\quad\quad\quad \textit{Divide by 6.}$$

Step 5

The width is 5 and the length is

$$L = 2W + 2 = 10 + 2 = 12.$$

The rectangle is 5 ft by 12 ft.

Step 6

12 is 2 more than twice 5 and
$$P = 2(12) + 2(5) = 34.$$

N4. *Step 2*

Let x = the number of TDs for Warner.
Then $x + 4$ = the number of TDs for Brees.

Copyright © 2012 Pearson Education, Inc. Publishing as Addison-Wesley.

Step 3
The sum of their TDs is 64, so an equation is

$$x + (x + 4) = 64.$$

Step 4
Solve the equation.

$$2x + 4 = 64$$
$$2x = 60 \quad \textit{Subtract 4.}$$
$$x = 30 \quad \textit{Divide by 2.}$$

Step 5
Warner had 30 TDs and Brees had
$30 + 4 = 34$ TDs.

Step 6
34 is 4 more than 30, and the sum of 30 and 34 is
64.

N5.   *Step 2*
Let $x =$ the number of Introductory Statistics
students in the fall of 1992. Since
$700\% = 700(0.01) = 7$, $7x =$ the number of
additional students in 2009.

Step 3

The number in 1992	plus	the increase	is	96.
↓	↓	↓	↓	↓
x	$+$	$7x$	$=$	96

Step 4
Solve the equation.

$$1x + 7x = 96 \quad \textit{Identity property}$$
$$8x = 96 \quad \textit{Combine like terms.}$$
$$x = 12 \quad \textit{Divide by 8.}$$

Step 5
There were 12 students in the fall of 1992.

Step 6
Check that the increase, $96 - 12 = 84$, is 700% of
12. $700\% \cdot 12 = 700(0.01)(12) = 84$, as
required.

N6.   Let $x =$ the amount invested at 3%.
Then $20,000 - x =$ the amount invested at 2.5%.

Use $I = prt$ with $t = 1$.

Make a table to organize the information.

Principal	Rate (as a decimal)	Interest
x	0.03	$0.03x$
$20,000 - x$	0.025	$0.025(20,000 - x)$
20,000	← Totals →	575

The last column gives the equation.

$$0.03x + 0.025(20,000 - x) = 575$$
$$0.03x + 500 - 0.025x = 575 \quad \textit{Distributive property}$$
$$0.005x + 500 = 575 \quad \textit{Combine terms.}$$
$$0.005x = 75 \quad \textit{Subtract 500.}$$
$$x = 15,000 \quad \textit{Divide by 0.005.}$$

The man invested \$15,000 at 3% and
$20,000 - \$15,000 = \$5000$ at 2.5%.

N7.   Let $x =$ the number of liters of 20% acid solution
needed. Make a table.

Liters of Solution	Percent (as a decimal)	Liters of Pure Acid
5	0.30	$0.30(5) = 1.5$
x	0.20	$0.20x$
$x + 5$	0.24	$0.24(x + 5)$

Acid in 30% solution	+	acid in 20% solution	=	acid in 24% solution.
1.5	+	$0.20x$	=	$0.24(x + 5)$

$$1.5 + 0.20x = 0.24x + 1.2$$
$$\textit{Distributive property}$$
$$0.3 = 0.04x$$
$$\textit{Subtract 0.20x and 1.2.}$$
$$x = 7.5 \quad \textit{Divide by 0.04.}$$

$7\frac{1}{2}$ liters of the 20% solution are needed.

Check   30% of 5 is 1.5 and 20% of 7.5 is 1.5;
$1.5 + 1.5 = 3.0$, which is the same as 24% of
$(5 + 7.5)$.

N8.   Let $x =$ the number of gallons of pure antifreeze.

Number of Liters	Percent (as a decimal)	Liters of Pure Antifreeze
x	$100\% = 1$	$1x$
3	$30\% = 0.30$	$0.30(3)$
$x + 3$	$40\% = 0.40$	$0.40(x + 3)$

The last column gives the equation.

$$1x + 0.30(3) = 0.40(x + 3)$$
$$1.0x + 0.9 = 0.4x + 1.2$$
$$0.6x + 0.9 = 1.2 \quad \textit{Subtract 0.4x.}$$
$$0.6x = 0.3 \quad \textit{Subtract 0.9.}$$
$$x = 0.5 \quad \textit{Divide by 0.6.}$$

$\frac{1}{2}$ gallon of pure antifreeze is needed.

2.3   Section Exercises

1.   **(a)** 15 more than a number $\underline{x + 15}$

    (b) 15 is more than a number. $\underline{15 > x}$

3.   **(a)** 8 less than a number $\underline{x - 8}$

    (b) 8 is less than a number. $\underline{8 < x}$

Copyright © 2012 Pearson Education, Inc. Publishing as Addison-Wesley.

5.   40% can be written as

$0.40 = 0.4 = \frac{40}{100} = \frac{4}{10} = \frac{2}{5}$, so "40% of a

number" can be written as $0.40x$, $0.4x$, or $\frac{2x}{5}$. We

see that "40% of a number" cannot be written as

$40x$, choice **D**.

7.   Twice a number, decreased by 13 $\underline{2x - 13}$

9.   12 increased by four times a number $\underline{12 + 4x}$

11.   The product of 8 and 16 less than a number

   $\underline{8(x - 16)}$

13.   The quotient of three times a number and 10 $\underline{\frac{3x}{10}}$

15.   The sentence "the sum of a number and 6 is -31"

can be translated as

$$x + 6 = -31.$$
$$x = -37 \quad \textit{Subtract 6.}$$

The number is -37.

17.   The sentence "if the product of a number and -4

is subtracted from the number, the result is 9 more

than the number" can be translated as

$$x - (-4x) = x + 9.$$
$$x + 4x = x + 9$$
$$4x = 9$$
$$x = \tfrac{9}{4}$$

The number is $\frac{9}{4}$.

19.   The sentence "when $\frac{2}{3}$ of a number is subtracted

from 14, the result is 10" can be translated as

$$14 - \tfrac{2}{3}x = 10.$$
$$42 - 2x = 30 \quad \textit{Multiply by 3.}$$
$$-2x = -12 \quad \textit{Subtract 42.}$$
$$x = 6 \quad \textit{Divide by } -2.$$

The number is 6.

21. $5(x + 3) - 8(2x - 6)$ is an *expression* because

there is no equals symbol.

$$5(x + 3) - 8(2x - 6)$$
$$= 5x + 15 - 16x + 48 \quad \textit{Distributive property}$$
$$= -11x + 63 \quad\quad\quad \textit{Combine like terms.}$$

23. $5(x + 3) - 8(2x - 6) = 12$ has an equals symbol,

so this represents an *equation*.

$$5(x + 3) - 8(2x - 6) = 12$$
$$5x + 15 - 16x + 48 = 12 \quad \textit{Dist. prop.}$$
$$-11x + 63 = 12 \quad \textit{Combine terms.}$$
$$-11x = -51 \quad \textit{Subtract 63.}$$
$$x = \tfrac{51}{11} \quad \textit{Divide by } -11.$$

The solution set is $\left\{\frac{51}{11}\right\}$.

25. $\frac{1}{2}x - \frac{1}{6}x + \frac{3}{2} - 8$ is an *expression* because there is

no equals symbol.

$$\frac{1}{2}x - \frac{1}{6}x + \frac{3}{2} - 8$$
$$= \tfrac{3}{6}x - \tfrac{1}{6}x + \tfrac{3}{2} - \tfrac{16}{2} \quad \textit{Common denominators}$$
$$= \tfrac{2}{6}x - \tfrac{13}{2} \quad\quad\quad \textit{Combine like terms.}$$
$$= \tfrac{1}{3}x - \tfrac{13}{2} \quad\quad\quad \textit{Reduce.}$$

27.   *Step 1*

We are asked to find <u>the number of patents each

corporation secured</u> .

Step 2

Let $x =$ the number of patents IBM secured.

Then $x - 667 =$ the number of <u>patents Samsung

secured</u> .

Step 3

A total of 7671 patents were secured, so

$$\underline{x} + \underline{x - 667} = 7671.$$

Step 4 $2x - 667 = 7671$
$$2x = 8338$$
$$x = \underline{4169}$$

Step 5

IBM secured <u>4169</u> patents and Samsung secured

$4169 - 667 = \underline{3502}$ patents.

Step 6

The number of Samsung patents was <u>667</u> fewer

than the number of <u>IBM patents</u> and the total

number of patents was $4169 + \underline{3502} = \underline{7671}$.

29.   *Step 2*

Let $W =$ the width of the base. Then $2W - 65$ is

the length of the base.

Step 3

The perimeter of the base is 860 feet. Using

$P = 2L + 2W$ gives us

$$2(2W - 65) + 2W = 860.$$

Step 4

$$4W - 130 + 2W = 860$$
$$6W - 130 = 860$$
$$6W = 990$$
$$W = \tfrac{990}{6} = 165$$

Step 5

The width of the base is 165 feet and the length of

the base is $2(165) - 65 = 265$ feet.

Step 6

$2L + 2W = 2(265) + 2(165) = 530 + 330 =$

860, which is the perimeter of the base.

Copyright © 2012 Pearson Education, Inc. Publishing as Addison-Wesley.

31. *Step 2*
Let x = the width of the painting. Then $x + 5.54$ = the height of the painting.

Step 3
The perimeter of the painting is 108.44 inches. Using $P = 2L + 2W$ gives us

$$2(x + 5.54) + 2x = 108.44.$$

Step 4

$$2x + 11.08 + 2x = 108.44$$
$$4x + 11.08 = 108.44$$
$$4x = 97.36$$
$$x = 24.34$$

Step 5
The width of the painting is 24.34 inches and the height is $24.34 + 5.54 = 29.88$ inches.

Step 6
29.88 is 5.54 more than 24.34 and $2(29.88) + 2(24.34) = 108.44$, as required.

33. *Step 2*
Let x = the length of the middle side. Then the shortest side is $x - 75$ and the longest side is $x + 375$.

Step 3
The perimeter of the Bermuda Triangle is 3075 miles. Using $P = a + b + c$ gives us

$$x + (x - 75) + (x + 375) = 3075.$$

Step 4

$$3x + 300 = 3075$$
$$3x = 2775 \quad \text{Subtract 300.}$$
$$x = 925 \quad \text{Divide by 3.}$$

Step 5
The length of the middle side is 925 miles. The length of the shortest side is $x - 75 = 925 - 75 = 850$ miles. The length of the longest side is $x + 375 = 925 + 375 = 1300$ miles.

Step 6
The answer checks since $925 + 850 + 1300 = 3075$ miles, which is the correct perimeter.

35. *Step 2*
Let x = the amount of revenue for Exxon Mobil. Then $x - 37.3$ is the amount of revenue for Wal-Mart (in billions).

Step 3
The total revenue was $848.5 billion, so

$$x + (x - 37.3) = 848.5.$$

Step 4

$$2x - 37.3 = 848.5$$
$$2x = 885.8 \quad \text{Add 37.3.}$$
$$x = 442.9 \quad \text{Divide by 2.}$$

Step 5
The amount of revenue for Exxon Mobil was $442.9 billion. The amount of revenue for Wal-Mart was $x - 37.3 = 442.9 - 37.3 = \$405.6$ billion.

Step 6
The answer checks since $442.9 + 405.6 = \$848.5$ billion, which is the correct total revenue.

37. *Step 2*
Let x = the height of the Eiffel Tower. Then $x - 880$ = the height of the Leaning Tower of Pisa.

Step 3
Together these heights are 1246 ft, so

$$x + (x - 880) = 1246.$$

Step 4

$$2x - 880 = 1246$$
$$2x = 2126$$
$$x = 1063$$

Step 5
The height of the Eiffel Tower is 1063 feet and the height of the Leaning Tower of Pisa is $1063 - 880 = 183$ feet.

Step 6
183 feet is 880 feet shorter than 1063 feet and the sum of 183 feet and 1063 feet is 1246 feet.

39. *Step 2*
Let x = number of electoral votes for McCain. Then $x + 192$ = number of electoral votes for Obama.

Step 3
There were 538 total electoral votes, so

$$x + (x + 192) = 538.$$

Step 4

$$2x + 192 = 538$$
$$2x = 346$$
$$x = 173$$

Step 5
McCain received 173 electoral votes, so Obama received $173 + 192 = 365$ electoral votes.

Step 6
365 is 192 more than 173 and the total is $173 + 365 = 538$.

41. Let x = the percent increase.

$$x = \frac{\text{amount of increase}}{\text{base number}}$$
$$= \frac{1,480,469 - 1,065,138}{1,065,138} = \frac{415,331}{1,065,138}$$
$$\approx 0.390 = 39.0\%$$

The percent increase was approximately 39.0%.

Copyright © 2012 Pearson Education, Inc. Publishing as Addison-Wesley.

43. Let x = the approximate cost in 2009.
Since x is 150% more than the 1995 cost,

$$x = 2811 + 1.5(2811)$$
$$= 2811 + 4216.5$$
$$= 7027.5.$$

To the nearest dollar, the cost was $7028.

45. Let x = the 2008 cost. Then

$$x - 3.8\%(x) = 42.91.$$
$$x - 3.8(0.01)(x) = 42.91$$
$$1x - 0.038x = 42.91$$
$$0.962x = 42.91$$
$$x = \tfrac{42.91}{0.962} \approx 44.60$$

The 2008 cost was $44.60.

47. Let x = the amount of the receipts excluding tax.
Since the sales tax is 9% of x, the total amount is

$$x + 0.09x = 2725$$
$$1x + 0.09x = 2725$$
$$1.09x = 2725$$
$$x = \tfrac{2725}{1.09} = 2500$$

Thus, the tax was $0.09(2500) = \$225$.

49. Let x = the amount invested at 3%. Then
$12,000 - x$ = the amount invested at 4%.
Complete the table. Use $I = prt$ with $t = 1$.

Principal	Rate (as a decimal)	Interest
x	0.03	$0.03x$
$12,000 - x$	0.04	$0.04(12,000 - x)$
12,000	← Totals →	440

The last column gives the equation.

$$\begin{array}{ccc} \text{Interest} & \text{interest} & \text{total} \\ \text{at 3\%} + & \text{at 4\%} = & \text{interest.} \end{array}$$
$$0.03x + 0.04(12,000 - x) = 440$$

$$3x + 4(12,000 - x) = 44,000 \quad \textit{Multiply by 100.}$$
$$3x + 48,000 - 4x = 44,000$$
$$-x = -4000$$
$$x = 4000$$

He should invest $4000 at 3% and
$12,000 - 4000 = \$8000$ at 4%.

Check $4000 @ 3% = $120 and
$8000 @ 4% = $320; $120 + $320 = $440.

51. Let x = the amount invested at 4.5%. Then
$2x - 1000$ = the amount invested at 3%.
Use $I = prt$ with $t = 1$. Make a table.

Principal	Rate (as a Decimal)	Interest
x	0.045	$0.045x$
$2x - 1000$	0.03	$0.03(2x - 1000)$
	Total →	1020

The last column gives the equation.

$$\begin{array}{ccc} \text{Interest} & \text{interest} & \text{total} \\ \text{at 4.5\%} + & \text{at 3\%} = & \text{interest.} \end{array}$$
$$0.045x + 0.03(2x - 1000) = 1020$$

$$45x + 30(2x - 1000) = 1,020,000 \quad \textit{Multiply by 1000.}$$
$$45x + 60x - 30,000 = 1,020,000$$
$$105x = 1,050,000$$
$$x = \frac{1,050,000}{105} = 10,000$$

She invested $10,000 at 4.5% and
$2x - 1000 = 2(10,000) - 1000 = \$19,000$ at 3%.

Check $19,000 is $1000 less than two times
$10,000. $10,000 @ 4.5% = $450 and
$19,000 @ 3% = $570; $450 + $570 = $1020.

53. Let x = the amount of additional money to be
invested at 3%.
Use $I = prt$ with $t = 1$. Make a table.
Use the fact that the total return on the two
investments is 4%.

Principal	Rate (as a decimal)	Interest
12,000	0.06	$0.06(12,000)$
x	0.03	$0.03x$
$12,000 + x$	0.04	$0.04(12,000 + x)$

The last column gives the equation.

$$\begin{array}{ccc} \text{Interest} & \text{interest} & \text{interest} \\ \text{at 6\%} + & \text{at 3\%} = & \text{at 4\%} . \end{array}$$
$$0.06(12,000) + 0.03x = 0.04(12,000 + x)$$

$$6(12,000) + 3x = 4(12,000 + x) \quad \textit{Multiply by 100.}$$
$$72,000 + 3x = 48,000 + 4x$$
$$24,000 = x$$

He should invest $24,000 at 3%.

Check $12,000 @ 6% = $720 and
$24,000 @ 3% = $720;
$720 + $720 = $1440, which is the same as
($12,000 + $24,000) @ 4%.

55. Let x = the number of liters of 10% acid solution
needed. Make a table.

Liters of Solution	Percent (as a decimal)	Liters of Pure Acid
10	0.04	$0.04(10) = 0.4$
x	0.10	$0.10x$
$x + 10$	0.06	$0.06(x + 10)$

Copyright © 2012 Pearson Education, Inc. Publishing as Addison-Wesley.

Acid in 4% solution $+$ acid in 10% solution $=$ acid in 6% solution.

$$0.4 + 0.10x = 0.06(x + 10)$$

$$0.4 + 0.10x = 0.06x + 0.6$$

Distributive property

$$0.04x = 0.2$$

Subtract 0.06x and 0.4.

$$x = 5 \qquad Divide\ by\ 0.04.$$

Five liters of the 10% solution are needed.

Check 4% of 10 is 0.4 and 10% of 5 is 0.5; $0.4 + 0.5 = 0.9$, which is the same as 6% of $(10 + 5)$.

57. Let x = the number of liters of the 20% alcohol solution. Make a table.

Liters of Solution	Percent (as a decimal)	Liters of Pure Alcohol
12	0.12	$0.12(12) = 1.44$
x	0.20	$0.20x$
$x + 12$	0.14	$0.14(x + 12)$

Alcohol in 12% solution $+$ alcohol in 20% solution $=$ alcohol in 14% solution.

$$1.44 + 0.20x = 0.14(x + 12)$$

$$144 + 20x = 14(x + 12) \quad Multiply\ by\ 100.$$
$$144 + 20x = 14x + 168 \quad Distributive\ property$$
$$6x = 24 \qquad Subtract\ 14x\ and\ 144.$$
$$x = 4 \qquad Divide\ by\ 6.$$

4L of 20% alcohol solution are needed.

Check 12% of 12 is 1.44 and 20% of 4 is 0.8; $1.44 + 0.8 = 2.24$, which is the same as 14% of $(12 + 4)$.

59. Let x = the amount of pure dye used (pure dye is 100% dye). Make a table.

Gallons of Solution	Percent (as a decimal)	Gallons of Pure Dye
x	1	$1x = x$
4	0.25	$0.25(4) = 1$
$x + 4$	0.40	$0.40(x + 4)$

Write the equation from the last column in the table.

$$x + 1 = 0.4(x + 4)$$
$$x + 1 = 0.4x + 1.6 \quad Distributive\ property$$
$$0.6x = 0.6 \qquad Subtract\ 0.4x\ and\ 1.$$
$$x = 1 \qquad Divide\ by\ 0.6.$$

One gallon of pure (100%) dye is needed.

Check 100% of 1 is 1 and 25% of 4 is 1; $1 + 1 = 2$, which is the same as 40% of $(1 + 4)$.

61. Let x = the amount of $6 per lb nuts. Make a table.

Pounds of nuts	Cost per lb	Total Cost
50	$2	$2(50) = 100$
x	$6	$6x$
$x + 50$	$5	$5(x + 50)$

The total value of the $2 per lb nuts and the $6 per lb nuts must equal the value of the $5 per lb nuts.

$$100 + 6x = 5(x + 50)$$
$$100 + 6x = 5x + 250$$
$$x = 150$$

He should use 150 lb of $6 nuts.

Check 50 pounds of the $2 per lb nuts are worth $100 and 150 pounds of the $6 per lb nuts are worth $900; $100 + $900 = 1000, which is the same as $(50 + 150)$ pounds worth $5 per lb.

63. We cannot expect the final mixture to be worth more than the more expensive of the two ingredients. Answers will vary.

♦ ♦ ♦ Relating Concepts 65–68 ♦ ♦ ♦

65. **(a)** Let x = the amount invested at 5%.
$800 - x$ = the amount invested at 10%.

(b) Let y = the amount of 5% acid used.
$800 - y$ = the amount of 10% acid used.

66. Organize the information in a table.

(a)

Principal	Percent (as a decimal)	Interest
x	0.05	$0.05x$
$800 - x$	0.10	$0.10(800 - x)$
800	0.0875	$0.0875(800)$

The amount of interest earned at 5% and 10% is found in the last column of the table, $0.05x$ and $0.10(800 - x)$.

(b)

Liters of Solution	Percent (as a decimal)	Liters of Pure Acid
y	0.05	$0.05y$
$800 - y$	0.10	$0.10(800 - y)$
800	0.0875	$0.0875(800)$

The amount of pure acid in the 5% and 10% mixtures is found in the last column of the table, $0.05y$ and $0.10(800 - y)$.

67. Refer to the tables for Exercise 66. In each case, the last column gives the equation.

(a) $0.05x + 0.10(800 - x) = 0.0875(800)$

(b) $0.05y + 0.10(800 - y) = 0.0875(800)$

Copyright © 2012 Pearson Education, Inc. Publishing as Addison-Wesley.

68. In both cases, multiply by 10,000 to clear the decimals.

(a)
$$0.05x + 0.10(800 - x) = 0.0875(800)$$
$$500x + 1000(800 - x) = 875(800)$$
$$500x + 800,000 - 1000x = 700,000$$
$$-500x = -100,000$$
$$x = 200$$

Jack invested $200 at 5% and
$800 - x = 800 - 200 = \$600$ at 10%.

(b)
$$0.05y + 0.10(800 - y) = 0.0875(800)$$
$$500y + 1000(800 - y) = 875(800)$$
$$500y + 800,000 - 1000y = 700,000$$
$$-500y = -100,000$$
$$y = 200$$

Jill used 200 L of 5% acid solution and
$800 - y = 800 - 200 = 600$ L of 10% acid solution.

(c) The processes used to solve Problems A and B were virtually the same. Aside from the variables chosen, the problem information was organized in similar tables and the equations solved were the same.

69. $d = rt; \quad r = 50, t = 4$
$$d = 50(4) = 200$$

71. $P = a + b + c; \quad b = 13, c = 14, P = 46$
$$46 = a + 13 + 14$$
$$46 = a + 27$$
$$19 = a$$

2.4 Further Applications of Linear Equations

2.4 Now Try Exercises

N1. Let x = the number of dimes.
Then $52 - x$ = the number of nickels.

Number of Coins	Denomination	Value
x	0.10	$0.10x$
$52 - x$	0.05	$0.05(52 - x)$
	Total →	3.70

Multiply the number of coins by the denominations, and add the results to get 3.70.

$$0.10x + 0.05(52 - x) = 3.70$$
$$10x + 5(52 - x) = 370 \quad \textit{Multiply by 100.}$$
$$10x + 260 - 5x = 370$$
$$5x = 110$$
$$x = 22$$

He has 22 dimes and $52 - 22 = 30$ nickels.

Check The number of coins is $22 + 30 = 52$ and the value of the coins is
$\$0.10(22) + \$0.05(30) = \$3.70$, as required.

N2. Let x = the amount of time needed for the trains to be 387.5 km apart.

Make a table. Use the formula $d = rt$, that is, find each distance by multiplying rate by time.

	Rate	Time	Distance
First Train	80	x	$80x$
Second Train	75	x	$75x$
Total			387.5

The total distance traveled is the sum of the distances traveled by each train, since they are traveling in opposite directions. This total is 387.5 km.

$$80x + 75x = 387.5$$
$$155x = 387.5$$
$$x = \tfrac{387.5}{155} = 2.5$$

The trains will be 387.5 km apart in 2.5 hr.

Check The first train traveled $80(2.5) = 200$ km. The second train traveled $75(2.5) = 187.5$ km, for a total of $200 + 187.5 = 387.5$, as required.

N3. Let x = the driving rate.
Then $x - 30$ = the bicycling rate.

Make a table. Use the formula $d = rt$, that is, find each distance by multiplying rate by time.

	Rate	Time	Distance
Car	x	$\frac{1}{2}$	$\frac{1}{2}x$
Bike	$x - 30$	$1\frac{1}{2} = \frac{3}{2}$	$\frac{3}{2}(x - 30)$

The distances are equal.

$$\tfrac{1}{2}x = \tfrac{3}{2}(x - 30)$$
$$1x = 3(x - 30) \quad \textit{Multiply by 2.}$$
$$x = 3x - 90$$
$$90 = 2x$$
$$45 = x$$

The distance he travels to work is
$$\tfrac{1}{2}x = \tfrac{1}{2}(45) = 22.5 \text{ miles.}$$

Check The distance he travels to work by bike is $\frac{3}{2}(x - 30) = \frac{3}{2}(45 - 30) = \frac{3}{2}(15) = 22.5$ miles, which is the same as we found above (by car).

N4. The sum of the three measures must equal 180°.

$$x + (x + 11) + (3x - 36) = 180$$
$$5x - 25 = 180$$
$$5x = 205$$
$$x = 41$$

Copyright © 2012 Pearson Education, Inc. Publishing as Addison-Wesley.

The angles measure $41°$, $41 + 11 = 52°$, and $3(41) - 36 = 87°$.

Check Since $41° + 52° + 87° = 180°$, the answers are correct.

2.4 Section Exercises

1. The total amount is
$$14(0.10) + 16(0.25) = 1.40 + 4.00$$
$$= \$5.40.$$

3. Use $d = rt$, or $r = \frac{d}{t}$. Substitute 300 for d and 10 for t.
$$r = \frac{300}{10} = 30$$
His rate was 30 mph.

5. The problem asks for the distance Jeff travels to the workplace, so we must multiply the rate, 10 mph, by the time, $\frac{3}{4}$ hr, to get the distance, 7.5 mi.

7. No, the answers must be whole numbers because they represent the number of coins.

9. Let $x =$ the number of pennies. Then x is also the number of dimes, and $44 - 2x$ is the number of quarters.

Number of Coins	Denomination	Value
x	0.01	$0.01x$
x	0.10	$0.10x$
$44 - 2x$	0.25	$0.25(44 - 2x)$
44	← Totals →	4.37

The sum of the values must equal the total value.
$$0.01x + 0.10x + 0.25(44 - 2x) = 4.37$$
$$x + 10x + 25(44 - 2x) = 437$$
Multiply both sides by 100.
$$x + 10x + 1100 - 50x = 437$$
$$-39x + 1100 = 437$$
$$-39x = -663$$
$$x = 17$$

There are 17 pennies, 17 dimes, and $44 - 2(17) = 10$ quarters.

Check The number of coins is $17 + 17 + 10 = 44$ and the value of the coins is $\$0.01(17) + \$0.10(17) + \$0.25(10) = \$4.37$, as required.

11. Let $x =$ the number of loonies. Then $37 - x$ is the number of toonies.

Number of Coins	Denomination	Value
x	1	$1x$
$37 - x$	2	$2(37 - x)$
37	← Totals →	51

The sum of the values must equal the total value.
$$1x + 2(37 - x) = 51$$
$$x + 74 - 2x = 51$$
$$-x + 74 = 51$$
$$23 = x$$

She has 23 loonies and $37 - 23 = 14$ toonies.

Check The number of coins is $23 + 14 = 37$ and the value of the coins is $\$1(23) + \$2(14) = \$51$, as required.

13. Let $x =$ the number of \$10 coins. Then $41 - x$ is the number of \$20 coins.

Number of Coins	Denomination	Value
x	10	$10x$
$41 - x$	20	$20(41 - x)$
41	← Totals →	540

The sum of the values must equal the total value.
$$10x + 20(41 - x) = 540$$
$$10x + 820 - 20x = 540$$
$$-10x = -280$$
$$x = 28$$

He has 28 \$10 coins and $41 - 28 = 13$ \$20 coins.

Check The number of coins is $28 + 13 = 41$ and the value of the coins is $\$10(28) + \$20(13) = \$540$, as required.

15. Let $x =$ the number of adult tickets sold. Then $1460 - x =$ the number of children and senior tickets sold.

Cost of Ticket	Number Sold	Amount Collected
\$18	x	$18x$
\$12	$1460 - x$	$12(1460 - x)$
Totals	1460	\$22,752

Write the equation from the last column of the table.
$$18x + 12(1460 - x) = 22{,}752$$
$$18x + 17{,}520 - 12x = 22{,}752$$
$$6x = 5232$$
$$x = 872$$

872 adult tickets were sold; $1460 - 872 = 588$ children and senior tickets were sold.

Check The number of tickets sold was $872 + 588 = 1460$ and the amount collected was $\$18(872) + \$12(588) = \$15{,}696 + \$7056 = \$22{,}752$, as required.

Copyright © 2012 Pearson Education, Inc. Publishing as Addison-Wesley.

17. $d = rt$, so

$$r = \frac{d}{t} = \frac{100}{12.54} \approx 7.97$$

Her rate was about 7.97 m/sec.

19. $d = rt$, so

$$r = \frac{d}{t} = \frac{400}{47.25} \approx 8.47$$

His rate was about 8.47 m/sec.

21. Let t = the time until they are 110 mi apart. Use the formula $d = rt$. Complete the table.

	Rate	Time	Distance
First Steamer	22	t	$22t$
Second Steamer	22	t	$22t$
			110

The total distance traveled is the sum of the distances traveled by each steamer, since they are traveling in opposite directions. This total is 110 mi.

$$22t + 22t = 110$$
$$44t = 110$$
$$t = \frac{110}{44} = \frac{5}{2} \quad \text{or} \quad 2\frac{1}{2}$$

It will take them $2\frac{1}{2}$ hr.

Check Each steamer traveled $22(2.5) = 55$ miles for a total of $2(55) = 110$ miles, as required.

23. Let t = Mulder's time. Then $t - \frac{1}{2}$ = Scully's time.

	Rate	Time	Distance
Mulder	65	t	$65t$
Scully	68	$t - \frac{1}{2}$	$68(t - \frac{1}{2})$

The distances are equal.

$$65t = 68(t - \frac{1}{2})$$
$$65t = 68t - 34$$
$$-3t = -34$$
$$t = \frac{34}{3} \quad \text{or} \quad 11\frac{1}{3}$$

Mulder's time will be $11\frac{1}{3}$ hr. Since he left at 8:30 A.M., $11\frac{1}{3}$ hr or 11 hr 20 min later is 7:50 P.M.

Check Mulder's distance was $65(\frac{34}{3}) = 736\frac{2}{3}$ miles. Scully's distance was $68(\frac{34}{3} - \frac{1}{2}) = 68(\frac{65}{6}) = 736\frac{2}{3}$, as required.

25. Let x = her average rate on Sunday. Then $x + 5$ = her average rate on Saturday.

	Rate	Time	Distance
Saturday	$x + 5$	3.6	$3.6(x + 5)$
Sunday	x	4	$4x$

The distances are equal.

$$3.6(x + 5) = 4x$$
$$3.6x + 18 = 4x$$
$$18 = 0.4x \qquad \textit{Subtract 3.6x.}$$
$$x = \frac{18}{0.4} = 45$$

Her average rate on Sunday was 45 mph.

Check On Sunday, 4 hours @ 45 mph = 180 miles. On Saturday, 3.6 hours @ 50 mph = 180 miles. The distances are equal.

27. Let x = Anne's time. Then $x + \frac{1}{2}$ = Johnny's time.

	Rate	Time	Distance
Anne	60	x	$60x$
Johnny	50	$x + \frac{1}{2}$	$50(x + \frac{1}{2})$

The total distance is 80.

$$60x + 50(x + \frac{1}{2}) = 80$$
$$60x + 50x + 25 = 80$$
$$110x = 55$$
$$x = \frac{55}{110} = \frac{1}{2}$$

They will meet $\frac{1}{2}$ hr after Anne leaves.

Check Anne travels $60(\frac{1}{2}) = 30$ miles. Johnny travels $50(\frac{1}{2} + \frac{1}{2}) = 50$ miles. The sum of the distances is 80 miles, as required.

29. The sum of the measures of the three angles of a triangle is 180°.

$$(x - 30) + (2x - 120) + (\frac{1}{2}x + 15) = 180$$
$$\frac{7}{2}x - 135 = 180$$
$$7x - 270 = 360 \qquad \textit{Multiply by 2.}$$
$$7x = 630$$
$$x = 90$$

With $x = 90$, the three angle measures become

$$(90 - 30)° = 60°,$$
$$[2(90) - 120]° = 60°,$$
$$\text{and} \quad [\frac{1}{2}(90) + 15]° = 60°.$$

31. The sum of the measures of the three angles of a triangle is 180°.

$$(3x + 7) + (9x - 4) + (4x + 1) = 180$$
$$16x + 4 = 180$$
$$16x = 176$$
$$x = 11$$

With $x = 11$, the three angle measures become

$$(3 \cdot 11 + 7)° = 40°,$$
$$(9 \cdot 11 - 4)° = 95°,$$
$$\text{and} \quad (4 \cdot 11 + 1)° = 45°.$$

Copyright © 2012 Pearson Education, Inc. Publishing as Addison-Wesley.

♦ ♦ ♦ Relating Concepts 33–36 ♦ ♦ ♦

33. The sum of the measures of the angles of a triangle is 180°.

$$x + 2x + 60 = 180$$
$$3x + 60 = 180$$
$$3x = 120$$
$$x = 40$$

The measures of the unknown angles are 40° and $2x = 80°$.

34. Two angles which form a straight line add to 180°, so $180° - 60° = 120°$. The measure of the unknown angle is 120°.

35. The sum of the measures of the unknown angles in Exercise 33 is $40° + 80° = 120°$. This is equal to the measure of the angle in Exercise 34.

36. The sum of the measures of angles ① and ② is equal to the measure of angle ③ .

37. Vertical angles have equal measure.

$$8x + 2 = 7x + 17$$
$$x = 15$$
$$8 \cdot 15 + 2 = 122 \quad \text{and} \quad 7 \cdot 15 + 17 = 122.$$

The angles are both 122°.

39. The sum of the two angles is 90°.

$$(5x - 1) + 2x = 90$$
$$7x - 1 = 90$$
$$7x = 91$$
$$x = 13$$

The measures of the two angles are $[5(13) - 1]° = 64°$ and $[2(13)]° = 26°$.

41. Let $x =$ the first consecutive integer. Then $x + 1$ will be the second consecutive integer, and $x + 2$ will be the third consecutive integer.

The sum of the first and twice the second is 17 more than twice the third.

$$x + 2(x + 1) = 2(x + 2) + 17$$
$$x + 2x + 2 = 2x + 4 + 17$$
$$3x + 2 = 2x + 21$$
$$x = 19$$

Since $x = 19$, $x + 1 = 20$, and $x + 2 = 21$. The three consecutive integers are 19, 20, and 21.

43. Let $x =$ my current age. Then $x + 1$ will be my age next year.
The sum of these ages will be 103 years.

$$x + (x + 1) = 103$$
$$2x + 1 = 103$$
$$2x = 102$$
$$x = 51$$

If my current age is 51, in 10 years I will be

$$51 + 10 = 61 \text{ years old.}$$

45. Let $x =$ the least even integer. Then $x + 2$ and $x + 4$ are the next two even consecutive integers. The sum of the least integer and middle integer is 26 more than the greatest integer.

$$x + (x + 2) = (x + 4) + 26$$
$$2x + 2 = x + 30$$
$$x = 28$$

The three consecutive even integers are 28, 30, and 32.

47. Let $x =$ the least odd integer. Then $x + 2$ and $x + 4$ are the next two odd consecutive integers. The sum of the least integer and middle integer is 19 more than the greatest integer.

$$x + (x + 2) = (x + 4) + 19$$
$$2x + 2 = x + 23$$
$$x = 21$$

The three consecutive odd integers are 21, 23, and 25.

49. $(4, \infty)$ is equivalent to $x > 4$.

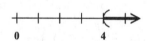

51. $(-2, 6)$ is equivalent to $-2 < x < 6$.

Summary Exercises on Solving Applied Problems

1. Let $x =$ the width of the rectangle. Then $x + 3$ is the length of the rectangle.

If the length were decreased by 2 inches and the width were increased by 1 inch, the perimeter would be 24 inches. Use the formula $P = 2L + 2W$, and substitute 24 for P, $(x + 3) - 2$ or $x + 1$ for L, and $x + 1$ for W.

$$P = 2L + 2W$$
$$24 = 2(x + 1) + 2(x + 1)$$
$$24 = 2x + 2 + 2x + 2$$
$$24 = 4x + 4$$
$$20 = 4x$$
$$5 = x$$

The width of the rectangle is 5 inches, and the length is $5 + 3 = 8$ inches.

Copyright © 2012 Pearson Education, Inc. Publishing as Addison-Wesley.

3. Let $x =$ the regular price of the item.
The sale price after a 46% (or 0.46) discount was $46.97, so an equation is

$$x - 0.46x = 46.97.$$
$$0.54x = 46.97$$
$$x = \frac{46.97}{0.54} \approx 86.98$$

To the nearest cent, the regular price was $86.98.

5. Let $x =$ the amount invested at 4%.
Then $2x$ is the amount invested at 5%.
Use $I = prt$ with $t = 1$ yr. Make a table.

Principal	Rate (as a decimal)	Interest
x	0.04	$0.04x$
$2x$	0.05	$0.05(2x) = 0.10x$
	Total →	112

The last column gives the equation.

$$\begin{array}{ccc} \text{Interest} & \text{interest} & \text{total} \\ \text{at 4\%} & \text{at 5\%} & \text{interest.} \end{array}$$
$$0.04x + 0.10x = 112$$

$$4x + 10x = 11{,}200 \quad \textit{Multiply by 100.}$$
$$14x = 11{,}200$$
$$x = 800$$

$800 is invested at 4% and 2($800) = $1600 at 5%.

Check $800 @ 4% = $32 and
$1600 @ 5% = $80; $32 + $80 = $112

7. Let $x =$ the number of points scored by Wade in the 2008–2009 season. Then $x - 136 =$ the number of points scored by James in the 2007–2008 season. The total number of points scored by both was 4636.

$$x + (x - 136) = 4636$$
$$2x - 136 = 4636$$
$$2x = 4772$$
$$x = 2386$$

Wade scored 2386 points and James scored $2386 - 136 = 2250$ points.

9. Let $t =$ the time it will take until John and Pat meet. Use $d = rt$ and make a table.

	Rate	Time	Distance
John	60	t	$60t$
Pat	28	t	$28t$

The total distance is 440 miles.

$$60t + 28t = 440$$
$$88t = 440$$
$$t = 5$$

It will take 5 hours for John and Pat to meet.

Check John traveled $60(5) = 300$ miles and Pat traveled $28(5) = 140$ miles; $300 + 140 = 440$, as required.

11. Let $x =$ the number of liters of the 5% drug solution.

Liters of Solution	Percent (as a decimal)	Liters of Pure Drug
20	0.10	$20(0.10) = 2$
x	0.05	$0.05x$
$20 + x$	0.08	$0.08(20 + x)$

$$\begin{array}{ccc} \text{Drug} & \text{drug} & \text{drug} \\ \text{in 10\%} + \text{in 5\%} & = & \text{in 8\%.} \end{array}$$
$$2 + 0.05x = 0.08(20 + x)$$

$$200 + 5x = 8(20 + x) \quad \textit{Multiply by 100.}$$
$$200 + 5x = 160 + 8x$$
$$40 = 3x$$
$$x = \frac{40}{3} \text{ or } 13\frac{1}{3}$$

The pharmacist should add $13\frac{1}{3}$ L.

Check 10% of 20 is 2 and 5% of $\frac{40}{3}$ is $\frac{2}{3}$; $2 + \frac{2}{3} = \frac{8}{3}$, which is the same as 8% of $(20 + \frac{40}{3})$.

13. Let $x =$ the number of $5 bills. Then $126 - x$ is the number of $10 bills.

Number of Bills	Denomination	Value
x	5	$5x$
$126 - x$	10	$10(126 - x)$
126	← Totals →	840

The sum of the values must equal the total value.

$$5x + 10(126 - x) = 840$$
$$5x + 1260 - 10x = 840$$
$$-5x = -420$$
$$x = 84$$

There are 84 $5 bills and $126 - 84 = 42$ $10 bills.

Check The number of bills is $84 + 42 = 126$ and the value of the bills is $5(84) + $10(42) = $840, as required.

15. The sum of the measures of the three angles of a triangle is 180°.

$$x + (6x - 50) + (x - 10) = 180$$
$$8x - 60 = 180$$
$$8x = 240$$
$$x = 30$$

With $x = 30$, the three angle measures become

$$(6 \cdot 30 - 50)° = 130°,$$
$$(30 - 10)° = 20°, \text{ and } 30°.$$

Copyright © 2012 Pearson Education, Inc. Publishing as Addison-Wesley.

17. Let $x =$ the least integer. Then $x + 1$ is the middle integer and $x + 2$ is the greatest integer.

"The sum of the least and greatest of three consecutive integers is 32 more than the middle integer" translates to

$$x + (x + 2) = 32 + (x + 1).$$
$$2x + 2 = x + 33$$
$$x = 31$$

The three consecutive integers are 31, 32, and 33.

Check The sum of the least and greatest integers is $31 + 33 = 64$, which is the same as 32 more than the middle integer.

19. Let $x =$ the length of the shortest side. Then $2x$ is the length of the middle side and $3x - 2$ is the length of the longest side.

The perimeter is 34 inches. Using $P = a + b + c$ gives us

$$x + 2x + (3x - 2) = 34.$$
$$6x - 2 = 34$$
$$6x = 36$$
$$x = 6$$

The lengths of the three sides are 6 inches, $2(6) = 12$ inches, and $3(6) - 2 = 16$ inches.

Check The sum of the lengths of the three sides is $6 + 12 + 16 = 34$ inches, as required.

2.5 Linear Inequalities in One Variable

2.5 Now Try Exercises

N1. $x - 10 > -7$
$\qquad x > 3 \qquad$ *Add 10.*

Check Substitute 3 for x in $x - 10 = -7$.

$$3 - 10 \overset{?}{=} -7$$
$$-7 = -7 \quad \textit{True}$$

This shows that 3 is the boundary point. Choose 0 and 7 as test points.

$$x - 10 > -7$$

Let x = 0.	*Let x = 7.*
$0 - 10 \overset{?}{>} -7$	$7 - 10 \overset{?}{>} -7$
$-10 > -7$ *False*	$-3 > -7$ *True*
0 is not in the	7 is in the
solution set.	solution set.

The check confirms that $(3, \infty)$ is the solution set.

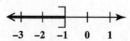

N2. $4x + 1 \geq 5x$
$\qquad 1 \geq x \qquad$ *Subtract 4x.*
$\qquad x \leq 1 \qquad$ *Equivalent inequality.*

Check Substitute 1 for x in $4x + 1 = 5x$.

$$4(1) + 1 \overset{?}{=} 5(1)$$
$$5 = 5 \qquad \textit{True}$$

So 1 satisfies the equality part of $\geq$. Choose 0 and 2 as test points.

$$4x + 1 \geq 5x$$

Let x = 0.

$$4(0) + 1 \overset{?}{\geq} 5(0)$$
$$1 \geq 0 \qquad\qquad \textit{True}$$

0 is in the solution set.

Let x = 2.

$$4(2) + 1 \overset{?}{\geq} 5(2)$$
$$9 \geq 10 \qquad\qquad \textit{False}$$

2 is not in the solution set.

The check confirms that $(-\infty, 1]$ is the solution set.

N3. (a) $8x \geq -40$
$\qquad \dfrac{8x}{8} \geq \dfrac{-40}{8} \qquad$ *Divide by 8 > 0; do not reverse the symbol.*
$\qquad x \geq -5$

Check that the solution set is the interval $[-5, \infty)$.

(b) $-20x > -60$
$\qquad \dfrac{-20x}{-20} < \dfrac{-60}{-20} \qquad$ *Divide by −20 < 0; reverse the symbol.*
$\qquad x < 3$

Check that the solution set is the interval $(-\infty, 3)$.

N4. $5 - 2(x - 4) \leq 11 - 4x$
$\qquad 5 - 2x + 8 \leq 11 - 4x$
$\qquad -2x + 13 \leq 11 - 4x$
$\qquad 2x + 13 \leq 11$
$\qquad 2x \leq -2$
$\qquad \dfrac{2x}{2} \leq \dfrac{-2}{2}$
$\qquad x \leq -1$

Check that the solution set is the interval $(-\infty, -1]$.

N5. $\frac{3}{4}(x-2) + \frac{1}{2} > \frac{1}{5}(x-8)$

$20\left[\frac{3}{4}(x-2) + \frac{1}{2}\right] > 20\left[\frac{1}{5}(x-8)\right]$ *Mult. by 20.*

$15(x-2) + 10 > 4(x-8)$

$15x - 30 + 10 > 4x - 32$

$15x - 20 > 4x - 32$

$11x - 20 > -32$ *Subtract 4x.*

$11x > -12$ *Add 20.*

$x > -\frac{12}{11}$ *Divide by 11.*

Check that the solution set is the interval $\left(-\frac{12}{11}, \infty\right)$.

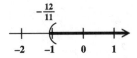

N6. $-1 < x - 2 < 3$

$1 < x < 5$ *Add 2 to each part.*

Check that the solution set is the interval $(1, 5)$.

N7. $-2 < -4x - 5 \le 7$

$3 < -4x \le 12$ *Add 5 to each part.*

$\dfrac{3}{-4} > \dfrac{-4x}{-4} \ge \dfrac{12}{-4}$ *Divide by -4.*

                 Reverse inequalities.

$-\dfrac{3}{4} > x \ge -3$ *Reduce.*

$-3 \le x < -\dfrac{3}{4}$ *Equivalent inequality.*

Check that the solution set is the interval $\left[-3, -\frac{3}{4}\right)$.

N8.   *Step 2*

Let $x = $ the number of months she belongs to the health club.

Step 3

She must pay $40, plus $35x$, to belong to the health club for x months, and this amount must be *no more than* $355.

Cost of belonging	is no more than	355 dollars.
↓	↓	↓
$40 + 35x$	$\le$	355

Step 4

$35x \le 315$ *Subtract 40.*

$x \le 9$ *Divide by 35.*

Step 5

She can belong to the health club for a maximum of 9 months. (She may belong for less time, as indicated by the inequality $x \le 9$.)

Step 6

If Sara belongs for 9 months, she will spend $40 + 35(9) = 355$, the maximum amount.

N9.   Let $x = $ the grade he must make on the fourth test.

To find the average of the four tests, add them and divide by 4. This average must be at least 90, that is, greater than or equal to 90.

$$\frac{82 + 97 + 93 + x}{4} \ge 90$$

$$\frac{272 + x}{4} \ge 90$$

$$272 + x \ge 360 \quad \text{\textit{Multiply by 4.}}$$

$$x \ge 88 \quad \text{\textit{Subtract 272.}}$$

Joel must score at least 88 on the fourth test.

Check $\dfrac{82 + 97 + 93 + 88}{4} = \dfrac{360}{4} = 90$

A score of 88 or more will give an average of at least 90, as required.

2.5   Section Exercises

1. $x \le 3$

In interval notation, this inequality is written $(-\infty, 3]$. The bracket indicates that 3 is included. The answer is choice **D**.

3. $x < 3$

In interval notation, this inequality is written $(-\infty, 3)$. The parenthesis indicates that 3 is not included. The graph of this inequality is shown in choice **B**.

5. $-3 \le x \le 3$

In interval notation, this inequality is written $[-3, 3]$. The brackets indicates that -3 and 3 are included. The answer is choice **F**.

7.   Since $4 > 0$, the student should not have reversed the direction of the inequality symbol when dividing by 4. We reverse the inequality symbol only when multiplying or dividing by a *negative* number. The solution set is $[-16, \infty)$.

9. $x - 4 \ge 12$

$x \ge 16$ *Add 4.*

Check that the solution set is the interval $[16, \infty)$.

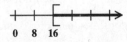

Copyright © 2012 Pearson Education, Inc. Publishing as Addison-Wesley.

11. $3k + 1 > 22$

$\quad 3k > 21 \quad$ *Subtract 1.*

$\quad k > 7 \quad$ *Divide by 3.*

Check that the solution set is the interval $(7, \infty)$.

13. $4x < -16$

$\quad x < -4 \quad$ *Divide by 4.*

Check that the solution set is the interval $(-\infty, -4)$.

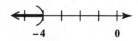

15. $\qquad -\frac{3}{4}x \geq 30$

Multiply both sides by $-\frac{4}{3}$ and reverse the inequality symbol.

$-\frac{4}{3}\left(-\frac{3}{4}x\right) \leq -\frac{4}{3}(30)$

$\qquad x \leq -40$

Check that the solution set is the interval $(-\infty, -40]$.

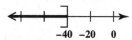

17. $\qquad -1.3x \geq -5.2$

Divide both sides by -1.3, and reverse the inequality symbol.

$\qquad \dfrac{-1.3x}{-1.3} \leq \dfrac{-5.2}{-1.3}$

$\qquad x \leq 4$

Check that the solution set is the interval $(-\infty, 4]$.

19. $5x + 2 \leq -48$

$\quad 5x \leq -50 \quad$ *Subtract 2.*

$\quad x \leq -10 \quad$ *Divide by 5.*

Check that the solution set is the interval $(-\infty, -10]$.

21. $\qquad \dfrac{5x - 6}{8} < 8$

$8\left(\dfrac{5x - 6}{8}\right) < 8 \cdot 8 \quad$ *Multiply by 8.*

$\qquad 5x - 6 < 64$

$\qquad 5x < 70 \quad$ *Add 6.*

$\qquad x < 14 \quad$ *Divide by 5.*

Check Let $x = 14$ in the *equation* $\dfrac{5x - 6}{8} = 8$.

$\dfrac{5(14) - 6}{8} \overset{?}{=} 8$

$\dfrac{64}{8} \overset{?}{=} 8$

$8 = 8 \quad$ *True*

This shows that 14 is the boundary point. Now test a number on each side of 14. We choose 0 and 20.

$$\dfrac{5x - 6}{8} < 8$$

Let $x = 0$.	*Let $x = 20$.*
$\dfrac{5(0) - 6}{8} \overset{?}{<} 8$	$\dfrac{5(20) - 6}{8} \overset{?}{<} 8$
$-\frac{6}{8} < 8 \quad$ *True*	$\frac{94}{8}$ (or $11\frac{6}{8}$) < 8 *False*
0 is in the solution set.	20 is not in the solution set.

The check confirms that $(-\infty, 14)$ is the solution set.

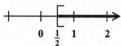

23. $\qquad \dfrac{2x - 5}{-4} > 5$

Multiply both sides by -4, and reverse the inequality symbol.

$-4\left(\dfrac{2x - 5}{-4}\right) < -4(5)$

$\qquad 2x - 5 < -20$

$\qquad 2x < -15 \qquad$ *Add 5.*

$\qquad x < -\frac{15}{2} \qquad$ *Divide by 2.*

Check that the solution set is the interval $\left(-\infty, -\frac{15}{2}\right)$.

25. $6x - 4 \geq -2x$

$\quad 8x - 4 \geq 0 \qquad$ *Add 2x.*

$\quad 8x \geq 4 \qquad$ *Add 4.*

$\quad x \geq \frac{4}{8} = \frac{1}{2} \quad$ *Divide by 8.*

Check that the solution set is the interval $\left[\frac{1}{2}, \infty\right)$.

27. $x - 2(x - 4) \leq 3x$

$\quad x - 2x + 8 \leq 3x$

$\quad -x + 8 \leq 3x$

$\quad 8 \leq 4x \qquad$ *Add x.*

$\quad 2 \leq x,$ or $x \geq 2$

Check that the solution set is the interval $[2, \infty)$.

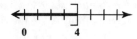

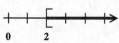

Copyright © 2012 Pearson Education, Inc. Publishing as Addison-Wesley.

29. $-(4+r)+2-3r < -14$

$\quad -4-r+2-3r < -14$ *Distributive property*

$\quad\quad -4r-2 < -14$ *Combine terms.*

$\quad\quad\quad -4r < -12$ *Add 2.*

Divide by -4, and reverse the inequality symbol.

$$r > 3$$

Check that the solution set is the interval $(3, \infty)$.

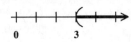

31. $-3(x-6) > 2x-2$

$\quad -3x+18 > 2x-2$ *Distributive property*

$\quad\quad -5x > -20$ *Subtract 2x and 18.*

Divide by -5, and reverse the inequality symbol.

$$x < 4$$

Check that the solution set is the interval $(-\infty, 4)$.

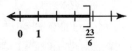

33. $\dfrac{2}{3}(3x-1) \geq \dfrac{3}{2}(2x-3)$

Multiply both sides by 6 to clear the fractions.

$6 \cdot \dfrac{2}{3}(3x-1) \geq 6 \cdot \dfrac{3}{2}(2x-3)$

$4(3x-1) \geq 9(2x-3)$

$12x - 4 \geq 18x - 27$ *Distributive property*

$-6x \geq -23$ *Subtract 18x; add 4.*

Divide by -6, and reverse the inequality symbol.

$$x \leq \dfrac{23}{6}$$

Check that the solution set is the interval $\left(-\infty, \dfrac{23}{6}\right]$.

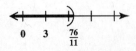

35. $-\dfrac{1}{4}(p+6) + \dfrac{3}{2}(2p-5) < 10$

Multiply each term by 4 to clear the fractions.

$-1(p+6) + 6(2p-5) < 40$

$-p-6+12p-30 < 40$

$11p - 36 < 40$

$11p < 76$

$p < \dfrac{76}{11}$

Check that the solution set is the interval $\left(-\infty, \dfrac{76}{11}\right)$.

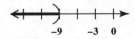

37. $3(2x-4) - 4x < 2x+3$

$6x - 12 - 4x < 2x+3$

$2x - 12 < 2x+3$

$-12 < 3$ *True*

The statement is true for all values of x. Therefore, the original inequality is true for any real number.

Check that the solution set is the interval $(-\infty, \infty)$.

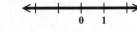

39. $8\left(\dfrac{1}{2}x+3\right) < 8\left(\dfrac{1}{2}x-1\right)$

$4x + 24 < 4x - 8$

$24 < -8$ *False*

This is a false statement, so the inequality is a contradiction.

Check that the solution set is $\emptyset$.

♦ ♦ ♦ Relating Concepts 41–45 ♦ ♦ ♦

41. $5(x+3) - 2(x-4) = 2(x+7)$

$5x + 15 - 2x + 8 = 2x + 14$

$3x + 23 = 2x + 14$

$x = -9$

Check that the solution set is $\{-9\}$.
The graph is the point -9 on a number line.

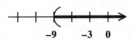

42. $5(x+3) - 2(x-4) > 2(x+7)$

$5x + 15 - 2x + 8 > 2x + 14$

$3x + 23 > 2x + 14$

$x > -9$

Check that the solution set is the interval $(-9, \infty)$.
The graph extends from -9 to the right on a number line; -9 is not included in the graph.

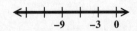

43. $5(x+3) - 2(x-4) < 2(x+7)$

$5x + 15 - 2x + 8 < 2x + 14$

$3x + 23 < 2x + 14$

$x < -9$

Check that the solution set is the interval $(-\infty, -9)$. The graph extends from -9 to the left on a number line; -9 is not included in the graph.

44. If we graph all the solution sets from Exercises 41–43; that is, $\{-9\}$, $(-9, \infty)$, and $(-\infty, -9)$, on the same number line, we will have graphed the set of all real numbers.

Copyright © 2012 Pearson Education, Inc. Publishing as Addison-Wesley.

45. The solution set of the given equation is the point -3 on a number line. The solution set of the first inequality extends from -3 to the right (toward ∞) on the same number line. Based on Exercises 41–43, the solution set of the second inequality should then extend from -3 to the left (toward $-\infty$) on the number line. Complete the statement with $\underline{(-\infty, -3)}$.

47. The goal is to isolate the variable x.

$$-4 < x - 5 \quad < 6$$
$$-4 + 5 < x - 5 + 5 < 6 + 5 \quad \textit{Add 5.}$$
$$1 < x \qquad\quad < 11$$

Check that the solution set is the interval $(1, 11)$.

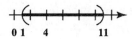

49.
$$-9 \le x + 5 \qquad \le 15$$
$$-9 - 5 \le x + 5 - 5 \le 15 - 5 \quad \textit{Subtract 5.}$$
$$-14 \le x \qquad\qquad \le 10$$

Check that the solution set is the interval $[-14, 10]$.

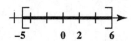

51.
$$-6 \le 2x + 4 \le 16$$
$$-10 \le 2x \qquad \le 12 \quad \textit{Subtract 4.}$$
$$-5 \le x \qquad \le 6 \quad \textit{Divide by 2.}$$

Check that the solution set is the interval $[-5, 6]$.

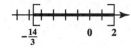

53.
$$-19 \le 3x - 5 \le 1$$
$$-14 \le 3x \qquad \le 6 \quad \textit{Add 5.}$$
$$-\tfrac{14}{3} \le x \qquad \le 2 \quad \textit{Divide by 3.}$$

Check that the solution set is the interval $[-\tfrac{14}{3}, 2]$.

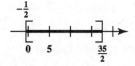

55. $-1 \le \dfrac{2x - 5}{6} \le 5$

$$-6 \le 2x - 5 \le 30 \quad \textit{Multiply by 6.}$$
$$-1 \le 2x \qquad \le 35 \quad \textit{Add 5.}$$
$$-\tfrac{1}{2} \le x \qquad \le \tfrac{35}{2} \quad \textit{Divide by 2.}$$

Check that the solution set is the interval $[-\tfrac{1}{2}, \tfrac{35}{2}]$.

57.
$$4 \le -9x + 5 < 8$$
$$-1 \le -9x \qquad < 3 \quad \textit{Subtract 5.}$$
$$\frac{1}{9} \ge x \qquad > -\frac{1}{3} \quad \begin{array}{l}\textit{Divide by } -9.\\ \textit{Reverse inequalities.}\end{array}$$

The last inequality may be written as

$$-\tfrac{1}{3} < x \le \tfrac{1}{9}.$$

Check that the solution set is the interval $(-\tfrac{1}{3}, \tfrac{1}{9}]$.

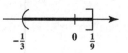

59. Six times a number is between -12 and 12.

$$-12 < 6x < 12$$
$$-2 < x < 2 \quad \textit{Divide by 6.}$$

This is the set of all numbers between -2 and 2—that is, $(-2, 2)$.

61. When 1 is added to twice a number, the result is greater than or equal to 7.

$$2x + 1 \ge 7$$
$$2x \ge 6 \quad \textit{Subtract 1.}$$
$$x \ge 3 \quad \textit{Divide by 2.}$$

This is the set of all numbers greater than or equal to 3—that is, $[3, \infty)$.

63. One third of a number is added to 6, giving a result of at least 3.

$$6 + \tfrac{1}{3}x \ge 3$$
$$\tfrac{1}{3}x \ge -3 \quad \textit{Subtract 6.}$$
$$x \ge -9 \quad \textit{Multiply by 3.}$$

This is the set of all numbers greater than or equal to -9—that is, $[-9, \infty)$.

65. Let $x =$ her score on the third test. Her average must be at least 84 (≥ 84). To find the average of three numbers, add them and divide by 3.

$$\frac{90 + 82 + x}{3} \ge 84$$
$$\frac{172 + x}{3} \ge 84 \quad \textit{Add.}$$
$$172 + x \ge 252 \quad \textit{Multiply by 3.}$$
$$x \ge 80 \quad \textit{Subtract 172.}$$

She must score at least 80 on her third test.

Copyright © 2012 Pearson Education, Inc. Publishing as Addison-Wesley.

67. Let x = the number of months. The cost of Plan A is $54.99x$ and the cost of Plan B is $49.99x + 129$. To determine the number of months that would be needed to make Plan B less expensive, solve the following inequality.

$$\text{Plan B (cost)} < \text{Plan A (cost)}$$

$$49.99x + 129 < 54.99x$$

$$129 < 5x \qquad \textit{Subtract 49.99x.}$$

$$5x > 129 \qquad \textit{Equivalent}$$

$$x > \tfrac{129}{5} \ \ [\, = 25.8\,] \qquad \textit{Divide by 5.}$$

It will take 26 months for Plan B to be the better deal.

69. Cost $C = 20x + 100$; Revenue $R = 24x$

The business will show a profit only when $R > C$. Substitute the given expressions for R and C.

$$R > C$$

$$24x > 20x + 100$$

$$4x > 100$$

$$x > 25$$

The company will show a profit upon selling 26 DVDs.

71. $\text{BMI} = \dfrac{704 \times (\text{weight in pounds})}{(\text{height in inches})^2}$

(a) Let the height equal 72.

$$19 \le \text{BMI} \le 25$$

$$19 \le \dfrac{704w}{72^2} \le 25$$

$$19(72^2) \le 704w \le 25(72^2)$$

$$\dfrac{19(72^2)}{704} \le w \le \dfrac{25(72^2)}{704}$$

$$(\approx 139.91) \le w \le (\approx 184.09)$$

According to the BMI formula, the healthy weight range (rounded to the nearest pound) for a person who is 72 inches tall is 140 to 184 pounds.

(b) Let the height equal 63.

$$19 \le \text{BMI} \le 25$$

$$19 \le \dfrac{704w}{63^2} \le 25$$

$$19(63^2) \le 704w \le 25(63^2)$$

$$\dfrac{19(63^2)}{704} \le w \le \dfrac{25(63^2)}{704}$$

$$(\approx 107.12) \le w \le (\approx 140.94)$$

According to the BMI formula, the healthy weight range (rounded to the nearest pound) for a person who is 63 inches tall is 107 to 141 pounds.

(c) Answers will vary.

73. (a) $x > 4$ is equivalent to $(4, \infty)$.

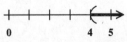

(b) $x < 5$ is equivalent to $(-\infty, 5)$.

(c) All numbers greater than 4 and less than 5 (that is, those satisfying $4 < x < 5$) belong to *both* sets.

2.6   Set Operations and Compound Inequalities

2.6 Now Try Exercises

N1. Let $A = \{2, 4, 6, 8\}$ and $B = \{0, 2, 6, 8\}$.

The set $A \cap B$, the intersection of A and B, contains those elements that belong to both A *and* B; that is, the numbers 2, 6, and 8. Therefore,

$$A \cap B = \{2, 6, 8\}.$$

N2. $x - 2 \le 5 \ \text{ and } \ x + 5 \ge 9$

Solve each inequality.

$$
\begin{array}{ccc}
x - 2 \le 5 & \text{and} & x + 5 \ge 9 \\
x - 2 + 2 \le 5 + 2 & \text{and} & x + 5 - 5 \ge 9 - 5 \\
x \le 7 & \text{and} & x \ge 4
\end{array}
$$

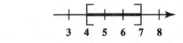

The values that satisfy both inequalities are the numbers between 4 and 7, including 4 and 7.

The solution set is $[4, 7]$.

N3.
$$
\begin{array}{ccc}
-4x - 1 < 7 & \text{and} & 3x + 4 \ge -5 \\
-4x < 8 & \text{and} & 3x \ge -9 \\
x > -2 & \text{and} & x \ge -3
\end{array}
$$

The overlap of the two graphs consists of the numbers that are greater than -2 and are also greater than or equal to -3; that is, the numbers greater than -2.

The solution set is $(-2, \infty)$.

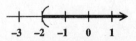

Copyright © 2012 Pearson Education, Inc. Publishing as Addison-Wesley.

N4. $x - 7 < -12$ and $2x + 1 > 5$
$$2x > 4$$
$$x < -5 \quad \text{and} \quad x > 2$$

The two graphs do not overlap. Therefore, there is no number that is both greater than 2 *and* less than -5, so the given compound inequality has no solution. The solution set is $\emptyset$.

N5. Let $A = \{5, 10, 15, 20\}$ and $B = \{5, 15, 25\}$.

The set $A \cup B$, the union of A and B, consists of all elements in either A *or* B (or both). Start by listing the elements of set A: 5, 10, 15, 20. Then list any additional elements from set B. In this case, the elements 5 and 15 are already listed, so the only additional element is 25. Therefore,

$$A \cup B = \{5, 10, 15, 20, 25\}.$$

N6. $-12x \leq -24$ or $x + 9 < 8$
$$x \geq 2 \qquad \text{or} \qquad x < -1$$

The graph of the solution set consists of all numbers greater than or equal to 2 *or* less than -1.

The solution set is $(-\infty, -1) \cup [2, \infty)$.

N7. $-x + 2 < 6$ or $6x - 8 \geq 10$
$$-x < 4 \qquad \text{or} \qquad 6x \geq 18$$
$$x > -4 \quad \text{or} \qquad x \geq 3$$

The solution set is all numbers that are either greater than -4 *or* greater than or equal to 3. All real numbers greater than -4 are included.

The solution set is $(-4, \infty)$.

N8. $8x - 4 \geq 20$ or $-2x + 1 > -9$
$$8x \geq 24 \quad \text{or} \qquad -2x > -10$$
$$x \geq 3 \quad \text{or} \qquad x < 5$$

The solution set is all numbers that are either greater than or equal to 3 *or* less than 5. All real numbers are included.

The solution set is $(-\infty, \infty)$.

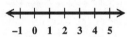

N9. **(a)** All films except *The Ten Commandments* had admissions greater than 140,000,000. *The Sound of Music, E.T.,* and *The Ten Commandments* had a gross income of less than \$1,200,000,000. Thus, the set of elements (films) that satisfy both sets is {*The Sound of Music, E.T.*}.

(b) *Star Wars, The Sound of Music, E.T.,* and *The Ten Commandments* had admissions less than 200,000,000. *The Sound of Music, E.T.,* and *The Ten Commandments* had a gross income of less than \$1,200,000,000. Thus, the set of elements (films) that satisfy either set is {*Star Wars, The Sound of Music, E.T., The Ten Commandments*}.

2.6 Section Exercises

1. This statement is *true*. The solution set of $x + 1 = 6$ is $\{5\}$. The solution set of $x + 1 > 6$ is $(5, \infty)$. The solution set of $x + 1 < 6$ is $(-\infty, 5)$. Taken together we have the set of real numbers. (See Section 2.5, Exercises 41–45, for a discussion of this concept.)

3. This statement is *false*. The union is $(-\infty, 7) \cup (7, \infty)$. The only real number that is *not* in the union is 7.

5. This statement is *false* since 0 is a rational number but not an irrational number. The sets of rational numbers and irrational numbers have no common elements so their intersection is $\emptyset$.

In Exercises 7–14, let $A = \{1, 2, 3, 4, 5, 6\}$, $B = \{1, 3, 5\}, C = \{1, 6\}$, and $D = \{4\}$.

7. The intersection of sets B and A contains only those elements in both sets B and A.

$$B \cap A = \{1, 3, 5\} \text{ or set } B$$

9. The intersection of sets A and D is the set of all elements in both set A and D. Therefore,

$$A \cap D = \{4\} \text{ or set } D.$$

Copyright © 2012 Pearson Education, Inc. Publishing as Addison-Wesley.

11. The intersection of set B and the set of no elements (empty set), $B \cap \emptyset$, is the set of no elements or $\emptyset$.

13. The union of sets A and B is the set of all elements that are in either set A or set B or both sets A and B. Since all numbers in set B are also in set A, the set $A \cup B$ will be the same as set A.

$$A \cup B = \{1, 2, 3, 4, 5, 6\} \text{ or set } A$$

15. The first graph represents the set $(-\infty, 2)$. The second graph represents the set $(-3, \infty)$. The intersection includes the elements common to both sets, that is, $(-3, 2)$.

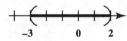

17. The first graph represents the set $(-\infty, 5]$. The second graph represents the set $(-\infty, 2]$. The intersection includes the elements common to both sets, that is, $(-\infty, 2]$.

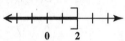

19. $x < 2$ and $x > -3$

The graph of the solution set will be all numbers that are both less than 2 and greater than -3. The solution set is $(-3, 2)$.

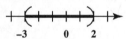

21. $x \leq 2$ and $x \leq 5$

The graph of the solution set will be all numbers that are both less than or equal to 2 and less than or equal to 5. The overlap is the numbers less than or equal to 2. The solution set is $(-\infty, 2]$.

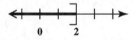

23. $x \leq 3$ and $x \geq 6$

The graph of the solution set will be all numbers that are both less than or equal to 3 and greater than or equal to 6. There are no such numbers. The solution set is $\emptyset$.

25. $x - 3 \leq 6$ and $x + 2 \geq 7$
   $x \leq 9$ and $x \geq 5$

The graph of the solution set is all numbers that are both less than or equal to 9 and greater than or equal to 5. This is the intersection. The elements common to both sets are the numbers between 5 and 9, including the endpoints. The solution set is $[5, 9]$.

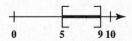

27. $-3x > 3$ and $x + 3 > 0$
   $x < -1$ and $x > -3$

The graph of the solution set is all numbers that are both less than -1 and greater than -3. This is the intersection. The elements common to both sets are the numbers between -3 and -1, not including the endpoints. The solution set is $(-3, -1)$.

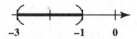

29. $3x - 4 \leq 8$ and $-4x + 1 \geq -15$
   $3x \leq 12$ and $-4x \geq -16$
   $x \leq 4$ and $x \leq 4$

Since both inequalities are identical, the graph of the solution set is the same as the graph of one of the inequalities. The solution set is $(-\infty, 4]$.

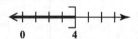

31. The first graph represents the set $(-\infty, 2]$. The second graph represents the set $[4, \infty)$. The union includes all elements in either set, or in both, that is, $(-\infty, 2] \cup [4, \infty)$.

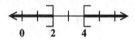

33. The first graph represents the set $[1, \infty)$. The second graph represents the set $(-\infty, 8]$. The union includes all elements in either set, or in both, that is, $(-\infty, \infty)$.

35. $x \leq 1$ or $x \leq 8$

The word "or" means to take the union of both sets. The graph of the solution set is all numbers that are either less than or equal to 1 *or* less than or equal to 8, or both. This is all numbers less than or equal to 8. The solution set is $(-\infty, 8]$.

37. $x \geq -2$ or $x \geq 5$

The graph of the solution set will be all numbers that are either greater than or equal to -2 or greater than or equal to 5. The solution set is $[-2, \infty)$.

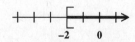

Copyright © 2012 Pearson Education, Inc. Publishing as Addison-Wesley.

39. $x \geq -2$ or $x \leq 4$

The graph of the solution set will be all numbers that are either greater than or equal to -2 or less than or equal to 4. This is the set of all real numbers. The solution set is $(-\infty, \infty)$.

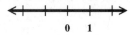

41. $x + 2 > 7$ or $1 - x > 6$
$$\qquad\qquad\qquad -x > 5$$
$\quad x > 5$ or $\quad x < -5$

The graph of the solution set is all numbers either greater than 5 or less than -5. This is the union. The solution set is $(-\infty, -5) \cup (5, \infty)$.

43. $x + 1 > 3$ or $-4x + 1 > 5$
$$\qquad\qquad\qquad -4x > 4$$
$\quad x > 2$ or $\quad\quad x < -1$

The graph of the solution set is all numbers either less than -1 or greater than 2. This is the union. The solution set is $(-\infty, -1) \cup (2, \infty)$.

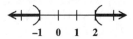

45. $4x + 1 \geq -7$ or $-2x + 3 \geq 5$
$\quad 4x \geq -8$ or $\quad -2x \geq 2$
$\quad\, x \geq -2$ or $\quad\quad\, x \leq -1$

The graph of the solution set is all numbers either greater than or equal to -2 or less than or equal to -1. This is the set of all real numbers. The solution set is $(-\infty, \infty)$.

47. $(-\infty, -1] \cap [-4, \infty)$

The intersection is the set of numbers less than or equal to -1 and greater than or equal to -4. The numbers common to both original sets are between, and including, -4 and -1. The simplest interval form is $[-4, -1]$.

49. $(-\infty, -6] \cap [-9, \infty)$

The intersection is the set of numbers less than or equal to -6 and greater than or equal to -9. The numbers common to both original sets are between, and including, -9 and -6. The simplest interval form is $[-9, -6]$.

51. $(-\infty, 3) \cup (-\infty, -2)$

The union is the set of numbers that are either less than 3 or less than -2, or both. This is all numbers less than 3. The simplest interval form is $(-\infty, 3)$.

53. $[3, 6] \cup (4, 9)$

The union is the set of numbers between, and including, 3 and 6, or between, but not including, 4 and 9. This is the set of numbers greater than or equal to 3 and less than 9. The simplest interval form is $[3, 9)$.

55. $x < -1$ and $x > -5$

The word "and" means to take the intersection of both sets. $x < -1$ and $x > -5$ is true only when
$$-5 < x < -1.$$

The graph of the solution set is all numbers greater than -5 *and* less than -1. This is all numbers between -5 and -1, not including -5 or -1. The solution set is $(-5, -1)$.

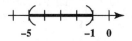

57. $x < 4$ or $x < -2$

The word "or" means to take the union of both sets. The graph of the solution set is all numbers that are either less than 4 *or* less than -2, or both. This is all numbers less than 4. The solution set is $(-\infty, 4)$.

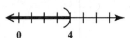

59. $-3x \leq -6$ or $-3x \geq 0$
$\quad\, x \geq 2$ or $\quad x \leq 0$

The word "or" means to take the union of both sets. The graph of the solution set is all numbers that are either greater than or equal to 2 *or* less than or equal to 0. The solution set is $(-\infty, 0] \cup [2, \infty)$.

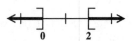

61. $x + 1 \geq 5$ and $x - 2 \leq 10$
$\quad\, x \geq 4$ and $\quad\, x \leq 12$

The word "and" means to take the intersection of both sets. The graph of the solution set is all numbers that are both greater than or equal to 4 *and* less than or equal to 12. This is all numbers between, and including, 4 and 12. The solution set is $[4, 12]$.

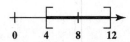

63. The set of expenses that are less than $6500 for public schools *and* are greater than $10,000 for private schools is {Tuition and fees}.

Copyright © 2012 Pearson Education, Inc. Publishing as Addison-Wesley.

65. The set of expenses that are less than $6500 for public schools *or* are greater than $10,000 for private schools is {Tuition and fees, Board rates, Dormitory charges}.

♦ ♦ ♦ Relating Concepts 67–72 ♦ ♦ ♦

67. Find "the yard can be fenced *and* the yard can be sodded."

A yard that can be fenced has $P \le 150$. Maria and Joe qualify.

A yard that can be sodded has $A \le 1400$. Again, Maria and Joe qualify.

Find the intersection. Maria's and Joe's yards are common to both sets, so Maria and Joe can have their yards both fenced and sodded.

68. Find "the yard can be fenced *and* the yard cannot be sodded."

A yard that can be fenced has $P \le 150$. Maria and Joe qualify.

A yard that cannot be sodded has $A > 1400$. Luigi and Than qualify.

Find the intersection. There are no yards common to both sets, so none of them qualify.

69. Find "the yard cannot be fenced *and* the yard can be sodded."

A yard that cannot be fenced has $P > 150$. Luigi and Than qualify.

A yard that can be sodded has $A \le 1400$. Maria and Joe qualify.

Find the intersection. There are no yards common to both sets, so none of them qualify.

70. Find "the yard cannot be fenced *and* the yard cannot be sodded."

A yard that cannot be fenced has $P > 150$. Luigi and Than qualify.

A yard that cannot be sodded has $A > 1400$. Again, Luigi and Than qualify.

Find the intersection. Luigi's and Than's yards are common to both sets, so Luigi and Than qualify.

71. Find "the yard can be fenced *or* the yard can be sodded." From Exercise 67, Maria's and Joe's yards qualify for both conditions, so the union is Maria and Joe.

72. Find "the yard cannot be fenced *or* the yard can be sodded." From Exercise 69, Luigi's and Than's yards cannot be fenced, and Maria's and Joe's yards can be sodded. The union includes all of them.

73.
$$2x - 4 \le 3x + 2$$
$$-x - 4 \le 2$$
$$-x \le 6$$
$$x \ge -6$$

The solution set is $[-6, \infty)$.

75.
$$-5 < 2x + 1 < 5$$
$$-6 < \quad 2x \quad < 4$$
$$-3 < \quad x \quad < 2$$

The solution set is $(-3, 2)$.

77.
$$-|6| - |-11| + (-4) = -6 - (11) + (-4)$$
$$= -17 + (-4)$$
$$= -21$$

79. The absolute value of 0 is 0, which is not a positive number, so the statement is *false*.

2.7   Absolute Value Equations and Inequalities

2.7   Now Try Exercises

N1. $|4x - 1| = 11$

$$4x - 1 = 11 \quad \text{or} \quad 4x - 1 = -11$$
$$4x = 12 \quad \text{or} \quad 4x = -10$$
$$x = 3 \quad \text{or} \quad x = -\tfrac{5}{2}$$

Check $x = 3$: $|11| = 11$ *True*

Check $x = -\tfrac{5}{2}$: $|-11| = 11$ *True*

The solution set is $\left\{ -\tfrac{5}{2}, 3 \right\}$.

N2. $|4x - 1| > 11$

$$4x - 1 > 11 \quad \text{or} \quad 4x - 1 < -11$$
$$4x > 12 \quad \text{or} \quad 4x < -10$$
$$x > 3 \quad \text{or} \quad x < -\tfrac{5}{2}$$

Check $x = -3, 0,$ and 6 in $|4x - 1| > 11$.

Check $x = -3$: $|-13| > 11$ *True*
Check $x = 0$: $|-1| > 11$ *False*
Check $x = 6$: $|23| > 11$ *True*

The solution set is $\left(-\infty, -\tfrac{5}{2} \right) \cup (3, \infty)$.

N3. $|4x - 1| < 11$

$$-11 < 4x - 1 < 11$$
$$-10 < \quad 4x \quad < 12$$
$$-\tfrac{5}{2} < \quad x \quad < 3$$

Check $x = -5, 0,$ and 5 in $|4x - 1| < 11$.

Check $x = -5$: $|-21| < 11$ *False*
Check $x = 0$: $|-1| < 11$ *True*
Check $x = 5$: $|19| < 11$ *False*

The solution set is $\left(-\tfrac{5}{2}, 3 \right)$.

Copyright © 2012 Pearson Education, Inc. Publishing as Addison-Wesley.

N4. $|10x - 2| - 2 = 12$

We first *isolate* the absolute value expression, that is, rewrite the equation so that the absolute value expression is alone on one side of the equals symbol.

$$|10x - 2| = 14$$

$$10x - 2 = 14 \quad \text{or} \quad 10x - 2 = -14$$
$$10x = 16 \quad \text{or} \quad 10x = -12$$
$$x = \tfrac{8}{5} \quad \text{or} \quad x = -\tfrac{6}{5}$$

Check $x = \tfrac{8}{5}$: $|14| - 2 = 12$ *True*
Check $x = -\tfrac{6}{5}$: $|-14| - 2 = 12$ *True*

The solution set is $\left\{-\tfrac{6}{5}, \tfrac{8}{5}\right\}$.

N5. **(a)** $|x - 1| - 4 \leq 2$
$$|x - 1| \leq 6 \quad \textit{Isolate.}$$

$$-6 \leq x - 1 \leq 6$$
$$-5 \leq x \quad\; \leq 7$$

The solution set is $[-5, 7]$.

(b) $|x - 1| - 4 \geq 2$
$$|x - 1| \geq 6 \quad \textit{Isolate.}$$

$$x - 1 \geq 6 \quad \text{or} \quad x - 1 \leq -6$$
$$x \geq 7 \quad \text{or} \quad x \leq -5$$

The solution set is $(-\infty, -5] \cup [7, \infty)$.

N6. $|3x - 4| = |5x + 12|$

$$3x - 4 = 5x + 12 \quad \text{or} \quad 3x - 4 = -(5x + 12)$$
$$-2x = 16 \quad \text{or} \quad 3x - 4 = -5x - 12$$
$$x = -8 \qquad\qquad 8x = -8$$
$$x = -1$$

Check $x = -8$: $|-28| = |-28|$ *True*
Check $x = -1$: $|-7| = |7|$ *True*

The solution set is $\{-8, -1\}$.

N7. **(a)** $|3x - 8| = -2$

Since the absolute value of an expression can never be negative, there are no solutions for this equation.

The solution set is $\emptyset$.

(b) $|7x + 12| = 0$

The expression $7x + 12$ will equal 0 *only* if $7x + 12 = 0$.
$$7x = -12 \qquad \textit{Subtract 12.}$$
$$x = -\tfrac{12}{7} \qquad \textit{Divide by 7.}$$

The solution set is $\left\{-\tfrac{12}{7}\right\}$.

N8. **(a)** $|x| > -10$

The absolute value of a number is always greater than or equal to 0. Therefore, the inequality is true for all real numbers.

The solution set is $(-\infty, \infty)$.

(b) $|4x + 1| + 5 < 4$
$$|4x + 1| < -1$$

There is no number whose absolute value is less than a negative number, so this inequality has no solution.

The solution set is $\emptyset$.

(c) $|x - 2| - 3 \leq -3$
$$|x - 2| \leq 0 \qquad \textit{Isolate abs. value.}$$

The value of $|x - 2|$ will never be less than 0. However, $|x - 2|$ will equal 0 when $x = 2$.

The solution set is $\{2\}$.

2.7 Section Exercises

1. $|x| = 5$ has two solutions, $x = 5$ or $x = -5$. The graph is Choice **E**.

$|x| < 5$ is written $-5 < x < 5$. Notice that -5 and 5 are not included. The graph is Choice **C**, which uses parentheses.

$|x| > 5$ is written $x < -5$ or $x > 5$. The graph is Choice **D**, which uses parentheses.

$|x| \leq 5$ is written $-5 \leq x \leq 5$. This time -5 and 5 are included. The graph is Choice **B**, which uses brackets.

$|x| \geq 5$ is written $x \leq -5$ or $x \geq 5$. The graph is Choice **A**, which uses brackets.

3.   **(a)** $|ax + b| = k, k = 0$
This means the distance from $ax + b$ to 0 is 0, so $ax + b = 0$, which has one solution.

(b) $|ax + b| = k, k > 0$
This means the distance from $ax + b$ to 0 is a positive number, so $ax + b = k$ or $ax + b = -k$. There are two solutions.

(c) $|ax + b| = k, k < 0$
This means the distance from $ax + b$ to 0 is a negative number, which is impossible because distance is always positive. There are no solutions.

5. $|x| = 12$
$$x = 12 \quad \text{or} \quad x = -12$$
The solution set is $\{-12, 12\}$.

7. $|4x| = 20$
$$4x = 20 \quad \text{or} \quad 4x = -20$$
$$x = 5 \quad \text{or} \quad x = -5$$
The solution set is $\{-5, 5\}$.

9. $|x - 3| = 9$
$$x - 3 = 9 \quad \text{or} \quad x - 3 = -9$$
$$x = 12 \quad \text{or} \quad x = -6$$
The solution set is $\{-6, 12\}$.

Copyright © 2012 Pearson Education, Inc. Publishing as Addison-Wesley.

11. $|2x - 1| = 11$

$2x - 1 = 11$ or $2x - 1 = -11$

$2x = 12$ $\qquad$ $2x = -10$

$x = 6$ or $\qquad$ $x = -5$

The solution set is $\{-5, 6\}$.

13. $|4x - 5| = 17$

$4x - 5 = 17$ or $4x - 5 = -17$

$4x = 22$ $\qquad$ $4x = -12$

$x = \frac{22}{4} = \frac{11}{2}$ or $\qquad$ $x = -3$

The solution set is $\left\{-3, \frac{11}{2}\right\}$.

15. $|2x + 5| = 14$

$2x + 5 = 14$ or $2x + 5 = -14$

$2x = 9$ $\qquad$ $2x = -19$

$x = \frac{9}{2}$ or $\qquad$ $x = -\frac{19}{2}$

The solution set is $\left\{-\frac{19}{2}, \frac{9}{2}\right\}$.

17. $\left|\frac{1}{2}x + 3\right| = 2$

$\frac{1}{2}x + 3 = 2$ or $\frac{1}{2}x + 3 = -2$

$\frac{1}{2}x = -1$ $\qquad$ $\frac{1}{2}x = -5$

$x = -2$ or $\qquad$ $x = -10$

The solution set is $\{-10, -2\}$.

19. $\left|1 + \frac{3}{4}x\right| = 7$

$1 + \frac{3}{4}x = 7$ or $1 + \frac{3}{4}x = -7$

Multiply each side by 4.

$4 + 3x = 28$ or $4 + 3x = -28$

$3x = 24$ $\qquad$ $3x = -32$

$x = 8$ or $\qquad$ $x = \frac{-32}{3}$

The solution set is $\left\{-\frac{32}{3}, 8\right\}$.

21. $|0.02x - 1| = 2.50$

$0.02x - 1 = 2.50$ or $0.02x - 1 = -2.50$

$0.02x = 3.50$ $\qquad$ $0.02x = -1.50$

$x = 50(3.5)$ $\qquad$ $x = 50(-1.5)$

$x = 175$ or $\qquad$ $x = -75$

The solution set is $\{-75, 175\}$.

23. $|x| > 3$ $\Leftrightarrow$ $x > 3$ or $x < -3$

The solution set is $(-\infty, -3) \cup (3, \infty)$.

25. $|x| \geq 4$ $\Leftrightarrow$ $x \geq 4$ or $x \leq -4$

The solution set is $(-\infty, -4] \cup [4, \infty)$.

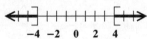

27. $|r + 5| \geq 20$

$r + 5 \leq -20$ or $r + 5 \geq 20$

$r \leq -25$ or $r \geq 15$

The solution set is $(-\infty, -25] \cup [15, \infty)$.

29. $|x + 2| > 10$

$x + 2 > 10$ or $x + 2 < -10$

$x > 8$ or $\qquad$ $x < -12$

The solution set is $(-\infty, -12) \cup (8, \infty)$.

31. $|3 - x| > 5$

$3 - x > 5$ or $3 - x < -5$

$-x > 2$ or $\qquad$ $-x < -8$

Multiply by -1, and reverse the inequality symbols.

$x < -2$ or $\qquad$ $x > 8$

The solution set is $(-\infty, -2) \cup (8, \infty)$.

33. $|-5x + 3| \geq 12$

$-5x + 3 \geq 12$ or $-5x + 3 \leq -12$

$-5x \geq 9$ $\qquad$ $-5x \leq -15$

$x \leq -\frac{9}{5}$ or $\qquad$ $x \geq 3$

The solution set is $\left(-\infty, -\frac{9}{5}\right] \cup [3, \infty)$.

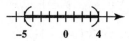

35. **(a)** $|2x + 1| < 9$

The graph of the solution set will be all numbers between -5 and 4, since the absolute value is less than 9.

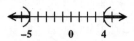

(b) $|2x + 1| > 9$

The graph of the solution set will be all numbers less than -5 or greater than 4, since the absolute value is greater than 9.

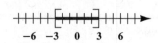

37. $|x| \leq 3 \Leftrightarrow -3 \leq x \leq 3$

The solution set is $[-3, 3]$.

39. $|x| < 4 \Leftrightarrow -4 < x < 4$

The solution set is $(-4, 4)$.

Copyright © 2012 Pearson Education, Inc. Publishing as Addison-Wesley.

41. $|r + 5| < 20$

$$-20 < r + 5 < 20$$
$$-25 < r \quad\quad < 15 \quad \textit{Subtract 5.}$$

The solution set is $(-25, 15)$.

43. $|x + 2| \leq 10$

$$-10 \leq x + 2 \leq 10$$
$$-12 \leq x \quad\quad \leq 8$$

The solution set is $[-12, 8]$.

45. $|3 - x| \leq 5$

$$-5 \leq 3 - x \leq 5$$
$$-8 \leq -x \quad \leq 2$$
$$8 \geq \quad x \quad \geq -2 \quad \begin{array}{l}\textit{Multiply by } -1, \\ \textit{reverse inequalities}\end{array}$$
$$-2 \leq \quad x \quad \leq 8 \quad \textit{Equivalent inequality}$$

The solution set is $[-2, 8]$.

47. $|-5x + 3| < 12$

$$-12 < -5x + 3 < 12$$
$$-15 < \quad -5x \quad < 9$$
$$3 > \quad x \quad > -\frac{9}{5} \quad \begin{array}{l}\textit{Divide by } -5, \\ \textit{reverse inequalities}\end{array}$$
$$-\frac{9}{5} < \quad x \quad < 3 \quad \textit{Equivalent inequality}$$

The solution set is $\left(-\frac{9}{5}, 3\right)$.

49. $|-4 + x| > 9$

$$-4 + x > 9 \quad \text{or} \quad -4 + x < -9$$
$$x > 13 \quad \text{or} \quad\quad x < -5$$

The solution set is $(-\infty, -5) \cup (13, \infty)$.

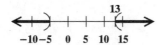

51. $|x + 5| > 20$

$$x + 5 > 20 \quad \text{or} \quad x + 5 < -20$$
$$x > 15 \quad \text{or} \quad\quad x < -25$$

The solution set is $(-\infty, -25) \cup (15, \infty)$.

53. $|7 + 2x| = 5$

$$7 + 2x = 5 \quad \text{or} \quad 7 + 2x = -5$$
$$2x = -2 \quad\quad\quad 2x = -12$$
$$x = -1 \quad \text{or} \quad\quad x = -6$$

The solution set is $\{-6, -1\}$.

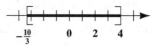

55. $|3x - 1| \leq 11$

$$-11 \leq 3x - 1 \leq 11$$
$$-10 \leq 3x \quad \leq 12$$
$$-\frac{10}{3} \leq x \quad \leq 4$$

The solution set is $\left[-\frac{10}{3}, 4\right]$.

57. $|-6x - 6| \leq 1$

$$-1 \leq -6x - 6 \leq 1$$
$$5 \leq \quad -6x \quad \leq 7$$
$$-\frac{5}{6} \geq \quad x \quad \geq -\frac{7}{6} \quad \begin{array}{l}\textit{Divide by } -6. \\ \textit{Reverse inequalities.}\end{array}$$
$$-\frac{7}{6} \leq \quad x \quad \leq -\frac{5}{6} \quad \textit{Equivalent inequality}$$

The solution set is $\left[-\frac{7}{6}, -\frac{5}{6}\right]$.

59. $|2x - 1| \geq 7$

$$2x - 1 \geq 7 \quad \text{or} \quad 2x - 1 \leq -7$$
$$2x \geq 8 \quad \text{or} \quad\quad 2x \leq -6$$
$$x \geq 4 \quad \text{or} \quad\quad x \leq -3$$

The solution set is $(-\infty, -3] \cup [4, \infty)$.

61. $|x + 2| = 3$

$$x + 2 = 3 \quad \text{or} \quad x + 2 = -3$$
$$x = 1 \quad \text{or} \quad\quad x = -5$$

The solution set is $\{-5, 1\}$.

63. $|x - 6| = 3$

$$x - 6 = 3 \quad \text{or} \quad x - 6 = -3$$
$$x = 9 \quad \text{or} \quad\quad x = 3$$

The solution set is $\{3, 9\}$.

65. $|2 - 0.2x| = 2$

$$2 - 0.2x = 2 \quad \text{or} \quad 2 - 0.2x = -2$$
$$-0.2x = 0 \quad\quad\quad -0.2x = -4$$
$$x = 0 \quad \text{or} \quad\quad x = 20$$

The solution set is $\{0, 20\}$.

Copyright © 2012 Pearson Education, Inc. Publishing as Addison-Wesley.

67. $|x| - 1 = 4$
$|x| = 5$

$x = 5$ or $x = -5$

The solution set is $\{-5, 5\}$.

69. $|x + 4| + 1 = 2$
$|x + 4| = 1$

$x + 4 = 1$ or $x + 4 = -1$
$x = -3$ or $x = -5$

The solution set is $\{-5, -3\}$.

71. $|2x + 1| + 3 > 8$
$|2x + 1| > 5$

$2x + 1 > 5$ or $2x + 1 < -5$
$2x > 4$ $\qquad$ $2x < -6$
$x > 2$ or $x < -3$

The solution set is $(-\infty, -3) \cup (2, \infty)$.

73. $|x + 5| - 6 \leq -1$
$|x + 5| \leq 5$
$-5 \leq x + 5 \leq 5$
$-10 \leq x \quad \leq 0$

The solution set is $[-10, 0]$.

75. $\left|\frac{1}{2}x + \frac{1}{3}\right| + \frac{1}{4} = \frac{3}{4}$
$\left|\frac{1}{2}x + \frac{1}{3}\right| = \frac{1}{2}$

$\frac{1}{2}x + \frac{1}{3} = \frac{1}{2}$ or $\frac{1}{2}x + \frac{1}{3} = -\frac{1}{2}$
$6\left(\frac{1}{2}x + \frac{1}{3}\right) = 6\left(\frac{1}{2}\right)$ $\quad$ $6\left(\frac{1}{2}x + \frac{1}{3}\right) = 6\left(-\frac{1}{2}\right)$
$3x + 2 = 3$ $\qquad$ $3x + 2 = -3$
$3x = 1$ $\qquad$ $3x = -5$
$x = \frac{1}{3}$ or $\qquad$ $x = -\frac{5}{3}$

The solution set is $\left\{-\frac{5}{3}, \frac{1}{3}\right\}$.

77. $|0.1x - 2.5| + 0.3 \geq 0.8$
$|0.1x - 2.5| \geq 0.5$

$0.1x - 2.5 \geq 0.5$ or $0.1x - 2.5 \leq -0.5$
$0.1x \geq 3$ $\quad$ or $\quad$ $0.1x \leq 2$
$x \geq 30$ $\quad$ or $\quad$ $x \leq 20$

The solution set is $(-\infty, 20] \cup [30, \infty)$.

79. $|3x + 1| = |2x + 4|$

$3x + 1 = 2x + 4$ or $3x + 1 = -(2x + 4)$
$\qquad$ $3x + 1 = -2x - 4$
$\qquad$ $5x = -5$
$x = 3$ or $\qquad$ $x = -1$

The solution set is $\{-1, 3\}$.

81. $\left|x - \frac{1}{2}\right| = \left|\frac{1}{2}x - 2\right|$

$x - \frac{1}{2} = \frac{1}{2}x - 2$ or $x - \frac{1}{2} = -\left(\frac{1}{2}x - 2\right)$
Multiply by 2. $\qquad$ $x - \frac{1}{2} = -\frac{1}{2}x + 2$
$\qquad$ *Multiply by 2.*
$2x - 1 = x - 4$ $\qquad$ $2x - 1 = -x + 4$
$\qquad$ $3x = 5$
$x = -3$ or $\qquad$ $x = \frac{5}{3}$

The solution set is $\left\{-3, \frac{5}{3}\right\}$.

83. $|6x| = |9x + 1|$

$6x = 9x + 1$ or $6x = -(9x + 1)$
$-3x = 1$ $\qquad$ $6x = -9x - 1$
$\qquad$ $15x = -1$
$x = -\frac{1}{3}$ or $\qquad$ $x = -\frac{1}{15}$

The solution set is $\left\{-\frac{1}{3}, -\frac{1}{15}\right\}$.

85. $|2x - 6| = |2x + 11|$

$2x - 6 = 2x + 11$ or $2x - 6 = -(2x + 11)$
$-6 = 11$ *False* $\qquad$ $2x - 6 = -2x - 11$
$\qquad$ $4x = -5$
No solution or $\qquad$ $x = -\frac{5}{4}$

The solution set is $\left\{-\frac{5}{4}\right\}$.

87. $|x| \geq -10$

The absolute value of a number is always greater than or equal to 0. Therefore, the inequality is true for all real numbers.

The solution set is $(-\infty, \infty)$.

89. $|12t - 3| = -8$

Since the absolute value of an expression can never be negative, there are no solutions for this equation.

The solution set is $\emptyset$.

91. $|4x + 1| = 0$

The expression $4x + 1$ will equal 0 *only* for the solution of the equation

$4x + 1 = 0.$
$4x = -1$
$x = \frac{-1}{4}$ or $-\frac{1}{4}$

The solution set is $\left\{-\frac{1}{4}\right\}$.

93. $|2x - 1| = -6$

Since the absolute value of an expression can never be negative, there are no solutions for this equation.

The solution set is $\emptyset$.

Copyright © 2012 Pearson Education, Inc. Publishing as Addison-Wesley.

95. $|x + 5| > -9$

Since the absolute value of an expression is always nonnegative (positive or zero), the inequality is true for any real number x.

The solution set is $(-\infty, \infty)$.

97. $|7x + 3| \le 0$

The absolute value of an expression is always nonnegative (positive or zero), so this inequality is true only when

$$7x + 3 = 0$$
$$7x = -3$$
$$x = -\tfrac{3}{7}.$$

The solution set is $\left\{ -\tfrac{3}{7} \right\}$.

99. $|5x - 2| = 0$

The expression $5x - 2$ will equal 0 *only* for the solution of the equation

$$5x - 2 = 0.$$
$$5x = 2$$
$$x = \tfrac{2}{5}$$

The solution set is $\left\{ \tfrac{2}{5} \right\}$.

101. $|x - 2| + 3 \ge 2$
$$|x - 2| \ge -1$$

Since the absolute value of an expression is always nonnegative (positive or zero), the inequality is true for any real number x.

The solution set is $(-\infty, \infty)$.

103. $|10x + 7| + 3 < 1$
$$|10x + 7| < -2$$

There is no number whose absolute value is less than -2, so this inequality has no solution.

The solution set is $\emptyset$.

105. Let x represent the calcium intake for a specific female. For x to be within 100 mg of 1000 mg, we must have

$$|x - 1000| \le 100.$$

$$-100 \le x - 1000 \le 100$$
$$900 \le \quad x \quad \le 1100$$

♦♦♦ Relating Concepts 107–110 ♦♦♦

107. Add the given heights with a calculator to get 8105. There are 10 numbers, so divide the sum by 10.

$$\frac{8105}{10} = 810.5$$

The average height is 810.5 ft.

108. $|x - k| < 50$

Substitute 810.5 for k and solve the inequality.

$$|x - 810.5| < 50$$

$$-50 < x - 810.5 < 50$$
$$760.5 < \quad x \quad < 860.5$$

The buildings with heights between 760.5 ft and 860.5 ft are Bank of America Center and Texaco Heritage Plaza.

109. $|x - k| < 95$

Substitute 810.5 for k and solve the inequality.

$$|x - 810.5| < 95$$

$$-95 < x - 810.5 < 95$$
$$715.5 < \quad x \quad < 905.5$$

The buildings with heights between 715.5 ft and 905.5 ft are Williams Tower, Bank of America Center, Texaco Heritage Plaza, Enterprise Plaza, Centerpoint Energy Plaza, Continental Center I, and Fulbright Tower.

110. **(a)** This would be the opposite of the inequality in Exercise 109, that is,

$$|x - 810.5| \ge 95.$$

(b) $|x - 810.5| \ge 95$

$$\begin{array}{lll} x - 810.5 \ge 95 & \text{or} & x - 810.5 \le -95 \\ x \ge 905.5 & \text{or} & x \le 715.5 \end{array}$$

(c) The buildings that are not within 95 ft of the average have height less than or equal to 715.5 or greater than or equal to 905.5. They are JPMorgan Chase Tower, Wells Fargo Plaza, and One Shell Plaza.

(d) The answer makes sense because it includes all the buildings *not* listed earlier which had heights within 95 ft of the average.

111. **(a)** $\quad 3x + 2y = 24$
$$3(0) + 2y = 24 \quad \textit{Let x = 0.}$$
$$0 + 2y = 24$$
$$2y = 24$$
$$y = 12$$

(b) $\quad -2x + 5y = 20$
$$-2(0) + 5y = 20 \quad \textit{Let x = 0.}$$
$$0 + 5y = 20$$
$$5y = 20$$
$$y = 4$$

113. **(a)** $\quad 3x + 2y = 24$
$$3(8) + 2y = 24 \quad \textit{Let x = 8.}$$
$$24 + 2y = 24$$
$$2y = 0$$
$$y = 0$$

Copyright © 2012 Pearson Education, Inc. Publishing as Addison-Wesley.

(b) $-2x + 5y = 20$

$-2(8) + 5y = 20$ *Let x = 8.*

$-16 + 5y = 20$

$5y = 36$

$y = \frac{36}{5}$

Summary Exercises on Solving Linear and Absolute Value Equations and Inequalities

1. $4x + 1 = 49$

$4x = 48$

$x = 12$

The solution set is $\{12\}$.

3. $6x - 9 = 12 + 3x$

$3x = 21$

$x = 7$

The solution set is $\{7\}$.

5. $|x + 3| = -4$

Since the absolute value of an expression is always nonnegative, there is no number that makes this statement true. Therefore, the solution set is $\emptyset$.

7. $8x + 2 \geq 5x$

$3x \geq -2$

$x \geq -\frac{2}{3}$

The solution set is $[-\frac{2}{3}, \infty)$.

9. $2x - 1 = -7$

$2x = -6$

$x = -3$

The solution set is $\{-3\}$.

11. $6x - 5 \leq 3x + 10$

$3x \leq 15$

$x \leq 5$

The solution set is $(-\infty, 5]$.

13. $9x - 3(x + 1) = 8x - 7$

$9x - 3x - 3 = 8x - 7$

$6x - 3 = 8x - 7$

$4 = 2x$

$2 = x$

The solution set is $\{2\}$.

15. $9x - 5 \geq 9x + 3$

$-5 \geq 3$ *False*

This is a false statement, so the inequality is a contradiction.

The solution set is $\emptyset$.

17. $|x| < 5.5$

$-5.5 < x < 5.5$

The solution set is $(-5.5, 5.5)$.

19. $\frac{2}{3}x + 8 = \frac{1}{4}x$

$8x + 96 = 3x$ *Multiply by 12.*

$5x = -96$

$x = -\frac{96}{5}$

The solution set is $\{-\frac{96}{5}\}$.

21. $\frac{1}{4}x < -6$

$4(\frac{1}{4}x) < 4(-6)$

$x < -24$

The solution set is $(-\infty, -24)$.

23. $\frac{3}{5}x - \frac{1}{10} = 2$

$6x - 1 = 20$ *Multiply by 10.*

$6x = 21$

$x = \frac{21}{6} = \frac{7}{2}$

The solution set is $\{\frac{7}{2}\}$.

25. $x + 9 + 7x = 4(3 + 2x) - 3$

$8x + 9 = 12 + 8x - 3$

$8x + 9 = 8x + 9$

$0 = 0$ *True*

The last statement is true for any real number x.

The solution set is $\{$all real numbers$\}$.

27. $|2x - 3| > 11$

$2x - 3 > 11$ or $2x - 3 < -11$

$2x > 14$ $\qquad\qquad$ $2x < -8$

$x > 7$ or $\qquad$ $x < -4$

The solution set is $(-\infty, -4) \cup (7, \infty)$.

29. $|5x + 1| \leq 0$

The expression $|5x + 1|$ is never less than 0 since an absolute value expression must be nonnegative. However, $|5x + 1| = 0$ if

$5x + 1 = 0$

$5x = -1$

$x = \frac{-1}{5} = -\frac{1}{5}$

The solution set is $\{-\frac{1}{5}\}$.

31. $-2 \leq 3x - 1 \leq 8$

$-1 \leq 3x \leq 9$

$-\frac{1}{3} \leq x \leq 3$

The solution set is $[-\frac{1}{3}, 3]$.

Copyright © 2012 Pearson Education, Inc. Publishing as Addison-Wesley.

33. $|7x - 1| = |5x + 3|$

$7x - 1 = 5x + 3$ or $7x - 1 = -(5x + 3)$

$2x = 4$ $\qquad$ $7x - 1 = -5x - 3$

$\qquad\qquad\qquad\qquad$ $12x = -2$

$x = 2$ $\quad$ or $\quad$ $x = \frac{-2}{12} = -\frac{1}{6}$

The solution set is $\left\{-\frac{1}{6}, 2\right\}$.

35. $|1 - 3x| \geq 4$

$1 - 3x \geq 4$ $\quad$ or $\quad$ $1 - 3x \leq -4$

$-3x \geq 3$ $\qquad\qquad$ $-3x \leq -5$

$x \leq -1$ $\quad$ or $\qquad$ $x \geq \frac{5}{3}$

The solution set is $(-\infty, -1] \cup [\frac{5}{3}, \infty)$.

37. $-(x + 4) + 2 = 3x + 8$

$-x - 4 + 2 = 3x + 8$

$-x - 2 = 3x + 8$

$-10 = 4x$

$x = \frac{-10}{4} = -\frac{5}{2}$

The solution set is $\left\{-\frac{5}{2}\right\}$.

39. $-6 \leq \frac{3}{2} - x \leq 6$

$-\frac{15}{2} \leq -x \quad \leq \frac{9}{2}$ $\quad$ *Subtract $\frac{3}{2}$.*

$\frac{15}{2} \geq x \quad \geq -\frac{9}{2}$ $\quad$ *Multiply by -1.*
$\qquad\qquad\qquad\qquad$ *Reverse inequalities.*

$-\frac{9}{2} \leq x \quad \leq \frac{15}{2}$ $\quad$ *Equivalent inequality*

The solution set is $[-\frac{9}{2}, \frac{15}{2}]$.

41. $|x - 1| \geq -6$

The absolute value of an expression is always nonnegative, so the inequality is true for any real number x.

The solution set is $(-\infty, \infty)$.

43. $8x - (1 - x) = 3(1 + 3x) - 4$

$8x - 1 + x = 3 + 9x - 4$

$9x - 1 = 9x - 1$ $\quad$ *True*

This is an identity.

The solution set is {all real numbers}.

45. $|x - 5| = |x + 9|$

$x - 5 = x + 9$ $\quad$ or $\qquad$ $x - 5 = -(x + 9)$

$-5 = 9$ $\;$ *False* $\qquad\qquad$ $x - 5 = -x - 9$

$\qquad\qquad\qquad\qquad$ $2x = -4$

$\quad$ *No solution* $\quad$ or $\qquad$ $x = -2$

The solution set is $\{-2\}$.

47. $2x + 1 > 5$ $\quad$ or $\quad$ $3x + 4 < 1$

$2x > 4$ $\qquad\qquad$ $3x < -3$

$x > 2$ $\quad$ or $\qquad$ $x < -1$

The solution set is $(-\infty, -1) \cup (2, \infty)$.

Chapter 2 Review Exercises

1. $-(8 + 3x) + 5 = 2x + 6$

$-8 - 3x + 5 = 2x + 6$

$-3x - 3 = 2x + 6$

$-5x = 9$

$x = -\frac{9}{5}$

The solution set is $\left\{-\frac{9}{5}\right\}$.

2. $-\frac{3}{4}x = -12$

$-3x = -48$ $\quad$ *Multiply by 4.*

$x = 16$

The solution set is $\{16\}$.

3. $\dfrac{2x + 1}{3} - \dfrac{x - 1}{4} = 0$

$4(2x + 1) - 3(x - 1) = 0$ $\quad$ *Multiply by 12.*

$8x + 4 - 3x + 3 = 0$

$5x + 7 = 0$

$5x = -7$

$x = -\frac{7}{5}$

The solution set is $\left\{-\frac{7}{5}\right\}$.

4. $5(2x - 3) = 6(x - 1) + 4x$

$10x - 15 = 6x - 6 + 4x$

$10x - 15 = 10x - 6$

$-15 = -6$ $\quad$ *False*

This is a false statement, so the equation is a contradiction.

The solution set is $\emptyset$.

5. $7x - 3(2x - 5) + 5 + 3x = 4x + 20$

$7x - 6x + 15 + 5 + 3x = 4x + 20$

$4x + 20 = 4x + 20$

$20 = 20$ $\quad$ *True*

This equation is an *identity*.

The solution set is {all real numbers}.

6. $8x - 4x - (x - 7) + 9x + 6 = 12x - 7$

$8x - 4x - x + 7 + 9x + 6 = 12x - 7$

$12x + 13 = 12x - 7$

$13 = -7$ $\quad$ *False*

This equation is a *contradiction*.

The solution set is $\emptyset$.

7. $-2x + 6(x - 1) + 3x - (4 - x) = -(x + 5) - 5$

$-2x + 6x - 6 + 3x - 4 + x = -x - 5 - 5$

$8x - 10 = -x - 10$

$9x = 0$

$x = 0$

This equation is a *conditional* equation.

The solution set is $\{0\}$.

8. Solve $V = LWH$ for L.

$$\frac{V}{WH} = \frac{LWH}{WH}$$
$$\frac{V}{WH} = L, \text{ or } L = \frac{V}{WH}$$

9. Solve $A = \frac{1}{2}h(b+B)$ for b.

$$2A = h(b+B) \quad \textit{Multiply by 2.}$$
$$2A = hb + hB \quad \textit{Distributive prop.}$$
$$2A - hB = hb \quad\quad \textit{Subtract hB.}$$
$$\frac{2A - hB}{h} = b \quad\quad\quad \textit{Divide by h.}$$

Another method:

$$2A = h(b+B) \quad \textit{Multiply by 2.}$$
$$\frac{2A}{h} = b + B \quad\quad \textit{Divide by h.}$$
$$\frac{2A}{h} - B = b, \quad\quad \textit{Subtract B.}$$

10. Solve $M = -\frac{1}{4}(x+3y)$ for x.

$$-4M = x + 3y \quad\quad \textit{Multiply by} -4.$$
$$x = -4M - 3y$$

11. Solve $P = \frac{3}{4}x - 12$ for x.

$$P + 12 = \frac{3}{4}x \quad\quad\quad \textit{Add 12.}$$
$$x = \frac{4}{3}(P + 12), \quad \textit{Multiply by 4/3.}$$
$$\text{or } x = \frac{4}{3}P + 16$$

12. Solve $-2x + 5 = 7$.

Begin by subtracting 5 from each side. Then divide each side by -2.

13. Use the formula $V = LWH$ and substitute 180 for V, 6 for L, and 5 for W.

$$180 = 6(5)H$$
$$180 = 30H$$
$$6 = H$$

The height is 6 feet.

14. Divide the amount of increase by the original amount (amounts in millions).

$$\frac{18.2 - 15.3}{15.3} = \frac{2.9}{15.3} \approx 0.190$$

The percent increase was about 19.0%.

15. Use the formula $I = prt$. Substitute $7800 for I, $30,000 for p, and 4 for t. Solve for r.

$$I = prt$$
$$\$7800 = (\$30,000)r(4)$$
$$7800 = 120,000r$$
$$r = \frac{7800}{120,000} = 0.065$$

The rate is 6.5%.

16. Use the formula $C = \frac{5}{9}(F - 32)$ and substitute 77 for F.

$$C = \frac{5}{9}(77 - 32) = \frac{5}{9}(45) = 25$$

The Celsius temperature is 25°.

17. The amount of money spent on Social Security in 2008 was about

$$0.207(\$2980 \text{ billion}) \approx \$617 \text{ billion}.$$

18. The amount of money spent on education and social services in 2008 was about

$$0.031(\$2980 \text{ billion}) \approx \$92.4 \text{ billion}$$

19. "One-third of a number, subtracted from 9" is written

$$9 - \frac{1}{3}x.$$

20. "The product of 4 and a number, divided by 9 more than the number" is written

$$\frac{4x}{x + 9}.$$

21. Let $x =$ the width of the rectangle.
Then $2x - 3 =$ the length of the rectangle.

Use the formula $P = 2L + 2W$ with $P = 42$.

$$42 = 2(2x - 3) + 2x$$
$$42 = 4x - 6 + 2x$$
$$48 = 6x$$
$$8 = x$$

The width is 8 meters and the length is $2(8) - 3 = 13$ meters.

22. Let $x =$ the length of each equal side. Then $2x - 15 =$ the length of the third side.

Use the formula $P = a + b + c$ with $P = 53$.

$$53 = x + x + (2x - 15)$$
$$53 = 4x - 15$$
$$68 = 4x$$
$$17 = x$$

The lengths of the three sides are 17 inches, 17 inches, and $2(17) - 15 = 19$ inches.

23. Let $x =$ the number of kilograms of peanut clusters. Then $3x$ is the number of kilograms of chocolate creams.
The clerk has a total of 48 kg.

$$x + 3x = 48$$
$$4x = 48$$
$$x = 12$$

The clerk has 12 kilograms of peanut clusters.

Copyright © 2012 Pearson Education, Inc. Publishing as Addison-Wesley.

24. Let x = the number of liters of the 20% solution. Make a table.

Liters of Solution	Percent (as a decimal)	Liters of Pure Chemical
x	0.20	$0.20x$
15	0.50	$0.50(15) = 7.5$
$x + 15$	0.30	$0.30(x + 15)$

The last column gives the equation.

$$0.20x + 7.5 = 0.30(x + 15)$$
$$0.20x + 7.5 = 0.30x + 4.5$$
$$3 = 0.10x$$
$$30 = x$$

30 L of the 20% solution should be used.

25. Let x = the number of liters of water.

Liters of Solution	Percent (as a decimal)	Liters of Pure Acid
30	0.40	$0.40(30) = 12$
x	0	$0(x) = 0$
$30 + x$	0.30	$0.30(30 + x)$

The last column gives the equation.

$$12 + 0 = 0.30(30 + x)$$
$$12 = 9 + 0.3x$$
$$3 = 0.3x$$
$$10 = x$$

10 L of water should be added.

26. Let x = the amount invested at 6%. Then $x - 4000$ = the amount invested at 4%.

Principal	Rate (as a decimal)	Interest
x	0.06	$0.06x$
$x - 4000$	0.04	$0.04(x - 4000)$
	Total →	$840

The last column gives the equation.

$$0.06x + 0.04(x - 4000) = 840$$
$$6x + 4(x - 4000) = 84{,}000 \quad \textit{Multiply by 100.}$$
$$6x + 4x - 16{,}000 = 84{,}000$$
$$10x = 100{,}000$$
$$x = 10{,}000$$

Eric should invest $10,000 at 6% and $10,000 − $4000 = $6000 at 4%.

27. Let x = the number of quarters. Then $2x - 1$ is the number of dimes.

Number of Coins	Denomination	Value
x	0.25	$0.25x$
$2x - 1$	0.10	$0.10(2x - 1)$
	Total →	3.50

The sum of the values equals the total value.

$$0.25x + 0.10(2x - 1) = 3.50$$
$$\textit{Multiply by 100.}$$
$$25x + 10(2x - 1) = 350$$
$$25x + 20x - 10 = 350$$
$$45x = 360$$
$$x = 8$$

There are 8 quarters and $2(8) - 1 = 15$ dimes.

Check $8(0.25) + 15(0.10) = 3.50$

28. Let x = the number of nickels. Then $19 - x$ is the number of dimes.

Number of Coins	Denomination	Value
x	0.05	$0.05x$
$19 - x$	0.10	$0.10(19 - x)$
	Total →	1.55

The sum of the values equals the total value.

$$0.05x + 0.10(19 - x) = 1.55$$
$$\textit{Multiply by 100.}$$
$$5x + 10(19 - x) = 155$$
$$5x + 190 - 10x = 155$$
$$-5x = -35$$
$$x = 7$$

He had 7 nickels and $19 - 7 = 12$ dimes.

Check $7(0.05) + 12(0.10) = 1.55$

29. Use the formula $d = rt$ or $r = \frac{d}{t}$. Here, d is about 400 mi and t is about 8 hr. Since $\frac{400}{8} = 50$, the best estimate is choice **A**.

30. Use the formula $d = rt$.

(a) Here, $r = 53$ mph and $t = 10$ hr.

$$d = 53(10) = 530$$

The distance is 530 miles.

(b) Here, $r = 164$ mph and $t = 2$ hr.

$$d = 164(2) = 328$$

The distance is 328 miles.

31. Let x = the time it takes for the trains to be 297 mi apart.

Use the formula $d = rt$.

	Rate	Time	Distance
Passenger Train	60	x	$60x$
Freight Train	75	x	$75x$
			297

The total distance traveled is the sum of the distances traveled by each train.

Copyright © 2012 Pearson Education, Inc. Publishing as Addison-Wesley.

$$60x + 75x = 297$$
$$135x = 297$$
$$x = 2.2$$

It will take the trains 2.2 hours before they are 297 miles apart.

32. Let x = the rate of the faster car and $x - 15$ = the rate of the slower car. Make a table.

	Rate	Time	Distance
Faster Car	x	2	$2x$
Slower Car	$x - 15$	2	$2(x - 15)$
			230

The total distance traveled is the sum of the distances traveled by each car.

$$2x + 2(x - 15) = 230$$
$$2x + 2x - 30 = 230$$
$$4x = 260$$
$$x = 65$$

The faster car travels at 65 km/hr, while the slower car travels at $65 - 15 = 50$ km/hr.

Check $2(65) + 2(50) = 230$

33. Let x = amount of time spent averaging 45 miles per hour. Then $4 - x$ = amount of time at 50 mph.

	Rate	Time	Distance
First Part	45	x	$45x$
Second Part	50	$4 - x$	$50(4 - x)$
Total			195

From the last column:

$$45x + 50(4 - x) = 195$$
$$45x + 200 - 50x = 195$$
$$-5x = -5$$
$$x = 1$$

The automobile averaged 45 mph for 1 hour.

Check 45 mph for 1 hour = 45 miles and 50 mph for 3 hours = 150 miles; $45 + 150 = 195$.

34. Let x = the average rate for the first hour. Then $x - 7$ = the average rate for the second hour. Using $d = rt$, the distance traveled for the first hour is $x(1)$, for the second hour is $(x - 7)(1)$, and for the whole trip, 85.

$$x + (x - 7) = 85$$
$$2x - 7 = 85$$
$$2x = 92$$
$$x = 46$$

The average rate for the first hour was 46 mph.

Check 46 mph for 1 hour = 46 miles and $46 - 7 = 39$ mph for 1 hour = 39 miles; $46 + 39 = 85$.

35. The sum of the angles in a triangle is $180°$.

$$(3x + 7) + (4x + 1) + (9x - 4) = 180$$
$$16x + 4 = 180$$
$$16x = 176$$
$$x = 11$$

The first angle is $3(11) + 7 = 40°$.
The second angle is $4(11) + 1 = 45°$.
The third angle is $9(11) - 4 = 95°$.

36. The marked angles are supplements which have a sum of $180°$.

$$(15x + 15) + (3x + 3) = 180$$
$$18x + 18 = 180$$
$$18x = 162$$
$$x = 9$$

The angle measures are
$15(9) + 15 = 150°$ and $3(9) + 3 = 30°$.

37. $-\dfrac{2}{3}x < 6$
$$-2x < 18 \qquad \textit{Multiply by 3.}$$
Divide by -2; reverse the inequality symbol.
$$x > -9$$

The solution set is $(-9, \infty)$.

38. $-5x - 4 \geq 11$
$$-5x \geq 15$$
Divide by -5; reverse the inequality symbol.
$$x \leq -3$$

The solution set is $(-\infty, -3]$.

39. $\dfrac{6x + 3}{-4} < -3$

Multiply by -4; reverse the inequality symbol.
$$6x + 3 > 12$$
$$6x > 9$$
$$x > \tfrac{9}{6} = \tfrac{3}{2}$$

The solution set is $(\tfrac{3}{2}, \infty)$.

40. $5 - (6 - 4x) \geq 2x - 7$
$$5 - 6 + 4x \geq 2x - 7$$
$$4x - 1 \geq 2x - 7$$
$$2x \geq -6$$
$$x \geq -3$$

The solution set is $[-3, \infty)$.

41. $8 \leq 3x - 1 < 14$
$$9 \leq 3x \quad < 15$$
$$3 \leq x \quad < 5$$

The solution set is $[3, 5)$.

Copyright © 2012 Pearson Education, Inc. Publishing as Addison-Wesley.

42. $\frac{5}{3}(x-2) + \frac{2}{5}(x+1) > 1$

$25(x-2) + 6(x+1) > 15$

Multiply by 15.

$25x - 50 + 6x + 6 > 15$

$31x - 44 > 15$

$31x > 59$

$x > \frac{59}{31}$

The solution set is $\left(\frac{59}{31}, \infty\right)$.

43. Let $x =$ the other dimension of the rectangle. One dimension of the rectangle is 22 and the perimeter can be no greater than 120.

$$P \le 120$$
$$2L + 2W \le 120$$
$$2(x) + 2(22) \le 120$$
$$2x + 44 \le 120$$
$$2x \le 76$$
$$x \le 38$$

The other dimension must be 38 meters or less.

44. Let $x =$ the number of tickets that can be purchased. The total cost of the tickets is $\$48x$. Including the $\$50$ discount and staying within the available $\$1600$, we have

$$48x - 50 \le 1600.$$
$$48x \le 1650$$
$$x \lesssim 34.4$$

The group can purchase 34 tickets or fewer (but at least 15).

45. Let $x =$ the student's score on the fifth test. The average of the five test scores must be at least 70. The inequality is

$$\frac{75 + 79 + 64 + 71 + x}{5} \ge 70.$$
$$75 + 79 + 64 + 71 + x \ge 350$$
$$289 + x \ge 350$$
$$x \ge 61$$

The student will pass algebra if any score greater than or equal to 61% on the fifth test is achieved.

46. The result, $-8 < -13$, is a false statement. There are no real numbers that make this inequality true. The solution set is $\emptyset$.

For Exercises 47–50, let $A = \{a, b, c, d\}$, $B = \{a, c, e, f\}$, and $C = \{a, e, f, g\}$.

47. $A \cap B = \{a, b, c, d\} \cap \{a, c, e, f\}$
$= \{a, c\}$

48. $A \cap C = \{a, b, c, d\} \cap \{a, e, f, g\}$
$= \{a\}$

49. $B \cup C = \{a, c, e, f\} \cup \{a, e, f, g\}$
$= \{a, c, e, f, g\}$

50. $A \cup C = \{a, b, c, d\} \cup \{a, e, f, g\}$
$= \{a, b, c, d, e, f, g\}$

51. $x > 6$ and $x < 9$

The graph of the solution set will be all numbers which are both greater than 6 and less than 9. The overlap is the numbers between 6 and 9, not including the endpoints.

The solution set is $(6, 9)$.

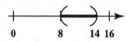

52. $x + 4 > 12$ and $x - 2 < 12$
$x > 8$ and $x < 14$

The graph of the solution set will be all numbers between 8 and 14, not including the endpoints.

The solution set is $(8, 14)$.

53. $x > 5$ or $x \le -3$

The graph of the solution set will be all numbers that are either greater than 5 or less than or equal to -3.

The solution set is $(-\infty, -3] \cup (5, \infty)$.

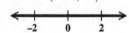

54. $x \ge -2$ or $x < 2$

The graph of the solution set will be all numbers that are either greater than or equal to -2 or less than 2. All real numbers satisfy these criteria.

The solution set is $(-\infty, \infty)$.

55. $x - 4 > 6$ and $x + 3 \le 10$
$x > 10$ and $x \le 7$

The graph of the solution set will be all numbers that are both greater than 10 and less than or equal to 7. There are no real numbers satisfying these criteria.

The solution set is $\emptyset$.

56. $-5x + 1 \ge 11$ or $3x + 5 \ge 26$
$-5x \ge 10$ $3x \ge 21$
$x \le -2$ or $x \ge 7$

The graph of the solution set will be all numbers that are either less than or equal to -2 or greater than or equal to 7.

The solution set is $(-\infty, -2] \cup [7, \infty)$.

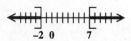

Copyright © 2012 Pearson Education, Inc. Publishing as Addison-Wesley.

57. $(-3, \infty) \cap (-\infty, 4)$

$(-3, \infty)$ includes all real numbers greater than -3.

$(-\infty, 4)$ includes all real numbers less than 4.
Find the intersection. The numbers common to both sets are greater than -3 and less than 4.

$$-3 < x < 4$$

The solution set is $(-3, 4)$.

58. $(-\infty, 6) \cap (-\infty, 2)$

$(-\infty, 6)$ includes all real numbers less than 6.

$(-\infty, 2)$ includes all real numbers less than 2.
Find the intersection. The numbers common to both sets are less than 2.

The solution set is $(-\infty, 2)$.

59. $(4, \infty) \cup (9, \infty)$

$(4, \infty)$ includes all real numbers greater than 4.

$(9, \infty)$ includes all real numbers greater than 9.
Find the union. The numbers in the first set, the second set, or in both sets are all the real numbers that are greater than 4.

The solution set is $(4, \infty)$.

60. $(1, 2) \cup (1, \infty)$

$(1, 2)$ includes the real numbers between 1 and 2, not including 1 and 2.

$(1, \infty)$ includes all real numbers greater than 1.
Find the union. The numbers in the first set, the second set, or in both sets are all real numbers greater than 1.

The solution set is $(1, \infty)$.

61. $|x| = 7 \Leftrightarrow x = 7 \text{ or } x = -7$

The solution set is $\{-7, 7\}$.

62. $|x + 2| = 9$

$x + 2 = 9 \quad \text{or} \quad x + 2 = -9$
$x = 7 \quad \text{or} \quad x = -11$

The solution set is $\{-11, 7\}$.

63. $|3x - 7| = 8$

$3x - 7 = 8 \quad \text{or} \quad 3x - 7 = -8$
$3x = 15 \qquad\qquad 3x = -1$
$x = 5 \quad \text{or} \quad x = -\frac{1}{3}$

The solution set is $\left\{-\frac{1}{3}, 5\right\}$.

64. $|x - 4| = -12$

Since the absolute value of an expression can never be negative, there are no solutions for this equation.

The solution set is $\emptyset$.

65. $|2x - 7| + 4 = 11$

$|2x - 7| = 7$

$2x - 7 = 7 \quad \text{or} \quad 2x - 7 = -7$
$2x = 14 \qquad\qquad 2x = 0$
$x = 7 \quad \text{or} \qquad x = 0$

The solution set is $\{0, 7\}$.

66. $|4x + 2| - 7 = -3$

$|4x + 2| = 4$

$4x + 2 = 4 \quad \text{or} \quad 4x + 2 = -4$
$4x = 2 \qquad\qquad 4x = -6$
$x = \frac{2}{4} \qquad\qquad x = -\frac{6}{4}$
$x = \frac{1}{2} \quad \text{or} \qquad x = -\frac{3}{2}$

The solution set is $\left\{-\frac{3}{2}, \frac{1}{2}\right\}$.

67. $|3x + 1| = |x + 2|$

$3x + 1 = x + 2 \quad \text{or} \quad 3x + 1 = -(x + 2)$
$2x = 1 \qquad\qquad 3x + 1 = -x - 2$
$\qquad\qquad\qquad\qquad 4x = -3$
$x = \frac{1}{2} \quad \text{or} \qquad x = -\frac{3}{4}$

The solution set is $\left\{-\frac{3}{4}, \frac{1}{2}\right\}$.

68. $|2x - 1| = |2x + 3|$

$2x - 1 = 2x + 3 \quad \text{or} \quad 2x - 1 = -(2x + 3)$
$-1 = 3 \quad \textit{False} \qquad 2x - 1 = -2x - 3$
$\qquad\qquad\qquad\qquad\qquad 4x = -2$
$\textit{No solution} \quad \text{or} \qquad x = -\frac{2}{4} = -\frac{1}{2}$

The solution set is $\left\{-\frac{1}{2}\right\}$.

69. $|x| < 14 \Leftrightarrow -14 < x < 14$

The solution set is $(-14, 14)$.

70. $|-x + 6| \leq 7$

$-7 \leq -x + 6 \leq 7$
$-13 \leq -x \qquad \leq 1 \quad \textit{Subtract 6.}$
$\qquad\qquad\qquad\qquad \textit{Multiply by } -1.$
$13 \geq x \qquad\quad \geq -1 \quad \textit{Reverse inequalities.}$
$-1 \leq x \qquad\quad \leq 13 \quad \textit{Equivalent inequality}$

The solution set is $[-1, 13]$.

71. $|2x + 5| \leq 1$

$-1 \leq 2x + 5 \leq 1$
$-6 \leq 2x \qquad \leq -4$
$-3 \leq x \qquad \leq -2$

The solution set is $[-3, -2]$.

72. $|x + 1| \geq -3$

Since the absolute value of an expression is always nonnegative (positive or zero), the inequality is *true* for any real number x.

The solution set is $(-\infty, \infty)$.

Copyright © 2012 Pearson Education, Inc. Publishing as Addison-Wesley.

73. **[2.5]** $5 - (6 - 4x) > 2x - 5$
$$5 - 6 + 4x > 2x - 5$$
$$-1 + 4x > 2x - 5$$
$$2x > -4$$
$$x > -2$$

The solution set is $(-2, \infty)$.

74. **[2.2]** Solve $ak + bt = 6r$ for k.

$$ak = 6r - bt$$
$$k = \frac{6r - bt}{a}$$

75. **[2.6]** $x < 3$ and $x \geq -2$

The real numbers that are common to both sets are the numbers greater than or equal to -2 and less than 3.

$$-2 \leq x < 3$$

The solution set is $[-2, 3)$.

76. **[2.1]** $\dfrac{4x + 2}{4} + \dfrac{3x - 1}{8} = \dfrac{x + 6}{16}$
Clear fractions by multiplying by the LCD, 16.
$$4(4x + 2) + 2(3x - 1) = x + 6$$
$$16x + 8 + 6x - 2 = x + 6$$
$$22x + 6 = x + 6$$
$$21x = 0$$
$$x = 0$$

The solution set is $\{0\}$.

77. **[2.7]** $|3x + 6| \geq 0$

The absolute value of an expression is always nonnegative, so the inequality is true for any real number k.

The solution set is $(-\infty, \infty)$.

78. **[2.5]** $-5x \geq -10$

$$x \leq 2 \qquad \begin{array}{l}\textit{Divide by } -5 < 0;\\ \textit{reverse the symbol.}\end{array}$$

The solution set is $(-\infty, 2]$.

79. **[2.2]** Use the formula $V = LWH$, and solve for H.

$$\frac{V}{LW} = \frac{LWH}{LW}$$
$$\frac{V}{LW} = H, \text{ or } H = \frac{V}{LW}$$

Substitute 1.5 for W, 5 for L, and 75 for V.

$$H = \frac{75}{5(1.5)} = \frac{75}{7.5} = 10$$

The height of the box is 10 ft.

80. **[2.4]** Let x = the first consecutive integer. Then $x + 1$ = the second consecutive integer and $x + 2$ = the third consecutive integer.
The sum of the first and third integers is 47 more than the second integer, so an equation is

$$x + (x + 2) = 47 + (x + 1).$$
$$2x + 2 = 48 + x$$
$$x = 46$$

Then $x + 1 = 47$, and $x + 2 = 48$.
The integers are 46, 47, and 48.

81. **[2.7]** $|3x + 2| + 4 = 9$
$$|3x + 2| = 5$$

$$\begin{array}{lll} 3x + 2 = 5 & \text{or} & 3x + 2 = -5 \\ 3x = 3 & & 3x = -7 \\ x = 1 & \text{or} & x = -\frac{7}{3} \end{array}$$

The solution set is $\left\{-\frac{7}{3}, 1\right\}$.

82. **[2.1]** $0.05x + 0.03(1200 - x) = 42$
$$5x + 3(1200 - x) = 4200$$
$$\textit{Multiply by 100.}$$
$$5x + 3600 - 3x = 4200$$
$$2x = 600$$
$$x = 300$$

The solution set is $\{300\}$.

83. **[2.7]** $|x + 3| \leq 13$

$$-13 \leq x + 3 \leq 13$$
$$-16 \leq \quad x \quad \leq 10$$

The solution set is $[-16, 10]$.

84. **[2.5]** $\frac{3}{4}(x - 2) - \frac{1}{3}(5 - 2x) < -2$
$$9(x - 2) - 4(5 - 2x) < -24$$
$$\textit{Multiply by 12.}$$
$$9x - 18 - 20 + 8x < -24$$
$$17x - 38 < -24$$
$$17x < 14$$
$$x < \tfrac{14}{17}$$

The solution set is $\left(-\infty, \frac{14}{17}\right)$.

85. **[2.5]** $-4 < 3 - 2x < 9$
$$-7 < -2x \quad < 6 \quad \textit{Subtract 3.}$$
$$\frac{7}{2} > x \qquad > -3 \quad \begin{array}{l}\textit{Divide by } -2.\\ \textit{Reverse inequalities.}\end{array}$$
$$-3 < x \qquad < \tfrac{7}{2} \quad \textit{Equivalent inequality}$$

The solution set is $\left(-3, \frac{7}{2}\right)$.

Copyright © 2012 Pearson Education, Inc. Publishing as Addison-Wesley.

86. **[2.5]** $-0.3x + 2.1(x - 4) \leq -6.6$

$$-3x + 21(x - 4) \leq -66$$

Multiply by 10.

$$-3x + 21x - 84 \leq -66$$
$$18x - 84 \leq -66$$
$$18x \leq 18$$
$$x \leq 1$$

The solution set is $(-\infty, 1]$.

87. **[2.4]** Let $x =$ the angle. Then $90 - x$ is its complement and $180 - x$ is its supplement. The complement of an angle measures $10°$ less than one-fifth of its supplement.

$$90 - x = \tfrac{1}{5}(180 - x) - 10$$
$$450 - 5x = 180 - x - 50 \quad \textit{Multiply by 5.}$$
$$450 - 5x = 130 - x$$
$$320 = 4x$$
$$80 = x$$

The measure of the angle is $80°$.

88. **[2.5]** Let $x =$ the employee's earnings during the fifth month. The average of the five months must be at least $1000.

$$\frac{900 + 1200 + 1040 + 760 + x}{5} \geq 1000$$
$$900 + 1200 + 1040 + 760 + x \geq 5000$$
$$3900 + x \geq 5000$$
$$x \geq 1100$$

Any amount greater than or equal to $1100 will qualify the employee for the pension plan.

89. **[2.7]** $|5x - 1| > 14$

$$5x - 1 > 14 \quad \text{or} \quad 5x - 1 < -14$$
$$5x > 15 \qquad\qquad 5x < -13$$
$$x > 3 \quad \text{or} \qquad x < -\tfrac{13}{5}$$

The solution set is $\left(-\infty, -\tfrac{13}{5}\right) \cup (3, \infty)$.

90. **[2.6]** $x \geq -2$ or $x < 4$

The solution set includes all numbers either greater than or equal to -2 or all numbers less than 4. This is the union and is the set of all real numbers.

The solution set is $(-\infty, \infty)$.

91. **[2.3]** Let $x =$ the number of liters of the 20% solution. Then $x + 10$ is the number of liters of the resulting 40% solution.

Liters of Solution	Percent (as a decimal)	Liters of Mixture
x	0.20	$0.20x$
10	0.50	$0.50(10) = 5$
$x + 10$	0.40	$0.40(x + 10)$

From the last column:

$$0.20x + 5 = 0.40(x + 10)$$
$$0.20x + 5 = 0.40x + 4$$
$$1 = 0.20x$$
$$5 = x$$

5 L of the 20% solution should be used.

92. **[2.7]** $|x - 1| = |2x + 3|$

$$x - 1 = 2x + 3 \quad \text{or} \quad x - 1 = -(2x + 3)$$
$$x - 1 = -2x - 3$$
$$3x = -2$$
$$-4 = x \qquad \text{or} \qquad x = -\tfrac{2}{3}$$

The solution set is $\left\{-4, -\tfrac{2}{3}\right\}$.

93. **[2.1]** $\dfrac{3x}{5} - \dfrac{x}{2} = 3$

$$6x - 5x = 30 \quad \textit{Multiply by 10.}$$
$$x = 30$$

The solution set is $\{30\}$.

94. **[2.7]** $|x + 3| \leq 1$

$$-1 \leq x + 3 \leq 1$$
$$-4 \leq x \qquad \leq -2$$

The solution set is $[-4, -2]$.

95. **[2.7]** $|3x - 7| = 4$

$$3x - 7 = 4 \quad \text{or} \quad 3x - 7 = -4$$
$$3x = 11 \qquad\qquad 3x = 3$$
$$x = \tfrac{11}{3} \quad \text{or} \qquad x = 1$$

The solution set is $\left\{1, \tfrac{11}{3}\right\}$.

96. **[2.1]** $5(2x - 7) = 2(5x + 3)$

$$10x - 35 = 10x + 6$$
$$-35 = 6 \quad \textit{False}$$

This equation is a *contradiction*.

The solution set is $\emptyset$.

Copyright © 2012 Pearson Education, Inc. Publishing as Addison-Wesley.

97. [2.7] (a) $|5x + 3| < k$

If $k < 0$, then $|5x + 3|$ would be less than a negative number. Since the absolute value of an expression is always nonnegative (positive or zero), the solution set is $\emptyset$.

(b) $|5x + 3| > k$

If $k < 0$, then $|5x + 3|$ would be greater than a negative number. Since the absolute value of an expression is always nonnegative (positive or zero), the solution set is the set of all real numbers, $(-\infty, \infty)$.

(c) $|5x + 3| = k$

If $k < 0$, then $|5x + 3|$ would be equal to a negative number. Since the absolute value of an expression is always nonnegative (positive or zero), the solution set is $\emptyset$.

98. [2.6] $x > 6$ and $x < 8$

The graph of the solution set is all numbers both greater than 6 *and* less than 8. This is the intersection. The elements common to both sets are the numbers between 6 and 8, not including the endpoints. The solution set is $(6, 8)$.

99. [2.6] $-5x + 1 \geq 11$ or $3x + 5 \geq 26$

$-5x \geq 10$ $3x \geq 21$

$x \leq -2$ or $x \geq 7$

The graph of the solution set is all numbers either less than or equal to -2 *or* greater than or equal to 7. This is the union. The solution set is $(-\infty, -2] \cup [7, \infty)$.

100. [2.6] (a) All states have less than 3 million female workers. The set of states with more than 3 million male workers is {Illinois}. Illinois is the only state in both sets, so the set of states with less than 3 million female workers *and* more than 3 million male workers is {Illinois}.

(b) The set of states with less than 1 million female workers *or* more than 2 million male workers is {Illinois, Maine, North Carolina, Oregon, Utah}.

(c) It is easy to see that the sum of the female and male workers for each state doesn't exceed 7 million, so the set of states with a total of more than 7 million civilian workers is { }, or $\emptyset$.

Chapter 2 Test

1. $3(2x - 2) - 4(x + 6) = 3x + 8 + x$

$6x - 6 - 4x - 24 = 4x + 8$

$2x - 30 = 4x + 8$

$-2x = 38$

$x = -19$

The solution set is $\{-19\}$.

2. $0.08x + 0.06(x + 9) = 1.24$

$8x + 6(x + 9) = 124$ *Multiply by 100.*

$8x + 6x + 54 = 124$

$14x + 54 = 124$

$14x = 70$

$x = 5$

The solution set is $\{5\}$.

3. $\dfrac{x + 6}{10} + \dfrac{x - 4}{15} = \dfrac{x + 2}{6}$

$3(x + 6) + 2(x - 4) = 5(x + 2)$ *Multiply by 30.*

$3x + 18 + 2x - 8 = 5x + 10$

$5x + 10 = 5x + 10$ *True*

This is an *identity*.

The solution set is {all real numbers}.

4. (a) $3x - (2 - x) + 4x + 2 = 8x + 3$

$3x - 2 + x + 4x + 2 = 8x + 3$

$8x = 8x + 3$

$0 = 3$ *False*

The false statement indicates that the equation is a *contradiction*.

The solution set is $\emptyset$.

(b) $\dfrac{x}{3} + 7 = \dfrac{5x}{6} - 2 - \dfrac{x}{2} + 9$

Multiply each side by the LCD, 6.

$2x + 42 = 5x - 12 - 3x + 54$

$2x + 42 = 2x + 42$

$0 = 0$ *True*

This equation is an *identity*.

The solution set is {all real numbers}.

(c) $-4(2x - 6) = 5x + 24 - 7x$

$-8x + 24 = -2x + 24$

$24 = 6x + 24$

$0 = 6x$

$0 = x$

This is a *conditional equation*.

The solution set is $\{0\}$.

Copyright © 2012 Pearson Education, Inc. Publishing as Addison-Wesley.

5.   Solve $V = \frac{1}{3}bh$ for h.

$$V = \tfrac{1}{3}bh$$
$$3V = bh \qquad \textit{Multiply by 3.}$$
$$\frac{3V}{b} = h \qquad \textit{Divide by b.}$$

6.   Solve $-16t^2 + vt - S = 0$ for v.

$$vt = S + 16t^2 \qquad \textit{Add S, } 16t^2.$$
$$v = \frac{S + 16t^2}{t} \qquad \textit{Divide by t.}$$

7.   Solve $d = rt$ for t and substitute 500 for d and 150.318 for r.

$$t = \frac{d}{r} = \frac{500}{150.318} \approx 3.326$$

His time was about 3.326 hr.

8.   Use $I = Prt$ and substitute \$2281.25 for I, \$36,500 for P, and 1 for t.

$$2281.25 = 36,500r(1)$$
$$r = \frac{2281.25}{36,500} = 0.0625$$

The rate of interest is 6.25%.

9. $\dfrac{27,232}{36,723} \approx 0.742$

74.2% were classified as post offices.

10. Let $x =$ the amount invested at 3%. Then $28,000 - x =$ the amount invested at 5%.

Principal	Rate (as a decimal)	Interest
x	0.03	0.03x
$28,000 - x$	0.05	$0.05(28,000 - x)$
\$28,000	← Totals →	\$1240

From the last column:

$$0.03x + 0.05(28,000 - x) = 1240$$
$$3x + 5(28,000 - x) = 124,000$$
$$\textit{Multiply by 100.}$$
$$3x + 140,000 - 5x = 124,000$$
$$-2x = -16,000$$
$$x = 8000$$

He invested \$8000 at 3% and
\$28,000 - \$8000 = \$20,000 at 5%.

11. Let $x =$ the rate of the slower car. Then $x + 15 =$ the rate of the faster car.

Use the formula $d = rt$.

	Rate	Time	Distance
Slower Car	x	6	$6x$
Faster Car	$x + 15$	6	$6(x + 15)$
			630

The total distance traveled is the sum of the distances traveled by each car.

$$6x + 6(x + 15) = 630$$
$$6x + 6x + 90 = 630$$
$$12x = 540$$
$$x = 45$$

The slower car traveled at 45 mph, while the faster car traveled at $45 + 15 = 60$ mph.

12. The sum of the three angle measures is $180°$.

$$(2x + 20) + x + x = 180$$
$$4x + 20 = 180$$
$$4x = 160$$
$$x = 40$$

The three angle measures are $40°$, $40°$, and $(2 \cdot 40 + 20)° = 100°$.

13.
$$4 - 6(x + 3) \le -2 - 3(x + 6) + 3x$$
$$4 - 6x - 18 \le -2 - 3x - 18 + 3x$$
$$-6x - 14 \le -20$$
$$-6x \le -6$$

Divide by -6, and reverse the inequality symbol.

$$x \ge 1$$

The solution set is $[1, \infty)$.

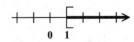

14. $-\dfrac{4}{7}x > -16$

$$-4x > -112 \qquad \textit{Multiply by 7.}$$

Divide by -4, and reverse the inequality symbol.

$$x < 28$$

The solution set is $(-\infty, 28)$.

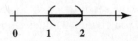

15. $-1 < 3x - 4 < 2$

$$3 < 3x \quad\; < 6 \quad \textit{Add 4.}$$
$$1 < x \quad\;\; < 2 \quad \textit{Divide by 3.}$$

The solution set is $(1, 2)$.

Copyright © 2012 Pearson Education, Inc. Publishing as Addison-Wesley.

16. $-6 \leq \frac{4}{3}x - 2 \leq 2$

$\quad -18 \leq 4x - 6 \leq 6 \quad$ *Multiply by 3.*

$\quad -12 \leq 4x \quad\;\; \leq 12 \quad$ *Add 6.*

$\quad\;\; -3 \leq x \quad\;\;\;\; \leq 3 \quad$ *Divide by 4.*

The solution set is $[-3, 3]$.

17. For each inequality, divide both sides by -3 and reverse the direction of the inequality symbol.

A. $\;\; -3x < 9$ **B.** $\;\; -3x > -9$

$\qquad x > -3 \qquad\qquad\qquad x < 3$

C. $\;\; -3x > 9$ **D.** $\;\; -3x < -9$

$\qquad x < -3 \qquad\qquad\qquad x > 3$

Thus, inequality **C** is equivalent to $x < -3$.

18. Let $x = $ the score on the fourth test.

$$\frac{83 + 76 + 79 + x}{4} \geq 80$$

$$\frac{238 + x}{4} \geq 80$$

$$238 + x \geq 320$$

$$x \geq 82$$

The minimum score must be 82 to guarantee a B.

19. **(a)** $A \cap B = \{1, 2, 5, 7\} \cap \{1, 5, 9, 12\}$

$\qquad\qquad = \{1, 5\}$

(b) $A \cup B = \{1, 2, 5, 7\} \cup \{1, 5, 9, 12\}$

$\qquad\qquad = \{1, 2, 5, 7, 9, 12\}$

20. $3x \geq 6 \quad$ and $\quad x - 4 < 5$

$\;\; x \geq 2 \quad$ and $\qquad x < 9$

The solution set is all numbers both greater than or equal to 2 *and* less than 9. This is the intersection. The numbers common to both sets are between 2 and 9, including 2 but not 9. The solution set is $[2, 9)$.

21. $-4x \leq -24 \quad$ or $\quad 4x - 2 < 10$

$\qquad\qquad\qquad\qquad\quad 4x < 12$

$\;\; x \geq 6 \qquad$ or $\qquad x < 3$

The solution set is all numbers less than 3 or greater than or equal to 6. This is the union. The solution set is $(-\infty, 3) \cup [6, \infty)$.

22. $|4x + 3| \leq 7$

$\quad -7 \leq 4x + 3 \leq 7$

$\quad -10 \leq 4x \quad\;\; \leq 4$

$\quad -\frac{10}{4} \leq x \quad\;\;\; \leq \frac{4}{4}$

$\quad\;\; -\frac{5}{2} \leq x \quad\;\;\; \leq 1$

The solution set is $\left[-\frac{5}{2}, 1\right]$.

23. $|5 - 6x| > 12$

$\quad 5 - 6x > 12 \quad$ or $\quad 5 - 6x < -12$

$\qquad -6x > 7 \qquad\qquad\quad -6x < -17$

$\qquad\;\; x < -\frac{7}{6} \quad$ or $\qquad\quad x > \frac{17}{6}$

The solution set is $\left(-\infty, -\frac{7}{6}\right) \cup \left(\frac{17}{6}, \infty\right)$.

24. $|7 - x| \leq -1$

Since the absolute value of an expression is always nonnegative (positive or zero), the inequality is *false* for any real number x.

The solution set is $\emptyset$.

25. $|-3x + 4| - 4 < -1$

$\qquad\;\; |-3x + 4| < 3$

$\quad -3 < -3x + 4 < 3$

$\quad -7 < -3x \quad\;\; < -1$

$\quad \frac{7}{3} > x \qquad\quad > \frac{1}{3} \quad$ *Reverse inequalities.*

$\quad \frac{1}{3} < x \qquad\quad < \frac{7}{3} \quad$ *Equivalent inequality*

The solution set is $\left(\frac{1}{3}, \frac{7}{3}\right)$.

26. $|3x - 2| + 1 = 8$

$\quad\;\; |3x - 2| = 7$

$\quad 3x - 2 = 7 \quad$ or $\quad 3x - 2 = -7$

$\qquad 3x = 9 \quad$ or $\qquad 3x = -5$

$\qquad\; x = \frac{9}{3} = 3 \quad$ or $\qquad\;\; x = -\frac{5}{3}$

The solution set is $\left\{-\frac{5}{3}, 3\right\}$.

27. $|3 - 5x| = |2x + 8|$

$\quad 3 - 5x = 2x + 8 \quad$ or $\quad 3 - 5x = -(2x + 8)$

$\qquad -7x = 5 \qquad\qquad\quad 3 - 5x = -2x - 8$

$\qquad\qquad\qquad\qquad\qquad\qquad -3x = -11$

$\qquad\;\; x = -\frac{5}{7} \quad$ or $\qquad\qquad x = \frac{11}{3}$

The solution set is $\left\{-\frac{5}{7}, \frac{11}{3}\right\}$.

28. **(a)** $|8x - 5| < k$

If $k < 0$, then $|8x - 5|$ would be less than a negative number. Since the absolute value of an expression is always nonnegative (positive or zero), the solution set is $\emptyset$.

(b) $|8x - 5| > k$

If $k < 0$, then $|8x - 5|$ would be greater than a negative number. Since the absolute value of an expression is always nonnegative (positive or zero), the solution set is the set of all real numbers, $(-\infty, \infty)$.

(c) $|8x - 5| = k$

If $k < 0$, then $|8x - 5|$ would be equal to a negative number. Since the absolute value of an expression is always nonnegative (positive or zero), the solution set is $\emptyset$.

Copyright © 2012 Pearson Education, Inc. Publishing as Addison-Wesley.

Cumulative Review Exercises
(Chapters 1–2)

Exercises 1–6 refer to set A.

Let $A = \left\{ -8, -\frac{2}{3}, -\sqrt{6}, 0, \frac{4}{5}, 9, \sqrt{36} \right\}$.

Simplify $\sqrt{36}$ to 6.

1.   The elements 9 and 6 are natural numbers.

2.   The elements 0, 9, and 6 are whole numbers.

3.   The elements -8, 0, 9, and 6 are integers.

4.   The elements $-8, -\frac{2}{3}, 0, \frac{4}{5}, 9,$ and 6 are rational numbers.

5.   The element $-\sqrt{6}$ is an irrational number.

6.   All the elements in set A are real numbers.

7. $\begin{aligned}[t] -\frac{4}{3} - \left(-\frac{2}{7} \right) &= -\frac{4}{3} + \frac{2}{7} \\ &= -\frac{28}{21} + \frac{6}{21} \\ &= -\frac{22}{21} \end{aligned}$

8. $\begin{aligned}[t] |-4| - |2| + |-6| &= 4 - 2 + 6 \\ &= 2 + 6 \\ &= 8 \end{aligned}$

9. $(-2)^4 + (-2)^3 = 16 + (-8) = 8$

10. $\sqrt{25} - 5(-1)^0 = 5 - 5(1) = 5 - 5 = 0$

11. $(-3)^5 = (-3)(-3)(-3)(-3)(-3) = -243$

12. $\left(\frac{6}{7} \right)^3 = \frac{6}{7} \cdot \frac{6}{7} \cdot \frac{6}{7} = \frac{216}{343}$

13. $\left(-\frac{2}{3} \right)^3 = \left(-\frac{2}{3} \right)\left(-\frac{2}{3} \right)\left(-\frac{2}{3} \right) = -\frac{8}{27}$

14. $-4^6 = -(4 \cdot 4 \cdot 4 \cdot 4 \cdot 4 \cdot 4) = -4096$

For Exercises 15–17, let $a = 2$, $b = -3$, and $c = 4$.

15. $\begin{aligned}[t] -3a + 2b - c &= -3(2) + 2(-3) - 4 \\ &= -6 - 6 - 4 \\ &= -16 \end{aligned}$

16. $\begin{aligned}[t] -8\left(a^2 + b^3 \right) &= -8\left[2^2 + (-3)^3 \right] \\ &= -8[4 + (-27)] \\ &= -8(-23) \\ &= 184 \end{aligned}$

17. $\begin{aligned}[t] \frac{3a^3 - b}{4 + 3c} &= \frac{3(2)^3 - (-3)}{4 + 3(4)} \\ &= \frac{3(8) - (-3)}{4 + 3(4)} \\ &= \frac{24 + 3}{4 + 12} \\ &= \frac{27}{16} \end{aligned}$

18. $\begin{aligned}[t] &-7r + 5 - 13r + 12 \\ &= -7r - 13r + 5 + 12 \\ &= (-7 - 13)r + (5 + 12) \\ &= -20r + 17 \end{aligned}$

19. $\begin{aligned}[t] &-(3k + 8) - 2(4k - 7) + 3(8k + 12) \\ &= -3k - 8 - 8k + 14 + 24k + 36 \\ &= -3k - 8k + 24k - 8 + 14 + 36 \\ &= 13k + 42 \end{aligned}$

20. $(a + b) + 4 = 4 + (a + b)$

The order of the terms $(a + b)$ and 4 have been reversed. This is an illustration of the commutative property.

21. $4x + 12x = (4 + 12)x$

The common variable, x, has been removed from each term. This is an illustration of the distributive property.

22. $\begin{aligned}[t] -4x + 7(2x + 3) &= 7x + 36 \\ -4x + 14x + 21 &= 7x + 36 \\ 10x + 21 &= 7x + 36 \\ 3x &= 15 \\ x &= 5 \end{aligned}$

The solution set is $\{5\}$.

23. $\begin{aligned}[t] -\tfrac{3}{5}x + \tfrac{2}{3}x &= 2 \\ 3(-3x) + 5(2x) &= 15(2) \quad \textit{Multiply by 15.} \\ -9x + 10x &= 30 \\ x &= 30 \end{aligned}$

The solution set is $\{30\}$.

24. $\begin{aligned}[t] 0.06x + 0.03(100 + x) &= 4.35 \\ 6x + 3(100 + x) &= 435 \quad \textit{Multiply by 100.} \\ 6x + 300 + 3x &= 435 \\ 9x &= 135 \\ x &= 15 \end{aligned}$

The solution set is $\{15\}$.

25.   Solve $P = a + b + c$ for b.

$$P - a - c = b, \ \text{ or } \ b = P - a - c$$

26. $\begin{aligned}[t] 4(2x - 6) + 3(x - 2) &= 11x + 1 \\ 8x - 24 + 3x - 6 &= 11x + 1 \\ 11x - 30 &= 11x + 1 \\ -30 &= 1 \quad \textit{False} \end{aligned}$

The solution set is $\emptyset$.

Copyright © 2012 Pearson Education, Inc. Publishing as Addison-Wesley.

27. $\frac{2}{3}x + \frac{5}{8}x = \frac{31}{24}x$

Multiply by the LCD, 24.

$8(2x) + 3(5x) = 31x$

$16x + 15x = 31x$

$31x = 31x$ *True*

The solution set is {all real numbers}.

28. $3 - 2(x + 7) \leq -x + 3$

$3 - 2x - 14 \leq -x + 3$

$-2x - 11 \leq -x + 3$

$-x \leq 14$

$x \geq -14$ *Multiply by -1.*
Reverse inequality.

The solution set is $[-14, \infty)$.

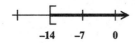

29. $-4 < 5 - 3x \leq 0$

$-9 < -3x \ \ \leq -5$ *Subtract 5.*

$3 > x \qquad \geq \dfrac{5}{3}$ *Divide by -3.*
Reverse inequalities.

$\dfrac{5}{3} \leq x \qquad < 3$ *Equivalent inequality*

The solution set is $[\frac{5}{3}, 3)$.

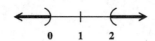

30. $2x + 1 > 5$ or $2 - x > 2$

$2x > 4 \qquad\qquad -x > 0$

$x > 2$ or $x < 0$

The solution set is $(-\infty, 0) \cup (2, \infty)$.

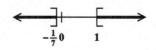

31. $|-7k + 3| \geq 4$

$-7k + 3 \geq 4$ or $-7k + 3 \leq -4$

$-7k \geq 1 \qquad\qquad -7k \leq -7$

$k \leq -\frac{1}{7}$ or $k \geq 1$

The solution set is $(-\infty, -\frac{1}{7}] \cup [1, \infty)$.

32. Let $x =$ the amount invested at 5%. Then $x + 2000 =$ the amount invested at 6%.

Use $I = prt$ and create a table.

Principal	Rate (as a decimal)	Interest
x	0.05	$0.05x$
$x + 2000$	0.06	$0.06(x + 2000)$
	Total →	$670

From the last column:

$0.05x + 0.06(x + 2000) = 670$

$5x + 6(x + 2000) = 67{,}000$ *Multiply by 100.*

$5x + 6x + 12{,}000 = 67{,}000$

$11x = 55{,}000$

$x = 5000$

$5000 was invested at 5% and $7000 at 6%.

33. Let $x =$ the amount of pure alcohol that should be added.

Liters of Solution	Percent (as a decimal)	Liters of Pure Alcohol
x	1.00	$1.00x$
7	0.10	$0.10(7) = 0.7$
$x + 7$	0.30	$0.30(x + 7)$

From the last column:

$1.00x + 0.7 = 0.30(x + 7)$

$10x + 7 = 3(x + 7)$ *Multiply by 10.*

$10x + 7 = 3x + 21$

$7x = 14$

$x = 2$

2 L of pure alcohol should be added to the solution.

34. Clark's rule:

$$\frac{\text{Weight of child in pounds}}{150} \times \frac{\text{adult}}{\text{dose}} = \frac{\text{child's}}{\text{dose}}$$

If the child weighs 55 lb and the adult dosage is 120 mg, then

$$\frac{55}{150} \times 120 = 44.$$

The child's dosage is 44 mg.

Copyright © 2012 Pearson Education, Inc. Publishing as Addison-Wesley.

35. (a) In 1975, there were 1756 daily newspapers.
In 2008, there were 1408 daily newspapers. The
difference is $1756 - 1408 = 348$, so the number
of daily newspapers decreased by 348 from 1975
to 2008.

(b) $\dfrac{348}{1756} \approx 0.198$, or 19.8%.

The number of daily newspapers decreased by
approximately 19.8% from 1975 to 2008.

Copyright © 2012 Pearson Education, Inc. Publishing as Addison-Wesley.

CHAPTER 3   GRAPHS, LINEAR EQUATIONS, AND FUNCTIONS

3.1   The Rectangular Coordinate System

3.1 Now Try Exercises

N1. $2x - y = 4$

To complete the ordered pairs, substitute the given value of x or y in the equation.
For $(0, \underline{\quad})$, let $x = 0$.

$$2x - y = 4$$
$$2(0) - y = 4$$
$$-y = 4$$
$$y = -4$$

The ordered pair is $(0, -4)$.

For $(\underline{\quad}, 0)$ let $y = 0$.

$$2x - y = 4$$
$$2x - 0 = 4$$
$$2x = 4$$
$$x = 2$$

The ordered pair is $(2, 0)$.

For $(4, \underline{\quad})$, let $x = 4$.

$$2x - y = 4$$
$$2(4) - y = 4$$
$$8 - y = 4$$
$$-y = -4$$
$$y = 4$$

The ordered pair is $(4, 4)$.

For $(\underline{\quad}, 2)$, let $y = 2$.

$$2x - y = 4$$
$$2x - 2 = 4$$
$$2x = 6$$
$$x = 3$$

The ordered pair is $(3, 2)$.
The completed table follows.

x	y
0	-4
2	0
4	4
3	2

N2. $x - 2y = 4$
To find the x-intercept, let $y = 0$.

$$x - 2y = 4$$
$$x - 2(0) = 4$$
$$x = 4$$

The x-intercept is $(4, 0)$.

To find the y-intercept, let $x = 0$.

$$x - 2y = 4$$
$$0 - 2y = 4$$
$$-2y = 4$$
$$y = -2$$

The y-intercept is $(0, -2)$.

Plot the intercepts, and draw the line through them.

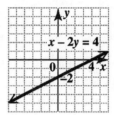

N3. $y = -2$

In standard form, the equation is $0x + y = -2$. Every value of x leads to $y = -2$, so the y-intercept is $(0, -2)$. There is no x-intercept. The graph is the horizontal line through $(0, -2)$.

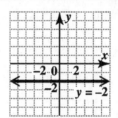

N4. $x + 3 = 0$

In standard form, the equation is $x + 0y = -3$. Every value of y leads to $x = -3$, the x-intercept is $(-3, 0)$. There is no y-intercept. The graph is the vertical line through $(-3, 0)$.

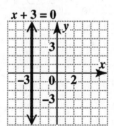

N5. $2x + 3y = 0$
To find the x-intercept, let $y = 0$.

$$2x + 3(0) = 0$$
$$2x = 0$$
$$x = 0$$

Since the x-intercept is $(0, 0)$, the y-intercept is also $(0, 0)$.

Find another point. Let $x = 3$.

Copyright © 2012 Pearson Education, Inc. Publishing as Addison-Wesley.

$$2(3) + 3y = 0$$
$$6 + 3y = 0$$
$$3y = -6$$
$$y = -2$$

This gives the ordered pair $(3, -2)$. Plot $(3, -2)$ and $(0, 0)$ and draw the line through them.

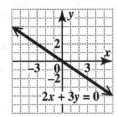

N6. By the midpoint formula, the midpoint of the segment with endpoints $(2, -5)$ and $(-4, 7)$ is

$$\left(\frac{2 + (-4)}{2}, \frac{-5 + 7}{2} \right) = \left(\frac{-2}{2}, \frac{2}{2} \right) = (-1, 1).$$

3.1 Section Exercises

1. **(a)** x represents the year; y represents the higher education aid in billions of dollars.

(b) The dot above the year 2007 appears to be at about 150, so the revenue in 2007 was \$150 billion.

(c) The ordered pair is $(2007, 150)$.

(d) In 1997, federal tax revenues were about \$75 billion.

3. The point with coordinates $(0, 0)$ is called the _origin_ of a rectangular coordinate system.

5. The x-intercept is the point where a line crosses the x-axis. To find the x-intercept of a line, we let _y_ equal 0 and solve for _x_.

The y-intercept is the point where a line crosses the y-axis. To find the y-intercept of a line, we let _x_ equal 0 and solve for _y_.

7. To graph a straight line, we must find a minimum of _two_ points. A third point is sometimes found to check the accuracy of the first two points.

9. **(a)** The point $(1, 6)$ is located in quadrant I, since the x- and y-coordinates are both positive.

(b) The point $(-4, -2)$ is located in quadrant III, since the x- and y-coordinates are both negative.

(c) The point $(-3, 6)$ is located in quadrant II, since the x-coordinate is negative and the y-coordinate is positive.

(d) The point $(7, -5)$ is located in quadrant IV, since the x-coordinate is positive and the y-coordinate is negative.

(e) The point $(-3, 0)$ is located on the x-axis, so it does not belong to any quadrant.

(f) The point $(0, -0.5)$ is located on the y-axis, so it does not belong to any quadrant.

11. **(a)** If $xy > 0$, then both x and y have the same sign.
(x, y) is in quadrant I if x and y are positive.
(x, y) is in quadrant III if x and y are negative.

(b) If $xy < 0$, then x and y have different signs.
(x, y) is in quadrant II if $x < 0$ and $y > 0$.
(x, y) is in quadrant IV if $x > 0$ and $y < 0$.

(c) If $\frac{x}{y} < 0$, then x and y have different signs.
(x, y) is in either quadrant II or IV. (See part (b).)

(d) If $\frac{x}{y} > 0$, then x and y have the same sign.
(x, y) is in either quadrant I or III. (See part (a).)

For Exercises 13–22, see the rectangular coordinate system after Exercise 21.

13. To plot $(2, 3)$, go two units from zero to the right along the x-axis, and then go three units up parallel to the y-axis.

15. To plot $(-3, -2)$, go three units from zero to the left along the x-axis, and then go two units down parallel to the y-axis.

17. To plot $(0, 5)$, do not move along the x-axis at all since the x-coordinate is 0. Move five units up along the y-axis.

19. To plot $(-2, 4)$, go two units from zero to the left along the x-axis, and then go four units up parallel to the y-axis.

21. To plot $(-2, 0)$, go two units to the left along the x-axis. Do not move up or down since the y-coordinate is 0.

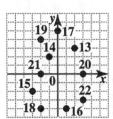

23. $y = x - 4$
(a) To complete the table, substitute the given values for x and y in the equation.
For $x = 0$: $y = x - 4$
$y = 0 - 4$
$y = -4$ **$(0, -4)$**

For $x = 1$: $y = x - 4$
$y = 1 - 4$
$y = -3$ **$(1, -3)$**

Copyright © 2012 Pearson Education, Inc. Publishing as Addison-Wesley.

In a similar manner, substitute $x = 2, 3$, and 4 to get $y = -2, -1$, and 0.

x	y
0	-4
1	-3
2	-2
3	-1
4	0

(b) Plot the ordered pairs and draw the line through them.

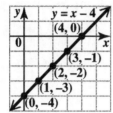

25. $x - y = 3$

(a) To complete the table, substitute the given values for x and y in the equation.

For $x = 0$: $x - y = 3$
$0 - y = 3$
$y = -3$ **$(0, -3)$**

For $y = 0$: $x - y = 3$
$x - 0 = 3$
$x = 3$ **$(3, 0)$**

For $x = 5$: $x - y = 3$
$5 - y = 3$
$-y = -2$
$y = 2$ **$(5, 2)$**

For $x = 2$: $x - y = 3$
$2 - y = 3$
$-y = 1$
$y = -1$ **$(2, -1)$**

(b) Plot the ordered pairs and draw the line through them.

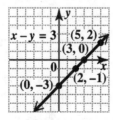

27. $x + 2y = 5$

(a) To complete the table, substitute the given values for x or y in the equation.

For $x = 0$: $x + 2y = 5$
$0 + 2y = 5$
$2y = 5$
$y = \frac{5}{2}$ **$(0, \frac{5}{2})$**

For $y = 0$: $x + 2y = 5$
$x + 2(0) = 5$
$x + 0 = 5$
$x = 5$ **$(5, 0)$**

For $x = 2$: $x + 2y = 5$
$2 + 2y = 5$
$2y = 3$
$y = \frac{3}{2}$ **$(2, \frac{3}{2})$**

For $y = 2$: $x + 2y = 5$
$x + 2(2) = 5$
$x + 4 = 5$
$x = 1$ **$(1, 2)$**

(b) Plot the ordered pairs and draw the line through them.

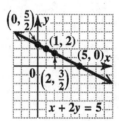

29. $4x - 5y = 20$

(a) For $x = 0$: $4x - 5y = 20$
$4(0) - 5y = 20$
$-5y = 20$
$y = -4$ **$(0, -4)$**

For $y = 0$: $4x - 5y = 20$
$4x - 5(0) = 20$
$4x = 20$
$x = 5$ **$(5, 0)$**

For $x = 2$: $4x - 5y = 20$
$4(2) - 5y = 20$
$8 - 5y = 20$
$-5y = 12$
$y = -\frac{12}{5}$ **$(2, -\frac{12}{5})$**

For $y = -3$: $4x - 5y = 20$
$4x - 5(-3) = 20$
$4x + 15 = 20$
$4x = 5$
$x = \frac{5}{4}$ **$(\frac{5}{4}, -3)$**

(b) Plot the ordered pairs and draw the line through them.

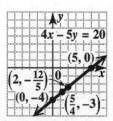

Copyright © 2012 Pearson Education, Inc. Publishing as Addison-Wesley.

31. $y = -2x + 3$

(a)

x	$-2x$	$y = -2x + 3$
0	0	3
1	-2	1
2	-4	-1
3	-6	-3

(b) Notice that as the value of x increases by 1, the value of y decreases by 2.

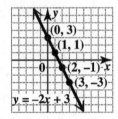

33. **(a)** The y-values are $-4, -3, -2, -1,$ and 0. They increase by <u>1</u> unit.

(b) The y-values are $3, 1, -1,$ and -3. They decrease by <u>2</u> units.

(c) It appears that the y-value increases (or decreases) by the value of the coefficient of x. So for $y = 2x + 4$, a conjecture is "For every increase in x by 1 unit, y increases by 2 units."

35. $2x + 3y = 12$

To find the x-intercept, let $y = 0$.

$$2x + 3y = 12$$
$$2x + 3(0) = 12$$
$$2x = 12$$
$$x = 6$$

The x-intercept is $(6, 0)$.
To find the y-intercept, let $x = 0$.

$$2x + 3y = 12$$
$$2(0) + 3y = 12$$
$$3y = 12$$
$$y = 4$$

The y-intercept is $(0, 4)$.
Plot the intercepts and draw the line through them.

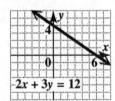

37. $x - 3y = 6$

To find the x-intercept, let $y = 0$.

$$x - 3y = 6$$
$$x - 3(0) = 6$$
$$x - 0 = 6$$
$$x = 6$$

The x-intercept is $(6, 0)$.
To find the y-intercept, let $x = 0$.

$$x - 3y = 6$$
$$0 - 3y = 6$$
$$-3y = 6$$
$$y = -2$$

The y-intercept is $(0, -2)$.
Plot the intercepts and draw the line through them.

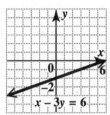

39. $5x + 6y = -10$

To find the x-intercept, let $y = 0$.

$$5x + 6(0) = -10$$
$$5x = -10$$
$$x = -2$$

The x-intercept is $(-2, 0)$.
To find the y-intercept, let $x = 0$.

$$5(0) + 6y = -10$$
$$6y = -10$$
$$y = -\frac{10}{6} = -\frac{5}{3}$$

The y-intercept is $(0, -\frac{5}{3})$.

Plot the intercepts and draw the line through them.

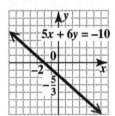

41. $\frac{2}{3}x - 3y = 7$

To find the x-intercept, let $y = 0$.

$$\frac{2}{3}x - 3(0) = 7$$
$$\frac{2}{3}x = 7$$
$$x = \frac{3}{2} \cdot 7 = \frac{21}{2}$$

The x-intercept is $(\frac{21}{2}, 0)$.
To find the y-intercept, let $x = 0$.

$$\frac{2}{3}(0) - 3y = 7$$
$$-3y = 7$$
$$y = -\frac{7}{3}$$

The y-intercept is $(0, -\frac{7}{3})$.
Plot the intercepts and draw the line through them.

Copyright © 2012 Pearson Education, Inc. Publishing as Addison-Wesley.

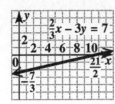

43. $y = 5$

This is a horizontal line. Every point has y-coordinate 5, so no point has y-coordinate 0. There is no x-intercept.

Since every point of the line has y-coordinate 5, the y-intercept is $(0, 5)$. Draw the horizontal line through $(0, 5)$.

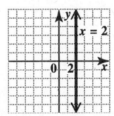

45. $x = 2$

This is a vertical line. Every point has x-coordinate 2, so the x-intercept is $(2, 0)$. Since every point of the line has x-coordinate 2, no point has x-coordinate 0. There is no y-intercept. Draw the vertical line through $(2, 0)$.

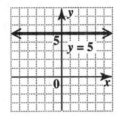

47. $x + 4 = 0$ $(x = -4)$

This is a vertical line. Every point has x-coordinate -4, so the x-intercept is $(-4, 0)$. Since every point of the line has x-coordinate -4, no point has x-coordinate 0. There is no y-intercept. Draw the vertical line through $(-4, 0)$.

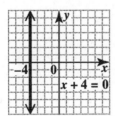

49. $y + 2 = 0$ $(y = -2)$

This is a horizontal line. Every point has y-coordinate -2, so no point has y-coordinate 0. There is no x-intercept.

Since every point of the line has y-coordinate -2, the y-intercept is $(0, -2)$. Draw the horizontal line through $(0, -2)$.

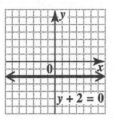

51. $x + 5y = 0$

To find the x-intercept, let $y = 0$.

$$x + 5y = 0$$
$$x + 5(0) = 0$$
$$x = 0$$

The x-intercept is $(0, 0)$, and since $x = 0$, this is also the y-intercept. Since the intercepts are the same, another point is needed to graph the line. Choose any number for y, say $y = -1$, and solve the equation for x.

$$x + 5y = 0$$
$$x + 5(-1) = 0$$
$$x = 5$$

This gives the ordered pair $(5, -1)$. Plot $(5, -1)$ and $(0, 0)$, and draw the line through them.

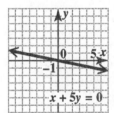

53. $2x = 3y$

If $x = 0$, then $y = 0$, so the x- and y-intercepts are $(0, 0)$. To get another point, let $x = 3$.

$$2(3) = 3y$$
$$2 = y$$

Plot $(3, 2)$ and $(0, 0)$, and draw the line through them.

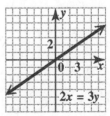

Copyright © 2012 Pearson Education, Inc. Publishing as Addison-Wesley.

55. $-\frac{2}{3}y = x$

If $x = 0$, then $y = 0$, so the x- and y-intercepts are $(0,0)$. To get another point, let $y = 3$.

$$-\frac{2}{3}(3) = x$$
$$-2 = x$$

Plot $(-2, 3)$ and $(0, 0)$, and draw the line through them.

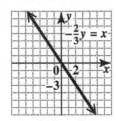

57. By the Midpoint Formula, the midpoint of the segment with endpoints $(-8, 4)$ and $(-2, -6)$ is

$$\left(\frac{-8 + (-2)}{2}, \frac{4 + (-6)}{2}\right) = \left(\frac{-10}{2}, \frac{-2}{2}\right) = (-5, -1).$$

59. By the Midpoint Formula, the midpoint of the segment with endpoints $(3, -6)$ and $(6, 3)$ is

$$\left(\frac{3 + 6}{2}, \frac{-6 + 3}{2}\right) = \left(\frac{9}{2}, \frac{-3}{2}\right) = \left(\frac{9}{2}, -\frac{3}{2}\right).$$

61. By the Midpoint Formula, the midpoint of the segment with endpoints $(-9, 3)$ and $(9, 8)$ is

$$\left(\frac{-9 + 9}{2}, \frac{3 + 8}{2}\right) = \left(\frac{0}{2}, \frac{11}{2}\right) = \left(0, \frac{11}{2}\right).$$

63. By the Midpoint Formula, the midpoint of the segment with endpoints $(2.5, 3.1)$ and $(1.7, -1.3)$ is

$$\left(\frac{2.5 + 1.7}{2}, \frac{3.1 + (-1.3)}{2}\right) = \left(\frac{4.2}{2}, \frac{1.8}{2}\right) = (2.1, 0.9).$$

65. By the Midpoint Formula, the midpoint of the segment with endpoints $\left(\frac{1}{2}, \frac{1}{3}\right)$ and $\left(\frac{3}{2}, \frac{5}{3}\right)$ is

$$\left(\frac{\frac{1}{2} + \frac{3}{2}}{2}, \frac{\frac{1}{3} + \frac{5}{3}}{2}\right) = \left(\frac{\frac{4}{2}}{2}, \frac{\frac{6}{3}}{2}\right) = \left(\frac{2}{2}, \frac{2}{2}\right) = (1, 1).$$

67. By the Midpoint Formula, the midpoint of the segment with endpoints $\left(-\frac{1}{3}, \frac{2}{7}\right)$ and $\left(-\frac{1}{2}, \frac{1}{14}\right)$ is

$$\left(\frac{-\frac{1}{3} + \left(-\frac{1}{2}\right)}{2}, \frac{\frac{2}{7} + \frac{1}{14}}{2}\right) = \left(\frac{-\frac{5}{6}}{2}, \frac{\frac{5}{14}}{2}\right) = \left(-\frac{5}{12}, \frac{5}{28}\right).$$

69. midpoint of $P(5, 8)$ and $Q(x, y) = M(8, 2)$

$$\left(\frac{5 + x}{2}, \frac{8 + y}{2}\right) = (8, 2)$$

The x- and y-coordinates must be equal.

$$\frac{5 + x}{2} = 8 \qquad \frac{8 + y}{2} = 2$$
$$5 + x = 16 \qquad 8 + y = 4$$
$$x = 11 \qquad y = -4$$

Thus, the endpoint Q is $(11, -4)$.

71. midpoint of $P(1.5, 1.25)$ and $Q(x, y) = M(3, 1)$

$$\left(\frac{1.5 + x}{2}, \frac{1.25 + y}{2}\right) = (3, 1)$$

The x- and y-coordinates must be equal.

$$\frac{1.5 + x}{2} = 3 \qquad \frac{1.25 + y}{2} = 1$$
$$1.5 + x = 6 \qquad 1.25 + y = 2$$
$$x = 4.5 \qquad y = 0.75$$

Thus, the endpoint Q is $(4.5, 0.75)$.

73. The graph goes through the point $(-2, 0)$ which satisfies only equations **B** and **C**. The graph also goes through the point $(0, 3)$ which satisfies only equations **A** and **B**. Therefore, the correct equation is **B**.

75. The screen on the right is more useful because it shows the intercepts.

77. We need to solve the given equation for y.

$$5x + 2y = -10$$
$$2y = -5x - 10 \qquad \textit{Subtract 5x.}$$
$$y = -\frac{5}{2}x - 5 \qquad \textit{Divide by 2.}$$

Graph $Y_1 = -2.5X - 5$ in a standard viewing window.

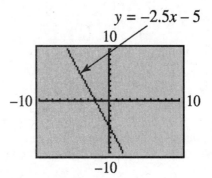

79. $\dfrac{6 - 2}{5 - 3} = \dfrac{4}{2} = 2$

81. $\dfrac{-5 - (-5)}{3 - 2} = \dfrac{-5 + 5}{1} = \dfrac{0}{1} = 0$

83. $4x - y = 10$
$$-y = -4x + 10 \qquad \textit{Subtract 4x.}$$
$$y = 4x - 10 \qquad \textit{Multiply by −1.}$$

85. $3x + 4y = 12$
$$4y = -3x + 12 \qquad \textit{Subtract 3x.}$$
$$y = \frac{12 - 3x}{4}, \qquad \textit{Divide by 4.}$$

or $y = -\frac{3}{4}x + 3$

Copyright © 2012 Pearson Education, Inc. Publishing as Addison-Wesley.

3.2   The Slope of a Line

3.2   Now Try Exercises

N1. If $(x_1, y_1) = (2, -6)$ and $(x_2, y_2) = (-3, 5)$, then

$$m = \frac{y_2 - y_1}{x_2 - x_1} = \frac{5 - (-6)}{-3 - 2} = \frac{11}{-5} = -\frac{11}{5}.$$

The slope is $-\frac{11}{5}$.

N2. To find the slope of the line with equation

$$3x - 7y = 21,$$

first find the intercepts. The x-intercept is $(7, 0)$, and the y-intercept is $(0, -3)$. The slope is then

$$m = \frac{-3 - 0}{0 - 7} = \frac{-3}{-7} = \frac{3}{7}.$$

The slope is $\frac{3}{7}$.

N3. **(a)** To find the slope of the line with equation

$$y - 6 = 0,$$

select two different points on the line, such as $(0, 6)$ and $(2, 6)$, and use the slope formula.

$$m = \frac{6 - 6}{2 - 0} = \frac{0}{2} = 0$$

The slope is 0.

(b) To find the slope of the line with equation

$$x = 4,$$

select two different points on the line, such as $(4, 0)$ and $(4, 3)$, and use the slope formula.

$$m = \frac{3 - 0}{4 - 4} = \frac{3}{0}$$

Since division by zero is undefined, the slope is *undefined*.

N4. Solve the equation for y.

$$5x - 4y = 7$$
$$-4y = -5x + 7 \quad \textit{Subtract 5x.}$$
$$y = \tfrac{5}{4}x - \tfrac{7}{4} \quad \textit{Divide by} -4.$$

The slope is given by the coefficient of x, so the slope is $\frac{5}{4}$.

N5. Through $(-4, 1)$; $m = -\frac{2}{3}$

Locate the point $(-4, 1)$ on the graph. Use the slope formula to find a second point on the line.

$$m = \frac{\text{change in } y}{\text{change in } x} = \frac{-2}{3}$$

From $(-4, 1)$, move *down* 2 units and then 3 units to the *right* to $(-1, -1)$. Draw the line through the two points.

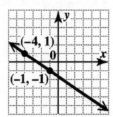

N6. Find the slope of each line.

The line through $(2, 5)$ and $(4, 8)$ has slope

$$m_1 = \frac{8 - 5}{4 - 2} = \frac{3}{2}.$$

The line through $(2, 0)$ and $(-1, -2)$ has slope

$$m_2 = \frac{-2 - 0}{-1 - 2} = \frac{-2}{-3} = \frac{2}{3}.$$

The slopes are *not* the same, so the lines are *not parallel*.

N7. Solve each equation for y.

$$x + 2y = 7 \qquad\qquad 2x - y = -4$$
$$2y = -x + 7 \qquad\quad -y = -2x - 4$$
$$y = -\tfrac{1}{2}x + \tfrac{7}{5} \qquad\quad y = 2x + 4$$
$$(\text{so } m_1 = -\tfrac{1}{2}) \qquad (\text{so } m_2 = 2)$$

Since $m_1 m_2 = (-\frac{1}{2})(2) = -1$, the lines are *perpendicular*.

N8. Solve each equation for y.

$$2x - y = 4 \qquad\qquad 2x + y = 6$$
$$-y = -2x + 4 \qquad\quad y = -2x + 6$$
$$y = 2x - 4$$
$$(\text{so } m_1 = 2) \qquad\qquad (\text{so } m_2 = -2)$$

Since $m_1 \neq m_2$, the lines are not parallel. Since $m_1 m_2 = 2(-2) = -4$, the lines are not perpendicular either. Therefore, the answer is *neither*.

N9. **(a)** $(x_1, y_1) = (2000, 690)$ and $(x_2, y_2) = (2002, 828)$.

$$\text{average rate of change} = \frac{y_2 - y_1}{x_2 - x_1}$$
$$= \frac{828 - 690}{2002 - 2000}$$
$$= \frac{138}{2} = 69$$

The average rate of change is 69 hr per yr.

(b) It is greater than 58, which is the average rate of change from 2000 to 2005 found in Example 9.

Copyright © 2012 Pearson Education, Inc. Publishing as Addison-Wesley.

N10. $(x_1, y_1) = (2000, 2838)$ and
$(x_2, y_2) = (2008, 1885)$.

$$\text{average rate of change} = \frac{y_2 - y_1}{x_2 - x_1}$$
$$= \frac{1885 - 2838}{2008 - 2000}$$
$$= \frac{-953}{8} = -119.125$$

Thus, the average rate of change in sales of digital camcorders in the United States from 2000 to 2008 was about $-\$119$ million per year.

3.2 Section Exercises

1. $\text{slope} = \dfrac{\text{change in vertical position}}{\text{change in horizontal position}}$
$$= \frac{30 \text{ feet}}{100 \text{ feet}}$$

Choices **A**, 0.3, **B**, $\frac{3}{10}$, **D**, $\frac{30}{100}$, and **G**, 30%, are all correct.

3. $\text{slope} = \dfrac{\text{change in vertical position}}{\text{change in horizontal position}}$
$$3 = \frac{15 \text{ feet}}{\text{change in horizontal position}}$$
$$3 \times \text{change} = 15$$
$$\text{change} = \frac{15}{3} = 5$$

So the change in horizontal position is 5 feet.

5. **(a)** Graph **C** indicates that sales leveled off during the second quarter.

(b) Graph **A** indicates that sales leveled off during the fourth quarter.

(c) Graph **D** indicates that sales rose sharply during the first quarter, and then fell to the original level during the second quarter.

(d) Graph **B** is the only graph that indicates that sales fell during the first two quarters.

7. To get to B from A, we must go up 2 units and move right 1 unit. Thus,

$$\text{slope of } AB = \frac{\text{rise}}{\text{run}} = \frac{2}{1} = 2.$$

9. $\text{slope of } CD = \dfrac{\text{rise}}{\text{run}} = \dfrac{-7}{0}$, which is undefined.

11. $\text{slope of } EF = \dfrac{\text{rise}}{\text{run}} = \dfrac{3}{3} = 1$

13. $\text{slope of } AF = \dfrac{\text{rise}}{\text{run}} = \dfrac{-3}{3} = -1$

15. $m = \dfrac{6 - 2}{5 - 3} = \dfrac{4}{2} = 2$

17. $m = \dfrac{4 - (-1)}{-3 - (-5)} = \dfrac{4 + 1}{-3 + 5} = \dfrac{5}{2}$

19. $m = \dfrac{-5 - (-5)}{3 - 2} = \dfrac{-5 + 5}{1} = \dfrac{0}{1} = 0$

21. $m = \dfrac{3 - 8}{-2 - (-2)} = \dfrac{-5}{-2 + 2} = \dfrac{-5}{0}$, which is *undefined.*

23. **(a)** "The line has positive slope" means that the line goes up from left to right. This is line B.

(b) "The line has negative slope" means that the line goes down from left to right. This is line C.

(c) "The line has slope 0" means that there is no vertical change; that is, the line is horizontal. This is line A.

(d) "The line has undefined slope" means that there is no horizontal change; that is, the line is vertical. This is line D.

25. **(a)** Let $(x_1, y_1) = (-2, -3)$ and $(x_2, y_2) = (-1, 5)$. Then
$$m = \frac{y_2 - y_1}{x_2 - x_1} = \frac{5 - (-3)}{-1 - (-2)} = \frac{8}{1} = 8.$$

The slope is 8.

(b) The slope is positive, so the line rises.

27. **(a)** Let $(x_1, y_1) = (-4, 1)$ and $(x_2, y_2) = (2, 6)$. Then
$$m = \frac{y_2 - y_1}{x_2 - x_1} = \frac{6 - 1}{2 - (-4)} = \frac{5}{6}.$$

The slope is $\frac{5}{6}$.

(b) The slope is positive, so the line rises.

29. **(a)** Let $(x_1, y_1) = (2, 4)$ and $(x_2, y_2) = (-4, 4)$. Then
$$m = \frac{y_2 - y_1}{x_2 - x_1} = \frac{4 - 4}{-4 - 2} = \frac{0}{-6} = 0.$$

The slope is 0.

(b) The slope is zero, so the line is horizontal.

31. **(a)** Let $(x_1, y_1) = (-2, 2)$ and $(x_2, y_2) = (4, -1)$. Then
$$m = \frac{y_2 - y_1}{x_2 - x_1} = \frac{-1 - 2}{4 - (-2)} = \frac{-3}{6} = -\frac{1}{2}.$$

The slope is $-\frac{1}{2}$.

(b) The slope is negative, so the line falls.

33. **(a)** Let $(x_1, y_1) = (5, -3)$ and $(x_2, y_2) = (5, 2)$. Then

$$m = \frac{y_2 - y_1}{x_2 - x_1} = \frac{2 - (-3)}{5 - 5} = \frac{5}{0}.$$

The slope is undefined.

(b) The slope is undefined, so the line is vertical.

35. (a) Let $(x_1, y_1) = (1.5, 2.6)$ and $(x_2, y_2) = (0.5, 3.6)$. Then

$$m = \frac{y_2 - y_1}{x_2 - x_1} = \frac{3.6 - 2.6}{0.5 - 1.5} = \frac{1}{-1} = -1.$$

The slope is -1.

(b) The slope is negative, so the line falls.

37. Let $(x_1, y_1) = \left(\frac{1}{6}, \frac{1}{2}\right)$ and $(x_2, y_2) = \left(\frac{5}{6}, \frac{9}{2}\right)$. Then

$$m = \frac{y_2 - y_1}{x_2 - x_1} = \frac{\frac{9}{2} - \frac{1}{2}}{\frac{5}{6} - \frac{1}{6}} = \frac{\frac{8}{2}}{\frac{4}{6}} = 4 \cdot \frac{3}{2} = 6.$$

The slope is 6.

39. Let $(x_1, y_1) = \left(-\frac{2}{9}, \frac{5}{18}\right)$ and $(x_2, y_2) = \left(\frac{1}{18}, -\frac{5}{9}\right)$. Then

$$m = \frac{y_2 - y_1}{x_2 - x_1} = \frac{-\frac{5}{9} - \frac{5}{18}}{\frac{1}{18} - \left(-\frac{2}{9}\right)} = \frac{-\frac{15}{18}}{\frac{5}{18}}$$
$$= -\frac{15}{18} \cdot \frac{18}{5} = -3.$$

The slope is -3.

41. The points shown on the line are $(-3, 3)$ and $(-1, -2)$. The slope is

$$m = \frac{-2 - 3}{-1 - (-3)} = \frac{-5}{2} = -\frac{5}{2}.$$

43. The points shown on the line are $(3, 3)$ and $(3, -3)$. The slope is

$$m = \frac{-3 - 3}{3 - 3} = \frac{-6}{0}, \text{ which is } \textit{undefined}.$$

45. To find the slope of

$$x + 2y = 4,$$

first find the intercepts. Replace y with 0 to find that the x-intercept is $(4, 0)$; replace x with 0 to find that the y-intercept is $(0, 2)$. The slope is then

$$m = \frac{2 - 0}{0 - 4} = -\frac{2}{4} = -\frac{1}{2}.$$

To sketch the graph, plot the intercepts and draw the line through them.

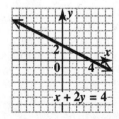

47. To find the slope of

$$5x - 2y = 10,$$

first find the intercepts. Replace y with 0 to find that the x-intercept is $(2, 0)$; replace x with 0 to find that the y-intercept is $(0, -5)$. The slope is then

$$m = \frac{-5 - 0}{0 - 2} = \frac{-5}{-2} = \frac{5}{2}.$$

To sketch the graph, plot the intercepts and draw the line through them.

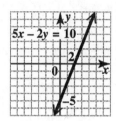

49. In the equation

$$y = 4x,$$

replace x with 0 and then x with 1 to get the ordered pairs $(0, 0)$ and $(1, 4)$, respectively. (There are other possibilities for ordered pairs.) The slope is then

$$m = \frac{4 - 0}{1 - 0} = \frac{4}{1} = 4.$$

To sketch the graph, plot the two points and draw the line through them.

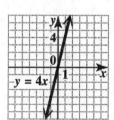

51. $x - 3 = 0 \quad (x = 3)$

The graph of $x = 3$ is the vertical line with x-intercept $(3, 0)$. The slope of a vertical line is undefined.

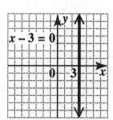

Copyright © 2012 Pearson Education, Inc. Publishing as Addison-Wesley.

53. $y = -5$

The graph of $y = -5$ is the horizontal line with y-intercept $(0, -5)$. The slope of a horizontal line is 0.

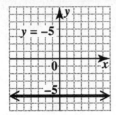

55. $2y = 3$ $\left(y = \frac{3}{2}\right)$

The graph of $y = \frac{3}{2}$ is the horizontal line with y-intercept $\left(0, \frac{3}{2}\right)$. The slope of a horizontal line is 0.

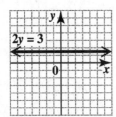

57. To graph the line through $(-4, 2)$ with slope $m = \frac{1}{2}$, locate $(-4, 2)$ on the graph. To find a second point, use the definition of slope.

$$m = \frac{\text{change in } y}{\text{change in } x} = \frac{1}{2}$$

From $(-4, 2)$, go up 1 unit. Then go 2 units to the right to get to $(-2, 3)$. Draw the line through $(-4, 2)$ and $(-2, 3)$.

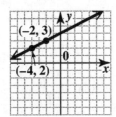

59. To graph the line through $(0, -2)$ with slope $m = -\frac{2}{3}$, locate the point $(0, -2)$ on the graph. To find a second point on the line, use the definition of slope, writing $-\frac{2}{3}$ as $\frac{-2}{3}$.

$$m = \frac{\text{change in } y}{\text{change in } x} = \frac{-2}{3}$$

From $(0, -2)$, move 2 units down and then 3 units to the right. Draw a line through this second point and $(0, -2)$. (Note that the slope could also be written as $\frac{2}{-3}$. In this case, move 2 units up and 3 units to the left to get another point on the same line.)

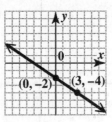

61. Locate $(-1, -2)$. Then use $m = 3 = \frac{3}{1}$ to go 3 units up and 1 unit right to $(0, 1)$. Draw the line through $(-1, -2)$ and $(0, 1)$.

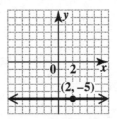

63. Locate $(2, -5)$. A slope of 0 means that the line is horizontal, so $y = -5$ at every point. Draw the horizontal line through $(2, -5)$.

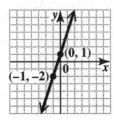

65. Locate $(-3, 1)$. Since the slope is undefined, the line is vertical. The x-value of every point is -3. Draw the vertical line through $(-3, 1)$.

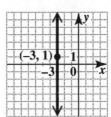

67. If a line has slope $-\frac{4}{9}$, then any line parallel to it has slope $-\frac{4}{9}$ (the slope must be the same), and any line perpendicular to it has slope $\frac{9}{4}$ (the slope must be the negative reciprocal).

69. The slope of the line through $(15, 9)$ and $(12, -7)$ is

$$m = \frac{-7 - 9}{12 - 15} = \frac{-16}{-3} = \frac{16}{3}.$$

The slope of the line through $(8, -4)$ and $(5, -20)$ is

$$m = \frac{-20 - (-4)}{5 - 8} = \frac{-16}{-3} = \frac{16}{3}.$$

Copyright © 2012 Pearson Education, Inc. Publishing as Addison-Wesley.

Since the slopes are equal, the two lines are *parallel*.

71. $x + 4y = 7$ and $4x - y = 3$
Solve the equations for y.

$$4y = -x + 7 \qquad\qquad -y = -4x + 3$$
$$y = -\tfrac{1}{4}x + \tfrac{7}{4} \qquad\qquad y = 4x - 3$$

The slopes, $-\tfrac{1}{4}$ and 4, are negative reciprocals of one another, so the lines are *perpendicular*.

73. $4x - 3y = 6$ and $3x - 4y = 2$
Solve the equations for y.

$$-3y = -4x + 6 \qquad -4y = -3x + 2$$
$$y = \tfrac{4}{3}x - 2 \qquad\quad y = \tfrac{3}{4}x - \tfrac{1}{2}$$

The slopes are $\tfrac{4}{3}$ and $\tfrac{3}{4}$. The lines are *neither* parallel nor perpendicular.

75. $x = 6$ and $6 - x = 8$

The second equation can be simplified as $x = -2$. Both lines are vertical lines, so they are *parallel*.

77. $4x + y = 0$ and $5x - 8 = 2y$
Solve the equations for y.

$$y = -4x \qquad \tfrac{5}{2}x - 4 = y$$

The slopes are -4 and $\tfrac{5}{2}$. The lines are *neither* parallel nor perpendicular.

79. $2x = y + 3$ and $2y + x = 3$
Solve the equations for y.

$$2x - 3 = y \qquad\qquad 2y = -x + 3$$
$$\qquad\qquad\qquad\qquad y = -\tfrac{1}{2}x + \tfrac{3}{2}$$

The slopes, 2 and $-\tfrac{1}{2}$, are negative reciprocals of one another, so the lines are *perpendicular*.

81. Use the points $(0, 20)$ and $(4, 4)$.

average rate of change

$$= \frac{\text{change in } y}{\text{change in } x} = \frac{4 - 20}{4 - 0} = \frac{-16}{4} = -4$$

The average rate of change is $-\$4000$ per year, that is, the value of the machine is decreasing $\$4000$ each year during these years.

83. We can see that there is no change in the percent of pay raise. Thus, the average rate of change is 0% per year, that is, the percent of pay raise is not changing—it is 3% each year during these years.

85. Let y be the vertical rise.
Since the slope is the vertical rise divided by the horizontal run,

$$0.13 = \frac{y}{150}.$$

Solving for y gives

$$y = 0.13(150) = 19.5.$$

The vertical rise could be a maximum of 19.5 ft.

87. **(a)** $m = \dfrac{270.9 - 207.3}{2008 - 2005} = \dfrac{63.6}{3} = 21.2$

(b) The number of subscribers increased by an average of 21.2 million each year from 2005 to 2008.

89. **(a)** Use $(2000, 443)$ and $(2007, 405)$.

$$m = \frac{405 - 443}{2007 - 2000} = \frac{-38}{7} \approx -5$$

The average rate of change is -5 theaters per year.

(b) The negative slope means that the number of drive-in theaters *decreased* by an average of 5 each year from 2000 to 2007.

91. Use $(2003, 1590)$ and $(2006, 5705)$.

average rate of change

$$= \frac{5705 - 1590}{2006 - 2003} = \frac{4115}{3} \approx 1371.67$$

The average rate of change in sales is about $\$1371.67$ million per year, that is, the sales of plasma TVs in the United States increased by an average of $\$1371.67$ million each year from 2003 to 2006.

93. Label the points as shown in the figure.

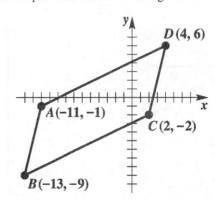

In order to determine whether $ABCD$ is a parallelogram, we need to show that the slope of $\overline{AB}$ equals the slope of $\overline{CD}$ and that the slope of $\overline{AD}$ equals the slope of $\overline{BC}$.

Slope of $\overline{AB} = \dfrac{-9 - (-1)}{-13 - (-11)} = \dfrac{-8}{-2} = 4$

Slope of $\overline{CD} = \dfrac{6 - (-2)}{4 - 2} = \dfrac{8}{2} = 4$

Slope of $\overline{AD} = \dfrac{6 - (-1)}{4 - (-11)} = \dfrac{7}{15}$

Slope of $\overline{BC} = \dfrac{-2 - (-9)}{2 - (-13)} = \dfrac{7}{15}$

Thus, the figure is a parallelogram.

Copyright © 2012 Pearson Education, Inc. Publishing as Addison-Wesley.

♦♦♦ Relating Concepts 95–100 ♦♦♦

95. For $A(3,1)$ and $B(6,2)$, the slope of $\overline{AB}$ is
$$m = \frac{2-1}{6-3} = \frac{1}{3}.$$

96. For $B(6,2)$ and $C(9,3)$, the slope of $\overline{BC}$ is
$$m = \frac{3-2}{9-6} = \frac{1}{3}.$$

97. For $A(3,1)$ and $C(9,3)$, the slope of $\overline{AC}$ is
$$m = \frac{3-1}{9-3} = \frac{2}{6} = \frac{1}{3}.$$

98. The slope of $\overline{AB}$ = slope of $\overline{BC}$
$$= \text{slope of } \overline{AC}$$
$$= \frac{1}{3}.$$

99. For $A(1,-2)$ and $B(3,-1)$, the slope of $\overline{AB}$ is
$$m = \frac{-1-(-2)}{3-1} = \frac{1}{2}.$$

For $B(3,-1)$ and $C(5,0)$, the slope of $\overline{BC}$ is
$$m = \frac{0-(-1)}{5-3} = \frac{1}{2}.$$

For $A(1,-2)$ and $C(5,0)$, the slope of $\overline{AC}$ is
$$m = \frac{0-(-2)}{5-1} = \frac{2}{4} = \frac{1}{2}.$$

Since the three slopes are the same, the three points are collinear.

100. For $A(0,6)$ and $B(4,-5)$, the slope of $\overline{AB}$ is
$$m = \frac{-5-6}{4-0} = \frac{-11}{4} = -\frac{11}{4}.$$

For $B(4,-5)$ and $C(-2,12)$, the slope of $\overline{BC}$ is
$$m = \frac{12-(-5)}{-2-4} = \frac{17}{-6} = -\frac{17}{6}.$$

Since these two slopes are not the same, the three points are not collinear.

101. $3x + 2y = 8$
$$2y = -3x + 8$$
$$y = -\frac{3}{2}x + 4$$

103. $y - 2 = 4(x+3)$
$$y - 2 = 4x + 12$$
$$y = 4x + 14$$

105. $y - (-1) = -\frac{1}{2}[x - (-2)]$
$$y + 1 = -\frac{1}{2}(x+2)$$
$$2(y+1) = -1(x+2)$$
$$2y + 2 = -x - 2$$
$$x + 2y = -4$$

3.3 Linear Equations in Two Variables

3.3 Now Try Exercises

N1. Slope $\frac{2}{3}$; y-intercept $(0,1)$

Here $m = \frac{2}{3}$ and $b = 1$. Substitute these values into the slope-intercept form.
$$y = mx + b$$
$$y = \frac{2}{3}x + 1$$

N2. $4x + 3y = 6$

Solve the equation for y.
$$3y = -4x + 6$$
$$y = -\frac{4}{3}x + 2$$

Plot the y-intercept $(0,2)$. The slope can be interpreted as either $\frac{-4}{3}$ or $\frac{4}{-3}$. Using $\frac{-4}{3}$, move from $(0,2)$ *down* 4 units and to the *right* 3 units to locate the point $(3,-2)$. Draw a line through the two points.

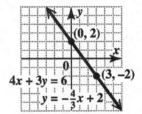

N3. Through $(5,-3)$; slope $= m = -\frac{1}{5}$

Use the point-slope form with $(x_1, y_1) = (5,-3)$ and $m = -\frac{1}{5}$.
$$y - y_1 = m(x - x_1)$$
$$y - (-3) = -\frac{1}{5}(x - 5)$$
$$y + 3 = -\frac{1}{5}(x - 5)$$
$$y + 3 = -\frac{1}{5}x + 1$$
$$y = -\frac{1}{5}x - 2 \quad \textit{Subtract 3.}$$

N4. **(a)** Through $(4,-4)$; m undefined

This is a vertical line since the slope is undefined. A vertical line through the point (a,b) has equation $x = a$. Here the x-coordinate is 4, so the equation is $x = 4$.

(b) Through $(4,-4)$; $m = 0$

Since the slope is 0, this is a horizontal line. A horizontal line through the point (a,b) has equation $y = b$. Here the y-coordinate is -4, so the equation is $y = -4$.

N5. Through $(3,-4)$ and $(-2,-1)$
$$m = \frac{-1-(-4)}{-2-3} = \frac{3}{-5} = -\frac{3}{5}$$

Let $(x_1, y_1) = (3,-4)$.

Copyright © 2012 Pearson Education, Inc. Publishing as Addison-Wesley.

$$y - y_1 = m(x - x_1)$$
$$y - (-4) = -\tfrac{3}{5}(x - 3)$$
$$y + 4 = -\tfrac{3}{5}(x - 3)$$
$$5y + 20 = -3x + 9 \qquad \textit{Multiply by 5.}$$
$$3x + 5y = -11 \qquad \textit{Standard form}$$

N6. (a) Through $(6, -1)$; parallel to the line $3x - 5y = 7$

Find the slope of the given line.

$$3x - 5y = 7$$
$$-5y = -3x + 7$$
$$y = \tfrac{3}{5}x - \tfrac{7}{5}$$

The slope is $\tfrac{3}{5}$, so a line parallel to it also has slope $\tfrac{3}{5}$. Use $m = \tfrac{3}{5}$ and $(x_1, y_1) = (6, -1)$ in the point-slope form.

$$y - y_1 = m(x - x_1)$$
$$y - (-1) = \tfrac{3}{5}(x - 6)$$
$$y + 1 = \tfrac{3}{5}x - \tfrac{18}{5}$$
$$y = \tfrac{3}{5}x - \tfrac{18}{5} - \tfrac{5}{5}$$
$$y = \tfrac{3}{5}x - \tfrac{23}{5}$$

(b) Through $(6, -1)$; perpendicular to $3x - 5y = 7$

The slope of $3x - 5y = 7$ is $\tfrac{3}{5}$. The negative reciprocal of $\tfrac{3}{5}$ is $-\tfrac{5}{3}$, so the slope of the line through $(6, -1)$ is $-\tfrac{5}{3}$.

$$y - y_1 = m(x - x_1)$$
$$y - (-1) = -\tfrac{5}{3}(x - 6)$$
$$y + 1 = -\tfrac{5}{3}x + 10$$
$$y = -\tfrac{5}{3}x + 9$$

N7. Since the price you pay is $85 per month plus a flat fee of $100, an equation for x months is

$$y = 85x + 100.$$

N8. (a) Use $(0, 2079)$ for 2004 and $(4, 2372)$ for 2008.

$$m = \frac{2372 - 2079}{4 - 0} = \frac{293}{4} = 73.25$$
$$y = mx + b$$
$$y = 73.25x + 2079$$

(b) For 2009, $x = 2009 - 2004 = 5$.

$$y = 73.25x + 2079$$
$$y = 73.25(5) + 2079 \quad \textit{Let x = 5.}$$
$$y = 366.25 + 2079$$
$$y = 2445.25$$

According to the model, average tuition and fees for in-state students at public two-year colleges in 2009 were about $2445.

N9. (a) Use $(2, 183)$ and $(6, 251)$.

$$m = \frac{251 - 183}{6 - 2} = \frac{68}{4} = 17$$

Use the point-slope form with $(x_1, y_1) = (2, 183)$.

$$y - y_1 = m(x - x_1)$$
$$y - 183 = 17(x - 2)$$
$$y - 183 = 17x - 34$$
$$y = 17x + 149$$

(b) For 2011, $x = 2011 - 2000 = 11$.

$$y = 17x + 149$$
$$y = 17(11) + 149 \quad \textit{Let x = 11.}$$
$$= 187 + 149$$
$$= 336$$

Using the equation from part (a), we estimate the retail spending on prescription drugs in 2011 to be $336 billion.

3.3 Section Exercises

1. Choice **A**, $3x - 2y = 5$, is in the form $Ax + By = C$ with $A \geq 0$ and integers A, B, and C having no common factor (except 1).

3. Choice **A**, $y = 6x + 2$, is in the form $y = mx + b$.

5.
$$y + 2 = -3(x - 4)$$
$$y + 2 = -3x + 12$$
$$3x + y = 10 \qquad \textit{Standard form}$$

7. $y = 2x + 3$
This line is in slope-intercept form with slope $m = 2$ and y-intercept $(0, b) = (0, 3)$. The only graph with positive slope and with a positive y-coordinate of its y-intercept is **A**.

9. $y = -2x - 3$
This line is in slope-intercept form with slope $m = -2$ and y-intercept $(0, b) = (0, -3)$. The only graph with negative slope and with a negative y-coordinate of its y-intercept is **C**.

11. $y = 2x$
This line has slope $m = 2$ and y-intercept $(0, b) = (0, 0)$. The only graph with positive slope and with y-intercept $(0, 0)$ is **H**.

13. $y = 3$
This line is a horizontal line with y-intercept $(0, 3)$. Its y-coordinate is positive. The only graph that has these characteristics is **B**.

15. $m = 5; b = 15$
Substitute these values in the slope-intercept form.

$$y = mx + b$$
$$y = 5x + 15$$

Copyright © 2012 Pearson Education, Inc. Publishing as Addison-Wesley.

17. $m = -\frac{2}{3}$; $b = \frac{4}{5}$

Substitute these values in the slope-intercept form.

$$y = mx + b$$
$$y = -\frac{2}{3}x + \frac{4}{5}$$

19. Slope 1; y-intercept $(0, -1)$

Here, $m = 1$ and $b = -1$. Substitute these values in the slope-intercept form.

$$y = mx + b$$
$$y = 1x - 1, \text{ or } y = x - 1$$

21. Slope $\frac{2}{5}$; y-intercept $(0, 5)$

Here, $m = \frac{2}{5}$ and $b = 5$. Substitute these values in the slope-intercept form.

$$y = mx + b$$
$$y = \frac{2}{5}x + 5$$

23. To get to the point $(3, 3)$ from the y-intercept $(0, 1)$, we must go up 2 units and to the right 3 units, so the slope is $\frac{2}{3}$. The slope-intercept form is

$$y = \frac{2}{3}x + 1.$$

25. To get to the point $(-3, 1)$ from the y-intercept $(0, -2)$, we must go up 3 units and to the left 3 units, so the slope is $\frac{3}{-3} = -1$. The slope-intercept form is

$$y = -1x - 2, \text{ or } y = -x - 2.$$

27. $-x + y = 4$

(a) Solve for y to get the equation in slope-intercept form.

$$-x + y = 4$$
$$y = x + 4$$

(b) The slope is the coefficient of x, 1.

(c) The y-intercept is the point $(0, b)$, or $(0, 4)$.

(d)

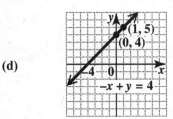

29. $6x + 5y = 30$

(a) Solve for y to get the equation in slope-intercept form.

$$6x + 5y = 30$$
$$5y = -6x + 30$$
$$y = -\frac{6}{5}x + 6$$

(b) The slope is the coefficient of x, $-\frac{6}{5}$.

(c) The y-intercept is the point $(0, b)$, or $(0, 6)$.

(d)

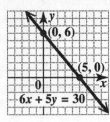

31. $4x - 5y = 20$

(a) Solve for y to get the equation in slope-intercept form.

$$4x - 5y = 20$$
$$-5y = -4x + 20$$
$$y = \frac{4}{5}x - 4$$

(b) The slope is the coefficient of x, $\frac{4}{5}$.

(c) The y-intercept is the point $(0, b)$, or $(0, -4)$.

(d)

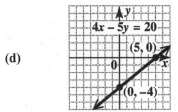

33. $x + 2y = -4$

(a) Solve for y to get the equation in slope-intercept form.

$$x + 2y = -4$$
$$2y = -x - 4$$
$$y = -\frac{1}{2}x - 2$$

(b) The slope is the coefficient of x, $-\frac{1}{2}$.

(c) The y-intercept is the point $(0, b)$, or $(0, -2)$.

(d)

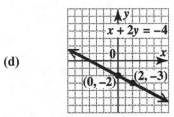

35. **(a)** Through $(5, 8)$; slope -2

Use the point-slope form with $(x_1, y_1) = (5, 8)$ and $m = -2$.

$$y - y_1 = m(x - x_1)$$
$$y - 8 = -2(x - 5)$$
$$y - 8 = -2x + 10$$
$$2x + y = 18$$

(b) Solve the last equation from part (a) for y.

$$2x + y = 18$$
$$y = -2x + 18$$

Copyright © 2012 Pearson Education, Inc. Publishing as Addison-Wesley.

37. **(a)** Through $(-2, 4)$; slope $-\frac{3}{4}$
Use the point-slope form with $(x_1, y_1) = (-2, 4)$
and $m = -\frac{3}{4}$.

$$y - y_1 = m(x - x_1)$$
$$y - 4 = -\frac{3}{4}[x - (-2)]$$
$$4(y - 4) = -3(x + 2)$$
$$4y - 16 = -3x - 6$$
$$3x + 4y = 10$$

(b) Solve the last equation from part (a) for y.

$$3x + 4y = 10$$
$$4y = -3x + 10$$
$$y = -\frac{3}{4}x + \frac{10}{4}$$
$$y = -\frac{3}{4}x + \frac{5}{2}$$

39. **(a)** Through $(-5, 4)$; slope $\frac{1}{2}$
Use the point-slope form with $(x_1, y_1) = (-5, 4)$
and $m = \frac{1}{2}$.

$$y - y_1 = m(x - x_1)$$
$$y - 4 = \frac{1}{2}[x - (-5)]$$
$$2(y - 4) = 1(x + 5)$$
$$2y - 8 = x + 5$$
$$-x + 2y = 13$$
$$x - 2y = -13$$

(b) Solve the last equation from part (a) for y.

$$x - 2y = -13$$
$$-2y = -x - 13$$
$$y = \frac{1}{2}x + \frac{13}{2}$$

41. **(a)** Through $(3, 0)$; slope 4
Use the point-slope form with $(x_1, y_1) = (3, 0)$
and $m = 4$.

$$y - y_1 = m(x - x_1)$$
$$y - 0 = 4(x - 3)$$
$$y = 4x - 12$$
$$-4x + y = -12$$
$$4x - y = 12$$

(b) Solve the last equation from part (a) for y.

$$4x - y = 12$$
$$-y = -4x + 12$$
$$y = 4x - 12$$

43. **(a)** Through $(2, 6.8)$; slope 1.4
Use the point-slope form with $(x_1, y_1) = (2, 6.8)$
and $m = 1.4$.

$$y - y_1 = m(x - x_1)$$
$$y - 6.8 = 1.4(x - 2)$$
$$y - 6.8 = \frac{7}{5}(x - 2)$$
$$5(y - 6.8) = 7(x - 2)$$

$$5y - 34 = 7x - 14$$
$$-7x + 5y = 20$$
$$7x - 5y = -20$$

(b) Solve the last equation from part (a) for y.

$$7x - 5y = -20$$
$$-5y = -7x - 20$$
$$y = \frac{7}{5}x + 4, \quad \text{or} \quad y = 1.4x + 4$$

45. Through $(9, 5)$; slope 0
A line with slope 0 is a horizontal line. A
horizontal line through the point (x, k) has
equation $y = k$. Here $k = 5$, so an equation is
$y = 5$.

47. Through $(9, 10)$; undefined slope
A vertical line has undefined slope and equation
$x = c$. Since the x-value in $(9, 10)$ is 9, the
equation is $x = 9$.

49. Through $\left(-\frac{3}{4}, -\frac{3}{2}\right)$; slope 0
The equation of this line is $y = -\frac{3}{2}$.

51. Through $(-7, 8)$; horizontal
A horizontal line through the point (x, k) has
equation $y = k$, so the equation is $y = 8$.

53. Through $(0.5, 0.2)$; vertical
A vertical line through the point (k, y) has
equation $x = k$. Here $k = 0.5$, so the equation is
$x = 0.5$.

55. **(a)** $(3, 4)$ and $(5, 8)$
Find the slope.

$$m = \frac{8 - 4}{5 - 3} = \frac{4}{2} = 2$$

Use the point-slope form with $(x_1, y_1) = (3, 4)$
and $m = 2$.

$$y - y_1 = m(x - x_1)$$
$$y - 4 = 2(x - 3)$$
$$y - 4 = 2x - 6$$
$$-2x + y = -2$$
$$2x - y = 2$$

(b) Solve the last equation from part (a) for y.

$$2x - y = 2$$
$$-y = -2x + 2$$
$$y = 2x - 2$$

Note: You could use *any* of the equations in part
(a) to solve for y. In this exercise, choosing

$$y - 4 = 2x - 6 \quad \text{or} \quad -2x + y = -2,$$

easily leads to the equation $y = 2x - 2$.

Copyright © 2012 Pearson Education, Inc. Publishing as Addison-Wesley.

57. (a) $(6, 1)$ and $(-2, 5)$
Find the slope.

$$m = \frac{5 - 1}{-2 - 6} = \frac{4}{-8} = -\frac{1}{2}$$

Use the point-slope form with $(x_1, y_1) = (6, 1)$
and $m = -\frac{1}{2}$.

$$y - y_1 = m(x - x_1)$$
$$y - 1 = -\frac{1}{2}(x - 6)$$
$$2(y - 1) = -1(x - 6)$$
$$2y - 2 = -x + 6$$
$$x + 2y = 8$$

(b) Solve the last equation from part (a) for y.

$$x + 2y = 8$$
$$2y = -x + 8$$
$$y = -\frac{1}{2}x + 4$$

59. (a) $(2, 5)$ and $(1, 5)$
Find the slope.

$$m = \frac{5 - 5}{1 - 2} = \frac{0}{-1} = 0$$

A line with slope 0 is horizontal. A horizontal line through the point (x, k) has equation $y = k$, so the equation is $y = 5$.

(b) $y = 5$ is already in the slope-intercept form.

61. (a) $(7, 6)$ and $(7, -8)$
Find the slope.

$$m = \frac{-8 - 6}{7 - 7} = \frac{-14}{0} \quad \textit{Undefined}$$

A line with undefined slope is a vertical line. The equation of a vertical line is $x = k$, where k is the common x-value. So the equation is $x = 7$.

(b) It is not possible to write $x = 7$ in slope-intercept form.

63. (a) $\left(\frac{1}{2}, -3\right)$ and $\left(-\frac{2}{3}, -3\right)$
Find the slope.

$$m = \frac{-3 - (-3)}{-\frac{2}{3} - \frac{1}{2}} = \frac{0}{-\frac{7}{6}} = 0$$

A line with slope 0 is horizontal. A horizontal line through the point (x, k) has equation $y = k$, so the equation is $y = -3$.

(b) $y = -3$ is already in the slope-intercept form.

65. (a) $\left(-\frac{2}{5}, \frac{2}{5}\right)$ and $\left(\frac{4}{3}, \frac{2}{3}\right)$
Find the slope.

$$m = \frac{\frac{2}{3} - \frac{2}{5}}{\frac{4}{3} - \left(-\frac{2}{5}\right)} = \frac{\frac{10 - 6}{15}}{\frac{20 + 6}{15}}$$

$$= \frac{\frac{4}{15}}{\frac{26}{15}} = \frac{4}{26} = \frac{2}{13}$$

Use the point-slope form with
$(x_1, y_1) = \left(-\frac{2}{5}, \frac{2}{5}\right)$ and $m = \frac{2}{13}$.

$$y - \frac{2}{5} = \frac{2}{13}\left[x - \left(-\frac{2}{5}\right)\right]$$
$$13\left(y - \frac{2}{5}\right) = 2\left(x + \frac{2}{5}\right)$$
$$13y - \frac{26}{5} = 2x + \frac{4}{5}$$
$$-2x + 13y = \frac{30}{5}$$
$$2x - 13y = -6$$

(b) Solve the last equation from part (a) for y.

$$2x - 13y = -6$$
$$-13y = -2x - 6$$
$$y = \frac{2}{13}x + \frac{6}{13}$$

67. (a) Through $(7, 2)$; parallel to the graph of the line having equation $3x - y = 8$
Find the slope of $3x - y = 8$.

$$-y = -3x + 8$$
$$y = 3x - 8$$

The slope is 3, so a line parallel to it also has slope 3. Use $m = 3$ and $(x_1, y_1) = (7, 2)$ in the point-slope form.

$$y - y_1 = m(x - x_1)$$
$$y - 2 = 3(x - 7)$$
$$y - 2 = 3x - 21$$
$$y = 3x - 19$$

(b)
$$y = 3x - 19$$
$$-3x + y = -19$$
$$3x - y = 19$$

69. (a) Through $(-2, -2)$; parallel to $-x + 2y = 10$
Find the slope of $-x + 2y = 10$.

$$2y = x + 10$$
$$y = \frac{1}{2}x + 5$$

The slope is $\frac{1}{2}$, so a line parallel to it also has slope $\frac{1}{2}$. Use $m = \frac{1}{2}$ and $(x_1, y_1) = (-2, -2)$ in the point-slope form.

$$y - y_1 = m(x - x_1)$$
$$y - (-2) = \frac{1}{2}[x - (-2)]$$
$$y + 2 = \frac{1}{2}(x + 2)$$
$$y + 2 = \frac{1}{2}x + 1$$
$$y = \frac{1}{2}x - 1$$

(b)
$$y = \frac{1}{2}x - 1$$
$$2y = x - 2 \quad \textit{Multiply by 2.}$$
$$-x + 2y = -2$$
$$x - 2y = 2$$

71. (a) Through $(8, 5)$; perpendicular to $2x - y = 7$
Find the slope of $2x - y = 7$.

$$-y = -2x + 7$$
$$y = 2x - 7$$

Copyright © 2012 Pearson Education, Inc. Publishing as Addison-Wesley.

The slope of the line is 2. Therefore, the slope of the line perpendicular to it is $-\frac{1}{2}$ since $2\left(-\frac{1}{2}\right) = -1$. Use $m = -\frac{1}{2}$ and $(x_1, y_1) = (8, 5)$ in the point-slope form.

$$y - y_1 = m(x - x_1)$$
$$y - 5 = -\frac{1}{2}(x - 8)$$
$$y - 5 = -\frac{1}{2}x + 4$$
$$y = -\frac{1}{2}x + 9$$

(b)
$$y = -\frac{1}{2}x + 9$$
$$2y = -x + 18 \quad \textit{Multiply by 2.}$$
$$x + 2y = 18$$

73. (a) Through $(-2, 7)$; perpendicular to $x = 9$
$x = 9$ is a vertical line so a line perpendicular to it will be a horizontal line. It goes through $(-2, 7)$, so its equation is

$$y = 7.$$

(b) $y = 7$ is already in standard form.

75. Distance $=$ (rate)(time), so

$$y = 45x.$$

x	$y = 45x$	Ordered Pair
0	$45(0) = 0$	$(0, 0)$
5	$45(5) = 225$	$(5, 225)$
10	$45(10) = 450$	$(10, 450)$

77. Total cost $=$ (cost/gal)(number of gallons), so

$$y = 3.10x.$$

x	$y = 3.10x$	Ordered Pair
0	$3.10(0) = 0$	$(0, 0)$
5	$3.10(5) = 15.50$	$(5, 15.50)$
10	$3.10(10) = 31.00$	$(10, 31.00)$

79. Total cost $=$ (cost/credit)(number of credits), so

$$y = 111x.$$

x	$y = 111x$	Ordered Pair
0	$111(0) = 0$	$(0, 0)$
5	$111(5) = 555$	$(5, 555)$
10	$111(10) = 1110$	$(10, 1110)$

81. (a) The fixed cost is $12, so that is the value of b. The variable cost is $112.50, so

$$y = mx + b = 112.50x + 12.$$

(b) If $x = 5$, $y = 112.50(5) + 12 = 574.50$. The ordered pair is $(5, 574.50)$. The cost of 5 tickets and a parking pass is $574.50.

(c) If $x = 2$, $y = 112.50(2) + 12 = 237$. The cost of 2 tickets and a parking pass is $237.

83. (a) The fixed cost is $99, so that is the value of b. The variable cost is $41, so

$$y = mx + b = 41x + 99.$$

(b) If $x = 5$, $y = 41(5) + 99 = 304$. The ordered pair is $(5, 304)$. The cost for a 5-month membership is $304.

(c) If $x = 12$, $y = 41(12) + 99 = 591$. The cost for the first year's membership is $591.

85. (a) The fixed cost is $36, so that is the value of b. The variable cost is $60, so

$$y = mx + b = 60x + 36.$$

(b) If $x = 5$, $y = 60(5) + 36 = 336$. The ordered pair is $(5, 336)$. The cost of the plan for 5 months is $336.

(c) For a 1-year contract, $x = 12$, so $y = 60(12) + 36 = 756$. The cost of the plan for 1 year is $756.

87. (a) The fixed cost is $30, so that is the value of b. The variable cost is $6, so

$$y = mx + b = 6x + 30.$$

(b) If $x = 5$, $y = 6(5) + 30 = 60$. The ordered pair is $(5, 60)$. It costs $60 to rent the saw for 5 days.

(c) $138 = 6x + 30 \quad \textit{Let y = 138.}$
$108 = 6x$
$$x = \frac{108}{6} = 18$$

The saw is rented for 18 days.

89. (a) Use $(0, 3921)$ and $(3, 7805)$.

$$m = \frac{7805 - 3921}{3 - 0} = \frac{3884}{3} \approx 1294.7$$

The equation is $y = 1294.7x + 3921$. The slope tells us that the sales of digital cameras in the United States increased by $1294.7 million per year from 2003 to 2006.

(b) The year 2007 corresponds to $x = 4$, so $y = 1294.7(4) + 3921 = \$9099.8$ million.

91. (a) Use $(3, 38.0)$ and $(7, 59.0)$.

$$m = \frac{59.0 - 38.0}{7 - 3} = \frac{21}{4} = 5.25$$

Now use the point-slope form.

$$y - 38 = 5.25(x - 3)$$
$$y - 38 = 5.25x - 15.75$$
$$y = 5.25x + 22.25$$

(b) The year 2005 corresponds to $x = 5$, so $y = 5.25(5) + 22.25 = 48.5$. This value is greater than the actual value.

♦ ♦ ♦ Relating Concepts 93–100 ♦ ♦ ♦

93. When $C = 0°$, $F = \underline{32°}$, and when $C = 100°$, $F = \underline{212°}$. These are the freezing and boiling temperatures for water.

Copyright © 2012 Pearson Education, Inc. Publishing as Addison-Wesley.

94. The two points of the form (C, F) would be $(0, 32)$ and $(100, 212)$.

95. $m = \dfrac{212 - 32}{100 - 0} = \dfrac{180}{100} = \dfrac{9}{5}$

96. Let $m = \frac{9}{5}$ and $(x_1, y_1) = (0, 32)$.

$$y - y_1 = m(x - x_1)$$
$$F - 32 = \frac{9}{5}(C - 0)$$
$$F - 32 = \frac{9}{5}C$$
$$F = \frac{9}{5}C + 32$$

97. $\qquad F = \frac{9}{5}C + 32$
$$F - 32 = \frac{9}{5}C$$
$$\frac{5}{9}(F - 32) = C$$

98. $F = \frac{9}{5}C + 32$

If $C = 30$, $\qquad F = \frac{9}{5}(30) + 32$
$$= 54 + 32 = 86.$$

Thus, when $C = 30°$, $F = 86°$.

99. $C = \frac{5}{9}(F - 32)$

When $F = 50$, $\qquad C = \frac{5}{9}(50 - 32)$
$$= \frac{5}{9}(18) = 10.$$

Thus, when $F = 50°$, $C = 10°$.

100. Let $F = C$ in the equation obtained in Exercise 96.

$$F = \frac{9}{5}C + 32$$
$$C = \frac{9}{5}C + 32 \qquad \textit{Let F = C.}$$
$$5C = 5(\tfrac{9}{5}C + 32) \quad \textit{Multiply by 5.}$$
$$5C = 9C + 160$$
$$-4C = 160 \qquad \textit{Subtract 9C.}$$
$$C = -40 \qquad \textit{Divide by } -4.$$

(The same result may be found by using either form of the equation obtained in Exercise 97.) The Celsius and Fahrenheit temperatures are equal $(F = C)$ at -40 degrees.

101. $2x + 5 < 9$
$$2x < 4$$
$$x < 2$$

The solution set is $(-\infty, 2)$.

103. $5 - 3x \geq 9$
$$-3x \geq 4$$
$$x \leq -\frac{4}{3}$$

The solution set is $(-\infty, -\frac{4}{3}]$.

Summary Exercises on Slopes and Equations of Lines

1. Solve the equation for y.

$$3x + 5y = 9$$
$$5y = -3x + 9$$
$$y = -\frac{3}{5}x + \frac{9}{5}$$

The slope is $-\frac{3}{5}$.

3. $y = 2x - 5$ is already in the slope-intercept form, so the slope is 2.

5. The graph of $x - 4 = 0$, or $x = 4$, is a vertical line with x-intercept $(4, 0)$. The slope of a vertical line is *undefined* because the denominator equals zero in the slope formula.

7. Through $(-2, 6)$ and $(4, 1)$

(a) The slope is

$$m = \frac{1 - 6}{4 - (-2)} = \frac{-5}{6} = -\frac{5}{6}.$$

Use the point-slope form.

$$y - y_1 = m(x - x_1)$$
$$y - 6 = -\frac{5}{6}[x - (-2)]$$
$$y - 6 = -\frac{5}{6}x - \frac{5}{3}$$
$$y = -\frac{5}{6}x - \frac{5}{3} + \frac{18}{3}$$
$$y = -\frac{5}{6}x + \frac{13}{3}$$

(b) $\qquad y = -\frac{5}{6}x + \frac{13}{3}$
$$6y = -5x + 26 \quad \textit{Multiply by 6.}$$
$$5x + 6y = 26$$

9. Through $(0, 0)$; perpendicular to $2x - 5y = 6$

(a) Find the slope of $2x - 5y = 6$.

$$-5y = -2x + 6$$
$$y = \frac{2}{5}x - \frac{6}{5}$$

The slope of the line is $\frac{2}{5}$. Therefore, the slope of the line perpendicular to it is $-\frac{5}{2}$ since $\frac{2}{5}(-\frac{5}{2}) = -1$. Use $m = -\frac{5}{2}$ and $(x_1, y_1) = (0, 0)$ in the point-slope form.

$$y - y_1 = m(x - x_1)$$
$$y - 0 = -\frac{5}{2}(x - 0)$$
$$y = -\frac{5}{2}x$$

(b) $\qquad y = -\frac{5}{2}x$
$$2y = -5x$$
$$5x + 2y = 0$$

Copyright © 2012 Pearson Education, Inc. Publishing as Addison-Wesley.

11. Through $\left(\frac{3}{4}, -\frac{7}{9}\right)$; perpendicular to $x = \frac{2}{3}$

(a) $x = \frac{2}{3}$ is a vertical line so a line perpendicular to it will be a horizontal line. It goes through $\left(\frac{3}{4}, -\frac{7}{9}\right)$, so its equation is

$$y = -\frac{7}{9}.$$

(b) $y = -\frac{7}{9}$

$9y = -7$ *Multiply by 9.*

13. Through $(-4, 2)$; parallel to the line through $(3, 9)$ and $(6, 11)$

(a) The slope of the line through $(3, 9)$ and $(6, 11)$ is

$$m = \frac{11 - 9}{6 - 3} = \frac{2}{3}.$$

Use the point-slope form with $(x_1, y_1) = (-4, 2)$ and $m = \frac{2}{3}$ (since the slope of the desired line must equal the slope of the given line).

$$y - y_1 = m(x - x_1)$$
$$y - 2 = \frac{2}{3}[x - (-4)]$$
$$y - 2 = \frac{2}{3}(x + 4)$$
$$y - 2 = \frac{2}{3}x + \frac{8}{3}$$
$$y = \frac{2}{3}x + \frac{8}{3} + \frac{6}{3}$$
$$y = \frac{2}{3}x + \frac{14}{3}$$

(b) $y = \frac{2}{3}x + \frac{14}{3}$

$3y = 2x + 14$

$-2x + 3y = 14$

$2x - 3y = -14$

15. Through $(4, -8)$ and $(-4, 12)$

(a) The slope is

$$m = \frac{12 - (-8)}{-4 - 4} = \frac{20}{-8} = -\frac{5}{2}.$$

Use the point-slope form.

$$y - y_1 = m(x - x_1)$$
$$y - (-8) = -\frac{5}{2}(x - 4)$$
$$y + 8 = -\frac{5}{2}x + 10$$
$$y = -\frac{5}{2}x + 2$$

(b) $y = -\frac{5}{2}x + 2$

$2y = -5x + 4$ *Multiply by 2.*

$5x + 2y = 4$

17. Through $(0, 3)$ and the midpoint of the segment with endpoints $(2, 8)$ and $(-4, 12)$

(a) The midpoint of the segment with endpoints $(2, 8)$ and $(-4, 12)$ is

$$\left(\frac{2 + (-4)}{2}, \frac{8 + 12}{2}\right) = \left(\frac{-2}{2}, \frac{20}{2}\right) = (-1, 10).$$

The slope of the line through $(0, 3)$ and $(-1, 10)$ is

$$m = \frac{10 - 3}{-1 - 0} = \frac{7}{-1} = -7.$$

Since we are given the y-intercept, $(0, 3)$, the slope-intercept form is

$$y = -7x + 3.$$

(b) $y = -7x + 3$

$7x + y = 3$

3.4   Linear Inequalities in Two Variables

3.4 Now Try Exercises

N1. $-x + 2y \geq 4$

Step 1
Graph the line, $-x + 2y = 4$, which has intercepts $(-4, 0)$ and $(0, 2)$, as a solid line since the inequality involves "$\leq$".

Step 2
Test $(0, 0)$.

$$x + y \geq 4$$
$$0 + 0 \overset{?}{\geq} 4$$
$$0 \geq 4 \qquad \textit{False}$$

Step 3
Since the result is false, shade the region that does not contain $(0, 0)$.

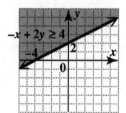

N2. $3x - y < 6$

Solve the inequality for y.

$$-y < -3x + 6 \quad \textit{Subtract 3x.}$$
$$y > 3x - 6 \quad \textit{Multiply by −1.}$$

Graph the boundary line, $y = 3x - 6$ [which has slope 3 and y-intercept $(0, -6)$], as a dashed line because the inequality symbol is $>$. Since the inequality is solved for y and the inequality symbol is $>$, we shade the half-plane above the boundary line.

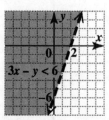

N3. $x + y < 3$ and $y \le 2$

Graph $x + y = 3$, which has intercepts $(3, 0)$ and $(0, 3)$, as a dashed line since the inequality involves " $<$ ". Test $(0, 0)$, which yields $0 < 3$, a true statement. Shade the region that includes $(0, 0)$.

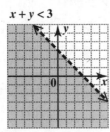

$x + y < 3$

Graph $y = 2$ as a solid horizontal line through $(0, 2)$. Shade the region below $y = 2$.

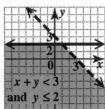

$y \le 2$

The graph of the intersection is the region common to both graphs.

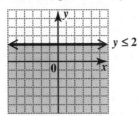

$x + y < 3$
and $y \le 2$

N4. $3x - 5y < 15$ or $x > 4$

Graph $3x - 5y = 15$ as a dashed line through its intercepts $(5, 0)$ and $(0, -3)$. Test $(0, 0)$, which yields $0 < 15$, a true statement. Shade the region that includes $(0, 0)$.

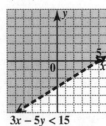

$3x - 5y < 15$

Graph $x = 4$ as a dashed vertical line through $(4, 0)$. Shade the region to the right of $x = 4$.

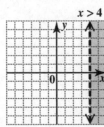

$x > 4$

The graph of the union is the region that includes all the points in both graphs.

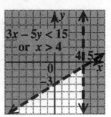

$3x - 5y < 15$
or $x > 4$

3.4 Section Exercises

1. The boundary of the graph of $y \le -x + 2$ will be a _solid_ line (since the inequality involves $\le$), and the shading will be _below_ the line (since the inequality sign is $\le$ or $<$).

3. The boundary of the graph of $y > -x + 2$ will be a _dashed_ line (since the inequality involves $>$), and the shading will be _above_ the line (since the inequality sign is $\ge$ or $>$).

5. The graph of $Ax + By = C$ divides the plane into two regions. In one of these regions, the ordered pairs satisfy $Ax + By < C$; in the other, they satisfy $Ax + By > C$.

7. $x + y \le 2$
Graph the line $x + y = 2$ by drawing a solid line (since the inequality involves $\le$) through the intercepts $(2, 0)$ and $(0, 2)$.
Test a point not on this line, such as $(0, 0)$.

$$x + y \le 2$$
$$0 + 0 \overset{?}{\le} 2$$
$$0 \le 2 \quad True$$

Shade the side of the line containing the test point $(0, 0)$.

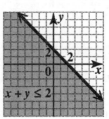

$x + y \le 2$

9. $4x - y < 4$
Graph the line $4x - y = 4$ by drawing a dashed line (since the inequality involves $<$) through the intercepts $(1, 0)$ and $(0, -4)$. Instead of using a test point, we will solve the inequality for y.

$$-y < -4x + 4$$
$$y > 4x - 4$$

Copyright © 2012 Pearson Education, Inc. Publishing as Addison-Wesley.

Since we have "$y >$" in the last inequality, shade the region *above* the boundary line.

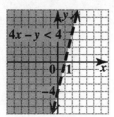

Shade the side of the line *not* containing the test point $(0, 0)$.

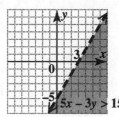

11. $x + 3y \geq -2$

Graph the solid line $x + 3y = -2$ (since the inequality involves $\geq$) through the intercepts $(-2, 0)$ and $(0, -\frac{2}{3})$.

Test a point not on this line such as $(0, 0)$.

$$0 + 3(0) \overset{?}{\geq} -2$$
$$0 \geq -2 \quad \textit{True}$$

Shade the side of the line containing the test point $(0, 0)$.

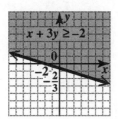

13. $2x + 3y \geq 6$

Graph the solid line $2x + 3y = 6$ (since the inequality involves $\geq$) through the intercepts $(3, 0)$ and $(0, 2)$.

Test a point not on this line such as $(0, 0)$.

$$2(0) + 3(0) \overset{?}{\geq} 6$$
$$0 \geq 6 \quad \textit{False}$$

Shade the side of the line *not* containing the test point $(0, 0)$.

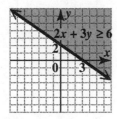

15. $5x - 3y > 15$

Graph the dashed line $5x - 3y = 15$ (since the inequality involves $>$) through the intercepts $(3, 0)$ and $(0, -5)$.

Test a point not on this line such as $(0, 0)$.

$$5(0) - 3(0) \overset{?}{>} 15$$
$$0 > 15 \quad \textit{False}$$

17. $x + y > 0$

Graph the line $x + y = 0$, which includes the points $(0, 0)$ and $(2, -2)$, as a dashed line (since the inequality involves $>$). Solving the inequality for y gives us

$$y > -x,$$

So shade the region above the boundary line.

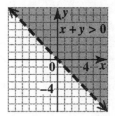

19. $x - 3y \leq 0$

Graph the solid line $x - 3y = 0$ through the points $(0, 0)$ and $(3, 1)$.

Solve the inequality for y.

$$-3y \leq -x$$
$$y \geq \tfrac{1}{3}x$$

Shade the region above the boundary line.

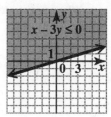

21. $y < x$

Graph the dashed line $y = x$ through $(0, 0)$ and $(2, 2)$. Since we have "$y <$" in the inequality, shade the region *below* the boundary line.

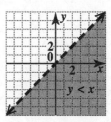

Copyright © 2012 Pearson Education, Inc. Publishing as Addison-Wesley.

23. $x + y \leq 1$ and $x \geq 1$

Graph the solid line $x + y = 1$ through $(0, 1)$ and $(1, 0)$. The inequality $x + y \leq 1$ can be written as $y \leq -x + 1$, so shade the region below the boundary line.

Graph the solid vertical line $x = 1$ through $(1, 0)$ and shade the region to the right. The required graph is the common shaded area as well as the portions of the lines that bound it.

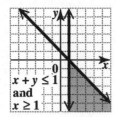

25. $2x - y \geq 2$ and $y < 4$

Graph the solid line $2x - y = 2$ through the intercepts $(1, 0)$ and $(0, -2)$. Test $(0, 0)$ to get $0 \geq 2$, a false statement. Shade the side of the line not containing $(0, 0)$. To graph $y < 4$ on the same axes, graph the dashed horizontal line through $(0, 4)$. Test $(0, 0)$ to get $0 < 4$, a true statement. Shade the side of the dashed line containing $(0, 0)$. The word "and" indicates the intersection of the two graphs. The final solution set consists of the region where the two shaded regions overlap.

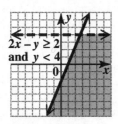

27. $x + y > -5$ and $y < -2$

Graph $x + y = -5$, which has intercepts $(-5, 0)$ and $(0, -5)$, as a dashed line. Test $(0, 0)$, which yields $0 > -5$, a true statement. Shade the region that includes $(0, 0)$.

Graph $y = -2$ as a dashed horizontal line. Shade the region below $y = -2$. The required graph of the intersection is the region common to both graphs.

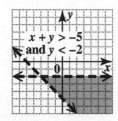

29. $|x| < 3$ can be rewritten as $-3 < x < 3$. The boundaries are the dashed vertical lines $x = -3$ and $x = 3$. Since x is between -3 and 3, the graph includes all points between the lines.

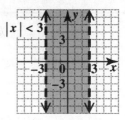

31. $|x + 1| < 2$ can be rewritten as

$$-2 < x + 1 < 2$$
$$-3 < x < 1.$$

The boundaries are the dashed vertical lines $x = -3$ and $x = 1$. Since x is between -3 and 1, the graph includes all points between the lines.

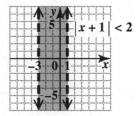

33. $x - y \geq 1$ or $y \geq 2$

Graph the solid line $x - y = 1$, which crosses the y-axis at -1 and the x-axis at 1. Use $(0, 0)$ as a test point, which yields $0 \geq 1$, a false statement. Shade the region that does not include $(0, 0)$. Now graph the solid line $y = 2$. Since the inequality is $y \geq 2$, shade above this line.

The required graph of the union includes all the shaded regions, that is, all the points that satisfy either inequality.

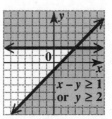

35. $x - 2 > y$ or $x < 1$

Graph $x - 2 = y$, which has intercepts $(2, 0)$ and $(0, -2)$, as a dashed line. Test $(0, 0)$, which yields $-2 > 0$, a false statement. Shade the region that does not include $(0, 0)$.

Graph $x = 1$ as a dashed vertical line. Shade the region to the left of $x = 1$.

The required graph of the union includes all the shaded regions, that is, all the points that satisfy either inequality.

Copyright © 2012 Pearson Education, Inc. Publishing as Addison-Wesley.

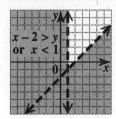

37. $3x + 2y < 6$ or $x - 2y > 2$

Graph $3x + 2y = 6$, which has intercepts $(2, 0)$ and $(0, 3)$, as a dashed line. Test $(0, 0)$, which yields $0 < 6$, a true statement. Shade the region that includes $(0, 0)$.

Graph $x - 2y = 2$, which has intercepts $(2, 0)$ and $(0, -1)$, as a dashed line. Test $(0, 0)$, which yields $0 > 2$, a false statement. Shade the region that does not include $(0, 0)$.

The required graph of the union includes all the shaded regions, that is, all the points that satisfy either inequality.

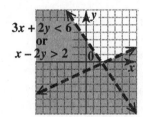

39. $y \leq 3x - 6$

The boundary line, $y = 3x - 6$, has slope 3 and y-intercept -6. This would be graph **B** or graph **C**. Since we want the region less than or equal to $3x - 6$, we want the region on or below the boundary line. The answer is graph **C**.

41. $y \leq -3x - 6$

The slope of the boundary line $y = -3x - 6$ is -3, and the y-intercept is -6. This would be graph **A** or graph **D**. The inequality sign is $\leq$, so we want the region on or below the boundary line. The answer is graph **A**.

43. (a) The x-intercept is $(-4, 0)$, so the solution set for $y = 0$ is $\{-4\}$.

(b) The solution set for $y < 0$ is $(-\infty, -4)$, since the graph is below the x-axis for these values of x.

(c) The solution set for $y > 0$ is $(-4, \infty)$, since the graph is above the x-axis for these values of x.

45. (a) The x-intercept is $(3.5, 0)$, so the solution set for $y = 0$ is $\{3.5\}$.

(b) The solution set for $y < 0$ is $(3.5, \infty)$, since the graph is below the x-axis for these values of x.

(c) The solution set for $y > 0$ is $(-\infty, 3.5)$, since the graph is above the x-axis for these values of x.

♦♦♦ Relating Concepts 47–52 ♦♦♦

47. "A factory can have *no more than* 200 workers on a shift, but must have *at least* 100" can be translated as $x \leq 200$ and $x \geq 100$. "Must manufacture *at least* 3000 units" can be translated as $y \geq 3000$.

48.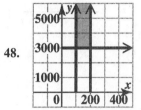

49. The total daily cost C consists of $50 per worker and $100 to manufacture 1 unit, so $C = 50x + 100y$.

50. Some examples of points in the shaded region are $(150, 4000)$, $(120, 3500)$, and $(180, 6000)$. Some examples of points on the boundary are $(100, 5000)$, $(150, 3000)$, and $(200, 4000)$. The corner points are $(100, 3000)$ and $(200, 3000)$.

51.

(x, y)	$50x + 100y = C$
$(150, 4000)$	$50(150) + 100(4000) = 407{,}500$
$(120, 3500)$	$50(120) + 100(3500) = 356{,}000$
$(180, 6000)$	$50(180) + 100(6000) = 609{,}000$
$(100, 5000)$	$50(100) + 100(5000) = 505{,}000$
$(150, 3000)$	$50(150) + 100(3000) = 307{,}500$
$(200, 4000)$	$50(200) + 100(4000) = 410{,}000$
$(100, 3000)$	$50(100) + 100(3000) = 305{,}000$
	(least value)
$(200, 3000)$	$50(200) + 100(3000) = 310{,}000$

52. The company should use 100 workers and manufacture 3000 units to achieve the least possible cost.

53. $x \geq 0$ written in interval notation is $[0, \infty)$.

55. $x < 1$ or $x > 1$
written in interval notation is $(-\infty, 1) \cup (1, \infty)$.

3.5 Introduction to Relations and Functions

3.5 Now Try Exercises

N1. (a) $\{(1, 5), (3, 5), (5, 5)\}$
The relation *is a function* because for each different x-value there is exactly one y-value. It is acceptable to have different x-values paired with the same y-value.

(b) $\{(-1, -3), (0, 2), (-1, 6)\}$
The first and last ordered pairs have the *same* x-value paired with *two different* y-values (-1 is paired with both -3 and 6), so this relation *is not a function.*

Copyright © 2012 Pearson Education, Inc. Publishing as Addison-Wesley.

N2. (a) $\{(2,2),(2,5),(4,8),(6,5)\}$
The first two ordered pairs have the *same* x-value paired with *two different* y-values (2 is paired with both 2 and 5), so this relation *is not a function.*

(b) The domain of this relation is the set of all first components, that is, $\{1, 10, 15\}$. The range of this relation is the set of all second components, that is, $\{\$1.39, \$10.39, \$20.85\}$. This relation *is a function* because for each different first component, there is exactly one second component.

N3. The arrowheads indicate that the graph extends indefinitely left and right, as well as upward. The domain includes all real numbers, written $(-\infty, \infty)$. Because there is a least y-value, -2, the range includes all numbers greater than or equal to -2, $[-2, \infty)$.

N4. A vertical line intersects the graph more than once, so the relation is *not* a function.

N5. (a) $y = 4x - 3$ is a function because each value of x corresponds to exactly one value of y. Its domain is the set of all real numbers, $(-\infty, \infty)$.

(b) $y = \sqrt{2x - 4}$ is a function because each value of x corresponds to exactly one value of y. Since the quantity under the radical must be nonnegative, the domain is the set of real numbers that satisfy the condition

$$2x - 4 \geq 0$$
$$2x \geq 4$$
$$x \geq 2.$$

Therefore, the domain is $[2, \infty)$.

(c) $y = \dfrac{1}{x - 2}$

Given any value of x in the domain, we find y by subtracting 2, and then dividing the result into 1. This process produces exactly one value of y for each value in the domain, so the given equation defines a function. The domain includes all real numbers except those that make the denominator 0. We find those numbers by setting the denominator equal to 0 and solving for x.

$$x - 2 = 0$$
$$x = 2$$

The domain includes all real numbers *except* 2, written as $(-\infty, 2) \cup (2, \infty)$.

(d) $y < 3x + 1$ is not a function because if $x = 0$, then $y < 1$. Thus, the x-value 0 corresponds to many y-values. Its domain is the set of all real numbers, $(-\infty, \infty)$.

3.5 Section Exercises

1. Answers will vary. A function is a set of ordered pairs in which each first component corresponds to exactly one second component. For example, $\{(0,1), (1,2), (2,3), (3,4), \dots\}$ is a function.

3. In an ordered pair of a relation, the first element is the independent variable.

5. $\{(0,2), (2,4), (4,6)\}$
We can represent this set of ordered pairs by plotting them on a graph.

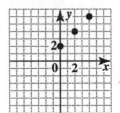

7. We can represent this diagram in table form.

x	y
-3	-4
-3	1
2	0

9. $\{(5,1), (3,2), (4,9), (7,6)\}$
The relation is a function since for each x-value, there is only one y-value.

The domain is the set of x-values: $\{5, 3, 4, 7\}$.
The range is the set of y-values: $\{1, 2, 9, 6\}$.

11. $\{(2,4), (0,2), (2,5)\}$
The relation is not a function since the x-value 2 has two different y-values associated with it, 4 and 5.

The domain is the set of x-values: $\{2, 0\}$.
The range is the set of y-values: $\{4, 2, 5\}$.

13. $\{(-3,1), (4,1), (-2,7)\}$
The relation is a function since for each x-value, there is only one y-value.

The domain is the set of x-values: $\{-3, 4, -2\}$.
The range is the set of y-values: $\{1, 7\}$.

15. $\{(1,1), (1,-1), (0,0), (2,4), (2,-4)\}$
The relation is not a function since the x-value 1 has two different y-values associated with it, 1 and -1. (A similar statement can be made for $x = 2$.)

The domain is the set of x-values: $\{1, 0, 2\}$.
The range is the set of y-values: $\{1, -1, 0, 4, -4\}$.

17. The relation can be described by the set of ordered pairs

$$\{(2,1), (5,1), (11,7), (17,20), (3,20)\}.$$

The relation is a function since for each x-value, there is only one y-value.

Copyright © 2012 Pearson Education, Inc. Publishing as Addison-Wesley.

The domain is the set of x-values: $\{2, 5, 11, 17, 3\}$.
The range is the set of y-values: $\{1, 7, 20\}$.

19. The relation can be described by the set of ordered pairs

$$\{(1, 5), (1, 2), (1, -1), (1, -4)\}.$$

The relation is not a function since the x-value 1 has four different y-values associated with it, 5, 2, -1, and -4.

The domain is the set of x-values: $\{1\}$.
The range is the set of y-values: $\{5, 2, -1, -4\}$.

21. The relation can be described by the set of ordered pairs

$$\{(4, -3), (2, -3), (0, -3), (-2, -3)\}.$$

The relation is a function since for each x-value, there is only one y-value.

The domain is the set of x-values: $\{4, 2, 0, -2\}$.
The range is the set of y-values: $\{-3\}$.

23. The relation can be described by the set of ordered pairs

$$\{(-2, 2), (0, 3), (3, 2)\}.$$

The relation is a function since for each x-value, there is only one y-value.

The domain is the set of x-values: $\{-2, 0, 3\}$.
The range is the set of y-values: $\{2, 3\}$.

25. Using the vertical line test, we find any vertical line will intersect the graph at most once. This indicates that the graph represents a function. This graph extends indefinitely to the left $(-\infty)$ and indefinitely to the right (∞). Therefore, the domain is $(-\infty, \infty)$. This graph extends indefinitely downward $(-\infty)$, and indefinitely upward (∞). Thus, the range is $(-\infty, \infty)$.

27. Since a vertical line, such as $x = -4$, intersects the graph in two points, the relation is not a function. The domain is $(-\infty, 0]$, and the range is $(-\infty, \infty)$.

29. Using the vertical line test, we find any vertical line will intersect the graph at most once. This indicates that the graph represents a function. This graph extends indefinitely to the left $(-\infty)$ and indefinitely to the right (∞). Therefore, the domain is $(-\infty, \infty)$. This graph extends indefinitely downward $(-\infty)$, and reaches a high point at $y = 4$. Therefore, the range is $(-\infty, 4]$.

31. Since a vertical line can intersect the graph of the relation in more than one point, the relation is not a function. The domain, the x-values of the points on the graph, is $[-4, 4]$. The range, the y-values of the points on the graph, is $[-3, 3]$.

33. $y = -6x$
Each value of x corresponds to one y-value. For example, if $x = 3$, then $y = -6(3) = -18$. Therefore, $y = -6x$ defines y as a function of x. Since any x-value, positive, negative, or zero, can be multiplied by -6, the domain is $(-\infty, \infty)$.

35. $y = 2x - 6$
For any value of x, there is exactly one value of y, so this equation defines a function. The domain is the set of all real numbers, $(-\infty, \infty)$.

37. $y = x^2$
Each value of x corresponds to one y-value. For example, if $x = 3$, then $y = 3^2 = 9$. Therefore, $y = x^2$ defines y as a function of x. Since any x-value, positive, negative, or zero, can be squared, the domain is $(-\infty, \infty)$.

39. $x = y^6$
The ordered pairs $(64, 2)$ and $(64, -2)$ both satisfy the equation. Since one value of x, 64, corresponds to two values of y, 2 and -2, the relation does not define a function. Because x is equal to the sixth power of y, the values of x must always be nonnegative. The domain is $[0, \infty)$.

41. $x + y < 4$
For a particular x-value, more than one y-value can be selected to satisfy $x + y < 4$. For example, if $x = 2$ and $y = 0$, then

$$2 + 0 < 4. \quad \textit{True}$$

Now, if $x = 2$ and $y = 1$, then

$$2 + 1 < 4. \quad \textit{Also true}$$

Therefore, $x + y < 4$ does not define y as a function of x.
The graph of $x + y < 4$ is equivalent to the graph of $y < -x + 4$, which consists of the shaded region below the dashed line $y = -x + 4$, which extends indefinitely from left to right. Therefore, the domain is $(-\infty, \infty)$.

43. $y = \sqrt{x}$
For any value of x, there is exactly one corresponding value for y, so this relation defines a function. Since the radicand must be a nonnegative number, x must always be nonnegative. The domain is $[0, \infty)$.

45. $y = \sqrt{x - 3}$
is a function because each value of x in the domain corresponds to exactly one value of y. Since the quantity under the radical must be nonnegative, the domain is the set of real numbers that satisfy the condition

$$x - 3 \geq 0$$
$$x \geq 3.$$

Therefore, the domain is $[3, \infty)$.

Copyright © 2012 Pearson Education, Inc. Publishing as Addison-Wesley.

47. $y = \sqrt{4x + 2}$
is a function because each value of x in the domain corresponds to exactly one value of y. Since the quantity under the radical must be nonnegative, the domain is the set of real numbers that satisfy the condition

$$4x + 2 \geq 0$$
$$4x \geq -2$$
$$x \geq -\tfrac{1}{2}.$$

Therefore, the domain is $[-\tfrac{1}{2}, \infty)$.

49. $y = \dfrac{x + 4}{5}$
Given any value of x, y is found by adding 4, then dividing the result by 5. This process produces exactly one value of y for each x-value in the domain, so the relation represents a function. The denominator is never 0, so the domain is $(-\infty, \infty)$.

51. $y = -\dfrac{2}{x}$
Given any value of x, y is found by dividing that value into 2, and negating that result. This process produces exactly one value of y for each x-value in the domain, so the relation represents a function. The domain includes all real numbers except those that make the denominator 0, namely 0. The domain is $(-\infty, 0) \cup (0, \infty)$.

53. $y = \dfrac{2}{x - 4}$
Given any value of x, y is found by subtracting 4, then dividing the result into 2. This process produces exactly one value of y for each x-value in the domain, so the relation represents a function. The domain includes all real numbers except those that make the denominator 0, namely 4. The domain is $(-\infty, 4) \cup (4, \infty)$.

55. $xy = 1$
Rewrite $xy = 1$ as $y = \tfrac{1}{x}$. Note that x can never equal 0, otherwise the denominator would equal 0. The domain is $(-\infty, 0) \cup (0, \infty)$.
Each nonzero x-value gives exactly one y-value. Therefore, $xy = 1$ defines y as a function of x.

57. **(a)** Each year corresponds to exactly one percentage, so the table defines a function.

(b) The domain is $\{2004, 2005, 2006, 2007, 2008\}$.
The range is $\{42.3, 42.8, 43.7, 43.8\}$.

(c) Answers will vary. Two possible answers are $(2005, 42.3)$ and $(2008, 43.8)$.

59. $y = -7x + 12$
$y = -7(3) + 12$ *Let x = 3.*
$y = -9$

61. $y = 3x - 8$
$y = 3(3) - 8$ *Let x = 3.*
$y = 1$

63. $2x - 4y = 7$
$-4y = -2x + 7$
$y = \tfrac{1}{2}x - \tfrac{7}{4}$

3.6 Function Notation and Linear Functions

3.6 Now Try Exercises

N1. $f(x) = 4x + 3$
$f(-2) = 4(-2) + 3$ *Replace x with −2.*
$\quad = -8 + 3$ *Multiply.*
$\quad = -5$ *Add.*

N2. $f(x) = 2x^2 - 4x + 1$

(a) $f(-2) = 2(-2)^2 - 4(-2) + 1$
$\quad = 8 + 8 + 1 = 17$

(b) $f(a) = 2a^2 - 4a + 1$

N3. $\quad g(x) = 8x - 5$
$g(a - 2) = 8(a - 2) - 5$
$\quad = 8a - 16 - 5$
$\quad = 8a - 21$

N4. **(a)** When $x = -1$, $y = 4$, so $f(-1) = 4$.

(b) $f(x) = x^2 - 12$
$f(-1) = (-1)^2 - 12 = 1 - 12 = -11$

N5. **(a)** When $x = -1$, $y = 0$, so $f(-1) = 0$.

(b) $f(x) = 2$ is equivalent to $y = 2$, and $y = 2$ when $x = 1$.

N6. $-4x^2 + y = 5$

Solving for y gives us $y = 4x^2 + 5$.

$f(x) = 4x^2 + 5$ *Replace y with f(x).*
$f(-3) = 4(-3)^2 + 5 = 36 + 5 = 41$
$f(h) = 4h^2 + 5$

N7. The graph of $g(x) = \tfrac{1}{3}x - 2$ is a line with slope $\tfrac{1}{3}$ and y-intercept -2.

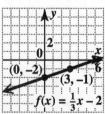

The domain and range are $(-\infty, \infty)$.

Copyright © 2012 Pearson Education, Inc. Publishing as Addison-Wesley.

3.6 Section Exercises

1. $f(3)$ is the value of the dependent variable when the independent variable is 3—choice **B**.

3. $f(x) = -3x + 4$
$f(0) = -3(0) + 4$
$= 0 + 4$
$= 4$

5. $g(x) = -x^2 + 4x + 1$
$g(-2) = -(-2)^2 + 4(-2) + 1$
$= -(4) - 8 + 1$
$= -11$

7. $f(x) = -3x + 4$
$f(\frac{1}{3}) = -3(\frac{1}{3}) + 4$
$= -1 + 4$
$= 3$

9. $g(x) = -x^2 + 4x + 1$
$g(0.5) = -(0.5)^2 + 4(0.5) + 1$
$= -0.25 + 2 + 1$
$= 2.75$

11. $f(x) = -3x + 4$
$f(p) = -3(p) + 4$
$= -3p + 4$

13. $f(x) = -3x + 4$
$f(-x) = -3(-x) + 4$
$= 3x + 4$

15. $f(x) = -3x + 4$
$f(x + 2) = -3(x + 2) + 4$
$= -3x - 6 + 4$
$= -3x - 2$

17. $g(x) = -x^2 + 4x + 1$
$g(\pi) = -\pi^2 + 4\pi + 1$

19. $f(x) = -3x + 4$
$f(x + h) = -3(x + h) + 4$
$= -3x - 3h + 4$

21. $f(x) = -3x + 4, g(x) = -x^2 + 4x + 1$
$f(4) - g(4)$
$= [-3(4) + 4] - [-(4)^2 + 4(4) + 1]$
$= [-8] - [1]$
$= -9$

23. $f = \{(-2, 2), (-1, -1), (2, -1)\}$

(a) When $x = 2, y = -1$, so $f(2) = -1$.

(b) When $x = -1, y = -1$, so $f(-1) = -1$.

25. (a) When $x = 2, y = 2$, so $f(2) = 2$.

(b) When $x = -1, y = 3$, so $f(-1) = 3$.

27. (a) When $x = 2, y = 15$, so $f(2) = 15$.

(b) When $x = -1, y = 10$, so $f(-1) = 10$.

29. (a) When $x = 2, y = 4$, so $f(2) = 4$.

(b) When $x = -1, y = 1$, so $f(-1) = 1$.

31. (a) The point $(2, 3)$ is on the graph of f, so $f(2) = 3$.

(b) The point $(-1, -3)$ is on the graph of f, so $f(-1) = -3$.

33. (a) The point $(2, -3)$ is on the graph of f, so $f(2) = -3$.

(b) The point $(-1, 2)$ is on the graph of f, so $f(-1) = 2$.

35. (a) $f(x) = 3$: when $y = 3, x = 2$

(b) $f(x) = -1$: when $y = -1, x = 0$

(c) $f(x) = -3$: when $y = -3, x = -1$

37. (a) Solve the equation for y.

$$x + 3y = 12$$
$$3y = 12 - x$$
$$y = \frac{12 - x}{3}$$

Since $y = f(x)$,

$$f(x) = \frac{12 - x}{3} = -\frac{1}{3}x + 4.$$

(b) $f(3) = \frac{12 - 3}{3} = \frac{9}{3} = 3$

39. (a) Solve the equation for y.

$$y + 2x^2 = 3$$
$$y = 3 - 2x^2$$

Since $y = f(x)$,

$$f(x) = 3 - 2x^2.$$

(b) $f(3) = 3 - 2(3)^2$
$= 3 - 2(9)$
$= -15$

41. (a) Solve the equation for y.

$$4x - 3y = 8$$
$$-3y = 8 - 4x$$
$$y = \frac{8 - 4x}{-3}$$

Since $y = f(x)$,

$$f(x) = \frac{8 - 4x}{-3} = \frac{4}{3}x - \frac{8}{3}.$$

(b) $f(3) = \frac{8 - 4(3)}{-3} = \frac{8 - 12}{-3}$
$= \frac{-4}{-3} = \frac{4}{3}$

Copyright © 2012 Pearson Education, Inc. Publishing as Addison-Wesley.

43. The equation $2x + y = 4$ has a straight <u>line</u> as its graph. To find y in $(3, y)$, let $x = 3$ in the equation.

$$2x + y = 4$$
$$2(3) + y = 4$$
$$6 + y = 4$$
$$y = -2$$

So one point that lies on the graph is $(3, \underline{-2})$. To use functional notation for $2x + y = 4$, solve for y to get

$$y = -2x + 4.$$

Replace y with $f(x)$ to get

$$f(x) = \underline{-2x + 4}.$$
$$f(3) = -2(3) + 4 = \underline{-2}$$

Because $y = -2$ when $x = 3$, the point $(\underline{3}, \underline{-2})$ lies on the graph of the function.

45. $f(x) = -2x + 5$
The graph will be a line. The intercepts are $(0, 5)$ and $\left(\frac{5}{2}, 0\right)$.
The domain is $(-\infty, \infty)$. The range is $(-\infty, \infty)$.

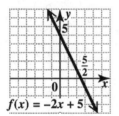

47. $h(x) = \frac{1}{2}x + 2$
The graph will be a line. The intercepts are $(0, 2)$ and $(-4, 0)$.
The domain is $(-\infty, \infty)$. The range is $(-\infty, \infty)$.

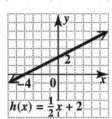

49. $G(x) = 2x$
This line includes the points $(0, 0), (1, 2)$, and $(2, 4)$. The domain is $(-\infty, \infty)$. The range is $(-\infty, \infty)$.

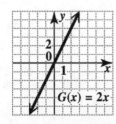

51. $g(x) = -4$
Using a y-intercept of $(0, -4)$ and a slope of $m = 0$, we graph the horizontal line. From the graph we see that the domain is $(-\infty, \infty)$. The range is $\{-4\}$.

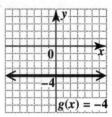

53. $f(x) = 0$
Draw the horizontal line through the point $(0, 0)$. On the horizontal line the value of x can be any real number, so the domain is $(-\infty, \infty)$. The range is $\{0\}$.

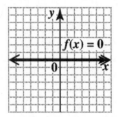

55. $f(x) = 0$, or $y = 0$, is the x-axis.

57. **(a)** $f(x) = 3.75x$
$$f(3) = 3.75(3)$$
$$= 11.25 \text{ (dollars)}$$

(b) 3 is the value of the independent variable, which represents a package weight of 3 pounds; $f(3)$ is the value of the dependent variable representing the cost to mail a 3-pound package.

(c) $3.75(5) = \$18.75$, the cost to mail a 5-lb package. Using function notation, we have $f(5) = 18.75$.

59. **(a)** Since the length of a man's femur is given, use the formula $h(r) = 69.09 + 2.24r$.

$$h(56) = 69.09 + 2.24(56) \quad \text{Let } r = 56.$$
$$= 194.53$$

The man is 194.53 cm tall.

(b) Use the formula $h(t) = 81.69 + 2.39t$.

$$h(40) = 81.69 + 2.39(40) \quad \text{Let } t = 40.$$
$$= 177.29$$

The man is 177.29 cm tall.

(c) Since the length of a woman's femur is given, use the formula $h(r) = 61.41 + 2.32r$.

$$h(50) = 61.41 + 2.32(50) \quad \text{Let } r = 50.$$
$$= 177.41$$

The woman is 177.41 cm tall.

Copyright © 2012 Pearson Education, Inc. Publishing as Addison-Wesley.

(d) Use the formula $h(t) = 72.57 + 2.53t$.

$$h(36) = 72.57 + 2.53(36) \quad \textit{Let } t = 36.$$
$$= 163.65$$

The woman is 163.65 cm tall.

61. **(a)** $f(x) = 12x + 100$

(b) $f(125) = 12(125) + 100$
$$= 1500 + 100 = 1600$$

The cost to print 125 t-shirts is $1600.

(c) $\qquad f(x) = 1000$
$$12x + 100 = 1000$$
$$12x = 900 \quad \textit{Subtract 100.}$$
$$x = 75 \quad \textit{Divide by 12.}$$

In function notation, $f(75) = 1000$.

The cost to print 75 t-shirts is $1000.

63. **(a)** $f(2) = 1.1$

(b) y is -2.5 when x is 5.
So, if $f(x) = -2.5$, then $x = 5$.

(c) Let $(x_1, y_1) = (0, 3.5)$ and
$(x_2, y_2) = (1, 2.3)$. Then

$$m = \frac{2.3 - 3.5}{1 - 0} = \frac{-1.2}{1} = -1.2.$$

The slope is -1.2.

(d) When $x = 0$, y is 3.5, so the y-intercept is $(0, 3.5)$.

(e) Use the slope-intercept form of the equation of a line and the information found in parts (c) and (d).

$$f(x) = mx + b$$
$$f(x) = -1.2x + 3.5$$

65. **(a)** The independent variable is t, the number of hours, and the possible values are in the set $[0, 100]$. The dependent variable is g, the number of gallons, and the possible values are in the set $[0, 3000]$.

(b) The graph rises for the first 25 hours, so the water level increases for 25 hours. The graph falls for $t = 50$ to $t = 75$, so the water level decreases for 25 hours.

(c) There are 2000 gallons in the pool when $t = 90$.

(d) $f(0)$ is the number of gallons in the pool at time $t = 0$. Here, $f(0) = 0$, which means the pool is empty at time 0.

(e) $f(25) = 3000$; After 25 hours, there are 3000 gallons of water in the pool.

67. $\qquad 6x + 5y = 2$
$$6(-3) + 5y = 2 \quad \textit{Let } x = -3.$$
$$-18 + 5y = 2$$
$$5y = 20$$
$$y = 4$$

69. $\qquad 1.5x + 2.5y = 5.5$
$$1.5(-3) + 2.5y = 5.5 \quad \textit{Let } x = -3.$$
$$-4.5 + 2.5y = 5.5$$
$$2.5y = 10$$
$$y = 4$$

71. $\qquad 2x - 4(x - 3) = 8$
$$2x - 4x + 12 = 8$$
$$-2x + 12 = 8$$
$$-2x = -4$$
$$x = 2$$

The solution set is $\{2\}$.

Chapter 3 Review Exercises

1. $\qquad 3x + 2y = 10$
For $x = 0$:

$$3(0) + 2y = 10$$
$$2y = 10$$
$$y = 5 \quad \mathbf{(0, 5)}$$

For $y = 0$:

$$3x + 2(0) = 10$$
$$3x = 10$$
$$x = \tfrac{10}{3} \quad \left(\tfrac{10}{3}, 0\right)$$

For $x = 2$:

$$3(2) + 2y = 10$$
$$6 + 2y = 10$$
$$2y = 4$$
$$y = 2 \quad \mathbf{(2, 2)}$$

For $y = -2$:

$$3x + 2(-2) = 10$$
$$3x - 4 = 10$$
$$3x = 14$$
$$x = \tfrac{14}{3} \quad \left(\tfrac{14}{3}, -2\right)$$

Plot the ordered pairs, and draw the line through them.

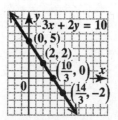

Copyright © 2012 Pearson Education, Inc. Publishing as Addison-Wesley.

2. $x - y = 8$
   For $x = 2$:

$$2 - y = 8$$
$$-y = 6$$
$$y = -6 \quad (2, -6)$$

   For $y = -3$:

$$x - (-3) = 8$$
$$x + 3 = 8$$
$$x = 5 \quad (5, -3)$$

   For $x = 3$:

$$3 - y = 8$$
$$-y = 5$$
$$y = -5 \quad (3, -5)$$

   For $y = -2$:

$$x - (-2) = 8$$
$$x + 2 = 8$$
$$x = 6 \quad (6, -2)$$

Plot the ordered pairs, and draw the line through them.

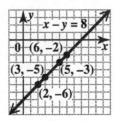

3. $4x - 3y = 12$
   To find the x-intercept, let $y = 0$.

$$4x - 3y = 12$$
$$4x - 3(0) = 12$$
$$4x = 12$$
$$x = 3$$

The x-intercept is $(3, 0)$.
To find the y-intercept, let $x = 0$.

$$4x - 3y = 12$$
$$4(0) - 3y = 12$$
$$-3y = 12$$
$$y = -4$$

The y-intercept is $(0, -4)$.
Plot the intercepts and draw the line through them.

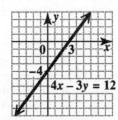

4. $5x + 7y = 28$
   To find the x-intercept, let $y = 0$.

$$5x + 7y = 28$$
$$5x + 7(0) = 28$$
$$5x = 28$$
$$x = \frac{28}{5}$$

The x-intercept is $\left(\frac{28}{5}, 0\right)$.
To find the y-intercept, let $x = 0$.

$$5x + 7y = 28$$
$$5(0) + 7y = 28$$
$$7y = 28$$
$$y = 4$$

The y-intercept is $(0, 4)$.
Plot the intercepts and draw the line through them.

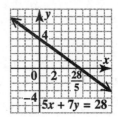

5. $2x + 5y = 20$
   To find the x-intercept, let $y = 0$.

$$2x + 5y = 20$$
$$2x + 5(0) = 20$$
$$2x = 20$$
$$x = 10$$

The x-intercept is $(10, 0)$.
To find the y-intercept, let $x = 0$.

$$2x + 5y = 20$$
$$2(0) + 5y = 20$$
$$5y = 20$$
$$y = 4$$

The y-intercept is $(0, 4)$.
Plot the intercepts and draw the line through them.

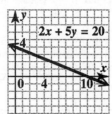

6. $x - 4y = 8$
   To find the x-intercept, let $y = 0$.

$$x - 4y = 8$$
$$x - 4(0) = 8$$
$$x = 8$$

Copyright © 2012 Pearson Education, Inc. Publishing as Addison-Wesley.

The x-intercept is $(8, 0)$.
To find the y-intercept, let $x = 0$.

$$0 - 4y = 8$$
$$-4y = 8$$
$$y = -2$$

The y-intercept is $(0, -2)$.
Plot the intercepts and draw the line through them.

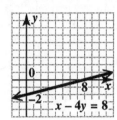

7. By the midpoint formula, the midpoint of the segment with endpoints $(-8, -12)$ and $(8, 16)$ is

$$\left(\frac{-8 + 8}{2}, \frac{-12 + 16}{2}\right) = \left(\frac{0}{2}, \frac{4}{2}\right) = (0, 2).$$

8. By the midpoint formula, the midpoint of the segment with endpoints $(0, -5)$ and $(-9, 8)$ is

$$\left(\frac{0 + (-9)}{2}, \frac{-5 + 8}{2}\right) = \left(\frac{-9}{2}, \frac{3}{2}\right) = \left(-\frac{9}{2}, \frac{3}{2}\right).$$

9. Through $(-1, 2)$ and $(4, -5)$

$$m = \frac{\text{change in } y}{\text{change in } x} = \frac{-5 - 2}{4 - (-1)} = \frac{-7}{5} = -\frac{7}{5}$$

10. Through $(0, 3)$ and $(-2, 4)$

Let $(x_1, y_1) = (0, 3)$ and $(x_2, y_2) = (-2, 4)$.

$$m = \frac{y_2 - y_1}{x_2 - x_1} = \frac{4 - 3}{-2 - 0} = \frac{1}{-2} = -\frac{1}{2}$$

11. The slope of $y = 2x + 3$ is 2, the coefficient of x.

12. $3x - 4y = 5$
Write the equation in slope-intercept form.

$$-4y = -3x + 5$$
$$y = \frac{3}{4}x - \frac{5}{4}$$

The slope is $\frac{3}{4}$.

13. $x = 5$ is a vertical line and has *undefined* slope.

14. Parallel to $3y = 2x + 5$
Write the equation in slope-intercept form.

$$3y = 2x + 5$$
$$y = \frac{2}{3}x + \frac{5}{3}$$

The slope of $3y = 2x + 5$ is $\frac{2}{3}$; all lines parallel to it will also have a slope of $\frac{2}{3}$.

15. Perpendicular to $3x - y = 4$
Solve for y.

$$y = 3x - 4$$

The slope is 3; the slope of a line perpendicular to it is $-\frac{1}{3}$ since

$$3\left(-\frac{1}{3}\right) = -1.$$

16. Through $(-1, 5)$ and $(-1, -4)$

$$m = \frac{\Delta y}{\Delta x} = \frac{-4 - 5}{-1 - (-1)} = \frac{-9}{0} \quad \textit{Undefined}$$

This is a vertical line; it has undefined slope.

17. Through $(3, -1)$ and $(-3, 1)$

$$m = \frac{\Delta y}{\Delta x} = \frac{1 - (-1)}{-3 - 3} = \frac{2}{-6} = -\frac{1}{3}.$$

18. The x-intercept is $(2, 0)$ and the y-intercept is $(0, 2)$. The slope is

$$m = \frac{\text{change in } y}{\text{change in } x} = \frac{2 - 0}{0 - 2} = \frac{2}{-2} = -1.$$

19. The line goes up from left to right, so it has positive slope.

20. The line goes down from left to right, so it has negative slope.

21. The line is vertical, so it has *undefined* slope.

22. The line is horizontal, so it has 0 slope.

23. To rise 1 foot, we must move 4 feet in the horizontal direction. To rise 3 feet, we must move $3(4) = 12$ feet in the horizontal direction.

24. Let $(x_1, y_1) = (1980, 21{,}000)$ and $(x_2, y_2) = (2007, 61{,}400)$. Then

$$m = \frac{\Delta y}{\Delta x} = \frac{61{,}400 - 21{,}000}{2007 - 1980} = \frac{40{,}400}{27}$$
$$\approx 1496.$$

The average rate of change is \$1496 per year (to the nearest dollar).

25. **(a)** Slope $-\frac{1}{3}$, y-intercept $(0, -1)$
Use the slope-intercept form with $m = -\frac{1}{3}$ and $b = -1$.

$$y = mx + b$$
$$y = -\frac{1}{3}x - 1$$

(b) $$y = -\frac{1}{3}x - 1$$
$$3y = -x - 3$$
$$x + 3y = -3$$

Copyright © 2012 Pearson Education, Inc. Publishing as Addison-Wesley.

26. **(a)** Slope 0, y-intercept $(0, -2)$

Use the slope-intercept form with $m = 0$ and $b = -2$.

$$y = mx + b$$
$$y = (0)x - 2$$
$$y = -2$$

(b) $y = -2$ is already in standard form.

27. **(a)** Slope $-\frac{4}{3}$, through $(2, 7)$

Use the point-slope form with $m = -\frac{4}{3}$ and $(x_1, y_1) = (2, 7)$.

$$y - y_1 = m(x - x_1)$$
$$y - 7 = -\frac{4}{3}(x - 2)$$
$$y - 7 = -\frac{4}{3}x + \frac{8}{3}$$
$$y = -\frac{4}{3}x + \frac{29}{3}$$

(b)
$$y = -\frac{4}{3}x + \frac{29}{3}$$
$$3y = -4x + 29$$
$$4x + 3y = 29$$

28. **(a)** Slope 3, through $(-1, 4)$

Use the point-slope form with $m = 3$ and $(x_1, y_1) = (-1, 4)$.

$$y - y_1 = m(x - x_1)$$
$$y - 4 = 3[x - (-1)]$$
$$y - 4 = 3(x + 1)$$
$$y - 4 = 3x + 3$$
$$y = 3x + 7$$

(b)
$$y = 3x + 7$$
$$-3x + y = 7$$
$$3x - y = -7$$

29. **(a)** Vertical, through $(2, 5)$

The equation of any vertical line is in the form $x = k$. Since the line goes through $(2, 5)$, the equation is $x = 2$. (Slope-intercept form is not possible.)

(b) $x = 2$ is already in standard form.

30. **(a)** Through $(2, -5)$ and $(1, 4)$

Find the slope.

$$m = \frac{\Delta y}{\Delta x} = \frac{4 - (-5)}{1 - 2} = \frac{9}{-1} = -9$$

Use the point-slope form with $m = -9$ and $(x_1, y_1) = (2, -5)$.

$$y - y_1 = m(x - x_1)$$
$$y - (-5) = -9(x - 2)$$
$$y + 5 = -9x + 18$$
$$y = -9x + 13$$

(b)
$$y = -9x + 13$$
$$9x + y = 13$$

31. **(a)** Through $(-3, -1)$ and $(2, 6)$

Find the slope.

$$m = \frac{\Delta y}{\Delta x} = \frac{6 - (-1)}{2 - (-3)} = \frac{7}{5}$$

Use the point-slope form with $m = \frac{7}{5}$ and $(x_1, y_1) = (2, 6)$.

$$y - y_1 = m(x - x_1)$$
$$y - 6 = \frac{7}{5}(x - 2)$$
$$y - 6 = \frac{7}{5}x - \frac{14}{5}$$
$$y = \frac{7}{5}x + \frac{16}{5}$$

(b)
$$y = \frac{7}{5}x + \frac{16}{5}$$
$$5y = 7x + 16$$
$$-7x + 5y = 16$$
$$7x - 5y = -16$$

32. **(a)** From Exercise 18, we have $m = -1$ and a y-intercept of $(0, 2)$. The slope-intercept form is

$$y = -1x + 2 \quad \text{or} \quad y = -x + 2.$$

(b)
$$y = -x + 2$$
$$x + y = 2$$

33. **(a)** Parallel to $4x - y = 3$ and through $(7, -1)$

Writing $4x - y = 3$ in slope-intercept form gives us $y = 4x - 3$, which has slope 4. Lines parallel to it will also have slope 4. The line with slope 4 through $(7, -1)$ is :

$$y - y_1 = m(x - x_1)$$
$$y - (-1) = 4(x - 7)$$
$$y + 1 = 4x - 28$$
$$y = 4x - 29$$

(b)
$$y = 4x - 29$$
$$-4x + y = -29$$
$$4x - y = 29$$

34. **(a)** Perpendicular to $2x - 5y = 7$ and through $(4, 3)$

Write the equation in slope-intercept form.

$$2x - 5y = 7$$
$$-5y = -2x + 7$$
$$y = \frac{2}{5}x - \frac{7}{5}$$

$y = \frac{2}{5}x - \frac{7}{5}$ has slope $\frac{2}{5}$ and is perpendicular to lines with slope $-\frac{5}{2}$.

The line with slope $-\frac{5}{2}$ through $(4, 3)$ is

$$y - y_1 = m(x - x_1)$$
$$y - 3 = -\frac{5}{2}(x - 4)$$
$$y - 3 = -\frac{5}{2}x + 10$$
$$y = -\frac{5}{2}x + 13$$

Copyright © 2012 Pearson Education, Inc. Publishing as Addison-Wesley.

(b)
$$y = -\tfrac{5}{2}x + 13$$
$$2y = -5x + 26$$
$$5x + 2y = 26$$

35. **(a)** The fixed cost is \$159, so that is the value of b. The variable cost is \$57, so

$$y = mx + b = 57x + 159.$$

The cost of a 1-year membership can be found by substituting 12 for x.

$$y = 57(12) + 159$$
$$= 684 + 159 = 843$$

The cost is \$843.

(b) As in part (a),

$$y = 47x + 159.$$
$$y = 47(12) + 159$$
$$= 564 + 159 = 723$$

The cost is \$723.

36. **(a)** Use $(3, 1839)$ and $(7, 2414)$.

$$m = \frac{\Delta y}{\Delta x} = \frac{2414 - 1839}{7 - 3} = \frac{575}{4} \approx 143.75$$

Use the point-slope form of a line.

$$y - y_1 = m(x - x_1)$$
$$y - 1839 = 143.75(x - 3)$$
$$y - 1839 = 143.75x - 431.25$$
$$y = 143.75x + 1407.75$$

The slope, 143.75, indicates that the revenue from skiing facilities increased by an average of \$143.75 million each year from 2003 to 2007.

(b) The year 2008 corresponds to $x = 8$.

$$y = 143.75(8) + 1407.75$$
$$= 1150 + 1407.75 = 2557.75$$

According to the equation from part (a), we estimate the revenue from skiing facilities to be \$2558 million (to the nearest million).

37. $3x - 2y \le 12$
Graph $3x - 2y = 12$ as a solid line through $(0, -6)$ and $(4, 0)$. Use $(0, 0)$ as a test point. Since $(0, 0)$ satisfies the inequality, shade the region on the side of the line containing $(0, 0)$.

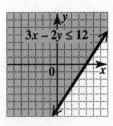

38. $5x - y > 6$
Graph $5x - y = 6$ as a dashed line through $(0, -6)$ and $(\tfrac{6}{5}, 0)$. Use $(0, 0)$ as a test point. Since $(0, 0)$ does not satisfy the inequality, shade the region on the side of the line that does not contain $(0, 0)$.

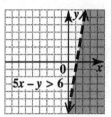

39. $2x + y \le 1$ and $x \ge 2y$
Graph $2x + y = 1$ as a solid line through $(\tfrac{1}{2}, 0)$ and $(0, 1)$, and shade the region on the side containing $(0, 0)$ since it satisfies the inequality. Next, graph $x = 2y$ as a solid line through $(0, 0)$ and $(2, 1)$, and shade the region on the side containing $(2, 0)$ since $2 > 2(0)$ or $2 > 0$ is true. The intersection is the region where the graphs overlap.

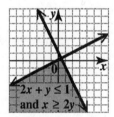

40. $x \ge 2$ or $y \ge 2$

Graph $x = 2$ as a solid vertical line through $(2, 0)$. Shade the region to the right of $x = 2$.

Graph $y = 2$ as a solid horizontal line through $(0, 2)$. Shade the region above $y = 2$. The graph of

$$x \ge 2 \quad \text{or} \quad y \ge 2$$

includes all the shaded regions.

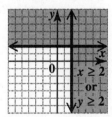

41. In $y < 4x + 3$, the "$<$" symbol indicates that the graph has a dashed boundary line and that the shading is below the line, so the correct choice is **D**.

42. $\{(-4, 2), (-4, -2), (1, 5), (1, -5)\}$
The domain, the set of x-values, is $\{-4, 1\}$.
The range, the set of y-values, is $\{2, -2, 5, -5\}$.
Since each x-value has more than one y-value, the relation is not a function.

Copyright © 2012 Pearson Education, Inc. Publishing as Addison-Wesley.

43. The relation can be described by the set of ordered pairs

$$\{(9, 32), (11, 47), (4, 47), (17, 69), (25, 14)\}.$$

The relation is a function since for each x-value, there is only one y-value.

The domain is the set of x-values: $\{9, 11, 4, 17, 25\}$.
The range is the set of y-values: $\{32, 47, 69, 14\}$.

44. The domain, the x-values of the points on the graph, is $[-4, 4]$. The range, the y-values of the points on the graph, is $[0, 2]$. Since a vertical line intersects the graph of the relation in at most one point, the relation is a function.

45. The x-values are negative or zero, so the domain is $(-\infty, 0]$. The y-values can be any real number, so the range is $(-\infty, \infty)$. A vertical line, such as $x = -3$, will intersect the graph twice, so by the vertical line test, the relation is not a function.

46. $y = 3x - 3$
For any value of x, there is exactly one value of y, so the equation defines a function, actually a linear function. The domain is the set of all real numbers, $(-\infty, \infty)$.

47. $y < x + 2$
For any value of x, there are many values of y. For example, $(1, 0)$ and $(1, 1)$ are both solutions of the inequality that have the same x-value but different y-values. The inequality does not define a function. The domain is the set of all real numbers, $(-\infty, \infty)$.

48. $y = |x|$
For any value of x, there is exactly one value of y, so the equation defines a function. The domain is the set of all real numbers, $(-\infty, \infty)$.

49. $y = \sqrt{4x + 7}$
Given any value of x, y is found by multiplying x by 4, adding 7, and taking the square root of the result. This process produces exactly one value of y for each x-value in the domain, so the equation defines a function. Since the radicand must be nonnegative,

$$4x + 7 \geq 0$$
$$4x \geq -7$$
$$x \geq -\tfrac{7}{4}.$$

The domain is $[-\tfrac{7}{4}, \infty)$.

50. $x = y^2$
The ordered pairs $(4, 2)$ and $(4, -2)$ both satisfy the equation. Since one value of x, 4, corresponds to two values of y, 2 and -2, the equation does not define a function. Because x is equal to the square of y, the values of x must always be nonnegative. The domain is $[0, \infty)$.

51. $y = \dfrac{7}{x - 6}$
Given any value of x, y is found by subtracting 6, then dividing the result into 7. This process produces exactly one value of y for each x-value in the domain, so the equation defines a function. The domain includes all real numbers except those that make the denominator 0, namely 6. The domain is $(-\infty, 6) \cup (6, \infty)$.

In Exercises 52–55, use

$$f(x) = -2x^2 + 3x - 6.$$

52. $f(0) = -2(0)^2 + 3(0) - 6 = -6$

53. $f(2.1) = -2(2.1)^2 + 3(2.1) - 6$
$$= -8.82 + 6.3 - 6 = -8.52$$

54. $f(-\tfrac{1}{2}) = -2(-\tfrac{1}{2})^2 + 3(-\tfrac{1}{2}) - 6$
$$= -\tfrac{1}{2} - \tfrac{3}{2} - 6 = -8$$

55. $f(k) = -2k^2 + 3k - 6$

56. $2x^2 - y = 0$
$$-y = -2x^2$$
$$y = 2x^2$$
Since $y = f(x)$,
$$f(x) = 2x^2,$$
and $f(3) = 2(3)^2 = 2(9) = 18$.

57. Solve for y in terms of x.
$$2x - 5y = 7$$
$$2x - 7 = 5y$$
$$\tfrac{2}{5}x - \tfrac{7}{5} = y$$

Thus, choice **C** is correct.

58. **(a)** For each year, there is exactly one life expectancy associated with the year, so the table defines a function.

(b) The domain is the set of years, that is, $\{1960, 1970, 1980, 1990, 2000, 2009\}$.
The range is the set of life expectancies, that is, $\{69.7, 70.8, 73.7, 75.4, 76.8, 78.1\}$.

(c) Answers will vary. Two possible answers are $(1960, 69.7)$ and $(2009, 78.1)$.

(d) $f(1980) = 73.7$. In 1980, life expectancy at birth was 73.7 yr.

(e) Since $f(2000) = 76.8$, $x = 2000$.

Copyright © 2012 Pearson Education, Inc. Publishing as Addison-Wesley.

♦♦♦ Relating Concepts 59–70 ♦♦♦

59. Because it falls from left to right, the slope is negative.

60. Use the points $(-1, 5)$ and $(3, -1)$.

$$m = \frac{\Delta y}{\Delta x} = \frac{-1 - 5}{3 - (-1)} = \frac{-6}{4} = -\frac{3}{2}$$

61. Since $m = -\frac{3}{2}$, the slope of any line parallel to this line is also $-\frac{3}{2}$, whereas the slope of any line perpendicular to this line is $\frac{2}{3}$ since $\frac{2}{3}$ is the negative reciprocal of $-\frac{3}{2}$.

62. $2y = -3x + 7$
To find the x-intercept, let $y = 0$.

$$2(0) = -3x + 7$$
$$3x = 7$$
$$x = \tfrac{7}{3}$$

The x-intercept is $\left(\frac{7}{3}, 0\right)$.

63. $2y = -3x + 7$
To find the y-intercept, let $x = 0$.

$$2y = -3(0) + 7$$
$$2y = 7$$
$$y = \tfrac{7}{2}$$

The y-intercept is $\left(0, \frac{7}{2}\right)$.

64. Solve $2y = -3x + 7$ for y.
$$y = -\tfrac{3}{2}x + \tfrac{7}{2}$$

Since $y = f(x)$,
$$f(x) = -\tfrac{3}{2}x + \tfrac{7}{2}.$$

65. $f(x) = -\frac{3}{2}x + \frac{7}{2}$
$$f(8) = -\tfrac{3}{2}(8) + \tfrac{7}{2}$$
$$= -\tfrac{24}{2} + \tfrac{7}{2} = -\tfrac{17}{2}$$

66. $f(x) = -\frac{3}{2}x + \frac{7}{2}$

$$-8 = -\tfrac{3}{2}x + \tfrac{7}{2} \quad \textit{Let f(x) = -8.}$$
$$-16 = -3x + 7 \quad \textit{Multiply by 2.}$$
$$-23 = -3x \quad \textit{Subtract 7.}$$
$$x = \tfrac{23}{3} \quad \textit{Divide by -3.}$$

67.
$$f(x) \geq 0$$
$$-\tfrac{3}{2}x + \tfrac{7}{2} \geq 0$$
$$-\tfrac{3}{2}x \geq -\tfrac{7}{2}$$
$$x \leq \left(-\tfrac{7}{2}\right)\left(-\tfrac{2}{3}\right)$$
$$x \leq \tfrac{7}{3}$$

68. $f(x) = 0$ is equivalent to $y = 0$, which is the equation we solved in Exercise 62 to find the x-intercept.
The solution set is $\left\{\frac{7}{3}\right\}$.

69. The graph is below the x-axis for $x > \frac{7}{3}$, so the solution set of $f(x) < 0$ is $\left(\frac{7}{3}, \infty\right)$.

70. The graph is above the x-axis for $x < \frac{7}{3}$, so the solution set of $f(x) > 0$ is $\left(-\infty, \frac{7}{3}\right)$.

Chapter 3 Test

1. $2x - 3y = 12$
For $x = 1$:

$$2(1) - 3y = 12$$
$$2 - 3y = 12$$
$$-3y = 10$$
$$y = -\tfrac{10}{3} \quad \left(1, -\tfrac{10}{3}\right)$$

For $x = 3$:

$$2(3) - 3y = 12$$
$$6 - 3y = 12$$
$$-3y = 6$$
$$y = -2 \quad (3, -2)$$

For $y = -4$:

$$2x - 3(-4) = 12$$
$$2x + 12 = 12$$
$$2x = 0$$
$$x = 0 \quad (0, -4)$$

2. $3x - 2y = 20$
To find the x-intercept, let $y = 0$.

$$3x - 2(0) = 20$$
$$3x = 20$$
$$x = \tfrac{20}{3}$$

The x-intercept is $\left(\frac{20}{3}, 0\right)$.
To find the y-intercept, let $x = 0$.

$$3(0) - 2y = 20$$
$$-2y = 20$$
$$y = -10$$

The y-intercept is $(0, -10)$.
Draw the line through these two points.

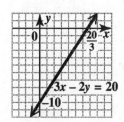

Copyright © 2012 Pearson Education, Inc. Publishing as Addison-Wesley.

3. The graph of $y = 5$ is the horizontal line with slope 0 and y-intercept $(0, 5)$. There is no x-intercept.

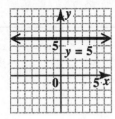

4. The graph of $x = 2$ is the vertical line with x-intercept at $(2, 0)$. There is no y-intercept.

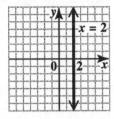

5. Through $(6, 4)$ and $(-4, -1)$

$$m = \frac{\Delta y}{\Delta x} = \frac{-1 - 4}{-4 - 6} = \frac{-5}{-10} = \frac{1}{2}$$

The slope of the line is $\frac{1}{2}$.

6. The graph of a line with undefined slope is the graph of a vertical line.

7. Find the slope of each line.

$$5x - y = 8$$
$$-y = -5x + 8$$
$$y = 5x - 8$$

The slope is 5.

$$5y = -x + 3$$
$$y = -\frac{1}{5}x + \frac{3}{5}$$

The slope is $-\frac{1}{5}$.
Since $5(-\frac{1}{5}) = -1$, the two slopes are negative reciprocals and the lines are perpendicular.

8. Find the slope of each line.

$$2y = 3x + 12$$
$$y = \frac{3}{2}x + 6$$

The slope is $\frac{3}{2}$.

$$3y = 2x - 5$$
$$y = \frac{2}{3}x - \frac{5}{3}$$

The slope is $\frac{2}{3}$.
The lines are neither parallel nor perpendicular.

9. Use the points $(1980, 119{,}000)$ and $(2008, 93{,}000)$.

average rate of change

$$= \frac{\text{change in } y}{\text{change in } x} = \frac{93{,}000 - 119{,}000}{2008 - 1980}$$

$$= \frac{-26{,}000}{28} \approx -929$$

The average rate of change is about -929 farms per year, that is, the number of farms decreased by about 929 each year from 1980 to 2008.

10. Through $(4, -1)$; $m = -5$

(a) Let $m = -5$ and $(x_1, y_1) = (4, -1)$ in the point-slope form.

$$y - y_1 = m(x - x_1)$$
$$y - (-1) = -5(x - 4)$$
$$y + 1 = -5x + 20$$
$$y = -5x + 19$$

(b) $\quad y = -5x + 19 \quad$ *From part (a)*
$\quad 5x + y = 19 \quad\quad$ *Standard form*

11. Through $(-3, 14)$; horizontal

(a) A horizontal line has equation $y = k$. Here $k = 14$, so the line has equation $y = 14$.

(b) $y = 14$ is already in standard form.

12. Through $(-2, 3)$ and $(6, -1)$

(a) First find the slope.

$$m = \frac{\Delta y}{\Delta x} = \frac{-1 - 3}{6 - (-2)} = \frac{-4}{8} = -\frac{1}{2}$$

Use $m = -\frac{1}{2}$ and $(x_1, y_1) = (-2, 3)$ in the point-slope form.

$$y - y_1 = m(x - x_1)$$
$$y - 3 = -\frac{1}{2}[x - (-2)]$$
$$y - 3 = -\frac{1}{2}(x + 2)$$
$$y - 3 = -\frac{1}{2}x - 1$$
$$y = -\frac{1}{2}x + 2$$

(b) $\quad y = -\frac{1}{2}x + 2 \quad$ *From part (a)*

$\quad \frac{1}{2}x + y = 2 \quad\quad$ *Variable terms on one side*

$\quad 2(\frac{1}{2}x + y) = 2(2) \quad$ *Multiply by 2.*

$\quad x + 2y = 4 \quad\quad$ *Standard form*

13. Through $(5, -6)$; vertical

(a) The equation of any vertical line is in the form $x = k$. Since the line goes through $(5, -6)$, the equation is $x = 5$. Writing $x = 5$ in slope-intercept form is *not possible* since there is no y-term.

(b) From part (a), the standard form is $x = 5$.

Copyright © 2012 Pearson Education, Inc. Publishing as Addison-Wesley.

14. Through $(-7, 2)$ and parallel to $3x + 5y = 6$

(a) To find the slope of $3x + 5y = 6$, write the equation in slope-intercept form by solving for y.

$$3x + 5y = 6$$
$$5y = -3x + 6$$
$$y = -\tfrac{3}{5}x + \tfrac{6}{5}$$

The slope is $-\tfrac{3}{5}$, so a line parallel to it also has slope $-\tfrac{3}{5}$. Let $m = -\tfrac{3}{5}$ and $(x_1, y_1) = (-7, 2)$ in the point-slope form.

$$y - y_1 = m(x - x_1)$$
$$y - 2 = -\tfrac{3}{5}[x - (-7)]$$
$$y - 2 = -\tfrac{3}{5}(x + 7)$$
$$y - 2 = -\tfrac{3}{5}x - \tfrac{21}{5}$$
$$y = -\tfrac{3}{5}x - \tfrac{11}{5}$$

(b) $y = -\tfrac{3}{5}x - \tfrac{11}{5}$ *From part (a)*

$\tfrac{3}{5}x + y = -\tfrac{11}{5}$ *Variable terms on one side*

$5(\tfrac{3}{5}x + y) = 5(-\tfrac{11}{5})$ *Multiply by 5.*

$3x + 5y = -11$ *Standard form*

15. Through $(-7, 2)$ and perpendicular to $y = 2x$

(a) Since $y = 2x$ is in slope-intercept form ($b = 0$), the slope, m, of $y = 2x$ is 2. A line perpendicular to it has a slope that is the negative reciprocal of 2, that is, $-\tfrac{1}{2}$. Let $m = -\tfrac{1}{2}$ and $(x_1, y_1) = (-7, 2)$ in the point-slope form.

$$y - y_1 = m(x - x_1)$$
$$y - 2 = -\tfrac{1}{2}(x + 7)$$
$$y - 2 = -\tfrac{1}{2}x - \tfrac{7}{2}$$
$$y = -\tfrac{1}{2}x - \tfrac{3}{2}$$

(b) $y = -\tfrac{1}{2}x - \tfrac{3}{2}$ *From part (a)*

$\tfrac{1}{2}x + y = -\tfrac{3}{2}$ *Variable terms on one side*

$2(\tfrac{1}{2}x + y) = 2(-\tfrac{3}{2})$ *Multiply by 2.*

$x + 2y = -3$ *Standard form*

16. Positive slope means that the line goes up from left to right. The only line that has positive slope and a negative y-coordinate for its y-intercept is choice **B**.

17. (a) Use the points $(1, 53{,}635)$ and $(7, 66{,}103)$.

$$m = \frac{\Delta y}{\Delta x} = \frac{66{,}103 - 53{,}635}{7 - 1} = \frac{12{,}468}{6}$$
$$= 2078$$

Use the point-slope form with $m = 2078$ and $(x_1, y_1) = (1, 53{,}635)$.

$$y - 53{,}635 = 2078(x - 1)$$
$$y - 53{,}635 = 2078x - 2078$$
$$y = 2078x + 51{,}557$$

(b) $y = 2078(5) + 51{,}557$ *Let x = 5.*
    $= \$61{,}947,$
which is more than the actual value.

18. $3x - 2y > 6$
Graph the line $3x - 2y = 6$, which has intercepts $(2, 0)$ and $(0, -3)$, as a dashed line since the inequality involves $>$. Test $(0, 0)$, which yields $0 > 6$, a false statement. Shade the region that does not include $(0, 0)$.

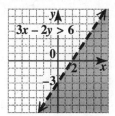

19. $y < 2x - 1$ and $x - y < 3$
First graph $y = 2x - 1$ as a dashed line through $(2, 3)$ and $(0, -1)$. Test $(0, 0)$, which yields $0 < -1$, a false statement. Shade the side of the line not containing $(0, 0)$.
Next, graph $x - y = 3$ as a dashed line through $(3, 0)$ and $(0, -3)$. Test $(0, 0)$, which yields $0 < 3$, a true statement. Shade the side of the line containing $(0, 0)$. The intersection is the region where the graphs overlap.

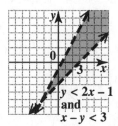

20. Choice **D** is the only graph that passes the vertical line test.

21. Choice **D** does not define a function, since its domain (input) element 0 is paired with two different range (output) elements, 1 and 2.

22. The x-values are greater than or equal to zero, so the domain is $[0, \infty)$. Since y can be any value, the range is $(-\infty, \infty)$.

23. The domain is the set of x-values: $\{0, -2, 4\}$.
The range is the set of y-values: $\{1, 3, 8\}$.

24. $f(x) = -x^2 + 2x - 1$

(a) $f(1) = -(1)^2 + 2(1) - 1$
    $= -1 + 2 - 1$
    $= 0$

(b) $f(a) = -a^2 + 2a - 1$

Copyright © 2012 Pearson Education, Inc. Publishing as Addison-Wesley.

25. $f(x) = \frac{2}{3}x - 1$

This function represents a line with y-intercept $(0, -1)$ and x-intercept $(\frac{3}{2}, 0)$.

Draw the line through these two points.

The domain is $(-\infty, \infty)$, and the range is $(-\infty, \infty)$.

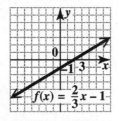

Cumulative Review Exercises (Chapters 1–3)

1. The absolute value of a negative number is a positive number and the additive inverse of the same negative number is the same positive number. For example, suppose the negative number is -5:

$$|-5| = -(-5) = 5$$
$$\text{and} \quad -(-5) = 5$$

The statement is *always true*.

2. The sum of two negative numbers is another negative number, so the statement is *never true*.

3. The statement is *sometimes true*. For example,

$$3 + (-3) = 0,$$
$$\text{but} \quad 3 + (-1) = 2 \neq 0.$$

4. $-|-2| - 4 + |-3| + 7 = -2 - 4 + 3 + 7$
$$= -6 + 3 + 7$$
$$= -3 + 7$$
$$= 4$$

5. $(-0.8)^2 = (-0.8)(-0.8) = 0.64$

6. $\sqrt{-64}$ is not a real number.

7. $-(-4m + 3) = -(-4m) - 3$
$$= 4m - 3$$

8. $3x^2 - 4x + 4 + 9x - x^2$
$$= 3x^2 - x^2 - 4x + 9x + 4$$
$$= 2x^2 + 5x + 4$$

9. $\dfrac{(4^2 - 4) - (-1)7}{4 + (-6)} = \dfrac{(16 - 4) - (-7)}{-2}$
$$= \dfrac{12 + 7}{-2} = -\dfrac{19}{2}$$

10. $-3 < x \leq 5$

This is the set of numbers between -3 and 5, not including -3 (use a parenthesis), but including 5 (use a bracket). In interval notation, the set is $(-3, 5]$.

For Exercises 11–12, let $p = -4$, $q = \frac{1}{2}$, and $r = 16$.

11. $-3(2q - 3p) = -3[2(\frac{1}{2}) - 3(-4)]$
$$= -3(1 + 12)$$
$$= -3(13)$$
$$= -39$$

12. $\dfrac{\sqrt{r}}{8p + 2r} = \dfrac{\sqrt{16}}{8(-4) + 2(16)}$
$$= \dfrac{4}{-32 + 32}$$
$$= \dfrac{4}{0}, \text{ which is } \textit{undefined.}$$

13. $2z - 5 + 3z = 2 - z$
$$5z - 5 = 2 - z$$
$$6z = 7$$
$$z = \frac{7}{6}$$

The solution set is $\left\{\frac{7}{6}\right\}$.

14. $\dfrac{3x - 1}{5} + \dfrac{x + 2}{2} = -\dfrac{3}{10}$

Multiply both sides by the LCD, 10.

$$10\left(\dfrac{3x - 1}{5} + \dfrac{x + 2}{2}\right) = 10\left(-\dfrac{3}{10}\right)$$
$$2(3x - 1) + 5(x + 2) = -3$$
$$6x - 2 + 5x + 10 = -3$$
$$11x + 8 = -3$$
$$11x = -11$$
$$x = -1$$

The solution set is $\{-1\}$.

15. Let x denote the side of the original square and $4x$ the perimeter. Now $x + 4$ is the side of the new square and $4(x + 4)$ is its perimeter.

"The perimeter would be 8 inches less than twice the perimeter of the original square " translates as

$$4(x + 4) = 2(4x) - 8.$$
$$4x + 16 = 8x - 8$$
$$24 = 4x$$
$$6 = x$$

The length of a side of the original square is 6 inches.

Copyright © 2012 Pearson Education, Inc. Publishing as Addison-Wesley.

16. Let $x =$ the time it takes for the planes to be 2100 miles apart.

Make a table. Use the formula $d = rt$.

	r	t	d
Eastbound Plane	550	x	$550x$
Westbound Plane	500	x	$500x$

The total distance is 2100 miles.

$$550x + 500x = 2100$$
$$1050x = 2100$$
$$x = 2$$

It will take 2 hr for the planes to be 2100 mi apart.

17. $-4 < 3 - 2k < 9$

$-7 < \ -2k \ < 6$

$\dfrac{7}{2} > \ \ k \ \ > -3$ *Divide by -2, reverse inequalities*

$-3 < \ \ k \ \ < \dfrac{7}{2}$ *Equivalent inequality*

The solution set is $\left(-3, \frac{7}{2}\right)$.

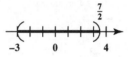

18. $-0.3x + 2.1(x - 4) \le -6.6$

$-3x + 21(x - 4) \le -66$

                Multiply by 10.

$-3x + 21x - 84 \le -66$

$18x - 84 \le -66$

$18x \le 18$

$x \le 1$

The solution set is $(-\infty, 1]$.

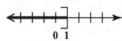

19. $\frac{1}{2}x > 3$ and $\frac{1}{3}x < \frac{8}{3}$

    $x > 6$ and $x < 8$

The graph of the solution set is all numbers both greater than 6 *and* less than 8. This is the intersection. The elements common to both sets are the numbers between 6 and 8, not including the endpoints. The solution set is $(6, 8)$.

20. $-5x + 1 \ge 11$ or $3x + 5 > 26$

     $-5x \ge 10$ $3x > 21$

      $x \le -2$ or $x > 7$

The graph of the solution set is all numbers either less than or equal to -2 *or* greater than 7. This is the union. The solution set is $(-\infty, -2] \cup (7, \infty)$.

21. $|2k - 7| + 4 = 11$

     $|2k - 7| = 7$

$2k - 7 = 7$ or $2k - 7 = -7$

$2k = 14$ $2k = 0$

$k = 7$ or $k = 0$

The solution set is $\{0, 7\}$.

22. $|3m + 6| \ge 0$

The absolute value of an expression is always nonnegative, so the inequality is true for any real number m.

The solution set is $(-\infty, \infty)$.

23. $3x + 5y = 12$

To find the x-intercept, let $y = 0$.

$$3x + 5(0) = 12$$
$$3x = 12$$
$$x = 4$$

The x-intercept is $(4, 0)$.

To find the y-intercept, let $x = 0$.

$$3(0) + 5y = 12$$
$$5y = 12$$
$$y = \tfrac{12}{5}$$

The y-intercept is $\left(0, \frac{12}{5}\right)$.

Plot the intercepts and draw the line through them.

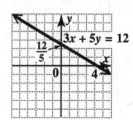

Copyright © 2012 Pearson Education, Inc. Publishing as Addison-Wesley.

24. $A(-2, 1)$ and $B(3, -5)$

(a) The slope of line AB is

$$m = \frac{\Delta y}{\Delta x} = \frac{-5 - 1}{3 - (-2)} = \frac{-6}{5} = -\frac{6}{5}.$$

(b) The slope of a line perpendicular to line AB is the negative reciprocal of $-\frac{6}{5}$, which is $\frac{5}{6}$.

25. $-2x + y < -6$

Graph the line $-2x + y = -6$, which has intercepts $(3, 0)$ and $(0, -6)$, as a dashed line since the inequality involves $<$. Test $(0, 0)$, which yields $0 < -6$, a false statement. Shade the region that does not include $(0, 0)$.

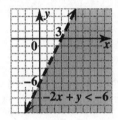

26. **(a)** Slope $-\frac{3}{4}$; y-intercept $(0, -1)$

To write an equation of this line, let $m = -\frac{3}{4}$ and $b = -1$ in the slope-intercept form.

$$y = mx + b$$
$$y = -\frac{3}{4}x - 1$$

(b)
$$y = -\frac{3}{4}x - 1$$
$$4y = -3x - 4$$
$$3x + 4y = -4$$

27. **(a)** Through $(4, -3)$ and $(1, 1)$

First find the slope of the line.

$$m = \frac{\Delta y}{\Delta x} = \frac{1 - (-3)}{1 - 4} = \frac{4}{-3} = -\frac{4}{3}$$

Now substitute $(x_1, y_1) = (4, -3)$ and $m = -\frac{4}{3}$ in the point-slope form. Then solve for y.

$$y - y_1 = m(x - x_1)$$
$$y - (-3) = -\frac{4}{3}(x - 4)$$
$$y + 3 = -\frac{4}{3}x + \frac{16}{3}$$
$$y = -\frac{4}{3}x + \frac{7}{3}$$

(b)
$$y = -\frac{4}{3}x + \frac{7}{3}$$
$$3y = -4x + 7$$
$$4x + 3y = 7$$

28. The domain of the relation consists of the elements in the leftmost figure; that is, $\{14, 91, 75, 23\}$.

The range of the relation consists of the elements in the rightmost figure; that is, $\{9, 70, 56, 5\}$.

Since the element 75 in the domain is paired with two different values, 70 and 56, in the range, the relation is not a function.

29. $f(x) = -4x + 10$

(a) The variable x can be any real number, so the domain is $(-\infty, \infty)$. The function is a non-constant linear function, so its range is $(-\infty, \infty)$.

(b) $f(-3) = -4(-3) + 10 = 12 + 10 = 22$

(c)
$$f(x) = 6$$
$$-4x + 10 = 6$$
$$-4x = -4 \quad \textit{Subtract 10.}$$
$$x = 1 \quad \textit{Divide by } -4.$$

30. Use $(2003, 46.8)$ and $(2008, 36.7)$.

$$m = \frac{\Delta y}{\Delta x} = \frac{36.7 - 46.8}{2008 - 2003} = \frac{-10.1}{5}$$
$$= -2.02$$

So the average rate of change is -2.02 per year; that is, the per capita consumption of potatoes in the United States decreased by an average of 2.02 pounds per year from 2003 to 2008.

Copyright © 2012 Pearson Education, Inc. Publishing as Addison-Wesley.

CHAPTER 4 SYSTEMS OF LINEAR EQUATIONS

4.1 Systems of Linear Equations in Two Variables

4.1 Now Try Exercises

N1. To determine whether $(2, 5)$ is a solution of the system

$$-x + 2y = 8$$
$$3x - 2y = 0,$$

replace x with 2 and y with 5 in each equation.

$-x + 2y = 8$ (1)	$3x - 2y = 0$ (2)
$-2 + 2(5) \overset{?}{=} 8$	$3(2) - 2(5) \overset{?}{=} 0$
$-2 + 10 \overset{?}{=} 8$	$6 - 10 \overset{?}{=} 0$
$8 = 8$ *True*	$-4 = 0$ *False*

The ordered pair $(2, 5)$ is not a solution of the system, since it does not make *both* equations true.

N2. $3x - 2y = 6$ (1)
$x - y = 1$ (2)

When the equations are graphed, the point of intersection appears to be $(4, 3)$. To check, substitute 4 for x and 3 for y in each equation of the system.

$3x - 2y = 6$ (1)	$x - y = 1$ (2)
$3(4) - 2(3) \overset{?}{=} 6$	$4 - 3 \overset{?}{=} 1$
$6 = 6$ *True*	$1 = 1$ *True*

Since $(4, 3)$ makes both equations true, the solution set is $\{(4, 3)\}$.

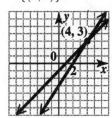

N3. $x = 3 + 2y$ (1)
$4x - 3y = 32$ (2)

Since equation (1) is solved for x, substitute $3 + 2y$ for x in equation (2).

$$4x - 3y = 32 \qquad (2)$$
$$4(3 + 2y) - 3y = 32$$
$$12 + 8y - 3y = 32$$
$$12 + 5y = 32$$
$$5y = 20$$
$$y = 4$$

Since $x = 3 + 2y$ and $y = 4$,

$$x = 3 + 2(4) = 11.$$

Check $(11, 4)$: $11 = 3 + 8$; $44 - 12 = 32$

The solution set is $\{(11, 4)\}$.

N4. $5x + y = 7$ (1)
$3x - 2y = 25$ (2)

Step 1
To use the substitution method, first solve one of the equations for x or y. Since the coefficient of y in equation (1) is 1, it is easiest to solve for y in this equation.

$$5x + y = 7 \qquad (1)$$
$$y = -5x + 7$$

Step 2
Substitute $-5x + 7$ for y in equation (2) and solve for x.

$$3x - 2y = 25 \qquad (2)$$
$$3x - 2(-5x + 7) = 25$$

Step 3

$$3x + 10x - 14 = 25$$
$$13x - 14 = 25$$
$$13x = 39$$
$$x = 3$$

Step 4
Since $y = -5x + 7$ and $x = 3$,

$$y = -5(3) + 7 = -15 + 7 = -8.$$

Step 5
Check $(3, -8)$: $15 - 8 = 7$; $9 + 16 = 25$

The solution set is $\{(3, -8)\}$.

N5. $\frac{1}{10}x - \frac{3}{5}y = \frac{2}{5}$ (1)
$-2x + 3y = 1$ (2)

Multiply equation (1) by the LCD, 10. The new system is

$$x - 6y = 4 \qquad (3)$$
$$-2x + 3y = 1. \qquad (2)$$

Solve equation (3) for x.

$$x = 6y + 4$$

Substitute $6y + 4$ for x in equation (2) and solve for y.

$$-2x + 3y = 1 \qquad (2)$$
$$-2(6y + 4) + 3y = 1$$
$$-12y - 8 + 3y = 1$$
$$-9y - 8 = 1$$
$$-9y = 9$$
$$y = -1$$

Since $x = 6y + 4$ and $y = -1$,

$$x = 6(-1) + 4 = -6 + 4 = -2.$$

Copyright © 2012 Pearson Education, Inc. Publishing as Addison-Wesley.

Check $(-2, -1)$: $-\frac{1}{5} + \frac{3}{5} = \frac{2}{5}$; $4 - 3 = 1$

The solution set is $\{(-2, -1)\}$.

N6. $8x - 2y = 5$ (1)
     $5x + 2y = -18$ (2)

Eliminate y by adding equations (1) and (2).

$$\begin{array}{rll} 8x - 2y = & 5 & (1) \\ 5x + 2y = & -18 & (2) \\ \hline 13x = & -13 & Add. \\ x = & -1 & Solve\ for\ x. \end{array}$$

To find y, replace x with -1 in either equation (1) or (2).

$$\begin{array}{rl} 8x - 2y = 5 & \quad (1) \\ 8(-1) - 2y = 5 & \\ -8 - 2y = 5 & \\ -2y = 13 & \\ y = -\frac{13}{2} & \end{array}$$

Check $\left(-1, -\frac{13}{2}\right)$: $-8 + 13 = 5$; $-5 - 13 = -18$

The solution set is $\left\{\left(-1, -\frac{13}{2}\right)\right\}$.

N7. $2x - 5y = 4$ (1)
     $5x + 2y = 10$ (2)

To eliminate y, multiply equation (1) by 2 and add the result to equation (2) multiplied by 5.

$$\begin{array}{rll} 4x - 10y = & 8 & 2 \times (1) \\ 25x + 10y = & 50 & 5 \times (2) \\ \hline 29x = & 58 & Add. \\ x = & 2 & Solve\ for\ x. \end{array}$$

To find y, substitute 2 for x in equation (1) or (2).

$$\begin{array}{rl} 5x + 2y = 10 & \quad (2) \\ 5(2) + 2y = 10 & \\ 10 + 2y = 10 & \\ 2y = 0 & \\ y = 0 & \end{array}$$

Check $(2, 0)$: $4 - 0 = 4$; $10 + 0 = 10$

The solution set is $\{(2, 0)\}$.

N8. $x - 3y = 7$ (1)
     $-3x + 9y = -21$ (2)

Multiply equation (1) by 3, and add the result to equation (2).

$$\begin{array}{rll} 3x - 9y = & 21 & 3 \times (1) \\ -3x + 9y = & -21 & (2) \\ \hline 0 = & 0 & True \end{array}$$

The equations are dependent.

The solution set is $\{(x, y) \mid x - 3y = 7\}$.

Equations (1) and (2) have the same graph.

N9. $-2x + 5y = 6$ (1)
     $6x - 15y = 4$ (2)

Multiply equation (1) by 3 and add the result to equation (2).

$$\begin{array}{rll} -6x + 15y = & 18 & 3 \times (1) \\ 6x - 15y = & 4 & (2) \\ \hline 0 = & 22 & False \end{array}$$

The system is inconsistent.

The solution set is $\emptyset$.

The graphs of the equations are parallel lines.

N10. (a) Solve each equation for y.

$$\begin{array}{l|l} 2x - 3y = 3 & 4x - 6y = -6 \\ -3y = -2x + 3 & -6y = -4x - 6 \\ y = \frac{2}{3}x - 1 & y = \frac{2}{3}x + 1 \end{array}$$

The system has no solution.

(b) Solve each equation for y.

$$\begin{array}{l|l} 5y = -x - 4 & -10y = 2x + 8 \\ y = -\frac{1}{5}x - \frac{4}{5} & y = -\frac{1}{5}x - \frac{4}{5} \end{array}$$

The system has infinitely many solutions.

4.1 Section Exercises

1. If $(4, -3)$ is a solution of a linear system in two variables, then substituting $\underline{4}$ for x and $\underline{-3}$ for y leads to true statements in *both* equations.

3. If the solution process leads to a false statement such as $0 = 3$, the solution set is $\underline{\emptyset}$.

5. If the two lines forming a system have the same slope and different y-intercepts, the system has $\underline{0}$ solutions. (The lines are parallel.)

7. **D**; The ordered pair solution must be in quadrant IV, since that is where the graphs of the equations intersect.

9. **(a)** $x - y = 0$ implies that $x = y$, so the coordinates of the solution must be equal—this limits our choice to **B** or **C**. Since $x + y = 6$, graph **B** is correct.

    (b) As in part (a), we must have **B** or **C**. Since $x + y = -6$, graph **C** is correct.

    (c) $x + y = 0$ implies that $x = -y$, so the coordinates of the solution must be opposites—this limits our choice to **A** or **D**. Since $x - y = -6$, graph **A** is correct.

    (d) As in part (c), we must have **A** or **D**. Since $x - y = 6$, graph **D** is correct.

Copyright © 2012 Pearson Education, Inc. Publishing as Addison-Wesley.

11. $x - y = 17$
$x + y = -1$

To decide if $(8, -9)$ is a solution, replace x with 8 and y with -9 in each equation of the system.

$$x - y = 17 \qquad\qquad x + y = -1$$
$$8 - (-9) \stackrel{?}{=} 17 \qquad 8 + (-9) \stackrel{?}{=} -1$$
$$17 = 17 \;\; \textit{True} \qquad -1 = -1 \;\; \textit{True}$$

Since $(8, -9)$ makes both equations true, $(8, -9)$ is a solution of the system.

13. $3x - 5y = -12$
$x - y = 1$

Replace x with -1 and y with 2.

$$3(-1) - 5(2) = -12$$
$$-3 - 10 \stackrel{?}{=} -12$$
$$-13 = -12 \;\; \textit{False}$$

It is not necessary to check if $(-1, 2)$ satisfies the second equation since it does not satisfy the first equation. Therefore, $(-1, 2)$ is *not* a solution of the system.

15. $x + y = -5$
$-2x + y = 1$

Graph the line $x + y = -5$ through its intercepts, $(-5, 0)$ and $(0, -5)$, and the line $-2x + y = 1$ through its intercepts, $(-\frac{1}{2}, 0)$ and $(0, 1)$. The lines appear to intersect at $(-2, -3)$.

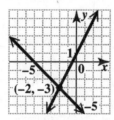

Check this ordered pair in the system.

$$(-2) + (-3) \stackrel{?}{=} -5$$
$$-5 = -5 \;\; \textit{True}$$

$$-2(-2) + (-3) \stackrel{?}{=} 1$$
$$4 + (-3) = 1 \;\; \textit{True}$$

The solution set is $\{(-2, -3)\}$.

17. $x + y = 4$
$2x - y = 2$

Graph the line $x + y = 4$ through its intercepts, $(4, 0)$ and $(0, 4)$, and the line $2x - y = 2$ through its intercepts, $(1, 0)$ and $(0, -2)$. The lines appear to intersect at $(2, 2)$.

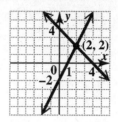

To check, substitute 2 for x and 2 for y in each equation of the system. Since $(2, 2)$ makes both equations true, the solution set of the system is $\{(2, 2)\}$.

19. $4x + y = 6 \quad (1)$
$y = 2x \quad (2)$

Equation (2) is already solved for y, so substitute $2x$ for y in equation (1).

$$4x + y = 6 \quad (1)$$
$$4x + 2x = 6 \quad \textit{Let y = 2x.}$$
$$6x = 6$$
$$x = 1$$

Substitute 1 for x in (2).

$$y = 2(1) = 2$$

The solution $(1, 2)$ checks.
The solution set is $\{(1, 2)\}$.

21. $-x - 4y = -14 \qquad (1)$
$y = 2x - 1 \qquad (2)$

Substitute $2x - 1$ for y in equation (1) and solve for x.

$$-x - 4y = -14 \qquad (1)$$
$$-x - 4(2x - 1) = -14$$
$$-x - 8x + 4 = -14$$
$$-9x = -18$$
$$x = 2$$

Substitute 2 for x in (2).

$$y = 2(2) - 1 = 3$$

The solution $(2, 3)$ checks.
The solution set is $\{(2, 3)\}$.

23. $3x - 4y = -22 \quad (1)$
$-3x + y = 0 \quad (2)$

Solve equation (2) for y to get

$$y = 3x.$$

Substitute $3x$ for y in equation (1).

$$3x - 4y = -22 \qquad (1)$$
$$3x - 4(3x) = -22 \qquad \textit{Let y = 3x.}$$
$$3x - 12x = -22$$
$$-9x = -22$$
$$x = \frac{-22}{-9} = \frac{22}{9}$$

Copyright © 2012 Pearson Education, Inc. Publishing as Addison-Wesley.

Substitute $\frac{22}{9}$ for x in $y = 3x$ to get

$$y = 3\left(\tfrac{22}{9}\right) = \tfrac{22}{3}.$$

The solution $\left(\frac{22}{9}, \frac{22}{3}\right)$ checks.
The solution set is $\left\{\left(\frac{22}{9}, \frac{22}{3}\right)\right\}$.

25. $5x - 4y = 9 \quad (1)$
      $3 - 2y = -x \quad (2)$

Solve equation (2) for x.

$$3 - 2y = -x \qquad (2)$$
$$2y - 3 = x \qquad (3) \qquad -1 \times (2)$$

Substitute $2y - 3$ for x in equation (1).

$$5x - 4y = 9 \qquad (1)$$
$$5(2y - 3) - 4y = 9 \qquad \textit{Let } x = 2y - 3.$$
$$10y - 15 - 4y = 9$$
$$6y - 15 = 9$$
$$6y = 24$$
$$y = 4$$

Substitute 4 for y in $x = 2y - 3$.

$$x = 2(4) - 3 = 5$$

The solution $(5, 4)$ checks.
The solution set is $\{(5, 4)\}$.

27. $4x - 5y = -11 \quad (1)$
    $x + 2y = 7 \qquad (2)$

Solve equation (2) for x.

$$x = -2y + 7$$

Substitute $-2y + 7$ for x in (1).

$$4x - 5y = -11 \qquad (1)$$
$$4(-2y + 7) - 5y = -11 \qquad \textit{Let } x = -2y + 7.$$
$$-8y + 28 - 5y = -11$$
$$-13y + 28 = -11$$
$$-13y = -39$$
$$y = 3$$

Substitute 3 for y in $x = -2y + 7$.

$$x = -2(3) + 7 = -6 + 7 = 1$$

The solution $(1, 3)$ checks.
The solution set is $\{(1, 3)\}$.

29. $x = 3y + 5 \quad (1)$
    $x = \frac{3}{2}y \qquad (2)$

Both equations are given in terms of x. Choose equation (2), and substitute $\frac{3}{2}y$ for x in equation (1).

$$x = 3y + 5 \qquad (1)$$
$$\tfrac{3}{2}y = 3y + 5 \qquad \textit{Let } x = \tfrac{3}{2}y.$$
$$3y = 6y + 10 \qquad \textit{Multiply by 2.}$$

$$-3y = 10$$
$$y = -\tfrac{10}{3}$$

Since $x = \frac{3}{2}y$ and $y = -\frac{10}{3}$,

$$x = \tfrac{3}{2}\left(-\tfrac{10}{3}\right) = -\tfrac{10}{2} = -5.$$

The solution $\left(-5, -\frac{10}{3}\right)$ checks.
The solution set is $\left\{\left(-5, -\frac{10}{3}\right)\right\}$.

31. $\frac{1}{2}x + \frac{1}{3}y = 3 \quad (1)$
    $-3x + y = 0 \quad (2)$

Multiply (1) by its LCD, 6, to eliminate fractions

$$3x + 2y = 18 \quad (3)$$

From (2), we have $y = 3x$, so substitute $3x$ for y in (3).

$$3x + 2(3x) = 18$$
$$3x + 6x = 18$$
$$9x = 18$$
$$x = 2$$

Substitute 2 for x in $y = 3x$.

$$y = 3(2) = 6$$

The solution $(2, 6)$ checks.
The solution set is $\{(2, 6)\}$.

33. $y = 2x \quad (1)$
   $4x - 2y = 0 \quad (2)$

From equation (1), substitute $2x$ for y in equation (2).

$$4x - 2y = 0 \qquad (2)$$
$$4x - 2(2x) = 0$$
$$4x - 4x = 0$$
$$0 = 0 \qquad \textit{True}$$

The equations are dependent, and the solution is the set of all points on the line.
The solution set is $\{(x, y) \mid y = 2x\}$.

35. $x = 5y \quad (1)$
   $5x - 25y = 5 \quad (2)$

From equation (1), substitute $5y$ for x in equation (2).

$$5x - 25y = 5 \qquad (2)$$
$$5(5y) - 25y = 5$$
$$25y - 25y = 5$$
$$0 = 5 \qquad \textit{False}$$

The system is *inconsistent*. Since the graphs of the equations are parallel lines, there are no ordered pairs that satisfy both equations.
The solution set is $\emptyset$.

Copyright © 2012 Pearson Education, Inc. Publishing as Addison-Wesley.

37.
$$y = 0.5x \quad (1)$$
$$1.5x - 0.5y = 5.0 \quad (2)$$

Multiply both equations by 2 to eliminate decimals. (We could just substitute $0.5x$ for y in the second equation.)

$$2y = x \quad (3) \quad 2 \times (1)$$
$$3x - y = 10 \quad (4) \quad 2 \times (2)$$

From (3), we have $x = 2y$, so substitute $2y$ for x in (4).

$$3(2y) - y = 10$$
$$6y - y = 10$$
$$5y = 10$$
$$y = 2$$

Substitute 2 for y in $x = 2y$.

$$x = 2(2) = 4$$

The solution $(4, 2)$ checks.
The solution set is $\{(4, 2)\}$.

39. The ordered pair $(0, 0)$ clearly satisfies the equations $Ax + By = 0$ and $Cx + Dy = 0$, so if there is a single solution, then it must be $(0, 0)$.

41.
$$-2x + 3y = -16 \quad (1)$$
$$2x - 5y = 24 \quad (2)$$

To eliminate x, add the equations.

$$-2x + 3y = -16 \quad (1)$$
$$\underline{2x - 5y = 24 \quad (2)}$$
$$-2y = 8$$
$$y = -4$$

To find x, substitute -4 for y in equation (2).

$$2x - 5y = 24 \quad (2)$$
$$2x - 5(-4) = 24$$
$$2x + 20 = 24$$
$$2x = 4$$
$$x = 2$$

The ordered pair $(2, -4)$ satisfies both equations, so it checks.
The solution set is $\{(2, -4)\}$.

43.
$$2x - 5y = 11 \quad (1)$$
$$3x + y = 8 \quad (2)$$

To eliminate y, multiply equation (2) by 5 and add the result to equation (1).

$$2x - 5y = 11 \quad (1)$$
$$\underline{15x + 5y = 40 \quad (3) \quad 5 \times (2)}$$
$$17x = 51$$
$$x = 3$$

To find y, substitute 3 for x in equation (2).

$$3x + y = 8 \quad (2)$$
$$3(3) + y = 8$$
$$9 + y = 8$$
$$y = -1$$

The ordered pair $(3, -1)$ satisfies both equations, so it checks.
The solution set is $\{(3, -1)\}$.

45.
$$3x + 4y = -6 \quad (1)$$
$$5x + 3y = 1 \quad (2)$$

To eliminate x, multiply equation (1) by 5 and equation (2) by -3. Then add the results.

$$15x + 20y = -30 \quad (3) \quad 5 \times (1)$$
$$\underline{-15x - 9y = -3 \quad (4) \quad -3 \times (2)}$$
$$11y = -33$$
$$y = -3$$

To find x, substitute -3 for y in equation (2).

$$5x + 3y = 1 \quad (2)$$
$$5x + 3(-3) = 1$$
$$5x - 9 = 1$$
$$5x = 10$$
$$x = 2$$

The ordered pair $(2, -3)$ satisfies both equations, so it checks.
The solution set is $\{(2, -3)\}$.

47.
$$7x + 2y = 6 \quad (1)$$
$$-14x - 4y = -12 \quad (2)$$

To eliminate y, multiply equation (1) by 2 and add the result to equation (2).

$$14x + 4y = 12 \quad (3) \quad 2 \times (1)$$
$$\underline{-14x - 4y = -12 \quad (2)}$$
$$0 = 0 \quad True$$

Multiplying equation (1) by -2 gives equation (2). The equations are dependent, and the solution is the set of all points on the line.

The solution set is $\{(x, y) \mid 7x + 2y = 6\}$.

49.
$$3x + 3y = 0 \quad (1)$$
$$4x + 2y = 3 \quad (2)$$

To eliminate y, multiply equation (1) by 2 and equation (2) by -3. Then add the results.

$$6x + 6y = 0 \quad (3) \quad 2 \times (1)$$
$$\underline{-12x - 6y = -9 \quad (4) \quad -3 \times (2)}$$
$$-6x = -9$$
$$x = \frac{-9}{-6} = \frac{3}{2}$$

To find y, substitute $\frac{3}{2}$ for x in equation (1).

Copyright © 2012 Pearson Education, Inc. Publishing as Addison-Wesley.

$$3x + 3y = 0 \quad (1)$$
$$3(\tfrac{3}{2}) + 3y = 0$$
$$\tfrac{9}{2} + 3y = 0$$
$$3y = -\tfrac{9}{2}$$
$$y = \tfrac{1}{3}(-\tfrac{9}{2}) = -\tfrac{3}{2}$$

The solution $(\tfrac{3}{2}, -\tfrac{3}{2})$ checks.
The solution set is $\left\{(\tfrac{3}{2}, -\tfrac{3}{2})\right\}$.

When you get a solution that has non-integer components, it is sometimes more difficult to check the problem than it was to solve it. A graphing calculator can be very helpful in this case. Just store the values for x and y in their respective memory locations, and then type the expressions as shown in the following screen. The results 0 and 3 (the right sides of the equations) indicate that we have found the correct solution.

```
3/2→X: -3/2→Y
                    -1.5
3X+3Y
                       0
4X+2Y
                       3
```

51. $5x - 5y = 3 \quad (1)$
$x - y = 12 \quad (2)$

To eliminate x, multiply equation (2) by -5 and add the result to equation (1).

$$
\begin{array}{rclll}
5x & - & 5y & = & 3 \quad (1)\\
-5x & + & 5y & = & -60 \quad (3) \qquad -5 \times (2)\\
\hline
& & 0 & = & -57 \quad \textit{False}
\end{array}
$$

The system is *inconsistent*. Since the graphs of the equations are parallel lines, there are no ordered pairs that satisfy both equations.

The solution set is $\emptyset$.

53. $x + y = 0 \quad (1)$
$2x - 2y = 0 \quad (2)$

To eliminate y, multiply equation (1) by 2 and add the result to equation (2).

$$
\begin{array}{rclll}
2x & + & 2y & = & 0 \quad (3) \qquad 2 \times (1)\\
2x & - & 2y & = & 0 \quad (2)\\
\hline
4x & & & = & 0\\
& & x & = & 0
\end{array}
$$

Substitute 0 for x in (1) to get $y = 0$.

The solution $(0, 0)$ checks.
The solution set is $\{(0, 0)\}$.

55. $x - \tfrac{1}{2}y = 2 \quad (1)$
$-x + \tfrac{2}{5}y = -\tfrac{8}{5} \quad (2)$

To eliminate x, add equations (1) and (2) to get

$$-\tfrac{1}{2}y + \tfrac{2}{5}y = 2 - \tfrac{8}{5}.$$

Multiply by the LCD, 10.

$$-5y + 4y = 20 - 16$$
$$-y = 4$$
$$y = -4$$

Substitute -4 for y in equation (1) to find x.

$$x - \tfrac{1}{2}(-4) = 2$$
$$x + 2 = 2$$
$$x = 0$$

The solution $(0, -4)$ checks.
The solution set is $\{(0, -4)\}$.

57. $\tfrac{1}{2}x + \tfrac{1}{3}y = \tfrac{49}{18} \quad (1)$
$\tfrac{1}{2}x + 2y = \tfrac{4}{3} \quad (2)$

Eliminate the fractions by multiplying equation (1) by 18 and equation (2) by 6.

$$
\begin{array}{rcll}
9x + 6y = 49 & (3) & 18 \times (1)\\
3x + 12y = 8 & (4) & 6 \times (2)
\end{array}
$$

Multiply equation (3) by -2 and add the resulting equations to eliminate y.

$$
\begin{array}{rclll}
-18x & - & 12y & = -98 \quad (5) & -2 \times (3)\\
3x & + & 12y & = 8 \quad (4)\\
\hline
-15x & & & = -90\\
& & x & = 6
\end{array}
$$

To find y, substitute 6 for x in equation (4).

$$3x + 12y = 8 \quad (4)$$
$$3(6) + 12y = 8$$
$$18 + 12y = 8$$
$$12y = -10$$
$$y = -\tfrac{5}{6}$$

The solution $(6, -\tfrac{5}{6})$ checks.
The solution set is $\left\{(6, -\tfrac{5}{6})\right\}$.

59. $3x + 7y = 4 \quad (1)$
$6x + 14y = 3 \quad (2)$

Write each equation in slope-intercept form by solving for y.

$$3x + 7y = 4 \qquad\qquad (1)$$
$$7y = -3x + 4$$
$$y = -\tfrac{3}{7}x + \tfrac{4}{7}$$

$$6x + 14y = 3 \qquad (2)$$
$$14y = -6x + 3$$
$$y = -\tfrac{6}{14}x + \tfrac{3}{14}$$
$$y = -\tfrac{3}{7}x + \tfrac{3}{14}$$

Since the equations have the same slope, $-\tfrac{3}{7}$, but different y-intercepts, $\tfrac{4}{7}$ and $\tfrac{3}{14}$, the lines when graphed are parallel. The system is inconsistent and has no solution.

61.
$$2x = -3y + 1 \quad (1)$$
$$6x = -9y + 3 \quad (2)$$

Write each equation in slope-intercept form by solving for y.

$$2x = -3y + 1 \qquad (1)$$
$$3y = -2x + 1$$
$$y = -\tfrac{2}{3}x + \tfrac{1}{3}$$

$$6x = -9y + 3 \qquad (2)$$
$$9y = -6x + 3$$
$$y = -\tfrac{6}{9}x + \tfrac{3}{9}$$
$$y = -\tfrac{2}{3}x + \tfrac{1}{3}$$

Since both equations are the same, the solution set is all points on the line $y = -\tfrac{2}{3}x + \tfrac{1}{3}$. The system has infinitely many solutions.

63.
$$3x + y = -7 \qquad (1)$$
$$x - y = -5 \qquad (2)$$

Add the equations to eliminate y.

$$
\begin{array}{rcr}
3x & + \; y \; = & -7 \\
x & - \; y \; = & -5 \\
\hline
4x & = & -12 \\
x & = & -3
\end{array}
$$

Substitute -3 for x in (2).

$$-3 - y = -5$$
$$-y = -2$$
$$y = 2$$

The solution $(-3, 2)$ checks.
The solution set is $\{(-3, 2)\}$.

65.
$$3x - 2y = 0 \qquad (1)$$
$$9x + 8y = 7 \qquad (2)$$

Multiply equation (1) by 4 and add the result to equation (2).

$$
\begin{array}{rclr}
12x & - \; 8y \; = & 0 & \quad 4 \times (1) \\
9x & + \; 8y \; = & 7 & \quad (2) \\
\hline
21x & = & 7 \\
x & = & \tfrac{1}{3}
\end{array}
$$

Substitute $\tfrac{1}{3}$ for x in equation (1).

$$3\left(\tfrac{1}{3}\right) - 2y = 0$$
$$1 - 2y = 0$$
$$-2y = -1$$
$$y = \tfrac{1}{2}$$

Check $\left(\tfrac{1}{3}, \tfrac{1}{2}\right)$: $1 - 1 = 0$; $3 + 4 = 7$
The solution set is $\left\{\left(\tfrac{1}{3}, \tfrac{1}{2}\right)\right\}$.

67.
$$2x + 3y = 10 \qquad (1)$$
$$-3x + y = 18 \qquad (2)$$

Solve equation (2) for y.

$$y = 3x + 18 \quad (3)$$

Substitute $3x + 18$ for y in (1).

$$2x + 3(3x + 18) = 10$$
$$2x + 9x + 54 = 10$$
$$11x + 54 = 10$$
$$11x = -44$$
$$x = -4$$

Substitute -4 for x in (3).

$$y = 3(-4) + 18 = 6$$

The solution $(-4, 6)$ checks.
The solution set is $\{(-4, 6)\}$.

69.
$$\tfrac{1}{2}x - \tfrac{1}{8}y = -\tfrac{1}{4} \qquad (1)$$
$$4x - y = -2 \qquad (2)$$

Multiply equation (1) by -8 and add to equation (2).

$$
\begin{array}{rclr}
-4x & + \; y \; = & 2 & \quad (3) \quad -8 \times (1) \\
4x & - \; y \; = & -2 & \quad (2) \\
\hline
0 & = & 0
\end{array}
$$

The equations are dependent.
The solution set is $\{(x, y) \mid 4x - y = -2\}$.

71.
$$0.3x + 0.2y = 0.4 \qquad (1)$$
$$0.5x + 0.4y = 0.7 \qquad (2)$$

To eliminate y, multiply equation (1) by -20 and equation (2) by 10, then add.

$$
\begin{array}{rclr}
-6x & - \; 4y \; = & -8 & \quad -20 \times (1) \\
5x & + \; 4y \; = & 7 & \quad (3) \quad 10 \times (2) \\
\hline
-x & = & -1 \\
x & = & 1
\end{array}
$$

Substitute 1 for x in equation (3).

$$5(1) + 4y = 7$$
$$4y = 2$$
$$y = \tfrac{1}{2}$$

Check $\left(1, \tfrac{1}{2}\right)$: $0.3 + 0.1 = 0.4$; $0.5 + 0.2 = 0.7$
The solution set is $\left\{\left(1, \tfrac{1}{2}\right)\right\}$.

Copyright © 2012 Pearson Education, Inc. Publishing as Addison-Wesley.

73. The table shows that when X = 3, $Y_1 = -4$ and $Y_2 = -4$. Since no other values of X in the table give the same values for Y_1 and Y_2, and since the functions are linear, the point $(3, -4)$ is the only point of intersection for the two graphs.

75. $y_1 = 3x - 5$
$y_2 = -4x + 2$

Both of the graphs in **B** have negative y-intercepts. But the y-intercept of y_2 is positive, so the graph in **B** is not acceptable. The slope of y_1 is positive, and y_1 has a negative y-intercept. The slope of y_2 is negative, and y_2 has a positive y-intercept. This fits graph **A**.

77. (a)
$$\begin{array}{r} x + y = 10 \quad (1) \\ 2x - y = 5 \quad (2) \\ \hline 3x = 15 \\ x = 5 \end{array}$$

Substitute 5 for x in equation (1).

$$\begin{aligned} x + y &= 10 \quad (1) \\ 5 + y &= 10 \\ y &= 5 \end{aligned}$$

The solution set is $\{(5, 5)\}$.

(b) Solve (1) and (2) for y.

$$\begin{aligned} x + y &= 10 \quad (1) \\ y &= -x + 10 \quad (3) \end{aligned}$$

$$\begin{aligned} 2x - y &= 5 \quad (2) \\ 2x - 5 &= y \quad (4) \end{aligned}$$

Now graph (3) and (4).

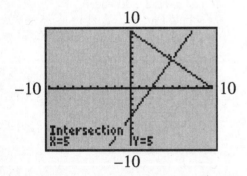

79. (a) The supply and demand graphs intersect at 4, so supply equals demand at a price of $4 per half-gallon.

(b) At a price of $4 per half-gallon, the supply and demand are both about 300 half-gallons.

(c) At a price of $2 per half-gallon, the supply is 200 half-gallons and the demand is 400 half-gallons.

81. (a) The graph for plasma flat panel displays is above the graph for front projection displays for years 2004–2008, so those are the years in which the sales for plasma flat panel displays exceeded the sales for front projection displays.

(b) LCD flat panel sales were greater than plasma flat panel sales between 2005 and 2006.

(c) The graph for LCD flat panel sales intersects the graph for plasma flat panel sales between 2005 and 2006. The sales value is about $4000 million.

(d) The ordered pair is $(2005, 4000)$.

(e) Sales of front projection displays were fairly constant. Sales of plasma flat panel displays increased from 2003 to 2006, but then declined. Sales of LCD flat panel displays increased over the whole period, fairly slowly at first, and then very rapidly.

83. The graphs intersect at about 1.5 (years since 2000). So the sales of digital cameras were less than the sales of digital camcorders for 2000, 2001, and the first half of 2002.

85.
$$\begin{aligned} 119x + y &= 2838 \quad (1) \\ 624x - y &= -1823 \quad (2) \end{aligned}$$

Adding the equations gives us $743x = 1015$, so $x = \frac{1015}{743} \approx 1.4$.

Solve equation (2) for y.

$$y = 624x + 1823 \quad (3)$$

Substitute $\frac{1015}{743}$ for x in (3).

$$y = 624\left(\tfrac{1015}{743}\right) + 1823 \approx 2675.4$$

Written as an ordered pair, the solution is $(1.4, 2675.4)$. (Values may vary slightly based on the method of solution used.)

87.
$$\begin{aligned} \frac{3}{x} + \frac{4}{y} &= \frac{5}{2} \quad (1) \\ \frac{5}{x} - \frac{3}{y} &= \frac{7}{4} \quad (2) \end{aligned}$$

If $p = \frac{1}{x}$ and $q = \frac{1}{y}$, equations (1) and (2) can be written as

$$\begin{aligned} 3p + 4q &= \tfrac{5}{2} \quad (3) \\ 5p - 3q &= \tfrac{7}{4}. \quad (4) \end{aligned}$$

To eliminate q, multiply equation (3) by 3 and equation (4) by 4. Then add the results.

$$\begin{array}{rll} 9p + 12q = \tfrac{15}{2} & (5) & 3 \times (3) \\ 20p - 12q = 7 & (6) & 4 \times (4) \\ \hline 29p = \tfrac{29}{2} & & \\ p = \tfrac{1}{2} & & \end{array}$$

Copyright © 2012 Pearson Education, Inc. Publishing as Addison-Wesley.

To find q, substitute $\frac{1}{2}$ for p in equation (3).

$$3p + 4q = \frac{5}{2} \qquad (3)$$
$$3\left(\tfrac{1}{2}\right) + 4q = \frac{5}{2} \qquad \text{Let } p = \tfrac{1}{2}.$$
$$\frac{3}{2} + 4q = \frac{5}{2}$$
$$4q = 1$$
$$q = \frac{1}{4}$$

Since $p = \frac{1}{x}$ and $p = \frac{1}{2}$, $\frac{1}{2} = \frac{1}{x}$ and $x = 2$.
Since $q = \frac{1}{y}$ and $q = \frac{1}{4}$, $\frac{1}{4} = \frac{1}{y}$ and $y = 4$.

The solution $(2, 4)$ checks.
The solution set is $\{(2, 4)\}$.

89.
$$\frac{2}{x} + \frac{3}{y} = \frac{11}{2} \quad (1)$$
$$-\frac{1}{x} + \frac{2}{y} = -1 \quad (2)$$

Let $p = \frac{1}{x}$ and $q = \frac{1}{y}$. Rewrite the system as

$$2p + 3q = \frac{11}{2} \qquad (3)$$
$$-p + 2q = -1. \qquad (4)$$

To eliminate p, multiply (3) by 2 and (4) by 4. Then add the results.

$$
\begin{array}{rrrll}
4p &+& 6q &=& 11 \quad (5) \quad 2 \times (3) \\
-4p &+& 8q &=& -4 \quad (6) \quad 4 \times (4) \\
\hline
&& 14q &=& 7 \\
&& q &=& \frac{1}{2}
\end{array}
$$

Substitute $\frac{1}{2}$ for q in (4).

$$-p + 2\left(\tfrac{1}{2}\right) = -1$$
$$-p + 1 = -1$$
$$-p = -2$$
$$p = 2$$

Since $p = \frac{1}{x}$ and $p = 2$,

$$2 = \frac{1}{x}, \text{ so } 2x = 1 \text{ and } x = \tfrac{1}{2}.$$

Since $q = \frac{1}{y}$ and $q = \frac{1}{2}$,

$$\frac{1}{2} = \frac{1}{y} \text{ and } y = 2.$$

The solution $\left(\frac{1}{2}, 2\right)$ checks.
The solution set is $\left\{\left(\frac{1}{2}, 2\right)\right\}$.

91.
$$ax + by = c \quad (1)$$
$$ax - 2by = c \quad (2)$$

To eliminate y, multiply (1) by 2 and add the result to equation (2).

$$
\begin{array}{rrrll}
2ax &+& 2by &=& 2c \quad (3) \quad 2 \times (1) \\
ax &-& 2by &=& c \quad (2) \\
\hline
3ax && &=& 3c \\
&& x &=& \dfrac{3c}{3a} = \dfrac{c}{a}
\end{array}
$$

Substitute $\frac{c}{a}$ for x in (1).

$$a\left(\tfrac{c}{a}\right) + by = c$$
$$c + by = c$$
$$by = 0$$
$$y = 0$$

The solution $\left(\frac{c}{a}, 0\right)$ checks.
The solution set is $\left\{\left(\frac{c}{a}, 0\right)\right\}$.

93.
$$2ax - y = 3 \quad (1)$$
$$y = 5ax \quad (2)$$

Substitute $5ax$ for y in (1).

$$2ax - 5ax = 3$$
$$-3ax = 3$$
$$x = \frac{3}{-3a} = -\frac{1}{a}$$

Substitute $-\frac{1}{a}$ for x in (2).

$$y = 5a\left(-\tfrac{1}{a}\right) = -5$$

The solution $\left(-\frac{1}{a}, -5\right)$ checks.
The solution set is $\left\{\left(-\frac{1}{a}, -5\right)\right\}$.

95.
$$4(2x - 3y + z) = 4(5)$$
$$8x - 12y + 4z = 20$$

97.
$$x + 2y + 3z = 9$$
$$1 + 2(-2) + 3z = 9 \qquad \text{Let } x = 1, y = -2.$$
$$1 - 4 + 3z = 9$$
$$3z = 12$$
$$z = 4$$

99. Multiplying the first equation by -3 will give us $-3x$, which when added to $3x$ in the second equation, gives us 0.

4.2 Systems of Linear Equations in Three Variables

4.2 Now Try Exercises

N1.
$$x - y + 2z = 1 \qquad (1)$$
$$3x + 2y + 7z = 8 \qquad (2)$$
$$-3x - 4y + 9z = -10 \qquad (3)$$

Step 1
Eliminate x by adding equations (2) and (3).

$$
\begin{array}{rrrrrll}
3x &+& 2y &+& 7z &=& 8 \quad (2) \\
-3x &-& 4y &+& 9z &=& -10 \quad (3) \\
\hline
&-& 2y &+& 16z &=& -2 \quad (4)
\end{array}
$$

Step 2
To eliminate x again, multiply equation (1) by 3 and add the result to equation (3).

$$
\begin{array}{rrrrrll}
3x &-& 3y &+& 6z &=& 3 \quad\quad 3 \times (1) \\
-3x &-& 4y &+& 9z &=& -10 \quad (3) \\
\hline
&-& 7y &+& 15z &=& -7 \quad (5)
\end{array}
$$

Copyright © 2012 Pearson Education, Inc. Publishing as Addison-Wesley.

Step 3
Use equations (4) and (5) to eliminate y. Multiply equation (4) by -7 and add the result to 2 times equation (5).

$$
\begin{array}{rrrl}
14y & - & 112z & = & 14 & -7 \times (4) \\
-14y & + & 30z & = & -14 & 2 \times (5) \\
\hline
 & - & 82z & = & 0 \\
 & & z & = & 0
\end{array}
$$

Step 4
Substitute 0 for z in equation (5) to find y.

$$
\begin{array}{rl}
-7y + 15z = -7 & (5) \\
-7y + 15(0) = -7 & \textit{Let z = 0.} \\
-7y = -7 \\
y = 1
\end{array}
$$

Step 5
Substitute 0 for z and 1 for y in equation (1) to find x.

$$
\begin{array}{rl}
x - y + 2z = 1 & (1) \\
x - 1 + 0 = 1 & \textit{Let y = 1, z = 0.} \\
x = 2
\end{array}
$$

Step 6
Check $(2, 1, 0)$

Equation (1): $2 - 1 + 0 = 1$ *True*
Equation (2): $6 + 2 + 0 = 8$ *True*
Equation (3): $-6 - 4 + 0 = -10$ *True*

The solution set is $\{(2, 1, 0)\}$.

N2. $\begin{array}{llll} 3x & & - z = -10 & (1) \\ & 4y & + 5z = 24 & (2) \\ x & - 6y & = -8 & (3) \end{array}$

Since equation (1) is missing y, eliminate y again from equations (2) and (3). Multiply equation (2) by 3 and add the result to 2 times equation (3).

$$
\begin{array}{rrrl}
 & 12y & + 15z & = & 72 & 3 \times (2) \\
2x & - 12y & & = & -16 & 2 \times (3) \\
\hline
2x & & + 15z & = & 56 & (4)
\end{array}
$$

Use equation (4) together with equation (1) to eliminate z. Multiply equation (1) by 15 and add it to equation (4).

$$
\begin{array}{rrrl}
45x & - 15z & = & -150 & 15 \times (1) \\
2x & + 15z & = & 56 & (4) \\
\hline
47x & & = & -94 \\
x & & = & -2
\end{array}
$$

Substitute -2 for x in equation (1) to find z.

$$
\begin{array}{rl}
3x - z = -10 & (1) \\
3(-2) - z = -10 & \textit{Let x = -2.} \\
-6 - z = -10
\end{array}
$$

$$
\begin{array}{r}
-z = -4 \\
z = 4
\end{array}
$$

Substitute 4 for z in equation (2) to find y.

$$
\begin{array}{rl}
4y + 5z = 24 & (2) \\
4y + 5(4) = 24 & \textit{Let z = 4.} \\
4y + 20 = 24 \\
4y = 4 \\
y = 1
\end{array}
$$

Check $(-2, 1, 4)$

Equation (1): $-6 - 4 = -10$ *True*
Equation (2): $4 + 20 = 24$ *True*
Equation (3): $-2 - 6 = -8$ *True*

The solution set is $\{(-2, 1, 4)\}$.

N3. $\begin{array}{rrl} x - & 5y + 2z = 4 & (1) \\ 3x + & y - z = 6 & (2) \\ -2x + & 10y - 4z = 7 & (3) \end{array}$

Multiply equation (1) by 2 and add the result to equation (3).

$$
\begin{array}{rrrrl}
2x & - 10y & + 4z & = & 8 & 2 \times (1) \\
-2x & + 10y & - 4z & = & 7 & (3) \\
\hline
 & & 0 & = & 15 & \textit{False}
\end{array}
$$

Since a false statement results, the system is inconsistent.

The solution set is $\emptyset$.

N4. $\begin{array}{rl} x - 3y + 2z = 10 & (1) \\ -2x + 6y - 4z = -20 & (2) \\ \frac{1}{2}x - \frac{3}{2}y + z = 5 & (3) \end{array}$

Since equation (2) is -2 times equation (1) and equation (3) is $\frac{1}{2}$ times equation (1), the three equations are dependent. All three have the same graph.

The solution set is $\{(x, y, z) \mid x - 3y + 2z = 10\}$.

N5. $\begin{array}{rl} x - 3y + 2z = 4 & (1) \\ \frac{1}{3}x - y + \frac{2}{3}z = 7 & (2) \\ \frac{1}{2}x - \frac{3}{2}y + z = 2 & (3) \end{array}$

Eliminate the fractions in equations (2) and (3).

$$
\begin{array}{rll}
x - 3y + 2z = 4 & (1) \\
x - 3y + 2z = 24 & (4) & 3 \times (2) \\
x - 3y + 2z = 4 & (5) & 2 \times (3)
\end{array}
$$

Equations (1) and (5) are dependent (they have the same graph). Equations (1) and (4) are not equivalent. Since they have the same coefficients, but a different constant term, their graphs have no points in common (the planes are parallel). Thus, the system is inconsistent and the solution set is $\emptyset$.

Copyright © 2012 Pearson Education, Inc. Publishing as Addison-Wesley.

4.2 Section Exercises

1. Substitute 1 for x, 2 for y, and 3 for z in $3x + 2y - z$, which is the left side of each equation.

$$3(1) + 2(2) - (3) = 3 + 4 - 3$$
$$= 7 - 3$$
$$= 4$$

Choice **B** is correct since its right side is 4.

3.
$$2x - 5y + 3z = -1 \quad (1)$$
$$x + 4y - 2z = 9 \quad (2)$$
$$x - 2y - 4z = -5 \quad (3)$$

Eliminate x by adding equation (1) to -2 times equation (2).

$$
\begin{array}{rcll}
2x - 5y + 3z &=& -1 & (1) \\
-2x - 8y + 4z &=& -18 & -2 \times (2) \\
\hline
-13y + 7z &=& -19 & (4)
\end{array}
$$

Now eliminate x by adding equation (2) to -1 times equation (3).

$$
\begin{array}{rcll}
x + 4y - 2z &=& 9 & (2) \\
-x + 2y + 4z &=& 5 & -1 \times (3) \\
\hline
6y + 2z &=& 14 & (5)
\end{array}
$$

Use equations (4) and (5) to eliminate z. Multiply equation (4) by -2 and add the result to 7 times equation (5).

$$
\begin{array}{rcll}
26y - 14z &=& 38 & -2 \times (4) \\
42y + 14z &=& 98 & 7 \times (5) \\
\hline
68y &=& 136 & \\
y &=& 2 &
\end{array}
$$

Substitute 2 for y in equation (5) to find z.

$$6y + 2z = 14 \quad (5)$$
$$6(2) + 2z = 14$$
$$12 + 2z = 14$$
$$2z = 2$$
$$z = 1$$

Substitute 2 for y and 1 for z in equation (3) to find x.

$$x - 2y - 4z = -5 \quad (3)$$
$$x - 2(2) - 4(1) = -5$$
$$x - 4 - 4 = -5$$
$$x - 8 = -5$$
$$x = 3$$

The solution $(3, 2, 1)$ checks in all three of the original equations.
The solution set is $\{(3, 2, 1)\}$.

5.
$$3x + 2y + z = 8 \quad (1)$$
$$2x - 3y + 2z = -16 \quad (2)$$
$$x + 4y - z = 20 \quad (3)$$

Eliminate z by adding equations (1) and (3).

$$
\begin{array}{rcll}
3x + 2y + z &=& 8 & (1) \\
x + 4y - z &=& 20 & (3) \\
\hline
4x + 6y &=& 28 & (4)
\end{array}
$$

To get another equation without z, multiply equation (3) by 2 and add the result to equation (2).

$$
\begin{array}{rcll}
2x + 8y - 2z &=& 40 & 2 \times (3) \\
2x - 3y + 2z &=& -16 & (2) \\
\hline
4x + 5y &=& 24 & (5)
\end{array}
$$

Use equations (4) and (5) to eliminate x. Multiply equation (4) by -1 and add the result to equation (5).

$$
\begin{array}{rcll}
-4x - 6y &=& -28 & -1 \times (4) \\
4x + 5y &=& 24 & (5) \\
\hline
-y &=& -4 & \\
y &=& 4 &
\end{array}
$$

Substitute 4 for y in equation (5) to find x.

$$4x + 5y = 24 \quad (5)$$
$$4x + 5(4) = 24$$
$$4x + 20 = 24$$
$$4x = 4$$
$$x = 1$$

Substitute 1 for x and 4 for y in equation (3) to find z.

$$x + 4y - z = 20 \quad (3)$$
$$1 + 4(4) - z = 20$$
$$1 + 16 - z = 20$$
$$17 - z = 20$$
$$-z = 3$$
$$z = -3$$

The solution $(1, 4, -3)$ checks in all three of the original equations.
The solution set is $\{(1, 4, -3)\}$.

7.
$$2x + 5y + 2z = 0 \quad (1)$$
$$4x - 7y - 3z = 1 \quad (2)$$
$$3x - 8y - 2z = -6 \quad (3)$$

Add equations (1) and (3) to eliminate z.

$$
\begin{array}{rcll}
2x + 5y + 2z &=& 0 & (1) \\
3x - 8y - 2z &=& -6 & (3) \\
\hline
5x - 3y &=& -6 & (4)
\end{array}
$$

Multiply equation (1) by 3 and equation (2) by 2. Then add the results to eliminate z again.

$$6x + 15y + 6z = 0 \qquad 3 \times (1)$$
$$\underline{8x - 14y - 6z = 2} \qquad 2 \times (2)$$
$$14x + y = 2 \quad (5)$$

Solve the system

$$5x - 3y = -6 \quad (4)$$
$$14x + y = 2. \quad (5)$$

Multiply equation (5) by 3 then add this result to (4).

$$5x - 3y = -6 \quad (4)$$
$$\underline{42x + 3y = 6} \qquad 3 \times (5)$$
$$47x = 0$$
$$x = 0$$

To find y, substitute $x = 0$ into equation (4).

$$5x - 3y = -6 \quad (4)$$
$$5(0) - 3y = -6$$
$$y = 2$$

To find z, substitute $x = 0$ and $y = 2$ in equation (1).

$$2x + 5y + 2z = 0 \quad (1)$$
$$2(0) + 5(2) + 2z = 0$$
$$10 + 2z = 0$$
$$2z = -10$$
$$z = -5$$

The solution set is $\{(0, 2, -5)\}$.

9.
$$x + 2y + z = 4 \quad (1)$$
$$2x + y - z = -1 \quad (2)$$
$$x - y - z = -2 \quad (3)$$

Add (1) and (2) to eliminate z.

$$x + 2y + z = 4 \quad (1)$$
$$\underline{2x + y - z = -1} \quad (2)$$
$$3x + 3y = 3 \quad (4)$$

Add (1) and (3) to eliminate z.

$$x + 2y + z = 4 \quad (1)$$
$$\underline{x - y - z = -2} \quad (3)$$
$$2x + y = 2 \quad (5)$$

Multiply equation (5) by -3.

$$-6x - 3y = -6 \quad (6) \ -3 \times (5)$$
$$\underline{3x + 3y = 3} \quad (4)$$
$$-3x = -3$$
$$x = 1$$

Substitute 1 for x in equation (5) and solve for y.

$$2(1) + y = 2$$
$$y = 0$$

Substitute 1 for x and 0 for y in (1).

$$(1) + 2(0) + z = 4$$
$$z = 3$$

The solution set is $\{(1, 0, 3)\}$.

11.
$$-x + 2y + 6z = 2 \quad (1)$$
$$3x + 2y + 6z = 6 \quad (2)$$
$$x + 4y - 3z = 1 \quad (3)$$

Eliminate y and z by adding equation (1) to -1 times equation (2).

$$-x + 2y + 6z = 2 \quad (1)$$
$$\underline{-3x - 2y - 6z = -6} \qquad -1 \times (2)$$
$$-4x = -4$$
$$x = 1$$

Eliminate z by adding equation (2) to 2 times equation (3).

$$3x + 2y + 6z = 6 \quad (2)$$
$$\underline{2x + 8y - 6z = 2} \qquad 2 \times (3)$$
$$5x + 10y = 8 \quad (4)$$

Substitute 1 for x in equation (4).

$$5x + 10y = 8 \quad (4)$$
$$5(1) + 10y = 8$$
$$10y = 3$$
$$y = \frac{3}{10}$$

Substitute 1 for x and $\frac{3}{10}$ for y in equation (1).

$$-x + 2y + 6z = 2 \quad (1)$$
$$-1 + 2\left(\tfrac{3}{10}\right) + 6z = 2$$
$$\tfrac{3}{5} + 6z = 3$$
$$6z = \tfrac{12}{5}$$
$$z = \tfrac{2}{5}$$

The solution set is $\left\{ \left(1, \frac{3}{10}, \frac{2}{5}\right) \right\}$.

13.
$$x + y - z = -2 \quad (1)$$
$$2x - y + z = -5 \quad (2)$$
$$-x + 2y - 3z = -4 \quad (3)$$

Eliminate y and z by adding equations (1) and (2).

$$x + y - z = -2 \quad (1)$$
$$\underline{2x - y + z = -5} \quad (2)$$
$$3x = -7$$
$$x = -\tfrac{7}{3}$$

To get another equation without y, multiply equation (2) by 2 and add the result to equation (3).

$$4x - 2y + 2z = -10 \qquad 2 \times (2)$$
$$\underline{-x + 2y - 3z = -4} \quad (3)$$
$$3x - z = -14 \quad (4)$$

Substitute $-\frac{7}{3}$ for x in equation (4) to find z.

Copyright © 2012 Pearson Education, Inc. Publishing as Addison-Wesley.

$$3x - z = -14 \quad (4)$$
$$3(-\tfrac{7}{3}) - z = -14$$
$$-7 - z = -14$$
$$-z = -7$$
$$z = 7$$

Substitute $-\tfrac{7}{3}$ for x and 7 for z in equation (1) to find y.

$$x + y - z = -2 \quad (1)$$
$$-\tfrac{7}{3} + y - 7 = -2$$
$$-7 + 3y - 21 = -6 \quad \textit{Multiply by 3.}$$
$$3y - 28 = -6$$
$$3y = 22$$
$$y = \tfrac{22}{3}$$

The solution set is $\left\{ \left(-\tfrac{7}{3}, \tfrac{22}{3}, 7\right) \right\}$.

A calculator check reduces the probability of making any arithmetic errors and is highly recommended. The following screen shows the substitution of the solution for x, y, and z along with the left sides of the three original equations. The evaluation of the three expressions, -2, -5, and -4 (the right sides of the three equations), indicates that we have found the correct solution.

```
-7/3→X:22/3→Y:7→
Z:X+Y-Z
                -2
2X-Y+Z
                -5
-X+2Y-3Z
                -4
```

15.
$$\tfrac{1}{3}x + \tfrac{1}{6}y - \tfrac{2}{3}z = -1 \quad (1)$$
$$-\tfrac{3}{4}x - \tfrac{1}{3}y - \tfrac{1}{4}z = 3 \quad (2)$$
$$\tfrac{1}{2}x + \tfrac{3}{2}y + \tfrac{3}{4}z = 21 \quad (3)$$

Eliminate all fractions.

$$2x + y - 4z = -6 \quad (4) \quad 6 \times (1)$$
$$-9x - 4y - 3z = 36 \quad (5) \quad 12 \times (2)$$
$$2x + 6y + 3z = 84 \quad (6) \quad 4 \times (3)$$

Multiply (4) by 4.

$$8x + 4y - 16z = -24$$
$$-9x - 4y - 3z = 36 \quad (5)$$
$$\overline{\quad -x \qquad - 19z = 12} \quad (7)$$

Multiply (4) by -6.

$$-12x - 6y + 24z = 36$$
$$2x + 6y + 3z = 84 \quad (6)$$
$$\overline{-10x \qquad + 27z = 120} \quad (8)$$

Multiply (7) by -10.

$$10x + 190z = -120$$
$$-10x + 27z = 120 \quad (8)$$
$$\overline{\quad 217z = 0}$$
$$z = 0$$

Substituting 0 for z in (7) gives us $x = -12$. Substitute -12 for x and 0 for z in (4).

$$2(-12) + y - 4(0) = -6$$
$$-24 + y = -6$$
$$y = 18$$

The solution set is $\{(-12, 18, 0)\}$.

17.
$$5.5x - 2.5y + 1.6z = 11.83 \quad (1)$$
$$2.2x + 5.0y - 0.1z = -5.97 \quad (2)$$
$$3.3x - 7.5y + 3.2z = 21.25 \quad (3)$$

Multiply the equations by 10 to eliminate the decimals for the coefficients of the variables.

$$55x - 25y + 16z = 118.3 \quad (4)$$
$$22x + 50y - z = -59.7 \quad (5)$$
$$33x - 75y + 32z = 212.5 \quad (6)$$

Multiply equation (4) by -3 and add the result to equation (6) to eliminate y.

$$-165x + 75y - 48z = -354.9 \quad -3 \times (4)$$
$$33x - 75y + 32z = 212.5 \quad (6)$$
$$\overline{-132x \qquad - 16z = -142.4} \quad (7)$$

Multiply equation (4) by 2 and add the result to equation (5) to eliminate y.

$$110x - 50y + 32z = 236.6 \quad 2 \times (4)$$
$$22x + 50y - z = -59.7 \quad (5)$$
$$\overline{132x \qquad + 31z = 176.9} \quad (8)$$

Now add equations (7) and (8) to eliminate x.

$$-132x - 16z = -142.4 \quad (7)$$
$$132x + 31z = 176.9 \quad (8)$$
$$\overline{\quad 15z = 34.5}$$
$$z = 2.3$$

Substitute 2.3 for z in equation (8) and solve for x.

$$132x + 31(2.3) = 176.9$$
$$132x + 71.3 = 176.9$$
$$132x = 105.6$$
$$x = 0.8$$

Use equation (5) to solve for y.

$$22x + 50y - z = -59.7$$
$$22(0.8) + 50y - 2.3 = -59.7$$
$$50y = -75$$
$$y = -1.5$$

The solution set is $\{(0.8, -1.5, 2.3)\}$.

Copyright © 2012 Pearson Education, Inc. Publishing as Addison-Wesley.

19.
$$\begin{aligned} 2x - 3y + 2z &= -1 \quad (1) \\ x + 2y + z &= 17 \quad (2) \\ 2y - z &= 7 \quad (3) \end{aligned}$$

Multiply equation (2) by -2, and add the result to equation (1).

$$\begin{array}{l} 2x - 3y + 2z = -1 \quad (1) \\ \underline{-2x - 4y - 2z = -34} \qquad -2 \times (2) \\ - 7y = -35 \\ y = 5 \end{array}$$

To find z, substitute 5 for y in equation (3).

$$\begin{aligned} 2y - z &= 7 \quad (3) \\ 2(5) - z &= 7 \\ 10 - z &= 7 \\ -z &= -3 \\ z &= 3 \end{aligned}$$

To find x, substitute $y = 5$ and $z = 3$ into equation (1).

$$\begin{aligned} 2x - 3y + 2z &= -1 \quad (1) \\ 2x - 3(5) + 2(3) &= -1 \\ 2x - 9 &= -1 \\ 2x &= 8 \\ x &= 4 \end{aligned}$$

The solution set is $\{(4, 5, 3)\}$.

21.
$$\begin{aligned} 4x + 2y - 3z &= 6 \quad (1) \\ x - 4y + z &= -4 \quad (2) \\ -x + 2z &= 2 \quad (3) \end{aligned}$$

Equation (3) is missing y. Eliminate y again by multiplying equation (1) by 2 and adding the result to equation (2).

$$\begin{array}{l} 8x + 4y - 6z = 12 \qquad 2 \times (1) \\ \underline{x - 4y + z = -4} \quad (2) \\ 9x - 5z = 8 \quad (4) \end{array}$$

Use equations (3) and (4) to eliminate x. Multiply equation (3) by 9 and add the result to equation (4).

$$\begin{array}{l} -9x + 18z = 18 \qquad 9 \times (3) \\ \underline{9x - 5z = 8} \quad (4) \\ 13z = 26 \\ z = 2 \end{array}$$

Substitute 2 for z in equation (3) to find x.

$$\begin{aligned} -x + 2z &= 2 \quad (3) \\ -x + 2(2) &= 2 \\ -x + 4 &= 2 \\ -x &= -2 \\ x &= 2 \end{aligned}$$

Substitute 2 for x and 2 for z in equation (2) to find y.

$$\begin{aligned} x - 4y + z &= -4 \quad (2) \\ 2 - 4y + 2 &= -4 \\ -4y + 4 &= -4 \\ -4y &= -8 \\ y &= 2 \end{aligned}$$

The solution set is $\{(2, 2, 2)\}$.

23.
$$\begin{aligned} 2x + y &= 6 \quad (1) \\ 3y - 2z &= -4 \quad (2) \\ 3x - 5z &= -7 \quad (3) \end{aligned}$$

To eliminate y, multiply equation (1) by -3 and add the result to equation (2).

$$\begin{array}{l} -6x - 3y = -18 \qquad -3 \times (1) \\ \underline{3y - 2z = -4} \quad (2) \\ -6x - 2z = -22 \quad (4) \end{array}$$

Since equation (3) does not have a y-term, we can multiply equation (3) by 2 and add the result to equation (4) to eliminate x and solve for z.

$$\begin{array}{l} 6x - 10z = -14 \qquad 2 \times (3) \\ \underline{-6x - 2z = -22} \quad (4) \\ - 12z = -36 \\ z = 3 \end{array}$$

To find x, substitute 3 for z into equation (3).

$$\begin{aligned} 3x - 5z &= -7 \quad (3) \\ 3x - 5(3) &= -7 \\ 3x &= 8 \\ x &= \tfrac{8}{3} \end{aligned}$$

To find y, substitute 3 for z into equation (2).

$$\begin{aligned} 3y - 2z &= -4 \quad (2) \\ 3y - 2(3) &= -4 \\ 3y &= 2 \\ y &= \tfrac{2}{3} \end{aligned}$$

The solution set is $\left\{ \left(\tfrac{8}{3}, \tfrac{2}{3}, 3 \right) \right\}$.

25.
$$\begin{aligned} -5x + 2y + z &= 5 \quad (1) \\ -3x - 2y - z &= 3 \quad (2) \\ -x + 6y &= 1 \quad (3) \end{aligned}$$

Add (1) and (2) to eliminate y and z.

$$\begin{aligned} -8x &= 8 \\ x &= -1 \end{aligned}$$

Substitute -1 for x in equation (3).

$$\begin{aligned} -x + 6y &= 1 \quad (3) \\ -(-1) + 6y &= 1 \\ 6y &= 0 \\ y &= 0 \end{aligned}$$

Substitute -1 for x and 0 for y in (1).

Copyright © 2012 Pearson Education, Inc. Publishing as Addison-Wesley.

$$-5x + 2y + z = 5 \quad (1)$$
$$-5(-1) + 2(0) + z = 5$$
$$5 + z = 5$$
$$z = 0$$

The solution set is $\{(-1, 0, 0)\}$.

27.
$$7x \qquad\; - 3z = -34 \quad (1)$$
$$2y + 4z = \;\;\; 20 \quad (2)$$
$$\tfrac{3}{4}x + \tfrac{1}{6}y \qquad\;\; = \;-2 \quad (3)$$

Eliminate fractions from equation (3) by multiplying by -12.

$$-9x - 2y = 24 \quad (4) \quad -12 \times (3)$$

Adding equations (2) and (4) gives us

$$-9x + 4z = 44. \quad (5)$$

Eliminate z by adding 4 times (1) to 3 times (5).

$$\begin{array}{rcll} 28x - 12z & = & -136 & 4 \times (1) \\ -27x + 12z & = & 132 & 3 \times (5) \\ \hline x & = & -4 & \end{array}$$

Substitute -4 for x in (5).

$$-9(-4) + 4z = 44$$
$$36 + 4z = 44$$
$$4z = 8$$
$$z = 2$$

Substitute 2 for z in (2).

$$2y + 4(2) = 20$$
$$2y + 8 = 20$$
$$2y = 12$$
$$y = 6$$

The solution set is $\{(-4, 6, 2)\}$.

29.
$$4x \qquad\;\; - z = -6 \quad (1)$$
$$\tfrac{3}{5}y + \tfrac{1}{2}z = \;\;\;\; 0 \quad (2)$$
$$\tfrac{1}{3}x \qquad\; + \tfrac{2}{3}z = -5 \quad (3)$$

Eliminate fractions first.

$$\begin{array}{rclll} 6y + 5z & = & 0 & (4) & 10 \times (2) \\ x \;\;\; + 2z & = & -15 & (5) & 3 \times (3) \end{array}$$

Eliminate z by adding (5) to 2 times (1).

$$\begin{array}{rcll} 8x - 2z & = & -12 & 2 \times (1) \\ x + 2z & = & -15 & (5) \\ \hline 9x & = & -27 & \\ x & = & -3 & \end{array}$$

Substitute -3 for x in (5).

$$x + 2z = -15 \quad (5)$$
$$-3 + 2z = -15$$
$$2z = -12$$
$$z = -6$$

Substitute -6 for z in (4).

$$6y + 5z = 0 \quad (4)$$
$$6y + 5(-6) = 0$$
$$6y = 30$$
$$y = 5$$

The solution set is $\{(-3, 5, -6)\}$.

31.
$$2x + 2y - 6z = \;\;\; 5 \quad (1)$$
$$-3x + \;\; y - \;\; z = -2 \quad (2)$$
$$-x - \;\; y + 3z = \;\;\; 4 \quad (3)$$

Multiply equation (3) by 2 and add the result to equation (1).

$$\begin{array}{rcll} 2x + 2y - 6z & = & 5 & (1) \\ -2x - 2y + 6z & = & 8 & 2 \times (3) \\ \hline 0 & = & 13 & \text{False} \end{array}$$

The solution set is $\emptyset$; inconsistent system.

33.
$$-5x + 5y - 20z = -40 \quad (1)$$
$$x - \;\; y + \;\; 4z = \;\;\; 8 \quad (2)$$
$$3x - 3y + 12z = \;\; 24 \quad (3)$$

Dividing equation (1) by -5 gives equation (2). Dividing equation (3) by 3 also gives equation (2). The resulting equations are the same, so the three equations are dependent.

The solution set is $\{(x, y, z) \,|\, x - y + 4z = 8\}$.

35.
$$x + 5y - 2z = -1 \quad (1)$$
$$-2x + 8y + z = -4 \quad (2)$$
$$3x - y + 5z = 19 \quad (3)$$

Eliminate x by adding (2) to 2 times (1).

$$\begin{array}{rcll} 2x + 10y - 4z & = & -2 & 2 \times (1) \\ -2x + \;\; 8y + \;\; z & = & -4 & (2) \\ \hline 18y - 3z & = & -6 & (4) \end{array}$$

Eliminate x by adding (3) to -3 times (1).

$$\begin{array}{rcll} -3x - 15y + 6z & = & 3 & -3 \times (1) \\ 3x - \;\;\; y + 5z & = & 19 & (3) \\ \hline -16y + 11z & = & 22 & (5) \end{array}$$

To eliminate z from equations (4) and (5), we could first divide equation (4) by 3 and then multiply the resulting equation by 11, or we could simply multiply equation (4) by $\frac{11}{3}$ and add it to equation (5).

$$\begin{array}{rcll} 66y - 11z & = & -22 & \frac{11}{3} \times (4) \\ -16y + 11z & = & 22 & (5) \\ \hline 50y & = & 0 & \\ y & = & 0 & \end{array}$$

Substitute 0 for y in equation (5).

Copyright © 2012 Pearson Education, Inc. Publishing as Addison-Wesley.

$$-16y + 11z = 22 \quad (5)$$
$$-16(0) + 11z = 22$$
$$11z = 22$$
$$z = 2$$

Substitute 0 for y and 2 for z in equation (1).

$$x + 5y - 2z = -1 \quad (1)$$
$$x + 5(0) - 2(2) = -1$$
$$x - 4 = -1$$
$$x = 3$$

The solution set is $\{(3, 0, 2)\}$.

37.
$$2x + y - z = 6 \quad (1)$$
$$4x + 2y - 2z = 12 \quad (2)$$
$$-x - \tfrac{1}{2}y + \tfrac{1}{2}z = -3 \quad (3)$$

Multiplying equation (1) by 2 gives equation (2). Multiplying equation (3) by -4 also gives equation (2). The resulting equations are the same, so the three equations are dependent.

The solution set is $\{(x, y, z) \mid 2x + y - z = 6\}$.

39.
$$x + y - 2z = 0 \quad (1)$$
$$3x - y + z = 0 \quad (2)$$
$$4x + 2y - z = 0 \quad (3)$$

Eliminate z by adding equations (2) and (3).

$$3x - y + z = 0 \quad (2)$$
$$\underline{4x + 2y - z = 0} \quad (3)$$
$$7x + y \qquad = 0 \quad (4)$$

To get another equation without z, multiply equation (2) by 2 and add the result to equation (1).

$$6x - 2y + 2z = 0 \qquad 2 \times (2)$$
$$\underline{x + y - 2z = 0} \quad (1)$$
$$7x - y \qquad = 0 \quad (5)$$

Add equations (4) and (5) to find x.

$$7x + y = 0 \quad (4)$$
$$\underline{7x - y = 0} \quad (5)$$
$$14x \qquad = 0$$
$$x = 0$$

Substitute 0 for x in equation (4) to find y.

$$7x + y = 0 \quad (4)$$
$$7(0) + y = 0$$
$$0 + y = 0$$
$$y = 0$$

Substitute 0 for x and 0 for y in equation (1) to find z.

$$x + y - 2z = 0 \quad (1)$$
$$0 + 0 - 2z = 0$$
$$-2z = 0$$
$$z = 0$$

The solution set is $\{(0, 0, 0)\}$.

41.
$$x - 2y + \tfrac{1}{3}z = 4 \quad (1)$$
$$3x - 6y + z = 12 \quad (2)$$
$$-6x + 12y - 2z = -3 \quad (3)$$

The coefficients of z are $\tfrac{1}{3}$, 1, and -2. We'll multiply the equations by values that will eliminate fractions and make the coefficients of z easy to compare.

$$-6x + 12y - 2z = -24 \quad (4) \qquad -6 \times (1)$$
$$-6x + 12y - 2z = -24 \quad (5) \qquad -2 \times (2)$$
$$-6x + 12y - 2z = -3 \quad (3)$$

We can now easily see that (4) and (5) are dependent equations (they have the same graph—in fact, their graph is the same plane). Equation (3) has the same coefficients, but a different constant term, so its graph is a plane parallel to the other plane—that is, there are no points in common. Thus, the system is inconsistent and the solution set is $\emptyset$.

43.
$$x + y + z - w = 5 \quad (1)$$
$$2x + y - z + w = 3 \quad (2)$$
$$x - 2y + 3z + w = 18 \quad (3)$$
$$-x - y + z + 2w = 8 \quad (4)$$

Eliminate w. Add equations (1) and (2).

$$x + y + z - w = 5 \quad (1)$$
$$\underline{2x + y - z + w = 3} \quad (2)$$
$$3x + 2y \qquad = 8 \quad (5)$$

Eliminate w again. Add equations (1) and (3).

$$x + y + z - w = 5 \quad (1)$$
$$\underline{x - 2y + 3z + w = 18} \quad (3)$$
$$2x - y + 4z \qquad = 23 \quad (6)$$

Eliminate w again. Multiply equation (2) by -2. Add the result to equation (4).

$$-4x - 2y + 2z - 2w = -6 \qquad -2 \times (2)$$
$$\underline{-x - y + z + 2w = 8} \quad (4)$$
$$-5x - 3y + 3z \qquad = 2 \quad (7)$$

Equations (5), (6), and (7) do not contain a w-term. Since (5) does not have a z-term, we will find another equation without a z-term.

Eliminate z. Multiply equation (6) by 3 and equation (7) by -4. Then add the results.

$$6x - 3y + 12z = 69 \qquad 3 \times (6)$$
$$\underline{20x + 12y - 12z = -8} \qquad -4 \times (7)$$
$$26x + 9y \qquad = 61 \quad (8)$$

Copyright © 2012 Pearson Education, Inc. Publishing as Addison-Wesley.

Eliminate y. Multiply equation (5) by 9 and equation (8) by -2. Then add the results.

$$
\begin{array}{rlr}
27x + 18y &= 72 & 9 \times (5) \\
-52x - 18y &= -122 & -2 \times (8) \\
\hline
-25x &= -50 & \\
x &= 2 &
\end{array}
$$

To find y, substitute $x = 2$ into equation (5).

$$
\begin{aligned}
3x + 2y &= 8 \qquad (5) \\
3(2) + 2y &= 8 \\
2y &= 2 \\
y &= 1
\end{aligned}
$$

To find z, substitute $x = 2$ and $y = 1$ into equation (6).

$$
\begin{aligned}
2x - y + 4z &= 23 \qquad (6) \\
2(2) - 1 + 4z &= 23 \\
4z &= 20 \\
z &= 5
\end{aligned}
$$

To find w, substitute $x = 2$, $y = 1$, and $z = 5$ into equation (1).

$$
\begin{aligned}
x + y + z - w &= 5 \qquad (1) \\
2 + 1 + 5 - w &= 5 \\
-w &= -3 \\
w &= 3
\end{aligned}
$$

The solution set is $\{(2, 1, 5, 3)\}$.

45.
$$
\begin{array}{rlr}
3x + y - z + w &= -3 & (1) \\
2x + 4y + z - w &= -7 & (2) \\
-2x + 3y - 5z + w &= 3 & (3) \\
5x + 4y - 5z + 2w &= -7 & (4)
\end{array}
$$

Eliminate w. Add equations (1) and (2).

$$
\begin{array}{rlr}
3x + y - z + w &= -3 & (1) \\
2x + 4y + z - w &= -7 & (2) \\
\hline
5x + 5y &= -10 & (5)
\end{array}
$$

Eliminate w again. Add equations (2) and (3).

$$
\begin{array}{rlr}
2x + 4y + z - w &= -7 & (2) \\
-2x + 3y - 5z + w &= 3 & (3) \\
\hline
7y - 4z &= -4 & (6)
\end{array}
$$

Eliminate w again. Multiply equation (2) by 2. Add the result to equation (4).

$$
\begin{array}{rlr}
4x + 8y + 2z - 2w &= -14 & 2 \times (2) \\
5x + 4y - 5z + 2w &= -7 & (4) \\
\hline
9x + 12y - 3z &= -21 & (7)
\end{array}
$$

Equations (5), (6), and (7) do not contain a w-term. Since (5) does not have a z-term, we will find another equation without a z-term.

Eliminate z. Multiply equation (6) by 3 and equation (7) by -4. Then add the results.

$$
\begin{array}{rlr}
21y - 12z &= -12 & 3 \times (6) \\
-36x - 48y + 12z &= 84 & -4 \times (7) \\
\hline
-36x - 27y &= 72 & (8)
\end{array}
$$

Eliminate y using (5) and (8).

$$
\begin{array}{rlr}
3x + 3y &= -6 & \frac{3}{5} \times (5) \\
-4x - 3y &= 8 & (8) \div 9 \\
\hline
-x &= 2 & \\
x &= -2 &
\end{array}
$$

Substitute -2 for x in (5).

$$
\begin{aligned}
5(-2) + 5y &= -10 \\
5y &= 0 \\
y &= 0
\end{aligned}
$$

Substitute 0 for y in (6).

$$
\begin{aligned}
7(0) - 4z &= -4 \\
-4z &= -4 \\
z &= 1
\end{aligned}
$$

Substitute -2 for x, 0 for y, and 1 for z in (1).

$$
\begin{aligned}
3(-2) + (0) - (1) + w &= -3 \\
-6 - 1 + w &= -3 \\
w &= 4
\end{aligned}
$$

The solution set is $\{(-2, 0, 1, 4)\}$.

47. Let $x =$ the length of the longest side.
Then $\frac{5}{6}x =$ the length of the shortest side,
and $x - 17 =$ the length of the medium side.

The perimeter is 323, so

$$
\begin{aligned}
x + \tfrac{5}{6}x + (x - 17) &= 323. \\
6x + 5x + 6(x - 17) &= 6(323) \qquad \textit{Multiply by 6.} \\
6x + 5x + 6x - 102 &= 1938 \\
17x &= 2040 \\
x &= 120
\end{aligned}
$$

Since $x = 120$, $\frac{5}{6}x = 100$, and $x - 17 = 103$. The lengths of the sides are 100 inches, 103 inches, and 120 inches.

49. Let $x =$ the least number.
Then $-3x =$ the greatest number,
and $-3x - 4 =$ the middle number.

The sum is 16, so

$$
\begin{aligned}
x + (-3x) + (-3x - 4) &= 16. \\
-5x - 4 &= 16 \\
-5x &= 20 \\
x &= -4
\end{aligned}
$$

Since $x = -4$, $-3x = 12$, and $-3x - 4 = 8$. The three numbers are -4, 8, and 12.

4.3   Applications of Systems of Linear Equations

4.3   Now Try Exercises

N1. *Step 2*
Let W = the width of the field
and L = the length of the field.

Step 3
Since the length is 10 feet more than twice its width,

$$L = 2W + 10. \quad (1)$$

The perimeter of a rectangle is given by

$$2W + 2L = P.$$

With perimeter $P = 620$ feet,

$$2W + 2L = 620. \quad (2)$$

Step 4
Substitute $2W + 10$ for L in equation (2).

$$2W + 2(2W + 10) = 620$$
$$2W + 4W + 20 = 620$$
$$6W = 600$$
$$W = 100$$

Substitute $W = 100$ into equation (1).

$$L = 2W + 10 = 200 + 10 = 210$$

Step 5
The length of the field is 210 feet and the width is 100 feet.

Step 6
Check   210 is 10 more than 2 times 100 and $2(210) + 2(100) = 620$, as required.

N2. *Step 2*
Let x = the cost of a general admission ticket, and y = the cost of a ticket for children under 48 inches tall.

Step 3
From the given information,

$$2x + 3y = 172.98 \quad (1)$$
$$x + 4y = 163.99. \quad (2)$$

Step 4
Multiply equation (2) by -2 and add to equation (1).

$$
\begin{array}{rl}
2x + 3y = & 172.98 \quad (1) \\
-2x - 8y = & -327.98 \quad\quad -2 \times (2) \\
\hline
- 5y = & -155.00 \\
y = & 31.00
\end{array}
$$

Let $y = 31.00$ in equation (2).

$$x + 4(31.00) = 163.99$$
$$x + 124.00 = 163.99$$
$$x = 39.99$$

Step 5
The cost of a general admission ticket is \$39.99 and the cost of a ticket for children under 48 inches tall is \$31.00.

Step 6
Check $2(39.99) + 3(31.00) = 172.98$
and $39.99 + 4(31.00) = 163.99$.

N3. *Step 2*
Let x = the amount of 15% acid solution and y = the amount of 25% acid solution.

Liters of Solution	Percent (as a decimal)	Liters of Pure Acid
x	$15\% = 0.15$	$0.15x$
y	$25\% = 0.25$	$0.25y$
30	$18\% = 0.18$	$0.18(30) = 5.4$

Step 3
Write the system from the columns.

$$x + y = 30 \quad (1)$$
$$0.15x + 0.25y = 5.4 \quad (2)$$

Step 4
Multiply (2) by 100 to clear the equation of decimals, and multiply (1) by -15.

$$
\begin{array}{rll}
-15x - 15y = & -450 & -15 \times (1) \\
15x + 25y = & 540 & 100 \times (2) \\
\hline
10y = & 90 & \\
y = & 9 &
\end{array}
$$

Substitute 9 for y in (1).

$$x + 9 = 30$$
$$x = 30 - 9 = 21$$

Step 5
21 L of 15% and 9 L of 25% should be mixed.

Step 6
Check $21 + 9 = 30$ and
$0.15(21) + 0.25(9) = 5.4$.

N4. *Step 2*
Let x = Vann's rate and y = Ivy's rate.

Make a table. Use $t = \frac{d}{r}$.

	Distance	Rate	Time
Vann	50	x	$\frac{50}{x}$
Ivy	40	y	$\frac{40}{y}$

Step 3
Since the times are the same,

Copyright © 2012 Pearson Education, Inc. Publishing as Addison-Wesley.

$$\frac{50}{x} = \frac{40}{y}$$
$$50y = 40x \quad \text{\textit{Multiply by xy.}}$$
$$-40x + 50y = 0. \quad (1)$$

Since Vann's rate, x, is 2 mph faster than Ivy's rate, y,

$$x = y + 2. \quad (2)$$

Step 4
Substitute $y + 2$ for x in equation (1) to find y.

$$-40x + 50y = 0 \quad (1)$$
$$-40(y + 2) + 50y = 0$$
$$-40y - 80 + 50y = 0$$
$$10y = 80$$
$$y = 8$$

Since $y = 8$ and $x = y + 2$, $x = 10$.

Step 5
Vann's rate is 10 mph, and Ivy's rate is 8 mph.

Step 6
Check 10 mph is 2 mph faster than 8 mph. It would take Vann 5 hours to travel 50 miles at 10 mph, which is the same amount of time it would take Ivy to travel 40 miles at 8 mph.

N5. Let x = the number of bottles of $2.00 Gatorade sold, y = the number of $1.50 pretzels sold, and z = the number of $1.00 candy items sold.

The number of pretzels was four times the number of candy items, so

$$y = 4z. \quad (1)$$

The number of Gatorades was 310 more than the number of pretzels, so

$$x = 310 + y. \quad (2)$$

Sales for these items totaled $1520, so

$$2x + 1.5y + 1z = 1520. \quad (3)$$

Substitute $310 + y$ for x in (3).

$$2(310 + y) + 1.5y + 1z = 1520 \quad (4)$$

Now substitute $4z$ for y in (4).

$$2(310 + 4z) + 1.5(4z) + 1z = 1520$$
$$620 + 8z + 6z + 1z = 1520$$
$$15z = 900$$
$$z = 60$$

From (1), $y = 4(60) = 240$.
From (2), $x = 310 + 240 = 550$.
Workers sold 60 candy items, 240 pretzels, and 550 bottles of Gatorade.

N6. Let x = the number of lone star quilts, y = the number of bandana quilts, and z = the number of log cabin quilts.

The time allocated for piecework is 74 hours, so

$$8x + 2y + 10z = 74. \quad (1)$$

The time allocated for machine quilting is 42 hours, so

$$4x + 2y + 5z = 42. \quad (2)$$

The time allocated for finishing is 24 hours, so

$$2x + 2y + 2z = 24, \text{ or equivalently,}$$
$$x + y + z = 12. \quad (3)$$

Solve the system of equations (1), (2), and (3). To eliminate z, multiply equation (3) by -10 and add the result to equation (1).

$$
\begin{array}{rrrrll}
-10x & - & 10y & - & 10z & = & -120 & \quad -10 \times (3) \\
8x & + & 2y & + & 10z & = & 74 & \quad (1) \\
\hline
-2x & - & 8y & & & = & -46 &
\end{array}
$$

Divide the last equation by -2 to get

$$x + 4y = 23. \quad (4)$$

To eliminate z again, multiply equation (3) by -5 and add the result to equation (2).

$$
\begin{array}{rrrrll}
-5x & - & 5y & - & 5z & = & -60 & \quad -5 \times (3) \\
4x & + & 2y & + & 5z & = & 42 & \quad (2) \\
\hline
-x & - & 3y & & & = & -18 & \quad (5)
\end{array}
$$

Add equations (4) and (5) to eliminate y.

$$
\begin{array}{rrrll}
x & + & 4y & = & 23 & \quad (4) \\
-x & - & 3y & = & -18 & \quad (5) \\
\hline
& & y & = & 5 &
\end{array}
$$

Substitute 5 for y in equation (4).

$$x + 4y = 23 \quad (4)$$
$$x + 4(5) = 23$$
$$x + 20 = 23$$
$$x = 3$$

Substitute 3 for x and 5 for y in equation (3).

$$x + y + z = 12 \quad (3)$$
$$3 + 5 + z = 12$$
$$8 + z = 12$$
$$z = 4$$

Each month, the quilting shop should make 3 lone star quilts, 5 bandana quilts, and 4 log cabin quilts.

Copyright © 2012 Pearson Education, Inc. Publishing as Addison-Wesley.

4.3 Section Exercises

1. *Step 2*

Let $x =$ the number of games that the Dodgers won and let $y =$ the number of games that they lost.

Step 3

They played 162 games, so

$$x + y = 162. \quad (1)$$

They won 28 more games than they lost, so

$$x = 28 + y. \quad (2)$$

Step 4

Substitute $28 + y$ for x in (1).

$$(28 + y) + y = 162$$
$$28 + 2y = 162$$
$$2y = 134$$
$$y = 67$$

Substitute 67 for y in (2).

$$x = 28 + 67 = 95$$

Step 5

The Dodgers win-loss record was 95 wins and 67 losses.

Step 6

95 is 28 more than 67 and the sum of 95 and 67 is 162.

3. Let $W =$ the width of the tennis court and $L =$ the length of the court.

Since the length is 42 ft more than the width,

$$L = W + 42. \quad (1)$$

The perimeter of a rectangle is given by

$$2W + 2L = P.$$

With perimeter $P = 228$ ft,

$$2W + 2L = 228. \quad (2)$$

Substitute $W + 42$ for L in equation (2).

$$2W + 2(W + 42) = 228$$
$$2W + 2W + 84 = 228$$
$$4W = 144$$
$$W = 36$$

Substitute $W = 36$ into equation (1).

$$L = W + 42 \quad (1)$$
$$L = 36 + 42$$
$$= 78$$

The length is 78 ft and the width is 36 ft.

5. Let $x =$ the revenue for Verizon and $y =$ the revenue for AT&T (both in billions of dollars).

The total revenue was $221.4 billion.

$$x + y = 221.4 \quad (1)$$

AT&T's revenue was $26.6 billion more than that of Verizon.

$$y = x + 26.6 \quad (2)$$

Substitute $x + 26.6$ for y in equation (1).

$$x + (x + 26.6) = 221.4$$
$$2x + 26.6 = 221.4$$
$$2x = 194.8$$
$$x = 97.4$$

Substitute 97.4 for x in equation (2).

$$y = x + 26.6 = 97.4 + 26.6 = 124.0$$

Verizon's revenue was $97.4 billion and AT&T's revenue was $124.0 billion.

7. From the figure in the text, the angles marked y and $3x + 10$ are supplementary, so

$$(3x + 10) + y = 180. \quad (1)$$

Also, the angles x and y are complementary, so

$$x + y = 90. \quad (2)$$

Solve equation (2) for y to get

$$y = 90 - x. \quad (3)$$

Substitute $90 - x$ for y in equation (1).

$$(3x + 10) + (90 - x) = 180$$
$$2x + 100 = 180$$
$$2x = 80$$
$$x = 40$$

Substitute $x = 40$ into equation (3) to get

$$y = 90 - x = 90 - 40 = 50.$$

The angles measure 40° and 50°.

9. Let $x =$ the hockey FCI and $y =$ the basketball FCI.

The sum is $580.16, so

$$x + y = 580.16. \quad (1)$$

The hockey FCI was $3.70 less than the basketball FCI, so

$$x = y - 3.70. \quad (2)$$

From (2), substitute $y - 3.70$ for x in (1).

$$(y - 3.70) + y = 580.16$$
$$2y - 3.70 = 580.16$$
$$2y = 583.86$$
$$y = 291.93$$

From (2),

$$x = y - 3.70 = 291.93 - 3.70 = 288.23.$$

The hockey FCI was $288.23 and the basketball FCI was $291.93.

11. Let x = the cost of a single Junior Roast Beef sandwich, and y = the cost of a single Big Montana sandwich.

15 Junior Roast Beef sandwiches and 10 Big Montana sandwiches cost $75.25.

$$15x + 10y = 75.25 \quad (1)$$

30 Junior Roast Beef sandwiches and 5 Big Montana sandwiches cost $84.65.

$$30x + 5y = 84.65 \quad (2)$$

Multiply equation (1) by -2 and add to equation (2).

$$
\begin{array}{rcl}
-30x - 20y &=& -150.50 \qquad -2 \times (1) \\
30x + 5y &=& 84.65 \qquad (2) \\
\hline
-15y &=& -65.85 \\
y &=& 4.39
\end{array}
$$

Substitute 4.39 for y in equation (1).

$$15x + 10y = 75.25$$
$$15x + 10(4.39) = 75.25$$
$$15x + 43.90 = 75.25$$
$$15x = 31.35$$
$$x = 2.09$$

A single Junior Roast Beef sandwich costs $2.09 and a single Big Montana sandwich costs $4.39.

13. Use the formula (rate of percent) • (base amount) = amount (percentage) of pure acid to compute parts (a) – (d).

(a) $0.10(60) = 6$ oz

(b) $0.25(60) = 15$ oz

(c) $0.40(60) = 24$ oz

(d) $0.50(60) = 30$ oz

15. The cost is the price per pound, $2.29, times the number of pounds, x, or $2.29x$.

17. Let x = the amount of 25% alcohol solution, and y = the amount of 35% alcohol solution.

Make a table. The percent times the amount of solution gives the amount of pure alcohol in the third column.

Gallons of Solution	Percent (as a decimal)	Gallons of Pure Alcohol
x	25% = 0.25	$0.25x$
y	35% = 0.35	$0.35y$
20	32% = 0.32	$0.32(20) = 6.4$

The third row gives the total amounts of solution and pure alcohol. From the columns in the table, write a system of equations.

$$x + y = 20 \quad (1)$$
$$0.25x + 0.35y = 6.4 \quad (2)$$

Solve the system. Multiply equation (1) by -25 and equation (2) by 100. Then add the results.

$$
\begin{array}{rcll}
-25x - 25y &=& -500 & -25 \times (1) \\
25x + 35y &=& 640 & 100 \times (2) \\
\hline
10y &=& 140 \\
y &=& 14
\end{array}
$$

Substitute $y = 14$ into equation (1).

$$x + y = 20 \quad (1)$$
$$x + 14 = 20$$
$$x = 6$$

Mix 6 gal of 25% solution and 14 gal of 35% solution.

19. Let x = the amount of pure acid and y = the amount of 10% acid.

Make a table.

Liters of Solution	Percent (as a decimal)	Liters of Pure Acid
x	100% = 1	$1.00x = x$
y	10% = 0.10	$0.10y$
54	20% = 0.20	$0.20(54) = 10.8$

Solve the following system.

$$x + y = 54 \quad (1)$$
$$x + 0.10y = 10.8 \quad (2)$$

Multiply equation (2) by 10 to clear the decimals.

$$10x + y = 108 \quad (3)$$

To eliminate y, multiply equation (1) by -1 and add the result to equation (3).

$$
\begin{array}{rcll}
-x - y &=& -54 & -1 \times (1) \\
10x + y &=& 108 & (3) \\
\hline
9x &=& 54 \\
x &=& 6
\end{array}
$$

Since $x = 6$,

$$x + y = 54 \quad (1)$$
$$6 + y = 54$$
$$y = 48.$$

Use 6 L of pure acid and 48 L of 10% acid.

Copyright © 2012 Pearson Education, Inc. Publishing as Addison-Wesley.

21. Complete the table.

	Number of Kilograms	Price per Kilogram	Value
Nuts	x	2.50	$2.50x$
Cereal	y	1.00	$1.00y$
Mixture	30	1.70	$1.70(30) = 51$

From the "Number of Kilograms" column,

$$x + y = 30. \quad (1)$$

From the "Value" column,

$$2.50x + 1.00y = 51. \quad (2)$$

Solve the system.

$$
\begin{array}{rl}
-10x - 10y = -300 & -10 \times (1) \\
\underline{25x + 10y = 510} & 10 \times (2) \\
15x = 210 & \\
x = 14 &
\end{array}
$$

From (1), $14 + y = 30$, so $y = 16$.
The party mix should be made from 14 kg of nuts and 16 kg of cereal.

23. From the "Principal" column in the text,

$$x + y = 3000. \quad (1)$$

From the "Interest" column in the text,

$$0.02x + 0.04y = 100. \quad (2)$$

Multiply equation (2) by 100 to clear the decimals.

$$2x + 4y = 10,000 \quad (3)$$

To eliminate x, multiply equation (1) by -2 and add the result to equation (3).

$$
\begin{array}{rl}
-2x - 2y = -6000 & -2 \times (1) \\
\underline{2x + 4y = 10,000} & (3) \\
2y = 4000 & \\
y = 2000 &
\end{array}
$$

From (1), $x + 2000 = 3000$, so $x = 1000$.
$1000 is invested at 2%, and $2000 is invested at 4%.

25. (a) The rate of the boat going upstream is *decreased* by the rate of the current, so it is $(10 - x)$ mph.

(b) The rate of the boat going downstream is *increased* by the rate of the current, so it is $(10 + x)$ mph.

27. Let x = the rate of the train
and y = the rate of the plane.

	r	t	d
Train	x	$\frac{150}{x}$	150
Plane	y	$\frac{400}{y}$	400

The times are equal, so

$$\frac{150}{x} = \frac{400}{y}. \quad (1)$$

The rate of the plane is 20 km per hour less than 3 times the rate of the train, so

$$3x - 20 = y. \quad (2)$$

Multiply (1) by xy (or use cross products).

$$150y = 400x \quad (3)$$

From (2), substitute $3x - 20$ for y in (3).

$$
\begin{array}{l}
150(3x - 20) = 400x \\
450x - 3000 = 400x \\
50x = 3000 \\
x = 60
\end{array}
$$

From (2), $y = 3(60) - 20 = 160$.
The rate of the train is 60 km/hr, and the rate of the plane is 160 km/hr.

29. Let x = the rate of the boat in still water
and y = the rate of the current.

Furthermore,

$$\text{rate upstream} = x - y$$

and $\text{rate downstream} = x + y$.

Use these rates and the information in the problem to make a table.

	r	t	d
Upstream	$x - y$	2	36
Downstream	$x + y$	1.5	36

From the table, use the formula $d = rt$ to write a system of equations.

$$
\begin{array}{l}
36 = 2(x - y) \\
36 = 1.5(x + y)
\end{array}
$$

Remove the parentheses and move the variables to the left side.

$$
\begin{array}{ll}
2x - 2y = 36 & (1) \\
1.5x + 1.5y = 36 & (2)
\end{array}
$$

Solve the system. Multiply equation (1) by -3 and equation (2) by 4. Then add the results.

$$
\begin{array}{rl}
-6x + 6y = -108 & -3 \times (1) \\
\underline{6x + 6y = 144} & 4 \times (2) \\
12y = 36 & \\
y = 3 &
\end{array}
$$

Substitute $y = 3$ into equation (1).

$$
\begin{array}{ll}
2x - 2y = 36 & (1) \\
2x - 2(3) = 36 & \\
2x = 42 & \\
x = 21 &
\end{array}
$$

The rate of the boat is 21 mph, and the rate of the current is 3 mph.

Copyright © 2012 Pearson Education, Inc. Publishing as Addison-Wesley.

31. Let $x =$ the number of pounds of the \$0.75-per-lb candy and $y =$ the number of pounds of the \$1.25-per-lb candy.

Make a table.

	Price per Pound	Number of Pounds	Value
Less Expensive Candy	\$0.75	x	\$0.75x
More Expensive Candy	\$1.25	y	\$1.25y
Mixture	\$0.96	9	\$0.96(9) = \$8.64

From the "Number of Pounds" column,

$$x + y = 9. \quad (1)$$

From the "Value" column,

$$0.75x + 1.25y = 8.64. \quad (2)$$

Solve the system.

$$
\begin{array}{rcll}
-75x - 75y &=& -675 & -75 \times (1) \\
75x + 125y &=& 864 & 100 \times (2) \\
\hline
50y &=& 189 & \\
y &=& 3.78 &
\end{array}
$$

From (1), $x + 3.78 = 9$, so $x = 5.22$. Mix 5.22 pounds of the \$0.75-per-lb candy with 3.78 pounds of the \$1.25-per-lb candy to obtain 9 pounds of a mixture that sells for \$0.96 per pound.

33. Let $x =$ the number of general admission tickets and $y =$ the number of student tickets.

Make a table.

Ticket	Number	Value of Tickets
General	x	$5 \cdot x = 5x$
Student	y	$4 \cdot y = 4y$
Totals	184	812

Solve the system.

$$
\begin{array}{rl}
x + y = 184 & (1) \\
5x + 4y = 812 & (2)
\end{array}
$$

To eliminate y, multiply equation (1) by -4 and add the result to equation (2).

$$
\begin{array}{rcll}
-4x - 4y &=& -736 & -4 \times (1) \\
5x + 4y &=& 812 & (2) \\
\hline
x &=& 76 &
\end{array}
$$

From (1), $76 + y = 184$, so $y = 108$.

76 general admission tickets and 108 student tickets were sold.

35. Let $x =$ the price for a citron and let $y =$ the price for a wood apple.

"9 citrons and 7 fragrant wood apples is 107" gives us

$$9x + 7y = 107. \quad (1)$$

"7 citrons and 9 fragrant wood apples is 101" gives us

$$7x + 9y = 101. \quad (2)$$

Multiply equation (1) by -7 and equation (2) by 9. Then add.

$$
\begin{array}{rcll}
-63x - 49y &=& -749 & -7 \times (1) \\
63x + 81y &=& 909 & 9 \times (2) \\
\hline
32y &=& 160 & \\
y &=& 5 &
\end{array}
$$

Substitute 5 for y in equation (1).

$$
\begin{array}{rcl}
9x + 7(5) &=& 107 \\
9x + 35 &=& 107 \\
9x &=& 72 \\
x &=& 8
\end{array}
$$

The prices are 8 for a citron and 5 for a wood apple.

37. Let $x =$ the measure of one angle, $y =$ the measure of another angle, and $z =$ the measure of the last angle.

Two equations are given, so

$$
\begin{array}{rl}
& z = x + 10 \\
\text{or} \quad -x + z = 10 & (1) \\
\text{and} \quad x + y = 100. & (2)
\end{array}
$$

Since the sum of the measures of the angles of a triangle is 180°, the third equation of the system is

$$x + y + z = 180. \quad (3)$$

Equation (1) is missing y. To eliminate y again, multiply equation (2) by -1 and add the result to equation (3).

$$
\begin{array}{rcll}
-x - y &=& -100 & -1 \times (2) \\
x + y + z &=& 180 & (3) \\
\hline
z &=& 80 &
\end{array}
$$

Since $z = 80$,

$$
\begin{array}{rcll}
-x + z &=& 10 & (1) \\
-x + 80 &=& 10 & \\
-x &=& -70 & \\
x &=& 70. &
\end{array}
$$

From (2), $70 + y = 100$, so $y = 30$. The measures of the angles are 70°, 30°, and 80°.

Copyright © 2012 Pearson Education, Inc. Publishing as Addison-Wesley.

39. Let $x =$ the measure of the first angle,
  $y =$ the measure of the second angle, and
  $z =$ the measure of the third angle.

The sum of the angles in a triangle equals $180°$, so

$$x + y + z = 180. \quad (1)$$

The measure of the second angle is $10°$ greater than 3 times that of the first angle, so

$$y = 3x + 10. \quad (2)$$

The third angle is equal to the sum of the other two, so

$$z = x + y. \quad (3)$$

Solve the system. Substitute z for $x + y$ in equation (1).

$$(x + y) + z = 180 \quad (1)$$
$$z + z = 180$$
$$2z = 180$$
$$z = 90$$

Substitute 90 for z and $3x + 10$ for y in equation (3).

$$z = x + y \quad (3)$$
$$90 = x + (3x + 10)$$
$$80 = 4x$$
$$20 = x$$

Substitute $x = 20$ and $z = 90$ into equation (3).

$$z = x + y \quad (3)$$
$$90 = 20 + y$$
$$70 = y$$

The three angles have measures of $20°$, $70°$, and $90°$.

41. Let $x =$ the length of the longest side,
  $y =$ the length of the middle side,
  and $z =$ the length of the shortest side.

Perimeter is the sum of the measures of the sides, so

$$x + y + z = 70. \quad (1)$$

The longest side is 4 cm less than the sum of the other sides, so

$$x = y + z - 4$$
or $x - y - z = -4.$ $\quad (2)$

Twice the shortest side is 9 cm less than the longest side, so

$$2z = x - 9$$
or $-x + 2z = -9$ $\quad (3)$

Add equations (1) and (2) to eliminate y and z.

$$\begin{array}{rcl} x + y + z &=& 70 \quad (1) \\ x - y - z &=& -4 \quad (2) \\ \hline 2x &=& 66 \\ x &=& 33 \end{array}$$

Substitute 33 for x in (3).

$$-x + 2z = -9 \quad (3)$$
$$-33 + 2z = -9$$
$$2z = 24$$
$$z = 12$$

Substitute 33 for x and 12 for z in (1).

$$x + y + z = 70 \quad (1)$$
$$33 + y + 12 = 70$$
$$y + 45 = 70$$
$$y = 25$$

The shortest side is 12 cm long, the middle side is 25 cm long, and the longest side is 33 cm long.

43. Let $x =$ the number of gold medals, $y =$ the number of silver medals, and $z =$ the number of bronze medals.

Russia earned 5 fewer gold medals than bronze, so

$$x = z - 5. \quad (1)$$

The number of silver medals earned was 35 less than twice the number of bronze medals, so

$$y = 2z - 35. \quad (2)$$

The total number of medals earned was 72, so

$$x + y + z = 72. \quad (3)$$

Substitute $z - 5$ for x and $2z - 35$ for y in (3).

$$(z - 5) + (2z - 35) + z = 72$$
$$4z - 40 = 72$$
$$4z = 112$$
$$z = 28$$

From (1), $x = 28 - 5 = 23.$
From (2), $y = 2(28) - 35 = 56 - 35 = 21.$

Russia earned 23 gold, 21 silver, and 28 bronze medals.

45. Let $x =$ the number of \$16 tickets, $y =$ the number of \$23 tickets, and $z =$ the number of VIP \$40 tickets.

Nine times as many \$16 tickets have been sold as VIP tickets, so

$$x = 9z. \quad (1)$$

The number of \$16 tickets was 55 more than the sum of the number of \$23 tickets and the number of VIP tickets, so

$$x = 55 + y + z. \quad (2)$$

Copyright © 2012 Pearson Education, Inc. Publishing as Addison-Wesley.

Sales of these tickets totaled $46,575.

$$16x + 23y + 40z = 46{,}575 \quad (3)$$

Since x is in terms of z in (1), we'll substitute $9z$ for x in (2) and then get y in terms of z.

$$9z = 55 + y + z$$
$$8z - 55 = y \quad (4)$$

Substitute $9z$ for x and $8z - 55$ for y in (3).

$$16(9z) + 23(8z - 55) + 40z = 46{,}575$$
$$144z + 184z - 1265 + 40z = 46{,}575$$
$$368z = 47{,}840$$
$$z = 130$$

From (1), $x = 9(130) = 1170$.
From (4), $y = 8(130) - 55 = 985$.

There were 1170 $16 tickets, 985 $23 tickets, and 130 $40 tickets sold.

47. Let $x =$ the number of T-shirts shipped
to bookstore A,
$y =$ the number of T-shirts shipped
to bookstore B,
and $z =$ the number of T-shirts shipped
to bookstore C.

Twice as many T-shirts were shipped to bookstore B as to bookstore A, so

$$y = 2x. \quad (1)$$

The number shipped to bookstore C was 40 less than the sum of the numbers shipped to the other two bookstores, so

$$z = x + y - 40. \quad (2)$$

The total number of T-shirts shipped was 800, so

$$x + y + z = 800. \quad (3)$$

Substitute $2x$ for y [from (1)] into equation (2) to get z in terms of x.

$$z = x + y - 40 \quad (2)$$
$$z = x + (2x) - 40$$
$$z = 3x - 40 \quad (4)$$

Substitute $2x$ for y and $3x - 40$ for z in (3).

$$x + (2x) + (3x - 40) = 800$$
$$6x - 40 = 800$$
$$6x = 840$$
$$x = 140$$

From (1), $y = 2(140) = 280$.
From (4), $z = 3(140) - 40 = 380$.
The number of T-shirts shipped to bookstores A, B, and C was 140, 280, and 380, respectively.

49. Let x, y, and z denote the number of kilograms of the first, second, and third chemicals, respectively. The mix must include 60% of the first and second chemicals, so

$$x + y = 0.60(750) = 450. \quad (1)$$

The second and third chemicals must be in a ratio of 4 to 3 by weight, so

$$y = \tfrac{4}{3}z. \quad (2)$$

The total is 750, so

$$x + y + z = 750. \quad (3)$$

From (1), $x = 450 - y$. $\quad (4)$

From (2), $z = \tfrac{3}{4}y$. $\quad (5)$

Substitute $450 - y$ for x and $\tfrac{3}{4}y$ for z in (3).

$$(450 - y) + y + \tfrac{3}{4}y = 750$$
$$\tfrac{3}{4}y = 300$$
$$y = 400$$

From (4), $x = 450 - 400 = 50$.
From (5), $z = \tfrac{3}{4}(400) = 300$.

Use 50 kg of the first chemical, 400 kg of the second chemical, and 300 kg of the third chemical to make the plant food.

51. *Step 2*
Let $x =$ the number of wins,
$y =$ the number of losses in regulation play,
and $z =$ the number of overtime losses.

Step 3
They played 82 games, so

$$x + y + z = 82. \quad (1)$$

The points from their wins and overtime losses totaled 116, so

$$2x + z = 116. \quad (2)$$

They had 9 more losses in regulation play than overtime losses, so

$$y = z + 9. \quad (3)$$

Step 4
Solving equation (2) for x gives us $x = 58 - \tfrac{1}{2}z$. Now substitute for x and y in (1).

$$x + y + z = 82 \quad (1)$$
$$(58 - \tfrac{1}{2}z) + (z + 9) + z = 82$$
$$1.5z + 67 = 82$$
$$1.5z = 15$$
$$z = 10$$

Since $y = z + 9$, $y = 10 + 9 = 19$.
Since $x = 58 - \tfrac{1}{2}z$, $x = 58 - \tfrac{1}{2}(10) = 53$.

Copyright © 2012 Pearson Education, Inc. Publishing as Addison-Wesley.

Step 5
The Bruins won 53 games, lost 19 games, and had 10 overtime losses.

Step 6
Adding 53, 19, and 10 gives 82 total games. The points from their wins and overtime losses totaled $2(53) + 10 = 106 + 10 = 116$, and there were 9 more losses in regulation play than overtime losses. The solution is correct.

53. (a) The additive inverse of -6 is $-(-6) = 6$.

(b) The multiplicative inverse (reciprocal) of -6 is $\frac{1}{-6} = -\frac{1}{6}$.

55. (a) The additive inverse of $\frac{7}{8}$ is $-\frac{7}{8}$.

(b) The multiplicative inverse (reciprocal) of $\frac{7}{8}$ is $\frac{1}{7/8} = \frac{8}{7}$.

4.4 Solving Systems of Linear Equations by Matrix Methods

4.4 Now Try Exercises

N1. $x + 3y = 3$
$2x - 3y = -12$

Write the augmented matrix for this system.

$$\begin{bmatrix} 1 & 3 & | & 3 \\ 2 & -3 & | & -12 \end{bmatrix}$$
$$\begin{bmatrix} 1 & 3 & | & 3 \\ 0 & -9 & | & -18 \end{bmatrix} \quad -2R_1 + R_2$$
$$\begin{bmatrix} 1 & 3 & | & 3 \\ 0 & 1 & | & 2 \end{bmatrix} \quad -\frac{1}{9}R_2$$

This matrix gives the system

$$x + 3y = 3$$
$$y = 2.$$

Substitute 2 for y in the first equation.

$$x + 3(2) = 3$$
$$x + 6 = 3$$
$$x = -3$$

The solution set is $\{(-3, 2)\}$.

N2. $x + y - 2z = -5$
$-x + 2y + z = -1$
$2x - y + 3z = 14$

Write the augmented matrix for this system.

$$\begin{bmatrix} 1 & 1 & -2 & | & -5 \\ -1 & 2 & 1 & | & -1 \\ 2 & -1 & 3 & | & 14 \end{bmatrix}$$
$$\begin{bmatrix} 1 & 1 & -2 & | & -5 \\ 0 & 3 & -1 & | & -6 \\ 2 & -1 & 3 & | & 14 \end{bmatrix} \quad R_1 + R_2$$
$$\begin{bmatrix} 1 & 1 & -2 & | & -5 \\ 0 & 3 & -1 & | & -6 \\ 0 & -3 & 7 & | & 24 \end{bmatrix} \quad -2R_1 + R_3$$
$$\begin{bmatrix} 1 & 1 & -2 & | & -5 \\ 0 & 1 & -\frac{1}{3} & | & -2 \\ 0 & -3 & 7 & | & 24 \end{bmatrix} \quad \frac{1}{3}R_2$$
$$\begin{bmatrix} 1 & 1 & -2 & | & -5 \\ 0 & 1 & -\frac{1}{3} & | & -2 \\ 0 & 0 & 6 & | & 18 \end{bmatrix} \quad 3R_2 + R_3$$
$$\begin{bmatrix} 1 & 1 & -2 & | & -5 \\ 0 & 1 & -\frac{1}{3} & | & -2 \\ 0 & 0 & 1 & | & 3 \end{bmatrix} \quad \frac{1}{6}R_3$$

This matrix gives the system

$$x + y - 2z = -5$$
$$y - \tfrac{1}{3}z = -2$$
$$z = 3.$$

Substitute 3 for z in the second equation.

$$y - \tfrac{1}{3}(3) = -2$$
$$y = -1$$

Substitute -1 for y and 3 for z in the first equation.

$$x + (-1) - 2(3) = -5$$
$$x - 7 = -5$$
$$x = 2$$

The solution set is $\{(2, -1, 3)\}$.

N3. (a) $3x - y = 8$
$-6x + 2y = 4$

Write the augmented matrix for this system.

$$\begin{bmatrix} 3 & -1 & | & 8 \\ -6 & 2 & | & 4 \end{bmatrix}$$
$$\begin{bmatrix} 3 & -1 & | & 8 \\ 0 & 0 & | & 20 \end{bmatrix} \quad 2R_1 + R_2$$

This matrix gives the system

$$3x - y = 8$$
$$0 = 20. \quad \textit{False}$$

The false statement indicates that the system is inconsistent and has no solution.

The solution set is $\emptyset$.

Copyright © 2012 Pearson Education, Inc. Publishing as Addison-Wesley.

(b) $\quad x + 2y = 7$
$\qquad -x - 2y = -7$

Write the augmented matrix for this system.

$$\begin{bmatrix} 1 & 2 & | & 7 \\ -1 & -2 & | & -7 \end{bmatrix}$$
$$\begin{bmatrix} 1 & 2 & | & 7 \\ 0 & 0 & | & 0 \end{bmatrix} \qquad R_1 + R_2$$

This matrix gives the system

$$x + 2y = 7$$
$$0 = 0. \quad True$$

The true statement indicates that the system has dependent equations.

The solution set is $\{(x, y) \mid x + 2y = 7\}$.

4.4 Section Exercises

1. $\begin{bmatrix} -2 & 3 & 1 \\ 0 & 5 & -3 \\ 1 & 4 & 8 \end{bmatrix}$

(a) The elements of the second row are 0, 5, and -3.

(b) The elements of the third column are 1, -3, and 8.

(c) The matrix is square since the number of rows (three) is the same as the number of columns.

(d) The matrix obtained by interchanging the first and third rows is

$$\begin{bmatrix} 1 & 4 & 8 \\ 0 & 5 & -3 \\ -2 & 3 & 1 \end{bmatrix}.$$

(e) The matrix obtained by multiplying the first row by $-\frac{1}{2}$ is

$$\begin{bmatrix} -2(-\frac{1}{2}) & 3(-\frac{1}{2}) & 1(-\frac{1}{2}) \\ 0 & 5 & -3 \\ 1 & 4 & 8 \end{bmatrix} = \begin{bmatrix} 1 & -\frac{3}{2} & -\frac{1}{2} \\ 0 & 5 & -3 \\ 1 & 4 & 8 \end{bmatrix}.$$

(f) The matrix obtained by multiplying the third row by 3 and adding to the first row is

$$\begin{bmatrix} -2+3(1) & 3+3(4) & 1+3(8) \\ 0 & 5 & -3 \\ 1 & 4 & 8 \end{bmatrix} = \begin{bmatrix} 1 & 15 & 25 \\ 0 & 5 & -3 \\ 1 & 4 & 8 \end{bmatrix}.$$

3. There are 3 rows and 2 columns, so the dimensions are written as 3×2.

5. There are 4 rows and 2 columns, so the dimensions are written as 4×2.

7. $\quad x + y = 5$
$\qquad x - y = 3$

Write the augmented matrix for this system.

$$\begin{bmatrix} 1 & 1 & | & 5 \\ 1 & -1 & | & 3 \end{bmatrix}$$
$$\begin{bmatrix} 1 & 1 & | & 5 \\ 0 & -2 & | & -2 \end{bmatrix} \quad -1R_1 + R_2$$
$$\begin{bmatrix} 1 & 1 & | & 5 \\ 0 & 1 & | & 1 \end{bmatrix} \qquad -\frac{1}{2}R_2$$

This matrix gives the system

$$x + y = 5$$
$$y = 1.$$

Substitute $y = 1$ in the first equation.

$$x + y = 5$$
$$x + 1 = 5$$
$$x = 4$$

The solution set is $\{(4, 1)\}$.

9. $\quad 2x + 4y = 6$
$\qquad 3x - y = 2$

Write the augmented matrix.

$$\begin{bmatrix} 2 & 4 & | & 6 \\ 3 & -1 & | & 2 \end{bmatrix}$$

The easiest way to get a 1 in the first row, first column position is to multiply the elements in the first row by $\frac{1}{2}$.

$$\begin{bmatrix} 1 & 2 & | & 3 \\ 3 & -1 & | & 2 \end{bmatrix} \quad \frac{1}{2}R_1$$

To get a 0 in row two, column 1, we need to subtract 3 from the 3 that is in that position. To do this we will multiply row 1 by -3 and add the result to row 2.

$$\begin{bmatrix} 1 & 2 & | & 3 \\ 0 & -7 & | & -7 \end{bmatrix} \quad -3R_1 + R_2$$
$$\begin{bmatrix} 1 & 2 & | & 3 \\ 0 & 1 & | & 1 \end{bmatrix} \qquad -\frac{1}{7}R_2$$

This matrix gives the system

$$x + 2y = 3$$
$$y = 1.$$

Substitute $y = 1$ in the first equation.

$$x + 2y = 3$$
$$x + 2(1) = 3$$
$$x = 1$$

The solution set is $\{(1, 1)\}$.

Copyright © 2012 Pearson Education, Inc. Publishing as Addison-Wesley.

11. $3x + 4y = 13$
$2x - 3y = -14$

Write the augmented matrix.

$$\begin{bmatrix} 3 & 4 & | & 13 \\ 2 & -3 & | & -14 \end{bmatrix}$$

$$\begin{bmatrix} 1 & 7 & | & 27 \\ 2 & -3 & | & -14 \end{bmatrix} \quad -1R_2 + R_1$$

$$\begin{bmatrix} 1 & 7 & | & 27 \\ 0 & -17 & | & -68 \end{bmatrix} \quad -2R_1 + R_2$$

$$\begin{bmatrix} 1 & 7 & | & 27 \\ 0 & 1 & | & 4 \end{bmatrix} \quad -\frac{1}{17}R_2$$

This matrix gives the system

$$x + 7y = 27$$
$$y = 4.$$

Substitute $y = 4$ in the first equation.

$$x + 7y = 27$$
$$x + 7(4) = 27$$
$$x + 28 = 27$$
$$x = -1$$

The solution set is $\{(-1, 4)\}$.

13. $-4x + 12y = 36$
$x - 3y = 9$

Write the augmented matrix.

$$\begin{bmatrix} -4 & 12 & | & 36 \\ 1 & -3 & | & 9 \end{bmatrix}$$

$$\begin{bmatrix} -1 & 3 & | & 9 \\ 1 & -3 & | & 9 \end{bmatrix} \quad \frac{1}{4}R_1$$

$$\begin{bmatrix} -1 & 3 & | & 9 \\ 0 & 0 & | & 18 \end{bmatrix} \quad R_1 + R_2$$

The corresponding system is

$$-x + 3y = 9$$
$$0 = 18 \quad \textit{False}$$

which is inconsistent and has no solution.

The solution set is $\emptyset$.

15. $2x + y = 4$
$4x + 2y = 8$

Write the augmented matrix.

$$\begin{bmatrix} 2 & 1 & | & 4 \\ 4 & 2 & | & 8 \end{bmatrix}$$

$$\begin{bmatrix} 1 & \frac{1}{2} & | & 2 \\ 4 & 2 & | & 8 \end{bmatrix} \quad \frac{1}{2}R_1$$

$$\begin{bmatrix} 1 & \frac{1}{2} & | & 2 \\ 0 & 0 & | & 0 \end{bmatrix} \quad -4R_1 + R_2$$

Row 2, $0 = 0$, indicates that the system has dependent equations.

The solution set is $\{(x, y) \mid 2x + y = 4\}$.

17. $-3x + 2y = 0$
$x - y = 0$

Write the augmented matrix.

$$\begin{bmatrix} -3 & 2 & | & 0 \\ 1 & -1 & | & 0 \end{bmatrix}$$

$$\begin{bmatrix} 1 & -1 & | & 0 \\ -3 & 2 & | & 0 \end{bmatrix} \quad R_1 \leftrightarrow R_2$$

We use $\leftrightarrow$ to represent the interchanging of two rows.

$$\begin{bmatrix} 1 & -1 & | & 0 \\ 0 & -1 & | & 0 \end{bmatrix} \quad 3R_1 + R_2$$

$$\begin{bmatrix} 1 & -1 & | & 0 \\ 0 & 1 & | & 0 \end{bmatrix} \quad -R_2$$

This matrix gives the system

$$x - y = 0$$
$$y = 0.$$

Substitute $y = 0$ in the first equation.

$$x - 0 = 0$$
$$x = 0$$

The solution set is $\{(0, 0)\}$.

19. $x + y - 3z = 1$
$2x - y + z = 9$
$3x + y - 4z = 8$

Write the augmented matrix.

$$\begin{bmatrix} 1 & 1 & -3 & | & 1 \\ 2 & -1 & 1 & | & 9 \\ 3 & 1 & -4 & | & 8 \end{bmatrix}$$

$$\begin{bmatrix} 1 & 1 & -3 & | & 1 \\ 0 & -3 & 7 & | & 7 \\ 0 & -2 & 5 & | & 5 \end{bmatrix} \quad \begin{matrix} -2R_1 + R_2 \\ -3R_1 + R_3 \end{matrix}$$

$$\begin{bmatrix} 1 & 1 & -3 & | & 1 \\ 0 & 1 & -\frac{7}{3} & | & -\frac{7}{3} \\ 0 & -2 & 5 & | & 5 \end{bmatrix} \quad -\frac{1}{3}R_2$$

$$\begin{bmatrix} 1 & 1 & -3 & | & 1 \\ 0 & 1 & -\frac{7}{3} & | & -\frac{7}{3} \\ 0 & 0 & \frac{1}{3} & | & \frac{1}{3} \end{bmatrix} \quad 2R_2 + R_3$$

$$\begin{bmatrix} 1 & 1 & -3 & | & 1 \\ 0 & 1 & -\frac{7}{3} & | & -\frac{7}{3} \\ 0 & 0 & 1 & | & 1 \end{bmatrix} \quad 3R_3$$

This matrix gives the system

$$x + y - 3z = 1$$
$$y - \frac{7}{3}z = -\frac{7}{3}$$
$$z = 1.$$

Substitute $z = 1$ in the second equation.

$$y - \frac{7}{3}z = -\frac{7}{3}$$
$$y - \frac{7}{3}(1) = -\frac{7}{3}$$
$$y = 0$$

Copyright © 2012 Pearson Education, Inc. Publishing as Addison-Wesley.

Substitute $y = 0$ and $z = 1$ in the first equation.

$$x + y - 3z = 1$$
$$x + 0 - 3(1) = 1$$
$$x - 3 = 1$$
$$x = 4$$

The solution set is $\{(4, 0, 1)\}$.

21. $\quad x + y - z = 6$
$\quad 2x - y + z = -9$
$\quad x - 2y + 3z = 1$

Write the augmented matrix.

$$\begin{bmatrix} 1 & 1 & -1 & | & 6 \\ 2 & -1 & 1 & | & -9 \\ 1 & -2 & 3 & | & 1 \end{bmatrix}$$

$$\begin{bmatrix} 1 & 1 & -1 & | & 6 \\ 0 & -3 & 3 & | & -21 \\ 0 & -3 & 4 & | & -5 \end{bmatrix} \begin{matrix} \\ -2R_1 + R_2 \\ -1R_1 + R_3 \end{matrix}$$

$$\begin{bmatrix} 1 & 1 & -1 & | & 6 \\ 0 & 1 & -1 & | & 7 \\ 0 & -3 & 4 & | & -5 \end{bmatrix} \begin{matrix} \\ -\frac{1}{3}R_2 \\ \\ \end{matrix}$$

$$\begin{bmatrix} 1 & 1 & -1 & | & 6 \\ 0 & 1 & -1 & | & 7 \\ 0 & 0 & 1 & | & 16 \end{bmatrix} \begin{matrix} \\ \\ 3R_2 + R_3 \end{matrix}$$

This matrix gives the system

$$x + y - z = 6$$
$$y - z = 7$$
$$z = 16.$$

Substitute $z = 16$ in the second equation.

$$y - 16 = 7$$
$$y = 23$$

Substitute $y = 23$ and $z = 16$ in the first equation.

$$x + 23 - 16 = 6$$
$$x + 7 = 6$$
$$x = -1$$

The solution set is $\{(-1, 23, 16)\}$.

23. $\quad x - y = 1$
$\quad\quad y - z = 6$
$\quad x + z = -1$

Write the augmented matrix.

$$\begin{bmatrix} 1 & -1 & 0 & | & 1 \\ 0 & 1 & -1 & | & 6 \\ 1 & 0 & 1 & | & -1 \end{bmatrix}$$

$$\begin{bmatrix} 1 & -1 & 0 & | & 1 \\ 0 & 1 & -1 & | & 6 \\ 0 & 1 & 1 & | & -2 \end{bmatrix} \begin{matrix} \\ \\ -1R_1 + R_3 \end{matrix}$$

$$\begin{bmatrix} 1 & -1 & 0 & | & 1 \\ 0 & 1 & -1 & | & 6 \\ 0 & 0 & 2 & | & -8 \end{bmatrix} \begin{matrix} \\ \\ -1R_2 + R_3 \end{matrix}$$

$$\begin{bmatrix} 1 & -1 & 0 & | & 1 \\ 0 & 1 & -1 & | & 6 \\ 0 & 0 & 1 & | & -4 \end{bmatrix} \begin{matrix} \\ \\ \frac{1}{2}R_3 \end{matrix}$$

This matrix gives the system

$$x - y = 1$$
$$y - z = 6$$
$$z = -4.$$

Substitute $z = -4$ in the second equation.

$$y - z = 6$$
$$y - (-4) = 6$$
$$y = 2$$

Substitute $y = 2$ in the first equation.

$$x - y = 1$$
$$x - 2 = 1$$
$$x = 3$$

The solution set is $\{(3, 2, -4)\}$.

25. $\quad x - 2y + z = 4$
$\quad 3x - 6y + 3z = 12$
$\quad -2x + 4y - 2z = -8$

Write the augmented matrix.

$$\begin{bmatrix} 1 & -2 & 1 & | & 4 \\ 3 & -6 & 3 & | & 12 \\ -2 & 4 & -2 & | & -8 \end{bmatrix}$$

$$\begin{bmatrix} 1 & -2 & 1 & | & 4 \\ 1 & -2 & 1 & | & 4 \\ -1 & 2 & -1 & | & -4 \end{bmatrix} \begin{matrix} \\ \frac{1}{3}R_2 \\ \frac{1}{2}R_3 \end{matrix}$$

$$\begin{bmatrix} 1 & -2 & 1 & | & 4 \\ 0 & 0 & 0 & | & 0 \\ 0 & 0 & 0 & | & 0 \end{bmatrix} \begin{matrix} \\ -1R_1 + R_2 \\ R_1 + R_3 \end{matrix}$$

This augmented matrix represents a system of dependent equations.
The solution set is $\{(x, y, z) \mid x - 2y + z = 4\}$.

27. $\quad x + 2y + 3z = -2$
$\quad 2x + 4y + 6z = -5$
$\quad x - y + 2z = 6$

Write the augmented matrix.

$$\begin{bmatrix} 1 & 2 & 3 & | & -2 \\ 2 & 4 & 6 & | & -5 \\ 1 & -1 & 2 & | & 6 \end{bmatrix}$$

$$\begin{bmatrix} 1 & 2 & 3 & | & -2 \\ 0 & 0 & 0 & | & -1 \\ 0 & -3 & -1 & | & 8 \end{bmatrix} \begin{matrix} \\ -2R_1 + R_2 \\ -R_1 + R_3 \end{matrix}$$

From the second row, $0 = -1$, we see that the system is inconsistent.
The solution set is $\emptyset$.

Copyright © 2012 Pearson Education, Inc. Publishing as Addison-Wesley.

29. $4x + y = 5$
$2x + y = 3$

Enter the augmented matrix as $[A]$.

$$\begin{bmatrix} 4 & 1 & 5 \\ 2 & 1 & 3 \end{bmatrix}$$

The TI-83/4 screen for A follows. (Use MATRX EDIT.)

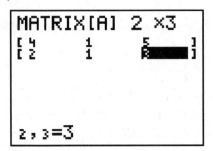

Now use the reduced row echelon form (rref) command to simplify the system. (Use MATRX MATH ALPHA B for rref and MATRX 1 for $[A]$.)

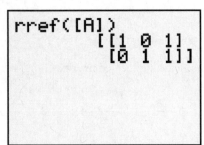

This matrix gives the system

$$1x + 0y = 1$$
$$0x + 1y = 1,$$

or simply, $x = 1$ and $y = 1$.
The solution set is $\{(1, 1)\}$.

31. $5x + y - 3z = -6$
$2x + 3y + z = 5$
$-3x - 2y + 4z = 3$

Enter the augmented matrix as $[C]$.

$$\begin{bmatrix} 5 & 1 & -3 & -6 \\ 2 & 3 & 1 & 5 \\ -3 & -2 & 4 & 3 \end{bmatrix}$$

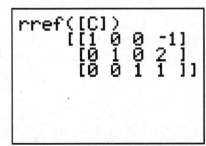

The solution set is $\{(-1, 2, 1)\}$.

33. $x + z = -3$
$y + z = 3$
$x + y = 8$

Enter the augmented matrix as $[E]$.

$$\begin{bmatrix} 1 & 0 & 1 & -3 \\ 0 & 1 & 1 & 3 \\ 1 & 1 & 0 & 8 \end{bmatrix}$$

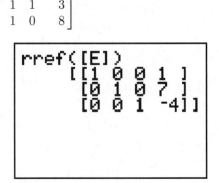

The solution set is $\{(1, 7, -4)\}$.

35. $2^6 = 2 \cdot 2 \cdot 2 \cdot 2 \cdot 2 \cdot 2 = 64$

37. $(-5)^4 = (-5)(-5)(-5)(-5) = 625$

39. $(\frac{3}{4})^4 = \frac{3}{4} \cdot \frac{3}{4} \cdot \frac{3}{4} \cdot \frac{3}{4}$
$= \frac{3 \cdot 3 \cdot 3 \cdot 3}{4 \cdot 4 \cdot 4 \cdot 4} = \frac{81}{256}$

Chapter 4 Review Exercises

1. **(a)** The graphs meet between the years of 1980 and 1985. Therefore, the number of degrees for men equal the number of degrees for women between 1980 and 1985.

(b) The number was just less than 500,000.

2. $x + 3y = 8$ (1)
$2x - y = 2$ (2)

Graph the two lines. They appear to intersect at the point $(2, 2)$. Check $(2, 2)$ in both equations.

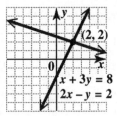

$$x + 3y = 8 \quad (1)$$
$$2 + 3(2) \overset{?}{=} 8$$
$$8 = 8 \quad \textit{True}$$

$$2x - y = 2 \quad (2)$$
$$2(2) - 2 \overset{?}{=} 2$$
$$2 = 2 \quad \textit{True}$$

The solution set is $\{(2, 2)\}$.

Copyright © 2012 Pearson Education, Inc. Publishing as Addison-Wesley.

3.   Checking the ordered pairs in choices **A**, **B**, and **C** yields true statements.

For choice **D**, we have:

$$3x + 2y = 6$$
$$3(3) + 2(-2) \overset{?}{=} 6$$
$$9 - 4 \overset{?}{=} 6$$
$$5 = 6 \qquad \textit{False}$$

So **D** is the answer.

4.   **(a)** The graphs of these two linear equations will intersect once.

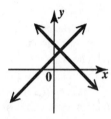

(b) The graphs of these two linear equations will not intersect. They are parallel lines.

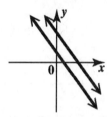

(c) The graphs of these two linear equations will be the same line.

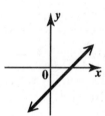

5. $3x + y = -4$ (1)
$x = \frac{2}{3}y$ (2)

Substitute $\frac{2}{3}y$ for x in equation (1) and solve for y.

$$3x + y = -4 \qquad (1)$$
$$3(\tfrac{2}{3}y) + y = 4$$
$$2y + y = -4$$
$$3y = -4$$
$$y = -\tfrac{4}{3}$$

Since $x = \frac{2}{3}y$ and $y = -\frac{4}{3}$,

$$x = \tfrac{2}{3}(-\tfrac{4}{3}) = -\tfrac{8}{9}.$$

The solution set is $\left\{\left(-\frac{8}{9}, -\frac{4}{3}\right)\right\}$.

6. $9x - y = -4$ (1)
$y = x + 4$ (2)

Substitute $x + 4$ for y in equation (1) and solve for x.

$$9x - y = -4 \qquad (1)$$
$$9x - (x + 4) = -4$$
$$9x - x - 4 = -4$$
$$8x = 0$$
$$x = 0$$

Since $x = 0$, $y = x + 4 = 0 + 4 = 4$.
The solution set is $\{(0, 4)\}$.

7. $-5x + 2y = -2$ (1)
$x + 6y = 26$ (2)

Solve equation (2) for x.

$$x = 26 - 6y$$

Substitute $26 - 6y$ for x in equation (1).

$$-5x + 2y = -2 \qquad (1)$$
$$-5(26 - 6y) + 2y = -2$$
$$-130 + 30y + 2y = -2$$
$$-130 + 32y = -2$$
$$32y = 128$$
$$y = 4$$

Since $x = 26 - 6y$ and $y = 4$,

$$x = 26 - 6(4) = 26 - 24 = 2.$$

The solution set is $\{(2, 4)\}$.

8. $5x + y = 12$ (1)
$2x - 2y = 0$ (2)
To eliminate y, multiply equation (1) by 2 and add the result to equation (2).

$$
\begin{array}{rcll}
10x + 2y &=& 24 & \quad 2 \times (1) \\
2x - 2y &=& 0 & \quad (2) \\
\hline
12x &=& 24 & \\
x &=& 2 &
\end{array}
$$

To find y, substitute 2 for x in equation (2).

$$2x - 2y = 0 \qquad (2)$$
$$2(2) - 2y = 0$$
$$4 = 2y$$
$$2 = y$$

The ordered pair $(2, 2)$ satisfies both equations, so it checks.
The solution set is $\{(2, 2)\}$.

9. $x - 4y = -4$ (1)
$3x + y = 1$ (2)

To eliminate y, multiply equation (2) by 4 and add the result to equation (1).

Copyright © 2012 Pearson Education, Inc. Publishing as Addison-Wesley.

$$x - 4y = -4 \quad (1)$$
$$\underline{12x + 4y = 4} \qquad 4 \times (2)$$
$$13x = 0$$
$$x = 0$$

Substitute 0 for x in equation (1) to find y.

$$x - 4y = -4 \quad (1)$$
$$0 - 4y = -4$$
$$y = 1$$

The solution set is $\{(0, 1)\}$.

10. $6x + 5y = 4 \quad (1)$
$-4x + 2y = 8 \quad (2)$

To eliminate x, multiply equation (1) by 2 and equation (2) by 3. Then add the results.

$$12x + 10y = 8 \quad 2 \times (1)$$
$$\underline{-12x + 6y = 24} \quad 3 \times (2)$$
$$16y = 32$$
$$y = 2$$

Since $y = 2$,

$$-4x + 2y = 8 \qquad (2)$$
$$-4x + 2(2) = 8$$
$$-4x + 4 = 8$$
$$-4x = 4$$
$$x = -1.$$

The solution set is $\{(-1, 2)\}$.

11. $\frac{1}{6}x + \frac{1}{6}y = -\frac{1}{2} \quad (1)$
$x - y = -9 \quad (2)$

Multiply equation (1) by 6 to clear the fractions. Add the result to equation (2) to eliminate y.

$$x + y = -3 \qquad 6 \times (1)$$
$$\underline{x - y = -9} \quad (2)$$
$$2x = -12$$
$$x = -6$$

Since $x = -6$,

$$x - y = -9 \qquad (2)$$
$$-6 - y = -9$$
$$-y = -3$$
$$y = 3.$$

The solution set is $\{(-6, 3)\}$.

12. $-3x + y = 6 \qquad (1)$
$ y = 6 + 3x \quad (2)$

Since equation (2) can be rewritten as $-3x + y = 6$, the two equations are the same, and hence, dependent.
The solution set is $\{(x, y) \mid -3x + y = 6\}$.

13. $5x - 4y = 2 \quad (1)$
$-10x + 8y = 7 \quad (2)$

Multiply equation (1) by 2 and add the result to equation (2).

$$10x - 8y = 4 \qquad 2 \times (1)$$
$$\underline{-10x + 8y = 7} \quad (2)$$
$$0 = 11 \quad False$$

Since a false statement results, the system is *inconsistent*. The solution set is $\emptyset$.

14. $y = 3x + 2$
$y = 3x - 4$

Since both equations are in slope-intercept form, their slopes and y-intercepts can be easily determined. The two lines have the same slope, 3, but the y-intercepts, $(0, 2)$ and $(0, -4)$, are different. Therefore, the lines are parallel, do not intersect, and have no common solution.

15. $2x + 3y - z = -16 \quad (1)$
$x + 2y + 2z = -3 \quad (2)$
$-3x + y + z = -5 \quad (3)$

To eliminate z, add equations (1) and (3).

$$2x + 3y - z = -16 \quad (1)$$
$$\underline{-3x + y + z = -5} \quad (3)$$
$$-x + 4y = -21 \quad (4)$$

To eliminate z again, multiply equation (1) by 2 and add the result to equation (2).

$$4x + 6y - 2z = -32 \qquad 2 \times (1)$$
$$\underline{x + 2y + 2z = -3} \quad (2)$$
$$5x + 8y = -35 \quad (5)$$

Use equations (4) and (5) to eliminate x. Multiply equation (4) by 5 and add the result to equation (5).

$$-5x + 20y = -105 \qquad 5 \times (4)$$
$$\underline{5x + 8y = -35} \quad (5)$$
$$28y = -140$$
$$y = -5$$

Substitute -5 for y in equation (4) to find x.

$$-x + 4y = -21 \qquad (4)$$
$$-x + 4(-5) = -21$$
$$-x - 20 = -21$$
$$-x = -1$$
$$x = 1$$

Substitute 1 for x and -5 for y in equation (2) to find z.

Copyright © 2012 Pearson Education, Inc. Publishing as Addison-Wesley.

$$x + 2y + 2z = -3 \quad (2)$$
$$1 + 2(-5) + 2z = -3$$
$$1 - 10 + 2z = -3$$
$$2z = 6$$
$$z = 3$$

The solution set is $\{(1, -5, 3)\}$.

16.
$$\begin{aligned} 4x - y \quad\quad &= 2 \quad (1) \\ 3y + z &= 9 \quad (2) \\ x \quad\quad + 2z &= 7 \quad (3) \end{aligned}$$

To eliminate y, multiply equation (1) by 3 and add the result to equation (2).

$$\begin{array}{ll} 12x - 3y \quad\quad = 6 & 3 \times (1) \\ \underline{\quad\quad 3y + z = 9} & (2) \\ 12x \quad\quad + z = 15 & (4) \end{array}$$

To eliminate z, multiply equation (4) by -2 and add the result to equation (3).

$$\begin{array}{ll} -24x - 2z = -30 & -2 \times (4) \\ \underline{\quad x + 2z = \quad 7} & (3) \\ -23x \quad\quad = -23 \\ \quad\quad x = 1 \end{array}$$

Substitute 1 for x in equation (3) to find z.

$$x + 2z = 7 \quad (3)$$
$$1 + 2z = 7$$
$$2z = 6$$
$$z = 3$$

Substitute 1 for x in equation (1) to find y.

$$4x - y = 2 \quad (1)$$
$$4(1) - y = 2$$
$$4 - y = 2$$
$$-y = -2$$
$$y = 2$$

The solution set is $\{(1, 2, 3)\}$.

17.
$$\begin{aligned} 3x - y - z &= -8 \quad (1) \\ 4x + 2y + 3z &= 15 \quad (2) \\ -6x + 2y + 2z &= 10 \quad (3) \end{aligned}$$

To eliminate y, multiply equation (1) by 2 and add the result to equation (3).

$$\begin{array}{ll} 6x - 2y - 2z = -16 & 2 \times (1) \\ \underline{-6x + 2y + 2z = \quad 10} & (3) \\ \quad\quad\quad\quad 0 = -6 & \textit{False} \end{array}$$

Since a false statement results, equations (1) and (3) have no common solution. The system is *inconsistent*. The solution set is $\emptyset$.

18. Let x = the width of the rink, and y = the length of the rink.

The length is 30 ft longer than twice the width.

$$y = 2x + 30 \quad (1)$$

The perimeter is 570 ft.

$$2x + 2y = 570 \quad (2)$$

Substitute $2x + 30$ for y in equation (2).

$$2x + 2y = 570 \quad (2)$$
$$2x + 2(2x + 30) = 570$$
$$2x + 4x + 60 = 570$$
$$6x = 510$$
$$x = 85$$

From (1), $y = 2(85) + 30 = 200$.

The width is 85 ft and the length is 200 ft.

19. Let x = the average price for a Yankees ticket and y = the average price for a Red Sox ticket.

From the given information, we get the following system of equations:

$$\begin{aligned} 2x + 3y &= 296.66 \quad (1) \\ 3x + 2y &= 319.39 \quad (2) \end{aligned}$$

To eliminate y, multiply equation (1) by 2 and add the result to -3 times equation (2).

$$\begin{array}{ll} 4x + 6y = \quad 593.32 & 2 \times (1) \\ \underline{-9x - 6y = -958.17} & -3 \times (2) \\ -5x \quad\quad = -364.85 \\ \quad x = \quad 72.97 \end{array}$$

Substitute 72.97 for x in equation (1).

$$2x + 3y = 296.66 \quad (1)$$
$$2(72.97) + 3y = 296.66$$
$$145.94 + 3y = 296.66$$
$$3y = 150.72$$
$$y = 50.24$$

The average price for a Yankees ticket was $72.97 and the average price for a Red Sox ticket was $50.24.

20. Let x = the speed of the plane and y = the speed of the wind.

Complete the chart.

	r	t	d
With Wind	$x + y$	1.75	$1.75(x + y)$
Against Wind	$x - y$	2	$2(x - y)$

The distance each way is 560 miles. From the chart,

$$1.75(x + y) = 560.$$

Divide by 1.75.

$$x + y = 320 \quad (1)$$

From the chart,

$$2(x - y) = 560$$
$$x - y = 280. \qquad (2)$$

Solve the system by adding equations (1) and (2) to eliminate y.

$$\begin{aligned} x + y &= 320 \quad (1) \\ x - y &= 280 \quad (2) \\ \hline 2x &= 600 \\ x &= 300 \end{aligned}$$

From (1), $300 + y = 320$, so $y = 20$.

The speed of the plane was 300 mph, and the speed of the wind was 20 mph.

21. Let $x =$ the amount of \$2-per-pound nuts and $y =$ the amount of \$1-per-pound chocolate candy.

Complete the chart.

	Number of Pounds	Price per Pound	Value
Nuts	x	2	$2x$
Chocolate	y	1	$1y = y$
Mixture	100	1.30	$1.30(100) = 130$

Solve the system formed from the second and fourth columns.

$$\begin{aligned} x + y &= 100 \quad (1) \\ 2x + y &= 130 \quad (2) \end{aligned}$$

Solve equation (1) for y.

$$y = 100 - x \quad (3)$$

Substitute $100 - x$ for y in equation (2).

$$2x + (100 - x) = 130$$
$$x = 30$$

From (3), $y = 100 - 30 = 70$.

She should use 30 lb of \$2-per-pound nuts and 70 lb of \$1-per-pound chocolate candy.

22. Let $x =$ the measure of the largest angle,
$y =$ the measure of the middle-sized angle,
and $z =$ the measure of the smallest angle.

Since the sum of the measures of the angles of a triangle is 180°,

$$x + y + z = 180. \quad (1)$$

Since the largest angle measures 10° less than the sum of the other two,

$$x = y + z - 10$$
$$\text{or} \quad x - y - z = -10. \qquad (2)$$

Since the measure of the middle-sized angle is the average of the other two,

$$y = \frac{x + z}{2}$$
$$2y = x + z$$
$$-x + 2y - z = 0. \qquad (3)$$

Solve the system.

$$\begin{aligned} x + y + z &= 180 \quad (1) \\ x - y - z &= -10 \quad (2) \\ -x + 2y - z &= 0 \quad (3) \end{aligned}$$

Add equations (1) and (2) to find x.

$$\begin{aligned} x + y + z &= 180 \quad (1) \\ x - y - z &= -10 \quad (2) \\ \hline 2x &= 170 \\ x &= 85 \end{aligned}$$

Add equations (1) and (3), to find y.

$$\begin{aligned} x + y + z &= 180 \quad (1) \\ -x + 2y - z &= 0 \quad (3) \\ \hline 3y &= 180 \\ y &= 60 \end{aligned}$$

Substitute 85 for x and 60 for y in equation (1) to find z.

$$\begin{aligned} x + y + z &= 180 \quad (1) \\ 85 + 60 + z &= 180 \\ 145 + z &= 180 \\ z &= 35 \end{aligned}$$

The three angle measures are 85°, 60°, and 35°.

23. Let $x =$ the value of sales at 10%,
$y =$ the value of sales at 6%,
and $z =$ the value of sales at 5%.

Since her total sales were \$280,000,

$$x + y + z = 280,000 \quad (1)$$

Since her commissions on the sales totaled \$17,000,

$$0.10x + 0.06y + 0.05z = 17,000.$$

Multiply by 100 to clear the decimals, so

$$10x + 6y + 5z = 1,700,000. \quad (2)$$

Since the 5% sale amounted to the sum of the other two sales,

$$z = x + y. \quad (3)$$

Solve the system.

$$\begin{aligned} x + y + z &= 280,000 \quad (1) \\ 10x + 6y + 5z &= 1,700,000 \quad (2) \\ z &= x + y \quad (3) \end{aligned}$$

Since equation (3) is given in terms of z, substitute $x + y$ for z in equations (1) and (2).

Copyright © 2012 Pearson Education, Inc. Publishing as Addison-Wesley.

$$x + y + z = 280,000 \quad (1)$$
$$x + y + (x + y) = 280,000$$
$$2x + 2y = 280,000$$
$$x + y = 140,000 \quad (4)$$

$$10x + 6y + 5z = 1,700,000 \quad (2)$$
$$10x + 6y + 5(x + y) = 1,700,000$$
$$10x + 6y + 5x + 5y = 1,700,000$$
$$15x + 11y = 1,700,000 \quad (5)$$

To eliminate x, multiply equation (4) by -11 and add the result to equation (5).

$$-11x - 11y = -1,540,000 \quad -11 \times (4)$$
$$\underline{15x + 11y = 1,700,000} \quad (5)$$
$$4x = 160,000$$
$$x = 40,000$$

From (4), $y = 100,000$.
From (3), $z = 40,000 + 100,000 = 140,000$.
He sold \$40,000 at 10%, \$100,000 at 6%, and \$140,000 at 5%.

24. Let x = the number of liters of 8% solution,
y = the number of liters of 10% solution,
and z = the number of liters of 20% solution.

Since the amount of the mixture will be 8 L,

$$x + y + z = 8. \quad (1)$$

Since the final solution will be 12.5% hydrogen peroxide,

$$0.08x + 0.10y + 0.20z = 0.125(8).$$

Multiply by 100 to clear the decimals.

$$8x + 10y + 20z = 100 \quad (2)$$

Since the amount of 8% solution used must be 2 L more than the amount of 20% solution,

$$x = z + 2. \quad (3)$$

Solve the system.

$$x + y + z = 8 \quad (1)$$
$$8x + 10y + 20z = 100 \quad (2)$$
$$x = z + 2 \quad (3)$$

Since equation (3) is given in terms of x, substitute $z + 2$ for x in equations (1) and (2).

$$x + y + z = 8 \quad (1)$$
$$(z + 2) + y + z = 8$$
$$y + 2z = 6 \quad (4)$$

$$8x + 10y + 20z = 100 \quad (2)$$
$$8(z + 2) + 10y + 20z = 100$$
$$8z + 16 + 10y + 20z = 100$$
$$10y + 28z = 84 \quad (5)$$

To eliminate y, multiply equation (4) by -10 and add the result to equation (5).

$$-10y - 20z = -60 \quad -10 \times (4)$$
$$\underline{10y + 28z = 84} \quad (5)$$
$$8z = 24$$
$$z = 3$$

From (3), $x = z + 2 = 3 + 2 = 5$.
From (4), $y = 6 - 2z = 6 - 2(3) = 0$.
Mix 5 L of 8% solution, none of 10% solution, and 3 L of 20% solution.

25. Let x = the number of home runs hit by Mantle,
y = the number of home runs hit by Maris,
and z = the number of home runs hit by Berra.

They combined for 137 home runs, so

$$x + y + z = 137. \quad (1)$$

Mantle hit 7 fewer than Maris, so

$$x = y - 7. \quad (2)$$

Maris hit 39 more than Berra, so

$$y = z + 39 \quad \text{or} \quad z = y - 39. \quad (3)$$

Substitute $y - 7$ for x and $y - 39$ for z in (1).

$$x + y + z = 137 \quad (1)$$
$$(y - 7) + y + (y - 39) = 137$$
$$3y - 46 = 137$$
$$3y = 183$$
$$y = 61$$

From (2), $x = y - 7 = 61 - 7 = 54$.
From (3), $z = y - 39 = 61 - 39 = 22$.
Mantle hit 54 home runs, Maris hit 61 home runs, and Berra hit 22 home runs.

26. $$2x + 5y = -4$$
$$4x - y = 14$$

Write the augmented matrix.

$$\begin{bmatrix} 2 & 5 & | & -4 \\ 4 & -1 & | & 14 \end{bmatrix}$$

$$\begin{bmatrix} 2 & 5 & | & -4 \\ 0 & -11 & | & 22 \end{bmatrix} \quad -2R_1 + R_2$$

$$\begin{bmatrix} 2 & 5 & | & -4 \\ 0 & 1 & | & -2 \end{bmatrix} \quad -\tfrac{1}{11}R_2$$

This matrix gives the system

$$2x + 5y = -4$$
$$y = -2.$$

Substitute $y = -2$ in the first equation.

$$2x + 5y = -4$$
$$2x + 5(-2) = -4$$
$$2x - 10 = -4$$
$$2x = 6$$
$$x = 3$$

The solution set is $\{(3, -2)\}$.

Copyright © 2012 Pearson Education, Inc. Publishing as Addison-Wesley.

27.
$$6x + 3y = 9$$
$$-7x + 2y = 17$$

Write the augmented matrix.

$$\begin{bmatrix} 6 & 3 & | & 9 \\ -7 & 2 & | & 17 \end{bmatrix}$$

$$\begin{bmatrix} 1 & -5 & | & -26 \\ -7 & 2 & | & 17 \end{bmatrix} \quad -R_1 - R_2$$

$$\begin{bmatrix} 1 & -5 & | & -26 \\ 0 & -33 & | & -165 \end{bmatrix} \quad 7R_1 + R_2$$

$$\begin{bmatrix} 1 & -5 & | & -26 \\ 0 & 1 & | & 5 \end{bmatrix} \quad -\frac{1}{33}R_2$$

This matrix gives the system

$$x - 5y = -26$$
$$y = 5.$$

Substitute $y = 5$ in the first equation.

$$x - 5y = -26$$
$$x - 5(5) = -26$$
$$x - 25 = -26$$
$$x = -1$$

The solution set is $\{(-1, 5)\}$.

28.
$$x + 2y - z = 1$$
$$3x + 4y + 2z = -2$$
$$-2x - y + z = -1$$

$$\begin{bmatrix} 1 & 2 & -1 & | & 1 \\ 3 & 4 & 2 & | & -2 \\ -2 & -1 & 1 & | & -1 \end{bmatrix}$$

$$\begin{bmatrix} 1 & 2 & -1 & | & 1 \\ 0 & -2 & 5 & | & -5 \\ 0 & 3 & -1 & | & 1 \end{bmatrix} \quad \begin{matrix} -3R_1 + R_2 \\ 2R_1 + R_3 \end{matrix}$$

$$\begin{bmatrix} 1 & 2 & -1 & | & 1 \\ 0 & 1 & 4 & | & -4 \\ 0 & 3 & -1 & | & 1 \end{bmatrix} \quad R_3 + R_2$$

$$\begin{bmatrix} 1 & 2 & -1 & | & 1 \\ 0 & 1 & 4 & | & -4 \\ 0 & 0 & -13 & | & 13 \end{bmatrix} \quad -3R_2 + R_3$$

$$\begin{bmatrix} 1 & 2 & -1 & | & 1 \\ 0 & 1 & 4 & | & -4 \\ 0 & 0 & 1 & | & -1 \end{bmatrix} \quad -\frac{1}{13}R_3$$

This matrix gives the system

$$x + 2y - z = 1$$
$$y + 4z = -4$$
$$z = -1.$$

Substitute $z = -1$ in the second equation.

$$y + 4z = -4$$
$$y + 4(-1) = -4$$
$$y = 0$$

Substitute $y = 0$ and $z = -1$ in the first equation.

$$x + 2y - z = 1$$
$$x + 2(0) - (-1) = 1$$
$$x + 1 = 1$$
$$x = 0$$

The solution set is $\{(0, 0, -1)\}$.

29.
$$x + 3y = 7$$
$$3x + z = 2$$
$$y - 2z = 4$$

$$\begin{bmatrix} 1 & 3 & 0 & | & 7 \\ 3 & 0 & 1 & | & 2 \\ 0 & 1 & -2 & | & 4 \end{bmatrix}$$

$$\begin{bmatrix} 1 & 3 & 0 & | & 7 \\ 0 & -9 & 1 & | & -19 \\ 0 & 1 & -2 & | & 4 \end{bmatrix} \quad -3R_1 + R_2$$

$$\begin{bmatrix} 1 & 3 & 0 & | & 7 \\ 0 & 1 & -2 & | & 4 \\ 0 & -9 & 1 & | & -19 \end{bmatrix} \quad R_2 \leftrightarrow R_3$$

We use $\leftrightarrow$ to represent the interchanging of two rows.

$$\begin{bmatrix} 1 & 3 & 0 & | & 7 \\ 0 & 1 & -2 & | & 4 \\ 0 & 0 & -17 & | & 17 \end{bmatrix} \quad 9R_2 + R_3$$

$$\begin{bmatrix} 1 & 3 & 0 & | & 7 \\ 0 & 1 & -2 & | & 4 \\ 0 & 0 & 1 & | & -1 \end{bmatrix} \quad -\frac{1}{17}R_3$$

This matrix gives the system

$$x + 3y = 7$$
$$y - 2z = 4$$
$$z = -1.$$

Substitute $z = -1$ in the second equation.

$$y - 2z = 4$$
$$y - 2(-1) = 4$$
$$y + 2 = 4$$
$$y = 2$$

Substitute $y = 2$ in the first equation.

$$x + 3y = 7$$
$$x + 3(2) = 7$$
$$x + 6 = 7$$
$$x = 1$$

The solution set is $\{(1, 2, -1)\}$.

Copyright © 2012 Pearson Education, Inc. Publishing as Addison-Wesley.

30. **[4.1]** System **B** would be easier to solve using the substitution method because the second equation is already solved for y.

31. **[4.1]** $\frac{2}{3}x + \frac{1}{6}y = \frac{19}{2}$ (1)

$\frac{1}{3}x - \frac{2}{9}y = 2$ (2)

Multiply equation (1) by 6 and equation (2) by 9 to clear the fractions.

$4x + y = 57$ (3) $6 \times (1)$
$3x - 2y = 18$ (4) $9 \times (2)$

To eliminate y, multiply equation (3) by 2 and add the result to equation (4).

$$\begin{array}{rcl} 8x + 2y &=& 114 \quad 2 \times (3) \\ 3x - 2y &=& 18 \quad (4) \\ \hline 11x &=& 132 \\ x &=& 12 \end{array}$$

Substitute 12 for x in equation (3) to find y.

$$\begin{array}{rcl} 4x + y &=& 57 \quad (3) \\ 4(12) + y &=& 57 \\ 48 + y &=& 57 \\ y &=& 9 \end{array}$$

The solution set is $\{(12, 9)\}$.

32. **[4.2]** $2x + 5y - z = 12$ (1)
$-x + y - 4z = -10$ (2)
$-8x - 20y + 4z = 31$ (3)

Multiply equation (1) by 4 and add the result to equation (3).

$$\begin{array}{rcl} 8x + 20y - 4z &=& 48 \quad 4 \times (1) \\ -8x - 20y + 4z &=& 31 \quad (3) \\ \hline 0 &=& 79 \; \textit{False} \end{array}$$

Since a false statement results, the system is *inconsistent*. The solution set is $\emptyset$.

33. **[4.1]** $x = 7y + 10$ (1)
$2x + 3y = 3$ (2)

Since equation (1) is given in terms of x, substitute $7y + 10$ for x in equation (2) and solve for y.

$$\begin{array}{rcl} 2(7y + 10) + 3y &=& 3 \\ 14y + 20 + 3y &=& 3 \\ 17y &=& -17 \\ y &=& -1 \end{array}$$

From (1), $x = 7(-1) + 10 = 3$.

The solution set is $\{(3, -1)\}$.

34. **[4.1]** $x + 4y = 17$ (1)
$-3x + 2y = -9$ (2)

To eliminate x, multiply equation (1) by 3 and add the result to equation (2).

$$\begin{array}{rcl} 3x + 12y &=& 51 \quad 3 \times (1) \\ -3x + 2y &=& -9 \quad (2) \\ \hline 14y &=& 42 \\ y &=& 3 \end{array}$$

Substitute 3 for y in equation (1) to find x.

$$\begin{array}{rcl} x + 4y &=& 17 \quad (1) \\ x + 4(3) &=& 17 \\ x + 12 &=& 17 \\ x &=& 5 \end{array}$$

The solution set is $\{(5, 3)\}$.

35. **[4.1]** $-7x + 3y = 12$ (1)
$5x + 2y = 8$ (2)

To eliminate y, multiply equation (1) by 2 and equation (2) by -3. Then add the results.

$$\begin{array}{rcl} -14x + 6y &=& 24 \quad 2 \times (1) \\ -15x - 6y &=& -24 \quad -3 \times (2) \\ \hline -29x &=& 0 \\ x &=& 0 \end{array}$$

Substitute 0 for x in equation (1) to find y.

$$\begin{array}{rcl} -7x + 3y &=& 12 \quad (1) \\ -7(0) + 3y &=& 12 \\ 3y &=& 12 \\ y &=& 4 \end{array}$$

The solution set is $\{(0, 4)\}$.

36. **[4.1]** $2x - 5y = 8$ (1)
$3x + 4y = 10$ (2)

To eliminate y, multiply equation (1) by 4 and equation (2) by 5 and add the results.

$$\begin{array}{rcl} 8x - 20y &=& 32 \quad 4 \times (1) \\ 15x + 20y &=& 50 \quad 5 \times (2) \\ \hline 23x &=& 82 \\ x &=& \frac{82}{23} \end{array}$$

Instead of substituting to find y, we'll choose different multipliers and eliminate x from the original system.

$$\begin{array}{rcl} 6x - 15y &=& 24 \quad 3 \times (1) \\ -6x - 8y &=& -20 \quad -2 \times (2) \\ \hline -23y &=& 4 \\ y &=& -\frac{4}{23} \end{array}$$

The solution set is $\left\{ \left(\frac{82}{23}, -\frac{4}{23} \right) \right\}$.

Copyright © 2012 Pearson Education, Inc. Publishing as Addison-Wesley.

37. **[4.3]** Let $x =$ the number of liters of 5% solution and $y =$ the number of liters of 10% solution.

Liters of Solution	Percent (as a decimal)	Liters of Pure Acid
x	0.05	$0.05x$
10	0.20	$0.20(10) = 2$
y	0.10	$0.10y$

Solve the system formed from the first and third columns.

$$x + 10 = y \qquad (1)$$
$$0.05x + 2 = 0.10y \qquad (2)$$

Multiply equation (2) by 100 to clear the decimals.

$$5x + 200 = 10y \quad (3)$$

Substitute $x + 10$ for y in equation (3) and solve for x.

$$5x + 200 = 10y \qquad (3)$$
$$5x + 200 = 10(x + 10)$$
$$5x + 200 = 10x + 100$$
$$100 = 5x$$
$$20 = x$$

He should use 20 L of 5% solution.

38. **[4.3]** Let x, y, and z denote the number of medals won by Germany, the United States, and Canada, respectively.

The total number of medals won was 93, so

$$x + y + z = 93. \quad (1)$$

Germany won seven fewer medals than the United States, so

$$x = y - 7. \quad (2)$$

Canada won eleven fewer medals than the United States, so

$$z = y - 11. \quad (3)$$

Substitute $y - 7$ for x and $y - 11$ for z in (1).

$$(y - 7) + y + (y - 11) = 93$$
$$3y - 18 = 93$$
$$3y = 111$$
$$y = 37$$

From (2), $x = 37 - 7 = 30$.
From (3), $z = 37 - 11 = 26$.

Germany won 30 medals, the United States won 37 medals, and Canada won 26 medals.

Chapter 4 Test

1. **(a)** Rising graphs indicate population growth, so Houston, Phoenix, and Dallas will experience population growth.

(b) Philadelphia's graph indicates that it will experience population decline.

(c) In the year 2000, the city populations from least to greatest are Dallas, Phoenix, Philadelphia, and Houston.

2. **(a)** The graphs for Dallas and Philadelphia intersect in the year 2010. The population for each city will be about 1.45 million.

(b) The graphs for Houston and Phoenix appear to intersect in the year 2025. The population for each city will be about 2.8 million. This can be represented by the ordered pair $(2025, 2.8)$.

3. When each equation of the system

$$x + y = 7$$
$$x - y = 5$$

is graphed, the point of intersection appears to be $(6, 1)$. To check, substitute 6 for x and 1 for y in each of the equations. Since $(6, 1)$ makes both equations true, the solution set of the system is $\{(6, 1)\}$.

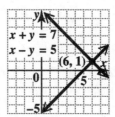

4.
$$2x - 3y = 24 \qquad (1)$$
$$y = -\tfrac{2}{3}x \qquad (2)$$

Since equation (2) is solved for y, substitute $-\tfrac{2}{3}x$ for y in equation (1) and solve for x.

$$2x - 3y = 24 \qquad (1)$$
$$2x - 3(-\tfrac{2}{3}x) = 24$$
$$2x + 2x = 24$$
$$4x = 24$$
$$x = 6$$

From (2), $y = -\tfrac{2}{3}(6) = -4$.

The solution set is $\{(6, -4)\}$.

5.
$$3x - y = -8 \quad (1)$$
$$2x + 6y = 3 \quad (2)$$

To eliminate y, multiply equation (1) by 6. Then add that equation and equation (2).

Copyright © 2012 Pearson Education, Inc. Publishing as Addison-Wesley.

$$18x - 6y = -48 \qquad 6 \times (1)$$
$$\underline{2x + 6y = 3 \qquad (2)}$$
$$20x = -45$$
$$x = -\tfrac{45}{20} = -\tfrac{9}{4}$$

To find y, substitute $-\tfrac{9}{4}$ for x in equation (2).
(We could also use elimination by adding $2 \times (1)$ and $-3 \times (2)$.)

$$2x + 6y = 3 \qquad (2)$$
$$2(-\tfrac{9}{4}) + 6y = 3$$
$$-\tfrac{9}{2} + 6y = 3$$
$$6y = \tfrac{15}{2}$$
$$y = \tfrac{15}{12} = \tfrac{5}{4}$$

The solution set is $\left\{-\tfrac{9}{4}, \tfrac{5}{4}\right\}$.

6.
$$12x - 5y = 8 \qquad (1)$$
$$3x = \tfrac{5}{4}y + 2$$
or
$$x = \tfrac{5}{12}y + \tfrac{2}{3} \qquad (2)$$

Substitute $\tfrac{5}{12}y + \tfrac{2}{3}$ for x in equation (1) and solve for y.

$$12x - 5y = 8 \qquad (1)$$
$$12(\tfrac{5}{12}y + \tfrac{2}{3}) - 5y = 8$$
$$5y + 8 - 5y = 8$$
$$8 = 8 \qquad \textit{True}$$

Equations (1) and (2) are dependent.
The solution set is $\{(x, y) \mid 12x - 5y = 8\}$.

7.
$$3x + y = 12 \qquad (1)$$
$$2x - y = 3 \qquad (2)$$

To eliminate y, add equations (1) and (2).

$$3x + y = 12 \qquad (1)$$
$$\underline{2x - y = 3 \qquad (2)}$$
$$5x = 15$$
$$x = 3$$

Substitute 3 for x in equation (1) to find y.

$$3x + y = 12 \qquad (1)$$
$$3(3) + y = 12$$
$$9 + y = 12$$
$$y = 3$$

The solution set is $\{(3, 3)\}$.

8.
$$-5x + 2y = -4 \qquad (1)$$
$$6x + 3y = -6 \qquad (2)$$

To eliminate x, multiply equation (1) by 6 and equation (2) by 5. Then add the results.

$$-30x + 12y = -24 \qquad 6 \times (1)$$
$$\underline{30x + 15y = -30 \qquad 5 \times (2)}$$
$$27y = -54$$
$$y = -2$$

Substitute -2 for y in equation (1) to find x.

$$-5x + 2y = -4 \qquad (1)$$
$$-5x + 2(-2) = -4$$
$$-5x - 4 = -4$$
$$-5x = 0$$
$$x = 0$$

The solution set is $\{(0, -2)\}$.

9.
$$3x + 4y = 8 \qquad (1)$$
$$8y = 7 - 6x$$
or
$$6x + 8y = 7 \qquad (2)$$

Multiply equation (1) by -2 and add the result to equation (2).

$$-6x - 8y = -16 \qquad -2 \times (1)$$
$$\underline{6x + 8y = 7 \qquad (2)}$$
$$0 = -9 \quad \textit{False}$$

Since a false statement results, the system is *inconsistent*. The solution set is $\emptyset$.

10.
$$3x + 5y + 3z = 2 \qquad (1)$$
$$6x + 5y + z = 0 \qquad (2)$$
$$3x + 10y - 2z = 6 \qquad (3)$$

To eliminate x, multiply equation (1) by -1 and add the result to equation (3).

$$-3x - 5y - 3z = -2 \qquad -1 \times (1)$$
$$\underline{3x + 10y - 2z = 6 \qquad (3)}$$
$$5y - 5z = 4 \qquad (4)$$

To eliminate x again, multiply equation (1) by -2 and add the result to equation (2).

$$-6x - 10y - 6z = -4 \qquad -2 \times (1)$$
$$\underline{6x + 5y + z = 0 \qquad (2)}$$
$$-5y - 5z = -4 \qquad (5)$$

To eliminate y, add equations (4) and (5).

$$5y - 5z = 4 \qquad (4)$$
$$\underline{-5y - 5z = -4 \qquad (5)}$$
$$-10z = 0$$
$$z = 0$$

Substitute 0 for z in equation (4) to find y.

$$5y - 5z = 4 \qquad (4)$$
$$5y - 5(0) = 4$$
$$5y - 0 = 4$$
$$5y = 4$$
$$y = \tfrac{4}{5}$$

Substitute $\tfrac{4}{5}$ for y and 0 for z in equation (1) to find x.

Copyright © 2012 Pearson Education, Inc. Publishing as Addison-Wesley.

$$3x + 5y + 3z = 2 \quad (1)$$
$$3x + 5(\tfrac{4}{5}) + 3(0) = 2$$
$$3x + 4 + 0 = 2$$
$$3x = -2$$
$$x = -\tfrac{2}{3}$$

The solution set is $\left\{\left(-\tfrac{2}{3}, \tfrac{4}{5}, 0\right)\right\}$.

11.
$$4x + y + z = 11 \quad (1)$$
$$x - y - z = 4 \quad (2)$$
$$y + 2z = 0 \quad (3)$$

To eliminate x, multiply equation (2) by -4 and add the result to equation (1).

$$4x + y + z = 11 \quad (1)$$
$$-4x + 4y + 4z = -16 \quad -4 \times (2)$$
$$\overline{\quad 5y + 5z = -5} \quad (4)$$

To eliminate y, divide equation (4) by -5 and add the result to equation (3).

$$-y - z = 1 \quad (4) \div (-5)$$
$$y + 2z = 0 \quad (3)$$
$$\overline{\quad z = 1}$$

From (3), $y + 2(1) = 0$, so $y = -2$.
From (2), $x - (-2) - 1 = 4$, so $x = 3$.

The solution set is $\{(3, -2, 1)\}$.

12. Let $x =$ the gross (in millions of dollars) for *Star Wars Episode IV: A New Hope*, and $y =$ the gross (in millions of dollars) for *Indiana Jones and the Kingdom of the Crystal Skull*.

Together the movies grossed $778.0 million, so

$$x + y = 778.0. \quad (1)$$

Indiana Jones and the Kingdom of the Crystal Skull grossed $144.0 million less than *Star Wars Episode IV: A New Hope*, so

$$y = x - 144.0. \quad (2)$$

Substitute $x - 144.0$ for y in equation (1).

$$x + y = 778.0 \quad (1)$$
$$x + (x - 144.0) = 778.0$$
$$2x = 922.0$$
$$x = 461.0$$

From (2), $y = 461.0 - 144.0 = 317.0$.

Star Wars Episode IV: A New Hope grossed $461.0 million and *Indiana Jones and the Kingdom of the Crystal Skull* grossed $317.0 million.

13. Let $x =$ the rate of the faster car and $y =$ the rate of the slower car.

Make a table.

	r	t	d
Faster Car	x	3.5	$3.5x$
Slower Car	y	3.5	$3.5y$

Since the slow car travels 30 mph slower than the fast car,

$$x - y = 30. \quad (1)$$

Since the cars travel a total of 420 miles,

$$3.5x + 3.5y = 420.$$

Multiply by 10 to clear the decimals.

$$35x + 35y = 4200 \quad (2)$$

To eliminate y, multiply equation (1) by 35 and add the result to equation (2).

$$35x - 35y = 1050 \quad 35 \times (1)$$
$$35x + 35y = 4200 \quad (2)$$
$$\overline{70x = 5250}$$
$$x = 75$$

Substitute 75 for x in equation (1) to find y.

$$x - y = 30 \quad (1)$$
$$75 - y = 30$$
$$-y = -45$$
$$y = 45$$

The faster car is traveling at 75 mph, and the slower car is traveling at 45 mph.

14. Let $x =$ the number of liters of 20% solution and $y =$ the number of liters of 50% solution.

Make a table.

Liters of Solution	Percent (as a decimal)	Liters of Pure Alcohol
x	0.20	$0.20x$
y	0.50	$0.50y$
12	0.40	$0.40(12) = 4.8$

Since 12 L of the mixture are needed,

$$x + y = 12. \quad (1)$$

Since the amount of pure alcohol in the 20% solution plus the amount of pure alcohol in the 50% solution must equal the amount of alcohol in the mixture,

$$0.2x + 0.5y = 4.8.$$

Multiply by 10 to clear the decimals.

$$2x + 5y = 48 \quad (2)$$

Multiply equation (1) by -2 and add the result to equation (2).

Copyright © 2012 Pearson Education, Inc. Publishing as Addison-Wesley.

$$-2x \; - \; 2y \; = \; -24 \qquad -2 \times (1)$$
$$\underline{\; 2x \; + \; 5y \; = \quad 48 \qquad (2)}$$
$$3y \; = \quad 24$$
$$y \; = \; 8$$

From (1), $x + 8 = 12$, so $x = 4$.
4 L of 20% solution and 8 L of 50% solution are needed.

15. Let $x =$ the price of an AC adaptor
and $y =$ the price of a rechargeable flashlight.

Since 7 AC adaptors and 2 rechargeable flashlights cost $86,

$$7x + 2y = 86. \quad (1)$$

Since 3 AC adaptors and 4 rechargeable flashlights cost $84,

$$3x + 4y = 84. \quad (2)$$

Solve the system.

$$7x \; + \; 2y \; = \; 86 \quad (1)$$
$$3x \; + \; 4y \; = \; 84 \quad (2)$$

To eliminate y, multiply equation (1) by -2 and add the result to equation (2).

$$-14x \; - \; 4y \; = \; -172 \qquad -2 \times (1)$$
$$\underline{\;\; 3x \; + \; 4y \; = \quad\;\; 84 \qquad (2)}$$
$$-11x \qquad\quad = \; -88$$
$$x \; = \; 8$$

Substitute 8 for x in equation (1) to find y.

$$7x + 2y = 86 \qquad (1)$$
$$7(8) + 2y = 86$$
$$56 + 2y = 86$$
$$2y = 30$$
$$y = 15$$

An AC adaptor costs $8, and a rechargeable flashlight costs $15.

16. Let $x =$ the amount of Orange Pekoe, $y =$ the amount of Irish Breakfast, and $z =$ the amount of Earl Grey.

The owner wants 100 oz of tea, so

$$x + y + z = 100. \quad (1)$$

An equation which relates the prices of the tea is

$$0.80x + 0.85y + 0.95z = 0.83(100).$$

Multiply by 100 to clear the decimals.

$$80x + 85y + 95z = 8300 \quad (2)$$

The mixture must contain twice as much Orange Pekoe as Irish Breakfast, so

$$x = 2y. \quad (3)$$

To eliminate z, multiply equation (1) by -95 and add the result to equation (2).

$$-95x \; - \; 95y \; - \; 95z \; = \; -9500 \qquad -95 \times (1)$$
$$\underline{\;\; 80x \; + \; 85y \; + \; 95z \; = \quad 8300 \qquad (2)}$$
$$-15x \; - \; 10y \qquad\qquad = \; -1200 \qquad (4)$$

Divide equation (4) by -5.

$$3x + 2y = 240 \quad (5)$$

Substitute $2y$ for x in equation (5) to find y.

$$3x + 2y = 240 \qquad (5)$$
$$3(2y) + 2y = 240$$
$$8y = 240$$
$$y = 30$$

From (3), $x = 2(30) = 60$.
Substitute 60 for x and 30 for y in equation (1) to find z.

$$x + y + z = 100 \qquad (1)$$
$$60 + 30 + z = 100$$
$$z = 10$$

He should use 60 oz of Orange Pekoe, 30 oz of Irish Breakfast, and 10 oz of Earl Grey.

17.
$$3x \; + \; 2y \; = \; 4$$
$$5x \; + \; 5y \; = \; 9$$

Write the augmented matrix.

$$\begin{bmatrix} 3 & 2 & | & 4 \\ 5 & 5 & | & 9 \end{bmatrix}$$

We could divide row 1 by 3, but to avoid working with fractions, we'll multiply row 1 by 2 and then multiply row 2 by -1 and add to row 1 to obtain a "1" in the first row.

$$\begin{bmatrix} 6 & 4 & | & 8 \\ 5 & 5 & | & 9 \end{bmatrix} \qquad 2R_1$$

$$\begin{bmatrix} 1 & -1 & | & -1 \\ 5 & 5 & | & 9 \end{bmatrix} \qquad -R_2 + R_1$$

$$\begin{bmatrix} 1 & -1 & | & -1 \\ 0 & 10 & | & 14 \end{bmatrix} \qquad -5R_1 + R_2$$

$$\begin{bmatrix} 1 & -1 & | & -1 \\ 0 & 1 & | & \frac{7}{5} \end{bmatrix} \qquad \frac{1}{10}R_2$$

This matrix gives the system

$$x - y = -1$$
$$y = \tfrac{7}{5}.$$

Substitute $y = \frac{7}{5}$ in the first equation.

$$x - \tfrac{7}{5} = -1$$
$$x = -1 + \tfrac{7}{5}$$
$$= -\tfrac{5}{5} + \tfrac{7}{5} = \tfrac{2}{5}$$

The solution set is $\left\{ \left(\frac{2}{5}, \frac{7}{5} \right) \right\}$.

Copyright © 2012 Pearson Education, Inc. Publishing as Addison-Wesley.

18.
$$x + 3y + 2z = 11$$
$$3x + 7y + 4z = 23$$
$$5x + 3y - 5z = -14$$

Write the augmented matrix.

$$\begin{bmatrix} 1 & 3 & 2 & | & 11 \\ 3 & 7 & 4 & | & 23 \\ 5 & 3 & -5 & | & -14 \end{bmatrix}$$

$$\begin{bmatrix} 1 & 3 & 2 & | & 11 \\ 0 & -2 & -2 & | & -10 \\ 0 & -12 & -15 & | & -69 \end{bmatrix} \begin{matrix} \\ -3R_1 + R_2 \\ -5R_1 + R_3 \end{matrix}$$

$$\begin{bmatrix} 1 & 3 & 2 & | & 11 \\ 0 & 1 & 1 & | & 5 \\ 0 & -12 & -15 & | & -69 \end{bmatrix} \begin{matrix} \\ -\frac{1}{2}R_2 \\ \\ \end{matrix}$$

$$\begin{bmatrix} 1 & 3 & 2 & | & 11 \\ 0 & 1 & 1 & | & 5 \\ 0 & 0 & -3 & | & -9 \end{bmatrix} \begin{matrix} \\ \\ 12R_2 + R_3 \end{matrix}$$

$$\begin{bmatrix} 1 & 3 & 2 & | & 11 \\ 0 & 1 & 1 & | & 5 \\ 0 & 0 & 1 & | & 3 \end{bmatrix} \begin{matrix} \\ \\ -\frac{1}{3}R_3 \end{matrix}$$

This matrix gives the system

$$x + 3y + 2z = 11$$
$$y + z = 5$$
$$z = 3.$$

Substitute $z = 3$ in the second equation.

$$y + z = 5$$
$$y + 3 = 5$$
$$y = 2$$

Substitute $y = 2$ and $z = 3$ in the first equation.

$$x + 3y + 2z = 11$$
$$x + 3(2) + 2(3) = 11$$
$$x + 6 + 6 = 11$$
$$x = -1$$

The solution set is $\{(-1, 2, 3)\}$.

Cumulative Review Exercises (Chapters 1–4)

1.
$$-\frac{3}{4} - \frac{2}{5} = -\frac{15}{20} - \frac{8}{20}$$
$$= \frac{-15 - 8}{20} = \frac{-23}{20} = -\frac{23}{20}$$

2.
$$\frac{8}{15} \div \left(-\frac{12}{5}\right) = \frac{8}{15}\left(-\frac{5}{12}\right)$$
$$= -\frac{2 \cdot 4 \cdot 5}{3 \cdot 5 \cdot 3 \cdot 4}$$
$$= -\frac{2}{3 \cdot 3} = -\frac{2}{9}$$

3. $(-3)^4 = (-3)(-3)(-3)(-3) = 81$

4. $-3^4 = -(3)(3)(3)(3) = -81$

5. $-(-3)^4 = -(-3)(-3)(-3)(-3) = -81$

6. $\sqrt{0.49} = 0.7$, since 0.7 is positive and $(0.7)^2 = 0.49$.

7. $-\sqrt{0.49} = -0.7$, since $(0.7)^2 = 0.49$ and the negative sign in front of the radical must be applied.

8. $\sqrt{-0.49}$ is not a real number because of the negative sign under the radical. No real number squared is negative.

In Exercises 9–10, let $x = -4$, $y = 3$, and $z = 6$.

9. $|2x| + 3y - z^3$
$$= |(2)(-4)| + 3(3) - (6)^3$$
$$= |-8| + 9 - 216$$
$$= 8 + 9 - 216$$
$$= -199$$

10. $-5(x^3 - y^3) = -5[(-4)^3 - (3)^3]$
$$= -5(-64 - 27)$$
$$= -5(-91)$$
$$= 455$$

11. The *commutative property* says that $3 \cdot 6 = 6 \cdot 3$, so that is the property that justifies the given statement.

12. $7(2x + 3) - 4(2x + 1) = 2(x + 1)$
$$14x + 21 - 8x - 4 = 2x + 2$$
$$6x + 17 = 2x + 2$$
$$4x = -15$$
$$x = -\frac{15}{4}$$

The solution set is $\left\{-\frac{15}{4}\right\}$.

13. $|6x - 8| = 4$

$$6x - 8 = 4 \quad \text{or} \quad 6x - 8 = -4$$
$$6x = 12 \qquad\qquad 6x = 4$$
$$x = 2 \quad \text{or} \quad x = \frac{4}{6} = \frac{2}{3}$$

The solution set is $\left\{\frac{2}{3}, 2\right\}$.

14. $ax + by = d$

To solve for x, get the term with x alone on one side of the equals symbol.

$$ax = d - by \quad \textit{Subtract the term by.}$$
$$x = \frac{d - by}{a} \quad \textit{Divide by the coefficient a.}$$

15. $0.04x + 0.06(x - 1) = 1.04$

Multiply both sides by 100 to clear the decimals.
$$4x + 6(x - 1) = 104$$
$$4x + 6x - 6 = 104$$
$$10x - 6 = 104$$
$$10x = 110$$
$$x = 11$$

The solution set is $\{11\}$.

Copyright © 2012 Pearson Education, Inc. Publishing as Addison-Wesley.

16. $\frac{2}{3}x + \frac{5}{12}x \le 20$

$12(\frac{2}{3}x + \frac{5}{12}x) \le 12(20)$ *Multiply by 12.*

$\qquad 8x + 5x \le 240$

$\qquad\qquad 13x \le 240$

$\qquad\qquad\quad x \le \frac{240}{13}$

The solution set is $(-\infty, \frac{240}{13}]$.

17. $|3x + 2| \le 4$

$-4 \le 3x + 2 \le 4$

$-6 \le 3x \qquad\;\; \le 2$ *Subtract 2.*

$-2 \le x \qquad\quad\;\; \le \frac{2}{3}$ *Divide by 3.*

The solution set is $[-2, \frac{2}{3}]$.

18. $|12t + 7| \ge 0$

The solution set is $(-\infty, \infty)$ since the absolute value of any number is greater than or equal to 0.

19. $2x + 3 > 5 \quad$ or $\quad x - 1 < 6$

$\quad\; 2x > 2$

$\qquad x > 1 \quad$ or $\qquad x < 7$

Any number is either greater than 1 or less than 7, so the solution set is $(-\infty, \infty)$.

20.   80.4% of 2500 is $0.804(2500) = 2010$

72.5% of 2500 is $0.725(2500) = 1812.5 \approx 1813$

$\dfrac{1570}{2500} = 0.628$ or 62.8%

$\dfrac{1430}{2500} = 0.572$ or 57.2%

Product or Company	Percent Recognition	Actual Number
Charmin	80.4%	2010
Wheaties	72.5%	1813
Budweiser	62.8%	1570
State Farm	57.2%	1430

21.   Let $x =$ the number of nickels,

$x + 1 =$ the number of dimes, and

$x + 6 =$ the number of pennies.

The total value is \$4.80, so

$0.05x + 0.10(x + 1) + 0.01(x + 6) = 4.80$.

Multiply both sides by 100 to clear the decimals.

$5x + 10(x + 1) + 1(x + 6) = 480$

$5x + 10x + 10 + x + 6 = 480$

$\qquad\qquad 16x + 16 = 480$

$\qquad\qquad\qquad 16x = 464$

$\qquad\qquad\qquad\quad x = 29$

Then, $x + 1 = 29 + 1 = 30$,

and $x + 6 = 29 + 6 = 35$.

There are 35 pennies, 29 nickels, and 30 dimes in the jar.

22.   Let $x =$ the measure of the equal angles and $2x - 4 =$ the measure of the third angle.

The sum of the measures of the angles in a triangle is 180°, so

$x + x + (2x - 4) = 180$

$\qquad\quad 4x - 4 = 180$

$\qquad\qquad 4x = 184$

$\qquad\qquad\; x = 46.$

So, $2x - 4 = 2(46) - 4 = 92 - 4 = 88$.

The measures of the angles are 46°, 46°, and 88°.

23.   A horizontal line through the point (x, k) has equation $y = k$. Since point A has coordinates $(-2, 6)$, $k = 6$. The equation of the horizontal line through A is $y = 6$.

24.   A vertical line through the point (k, y) has equation $x = k$. Since point B has coordinates $(4, -2)$, $k = 4$. The equation of the vertical line through B is $x = 4$.

25.   Let $(x_1, y_1) = (-2, 6)$ and $(x_2, y_2) = (4, -2)$. Then,

$$m = \frac{y_2 - y_1}{x_2 - x_1} = \frac{-2 - 6}{4 - (-2)} = \frac{-8}{6} = -\frac{4}{3}.$$

The slope is $-\frac{4}{3}$.

26.   Perpendicular lines have slopes that are negative reciprocals of each other. The slope of line AB is $-\frac{4}{3}$ (from Exercise 25). The negative reciprocal of $-\frac{4}{3}$ is $\frac{3}{4}$, so the slope of a line perpendicular to line AB is $\frac{3}{4}$.

27.   Let $m = -\frac{4}{3}$ and $(x_1, y_1) = (4, -2)$ in the point-slope form.

$$y - y_1 = m(x - x_1)$$
$$y - (-2) = -\frac{4}{3}(x - 4)$$
$$y + 2 = -\frac{4}{3}x + \frac{16}{3}$$

Multiply by 3 to clear the fractions, and then write the equation in standard form, $Ax + By = C$.

$$3y + 6 = -4x + 16$$
$$4x + 3y = 10$$

28.   First locate the point $(-1, -3)$ on a graph. Then use the definition of slope to find a second point on the line.

$$m = \frac{\text{change in } y}{\text{change in } x} = \frac{2}{3}$$

From $(-1, -3)$, move 2 units up and 3 units to the right. The line through $(-1, -3)$ and the new point, $(2, -1)$, is the graph.

Copyright © 2012 Pearson Education, Inc. Publishing as Addison-Wesley.

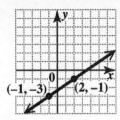

(−1, −3)   (2, −1)

29. $-3x - 2y \le 6$

Graph the line $-3x - 2y = 6$ through its intercepts, $(-2, 0)$ and $(0, -3)$, as a solid line, since the inequality involves $\le$.
To determine the region that belongs to the graph, test $(0, 0)$.

$$-3x - 2y \le 6$$
$$-3(0) - 2(0) \overset{?}{\le} 6$$
$$0 \le 6 \qquad True$$

Since the result is true, shade the region that includes $(0, 0)$.

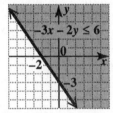

30. $f(x) = x^2 + 3x - 6$

(a) $f(-3) = (-3)^2 + 3(-3) - 6$
$$= 9 - 9 - 6 = -6$$

(b) $f(a) = (a)^2 + 3(a) - 6$
$$= a^2 + 3a - 6$$

31.
$$-2x + 3y = -15 \quad (1)$$
$$4x - y = 15 \quad (2)$$

To eliminate x, multiply equation (1) by 2 and add the result to equation (2).

$$-4x + 6y = -30 \qquad 2 \times (1)$$
$$\underline{4x - y = 15 \quad (2)}$$
$$5y = -15$$
$$y = -3$$

Substitute -3 for y in equation (2) to find x.
$$4x - y = 15 \qquad (2)$$
$$4x - (-3) = 15$$
$$4x + 3 = 15$$
$$4x = 12$$
$$x = 3$$

The solution set is $\{(3, -3)\}$.

32.
$$x - 3y = 7 \qquad (1)$$
$$2x - 6y = 14 \qquad (2)$$

Multiplying equation (1) by 2 gives equation (2). Since those equations are the same, the two equations are dependent.
The solution set is $\{(x, y) \mid x - 3y = 7\}$.

33.
$$x + y + z = 10 \quad (1)$$
$$x - y - z = 0 \quad (2)$$
$$-x + y - z = -4 \quad (3)$$

Add equations (1) and (2) to eliminate y and z. The result is

$$2x = 10$$
$$x = 5.$$

Add equations (2) and (3) to eliminate x and y. The result is

$$-2z = -4$$
$$z = 2.$$

Substitute 5 for x and 2 for z in equation (1) to find y.

$$x + y + z = 10 \qquad (1)$$
$$5 + y + 2 = 10$$
$$y + 7 = 10$$
$$y = 3$$

The solution set is $\{(5, 3, 2)\}$.

34. Let $x =$ the average cost of the original Tickle Me Elmo
and $y =$ the recommended cost of T.M.X.

The original's cost was \$12.37 less than T.M.X.'s, so

$$x = y - 12.37. \quad (1)$$

Together they cost \$67.63, so

$$x + y = 67.63. \quad (2)$$

From (1), substitute $y - 12.37$ for x in equation (2).

$$(y - 12.37) + y = 67.63$$
$$2y - 12.37 = 67.63$$
$$2y = 80$$
$$y = 40$$

From (1), $x = y - 12.37 = 40 - 12.37 = 27.63$. The average cost of the original Elmo was \$27.63 and the recommended cost of a T.M.X. is \$40.

35. (a) The lines intersect at $(8, 3000)$, so the cost equals the revenue at $x = 8$ (which is 800 items). The revenue is \$3000.

(b) On the sale of 1100 parts ($x = 11$), the revenue is about \$4100 and the cost is about \$3700.

$$\text{Profit} = \text{Revenue} - \text{Cost}$$
$$\approx 4100 - 3700$$
$$= 400$$

The profit is about \$400.

Copyright © 2012 Pearson Education, Inc. Publishing as Addison-Wesley.

CHAPTER 5   EXPONENTS, POLYNOMIALS, AND POLYNOMIAL FUNCTIONS

5.1   Integer Exponents and Scientific Notation

5.1   Now Try Exercises

N1. Apply the product rule for exponents, if possible.

(a) $8^5 \cdot 8^4 = 8^{5+4} = 8^9$

(b) $(5x^4y^7)(-7xy^3) = -5(7)x^4xy^7y^3$
$$= -35x^{4+1}y^{7+3}$$
$$= -35x^5y^{10}$$

(c) $p^2 \cdot q^2$ cannot be simplified further because the bases p and q are not the same. The product rule does not apply.

N2. **(a)** $5^0 = 1$

Any real number (except 0) raised to the power zero is equal to 1.

(b) Since $x \neq 0$, $5x \neq 0$, and $(-5x)^0 = 1$.

(c) $-5^0 = -(5^0) = -1$

(d) $10^0 - 9^0 = 1 - 1 = 0$

N3. **(a)** $9^{-4} = \dfrac{1}{9^4}$

(b) $(3y)^{-6} = \dfrac{1}{(3y)^6}, y \neq 0$

(c) $-4k^{-3} = -4\left(\dfrac{1}{k^3}\right) = -\dfrac{4}{k^3}, k \neq 0$

(d) $4^{-1} + 6^{-1} = \dfrac{1}{4} + \dfrac{1}{6} = \dfrac{3}{12} + \dfrac{2}{12} = \dfrac{5}{12}$

N4. **(a)** $\dfrac{1}{5^{-3}} = \dfrac{1}{\frac{1}{5^3}} = 1 \div \dfrac{1}{5^3} = 1 \cdot \dfrac{5^3}{1} = 5^3 = 125$

(b) $\dfrac{10^{-2}}{2^{-5}} = \dfrac{\frac{1}{10^2}}{\frac{1}{2^5}} = \dfrac{1}{10^2} \div \dfrac{1}{2^5} = \dfrac{1}{10^2} \cdot \dfrac{2^5}{1}$
$$= \dfrac{1}{100} \cdot \dfrac{32}{1} = \dfrac{32}{100} = \dfrac{8 \cdot 4}{25 \cdot 4} = \dfrac{8}{25}$$

N5. Apply the quotient rule for exponents, if possible.

(a) $\dfrac{t^8}{t^2} = t^{8-2} = t^6, t \neq 0$

(b) $\dfrac{4^5}{4^{-2}} = 4^{5-(-2)} = 4^7$

(c) $\dfrac{m^4}{n^3}, n \neq 0$, cannot be simplified because the bases m and n are different. The quotient rule does not apply.

N6. **(a)** $(-2m^3)^4 = (-2)^4(m^3)^4 = 16m^{3 \cdot 4} = 16m^{12}$

(b) $\left(\dfrac{3x^2}{y^3}\right)^3 = \dfrac{3^3(x^2)^3}{(y^3)^3} = \dfrac{27x^{2 \cdot 3}}{y^{3 \cdot 3}} = \dfrac{27x^6}{y^9}$, $y \neq 0$

N7. $\left(\dfrac{5}{3}\right)^{-3} = \left(\dfrac{3}{5}\right)^3 = \dfrac{3^3}{5^3} = \dfrac{27}{125}$

N8. **(a)** $x^{-8} \cdot x \cdot x^4 = x^{-8+1+4} = x^{-3} = \dfrac{1}{x^3}$

(b) $(5^{-3})^2 = 5^{-3(2)} = 5^{-6} = \dfrac{1}{5^6}$

(c) $\dfrac{p^2q^{-4}}{p^{-2}q^{-1}} = \dfrac{p^2}{p^{-2}} \cdot \dfrac{q^{-4}}{q^{-1}}$
$$= p^{2-(-2)}q^{-4-(-1)}$$
$$= p^4q^{-3} = \dfrac{p^4}{q^3}$$

(d) $\left(\dfrac{2x^2}{y^2}\right)^3\left(\dfrac{-5x^{-2}}{y}\right)^{-2}$
$$= \dfrac{2^3x^6}{y^6} \cdot \dfrac{(-5)^{-2}x^4}{y^{-2}} \quad \textit{Combination of rules}$$
$$= \dfrac{2^3x^6(-5)^{-2}x^4}{y^6y^{-2}}$$
$$= \dfrac{8x^{10}}{(-5)^2y^4} = \dfrac{8x^{10}}{25y^4}$$

N9. **(a)** $7{,}560{,}000{,}000 = 7\wedge560{,}000{,}000.$

Place a caret after the first nonzero digit, 7. Count 9 places from the decimal point (understood to be after the last 0) to the caret. Use a positive exponent on 10 since $7{,}560{,}000{,}000 > 7.56$.

$$7{,}560{,}000{,}000 = 7.56 \times 10^9$$

(b) $-0.000\,000\,245 = 0.000\,000\,2\wedge45$

Count 7 places. Use a negative exponent on 10 since $0.000\,000\,245 < 2.45$.

$$-0.000\,000\,245 = -2.45 \times 10^{-7}$$

N10. **(a)** $4.45 \times 10^{10} = 4.4{,}500{,}000{,}000.$
$$= 44{,}500{,}000{,}000$$

Move the decimal 10 places to the right.

(b) $-5.9 \times 10^{-5} = -0.00005.9 = -0.000\,059$

Move the decimal 5 places to the left.

Copyright © 2012 Pearson Education, Inc. Publishing as Addison-Wesley.

N11. $\dfrac{0.00063 \times 400{,}000}{1400 \times 0.000003} = \dfrac{6.3 \times 10^{-4} \times 4 \times 10^{5}}{1.4 \times 10^{3} \times 3 \times 10^{-6}}$

$= \dfrac{6.3 \times 4 \times 10^{-4} \times 10^{5}}{1.4 \times 3 \times 10^{3} \times 10^{-6}}$

$= \dfrac{6.3 \times 4 \times 10^{1}}{1.4 \times 3 \times 10^{-3}}$

$= \dfrac{6.3 \times 4}{1.4 \times 3} \times 10^{4}$

$= 6 \times 10^{4} = 60{,}000$

N12. $d = rt$, so

$t = \dfrac{d}{r}$ [Note that $30{,}000{,}000{,}000 = 3 \times 10^{10}$.]

$= \dfrac{1.2 \times 10^{15}}{3 \times 10^{10}}$

$= \dfrac{1.2}{3} \times 10^{15-10}$

$= 0.4 \times 10^{5} = 4.0 \times 10^{4}$, or $40{,}000$ sec.

It would take 4.0×10^{4} seconds.

5.1 Section Exercises

1. $(ab)^2 = a^2 b^2$ by a power rule. Since $a^2 b^2 \neq ab^2$, the expression $(ab)^2 = ab^2$ has been simplified incorrectly. The exponent should apply to both a and b.

3. $\left(\dfrac{3}{a}\right)^4 = \dfrac{3^4}{a^4}$, $a \neq 0$

Since $\dfrac{3^4}{a^4} \neq \dfrac{3^4}{a}$, the expression

$$\left(\dfrac{3}{a}\right)^4 = \dfrac{3^4}{a}$$

has been simplified incorrectly.

5. $x^3 \cdot x^4 = x^{3+4} = x^7$ is correct.

7. Your friend multiplied the bases, which is incorrect. Instead, keep the same base and add the exponents.

$$4^5 \cdot 4^2 = 4^{5+2} = 4^7$$

9. $13^4 \cdot 13^8 = 13^{4+8} = 13^{12}$

11. $8^9 \cdot 8 = 8^9 \cdot 8^1 = 8^{9+1} = 8^{10}$

13. $x^3 \cdot x^5 \cdot x^9 = x^{3+5+9} = x^{17}$

15. $(-3w^5)(9w^3) = (-3)(9)w^{5+3} = -27w^8$

17. $(2x^2 y^5)(9xy^3) = (2)(9)x^{2+1}y^{5+3} = 18x^3 y^8$

19. $r^2 \cdot s^4$ cannot be simplified because the product rule does not apply.

21. (a) $9^0 = 1$ **(B)**

(b) $-9^0 = -(9^0) = -(1) = -1$ **(C)**

(c) $(-9)^0 = 1$ **(B)**

(d) $-(-9)^0 = -1$ **(C)**

23. $15^0 = 1$, since $a^0 = 1$ for any nonzero base a.

25. $-8^0 = -(8^0) = -(1) = -1$

27. $(-25)^0 = 1$ since -25 is in parentheses.

29. $3^0 + (-3)^0 = 1 + 1 = 2$

31. $-3^0 + 3^0 = -1 + 1 = 0$

33. $-4^0 - m^0 = -1 - 1 = -2$

35. (a) $5^{-2} = \dfrac{1}{5^2} = \dfrac{1}{25}$ **(B)**

(b) $-5^{-2} = -\dfrac{1}{5^2} = -\dfrac{1}{25}$ **(D)**

(c) $(-5)^{-2} = \dfrac{1}{(-5)^2} = \dfrac{1}{25}$ **(B)**

(d) $-(-5)^{-2} = -\dfrac{1}{(-5)^2} = -\dfrac{1}{25}$ **(D)**

37. $5^{-4} = \dfrac{1}{5^4}$, or $\dfrac{1}{625}$

39. $9^{-1} = \dfrac{1}{9^1} = \dfrac{1}{9}$

41. $(4x)^{-2} = \dfrac{1}{(4x)^2} = \dfrac{1}{4^2 x^2} = \dfrac{1}{16x^2}$

43. $4x^{-2} = \dfrac{4}{x^2}$

45. $-a^{-3} = -\dfrac{1}{a^3}$

47. $(-a)^{-4} = \dfrac{1}{(-a)^4} = \dfrac{1}{a^4}$

49. $5^{-1} + 6^{-1} = \dfrac{1}{5} + \dfrac{1}{6} = \dfrac{6}{30} + \dfrac{5}{30} = \dfrac{11}{30}$

51. $8^{-1} - 3^{-1} = \dfrac{1}{8} - \dfrac{1}{3} = \dfrac{3}{24} - \dfrac{8}{24} = -\dfrac{5}{24}$

53. $\dfrac{1}{4^{-2}} = 4^2 = 16$

55. $\dfrac{2^{-2}}{3^{-3}} = \dfrac{3^3}{2^2} = \dfrac{27}{4}$

57. $\left(\dfrac{2}{3}\right)^{-3} = \left(\dfrac{3}{2}\right)^3 = \dfrac{3^3}{2^3} = \dfrac{27}{8}$

59. $\left(\dfrac{4}{5}\right)^{-2} = \left(\dfrac{5}{4}\right)^2 = \dfrac{5^2}{4^2} = \dfrac{25}{16}$

61. (a) $\left(\dfrac{1}{3}\right)^{-1} = \left(\dfrac{3}{1}\right)^1 = 3$ **(B)**

(b) $\left(-\dfrac{1}{3}\right)^{-1} = \left(-\dfrac{3}{1}\right)^1 = -3$ **(D)**

Copyright © 2012 Pearson Education, Inc. Publishing as Addison-Wesley.

(c) $-\left(\dfrac{1}{3}\right)^{-1} = -\left(\dfrac{3}{1}\right)^{1} = -3$ **(D)**

(d) $-\left(-\dfrac{1}{3}\right)^{-1} = -\left(-\dfrac{3}{1}\right)^{1} = -(-3) = 3$ **(B)**

63. $\dfrac{4^8}{4^6} = 4^{8-6} = 4^2$, or 16

65. $\dfrac{x^{12}}{x^8} = x^{12-8} = x^4$

67. $\dfrac{r^7}{r^{10}} = r^{7-10} = r^{-3} = \dfrac{1}{r^3}$

69. $\dfrac{6^4}{6^{-2}} = 6^{4-(-2)} = 6^{4+2} = 6^6$

71. $\dfrac{6^{-3}}{6^7} = 6^{-3-7} = 6^{-10} = \dfrac{1}{6^{10}}$

73. $\dfrac{7}{7^{-1}} = 7^{1-(-1)} = 7^2$, or 49

75. $\dfrac{r^{-3}}{r^{-6}} = r^{-3-(-6)} = r^{-3+6} = r^3$

77. $\dfrac{x^3}{y^2}$ cannot be simplified because the quotient rule does not apply.

79. $(x^3)^6 = x^{3\cdot 6} = x^{18}$

81. $\left(\dfrac{3}{5}\right)^3 = \dfrac{3^3}{5^3} = \dfrac{27}{125}$

83. $(4t)^3 = 4^3 t^3 = 64t^3$

85. $(-6x^2)^3 = (-6)^3 x^{2\cdot 3} = -216x^6$

87. $\left(\dfrac{-4m^2}{t}\right)^3 = \dfrac{(-4)^3 m^{2\cdot 3}}{t^3} = \dfrac{-64m^6}{t^3} = -\dfrac{64m^6}{t^3}$

89. $\left(\dfrac{-s^3}{t^5}\right)^4 = \dfrac{(-1)^4 s^{3\cdot 4}}{t^{5\cdot 4}} = \dfrac{s^{12}}{t^{20}}$

91. $3^5 \cdot 3^{-6} = 3^{5+(-6)} = 3^{-1} = \dfrac{1}{3^1} = \dfrac{1}{3}$

93. $a^{-3} a^2 a^{-4} = a^{-3+2+(-4)} = a^{-5} = \dfrac{1}{a^5}$

95. $(k^2)^{-3} k^4 = k^{2(-3)} k^4$
$= k^{-6} k^4$
$= k^{-6+4}$
$= k^{-2}$
$= \dfrac{1}{k^2}$

97. $-4r^{-2}(r^4)^2 = -4r^{-2}(r^8)$
$= -4r^{-2+8}$
$= -4r^6$

99. $(5a^{-1})^4 (a^2)^{-3} = 5^4 a^{-1\cdot 4} a^{2(-3)}$
$= 5^4 a^{-4} a^{-6}$
$= 5^4 a^{-4-6}$
$= 5^4 a^{-10}$
$= \dfrac{5^4}{a^{10}} = \dfrac{625}{a^{10}}$

101. $(z^{-4} x^3)^{-1} = z^{-4(-1)} x^{3(-1)}$
$= z^4 x^{-3}$
$= \dfrac{z^4}{x^3}$

103. $7k^2(-2k)(4k^{-5})^0 = 7(-2)k^2 k \cdot 1$
$= -14k^{2+1}$
$= -14k^3$

105. $\dfrac{(p^{-2})^0}{5p^{-4}} = \dfrac{1\cdot p^4}{5} = \dfrac{p^4}{5}$

107. $\dfrac{(3pq)q^2}{6p^2 q^4} = \dfrac{3p^1 q^{1+2}}{6p^2 q^4} = \dfrac{3pq^3}{6p^2 q^4}$
$= \dfrac{1}{2} p^{1-2} q^{3-4}$
$= \dfrac{1}{2} p^{-1} q^{-1} = \dfrac{1}{2} \cdot \dfrac{1}{p^1} \cdot \dfrac{1}{q^1}$
$= \dfrac{1}{2pq}$

109. $\dfrac{4a^5(a^{-1})^3}{(a^{-2})^{-2}} = \dfrac{4a^5 a^{-1\cdot 3}}{a^{-2(-2)}} = \dfrac{4a^5 a^{-3}}{a^4}$
$= 4a^{5-3-4}$
$= 4a^{-2}$
$= \dfrac{4}{a^2}$

111. The first step may be the most confusing.

$$(-y^{-4})^2 = (-1 \cdot y^{-4})^2$$
$$= (-1)^2 (y^{-4})^2$$
$$= 1 \cdot y^{-4(2)} = y^{-8}$$

It is not necessary to include these steps once the concept of squaring a negative is committed to memory.

$$\dfrac{(-y^{-4})^2}{6(y^{-5})^{-1}} = \dfrac{y^{-4(2)}}{6y^{-5(-1)}}$$
$$= \dfrac{y^{-8}}{6y^5}$$
$$= \dfrac{1}{6y^5 y^8}$$
$$= \dfrac{1}{6y^{13}}$$

Copyright © 2012 Pearson Education, Inc. Publishing as Addison-Wesley.

113. $\dfrac{(2k)^2 m^{-5}}{(km)^{-3}} = \dfrac{2^2 k^2 m^{-5}}{k^{-3} m^{-3}}$

$= 2^2 k^{2-(-3)} m^{-5-(-3)}$

$= 2^2 k^5 m^{-2}$

$= \dfrac{2^2 k^5}{m^2} = \dfrac{4k^5}{m^2}$

115. $\dfrac{(2k)^2 k^3}{k^{-1} k^{-5}} (5k^{-2})^{-3} = \dfrac{2^2 k^2 k^3}{k^{-1} k^{-5}} \left(5^{-3} k^{-2(-3)}\right)$

$= \dfrac{2^2 k^5}{k^{-6}} \left(5^{-3} k^6\right)$

$= 2^2 k^{11} \left(5^{-3} k^6\right)$

$= 2^2 \cdot 5^{-3} k^{11+6}$

$= \dfrac{2^2 k^{17}}{5^3} = \dfrac{4k^{17}}{125}$

117. $\left(\dfrac{3k^{-2}}{k^4}\right)^{-1} \cdot \dfrac{2}{k} = \dfrac{(3k^{-2})^{-1}}{(k^4)^{-1}} \cdot \dfrac{2}{k}$

$= \dfrac{3^{-1} k^2 \cdot 2}{k^{-4} k^1}$

$= \dfrac{3^{-1} \cdot 2k^2}{k^{-3}}$

$= \dfrac{2k^2 k^3}{3} = \dfrac{2k^5}{3}$

119. $\left(\dfrac{2p}{q^2}\right)^3 \left(\dfrac{3p^4}{q^{-4}}\right)^{-1} = \dfrac{(2p)^3}{(q^2)^3} \cdot \dfrac{(3p^4)^{-1}}{(q^{-4})^{-1}}$

$= \dfrac{2^3 p^3}{q^6} \cdot \dfrac{3^{-1} p^{-4}}{q^4}$

$= \dfrac{2^3 p^{3-4}}{3^1 q^{6+4}}$

$= \dfrac{8p^{-1}}{3q^{10}} = \dfrac{8}{3pq^{10}}$

121. $\dfrac{2^2 y^4 (y^{-3})^{-1}}{2^5 y^{-2}} = \dfrac{y^4 y^3 y^2}{2^{5-2}}$

$= \dfrac{y^{4+3+2}}{2^3} = \dfrac{y^9}{8}$

123. $\left(\dfrac{5m^4 n^{-3}}{m^{-5} n^2}\right)^{-2} = (5m^9 n^{-5})^{-2}$

$= 5^{-2} (m^9)^{-2} (n^{-5})^{-2}$

$= \dfrac{1}{5^2} m^{-18} n^{10}$

$= \dfrac{n^{10}}{25m^{18}}$

125. $\left(\dfrac{-3x^4 y^6}{15x^{-6} y^7}\right)^{-3} = \left(\dfrac{x^{10} y^{-1}}{-5}\right)^{-3}$

$= \left(\dfrac{-5}{x^{10} y^{-1}}\right)^3$

$= \dfrac{-125}{x^{30} y^{-3}}$

$= -\dfrac{125 y^3}{x^{30}}$

127. $\dfrac{(2m^2 p^3)^2 (4m^2 p)^{-2}}{(-3mp^4)^{-1} (2m^3 p^4)^3}$

$= \dfrac{2^2 m^4 p^6 4^{-2} m^{-4} p^{-2}}{(-3)^{-1} m^{-1} p^{-4} 2^3 m^9 p^{12}}$ *Power rule*

$= \dfrac{4(-3)^1 (m^{4-4})(p^{6-2})}{4^2 \cdot 2^3 (m^{-1+9})(p^{-4+12})}$ *Product rule*

$= \dfrac{-3m^0 p^4}{4 \cdot 2^3 m^8 p^8}$

$= \dfrac{-3(p^{4-8})}{32m^8}$ *Quotient rule*

$= \dfrac{-3p^{-4}}{32m^8} = -\dfrac{3}{32m^8 p^4}$

129. $\dfrac{(-3y^3 x^3)(-4y^4 x^2)(x^2)^{-4}}{18x^3 y^2 (y^3)^3 (x^3)^{-2}}$

$= \dfrac{(-3y^3 x^3)(-4y^4 x^2)(x^{-8})}{18x^3 y^2 y^9 x^{-6}}$ *Power rule*

$= \dfrac{12x^{3+2-8} y^{3+4}}{18x^{3-6} y^{2+9}}$ *Product rule*

$= \dfrac{2x^{-3} y^7}{3x^{-3} y^{11}}$

$= \dfrac{2x^{-3-(-3)} y^{7-11}}{3}$ *Quotient rule*

$= \dfrac{2y^{-4}}{3}$

$= \dfrac{2}{3y^4}$

131. $\left(\dfrac{p^2 q^{-1}}{2p^{-2}}\right)^2 \cdot \left(\dfrac{p^3 \cdot 4q^{-2}}{3q^{-5}}\right)^{-1} \cdot \left(\dfrac{pq^{-5}}{q^{-2}}\right)^3$

$= \dfrac{p^4 q^{-2} p^{-3} 4^{-1} q^2 p^3 q^{-15}}{2^2 p^{-4} 3^{-1} q^5 q^{-6}}$ *Power rule*

$= \dfrac{4^{-1} p^{4-3+3} q^{-2+2-15}}{2^2 3^{-1} p^{-4} q^{5-6}}$ *Product rule*

$= \dfrac{3p^4 q^{-15}}{2^2 \cdot 4p^{-4} q^{-1}}$

$= \dfrac{3p^{4-(-4)} q^{-15-(-1)}}{4 \cdot 4}$ *Quotient rule*

$= \dfrac{3p^8 q^{-14}}{16}$

$= \dfrac{3p^8}{16q^{14}}$

133. $530 = 5\wedge 30.$ ← Decimal point

Count 2 places.

Since the number 5.3 is to be made larger, the exponent on 10 is positive.

$$530 = 5.3 \times 10^2$$

Copyright © 2012 Pearson Education, Inc. Publishing as Addison-Wesley.

135. $0.830 = 0.8_\wedge 30$

Count 1 place.

Since the number 8.3 is to be made smaller, the exponent on 10 is negative.

$$0.830 = 8.3 \times 10^{-1}$$

137. $0.000\,006\,92 = 0.0\,0\,0\,0\,0\,6_\wedge 92$

Count 6 places.

Since the number 6.92 is to be made smaller, the exponent on 10 is negative.

$$0.00000692 = 6.92 \times 10^{-6}$$

139. $-38,500 = -3_\wedge 8\,5\,0\,0.$

Count 4 places.

Since the number 3.85 is to be made larger, the exponent on 10 is positive. Also, affix a negative sign in front of the number.

$$-38,500 = -3.85 \times 10^4$$

141. $7.2 \times 10^4 = 72,000$

Move the decimal point 4 places to the *right* because of the *positive* exponent. Attach extra zeros.

143. $2.54 \times 10^{-3} = 0.002\,54$

Since the exponent is *negative*, move the decimal point 3 places to the *left*.

145. $-6 \times 10^4 = -60,000$

Move the decimal point 4 places to the *right* because of the *positive* exponent. Attach extra zeros.

147. $1.2 \times 10^{-5} = 0.000\,012$

Since the exponent is *negative*, move the decimal point 5 places to the *left*.

149.
$$\frac{12 \times 10^4}{2 \times 10^6} = \frac{12}{2} \times \frac{10^4}{10^6}$$
$$= 6 \times 10^{4-6}$$
$$= 6 \times 10^{-2}$$
$$= 0.06$$

151.
$$\frac{3 \times 10^{-2}}{12 \times 10^3} = \frac{3 \times 10^{-2}}{1.2 \times 10^4}$$
$$= \frac{3}{1.2} \times \frac{10^{-2}}{10^4}$$
$$= 2.5 \times 10^{-6}$$
$$= 0.000\,002\,5$$

153.
$$\frac{0.05 \times 1600}{0.0004} = \frac{5 \times 10^{-2} \times 1.6 \times 10^3}{4 \times 10^{-4}}$$
$$= \frac{5(1.6)}{4} \times \frac{10^{-2} \times 10^3}{10^{-4}}$$
$$= 2 \times 10^{-2+3-(-4)}$$
$$= 2 \times 10^5$$
$$= 200,000$$

155.
$$\frac{20,000 \times 0.018}{300 \times 0.0004} = \frac{2 \times 10^4 \times 1.8 \times 10^{-2}}{3 \times 10^2 \times 4 \times 10^{-4}}$$
$$= \frac{2 \times 1.8}{3 \times 4} \times \frac{10^4 \times 10^{-2}}{10^2 \times 10^{-4}}$$
$$= 0.3 \times 10^{4-2-2-(-4)}$$
$$= 0.3 \times 10^4$$
$$= 3000$$

157.
$$\$1,000,000,000 = \$1 \times 10^9$$
$$\$1,000,000,000,000 = \$1 \times 10^{12}$$
$$\$3,100,000,000,000 = \$3.1 \times 10^{12}$$
$$210,385 = 2.10385 \times 10^5$$

159. (a) 308.4 million
$$= 308,400,000$$
$$= 3.084 \times 10^8$$

(b) $\$1,000,000,000,000 = \$1 \times 10^{12}$

(c) Divide the amount by the number of people to determine how much each person would have to contribute.

$$\frac{\$1 \times 10^{12}}{3.084 \times 10^8} \approx \$0.3243 \times 10^4$$
$$= \$3243$$

In 2010, each person in the United States would have had to contribute about $3243 in order to make someone a trillionaire.

161. Divide the amount by the number of people.

$$\frac{\$11.5 \times 10^{12}}{3.07 \times 10^8} \approx \$3.7459 \times 10^4$$
$$= \$37,459$$

In 2009, the national debt was about $37,459 for each person in the United States.

163. Since $d = rt$, $t = \frac{d}{r}$. Divide the distance traveled by the speed of light.

$$\frac{9 \times 10^{12}}{3 \times 10^{10}} = \frac{9}{3} \times 10^{12-10} = 3 \times 10^2$$

It will take 300 seconds.

Copyright © 2012 Pearson Education, Inc. Publishing as Addison-Wesley.

165. First find the number of seconds in a year.

$$1 \text{ year} = 365 \text{ days}$$
$$= 365(24) \text{ hr}$$
$$= 8760 \text{ hr}$$
$$= 8760(60) \text{ min}$$
$$= 525{,}600 \text{ min}$$
$$= 525{,}600(60) \text{ sec}$$
$$= 31{,}536{,}000 \text{ sec}$$
$$= 3.1536 \times 10^7 \text{ sec}$$

Now use $d = rt$ and multiply the rate light travels by the number of seconds in a year.

$$1 \text{ light year} = (1.86 \times 10^5 \text{ mi/sec})$$
$$\cdot (3.1536 \times 10^7 \text{ sec})$$
$$= 1.86(3.1536) \times 10^{5+7}$$
$$\approx 5.87 \times 10^{12} \text{ mi}$$

There are about 5.87×10^{12} miles in a light year.

167. The density D is the population P divided by the area $\mathcal{A}$.

$$D = \frac{P}{\mathcal{A}}$$

We want to find the area, so solve the formula for $\mathcal{A}$.

$$D\mathcal{A} = P$$
$$\mathcal{A} = \frac{P}{D}$$
$$= \frac{4.92 \times 10^5}{493}$$
$$= \frac{4.92 \times 10^5}{4.93 \times 10^2}$$
$$\approx 0.998 \times 10^{5-2}$$
$$= 0.998 \times 10^3 = 998$$

The area is 998 square miles.

169. $(1.5 \,\text{E}\, 12) * (5 \,\text{E}\, ^-3)$
$$= (1.5 \times 10^{12})(5 \times 10^{-3})$$
$$= (1.5 \times 5)(10^{12-3})$$
$$= 7.5 \times 10^9$$

171. $(8.4 \,\text{E}\, 14)/(2.1 \,\text{E}\, ^-3)$
$$= \frac{8.4 \times 10^{14}}{2.1 \times 10^{-3}}$$
$$= \frac{8.4}{2.1} \times \frac{10^{14}}{10^{-3}}$$
$$= 4 \times 10^{14-(-3)}$$
$$= 4 \times 10^{17}$$

173. $9x + 5x - x + 8x - 12x$
$$= (9 + 5 - 1 + 8 - 12)x$$
$$= 9x$$

175. $3(5 + q) - 2(6 - q)$
$$= 15 + 3q - 12 + 2q$$
$$= (15 - 12) + (3q + 2q)$$
$$= 3 + 5q$$

5.2 Adding and Subtracting Polynomials

5.2 Now Try Exercises

N1. $-2x^3 - 2x^5 + 4x^2 + 7 - x$ is written in descending powers as $-2x^5 - 2x^3 + 4x^2 - x + 7$. The leading term is $-2x^5$. The leading coefficient is -2.

N2. $-4x^3 + 10x^5 + 7$ has three terms, so it is a trinomial, and its degree is 5 (from $10x^5$).

N3. **(a)** $3p^3 - 2q + p^3 - 5q$
$$= 3p^3 + p^3 - 2q - 5q$$
$$= 4p^3 - 7q$$

(b) $-x^2t + 4x^2t + 3xt^2 - 7xt^2$
$$= 3x^2t - 4xt^2$$

N4. Add using the horizontal method.

$(7x^2 - 9x + 4) + (x^3 - 3x^2 - 5)$
$$= x^3 + 7x^2 - 3x^2 - 9x + 4 - 5$$
$$= x^3 + 4x^2 - 9x - 1$$

N5. Subtract using the vertical method.

$$\begin{array}{rrrr} 2y^2 & - 7y & - & 4 \\ 8y^2 & - 2y & + & 10 \end{array}$$

Change all the signs in the second polynomial, and add.

$$\begin{array}{rrrr} 2y^2 & - 7y & - & 4 \\ -8y^2 & + 2y & - & 10 \\ \hline -6y^2 & - 5y & - & 14 \end{array}$$

5.2 Section Exercises

1. In $7z$, the coefficient is 7 and, since $7z = 7z^1$, the degree is 1.

3. In $-15p^2$, the coefficient is -15 and the degree is 2.

5. In x^4, since $x^4 = 1x^4$, the coefficient is 1 and the degree is 4.

7. In $\frac{t}{6} = \frac{1}{6}t$, the coefficient is $\frac{1}{6}$ and since $\frac{t}{6} = \frac{1}{6}t^1$, the degree is 1.

9. In $8 = 8x^0$, the coefficient is 8 and the degree is 0.

11. In $-x^3 = -1x^3$, the coefficient is -1 and the degree is 3.

Copyright © 2012 Pearson Education, Inc. Publishing as Addison-Wesley.

13. $2x^3 + x - 3x^2 + 4$

The polynomial is written in descending powers of the variable if the exponents on the terms of the polynomial decrease from left to right.

$$2x^3 - 3x^2 + x + 4$$

The leading term is $2x^3$. The leading coefficient is 2.

15. $4p^3 - 8p^5 + p^7 = p^7 - 8p^5 + 4p^3$
The leading term is p^7. The leading coefficient is 1.

17. $10 - m^3 - 3m^4 = -3m^4 - m^3 + 10$
The leading term is $-3m^4$. The leading coefficient is -3.

19. 25 is one term, so it's a *monomial*. 25 is a nonzero constant, so it has degree zero.

21. $7m - 22$ has two terms, so it's a *binomial*. The exponent on m is 1, so $7m - 22$ has degree 1.

23. $-7y^6 + 11y^8$ is a *binomial* of degree 8.

25. $-mn^5$ is one term, so it's a *monomial*. Since $m = m^1$ and $1 + 5 = 6$, the degree is 6.

27. $-5m^3 + 6m - 9m^2$ has three terms, so it's a *trinomial*. The greatest exponent is 3, so the degree is 3.

29. $-6p^4q - 3p^3q^2 + 2pq^3 - q^4$ has four terms, so it is classified as *none of these*. The greatest sum of exponents on any term is 5, so the polynomial has degree 5.

31. Only choice **A** is a trinomial (it has three terms) with descending powers and having degree 6.

33. $5z^4 + 3z^4 = (5 + 3)z^4 = 8z^4$

35. $-m^3 + 2m^3 + 6m^3 = (-1 + 2 + 6)m^3 = 7m^3$

37. $x + x + x + x + x$
$$= (1 + 1 + 1 + 1 + 1)x$$
$$= 5x$$

39. $m^4 - 3m^2 + m$ is *already simplified* since there are no like terms to be combined.

41. $5t + 4s - 6t + 9s = (5t - 6t) + (4s + 9s)$
$$= -t + 13s$$

43. $2k + 3k^2 + 5k^2 - 7$
$$= (3k^2 + 5k^2) + 2k - 7$$
$$= (3 + 5)k^2 + 2k - 7$$
$$= 8k^2 + 2k - 7$$

45. $n^4 - 2n^3 + n^2 - 3n^4 + n^3$
$$= n^4 - 3n^4 - 2n^3 + n^3 + n^2$$
$$= (1 - 3)n^4 + (-2 + 1)n^3 + n^2$$
$$= -2n^4 - n^3 + n^2$$

47. $3ab^2 + 7a^2b - 5ab^2 + 13a^2b$
$$= (3 - 5)ab^2 + (7 + 13)a^2b$$
$$= -2ab^2 + 20a^2b$$

49. $4 - (2 + 3m) + 6m + 9$
$$= 4 - 2 - 3m + 6m + 9$$
$$= (-3 + 6)m + (4 - 2 + 9)$$
$$= 3m + 11$$

51. $6 + 3p - (2p + 1) - (2p + 9)$
$$= 6 + 3p - 2p - 1 - 2p - 9$$
$$= (3 - 2 - 2)p + (6 - 1 - 9)$$
$$= -p - 4$$

53. $(5x^2 + 7x - 4) + (3x^2 - 6x + 2)$
$$= 5x^2 + 3x^2 + 7x - 6x - 4 + 2$$
$$= 8x^2 + x - 2$$

55. $(6t^2 - 4t^4 - t) + (3t^4 - 4t^2 + 5)$
$$= -4t^4 + 3t^4 + 6t^2 - 4t^2 - t + 5$$
$$= (-4 + 3)t^4 + (6 - 4)t^2 - t + 5$$
$$= -t^4 + 2t^2 - t + 5$$

57. $(y^3 + 3y + 2) + (4y^3 - 3y^2 + 2y - 1)$
$$= y^3 + 4y^3 - 3y^2 + 3y + 2y + 2 - 1$$
$$= (1 + 4)y^3 - 3y^2 + (3 + 2)y + (2 - 1)$$
$$= 5y^3 - 3y^2 + 5y + 1$$

59. $(3r + 8) - (2r - 5)$

Change all signs in the second polynomial and add.

$$= (3r + 8) + (-2r + 5)$$
$$= 3r + 8 - 2r + 5$$
$$= 3r - 2r + 8 + 5$$
$$= r + 13$$

61. $(2a^2 + 3a - 1) - (4a^2 + 5a + 6)$
$$= (2a^2 + 3a - 1) + (-4a^2 - 5a - 6)$$
$$= 2a^2 - 4a^2 + 3a - 5a - 1 - 6$$
$$= -2a^2 - 2a - 7$$

63. $(z^5 + 3z^2 + 2z) - (4z^5 + 2z^2 - 5z)$
$$= z^5 + 3z^2 + 2z - 4z^5 - 2z^2 + 5z$$
$$= z^5 - 4z^5 + 3z^2 - 2z^2 + 2z + 5z$$
$$= -3z^5 + z^2 + 7z$$

65.
$$\begin{array}{r} 21p - 8 \\ -9p + 4 \\ \hline 12p - 4 \end{array}$$ *Add vertically.*

67.
$$\begin{array}{r} -12p^2 + 4p - 1 \\ 3p^2 + 7p - 8 \\ \hline -9p^2 + 11p - 9 \end{array}$$ *Add vertically.*

Copyright © 2012 Pearson Education, Inc. Publishing as Addison-Wesley.

69. Subtract.

$$12a + 15$$
$$\underline{7a - 3}$$

Change all the signs in the second polynomial, and add.

$$12a + 15$$
$$\underline{-7a + 3}$$
$$5a + 18$$

71. Subtract.

$$6m^2 - 11m + 5$$
$$\underline{-8m^2 + 2m - 1}$$

Change all the signs in the second polynomial, and add.

$$6m^2 - 11m + 5$$
$$\underline{8m^2 - 2m + 1}$$
$$14m^2 - 13m + 6$$

73. Add column by column to obtain the result on the bottom line.

$$12z^2 - 11z + 8$$
$$5z^2 + 16z - 2$$
$$\underline{-4z^2 + 5z - 9}$$
$$13z^2 + 10z - 3$$

75.
$$6y^3 - 9y^2 + 8$$
$$\underline{4y^3 + 2y^2 + 5y}$$
$$10y^3 - 7y^2 + 5y + 8$$

77. Subtract.

$$-5a^4 + 8a^2 - 9$$
$$\underline{6a^3 - a^2 + 2}$$

Change all the signs in the second polynomial, and add.

$$-5a^4 + 8a^2 - 9$$
$$\underline{ - 6a^3 + a^2 - 2}$$
$$-5a^4 - 6a^3 + 9a^2 - 11$$

79. $7y^2 - 6y + 5 - (4y^2 - 2y + 3)$
$$= 7y^2 - 6y + 5 - 4y^2 + 2y - 3$$
$$= 7y^2 - 4y^2 - 6y + 2y + 5 - 3$$
$$= 3y^2 - 4y + 2$$

81. Simplify the expression in brackets first.

$(3m^2 - 5n^2 + 2n) + (-3m^2) + 4n^2$
$$= 3m^2 - 5n^2 + 2n - 3m^2 + 4n^2$$
$$= 3m^2 - 3m^2 - 5n^2 + 4n^2 + 2n$$
$$= -n^2 + 2n$$

Now perform the subtraction.

$(-4m^2 + 3n^2 - 5n) - (-n^2 + 2n)$
$$= -4m^2 + 3n^2 - 5n + n^2 - 2n$$
$$= -4m^2 + 3n^2 + n^2 - 5n - 2n$$
$$= -4m^2 + 4n^2 - 7n$$

83. $[-(y^4 - y^2 + 1) - (y^4 + 2y^2 + 1)]$
$$ + (3y^4 - 3y^2 - 2)$$
$$= -y^4 + y^2 - 1 - y^4 - 2y^2 - 1$$
$$ + 3y^4 - 3y^2 - 2$$
$$= -y^4 - y^4 + 3y^4 + y^2 - 2y^2 - 3y^2$$
$$ - 1 - 1 - 2$$
$$= y^4 - 4y^2 - 4$$

85. $-[3z^2 + 5z - (2z^2 - 6z)]$
$$ + [(8z^2 - [5z - z^2]) + 2z^2]$$
$$= -(3z^2 + 5z - 2z^2 + 6z)$$
$$ + (8z^2 - 5z + z^2 + 2z^2)$$
$$= -(3z^2 - 2z^2 + 5z + 6z)$$
$$ + (8z^2 + z^2 + 2z^2 - 5z)$$
$$= -(z^2 + 11z) + (11z^2 - 5z)$$
$$= -z^2 - 11z + 11z^2 - 5z$$
$$= -z^2 + 11z^2 - 11z - 5z$$
$$= 10z^2 - 16z$$

87. Using the vertical line test, we find any vertical line will intersect the graph at most once. This indicates that the graph represents a function. The domain is the set of x-values: $(-\infty, \infty)$. The range is the set of y-values: $(-\infty, \infty)$.

89. Since a vertical line can intersect the graph of the relation in more than one point, the relation is not a function. The domain is the set of x-values: $[0, \infty)$. The range is the set of y-values: $(-\infty, \infty)$.

91. $f(x) = x^2 + 2$

(a) $f(-1) = (-1)^2 + 2$
$$= 1 + 2 = 3$$

(b) $f(2) = 2^2 + 2$
$$= 4 + 2 = 6$$

5.3   Polynomial Functions, Graphs, and Composition

5.3   Now Try Exercises

N1. $f(x) = x^3 - 2x^2 + 7$

$$f(-3) = (-3)^3 - 2(-3)^2 + 7$$
$$= -27 - 2(9) + 7$$
$$= -38$$

Copyright © 2012 Pearson Education, Inc. Publishing as Addison-Wesley.

N2. $P(x) = -0.01774x^2 + 0.7871x + 41.26$

The year 2002 corresponds to $2002 - 1900 = 12$.

$$P(12) = -0.01774(12)^2 + 0.7871(12) + 41.26$$
$$= 48.15064 \quad \text{Let } x = 12.$$

Thus, in 2002 about 48.2 million students were enrolled in public schools in the United States.

N3. $f(x) = x^3 - 3x^2 + 4,\ g(x) = -2x^3 + x^2 - 12$

(a) $(f + g)(x)$
$$= f(x) + g(x)$$
$$= (x^3 - 3x^2 + 4) + (-2x^3 + x^2 - 12)$$
$$= x^3 - 3x^2 + 4 - 2x^3 + x^2 - 12$$
$$= -x^3 - 2x^2 - 8$$

(b) $(f - g)(x)$
$$= f(x) - g(x)$$
$$= (x^3 - 3x^2 + 4) - (-2x^3 + x^2 - 12)$$
$$= x^3 - 3x^2 + 4 + 2x^3 - x^2 + 12$$
$$= 3x^3 - 4x^2 + 16$$

N4. $f(x) = x^2 - 4,\ g(x) = -6x^2$

(a) $(f + g)(x)$
$$= f(x) + g(x)$$
$$= (x^2 - 4) + (-6x^2)$$
$$= -5x^2 - 4$$

(b) $(f - g)(x)$
$$= f(x) - g(x)$$
$$= (x^2 - 4) - (-6x^2)$$
$$= 7x^2 - 4$$
$$(f - g)(-4)$$
$$= 7(-4)^2 - 4$$
$$= 7(16) - 4$$
$$= 108$$

N5. $f(x) = 3x + 7$ and $g(x) = x - 2$

$$(f \circ g)(7) = f(g(7)) \quad \textit{Definition}$$
$$= f(7 - 2) \quad \textit{g(x) = x - 2}$$
$$= f(5) \quad \textit{Subtract.}$$
$$= 3(5) + 7 \quad \textit{f(x) = 3x + 7}$$
$$= 22$$

N6. $f(x) = x - 5$ and $g(x) = -x^2 + 2$

(a) $(g \circ f)(-1) = g(f(-1)) \quad \textit{Definition}$
$$= g(-1 - 5) \quad \textit{f(x) = x - 5}$$
$$= f(-6) \quad \textit{Subtract.}$$
$$= -(-6)^2 + 2 \quad \textit{g(x) = -x^2 + 2}$$
$$= -34$$

(b) $(f \circ g)(x) = f(g(x)) \quad \textit{Definition}$
$$= f(-x^2 + 2) \quad \textit{g(x) = -x^2 + 2}$$
$$= (-x^2 + 2) - 5 \quad \textit{f(x) = x - 5}$$
$$= -x^2 - 3$$

N7.

x	$f(x) = x^2 - 4$
-2	0
-1	-3
0	-4
1	-3
2	0

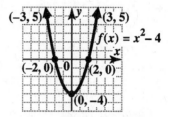

Any value of x can be used, so the domain is $(-\infty, \infty)$. The minimum y-value is -4 and there is no maximum y-value, so the range is $[-4, \infty)$.

5.3 Section Exercises

1. $f(x) = 6x - 4$

(a) $f(-1) = 6(-1) - 4$
$$= -6 - 4 = -10$$

(b) $f(2) = 6(2) - 4$
$$= 12 - 4 = 8$$

3. $f(x) = x^2 - 3x + 4$

(a) $f(-1) = (-1)^2 - 3(-1) + 4$
$$= 1 + 3 + 4 = 8$$

(b) $f(2) = (2)^2 - 3(2) + 4$
$$= 4 - 6 + 4 = 2$$

5. $f(x) = 5x^4 - 3x^2 + 6$

(a) $f(-1) = 5(-1)^4 - 3(-1)^2 + 6$
$$= 5 \cdot 1 - 3 \cdot 1 + 6$$
$$= 5 - 3 + 6 = 8$$

(b) $f(2) = 5(2)^4 - 3(2)^2 + 6$
$$= 5 \cdot 16 - 3 \cdot 4 + 6$$
$$= 80 - 12 + 6 = 74$$

7. $f(x) = -x^2 + 2x^3 - 8$

(a) $f(-1) = -(-1)^2 + 2(-1)^3 - 8$
$$= -(1) + 2(-1) - 8$$
$$= -1 - 2 - 8 = -11$$

(b) $f(2) = -(2)^2 + 2(2)^3 - 8$
$$= -(4) + 2 \cdot 8 - 8$$
$$= -4 + 16 - 8 = 4$$

9. $P(x) = 1667x^2 + 22.78x + 4300$

(a) $x = 2000 - 2000 = 0$

$$P(0) = 1667(0)^2 + 22.78(0) + 4300$$
$$= 4300 \text{ thousand pounds}$$

Copyright © 2012 Pearson Education, Inc. Publishing as Addison-Wesley.

(b) $x = 2003 - 2000 = 3$

$P(3) = 1667(3)^2 + 22.78(3) + 4300$
$= 19{,}371.34 \approx 19{,}371$ thousand pounds

(c) $x = 2006 - 2000 = 6$

$P(6) = 1667(6)^2 + 22.78(6) + 4300$
$= 64{,}448.68 \approx 64{,}449$ thousand pounds

11. $P(x) =$
$\quad -0.00189x^3 + 0.1193x^2 + 2.027x + 28.19$

(a) $x = 1985 - 1985 = 0$

$P(0) = -0.00189(0)^3 + 0.1193(0)^2$
$\quad + 2.027(0) + 28.19$
$= 28.19 \qquad (\$28.2 \text{ billion})$

(b) $x = 2000 - 1985 = 15$

$P(15) = -0.00189(15)^3 + 0.1193(15)^2$
$\quad + 2.027(15) + 28.19$
$= 79.05875 \qquad (\$79.1 \text{ billion})$

(c) $x = 2006 - 1985 = 21$

$P(21) = -0.00189(21)^3 + 0.1193(21)^2$
$\quad + 2.027(21) + 28.19$
$= 105.86501 \qquad (\$105.9 \text{ billion})$

13. **(a)** $(f + g)(x) = f(x) + g(x)$
$\quad = (5x - 10) + (3x + 7)$
$\quad = 8x - 3$

(b) $(f - g)(x) = f(x) - g(x)$
$\quad = (5x - 10) - (3x + 7)$
$\quad = (5x - 10) + (-3x - 7)$
$\quad = 2x - 17$

15. **(a)** $(f + g)(x)$
$\quad = f(x) + g(x)$
$\quad = (4x^2 + 8x - 3) + (-5x^2 + 4x - 9)$
$\quad = -x^2 + 12x - 12$

(b) $(f - g)(x)$
$\quad = f(x) - g(x)$
$\quad = (4x^2 + 8x - 3) - (-5x^2 + 4x - 9)$
$\quad = (4x^2 + 8x - 3) + (5x^2 - 4x + 9)$
$\quad = 9x^2 + 4x + 6$

For Exercises 17–32, let $f(x) = x^2 - 9$, $g(x) = 2x$, and $h(x) = x - 3$.

17. $(f + g)(x) = f(x) + g(x)$
$\quad = (x^2 - 9) + (2x)$
$\quad = x^2 + 2x - 9$

19. $(f + g)(3) = f(3) + g(3)$
$\quad = (3^2 - 9) + 2(3)$
$\quad = 0 + 6 = 6$

Alternatively, we could evaluate the polynomial in Exercise 17, $x^2 + 2x - 9$, using $x = 3$.

21. $(f - h)(x) = f(x) - h(x)$
$\quad = (x^2 - 9) - (x - 3)$
$\quad = x^2 - 9 - x + 3$
$\quad = x^2 - x - 6$

23. $(f - h)(-3) = f(-3) - h(-3)$
$\quad = [(-3)^2 - 9] - [(-3) - 3]$
$\quad = (9 - 9) - (-6)$
$\quad = 0 + 6 = 6$

25. $(g + h)(-10) = g(-10) + h(-10)$
$\quad = 2(-10) + [(-10) - 3]$
$\quad = -20 + (-13)$
$\quad = -33$

27. $(g - h)(-3) = g(-3) - h(-3)$
$\quad = 2(-3) - [(-3) - 3]$
$\quad = -6 - (-6)$
$\quad = -6 + 6 = 0$

29. $(g + h)\left(\frac{1}{4}\right) = g\left(\frac{1}{4}\right) + h\left(\frac{1}{4}\right)$
$\quad = 2\left(\frac{1}{4}\right) + \left(\frac{1}{4} - 3\right)$
$\quad = \frac{1}{2} + \left(-\frac{11}{4}\right)$
$\quad = -\frac{9}{4}$

31. $(g + h)\left(-\frac{1}{2}\right) = g\left(-\frac{1}{2}\right) + h\left(-\frac{1}{2}\right)$
$\quad = 2\left(-\frac{1}{2}\right) + \left(-\frac{1}{2} - 3\right)$
$\quad = -1 + \left(-\frac{7}{2}\right)$
$\quad = -\frac{9}{2}$

33. Answers will vary. Let $f(x) = x^3$ and $g(x) = x^4$.

$\quad (f - g)(x) = f(x) - g(x)$
$\quad\quad = x^3 - x^4$
$\quad (g - f)(x) = g(x) - f(x)$
$\quad\quad = x^4 - x^3$

Because the two differences are not equal, subtraction of polynomial functions is not commutative.

For Exercises 35–50,
$f(x) = x^2 + 4$, $g(x) = 2x + 3$, and $h(x) = x - 5$.

35. $(h \circ g)(4) = h(g(4))$
$\quad = h(2 \cdot 4 + 3)$
$\quad = h(11)$
$\quad = 11 - 5 = 6$

37. $(g \circ f)(6) = g(f(6))$
$\quad = g(6^2 + 4)$
$\quad = g(40)$
$\quad = 2 \cdot 40 + 3 = 83$

39. $(f \circ h)(-2) = f(h(-2))$
$\quad = f(-2 - 5)$
$\quad = f(-7)$
$\quad = (-7)^2 + 4 = 53$

Copyright © 2012 Pearson Education, Inc. Publishing as Addison-Wesley.

41. $(f \circ g)(0) = f(g(0))$
$$= f(2(0) + 3)$$
$$= f(3)$$
$$= 3^2 + 4$$
$$= 13$$

43. $(g \circ f)(x) = g(f(x))$
$$= g(x^2 + 4)$$
$$= 2(x^2 + 4) + 3$$
$$= 2x^2 + 8 + 3$$
$$= 2x^2 + 11$$

45. $(h \circ g)(x) = h(g(x))$
$$= h(2x + 3)$$
$$= 2x + 3 - 5$$
$$= 2x - 2$$

47. $(f \circ h)\left(\frac{1}{2}\right) = f\left(h\left(\frac{1}{2}\right)\right)$
$$= f\left(\frac{1}{2} - 5\right)$$
$$= f\left(-\frac{9}{2}\right)$$
$$= \left(-\frac{9}{2}\right)^2 + 4$$
$$= \frac{81}{4} + \frac{16}{4} = \frac{97}{4}$$

49. $(f \circ g)\left(-\frac{1}{2}\right) = f\left(g\left(-\frac{1}{2}\right)\right)$
$$= f\left[2\left(-\frac{1}{2}\right) + 3\right]$$
$$= f(2)$$
$$= 2^2 + 4$$
$$= 4 + 4 = 8$$

51. $f(x) = 12x, g(x) = 5280x$

$(f \circ g)(x) = f(g(x))$
$$= f(5280x)$$
$$= 12(5280x)$$
$$= 63,360x$$

$(f \circ g)(x)$ computes the number of inches in x mi.

53. $r(t) = 2t, \mathcal{A}(r) = \pi r^2$

$(\mathcal{A} \circ r)(t) = \mathcal{A}(r(t))$
$$= \mathcal{A}(2t)$$
$$= \pi(2t)^2$$
$$= 4\pi t^2$$

This is the area of the circular layer as a function of time.

55.

x	$f(x) = -2x + 1$
-2	$-2(-2) + 1 = 5$
-1	$-2(-1) + 1 = 3$
0	$-2(0) + 1 = 1$
1	$-2(1) + 1 = -1$
2	$-2(2) + 1 = -3$

This is a linear function, so plot the points and draw a line through them.

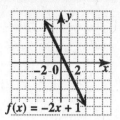

$f(x) = -2x + 1$

Any x-value can be used, so the domain is $(-\infty, \infty)$. From the graph, we see that any y-value can be obtained from the function, so the range is $(-\infty, \infty)$.

57.

x	$f(x) = -3x^2$
-2	$-3(-2)^2 = -12$
-1	$-3(-1)^2 = -3$
0	$-3(0)^2 = 0$
1	$-3(1)^2 = -3$
2	$-3(2)^2 = -12$

Since the greatest exponent is 2, the graph of f is a parabola.

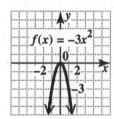

$f(x) = -3x^2$

Any x-value can be used, so the domain is $(-\infty, \infty)$. From the graph, we see that the y-values are at most 0, so the range is $(-\infty, 0]$.

59.

x	$f(x) = x^3 + 1$
-2	$(-2)^3 + 1 = -7$
-1	$(-1)^3 + 1 = 0$
0	$(0)^3 + 1 = 1$
1	$(1)^3 + 1 = 2$
2	$(2)^3 + 1 = 9$

The greatest exponent is 3, so the graph of f is s-shaped.

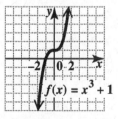

$f(x) = x^3 + 1$

Any x-value can be used, so the domain is $(-\infty, \infty)$. From the graph, we see that any y-value can be obtained from the function, so the range is $(-\infty, \infty)$.

61. $3m^3(4m^2) = 3(4)m^3m^2$
$$= 12m^{3+2}$$
$$= 12m^5$$

Copyright © 2012 Pearson Education, Inc. Publishing as Addison-Wesley.

63. $-3b^5(2a^3b^4) = -3(2)a^3b^5b^4$
$$= -6a^3b^{5+4}$$
$$= -6a^3b^9$$

65. $12x^2y(5xy^3) = 12(5)x^2xyy^3$
$$= 60x^{2+1}y^{1+3}$$
$$= 60x^3y^4$$

5.4   Multiplying Polynomials

5.4 Now Try Exercises

N1. $-3s^2t(15s^3t^4) = -3(15)s^2 \cdot s^3 \cdot t^1 \cdot t^4$
$$= -45s^{2+3}t^{1+4}$$
$$= -45s^5t^5$$

N2. (a) $3k^3(-2k^5 + 3k^2 - 4)$
$$= 3k^3(-2k^5) + 3k^3(3k^2) + 3k^3(-4)$$
$$= -6k^8 + 9k^5 - 12k^3$$

(b) $5x(2x - 1)(x + 4)$
$$= 5x[(2x - 1)(x) + (2x - 1)(4)]$$
$$= 5x[2x^2 - x + 8x - 4]$$
$$= 5x[2x^2 + 7x - 4]$$
$$= 10x^3 + 35x^2 - 20x$$

N3.
$$
\begin{array}{r}
3t^2 - 5t + 4 \\
t - 3 \\
\hline
-9t^2 + 15t - 12 \\
3t^3 - 5t^2 + 4t \\
\hline
3t^3 - 14t^2 + 19t - 12
\end{array}
$$

N4. $(3p - k)(5p + 4k)$
$$\ \mathbf{F}\qquad \mathbf{O}\qquad \mathbf{I}\qquad \mathbf{L}$$
$$= 15p^2 + 12kp - 5kp - 4k^2$$
$$= 15p^2 + 7kp - 4k^2$$

N5. Use $(x + y)(x - y) = x^2 - y^2$.

(a) $(3x - 7y)(3x + 7y) = (3x)^2 - (7y)^2$
$$= 3^2x^2 - 7^2y^2$$
$$= 9x^2 - 49y^2$$

(b) $5k(2k - 3)(2k + 3)$
$$= 5k[(2k)^2 - 3^2]$$
$$= 5k[4k^2 - 9]$$
$$= 20k^3 - 45k$$

N6. Use $(x + y)^2 = x^2 + 2xy + y^2$
or $(x - y)^2 = x^2 - 2xy + y^2$.

(a) $(y - 10)^2 = y^2 - 2 \cdot y \cdot 10 + 10^2$
$$= y^2 - 20y + 100$$

(b) $(4x + 5y)^2 = (4x)^2 + 2(4x)(5y) + (5y)^2$
$$= 16x^2 + 40xy + 25y^2$$

N7. (a) $[(4x - y) + 2][(4x - y) - 2]$
This is the product of a sum and a difference.
$$= (4x - y)^2 - 2^2$$
The first term is the square of a binomial.
$$= (4x)^2 - 2(4x)(y) + y^2 - 2^2$$
$$= 16x^2 - 8xy + y^2 - 4$$

(b) $(y - 3)^4$
$$= (y - 3)^2(y - 3)^2$$
$$= (y^2 - 6y + 9)(y^2 - 6y + 9)$$
$$= y^4 - 6y^3 + 9y^2 - 6y^3 + 36y^2 - 54y$$
$$\ + 9y^2 - 54y + 81$$
$$= y^4 - 12y^3 + 54y^2 - 108y + 81$$

N8. $f(x) = 3x^2 - 1, \ g(x) = 8x + 7$

$$(fg)(x) = f(x) \cdot g(x)$$
$$= (3x^2 - 1)(8x + 7)$$
$$= 24x^3 + 21x^2 - 8x - 7$$

Using the last result,

$$(fg)(-2)$$
$$= 24(-2)^3 + 21(-2)^2 - 8(-2) - 7$$
$$= -192 + 84 + 16 - 7$$
$$= -99.$$

5.4 Section Exercises

1. $(2x - 5)(3x + 4)$
$$\ \mathbf{F}\qquad \mathbf{O}\qquad \mathbf{I}\qquad \mathbf{L}$$
$$= 2x(3x) + 2x(4) + (-5)(3x) + (-5)(4)$$
$$= 6x^2 + 8x - 15x - 20$$
$$= 6x^2 - 7x - 20 \quad \text{(Choice } \mathbf{C})$$

3. $(2x - 5)(3x - 4)$
$$\ \mathbf{F}\qquad \mathbf{O}\qquad \mathbf{I}\qquad \mathbf{L}$$
$$= 2x(3x) + 2x(-4) + (-5)(3x) + (-5)(-4)$$
$$= 6x^2 - 8x - 15x + 20$$
$$= 6x^2 - 23x + 20 \quad \text{(Choice } \mathbf{D})$$

5. $-8m^3(3m^2) = -8(3)m^{3+2} = -24m^5$

7. $14x^2y^3(-2x^5y) = 14(-2)x^{2+5}y^{3+1}$
$$= -28x^7y^4$$

9. $3x(-2x + 5) = 3x(-2x) + 3x(5) = -6x^2 + 15x$

11. $-q^3(2 + 3q) = -q^3(2) - q^3(3q)$
$$= -2q^3 - 3q^4$$

13. $6k^2(3k^2 + 2k + 1)$
$$= 6k^2(3k^2) + 6k^2(2k) + 6k^2(1)$$
$$= 18k^4 + 12k^3 + 6k^2$$

Copyright © 2012 Pearson Education, Inc. Publishing as Addison-Wesley.

15. $(2t+3)(3t^2 - 4t - 1)$
$$= 2t(3t^2 - 4t - 1) + 3(3t^2 - 4t - 1)$$
$$= 2t(3t^2) + 2t(-4t) + 2t(-1)$$
$$+ 3(3t^2) + 3(-4t) + 3(-1)$$
$$= 6t^3 - 8t^2 - 2t + 9t^2 - 12t - 3$$
$$= 6t^3 - 8t^2 + 9t^2 - 2t - 12t - 3$$
$$= 6t^3 + t^2 - 14t - 3$$

17. $m(m+5)(m-8)$
$$= m(m^2 - 8m + 5m - 40)$$
$$= m(m^2 - 3m - 40)$$
$$= m(m^2) + m(-3m) + m(-40)$$
$$= m^3 - 3m^2 - 40m$$

19. $4z(2z+1)(3z-4)$
$$= 4z(6z^2 - 8z + 3z - 4)$$
$$= 4z(6z^2 - 5z - 4)$$
$$= 4z(6z^2) + 4z(-5z) + 4z(-4)$$
$$= 24z^3 - 20z^2 - 16z$$

21. $4x^3(x-3)(x+2)$
$$= 4x^3(x^2 + 2x - 3x - 6)$$
$$= 4x^3(x^2 - x - 6)$$
$$= 4x^3(x^2) + 4x^3(-x) + 4x^3(-6)$$
$$= 4x^5 - 4x^4 - 24x^3$$

23. $(2y+3)(3y-4)$
Rewrite vertically and multiply.

$$
\begin{array}{r}
2y + 3 \\
3y - 4 \\
\hline
-8y - 12 \leftarrow -4(2y+3) \\
6y^2 + 9y \qquad \leftarrow 3y(2y+3) \\
\hline
6y^2 + y - 12 \qquad \text{Combine} \\
\text{like terms.}
\end{array}
$$

25.
$$
\begin{array}{r}
-b^2 + 3b + 3 \\
2b + 4 \\
\hline
-4b^2 + 12b + 12 \\
-2b^3 + 6b^2 + 6b \\
\hline
-2b^3 + 2b^2 + 18b + 12
\end{array}
$$

27.
$$
\begin{array}{r}
5m - 3n \\
5m + 3n \\
\hline
15mn - 9n^2 \\
25m^2 - 15mn \\
\hline
25m^2 \qquad - 9n^2
\end{array}
$$

29.
$$
\begin{array}{r}
2z^3 - 5z^2 + 8z - 1 \\
4z + 3 \\
\hline
6z^3 - 15z^2 + 24z - 3 \\
8z^4 - 20z^3 + 32z^2 - 4z \\
\hline
8z^4 - 14z^3 + 17z^2 + 20z - 3
\end{array}
$$

31.
$$
\begin{array}{r}
2p^2 + 3p + 6 \\
3p^2 - 4p - 1 \\
\hline
-2p^2 - 3p - 6 \\
-8p^3 - 12p^2 - 24p \\
6p^4 + 9p^3 + 18p^2 \\
\hline
6p^4 + p^3 + 4p^2 - 27p - 6
\end{array}
$$

33. $(m+5)(m-8)$
$$\text{F} \qquad \text{O} \quad \text{I} \qquad \text{L}$$
$$= m^2 - 8m + 5m - 40$$
$$= m^2 - 3m - 40$$

35. $(4k+3)(3k-2)$
$$\text{F} \qquad \text{O} \quad \text{I} \qquad \text{L}$$
$$= 12k^2 - 8k + 9k - 6$$
$$= 12k^2 + k - 6$$

37. $(z-w)(3z+4w)$
$$\text{F} \qquad \text{O} \quad \text{I} \qquad \text{L}$$
$$= 3z^2 + 4zw - 3zw - 4w^2$$
$$= 3z^2 + zw - 4w^2$$

39. $(6c-d)(2c+3d)$
$$\text{F} \qquad \text{O} \quad \text{I} \qquad \text{L}$$
$$= 12c^2 + 18cd - 2cd - 3d^2$$
$$= 12c^2 + 16cd - 3d^2$$

41. A description of the FOIL method is as follows: The product of two binomials is the sum of the product of the first terms, the product of the outer terms, the product of the inner terms, and the product of the last terms.

43. Use the formula for the product of the sum and difference of two terms.
$$(x+9)(x-9) = x^2 - 9^2$$
$$= x^2 - 81$$

45. Use the formula for the product of the sum and difference of two terms.
$$(2p-3)(2p+3) = (2p)^2 - 3^2$$
$$= 4p^2 - 9$$

47. $(5m-1)(5m+1) = (5m)^2 - 1^2$
$$= 25m^2 - 1$$

49. $(3a+2c)(3a-2c) = (3a)^2 - (2c)^2$
$$= 9a^2 - 4c^2$$

51. $(4m+7n^2)(4m-7n^2) = (4m)^2 - (7n^2)^2$
$$= 16m^2 - 49n^4$$

Copyright © 2012 Pearson Education, Inc. Publishing as Addison-Wesley.

53. $3y(5y^3 + 2)(5y^3 - 2)$
First multiply $(5y^3 + 2)(5y^3 - 2)$.

$$(5y^3 + 2)(5y^3 - 2) = (5y^3)^2 - (2)^2$$
$$= 25y^6 - 4$$

Now multiply the last polynomial times $3y$.

$$3y(25y^6 - 4) = 75y^7 - 12y$$

55. Use the formula for the square of a binomial.

$$(y - 5)^2 = y^2 - 2 \cdot y \cdot 5 + 5^2$$
$$= y^2 - 10y + 25$$

57. Use the formula for the square of a binomial.

$$(x + 1)^2 = x^2 + 2 \cdot x \cdot 1 + 1^2$$
$$= x^2 + 2x + 1$$

59. $(2p + 7)^2 = (2p)^2 + 2(2p)(7) + 7^2$
$$= 4p^2 + 28p + 49$$

61. $(4n - 3m)^2$
$$= (4n)^2 - 2(4n)(3m) + (3m)^2$$
$$= 16n^2 - 24nm + 9m^2$$

63. To find the product $101 \cdot 99$ using the special product rule

$$(x + y)(x - y) = x^2 - y^2,$$

let $x = 100$ and $y = 1$. Then

$$101 \cdot 99 = (100 + 1)(100 - 1)$$
$$= 100^2 - 1^2$$
$$= 10{,}000 - 1$$
$$= 9999.$$

65. $(0.2x + 1.3)(0.5x - 0.1)$
$$\quad \mathbf{F} \qquad \mathbf{O} \qquad \mathbf{I} \qquad \mathbf{L}$$
$$= 0.1x^2 - 0.02x + 0.65x - 0.13$$
$$= 0.1x^2 + 0.63x - 0.13$$

67. $\left(3w + \frac{1}{4}z\right)(w - 2z)$
$$\quad \mathbf{F} \qquad \mathbf{O} \qquad \mathbf{I} \qquad \mathbf{L}$$
$$= 3w^2 - 6wz + \frac{1}{4}wz - \frac{1}{2}z^2$$
$$= 3w^2 - \frac{23}{4}wz - \frac{1}{2}z^2$$

69. $\left(4x - \frac{2}{3}\right)\left(4x + \frac{2}{3}\right) = (4x)^2 - \left(\frac{2}{3}\right)^2$
$$= 16x^2 - \frac{4}{9}$$

71. $\left(k - \frac{5}{7}p\right)^2 = k^2 - 2(k)\left(\frac{5}{7}p\right) + \left(\frac{5}{7}p\right)^2$
$$= k^2 - \frac{10}{7}kp + \frac{25}{49}p^2$$

73. $(0.2x - 1.4y)^2$
$$= (0.2x)^2 - 2(0.2x)(1.4y) + (1.4y)^2$$
$$= 0.04x^2 - 0.56xy + 1.96y^2$$

75. $[(5x + 1) + 6y]^2$
$$= (5x + 1)^2 + 2(5x + 1)(6y) + (6y)^2$$
$$\qquad\qquad \textit{Square of a binomial}$$
$$= (5x + 1)^2 + 12y(5x + 1) + 36y^2$$
$$= \left[(5x)^2 + 2(5x)(1) + 1^2\right]$$
$$\quad + 60xy + 12y + 36y^2$$
$$\qquad\qquad \textit{Square of a binomial}$$
$$= 25x^2 + 10x + 1 + 60xy + 12y + 36y^2$$

77. $[(2a + b) - 3]^2$
$$= (2a + b)^2 - 2(2a + b)(3) + 3^2$$
$$= \left[(2a)^2 + 2(2a)(b) + b^2\right]$$
$$\quad - 6(2a + b) + 3^2$$
$$= 4a^2 + 4ab + b^2 - 12a - 6b + 9$$

79. $[(2a + b) - 3][(2a + b) + 3]$
$$= (2a + b)^2 - 3^2$$
$$\qquad\qquad \textit{Product of the sum and}$$
$$\qquad\qquad \textit{difference of two terms}$$
$$= \left[(2a)^2 + 2(2a)(b) + b^2\right] - 9$$
$$\qquad\qquad \textit{Square of a binomial}$$
$$= 4a^2 + 4ab + b^2 - 9$$

81. $[(2h - k) + j][(2h - k) - j]$
$$= (2h - k)^2 - j^2$$
$$= (2h)^2 - 2(2h)(k) + k^2 - j^2$$
$$= 4h^2 - 4hk + k^2 - j^2$$

83. $(y + 2)^3 = (y + 2)^2(y + 2)$
$$= [y^2 + 2(y)(2) + 2^2](y + 2)$$
$$= (y^2 + 4y + 4)(y + 2)$$

$$
\begin{array}{r}
y^2 + 4y + 4 \\
y + 2 \\
\hline
2y^2 + 8y + 8 \\
y^3 + 4y^2 + 4y \\
\hline
y^3 + 6y^2 + 12y + 8
\end{array}
$$

85. $(5r - s)^3$
$$= (5r - s)^2(5r - s)$$
$$= (25r^2 - 10rs + s^2)(5r - s)$$
$$= 125r^3 - 50r^2s + 5rs^2 - 25r^2s + 10rs^2 - s^3$$
$$= 125r^3 - 75r^2s + 15rs^2 - s^3$$

87. $(q - 2)^4 = (q - 2)^2(q - 2)^2$
$$= (q^2 - 4q + 4)(q^2 - 4q + 4)$$

$$
\begin{array}{r}
q^2 - 4q + 4 \\
q^2 - 4q + 4 \\
\hline
4q^2 - 16q + 16 \\
- 4q^3 + 16q^2 - 16q \\
q^4 - 4q^3 + 4q^2 \\
\hline
q^4 - 8q^3 + 24q^2 - 32q + 16
\end{array}
$$

Copyright © 2012 Pearson Education, Inc. Publishing as Addison-Wesley.

89. $(2a + b)(3a^2 + 2ab + b^2)$
Rewrite vertically and multiply.

$$
\begin{array}{r}
3a^2 + 2ab + b^2 \\
2a + b \\
\hline
3a^2b + 2ab^2 + b^3 \\
6a^3 + 4a^2b + 2ab^2 \\
\hline
6a^3 + 7a^2b + 4ab^2 + b^3
\end{array}
$$

91. $(4z - x)(z^3 - 4z^2x + 2zx^2 - x^3)$
Rewrite vertically and multiply.

$$
\begin{array}{r}
z^3 - 4z^2x + 2zx^2 - x^3 \\
4z - x \\
\hline
- z^3x + 4z^2x^2 - 2zx^3 + x^4 \\
4z^4 - 16z^3x + 8z^2x^2 - 4zx^3 \\
\hline
4z^4 - 17z^3x + 12z^2x^2 - 6zx^3 + x^4
\end{array}
$$

93. $(m^2 - 2mp + p^2)(m^2 + 2mp - p^2)$
Rewrite vertically and multiply.

$$
\begin{array}{r}
m^2 - 2mp + p^2 \\
m^2 + 2mp - p^2 \\
\hline
- m^2p^2 + 2mp^3 - p^4 \\
2m^3p - 4m^2p^2 + 2mp^3 \\
m^4 - 2m^3p + m^2p^2 \\
\hline
m^4 \qquad\qquad - 4m^2p^2 + 4mp^3 - p^4
\end{array}
$$

95. $ab(a + b)(a + 2b)(a - 3b)$
First multiply $(a + b)(a + 2b)(a - 3b)$.

$$
\begin{aligned}
(a + b)&(a + 2b)(a - 3b) \\
&= [(a + b)(a + 2b)](a - 3b) \\
&= [a^2 + 3ab + 2b^2](a - 3b)
\end{aligned}
$$

$$
\begin{array}{r}
a^2 + 3ab + 2b^2 \\
a - 3b \\
\hline
- 3a^2b - 9ab^2 - 6b^3 \\
a^3 + 3a^2b + 2ab^2 \\
\hline
a^3 \qquad\qquad - 7ab^2 - 6b^3
\end{array}
$$

Now multiply the last polynomial times ab.

$$
\begin{aligned}
ab(a^3 &- 7ab^2 - 6b^3) \\
&= a^4b - 7a^2b^3 - 6ab^4
\end{aligned}
$$

In Exercises 97–100, substitute 3 for x and 4 for y.

97. $(x + y)^2 = (3 + 4)^2 = 7^2 = 49$
$x^2 + y^2 = 3^2 + 4^2 = 9 + 16 = 25$
Since $49 \neq 25$, $(x + y)^2 \neq x^2 + y^2$.

99. $(x + y)^4 = (3 + 4)^4 = 7^4 = 2401$
$x^4 + y^4 = 3^4 + 4^4 = 81 + 256 = 337$
Since $2401 \neq 337$, $(x + y)^4 \neq x^4 + y^4$.

101. The formula for the area of a triangle is $A = \frac{1}{2}bh$.
Use $b = 3x + 2y$ and $h = 3x - 2y$.

$$
\begin{aligned}
A &= \tfrac{1}{2}(3x + 2y)(3x - 2y) \\
&= \tfrac{1}{2}(9x^2 - 4y^2) \\
&= \tfrac{9}{2}x^2 - 2y^2
\end{aligned}
$$

103. The formula for the area of a parallelogram is
$A = bh$. Use $b = 5x + 6$ and $h = 3x - 4$.

$$
\begin{aligned}
A &= (5x + 6)(3x - 4) \\
&= 15x^2 - 20x + 18x - 24 \\
&= 15x^2 - 2x - 24
\end{aligned}
$$

♦♦♦ Relating Concepts 105–112 ♦♦♦

105. The length of each side of the entire square is a.
To find the length of each side of the blue square,
subtract b, the length of a side of the green
rectangle, that is, $a - b$.

106. The formula for the area A of a square with side s
is $A = s^2$. Since $s = a - b$, the formula for the
area of the blue square would be $A = (a - b)^2$.

107. The formula for the area of a rectangle is
$A = LW$. The green rectangle has length $a - b$
and width b, so each green rectangle has an area of
$\underline{(a - b)b}$ or $ab - b^2$.

Since there are two green rectangles, the total area
in green is represented by the polynomial
$\underline{2(ab - b^2)}$ or $2ab - 2b^2$.

108. The length of each side of the yellow square is b,
so the yellow square has an area of $\underline{b^2}$.

109. The area of the entire colored region is represented
by $\underline{a^2}$, because each side of the entire colored
region has length $\underline{a}$.

110. Using the results from Exercises 105–109, the area
of the blue square equals

$$a^2 - (2ab - 2b^2) - b^2 = a^2 - 2ab + b^2.$$

111. (a) Both expressions for the area of the blue
square must be equal to each other; that is,
$(a - b)^2$ from Exercise 106 and $a^2 - 2ab + b^2$
from Exercise 110.

(b) From Exercise 106, the area of the blue square
is $(a - b)^2$. From Exercise 110, the area of the
blue square is $a^2 - 2ab + b^2$. Since these
expressions must be equal

$$(a - b)^2 = a^2 - 2ab + b^2.$$

This equation reinforces the special product for the
square of a binomial difference.

Copyright © 2012 Pearson Education, Inc. Publishing as Addison-Wesley.

112.

	Area: a^2	Area: ab

(diagram with sides a and b)

The large square is made up of two smaller squares and two congruent rectangles. The sum of the areas is $a^2 + 2ab + b^2$. Since $(a+b)^2$ must represent the same quantity, they must be equal. Thus,

$$(a+b)^2 = a^2 + 2ab + b^2.$$

113. $f(x) = 2x, g(x) = 5x - 1$

$$\begin{aligned}(fg)(x) &= f(x) \cdot g(x) \\ &= 2x(5x - 1) \\ &= 10x^2 - 2x\end{aligned}$$

115. $f(x) = x + 1, g(x) = 2x - 3$

$$\begin{aligned}(fg)(x) &= f(x) \cdot g(x) \\ &= (x+1)(2x - 3) \\ &= 2x^2 - 3x + 2x - 3 \\ &= 2x^2 - x - 3\end{aligned}$$

117. $f(x) = 2x - 3, g(x) = 4x^2 + 6x + 9$

$$\begin{aligned}(fg)(x) &= f(x) \cdot g(x) \\ &= (2x - 3)(4x^2 + 6x + 9)\end{aligned}$$

Multiply vertically.

$$\begin{array}{r} 4x^2 + 6x + 9 \\ 2x - 3 \\ \hline -12x^2 - 18x - 27 \\ 8x^3 + 12x^2 + 18x \\ \hline 8x^3 - 27 \end{array}$$

For Exercises 119–130, let $f(x) = x^2 - 9$, $g(x) = 2x$, and $h(x) = x - 3$.

119. $(fg)(x) = f(x) \cdot g(x)$
$= (x^2 - 9)(2x)$
$= 2x^3 - 18x$

121. $(fg)(2) = f(2) \cdot g(2)$
$= (2^2 - 9)[2(2)]$
$= -5 \cdot 4 = -20$

123. $(gh)(x) = g(x) \cdot h(x)$
$= (2x)(x - 3)$
$= 2x^2 - 6x$

125. $(gh)(-3) = g(-3) \cdot h(-3)$
$= [2(-3)] \cdot [(-3) - 3]$
$= -6(-6) = 36$

127. $(fg)\left(-\frac{1}{2}\right) = f\left(-\frac{1}{2}\right) \cdot g\left(-\frac{1}{2}\right)$
$= \left[\left(-\frac{1}{2}\right)^2 - 9\right] \cdot \left[2\left(-\frac{1}{2}\right)\right]$
$= \left(\frac{1}{4} - \frac{36}{4}\right)(-1)$
$= -\frac{35}{4}(-1) = \frac{35}{4}$

129. $(fh)\left(-\frac{1}{4}\right) = f\left(-\frac{1}{4}\right) \cdot h\left(-\frac{1}{4}\right)$
$= \left[\left(-\frac{1}{4}\right)^2 - 9\right] \cdot \left[-\frac{1}{4} - 3\right]$
$= \left(\frac{1}{16} - \frac{144}{16}\right)\left(-\frac{1}{4} - \frac{12}{4}\right)$
$= -\frac{143}{16}\left(-\frac{13}{4}\right) = \frac{1859}{64}$

131. $\dfrac{12p^7}{6p^3} = \dfrac{12}{6}p^{7-3} = 2p^4$

133. $\dfrac{-8a^3b^7}{6a^5b} = \dfrac{-8}{6}a^{3-5}b^{7-1} = \dfrac{-4}{3}a^{-2}b^6 = \dfrac{-4b^6}{3a^2}$

135. Subtract.

$$\begin{array}{r} -3a^2 + 4a - 5 \\ 5a^2 + 3a - 9 \\ \hline \end{array}$$

Change all the signs in the second row, then add.

$$\begin{array}{r} -3a^2 + 4a - 5 \\ -5a^2 - 3a + 9 \\ \hline -8a^2 + a + 4 \end{array}$$

5.5 Dividing Polynomials

5.5 Now Try Exercises

N1. $\dfrac{81y^4 - 54y^3 + 18y}{9y^3} = \dfrac{81y^4}{9y^3} - \dfrac{54y^3}{9y^3} + \dfrac{18y}{9y^3}$

$$= 9y - 6 + \dfrac{2}{y^2}$$

N2. $\dfrac{3x^2 - 4x - 15}{x - 3}$

$$\begin{array}{r} 3x + 5 \\ x - 3 \overline{\smash{\big)}\, 3x^2 - 4x - 15} \\ \underline{3x^2 - 9x} \\ 5x - 15 \\ \underline{5x - 15} \\ 0 \end{array}$$

First step: $\dfrac{3x^2}{x} = 3x$

Check: $(x - 3)(3x + 5) = 3x^2 - 4x - 15$

Answer: $3x + 5$

Copyright © 2012 Pearson Education, Inc. Publishing as Addison-Wesley.

N3. Divide $2x^3 - 12x - 10$ by $x - 4$.

Add a term with 0 coefficient as a placeholder for the missing x^2-term.

$$
\begin{array}{r}
2x^2 + 8x + 20 \\
x - 4 \overline{\smash{\big)}\, 2x^3 + 0x^2 - 12x - 10} \\
\underline{2x^3 - 8x^2} \\
8x^2 - 12x \\
\underline{8x^2 - 32x} \\
20x - 10 \\
\underline{20x - 80} \\
70
\end{array}
$$
$\qquad\qquad\qquad$ *Remainder*

Check: $(x - 4)(2x^2 + 8x + 20) + 70$
$\qquad = (2x^3 - 12x - 80) + 70$
$\qquad = 2x^3 - 12x - 10$

Answer: $2x^2 + 8x + 20 + \dfrac{70}{x - 4}$

N4. Divide $2x^4 + 8x^3 + 2x^2 - 5x - 3$ by $2x^2 - 2$.

$$
\begin{array}{r}
x^2 + 4x + 2 \\
2x^2 - 2 \overline{\smash{\big)}\, 2x^4 + 8x^3 + 2x^2 - 5x - 3} \\
\underline{2x^4 - 2x^2} \\
8x^3 + 4x^2 \\
\underline{8x^3 - 8x} \\
4x^2 + 3x - 3 \\
\underline{4x^2 - 4} \\
3x + 1
\end{array}
$$
$\qquad\qquad\qquad$ *Remainder*

Answer: $x^2 + 4x + 2 + \dfrac{3x + 1}{2x^2 - 2}$

N5. Divide $6m^3 - 8m^2 - 5m - 6$ by $3m - 6$.

$$
\begin{array}{r}
2m^2 + \frac{4}{3}m + 1 \\
3m - 6 \overline{\smash{\big)}\, 6m^3 - 8m^2 - 5m - 6} \\
\underline{6m^3 - 12m^2} \\
4m^2 - 5m \\
\underline{4m^2 - 8m} \\
3m - 6 \\
\underline{3m - 6} \\
0
\end{array}
$$

Answer: $2m^2 + \frac{4}{3}m + 1$

N6. $f(x) = 8x^2 + 2x - 3$, $g(x) = 2x - 1$

$$\left(\frac{f}{g}\right)(x) = \frac{f(x)}{g(x)} = \frac{8x^2 + 2x - 3}{2x - 1}$$

$$
\begin{array}{r}
4x + 3 \\
2x - 1 \overline{\smash{\big)}\, 8x^2 + 2x - 3} \\
\underline{8x^2 - 4x} \\
6x - 3 \\
\underline{6x - 3} \\
0
\end{array}
$$
First step: $\dfrac{8x^2}{2x} = 4x$

We conclude that $\left(\frac{f}{g}\right)(x) = 4x + 3$, provided the denominator, $2x - 1$, is *not* equal to zero; that is, $x \neq \frac{1}{2}$.

Using the above result, $\left(\frac{f}{g}\right)(x) = 4x + 3$, we have

$$\left(\frac{f}{g}\right)(8) = 32 + 3 = 35.$$

5.5 Section Exercises

1. We find the quotient of two monomials by using the _quotient_ rule for _exponents_.

3. If a polynomial in a division problem has a missing term, insert a term with coefficient equal to _0_ as a placeholder.

5. $\dfrac{15x^3 - 10x^2 + 5}{5} = \dfrac{15x^3}{5} - \dfrac{10x^2}{5} + \dfrac{5}{5}$
$\qquad\qquad\qquad\quad = 3x^3 - 2x^2 + 1$

7. $\dfrac{9y^2 + 12y - 15}{3y} = \dfrac{9y^2}{3y} + \dfrac{12y}{3y} - \dfrac{15}{3y}$
$\qquad\qquad\qquad = 3y + 4 - \dfrac{5}{y}$

9. $\dfrac{15m^3 + 25m^2 + 30m}{5m^3}$
$\quad = \dfrac{15m^3}{5m^3} + \dfrac{25m^2}{5m^3} + \dfrac{30m}{5m^3}$
$\quad = 3 + \dfrac{5}{m} + \dfrac{6}{m^2}$

11. $\dfrac{4m^2n^2 - 21mn^3 + 18mn^2}{14m^2n^3}$
$\quad = \dfrac{4m^2n^2}{14m^2n^3} - \dfrac{21mn^3}{14m^2n^3} + \dfrac{18mn^2}{14m^2n^3}$
$\quad = \dfrac{2}{7n} - \dfrac{3}{2m} + \dfrac{9}{7mn}$

13. $\dfrac{8wxy^2 + 3wx^2y + 12w^2xy}{4wx^2y}$
$\quad = \dfrac{8wxy^2}{4wx^2y} + \dfrac{3wx^2y}{4wx^2y} + \dfrac{12w^2xy}{4wx^2y}$
$\quad = \dfrac{2y}{x} + \dfrac{3}{4} + \dfrac{3w}{x}$

15.
$$
\begin{array}{r}
r^2 - 7r + 6 \\
3r - 1 \overline{\smash{\big)}\, 3r^3 - 22r^2 + 25r - 6} \\
\underline{3r^3 - r^2} \\
-21r^2 + 25r \\
\underline{-21r^2 + 7r} \\
18r - 6 \\
\underline{18r - 6} \\
0
\end{array}
$$

Answer: $r^2 - 7r + 6$

Copyright © 2012 Pearson Education, Inc. Publishing as Addison-Wesley.

17. $\dfrac{y^2 + y - 20}{y + 5}$

$$
\begin{array}{r}
y - 4 \\
y + 5 \overline{\smash{)}\, y^2 + y - 20} \\
\underline{y^2 + 5y} \\
-4y - 20 \\
\underline{-4y - 20} \\
0
\end{array}
$$

Answer: $y - 4$

19. $\dfrac{q^2 + 4q - 32}{q - 4}$

$$
\begin{array}{r}
q + 8 \\
q - 4 \overline{\smash{)}\, q^2 + 4q - 32} \\
\underline{q^2 - 4q} \\
8q - 32 \\
\underline{8q - 32} \\
0
\end{array}
$$

Answer: $q + 8$

21. $\dfrac{3t^2 + 17t + 10}{3t + 2}$

$$
\begin{array}{r}
t + 5 \\
3t + 2 \overline{\smash{)}\, 3t^2 + 17t + 10} \\
\underline{3t^2 + 2t} \\
15t + 10 \\
\underline{15t + 10} \\
0
\end{array}
$$

Answer: $t + 5$

23. $\dfrac{p^2 + 2p + 20}{p + 6}$

$$
\begin{array}{r}
p - 4 \\
p + 6 \overline{\smash{)}\, p^2 + 2p + 20} \\
\underline{p^2 + 6p} \\
-4p + 20 \\
\underline{-4p - 24} \\
44
\end{array}
$$

Remainder

Answer: $p - 4 + \dfrac{44}{p + 6}$

25. $\dfrac{3m^3 + 5m^2 - 5m + 1}{3m - 1}$

$$
\begin{array}{r}
m^2 + 2m - 1 \\
3m - 1 \overline{\smash{)}\, 3m^3 + 5m^2 - 5m + 1} \\
\underline{3m^3 - m^2} \\
6m^2 - 5m \\
\underline{6m^2 - 2m} \\
-3m + 1 \\
\underline{-3m + 1} \\
0
\end{array}
$$

Answer: $m^2 + 2m - 1$

27. $\dfrac{m^3 - 2m^2 - 9}{m - 3}$

$$
\begin{array}{r}
m^2 + m + 3 \\
m - 3 \overline{\smash{)}\, m^3 - 2m^2 + 0m - 9} \\
\underline{m^3 - 3m^2} \\
m^2 + 0m \\
\underline{m^2 - 3m} \\
3m - 9 \\
\underline{3m - 9} \\
0
\end{array}
$$

Answer: $m^2 + m + 3$

29.

$$
\begin{array}{r}
z^2 + 3 \\
2z - 5 \overline{\smash{)}\, 2z^3 - 5z^2 + 6z - 15} \\
\underline{2z^3 - 5z^2} \\
6z - 15 \\
\underline{6z - 15} \\
0
\end{array}
$$

Answer: $z^2 + 3$

31.

$$
\begin{array}{r}
x^2 + 2x - 3 \\
4x + 1 \overline{\smash{)}\, 4x^3 + 9x^2 - 10x + 3} \\
\underline{4x^3 + x^2} \\
8x^2 - 10x \\
\underline{8x^2 + 2x} \\
-12x + 3 \\
\underline{-12x - 3} \\
6
\end{array}
$$

Remainder

Answer: $x^2 + 2x - 3 + \dfrac{6}{4x + 1}$

33.

$$
\begin{array}{r}
2x - 5 \\
3x^2 - 2x + 4 \overline{\smash{)}\, 6x^3 - 19x^2 + 14x - 15} \\
\underline{6x^3 - 4x^2 + 8x} \\
-15x^2 + 6x - 15 \\
\underline{-15x^2 + 10x - 20} \\
-4x + 5
\end{array}
$$

Remainder

Answer: $2x - 5 + \dfrac{-4x + 5}{3x^2 - 2x + 4}$

35.

$$
\begin{array}{r}
x^2 + x + 3 \\
x - 1 \overline{\smash{)}\, x^3 + 0x^2 + 2x - 3} \\
\underline{x^3 - x^2} \\
x^2 + 2x \\
\underline{x^2 - x} \\
3x - 3 \\
\underline{3x - 3} \\
0
\end{array}
$$

Answer: $x^2 + x + 3$

Copyright © 2012 Pearson Education, Inc. Publishing as Addison-Wesley.

37.

$$
\begin{array}{r}
2x^2 - x - 5 \\
x - 5 \overline{\smash{\big)}\ 2x^3 - 11x^2 + 0x + 25} \\
\underline{2x^3 - 10x^2} \\
-x^2 + 0x \\
\underline{-x^2 + 5x} \\
-5x + 25 \\
\underline{-5x + 25} \\
0
\end{array}
$$

Answer: $2x^2 - x - 5$

39.

$$
\begin{array}{r}
3x^2 + 6x + 11 \\
x - 2 \overline{\smash{\big)}\ 3x^3 + 0x^2 - x + 4} \\
\underline{3x^3 - 6x^2} \\
6x^2 - x \\
\underline{6x^2 - 12x} \\
11x + 4 \\
\underline{11x - 22} \\
26
\end{array}
$$

Answer: $3x^2 + 6x + 11 + \dfrac{26}{x - 2}$

41.

$$
\begin{array}{r}
2k^2 + 3k - 1 \\
2k^2 + 1 \overline{\smash{\big)}\ 4k^4 + 6k^3 + 0k^2 + 3k - 1} \\
\underline{4k^4 + 2k^2} \\
6k^3 - 2k^2 + 3k \\
\underline{6k^3 + 3k} \\
-2k^2 - 1 \\
\underline{-2k^2 - 1} \\
0
\end{array}
$$

Answer: $2k^2 + 3k - 1$

43.

$$
\begin{array}{r}
2y^2 + 2 \\
3y^2 + 2y - 3 \overline{\smash{\big)}\ 6y^4 + 4y^3 + 0y^2 + 4y - 6} \\
\underline{6y^4 + 4y^3 - 6y^2} \\
6y^2 + 4y - 6 \\
\underline{6y^2 + 4y - 6} \\
0
\end{array}
$$

Answer: $2y^2 + 2$

45.

$$
\begin{array}{r}
x^2 - 4x + 2 \\
x^2 + 3 \overline{\smash{\big)}\ x^4 - 4x^3 + 5x^2 - 3x + 2} \\
\underline{x^4 + 3x^2} \\
-4x^3 + 2x^2 - 3x \\
\underline{-4x^3 - 12x} \\
2x^2 + 9x + 2 \\
\underline{2x^2 + 6} \\
9x - 4
\end{array}
$$

Answer: $x^2 - 4x + 2 + \dfrac{9x - 4}{x^2 + 3}$

47.

$$
\begin{array}{r}
p^2 + \frac{5}{2}p + 2 \\
2p + 2 \overline{\smash{\big)}\ 2p^3 + 7p^2 + 9p + 3} \\
\underline{2p^3 + 2p^2} \\
5p^2 + 9p \\
\underline{5p^2 + 5p} \\
4p + 3 \\
\underline{4p + 4} \\
-1
\end{array}
$$

Answer: $p^2 + \dfrac{5}{2}p + 2 + \dfrac{-1}{2p + 2}$

49.

$$
\begin{array}{r}
\frac{3}{2}a - 10 \\
2a + 6 \overline{\smash{\big)}\ 3a^2 - 11a + 17} \\
\underline{3a^2 + 9a} \\
-20a + 17 \\
\underline{-20a - 60} \\
77
\end{array}
$$

Answer: $\dfrac{3}{2}a - 10 + \dfrac{77}{2a + 6}$

51.

$$
\begin{array}{r}
p^2 + p + 1 \\
p - 1 \overline{\smash{\big)}\ p^3 + 0p^2 + 0p - 1} \\
\underline{p^3 - p^2} \\
p^2 + 0p \\
\underline{p^2 - p} \\
p - 1 \\
\underline{p - 1} \\
0
\end{array}
$$

Answer: $p^2 + p + 1$

53. To start: $\dfrac{2x^2}{3x} = \dfrac{2}{3}x$

$$
\begin{array}{r}
\frac{2}{3}x - 1 \\
3x + 1 \overline{\smash{\big)}\ 2x^2 - \frac{7}{3}x - 1} \\
\underline{2x^2 + \frac{2}{3}x} \\
-3x - 1 \\
\underline{-3x - 1} \\
0
\end{array}
$$

Answer: $\frac{2}{3}x - 1$

55.

$$
\begin{array}{r}
\frac{3}{4}a - 2 \\
4a + 3 \overline{\smash{\big)}\ 3a^2 - \frac{23}{4}a - 5} \\
\underline{3a^2 + \frac{9}{4}a} \\
-8a - 5 \\
\underline{-8a - 6} \\
1
\end{array}
$$
Remainder

Answer: $\dfrac{3}{4}a - 2 + \dfrac{1}{4a + 3}$

Copyright © 2012 Pearson Education, Inc. Publishing as Addison-Wesley.

57. The volume of a box is the product of the height, length, and width. Use the formula $V = LWH$.

$$V = LWH$$
$$\frac{V}{LH} = W$$

Here,
$$L \cdot H = (p+4)p = p^2 + 4p, \text{ so}$$
$$W = \frac{V}{LH} = \frac{2p^3 + 15p^2 + 28p}{p^2 + 4p}.$$

$$
\begin{array}{r}
2p + 7 \\
p^2 + 4p \overline{\smash{\big)}\, 2p^3 + 15p^2 + 28p} \\
\underline{2p^3 + 8p^2 } \\
7p^2 + 28p \\
\underline{7p^2 + 28p} \\
0
\end{array}
$$

The width is $2p + 7$.

59.
$$P(x) = x^3 - 4x^2 + 3x - 5$$
$$P(-1) = (-1)^3 - 4(-1)^2 + 3(-1) - 5$$
$$= -1 - 4 - 3 - 5 = -13$$

Now divide the given polynomial by $x + 1$.

$$
\begin{array}{r}
x^2 - 5x + 8 \\
x + 1 \overline{\smash{\big)}\, x^3 - 4x^2 + 3x - 5} \\
\underline{x^3 + x^2 } \\
-5x^2 + 3x \\
\underline{-5x^2 - 5x } \\
8x - 5 \\
\underline{8x + 8} \\
-13
\end{array}
$$
Remainder

The remainder in the division is the same as $P(-1)$, that is, -13. This suggests that if a polynomial is divided by $x - r$, in this case $x - (-1)$ or $x + 1$, then the remainder is equal to $P(r)$, in this case $P(-1)$.

61.
$$\left(\frac{f}{g}\right)(x) = \frac{f(x)}{g(x)} = \frac{10x^2 - 2x}{2x}$$
$$= \frac{10x^2}{2x} - \frac{2x}{2x}$$
$$= 5x - 1$$

The x-values that are not in the domain of the quotient function are found by solving $g(x) = 0$.

$$2x = 0$$
$$x = 0$$

Exclude $x = 0$ from the domain.

63.
$$
\begin{array}{r}
2x - 3 \\
x + 1 \overline{\smash{\big)}\, 2x^2 - x - 3} \\
\underline{2x^2 + 2x } \\
-3x - 3 \\
\underline{-3x - 3} \\
0
\end{array}
$$

Quotient: $2x - 3$

$$g(x) = 0$$
$$x + 1 = 0$$
$$x = -1$$

Exclude $x = -1$ from the domain.

65.
$$
\begin{array}{r}
4x^2 + 6x + 9 \\
2x - 3 \overline{\smash{\big)}\, 8x^3 + 0x^2 + 0x - 27} \\
\underline{8x^3 - 12x^2 } \\
12x^2 + 0x \\
\underline{12x^2 - 18x } \\
18x - 27 \\
\underline{18x - 27} \\
0
\end{array}
$$

Quotient: $4x^2 + 6x + 9$

$$g(x) = 0$$
$$2x - 3 = 0$$
$$2x = 3$$
$$x = \tfrac{3}{2}$$

Exclude $x = \frac{3}{2}$ from the domain.

For Exercises 67–78, let $f(x) = x^2 - 9$, $g(x) = 2x$, and $h(x) = x - 3$.

67.
$$\left(\frac{f}{g}\right)(x) = \frac{f(x)}{g(x)} = \frac{x^2 - 9}{2x}$$

We must exclude any values of x that make the denominator equal to zero, that is, $x \neq 0$.

69.
$$\left(\frac{f}{g}\right)(2) = \frac{f(2)}{g(2)} = \frac{2^2 - 9}{2(2)} = \frac{-5}{4} = -\frac{5}{4}$$

71.
$$\left(\frac{h}{g}\right)(x) = \frac{h(x)}{g(x)} = \frac{x - 3}{2x}, \; x \neq 0$$

73.
$$\left(\frac{h}{g}\right)(3) = \frac{h(3)}{g(3)} = \frac{(3) - 3}{2(3)} = \frac{0}{6} = 0$$

75.
$$\left(\frac{f}{g}\right)\left(\frac{1}{2}\right) = \frac{f\left(\frac{1}{2}\right)}{g\left(\frac{1}{2}\right)} = \frac{\left(\frac{1}{2}\right)^2 - 9}{2\left(\frac{1}{2}\right)} = \frac{\frac{1}{4} - \frac{36}{4}}{1}$$
$$= -\frac{35}{4}$$

77.
$$\left(\frac{h}{g}\right)\left(-\frac{1}{2}\right) = \frac{h\left(-\frac{1}{2}\right)}{g\left(-\frac{1}{2}\right)} = \frac{-\frac{1}{2} - 3}{2\left(-\frac{1}{2}\right)} = \frac{-\frac{7}{2}}{-1} = \frac{7}{2}$$

79.
$$8(y - 5) = 8y - 8 \cdot 5$$
$$= 8y - 40$$

Copyright © 2012 Pearson Education, Inc. Publishing as Addison-Wesley.

81. $4p(2p+1) = 4p(2p) + 4p(1)$
$= 8p^2 + 4p$

83. $7(2x) - 7(3z) = 7(2x - 3z)$

Chapter 5 Review Exercises

1. $4^3 = 4 \cdot 4 \cdot 4 = 64$

2. $\left(\frac{1}{3}\right)^4 = \frac{1}{3} \cdot \frac{1}{3} \cdot \frac{1}{3} \cdot \frac{1}{3} = \frac{1}{81}$

3. $(-5)^3 = (-5)(-5)(-5) = -125$

4. $\dfrac{2}{(-3)^{-2}} = \dfrac{2}{\dfrac{1}{(-3)^2}} = \dfrac{2}{1} \cdot \dfrac{(-3)^2}{1}$
$= 2 \cdot (-3)^2$
$= 2 \cdot (-3)(-3)$
$= 18$

5. $\left(\frac{2}{3}\right)^{-4} = \left(\frac{3}{2}\right)^4$
$= \frac{3}{2} \cdot \frac{3}{2} \cdot \frac{3}{2} \cdot \frac{3}{2}$
$= \frac{81}{16}$

6. $\left(\frac{5}{4}\right)^{-2} = \left(\frac{4}{5}\right)^2 = \frac{4}{5} \cdot \frac{4}{5} = \frac{16}{25}$

7. $5^{-1} - 6^{-1} = \frac{1}{5} - \frac{1}{6} = \frac{6}{30} - \frac{5}{30} = \frac{1}{30}$

8. $2^{-1} + 4^{-1} = \frac{1}{2} + \frac{1}{4} = \frac{2}{4} + \frac{1}{4} = \frac{3}{4}$

9. $-3^0 + 3^0 = -1 + 1 = 0$

10. $(3^{-4})^2 = 3^{(-4) \cdot 2} = 3^{-8} = \dfrac{1}{3^8}$

11. $(x^{-4})^{-2} = x^{-4(-2)} = x^8$

12. $(xy^{-3})^{-2} = x^{1(-2)} y^{(-3)(-2)}$
$= x^{-2} y^6$
$= \frac{1}{x^2} \cdot y^6 = \frac{y^6}{x^2}$

13. $(z^{-3})^3 z^{-6} = z^{-9} z^{-6}$
$= z^{-9+(-6)}$
$= z^{-15} = \dfrac{1}{z^{15}}$

14. $(5m^{-3})^2 (m^4)^{-3} = 5^2 m^{-6} m^{-12}$
$= 25 m^{-6-12}$
$= 25 m^{-18} = \dfrac{25}{m^{18}}$

15. $\dfrac{(3r)^2 r^4}{r^{-2} r^{-3}} (9r^{-3})^{-2} = \dfrac{3^2 r^2 r^4 9^{-2} r^6}{r^{-2} r^{-3}}$
$= \dfrac{9^1 9^{-2} r^{2+4+6}}{r^{-2-3}}$
$= \dfrac{9^{-1} r^{12}}{r^{-5}}$
$= \dfrac{r^{12-(-5)}}{9} = \dfrac{r^{17}}{9}$

16. $\left(\dfrac{5z^{-3}}{z^{-1}}\right) \dfrac{5}{z^2} = \dfrac{25z^{-3}}{z^{-1+2}}$
$= \dfrac{25}{z^3 z^1}$
$= \dfrac{25}{z^{3+1}} = \dfrac{25}{z^4}$

17. $\left(\dfrac{6m^{-4}}{m^{-9}}\right)^{-1} \left(\dfrac{m^{-2}}{16}\right) = \dfrac{6^{-1} m^4}{m^9} \cdot \dfrac{m^{-2}}{16}$
$= \dfrac{1}{6 \cdot 16} m^{4+(-2)-9}$
$= \dfrac{1}{96} m^{-7} = \dfrac{1}{96 m^7}$

18. $\left(\dfrac{3r^5}{5r^{-3}}\right)^{-2} \left(\dfrac{9r^{-1}}{2r^{-5}}\right)^3$
$= \dfrac{3^{-2} r^{-10}}{5^{-2} r^6} \cdot \dfrac{9^3 r^{-3}}{2^3 r^{-15}}$
$= \dfrac{5^2}{3^2} r^{-10-6} \cdot \dfrac{9^3}{2^3} r^{-3-(-15)}$
$= \dfrac{25}{9} r^{-16} \cdot \dfrac{729}{8} r^{12}$
$= \dfrac{(25)(729)}{(9)(8)} r^{-16+12}$
$= \dfrac{2025}{8} r^{-4} = \dfrac{2025}{8r^4}$

19. $(-3x^4 y^3)(4x^{-2} y^5) = -3(4) x^4 x^{-2} y^3 y^5$
$= -12 x^{4-2} y^{3+5}$
$= -12 x^2 y^8$

20. $\dfrac{6m^{-4} n^3}{-3mn^2} = -2m^{-4-1} n^{3-2}$
$= -2m^{-5} n^1$
$= -\dfrac{2n}{m^5}, \ \ \text{or} \ \ \dfrac{-2n}{m^5}$

21. $\dfrac{(5p^{-2}q)(4p^5 q^{-3})}{2p^{-5} q^5} = \dfrac{20p^{-2+5} q^{1-3}}{2p^{-5} q^5}$
$= \dfrac{10p^3 q^{-2}}{p^{-5} q^5}$
$= 10p^{3-(-5)} q^{-2-5}$
$= 10p^8 q^{-7}$
$= \dfrac{10p^8}{q^7}$

22. In $(-6)^0$, the base is -6 and the expression simplifies to 1. In -6^0, the base is 6 and the expression simplifies to -1.

23. Yes, $\left(\dfrac{a}{b}\right)^{-1} = \dfrac{a^{-1}}{b^{-1}}$ for all $a, b \neq 0$.

24. No, $(ab)^{-1} \neq ab^{-1}$ for all $a, b \neq 0$.
For example, let $a = 3$ and $b = 4$.
Then $(ab)^{-1} = (3 \cdot 4)^{-1} = 12^{-1} = \frac{1}{12}$,
while $ab^{-1} = 3 \cdot 4^{-1} = 3 \cdot \frac{1}{4} = \frac{3}{4} \cdot \frac{1}{12} \neq \frac{3}{4}$.
Remember, in general, $(ab)^{-1} = a^{-1} b^{-1}$.

Copyright © 2012 Pearson Education, Inc. Publishing as Addison-Wesley.

25. $13,450 = 1{\wedge}3\,4\,5\,0.$

Place a caret to the right of the first nonzero digit. Count 4 places.

Since the number 1.345 is to be made larger, the exponent on 10 is positive.
$$13,450 = 1.345 \times 10^4$$

26. $0.000\,000\,076\,5 = 0.0\,0\,0\,0\,0\,0\,0\,7{\wedge}65$

Count 8 places

Since the number 7.65 is to be made smaller, the exponent on 10 is negative.
$$0.000\,000\,076\,5 = 7.65 \times 10^{-8}$$

27. $0.138 = 0.1\,{\wedge}38$

Count 1 place.

Since the number 1.38 is to be made smaller, the exponent on 10 is negative.
$$0.138 = 1.38 \times 10^{-1}$$

28. $304,060,000 = 3.0406 \times 10^8$
$$92,000 = 9.2 \times 10^4$$
$$100 = 1 \times 10^2$$

29. $1.21 \times 10^6 = 1,210,000$

Move the decimal point 6 places to the right because the exponent is positive. Attach extra zeros.

30. $5.8 \times 10^{-3} = 0.0058$

Move the decimal point 3 places to the left because the exponent is negative.

31. $\dfrac{16 \times 10^4}{8 \times 10^8} = \dfrac{16}{8} \times 10^{4-8}$
$$= 2 \times 10^{-4} \text{ or } 0.0002$$

32. $\dfrac{6 \times 10^{-2}}{4 \times 10^{-5}} = \dfrac{6}{4} \times 10^{-2-(-5)}$
$$= 1.5 \times 10^3 \text{ or } 1500$$

33. $\dfrac{0.000\,000\,016\,4}{0.0004} = \dfrac{1.64 \times 10^{-8}}{4 \times 10^{-4}}$
$$= \dfrac{1.64}{4} \times 10^{-8-(-4)}$$
$$= 0.41 \times 10^{-4}$$
$$= 4.1 \times 10^{-5} \text{ or } 0.000\,041$$

34. $\dfrac{0.0009 \times 12,000,000}{400,000}$
$$= \dfrac{9 \times 10^{-4} \times 1.2 \times 10^7}{4 \times 10^5}$$
$$= \dfrac{9 \times 1.2}{4} \times \dfrac{10^{-4} \times 10^7}{10^5}$$
$$= \dfrac{10.8}{4} \times 10^{-4+7-5}$$
$$= 2.7 \times 10^{-2} \text{ or } 0.027$$

35. (a) The distance from Mercury to the sun is 3.6×10^7 mi and the distance from Venus to the sun is 6.7×10^7 mi, so the distance between Mercury and Venus in miles is
$$(6.7 \times 10^7) - (3.6 \times 10^7)$$
$$= (6.7 - 3.6) \times 10^7$$
$$= 3.1 \times 10^7.$$

Use $d = rt$, or $\frac{d}{r} = t$, where $d = 3.1 \times 10^7$ and $r = 1.55 \times 10^3$.
$$\dfrac{3.1 \times 10^7}{1.55 \times 10^3} = \dfrac{3.1}{1.55} \times 10^{7-3}$$
$$= 2 \times 10^4$$
$$= 20,000$$

It would take 20,000 hr.

(b) From part (a), it takes 20,000 hr for a spacecraft to travel from Venus to Mercury. Convert this to days (24 hours = 1 day).
$$20,000 \text{ hr} = \dfrac{20,000}{24} \text{ days}$$
$$\approx 833 \text{ days}$$

It would take about 833 days.

36. The coefficient of $14p^5$ is 14.

37. The coefficient of $-z = -1z$ is -1.

38. The coefficient of $\frac{x}{10} = \frac{1}{10}x$ is $\frac{1}{10}$.

39. The coefficient of $504p^3r^5$ is 504.

40. $9k + 11k^3 - 3k^2$

(a) In descending powers of k, the polynomial is
$$11k^3 - 3k^2 + 9k.$$

(b) The polynomial is a trinomial since it has three terms.

(c) The degree of the polynomial is 3 since the highest power of k is 3.

41. $14m^6 + 9m^7$

(a) In descending powers of m, the polynomial is
$$9m^7 + 14m^6.$$

(b) The polynomial is a binomial since it has two terms.

(c) The degree of the polynomial is 7 since the highest power of m is 7.

42. $-5y^4 + 3y^3 + 7y^2 - 2y$

(a) The polynomial is already written in descending powers of y.

(b) The polynomial has four terms, so it is none of these choices.

Copyright © 2012 Pearson Education, Inc. Publishing as Addison-Wesley.

(c) The degree of the polynomial is 4 since the highest power of y is 4.

43. $-7q^5r^3$

(a) The polynomial is already written in descending powers.

(b) The polynomial is a monomial since it has just one term.

(c) The degree is $5 + 3 = 8$, the sum of the exponents of this term.

44. One example of a polynomial in the variable x that has degree 5, is lacking a third-degree term, and is written in descending powers of the variable is

$$x^5 + 2x^4 - x^2 + x + 2.$$

45. Add by columns.

$$\begin{array}{r} 3x^2 - 5x + 6 \\ -4x^2 + 2x - 5 \\ \hline -1x^2 - 3x + 1 \end{array} \text{ or } -x^2 - 3x + 1$$

46. Subtract.

$$\begin{array}{r} -5y^3 \quad + 8y - 3 \\ 4y^2 + 2y + 9 \\ \hline \end{array}$$

Change the signs in the second polynomial and add.

$$\begin{array}{r} -5y^3 \quad + 8y - 3 \\ - 4y^2 - 2y - 9 \\ \hline -5y^3 - 4y^2 + 6y - 12 \end{array}$$

47. $(4a^3 - 9a + 15) - (-2a^3 + 4a^2 + 7a)$
$= 4a^3 - 9a + 15 + 2a^3 - 4a^2 - 7a$
$= 4a^3 + 2a^3 - 4a^2 - 9a - 7a + 15$
$= 6a^3 - 4a^2 - 16a + 15$

48. $(3y^2 + 2y - 1) + (5y^2 - 11y + 6)$
$= 3y^2 + 5y^2 + 2y - 11y - 1 + 6$
$= 8y^2 - 9y + 5$

49. To find the perimeter, add the measures of the three sides.

$(4x^2 + 2) + (6x^2 + 5x + 2) + (2x^2 + 3x + 1)$
$= 4x^2 + 6x^2 + 2x^2 + 5x + 3x + 2 + 2 + 1$
$= 12x^2 + 8x + 5$

The perimeter is $12x^2 + 8x + 5$.

50. $f(x) = -2x^2 + 5x + 7$

(a) $f(-2) = -2(-2)^2 + 5(-2) + 7$
$= -2(4) - 10 + 7$
$= -8 - 10 + 7$
$= -18 + 7 = -11$

(b) $f(3) = -2(3)^2 + 5(3) + 7$
$= -2(9) + 15 + 7$
$= -18 + 15 + 7$
$= -3 + 7 = 4$

51. $f(x) = 2x + 3, g(x) = 5x^2 - 3x + 2$

(a) $(f + g)(x) = f(x) + g(x)$
$= (2x + 3) + (5x^2 - 3x + 2)$
$= 5x^2 + 2x - 3x + 3 + 2$
$= 5x^2 - x + 5$

(b) $(f - g)(x) = f(x) - g(x)$
$= (2x + 3) - (5x^2 - 3x + 2)$
$= 2x + 3 - 5x^2 + 3x - 2$
$= -5x^2 + 5x + 1$

(c) $(f + g)(-1)$
$= f(-1) + g(-1)$
$= [2(-1) + 3] + [5(-1)^2 - 3(-1) + 2]$
$= [1] + [10]$
$= 11$

(d) $(f - g)(-1) = f(-1) - g(-1)$
$= 1 - 10$ *from part (c)*
$= -9$

52. $f(x) = 3x^2 + 2x - 1, g(x) = 5x + 7$

(a) $(g \circ f)(3) = g(f(3))$
$= g(3 \cdot 3^2 + 2 \cdot 3 - 1)$
$= g(32)$
$= 5 \cdot 32 + 7$
$= 167$

(b) $(f \circ g)(3) = f(g(3))$
$= f(5 \cdot 3 + 7)$
$= f(22)$
$= 3 \cdot 22^2 + 2 \cdot 22 - 1$
$= 1495$

(c) $(f \circ g)(-2) = f(g(-2))$
$= f[5(-2) + 7]$
$= f(-3)$
$= 3(-3)^2 + 2(-3) - 1$
$= 20$

(d) $(g \circ f)(-2) = g(f(-2))$
$= g[3(-2)^2 + 2(-2) - 1]$
$= g(7)$
$= 5 \cdot 7 + 7$
$= 42$

Copyright © 2012 Pearson Education, Inc. Publishing as Addison-Wesley.

(e) $(f \circ g)(x)$
$$= f(g(x))$$
$$= f(5x + 7)$$
$$= 3(5x + 7)^2 + 2(5x + 7) - 1$$
$$= 3(25x^2 + 70x + 49) + 10x + 14 - 1$$
$$= 75x^2 + 210x + 147 + 10x + 13$$
$$= 75x^2 + 220x + 160$$

(f) $(g \circ f)(x) = g(f(x))$
$$= g(3x^2 + 2x - 1)$$
$$= 5(3x^2 + 2x - 1) + 7$$
$$= 15x^2 + 10x - 5 + 7$$
$$= 15x^2 + 10x + 2$$

53. $f(x) = ax^3 + bx^2 + cx + d$, where $a = -21.307$, $b = 603.07$, $c = -1576.7$, and $d = 94{,}319$.

(a) $x = 1990 - 1990 = 0$

$$f(0) = a(0)^3 + b(0)^2 + c(0) + d$$
$$= 94{,}319 \text{ twin births}$$

(b) $x = 2000 - 1990 = 10$

$$f(10) = a(10)^3 + b(10)^2 + c(10) + d$$
$$= 117{,}552 \text{ twin births}$$

(c) $x = 2006 - 1990 = 16$

$$f(16) = a(16)^3 + b(16)^2 + c(16) + d$$
$$= 136{,}204.248 \approx 136{,}204 \text{ twin births}$$

54.

x	$f(x) = -2x + 5$
-2	$-2(-2) + 5 = 9$
-1	$-2(-1) + 5 = 7$
0	$-2(0) + 5 = 5$
1	$-2(1) + 5 = 3$
2	$-2(2) + 5 = 1$

This is a linear function, so plot the points and draw a line through them.

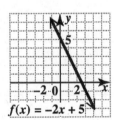

Any x-value can be used, so the domain is $(-\infty, \infty)$. From the graph, we see that any y-value can be obtained from the function, so the range is $(-\infty, \infty)$.

55.

x	$f(x) = x^2 - 6$
-2	$(-2)^2 - 6 = -2$
-1	$(-1)^2 - 6 = -5$
0	$(0)^2 - 6 = -6$
1	$(1)^2 - 6 = -5$
2	$(2)^2 - 6 = -2$

Since the greatest exponent is 2, the graph of f is a parabola.

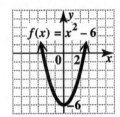

Any x-value can be used, so the domain is $(-\infty, \infty)$. From the graph, we see that the y-values are at least -6, so the range is $[-6, \infty)$.

56.

x	$f(x) = -x^3 + 1$
-2	$-(-2)^3 + 1 = 9$
-1	$-(-1)^3 + 1 = 2$
0	$-(0)^3 + 1 = 1$
1	$-(1)^3 + 1 = 0$
2	$-(2)^3 + 1 = -7$

The greatest exponent is 3, so the graph of f is s-shaped.

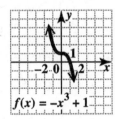

Any x-value can be used, so the domain is $(-\infty, \infty)$. From the graph, we see that any y-value can be obtained from the function, so the range is $(-\infty, \infty)$.

57. $-6k(2k^2 + 7) = -6k(2k^2) - 6k(7)$
$$= -12k^3 - 42k$$

58. $(3m - 2)(5m + 1)$

$$\begin{array}{cccc} \mathbf{F} & \mathbf{O} & \mathbf{I} & \mathbf{L} \end{array}$$
$$= 15m^2 + 3m - 10m - 2$$
$$= 15m^2 - 7m - 2$$

59. $(3w - 2t)(2w - 3t)$

$$\begin{array}{cccc} \mathbf{F} & \mathbf{O} & \mathbf{I} & \mathbf{L} \end{array}$$
$$= 6w^2 - 9wt - 4wt + 6t^2$$
$$= 6w^2 - 13wt + 6t^2$$

Copyright © 2012 Pearson Education, Inc. Publishing as Addison-Wesley.

60. $(2p^2 + 6p)(5p^2 - 4)$

$\qquad$ **F** $\quad$ **O** $\quad$ **I** $\quad$ **L**

$= 10p^4 - 8p^2 + 30p^3 - 24p$

$= 10p^4 + 30p^3 - 8p^2 - 24p$

61. $(3q^2 + 2q - 4)(q - 5)$

$= (3q^2 + 2q - 4)(q) + (3q^2 + 2q - 4)(-5)$

$= 3q^3 + 2q^2 - 4q - 15q^2 - 10q + 20$

$= 3q^3 + 2q^2 - 15q^2 - 4q - 10q + 20$

$= 3q^3 - 13q^2 - 14q + 20$

62. $(6r^2 - 1)(6r^2 + 1) = (6r^2)^2 - 1^2$

$= 36r^4 - 1$

63. $(4m + 3)^2 = (4m)^2 + 2(4m)(3) + 3^2$

$= 16m^2 + 24m + 9$

64. $(3t + 2)^2 = (3t)^2 + 2(3t)(2) + 2^2$

$= 9t^2 + 12t + 4$

Now multiply the last polynomial by t.

$t(3t + 2)^2 = 9t^3 + 12t^2 + 4t$

65. $\dfrac{4y^3 - 12y^2 + 5y}{4y} = \dfrac{4y^3}{4y} - \dfrac{12y^2}{4y} + \dfrac{5y}{4y}$

$= y^2 - 3y + \tfrac{5}{4}$

66. $\dfrac{x^3 - 9x^2 + 26x - 30}{x - 5}$

$$
\begin{array}{r}
x^2 - 4x + 6 \\
x - 5 \overline{\smash{)}\, x^3 - 9x^2 + 26x - 30} \\
\underline{x^3 - 5x^2} \\
-4x^2 + 26x \\
\underline{-4x^2 + 20x} \\
6x - 30 \\
\underline{6x - 30} \\
0
\end{array}
$$

Answer: $x^2 - 4x + 6$

67. $\dfrac{2p^3 + 9p^2 + 27}{2p - 3}$

$$
\begin{array}{r}
p^2 + 6p + 9 \\
2p - 3 \overline{\smash{)}\, 2p^3 + 9p^2 + 0p + 27} \\
\underline{2p^3 - 3p^2} \\
12p^2 + 0p \\
\underline{12p^2 - 18p} \\
18p + 27 \\
\underline{18p - 27} \\
54 \quad \textit{Remainder}
\end{array}
$$

Answer: $p^2 + 6p + 9 + \dfrac{54}{2p - 3}$

68. $\dfrac{5p^4 + 15p^3 - 33p^2 - 9p + 18}{5p^2 - 3}$

$$
\begin{array}{r}
p^2 + 3p - 6 \\
5p^2 - 3 \overline{\smash{)}\, 5p^4 + 15p^3 - 33p^2 - 9p + 18} \\
\underline{5p^4 \qquad - 3p^2} \\
15p^3 - 30p^2 - 9p \\
\underline{15p^3 \qquad - 9p} \\
-30p^2 \qquad + 18 \\
\underline{-30p^2 \qquad + 18} \\
0
\end{array}
$$

Answer: $p^2 + 3p - 6$

69. **[5.1]** **(a)** $4^{-2} = \dfrac{1}{4^2} = \dfrac{1}{16}$ **(A)**

(b) $-4^2 = -(4^2) = -16$ **(G)**

(c) $4^0 = 1$ **(C)**

(d) $(-4)^0 = 1$ **(C)**

(e) $(-4)^{-2} = \dfrac{1}{(-4)^2} = \dfrac{1}{16}$ **(A)**

(f) $-4^0 = -(4^0) = -1$ **(E)**

(g) $-4^0 + 4^0 = -1 + 1 = 0$ **(B)**

(h) $-4^0 - 4^0 = (-1) - 1 = -2$ **(H)**

(i) $4^{-2} + 4^{-1} = \dfrac{1}{4^2} + \dfrac{1}{4} = \dfrac{1}{16} + \dfrac{4}{16} = \dfrac{5}{16}$ **(F)**

(j) $4^2 = 16$ **(I)**

70. **[5.1]** $\dfrac{6^{-1}y^3(y^2)^{-2}}{6y^{-4}(y^{-1})} = \dfrac{y^3 y^{-4}}{6^1 \cdot 6y^{-4}y^{-1}}$

$= \dfrac{y^{3-4-(-4-1)}}{36}$

$= \dfrac{y^{-1+5}}{36} = \dfrac{y^4}{36}$

71. **[5.1]** $5^{-3} = \dfrac{1}{5^3} = \dfrac{1}{5 \cdot 5 \cdot 5} = \dfrac{1}{125}$

72. **[5.1]** $-(-3)^2 = -(9) = -9$

73. **[5.4]** $7p^5(3p^4 + p^3 + 2p^2)$

$= 7p^5(3p^4) + 7p^5(p^3) + 7p^5(2p^2)$

$= 21p^9 + 7p^8 + 14p^7$

74. **[5.4]** $(2x - 9)^2 = (2x)^2 - 2(2x)(9) + 9^2$

$= 4x^2 - 36x + 81$

Copyright © 2012 Pearson Education, Inc. Publishing as Addison-Wesley.

75. [5.1] $\dfrac{(-z^{-2})^3}{5(z^{-3})^{-1}} = \dfrac{(-1)^3 z^{-2(3)}}{5z^{-3(-1)}}$

$= \dfrac{-z^{-6}}{5z^3}$

$= \dfrac{-z^{-6-3}}{5}$

$= \dfrac{-z^{-9}}{5} = -\dfrac{1}{5z^9}$

76. [5.1] $(y^6)^{-5}(2y^{-3})^{-4}$

$= y^{-30}(2)^{-4}y^{12}$

$= \dfrac{y^{-30+12}}{2^4}$

$= \dfrac{y^{-18}}{2^4} = \dfrac{1}{16y^{18}}$

77. [5.5]

$$
\begin{array}{r}
8x + 1 \\
x - 3 \overline{\smash{\big)}\, 8x^2 - 23x + 2} \\
\underline{8x^2 - 24x} \\
x + 2 \\
\underline{x - 3} \\
5
\end{array}
$$

Answer: $8x + 1 + \dfrac{5}{x - 3}$

78. [5.1] $\dfrac{(5z^2x^3)^2(2zx^2)^{-1}}{(-10zx^{-3})^{-2}(3z^{-1}x^{-4})^2}$

$= \dfrac{5^2 z^4 x^6 2^{-1} z^{-1} x^{-2}}{(-10)^{-2} z^{-2} x^6 3^2 z^{-2} x^{-8}}$

$= \dfrac{25(-10)^2 z^3 x^4}{2 \cdot 9 z^{-4} x^{-2}}$

$= \dfrac{25(100) z^7 x^6}{2 \cdot 9}$

$= \dfrac{1250 z^7 x^6}{9}$

79. [5.4] $[(3m - 5n) + p][(3m - 5n) - p]$

$= (3m - 5n)^2 - (p)^2$

$= (3m)^2 - 2(3m)(5n) + (5n)^2 - p^2$

$= 9m^2 - 30mn + 25n^2 - p^2$

80. [5.5] $\dfrac{20y^3x^3 + 15y^4x + 25yx^4}{10yx^2}$

$= \dfrac{20y^3x^3}{10yx^2} + \dfrac{15y^4x}{10yx^2} + \dfrac{25yx^4}{10yx^2}$

$= 2y^2x + \dfrac{3y^3}{2x} + \dfrac{5x^2}{2}$

81. [5.2] $(2k - 1) - (3k^2 - 2k + 6)$

$= 2k - 1 - 3k^2 + 2k - 6$

$= -3k^2 + 2k + 2k - 1 - 6$

$= -3k^2 + 4k - 7$

82. [5.1] See the solution for Exercise 167 in Section 5.1.

$A = \dfrac{P}{D}$

$= \dfrac{4.2134 \times 10^6}{40.76}$

$= \dfrac{4.2134 \times 10^6}{4.076 \times 10^1}$

$\approx 1.03371 \times 10^5$

$= 103{,}371$

The area is approximately 103,371 mi^2.

Chapter 5 Test

1. **(a)** $7^{-2} = \dfrac{1}{7^2} = \dfrac{1}{49}$ **(C)**

  (b) $7^0 = 1$ **(A)**

  (c) $-7^0 = -(1) = -1$ **(D)**

  (d) $(-7)^0 = 1$ **(A)**

  (e) $-7^2 = -49$ **(E)**

  (f) $7^{-1} + 2^{-1} = \dfrac{1}{7} + \dfrac{1}{2}$

$= \dfrac{2}{14} + \dfrac{7}{14} = \dfrac{9}{14}$ **(F)**

  (g) $(7 + 2)^{-1} = 9^{-1} = \dfrac{1}{9}$ **(B)**

  (h) $\dfrac{7^{-1}}{2^{-1}} = \dfrac{2^1}{7^1} = \dfrac{2}{7}$ **(G)**

  (i) $7^2 = 49$ **(I)**

  (j) $(-7)^{-2} = \dfrac{1}{(-7)^2} = \dfrac{1}{49}$ **(C)**

2. $(3x^{-2}y^3)^{-2}(4x^3y^{-4})$

$= 3^{-2}x^{-2(-2)}y^{3(-2)}4x^3y^{-4}$

$= 3^{-2}x^4y^{-6}4x^3y^{-4}$

$= \dfrac{4x^{4+3}y^{-6-4}}{3^2}$

$= \dfrac{4x^7y^{-10}}{9} = \dfrac{4x^7}{9y^{10}}$

3. $\dfrac{36r^{-4}(r^2)^{-3}}{6r^4} = \dfrac{36r^{-4}r^{2(-3)}}{6r^4}$

$= \dfrac{6r^{-4}r^{-6}}{r^4}$

$= \dfrac{6r^{-10}}{r^4} = \dfrac{6}{r^{14}}$

Copyright © 2012 Pearson Education, Inc. Publishing as Addison-Wesley.

4. $\left(\dfrac{4p^2}{q^4}\right)^3 \left(\dfrac{6p^8}{q^{-8}}\right)^{-2}$

$= \dfrac{4^3 p^6}{q^{12}} \cdot \dfrac{6^{-2} p^{-16}}{q^{16}}$

$= \dfrac{4^3 p^{-10}}{6^2 q^{28}}$

$= \dfrac{64}{36 p^{10} q^{28}} = \dfrac{16}{9 p^{10} q^{28}}$

5. $\left(-2x^4 y^{-3}\right)^0 \left(-4x^{-3} y^{-8}\right)^2$

$= 1(-4)^2 x^{-6} y^{-16}$

$= \dfrac{16}{x^6 y^{16}}$

6. $9.1 \times 10^{-7} = 0.000\,000\,91$

Move the decimal point 7 places to the left because the exponent is negative.

7. $\dfrac{2{,}500{,}000 \times 0.00003}{0.05 \times 5{,}000{,}000}$

$= \dfrac{(2.5 \times 10^6)(3 \times 10^{-5})}{(5 \times 10^{-2})(5 \times 10^6)}$

$= \dfrac{7.5 \times 10^1}{25 \times 10^4}$

$= 0.3 \times 10^{1-4}$

$= 0.3 \times 10^{-3}$

$= 3 \times 10^{-4}, \text{ or } 0.0003$

8. $f(x) = -2x^2 + 5x - 6, \; g(x) = 7x - 3$

(a) $f(x) = -2x^2 + 5x - 6$

$f(4) = -2(4)^2 + 5(4) - 6$

$= -2 \cdot 16 + 20 - 6$

$= -32 + 20 - 6$

$= -12 - 6 = -18$

(b) $(f + g)(x) = f(x) + g(x)$

$= (-2x^2 + 5x - 6) + (7x - 3)$

$= -2x^2 + 12x - 9$

(c) $(f - g)(x) = f(x) - g(x)$

$= (-2x^2 + 5x - 6) - (7x - 3)$

$= -2x^2 + 5x - 6 - 7x + 3$

$= -2x^2 - 2x - 3$

(d) Using the answer in part (c), we have

$(f - g)(-2) = -2(-2)^2 - 2(-2) - 3$

$= -8 + 4 - 3 = -7.$

9. $f(x) = 3x + 5, \; g(x) = x^2 + 2$

(a) $(f \circ g)(-2) = f(g(-2))$

$= f\left((-2)^2 + 2\right)$

$= f(6)$

$= 3 \cdot 6 + 5$

$= 23$

(b) $(f \circ g)(x) = f(g(x))$

$= f(x^2 + 2)$

$= 3(x^2 + 2) + 5$

$= 3x^2 + 6 + 5$

$= 3x^2 + 11$

(c) $(g \circ f)(x) = g(f(x))$

$= g(3x + 5)$

$= (3x + 5)^2 + 2$

$= 9x^2 + 30x + 25 + 2$

$= 9x^2 + 30x + 27$

10.

x	$f(x) = -2x^2 + 3$
-2	$-2(-2)^2 + 3 = -5$
-1	$-2(-1)^2 + 3 = 1$
0	$-2(0)^2 + 3 = 3$
1	$-2(1)^2 + 3 = 1$
2	$-2(2)^2 + 3 = -5$

Since the greatest exponent is 2, the graph of f is a parabola.

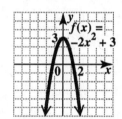

11.

x	$f(x) = -x^3 + 3$
-2	$-(-2)^3 + 3 = 11$
-1	$-(-1)^3 + 3 = 4$
0	$-(0)^3 + 3 = 3$
1	$-(1)^3 + 3 = 2$
2	$-(2)^3 + 3 = -5$

Since the greatest exponent is 3, the graph of f is a cubic (s-shaped).

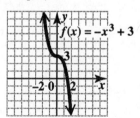

12. $f(x) = 0.139x^2 + 14.16x + 465.9$

$f(x)$ represents the number of medical doctors, in thousands, in the United States.

	Year	x	$f(x)$
(a)	1980	0	$465.9 \approx 466$
(b)	1995	15	$709.575 \approx 710$
(c)	2005	25	$906.775 \approx 907$

Copyright © 2012 Pearson Education, Inc. Publishing as Addison-Wesley.

13. $(4x^3 - 3x^2 + 2x - 5)$
$\quad - (3x^3 + 11x + 8) + (x^2 - x)$
$= 4x^3 - 3x^2 + 2x - 5 - 3x^3 - 11x$
$\quad - 8 + x^2 - x$
$= x^3 - 2x^2 - 10x - 13$

14. $(5x - 3)(2x + 1)$

$\quad\quad\textbf{F}\quad\quad\textbf{O}\quad\quad\textbf{I}\quad\textbf{L}$
$= 10x^2 + 5x - 6x - 3$
$= 10x^2 - x - 3$

15. $(2m - 5)(3m^2 + 4m - 5)$
$= 2m(3m^2 + 4m - 5)$
$\quad + (-5)(3m^2 + 4m - 5)$
$= 6m^3 + 8m^2 - 10m$
$\quad - 15m^2 - 20m + 25$
$= 6m^3 - 7m^2 - 30m + 25$

16. $(6x + y)(6x - y) = (6x)^2 - y^2$
$\quad\quad\quad\quad\quad\quad\quad = 36x^2 - y^2$

17. $(3k + q)^2 = (3k)^2 + 2(3k)(q) + q^2$
$\quad\quad\quad\quad\quad = 9k^2 + 6kq + q^2$

18. $[2y + (3z - x)][2y - (3z - x)]$
$= (2y)^2 - (3z - x)^2$
$= 4y^2 - (9z^2 - 6zx + x^2)$
$= 4y^2 - 9z^2 + 6zx - x^2$

19. $\dfrac{16p^3 - 32p^2 + 24p}{4p^2}$

$= \dfrac{16p^3}{4p^2} - \dfrac{32p^2}{4p^2} + \dfrac{24p}{4p^2}$

$= 4p - 8 + \dfrac{6}{p}$

20. $(x^3 + 3x^2 - 4) \div (x - 1)$

Insert $0x$ for the missing x-term.

$$
\begin{array}{r}
x^2 + 4x + 4 \\
x - 1\overline{)\,x^3 + 3x^2 + 0x - 4} \\
\underline{x^3 - x^2} \\
4x^2 + 0x \\
\underline{4x^2 - 4x} \\
4x - 4 \\
\underline{4x - 4} \\
0
\end{array}
$$

Answer: $x^2 + 4x + 4$

21. $f(x) = x^2 + 3x + 2,\ g(x) = x + 1$

(a) $(fg)(x) = f(x) \cdot g(x)$
$\quad\quad\quad = (x^2 + 3x + 2)(x + 1)$
$\quad\quad\quad = (x^2 + 3x + 2)(x)$
$\quad\quad\quad\quad + (x^2 + 3x + 2)(1)$
$\quad\quad\quad = x^3 + 3x^2 + 2x + x^2 + 3x + 2$
$\quad\quad\quad = x^3 + 4x^2 + 5x + 2$

(b) $(fg)(-2) = f(-2) \cdot g(-2)$
$= [(-2)^2 + 3(-2) + 2] \cdot [(-2) + 1]$
$= [4 - 6 + 2] \cdot [-1]$
$= 0(-1) = 0$

Alternatively, we could have substituted -2 for x into our answer from part (a).

22. **(a)** $\left(\dfrac{f}{g}\right)(x) = \dfrac{f(x)}{g(x)} = \dfrac{x^2 + 3x + 2}{x + 1}$

$$
\begin{array}{r}
x + 2 \\
x + 1\overline{)\,x^2 + 3x + 2} \\
\underline{x^2 + x} \\
2x + 2 \\
\underline{2x + 2} \\
0
\end{array}
$$

Thus, $\left(\dfrac{f}{g}\right)(x) = x + 2$ if $x + 1 \neq 0$, that is,
$x \neq -1$.

(b) Using our answer from part (a),

$\quad\quad \left(\dfrac{f}{g}\right)(-2) = (-2) + 2 = 0.$

Cumulative Review Exercises
(Chapters 1–5)

1. **(a)** 34 is a natural number, so it is also a whole number, an integer, a rational number, and a real number.   **A, B, C, D, F**

(b) 0 is a whole number, so it is also an integer, a rational number, and a real number.   **B, C, D, F**

(c) 2.16 is a rational number, so it is also a real number.   **D, F**

(d) $-\sqrt{36} = -6$ is an integer, so it is also a rational number and a real number.   **C, D, F**

(e) $\sqrt{13}$ is an irrational number, so it is also a real number.   **E, F**

(f) $-\frac{4}{5}$ is a rational number, so it is also a real number.   **D, F**

2. $9 \cdot 4 - 16 \div 4 = (9 \cdot 4) - (16 \div 4) = 36 - 4 = 32$

3. $\left(\frac{1}{3}\right)^2 - \left(\frac{1}{2}\right)^3 = \frac{1}{9} - \frac{1}{8}$
$\quad\quad\quad\quad\quad\quad = \frac{8}{72} - \frac{9}{72} = -\frac{1}{72}$

4. $-|8 - 13| - |-4| + |-9| = -|-5| - 4 + 9$
$\quad\quad\quad\quad\quad\quad\quad\quad = -5 - 4 + 9$
$\quad\quad\quad\quad\quad\quad\quad\quad = -9 + 9 = 0$

5. $-5(8 - 2z) + 4(7 - z) = 7(8 + z) - 3$
$\quad -40 + 10z + 28 - 4z = 56 + 7z - 3$
$\quad\quad\quad\quad\quad\quad\quad\quad\quad\quad\quad$ *Distributive property*
$\quad\quad\quad\quad 6z - 12 = 7z + 53\quad$ *Combine like terms.*
$\quad\quad\quad\quad\quad\quad -65 = z\quad$ *Subtract 6z; 53.*

Thus, the solution set is $\{-65\}$.

Copyright © 2012 Pearson Education, Inc. Publishing as Addison-Wesley.

6. $3(x+2) - 5(x+2) = -2x - 4$
$3x + 6 - 5x - 10 = -2x - 4$
$-2x - 4 = -2x - 4$

The last statement is true for all real numbers, so the solution set is {all real numbers}.

7. Solve $A = p + prt$ for t.

$$A = p + prt$$
$$A - p = prt \qquad \textit{Subtract p.}$$
$$\frac{A - p}{pr} = t \qquad \textit{Divide by pr.}$$

8. $2(m+5) - 3m + 1 > 5$
$2m + 10 - 3m + 1 > 5$
$-m + 11 > 5$
$-m > -6$
$m < 6$

The solution set is $(-\infty, 6)$.

9. $|3x - 1| = 2$

$$3x - 1 = 2 \quad \text{or} \quad 3x - 1 = -2$$
$$3x = 3 \qquad\qquad 3x = -1$$
$$x = 1 \quad \text{or} \qquad x = -\tfrac{1}{3}$$

The solution set is $\left\{-\tfrac{1}{3}, 1\right\}$.

10. $|3z + 1| \geq 7$

$$3z + 1 \geq 7 \quad \text{or} \quad 3z + 1 \leq -7$$
$$3z \geq 6 \qquad\qquad 3z \leq -8$$
$$z \geq 2 \quad \text{or} \qquad z \leq -\tfrac{8}{3}$$

The solution set is $\left(-\infty, -\tfrac{8}{3}\right] \cup [2, \infty)$.

11. Personal computer: $\frac{480}{1500} = 32\%$

Pacemaker:

$$26\% \text{ of } 1500 = 0.26(1500) = 390$$

Wireless communication:

$$18\% \text{ of } 1500 = 0.18(1500) = 270$$

Television: $\frac{150}{1500} = 10\%$

12. The sum of the measures of the angles of any triangle is $180°$, so

$$(x + 15) + (6x + 10) + (x - 5) = 180.$$

Solve this equation.

$$8x + 20 = 180$$
$$8x = 160$$
$$x = 20$$

Substitute 20 for x to find the measures of the angles.

$x - 5 = 20 - 5 = 15$
$x + 15 = 20 + 15 = 35$
$6x + 10 = 6(20) + 10 = 130$

The measures of the angles of the triangle are $15°$, $35°$, and $130°$.

13. Through $(-4, 5)$ and $(2, -3)$

Use the definition of slope with $x_1 = -4$, $y_1 = 5$, $x_2 = 2$, and $y_2 = -3$.

$$m = \frac{y_2 - y_1}{x_2 - x_1} = \frac{-3 - 5}{2 - (-4)} = \frac{-8}{6} = -\frac{4}{3}$$

14. Horizontal, through $(4, 5)$

The slope of every horizontal line is 0.

15. Through $(4, -1)$, $m = -4$

(a) Use the point-slope form with $x_1 = 4$, $y_1 = -1$, and $m = -4$.

$$y - y_1 = m(x - x_1)$$
$$y - (-1) = -4(x - 4)$$
$$y + 1 = -4x + 16$$
$$y = -4x + 15$$

(b) The standard form is $Ax + By = C$.

$$y = -4x + 15$$
$$4x + y = 15$$

16. Through $(0, 0)$ and $(1, 4)$

(a) Find the slope.

$$m = \frac{4 - 0}{1 - 0} = 4$$

Because the slope is 4 and the y-intercept is 0, the equation of the line in slope-intercept form is

$$y = 4x.$$

(b) The standard form is $Ax + By = C$.

$$y = 4x$$
$$0 = 4x - y, \quad \text{or} \quad 4x - y = 0$$

17. $-3x + 4y = 12$

If $y = 0$, $x = -4$, so the x-intercept is $(-4, 0)$.

If $x = 0$, $y = 3$, so the y-intercept is $(0, 3)$.

Draw a line through these intercepts. A third point may be used as a check.

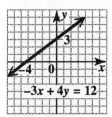

Copyright © 2012 Pearson Education, Inc. Publishing as Addison-Wesley.

18. $y \le 2x - 6$

Graph the boundary, $y = 2x - 6$, as a solid line through the intercepts $(3, 0)$ and $(0, -6)$. A third point such as $(1, -4)$ can be used as a check. Using $(0, 0)$ as a test point results in the false inequality $0 \le -6$, so shade the region *not* containing the origin. This is the region below the line. The solid line shows that the boundary is part of the graph.

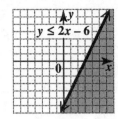

19. $3x + 2y < 0$

Graph the boundary, $3x + 2y = 0$, as a dashed line through $(0, 0)$, $(-2, 3)$, and $(2, -3)$. Choose a test point not on the line. Using $(1, 1)$ results in the false statement $5 < 0$, so shade the region *not* containing $(1, 1)$. This is the region below the line. The dashed line shows that the boundary is not part of the graph.

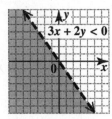

20. **(a)** $m = \dfrac{261{,}122 - 322{,}486}{5 - 0} = \dfrac{-61{,}364}{5}$

$= -12{,}272.8$

The average rate of change is $-12{,}272.8$ thousand pounds per year, that is, the number of pounds of shrimp caught in the United States decreased an average of $-12{,}272.8$ thousand pounds per year during the selected years.

(b) $y = mx + b$

$y = -12{,}272.8x + 322{,}486$

(c) For the year 2003, $x = 2003 - 2000 = 3$.

$y = -12{,}272.8(3) + 322{,}486$

$= 285{,}667.6 \approx 285{,}668$

The model approximates about 285,668 thousand pounds of shrimp caught in 2003.

21. $\{(-4, -2), (-1, 0), (2, 0), (5, 2)\}$

The domain is the set of first components, that is, $\{-4, -1, 2, 5\}$.
The range is the set of second components, that is, $\{-2, 0, 2\}$.
The relation is a function since each first component is paired with a unique second component.

22. $g(x) = -x^2 - 2x + 6$

$g(3) = -3^2 - 2(3) + 6$

$= -9 - 6 + 6$

$= -9$

23. $\begin{array}{ll} 3x - 4y = 1 & (1) \\ 2x + 3y = 12 & (2) \end{array}$

To eliminate y, multiply equation (1) by 3 and equation (2) by 4. Then add the results.

$$\begin{array}{rcll} 9x - 12y & = & 3 & 3 \times (1) \\ 8x + 12y & = & 48 & 4 \times (2) \\ \hline 17x & = & 51 & \\ x & = & 3 & \end{array}$$

Since $x = 3$,

$$\begin{array}{ll} 3x - 4y = 1 & (1) \\ 3(3) - 4y = 1 & \\ 9 - 4y = 1 & \\ -4y = -8 & \\ y = 2 & \end{array}$$

The solution set is $\{(3, 2)\}$.

24. $\begin{array}{ll} 3x - 2y = 4 & (1) \\ -6x + 4y = 7 & (2) \end{array}$

Multiply equation (1) by 2 and add the result to equation (2).

$$\begin{array}{rcll} 6x - 4y & = & 8 & 2 \times (1) \\ -6x + 4y & = & 7 & (2) \\ \hline 0 & = & 15 & \textit{False} \end{array}$$

Since a false statement results, the system is *inconsistent*. The solution set is $\emptyset$.

Copyright © 2012 Pearson Education, Inc. Publishing as Addison-Wesley.

25.
$$x + 3y - 6z = 7 \quad (1)$$
$$2x - y + z = 1 \quad (2)$$
$$x + 2y + 2z = -1 \quad (3)$$

To eliminate x, multiply equation (1) by -2 and add the result to equation (2).

$$
\begin{array}{rr}
-2x - 6y + 12z = -14 & -2 \times (1) \\
\underline{2x - y + z = 1} & (2) \\
-7y + 13z = -13 & (4)
\end{array}
$$

To eliminate x again, multiply equation (3) by -2 and add the result to equation (2).

$$
\begin{array}{rr}
2x - y + z = 1 & (2) \\
\underline{-2x - 4y - 4z = 2} & -2 \times (3) \\
-5y - 3z = 3 & (5)
\end{array}
$$

Use equations (4) and (5) to eliminate z. Multiply equation (4) by 3 and add the result to 13 times equation (5).

$$
\begin{array}{rr}
-21y + 39z = -39 & 3 \times (4) \\
\underline{-65y - 39z = 39} & 13 \times (5) \\
-86y = 0 & \\
y = 0 &
\end{array}
$$

From (5), $-3z = 3$, so $z = -1$.
From (3), $x - 2 = -1$, so $x = 1$.

The solution set is $\{(1, 0, -1)\}$.

26. The length L of the rectangular flag measured 12 feet more than its width W, so

$$L = W + 12. \quad (1)$$

The perimeter is 144 feet.

$$P = 2L + 2W \quad (2)$$

Substitute $W + 12$ for L into equation (2).

$$144 = 2(W + 12) + 2W$$
$$144 = 2W + 24 + 2W$$
$$120 = 4W$$
$$30 = W$$

From (1), $L = 30 + 12 = 42$.

The length is 42 feet and the width is 30 feet.

27. Make a chart.

Number of Liters of Solution	Percent (as a decimal)	Pure Liters of Alcohol
x	0.15	$0.15x$
y	0.30	$0.30y$
9	0.20	$0.20(9) = 1.8$

From the first and third columns, we have the following system:

$$x + y = 9 \quad (1)$$
$$0.15x + 0.30y = 1.8 \quad (2)$$

To eliminate x, multiply equation (1) by -15 and add the result to 100 times equation (2).

$$
\begin{array}{r}
-15x - 15y = -135 \\
\underline{15x + 30y = 180} \\
15y = 45 \\
y = 3
\end{array}
$$

From (1), $x + 3 = 9$, so $x = 6$.

Agbe should use 6 L of 15% solution and 3 L of 30% solution.

28. $\left(\dfrac{2m^3n}{p^2}\right)^3 = \dfrac{2^3(m^3)^3 n^3}{(p^2)^3} = \dfrac{8m^9n^3}{p^6}$

29. $\dfrac{x^{-6}y^3z^{-1}}{x^7y^{-4}z} = \dfrac{y^4y^3}{x^6x^7z^1z} = \dfrac{y^7}{x^{13}z^2}$

30. $(2m^{-2}n^3)^{-3}$

$$= 2^{-3}(m^{-2})^{-3}(n^3)^{-3}$$
$$= 2^{-3}m^{(-2)(-3)}n^{3(-3)}$$
$$= 2^{-3}m^6n^{-9}$$
$$= \dfrac{m^6}{2^3n^9} = \dfrac{m^6}{8n^9}$$

31. $(3x^2 - 8x + 1) - (x^2 - 3x - 9)$

$$= 3x^2 - 8x + 1 - x^2 + 3x + 9$$
$$\hspace{3cm} \textit{Distributive property}$$
$$= (3x^2 - x^2) + (-8x + 3x) + (1 + 9)$$
$$\hspace{3cm} \textit{Combine like terms.}$$
$$= 2x^2 - 5x + 10$$

Copyright © 2012 Pearson Education, Inc. Publishing as Addison-Wesley.

32. $(x + 2y)(x^2 - 2xy + 4y^2)$

Multiply vertically.

$$
\begin{array}{r}
x^2 \;-\; 2xy \;+\; 4y^2 \\
x \;+\; 2y \\
\hline
2x^2y \;-\; 4xy^2 \;+\; 8y^3 \\
x^3 \;-\; 2x^2y \;+\; 4xy^2 \\
\hline
x^3 \qquad\qquad\qquad\; +\; 8y^3
\end{array}
$$

Thus,

$$(x + 2y)(x^2 - 2xy + 4y^2) = x^3 + 8y^3.$$

33. $(3x + 2y)(5x - y)$

$$
\begin{array}{cccc}
\mathbf{F} & \mathbf{O} & \mathbf{I} & \mathbf{L}
\end{array}
$$
$$= 3x(5x) + 3x(-y) + 2y(5x) + 2y(-y)$$
$$= 15x^2 - 3xy + 10xy - 2y^2$$
$$= 15x^2 + 7xy - 2y^2$$

34. $\dfrac{16x^3y^5 - 8x^2y^2 + 4}{4x^2y}$

$$= \dfrac{16x^3y^5}{4x^2y} - \dfrac{8x^2y^2}{4x^2y} + \dfrac{4}{4x^2y}$$

$$= 4xy^4 - 2y + \dfrac{1}{x^2y}$$

35. $\dfrac{m^3 - 3m^2 + 5m - 3}{m - 1}$

$$
\require{enclose}
\begin{array}{r}
m^2 \;-\; 2m \;+\; 3 \\
m - 1 \enclose{longdiv}{m^3 \;-\; 3m^2 + 5m - 3} \\
\underline{m^3 \;-\; m^2 } \\
-2m^2 \;+\; 5m \\
\underline{-2m^2 \;+\; 2m } \\
3m \;-\; 3 \\
\underline{3m \;-\; 3} \\
0
\end{array}
$$

The remainder is 0. The answer is the quotient,

$$m^2 - 2m + 3.$$

Copyright © 2012 Pearson Education, Inc. Publishing as Addison-Wesley.

CHAPTER 6   FACTORING

6.1   Greatest Common Factors and Factoring by Grouping

6.1 Now Try Exercises

N1. **(a)** $54m - 45 = 9 \cdot 6m - 9 \cdot 5 \quad GCF = 9$
$$= 9(6m - 5)$$

Check: $9(6m - 5) = 9(6m) - 9(5)$
$$= 54m - 45$$

(b) $2k - 7$ *cannot be factored* because $2k$ and 7 do not have a common factor other than 1.

N2. $20m^3n^3 - 15m^3n^2 + 10m^2n$

The numerical part of the GCF is 5, the largest number that divides into 20, 15, and 10. For the variable parts, m^3, m^3, and m^2, use the least exponent that appears on m; here the least exponent is 2. For the variable parts, n^3, n^2, and n^1, use the least exponent that appears on n; here the least exponent is 1. The GCF is $5m^2n$.

$$20m^3n^3 - 15m^3n^2 + 10m^2n$$
$$= 5m^2n(4mn^2) + 5m^2n(-3mn) + 5m^2n(2)$$
$$= 5m^2n(4mn^2 - 3mn + 2)$$

N3. **(a)** $(3x - 2)(x + 1) + (3x - 2)(2x - 5)$

The greatest common factor is $(3x - 2)$.

$$= (3x - 2)[(x + 1) + (2x - 5)]$$
$$= (3x - 2)(x + 1 + 2x - 5)$$
$$= (3x - 2)(3x - 4)$$

(b) $z(a - b)^3 - 2w(a - b)^2 \quad GCF = (a - b)^2$
$$= (a - b)^2[z(a - b) - 2w]$$
$$= (a - b)^2(za - zb - 2w)$$

N4. $-4y^5 - 3y^3 + 8y$

Factor with GCF $= y$.

$$-4y^5 - 3y^3 + 8y = y(-4y^4) + y(-3y^2) + y(8)$$
$$= y(-4y^4 - 3y^2 + 8)$$

Factor with GCF $= -y$.

$$-4y^5 - 3y^3 + 8y$$
$$= -y(4y^4) + (-y)(3y^2) + (-y)(-8)$$
$$= -y(4y^4 + 3y^2 - 8)$$

N5. $3m - 3n + xm - xn$
$$= (3m - 3n) + (xm - xn)$$
$$= 3(m - n) + x(m - n)$$
$$= (m - n)(3 + x)$$

N6. $ab - 7a - 5b + 35 = (ab - 7a) - (5b - 35)$
$$= a(b - 7) - 5(b - 7)$$
$$= (b - 7)(a - 5)$$

N7. $3ax - 6xy - a + 2y$
$$= (3ax - 6xy) + (-a + 2y)$$
$$= 3x(a - 2y) + (-1)(a - 2y)$$
$$= (a - 2y)(3x - 1)$$

N8. $2kp^2 + 6 - 3p^2 - 4k$

Rearrange the terms so that there is a common factor in the first two terms and a common factor in the last two terms.

$$= 2kp^2 - 3p^2 + 6 - 4k$$

Group the first two terms and the last two terms.

$$= (2kp^2 - 3p^2) + (6 - 4k)$$
$$= p^2(2k - 3) - 2(2k - 3)$$
$$= (2k - 3)(p^2 - 2)$$

6.1 Section Exercises

1. $12m - 60 = 12 \cdot m - 12 \cdot 5$
$$= 12(m - 5)$$

3. $4 + 20z = 4 \cdot 1 + 4 \cdot 5z$
$$= 4(1 + 5z)$$

5. $8y - 15$ *cannot be factored.*

7. $8k^3 + 24k = 8k \cdot k^2 + 8k \cdot 3$
$$= 8k(k^2 + 3)$$

9. $-4p^3q^4 - 2p^2q^5 = -2p^2q^4 \cdot 2p - 2p^2q^4 \cdot q$
$$= -2p^2q^4(2p + q)$$

11. $7x^3 + 35x^4 - 14x^5$
$$= 7x^3(1 + 5x - 2x^2)$$

13. $10t^5 - 2t^3 - 4t^4$
$$= 2t^3(5t^2 - 1 - 2t)$$

15. $15a^2c^3 - 25ac^2 + 5ac$
$$= 5ac(3ac^2 - 5c + 1)$$

17. $16z^2n^6 + 64zn^7 - 32z^3n^3$
$$= 16zn^3(zn^3 + 4n^4 - 2z^2)$$

19. $14a^3b^2 + 7a^2b - 21a^5b^3 + 42ab^4$
$$= 7ab(2a^2b + a - 3a^4b^2 + 6b^3)$$

21. $(m - 4)(m + 2) + (m - 4)(m + 3)$
The GCF is $(m - 4)$.
$$= (m - 4)[(m + 2) + (m + 3)]$$
$$= (m - 4)(m + 2 + m + 3)$$
$$= (m - 4)(2m + 5)$$

23. $(2z - 1)(z + 6) - (2z - 1)(z - 5)$
$$= (2z - 1)[(z + 6) - (z - 5)]$$
$$= (2z - 1)[z + 6 - z + 5]$$
$$= (2z - 1)(11)$$
$$= 11(2z - 1)$$

Copyright © 2012 Pearson Education, Inc. Publishing as Addison-Wesley.

25. $5(2 - x)^2 - 2(2 - x)^3$
$= (2 - x)^2[5 - 2(2 - x)]$
$= (2 - x)^2(5 - 4 + 2x)$
$= (2 - x)^2(1 + 2x)$

27. $4(3 - x)^2 - (3 - x)^3 + 3(3 - x)$
$= (3 - x)[4(3 - x) - (3 - x)^2 + 3]$
$= (3 - x)[12 - 4x - (9 - 6x + x^2) + 3]$
$= (3 - x)[12 - 4x - 9 + 6x - x^2 + 3]$
$= (3 - x)(6 + 2x - x^2)$

29. $15(2z + 1)^3 + 10(2z + 1)^2 - 25(2z + 1)$
The GCF is $5(2z + 1)$.
$= 5(2z + 1)$
    $\cdot [3(2z + 1)^2 + 2(2z + 1) - 5]$
$= 5(2z + 1)$
    $\cdot [3(4z^2 + 4z + 1) + 4z + 2 - 5]$
$= 5(2z + 1)[12z^2 + 12z + 3 + 4z + 2 - 5]$
$= 5(2z + 1)[12z^2 + 16z]$
$= 5(2z + 1)[4z(3z + 4)]$
$= 20z(2z + 1)(3z + 4)$

31. $5(m + p)^3 - 10(m + p)^2 - 15(m + p)^4$
The GCF is $5(m + p)^2$.
$= 5(m + p)^2[(m + p) - 2 - 3(m + p)^2]$
$= 5(m + p)^2$
    $\cdot [m + p - 2 - 3(m^2 + 2mp + p^2)]$
$= 5(m + p)^2$
    $\cdot (m + p - 2 - 3m^2 - 6mp - 3p^2)$

33. $-r^3 + 3r^2 + 5r$
Factor out r.
$= r(-r^2 + 3r + 5)$
Factor out $-r$.
$= -r(r^2 - 3r - 5)$

35. $-12s^5 + 48s^4$
Factor out $12s^4$.
$= 12s^4(-s + 4)$
Factor out $-12s^4$.
$= -12s^4(s - 4)$

37. $-2x^5 + 6x^3 + 4x^2$
Factor out $2x^2$.
$= 2x^2(-x^3 + 3x + 2)$
Factor out $-2x^2$.
$= -2x^2(x^3 - 3x - 2)$

39. $mx + qx + my + qy$
$= (mx + qx) + (my + qy)$
$= x(m + q) + y(m + q)$
$= (m + q)(x + y)$

41. $10m + 2n + 5mk + nk$
$= (10m + 2n) + (5mk + nk)$

$= 2(5m + n) + k(5m + n)$
$= (5m + n)(2 + k)$

43. $4 - 2q - 6p + 3pq$
$= (4 - 2q) + (-6p + 3pq)$
$= 2(2 - q) - 3p(2 - q)$
$= (2 - q)(2 - 3p)$

45. $p^2 - 4zq + pq - 4pz$
$= (p^2 + pq) + (-4zq - 4pz)$
$= p(p + q) - 4z(q + p)$
$= (p + q)(p - 4z)$

47. $2xy + 3y + 2x + 3$
$= (2xy + 3y) + (2x + 3)$
$= y(2x + 3) + 1(2x + 3)$
$= (2x + 3)(y + 1)$

49. $m^3 + 4m^2 - 6m - 24$
$= (m^3 + 4m^2) + (-6m - 24)$
$= m^2(m + 4) - 6(m + 4)$
$= (m + 4)(m^2 - 6)$

51. $-3a^3 - 3ab^2 + 2a^2b + 2b^3$
$= (-3a^3 - 3ab^2) + (2a^2b + 2b^3)$
$= -3a(a^2 + b^2) + 2b(a^2 + b^2)$
$= (a^2 + b^2)(-3a + 2b)$

53. $4 + xy - 2y - 2x$
$= xy - 2x - 2y + 4$
$= (xy - 2x) + (-2y + 4)$
$= x(y - 2) - 2(y - 2)$
$= (y - 2)(x - 2)$

55. $8 + 9y^4 - 6y^3 - 12y$
$= 9y^4 - 6y^3 - 12y + 8$
$= (9y^4 - 6y^3) + (-12y + 8)$
$= 3y^3(3y - 2) - 4(3y - 2)$
$= (3y - 2)(3y^3 - 4)$

57. $1 - a + ab - b$
$= 1 - a - b + ab$
$= (1 - a) + (-b + ab)$
$= 1(1 - a) - b(1 - a)$
$= (1 - a)(1 - b)$

59. $3m^{-5} + m^{-3}$

Factor out m^{-5} since -5 is the smaller exponent.

$= m^{-5}(3) + m^{-5}(m^{-3 - (-5)})$
$= m^{-5}(3 + m^2)$ or $\dfrac{3 + m^2}{m^5}$

61. $3p^{-3} + 2p^{-2}$

Factor out p^{-3} since -3 is the smaller exponent.

$= p^{-3}(3) + p^{-3}(2p^{-2 - (-3)})$
$= p^{-3}(3 + 2p)$ or $\dfrac{3 + 2p}{p^3}$

Copyright © 2012 Pearson Education, Inc. Publishing as Addison-Wesley.

63. The directions said that the student was to factor the polynomial *completely*. The completely factored form is $4xy^3(xy^2 - 2)$.

65. Factoring a polynomial involves writing the polynomial as the product of two or more simpler polynomials. Here, choice **A** is the sum of two polynomials; choices **B** and **D** are differences of two polynomials. Only choice **C** involves the product of two polynomials and is an example of a polynomial in factored form. The correct answer is choice **C**.

67. $(k + 7)(k - 1) = k^2 - k + 7k - 7$
$$= k^2 + 6k - 7$$

69. $(5x - 2t)(5x + 2t) = (5x)^2 - (2t)^2$
$$= 25x^2 - 4t^2$$

71. $(3y^3 - 4)(2y^3 + 3) = 6y^6 + 9y^3 - 8y^3 - 12$
$$= 6y^6 + y^3 - 12$$

6.2 Factoring Trinomials

6.2 Now Try Exercises

N1. **(a)** $t^2 - t - 30$

Step 1 Find pairs of integers whose product is −30.	Step 2 Write sums of those integers.
−30(1)	−30 + 1 = −29
30(−1)	30 + (−1) = 29
−15(2)	−15 + 2 = −13
15(−2)	15 + (−2) = 13
−10(3)	−10 + 3 = −7
10(−3)	10 + (−3) = 7
−6(5)	−6 + 5 = −1 ←
6(−5)	6 + (−5) = 1

The integers −6 and 5 have the necessary product, −30, and sum, −1, so

$t^2 - t - 30$ factors as $(t - 6)(t + 5)$.

Check $(t - 6)(t + 5) = t^2 - t - 30$

(b) $w^2 + 12w + 32$

Step 1 Find pairs of integers whose product is 32.	Step 2 Write sums of those integers.
32(1)	32 + 1 = 33
−32(−1)	−32 + (−1) = −33
16(2)	16 + 2 = 18
−16(−2)	−16 + (−2) = −18
8(4)	8 + 4 = 12 ←
−8(−4)	−8 + (−4) = −12

The integers 8 and 4 have the necessary product, 32, and sum, 12, so

$w^2 + 12w + 32$ factors as $(w + 8)(w + 4)$.

Check $(w + 8)(w + 4) = w^2 + 12w + 32$

N2. $m^2 + 12m - 11$

Look for two integers whose product is −11 and whose sum is 12. Only two pairs of integers, 11 and −1 and −11 and 1, give a product of −11. Neither of these pairs has a sum of 12, so $m^2 + 12m - 11$ cannot be factored with integer coefficients and is *prime*.

N3. $a^2 + ab - 20b^2$

Look for two expressions whose product is $-20b^2$ and whose sum is b. The expressions $5b$ and $-4b$ have the necessary product and sum, so

$a^2 + ab - 20b^2$ factors as $(a + 5b)(a - 4b)$.

N4. $5w^3 - 40w^2 + 60w$

Start by factoring out the GCF, $5w$.

$5w^3 - 40w^2 + 60w = 5w(w^2 - 8w + 12)$

To factor $w^2 - 8w + 12$, look for two integers whose product is 12 and whose sum is −8. The necessary integers are −6 and −2. Remember to write the common factor $5w$ as part of the answer.

$$5w^3 - 40w^2 + 60w = 5w(w - 6)(w - 2)$$

N5. $8y^2 - 10y - 3$

Since $a = 8$, $b = -10$, and $c = -3$, the product ac is $8(-3) = -24$. The two integers whose product is −24 and whose sum is −10, are −12 and 2. Write $-10y$ as $-12y + 2y$ and then factor by grouping.

$$8y^2 - 10y - 3 = 8y^2 - 12y + 2y - 3$$
$$= (8y^2 - 12y) + (2y - 3)$$
$$= 4y(2y - 3) + 1(2y - 3)$$
$$= (2y - 3)(4y + 1)$$

N6. Use trial and error to factor the trinomial

$10r^2 + 19r + 6$ as $(2r + 3)(5r + 2)$.

N7. $12a^2 - 19ab - 21b^2$

Try some possibilities.

$(2a - 7b)(6a + 3b) = 12a^2 - 36ab - 21b^2$ *No*
$(2a + 7b)(6a - 3b) = 12a^2 + 36ab - 21b^2$ *No*
$(4a - 3b)(3a + 7b) = 12a^2 + 19ab - 21b^2$ *No*
$(4a + 3b)(3a - 7b) = 12a^2 - 19ab - 21b^2$ *Yes*

$12a^2 - 19ab - 21b^2$ factors as $(4a + 3b)(3a - 7b)$

Copyright © 2012 Pearson Education, Inc. Publishing as Addison-Wesley.

N8. $-8x^2 + 22x - 15$

First factor out -1, then factor the trinomial.

$$-8x^2 + 22x - 15 = -1(8x^2 - 22x + 15)$$
$$= -1(2x - 3)(4x - 5)$$
$$= -(4x - 5)(2x - 3)$$

N9. $12y^3 + 33y^2 - 9y$

First, factor out the GCF, $3y$.

$$12y^3 + 33y^2 - 9y = 3y(4y^2 + 11y - 3)$$

Look for two integers whose product is $4(-3) = -12$ and whose sum is 11. The integers are -1 and 12.

$$12y^3 + 33y^2 - 9y$$
$$= 3y(4y^2 - y + 12y - 3)$$
$$= 3y[y(4y - 1) + 3(4y - 1)]$$
$$= 3y[(4y - 1)(y + 3)]$$
$$= 3y(4y - 1)(y + 3)$$

N10. $3(a + 2)^2 - 11(a + 2) - 4$
$$= 3x^2 - 11x - 4 \qquad Let\ x = a + 2.$$
$$= (3x + 1)(x - 4)$$

Now replace x with $a + 2$.

$$3(a + 2)^2 - 11(a + 2) - 4$$
$$= [3(a + 2) + 1][(a + 2) - 4]$$
$$= (3a + 6 + 1)(a + 2 - 4)$$
$$= (3a + 7)(a - 2)$$

N11. $6x^4 + 11x^2 + 3$
$$= 6(x^2)^2 + 11x^2 + 3$$
$$= 6y^2 + 11y + 3 \qquad Let\ y = x^2.$$
$$= (3y + 1)(2y + 3) \qquad Factor.$$
$$= (3x^2 + 1)(2x^2 + 3) \quad y = x^2$$

6.2   Section Exercises

1.   **D** is not valid.

$$(8x)(4x) = 32x^2 \neq 12x^2$$

3.   **B** is not a factored form.

$$(-x - 10)(x + 6) = -x^2 - 16x - 60$$
$$\neq -x^2 + 16x - 60$$

5.   To factor $y^2 + 7y - 30$, we need two integer factors whose sum is 7 (coefficient of the middle term) and whose product is -30 (the last term). Since $-3 + 10 = 7$ and $-3 \cdot 10 = -30$, we have

$$y^2 + 7y - 30 = (y - 3)(y + 10).$$

7. $p^2 + 15p + 56$

Two integer factors whose product is 56 and whose sum is 15 are 8 and 7.

$$p^2 + 15p + 56 = (p + 8)(p + 7)$$

9. $m^2 - 11m + 60$

To factor $m^2 - 11m + 60$, we need two integer factors whose product is 60 (the last term) and whose sum is -11 (coefficient of the middle term).

Factors	Sum
$-1, -60$	-61
$-2, -30$	-32
$-3, -20$	-23
$-4, -15$	-19
$-5, -12$	-17
$-6, -10$	-16

No sum is -11, so the trinomial is prime.

11. $a^2 - 2ab - 35b^2$

Two integer factors whose product is -35 and whose sum is -2 are 5 and -7.

$$a^2 - 2ab - 35b^2 = (a + 5b)(a - 7b)$$

13. $a^2 - 9ab + 18b^2$

Two integer factors whose product is 18 and whose sum is -9 are -6 and -3.

$$a^2 - 9ab + 18b^2 = (a - 6b)(a - 3b)$$

15. $x^2y^2 + 12xy + 18$

There are no integer factors of 18 that add up to 12, so this trinomial is prime.

17. $-6m^2 - 13m + 15 = -1(6m^2 + 13m - 15)$
We'll try to factor $6m^2 + 13m - 15$.
Multiply the first and last coefficients to get $6(-15) = -90$.
Two integer factors whose product is -90 and whose sum is 13 are -5 and 18.
Rewrite the trinomial in a form that can be factored by grouping.

$$6m^2 + (+13m) - 15$$
$$= 6m^2 + (-5m + 18m) - 15$$
$$= (6m^2 - 5m) + (18m - 15)$$
$$= m(6m - 5) + 3(6m - 5)$$
$$= (6m - 5)(m + 3)$$

Thus, the final factored form is

$$-1(6m - 5)(m + 3).$$

Note: These exercises can be worked using the alternative method of repeated combinations and FOIL or the grouping method.

Copyright © 2012 Pearson Education, Inc. Publishing as Addison-Wesley.

19. $10x^2 + 3x - 18$

Two integer factors whose product is $(10)(-18) = -180$ and whose sum is 3 are 15 and -12.
Rewrite the trinomial in a form that can be factored by grouping.

$$
\begin{aligned}
10x^2 &+ 3x - 18 \\
&= 10x^2 + 15x - 12x - 18 \\
&= 5x(2x + 3) - 6(2x + 3) \\
&= (2x + 3)(5x - 6)
\end{aligned}
$$

21. $20k^2 + 47k + 24$

Two integer factors whose product is $(20)(24) = 480$ and whose sum is 47 are 15 and 32.
Rewrite the trinomial in a form that can be factored by grouping.

$$
\begin{aligned}
20k^2 &+ 47k + 24 \\
&= 20k^2 + 15k + 32k + 24 \\
&= 5k(4k + 3) + 8(4k + 3) \\
&= (4k + 3)(5k + 8)
\end{aligned}
$$

23. $15a^2 - 22ab + 8b^2$

Two integer factors whose product is $(15)(8) = 120$ and whose sum is -22 are -10 and -12.
Rewrite the trinomial in a form that can be factored by grouping.

$$
\begin{aligned}
15a^2 &- 22ab + 8b^2 \\
&= 15a^2 - 10ab - 12ab + 8b^2 \\
&= 5a(3a - 2b) - 4b(3a - 2b) \\
&= (3a - 2b)(5a - 4b)
\end{aligned}
$$

25. $36m^2 - 60m + 25$

Use the alternative method and write $36m^2$ as $6m \cdot 6m$ and 25 as $5 \cdot 5$. Use these factors in the binomial factors to obtain

$$
\begin{aligned}
36m^2 - 60m + 25 &= (6m - 5)(6m - 5) \\
&= (6m - 5)^2.
\end{aligned}
$$

27. $40x^2 + xy + 6y^2$

There are no integer factors of $(40)(6) = 240$ that add up to 1, so this trinomial is prime.

29. $6x^2z^2 + 5xz - 4$

Two integer factors whose product is $(6)(-4) = -24$ and whose sum is 5 are 8 and -3.
Rewrite the trinomial in a form that can be factored by grouping.

$$
\begin{aligned}
6x^2z^2 &+ 5xz - 4 \\
&= 6x^2z^2 + 8xz - 3xz - 4 \\
&= 2xz(3xz + 4) - 1(3xz + 4) \\
&= (3xz + 4)(2xz - 1)
\end{aligned}
$$

31. $24x^2 + 42x + 15$

Always factor out the GCF first.

$$
24x^2 + 42x + 15 = 3(8x^2 + 14x + 5)
$$

Now factor $8x^2 + 14x + 5$ by trial and error.

$$
8x^2 + 14x + 5 = (4x + 5)(2x + 1)
$$

The final factored form is

$$
3(4x + 5)(2x + 1).
$$

33. $-15a^2 - 70a + 120$

$$
\begin{aligned}
&= -5(3a^2 + 14a - 24) \\
&= -5(a + 6)(3a - 4)
\end{aligned}
$$

35. $-11x^3 + 110x^2 - 264x$

$$
\begin{aligned}
&= -11x(x^2 - 10x + 24) \\
&= -11x(x - 6)(x - 4)
\end{aligned}
$$

37. $2x^3y^3 - 48x^2y^4 + 288xy^5$

$$
\begin{aligned}
&= 2xy^3(x^2 - 24xy + 144y^2) \\
&= 2xy^3(x - 12y)(x - 12y)
\end{aligned}
$$
or $\quad 2xy^3(x - 12y)^2$

39. $6a^3 + 12a^2 - 90a$

$$
\begin{aligned}
&= 6a(a^2 + 2a - 15) \\
&= 6a(a - 3)(a + 5)
\end{aligned}
$$

41. $13y^3 + 39y^2 - 52y$

$$
\begin{aligned}
&= 13y(y^2 + 3y - 4) \\
&= 13y(y + 4)(y - 1)
\end{aligned}
$$

43. $12p^3 - 12p^2 + 3p$

$$
\begin{aligned}
&= 3p(4p^2 - 4p + 1) \\
&= 3p(2p - 1)(2p - 1)
\end{aligned}
$$
or $\quad 3p(2p - 1)^2$

45. There is a GCF of 2. She did not factor the polynomial *completely*. The factor $(4x + 10)$ can be factored further as $2(2x + 5)$, giving the final form as $2(2x + 5)(x - 2)$.

47. $12p^6 - 32p^3r + 5r^2$

Let $x = p^3$ and then factor the trinomial.

$$
12x^2 - 32xr + 5r^2 = (6x - r)(2x - 5r)
$$

Now replace x with p^3 to get

$$
12p^6 - 32p^3r + 5r^2 = (6p^3 - r)(2p^3 - 5r).
$$

49. $10(k + 1)^2 - 7(k + 1) + 1$

Let $x = k + 1$ and then factor the trinomial.

$$
10x^2 - 7x + 1 = (5x - 1)(2x - 1)
$$

Now replace x with $k + 1$.

$$
\begin{aligned}
10(k + 1)^2 &- 7(k + 1) + 1 \\
&= [5(k + 1) - 1][2(k + 1) - 1] \\
&= (5k + 5 - 1)(2k + 2 - 1) \\
&= (5k + 4)(2k + 1)
\end{aligned}
$$

Copyright © 2012 Pearson Education, Inc. Publishing as Addison-Wesley.

51. $3(m+p)^2 - 7(m+p) - 20$

Let $x = m + p$ and then factor the trinomial.

$$3x^2 - 7x - 20 = (3x + 5)(x - 4)$$

Now replace x with $m + p$.

$$3(m+p)^2 - 7(m+p) - 20$$
$$= [3(m+p) + 5][(m+p) - 4]$$
$$= (3m + 3p + 5)(m + p - 4)$$

53. $a^2(a+b)^2 - ab(a+b)^2 - 6b^2(a+b)^2$

Factor out the GCF, $(a+b)^2$.
$$= (a+b)^2(a^2 - ab - 6b^2)$$
Factor the trinomial.
$$= (a+b)^2(a - 3b)(a + 2b)$$

55. $p^2(p+q) + 4pq(p+q) + 3q^2(p+q)$

Factor out the GCF, $p + q$.
$$= (p+q)(p^2 + 4pq + 3q^2)$$
Factor the trinomial.
$$= (p+q)(p+q)(p+3q)$$
$$= (p+q)^2(p+3q)$$

57. $z^2(z - x) - zx(x - z) - 2x^2(z - x)$

Factor out -1 from the middle term:
$x - z = -1(z - x)$.

$$= z^2(z - x) + zx(z - x) - 2x^2(z - x)$$
Factor out the GCF, $z - x$.
$$= (z - x)(z^2 + zx - 2x^2)$$
Factor the trinomial.
$$= (z - x)(z + 2x)(z - x)$$
$$= (z - x)^2(z + 2x)$$

59. In $p^4 - 10p^2 + 16$, let $x = p^2$ to obtain

$$x^2 - 10x + 16 = (x - 8)(x - 2).$$

Replace x with p^2.

$$p^4 - 10p^2 + 16 = (p^2 - 8)(p^2 - 2)$$

61. In $2x^4 - 9x^2 - 18$, let $y = x^2$ to obtain

$$2y^2 - 9y - 18 = (2y + 3)(y - 6).$$

Replace y with x^2.

$$2x^4 - 9x^2 - 18 = (2x^2 + 3)(x^2 - 6)$$

63. In $16x^4 + 16x^2 + 3$, let $m = x^2$ to obtain

$$16m^2 + 16m + 3 = (4m + 3)(4m + 1).$$

Replace m with x^2.

$$16x^4 + 16x^2 + 3 = (4x^2 + 3)(4x^2 + 1)$$

65. $(3x - 5)(3x + 5) = (3x)^2 - 5^2$
$$= 9x^2 - 25$$

67. $(p + 3q)^2 = p^2 + 2 \cdot p \cdot 3q + (3q)^2$
$$= p^2 + 6pq + 9q^2$$

69. $(y + 3)(y^2 - 3y + 9)$
$$= y(y^2 - 3y + 9) + 3(y^2 - 3y + 9)$$
$$= y^3 - 3y^2 + 9y + 3y^2 - 9y + 27$$
$$= y^3 + 27$$

6.3 Special Factoring

6.3 Now Try Exercises

N1. Use the difference of squares formula,
$$x^2 - y^2 = (x + y)(x - y).$$

(a) $4m^2 - 25n^2 = (2m)^2 - (5n)^2$
$$= (2m + 5n)(2m - 5n)$$

(b) $9x^2 - 729 = 9(x^2 - 81)$ GCF $= 9$
$$= 9(x^2 - 9^2)$$
$$= 9(x + 9)(x - 9)$$

(c) $(a + b)^2 - 25$
$$= (a + b)^2 - 5^2$$
$$= (a + b + 5)(a + b - 5)$$

(d) $v^4 - 1 = (v^2)^2 - 1^2$
$$= (v^2 + 1)(v^2 - 1)$$
$$= (v^2 + 1)(v^2 - 1^2)$$
$$= (v^2 + 1)(v + 1)(v - 1)$$

N2. **(a)** $a^2 + 12a + 36 = a^2 + 12a + 6^2$
$$= (a + 6)^2$$

Check $2(a)(6) = 12a$, which is the middle term.

Thus, $a^2 + 12a + 36 = (a + 6)^2$.

(b) $16x^2 - 56xy + 49y^2$
$$= (4x)^2 - 56xy + (7y)^2$$
$$= (4x - 7y)^2$$

Check $2(4x)(-7y) = -56xy$, which is the middle term.

Thus, $16x^2 - 56xy + 49y^2 = (4x - 7y)^2$.

(c) $y^2 - 16y + 64 - z^2$

Group the first three terms.
$$= (y^2 - 16y + 64) - z^2$$
Factor the perfect square trinomial.
$$= (y - 8)^2 - z^2$$
This is the difference of two squares.
$$= [(y - 8) + z][(y - 8) - z]$$
$$= (y - 8 + z)(y - 8 - z)$$

N3. Use the difference of cubes formula,
$$x^3 - y^3 = (x - y)(x^2 + xy + y^2).$$

(a) $t^3 - 1 = t^3 - 1^3$
$$= (t - 1)(t^2 + 1t + 1^2)$$
$$= (t - 1)(t^2 + t + 1)$$

Copyright © 2012 Pearson Education, Inc. Publishing as Addison-Wesley.

(b) $125a^3 - 8b^3$

$= (5a)^3 - (2b)^3$

$= (5a - 2b)[(5a)^2 + 5a(2b) + (2b)^2]$

$= (5a - 2b)(25a^2 + 10ab + 4b^2)$

N4. Use the sum of cubes formula,

$$x^3 + y^3 = (x + y)(x^2 - xy + y^2).$$

(a) $1000 + z^3 = 10^3 + z^3$

$= (10 + z)[10^2 - 10z + z^2]$

$= (10 + z)(100 - 10z + z^2)$

(b) $81a^6 + 3b^3$

$= 3(27a^6 + b^3)$ *GCF = 3*

$= 3[(3a^2)^3 + b^3)]$

$= 3(3a^2 + b)[(3a^2)^2 - 3a^2b + b^2)]$

$= 3(3a^2 + b)(9a^4 - 3a^2b + b^2)$

(c) $(x - 3)^3 + y^3$

$= [(x - 3) + y][(x - 3)^2 - (x - 3)y + y^2]$

$= (x - 3 + y)(x^2 - 6x + 9 - xy + 3y + y^2)$

6.3 Section Exercises

1. **A.** Yes, 64 and k^2 are squares.

B. No, $2x^2$ is not a square.

C. No, $k^2 + 9$ is a *sum* of squares.

D. Yes, $4z^2$ and 49 are squares.

Thus, the binomials that are differences of squares are **A** and **D**.

3. **A.** Since $x^2 - 8x - 16$ has a negative third term, it is not a perfect square trinomial.

B. $4m^2 + 20m + 25 = (2m)^2 + 20m + 5^2$

$2(2m)(5) = 20m$, the middle term. Therefore, this trinomial is a perfect square.

C. $9z^4 + 30z^2 + 25 = (3z^2)^2 + 30z^2 + 5^2$

$2(3z^2)(5) = 30z^2$, the middle term. Therefore, this trinomial is a perfect square.

D. $25p^2 - 45p + 81 = (5p)^2 - 45p + (-9)^2$

$2(5p)(-9) = -90p$, which is not the middle term. Therefore, this trinomial is not a perfect square.

Thus, the perfect square trinomials are **B** and **C**.

5. The sum of two squares can be factored if the binomial terms have a common factor greater than 1. For example,

$$4x^2 + 64 = 4(x^2 + 16).$$

7. $p^2 - 16 = p^2 - 4^2$

$= (p + 4)(p - 4)$

9. $25x^2 - 4 = (5x)^2 - 2^2$

$= (5x + 2)(5x - 2)$

11. $18a^2 - 98b^2 = 2(9a^2 - 49b^2)$

$= 2[(3a)^2 - (7b)^2]$

$= 2[(3a + 7b)(3a - 7b)]$

$= 2(3a + 7b)(3a - 7b)$

13. $64m^4 - 4y^4$

$= 4(16m^4 - y^4)$

$= [4(4m^2)^2 - (y^2)^2]$

Factor the difference of squares.

$= 4(4m^2 + y^2)(4m^2 - y^2)$

$= 4(4m^2 + y^2)[(2m)^2 - y^2]$

Factor the difference of squares again.

$= 4(4m^2 + y^2)(2m + y)(2m - y)$

15. $(y + z)^2 - 81$

$= (y + z)^2 - 9^2$

$= [(y + z) + 9][(y + z) - 9]$

$= (y + z + 9)(y + z - 9)$

17. $16 - (x + 3y)^2$

$= 4^2 - z^2$ *Let $z = (x + 3y)$.*

$= (4 + z)(4 - z)$

Substitute $x + 3y$ for z.

$= [4 + (x + 3y)][4 - (x + 3y)]$

$= (4 + x + 3y)(4 - x - 3y)$

19. $p^4 - 256 = (p^2)^2 - 16^2$

$= (p^2 + 16)(p^2 - 16)$

$= (p^2 + 16)(p^2 - 4^2)$

$= (p^2 + 16)(p + 4)(p - 4)$

21. $k^2 - 6k + 9 = (k)^2 - 2(k)(3) + 3^2$

$= (k - 3)^2$

23. $4z^2 + 4zw + w^2$

$= (2z)^2 + 2(2z)(w) + w^2$

$= (2z + w)^2$

25. $16m^2 - 8m + 1 - n^2$

Group the first three terms.

$= (16m^2 - 8m + 1) - n^2$

$= [(4m)^2 - 2(4m)(1) + 1^2] - n^2$

$= (4m - 1)^2 - n^2$

$= [(4m - 1) + n][(4m - 1) - n]$

$= (4m - 1 + n)(4m - 1 - n)$

27. $4r^2 - 12r + 9 - s^2$

Group the first three terms.

$= (4r^2 - 12r + 9) - s^2$

$= [(2r)^2 - 2(2r)(3) + 3^2] - s^2$

$= (2r - 3)^2 - s^2$

$= [(2r - 3) + s][(2r - 3) - s]$

$= (2r - 3 + s)(2r - 3 - s)$

Copyright © 2012 Pearson Education, Inc. Publishing as Addison-Wesley.

29. $x^2 - y^2 + 2y - 1$

Group the last three terms.

$= x^2 - (y^2 - 2y + 1)$

$= x^2 - (y - 1)^2$

$= [x + (y - 1)][x - (y - 1)]$

$= (x + y - 1)(x - y + 1)$

31. $98m^2 + 84mn + 18n^2$

$= 2(49m^2 + 42mn + 9n^2)$

$= 2[(7m)^2 + 2(7m)(3n) + (3n)^2]$

$= 2(7m + 3n)^2$

33. $(p + q)^2 + 2(p + q) + 1$

$= x^2 + 2x + 1$ *Let* $x = p + q$.

$= (x + 1)^2$

$= (p + q + 1)^2$ *Resubstitute.*

35. $(a - b)^2 + 8(a - b) + 16$

$= (a - b)^2 + 2(a - b)(4) + 4^2$

$= [(a - b) + 4]^2$

$= (a - b + 4)^2$

37. $x^3 - 27 = x^3 - 3^3$

$= (x - 3)(x^2 + x \cdot 3 + 3^2)$

$= (x - 3)(x^2 + 3x + 9)$

39. $216 - t^3 = 6^3 - t^3$

$= (6 - t)(6^2 + 6 \cdot t + t^2)$

$= (6 - t)(36 + 6t + t^2)$

41. $x^3 + 64 = x^3 + 4^3$

$= (x + 4)(x^2 - x \cdot 4 + 4^2)$

$= (x + 4)(x^2 - 4x + 16)$

43. $1000 + y^3 = 10^3 + y^3$

$= (10 + y)(10^2 - 10 \cdot y + y^2)$

$= (10 + y)(100 - 10y + y^2)$

45. $8x^3 + 1 = (2x)^3 + 1^3$

$= (2x + 1)[(2x)^2 - 2x \cdot 1 + 1^2]$

$= (2x + 1)(4x^2 - 2x + 1)$

47. $125x^3 - 216 = (5x)^3 - 6^3$

$= (5x - 6)[(5x)^2 + 5x \cdot 6 + 6^2]$

$= (5x - 6)(25x^2 + 30x + 36)$

49. $x^3 - 8y^3 = x^3 - (2y)^3$

$= (x - 2y)[x^2 + x \cdot 2y + (2y)^2]$

$= (x - 2y)(x^2 + 2xy + 4y^2)$

51. $64g^3 - 27h^3$

$= (4g)^3 - (3h)^3$

$= (4g - 3h)[(4g)^2 + (4g)(3h) + (3h)^2]$

$= (4g - 3h)(16g^2 + 12gh + 9h^2)$

53. $343p^3 + 125q^3$

$= (7p)^3 + (5q)^3$

$= (7p + 5q)[(7p)^2 - (7p)(5q) + (5q)^2]$

$= (7p + 5q)(49p^2 - 35pq + 25q^2)$

55. $24n^3 + 81p^3$

$= 3(8n^3 + 27p^3)$

$= 3[(2n)^3 + (3p)^3]$

$= 3[2n + 3p][(2n)^2 - (2n)(3p) + (3p)^2]$

$= 3(2n + 3p)(4n^2 - 6np + 9p^2)$

57. $(y + z)^3 + 64$

$= (y + z)^3 + 4^3$

$= [(y + z) + 4][(y + z)^2 - (y + z)(4) + 4^2]$

$= (y + z + 4)(y^2 + 2yz + z^2 - 4y - 4z + 16)$

59. $m^6 - 125 = (m^2)^3 - (5)^3$

$= (m^2 - 5)[(m^2)^2 + (m^2)(5) + 5^2]$

$= (m^2 - 5)(m^4 + 5m^2 + 25)$

61. $27 - 1000x^9$

$= 3^3 - (10x^3)^3$

$= (3 - 10x^3)[3^2 + (3)(10x^3) + (10x^3)^2]$

$= (3 - 10x^3)(9 + 30x^3 + 100x^6)$

63. $125y^6 + z^3$

$= (5y^2)^3 + z^3$

$= (5y^2 + z)[(5y^2)^2 - (5y^2)(z) + z^2]$

$= (5y^2 + z)(25y^4 - 5y^2z + z^2)$

♦♦♦ Relating Concepts 65–70 ♦♦♦

65. $x^6 - y^6$

$= (x^3)^2 - (y^3)^2$

$= (x^3 + y^3)(x^3 - y^3)$

$= [(x + y)(x^2 - xy + y^2)]$

    $\cdot [(x - y)(x^2 + xy + y^2)]$

$= (x + y)(x^2 - xy + y^2)(x - y)$

    $\cdot (x^2 + xy + y^2)$

66. $x^6 - y^6 = (x - y)(x + y)$

    $\cdot \underline{(x^2 + xy + y^2)(x^2 - xy + y^2)}$

67. $x^6 - y^6$

$= (x^2)^3 - (y^2)^3$

$= (x^2 - y^2)(x^4 + x^2y^2 + y^4)$

$= (x + y)(x - y)(x^4 + x^2y^2 + y^4)$

68. $x^6 - y^6 = (x - y)(x + y)$

    $\cdot \underline{(x^4 + x^2y^2 + y^4)}$

69. The product written on the blank in Exercise 66 must equal the product written on the blank in Exercise 68. To verify this, multiply the two factors written in Exercise 66.

Copyright © 2012 Pearson Education, Inc. Publishing as Addison-Wesley.

$$(x^2 + xy + y^2)(x^2 - xy + y^2)$$
$$= x^2(x^2 - xy + y^2)$$
$$\quad + xy(x^2 - xy + y^2)$$
$$\quad + y^2(x^2 - xy + y^2)$$
$$= x^4 - x^3y + x^2y^2 + x^3y - x^2y^2$$
$$\quad + xy^3 + x^2y^2 - xy^3 + y^4$$
$$= x^4 + x^2y^2 + y^4$$

They are equal.

70. Start by factoring as the difference of squares since doing so resulted in the complete factorization more directly.

71. $125p^3 + 25p^2 + 8q^3 - 4q^2$
$$= (125p^3 + 8q^3) + (25p^2 - 4q^2)$$
$$= [(5p)^3 + (2q)^3] + [(5p)^2 - (2q)^2]$$
Factor within groups.
$$= [(5p + 2q)(25p^2 - 10pq + 4q^2)]$$
$$\quad + [(5p + 2q)(5p - 2q)]$$
Factor out the GCF, $5p + 2q$.
$$= (5p + 2q)$$
$$\quad \cdot [(25p^2 - 10pq + 4q^2) + (5p - 2q)]$$
$$= (5p + 2q)$$
$$\quad \cdot (25p^2 - 10pq + 4q^2 + 5p - 2q)$$

73. $27a^3 + 15a - 64b^3 - 20b$
$$= (27a^3 - 64b^3) + (15a - 20b)$$
$$= [(3a)^3 - (4b)^3] + (15a - 20b)$$
Factor within groups.
$$= (3a - 4b)(9a^2 + 12ab + 16b^2)$$
$$\quad + 5(3a - 4b)$$
Factor out the GCF, $3a - 4b$.
$$= (3a - 4b)[(9a^2 + 12ab + 16b^2) + 5]$$
$$= (3a - 4b)(9a^2 + 12ab + 16b^2 + 5)$$

75. $8t^4 - 24t^3 + t - 3$
$$= (8t^4 - 24t^3) + (t - 3)$$
$$= 8t^3(t - 3) + 1(t - 3)$$
Factor out the GCF, $t - 3$.
$$= (t - 3)(8t^3 + 1)$$
Factor the sum of cubes.
$$= (t - 3)(2t + 1)(4t^2 - 2t + 1)$$

77. $64m^2 - 512m^3 - 81n^2 + 729n^3$
$$= (64m^2 - 81n^2) - (512m^3 - 729n^3)$$
$$= [(8m)^2 - (9n)^2] - [(8m)^3 - (9n)^3]$$
Factor within groups.
$$= (8m + 9n)(8m - 9n)$$
$$\quad - (8m - 9n)(64m^2 + 72mn + 81n^2)$$
Factor out the GCF, $8m - 9n$.
$$= (8m - 9n)$$
$$\quad \cdot [(8m + 9n) - (64m^2 + 72mn + 81n^2)]$$
$$= (8m - 9n)$$
$$\quad \cdot (8m + 9n - 64m^2 - 72mn - 81n^2)$$

79. $2ax + ay - 2bx - by$
$$= (2ax + ay) + (-2bx - by)$$
$$= a(2x + y) - b(2x + y)$$
$$= (2x + y)(a - b)$$

81. $p^2 + 4p - 21$

By trial and error,
$$p^2 + 4p - 21 = (p + 7)(p - 3).$$

6.4 A General Approach to Factoring

6.4 Now Try Exercises

N1. (a) $21x^3y^2 - 27x^2y^4$
$$= 3x^2y^2(7x - 9y^2) \quad \textit{GCF} = 3x^2y^2$$

(b) $8y(m - n) - 5(m - n)$
$$= (m - n)(8y - 5) \quad \textit{GCF} = m - n$$

N2. (a) $4a^2 - 49b^2$
$$= (2a)^2 - (7b)^2 \qquad \textit{Difference of squares}$$
$$= (2a + 7b)(2a - 7b)$$

(b) $9x^2 + 100$ is a *sum* of squares and cannot be factored. The binomial is prime.

(c) $27v^3 - 1000$
$$= (3v)^3 - 10^3 \qquad \textit{Difference of cubes}$$
$$= (3v - 10)[(3v)^2 + 3v \cdot 10 + 10^2]$$
$$= (3v - 10)(9v^2 + 30v + 100)$$

N3. (a) $25x^2 - 90x + 81$
$$= (5x)^2 - 2(5x)(9) + 9^2$$
$$\textit{Perfect square trinomial}$$
$$= (5x - 9)^2$$

(b) $7x^2 - 7xy - 84y^2 = 7(x^2 - xy - 12y^2)$

Two integer factors whose product is -12 and whose sum is -1 are -4 and 3.

$$= 7[x^2 - 4xy + 3xy - 12y^2]$$
$$= 7[x(x - 4y) + 3y(x - 4y)]$$
$$= 7[(x - 4y)(x + 3y)]$$
$$= 7(x - 4y)(x + 3y)$$

(c) $12m^2 + 5m - 28$

Two integer factors whose product is $12(-28) = -336$ and whose sum is 5 are 21 and -16.

$$= 12m^2 + 21m - 16m - 28$$
$$= 3m(4m + 7) - 4(4m + 7)$$
$$= (4m + 7)(3m - 4)$$

Copyright © 2012 Pearson Education, Inc. Publishing as Addison-Wesley.

N4. (a) $5a^3 + 5a^2b - ab^2 - b^3$
$$= (5a^3 + 5a^2b) - (ab^2 + b^3)$$
$$= 5a^2(a+b) - b^2(a+b)$$
$$= (a+b)(5a^2 - b^2)$$

(b) $9u^2 - 48u + 64 - v^2$
$$= (9u^2 - 48u + 64) - v^2$$
$$= (3u - 8)^2 - v^2$$
$$= [(3u - 8) + v][(3u - 8) - v]$$
$$= (3u - 8 + v)(3u - 8 - v)$$

(c) $x^3 - 9y^2 - 27y^3 + x^2$
$$= (x^3 - 27y^3) + (x^2 - 9y^2)$$
$$= [x^3 - (3y)^3] + [x^2 - (3y)^2]$$
$$= \{(x - 3y)[x^2 + x \cdot 3y + (3y)^2]\}$$
$$+ [(x + 3y)(x - 3y)]$$
$$= [(x - 3y)(x^2 + 3xy + 9y^2]$$
$$+ [(x + 3y)(x - 3y)]$$
$$= (x - 3y)(x^2 + 3xy + 9y^2 + x + 3y)$$

6.4 Section Exercises

1. $100a^2 - 9b^2$
$$= (10a)^2 - (3b)^2 \qquad \textit{Difference of squares}$$
$$= (10a + 3b)(10a - 3b)$$

3. $3p^4 - 3p^3 - 90p^2$
$$= 3p^2(p^2 - p - 30)$$
$$= 3p^2(p - 6)(p + 5)$$

5. $3a^2pq + 3abpq - 90b^2pq$
$$= 3pq(a^2 + ab - 30b^2)$$
$$= 3pq(a + 6b)(a - 5b)$$

7. $225p^2 + 256$ is a *sum* of squares and cannot be factored. The binomial is prime.

9. $6b^2 - 17b - 3$
Two integer factors whose product is $(6)(-3) = -18$ and whose sum is -17 are -18 and 1.
$$= 6b^2 - 18b + b - 3$$
$$= 6b(b - 3) + 1(b - 3)$$
$$= (b - 3)(6b + 1)$$

11. $x^3 - 1000$
$$= (x)^3 - 10^3 \quad \textit{Difference of cubes}$$
$$= (x - 10)(x^2 + 10x + 100)$$

13. $4(p + 2) + m(p + 2) = (p + 2)(4 + m)$

15. $9m^2 - 45m + 18m^3$
Factor out the GCF, $9m$.
$$= 9m(m - 5 + 2m^2) \text{ or } 9m(2m^2 + m - 5)$$
There is no pair of integers with a product of -10 and a sum of 1, so this cannot be factored further.

17. $54m^3 - 2000$
Factor out the GCF, 2.
$2(27m^3 - 1000)$
$$= 2[(3m)^3 - 10^3] \qquad \textit{Difference of cubes}$$
$$= 2(3m - 10)$$
$$\cdot [(3m)^2 + (3m)(10) + 10^2]$$
$$= 2(3m - 10)(9m^2 + 30m + 100)$$

19. $9m^2 - 30mn + 25n^2$
$$= (3m)^2 - 2(3m)(5n) + (5n)^2$$
$$\textit{Perfect square trinomial}$$
$$= (3m - 5n)^2$$

21. $kq - 9q + kr - 9r$
$$= q(k - 9) + r(k - 9)$$
$$= (k - 9)(q + r)$$

23. $16z^3x^2 - 32z^2x = 16z^2x(zx - 2)$

25. $x^2 + 2x - 35$
The integers 7 and -5 have a product of -35 and a sum of 2.
$$= (x + 7)(x - 5)$$

27. $625 - x^4$
$$= 25^2 - (x^2)^2 \qquad \textit{Difference of squares}$$
$$= (25 + x^2)(25 - x^2)$$
$$= (25 + x^2)(5 + x)(5 - x) \qquad \textit{Difference of squares again}$$

29. $p^3 + 1 = p^3 + 1^3 \qquad \textit{Sum of cubes}$
$$= (p + 1)(p^2 - p \cdot 1 + 1^2)$$
$$= (p + 1)(p^2 - p + 1)$$

31. $64m^2 - 625$
$$= (8m)^2 - 25^2 \qquad \textit{Difference of squares}$$
$$= (8m + 25)(8m - 25)$$

33. $12z^3 - 6z^2 + 18z$
Factor out the GCF, $6z$.
$$6z(2z^2 - z + 3)$$
Further factoring is not possible. There is no pair of integers whose product is 6 and whose sum is -1.

35. $256b^2 - 400c^2$
$$= 16(16b^2 - 25c^2)$$
$$= 16[(4b)^2 - (5c)^2] \qquad \textit{Difference of squares}$$
$$= 16(4b + 5c)(4b - 5c)$$

Copyright © 2012 Pearson Education, Inc. Publishing as Addison-Wesley.

37. $512 + 1000z^3$
$$= 8(64 + 125z^3)$$
$$= 8[4^3 + (5z)^3] \qquad \textit{Sum of cubes}$$
$$= 8[4 + 5z][4^2 - (4)(5z) + (5z)^2]$$
$$= 8(4 + 5z)(16 - 20z + 25z^2)$$

39. $10r^2 + 23rs - 5s^2$
Two integer factors whose product is $(10)(-5) = -50$ and whose sum is 23 are 25 and -2.
$$= 10r^2 + 25rs - 2rs - 5s^2$$
$$= 5r(2r + 5s) - s(2r + 5s)$$
$$= (2r + 5s)(5r - s)$$

41. $24p^3q + 52p^2q^2 + 20pq^3$
$$= 4pq(6p^2 + 13pq + 5q^2)$$
Two integer factors whose product is $(6)(5) = 30$ and whose sum is 13 are 10 and 3.
$$= 4pq(6p^2 + 10pq + 3pq + 5q^2)$$
$$= 4pq[2p(3p + 5q) + q(3p + 5q)]$$
$$= 4pq(3p + 5q)(2p + q)$$

43. $48k^4 - 243$
$$= 3(16k^4 - 81)$$
$$= 3[(4k^2)^2 - 9^2]$$
$$= 3(4k^2 + 9)(4k^2 - 9)$$
$$= 3(4k^2 + 9)[(2k)^2 - 3^2]$$
$$= 3(4k^2 + 9)(2k + 3)(2k - 3)$$

45. $m^3 + m^2 - n^3 - n^2$
$$= (m^3 - n^3) + (m^2 - n^2)$$
$\qquad$ *Difference* $\qquad$ *Difference*
$\qquad$ *of cubes;* $\qquad$ *of squares*
$$= (m - n)(m^2 + mn + n^2)$$
$$\quad + (m + n)(m - n)$$
$$= (m - n)[(m^2 + mn + n^2) + (m + n)]$$
$$= (m - n)(m^2 + mn + n^2 + m + n)$$

47. $x^2 - 4m^2 - 4mn - n^2$
$$= x^2 - (4m^2 + 4mn + n^2)$$
$$= x^2 - [(2m)^2 + 2(2m)n + n^2]$$
$\qquad\qquad$ *Perfect square trinomial*
$$= x^2 - (2m + n)^2$$
$\qquad\qquad$ *Difference of squares*
$$= [x + (2m + n)][x - (2m + n)]$$
$$= (x + 2m + n)(x - 2m - n)$$

49. $18p^5 - 24p^3 + 12p^6$
Factor out the GCF, $6p^3$.
$$6p^3(3p^2 - 4 + 2p^3)$$
Further factoring is not possible.

51. $2x^2 - 2x - 40 = 2(x^2 - x - 20)$
$$= 2(x + 4)(x - 5)$$

53. $(2m + n)^2 - (2m - n)^2$
$\qquad\qquad$ *Difference of squares*
$$= [(2m + n) + (2m - n)]$$
$$\quad \cdot [(2m + n) - (2m - n)]$$
$$= (2m + n + 2m - n)(2m + n - 2m + n)$$
$$= 4m(2n)$$
$$= 8mn$$

55. $50p^2 - 162$
$$= 2(25p^2 - 81)$$
$$= 2[(5p)^2 - 9^2]$$
$$= 2(5p + 9)(5p - 9)$$

57. $12m^2rx + 4mnrx + 40n^2rx$
Factor out the GCF, $4rx$.
$$= 4rx(3m^2 + mn + 10n^2)$$

59. $21a^2 - 5ab - 4b^2$
Two integer factors whose product is $(21)(-4) = -84$ and whose sum is -5 are -12 and 7.
$$= 21a^2 - 12ab + 7ab - 4b^2$$
$$= 3a(7a - 4b) + b(7a - 4b)$$
$$= (7a - 4b)(3a + b)$$

61. $x^2 - y^2 - 4$ cannot be factored. The polynomial is *prime*.

63. $(p + 8q)^2 - 10(p + 8q) + 25$
$$= x^2 - 10x + 25 \qquad \textit{Let x = p + 8q.}$$
$$= (x - 5)^2$$
$$= [(p + 8q) - 5]^2 \qquad \textit{Resubstitute.}$$
$$= (p + 8q - 5)^2$$

65. $21m^4 - 32m^2 - 5$
$$= 21x^2 - 32x - 5 \qquad \textit{Let x = m^2.}$$
Two integer factors whose product is $(21)(-5) = -105$ and whose sum is -32 are -35 and 3.
$$= 21x^2 - 35x + 3x - 5$$
$$= 7x(3x - 5) + 1(3x - 5)$$
$$= (3x - 5)(7x + 1)$$
$$= (3m^2 - 5)(7m^2 + 1) \qquad \textit{Resubstitute.}$$

67. $(r + 2t)^3 + (r - 3t)^3$
Let $x = r + 2t$ and $y = r - 3t$.
$$= x^3 + y^3 \qquad\qquad \textit{Sum of cubes}$$
$$= (x + y)(x^2 - xy + y^2)$$
Substitute $r + 2t$ for x and $r - 3t$ for y.
$$= [(r + 2t) + (r - 3t)]$$
$$\quad \cdot [(r + 2t)^2 - (r + 2t)(r - 3t)$$
$$\qquad + (r - 3t)^2]$$
$$= (2r - t)$$
$$\quad \cdot (r^2 + 4rt + 4t^2 - r^2 + rt + 6t^2$$
$$\qquad + r^2 - 6rt + 9t^2)$$
$$= (2r - t)(r^2 - rt + 19t^2)$$

Copyright © 2012 Pearson Education, Inc. Publishing as Addison-Wesley.

69. $x^5 + 3x^4 - x - 3$
$$= x^4(x + 3) - 1(x + 3)$$
$$= (x + 3)(x^4 - 1)$$
$$= (x + 3)[(x^2)^2 - 1^2]$$
$$= (x + 3)(x^2 + 1)(x^2 - 1)$$
$$= (x + 3)(x^2 + 1)[x^2 - 1^2]$$
$$= (x + 3)(x^2 + 1)(x + 1)(x - 1)$$

71. $m^2 - 4m + 4 - n^2 + 6n - 9$
$$= (m^2 - 4m + 4) - (n^2 - 6n + 9)$$
$$= [m^2 - 2(2)m + 2^2]$$
$$\quad - [n^2 - 2(3)n + 3^2]$$
Perfect square trinomials
$$= (m - 2)^2 - (n - 3)^2$$
$$= [(m - 2) + (n - 3)]$$
$$\quad \cdot [(m - 2) - (n - 3)]$$
Difference of two squares
$$= (m - 2 + n - 3)(m - 2 - n + 3)$$
$$= (m + n - 5)(m - n + 1)$$

73. $3x + 2 = 0$
$$3x = -2$$
$$x = \frac{-2}{3} = -\frac{2}{3}$$
The solution set is $\left\{-\frac{2}{3}\right\}$.

75. $5x = 0$
$$x = \frac{0}{5} = 0$$
The solution set is $\{0\}$.

77. $\frac{1}{2}x + 5 = 0$
$$\frac{1}{2}x = -5$$
$$2\left(\frac{1}{2}x\right) = 2(-5)$$
$$x = -10$$
The solution set is $\{-10\}$.

6.5 Solving Equations by Factoring

6.5 Now Try Exercises

N1. $(x + 5)(4x - 7) = 0$
$$x + 5 = 0 \quad \text{or} \quad 4x - 7 = 0$$
$$\qquad\qquad\qquad 4x = 7$$
$$x = -5 \quad \text{or} \qquad x = \frac{7}{4}$$
Check $x = -5$: $\quad 0(-27) = 0 \quad$ *True*
Check $x = \frac{7}{4}$: $\qquad \frac{27}{4}(0) = 0 \quad$ *True*
The solution set is $\left\{-5, \frac{27}{4}\right\}$.

N2. **(a)** *Step 1 Standard form*
$$7x = 3 - 6x^2$$
$$6x^2 + 7x - 3 = 0$$

Step 2 Factor.
$$(2x + 3)(3x - 1) = 0$$

Step 3 Zero-factor property
$$2x + 3 = 0 \quad \text{or} \quad 3x - 1 = 0$$

Step 4 Solve each equation.
$$2x = -3 \quad \text{or} \quad 3x = 1$$
$$x = -\frac{3}{2} \qquad x = \frac{1}{3}$$

Step 5
Check each solution in the original equation.

Check $x = -\frac{3}{2}$: $-\frac{21}{2} = \frac{6}{2} - \frac{27}{2}$ *True*
Check $x = \frac{1}{3}$: $\quad \frac{7}{3} = \frac{9}{3} - \frac{2}{3}$ *True*

The solution set is $\left\{-\frac{3}{2}, \frac{1}{3}\right\}$.

(b) $16x^2 + 40x + 25 = 0$ *Standard form*
$$(4x + 5)^2 = 0 \qquad \text{Factor.}$$
$$4x + 5 = 0 \qquad \text{Zero-factor prop.}$$
$$4x = -5 \qquad \text{Subtract 5.}$$
$$x = -\frac{5}{4} \qquad \text{Divide by 4.}$$

Check $x = -\frac{5}{4}$: $16\left(\frac{25}{16}\right) - 50 + 25 = 0 \quad$ *True*
The solution set is $\left\{-\frac{5}{4}\right\}$.

N3. $3x^2 + 12x = 0$
$$3x(x + 4) = 0$$
$$3x = 0 \quad \text{or} \quad x + 4 = 0$$
$$x = 0 \qquad\qquad x = -4$$

Check $x = 0$: $\qquad 0 + 0 = 0 \quad$ *True*
Check $x = -4$: $\quad 48 - 48 = 0 \quad$ *True*

The solution set is $\{-4, 0\}$.

N4. $\qquad 4x^2 - 100 = 0$
$$4(x^2 - 25) = 0 \quad \text{Factor out 4.}$$
$$4(x + 5)(x - 5) = 0 \quad \text{Difference of squares}$$
$$x + 5 = 0 \quad \text{or} \quad x - 5 = 0 \quad \begin{matrix}\text{Zero-factor}\\\text{property}\end{matrix}$$
$$x = -5 \quad \text{or} \qquad x = 5$$

Check $x = \pm 5$: $\quad 4(25) - 100 = 0 \quad$ *True*

The solution set is $\{-5, 5\}$.

N5. $(x + 3)(2x - 1) = 4(x + 4) - 4$
$$2x^2 + 5x - 3 = 4x + 16 - 4$$
$$2x^2 + x - 15 = 0$$
$$(x + 3)(2x - 5) = 0$$
$$x + 3 = 0 \quad \text{or} \quad 2x - 5 = 0$$
$$x = -3 \quad \text{or} \qquad x = \frac{5}{2}$$

Check $x = -3$: $\qquad 0 = 4 - 4 \quad$ *True*
Check $x = \frac{5}{2}$: $\quad \frac{11}{2}(4) = 26 - 4 \quad$ *True*

The solution set is $\left\{-3, \frac{5}{2}\right\}$.

Copyright © 2012 Pearson Education, Inc. Publishing as Addison-Wesley.

N6.
$$12x = 2x^3 + 5x^2$$
$$2x^3 + 5x^2 - 12x = 0$$
$$x(2x^2 + 5x - 12) = 0$$
$$x(x + 4)(2x - 3) = 0$$

$$x = 0 \quad \text{or} \quad x + 4 = 0 \quad \text{or} \quad 2x - 3 = 0$$
$$x = -4 \quad \text{or} \quad x = \tfrac{3}{2}$$

Check $x = 0$: $0 = 0$ *True*
Check $x = -4$: $-48 = -128 + 80$ *True*
Check $x = \tfrac{3}{2}$: $18 = \tfrac{27}{4} + \tfrac{45}{4}$ *True*

The solution set is $\left\{-4, 0, \tfrac{3}{2}\right\}$.

N7. Let $x =$ the base of the triangle. Then $2x - 1 =$ the height.

Use the formula $\tfrac{1}{2}bh = \mathcal{A}$, where $b = x$, $h = 2x - 1$, and $\mathcal{A} = 14$.

$$\tfrac{1}{2}x(2x - 1) = 14$$
$$x(2x - 1) = 28$$
$$2x^2 - x - 28 = 0$$
$$(2x + 7)(x - 4) = 0$$

$$2x + 7 = 0 \quad \text{or} \quad x - 4 = 0$$
$$x = -\tfrac{7}{2} \quad \text{or} \quad x = 4$$

The triangle cannot have a negative base, so reject $x = -\tfrac{7}{2}$. The base is 4 ft and the height is $2(4) - 1 = 7$ ft. The area of the triangle is $\tfrac{1}{2}(4)(7) = 14$ ft^2, as required.

N8.
$$h(t) = -16t^2 + 128t$$
$$192 = -16t^2 + 128t$$
$$16t^2 - 128t + 192 = 0$$
$$t^2 - 8t + 12 = 0 \quad \textit{Divide by 16.}$$
$$(t - 2)(t - 6) = 0 \quad \textit{Factor.}$$

$$t - 2 = 0 \quad \text{or} \quad t - 6 = 0$$
$$t = 2 \quad \text{or} \quad t = 6$$

The rocket will reach a height of 192 feet after 2 seconds (on the way up) and again after 6 seconds (on the way down).

N9. Solve $\mathcal{A} = 2HW + 2LW + 2LH$ for H.
$$\mathcal{A} - 2LW = 2HW + 2LH$$
       Get the H-terms on one side.
$$\mathcal{A} - 2LW = H(2W + 2L)$$
       Factor out H.
$$\frac{\mathcal{A} - 2LW}{2W + 2L} = \frac{H(2W + 2L)}{2W + 2L}$$
       Divide by 2W + 2L.
$$\frac{\mathcal{A} - 2LW}{2W + 2L} = H, \text{ or } H = \frac{\mathcal{A} - 2LW}{2W + 2L}$$

6.5 Section Exercises

1. First rewrite the equation so that one side is 0. Factor the other side and set each factor equal to 0. The solutions of these linear equations are solutions of the quadratic equation.

In the exercises in this section, check all solutions to the equations by substituting them back in the original equations.

3. $(x + 10)(x - 5) = 0$
$$x + 10 = 0 \quad \text{or} \quad x - 5 = 0$$
$$x = -10 \quad \text{or} \quad x = 5$$

Check $x = -10$: $0(-15) = 0$ *True*
Check $x = 5$: $15(0) = 0$ *True*

The solution set is $\{-10, 5\}$.

5. $(2k - 5)(3k + 8) = 0$
$$2k - 5 = 0 \quad \text{or} \quad 3k + 8 = 0$$
$$2k = 5 \qquad\qquad 3k = -8$$
$$k = \tfrac{5}{2} \quad \text{or} \quad k = -\tfrac{8}{3}$$

The solution set is $\left\{-\tfrac{8}{3}, \tfrac{5}{2}\right\}$.

7. $x^2 - 3x - 10 = 0$
$(x + 2)(x - 5) = 0$
Use the zero-factor property.
$$x + 2 = 0 \quad \text{or} \quad x - 5 = 0$$
$$x = -2 \quad \text{or} \quad x = 5$$

The solution set is $\{-2, 5\}$.

9. $x^2 + 9x + 18 = 0$
$(x + 6)(x + 3) = 0$
$$x + 6 = 0 \quad \text{or} \quad x + 3 = 0$$
$$x = -6 \quad \text{or} \quad x = -3$$

The solution set is $\{-6, -3\}$.

11. $$2x^2 = 7x + 4$$
Get 0 on one side.
$$2x^2 - 7x - 4 = 0$$
$$(2x + 1)(x - 4) = 0$$
$$2x + 1 = 0 \quad \text{or} \quad x - 4 = 0$$
$$2x = -1 \qquad\qquad x = 4$$
$$x = -\tfrac{1}{2}$$

The solution set is $\left\{-\tfrac{1}{2}, 4\right\}$.

13. $$15x^2 - 7x = 4$$
$$15x^2 - 7x - 4 = 0$$
$$(3x + 1)(5x - 4) = 0$$
$$3x + 1 = 0 \quad \text{or} \quad 5x - 4 = 0$$
$$3x = -1 \qquad\qquad 5x = 4$$
$$x = -\tfrac{1}{3} \quad \text{or} \quad x = \tfrac{4}{5}$$

The solution set is $\left\{-\tfrac{1}{3}, \tfrac{4}{5}\right\}$.

Copyright © 2012 Pearson Education, Inc. Publishing as Addison-Wesley.

15. $2x^2 - 12 - 4x = x^2 - 3x$

$x^2 - x - 12 = 0$

$(x + 3)(x - 4) = 0$

$x + 3 = 0$ or $x - 4 = 0$

$x = -3$ or $x = 4$

The solution set is $\{-3, 4\}$.

17. $\qquad\qquad (5x + 1)(x + 3) = -2(5x + 1)$

$(5x + 1)(x + 3) + 2(5x + 1) = 0$

$(5x + 1)[(x + 3) + 2] = 0$

$(5x + 1)(x + 5) = 0$

$5x + 1 = 0$ or $x + 5 = 0$

$5x = -1$ $x = -5$

$x = -\frac{1}{5}$

The solution set is $\left\{-5, -\frac{1}{5}\right\}$.

19. $4p^2 + 16p = 0$

$4p(p + 4) = 0$

$4p = 0$ or $p + 4 = 0$

$p = 0$ $p = -4$

The solution set is $\{-4, 0\}$.

21. $6x^2 - 36x = 0$

$6x(x - 6) = 0$

$6x = 0$ or $x - 6 = 0$

$x = 0$ or $x = 6$

The solution set is $\{0, 6\}$.

23. $\qquad 4p^2 - 16 = 0$

$4(p^2 - 4) = 0$

$4(p + 2)(p - 2) = 0$

$p + 2 = 0$ or $p - 2 = 0$

$p = -2$ or $p = 2$

The solution set is $\{-2, 2\}$.

25. $\qquad -3x^2 + 27 = 0$

$-3(x^2 - 9) = 0$

$-3(x + 3)(x - 3) = 0$

Note that the leading -3 does not affect the solution set of the equation.

$x + 3 = 0$ or $x - 3 = 0$

$x = -3$ or $x = 3$

The solution set is $\{-3, 3\}$.

27. $-x^2 = 9 - 6x$

$0 = x^2 - 6x + 9$

$0 = (x - 3)(x - 3)$

$0 = x - 3$

$3 = x$

The solution set is $\{3\}$.

29. $9x^2 + 24x + 16 = 0$

$(3x + 4)(3x + 4) = 0$

$3x + 4 = 0$

$3x = -4$

$x = -\frac{4}{3}$

The solution set is $\left\{-\frac{4}{3}\right\}$.

31. $(x - 3)(x + 5) = -7$

Multiply the factors, and then add 7 on both sides of the equation to get 0 on the right.

$x^2 + 5x - 3x - 15 = -7$

$x^2 + 2x - 8 = 0$

Now factor the polynomial.

$(x + 4)(x - 2) = 0$

$x + 4 = 0$ or $x - 2 = 0$

$x = -4$ or $x = 2$

The solution set is $\{-4, 2\}$.

33. $(2x + 1)(x - 3) = 6x + 3$

$2x^2 - 6x + x - 3 = 6x + 3$

$2x^2 - 5x - 3 = 6x + 3$

$2x^2 - 11x - 6 = 0$

$(2x + 1)(x - 6) = 0$

$2x + 1 = 0$ or $x - 6 = 0$

$2x = -1$ $x = 6$

$x = -\frac{1}{2}$

The solution set is $\left\{-\frac{1}{2}, 6\right\}$.

35. $(x + 3)(x - 6) = (2x + 2)(x - 6)$

$x^2 - 3x - 18 = 2x^2 - 10x - 12$

$0 = x^2 - 7x + 6$

$0 = (x - 1)(x - 6)$

$x - 1 = 0$ or $x - 6 = 0$

$x = 1$ or $x = 6$

The solution set is $\{1, 6\}$.

37. $2x^3 - 9x^2 - 5x = 0$

$x(2x^2 - 9x - 5) = 0$

$x(2x + 1)(x - 5) = 0$

$x = 0$ or $2x + 1 = 0$ or $x - 5 = 0$

$2x = -1$ $x = 5$

$x = -\frac{1}{2}$

The solution set is $\left\{-\frac{1}{2}, 0, 5\right\}$.

39. $\qquad\quad x^3 - 2x^2 = 3x$

$x^3 - 2x^2 - 3x = 0$

$x(x^2 - 2x - 3) = 0$

$x(x - 3)(x + 1) = 0$

$x = 0$ or $x - 3 = 0$ or $x + 1 = 0$

or $x = 3$ $x = -1$

The solution set is $\{-1, 0, 3\}$.

Copyright © 2012 Pearson Education, Inc. Publishing as Addison-Wesley.

41.
$$9x^3 = 16x$$
$$9x^3 - 16x = 0$$
$$x(9x^2 - 16) = 0$$
$$x(3x + 4)(3x - 4) = 0$$

$$x = 0 \quad \text{or} \quad 3x + 4 = 0 \quad \text{or} \quad 3x - 4 = 0$$
$$3x = -4 \qquad\qquad 3x = 4$$
$$x = -\tfrac{4}{3} \quad \text{or} \qquad x = \tfrac{4}{3}$$

The solution set is $\left\{-\tfrac{4}{3}, 0, \tfrac{4}{3}\right\}$.

43.
$$2x^3 + 5x^2 - 2x - 5 = 0$$
Factor by grouping.
$$(2x^3 - 2x) + (5x^2 - 5) = 0$$
$$2x(x^2 - 1) + 5(x^2 - 1) = 0$$
$$(x^2 - 1)(2x + 5) = 0$$
$$(x + 1)(x - 1)(2x + 5) = 0$$

$$x + 1 = 0 \quad \text{or} \quad x - 1 = 0 \quad \text{or} \quad 2x + 5 = 0$$
$$x = -1 \quad \text{or} \qquad x = 1 \quad \text{or} \qquad x = -\tfrac{5}{2}$$

The solution set is $\left\{-\tfrac{5}{2}, -1, 1\right\}$.

45.
$$x^3 - 6x^2 - 9x + 54 = 0$$
Factor by grouping.
$$(x^3 - 6x^2) + (-9x + 54) = 0$$
$$x^2(x - 6) - 9(x - 6) = 0$$
$$(x - 6)(x^2 - 9) = 0$$
$$(x - 6)(x + 3)(x - 3) = 0$$

$$x - 6 = 0 \quad \text{or} \quad x + 3 = 0 \quad \text{or} \quad x - 3 = 0$$
$$x = 6 \qquad\qquad x = -3 \qquad\qquad x = 3$$

The solution set is $\{-3, 3, 6\}$.

47. By dividing each side by a variable expression, she "lost" the solution 0. The solution set is $\left\{-\tfrac{4}{3}, 0, \tfrac{4}{3}\right\}$.

49. $2(x - 1)^2 - 7(x - 1) - 15 = 0$
Let $y = x - 1$.
$$2y^2 - 7y - 15 = 0$$
$$(2y + 3)(y - 5) = 0$$

$$2y + 3 = 0 \quad \text{or} \quad y - 5 = 0$$
$$2y = -3 \qquad\qquad y = 5$$
$$y = -\tfrac{3}{2}$$

Substitute $x - 1$ for y.

$$x - 1 = -\tfrac{3}{2} \quad \text{or} \quad x - 1 = 5$$
$$x = -\tfrac{1}{2} \quad \text{or} \qquad x = 6$$

The solution set is $\left\{-\tfrac{1}{2}, 6\right\}$.

51. $5(3x - 1)^2 + 3 = -16(3x - 1)$
Let $y = 3x - 1$.
$$5y^2 + 3 = -16y$$
$$5y^2 + 16y + 3 = 0$$
$$(y + 3)(5y + 1) = 0$$

$$y + 3 = 0 \quad \text{or} \quad 5y + 1 = 0$$
$$y = -3 \qquad\qquad 5y = -1$$
$$y = -\tfrac{1}{5}$$

Substitute $3x - 1$ for y.

$$3x - 1 = -3 \quad \text{or} \quad 3x - 1 = -\tfrac{1}{5}$$
$$3x = -2 \qquad\qquad 3x = \tfrac{4}{5}$$
$$x = -\tfrac{2}{3} \qquad\qquad x = \tfrac{4}{15}$$

The solution set is $\left\{-\tfrac{2}{3}, \tfrac{4}{15}\right\}$.

53.
$$(2x - 3)^2 = 16x^2$$
$$4x^2 - 12x + 9 = 16x^2$$
$$-12x^2 - 12x + 9 = 0$$
$$4x^2 + 4x - 3 = 0 \qquad \textit{Divide by } -3.$$
$$(2x + 3)(2x - 1) = 0$$

$$2x + 3 = 0 \quad \text{or} \quad 2x - 1 = 0$$
$$2x = -3 \qquad\qquad 2x = 1$$
$$x = -\tfrac{3}{2} \quad \text{or} \qquad x = \tfrac{1}{2}$$

The solution set is $\left\{-\tfrac{3}{2}, \tfrac{1}{2}\right\}$.

55. Let $x =$ the width of the garden.
Then $x + 4 =$ the length of the garden.

The area of the rectangular-shaped garden is 320 ft^2, so use the formula $A = LW$ and substitute 320 for A, $x + 4$ for L, and x for W.

$$A = LW$$
$$320 = (x + 4)x$$
$$320 = x^2 + 4x$$
$$0 = x^2 + 4x - 320$$
$$0 = (x - 16)(x + 20)$$

$$x - 16 = 0 \quad \text{or} \quad x + 20 = 0$$
$$x = 16 \quad \text{or} \qquad x = -20$$

A rectangle cannot have a width that is a negative measure, so reject -20 as a solution. The only possible solution is 16.

The width of the garden is 16 feet, and the length is $16 + 4 = 20$ feet.

57. Let $h =$ the height of the parallelogram.
Then $h + 7 =$ the base of the parallelogram.

The area is 60 ft^2, so use the formula $A = bh$ and substitute 60 for A, and $h + 7$ for b.

$$A = bh$$
$$60 = (h + 7)h$$
$$60 = h^2 + 7h$$
$$0 = h^2 + 7h - 60$$
$$0 = (h + 12)(h - 5)$$

$$h + 12 = 0 \quad \text{or} \quad h - 5 = 0$$
$$h = -12 \quad \text{or} \qquad h = 5$$

Copyright © 2012 Pearson Education, Inc. Publishing as Addison-Wesley.

A parallelogram cannot have a height that is negative, so reject -12 as a solution. The only possible solution is 5.
The height of the parallelogram is 5 feet and the base is $5 + 7 = 12$ feet.

59. Let $L =$ the length of the rectangular area and $W =$ the width.

Use the formula for perimeter, $P = 2L + 2W$, and solve for W in terms of L. The perimeter is 300 ft.

$$300 = 2L + 2W$$
$$300 - 2L = 2W$$
$$150 - L = W$$

Now use the formula for area, $\mathcal{A} = LW$, substitute 5000 for $\mathcal{A}$, and solve for L.

$$5000 = L(150 - L)$$
$$5000 = 150L - L^2$$
$$L^2 - 150L + 5000 = 0$$
$$(L - 50)(L - 100) = 0$$

$$L - 50 = 0 \quad \text{or} \quad L - 100 = 0$$
$$L = 50 \quad \text{or} \qquad L = 100$$

When $L = 50$, $W = 150 - 50 = 100$.
When $L = 100$, $W = 150 - 100 = 50$.
The dimensions should be 50 feet by 100 feet.

61. Let x and $x + 1$ denote the two consecutive integers.

The sum of their squares is 61, so

$$x^2 + (x + 1)^2 = 61.$$
$$x^2 + x^2 + 2x + 1 = 61$$
$$2x^2 + 2x - 60 = 0 \quad \textit{Divide by 2.}$$
$$x^2 + x - 30 = 0$$
$$(x + 6)(x - 5) = 0$$

$$x + 6 = 0 \quad \text{or} \quad x - 5 = 0$$
$$x = -6 \quad \text{or} \qquad x = 5$$

If $x = -6$, then $x + 1 = -5$.
If $x = 5$, then $x + 1 = 6$.

The two possible pairs of consecutive integers are -6 and -5 or 5 and 6.

63. Let $w =$ the width of the cardboard.
Then $w + 6 =$ the length of the cardboard.

If squares that measure 2 inches are cut from each corner of the cardboard, then the width becomes $w - 4$ and the length becomes $(w + 6) - 4 = w + 2$. Use the formula $V = LWH$ and substitute 110 for V, $w + 2$ for L, $w - 4$ for W, and 2 for H.

$$V = LWH$$
$$110 = (w + 2)(w - 4)2$$
$$110 = (w^2 - 2w - 8)2$$
$$55 = w^2 - 2w - 8$$
$$0 = w^2 - 2w - 63$$
$$0 = (w - 9)(w + 7)$$

$$w - 9 = 0 \quad \text{or} \quad w + 7 = 0$$
$$w = 9 \quad \text{or} \qquad w = -7$$

A box cannot have a negative width, so reject -7 as a solution. The only possible solution is 9.
The piece of cardboard has width 9 inches and length $9 + 6 = 15$ inches.

65. When the object hits the ground, its height is given by $f(t) = 0$. Let $f(t) = 0$ in the given equation, and solve for t.

$$-16t^2 + 64t + 80 = f(t)$$
$$-16t^2 + 64t + 80 = 0$$
$$t^2 - 4t - 5 = 0 \quad \textit{Divide by -16.}$$
$$(t - 5)(t + 1) = 0$$

$$t - 5 = 0 \quad \text{or} \quad t + 1 = 0$$
$$t = 5 \quad \text{or} \qquad t = -1$$

Time cannot be negative, so reject -1 as a solution. The only possible solution is 5.
The object will hit the ground 5 seconds after it is thrown.

67. Use $f(t) = -16t^2 + 625$ with $f(t) = 0$.

$$0 = -16t^2 + 625$$
$$0 = 16t^2 - 625 \quad \textit{Multiply by -1.}$$
$$0 = (4t + 25)(4t - 25)$$

$$4t + 25 = 0 \quad \text{or} \quad 4t - 25 = 0$$
$$4t = -25 \qquad\qquad 4t = 25$$
$$t = -\frac{25}{4} \quad \text{or} \qquad t = \frac{25}{4}$$

Time cannot be negative, so reject $-\frac{25}{4}$ as a solution. The only possible solution is $\frac{25}{4}$ or $6\frac{1}{4}$.
The ball will hit the ground after $6\frac{1}{4}$ seconds.

69. Solve $2k + ar = r - 3y$ for r.

Get the "r-terms" on one side and the other terms on the other side.

$$ar - r = -2k - 3y$$
$$(a - 1)r = -2k - 3y \quad \textit{Distributive property}$$
$$r = \frac{-2k - 3y}{a - 1} \quad \textit{Divide by $a - 1$.}$$

The answer can also be written as

$$r = \frac{2k + 3y}{1 - a},$$

which would occur if you took the r-terms to the right side in your first step.

Copyright © 2012 Pearson Education, Inc. Publishing as Addison-Wesley.

71. Solve $w = \dfrac{3y - x}{y}$ for y.

$$wy = 3y - x \qquad \text{Multiply by } y.$$

$$x = 3y - wy \qquad \begin{array}{l}\text{Get } y\text{-terms} \\ \text{on one side.}\end{array}$$

$$x = (3 - w)y \qquad \begin{array}{l}\text{Distributive} \\ \text{property}\end{array}$$

$$\frac{x}{3 - w} = y \qquad \text{Divide by } 3 - w.$$

Equivalently, we have

$$y = \frac{-x}{w - 3}.$$

73. The formula is not solved for L because L should not appear on each side of the final equation. Solve for L as follows:

$$\mathcal{A} = 2HW + 2LW + 2LH$$
$$\mathcal{A} - 2HW = 2LW + 2LH$$
$$\mathcal{A} - 2HW = L(2W + 2H)$$
$$\frac{\mathcal{A} - 2HW}{2W + 2H} = L$$

75.
$$2x^2 - 7x - 4 = 0$$
$$(2x + 1)(x - 4) = 0$$

$$2x + 1 = 0 \quad \text{or} \quad x - 4 = 0$$
$$2x = -1 \qquad\qquad x = 4$$
$$x = -\tfrac{1}{2}$$

These values correspond to the zeros shown on the screens.

The solution set is $\{-0.5, 4\}$.

♦ ♦ ♦ Relating Concepts 77–82 ♦ ♦ ♦

77. With $v_0 = 80$ and $s_0 = 100$, the function $f(x) = -16x^2 + v_0 x + s_0$ becomes $f(x) = -16x^2 + 80x + 100$.

78.

$$f(x) = -16x^2 + 80x + 100$$

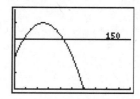

79. The maximum height appears to be 200 feet after about 2.5 seconds.

80. The graph intersects the x-axis at about 6, which means the ball reaches the ground in about 6 seconds.

Check $f(6) = -576 + 480 + 100 = 4$, which is close to 0, so the actual time must be slightly greater than 6 seconds.

81. The graph of f and $g(x) = 150$ indicates that the height of the ball is greater than 150 during the time interval $(0.7, 4.3)$.

Check $f(0.7) = -7.84 + 56 + 100 = 148.16$
Check $f(4.3) = -295.84 + 344 + 100 = 148.16$

The values are close to 150, which confirms our estimate of $(0.7, 4.3)$.

82. With $v_0 = 100$ and $s_0 = 0$, the function $f(x) = -16x^2 + v_0 x + s_0$ becomes $f(x) = -16x^2 + 100x$.

83. $\dfrac{12p^2}{3p} = \dfrac{12}{3} p^{2-1} = 4p$

85. $\dfrac{-27m^2n^5}{36m^6n^8} = -\dfrac{3(9)}{4(9)} m^{2-6}n^{5-8} = -\dfrac{3}{4}m^{-4}n^{-3}$

$$= -\dfrac{3}{4m^4n^3}$$

87. $\dfrac{12}{25} = \dfrac{?}{75} \qquad 25(3) = 75$

$$\frac{12}{25} = \frac{12(3)}{25(3)} = \frac{36}{75}$$

Chapter 6 Review Exercises

1. $12p^2 - 6p = 6p(2p - 1)$

2. $21x^2 + 35x = 7x(3x + 5)$

3. $12q^2b + 8qb^2 - 20q^3b^2$
   $= 4qb(3q + 2b - 5q^2b)$

4. $6r^3t - 30r^2t^2 + 18rt^3$
   $= 6rt(r^2 - 5rt + 3t^2)$

5. $(x + 3)(4x - 1) - (x + 3)(3x + 2)$

The GCF is $(x + 3)$.

$= (x + 3)[(4x - 1) - (3x + 2)]$
$= (x + 3)(4x - 1 - 3x - 2)$
$= (x + 3)(x - 3)$

6. $(z + 1)(z - 4) + (z + 1)(2z + 3)$
   $= (z + 1)[(z - 4) + (2z + 3)]$
   $= (z + 1)(3z - 1)$

Copyright © 2012 Pearson Education, Inc. Publishing as Addison-Wesley.

7. $4m + nq + mn + 4q$

Rearrange the terms.

$$= 4m + 4q + mn + nq$$
$$= 4(m + q) + n(m + q)$$
$$= (m + q)(4 + n)$$

8. $x^2 + 5y + 5x + xy$
$$= x^2 + xy + 5x + 5y$$
$$= x(x + y) + 5(x + y)$$
$$= (x + y)(x + 5)$$

9. $2m + 6 - am - 3a$
$$= 2(m + 3) - a(m + 3)$$
$$= (m + 3)(2 - a)$$

10. $x^2 + 3x - 3y - xy$
$$= (x^2 + 3x) + (-3y - xy)$$
$$= x(x + 3) - y(3 + x)$$
$$= x(x + 3) - y(x + 3)$$
$$= (x + 3)(x - y)$$

11. $3p^2 - p - 4$

Two integer factors whose product is $(3)(-4) = -12$ and whose sum is -1 are -4 and 3.
$$= 3p^2 - 4p + 3p - 4$$
$$= p(3p - 4) + 1(3p - 4)$$
$$= (3p - 4)(p + 1)$$

12. $6k^2 + 11k - 10$

Two integer factors whose product is $(6)(-10) = -60$ and whose sum is 11 are 15 and -4.
$$= 6k^2 + 15k - 4k - 10$$
$$= 3k(2k + 5) - 2(2k + 5)$$
$$= (2k + 5)(3k - 2)$$

13. $12r^2 - 5r - 3$

Two integer factors whose product is $(12)(-3) = -36$ and whose sum is -5 are -9 and 4.
$$= 12r^2 - 9r + 4r - 3$$
$$= 3r(4r - 3) + 1(4r - 3)$$
$$= (4r - 3)(3r + 1)$$

14. $10m^2 + 37m + 30$

Two integer factors whose product is $(10)(30) = 300$ and whose sum is 37 are 12 and 25.
$$= 10m^2 + 12m + 25m + 30$$
$$= 2m(5m + 6) + 5(5m + 6)$$
$$= (5m + 6)(2m + 5)$$

15. $10k^2 - 11kh + 3h^2$

Two integer factors whose product is $(10)(3) = 30$ and whose sum is -11 are -6 and -5.
$$= 10k^2 - 6kh - 5kh + 3h^2$$
$$= 2k(5k - 3h) - h(5k - 3h)$$
$$= (5k - 3h)(2k - h)$$

16. $9x^2 + 4xy - 2y^2$

There are no integers that have a product of $9(-2) = -18$ and a sum of 4. Therefore, the trinomial cannot be factored and is *prime*.

17. $24x - 2x^2 - 2x^3$
$$= 2x(12 - x - x^2)$$
$$= 2x(4 + x)(3 - x)$$

18. $6b^3 - 9b^2 - 15b$
$$= 3b(2b^2 - 3b - 5)$$
$$= 3b(2b - 5)(b + 1)$$

19. $y^4 + 2y^2 - 8$
$$= (y^2)^2 + 2y^2 - 8$$
$$= (y^2 + 4)(y^2 - 2)$$

20. $2k^4 - 5k^2 - 3$
$$= 2(k^2)^2 - 5k^2 - 3$$
$$= (2k^2 + 1)(k^2 - 3)$$

21. $p^2(p + 2)^2 + p(p + 2)^2 - 6(p + 2)^2$
Factor out $(p + 2)^2$.
$$= (p + 2)^2(p^2 + p - 6)$$
$$= (p + 2)^2(p + 3)(p - 2)$$

22. $3(r + 5)^2 - 11(r + 5) - 4$
$$= 3x^2 - 11x - 4 \qquad \textit{Let } x = r + 5.$$
$$= (3x + 1)(x - 4)$$
$$= [3(r + 5) + 1][(r + 5) - 4] \quad \textit{Resubstitute.}$$
$$= (3r + 15 + 1)(r + 1)$$
$$= (3r + 16)(r + 1)$$

23. The student's answer is
$$x^2y^2 - 6x^2 + 5y^2 - 30$$
$$= x^2(y^2 - 6) + 5(y^2 - 6).$$

This is incorrect because the polynomial still has two terms, so it is not factored. The correct answer is
$$(y^2 - 6)(x^2 + 5).$$

24. Since area equals length times width, factor the polynomial given for the area.
$$4p^2 + 3p - 1 = (4p - 1)(p + 1)$$

Since the length is given as $4p - 1$, the width is $p + 1$.

Copyright © 2012 Pearson Education, Inc. Publishing as Addison-Wesley.

25. $16x^2 - 25 = (4x)^2 - 5^2$ *Difference of squares*

$= (4x + 5)(4x - 5)$

26. $9t^2 - 49 = (3t)^2 - 7^2$

$= (3t + 7)(3t - 7)$

27. $36m^2 - 25n^2 = (6m)^2 - (5n)^2$

$= (6m + 5n)(6m - 5n)$

28. $x^2 + 14x + 49 = x^2 + 2(x)(7) + 7^2$

Perfect square trinomial

$= (x + 7)^2$

29. $9k^2 - 12k + 4 = (3k)^2 - 2(3k)(2) + 2^2$

$= (3k - 2)^2$

30. $r^3 + 27 = r^3 + 3^3$ *Sum of cubes*

$= (r + 3)(r^2 - 3r + 9)$

31. $125x^3 - 1 = (5x)^3 - 1^3$ *Difference of cubes*

$= (5x - 1)(25x^2 + 5x + 1)$

32. $m^6 - 1 = (m^3)^2 - 1^2$

Difference of squares

$= (m^3 + 1)(m^3 - 1)$

$= (m^3 + 1^3)(m^3 - 1^3)$

Sum of   Difference
cubes;   of cubes

$= (m + 1)(m^2 - m + 1)$

$\cdot (m - 1)(m^2 + m + 1)$

33. $x^8 - 1 = (x^4)^2 - 1^2$

Difference of squares

$= (x^4 + 1)(x^4 - 1)$

$= (x^4 + 1)[(x^2)^2 - 1^2]$

Difference of squares again

$= (x^4 + 1)(x^2 + 1)(x^2 - 1)$

Difference of squares again

$= (x^4 + 1)(x^2 + 1)(x + 1)(x - 1)$

34. $x^2 + 6x + 9 - 25y^2$

$= (x^2 + 6x + 9) - 25y^2$

$= [x^2 + 2(x)(3) + 3^2] - 25y^2$

Perfect square trinomial

$= (x + 3)^2 - (5y)^2$

Difference of squares

$= [(x + 3) + 5y][(x + 3) - 5y]$

$= (x + 3 + 5y)(x + 3 - 5y)$

35. $(a + b)^3 - (a - b)^3$

Difference of cubes

$= [(a + b) - (a - b)]$

$\cdot [(a + b)^2 + (a + b)(a - b)$

$+ (a - b)^2]$

$= [2b](a^2 + 2ab + b^2 + a^2 - b^2$

$+ a^2 - 2ab + b^2)$

$= 2b(3a^2 + b^2)$

36. $x^5 - x^3 - 8x^2 + 8$

$= x^3(x^2 - 1) - 8(x^2 - 1)$

$= (x^2 - 1)(x^3 - 8)$

$= (x^2 - 1^2)(x^3 - 2^3)$

Difference   Difference
of squares;   of cubes

$= (x + 1)(x - 1)(x - 2)(x^2 + 2x + 4)$

37. $x^2 - 8x + 16 = 0$

$(x - 4)(x - 4) = 0$

$x - 4 = 0$

$x = 4$

The solution set is $\{4\}$.

38. $(5x + 2)(x + 1) = 0$

$5x + 2 = 0$ or $x + 1 = 0$

$5x = -2$ $x = -1$

$x = -\frac{2}{5}$

The solution set is $\left\{-1, -\frac{2}{5}\right\}$.

39. $x^2 - 5x + 6 = 0$

$(x - 2)(x - 3) = 0$

$x - 2 = 0$ or $x - 3 = 0$

$x = 2$ or $x = 3$

The solution set is $\{2, 3\}$.

40. $x^2 + 2x = 8$

$x^2 + 2x - 8 = 0$

$(x + 4)(x - 2) = 0$

$x + 4 = 0$ or $x - 2 = 0$

$x = -4$ or $x = 2$

The solution set is $\{-4, 2\}$.

41. $6x^2 = 5x + 50$

$6x^2 - 5x - 50 = 0$

$(3x - 10)(2x + 5) = 0$

$3x - 10 = 0$ or $2x + 5 = 0$

$3x = 10$ $2x = -5$

$x = \frac{10}{3}$ or $x = -\frac{5}{2}$

The solution set is $\left\{-\frac{5}{2}, \frac{10}{3}\right\}$.

Copyright © 2012 Pearson Education, Inc. Publishing as Addison-Wesley.

42.
$$6x^2 + 7x = 3$$
$$6x^2 + 7x - 3 = 0$$
$$(2x + 3)(3x - 1) = 0$$

$2x + 3 = 0$ or $3x - 1 = 0$
$2x = -3$ $3x = 1$
$x = -\frac{3}{2}$ or $x = \frac{1}{3}$

The solution set is $\left\{-\frac{3}{2}, \frac{1}{3}\right\}$.

43.
$$8x^2 + 14x + 3 = 0$$
$$(2x + 3)(4x + 1) = 0$$

$2x + 3 = 0$ or $4x + 1 = 0$
$2x = -3$ $4x = -1$
$x = -\frac{3}{2}$ or $x = -\frac{1}{4}$

The solution set is $\left\{-\frac{3}{2}, -\frac{1}{4}\right\}$.

44.
$$-4x^2 + 36 = 0$$
$$x^2 - 9 = 0 \quad \textit{Divide by -4.}$$
$$(x + 3)(x - 3) = 0$$

$x + 3 = 0$ or $x - 3 = 0$
$x = -3$ or $x = 3$

The solution set is $\{-3, 3\}$.

45.
$$6x^2 + 9x = 0$$
$$3x(2x + 3) = 0$$

$3x = 0$ or $2x + 3 = 0$
$x = 0$ $2x = -3$
$x = -\frac{3}{2}$

The solution set is $\left\{-\frac{3}{2}, 0\right\}$.

46.
$$(2x + 1)(x - 2) = -3$$
$$2x^2 - 3x - 2 = -3$$
$$2x^2 - 3x + 1 = 0$$
$$(2x - 1)(x - 1) = 0$$

$2x - 1 = 0$ or $x - 1 = 0$
$2x = 1$ $x = 1$
$x = \frac{1}{2}$

The solution set is $\left\{\frac{1}{2}, 1\right\}$.

47.
$$(x + 2)(x - 2) = (x - 2)(x + 3) - 2$$
$$x^2 - 4 = x^2 + x - 6 - 2$$
$$-4 = x - 8$$
$$4 = x$$

The solution set is $\{4\}$.

48.
$$2x^3 - x^2 - 28x = 0$$
$$x(2x^2 - x - 28) = 0$$
$$x(2x + 7)(x - 4) = 0$$

$x = 0$ or $2x + 7 = 0$ or $x - 4 = 0$
$2x = -7$ $x = 4$
$x = -\frac{7}{2}$

The solution set is $\left\{-\frac{7}{2}, 0, 4\right\}$.

49.
$$-x^3 - 3x^2 + 4x + 12 = 0$$
$$x^3 + 3x^2 - 4x - 12 = 0 \quad \textit{Multiply by -1.}$$
$$x^2(x + 3) - 4(x + 3) = 0$$
$$(x + 3)(x^2 - 4) = 0$$
$$(x + 3)(x + 2)(x - 2) = 0$$

$x + 3 = 0$ or $x + 2 = 0$ or $x - 2 = 0$
$x = -3$ or $x = -2$ or $x = 2$

The solution set is $\{-3, -2, 2\}$.

50.
$$(x + 2)(5x^2 - 9x - 18) = 0$$
$$(x + 2)(5x + 6)(x - 3) = 0$$

$x + 2 = 0$ or $5x + 6 = 0$ or $x - 3 = 0$
$x = -2$ $5x = -6$ $x = 3$
$x = -\frac{6}{5}$

The solution set is $\left\{-2, -\frac{6}{5}, 3\right\}$.

51. Let x be the length of the shorter side. Then the length of the longer side will be $2x + 1$. The area is 10.5 ft^2. Use the formula for area of a triangle, $A = \frac{1}{2}bh$.

$$\frac{1}{2}(2x + 1)(x) = 10.5$$
$$x(2x + 1) = 21 \quad \textit{Multiply by 2.}$$
$$2x^2 + x = 21$$
$$2x^2 + x - 21 = 0$$
$$(2x + 7)(x - 3) = 0$$

$2x + 7 = 0$ or $x - 3 = 0$
$2x = -7$ $x = 3$
$x = -\frac{7}{2}$

The side cannot have a negative length, so reject $x = -\frac{7}{2}$.
The length of the shorter side is 3 feet.

52. Let w be the width of the lot. Then $w + 20$ will be the length of the lot. The area is 2400 ft^2. Use the formula $LW = A$.

$$(w + 20)w = 2400$$
$$w^2 + 20w = 2400$$
$$w^2 + 20w - 2400 = 0$$
$$(w - 40)(w + 60) = 0$$

$w - 40 = 0$ or $w + 60 = 0$
$w = 40$ or $w = -60$

The lot cannot have a negative width, so reject $w = -60$.
The width of the lot is 40 feet, and the length is $40 + 20 = 60$ feet.

Copyright © 2012 Pearson Education, Inc. Publishing as Addison-Wesley.

53. The height is 0 when the rock returns to the ground.

$$f(t) = -16t^2 + 256t$$
$$0 = -16t^2 + 256t$$
$$0 = -16t(t - 16)$$

$$-16t = 0 \quad \text{or} \quad t - 16 = 0$$
$$t = 0 \quad \text{or} \quad t = 16$$

The rock is on the ground when $t = 0$. It will return to the ground again after 16 seconds.

54. $f(t) = -16t^2 + 256t$
$$240 = -16t^2 + 256t \qquad \textit{Let f(t) = 240.}$$
$$0 = -16t^2 + 256t - 240$$
$$0 = -16(t^2 - 16t + 15)$$
$$0 = -16(t - 15)(t - 1)$$

$$t - 15 = 0 \quad \text{or} \quad t - 1 = 0$$
$$t = 15 \quad \text{or} \quad t = 1$$

The rock will be 240 ft above the ground after 1 second and again after 15 seconds.

55. The question in Exercise 54 has two answers because the rock will be 240 ft above the ground after 1 second on the way up and again after 15 seconds on the way back down.

56. $f(t) = -16t^2 + 256t$
$$1024 = -16t^2 + 256t \qquad \textit{Let f(t) = 1024.}$$
$$0 = -16t^2 + 256t - 1024$$
$$0 = -16(t^2 - 16t + 64)$$
$$0 = -16(t - 8)^2$$

$$t - 8 = 0$$
$$t = 8$$

The rock will be 1024 ft above the ground after 8 seconds.

57. Solve $3s + bk = k - 2t$ for k.

$$3s + 2t = k - bk$$
$$3s + 2t = k(1 - b)$$
$$k = \frac{3s + 2t}{1 - b}, \text{ or } p = \frac{-3s - 2t}{b - 1}$$

58. Solve $z = \dfrac{3w + 7}{w}$ for w.

$$zw = 3w + 7 \qquad \textit{Multiply by w.}$$
$$zw - 3w = 7 \qquad \textit{Get y-terms on one side.}$$
$$w(z - 3) = 7 \qquad \textit{Distributive property}$$
$$w = \frac{7}{z - 3} \qquad \textit{Divide by z - 3.}$$

Equivalently, we have

$$w = \frac{-7}{3 - z}.$$

59. **[6.3]** $16 - 81k^2 = 4^2 - (9k)^2$
$$= (4 + 9k)(4 - 9k)$$

60. **[6.2]** $30a + am - am^2$
$$= a(30 + m - m^2)$$
$$= a(6 - m)(5 + m)$$

61. **[6.2]** $9x^2 + 13xy - 3y^2$ is *prime* since it cannot be factored further.

62. **[6.3]** $8 - a^3$
$$= 2^3 - a^3$$
$$= (2 - a)(2^2 + 2a + a^2)$$
$$= (2 - a)(4 + 2a + a^2)$$

63. **[6.3]** $25z^2 - 30zm + 9m^2$
$$= (5z)^2 - 2(5z)(3m) + (3m)^2$$
$$= (5z - 3m)^2$$

64. **[6.1]** $15y^3 + 20y^2 = 5y^2(3y + 4)$

65. **[6.5]** $$5x^2 - 17x = 12$$
$$5x^2 - 17x - 12 = 0$$
$$(5x + 3)(x - 4) = 0$$

$$5x + 3 = 0 \quad \text{or} \quad x - 4 = 0$$
$$5x = -3 \qquad\qquad x = 4$$
$$x = -\tfrac{3}{5}$$

The solution set is $\left\{-\tfrac{3}{5}, 4\right\}$.

66. **[6.5]** $$x^3 - x = 0$$
$$x(x^2 - 1) = 0$$
$$x(x + 1)(x - 1) = 0$$

$$x = 0 \quad \text{or} \quad x + 1 = 0 \quad \text{or} \quad x - 1 = 0$$
$$x = -1 \qquad\qquad x = 1$$

The solution set is $\{-1, 0, 1\}$.

67. **[6.5]** Let x be the width of the frame. Then $x + 2$ will be the length of the frame. The area is 48 in². Use the formula $LW = A$.

$$(x + 2)x = 48$$
$$x^2 + 2x = 48$$
$$x^2 + 2x - 48 = 0$$
$$(x + 8)(x - 6) = 0$$

$$x + 8 = 0 \quad \text{or} \quad x - 6 = 0$$
$$x = -8 \quad \text{or} \quad x = 6$$

The frame cannot have a negative width, so reject $x = -8$.
The width of the frame is 6 inches.

Copyright © 2012 Pearson Education, Inc. Publishing as Addison-Wesley.

68. **[6.5]** Let x be the width of the floor. Then $x + 85$ will be the length of the floor. The area is 2750 ft^2. Use the formula $LW = A$.

$$(x + 85)x = 2750$$
$$x^2 + 85x = 2750$$
$$x^2 + 85x - 2750 = 0$$
$$(x + 110)(x - 25) = 0$$

$$x + 110 = 0 \qquad \text{or} \quad x - 25 = 0$$
$$x = -110 \quad \text{or} \qquad x = 25$$

The floor cannot have a negative width, so reject $x = -110$.
The width of the floor was 25 feet and the length was $25 + 85 = 110$ feet.

Chapter 6 Test

1. $11z^2 - 44z = 11z(z - 4)$

2. $10x^2y^5 - 5x^2y^3 - 25x^5y^3$

Factor out the GCF, $5x^2y^3$.

$$= 5x^2y^3(2y^2 - 1 - 5x^3)$$

3. $3x + by + bx + 3y$
$$= 3x + 3y + bx + by$$
$$= 3(x + y) + b(x + y)$$
$$= (x + y)(3 + b)$$

4. $-2x^2 - x + 36 = -1(2x^2 + x - 36)$

Two integer factors whose product is $(2)(-36) = -72$ and whose sum is 1 are 9 and -8.

$$2x^2 + x - 36 = 2x^2 + 9x - 8x - 36$$
$$= x(2x + 9) - 4(2x + 9)$$
$$= (2x + 9)(x - 4)$$

Thus, the final factored form is

$$-(2x + 9)(x - 4).$$

5. $6x^2 + 11x - 35$

Two integer factors whose product is $(6)(-35) = -210$ and whose sum is 11 are 21 and -10.

$$= 6x^2 + 21x - 10x - 35$$
$$= 3x(2x + 7) - 5(2x + 7)$$
$$= (2x + 7)(3x - 5)$$

6. $4p^2 + 3pq - q^2$

Two integer factors whose product is $(4)(-1) = -4$ and whose sum is 3 are 4 and -1.

$$= 4p^2 + 4pq - pq - q^2$$
$$= 4p(p + q) - q(p + q)$$
$$= (p + q)(4p - q)$$

7. $16a^2 + 40ab + 25b^2$
$$= (4a)^2 + 2(4a)(5b) + (5b)^2$$
$$= (4a + 5b)^2$$

8. $x^2 + 2x + 1 - 4z^2$
$$= (x^2 + 2x + 1) - 4z^2$$
$$= (x + 1)^2 - (2z)^2$$
$$= [(x + 1) + 2z][(x + 1) - 2z]$$
$$= (x + 1 + 2z)(x + 1 - 2z)$$

9. $a^3 + 2a^2 - ab^2 - 2b^2$
$$= a^2(a + 2) - b^2(a + 2)$$
$$= (a + 2)(a^2 - b^2)$$
$$= (a + 2)(a + b)(a - b)$$

10. $9k^2 - 121j^2 = (3k)^2 - (11j)^2$
$$= (3k + 11j)(3k - 11j)$$

11. $y^3 - 216 = y^3 - 6^3$
$$= (y - 6)(y^2 + 6y + 6^2)$$
$$= (y - 6)(y^2 + 6y + 36)$$

12. $6k^4 - k^2 - 35 = 6(k^2)^2 - k^2 - 35$

Two integer factors whose product is $(6)(-35) = -210$ and whose sum is -1 are -15 and 14.

$$= 6(k^2)^2 - 15k^2 + 14k^2 - 35$$
$$= 3k^2(2k^2 - 5) + 7(2k^2 - 5)$$
$$= (2k^2 - 5)(3k^2 + 7)$$

13. $27x^6 + 1$
$$= (3x^2)^3 + (1)^3$$
$$= (3x^2 + 1)[(3x^2)^2 - (3x^2)(1) + 1^2]$$
$$= (3x^2 + 1)(9x^4 - 3x^2 + 1)$$

14. **A.** $(3 - x)(x + 4)$
$$= 3x + 12 - x^2 - 4x$$
$$= -x^2 - x + 12$$

B. $-(x - 3)(x + 4)$
$$= -(x^2 + 4x - 3x - 12)$$
$$= -(x^2 + x - 12)$$
$$= -x^2 - x + 12$$

C. $(-x + 3)(x + 4)$
$$= -x^2 - 4x + 3x + 12$$
$$= -x^2 - x + 12$$

D. $(x - 3)(-x + 4)$
$$= -x^2 + 4x + 3x - 12$$
$$= -x^2 + 7x - 12$$

Therefore, only **D** is *not* a factored form of $-x^2 - x + 12$.

15.
$$3x^2 + 8x = -4$$
$$3x^2 + 8x + 4 = 0$$
$$(x + 2)(3x + 2) = 0$$

$$x + 2 = 0 \quad \text{or} \quad 3x + 2 = 0$$
$$x = -2 \qquad\qquad 3x = -2$$
$$x = -\tfrac{2}{3}$$

The solution set is $\left\{-2, -\tfrac{2}{3}\right\}$.

16. $\quad 3x^2 - 5x = 0$
$$x(3x - 5) = 0$$

$$x = 0 \quad \text{or} \quad 3x - 5 = 0$$
$$3x = 5$$
$$x = \tfrac{5}{3}$$

The solution set is $\left\{0, \tfrac{5}{3}\right\}$.

17.
$$5m(m - 1) = 2(1 - m)$$
$$5m^2 - 5m = 2 - 2m$$
$$5m^2 - 3m - 2 = 0$$
$$(5m + 2)(m - 1) = 0$$

$$5m + 2 = 0 \quad \text{or} \quad m - 1 = 0$$
$$5m = -2 \qquad\qquad m = 1$$
$$m = -\tfrac{2}{5}$$

The solution set is $\left\{-\tfrac{2}{5}, 1\right\}$.

18. Solve $ar + 2 = 3r - 6t$ for r.

$$ar - 3r = -2 - 6t$$
$$r(a - 3) = -2 - 6t$$
$$r = \frac{-2 - 6t}{a - 3}, \quad \text{or} \quad r = \frac{2 + 6t}{3 - a}$$

19. Using $A = LW$, substitute 40 for A, $x + 7$ for L, and $2x + 3$ for W.

$$A = LW$$
$$40 = (x + 7)(2x + 3)$$
$$40 = 2x^2 + 3x + 14x + 21$$
$$40 = 2x^2 + 17x + 21$$
$$0 = 2x^2 + 17x - 19$$
$$0 = (2x + 19)(x - 1)$$

$$2x + 19 = 0 \quad \text{or} \quad x - 1 = 0$$
$$2x = -19 \qquad\qquad x = 1$$
$$x = -\tfrac{19}{2}$$

Length and width will be negative if $x = -\tfrac{19}{2}$, so reject it as a possible solution. If $x = 1$, then

$$x + 7 = 1 + 7 = 8$$
and
$$2x + 3 = 2(1) + 3 = 5.$$

The length is 8 inches, and the width is 5 inches.

20. Substitute 128 for $f(t)$ in the equation.

$$f(t) = -16t^2 + 96t$$
$$128 = -16t^2 + 96t$$
$$16t^2 - 96t + 128 = 0$$
$$16(t^2 - 6t + 8) = 0$$
$$16(t - 4)(t - 2) = 0$$

$$t - 4 = 0 \quad \text{or} \quad t - 2 = 0$$
$$t = 4 \quad \text{or} \qquad t = 2$$

The ball is 128 ft high at 2 seconds (on the way up) and again at 4 seconds (on the way down).

Cumulative Review Exercises (Chapters 1–6)

1. $\quad -2(m - 3) = -2(m) - 2(-3) = -2m + 6$

2. $\quad 3x^2 - 4x + 4 + 9x - x^2$
$$= 3x^2 - x^2 - 4x + 9x + 4$$
$$= 2x^2 + 5x + 4$$

For Exercises 3–4, let $p = -4, q = -2$, and $r = 5$.

3.
$$\frac{5p + 6r^2}{p^2 + q - 1} = \frac{5(-4) + 6(5)^2}{(-4)^2 + (-2) - 1}$$
$$= \frac{-20 + 6(25)}{16 - 2 - 1}$$
$$= \frac{-20 + 150}{13}$$
$$= \tfrac{130}{13} = 10$$

4.
$$\frac{\sqrt{r}}{-p + 2q} = \frac{\sqrt{5}}{-(-4) + 2(-2)}$$
$$= \frac{\sqrt{5}}{4 - 4} = \frac{\sqrt{5}}{0}$$

This is *undefined* since the denominator is zero.

5. $\quad 2x - 5 + 3x = 4 - (x + 2)$
$$5x - 5 = 4 - x - 2$$
$$5x - 5 = 2 - x$$
$$6x = 7$$
$$x = \tfrac{7}{6}$$

The solution set is $\left\{\tfrac{7}{6}\right\}$.

6.
$$\frac{3x - 1}{5} + \frac{x + 2}{2} = -\frac{3}{10}$$
$$2(3x - 1) + 5(x + 2) = -3 \qquad \textit{Multiply by 10.}$$
$$6x - 2 + 5x + 10 = -3$$
$$11x + 8 = -3$$
$$11x = -11$$
$$x = -1$$

The solution set is $\{-1\}$.

Copyright © 2012 Pearson Education, Inc. Publishing as Addison-Wesley.

7. $3 - 2(x + 3) < 4x$

$3 - 2x - 6 < 4x$

$-2x - 3 < 4x$

$-6x < 3$

$x > -\frac{3}{6} \text{ or } -\frac{1}{2}$

The solution set is $\left(-\frac{1}{2}, \infty\right)$.

8. $2x + 4 < 10 \quad \text{and} \quad 3x - 1 > 5$

$2x < 6 \qquad\qquad 3x > 6$

$x < 3 \quad \text{and} \qquad x > 2$

The overlap of these inequalities is the set of all numbers between 2 and 3.

The solution set is $(2, 3)$.

9. $2x + 4 > 10 \quad \text{or} \quad 3x - 1 < 5$

$2x > 6 \qquad\qquad 3x < 6$

$x > 3 \quad \text{or} \qquad x < 2$

The solution set is the set of numbers that are either greater than 3 or less than 2.

The solution set is $(-\infty, 2) \cup (3, \infty)$.

10. $|5x + 3| - 10 = 3$

$|5x + 3| = 13$

$5x + 3 = 13 \quad \text{or} \quad 5x + 3 = -13$

$5x = 10 \qquad\qquad 5x = -16$

$x = 2 \quad \text{or} \qquad x = -\frac{16}{5}$

The solution set is $\left\{-\frac{16}{5}, 2\right\}$.

11. $|x + 2| < 9$

$-9 < x + 2 < 9$

$-11 < x < 7$

The solution set is $(-11, 7)$.

12. $|2x - 5| \geq 9$

$2x - 5 \geq 9 \quad \text{or} \quad 2x - 5 \leq -9$

$2x \geq 14 \qquad\qquad 2x \leq -4$

$x \geq 7 \quad \text{or} \qquad x \leq -2$

The solution set is $(-\infty, -2] \cup [7, \infty)$.

13. Let x be the time it takes for the planes to be 2100 mi apart. Use the formula $d = rt$ to complete the table.

Plane	r	t	d
Eastbound	550	x	$550x$
Westbound	500	x	$500x$

The total distance is 2100 mi.

$550x + 500x = 2100$

$1050x = 2100$

$x = 2$

It will take 2 hours for the planes to be 2100 miles apart.

14. $4x + 2y = -8$

Draw a line through the x- and y-intercepts, $(-2, 0)$ and $(0, -4)$, respectively.

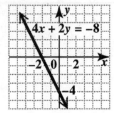

15. The slope m of the line through $(-4, 8)$ and $(-2, 6)$ is

$$m = \frac{6 - 8}{-2 - (-4)} = \frac{-2}{2} = -1.$$

16. $y = -3$ is an equation of a horizontal line. Its slope is 0.

17. $f(x) = 2x + 7$

$f(-4) = 2(-4) + 7$

$= -8 + 7 = -1$

18. To find the x-intercept of the graph of $f(x) = 2x + 7$, let $f(x) = 0$ (which is the same as letting $y = 0$) and solve for x.

$$0 = 2x + 7$$

$$-7 = 2x$$

$$-\frac{7}{2} = x$$

The x-intercept is $\left(-\frac{7}{2}, 0\right)$.

19. $f(0) = 7$, so the y-intercept is $(0, 7)$.

20. $3x - 2y = -7 \quad (1)$

$2x + 3y = 17 \quad (2)$

To eliminate y, multiply (1) by 3 and (2) by 2, and then add the resulting equations.

$$
\begin{array}{rcll}
9x - 6y &=& -21 & 3 \times (1) \\
4x + 6y &=& 34 & 2 \times (2) \\
\hline
13x &=& 13 & \\
x &=& 1 &
\end{array}
$$

Substitute 1 for x in (1).

$$
\begin{array}{rcll}
3x - 2y &=& -7 & (1) \\
3(1) - 2y &=& -7 & \\
-2y &=& -10 & \\
y &=& 5 &
\end{array}
$$

The solution set is $\{(1, 5)\}$.

Copyright © 2012 Pearson Education, Inc. Publishing as Addison-Wesley.

21.
$$\begin{aligned} 2x + 3y - 6z &= 5 \quad (1) \\ 8x - y + 3z &= 7 \quad (2) \\ 3x + 4y - 3z &= 7 \quad (3) \end{aligned}$$

To eliminate z, add (2) and (3).

$$\begin{aligned} 8x - y + 3z &= 7 \quad (2) \\ 3x + 4y - 3z &= 7 \quad (3) \\ \hline 11x + 3y &= 14 \quad (4) \end{aligned}$$

To eliminate z again, multiply (2) by 2 and add the result to (1).

$$\begin{aligned} 2x + 3y - 6z &= 5 \quad (1) \\ 16x - 2y + 6z &= 14 \quad 2 \times (2) \\ \hline 18x + y &= 19 \quad (5) \end{aligned}$$

To eliminate y, multiply (5) by -3 and add the result to (4).

$$\begin{aligned} 11x + 3y &= 14 \quad (4) \\ -54x - 3y &= -57 \quad -3 \times (5) \\ \hline -43x &= -43 \\ x &= 1 \end{aligned}$$

From (5), $18(1) + y = 19$, so $y = 1$.
To find z, let $x = 1$ and $y = 1$ in (1).

$$\begin{aligned} 2x + 3y - 6z &= 5 \quad (1) \\ 2(1) + 3(1) - 6z &= 5 \\ 5 - 6z &= 5 \\ -6z &= 0 \\ z &= 0 \end{aligned}$$

The solution set is $\{(1, 1, 0)\}$.

22. $(3x^2y^{-1})^{-2}(2x^{-3}y)^{-1}$
$$= 3^{-2}x^{-4}y^2 \cdot 2^{-1}x^3y^{-1}$$
$$= 3^{-2}2^{-1}x^{-1}y$$
$$= \frac{y}{3^2 \cdot 2x} = \frac{y}{18x}$$

23. $(7x + 3y)^2 = (7x)^2 + 2(7x)(3y) + (3y)^2$
$$= 49x^2 + 42xy + 9y^2$$

24. $(3x^3 + 4x^2 - 7) - (2x^3 - 8x^2 + 3x)$
$$= 3x^3 + 4x^2 - 7 - 2x^3 + 8x^2 - 3x$$
$$= x^3 + 12x^2 - 3x - 7$$

25. $16w^2 + 50wz - 21z^2$

Two integer factors whose product is $(16)(-21) = -336$ and whose sum is 50 are 56 and -6.

$$= 16w^2 + 56wz - 6wz - 21z^2$$
$$= 8w(2w + 7z) - 3z(2w + 7z)$$
$$= (2w + 7z)(8w - 3z)$$

26. $4x^2 - 4x + 1 - y^2$
Group the first three terms.
$$= (4x^2 - 4x + 1) - y^2$$
$$= (2x - 1)^2 - y^2$$
$$= [(2x - 1) + y][(2x - 1) - y]$$
$$= (2x - 1 + y)(2x - 1 - y)$$

27. $100x^4 - 81 = (10x^2)^2 - 9^2$
$$= (10x^2 + 9)(10x^2 - 9)$$

28. $8p^3 + 27$
$$= (2p)^3 + 3^3$$
$$= (2p + 3)[(2p)^2 - (2p)(3) + 3^2]$$
$$= (2p + 3)(4p^2 - 6p + 9)$$

29. $(x - 1)(2x + 3)(x + 4) = 0$

$$x - 1 = 0 \quad \text{or} \quad 2x + 3 = 0 \quad \text{or} \quad x + 4 = 0$$
$$x = 1 \qquad 2x = -3 \qquad x = -4$$
$$x = -\tfrac{3}{2}$$

The solution set is $\{-4, -\tfrac{3}{2}, 1\}$.

30.
$$\begin{aligned} 9x^2 &= 6x - 1 \\ 9x^2 - 6x + 1 &= 0 \\ (3x - 1)^2 &= 0 \\ 3x - 1 &= 0 \\ 3x &= 1 \\ x &= \tfrac{1}{3} \end{aligned}$$

The solution set is $\{\tfrac{1}{3}\}$.

31. Let x be the length of the base. Then $x + 3$ will be the height. The area is 14 square feet. Use the formula $A = \frac{1}{2}bh$, and substitute 14 for A, x for b, and $x + 3$ for h.

$$\begin{aligned} \tfrac{1}{2}bh &= A \\ \tfrac{1}{2}(x)(x + 3) &= 14 \\ x(x + 3) &= 28 \quad \textit{Multiply by 2.} \\ x^2 + 3x &= 28 \\ x^2 + 3x - 28 &= 0 \\ (x + 7)(x - 4) &= 0 \end{aligned}$$

$$x + 7 = 0 \quad \text{or} \quad x - 4 = 0$$
$$x = -7 \quad \text{or} \quad x = 4$$

The length cannot be negative, so reject -7 as a solution. The only possible solution is 4. The base is 4 feet long.

Copyright © 2012 Pearson Education, Inc. Publishing as Addison-Wesley.

32. Let x be the distance between the longer sides. (This is actually the width.) Then $x + 2$ will be the length of the longer side. The area of the rectangle is 288 in². Use the formula $LW = \mathcal{A}$. Substitute 288 for $\mathcal{A}$, $x + 2$ for L, and x for W.

$$(x + 2)x = 288$$
$$x^2 + 2x = 288$$
$$x^2 + 2x - 288 = 0$$
$$(x + 18)(x - 16) = 0$$

$$x + 18 = 0 \quad \text{or} \quad x - 16 = 0$$
$$x = -18 \quad \text{or} \quad x = 16$$

The distance cannot be negative, so reject -18 as a solution. The only possible solution is 16. The distance between the longer sides is 16 inches, and the length of the longer sides is $16 + 2 = 18$ inches.

Copyright © 2012 Pearson Education, Inc. Publishing as Addison-Wesley.

CHAPTER 7 RATIONAL EXPRESSIONS AND FUNCTIONS

7.1 Rational Expressions and Functions; Multiplying and Dividing

7.1 Now Try Exercises

N1. **(a)** $f(x) = \dfrac{2x - 1}{x^2 - 4x - 5}$

Set the denominator equal to zero, and solve the equation.

$$x^2 - 4x - 5 = 0$$
$$(x - 5)(x + 1) = 0$$
$$x - 5 = 0 \quad \text{or} \quad x + 1 = 0$$
$$x = 5 \quad \text{or} \quad x = -1$$

Both 5 and -1 make the function undefined.

Domain: $\{x \mid x \neq -1, 5\}$

(b) $f(x) = \dfrac{15}{2x^2 + 1}$

The denominator, $2x^2 + 1$, can never be 0, so the domain includes all real numbers, written in set-builder notation as $\{x \mid x \text{ is a real number}\}$.

N2. **(a)** $\dfrac{3a^2 - 7a + 2}{a^2 + 2a - 8} = \dfrac{(3a - 1)(a - 2)}{(a + 4)(a - 2)}$ *Factor.*

$$= \dfrac{3a - 1}{a + 4} \cdot 1$$

$$= \dfrac{3a - 1}{a + 4} \qquad \text{\textit{Lowest terms}}$$

(b) $\dfrac{t^3 + 8}{t + 2}$

$$= \dfrac{(t + 2)(t^2 - 2t + 4)}{t + 2} \qquad \begin{array}{l}\textit{Factor the sum}\\\textit{of cubes.}\end{array}$$

$$= t^2 - 2t + 4 \qquad \textit{Lowest terms}$$

(c) $\dfrac{am - bm + an - bn}{am + bm + an + bn}$

$$= \dfrac{(am - bm) + (an - bn)}{(am + bm) + (an + bn)} \qquad \begin{array}{l}\textit{Group}\\\textit{the terms.}\end{array}$$

$$= \dfrac{m(a - b) + n(a - b)}{m(a + b) + n(a + b)} \qquad \begin{array}{l}\textit{Factor within}\\\textit{the groups.}\end{array}$$

$$= \dfrac{(a - b)(m + n)}{(a + b)(m + n)} \qquad \begin{array}{l}\textit{Factor by}\\\textit{grouping.}\end{array}$$

$$= \dfrac{a - b}{a + b} \qquad \textit{Lowest terms}$$

N3. **(a)** $\dfrac{a - 10}{10 - a} = \dfrac{a - 10}{-1(a - 10)} = \dfrac{1}{-1} = -1$

(b) $\dfrac{81 - y^2}{y - 9}$

$$= \dfrac{(9 + y)(9 - y)}{y - 9} \qquad \textit{Factor.}$$

$$= \dfrac{-1(9 + y)(y - 9)}{y - 9}$$

$$= -1(9 + y), \text{ or } -9 - y \qquad \begin{array}{l}\textit{Fundamental}\\\textit{property}\end{array}$$

N4. **(a)** $\dfrac{8t^2}{t^2 - 4} \cdot \dfrac{3t + 6}{9t}$

$$= \dfrac{8t^2}{(t + 2)(t - 2)} \cdot \dfrac{3(t + 2)}{9t}$$

$$= \dfrac{8t}{3(t - 2)}$$

(b) $\dfrac{m^2 + 2m - 15}{m^2 - 5m + 6} \cdot \dfrac{m^2 - 4}{m^2 + 5m}$

$$= \dfrac{(m + 5)(m - 3)}{(m - 2)(m - 3)} \cdot \dfrac{(m + 2)(m - 2)}{m(m + 5)}$$

$$= \dfrac{m + 2}{m}$$

N5. **(a)** $\dfrac{2p^2q}{3pq^4} \div \dfrac{pq}{6p^2q^2} = \dfrac{2p^2q}{3pq^4} \cdot \dfrac{6p^2q^2}{pq}$

$$= \dfrac{12p^4q^3}{3p^2q^5}$$

$$= \dfrac{4p^2}{q^2}$$

(b) $\dfrac{3k^2 + 5k - 2}{9k^2 - 1} \div \dfrac{4k^2 + 8k}{k^2 - 7k}$

$$= \dfrac{3k^2 + 5k - 2}{9k^2 - 1} \cdot \dfrac{k^2 - 7k}{4k^2 + 8k}$$

$$= \dfrac{(3k - 1)(k + 2)}{(3k + 1)(3k - 1)} \cdot \dfrac{k(k - 7)}{4k(k + 2)}$$

$$= \dfrac{k - 7}{4(3k + 1)}$$

7.1 Section Exercises

1. $f(x) = \dfrac{x}{x - 7}$

Set the denominator equal to zero, and solve the equation.

$$x - 7 = 0$$
$$x = 7$$

The number 7 makes the rational expression undefined, so 7 is not in the domain of the function. In set-builder notation, the domain is $\{x \mid x \neq 7\}$.

Copyright © 2012 Pearson Education, Inc. Publishing as Addison-Wesley.

3. $f(x) = \dfrac{6x-5}{7x+1}$

Set the denominator equal to zero, and solve the equation.

$$7x + 1 = 0$$
$$7x = -1$$
$$x = -\tfrac{1}{7}$$

The number $-\tfrac{1}{7}$ makes the rational expression undefined, so $-\tfrac{1}{7}$ is not in the domain of the function. In set-builder notation, the domain is $\left\{x \mid x \neq -\tfrac{1}{7}\right\}$.

5. $f(x) = \dfrac{12x+3}{x}$ is undefined when the denominator x equals 0. So $x = 0$ makes the rational expression undefined and 0 is not in the domain of the function. In set-builder notation, the domain is $\{x \mid x \neq 0\}$.

7. $f(x) = \dfrac{3x+1}{2x^2 + x - 6}$

Set the denominator equal to zero and solve.

$$2x^2 + x - 6 = 0$$
$$(x+2)(2x-3) = 0$$

$$x + 2 = 0 \quad \text{or} \quad 2x - 3 = 0$$
$$x = -2 \quad \text{or} \quad 2x = 3$$
$$x = \tfrac{3}{2}$$

The numbers -2 and $\tfrac{3}{2}$ are not in the domain of the function. In set-builder notation, the domain is $\left\{x \mid x \neq -2, \tfrac{3}{2}\right\}$.

9. $f(x) = \dfrac{x+2}{14}$ is never undefined, so all numbers are in the domain of the function. In set-builder notation, the domain is $\{x \mid x \text{ is a real number}\}$.

11. $f(x) = \dfrac{2x^2 - 3x + 4}{3x^2 + 8}$

Set the denominator equal to zero and solve.

$$3x^2 + 8 = 0$$
$$3x^2 = -8$$
$$x^2 = -\tfrac{8}{3}$$

The square of any real number x is positive or zero, so this equation has no solution. There are no real numbers which make this rational expression undefined, so all numbers are in the domain of the function. In set-builder notation, the domain is $\{x \mid x \text{ is a real number}\}$.

13. $\dfrac{4}{21} \cdot \dfrac{7}{10} = \dfrac{2 \cdot 2}{3 \cdot 7} \cdot \dfrac{7}{2 \cdot 5} = \dfrac{2}{3 \cdot 5} = \dfrac{2}{15}$

15. $\dfrac{3}{8} \div \dfrac{5}{12} = \dfrac{3}{8} \cdot \dfrac{12}{5}$

$$= \dfrac{3}{2 \cdot 2 \cdot 2} \cdot \dfrac{2 \cdot 2 \cdot 3}{5} = \dfrac{3 \cdot 3}{2 \cdot 5} = \dfrac{9}{10}$$

17. $\dfrac{2}{3} \div \dfrac{8}{9} = \dfrac{2}{3} \cdot \dfrac{9}{8}$

$$= \dfrac{2}{3} \cdot \dfrac{3 \cdot 3}{2 \cdot 2 \cdot 2} = \dfrac{3}{2 \cdot 2} = \dfrac{3}{4}$$

19. (a) $\dfrac{x-3}{x+4} = \dfrac{(-1)(x-3)}{(-1)(x+4)}$

$$= \dfrac{-x+3}{-x-4}, \text{ or } \dfrac{3-x}{-x-4} \quad \textbf{(C)}$$

(b) $\dfrac{x+3}{x-4} = \dfrac{(-1)(x+3)}{(-1)(x-4)}$

$$= \dfrac{-x-3}{-x+4}, \text{ or } \dfrac{-x-3}{4-x} \quad \textbf{(A)}$$

(c) $\dfrac{x-3}{x-4} = \dfrac{(-1)(x-3)}{(-1)(x-4)}$

$$= \dfrac{-x+3}{-x+4} \quad \textbf{(D)}$$

(d) $\dfrac{x+3}{x+4} = \dfrac{(-1)(x+3)}{(-1)(x+4)}$

$$= \dfrac{-x-3}{-x-4} \quad \textbf{(B)}$$

(e) $\dfrac{3-x}{x+4} = \dfrac{(-1)(3-x)}{(-1)(x+4)}$

$$= \dfrac{-3+x}{-x-4}, \text{ or } \dfrac{x-3}{-x-4} \quad \textbf{(E)}$$

(f) $\dfrac{x+3}{4-x} = \dfrac{(-1)(x+3)}{(-1)(4-x)}$

$$= \dfrac{-x-3}{-4+x}, \text{ or } \dfrac{-x-3}{x-4} \quad \textbf{(F)}$$

21. $\dfrac{x^2 + 4x}{x+4}$

The two terms in the numerator are x^2 and $4x$. The two terms in the denominator are x and 4.

To express the rational expression in lowest terms, factor the numerator and denominator and replace the quotient of common factors with 1.

$$\dfrac{x^2 + 4x}{x+4} = \dfrac{x(x+4)}{x+4}$$
$$= x \cdot 1 = x$$

23. A. $\dfrac{3-x}{x-4} \cdot \dfrac{-1(x-3)}{-1(4-x)} = \dfrac{x-3}{4-x}$

B. $\dfrac{x+3}{4+x}$ cannot be transformed to equal $\dfrac{x-3}{4-x}$.

C. $-\dfrac{3-x}{4-x} = \dfrac{-(3-x)}{4-x} = \dfrac{x-3}{4-x}$

D. $-\dfrac{x-3}{x-4} = \dfrac{x-3}{-(x-4)} = \dfrac{x-3}{4-x}$

Only the expression in **B** is *not* equivalent to $\dfrac{x-3}{4-x}$.

Copyright © 2012 Pearson Education, Inc. Publishing as Addison-Wesley.

25. $\dfrac{x^2(x+1)}{x(x+1)} = \dfrac{x}{1} \cdot \dfrac{x(x+1)}{x(x+1)} = \dfrac{x}{1} \cdot 1 = x$

27. $\dfrac{(x+4)(x-3)}{(x+5)(x+4)} = \dfrac{x-3}{x+5} \cdot \dfrac{x+4}{x+4} = \dfrac{x-3}{x+5}$

29. $\dfrac{4x(x+3)}{8x^2(x-3)} = \dfrac{(x+3) \cdot 4x}{2x(x-3) \cdot 4x}$

$= \dfrac{x+3}{2x(x-3)}$

31. $\dfrac{3x+7}{3}$ Since the numerator and denominator have no common factors, the expression is already in lowest terms.

33. $\dfrac{6m+18}{7m+21} = \dfrac{6(m+3)}{7(m+3)} = \dfrac{6}{7}$

35. $\dfrac{3z^2+z}{18z+6} = \dfrac{z(3z+1)}{6(3z+1)} = \dfrac{z}{6}$

37. $\dfrac{t^2-9}{3t+9} = \dfrac{(t+3)(t-3)}{3(t+3)} = \dfrac{t-3}{3}$

39. $\dfrac{2t+6}{t^2-9} = \dfrac{2(t+3)}{(t-3)(t+3)} = \dfrac{2}{t-3}$

41. $\dfrac{x^2+2x-15}{x^2+6x+5} = \dfrac{(x+5)(x-3)}{(x+5)(x+1)}$

$= \dfrac{x-3}{x+1}$

43. $\dfrac{8x^2-10x-3}{8x^2-6x-9} = \dfrac{(4x+1)(2x-3)}{(4x+3)(2x-3)}$

$= \dfrac{4x+1}{4x+3}$

45. $\dfrac{a^3+b^3}{a+b} = \dfrac{(a+b)(a^2-ab+b^2)}{a+b}$

$= a^2 - ab + b^2$

47. $\dfrac{2c^2+2cd-60d^2}{2c^2-12cd+10d^2}$

$= \dfrac{2(c^2+cd-30d^2)}{2(c^2-6cd+5d^2)}$

$= \dfrac{2(c+6d)(c-5d)}{2(c-d)(c-5d)}$

$= \dfrac{c+6d}{c-d}$

49. $\dfrac{ac-ad+bc-bd}{ac-ad-bc+bd}$

$= \dfrac{a(c-d)+b(c-d)}{a(c-d)-b(c-d)}$

$= \dfrac{(c-d)(a+b)}{(c-d)(a-b)}$ *Factor by grouping.*

$= \dfrac{a+b}{a-b}$

51. $\dfrac{7-b}{b-7} = \dfrac{-1(b-7)}{b-7} = -1$

In Exercises 53 and 55, there are other acceptable ways to express each answer.

53. $\dfrac{x^2-y^2}{y-x} = \dfrac{(x-y)(x+y)}{y-x}$

$= \dfrac{-1(y-x)(x+y)}{y-x}$

$= -(x+y)$

55. $\dfrac{(a-3)(x+y)}{(3-a)(x-y)} = \dfrac{(a-3)(x+y)}{-1(a-3)(x-y)}$

$= \dfrac{x+y}{-1(x-y)}$

$= -\dfrac{x+y}{x-y}$

57. $\dfrac{5k-10}{20-10k} = \dfrac{-5(2-k)}{10(2-k)} = \dfrac{-5}{10} = -\dfrac{1}{2}$

59. $\dfrac{a^2-b^2}{a^2+b^2} = \dfrac{(a+b)(a-b)}{a^2+b^2}$

The numerator and denominator have no common factors except 1, so the original expression is already in lowest terms.

61. $\dfrac{x^3}{3y} \cdot \dfrac{9y^2}{x^5} = \dfrac{3y \cdot 3x^3 y}{x^2 \cdot 3x^3 y} = \dfrac{3y}{x^2} \cdot 1 = \dfrac{3y}{x^2}$

63. $\dfrac{5a^4b^2}{16a^2b} \div \dfrac{25a^2b}{60a^3b^2}$

Multiply by the reciprocal of the divisor.

$= \dfrac{5a^4b^2}{16a^2b} \cdot \dfrac{60a^3b^2}{25a^2b}$

$= \dfrac{300a^7b^4}{400a^4b^2}$

$= \dfrac{100a^4b^2 \cdot 3a^3b^2}{100a^4b^2 \cdot 4}$

$= \dfrac{3a^3b^2}{4}$

65. $\dfrac{(-3mn)^2 \cdot 64(m^2n)^3}{16m^2n^4(mn^2)^3} \div \dfrac{24(m^2n^2)^4}{(3m^2n^3)^2}$

$= \dfrac{9m^2n^2 \cdot 64m^6n^3}{16m^2n^4 \cdot m^3n^6} \cdot \dfrac{9m^4n^6}{24m^8n^8}$

$= \dfrac{9 \cdot 4 \cdot 9m^{12}n^{11}}{3 \cdot 4 \cdot 2m^{13}n^{18}}$

$= \dfrac{3 \cdot 9}{2m^1n^7} = \dfrac{27}{2mn^7}$

67. $\dfrac{(x+2)(x+1)}{(x+3)(x-2)} \cdot \dfrac{(x+3)(x+4)}{(x+2)(x+1)}$

$= \dfrac{(x+1)(x+2)(x+3)}{(x+1)(x+2)(x+3)} \cdot \dfrac{x+4}{x-2} = \dfrac{x+4}{x-2}$

Copyright © 2012 Pearson Education, Inc. Publishing as Addison-Wesley.

69. $\dfrac{(2x+3)(x-4)}{(x+8)(x-4)} \div \dfrac{(x-4)(x+2)}{(x-4)(x+8)}$

$= \dfrac{2x+3}{x+8} \div \dfrac{x+2}{x+8}$

$= \dfrac{2x+3}{x+8} \cdot \dfrac{x+8}{x+2}$

$= \dfrac{2x+3}{x+2}$

71. $\dfrac{4x}{8x+4} \cdot \dfrac{14x+7}{6} = \dfrac{4x \cdot 7(2x+1)}{4(2x+1)\cdot 6}$

$= \dfrac{7x}{6}$

73. $\dfrac{p^2-25}{4p} \cdot \dfrac{2}{5-p} = \dfrac{(p+5)(p-5)2}{2 \cdot 2p(-1)(p-5)}$

$= \dfrac{p+5}{(2p)(-1)} = -\dfrac{p+5}{2p}$

There are several other ways to express the answer.

75. $(7k+7) \div \dfrac{4k+4}{5}$

$= \dfrac{7(k+1)}{1} \cdot \dfrac{5}{4(k+1)}$

$= \dfrac{7\cdot 5}{4} = \dfrac{35}{4}$

77. $(z^2-1) \cdot \dfrac{1}{1-z}$

$= \dfrac{(z+1)(z-1)}{1} \cdot \dfrac{1}{-1(z-1)}$

$= \dfrac{z+1}{-1} = -(z+1)$ or $-z-1$

79. $\dfrac{4x-20}{5x} \div \dfrac{2x-10}{7x^3}$

$= \dfrac{4(x-5)}{5x} \cdot \dfrac{7x^3}{2(x-5)}$

$= \dfrac{2 \cdot 7x^2}{5}$

$= \dfrac{14x^2}{5}$

81. $\dfrac{12x-10y}{3x+2y} \cdot \dfrac{6x+4y}{10y-12x}$

$= \dfrac{2(6x-5y)\cdot 2(3x+2y)}{(3x+2y)\cdot 2(5y-6x)}$

$= \dfrac{2(-1)(5y-6x)}{(5y-6x)} = -2$

83. $\dfrac{x^2-25}{x^2+x-20} \cdot \dfrac{x^2+7x+12}{x^2-2x-15}$

$= \dfrac{(x-5)(x+5)}{(x+5)(x-4)} \cdot \dfrac{(x+3)(x+4)}{(x-5)(x+3)}$

$= \dfrac{x+4}{x-4}$

85. $\dfrac{a^3-b^3}{a^2-b^2} \div \dfrac{2a-2b}{2a+2b}$

$= \dfrac{(a-b)(a^2+ab+b^2)}{(a+b)(a-b)} \cdot \dfrac{2(a+b)}{2(a-b)}$

$= \dfrac{a^2+ab+b^2}{a-b}$

87. $\dfrac{8x^3-27}{2x^2-18} \cdot \dfrac{2x+6}{8x^2+12x+18}$

$= \dfrac{(2x)^3-3^3}{2(x^2-9)} \cdot \dfrac{2(x+3)}{2(4x^2+6x+9)}$

$= \dfrac{(2x-3)(4x^2+6x+9)}{2(x+3)(x-3)} \cdot \dfrac{2(x+3)}{2(4x^2+6x+9)}$

$= \dfrac{2x-3}{2(x-3)}$

89. $\dfrac{a^3-8b^3}{a^2-ab-6b^2} \cdot \dfrac{a^2+ab-12b^2}{a^2+2ab-8b^2}$

$= \dfrac{(a-2b)(a^2+2ab+4b^2)}{(a-3b)(a+2b)} \cdot \dfrac{(a-3b)(a+4b)}{(a-2b)(a+4b)}$

$= \dfrac{a^2+2ab+4b^2}{a+2b}$

91. $\dfrac{6x^2+5x-6}{12x^2-11x+2} \div \dfrac{4x^2-12x+9}{8x^2-14x+3}$

$= \dfrac{(3x-2)(2x+3)}{(3x-2)(4x-1)} \cdot \dfrac{(2x-3)(4x-1)}{(2x-3)(2x-3)}$

$= \dfrac{2x+3}{2x-3}$

93. $\dfrac{3k^2+17kp+10p^2}{6k^2+13kp-5p^2} \div \dfrac{6k^2+kp-2p^2}{6k^2-5kp+p^2}$

$= \dfrac{(3k+2p)(k+5p)}{(3k-p)(2k+5p)} \cdot \dfrac{(3k-p)(2k-p)}{(3k+2p)(2k-p)}$

$= \dfrac{k+5p}{2k+5p}$

95. $\left(\dfrac{6k^2-13k-5}{k^2+7k} \div \dfrac{2k-5}{k^3+6k^2-7k} \right)$

$\cdot \dfrac{k^2-5k+6}{3k^2-8k-3}$

Factor k from the denominator of the divisor; multiply by the reciprocal.

$= \left[\dfrac{6k^2-13k-5}{k^2+7k} \cdot \dfrac{k(k^2+6k-7)}{2k-5} \right]$

$\cdot \dfrac{k^2-5k+6}{3k^2-8k-3}$

$= \left[\dfrac{(3k+1)(2k-5)}{k(k+7)} \cdot \dfrac{k(k+7)(k-1)}{2k-5} \right]$

$\cdot \dfrac{(k-2)(k-3)}{(3k+1)(k-3)}$

$= (k-1)(k-2)$

Copyright © 2012 Pearson Education, Inc. Publishing as Addison-Wesley.

97. $-\dfrac{3}{4} + \dfrac{1}{12} = -\dfrac{9}{12} + \dfrac{1}{12}$ *LCD = 12*

$\qquad = \dfrac{-9+1}{12} = \dfrac{-8}{12} = -\dfrac{2}{3}$

99. $\dfrac{4}{7} + \dfrac{1}{3} - \dfrac{1}{2} = \dfrac{24}{42} + \dfrac{14}{42} - \dfrac{21}{42}$ *LCD = 42*

$\qquad = \dfrac{24 + 14 - 21}{42} = \dfrac{17}{42}$

7.2 Adding and Subtracting Rational Expressions

7.2 Now Try Exercises

N1. (a) $\dfrac{5}{3x} + \dfrac{2}{3x} = \dfrac{5+2}{3x} = \dfrac{7}{3x}$

(b) $\dfrac{x^2}{x-3} - \dfrac{9}{x-3} = \dfrac{x^2-9}{x-3}$

$\qquad = \dfrac{(x+3)(x-3)}{x-3} = x+3$

(c) $\dfrac{2}{x^2+x-2} + \dfrac{x}{x^2+x-2}$

$\qquad = \dfrac{2+x}{x^2+x-2}$

$\qquad = \dfrac{x+2}{(x+2)(x-1)}$

$\qquad = \dfrac{1}{x-1}$

N2. (a) $15m^3n,\ 10m^2n$

Factor the denominators.

$$15m^3n = 3 \cdot 5 \cdot m^3 \cdot n$$
$$10m^2n = 2 \cdot 5 \cdot m^2 \cdot n$$

The least common denominator (LCD) is the product of all the different factors, with each factor raised to the greatest power in any denominator.

$$LCD = 2 \cdot 3 \cdot 5 \cdot m^3 \cdot n$$
$$= 30m^3n$$

(b) $t,\ t-8$

Each denominator is already factored.

$$LCD = t(t-8)$$

(c) $3x^2 + 9x - 30,\ x^2 - 4,\ x^2 + 10x + 25$

Factor the denominators.

$$3x^2 + 9x - 30 = 3(x+5)(x-2)$$
$$x^2 - 4 = (x+2)(x-2)$$
$$x^2 + 10x + 25 = (x+5)^2$$

$$LCD = 3(x-2)(x+2)(x+5)^2$$

N3. (a) $\dfrac{2}{3x} + \dfrac{7}{4x}$

The LCD for $3x$ and $4x$ is $12x$. To write $\dfrac{2}{3x}$ with a denominator of $12x$, multiply by $\frac{4}{4}$.

$$\dfrac{2}{3x} + \dfrac{7}{4x} = \dfrac{2\cdot 4}{3x\cdot 4} + \dfrac{7\cdot 3}{4x\cdot 3}$$

$$= \dfrac{8}{12x} + \dfrac{21}{12x}$$

$$= \dfrac{8+21}{12x}$$

$$= \dfrac{29}{12x}$$

(b) $\dfrac{3}{z} - \dfrac{6}{z-5}$

The LCD is $z(z-5)$. Rewrite each rational expression with this denominator.

$$\dfrac{3}{z} - \dfrac{6}{z-5} = \dfrac{3(z-5)}{z(z-5)} - \dfrac{(6)z}{(z-5)z}$$

$$= \dfrac{3(z-5)}{z(z-5)} - \dfrac{6z}{z(z-5)}$$

$$= \dfrac{3z-15-6z}{z(z-5)}$$

$$= \dfrac{-3z-15}{z(z-5)}$$

N4. (a) $\dfrac{18x+7}{4x+5} - \dfrac{2x-13}{4x+5}$

The denominator is already the same for both rational expressions, so write them as a single expression. Be careful to apply the subtraction sign to both terms of the second expression.

$$= \dfrac{18x+7-(2x-13)}{4x+5}$$

$$= \dfrac{18x+7-2x+13}{4x+5}$$

$$= \dfrac{16x+20}{4x+5}$$

$$= \dfrac{4(4x+5)}{4x+5}$$

$$= 4$$

(b) $\dfrac{5}{x-3} - \dfrac{5}{x+3}$

The LCD is $(x-3)(x+3)$.

$$= \dfrac{5(x+3)}{(x-3)(x+3)} - \dfrac{5(x-3)}{(x-3)(x+3)}$$

$$= \dfrac{5(x+3) - 5(x-3)}{(x-3)(x+3)}$$

$$= \dfrac{5x+15 - 5x+15}{(x-3)(x+3)}$$

$$= \dfrac{30}{(x-3)(x+3)}$$

Copyright © 2012 Pearson Education, Inc. Publishing as Addison-Wesley.

N5. $\dfrac{t}{t-9} + \dfrac{2}{9-t}$

To get a common denominator of $t - 9$, multiply both the numerator and denominator of the second expression by -1.

$$= \frac{t}{t-9} + \frac{2(-1)}{(9-t)(-1)}$$

$$= \frac{t}{t-9} + \frac{-2}{t-9}$$

$$= \frac{t+(-2)}{t-9}$$

$$= \frac{t-2}{t-9}$$

An equivalent answer is

$$\frac{-1(t-2)}{-1(t-9)} = \frac{-t+2}{-t+9} = \frac{2-t}{9-t}.$$

N6. $\dfrac{6}{y} - \dfrac{1}{y-3} + \dfrac{3}{y^2 - 3y}$

$$= \frac{6}{y} - \frac{1}{y-3} + \frac{3}{y(y-3)}$$

Factor denominators.

$$= \frac{6(y-3)}{y(y-3)} - \frac{1y}{(y-3)y} + \frac{3}{y(y-3)}$$

LCD = y(y − 3)

$$= \frac{6(y-3) - y + 3}{y(y-3)}$$

$$= \frac{6y - 18 - y + 3}{y(y-3)}$$

$$= \frac{5y - 15}{y(y-3)} = \frac{5(y-3)}{y(y-3)} = \frac{5}{y}$$

N7. $\dfrac{t-1}{t^2 - 2t - 8} - \dfrac{2t+3}{t^2 + 3t + 2}$

$$= \frac{t-1}{(t-4)(t+2)} - \frac{2t+3}{(t+1)(t+2)}$$

The LCD is $(t-4)(t+2)(t+1)$.

$$= \frac{(t-1)(t+1)}{(t-4)(t+2)(t+1)}$$

$$\quad - \frac{(2t+3)(t-4)}{(t-4)(t+2)(t+1)}$$

$$= \frac{(t-1)(t+1) - (2t+3)(t-4)}{(t-4)(t+2)(t+1)}$$

$$= \frac{t^2 - 1 - (2t^2 - 5t - 12)}{(t-4)(t+2)(t+1)}$$

$$= \frac{t^2 - 1 - 2t^2 + 5t + 12}{(t-4)(t+2)(t+1)}$$

$$= \frac{-t^2 + 5t + 11}{(t-4)(t+2)(t+1)}$$

N8. $\dfrac{2}{m^2 - 6m + 9} + \dfrac{4}{m^2 + m - 12}$

$$= \frac{2}{(m-3)(m-3)} + \frac{4}{(m+4)(m-3)}$$

The LCD is $(m-3)^2(m+4)$.

$$= \frac{2(m+4)}{(m-3)^2(m+4)} + \frac{4(m-3)}{(m+4)(m-3)^2}$$

$$= \frac{2(m+4) + 4(m-3)}{(m-3)^2(m+4)}$$

$$= \frac{2m + 8 + 4m - 12}{(m-3)^2(m+4)}$$

$$= \frac{6m - 4}{(m-3)^2(m+4)}$$

7.2 Section Exercises

1. $\dfrac{8}{15} + \dfrac{4}{15} = \dfrac{8+4}{15} = \dfrac{12}{15} = \dfrac{3 \cdot 4}{3 \cdot 5} = \dfrac{4}{5}$

3. $\dfrac{5}{6} - \dfrac{8}{9} = \dfrac{15}{18} - \dfrac{16}{18} = \dfrac{15 - 16}{18} = \dfrac{-1}{18} = -\dfrac{1}{18}$

5. $\dfrac{5}{18} + \dfrac{7}{12} = \dfrac{10}{36} + \dfrac{21}{36} = \dfrac{10 + 21}{36} = \dfrac{31}{36}$

7. $\dfrac{7}{t} + \dfrac{2}{t} = \dfrac{7+2}{t} = \dfrac{9}{t}$

9. $\dfrac{6x}{7} + \dfrac{y}{7} = \dfrac{6x + y}{7}$

11. $\dfrac{11}{5x} - \dfrac{1}{5x} = \dfrac{11 - 1}{5x} = \dfrac{10}{5x} = \dfrac{2}{x}$

13. $\dfrac{9}{4x^3} - \dfrac{17}{4x^3} = \dfrac{9 - 17}{4x^3} = \dfrac{-8}{4x^3} = -\dfrac{2}{x^3}$

15. $\dfrac{5x+4}{6x+5} + \dfrac{x+1}{6x+5} = \dfrac{5x+4+x+1}{6x+5}$

$$= \frac{6x+5}{6x+5} = 1$$

17. $\dfrac{x^2}{x+5} - \dfrac{25}{x+5} = \dfrac{x^2 - 25}{x+5}$

$$= \frac{(x+5)(x-5)}{x+5}$$

$$= x - 5$$

19. $\dfrac{-3p+7}{p^2 + 7p + 12} + \dfrac{8p+13}{p^2 + 7p + 12}$

$$= \frac{-3p + 7 + 8p + 13}{p^2 + 7p + 12}$$

$$= \frac{5p + 20}{(p+3)(p+4)}$$

$$= \frac{5(p+4)}{(p+3)(p+4)} = \frac{5}{p+3}$$

21. $\dfrac{a^3}{a^2 + ab + b^2} - \dfrac{b^3}{a^2 + ab + b^2}$

$$= \frac{a^3 - b^3}{a^2 + ab + b^2}$$

$$= \frac{(a-b)(a^2 + ab + b^2)}{a^2 + ab + b^2}$$

$$= a - b$$

Copyright © 2012 Pearson Education, Inc. Publishing as Addison-Wesley.

23. $18x^2y^3,\ 24x^4y^5$

Factor each denominator.

$$18x^2y^3 = 2 \cdot 3 \cdot 3 \cdot x^2 \cdot y^3$$
$$= 2 \cdot 3^2 \cdot x^2 \cdot y^3$$
$$24x^4y^5 = 2 \cdot 2 \cdot 2 \cdot 3 \cdot x^4 \cdot y^5$$
$$= 2^3 \cdot 3 \cdot x^4 \cdot y^5$$

The least common denominator (LCD) is the product of all the different factors, with each factor raised to the greatest power in any denominator.

$$\text{LCD} = 2^3 \cdot 3^2 \cdot x^4 \cdot y^5$$
$$= 8 \cdot 9 \cdot x^4 \cdot y^5$$
$$= 72x^4y^5$$

The LCD is $72x^4y^5$.

25. $z - 2,\ z$

Both $z - 2$ and z have only 1 and themselves for factors.

$$\text{LCD} = z(z - 2)$$

27. $2y + 8,\ y + 4$

Factor each denominator.

$$2y + 8 = 2(y + 4)$$

The second denominator, $y + 4$, is already factored. The LCD is

$$2(y + 4).$$

29. $x^2 - 81,\ x^2 + 18x + 81$

Factor each denominator.

$$x^2 - 81 = (x + 9)(x - 9)$$
$$x^2 + 18x + 81 = (x + 9)(x + 9)$$

$$\text{LCD} = (x + 9)^2(x - 9)$$

31. $m + n,\ m - n,\ m^2 - n^2$

Both $m + n$ and $m - n$ have only 1 and themselves for factors, while $m^2 - n^2 = (m + n)(m - n)$.

$$\text{LCD} = (m + n)(m - n)$$

33. $x^2 - 3x - 4,\ x + x^2$

Factor each denominator.

$$x^2 - 3x - 4 = (x - 4)(x + 1)$$
$$x + x^2 = x(1 + x) = x(x + 1)$$

The LCD is $x(x - 4)(x + 1)$.

35. $2t^2 + 7t - 15,\ t^2 + 3t - 10$

Factor each denominator.

$$2t^2 + 7t - 15 = (2t - 3)(t + 5)$$
$$t^2 + 3t - 10 = (t + 5)(t - 2)$$

The LCD is $(2t - 3)(t + 5)(t - 2)$.

37. $2y + 6,\ y^2 - 9,\ y$

Factor each denominator.

$$2y + 6 = 2(y + 3)$$
$$y^2 - 9 = (y + 3)(y - 3)$$

Remember the factor y from the third denominator. The LCD is

$$2y(y + 3)(y - 3).$$

39. $2x - 6,\ x^2 - x - 6,\ (x + 2)^2$

Factor each denominator.

$$2x - 6 = 2(x - 3)$$
$$x^2 - x - 6 = (x - 3)(x + 2)$$
$$(x + 2)^2 \text{ is already factored.}$$

$$\text{LCD} = 2(x + 2)^2(x - 3)$$

41. The first step is incorrect. The third term in the numerator should be $+1$, because the subtraction sign should be distributed to both $4x$ and -1. The correct solution follows.

$$\frac{x}{x + 2} - \frac{4x - 1}{x + 2} = \frac{x - (4x - 1)}{x + 2}$$
$$= \frac{x - 4x + 1}{x + 2}$$
$$= \frac{-3x + 1}{x + 2}$$

43. $\dfrac{8}{t} + \dfrac{7}{3t}$ The LCD is $3t$.

$$\frac{8}{t} + \frac{7}{3t} = \frac{8 \cdot 3}{t \cdot 3} + \frac{7}{3t}$$
$$= \frac{24 + 7}{3t} = \frac{31}{3t}$$

45. $\dfrac{5}{12x^2y} - \dfrac{11}{6xy}$ The LCD is $12x^2y$.

$$\frac{5}{12x^2y} - \frac{11}{6xy} = \frac{5}{12x^2y} - \frac{11 \cdot 2x}{6xy \cdot 2x}$$
$$= \frac{5}{12x^2y} - \frac{22x}{12x^2y}$$
$$= \frac{5 - 22x}{12x^2y}$$

47. $\dfrac{4}{15a^4b^5} + \dfrac{3}{20a^2b^6}$ $\text{LCD} = 60a^4b^6$

$$= \frac{4 \cdot 4b}{15a^4b^5 \cdot 4b} + \frac{3 \cdot 3a^2}{20a^2b^6 \cdot 3a^2}$$
$$= \frac{16b + 9a^2}{60a^4b^6}$$

Copyright © 2012 Pearson Education, Inc. Publishing as Addison-Wesley.

49. $\dfrac{2r}{7p^3q^4} + \dfrac{3s}{14p^4q}$ $\text{LCD} = 14p^4q^4$

$= \dfrac{2r \cdot 2p}{7p^3q^4 \cdot 2p} + \dfrac{3s \cdot q^3}{14p^4q \cdot q^3}$

$= \dfrac{4pr + 3sq^3}{14p^4q^4}$

51. $\dfrac{1}{a^3b^2} - \dfrac{2}{a^4b} + \dfrac{3}{a^5b^7}$ $\text{LCD} = a^5b^7$

$= \dfrac{1 \cdot a^2b^5}{a^3b^2 \cdot a^2b^5} - \dfrac{2 \cdot ab^6}{a^4b \cdot ab^6} + \dfrac{3}{a^5b^7}$

$= \dfrac{a^2b^5 - 2ab^6 + 3}{a^5b^7}$

53. $\dfrac{1}{x-1} - \dfrac{1}{x}$ $\text{LCD} = x(x-1)$

$= \dfrac{1 \cdot x}{(x-1)x} - \dfrac{1 \cdot (x-1)}{x(x-1)}$

$= \dfrac{x - (x-1)}{x(x-1)}$

$= \dfrac{x - x + 1}{x(x-1)}$

$= \dfrac{1}{x(x-1)}$

55. $\dfrac{3a}{a+1} + \dfrac{2a}{a-3}$ $\text{LCD} = (a+1)(a-3)$

$= \dfrac{3a(a-3)}{(a+1)(a-3)} + \dfrac{2a(a+1)}{(a-3)(a+1)}$

$= \dfrac{3a(a-3) + 2a(a+1)}{(a+1)(a-3)}$

$= \dfrac{3a^2 - 9a + 2a^2 + 2a}{(a+1)(a-3)}$

$= \dfrac{5a^2 - 7a}{(a+1)(a-3)}$

57. $\dfrac{17y+3}{9y+7} - \dfrac{-10y-18}{9y+7}$

$= \dfrac{17y + 3 - (-10y - 18)}{9y+7}$

$= \dfrac{17y + 3 + 10y + 18}{9y+7}$

$= \dfrac{27y + 21}{9y+7}$

$= \dfrac{3(9y+7)}{9y+7}$

$= 3$

59. $\dfrac{11x-13}{2x-3} - \dfrac{3x-1}{2x-3}$

$= \dfrac{11x - 13 - (3x - 1)}{2x-3}$

$= \dfrac{11x - 13 - 3x + 1}{2x-3}$

$= \dfrac{8x - 12}{2x-3}$

$= \dfrac{4(2x-3)}{2x-3} = 4$

61. $\dfrac{2}{4-x} + \dfrac{5}{x-4}$

To get a common denominator of $x - 4$, multiply both the numerator and denominator of the first expression by -1.

$= \dfrac{(2)(-1)}{(4-x)(-1)} + \dfrac{5}{x-4}$

$= \dfrac{-2}{x-4} + \dfrac{5}{x-4}$

$= \dfrac{-2 + 5}{x-4}$

$= \dfrac{3}{x-4}$

If you chose $4 - x$ for the LCD, then you should have obtained the equivalent answer, $\dfrac{-3}{4-x}$.

63. $\dfrac{w}{w-z} - \dfrac{z}{z-w}$

$w - z$ and $z - w$ are opposites, so factor out -1 from $z - w$ to get a common denominator.

$= \dfrac{w}{w-z} - \dfrac{z}{-1(w-z)}$

$= \dfrac{w}{w-z} + \dfrac{z}{w-z}$

$= \dfrac{w+z}{w-z}$,   or $\dfrac{-w-z}{z-w}$

65. $\dfrac{1}{x+1} - \dfrac{1}{x-1}$ $\text{LCD} = (x+1)(x-1)$

$= \dfrac{1(x-1)}{(x+1)(x-1)} - \dfrac{1(x+1)}{(x-1)(x+1)}$

$= \dfrac{x - 1 - x - 1}{(x+1)(x-1)}$

$= \dfrac{-2}{(x+1)(x-1)}$

67. $\dfrac{4x}{x-1} - \dfrac{2}{x+1} - \dfrac{4}{x^2-1}$

$x^2 - 1 = (x+1)(x-1)$, the LCD.

$= \dfrac{4x(x+1)}{(x-1)(x+1)} - \dfrac{2(x-1)}{(x+1)(x-1)}$

$\quad - \dfrac{4}{(x+1)(x-1)}$

$= \dfrac{4x(x+1) - 2(x-1) - 4}{(x+1)(x-1)}$

$= \dfrac{4x^2 + 4x - 2x + 2 - 4}{(x-1)(x+1)}$

$= \dfrac{4x^2 + 2x - 2}{(x-1)(x+1)}$

$= \dfrac{2(2x^2 + x - 1)}{(x-1)(x+1)}$

$= \dfrac{2(2x-1)(x+1)}{(x-1)(x+1)}$

$= \dfrac{2(2x-1)}{x-1}$

Copyright © 2012 Pearson Education, Inc. Publishing as Addison-Wesley.

69. $\dfrac{15}{y^2+3y}+\dfrac{2}{y}+\dfrac{5}{y+3}$

$y^2+3y=y(y+3)$, the LCD.

$=\dfrac{15}{y(y+3)}+\dfrac{2(y+3)}{y(y+3)}+\dfrac{5y}{(y+3)y}$

$=\dfrac{15+2(y+3)+5y}{y(y+3)}$

$=\dfrac{15+2y+6+5y}{y(y+3)}$

$=\dfrac{7y+21}{y(y+3)}$

$=\dfrac{7(y+3)}{y(y+3)}=\dfrac{7}{y}$

71. $\dfrac{5}{x-2}+\dfrac{1}{x}+\dfrac{2}{x^2-2x}$

$x^2-2x=x(x-2)$, the LCD.

$\dfrac{5}{x-2}+\dfrac{1}{x}+\dfrac{2}{x^2-2x}$

$=\dfrac{5x}{(x-2)x}+\dfrac{1(x-2)}{x(x-2)}+\dfrac{2}{x(x-2)}$

$=\dfrac{5x+x-2+2}{x(x-2)}$

$=\dfrac{6x}{x(x-2)}=\dfrac{6}{x-2}$

73. $\dfrac{3x}{x+1}+\dfrac{4}{x-1}-\dfrac{6}{x^2-1}$

$x^2-1=(x+1)(x-1)$, the LCD.

$=\dfrac{3x(x-1)}{(x+1)(x-1)}+\dfrac{4(x+1)}{(x-1)(x+1)}$

$\qquad -\dfrac{6}{(x+1)(x-1)}$

$=\dfrac{3x(x-1)+4(x+1)-6}{(x+1)(x-1)}$

$=\dfrac{3x^2-3x+4x+4-6}{(x+1)(x-1)}$

$=\dfrac{3x^2+x-2}{(x+1)(x-1)}$

$=\dfrac{(3x-2)(x+1)}{(x+1)(x-1)}$

$=\dfrac{3x-2}{x-1}$

75. $\dfrac{4}{x+1}+\dfrac{1}{x^2-x+1}-\dfrac{12}{x^3+1}$

$x^3+1=(x+1)(x^2-x+1)$, the LCD.

$=\dfrac{4(x^2-x+1)}{(x+1)(x^2-x+1)}$

$\quad +\dfrac{1\cdot(x+1)}{(x^2-x+1)(x+1)}$

$\quad -\dfrac{12}{(x+1)(x^2-x+1)}$

$=\dfrac{4(x^2-x+1)+(x+1)-12}{(x+1)(x^2-x+1)}$

$=\dfrac{4x^2-4x+4+x+1-12}{(x+1)(x^2-x+1)}$

$=\dfrac{4x^2-3x-7}{(x+1)(x^2-x+1)}$

$=\dfrac{(4x-7)(x+1)}{(x+1)(x^2-x+1)}$

$=\dfrac{4x-7}{x^2-x+1}$

77. $\dfrac{2x+4}{x+3}+\dfrac{3}{x}-\dfrac{6}{x^2+3x}$

$x^2+3x=x(x+3)$, the LCD.

$=\dfrac{(2x+4)x}{(x+3)x}+\dfrac{3(x+3)}{x(x+3)}-\dfrac{6}{x(x+3)}$

$=\dfrac{(2x+4)x+3(x+3)-6}{x(x+3)}$

$=\dfrac{2x^2+4x+3x+9-6}{x(x+3)}$

$=\dfrac{2x^2+7x+3}{x(x+3)}$

$=\dfrac{(2x+1)(x+3)}{x(x+3)}=\dfrac{2x+1}{x}$

79. $\dfrac{3}{(p-2)^2}-\dfrac{5}{p-2}+4$

$=\dfrac{3}{(p-2)^2}-\dfrac{5(p-2)}{(p-2)^2}+\dfrac{4(p-2)^2}{(p-2)^2}$

$\qquad\qquad LCD=(p-2)^2$

$=\dfrac{3-5(p-2)+4(p^2-4p+4)}{(p-2)^2}$

$=\dfrac{3-5p+10+4p^2-16p+16}{(p-2)^2}$

$=\dfrac{4p^2-21p+29}{(p-2)^2}$

81. $\dfrac{3}{x^2-5x+6}-\dfrac{2}{x^2-4x+4}$

$=\dfrac{3}{(x-2)(x-3)}-\dfrac{2}{(x-2)(x-2)}$

$=\dfrac{3(x-2)}{(x-2)(x-3)(x-2)}$

$\quad -\dfrac{2(x-3)}{(x-2)(x-2)(x-3)}$

$\qquad\qquad LCD=(x-2)^2(x-3)$

$=\dfrac{3x-6-2x+6}{(x-2)^2(x-3)}$

$=\dfrac{x}{(x-2)^2(x-3)}$

Copyright © 2012 Pearson Education, Inc. Publishing as Addison-Wesley.

83. $\dfrac{5x}{x^2 + xy - 2y^2} - \dfrac{3x}{x^2 + 5xy - 6y^2}$

Factor each denominator.

$$x^2 + xy - 2y^2 = (x + 2y)(x - y)$$
$$x^2 + 5xy - 6y^2 = (x + 6y)(x - y)$$

The LCD is $(x + 2y)(x - y)(x + 6y)$.

$$\dfrac{5x}{(x + 2y)(x - y)} - \dfrac{3x}{(x + 6y)(x - y)}$$

$$= \dfrac{(5x)(x + 6y)}{(x + 2y)(x - y)(x + 6y)}$$

$$- \dfrac{(3x)(x + 2y)}{(x + 6y)(x - y)(x + 2y)}$$

$$= \dfrac{(5x)(x + 6y) - (3x)(x + 2y)}{(x + 6y)(x - y)(x + 2y)}$$

$$= \dfrac{5x^2 + 30xy - (3x^2 + 6xy)}{(x + 2y)(x - y)(x + 6y)}$$

$$= \dfrac{2x^2 + 24xy}{(x + 2y)(x - y)(x + 6y)}$$

$$= \dfrac{2x(x + 12y)}{(x + 2y)(x - y)(x + 6y)}$$

85. $\dfrac{5x - y}{x^2 + xy - 2y^2} - \dfrac{3x + 2y}{x^2 + 5xy - 6y^2}$

Factor each denominator.

$$x^2 + xy - 2y^2 = (x + 2y)(x - y)$$
$$x^2 + 5xy - 6y^2 = (x + 6y)(x - y)$$

The LCD is $(x + 2y)(x - y)(x + 6y)$.

$$\dfrac{5x - y}{(x + 2y)(x - y)} - \dfrac{3x + 2y}{(x + 6y)(x - y)}$$

$$= \dfrac{(5x - y)(x + 6y)}{(x + 2y)(x - y)(x + 6y)}$$

$$- \dfrac{(3x + 2y)(x + 2y)}{(x + 6y)(x - y)(x + 2y)}$$

$$= \dfrac{(5x - y)(x + 6y) - (3x + 2y)(x + 2y)}{(x + 6y)(x - y)(x + 2y)}$$

$$= \dfrac{5x^2 + 29xy - 6y^2 - (3x^2 + 8xy + 4y^2)}{(x + 2y)(x - y)(x + 6y)}$$

$$= \dfrac{2x^2 + 21xy - 10y^2}{(x + 2y)(x - y)(x + 6y)}$$

87. $\dfrac{r + s}{3r^2 + 2rs - s^2} - \dfrac{s - r}{6r^2 - 5rs + s^2}$

Factor each denominator.

$$3r^2 + 2rs - s^2 = (3r - s)(r + s)$$
$$6r^2 - 5rs + s^2 = (3r - s)(2r - s)$$

The LCD is $(3r - s)(r + s)(2r - s)$.

$\dfrac{r + s}{3r^2 + 2rs - s^2} - \dfrac{s - r}{6r^2 - 5rs + s^2}$

$$= \dfrac{(r + s)(2r - s)}{(3r - s)(r + s)(2r - s)}$$

$$- \dfrac{(s - r)(r + s)}{(3r - s)(2r - s)(r + s)}$$

$$= \dfrac{(r + s)(2r - s) - (s - r)(s + r)}{(3r - s)(r + s)(2r - s)}$$

$$= \dfrac{2r^2 + rs - s^2 - (s^2 - r^2)}{(3r - s)(r + s)(2r - s)}$$

$$= \dfrac{2r^2 + rs - s^2 - s^2 + r^2}{(3r - s)(r + s)(2r - s)}$$

$$= \dfrac{3r^2 + rs - 2s^2}{(3r - s)(r + s)(2r - s)}$$

$$= \dfrac{(3r - 2s)(r + s)}{(3r - s)(r + s)(2r - s)}$$

$$= \dfrac{3r - 2s}{(3r - s)(2r - s)}$$

89. $\dfrac{3}{x^2 + 4x + 4} + \dfrac{7}{x^2 + 5x + 6}$

$$= \dfrac{3}{(x + 2)^2} + \dfrac{7}{(x + 2)(x + 3)}$$

$$LCD = (x + 2)^2(x + 3)$$

$$= \dfrac{3(x + 3)}{(x + 2)^2(x + 3)} + \dfrac{7(x + 2)}{(x + 2)^2(x + 3)}$$

$$= \dfrac{3x + 9 + 7x + 14}{(x + 2)^2(x + 3)}$$

$$= \dfrac{10x + 23}{(x + 2)^2(x + 3)}$$

91. (a) $c(x) = \dfrac{1010}{49(101 - x)} - \dfrac{10}{49}$

$$= \dfrac{1010}{49(101 - x)} - \dfrac{10(101 - x)}{49(101 - x)}$$

$$= \dfrac{1010 - 1010 + 10x}{49(101 - x)}$$

$$= \dfrac{10x}{49(101 - x)}$$

(b) $c(95) = \dfrac{10(95)}{49(101 - 95)}$

$$= \dfrac{950}{294} \approx 3.23$$

It would cost approximately 3.23 thousand dollars to win 95 points.

93. $\dfrac{\frac{2}{3}}{4} = \dfrac{\frac{2}{3}}{\frac{4}{1}} = \dfrac{2}{3} \div \dfrac{4}{1} = \dfrac{\overset{1}{\cancel{2}}}{3} \cdot \dfrac{1}{\underset{2}{\cancel{4}}} = \dfrac{1}{6}$

95. $\dfrac{\frac{3}{4}}{\frac{5}{12}} = \dfrac{3}{4} \div \dfrac{5}{12} = \dfrac{3}{\underset{1}{\cancel{4}}} \cdot \dfrac{\overset{3}{\cancel{12}}}{5} = \dfrac{9}{5}$

Copyright © 2012 Pearson Education, Inc. Publishing as Addison-Wesley.

7.3 Complex Fractions

7.3 Now Try Exercises

N1. **(a)** $\dfrac{\dfrac{t+4}{3t}}{\dfrac{2t+1}{9t}}$

Both the numerator and denominator are already simplified.

$= \dfrac{t+4}{3t} \div \dfrac{2t+1}{9t}$

$= \dfrac{t+4}{3t} \cdot \dfrac{9t}{2t+1}$ *Multiply by the reciprocal of the divisor.*

$= \dfrac{3(t+4)}{2t+1}$

(b) $\dfrac{5 - \dfrac{2}{y}}{4 + \dfrac{1}{y}}$

$= \dfrac{\dfrac{5y}{y} - \dfrac{2}{y}}{\dfrac{4y}{y} + \dfrac{1}{y}}$ *Simplify the numerator and denominator.*

$= \dfrac{\dfrac{5y-2}{y}}{\dfrac{4y+1}{y}}$

$= \dfrac{5y-2}{y} \cdot \dfrac{y}{4y+1}$ *Multiply by the reciprocal of the divisor.*

$= \dfrac{5y-2}{4y+1}$

N2. **(a)** $\dfrac{5 - \dfrac{2}{y}}{4 + \dfrac{1}{y}}$

The LCD of $\frac{2}{y}$ and $\frac{1}{y}$ is y. Multiply the numerator and denominator by y.

$= \dfrac{\left(5 - \dfrac{2}{y}\right) \cdot y}{\left(4 + \dfrac{1}{y}\right) \cdot y}$

$= \dfrac{5 \cdot y - \dfrac{2}{y} \cdot y}{4 \cdot y + \dfrac{1}{y} \cdot y}$ *Distributive property*

$= \dfrac{5y-2}{4y+1}$ *Simplify.*

(b) $\dfrac{x + \dfrac{5}{x}}{4x - \dfrac{1}{x-2}}$

Multiply the numerator and denominator by $x(x-2)$, the LCD of all the fractions.

$= \dfrac{\left(x + \dfrac{5}{x}\right) \cdot x(x-2)}{\left(4x - \dfrac{1}{x-2}\right) \cdot x(x-2)}$

$= \dfrac{x \cdot x(x-2) + \dfrac{5}{x} \cdot x(x-2)}{4x \cdot x(x-2) - \dfrac{1}{x-2} \cdot x(x-2)}$

$= \dfrac{x^2(x-2) + 5(x-2)}{4x^2(x-2) - x}$

$= \dfrac{x^3 - 2x^2 + 5x - 10}{4x^3 - 8x^2 - x}$

N3. **(a)** **Method 1**

$\dfrac{\dfrac{1}{p-6}}{\dfrac{5}{p^2-36}} = \dfrac{\dfrac{1}{p-6}}{\dfrac{5}{(p+6)(p-6)}}$

$= \dfrac{1}{p-6} \div \dfrac{5}{(p+6)(p-6)}$

$= \dfrac{1}{p-6} \cdot \dfrac{(p+6)(p-6)}{5}$

$= \dfrac{p+6}{5}$

Method 2

$\dfrac{\dfrac{1}{p-6}}{\dfrac{5}{p^2-36}} = \dfrac{\dfrac{1}{p-6}}{\dfrac{5}{(p+6)(p-6)}}$

The LCD of the numerator and denominator is $(p+6)(p-6)$.

$= \dfrac{\dfrac{1}{p-6} \cdot (p+6)(p-6)}{\dfrac{5}{(p+6)(p-6)} \cdot (p+6)(p-6)}$

$= \dfrac{p+6}{5}$

(b) **Method 1**

$\dfrac{\dfrac{1}{m^2} - \dfrac{1}{n^2}}{\dfrac{1}{m} + \dfrac{1}{n}} = \dfrac{\dfrac{n^2}{m^2n^2} - \dfrac{m^2}{m^2n^2}}{\dfrac{n}{mn} + \dfrac{m}{mn}}$ $LCD = m^2n^2$
$LCD = mn$

$= \dfrac{\dfrac{n^2 - m^2}{m^2n^2}}{\dfrac{n+m}{mn}}$

Copyright © 2012 Pearson Education, Inc. Publishing as Addison-Wesley.

$$= \frac{\dfrac{(n+m)(n-m)}{m^2 n^2}}{\dfrac{n+m}{mn}}$$

$$= \frac{(n+m)(n-m)}{m^2 n^2} \div \frac{n+m}{mn}$$

$$= \frac{(n+m)(n-m)}{m^2 n^2} \cdot \frac{mn}{n+m}$$

$$= \frac{n-m}{mn}$$

Method 2

$$\frac{\dfrac{1}{m^2} - \dfrac{1}{n^2}}{\dfrac{1}{m} + \dfrac{1}{n}}$$

The LCD of the numerator and the denominator is $m^2 n^2$.

$$= \frac{\left(\dfrac{1}{m^2} - \dfrac{1}{n^2}\right) \cdot m^2 n^2}{\left(\dfrac{1}{m} + \dfrac{1}{n}\right) \cdot m^2 n^2}$$

$$= \frac{n^2 - m^2}{mn^2 + m^2 n}$$

$$= \frac{(n+m)(n-m)}{mn(n+m)}$$

$$= \frac{n-m}{mn}$$

N4. $\dfrac{2y^{-1} - 3y^{-2}}{y^{-2} + 3x^{-1}} = \dfrac{\dfrac{2}{y} - \dfrac{3}{y^2}}{\dfrac{1}{y^2} + \dfrac{3}{x}}$ $LCD = xy^2$

$$= \frac{xy^2 \left(\dfrac{2}{y} - \dfrac{3}{y^2}\right)}{xy^2 \cdot \left(\dfrac{1}{y^2} + \dfrac{3}{x}\right)}$$

$$= \frac{xy^2 \cdot \dfrac{2}{y} - xy^2 \cdot \dfrac{3}{y^2}}{xy^2 \cdot \dfrac{1}{y^2} + xy^2 \cdot \dfrac{3}{x}}$$

$$= \frac{2xy - 3x}{x + 3y^2}$$

7.3 Section Exercises

1. $\dfrac{\dfrac{5}{9} - \dfrac{1}{3}}{\dfrac{2}{3} + \dfrac{1}{6}} = \dfrac{\dfrac{5}{9} - \dfrac{3}{9}}{\dfrac{4}{6} + \dfrac{1}{6}} = \dfrac{\dfrac{2}{9}}{\dfrac{5}{6}} = \dfrac{2}{9} \div \dfrac{5}{6} = \dfrac{2}{\overset{3}{\cancel{9}}} \cdot \dfrac{\overset{2}{\cancel{6}}}{5} = \dfrac{4}{15}$

3. $\dfrac{2 - \dfrac{1}{4}}{\dfrac{5}{4} + 3} = \dfrac{\dfrac{8}{4} - \dfrac{1}{4}}{\dfrac{5}{4} + \dfrac{12}{4}} = \dfrac{\dfrac{7}{4}}{\dfrac{17}{4}} = \dfrac{7}{4} \div \dfrac{17}{4} = \dfrac{7}{\overset{}{\underset{1}{\cancel{4}}}} \cdot \dfrac{\overset{1}{\cancel{4}}}{17} = \dfrac{7}{17}$

5. $\dfrac{\dfrac{12}{x-1}}{\dfrac{6}{x}} = \dfrac{12}{x-1} \div \dfrac{6}{x}$

Multiply by the reciprocal of the divisor.

$$= \frac{12}{x-1} \cdot \frac{x}{6}$$

$$= \frac{2x}{x-1}$$

7. $\dfrac{\dfrac{k+1}{2k}}{\dfrac{3k-1}{4k}} = \dfrac{k+1}{2k} \cdot \dfrac{4k}{3k-1}$

$$= \frac{4k(k+1)}{2k(3k-1)}$$

$$= \frac{2(k+1)}{3k-1}$$

9. $\dfrac{\dfrac{4z^2 x^4}{9}}{\dfrac{12x^2 z^5}{15}} = \dfrac{\dfrac{4z^2 x^4}{9}}{\dfrac{4x^2 z^5}{5}}$

$$= \frac{4z^2 x^4}{9} \div \frac{4x^2 z^5}{5}$$

$$= \frac{4z^2 x^4}{9} \cdot \frac{5}{4x^2 z^5}$$

$$= \frac{5z^2 x^4}{9x^2 z^5} = \frac{5x^2}{9z^3}$$

11. $\dfrac{6 + \dfrac{1}{x}}{7 - \dfrac{3}{x}}$

Multiply the numerator and denominator by x, the LCD of all the fractions.

$$= \frac{x\left(6 + \dfrac{1}{x}\right)}{x\left(7 - \dfrac{3}{x}\right)}$$

$$= \frac{x \cdot 6 + x \cdot \dfrac{1}{x}}{x \cdot 7 + x\left(-\dfrac{3}{x}\right)}$$

$$= \frac{6x + 1}{7x - 3}$$

13. $\dfrac{\dfrac{3}{x} + \dfrac{3}{y}}{\dfrac{3}{x} - \dfrac{3}{y}}$

Multiply the numerator and denominator by xy, the LCD of all the fractions.

$$= \frac{\left(\dfrac{3}{x} + \dfrac{3}{y}\right) xy}{\left(\dfrac{3}{x} - \dfrac{3}{y}\right) xy}$$

$$= \frac{\dfrac{3}{x} \cdot xy + \dfrac{3}{y} \cdot xy}{\dfrac{3}{x} \cdot xy - \dfrac{3}{y} \cdot xy}$$

$$= \frac{3y + 3x}{3y - 3x}$$

$$= \frac{3(y+x)}{3(y-x)} = \frac{y+x}{y-x}$$

Copyright © 2012 Pearson Education, Inc. Publishing as Addison-Wesley.

15. $\dfrac{\dfrac{8x - 24y}{10}}{\dfrac{x - 3y}{5x}} = \dfrac{8x - 24y}{10} \cdot \dfrac{5x}{x - 3y}$

$= \dfrac{8(x - 3y)5x}{10(x - 3y)}$

$= \dfrac{40x}{10} = 4x$

17. $\dfrac{\dfrac{x^2 - 16y^2}{xy}}{\dfrac{1}{y} - \dfrac{4}{x}}$

Multiply the numerator and denominator by xy, the LCD of all the fractions.

$= \dfrac{\left(\dfrac{x^2 - 16y^2}{xy}\right)xy}{\left(\dfrac{1}{y} - \dfrac{4}{x}\right)xy}$

$= \dfrac{x^2 - 16y^2}{\dfrac{1}{y} \cdot xy - \dfrac{4}{x} \cdot xy}$

$= \dfrac{x^2 - 16y^2}{x - 4y}$

$= \dfrac{(x + 4y)(x - 4y)}{x - 4y}$

$= x + 4y$

19. $\dfrac{\dfrac{6}{y - 4}}{\dfrac{12}{y^2 - 16}} = \dfrac{6}{y - 4} \cdot \dfrac{y^2 - 16}{12}$

$= \dfrac{6}{y - 4} \cdot \dfrac{(y + 4)(y - 4)}{12}$

$= \dfrac{y + 4}{2}$

21. $\dfrac{\dfrac{1}{b^2} - \dfrac{1}{a^2}}{\dfrac{1}{b} - \dfrac{1}{a}} = \dfrac{\dfrac{a^2 - b^2}{b^2a^2}}{\dfrac{a - b}{ba}}$

$= \dfrac{a^2 - b^2}{b^2a^2} \cdot \dfrac{ba}{a - b}$

$= \dfrac{(a + b)(a - b)}{b^2a^2} \cdot \dfrac{ba}{a - b}$

$= \dfrac{a + b}{ab}$

23. $\dfrac{x + y}{\dfrac{1}{y} + \dfrac{1}{x}} = \dfrac{x + y}{\dfrac{x + y}{yx}}$

$= \dfrac{x + y}{1} \cdot \dfrac{yx}{x + y}$

$= xy$

25. $\dfrac{y - \dfrac{y - 3}{3}}{\dfrac{4}{9} + \dfrac{2}{3y}}$

Multiply the numerator and denominator by $9y$, the LCD of all the fractions.

$= \dfrac{9y\left(y - \dfrac{y - 3}{3}\right)}{9y\left(\dfrac{4}{9} + \dfrac{2}{3y}\right)}$

$= \dfrac{9y^2 - 3y(y - 3)}{4y + 6}$

$= \dfrac{9y^2 - 3y^2 + 9y}{4y + 6}$

$= \dfrac{6y^2 + 9y}{4y + 6}$

$= \dfrac{3y(2y + 3)}{2(2y + 3)} = \dfrac{3y}{2}$

27. $\dfrac{\dfrac{x + 2}{x} + \dfrac{1}{x + 2}}{\dfrac{5}{x} + \dfrac{x}{x + 2}}$

Multiply the numerator and denominator by $x(x + 2)$, the LCD of all the fractions.

$= \dfrac{x(x + 2)\left(\dfrac{x + 2}{x} + \dfrac{1}{x + 2}\right)}{x(x + 2)\left(\dfrac{5}{x} + \dfrac{x}{x + 2}\right)}$

$= \dfrac{x(x + 2)\left(\dfrac{x + 2}{x}\right) + x(x + 2)\left(\frac{1}{x+2}\right)}{x(x + 2)\left(\dfrac{5}{x}\right) + x(x + 2)\left(\frac{x}{x+2}\right)}$

$= \dfrac{(x + 2)(x + 2) + x}{5(x + 2) + x^2}$

$= \dfrac{x^2 + 4x + 4 + x}{5x + 10 + x^2}$

$= \dfrac{x^2 + 5x + 4}{x^2 + 5x + 10}$

◆◆◆ Relating Concepts 29–34 ◆◆◆

29. To add the fractions in the numerator, use the LCD $m(m - 1)$.

$\dfrac{4}{m} + \dfrac{m + 2}{m - 1} = \dfrac{4(m - 1)}{m(m - 1)} + \dfrac{m(m + 2)}{m(m - 1)}$

$= \dfrac{4m - 4 + m^2 + 2m}{m(m - 1)}$

$= \dfrac{m^2 + 6m - 4}{m(m - 1)}$

Copyright © 2012 Pearson Education, Inc. Publishing as Addison-Wesley.

30. To subtract the fractions in the denominator, use the same LCD, $m(m-1)$.

$$\frac{m+2}{m} - \frac{2}{m-1}$$

$$= \frac{(m+2)(m-1)}{m(m-1)} - \frac{m \cdot 2}{m(m-1)}$$

$$= \frac{m^2 + m - 2 - 2m}{m(m-1)}$$

$$= \frac{m^2 - m - 2}{m(m-1)}$$

31.

$$\begin{array}{cc} \text{Exercise 29} & \text{Exercise 30} \\ \text{answer} & \text{answer} \\ \downarrow & \downarrow \end{array}$$

$$\frac{m^2 + 6m - 4}{m(m-1)} \div \frac{m^2 - m - 2}{m(m-1)}$$

Multiply by the reciprocal.

$$= \frac{m^2 + 6m - 4}{m(m-1)} \cdot \frac{m(m-1)}{m^2 - m - 2}$$

$$= \frac{m^2 + 6m - 4}{m^2 - m - 2}$$

32. The LCD of all the denominators in the complex fraction is $m(m-1)$.

33.

$$\frac{\left(\dfrac{4}{m} + \dfrac{m+2}{m-1}\right) \cdot m(m-1)}{\left(\dfrac{m+2}{m} - \dfrac{2}{m-1}\right) \cdot m(m-1)}$$

$$= \frac{4(m-1) + m(m+2)}{(m+2)(m-1) - 2m}$$

$$= \frac{4m - 4 + m^2 + 2m}{m^2 + m - 2 - 2m}$$

$$= \frac{m^2 + 6m - 4}{m^2 - m - 2}$$

34. Answers will vary. Because of the complicated nature of the numerator and denominator of the complex fraction, using Method 1 takes much longer to simplify the complex fraction. Method 2 is a simpler, more direct means of simplifying and is most likely the preferred method.

35.

$$\frac{1}{x^{-2} + y^{-2}} = \frac{1}{\dfrac{1}{x^2} + \dfrac{1}{y^2}}$$

$$= \frac{x^2 y^2 (1)}{x^2 y^2 \left(\dfrac{1}{x^2} + \dfrac{1}{y^2}\right)} \qquad LCD = x^2 y^2$$

$$= \frac{x^2 y^2}{y^2 + x^2}$$

37. $\dfrac{x^{-2} + y^{-2}}{x^{-1} + y^{-1}}$

$$= \frac{\dfrac{1}{x^2} + \dfrac{1}{y^2}}{\dfrac{1}{x} + \dfrac{1}{y}}$$

Multiply the numerator and denominator by $x^2 y^2$, the LCD of all the fractions.

$$= \frac{x^2 y^2 \left(\dfrac{1}{x^2} + \dfrac{1}{y^2}\right)}{x^2 y^2 \left(\dfrac{1}{x} + \dfrac{1}{y}\right)}$$

$$= \frac{x^2 y^2 \cdot \dfrac{1}{x^2} + x^2 y^2 \cdot \dfrac{1}{y^2}}{x^2 y^2 \cdot \dfrac{1}{x} + x^2 y^2 \cdot \dfrac{1}{y}}$$

$$= \frac{y^2 + x^2}{xy^2 + x^2 y}, \quad \text{or} \quad \frac{y^2 + x^2}{xy(y + x)}$$

39. $\dfrac{x^{-1} + 2y^{-1}}{2y + 4x} = \dfrac{\dfrac{1}{x} + \dfrac{2}{y}}{2y + 4x}$

Multiply the numerator and denominator by xy, the LCD of all the fractions.

$$= \frac{xy \left(\dfrac{1}{x} + \dfrac{2}{y}\right)}{xy(2y + 4x)}$$

$$= \frac{y + 2x}{2xy(y + 2x)}$$

$$= \frac{1}{2xy}$$

41. (a) $\dfrac{\dfrac{3}{mp} - \dfrac{4}{p} + \dfrac{8}{m}}{2m^{-1} - 3p^{-1}}$

$$= \frac{\dfrac{3}{mp} - \dfrac{4}{p} + \dfrac{8}{m}}{\dfrac{2}{m} - \dfrac{3}{p}}$$

(b) $2m^{-1} = \dfrac{2}{m}$, not $\dfrac{1}{2m}$, since the exponent applies only to m, not to 2. Likewise, $3p^{-1} = \dfrac{3}{p}$, not $\dfrac{1}{3p}$.

(c) $\dfrac{\dfrac{3}{mp} - \dfrac{4}{p} + \dfrac{8}{m}}{2m^{-1} - 3p^{-1}} = \dfrac{\dfrac{3}{mp} - \dfrac{4}{p} + \dfrac{8}{m}}{\dfrac{2}{m} - \dfrac{3}{p}}$

Multiply the numerator and denominator by mp, the LCD of all the fractions.

Copyright © 2012 Pearson Education, Inc. Publishing as Addison-Wesley.

$$= \frac{mp\left(\dfrac{3}{mp} - \dfrac{4}{p} + \dfrac{8}{m}\right)}{mp\left(\dfrac{2}{m} - \dfrac{3}{p}\right)}$$

$$= \frac{mp \cdot \dfrac{3}{mp} - mp \cdot \dfrac{4}{p} + mp \cdot \dfrac{8}{m}}{mp \cdot \dfrac{2}{m} - mp \cdot \dfrac{3}{p}}$$

$$= \frac{3 - 4m + 8p}{2p - 3m}$$

43.
$$\tfrac{1}{2}x + \tfrac{1}{4}x = -9$$
$$4\left(\tfrac{1}{2}x + \tfrac{1}{4}x\right) = 4(-9)$$
$$2x + x = -36$$
$$3x = -36$$
$$x = -12$$

The solution set is $\{-12\}$.

45.
$$\frac{x-6}{5} = \frac{x+4}{10}$$
$$10\left(\frac{x-6}{5}\right) = 10\left(\frac{x+4}{10}\right)$$
$$2(x-6) = x+4$$
$$2x - 12 = x + 4$$
$$x = 16$$

The solution set is $\{16\}$.

47. $f(x) = \dfrac{1}{x}$

$x = 0$ is not in the domain.
The domain is $\{x \mid x \neq 0\}$.

7.4 Equations with Rational Expressions and Graphs

7.4 Now Try Exercises

N1. (a) $\dfrac{1}{3x} - \dfrac{3}{4x} = \dfrac{1}{3}$

Find the domain of each term.

For $\dfrac{1}{3x}$, $3x \neq 0$, so the domain is $\{x \mid x \neq 0\}$.

For $\dfrac{3}{4x}$, $4x \neq 0$, so the domain is $\{x \mid x \neq 0\}$.

The domain of $\tfrac{1}{3}$ is $(-\infty, \infty)$.

The intersection of these three domains is $\{x \mid x \neq 0\}$.

(b) $\dfrac{1}{x-7} + \dfrac{2}{x+7} = \dfrac{14}{x^2 - 49}$

Find the domain of each term.

For $\dfrac{1}{x-7}$, $x - 7 \neq 0$, so the domain is $\{x \mid x \neq 7\}$.

For $\dfrac{2}{x+7}$, $x + 7 \neq 0$, so the domain is $\{x \mid x \neq -7\}$.

For $\dfrac{14}{x^2 - 49}$, $x^2 - 49 = (x+7)(x-7) \neq 0$, so the domain is $\{x \mid x \neq \pm 7\}$.

The intersection of these three domains is $\{x \mid x \neq \pm 7\}$.

N2. $\dfrac{1}{3x} - \dfrac{3}{4x} = \dfrac{1}{3}$

Step 1
The domain is $\{x \mid x \neq 0\}$.

Step 2
Multiply each side by the LCD, $12x$.

$$12x\left(\frac{1}{3x} - \frac{3}{4x}\right) = 12x\left(\frac{1}{3}\right)$$

Step 3
$$12x\left(\frac{1}{3x}\right) - 12x\left(\frac{3}{4x}\right) = 12x\left(\frac{1}{3}\right)$$
$$4 - 9 = 4x$$
$$-5 = 4x$$
$$x = -\tfrac{5}{4}$$

Step 4
Check $x = -\tfrac{5}{4}$: $\quad -\tfrac{4}{15} + \tfrac{9}{15} = \tfrac{5}{15}$ *True*

The solution set is $\left\{-\tfrac{5}{4}\right\}$.

N3. $\dfrac{1}{t-7} + \dfrac{2}{t+7} = \dfrac{14}{t^2 - 49}$

The domain is $\{t \mid t \neq \pm 7\}$.

Multiply each side by the LCD, $(t+7)(t-7)$.

$$(t+7)(t-7)\left(\frac{1}{t-7} + \frac{2}{t+7}\right)$$
$$= (t+7)(t-7)\left(\frac{14}{t^2 - 49}\right)$$
$$(t+7)(t-7)\left(\frac{1}{t-7}\right) + (t+7)(t-7)\left(\frac{2}{t+7}\right)$$
$$= (t+7)(t-7)\left(\frac{14}{(t+7)(t-7)}\right)$$
$$1(t+7) + 2(t-7) = 14$$
$$t + 7 + 2t - 14 = 14$$
$$3t - 7 = 14$$
$$3t = 21$$
$$t = 7$$

Since 7 is not in the domain, the solution set is $\emptyset$.

Copyright © 2012 Pearson Education, Inc. Publishing as Addison-Wesley.

N4. $\dfrac{2}{t^2 - 2t - 8} - \dfrac{4}{t^2 + 6t + 8} = \dfrac{2}{t^2 - 16}$

Factor the denominators and determine the domain.

$t^2 - 2t - 8 = (t + 2)(t - 4)$, so $t \neq -2, 4$.
$t^2 + 6t + 8 = (t + 4)(t + 2)$, so $t \neq -4, -2$.
$t^2 - 16 = (t + 4)(t - 4)$, so $t \neq \pm 4$.

The domain is $\{x \mid x \neq -2, \pm 4\}$.

Multiply each side by the LCD, $(t + 2)(t + 4)(t - 4)$.

$(t + 2)(t + 4)(t - 4)\left[\dfrac{2}{t^2 - 2t - 8} - \dfrac{4}{t^2 + 6t + 8}\right]$

$= (t + 2)(t + 4)(t - 4)\dfrac{2}{t^2 - 16}$

$2(t + 4) - 4(t - 4) = 2(t + 2)$
$2t + 8 - 4t + 16 = 2t + 4$
$-2t + 24 = 2t + 4$
$-4t = -20$
$t = 5$

The solution set is $\{5\}$.

N5. $\dfrac{2x}{x - 2} = \dfrac{-3}{x} + \dfrac{4}{x - 2}$

The domain is $\{x \mid x \neq 0, 2\}$.

Multiply each side by the LCD, $x(x - 2)$.

$x(x - 2)\left(\dfrac{2x}{x - 2}\right)$

$= x(x - 2)\left(\dfrac{-3}{x} + \dfrac{4}{x - 2}\right)$

$x(2x) = -3(x - 2) + 4x$
$2x^2 = -3x + 6 + 4x$
$2x^2 - x - 6 = 0$
$(2x + 3)(x - 2) = 0$

$2x + 3 = 0$ or $x - 2 = 0$
$x = -\frac{3}{2}$ or $x = 2$

Because 2 is not in the domain of the equation, it is not a solution.

Check $x = -\frac{3}{2}$: $\frac{6}{7} = 2 - \frac{8}{7}$ *True*

The solution set is $\left\{-\frac{3}{2}\right\}$.

N6. $f(x) = \dfrac{1}{x + 1}$

x	-11	-6	-2	0	4	9
y	$-\frac{1}{10}$	$-\frac{1}{5}$	-1	1	$\frac{1}{5}$	$\frac{1}{10}$

Set the denominator equal to zero and solve for x.

$x + 1 = 0$
$x = -1$

Since $x \neq -1$, the equation of the vertical asymptote is $x = -1$. As the values of x get larger, the values of y get closer to 0, so $y = 0$ is the equation of the horizontal asymptote.

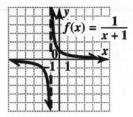

7.4 Section Exercises

1. (a) $\dfrac{1}{3x} + \dfrac{1}{2x} = \dfrac{x}{3}$

Only 0 would make any of the denominators equal to 0, so 0 would have to be rejected as a potential solution.

(b) The domain is $\{x \mid x \neq 0\}$.

3. (a) $\dfrac{1}{x + 1} - \dfrac{1}{x - 2} = 0$

Set each denominator equal to 0 and solve.

$x + 1 = 0$ or $x - 2 = 0$
$x = -1$ or $x = 2$

Solutions of -1 and 2 would be rejected since these values would make a denominator of the original equation equal to 0.

(b) The domain is $\{x \mid x \neq -1, 2\}$.

5. (a) $\dfrac{1}{x^2 - 16} - \dfrac{2}{x - 4} = \dfrac{1}{x + 4}$

$\dfrac{1}{(x + 4)(x - 4)} - \dfrac{2}{x - 4} = \dfrac{1}{x + 4}$

$x^2 - 16$ equals 0 if $x = \pm 4$.
$x - 4$ equals 0 if $x = 4$.
$x + 4$ equals 0 if $x = -4$.

So ± 4 would have to be rejected as potential solutions.

(b) The domain is $\{x \mid x \neq \pm 4\}$.

7. (a) $\dfrac{2}{x^2 - x} + \dfrac{1}{x + 3} = \dfrac{4}{x - 2}$

$x^2 - x = x(x - 1)$ is 0 if $x = 0$ or $x = 1$.
$x + 3$ is 0 if $x = -3$.
$x - 2$ is 0 if $x = 2$.

So 0, 1, -3, and 2 would have to be rejected as potential solutions.

(b) The domain is $\{x \mid x \neq 0, 1, -3, 2\}$.

Copyright © 2012 Pearson Education, Inc. Publishing as Addison-Wesley.

9. (a) $\dfrac{6}{4x+7} - \dfrac{3}{x} = \dfrac{5}{6x-13}$

$4x + 7$ is 0 if $x = -\frac{7}{4}$.

x is 0 if $x = 0$.

$6x - 13$ is 0 if $x = \frac{13}{6}$.

So $-\frac{7}{4}, 0,$ and $\frac{13}{6}$ would have to be rejected as potential solutions.

(b) The domain is $\left\{ x \mid x \neq -\frac{7}{4}, 0, \frac{13}{6} \right\}$.

11. (a) $\dfrac{3x+1}{x-4} = \dfrac{6x+5}{2x-7}$

$x - 4$ is 0 if $x = 4$.

$2x - 7$ is 0 if $x = \frac{7}{2}$.

So 4 and $\frac{7}{2}$ would have to be rejected as potential solutions.

(b) The domain is $\left\{ x \mid x \neq 4, \frac{7}{2} \right\}$.

13. No, there is no possibility that the proposed solution will be rejected, because there are no variables in the denominators in the original equation.

In Exercises 15–46, check each potential solution in the original equation.

15. $\dfrac{3}{4x} = \dfrac{5}{2x} - \dfrac{7}{4}$

Multiply by the LCD, $4x$. $(x \neq 0)$

$4x \left(\dfrac{3}{4x} \right) = 4x \left(\dfrac{5}{2x} - \dfrac{7}{4} \right)$

$3 = 2(5) - x(7)$

$3 = 10 - 7x$

$7x = 7$

$x = 1$

Check $x = 1$: $\frac{3}{4} = \frac{10}{4} - \frac{7}{4}$ *True*

The solution set is $\{1\}$.

17. $x - \dfrac{24}{x} = -2$

Multiply by the LCD, x. $(x \neq 0)$

$x \left(x - \dfrac{24}{x} \right) = -2 \cdot x$

$x^2 - 24 = -2x$

$x^2 + 2x - 24 = 0$

$(x+6)(x-4) = 0$

$x + 6 = 0$ or $x - 4 = 0$

$x = -6$ or $x = 4$

Check $x = -6$: $-6 + 4 = -2$ *True*

Check $x = 4$: $4 - 6 = -2$ *True*

The solution set is $\{-6, 4\}$.

19. $\dfrac{x}{4} - \dfrac{21}{4x} = -1$

Multiply by the LCD, $4x$. $(x \neq 0)$

$x^2 - 21 = -4x$

$x^2 + 4x - 21 = 0$

$(x+7)(x-3) = 0$

$x + 7 = 0$ or $x - 3 = 0$

$x = -7$ or $x = 3$

Check $x = -7$: $-\frac{7}{4} + \frac{3}{4} = -1$ *True*

Check $x = 3$: $\frac{3}{4} - \frac{7}{4} = -1$ *True*

The solution set is $\{-7, 3\}$.

21. $\dfrac{x-4}{x+6} = \dfrac{2x+3}{2x-1}$

Multiply by the LCD, $(x+6)(2x-1)$.

Note that $x \neq -6$ and $x \neq \frac{1}{2}$.

$(x+6)(2x-1)\left(\dfrac{x-4}{x+6} \right) = (x+6)(2x-1)\left(\dfrac{2x+3}{2x-1} \right)$

$(2x-1)(x-4) = (x+6)(2x+3)$

$2x^2 - 9x + 4 = 2x^2 + 15x + 18$

$-24x = 14$

$x = \dfrac{14}{-24} = -\dfrac{7}{12}$

A calculator check is suggested.

```
-7/12→X:(X-4)/(X
+6)►Frac
             -11/13
(2X+3)/(2X-1)►Fr
ac
             -11/13
```

Check $x = -\frac{7}{12}$: $-\frac{11}{13} = -\frac{11}{13}$ *True*

The solution set is $\left\{ -\frac{7}{12} \right\}$.

23. $\dfrac{3x+1}{x-4} = \dfrac{6x+5}{2x-7}$

Multiply by the LCD, $(x-4)(2x-7)$.

Note that $x \neq 4$ and $x \neq \frac{7}{2}$.

$(x-4)(2x-7)\left(\dfrac{3x+1}{x-4} \right) = (x-4)(2x-7)\left(\dfrac{6x+5}{2x-7} \right)$

$(2x-7)(3x+1) = (x-4)(6x+5)$

$6x^2 - 19x - 7 = 6x^2 - 19x - 20$

$-7 = -20$ *False*

The false statement indicates that the original equation has no solution.

The solution set is $\emptyset$.

Copyright © 2012 Pearson Education, Inc. Publishing as Addison-Wesley.

25.
$$\frac{1}{y-1} + \frac{5}{12} = \frac{-2}{3y-3}$$
$$\frac{1}{y-1} + \frac{5}{12} = \frac{-2}{3(y-1)}$$

Multiply by the LCD, $12(y-1)$. $(y \neq 1)$
$$12(y-1)\left(\frac{1}{y-1} + \frac{5}{12}\right) = 12(y-1)\left(\frac{-2}{3(y-1)}\right)$$
$$12 + 5(y-1) = -8$$
$$12 + 5y - 5 = -8$$
$$5y + 7 = -8$$
$$5y = -15$$
$$y = -3$$

Check $y = -3$: $-\frac{3}{12} + \frac{5}{12} = \frac{2}{12}$ *True*

The solution set is $\{-3\}$.

27.
$$\frac{7}{6x+3} - \frac{1}{3} = \frac{2}{2x+1}$$
$$\frac{7}{3(2x+1)} - \frac{1}{3} = \frac{2}{2x+1}$$

Multiply by the LCD, $3(2x+1)$. $\left(x \neq -\frac{1}{2}\right)$
$$3(2x+1)\left(\frac{7}{6x+3} - \frac{1}{3}\right) = 3(2x+1)\left(\frac{2}{2x+1}\right)$$
$$7 - 1(2x+1) = 3(2)$$
$$7 - 2x - 1 = 6$$
$$-2x = 0$$
$$x = 0$$

Check $x = 0$: $\frac{7}{3} - \frac{1}{3} = 2$ *True*

The solution set is $\{0\}$.

29.
$$\frac{6}{x-4} + \frac{5}{x} = \frac{-20}{x^2 - 4x}$$
$$\frac{6}{x-4} + \frac{5}{x} = \frac{-20}{x(x-4)}$$

Multiply by the LCD, $x(x-4)$. $(x \neq 0, 4)$
$$x(x-4)\left(\frac{6}{x-4} + \frac{5}{x}\right) = x(x-4)\left(\frac{-20}{x(x-4)}\right)$$
$$6x + 5(x-4) = -20$$
$$6x + 5x - 20 = -20$$
$$11x = 0$$
$$x = 0$$

Since 0 cannot appear in the denominator, this equation has no solution.

The solution set is $\emptyset$.

31.
$$\frac{3}{x+2} - \frac{2}{x^2 - 4} = \frac{1}{x-2}$$
$$\frac{3}{x+2} - \frac{2}{(x+2)(x-2)} = \frac{1}{x-2}$$
Multiply by the LCD, $(x+2)(x-2)$.
$(x \neq -2, 2)$

$$(x+2)(x-2)\left(\frac{3}{x+2} - \frac{2}{(x+2)(x-2)}\right)$$
$$= (x+2)(x-2)\left(\frac{1}{x-2}\right)$$
$$3(x-2) - 2 = x+2$$
$$3x - 6 - 2 = x+2$$
$$3x - 8 = x+2$$
$$2x = 10$$
$$x = 5$$

Check $x = 5$: $\frac{9}{21} - \frac{2}{21} = \frac{1}{3}$ *True*

The solution set is $\{5\}$.

33.
$$\frac{1}{y+2} + \frac{3}{y+7} = \frac{5}{y^2 + 9y + 14}$$
$$\frac{1}{y+2} + \frac{3}{y+7} = \frac{5}{(y+2)(y+7)}$$
Multiply by the LCD, $(y+2)(y+7)$. $(y \neq -2, -7)$

$$(y+2)(y+7)\left(\frac{1}{y+2} + \frac{3}{y+7}\right)$$
$$= (y+2)(y+7)\left(\frac{5}{(y+2)(y+7)}\right)$$
$$(y+7) + 3(y+2) = 5$$
$$y + 7 + 3y + 6 = 5$$
$$4y + 13 = 5$$
$$4y = -8$$
$$y = -2$$

But y cannot equal -2 because that would make the denominator $y + 2$ equal to 0. Since division by 0 is undefined, the equation has no solution.

The solution set is $\emptyset$.

35.
$$\frac{9}{x} + \frac{4}{6x-3} = \frac{2}{6x-3}$$
Multiply by the LCD, $x(6x-3)$. $\left(x \neq 0, \frac{1}{2}\right)$
$$x(6x-3)\left(\frac{9}{x} + \frac{4}{6x-3}\right) = x(6x-3)\left(\frac{2}{6x-3}\right)$$
$$9(6x-3) + 4x = 2x$$
$$54x - 27 + 4x = 2x$$
$$56x = 27$$
$$x = \frac{27}{56}$$

Check $x = \frac{27}{56}$: $\frac{56}{3} + \left(-\frac{112}{3}\right) = -\frac{56}{3}$ *True*

The solution set is $\left\{\frac{27}{56}\right\}$.

37.
$$\frac{1}{x-2} + \frac{1}{4} = \frac{1}{4(x^2 - 4)}$$
$$\frac{1}{x-2} + \frac{1}{4} = \frac{1}{4(x+2)(x-2)}$$
Multiply by the LCD, $4(x+2)(x-2)$.
$(x \neq -2, 2)$

$$4(x+2)(x-2)\left(\frac{1}{x-2}+\frac{1}{4}\right)$$
$$= 4(x+2)(x-2)\left[\frac{1}{4(x^2-4)}\right]$$
$$4(x+2)+(x+2)(x-2)=1$$
$$4x+8+x^2-4=1$$
$$x^2+4x+3=0$$
$$(x+3)(x+1)=0$$
$$x+3=0 \quad \text{or} \quad x+1=0$$
$$x=-3 \quad \text{or} \quad x=-1$$

Check $x=-3$: $-\frac{1}{5}+\frac{1}{4}=\frac{1}{20}$ *True*
Check $x=-1$: $-\frac{1}{3}+\frac{1}{4}=-\frac{1}{12}$ *True*
The solution set is $\{-3,-1\}$.

39.
$$\frac{6}{w+3}+\frac{-7}{w-5}=\frac{-48}{w^2-2w-15}$$
$$\frac{6}{w+3}+\frac{-7}{w-5}=\frac{-48}{(w+3)(w-5)}$$
Multiply by the LCD, $(w+3)(w-5)$.
$(w \neq -3, 5)$
$$(w+3)(w-5)\left(\frac{6}{w+3}+\frac{-7}{w-5}\right)$$
$$= (w+3)(w-5)\left[\frac{-48}{(w+3)(w-5)}\right]$$
$$6(w-5)-7(w+3)=-48$$
$$6w-30-7w-21=-48$$
$$-w-51=-48$$
$$-w=3$$
$$w=-3$$

But w cannot equal -3 because that would make the denominator $w+3$ equal to 0. Since division by 0 is undefined, the equation has no solution.

The solution set is $\emptyset$.

41.
$$\frac{x}{x-3}+\frac{4}{x+3}=\frac{18}{x^2-9}$$
$$\frac{x}{x-3}+\frac{4}{x+3}=\frac{18}{(x-3)(x+3)}$$
Multiply by the LCD, $(x-3)(x+3)$. $(x \neq 3, -3)$
$$(x-3)(x+3)\left(\frac{x}{x-3}+\frac{4}{x+3}\right)$$
$$= (x-3)(x+3)\left(\frac{18}{(x-3)(x+3)}\right)$$
$$x(x+3)+4(x-3)=18$$
$$x^2+3x+4x-12=18$$
$$x^2+7x-30=0$$
$$(x-3)(x+10)=0$$
$$x-3=0 \quad \text{or} \quad x+10=0$$
$$x=3 \quad \text{or} \quad x=-10$$

But $x \neq 3$ since a denominator of 0 results. The only solution to check is -10.

Check $x=-10$: $\frac{10}{13}+\left(-\frac{4}{7}\right)=\frac{18}{91}$ *True*

The solution set is $\{-10\}$.

43.
$$\frac{1}{x+4}+\frac{x}{x-4}=\frac{-8}{x^2-16}$$
$$\frac{1}{x+4}+\frac{x}{x-4}=\frac{-8}{(x+4)(x-4)}$$
Multiply by the LCD, $(x+4)(x-4)$. $(x \neq -4, 4)$
$$(x+4)(x-4)\left(\frac{1}{x+4}+\frac{x}{x-4}\right)$$
$$= (x+4)(x-4)\left(\frac{-8}{(x+4)(x-4)}\right)$$
$$(x-4)+x(x+4)=-8$$
$$x-4+x^2+4x=-8$$
$$x^2+5x+4=0$$
$$(x+4)(x+1)=0$$

$$x+4=0 \quad \text{or} \quad x+1=0$$
$$x=-4 \qquad\qquad x=-1$$

But x cannot equal -4, so we only need to check -1.
Check $x=-1$: $\frac{1}{3}+\frac{1}{5}=\frac{8}{15}$ *True*

The solution set is $\{-1\}$.

45.
$$\frac{2}{k^2+k-6}+\frac{1}{k^2-k-2}=\frac{4}{k^2+4k+3}$$
$$\frac{2}{(k+3)(k-2)}+\frac{1}{(k-2)(k+1)}=\frac{4}{(k+1)(k+3)}$$
Multiply by the LCD, $(k+3)(k-2)(k+1)$.
$(k \neq -3, 2, -1)$
$$2(k+1)+1(k+3)=4(k-2)$$
$$2k+2+k+3=4k-8$$
$$13=k$$

Check $k=13$: $\frac{1}{88}+\frac{1}{154}=\frac{1}{56}$ *True*

The solution set is $\{13\}$.

47.
$$\frac{5x+14}{x^2-9}=\frac{-2x^2-5x+2}{x^2-9}+\frac{2x+4}{x-3}$$
$$\frac{5x+14}{(x+3)(x-3)}$$
$$= \frac{-2x^2-5x+2}{(x+3)(x-3)}+\frac{2x+4}{x-3}$$
Multiply by the LCD, $(x+3)(x-3)$.
$(x \neq -3, 3)$
$$5x+14=-2x^2-5x+2+(2x+4)(x+3)$$
$$5x+14=-2x^2-5x+2+2x^2+10x+12$$
$$5x+14=5x+14 \quad \text{True}$$

This equation is true for every real number value of x, but we have already determined that $x \neq -3$ or $x \neq 3$. So every real number except -3 and 3 is a solution.

The solution set is $\{x \mid x \neq \pm 3\}$ or $(-\infty, -3) \cup (-3, 3) \cup (3, \infty)$.

Copyright © 2012 Pearson Education, Inc. Publishing as Addison-Wesley.

49. $f(x) = \dfrac{2}{x}$ is not defined when $x = 0$, so an equation of the vertical asymptote is $x = 0$. The graph approaches the x-axis as x gets very large (positive or negative), so $y = 0$ is the horizontal asymptote.

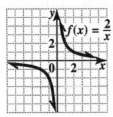

51. $g(x) = -\dfrac{1}{x}$ is not defined when $x = 0$, so an equation of the vertical asymptote is $x = 0$. The graph approaches the x-axis as x gets very large (positive or negative), so $y = 0$ is the horizontal asymptote.

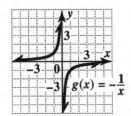

53. $f(x) = \dfrac{1}{x - 2}$ is not defined when $x = 2$, so an equation of the vertical asymptote is $x = 2$. The graph approaches the x-axis as x gets very large (positive or negative), so $y = 0$ is the horizontal asymptote.

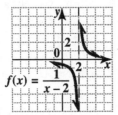

55. $w(x) = \dfrac{x^2}{2(1 - x)}$

(a) $w(0.1) = \dfrac{(0.1)^2}{2(1 - 0.1)}$

$= \dfrac{0.01}{2(0.9)} \approx 0.006$

To the nearest tenth, $w(0.1)$ is 0.

(b) $w(0.8) = \dfrac{(0.8)^2}{2(1 - 0.8)}$

$= \dfrac{0.64}{2(0.2)} = 1.6$

(c) $w(0.9) = \dfrac{(0.9)^2}{2(1 - 0.9)}$

$= \dfrac{0.81}{2(0.1)} = 4.05 \approx 4.1$

(d) Based on the answers in (a), (b), and (c), we see that as the traffic intensity increases, the waiting time also increases.

57. **(a)** $F(r) = \dfrac{225{,}000}{r}$

$450 = \dfrac{225{,}000}{r}$ *Let $F(r) = 450$.*

$450r = 225{,}000$

$r = \dfrac{225{,}000}{450} = 500$

The radius must be 500 feet.

(b) As r increases, the fraction $\frac{225{,}000}{r}$ gets smaller, so the force, $f(r)$, decreases.

59. The number of solutions of the equation $f(x) = 0$ is the same as the number of x-intercepts, four.

61. Solve $d = rt$ for t.

$\dfrac{d}{r} = \dfrac{rt}{r}$ *Divide by r.*

$\dfrac{d}{r} = t$

63.   Solve $P = a + b + c$ for c.

$P - a = b + c$ *Subtract a.*

$P - a - b = c$ *Subtract b.*

Summary Exercises on Rational Expressions and Equations

1. $\dfrac{x}{2} - \dfrac{x}{4} = 5$

There is an equals sign, so this is an *equation*.

$4\left(\dfrac{x}{2} - \dfrac{x}{4}\right) = 4(5)$ *Multiply by 4.*

$2x - x = 20$

$x = 20$

Check $x = 20$: $10 - 5 = 5$ *True*

The solution set is $\{20\}$.

3. $\dfrac{6}{7x} - \dfrac{4}{x}$

No equals sign appears so this is an *expression*.

$= \dfrac{6}{7x} - \dfrac{4 \cdot 7}{x \cdot 7}$ *LCD $= 7x$*

$= \dfrac{6 - 28}{7x} = \dfrac{-22}{7x}$

Copyright © 2012 Pearson Education, Inc. Publishing as Addison-Wesley.

5. $\dfrac{5}{7t} = \dfrac{52}{7} - \dfrac{3}{t}$

There is an equals sign, so this is an *equation*.

Multiply by the LCD, $7t$. ($t \neq 0$)

$7t\left(\dfrac{5}{7t}\right) = 7t\left(\dfrac{52}{7} - \dfrac{3}{t}\right)$

$5 = 52t - 3(7)$

$5 = 52t - 21$

$26 = 52t$

$t = \dfrac{26}{52} = \dfrac{1}{2}$

Check $t = \dfrac{1}{2}$: $\quad \dfrac{10}{7} = \dfrac{52}{7} - 6 \quad$ *True*

The solution set is $\left\{\dfrac{1}{2}\right\}$.

7. $\dfrac{7}{6x} + \dfrac{5}{8x}$

No equals sign appears so this is an *expression*.

$= \dfrac{7}{3 \cdot 2x} + \dfrac{5}{4 \cdot 2x}$

$= \dfrac{7(4)}{3 \cdot 2x \cdot 4} + \dfrac{5(3)}{4 \cdot 2x \cdot 3} \qquad LCD = 24x$

$= \dfrac{28 + 15}{3 \cdot 4 \cdot 2x} = \dfrac{43}{24x}$

9. $\dfrac{\dfrac{6}{x+1} - \dfrac{1}{x}}{\dfrac{2}{x} - \dfrac{4}{x+1}}$

No equals sign appears so this is an *expression*.
Multiply the numerator and denominator by the LCD of all the fractions, $x(x + 1)$.

$= \dfrac{6(x) - 1(x+1)}{2(x+1) - 4(x)}$

$= \dfrac{6x - x - 1}{2x + 2 - 4x}$

$= \dfrac{5x - 1}{-2x + 2}, \quad \text{or} \quad \dfrac{5x - 1}{-2(x - 1)}$

11. $\dfrac{x}{x+y} + \dfrac{2y}{x-y}$

No equals sign appears so this is an *expression*.

$\dfrac{x(x-y)}{(x+y)(x-y)} + \dfrac{2y(x+y)}{(x-y)(x+y)}$

$\qquad\qquad LCD = (x+y)(x-y)$

$= \dfrac{x^2 - xy + 2xy + 2y^2}{(x+y)(x-y)}$

$= \dfrac{x^2 + xy + 2y^2}{(x+y)(x-y)}$

13. $\dfrac{x-2}{9} \cdot \dfrac{5}{8 - 4x}$

No equals sign appears so this is an *expression*.

$= \dfrac{x-2}{9} \cdot \dfrac{5}{-4(x-2)}$

$= \dfrac{5}{-36} = -\dfrac{5}{36}$

15. $\dfrac{b^2 + b - 6}{b^2 + 2b - 8} \cdot \dfrac{b^2 + 8b + 16}{3b + 12}$

No equals sign appears so this is an *expression*.

$= \dfrac{(b+3)(b-2)}{(b+4)(b-2)} \cdot \dfrac{(b+4)(b+4)}{3(b+4)}$

$= \dfrac{b+3}{3}$

17. $\dfrac{5}{x^2 - 2x} - \dfrac{3}{x^2 - 4}$

No equals sign appears so this is an *expression*.

$= \dfrac{5}{x(x-2)} - \dfrac{3}{(x+2)(x-2)}$

$= \dfrac{5(x+2)}{x(x-2)(x+2)} - \dfrac{3x}{(x-2)(x+2)x}$

$= \dfrac{5x + 10 - 3x}{x(x-2)(x+2)}$

$= \dfrac{2x + 10}{x(x-2)(x+2)}$

19. $\dfrac{\dfrac{5}{x} - \dfrac{3}{y}}{\dfrac{9x^2 - 25y^2}{x^2 y}}$

No equals sign appears so this is an *expression*.
Multiply the numerator and denominator by the LCD of all the fractions, $x^2 y$.

$= \dfrac{x^2 y\left(\dfrac{5}{x} - \dfrac{3}{y}\right)}{x^2 y\left(\dfrac{9x^2 - 25y^2}{x^2 y}\right)}$

$= \dfrac{5xy - 3x^2}{9x^2 - 25y^2}$

$= \dfrac{-x(3x - 5y)}{(3x + 5y)(3x - 5y)}$

$= \dfrac{-x}{3x + 5y}$

Copyright © 2012 Pearson Education, Inc. Publishing as Addison-Wesley.

21. $\dfrac{4y^2 - 13y + 3}{2y^2 - 9y + 9} \div \dfrac{4y^2 + 11y - 3}{6y^2 - 5y - 6}$

No equals sign appears so this is an *expression*.

$= \dfrac{(4y - 1)(y - 3)}{(2y - 3)(y - 3)} \cdot \dfrac{(2y - 3)(3y + 2)}{(4y - 1)(y + 3)}$

$= \dfrac{3y + 2}{y + 3}$

23. $\dfrac{3r}{r - 2} = 1 + \dfrac{6}{r - 2}$

There is an equals sign, so this is an *equation*.

Multiply by the LCD, $r - 2$. $(r \neq 2)$

$3r = r - 2 + 6$

$2r = 4$

$r = 2$

But $r \neq 2$.

The solution set is $\emptyset$.

25. $\dfrac{-1}{3 - x} - \dfrac{2}{x - 3} = \dfrac{-1}{-(x - 3)} - \dfrac{2}{x - 3}$

No equals sign appears so this is an *expression*.

$= \dfrac{1}{x - 3} - \dfrac{2}{x - 3}$

$= \dfrac{-1}{x - 3}$, or $\dfrac{-1}{-(3 - x)} = \dfrac{1}{3 - x}$

27. $\dfrac{2}{y + 1} - \dfrac{3}{y^2 - y - 2} = \dfrac{3}{y - 2}$

$\dfrac{2}{y + 1} - \dfrac{3}{(y - 2)(y + 1)} = \dfrac{3}{y - 2}$

There is an equals sign, so this is an *equation*.

Multiply by the LCD, $(y - 2)(y + 1)$.

$(y \neq -1, 2)$

$2(y - 2) - 3 = 3(y + 1)$

$2y - 4 - 3 = 3y + 3$

$2y - 7 = 3y + 3$

$-10 = y$

Check $y = -10$: $-\dfrac{2}{9} - \dfrac{1}{36} = -\dfrac{1}{4}$ *True*

The solution set is $\{-10\}$.

29. $\dfrac{3}{y - 3} - \dfrac{3}{y^2 - 5y + 6} = \dfrac{2}{y - 2}$

$\dfrac{3}{y - 3} - \dfrac{3}{(y - 3)(y - 2)} = \dfrac{2}{y - 2}$

There is an equals sign, so this is an *equation*.

Multiply by the LCD, $(y - 3)(y - 2)$. $(y \neq 2, 3)$

$3(y - 2) - 3 = 2(y - 3)$

$3y - 6 - 3 = 2y - 6$

$3y - 9 = 2y - 6$

$y = 3$

But $y \neq 3$.

The solution set is $\emptyset$.

7.5   Applications of Rational Expressions

7.5   Now Try Exercises

N1. $\dfrac{1}{f} = \dfrac{1}{p} + \dfrac{1}{q}$

$\dfrac{1}{f} = \dfrac{1}{50} + \dfrac{1}{10}$ *Let p = 50, q = 10.*

Multiply by the LCD, $50f$.

$50f \cdot \dfrac{1}{f} = 50f \left(\dfrac{1}{50} + \dfrac{1}{10} \right)$

$50f \cdot \dfrac{1}{f} = 50f \left(\dfrac{1}{50} \right) + 50f \left(\dfrac{1}{10} \right)$

$50 = f + 5f$

$50 = 6f$

$f = \dfrac{50}{6} = \dfrac{25}{3}$

The focal length is $\dfrac{25}{3}$ cm.

N2. Solve $\dfrac{1}{m} - \dfrac{2}{n} = 5$ for m.

Multiply by the LCD, mn.

$mn \left(\dfrac{1}{m} - \dfrac{2}{n} \right) = mn(5)$

$mn \left(\dfrac{1}{m} \right) - mn \left(\dfrac{2}{n} \right) = mn(5)$

$n - 2m = 5mn$

Get the m terms on one side.

$n = 2m + 5mn$

$n = m(2 + 5n)$ *Factor out m.*

$m = \dfrac{n}{2 + 5n}$,   or $\dfrac{n}{5n + 2}$

N3. Solve $\dfrac{NR}{N - n} = t$ for N.

Multiply by the LCD, $N - n$.

$\dfrac{NR}{N - n}(N - n) = t(N - n)$

$NR = tN - tn$

Get the N terms on one side.

$NR - tN = -tn$

$N(R - t) = -tn$ *Factor out N.*

$N = \dfrac{-tn}{R - t}$,   or $\dfrac{nt}{t - R}$

Copyright © 2012 Pearson Education, Inc. Publishing as Addison-Wesley.

N4. *Step 2*
Let $x =$ the number (in millions) of Americans who lived in poverty in 2008.

Step 3
Set up a proportion.

$$\frac{13}{100} = \frac{x}{302}$$

Step 4
To solve the equation, multiply by the LCD.

$$30{,}200\left(\frac{13}{100}\right) = 30{,}200\left(\frac{x}{302}\right)$$
$$302(13) = 100(x)$$
$$3926 = 100x$$
$$x = 39.26$$

Step 5
There were 39.26 million Americans who lived in poverty in 2008.

Step 6
The ratio of 39.26 to 302 equals $\frac{13}{100}$.

N5. *Step 2*
Let $x =$ the additional number of gallons of gasoline needed.

Step 3
He knows that he can drive 500 miles with 28 gallons of gasoline. He wants to drive 400 miles using $(10 + x)$ gallons of gasoline. Set up a proportion.

$$\frac{500}{28} = \frac{400}{10 + x}$$
$$\frac{125}{7} = \frac{400}{10 + x} \qquad \text{\textit{Reduce.}}$$

Step 4
Find the cross products and solve for x.

$$125(10 + x) = 7(400)$$
$$1250 + 125x = 2800 \qquad \text{\textit{Distributive property}}$$
$$125x = 1550 \qquad \text{\textit{Subtract 1250.}}$$
$$x = \frac{1550}{125} \qquad \text{\textit{Divide by 125.}}$$
$$x = 12.4$$

Step 5
He will need about 12.4 more gallons of gasoline.

Step 6
Check The 10 gallons plus the 12.4 gallons equals 22.4 gallons. We'll check the rates (miles/gallon). Note that we could also use gallons/mile.

$$\frac{500}{28} \approx 17.86 \text{ mpg} \qquad \frac{400}{22.4} \approx 17.86 \text{ mpg}$$

The rates are approximately equal, so the solution is correct.

N6. **(a)** Let $x =$ the rate of the current.

Use $d = rt$, or $t = \frac{d}{r}$, to complete the table.

	d	r	t
Against Current	36	$20 - x$	$\dfrac{36}{20 - x}$
With Current	44	$20 + x$	$\dfrac{44}{20 + x}$

(b) Because the time against the current equals the time with the current,

$$\frac{36}{20 - x} = \frac{44}{20 + x}.$$

Multiply by the LCD, $(20 - x)(20 + x)$.

$$\frac{36}{20 - x}(20 - x)(20 + x)$$
$$= \frac{44}{20 + x}(20 - x)(20 + x)$$
$$36(20 + x) = 44(20 - x)$$
$$720 + 36x = 880 - 44x$$
$$80x = 160$$
$$x = 2$$

The rate of the current is 2 mph.

N7. Let $x =$ the driving rate for Pat.
Then $x + 5 =$ the driving rate for James.

Use $d = rt$, or $t = \frac{d}{r}$, to complete the table.

	d	r	t
James	130	$x + 5$	$\dfrac{130}{x + 5}$
Pat	$310 - 130 = 180$	x	$\dfrac{180}{x}$

Now write an equation.

Time for James	plus	time for Pat	equals	5 hr.
↓	↓	↓	↓	↓
$\dfrac{130}{x + 5}$	$+$	$\dfrac{180}{x}$	$=$	5

Multiply by the LCD, $x(x + 5)$.

$$x(x + 5)\left(\frac{130}{x + 5} + \frac{180}{x}\right) = 5x(x + 5)$$
$$130x + 180(x + 5) = 5x(x + 5)$$
$$130x + 180x + 900 = 5x^2 + 25x$$
$$0 = 5x^2 - 285x - 900$$
$$0 = x^2 - 57x - 180$$
$$0 = (x - 60)(x + 3)$$
$$x = 60 \quad \text{or} \quad -3$$

Reject -3 for a rate, so Pat drove 60 miles per hour and James drove 65 miles per hour.

Copyright © 2012 Pearson Education, Inc. Publishing as Addison-Wesley.

N8. Let x = the time it will take them working together.

Make a table.

	Rate	Time Working Together	Fractional Part of the Job Done
Gina	$\frac{1}{2}$	x	$\frac{1}{2}x$
Matt	$\frac{1}{3}$	x	$\frac{1}{3}x$

Part done by Gina	plus	part done by Matt	equals	1 whole job.
↓	↓	↓	↓	↓
$\frac{1}{2}x$	$+$	$\frac{1}{3}x$	$=$	1

Multiply by the LCD, 6.

$$6\left(\tfrac{1}{2}x + \tfrac{1}{3}x\right) = 6 \cdot 1$$
$$3x + 2x = 6$$
$$5x = 6$$
$$x = \tfrac{6}{5}$$

It will take them $\frac{6}{5}$ hr, or 1 hr, 12 minutes working together.

7.5 Section Exercises

1. **A.** $b = \frac{p}{r}$ is the same as $p = br$.

B. $r = \frac{b}{p}$ is the same as $b = pr$.

C. $b = \frac{r}{p}$ is the same as $r = bp$.

D. $p = \frac{r}{b}$ is the same as $r = bp$.

Choice **A** is correct.

3. **A.** $a = mF$ is the same as $m = \frac{a}{F}$.

B. $F = \frac{m}{a}$ is the same as $m = Fa$.

C. $F = \frac{a}{m}$ is the same as $Fm = a$, which is the same as $m = \frac{a}{F}$.

D. $F = ma$ is the same as $m = \frac{F}{a}$.

Choice **D** is correct.

5.
$$\frac{1}{a} = \frac{1}{b} + \frac{1}{c}$$
$$\frac{1}{8} = \frac{1}{b} + \frac{1}{12} \qquad \textit{Let } a = 8, c = 12.$$
$$24b\left(\frac{1}{8}\right) = 24b\left(\frac{1}{b} + \frac{1}{12}\right) \quad \textit{Multiply by 24b.}$$
$$3b = 24 + 2b$$
$$b = 24$$

7.
$$c = \frac{100b}{L}$$
$$80 = \frac{100 \cdot 5}{L} \qquad \textit{Let } c = 80, b = 5.$$
$$80L = 500 \qquad \textit{Multiply by L.}$$
$$L = \tfrac{500}{80} = \tfrac{25}{4}, \text{ or } 6.25$$

9. Solve $F = \dfrac{GMm}{d^2}$ for G.
$$Fd^2 = GMm \qquad \textit{Multiply by } d^2.$$
$$\frac{Fd^2}{Mm} = G \qquad \textit{Divide by } Mm.$$

11. Solve $\dfrac{1}{a} = \dfrac{1}{b} + \dfrac{1}{c}$ for a.
Multiply by the LCD, abc.
$$abc\left(\frac{1}{a}\right) = abc\left(\frac{1}{b} + \frac{1}{c}\right)$$
$$bc = ac + ab$$
$$bc = a(c + b) \qquad \textit{Factor out a.}$$
$$\frac{bc}{c + b} = a \qquad \textit{Divide by } c + b.$$

13. Solve $\dfrac{PV}{T} = \dfrac{pv}{t}$ for v.
$$\frac{PVt}{T} = pv \qquad \textit{Multiply by t.}$$
$$\frac{PVt}{pT} = v \qquad \textit{Divide by p.}$$

15. Solve $I = \dfrac{nE}{R + nr}$ for r.
$$I(R + nr) = nE \qquad \textit{Multiply by } R + nr.$$
$$IR + Inr = nE$$
$$Inr = nE - IR$$
$$r = \frac{nE - IR}{In}, \quad \text{ or } \quad r = \frac{IR - nE}{-In}$$

17. Solve $\mathcal{A} = \dfrac{1}{2}h(b + B)$ for b.
$$\frac{2}{h}(\mathcal{A}) = \frac{2}{h}\left[\frac{1}{2}h(b + B)\right] \qquad \textit{Multiply by } \tfrac{2}{h}.$$
$$\frac{2\mathcal{A}}{h} = b + B$$
$$\frac{2\mathcal{A}}{h} - B = b, \quad \text{ or } \quad b = \frac{2\mathcal{A} - hB}{h}$$

19. Solve $\dfrac{E}{e} = \dfrac{R + r}{r}$ for r.
$$Er = e(R + r) \qquad \textit{Multiply by er.}$$
$$Er = eR + er$$
$$Er - er = eR \qquad \textit{Subtract er.}$$
$$r(E - e) = eR$$
$$r = \frac{eR}{E - e} \qquad \textit{Divide by } E - e.$$

21. To solve the equation $m = \dfrac{ab}{a - b}$ for a, the first step is to multiply both sides of the equation by the LCD, $a - b$.

Copyright © 2012 Pearson Education, Inc. Publishing as Addison-Wesley.

23. Let $x = $ the number of girls in the class.
Write a proportion using ratios of girl students to total students.

$$\frac{3}{4} = \frac{x}{28}$$
$$28\left(\tfrac{3}{4}\right) = 28\left(\tfrac{x}{28}\right) \quad \textit{Multiply by 28.}$$
$$21 = x$$

There are 21 girls and $28 - 21 = 7$ boys in the class.

25. Marin's rate $= \dfrac{1 \text{ job}}{\text{time to complete 1 job}}$

$$= \frac{1 \text{ job}}{3 \text{ hours}}$$
$$= \tfrac{1}{3} \text{ job per hour}$$

27. Let $x = $ the amount of liquid precipitation that would produce 31.5 inches of snow.
Write a proportion using ratios of snow to liquid precipitation.

$$\frac{31.5}{x} = \frac{18}{1}$$
$$x\left(\tfrac{31.5}{x}\right) = x(18) \qquad \textit{Multiply by x.}$$
$$31.5 = 18x$$
$$x = \frac{31.5}{18} = 1.75$$

The amount of liquid precipitation that would produce 31.5 inches of snow is 1.75 inches.

29. Let $x = $ the distance between Chicago and El Paso on the map (in inches).

Write a proportion with one ratio involving map distances and the other involving actual distances.

$$\frac{x \text{ inches}}{4.125 \text{ inches}} = \frac{1606 \text{ miles}}{1238 \text{ miles}}$$
$$1238x = 4.125(1606)$$
$$1238x = 6624.75$$
$$x \approx 5.351 \approx 5.4$$

The distance on the map between Chicago and El Paso would be about 5.4 inches.

31. Let $x = $ the distance between Madrid and Rio de Janeiro on the map (in inches).

Write a proportion with one ratio involving map distances and the other involving actual distances.

$$\frac{x \text{ inches}}{8.5 \text{ inches}} = \frac{5045 \text{ miles}}{5619 \text{ miles}}$$
$$5619x = 8.5(5045)$$
$$5619x = 42{,}882.5$$
$$x \approx 7.632 \approx 7.6$$

The distance on the map between Madrid and Rio de Janeiro would be about 7.6 inches.

33. Let $x = $ the number of teachers.
Write a proportion using ratios of teachers to students.

$$\frac{1}{15} = \frac{x}{846}$$

Multiply each side by $846 \cdot 15$.

$$846 \cdot 15 \cdot \frac{1}{15} = 846 \cdot 15 \cdot \frac{x}{846}$$
$$846 = 15x$$
$$x = \tfrac{846}{15} = 56.4$$

There would be 56 teachers.

35. Let $x = $ the number of deer in the forest preserve.
Write and solve a proportion.

$$\frac{\text{total in forest}}{\text{tagged in forest}} = \frac{\text{total in sample}}{\text{tagged in sample}}$$
$$\frac{x}{42} = \frac{75}{15}$$
$$\frac{x}{42} = \frac{5}{1} \qquad \textit{Reduce.}$$
$$x = 5(42)$$
$$x = 210$$

There are approximately 210 deer in the forest preserve.

37. Let $x = $ the number of fish in the lake.
Write and solve a proportion.

$$\frac{\text{total in lake}}{\text{tagged in lake}} = \frac{\text{total in sample}}{\text{tagged in sample}}$$
$$\frac{x}{500} = \frac{400}{8}$$
$$\frac{x}{500} = \frac{50}{1} \qquad \textit{Reduce.}$$
$$x = 500(50)$$
$$= 25{,}000$$

There are approximately 25,000 fish in the lake.

39. *Step 2*
Let $x = $ the additional number of gallons of gasoline needed.

Step 3
He knows that he can drive 156 miles with 5 gallons of gasoline. He wants to drive 300 miles using $(3 + x)$ gallons of gasoline. Set up a proportion.

$$\frac{156}{5} = \frac{300}{3 + x}$$

Step 4
Find the cross products and solve for x.

Copyright © 2012 Pearson Education, Inc. Publishing as Addison-Wesley.

$$156(3 + x) = 5(300)$$
$$468 + 156x = 1500$$
$$156x = 1032$$
$$x = \frac{1032}{156}$$
$$x \approx 6.6$$

Step 5
He will need about 6.6 more gallons of gasoline.

Step 6
Check The 3 gallons plus the 6.6 gallons equals 9.6 gallons. We'll check the rates (miles/gallon). Note that we could also use gallons/mile.

$$\frac{156}{5} = 31.2 \text{ mpg} \qquad \frac{300}{9.6} = 31.25 \text{ mpg}$$

The rates are approximately equal, so the solution is correct. Note that we could have used our exact value for x to get the exact rate

$$\frac{300}{3 + (1032/156)} = 31.2.$$

41. Since $\frac{4}{6} = \frac{6}{9} = \frac{2}{3}$, use the proportion

$$\frac{2}{3} = \frac{2x + 1}{2x + 5}.$$
$$2(2x + 5) = 3(2x + 1)$$
$$4x + 10 = 6x + 3$$
$$7 = 2x$$
$$\frac{7}{2} = x$$

Since $x = \frac{7}{2}$,

$$AC = 2x + 1 = 2(\tfrac{7}{2}) + 1 = 8$$
$$\text{and} \quad DF = 2x + 5 = 2(\tfrac{7}{2}) + 5 = 12.$$

43. Let x represent the amount to administer in milliliters.

$$\frac{100 \text{ mg}}{2 \text{ mL}} = \frac{120 \text{ mg}}{x \text{ mL}}$$

Multiply each side by the LCD, $2x$.

$$2x\left(\frac{100}{2}\right) = 2x\left(\frac{120}{x}\right)$$
$$100x = 240$$
$$x = 2.4$$

The correct dose is 2.4 mL.

45. *Step 2*
Let x represent the rate of the current of the river. The boat goes 12 mph, so the downstream rate is $12 + x$ and the upstream rate is $12 - x$.

Use $t = \frac{d}{r}$ and make a table.

	Distance	Rate	Time
Downstream	10	$12 + x$	$\dfrac{10}{12 + x}$
Upstream	6	$12 - x$	$\dfrac{6}{12 - x}$

Step 3
Because the time upstream equals the time downstream,

$$\frac{6}{12 - x} = \frac{10}{12 + x}$$

Step 4
Multiply by the LCD, $(12 - x)(12 + x)$.

$$(12 - x)(12 + x)\left(\frac{6}{12 - x}\right)$$
$$= (12 - x)(12 + x)\left(\frac{10}{12 + x}\right)$$
$$6(12 + x) = 10(12 - x)$$
$$72 + 6x = 120 - 10x$$
$$16x = 48$$
$$x = 3$$

Step 5
The rate of the current of the river is 3 mph.

Step 6
Check The rate downstream is $12 + 3 = 15$ mph, so she can go 10 miles in $\frac{10}{15} = \frac{2}{3}$ hour. The rate upstream is $12 - 3 = 9$ mph, so she can go 6 miles in $\frac{6}{9} = \frac{2}{3}$ hour. The times are the same, as required.

47. *Step 2*
Find the distance from Montpelier to Columbia. Let x represent that distance.

Complete the table.

	d	r	t
Actual Trip	x	51	$\dfrac{x}{51}$
Alternative Trip	x	60	$\dfrac{x}{60}$

Step 3
At 60 mph, his time at 51 mph would be decreased 3 hr.

$$\frac{x}{60} = \frac{x}{51} - 3$$

Step 4
Multiply by 1020.

$$17x = 20x - 3060$$
$$3060 = 3x$$
$$1020 = x$$

Step 5
The distance from Montpelier to Columbia is 1020 miles.

Step 6
Check 1020 miles at 51 mph takes $\frac{1020}{51}$ or 20 hours; 1020 miles at 60 mph takes $\frac{1020}{60}$ or 17 hours; $17 = 20 - 3$ as required.

Copyright © 2012 Pearson Education, Inc. Publishing as Addison-Wesley.

49. *Step 2*
Let x = the one-way distance.

Use $t = \frac{d}{r}$ and make a table.

	d	r	t
Trip Going East	x	500	$\frac{x}{500}$
Return Trip	x	350	$\frac{x}{350}$

Step 3
The total flying time in both directions was 8.5 hours, so

$$\frac{x}{500} + \frac{x}{350} = 8.5.$$

Step 4
Multiply by the LCD, 3500.

$$7x + 10x = 29{,}750$$
$$17x = 29{,}750$$
$$x = 1750$$

Step 5
The one-way distance was 1750 miles.

Step 6
Check The time for the trip going east was $\frac{1750}{500} = 3.5$ hours. The time for the return trip was $\frac{1750}{350} = 5$ hours. The total time was $3.5 + 5 = 8.5$ hours.

51. *Step 2*
Let x = the distance on the first part of the trip.

Use $t = \frac{d}{r}$ and make a table.

	d	r	t
First Part	x	60	$\frac{x}{60}$
Second Part	$x + 10$	50	$\frac{x+10}{50}$

Step 3
From the problem, the equation is stated in words:
Time for the second part = time for the first part + $\frac{1}{2}$ (Note that 30 min = $\frac{1}{2}$ hr.). Use the times given in the table to write the equation.

$$\frac{x+10}{50} = \frac{x}{60} + \frac{1}{2}$$

Step 4
Multiply by the LCD, 300.

$$300\left(\frac{x+10}{50}\right) = 300\left(\frac{x}{60} + \frac{1}{2}\right)$$
$$6(x+10) = 5x + 150$$
$$6x + 60 = 5x + 150$$
$$x = 90$$

Step 5
The distance for both parts of the trip is given by

$$x + (x + 10) = 90 + (90 + 10) = 190.$$

The distance is 190 miles.

Step 6
Check 90 miles at 60 mph takes $\frac{90}{60}$ or $1\frac{1}{2}$ hours; 100 miles at 50 mph takes 2 hours. The second part of the trip takes $\frac{1}{2}$ hour more than the first part, as required.

53. Let x = the time it would take them working together.

Complete the table.

Worker	Rate	Time Working Together	Fractional Part of the Job Done
Butch	$\frac{1}{15}$	x	$\frac{1}{15}x$
Peggy	$\frac{1}{12}$	x	$\frac{1}{12}x$

$$\begin{array}{ccccc}
\text{Part done} & & \text{part done} & & \text{1 whole} \\
\text{by Butch} & + & \text{by Peggy} & = & \text{job.} \\
\frac{1}{15}x & + & \frac{1}{12}x & = & 1
\end{array}$$

$$60\left(\tfrac{1}{15}x + \tfrac{1}{12}x\right) = 60 \cdot 1 \quad \textit{Multiply by 60.}$$
$$4x + 5x = 60$$
$$9x = 60$$
$$x = \frac{60}{9} = \frac{20}{3} \quad \text{or} \quad 6\frac{2}{3}$$

Together they could do the job in $\frac{20}{3}$ or $6\frac{2}{3}$ minutes.

55. Let x = the time it would take Kuba working alone.

Worker	Rate	Time Working Together	Fractional Part of the Job Done
Jerry	$\frac{1}{20}$	12	$\frac{1}{20}(12) = \frac{3}{5}$
Kuba	$\frac{1}{x}$	12	$\frac{1}{x}(12) = \frac{12}{x}$

$$\begin{array}{ccccc}
\text{Part done} & & \text{part done} & & \text{1 whole} \\
\text{by Jerry} & + & \text{by Kuba} & = & \text{job.} \\
\frac{3}{5} & + & \frac{12}{x} & = & 1
\end{array}$$

$$5x\left(\tfrac{3}{5} + \tfrac{12}{x}\right) = 5x \cdot 1 \quad \textit{Multiply by 5x.}$$
$$3x + 60 = 5x$$
$$60 = 2x$$
$$30 = x$$

It would take Kuba 30 hours to do the job alone.

Copyright © 2012 Pearson Education, Inc. Publishing as Addison-Wesley.

57. Let $x =$ the time needed for Dixie and Trixie to do the job together.
Make a table.

Worker	Rate	Time Working Together	Fractional Part of the Job Done
Dixie	$\frac{1}{3}$	x	$\frac{1}{3}x$
Trixie	$\frac{1}{6}$	x	$\frac{1}{6}x$

Since Dixie has been painting for one hour, $\frac{1}{3}$ of the room is already painted. Thus, the sum of the fractional parts equals $\frac{2}{3}$, not 1.

$$\frac{1}{3}x + \frac{1}{6}x = \frac{2}{3}$$
$$6\left(\frac{1}{3}x + \frac{1}{6}x\right) = 6 \cdot \frac{2}{3} \quad \textit{Multiply by 6.}$$
$$2x + x = 4$$
$$3x = 4$$
$$x = \frac{4}{3}$$

In summary, Dixie will paint for 1 hour alone, Dixie and Trixie will paint for $1\frac{1}{3}$ hours together, so it will take $2\frac{1}{3}$ hours, after Dixie starts, to finish painting the room.

59. Let $x =$ the time it will take to fill the vat if both pipes are open.

	Rate	Time to Fill the Vat	Fractional Part of the Job Done
Inlet Pipe	$\frac{1}{10}$	x	$\frac{1}{10}x$
Outlet Pipe	$-\frac{1}{20}$	x	$-\frac{1}{20}x$

Notice that the rate of the outlet pipe is negative because it will empty the vat, not fill it.

Part done with the inlet pipe open $+$ Part done with the outlet pipe open $=$ equals 1 whole job.

$$\frac{1}{10}x \quad + \quad \left(-\frac{1}{20}x\right) \quad = \quad 1$$
$$20\left(\frac{1}{10}x - \frac{1}{20}x\right) = 20 \cdot 1 \quad \textit{Multiply by 20.}$$
$$2x - x = 20$$
$$x = 20$$

It will take 20 hours to fill the vat.

61. Let $x =$ the time from Mimi's arrival home to the time the place is a shambles.

	Rate	Time to Mess up House	Fractional Part of the Job Done
Hortense and Mort	$-\frac{1}{7}$	x	$-\frac{1}{7}x$
Mimi	$\frac{1}{2}$	x	$\frac{1}{2}x$

Notice that Hortense and Mort's rate is negative since they are opposing the messing up by cleaning the house.

Part done by Hortense and Mort $+$ Part done by Mimi $=$ equals 1 whole job of messing up.

$$-\frac{1}{7}x \quad + \quad \frac{1}{2}x \quad = \quad 1$$
$$14\left(-\frac{1}{7}x + \frac{1}{2}x\right) = 14 \cdot 1 \quad \textit{Multiply by 14.}$$
$$-2x + 7x = 14$$
$$5x = 14$$
$$x = \frac{14}{5} \quad \text{or} \quad 2\frac{4}{5}$$

It would take $\frac{14}{5}$ or $2\frac{4}{5}$ hours after Mimi got home for the house to be a shambles.

63.
$$y = kx$$
$$1 = k(3) \quad \textit{Let } y = 1, x = 3.$$
$$\frac{1}{3} = k$$

65.
$$y = \frac{k}{x}$$
$$1 = \frac{k}{3} \quad \textit{Let } y = 1, x = 3.$$
$$1(3) = k$$
$$3 = k$$

7.6 Variation

7.6 Now Try Exercises

N1. Let C represent the cost of d dozen eggs. C varies directly as d, so

$$C = kd,$$

where k represents the cost of one dozen eggs. Since $C = 20$ when $d = 8$,

$$20 = k \cdot 8$$
$$k = \frac{20}{8} = 2.50$$

One dozen eggs costs $2.50, and C and d are related by

$$C = 2.50d.$$

N2. Let A represent the area of a parallelogram with base b and constant height. Use $A = kb$ with $A = 20$ and $b = 4$ to find k.

$$A = kb$$
$$20 = k(4)$$
$$\frac{20}{4} = k$$
$$5 = k$$

So $A = 5b$.

Let $b = 7$. Find A.

$$A = 5(7) = 35$$

The area is 35 cm^2.

Copyright © 2012 Pearson Education, Inc. Publishing as Addison-Wesley.

N3. y varies directly as the square of x, so
$$y = kx^2.$$
$y = 200$ when $x = 5$, so
$$200 = k(5)^2.$$
$$200 = k(25)$$
$$8 = k$$
Thus, $y = 8x^2$. When $x = 7$,
$$y = 8(7)^2 = 8(49) = 392.$$

N4. Let h represent the height of a triangle with base b and constant area. h varies inversely as b, so
$$h = \frac{k}{b},$$
for some constant k. Since $h = 7$ when $b = 8$,
$$7 = \frac{k}{8}.$$
$$56 = k$$
Thus, $h = \frac{56}{b}$. When $b = 14$,
$$h = \frac{56}{14} = 4.$$
The height is 4 centimeters.

N5. Let w represent the weight of an object above Earth at a distance d from the center of Earth. w varies inversely as the square of d, so
$$w = \frac{k}{d^2}.$$
$w = 150$ when $d = 3960$, so
$$150 = \frac{k}{3960^2}.$$
$$150(3960^2) = k$$
Thus, $w = \frac{150(3960^2)}{d^2}$.
When $d = 3960 + 1000 = 4960$,
$$w = \frac{150(3960^2)}{4960^2} \approx 95.6.$$
The object weighs about 96 pounds.

N6. Let V represent the volume of a right pyramid with height h and base area $\mathcal{A}$. V varies jointly as h and $\mathcal{A}$, so
$$V = kh\mathcal{A}.$$
$V = 100$ when $h = 10$ and $\mathcal{A} = 30$, so
$$100 = k(10)(30).$$
$$100 = k(300)$$
$$k = \frac{100}{300} = \frac{1}{3}$$
Thus, $V = \frac{1}{3}h\mathcal{A}$. When $h = 20$ and $\mathcal{A} = 90$,
$$V = \frac{1}{3}(20)(90) = 20(30) = 600.$$
The volume is 600 cubic feet.

N7. Let S represent the sample size, v the variance, and e the maximum error of the estimate. S varies directly as v and inversely as e^2, so
$$S = \frac{kv}{e^2}.$$
$S = 200$ when $v = 25$ and $e = 0.5$, so
$$200 = \frac{k(25)}{(0.5)^2}$$
$$200 = 100k$$
$$k = \frac{200}{100} = 2$$
Thus, $S = \frac{2v}{e^2}$. When $v = 25$ and $e = 0.1$,
$$S = \frac{2(25)}{(0.1)^2} = 5000.$$

7.6 Section Exercises

1. This suggests *direct* variation.
   As the number of tickets you buy increases, so does the probability that you will win.

3. This suggests *direct* variation.
   As the pressure put on the accelerator increases, so does the speed of the car.

5. This suggests *inverse* variation.
   As your age gets larger, the probability that you believe in Santa Claus gets smaller.

7. This suggests *inverse* variation.
   As the number of days gets smaller, the number of home runs increases.

9. The equation $y = \frac{3}{x}$ represents *inverse* variation.
   y varies inversely as x because x is in the denominator.

11. The equation $y = 10x^2$ represents *direct* variation.
    The number 10 is the constant of variation, and y varies directly as the square of x.

13. The equation $y = 3xz^4$ represents *joint* variation.
    y varies directly as x and z^4.

15. The equation $y = \frac{4x}{wz}$ represents *combined* variation. In the numerator, 4 is the constant of variation, and y varies directly as x. In the denominator, y varies inversely as w and z.

17. For $k > 0$, if y varies directly as x (then $y = kx$), when x increases, y _increases_, and when x decreases, y _decreases_.

19. $P = 4s$ The perimeter of a square varies directly as the length of its side.

21. $S = 4\pi r^2$ The surface area of a sphere varies directly as the square of its radius.

Copyright © 2012 Pearson Education, Inc. Publishing as Addison-Wesley.

23. $A = \frac{1}{2}bh$ The area of a triangle varies jointly as the length of its base and its height.

25. For $P = 4s$, $d = 2r$, $S = 4\pi r^2$, $V = \frac{4}{3}\pi r^3$, $A = \frac{1}{2}bh$, and $V = \frac{1}{3}\pi r^2 h$, the constants of variation are 4, 2, 4π, $\frac{4}{3}\pi$, $\frac{1}{2}$, and $\frac{1}{3}\pi$, respectively.

27. "x varies directly as y" means

$$x = ky$$

for some constant k. Let $x = 9$ and $y = 3$ to find k.

$$x = ky$$
$$9 = k(3)$$
$$k = \frac{9}{3} = 3$$

So $x = 3y$. To find x when $y = 12$, substitute 12 for y in the equation.

$$x = 3y$$
$$x = 3(12)$$
$$x = 36$$

29. "a varies directly as the square of b" means

$$a = kb^2$$

for some constant k. Substitute $a = 4$ and $b = 3$ in the equation and solve for k.

$$a = kb^2$$
$$4 = k(3)^2$$
$$k = \frac{4}{9}$$

So $a = \frac{4}{9}b^2$. To find a when $b = 2$, substitute 2 for b in the equation.

$$a = \frac{4}{9}b^2$$
$$a = \frac{4}{9}(2)^2$$
$$a = \frac{4 \cdot 4}{9} = \frac{16}{9}$$

31. "z varies inversely as w" means

$$z = \frac{k}{w}$$

for some constant k. Since $z = 10$ when $w = 0.5$, substitute these values in the equation and solve for k.

$$z = \frac{k}{w}$$
$$10 = \frac{k}{0.5}$$
$$k = 10(0.5) = 5$$

So $z = \frac{5}{w}$. To find z when $w = 8$, substitute 8 for w in the equation.

$$z = \frac{5}{w}$$
$$z = \frac{5}{8} \text{ or } 0.625$$

33. "m varies inversely as p^2" means

$$m = \frac{k}{p^2}$$

for some constant k. Since $m = 20$ when $p = 2$, substitute these values in the equation and solve for k.

$$m = \frac{k}{p^2}$$
$$20 = \frac{k}{2^2}$$
$$k = 20(4) = 80$$

So $m = \frac{80}{p^2}$. Now let $p = 5$.

$$m = \frac{80}{p^2}$$
$$m = \frac{80}{5^2} = \frac{16}{5}$$

35. "p varies jointly as q and r^2" means

$$p = kqr^2$$

for some constant k. Given that $p = 200$ when $q = 2$ and $r = 3$, solve for k.

$$p = kqr^2$$
$$200 = k(2)(3)^2$$
$$200 = 18k$$
$$k = \frac{200}{18} = \frac{100}{9}$$

So $p = \frac{100}{9}qr^2$. Now let $q = 5$ and $r = 2$.

$$p = \frac{100}{9}qr^2$$
$$p = \frac{100}{9}(5)(2)^2$$
$$= \frac{100}{9}(20)$$
$$= \frac{2000}{9} \text{ or } 222\frac{2}{9}$$

37. Let $x =$ the number of gallons he bought and let $C =$ the cost.
C varies directly as x, so

$$C = kx.$$

Since $C = 43.79$ when $x = 15$,

$$43.79 = k(15)$$
$$k = \frac{43.79}{15} \approx 2.919.$$

The price per gallon is \$2.919, or \$$2.91\frac{9}{10}$.

39. Let $y =$ the weight of an object on earth and $x =$ the weight of the object on the moon.
y varies directly as x, so

$$y = kx$$

for some constant k. Since $y = 200$ when $x = 32$, substitute these values in the equation and solve for k.

$$y = kx$$
$$200 = k(32)$$
$$k = \frac{200}{32} = 6.25$$

So $y = 6.25x$. To find x when $y = 50$, substitute 50 for y in the equation

$$y = 6.25x.$$
$$50 = 6.25x$$
$$x = \frac{50}{6.25} = 8$$

The dog would weigh 8 lb on the moon.

41. Let V = the volume of the can and let h = the height of the can. V varies directly as h, so

$$V = kh.$$

Since $V = 300$ when $h = 10.62$,

$$300 = k(10.62)$$
$$k = \frac{300}{10.62} \approx 28.25.$$

So $V = 28.25h$. Now let $h = 15.92$.

$$V = 28.25h$$
$$V = 28.25(15.92) = 449.74$$

The volume is about 450 cubic centimeters.

43. Let d = the distance and t = the time. d varies directly as the square of t, so $d = kt^2$. Let $d = -576$ and $t = 6$. (You could also use $d = 576$, but the negative sign indicates the direction of the body.)

$$-576 = k(6)^2$$
$$-576 = 36k$$
$$-16 = k$$

So $d = -16t^2$. Now let $t = 4$.

$$d = -16(4)^2 = -256$$

The object fell 256 feet in the first 4 seconds.

45. Let s = the rate and t = the time. The rate varies inversely with time, so there is a constant k such that $s = k/t$. Find the value of k by replacing s with 160 and t with $\frac{1}{2}$.

$$s = \frac{k}{t}$$
$$160 = \frac{k}{\frac{1}{2}}$$
$$k = 160(\tfrac{1}{2}) = 80$$

So $s = \frac{80}{t}$. Now let $t = \frac{3}{4}$.

$$s = \frac{80}{\frac{3}{4}}$$
$$s = \frac{80}{1} \cdot \frac{4}{3} = \frac{320}{3} \quad \text{or} \quad 106\frac{2}{3}$$

A rate of $106\frac{2}{3}$ miles per hour is needed to go the same distance in $\frac{3}{4}$ minute.

47. Let f = the frequency of a string in cycles per second and s = the length in feet. f varies inversely as s, so

$$f = \frac{k}{s}$$

for some constant k. Since $f = 250$ when $s = 2$, substitute these values in the equation and solve for k.

$$f = \frac{k}{s}$$
$$250 = \frac{k}{2}$$
$$k = 250(2) = 500$$

So $f = \frac{500}{s}$. Now let $s = 5$.

$$f = \frac{500}{5} = 100.$$

The string would have a frequency of 100 cycles per second.

49. Let I = the illumination produced by a light source and d = the distance from the source. I varies inversely as d^2, so

$$I = \frac{k}{d^2}$$

for some constant k. Since $I = 768$ when $d = 1$, substitute these values in the equation and solve for k.

$$I = \frac{k}{d^2}$$
$$768 = \frac{k}{1^2}$$
$$768 = k$$

So $I = \frac{768}{d^2}$. Now let $d = 6$.

$$I = \frac{768}{d^2}$$
$$I = \frac{768}{6^2} = \frac{768}{36} = \frac{64}{3} \quad \text{or} \quad 21\frac{1}{3}$$

The illumination produced by the light source is $21\frac{1}{3}$ foot-candles.

51. Let I = the simple interest, P the principal, and t the time.
Since I varies jointly as the principal and time, there is a constant k such that $I = kPt$. Find k by replacing I with 280, P with 2000, and t with 4.

$$I = kPt$$
$$280 = k(2000)(4)$$
$$k = \frac{280}{8000} = 0.035$$

So $I = 0.035Pt$. Now let $t = 6$.

$$I = 0.035(2000)(6)$$
$$= 420$$

The interest would be $420.

Copyright © 2012 Pearson Education, Inc. Publishing as Addison-Wesley.

53. The weight W of a bass varies jointly as its girth G and the square of its length L, so

$$W = kGL^2$$

for some constant k. Substitute 22.7 for W, 21 for G, and 36 for L.

$$22.7 = k(21)(36)^2$$
$$k = \frac{22.7}{27{,}216} \approx 0.000834$$

So $W = 0.000834GL^2$. Now let $G = 18$ and $L = 28$.

$$W = 0.000834GL^2$$
$$= 0.000834(18)(28)^2$$
$$\approx 11.8$$

The bass would weigh about 11.8 pounds.

55. Let $F =$ the force, $w =$ the weight of the car, $s =$ the speed, and $r =$ the radius.
The force varies inversely as the radius and jointly as the weight and the square of the speed, so

$$F = \frac{kws^2}{r}.$$

Let $F = 242$, $w = 2000$, $r = 500$, and $s = 30$.

$$242 = \frac{k(2000)(30)^2}{500}$$
$$k = \frac{242(500)}{2000(900)} = \frac{121}{1800}$$

So $F = \frac{121ws^2}{1800r}$.
Let $r = 750$, $s = 50$, and $w = 2000$.

$$F = \frac{121(2000)(50)^2}{1800(750)} \approx 448.1$$

Approximately 448.1 pounds of force would be needed.

57. Let $N =$ the number of long distance calls,
$p_1 =$ the population of City 1,
$p_2 =$ the population of City 2,
and $d =$ the distance between them.

$$N = \frac{kp_1p_2}{d}$$

Let $N = 80{,}000$, $p_1 = 70{,}000$, $p_2 = 100{,}000$, and $d = 400$.

$$80{,}000 = \frac{k(70{,}000)(100{,}000)}{400}$$
$$80{,}000 = 17{,}500{,}000k$$
$$k = \frac{80{,}000}{17{,}500{,}000} = \frac{4}{875}$$
$$N = \frac{4}{875}\left(\frac{p_1p_2}{d}\right)$$

Let $p_1 = 50{,}000$, $p_2 = 75{,}000$, and $d = 250$.

$$N = \frac{4}{875}\left(\frac{50{,}000 \cdot 75{,}000}{250}\right)$$
$$= \frac{480{,}000}{7} = 68{,}571\tfrac{3}{7}$$

Rounded to the nearest hundred, there are about 68,600 calls.

59. Use the BMI from Example 7, with $k = 694$.

$$B = \frac{694w}{h^2}$$

Substitute your weight in pounds for w and your height in inches for h to determine your BMI. Answers will vary. A BMI from 19 to 25 is considered desirable.

61. $\sqrt{64} = 8$ since 8 is positive and $8^2 = 64$.

63. $-\sqrt{16} = -4$ since the negative sign is outside the radical symbol.

65. $\sqrt{-25}$ is not a real number.

Chapter 7 Review Exercises

1. **(a)** $f(x) = \dfrac{-7}{3x + 18}$

Set the denominator equal to zero and solve.

$$3x + 18 = 0$$
$$3x = -18$$
$$x = -6$$

The number -6 makes the expression undefined, so it is excluded from the domain.

(b) The domain is $\{x \mid x \neq -6\}$.

2. **(a)** $f(x) = \dfrac{5x + 17}{x^2 - 7x + 10}$

Set the denominator equal to zero and solve.

$$x^2 - 7x + 10 = 0$$
$$(x - 5)(x - 2) = 0$$
$$x - 5 = 0 \quad \text{or} \quad x - 2 = 0$$
$$x = 5 \quad \text{or} \quad x = 2$$

The numbers 2 and 5 make the expression undefined, so they are excluded from the domain.

(b) The domain is $\{x \mid x \neq 2, 5\}$.

3. **(a)** $f(x) = \dfrac{9}{x^2 - 18x + 81}$

Set the denominator equal to zero and solve.

$$x^2 - 18x + 81 = 0$$
$$(x - 9)^2 = 0$$
$$x - 9 = 0$$
$$x = 9$$

Copyright © 2012 Pearson Education, Inc. Publishing as Addison-Wesley.

The number 9 makes the expression undefined, so it is excluded from the domain.

(b) The domain is $\{x \mid x \neq 9\}$.

4. $\dfrac{12x^2 + 6x}{24x + 12} = \dfrac{6x(2x+1)}{12(2x+1)} = \dfrac{x}{2}$

5. $\dfrac{25m^2 - n^2}{25m^2 - 10mn + n^2} = \dfrac{(5m+n)(5m-n)}{(5m-n)(5m-n)}$
$$= \dfrac{5m+n}{5m-n}$$

6. $\dfrac{r-2}{4-r^2} = \dfrac{r-2}{(2+r)(2-r)}$
$$= \dfrac{(-1)(2-r)}{(2+r)(2-r)}$$
$$= \dfrac{-1}{2+r}$$

7. The reciprocal of a rational expression is another rational expression such that the two rational expressions have a product of 1.

8. $\dfrac{(2y+3)^2}{5y} \cdot \dfrac{15y^3}{4y^2 - 9}$
$$= \dfrac{15y^3(2y+3)^2}{5y(2y+3)(2y-3)}$$
$$= \dfrac{3y^2(2y+3)}{2y-3}$$

9. $\dfrac{w^2 - 16}{w} \cdot \dfrac{3}{4-w}$
$$= \dfrac{(w-4)(w+4)}{w} \cdot \dfrac{3}{4-w}$$
$$= \dfrac{(-1)(4-w)(w+4)}{w} \cdot \dfrac{3}{4-w}$$
$$= \dfrac{-3(w+4)}{w}$$

10. $\dfrac{z^2 - z - 6}{z-6} \cdot \dfrac{z^2 - 6z}{z^2 + 2z - 15}$
$$= \dfrac{(z-3)(z+2)}{z-6} \cdot \dfrac{z(z-6)}{(z-3)(z+5)}$$
$$= \dfrac{z(z+2)}{z+5}$$

11. $\dfrac{m^3 - n^3}{m^2 - n^2} \div \dfrac{m^2 + mn + n^2}{m+n}$
Multiply by the reciprocal.
$$= \dfrac{m^3 - n^3}{m^2 - n^2} \cdot \dfrac{m+n}{m^2 + mn + n^2}$$
$$= \dfrac{(m-n)(m^2 + mn + n^2)}{(m-n)(m+n)} \cdot \dfrac{m+n}{m^2 + mn + n^2}$$
$$= 1$$

12. $32b^3,\ 24b^5$

Factor each denominator.
$$32b^3 = 2 \cdot 2 \cdot 2 \cdot 2 \cdot 2 \cdot b^3 = 2^5 \cdot b^3$$
$$24b^5 = 2 \cdot 2 \cdot 2 \cdot 3 \cdot b^5 = 2^3 \cdot 3 \cdot b^5$$
$$\text{LCD} = 2^5 \cdot 3 \cdot b^5 = 96b^5$$

13. $9r^2,\ 3r + 1,\ 9$

Factor each denominator.
$$9r^2 = 3^2 \cdot r^2$$

The second denominator is already in factored form. The third denominator is $9 = 3^2$.
The LCD is
$$3^2 \cdot r^2 \cdot (3r+1) \quad \text{or} \quad 9r^2(3r+1).$$

14. $6x^2 + 13x - 5,\ 9x^2 + 9x - 4$

Factor each denominator.
$$6x^2 + 13x - 5 = (3x-1)(2x+5)$$
$$9x^2 + 9x - 4 = (3x-1)(3x+4)$$

The LCD is $(3x-1)(2x+5)(3x+4)$.

15. $3x - 12,\ x^2 - 2x - 8,\ x^2 - 8x + 16$

Factor each denominator.
$$3x - 12 = 3(x-4)$$
$$x^2 - 2x - 8 = (x-4)(x+2)$$
$$x^2 - 8x + 16 = (x-4)^2$$

The LCD is $3(x-4)^2(x+2)$.

16. $\dfrac{5}{3x^6y^5} - \dfrac{8}{9x^4y^7}$
The LCD is $9x^6y^7$.
$$= \dfrac{5 \cdot 3y^2}{3x^6y^5 \cdot 3y^2} - \dfrac{8 \cdot x^2}{9x^4y^7 \cdot x^2}$$
$$= \dfrac{15y^2}{9x^6y^7} - \dfrac{8x^2}{9x^6y^7}$$
$$= \dfrac{15y^2 - 8x^2}{9x^6y^7}$$

17. $\dfrac{5y+13}{y+1} - \dfrac{1-7y}{y+1}$
$$= \dfrac{5y+13 - (1-7y)}{y+1}$$
$$= \dfrac{5y+13 - 1 + 7y}{y+1}$$
$$= \dfrac{12y+12}{y+1}$$
$$= \dfrac{12(y+1)}{y+1} = 12$$

Copyright © 2012 Pearson Education, Inc. Publishing as Addison-Wesley.

18. $\dfrac{6}{5a+10}+\dfrac{7}{6a+12}$

$=\dfrac{6}{5(a+2)}+\dfrac{7}{6(a+2)}$

The LCD is $30(a+2)$.

$=\dfrac{6\cdot 6}{5(a+2)\cdot 6}+\dfrac{7\cdot 5}{6(a+2)\cdot 5}$

$=\dfrac{36}{30(a+2)}+\dfrac{35}{30(a+2)}$

$=\dfrac{36+35}{30(a+2)}=\dfrac{71}{30(a+2)}$

19. $\dfrac{3r}{10r^2-3rs-s^2}+\dfrac{2r}{2r^2+rs-s^2}$

$=\dfrac{3r}{(5r+s)(2r-s)}+\dfrac{2r}{(2r-s)(r+s)}$

The LCD is $(5r+s)(2r-s)(r+s)$.

$=\dfrac{3r(r+s)}{(5r+s)(2r-s)(r+s)}$

$\quad+\dfrac{2r(5r+s)}{(2r-s)(r+s)(5r+s)}$

$=\dfrac{3r^2+3rs+10r^2+2rs}{(5r+s)(2r-s)(r+s)}$

$=\dfrac{13r^2+5rs}{(5r+s)(2r-s)(r+s)}$

20. $\dfrac{1}{y-x}=\dfrac{1}{y-x}\cdot\dfrac{-1}{-1}=\dfrac{-1}{-1(y-x)}=\dfrac{-1}{x-y}$

Both students got the correct answer. The two expressions obtained are equivalent.

21. $\dfrac{\frac{3}{t}+2}{\frac{4}{t}-7}$

Multiply the numerator and denominator by the LCD of all the fractions, t.

$=\dfrac{t\left(\frac{3}{t}+2\right)}{t\left(\frac{4}{t}-7\right)}=\dfrac{3+2t}{4-7t}$

22. $\dfrac{\frac{2}{m-3n}}{\frac{1}{3n-m}}=\dfrac{2}{m-3n}\div\dfrac{1}{3n-m}$

$=\dfrac{2}{m-3n}\cdot\dfrac{3n-m}{1}$

$=\dfrac{2}{m-3n}\cdot\dfrac{-1(m-3n)}{1}$

$=\dfrac{2\cdot(-1)}{1}=-2$

23. $\dfrac{\frac{3}{p}-\frac{2}{q}}{\frac{9q^2-4p^2}{qp}}$

Multiply the numerator and denominator by the LCD of all the fractions, qp.

$=\dfrac{qp\left(\frac{3}{p}-\frac{2}{q}\right)}{qp\left(\frac{9q^2-4p^2}{qp}\right)}$

$=\dfrac{3q-2p}{9q^2-4p^2}$

$=\dfrac{3q-2p}{(3q+2p)(3q-2p)}$

$=\dfrac{1}{3q+2p}$

24. $\dfrac{x^{-2}-y^{-2}}{x^{-1}-y^{-1}}=\dfrac{\frac{1}{x^2}-\frac{1}{y^2}}{\frac{1}{x}-\frac{1}{y}}$

Multiply the numerator and denominator by the LCD of all the fractions, x^2y^2.

$=\dfrac{x^2y^2\left(\frac{1}{x^2}-\frac{1}{y^2}\right)}{x^2y^2\left(\frac{1}{x}-\frac{1}{y}\right)}$

$=\dfrac{y^2-x^2}{xy^2-x^2y}$

$=\dfrac{(y+x)(y-x)}{xy(y-x)}$

$=\dfrac{y+x}{xy}$

25. $\dfrac{1}{t+4}+\dfrac{1}{2}=\dfrac{3}{2t+8}$

$\dfrac{1}{t+4}+\dfrac{1}{2}=\dfrac{3}{2(t+4)}$

Multiply by the LCD, $2(t+4)$. $(t\neq -4)$

$2(t+4)\left(\dfrac{1}{t+4}+\dfrac{1}{2}\right)=2(t+4)\left(\dfrac{3}{2(t+4)}\right)$

$2+(t+4)=3$

$t+6=3$

$t=-3$

Check $t=-3$: $1+\frac{1}{2}=\frac{3}{2}$ *True*

The solution set is $\{-3\}$.

Copyright © 2012 Pearson Education, Inc. Publishing as Addison-Wesley.

26.
$$\frac{-5m}{m+1} + \frac{m}{3m+3} = \frac{56}{6m+6}$$
$$\frac{-5m}{m+1} + \frac{m}{3(m+1)} = \frac{56}{6(m+1)}$$
$$\frac{-5m}{m+1} + \frac{m}{3(m+1)} = \frac{28}{3(m+1)}$$
Multiply by the LCD, $3(m+1)$. $(m \neq -1)$
$$3(m+1)\left(\frac{-5m}{m+1} + \frac{m}{3(m+1)}\right)$$
$$= 3(m+1)\left(\frac{28}{3(m+1)}\right)$$
$$-15m + m = 28$$
$$-14m = 28$$
$$m = -2$$
Check $m = -2$: $-10 + \frac{2}{3} = -\frac{56}{6}$ *True*
The solution set is $\{-2\}$.

27.
$$\frac{2}{k-1} - \frac{4k+1}{k^2-1} = \frac{-1}{k+1}$$
$$\frac{2}{k-1} - \frac{4k+1}{(k+1)(k-1)} = \frac{-1}{k+1}$$
Multiply by the LCD, $(k+1)(k-1)$.
$$(k+1)(k-1)\left(\frac{2}{k-1} - \frac{4k+1}{(k+1)(k-1)}\right)$$
$$= (k+1)(k-1)\left(\frac{-1}{k+1}\right)$$
$$2(k+1) - (4k+1) = -1(k-1)$$
$$2k+2-4k-1 = -k+1$$
$$-2k+1 = -k+1$$
$$0 = k$$
Check $k = 0$: $-2 + 1 = -1$ *True*
The solution set is $\{0\}$.

28.
$$\frac{5}{x+2} + \frac{3}{x+3} = \frac{x}{x^2+5x+6}$$
$$\frac{5}{x+2} + \frac{3}{x+3} = \frac{x}{(x+2)(x+3)}$$
Multiply by the LCD, $(x+2)(x+3)$.
$(x \neq -3, -2)$
$$(x+2)(x+3)\left(\frac{5}{x+2} + \frac{3}{x+3}\right)$$
$$= (x+2)(x+3)\left(\frac{x}{(x+2)(x+3)}\right)$$
$$5(x+3) + 3(x+2) = x$$
$$5x+15+3x+6 = x$$
$$8x+21 = x$$
$$7x = -21$$
$$x = -3$$
Substituting -3 in the original equation results in division by 0, so -3 is not a solution.
The solution set is $\emptyset$.

29. (a) $\dfrac{x+3}{x+3} = 1$

The left side of the equation is equal to 1 except for $x = -3$, when it is undefined. Hence, the solution set is $\{x \mid x \neq -3\}$.

(b) The solution set is not $\{$all real numbers$\}$ because -3 is not in the domain.

30. In simplifying the expression, we are combining terms to get a single fraction with a denominator of $6x$. In solving the equation, we are finding a value for x that makes the equation true.

31. The graph in choice **C** has a vertical and a horizontal asymptote. The equations are $x = 0$ and $y = 0$, respectively.

32. $\dfrac{1}{A} = \dfrac{1}{B} + \dfrac{1}{C}$
Let $B = 30$ and $C = 10$.
$$\frac{1}{A} = \frac{1}{30} + \frac{1}{10}$$
To solve for A, multiply both sides by the LCD, $30A$.
$$30A\left(\frac{1}{A}\right) = 30A\left(\frac{1}{30} + \frac{1}{10}\right)$$
$$30 = A + 3A$$
$$30 = 4A$$
$$A = \frac{30}{4} = \frac{15}{2}$$

33. Solve $F = \dfrac{GMm}{d^2}$ for m.
$$Fd^2 = GMm \qquad \textit{Multiply by } d^2.$$
$$\frac{Fd^2}{GM} = m \qquad \textit{Divide by } GM.$$

34. Solve $\mu = \dfrac{Mv}{M+m}$ for M.
$$\mu(M+m) = Mv \qquad \textit{Multiply by } M + m.$$
$$\mu M + \mu m = Mv$$
$$\mu m = Mv - \mu M$$
$$m\mu = M(v - \mu)$$
$$M = \frac{m\mu}{v - \mu}$$

35. Let $x =$ the number of passenger-kilometers per day provided by high-speed trains.
Write a proportion.
$$\frac{x}{15,000} = \frac{23,200}{58,000}$$
$$x = \frac{23,200(15,000)}{58,000}$$
$$= 6000$$

The high-speed train would provide 6000 passenger-km per day in that region.

Copyright © 2012 Pearson Education, Inc. Publishing as Addison-Wesley.

36. Let $x =$ the rate of the boat in still water.

Use $d = rt$, or $t = \frac{d}{r}$, to make a table.

	Distance	Rate	Time
Upstream	24	$x - 4$	$\dfrac{24}{x-4}$
Downstream	40	$x + 4$	$\dfrac{40}{x+4}$

Because the times are equal,

$$\frac{40}{x+4} = \frac{24}{x-4}.$$

Multiply by the LCD, $(x + 4)(x - 4)$. $(x \neq -4, 4)$

$$(x+4)(x-4)\left(\frac{40}{x+4}\right) = (x+4)(x-4)\left(\frac{24}{x-4}\right)$$
$$40(x - 4) = 24(x + 4)$$
$$40x - 160 = 24x + 96$$
$$16x = 256$$
$$x = 16$$

The rate of the boat in still water is 16 km per hr.

37. Let $x =$ the time it takes to fill the sink with both taps open.

Make a table.

	Rate	Time Working Together	Fractional Part of the Job Done
Cold	$\dfrac{1}{8}$	x	$\dfrac{x}{8}$
Hot	$\dfrac{1}{12}$	x	$\dfrac{x}{12}$

Part done by cold	plus	part done by hot	equals	1 whole job.
$\dfrac{x}{8}$	$+$	$\dfrac{x}{12}$	$=$	1

$$24\left(\frac{x}{8} + \frac{x}{12}\right) = 24 \cdot 1 \quad \textit{Multiply by 24.}$$
$$3x + 2x = 24$$
$$5x = 24$$
$$x = \frac{24}{5} \quad \text{or} \quad 4\tfrac{4}{5}$$

The sink will be filled in $\frac{24}{5}$ or $4\frac{4}{5}$ minutes.

38. Let $x =$ the time to do the job working together. Make a table.

Worker	Rate	Time Working Together	Fractional Part of the Job Done
Jane	$\dfrac{1}{9}$	x	$\dfrac{x}{9}$
Jessica	$\dfrac{1}{6}$	x	$\dfrac{x}{6}$

Part done by Jane	plus	part done by Jessica	equals	1 whole job.
$\dfrac{x}{9}$	$+$	$\dfrac{x}{6}$	$=$	1

$$36\left(\frac{x}{9} + \frac{x}{6}\right) = 36 \cdot 1 \quad \textit{Multiply by 36.}$$
$$4x + 6x = 36$$
$$10x = 36$$
$$x = \frac{36}{10}$$
$$x = \frac{18}{5} \quad \text{or} \quad 3\tfrac{3}{5}$$

Working together, they can do the job in $\frac{18}{5}$ or $3\frac{3}{5}$ hours.

39. If y varies inversely as x, then $y = \frac{k}{x}$, for some constant k. This form fits choice **C**.

40. Let $v =$ the viewing distance and $e =$ the amount of enlargement. v varies directly as e, so

$$v = ke$$

for some constant k. Since $v = 250$ when $e = 5$, substitute these values in the equation and solve for k.

$$v = ke$$
$$250 = k(5)$$
$$50 = k$$

So $v = 50e$. Now let $e = 8.6$.

$$v = 50(8.6) = 430$$

It should be viewed from 430 mm.

41. The frequency f of a vibrating guitar string varies inversely as its length L, so

$$f = \frac{k}{L}$$

for some constant k. Substitute 0.65 for L and 4.3 for f.

$$4.3 = \frac{k}{0.65}$$
$$k = 4.3(0.65) = 2.795$$

So $f = \frac{2.795}{L}$. Now let $L = 0.5$.

$$f = \frac{2.795}{0.5} = 5.59$$

The frequency would be 5.59 vibrations per second.

42. The volume V of a rectangular box of a given height is proportional to its width W and length L, so

$$V = kWL$$

for some constant k. Substitute 2 for W, 4 for L, and 12 for V.

Copyright © 2012 Pearson Education, Inc. Publishing as Addison-Wesley.

$$12 = k(2)(4)$$
$$k = \tfrac{12}{8} = \tfrac{3}{2}$$

So $V = \tfrac{3}{2}WL$. Now let $W = 3$ and $L = 5$.

$$V = \tfrac{3}{2}(3)(5) = \tfrac{45}{2}$$

The volume is 22.5 cubic feet.

43. **[7.1]** $\dfrac{x + 2y}{x^2 - 4y^2} = \dfrac{x + 2y}{(x + 2y)(x - 2y)}$

$$= \dfrac{1}{x - 2y}$$

44. **[7.1]** $\dfrac{x^2 + 2x - 15}{x^2 - x - 6} = \dfrac{(x + 5)(x - 3)}{(x - 3)(x + 2)}$

$$= \dfrac{x + 5}{x + 2}$$

45. **[7.2]** $\dfrac{2}{m} + \dfrac{5}{3m^2}$

The LCD is $3m^2$.

$$= \dfrac{2 \cdot 3m}{m \cdot 3m} + \dfrac{5}{3m^2}$$

$$= \dfrac{6m}{3m^2} + \dfrac{5}{3m^2} = \dfrac{6m + 5}{3m^2}$$

46. **[7.2]** $\dfrac{9}{3 - x} - \dfrac{2}{x - 3}$

$$= \dfrac{9}{3 - x} - \dfrac{2(-1)}{(x - 3)(-1)}$$

$$= \dfrac{9}{3 - x} - \dfrac{-2}{3 - x}$$

$$= \dfrac{9 - (-2)}{3 - x}$$

$$= \dfrac{11}{3 - x}, \text{ or } \dfrac{-11}{x - 3}$$

47. **[7.3]** $\dfrac{\dfrac{-3}{x} + \dfrac{x}{2}}{1 + \dfrac{x + 1}{x}}$

Multiply the numerator and denominator by the LCD of all the fractions, $2x$.

$$= \dfrac{2x\left(\dfrac{-3}{x} + \dfrac{x}{2}\right)}{2x\left(1 + \dfrac{x + 1}{x}\right)}$$

$$= \dfrac{-6 + x^2}{2x + 2(x + 1)}$$

$$= \dfrac{x^2 - 6}{2x + 2x + 2}$$

$$= \dfrac{x^2 - 6}{4x + 2} = \dfrac{x^2 - 6}{2(2x + 1)}$$

48. **[7.3]** $\dfrac{\dfrac{3}{x} - 5}{6 + \dfrac{1}{x}}$

Multiply the numerator and denominator by the LCD of all the fractions, x.

$$= \dfrac{x\left(\dfrac{3}{x} - 5\right)}{x\left(6 + \dfrac{1}{x}\right)}$$

$$= \dfrac{3 - 5x}{6x + 1}$$

49. **[7.1]** $\dfrac{4y + 16}{30} \div \dfrac{2y + 8}{5}$

Multiply by the reciprocal.

$$= \dfrac{4y + 16}{30} \cdot \dfrac{5}{2y + 8}$$

$$= \dfrac{4(y + 4)}{30} \cdot \dfrac{5}{2(y + 4)}$$

$$= \dfrac{4 \cdot 5}{2 \cdot 30} = \dfrac{2}{6} = \dfrac{1}{3}$$

50. **[7.3]** $\dfrac{t^{-2} + s^{-2}}{t^{-1} - s^{-1}} = \dfrac{\dfrac{1}{t^2} + \dfrac{1}{s^2}}{\dfrac{1}{t} - \dfrac{1}{s}}$

Multiply the numerator and denominator by the LCD of all the fractions, $t^2 s^2$.

$$= \dfrac{t^2 s^2\left(\dfrac{1}{t^2} + \dfrac{1}{s^2}\right)}{t^2 s^2\left(\dfrac{1}{t} - \dfrac{1}{s}\right)}$$

$$= \dfrac{s^2 + t^2}{ts^2 - t^2 s}$$

$$= \dfrac{s^2 + t^2}{st(s - t)}$$

51. **[7.1]** $\dfrac{k^2 - 6k + 9}{1 - 216k^3} \cdot \dfrac{6k^2 + 17k - 3}{9 - k^2}$

Factor $1 - 216k^3$ as the difference of cubes, $1^3 - (6k)^3$.

$$= \dfrac{(k - 3)(k - 3)}{(1 - 6k)(1 + 6k + 36k^2)}$$

$$\cdot \dfrac{(6k - 1)(k + 3)}{(3 - k)(3 + k)}$$

$$= \dfrac{(k - 3)(k - 3)}{(-1)(6k - 1)(1 + 6k + 36k^2)}$$

$$\cdot \dfrac{(6k - 1)(k + 3)}{(-1)(k - 3)(k + 3)}$$

$$= \dfrac{k - 3}{1 + 6k + 36k^2} \text{ or } \dfrac{k - 3}{36k^2 + 6k + 1}$$

Copyright © 2012 Pearson Education, Inc. Publishing as Addison-Wesley.

52. **[7.1]** $\dfrac{9x^2 + 46x + 5}{3x^2 - 2x - 1} \div \dfrac{x^2 + 11x + 30}{x^3 + 5x^2 - 6x}$

Multiply by the reciprocal.

$= \dfrac{9x^2 + 46x + 5}{3x^2 - 2x - 1} \cdot \dfrac{x(x^2 + 5x - 6)}{x^2 + 11x + 30}$

$= \dfrac{(9x + 1)(x + 5)}{(3x + 1)(x - 1)} \cdot \dfrac{x(x + 6)(x - 1)}{(x + 6)(x + 5)}$

$= \dfrac{x(9x + 1)}{3x + 1}$

53. **[7.2]** $\dfrac{4a}{a^2 - ab - 2b^2} - \dfrac{6b - a}{a^2 + 4ab + 3b^2}$

$= \dfrac{4a}{(a - 2b)(a + b)} - \dfrac{6b - a}{(a + 3b)(a + b)}$

The LCD is $(a + 3b)(a - 2b)(a + b)$.

$= \dfrac{4a(a + 3b)}{(a - 2b)(a + b)(a + 3b)}$

$- \dfrac{(6b - a)(a - 2b)}{(a + 3b)(a + b)(a - 2b)}$

$= \dfrac{4a(a + 3b) - (6b - a)(a - 2b)}{(a + 3b)(a + b)(a - 2b)}$

$= \dfrac{4a^2 + 12ab - (6ab - 12b^2 - a^2 + 2ab)}{(a + 3b)(a + b)(a - 2b)}$

$= \dfrac{4a^2 + 12ab - 6ab + 12b^2 + a^2 - 2ab}{(a + 3b)(a + b)(a - 2b)}$

$= \dfrac{5a^2 + 4ab + 12b^2}{(a + 3b)(a + b)(a - 2b)}$

54. **[7.2]** $\dfrac{a}{b} + \dfrac{b}{c} + \dfrac{c}{d}$

The LCD is bcd.

$= \dfrac{a \cdot cd}{b \cdot cd} + \dfrac{b \cdot bd}{c \cdot bd} + \dfrac{c \cdot bc}{d \cdot bc}$

$= \dfrac{acd + b^2 d + bc^2}{bcd}$

55. **[7.4]** $\dfrac{x + 3}{x^2 - 5x + 4} - \dfrac{1}{x} = \dfrac{2}{x^2 - 4x}$

$\dfrac{x + 3}{(x - 4)(x - 1)} - \dfrac{1}{x} = \dfrac{2}{x(x - 4)}$

Multiply by the LCD, $x(x - 4)(x - 1)$.

$(x \neq 0, 1, 4)$

$x(x - 4)(x - 1)\left(\dfrac{x + 3}{(x - 4)(x - 1)} - \dfrac{1}{x} \right)$

$= x(x - 4)(x - 1) \cdot \left(\dfrac{2}{x(x - 4)} \right)$

$x(x + 3) - (x - 4)(x - 1) = 2(x - 1)$

$x^2 + 3x - (x^2 - 5x + 4) = 2x - 2$

$x^2 + 3x - x^2 + 5x - 4 = 2x - 2$

$8x - 4 = 2x - 2$

$6x = 2$

$x = \dfrac{1}{3}$

Check $x = \dfrac{1}{3}$: $\dfrac{15}{11} - 3 = -\dfrac{18}{11}$ *True*

The solution set is $\left\{ \dfrac{1}{3} \right\}$.

56. **[7.5]** Solve $A = \dfrac{Rr}{R + r}$ for r.

$A(R + r) = Rr$

$AR + Ar = Rr$

$AR = Rr - Ar$

$AR = (R - A)r$

$\dfrac{AR}{R - A} = r$, or $r = \dfrac{-AR}{A - R}$

57. **[7.4]** $1 - \dfrac{5}{r} = \dfrac{-4}{r^2}$

Multiply by the LCD, r^2. $(r \neq 0)$

$r^2\left(1 - \dfrac{5}{r} \right) = r^2\left(\dfrac{-4}{r^2} \right)$

$r^2 - 5r = -4$

$r^2 - 5r + 4 = 0$

$(r - 4)(r - 1) = 0$

$r - 4 = 0$ or $r - 1 = 0$

$r = 4$ or $r = 1$

Check $r = 1$: $1 - 5 = -4$ *True*

Check $r = 4$: $1 - \dfrac{5}{4} = -\dfrac{1}{4}$ *True*

The solution set is $\{1, 4\}$.

58. **[7.4]** $\dfrac{3x}{x - 4} + \dfrac{2}{x} = \dfrac{48}{x^2 - 4x}$

$\dfrac{3x}{x - 4} + \dfrac{2}{x} = \dfrac{48}{x(x - 4)}$

Multiply by the LCD, $x(x - 4)$. $(x \neq 0, 4)$

$x(x - 4)\left(\dfrac{3x}{x - 4} + \dfrac{2}{x} \right) = x(x - 4)\left(\dfrac{48}{x(x - 4)} \right)$

$3x^2 + 2(x - 4) = 48$

$3x^2 + 2x - 8 = 48$

$3x^2 + 2x - 56 = 0$

$(3x + 14)(x - 4) = 0$

$3x + 14 = 0$ or $x - 4 = 0$

$3x = -14$ or $x = 4$

$x = -\dfrac{14}{3}$

The number 4 is not allowed as a solution because substituting it in the original equation results in division by 0.

Check $x = -\dfrac{14}{3}$: $\dfrac{21}{13} - \dfrac{3}{7} = \dfrac{108}{91}$ *True*

The solution set is $\left\{ -\dfrac{14}{3} \right\}$.

59. **[7.5]** $f(x) = \dfrac{337}{x}$

(a) $f(40.5) = \dfrac{337}{40.5}$ *Let x = 40.5.*

≈ 8.32 mm

Copyright © 2012 Pearson Education, Inc. Publishing as Addison-Wesley.

(b) $7.51 = \dfrac{337}{x}$ *Let f(x) = 7.51.*

$$7.51x = 337$$

$$x = \dfrac{337}{7.51} \approx 44.9 \text{ diopters}$$

60. **[7.5]** Let $x =$ the time to fill the tub working together. Make a table.

	Rate	Time Working Together	Fractional Part of the Job Done
Hot	$\dfrac{1}{20}$	x	$\dfrac{x}{20}$
Cold	$\dfrac{1}{15}$	x	$\dfrac{x}{15}$

Part done by hot	plus	part done by cold	equals	1 whole job.
$\dfrac{x}{20}$	$+$	$\dfrac{x}{15}$	$=$	1

$$60\left(\dfrac{x}{20} + \dfrac{x}{15}\right) = 60 \cdot 1 \quad \textit{Multiply by 60.}$$

$$3x + 4x = 60$$

$$7x = 60$$

$$x = \tfrac{60}{7} \quad \text{or} \quad 8\tfrac{4}{7}$$

Working together, the tub can be filled in $\tfrac{60}{7}$ or $8\tfrac{4}{7}$ minutes.

61. **[7.5]** Let $x =$ the cost of 13 gallons. Write a proportion.

$$\dfrac{x}{13} = \dfrac{8.37}{3}$$

$$x = \dfrac{13(8.37)}{3}$$

$$= 36.27$$

The cost of 13 gallons of unleaded gasoline is $36.27.

62. **[7.6]** The area A of a triangle varies jointly as the lengths of the base b and height h, so

$$A = kbh.$$

When $b = 10$ and $h = 4$, $A = 20$.

$$20 = k(10)(4)$$

$$k = \tfrac{20}{40} = \tfrac{1}{2}$$

Thus, $A = \tfrac{1}{2}bh$. When $b = 3$ and $h = 8$,

$$A = \tfrac{1}{2}(3)(8) = 12.$$

The area of the triangle is 12 square feet.

63. **[7.5]** Let $x =$ the distance between his office and home. Make a table using the information in the problem and the formula $t = \tfrac{d}{r}$.

	d	r	t
Bike	x	12	$\dfrac{x}{12}$
Car	x	36	$\dfrac{x}{36}$

From the problem, the equation is stated in words: Car driving time = Bike riding time less $\tfrac{1}{4}$ hour.

Use the car time and the bike time given in the table to write the equation.

$$\dfrac{x}{36} = \dfrac{x}{12} - \dfrac{1}{4}$$

$$36\left(\dfrac{x}{36}\right) = 36\left(\dfrac{x}{12}\right) - 36\left(\dfrac{1}{4}\right)$$

$$x = 3x - 9$$

$$-2x = -9$$

$$x = 4\tfrac{1}{2} \text{ miles}$$

His office is $4\tfrac{1}{2}$ miles from home.

64. **[7.5]** *Step 2*

Let $x =$ the distance from San Francisco to the secret rendezvous.

Make a table.

	d	r	t
First Trip	x	200	$\dfrac{x}{200}$
Return Trip	x	300	$\dfrac{x}{300}$

Step 3

Time there	plus	time back	equals	4 hr.
$\dfrac{x}{200}$	$+$	$\dfrac{x}{300}$	$=$	4

Step 4

Multiply by the LCD, 600.

$$600\left(\dfrac{x}{200} + \dfrac{x}{300}\right) = 600(4)$$

$$3x + 2x = 2400$$

$$5x = 2400$$

$$x = 480$$

Step 5

The distance is 480 miles.

Step 6

Check 480 miles at 200 mph takes $\tfrac{480}{200}$ or 2.4 hours; 480 miles at 300 mph takes $\tfrac{480}{300}$ or 1.6 hours. The total time is 4 hours, as required.

Copyright © 2012 Pearson Education, Inc. Publishing as Addison-Wesley.

65. *[7.5] Step 2*
Let x = the distance to the fishing hole.

Make a table.

	d	r	t
Old Highway	x	30	$\dfrac{x}{30}$
Interstate	x	50	$\dfrac{x}{50}$

Step 3
The time on the interstate is 2 hr less than the time on the old highway.

$$\frac{x}{50} = \frac{x}{30} - 2$$

Step 4
Multiply by the LCD, 150.

$$3x = 5x - 300$$
$$300 = 2x$$
$$150 = x$$

Step 5
The distance to the fishing hole is 150 miles.

Step 6
Check 150 miles at 30 mph takes $\frac{150}{30}$ or 5 hours; 150 miles at 50 mph takes $\frac{150}{50}$ or 3 hours. The interstate time is 2 hours less than the highway time, as required.

Chapter 7 Test

1. $f(x) = \dfrac{x+3}{3x^2 + 2x - 8}$

Set the denominator equal to zero and solve.

$$3x^2 + 2x - 8 = 0$$
$$(3x - 4)(x + 2) = 0$$

$$3x - 4 = 0 \quad \text{or} \quad x + 2 = 0$$
$$3x = 4$$
$$x = \tfrac{4}{3} \quad \text{or} \quad x = -2$$

The numbers -2 and $\frac{4}{3}$ make the rational expression undefined and are excluded from the domain of f, which can be written in set-builder notation as $\left\{ x \mid x \neq -2, \frac{4}{3} \right\}$.

2. $\dfrac{6x^2 - 13x - 5}{9x^3 - x} = \dfrac{(3x+1)(2x-5)}{x(9x^2 - 1)}$

$$= \frac{(3x+1)(2x-5)}{x(3x+1)(3x-1)}$$

$$= \frac{2x-5}{x(3x-1)}$$

3. $\dfrac{(x+3)^2}{4} \cdot \dfrac{6}{2x+6} = \dfrac{(x+3)^2}{4} \cdot \dfrac{2 \cdot 3}{2(x+3)}$

$$= \frac{3(x+3)}{4}$$

4. $\dfrac{y^2 - 16}{y^2 - 25} \cdot \dfrac{y^2 + 2y - 15}{y^2 - 7y + 12}$

$$= \frac{(y+4)(y-4)}{(y+5)(y-5)} \cdot \frac{(y+5)(y-3)}{(y-4)(y-3)}$$

$$= \frac{y+4}{y-5}$$

5. $\dfrac{3-t}{5} \div \dfrac{t-3}{10}$

Multiply by the reciprocal of the divisor.

$$= \frac{3-t}{5} \cdot \frac{10}{t-3}$$

$$= \frac{-1(t-3) \cdot 10}{5(t-3)}$$

$$= -1(2) = -2$$

6. $\dfrac{x^2 - 9}{x^3 + 3x^2} \div \dfrac{x^2 + x - 12}{x^3 + 9x^2 + 20x}$

Multiply by the reciprocal.

$$= \frac{x^2 - 9}{x^3 + 3x^2} \cdot \frac{x(x^2 + 9x + 20)}{x^2 + x - 12}$$

$$= \frac{(x+3)(x-3)}{x^2(x+3)} \cdot \frac{x(x+5)(x+4)}{(x+4)(x-3)}$$

$$= \frac{x+5}{x}$$

7. $t^2 + t - 6, t^2 + 3t, t^2$

Factor each denominator.

$$t^2 + t - 6 = (t+3)(t-2)$$
$$t^2 + 3t = t(t+3)$$

The third denominator is already in factored form. The LCD is

$$t^2(t+3)(t-2).$$

8. $\dfrac{7}{6t^2} - \dfrac{1}{3t}$

The LCD is $6t^2$.

$$= \frac{7}{6t^2} - \frac{1 \cdot 2t}{3t \cdot 2t}$$

$$= \frac{7}{6t^2} - \frac{2t}{6t^2} = \frac{7 - 2t}{6t^2}$$

9. $\dfrac{3}{7a^4 b^3} + \dfrac{5}{21a^5 b^2}$

The LCD is $21a^5 b^3$.

$$= \frac{3 \cdot 3a}{7a^4 b^3 \cdot 3a} + \frac{5 \cdot b}{21a^5 b^2 \cdot b}$$

$$= \frac{9a}{21a^5 b^3} + \frac{5b}{21a^5 b^3}$$

$$= \frac{9a + 5b}{21a^5 b^3}$$

10. $\dfrac{9}{x^2 - 6x + 9} + \dfrac{2}{x^2 - 9}$

Factor each denominator.

$$x^2 - 6x + 9 = (x-3)^2$$
$$x^2 - 9 = (x+3)(x-3)$$

Copyright © 2012 Pearson Education, Inc. Publishing as Addison-Wesley.

The LCD is $(x+3)(x-3)^2$.

$$\frac{9}{(x-3)^2} + \frac{2}{(x+3)(x-3)}$$

$$= \frac{9(x+3)}{(x-3)^2(x+3)} + \frac{2(x-3)}{(x+3)(x-3)(x-3)}$$

$$= \frac{9(x+3) + 2(x-3)}{(x+3)(x-3)^2}$$

$$= \frac{9x + 27 + 2x - 6}{(x+3)(x-3)^2}$$

$$= \frac{11x + 21}{(x+3)(x-3)^2}$$

11. $\frac{6}{x+4} + \frac{1}{x+2} - \frac{3x}{x^2 + 6x + 8}$

$$= \frac{6}{x+4} + \frac{1}{x+2} - \frac{3x}{(x+4)(x+2)}$$

The LCD is $(x+4)(x+2)$.

$$= \frac{6(x+2)}{(x+4)(x+2)} + \frac{1(x+4)}{(x+2)(x+4)}$$

$$- \frac{3x}{(x+4)(x+2)}$$

$$= \frac{6(x+2) + x + 4 - 3x}{(x+4)(x+2)}$$

$$= \frac{6x + 12 + x + 4 - 3x}{(x+4)(x+2)}$$

$$= \frac{4x + 16}{(x+4)(x+2)}$$

$$= \frac{4(x+4)}{(x+4)(x+2)} = \frac{4}{x+2}$$

12. $\dfrac{\dfrac{12}{r+4}}{\dfrac{11}{6r+24}} = \dfrac{12}{r+4} \div \dfrac{11}{6r+24}$

Multiply by the reciprocal.

$$= \frac{12}{r+4} \cdot \frac{6r+24}{11}$$

$$= \frac{12}{r+4} \cdot \frac{6(r+4)}{11} = \frac{72}{11}$$

13. $\dfrac{\dfrac{1}{a} - \dfrac{1}{b}}{\dfrac{a}{b} - \dfrac{b}{a}}$

Multiply the numerator and the denominator by the LCD of all the fractions, ab.

$$= \frac{ab\left(\frac{1}{a} - \frac{1}{b}\right)}{ab\left(\frac{a}{b} - \frac{b}{a}\right)} = \frac{b - a}{a^2 - b^2}$$

$$= \frac{b - a}{(a-b)(a+b)} = \frac{(-1)(a-b)}{(a-b)(a+b)}$$

$$= \frac{-1}{a+b} \quad \text{or} \quad -\frac{1}{a+b}$$

14. $\dfrac{2x^{-2} + y^{-2}}{x^{-1} - y^{-1}} = \dfrac{\dfrac{2}{x^2} + \dfrac{1}{y^2}}{\dfrac{1}{x} - \dfrac{1}{y}}$

Multiply the numerator and denominator by the LCD of all the fractions, x^2y^2.

$$= \frac{x^2y^2\left(\frac{2}{x^2} + \frac{1}{y^2}\right)}{x^2y^2\left(\frac{1}{x} - \frac{1}{y}\right)}$$

$$= \frac{2y^2 + x^2}{xy^2 - x^2y}$$

$$= \frac{2y^2 + x^2}{xy(y - x)}$$

15. **(a)** Simplify this expression.

$$\frac{2x}{3} + \frac{x}{4} - \frac{11}{2}$$

The LCD is 12.

$$= \frac{2x \cdot 4}{3 \cdot 4} + \frac{x \cdot 3}{4 \cdot 3} - \frac{11 \cdot 6}{2 \cdot 6}$$

$$= \frac{8x}{12} + \frac{3x}{12} - \frac{66}{12}$$

$$= \frac{8x + 3x - 66}{12}$$

$$= \frac{11x - 66}{12} = \frac{11(x - 6)}{12}$$

(b) Solve this equation.

$$\frac{2x}{3} + \frac{x}{4} = \frac{11}{2}$$

Multiply by the LCD, 12.

$$12\left(\frac{2x}{3} + \frac{x}{4}\right) = 12\left(\frac{11}{2}\right)$$

$$8x + 3x = 66$$

$$11x = 66$$

$$x = 6$$

Check $x = 6$: $4 + \frac{3}{2} = \frac{11}{2}$ *True*

The solution set is $\{6\}$.

16. $\dfrac{1}{x} - \dfrac{4}{3x} = \dfrac{1}{x - 2}$

Multiply by the LCD, $3x(x-2)$. ($x \neq 0, 2$)

$$3x(x-2)\left(\frac{1}{x} - \frac{4}{3x}\right) = 3x(x-2)\left(\frac{1}{x-2}\right)$$

$$3(x-2) - 4(x-2) = 3x$$

$$3x - 6 - 4x + 8 = 3x$$

$$-x + 2 = 3x$$

$$-4x = -2$$

$$x = \frac{1}{2}$$

Check $x = \frac{1}{2}$: $2 - \frac{8}{3} = -\frac{2}{3}$ *True*

The solution set is $\left\{\frac{1}{2}\right\}$.

Copyright © 2012 Pearson Education, Inc. Publishing as Addison-Wesley.

17.

$$\frac{y}{y+2} - \frac{1}{y-2} = \frac{8}{y^2 - 4}$$

$$\frac{y}{y+2} - \frac{1}{y-2} = \frac{8}{(y+2)(y-2)}$$

Multiply by the LCD, $(y+2)(y-2)$.

$(y \neq -2, 2)$

$$(y+2)(y-2)\left(\frac{y}{y+2} - \frac{1}{y-2}\right)$$

$$= (y+2)(y-2)\left(\frac{8}{(y+2)(y-2)}\right)$$

$$y(y-2) - 1(y+2) = 8$$

$$y^2 - 2y - y - 2 = 8$$

$$y^2 - 3y - 10 = 0$$

$$(y-5)(y+2) = 0$$

$$y - 5 = 0 \quad \text{or} \quad y + 2 = 0$$

$$y = 5 \quad \text{or} \quad y = -2$$

The number -2 is not allowed as a solution because substituting it in the original equation results in division by 0.

Check $y = 5$: $\frac{5}{7} - \frac{1}{3} = \frac{8}{21}$ *True*

The solution set is $\{5\}$.

18.   Solve $S = \frac{n}{2}(a + \ell)$ for ℓ.

To solve for ℓ, multiply both sides by $\frac{2}{n}$.

$$\frac{2}{n} \cdot S = \frac{2}{n} \cdot \frac{n}{2}(a + \ell)$$

$$\frac{2S}{n} = a + \ell$$

$$\frac{2S}{n} - a = \ell, \text{ or } \ell = \frac{2S - na}{n}$$

19. $f(x) = \dfrac{-2}{x+1}$ is not defined when $x = -1$, so an equation of the vertical asymptote is $x = -1$. As the values of x get larger, the values of y get closer to 0, so $y = 0$ is the equation of the horizontal asymptote.

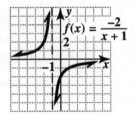

20.   Let $x =$ the time to do the job working together. Make a table.

Worker	Rate	Time Working Together	Fractional Part of the Job Done
Chris	$\frac{1}{9}$	x	$\frac{x}{9}$
Dana	$\frac{1}{5}$	x	$\frac{x}{5}$

Part done by Chris	plus	part done by Dana	equals	1 whole job.
$\frac{x}{9}$	$+$	$\frac{x}{5}$	$=$	1

$$45\left(\frac{x}{9} + \frac{x}{5}\right) = 45 \cdot 1 \quad \textit{Multiply by 45.}$$

$$5x + 9x = 45$$

$$14x = 45$$

$$x = \tfrac{45}{14} \quad \text{or} \quad 3\tfrac{3}{14}$$

Working together, they can do the job in $\frac{45}{14}$ or $3\frac{3}{14}$ hours.

21.   Let $x =$ the rate of the boat in still water. Use $d = rt$, or $t = \frac{d}{r}$, to make a table.

	d	r	t
Downstream	36	$x+3$	$\frac{36}{x+3}$
Upstream	24	$x-3$	$\frac{24}{x-3}$

Because the times are equal,

$$\frac{36}{x+3} = \frac{24}{x-3}.$$

Multiply by the LCD, $(x+3)(x-3)$. $(x \neq -3, 3)$

$$(x+3)(x-3)\left(\frac{36}{x+3}\right) = (x+3)(x-3)\left(\frac{24}{x-3}\right)$$

$$36(x-3) = 24(x+3)$$

$$36x - 108 = 24x + 72$$

$$12x = 180$$

$$x = 15$$

The rate of the boat in still water is 15 mph.

22.   Let $x =$ the number of fish in West Okoboji Lake. Write a proportion.

$$\frac{x}{600} = \frac{800}{10}$$

Multiply by the LCD, 600.

$$600\left(\frac{x}{600}\right) = 600\left(\frac{800}{10}\right)$$

$$x = 60 \cdot 800$$

$$x = 48{,}000$$

There are about 48,000 fish in the lake.

Copyright © 2012 Pearson Education, Inc. Publishing as Addison-Wesley.

23. $g(x) = \dfrac{5x}{2+x}$

(a) $\qquad 3 = \dfrac{5x}{2+x} \quad$ *Let g(x) = 3.*

$3(2+x) = 5x$

$6 + 3x = 5x$

$6 = 2x$

$x = 3 \text{ units}$

(b) If no food is available, then $x = 0$.

$$g(0) = \frac{5(0)}{2+0} = \frac{0}{2} = 0$$

The growth rate is 0.

24. The current I is inversely proportional to the resistance R, so

$$I = \frac{k}{R}$$

for some constant k. Let $I = 80$ and $R = 30$. Find k.

$$80 = \frac{k}{30}$$

$$k = 80(30) = 2400$$

So $I = \frac{2400}{R}$. Now let $R = 12$.

$$I = \frac{2400}{12} = 200$$

The current is 200 amperes.

25. The force F of the wind blowing on a vertical surface varies jointly as the area A of the surface and the square of the velocity V, so

$$F = kAV^2$$

for some constant k. Let $F = 50$, $A = 500$, and $V = 40$. Find k.

$$50 = k(500)(40)^2$$

$$k = \frac{50}{500(1600)} = \frac{1}{16{,}000}$$

So $F = \frac{1}{16{,}000}AV^2$. Now let $A = 2$ and $V = 80$.

$$F = \frac{1}{16{,}000}(2)(80)^2 = 0.8$$

The force of the wind is 0.8 pounds.

Cumulative Review Exercises (Chapters 1–7)

1. $|2x| + 3y - z^3$

Let $x = -4$, $y = 3$, and $z = 6$.

$= |(2)(-4)| + 3(3) - (6)^3$

$= |-8| + 9 - 216$

$= 8 + 9 - 216 = -199$

2. $7(2x + 3) - 4(2x + 1) = 2(x + 1)$

$14x + 21 - 8x - 4 = 2x + 2$

$6x + 17 = 2x + 2$

$4x = -15$

$x = -\frac{15}{4}$

The solution set is $\left\{ -\frac{15}{4} \right\}$.

3. $|6x - 8| - 4 = 0$

$|6x - 8| = 4$

$6x - 8 = 4 \quad$ or $\quad 6x - 8 = -4$

$6x = 12 \qquad\qquad 6x = 4$

$x = 2 \quad$ or $\quad x = \frac{4}{6} = \frac{2}{3}$

The solution set is $\left\{ \frac{2}{3}, 2 \right\}$.

4. $\frac{2}{3}y + \frac{5}{12}y \le 20$

Multiply both sides by 12.

$12(\frac{2}{3}y + \frac{5}{12}y) \le 12(20)$

$8y + 5y \le 240$

$13y \le 240$

$y \le \frac{240}{13}$

The solution set is $(-\infty, \frac{240}{13}]$.

5. $|3x + 2| \ge 4$

$3x + 2 \ge 4 \quad$ or $\quad 3x + 2 \le -4$

$3x \ge 2 \quad$ or $\quad 3x \le -6$

$x \ge \frac{2}{3} \quad$ or $\quad x \le -2$

The solution set is $(-\infty, -2] \cup [\frac{2}{3}, \infty)$.

6. Let $x = $ the amount of money invested at 4% and $2x = $ the amount of money invested at 3%.

Use $I = prt$ with the time, t, equal to 1 yr.

$0.04x + 0.03(2x) = 400$

$0.04x + 0.06x = 400$

$0.10x = 400$

Multiply by 10 to clear the decimal.

$x = 4000$

Then $2x = 2(4000) = 8000$.

He invested \$4000 at 4% and \$8000 at 3%.

7. Let $h = $ the height of the triangle. Use the formula $A = \frac{1}{2}bh$. Here, $A = 42$ and $b = 14$.

$A = \frac{1}{2}bh$

$42 = \frac{1}{2}(14)h$

$42 = 7h$

$6 = h$

The height is 6 meters.

Copyright © 2012 Pearson Education, Inc. Publishing as Addison-Wesley.

8. $-4x + 2y = 8$
To find the x-intercept, let $y = 0$.

$$-4x + 2(0) = 8$$
$$-4x = 8$$
$$x = -2$$

The x-intercept is $(-2, 0)$.
To find the y-intercept, let $x = 0$.

$$-4(0) + 2y = 8$$
$$2y = 8$$
$$y = 4$$

The y-intercept is $(0, 4)$. Plot the intercepts, and draw the line through them.

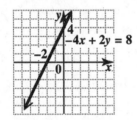

9. Through $(-5, 8)$ and $(-1, 2)$
Let $(x_1, y_1) = (-5, 8)$ and $(x_2, y_2) = (-1, 2)$.
Then,

$$m = \frac{y_2 - y_1}{x_2 - x_1} = \frac{2 - 8}{-1 - (-5)} = \frac{-6}{4} = -\frac{3}{2}.$$

The slope is $-\frac{3}{2}$.

10. Perpendicular to $4x - 3y = 12$

Solve for y to write the equation in slope-intercept form and find the slope.

$$4x - 3y = 12$$
$$-3y = -4x + 12$$
$$y = \frac{4}{3}x - 4$$

The slope is $\frac{4}{3}$. Perpendicular lines have slopes that are negative reciprocals of each other. The negative reciprocal of $\frac{4}{3}$ is $-\frac{3}{4}$. The slope of a line perpendicular to the given line is $-\frac{3}{4}$.

11. Use $(x_1, y_1) = (-5, 8)$ and $m = -\frac{3}{2}$ in the point-slope form.

$$y - y_1 = m(x - x_1)$$
$$y - 8 = -\frac{3}{2}[x - (-5)]$$
$$y - 8 = -\frac{3}{2}(x + 5)$$
$$y - 8 = -\frac{3}{2}x - \frac{15}{2}$$
$$y = -\frac{3}{2}x + \frac{1}{2}$$

12. $2x + 5y > 10$
Graph the line $2x + 5y = 10$ by drawing a dashed line (since the inequality involves $>$) through the intercepts $(5, 0)$ and $(0, 2)$.
Test a point not on this line, such as $(0, 0)$.

$$2x + 5y > 10$$
$$2(0) + 5(0) > 10$$
$$0 > 10 \quad \textit{False}$$

Shade the side of the line not containing $(0, 0)$.

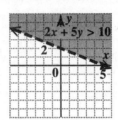

13. $x - y \geq 3$ and $3x + 4y \leq 12$
Graph the solid line $x - y = 3$ through $(3, 0)$ and $(0, -3)$. The inequality $x - y \geq 3$ can be written as $y \leq x - 3$, so shade the region below the boundary line.
Graph the solid line $3x + 4y = 12$ through $(4, 0)$ and $(0, 3)$. The inequality $3x + 4y \leq 12$ can be written as $y \leq -\frac{3}{4}x + 3$, so shade the region below the boundary line.
The required graph is the common shaded area as well as the portions of the lines that bound it.

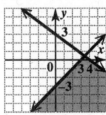

$x - y \geq 3$ and
$3x + 4y \leq 12$

14. $y = -\sqrt{x + 2}$
The relation is a function since each input value corresponds to exactly one output value. The radicand must be nonnegative; that is, $x + 2 \geq 0$, or $x \geq -2$. The domain is $[-2, \infty)$. The values of $\sqrt{x + 2}$ are nonnegative, so the values of $-\sqrt{x + 2}$ are nonpositive. Thus, the range is $(-\infty, 0]$.

15. **(a)** Solve the equation for y.

$$5x - 3y = 8$$
$$5x - 8 = 3y$$
$$\frac{5x - 8}{3} = y$$

So $f(x) = \dfrac{5x - 8}{3}$ or $f(x) = \dfrac{5}{3}x - \dfrac{8}{3}$.

Copyright © 2012 Pearson Education, Inc. Publishing as Addison-Wesley.

(b) $f(1) = \dfrac{5(1) - 8}{3} = \dfrac{-3}{3} = -1$

16. $4x - y = -7$ (1)
$5x + 2y = 1$ (2)

To eliminate y, multiply equation (1) by 2 and add the result to (2).

$$
\begin{array}{rlr}
8x - 2y & = -14 & \quad 2 \times (1) \\
5x + 2y & = 1 & \quad (2) \\
\hline
13x & = -13 & \\
x & = -1 &
\end{array}
$$

To find y, substitute -1 for x in (1).

$$
\begin{aligned}
4x - y &= -7 \qquad (1) \\
4(-1) - y &= -7 \\
-4 - y &= -7 \\
-y &= -3 \\
y &= 3
\end{aligned}
$$

The solution set is $\{(-1, 3)\}$.

17. $x + y - 2z = -1$ (1)
$2x - y + z = -6$ (2)
$3x + 2y - 3z = -3$ (3)

Add (1) and (2) to eliminate y.

$$
\begin{array}{rlr}
x + y - 2z & = -1 & (1) \\
2x - y + z & = -6 & (2) \\
\hline
3x - z & = -7 & (4)
\end{array}
$$

Multiply (2) by 2 and add it to (3) to eliminate y.

$$
\begin{array}{rlr}
4x - 2y + 2z & = -12 & \quad 2 \times (2) \\
3x + 2y - 3z & = -3 & (3) \\
\hline
7x - z & = -15 & (5)
\end{array}
$$

Multiply (4) by -1 and add to (5) to eliminate z.

$$
\begin{array}{rlr}
-3x + z & = 7 & \quad -1 \times (4) \\
7x - z & = -15 & (5) \\
\hline
4x & = -8 & \\
x & = -2 &
\end{array}
$$

Substitute -2 for x into (4) and solve for z.

$$
\begin{aligned}
3(-2) - z &= -7 \\
-6 - z &= -7 \\
-z &= -1 \\
z &= 1
\end{aligned}
$$

Substitute -2 for x and 1 for z into (1) and solve for y.

$$
\begin{aligned}
-2 + y - 2(1) &= -1 \\
-2 + y - 2 &= -1 \\
y - 4 &= -1 \\
y &= 3
\end{aligned}
$$

The solution set is $\{(-2, 3, 1)\}$.

18. $x + 2y + z = 5$ (1)
$x - y + z = 3$ (2)
$2x + 4y + 2z = 11$ (3)

Multiply (1) by -1 and add to (2).

$$
\begin{array}{rlr}
-x - 2y - z & = -5 & \quad -1 \times (1) \\
x - y + z & = 3 & (2) \\
\hline
- 3y & = -2 & \\
y & = \dfrac{2}{3} &
\end{array}
$$

Substitute $\frac{2}{3}$ for y in (1) and (3), then add (1) and (3).

$$
\begin{aligned}
x + 2\left(\tfrac{2}{3}\right) + z &= 5 \qquad \text{Let } y = \tfrac{2}{3} \text{ in (1).} \\
x + \tfrac{4}{3} + z &= 5 \\
x + z &= \tfrac{11}{3} \qquad (4)
\end{aligned}
$$

$$
\begin{aligned}
2x + 4\left(\tfrac{2}{3}\right) + 2z &= 11 \qquad \text{Let } y = \tfrac{2}{3} \text{ in (3).} \\
2x + \tfrac{8}{3} + 2z &= 11 \\
2x + 2z &= \tfrac{25}{3} \qquad (5)
\end{aligned}
$$

Multiply (4) by -2 and add to (5).

$$
\begin{array}{rlr}
-2x - 2z & = -\tfrac{22}{3} & \quad -2 \times (4) \\
2x + 2z & = \tfrac{25}{3} & (5) \\
\hline
0 & = 1 & \quad \textit{False}
\end{array}
$$

Since this statement is false, the solution is $\emptyset$.

19.   Let $x =$ the average rate of the automobile.
Then $x + 558 =$ the average rate of the airplane.
Use $d = rt$, or $t = \frac{d}{r}$, to make a table.

	d	r	t
Automobile	7	x	$\dfrac{7}{x}$
Airplane	100	$x + 558$	$\dfrac{100}{x + 558}$

Because the times are equal,

$$
\frac{7}{x} = \frac{100}{x + 558}.
$$

Multiply each side by the LCD, $x(x + 558)$.

$$
\begin{aligned}
x(x + 558)\left(\frac{7}{x}\right) &= x(x + 558)\left(\frac{100}{x + 558}\right) \\
7(x + 558) &= 100x \\
7x + 3906 &= 100x \\
3906 &= 93x \\
42 &= x
\end{aligned}
$$

Since $x = 42$, $x + 558 = 600$.

The average rate of the automobile is 42 km per hr and that of the airplane is 600 km per hr.

Copyright © 2012 Pearson Education, Inc. Publishing as Addison-Wesley.

20. $\left(\dfrac{m^{-4}n^2}{m^2n^{-3}}\right) \cdot \left(\dfrac{m^5n^{-1}}{m^{-2}n^5}\right)$

$= \left(\dfrac{n^2n^3}{m^2m^4}\right) \cdot \left(\dfrac{m^5m^2}{n^1n^5}\right)$

$= \left(\dfrac{n^5}{m^6}\right) \cdot \left(\dfrac{m^7}{n^6}\right)$

$= \dfrac{n^5m^7}{n^6m^6} = \dfrac{m}{n}$

21. $(3y^2 - 2y + 6) - (-y^2 + 5y + 12)$

$= 3y^2 - 2y + 6 + y^2 - 5y - 12$

$= 4y^2 - 7y - 6$

22. $(4f + 3)(3f - 1) = 12f^2 - 4f + 9f - 3$

$= 12f^2 + 5f - 3$

23. $\left(\frac{1}{4}x + 5\right)^2$

Use the formula for the square of a binomial,
$(a + b)^2 = a^2 + 2ab + b^2$.

$\left(\frac{1}{4}x + 5\right)^2 = \left(\frac{1}{4}x\right)^2 + 2\left(\frac{1}{4}x\right)(5) + 5^2$

$= \frac{1}{16}x^2 + \frac{5}{2}x + 25$

24. $(3x^3 + 13x^2 - 17x - 7) \div (3x + 1)$

$$
\begin{array}{r}
x^2 + 4x - 7 \\
3x + 1 \overline{)3x^3 + 13x^2 - 17x - 7} \\
\underline{3x^3 + x^2} \\
12x^2 - 17x \\
\underline{12x^2 + 4x} \\
-21x - 7 \\
\underline{-21x - 7} \\
0
\end{array}
$$

Answer: $x^2 + 4x - 7$

25. **(a)** $(f + g)(x) = f(x) + g(x)$

$= (x^2 + 2x - 3)$

$\qquad + (2x^3 - 3x^2 + 4x - 1)$

$= 2x^3 - 2x^2 + 6x - 4$

(b) $(g - f)(x) = g(x) - f(x)$

$= (2x^3 - 3x^2 + 4x - 1)$

$\qquad - (x^2 + 2x - 3)$

$= 2x^3 - 3x^2 + 4x - 1$

$\qquad - x^2 - 2x + 3$

$= 2x^3 - 4x^2 + 2x + 2$

(c) Using part (a),

$(f + g)(-1)$

$= 2(-1)^3 - 2(-1)^2 + 6(-1) - 4$

$= -2 - 2 - 6 - 4$

$= -14$

(d) $(f \circ h)(x) = f(h(x))$

$= f(x^2)$

$= (x^2)^2 + 2(x^2) - 3$

$= x^4 + 2x^2 - 3$

26. $2x^2 - 13x - 45 = (2x + 5)(x - 9)$

27. $100t^4 - 25 = 25(4t^4 - 1)$

$= 25[(2t^2)^2 - 1^2]$

$= 25(2t^2 + 1)(2t^2 - 1)$

28. Use the sum of cubes formula,

$$x^3 + y^3 = (x + y)(x^2 - xy + y^2).$$

$8p^3 + 125 = (2p)^3 + 5^3$

$= (2p + 5)[(2p)^2 - (2p)(5) + 5^2]$

$= (2p + 5)(4p^2 - 10p + 25)$

29.
$$3x^2 + 4x = 7$$

$$3x^2 + 4x - 7 = 0$$

$$(3x + 7)(x - 1) = 0$$

$3x + 7 = 0 \qquad$ or $\quad x - 1 = 0$

$3x = -7$

$x = -\frac{7}{3}$ or $\qquad x = 1$

The solution set is $\left\{-\frac{7}{3}, 1\right\}$.

30. $\dfrac{y^2 - 16}{y^2 - 8y + 16} = \dfrac{(y + 4)(y - 4)}{(y - 4)(y - 4)}$

$= \dfrac{y + 4}{y - 4}$

31. $\dfrac{2a^2}{a + b} \cdot \dfrac{a - b}{4a} = \dfrac{2a^2(a - b)}{4a(a + b)}$

$= \dfrac{a(a - b)}{2(a + b)}$

32. $\dfrac{x^2 - 9}{2x + 4} \div \dfrac{x^3 - 27}{4}$

$= \dfrac{x^2 - 9}{2x + 4} \cdot \dfrac{4}{x^3 - 27}$

$= \dfrac{(x + 3)(x - 3)}{2(x + 2)} \cdot \dfrac{4}{(x - 3)(x^2 + 3x + 9)}$

$= \dfrac{2(x + 3)}{(x + 2)(x^2 + 3x + 9)}$

33. $\dfrac{x + 4}{x - 2} + \dfrac{2x - 10}{x - 2} = \dfrac{x + 4 + 2x - 10}{x - 2}$

$= \dfrac{3x - 6}{x - 2}$

$= \dfrac{3(x - 2)}{x - 2} = 3$

Copyright © 2012 Pearson Education, Inc. Publishing as Addison-Wesley.

34.

$$\frac{-3x}{x+1} + \frac{4x+1}{x} = \frac{-3}{x^2+x}$$
$$\frac{-3x}{x+1} + \frac{4x+1}{x} = \frac{-3}{x(x+1)}$$

Multiply by the LCD, $x(x+1)$. ($x \neq -1, 0$)

$$x(x+1)\left(\frac{-3x}{x+1} + \frac{4x+1}{x}\right)$$
$$= x(x+1)\left(\frac{-3}{x(x+1)}\right)$$

$$x(-3x) + (x+1)(4x+1) = -3$$
$$-3x^2 + 4x^2 + x + 4x + 1 = -3$$
$$x^2 + 5x + 4 = 0$$
$$(x+4)(x+1) = 0$$

$$x + 4 = 0 \quad \text{or} \quad x + 1 = 0$$
$$x = -4 \quad \text{or} \quad x = -1$$

The number -1 is not allowed as a solution because substituting it in the original equation results in division by 0.

Check $x = -4$: $-4 + \frac{15}{4} = -\frac{1}{4}$ *True*

The solution set is $\{-4\}$.

35. Let $x =$ the time to complete the job working together.
Make a table.

	Rate	Time Working Together	Fractional Part of the Job Done
Machine A	$\frac{1}{2}$	x	$\frac{x}{2}$
Machine B	$\frac{1}{3}$	x	$\frac{x}{3}$

Part done by A	plus	part done by B	equals	1 whole job.
$\frac{x}{2}$	$+$	$\frac{x}{3}$	$=$	1

$$6\left(\frac{x}{2} + \frac{x}{3}\right) = 6 \cdot 1 \quad \text{Multiply by 6.}$$
$$3x + 2x = 6$$
$$5x = 6$$
$$x = \frac{6}{5} \quad \text{or} \quad 1\frac{1}{5}$$

Working together, the machines can complete the job in $\frac{6}{5}$ or $1\frac{1}{5}$ hours.

36. Let $C =$ the cost of a pizza and $r =$ the radius of the pizza. C varies directly as r^2, so

$$C = kr^2$$

for some constant k. Since $C = 6$ when $r = 7$, substitute these values in the equation and solve for k.

$$C = kr^2$$
$$6 = k(7)^2$$
$$6 = 49k$$
$$\frac{6}{49} = k$$

So $C = \frac{6}{49}r^2$. Now let $r = 9$.

$$C = \frac{6}{49}(9)^2 = \frac{6}{49}(81) = \frac{486}{49} \approx 9.92.$$

A pizza with a 9-inch radius should cost $9.92.

Copyright © 2012 Pearson Education, Inc. Publishing as Addison-Wesley.

Copyright © 2012 Pearson Education, Inc. Publishing as Addison-Wesley.

CHAPTER 8   ROOTS, RADICALS, AND ROOT FUNCTIONS

8.1   Radical Expressions and Graphs

8.1   Now Try Exercises

N1. (a) $\sqrt[3]{1000} = 10$, because $10^3 = 1000$.

(b) $\sqrt[4]{625} = 5$, because $5^4 = 625$.

(c) $\sqrt[4]{\frac{1}{256}} = \frac{1}{4}$, because $\left(\frac{1}{4}\right)^4 = \frac{1}{256}$.

(d) $\sqrt[3]{0.027} = 0.3$, because $(0.3)^3 = 0.027$.

N2. (a) $\sqrt{25} = 5$, because $5^2 = 25$, so $-\sqrt{25} = -5$. The negative sign outside the radical makes the answer negative.

(b) $\sqrt[4]{-625}$ is *not a real number*, since the index, 4, is even and the radicand, -625, is negative.

(c) $\sqrt[5]{-32} = -2$, because $(-2)^5 = -32$.

(d) $-\sqrt[3]{64} = -\left(\sqrt[3]{64}\right) = -(4) = -4$

N3. (a) $f(x) = \sqrt{x+1}$

For the radicand to be nonnegative, we must have $x + 1 \geq 0$, or, equivalently $x \geq -1$. Therefore, the domain is $[-1, \infty)$. The function values are at least 0 since $\sqrt{x+1} \geq 0$, so the range is $[0, \infty)$.

x	$f(x) = \sqrt{x+1}$
-1	$\sqrt{-1+1} = 0$
0	$\sqrt{0+1} = 1$
3	$\sqrt{3+1} = 2$
8	$\sqrt{8+1} = 3$

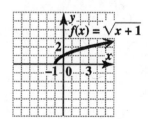

(b) $f(x) = \sqrt[3]{x} - 1$

Since we can take the cube root of any real number, the domain is $(-\infty, \infty)$.
The result of a cube root can be any real number, so the range is $(-\infty, \infty)$ (subtracting 1 doesn't change the range).

x	$f(x) = \sqrt[3]{x} - 1$
-8	$\sqrt[3]{-8} - 1 = -3$
-1	$\sqrt[3]{-1} - 1 = -2$
0	$\sqrt[3]{0} - 1 = -1$
1	$\sqrt[3]{1} - 1 = 0$
8	$\sqrt[3]{8} - 1 = 1$

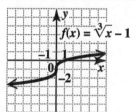

N4. (a) $\sqrt{11^2} = |11| = 11$

(b) $\sqrt{(-11)^2} = |-11| = 11$

(c) $\sqrt{z^2} = |z|$

(d) $\sqrt{(-z)^2} = |-z| = |z|$

N5. (a) $\sqrt[8]{(-2)^8} = |-2| = 2$ (n is even)

(b) $\sqrt[3]{(-9)^3} = -9$ (n is odd)

(c) $-\sqrt[4]{(-10)^4} = -|-10| = -10$ (n is even)

(d) $-\sqrt{m^8} = -\sqrt{(m^4)^2} = -m^4$ (n is even)
No absolute value bars are needed here because m^4 is nonnegative for any real number value of m.

(e) $\sqrt[3]{x^{18}} = \sqrt[3]{(x^6)^3} = x^6$

(f) $\sqrt[4]{t^{20}} = \sqrt[4]{(t^5)^4} = |t^5|$
Since t^5 could be negative, we need the absolute value bars.

N6. Use a calculator, and round to three decimal places.

(a) $-\sqrt{92} \approx -9.592$

(b) $\sqrt[4]{39} \approx 2.499$

(c) $\sqrt[5]{33} \approx 2.012$

N7. $f = \dfrac{1}{2\pi\sqrt{LC}}$

$= \dfrac{1}{2\pi\sqrt{(7 \times 10^{-5})(3 \times 10^{-9})}}$

$\approx 347{,}305$

or about 347,000 cycles per second.

8.1   Section Exercises

1. $-\sqrt{25} = -(5) = -5$ **(E)**

3. $\sqrt[3]{-27} = -3$, because $(-3)^3 = -27$. **(D)**

5. $\sqrt[4]{81} = 3$, because $3^4 = 81$. **(A)**

Copyright © 2012 Pearson Education, Inc. Publishing as Addison-Wesley.

7. $\sqrt{81} = 9$, because $9^2 = 81$. Notice that -9 is not an answer because the symbol $\sqrt{81}$ means the *nonnegative* square root of 81. However, the negative in front of the radical does lead to a negative answer since

$$-\sqrt{81} = -(9) = -9.$$

9. $\sqrt[3]{216} = 6$, because $6^3 = 216$.

11. $\sqrt[3]{-64} = -4$, because $(-4)^3 = -64$.

13. $\sqrt[3]{512} = 8$, because $8^3 = 512$, so $-\sqrt[3]{512} = -8$.

15. $\sqrt[4]{1296} = 6$, because $6^4 = 1296$.

17. $\sqrt[4]{16} = 2$, because $2^4 = 16$, so $-\sqrt[4]{16} = -2$.

19. $\sqrt[4]{-625}$ is not a real number since no real number to the fourth power equals -625. Any real number raised to the fourth power is 0 or positive.

21. $\sqrt[6]{64} = 2$, because $2^6 = 64$.

23. $\sqrt[6]{-32}$ is not a real number, because the index, 6, is even and the radicand, -32, is negative.

25. $\sqrt{\frac{64}{81}} = \frac{8}{9}$, because $\left(\frac{8}{9}\right)^2 = \frac{64}{81}$.

27. $\sqrt[3]{\frac{64}{27}} = \frac{4}{3}$, because $\left(\frac{4}{3}\right)^3 = \frac{64}{27}$.

29. $\sqrt[6]{\frac{1}{64}} = \frac{1}{2}$, because $\left(\frac{1}{2}\right)^6 = \frac{1}{64}$, so $-\sqrt[6]{\frac{1}{64}} = -\frac{1}{2}$.

31. $\sqrt[3]{-27} = -3$, because $(-3)^3 = -27$, so $-\sqrt[3]{-27} = -(-3) = 3$.

33. $\sqrt{0.25} = 0.5$, because $(0.5)^2 = 0.25$.

35. $\sqrt{0.49} = 0.7$, because $(0.7)^2 = 0.49$, so $-\sqrt{0.49} = -0.7$.

37. $\sqrt[3]{0.001} = 0.1$, because $(0.1)^3 = 0.001$.

39. **(a)** If $a > 0$, then $\sqrt{-a}$ is not a real number, so $-\sqrt{-a}$ is *not a real number*.

(b) If $a < 0$, then $\sqrt{-a}$ is a positive number and then $-\sqrt{-a}$ is a *negative* number.

(c) If $a = 0$, then $\sqrt{-a} = \sqrt{0} = 0$, so $-\sqrt{-a} = 0$.

41. $f(x) = \sqrt{x+3}$

For the radicand to be nonnegative, we must have

$$x + 3 \geq 0 \quad \text{or} \quad x \geq -3.$$

Thus, the domain is $[-3, \infty)$.
The function values are positive or zero (the result of the radical), so the range is $[0, \infty)$.

x	$f(x) = \sqrt{x+3}$
-3	$\sqrt{-3+3} = 0$
-2	$\sqrt{-2+3} = 1$
1	$\sqrt{1+3} = 2$

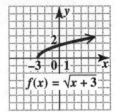

43. $f(x) = \sqrt{x} - 2$

For the radicand to be nonnegative, we must have $x \geq 0$. Note that the "-2" does not affect the domain, which is $[0, \infty)$.

The result of the radical is positive or zero, but the function values are 2 less than those values, so the range is $[-2, \infty)$.

x	$f(x) = \sqrt{x} - 2$
0	$\sqrt{0} - 2 = -2$
1	$\sqrt{1} - 2 = -1$
4	$\sqrt{4} - 2 = 0$

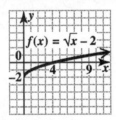

45. $f(x) = \sqrt[3]{x} - 3$

Since we can take the cube root of any real number, the domain is $(-\infty, \infty)$.
The result of a cube root can be any real number, so the range is $(-\infty, \infty)$. (The "-3" does not affect the range.)

x	$f(x) = \sqrt[3]{x} - 3$
-8	$\sqrt[3]{-8} - 3 = -5$
-1	$\sqrt[3]{-1} - 3 = -4$
0	$\sqrt[3]{0} - 3 = -3$
1	$\sqrt[3]{1} - 3 = -2$
8	$\sqrt[3]{8} - 3 = -1$

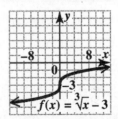

Copyright © 2012 Pearson Education, Inc. Publishing as Addison-Wesley.

47. $f(x) = \sqrt[3]{x} - 3$

Both the domain and range are $(-\infty, \infty)$.

x	$f(x) = \sqrt[3]{x} - 3$
-5	$\sqrt[3]{-5} - 3 = -2$
2	$\sqrt[3]{2} - 3 = -1$
3	$\sqrt[3]{3} - 3 = 0$
4	$\sqrt[3]{4} - 3 = 1$
11	$\sqrt[3]{11} - 3 = 2$

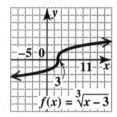

49. $\sqrt{12^2} = |12| = 12$

51. $\sqrt{(-10)^2} = |-10| = -(-10) = 10$

53. Since 6 is an even positive integer, $\sqrt[6]{a^6} = |a|$, so $\sqrt[6]{(-2)^6} = |-2| = 2$.

55. Since 5 is odd, $\sqrt[5]{a^5} = a$, so $\sqrt[5]{(-9)^5} = -9$.

57. $\sqrt[6]{(-5)^6} = |-5| = 5$, so $-\sqrt[6]{(-5)^6} = -5$.

59. Since the index is even, $\sqrt{x^2} = |x|$.

61. $\sqrt{(-z)^2} = |-z| = |z|$

63. Since the index is odd, $\sqrt[3]{x^3} = x$.

65. $\sqrt[3]{x^{15}} = \sqrt[3]{(x^5)^3} = x^5$ (3 is odd)

67. $\sqrt[6]{x^{30}} = \sqrt[6]{(x^5)^6} = |x^5|$, or $|x|^5$ (6 is even)

69. $\sqrt{123.5} \approx \sqrt{121} = 11$, because $11^2 = 121$. **(C)**

In Exercises 71–82, use a calculator and round to three decimal places

71. $\sqrt{9483} \approx 97.381$

73. $\sqrt{284.361} \approx 16.863$

75. $-\sqrt{82} \approx -9.055$

77. $\sqrt[3]{423} \approx 7.507$

79. $\sqrt[4]{100} \approx 3.162$

81. $\sqrt[5]{23.8} \approx 1.885$

83. The length $\sqrt{98}$ is closer to $\sqrt{100} = 10$ than to $\sqrt{81} = 9$. The width $\sqrt{26}$ is closer to $\sqrt{25} = 5$ than to $\sqrt{36} = 6$. Use the estimates $L = 10$ and $W = 5$ in $A = LW$ to find an estimate of the area.

$$A \approx 10 \cdot 5 = 50$$

Choice **A** is the best estimate.

85. $f = \dfrac{1}{2\pi\sqrt{LC}}$

$= \dfrac{1}{2\pi\sqrt{(7.237 \times 10^{-5})(2.5 \times 10^{-10})}}$

$\approx 1{,}183{,}235$

or about 1,183,000 cycles per second.

87. Since $H = 44 + 6 = 50$ ft, substitute 50 for H in the formula.

$$D = \sqrt{2H} = \sqrt{2 \cdot 50} = \sqrt{100} = 10$$

She will be able to see about 10 miles.

89. Let $a = 850$ mi, $b = 925$ mi, and $c = 1300$ mi. First find s.

$$s = \tfrac{1}{2}(a + b + c)$$
$$= \tfrac{1}{2}(850 + 925 + 1300)$$
$$= \tfrac{3075}{2} = 1537.5$$

Now find the area using Heron's formula.

$$A = \sqrt{s(s-a)(s-b)(s-c)}$$
$$= \sqrt{1537.5(687.5)(612.5)(237.5)}$$
$$\approx 392{,}128.8$$

The area of the Bermuda Triangle is about 392,000 square miles.

91. **(a)** $I = \sqrt{\dfrac{2P}{L}} = \sqrt{\dfrac{2(120)}{80}}$ *Let P = 120, L = 80.*

$= \sqrt{3} \approx 1.732$ amps

(b) $I = \sqrt{\dfrac{2P}{L}} = \sqrt{\dfrac{2(100)}{40}}$ *Let P = 100, L = 40.*

$= \sqrt{5} \approx 2.236$ amps

93. $x^5 \cdot x^{-1} \cdot x^{-3} = x^{5+(-1)+(-3)} = x^1$, or x

95. $\left(\dfrac{2}{3}\right)^{-3} = \left(\dfrac{3}{2}\right)^3 = \dfrac{3^3}{2^3}$, or $\dfrac{27}{8}$

8.2   Rational Exponents

8.2 Now Try Exercises

N1. **(a)** $81^{1/2} = \sqrt[2]{81} = \sqrt{81} = 9$ $(9^2 = 81)$

(b) $125^{1/3} = \sqrt[3]{125} = 5$ $(5^3 = 125)$

(c) $-625^{1/4} = -\sqrt[4]{625} = -5$ $(5^4 = 625)$

(d) $(-625)^{1/4} = \sqrt[4]{-625}$ is *not a real number* because the radicand, -625, is negative and the index, 4, is even.

(e) $(-125)^{1/3} = \sqrt[3]{-125} = -5$ $[(-5)^3 = -125]$

(f) $\left(\tfrac{1}{16}\right)^{1/4} = \sqrt[4]{\tfrac{1}{16}} = \tfrac{1}{2}$ $[(\tfrac{1}{2})^4 = \tfrac{1}{16}]$

N2. **(a)** $32^{2/5} = (32^{1/5})^2 = \left(\sqrt[5]{32}\right)^2 = 2^2 = 4$

(b) $8^{5/3} = (8^{1/3})^5 = \left(\sqrt[3]{8}\right)^5 = 2^5 = 32$

Copyright © 2012 Pearson Education, Inc. Publishing as Addison-Wesley.

(c) $-100^{3/2} = -(100^{1/2})^3 = -(\sqrt{100})^3$

$\qquad = -10^3 = -1000$

(d) $(-121)^{3/2}$ is *not a real number*, since $(-121)^{1/2}$, or $\sqrt{-121}$, is not a real number.

(e) $(-125)^{4/3} = [(-125)^{1/3}]^4 = (\sqrt[3]{-125})^4$

$\qquad = (-5)^4 = 625$

N3. **(a)** $243^{-3/5} = \dfrac{1}{243^{3/5}} = \dfrac{1}{(243^{1/5})^3} = \dfrac{1}{(\sqrt[5]{243})^3}$

$\qquad = \dfrac{1}{3^3} = \dfrac{1}{27}$

(b) $4^{-5/2} = \dfrac{1}{4^{5/2}} = \dfrac{1}{(4^{1/2})^5} = \dfrac{1}{(\sqrt{4})^5}$

$\qquad = \dfrac{1}{2^5} = \dfrac{1}{32}$

(c) $\left(\dfrac{216}{125}\right)^{-2/3} = \left(\dfrac{125}{216}\right)^{2/3} = \left(\sqrt[3]{\dfrac{125}{216}}\right)^2$

$\qquad = \left(\dfrac{5}{6}\right)^2 = \dfrac{25}{36}$

N4. **(a)** $21^{1/2} = (\sqrt[2]{21})^1 = \sqrt{21}$

(b) $17^{5/4} = (\sqrt[4]{17})^5$

(c) $4t^{3/5} + (4t)^{2/3} = 4(\sqrt[5]{t})^3 + (\sqrt[3]{4t})^2$

(d) $w^{-2/5} = \dfrac{1}{w^{2/5}} = \dfrac{1}{(\sqrt[5]{w})^2}$

(e) $(a^2 - b^2)^{1/4} = \sqrt[4]{a^2 - b^2}$

(f) $\sqrt[3]{15} = 15^{1/3}$

(g) $\sqrt[4]{4^2} = 4^{2/4} = 4^{1/2} = \sqrt{4} = 2$

(h) $\sqrt[4]{x^4} = x$, since x is assumed to be positive.

N5. **(a)** $5^{1/4} \cdot 5^{2/3} = 5^{1/4+2/3} = 5^{3/12+8/12} = 5^{11/12}$

(b) $\dfrac{9^{3/5}}{9^{7/5}} = 9^{3/5-7/5} = 9^{-4/5} = \dfrac{1}{9^{4/5}}$

(c) $\dfrac{(r^{2/3}t^{1/4})^8}{t} = \dfrac{(r^{2/3})^8(t^{1/4})^8}{t}$

$\qquad = \dfrac{r^{(2/3)8}t^{(1/4)8}}{t}$

$\qquad = \dfrac{r^{16/3}t^2}{t}$

$\qquad = r^{16/3}t$

(d) $\left(\dfrac{2x^{1/2}y^{-2/3}}{x^{-3/5}y^{-1/5}}\right)^{-3}$

$\qquad = (2x^{1/2-(-3/5)}y^{-2/3-(-1/5)})^{-3}$

$\qquad = (2x^{5/10+6/10}y^{-10/15+3/15})^{-3}$

$\qquad = (2x^{11/10}y^{-7/15})^{-3}$

$\qquad = (2)^{-3}(x^{11/10})^{-3}(y^{-7/15})^{-3}$

$\qquad = \dfrac{1}{2^3}x^{-33/10}y^{21/15}$

$\qquad = \dfrac{y^{7/5}}{8x^{33/10}}$

(e) $y^{2/3}(y^{1/3} + y^{5/3}) = y^{2/3} \cdot y^{1/3} + y^{2/3} \cdot y^{5/3}$

$\qquad = y^{2/3+1/3} + y^{2/3+5/3}$

$\qquad = y^{3/3} + y^{7/3}$

$\qquad = y + y^{7/3}$

N6. **(a)** $\sqrt[5]{y^3} \cdot \sqrt[3]{y} = y^{3/5} \cdot y^{1/3}$

$\qquad = y^{3/5+1/3}$

$\qquad = y^{9/15+5/15}$

$\qquad = y^{14/15}$

(b) $\dfrac{\sqrt[4]{y^3}}{\sqrt{y^5}} = \dfrac{y^{3/4}}{y^{5/2}} = y^{3/4-5/2}$

$\qquad = y^{3/4-10/4} = y^{-7/4} = \dfrac{1}{y^{7/4}}$

(c) $\sqrt{\sqrt[3]{y}} = \sqrt{y^{1/3}} = (y^{1/3})^{1/2}$

$\qquad = y^{(1/3)(1/2)} = y^{1/6}$

8.2   Section Exercises

1. $3^{1/2} = \sqrt[2]{3^1} = \sqrt{3}$ **(C)**

3. $-16^{1/2} = -\sqrt{16} = -(4) = -4$ **(A)**

5. $(-32)^{1/5} = \sqrt[5]{-32} = -2$ **(H)**

7. $4^{3/2} = (4^{1/2})^3 = 2^3 = 8$ **(B)**

9. $-6^{2/4} = -(6^{1/2}) = -\sqrt{6}$ **(D)**

11. $169^{1/2} = \sqrt{169} = 13$

13. $729^{1/3} = \sqrt[3]{729} = 9$

15. $16^{1/4} = \sqrt[4]{16} = 2$

17. $\left(\dfrac{64}{81}\right)^{1/2} = \sqrt{\dfrac{64}{81}} = \dfrac{8}{9}$

19. $(-27)^{1/3} = \sqrt[3]{-27} = -3$

21. $(-144)^{1/2}$ is not a real number, because no real number squared equals -144.

23. $100^{3/2} = (100^{1/2})^3 = 10^3 = 1000$

25. $81^{3/4} = (81^{1/4})^3 = (\sqrt[4]{81})^3 = 3^3 = 27$

27. $-16^{5/2} = -(16^{1/2})^5 = -(4)^5 = -1024$

29. $(-8)^{4/3} = [(-8)^{1/3}]^4 = (\sqrt[3]{-8})^4$

$\qquad = (-2)^4 = 16$

31. $32^{-3/5} = \dfrac{1}{32^{3/5}} = \dfrac{1}{(32^{1/5})^3} = \dfrac{1}{(\sqrt[5]{32})^3}$

$\qquad = \dfrac{1}{2^3} = \dfrac{1}{8}$

Copyright © 2012 Pearson Education, Inc. Publishing as Addison-Wesley.

33. $64^{-3/2} = \dfrac{1}{64^{3/2}} = \dfrac{1}{(64^{1/2})^3} = \dfrac{1}{8^3} = \dfrac{1}{512}$

35. $\left(\dfrac{125}{27}\right)^{-2/3} = \left(\dfrac{27}{125}\right)^{2/3}$
$= \left[\left(\dfrac{27}{125}\right)^{1/3}\right]^2$
$= \left(\dfrac{3}{5}\right)^2 = \dfrac{9}{25}$

37. $\left(\dfrac{16}{81}\right)^{-3/4} = \left(\dfrac{81}{16}\right)^{3/4}$
$= \left[\left(\dfrac{81}{16}\right)^{1/4}\right]^3$
$= \left(\dfrac{3}{2}\right)^3 = \dfrac{27}{8}$

39. $10^{1/2} = \left(\sqrt[2]{10}\right)^1 = \sqrt{10}$

41. $8^{3/4} = (8^{1/4})^3 = \left(\sqrt[4]{8}\right)^3$

43. $(9q)^{5/8} - (2x)^{2/3}$
$= \left[(9q)^{1/8}\right]^5 - \left[(2x)^{1/3}\right]^2$
$= \left(\sqrt[8]{9q}\right)^5 - \left(\sqrt[3]{2x}\right)^2$

45. $(2m)^{-3/2} = \left[(2m)^{1/2}\right]^{-3}$
$= \left(\sqrt{2m}\right)^{-3}$
$= \dfrac{1}{\left(\sqrt{2m}\right)^3}$

47. $(2y+x)^{2/3} = \left[(2y+x)^{1/3}\right]^2$
$= \left(\sqrt[3]{2y+x}\right)^2$

49. $\left(3m^4 + 2k^2\right)^{-2/3}$
$= \dfrac{1}{(3m^4+2k^2)^{2/3}}$
$= \dfrac{1}{\left[(3m^4+2k^2)^{1/3}\right]^2}$
$= \dfrac{1}{\left(\sqrt[3]{3m^4+2k^2}\right)^2}$

51. $\sqrt{2^{12}} = (2^{12})^{1/2} = 2^{12/2} = 2^6 = 64$

53. $\sqrt[3]{4^9} = 4^{9/3} = 4^3 = 64$

55. $\sqrt{x^{20}} = x^{20/2} = x^{10}$

57. $\sqrt[3]{x} \cdot \sqrt{x} = x^{1/3} \cdot x^{1/2}$
$= x^{1/3+1/2} = x^{2/6+3/6}$
$= x^{5/6} = \sqrt[6]{x^5}$

59. $\dfrac{\sqrt[3]{t^4}}{\sqrt[5]{t^4}} = \dfrac{t^{4/3}}{t^{4/5}} = \dfrac{t^{20/15}}{t^{12/15}} = t^{20/15-12/15}$
$= t^{8/15} = \sqrt[15]{t^8}$

61. $3^{1/2} \cdot 3^{3/2} = 3^{1/2+3/2} = 3^{4/2} = 3^2 = 9$

63. $\dfrac{64^{5/3}}{64^{4/3}} = 64^{5/3-4/3} = 64^{1/3} = \sqrt[3]{64} = 4$

65. $y^{7/3} \cdot y^{-4/3} = y^{7/3+(-4/3)}$
$= y^{3/3} = y^1 = y$

67. $x^{2/3} \cdot x^{-1/4} = x^{2/3+(-1/4)}$
$= x^{8/12-3/12} = x^{5/12}$

69. $\dfrac{k^{1/3}}{k^{2/3} \cdot k^{-1}} = \dfrac{k^{1/3}}{k^{-1/3}} = k^{1/3-(-1/3)} = k^{2/3}$

71. $\dfrac{(x^{1/4}y^{2/5})^{20}}{x^2} = \dfrac{x^{(1/4)\cdot 20}y^{(2/5)\cdot 20}}{x^2}$
$= \dfrac{x^5 y^8}{x^2} = x^3 y^8$

73. $\dfrac{(x^{2/3})^2}{(x^2)^{7/3}} = \dfrac{x^{4/3}}{x^{14/3}}$
$= x^{4/3-14/3} = x^{-10/3} = \dfrac{1}{x^{10/3}}$

75. $\dfrac{m^{3/4}n^{-1/4}}{(m^2 n)^{1/2}} = \dfrac{m^{3/4}n^{-1/4}}{m^1 n^{1/2}}$
$= m^{3/4-1}n^{-1/4-1/2}$
$= m^{3/4-4/4}n^{-1/4-2/4}$
$= m^{-1/4}n^{-3/4}$
$= \dfrac{1}{m^{1/4}n^{3/4}}$

77. $\dfrac{p^{1/5}p^{7/10}p^{1/2}}{(p^3)^{-1/5}} = \dfrac{p^{2/10+7/10+5/10}}{p^{-3/5}}$
$= \dfrac{p^{14/10}}{p^{-6/10}}$
$= p^{14/10-(-6/10)}$
$= p^{20/10} = p^2$

79. $\left(\dfrac{b^{-3/2}}{c^{-5/3}}\right)^2 \left(b^{-1/4}c^{-1/3}\right)^{-1}$
$= \left(\dfrac{c^{5/3}}{b^{3/2}}\right)^2 \left(b^{1/4}c^{1/3}\right)$
$= \dfrac{c^{10/3}}{b^3}\left(b^{1/4}c^{1/3}\right)$
$= \dfrac{c^{10/3}b^{1/4}c^{1/3}}{b^3}$
$= c^{10/3+1/3}b^{1/4-3}$
$= c^{11/3}b^{-11/4}$
$= \dfrac{c^{11/3}}{b^{11/4}}$

81. $\left(\dfrac{p^{-1/4}q^{-3/2}}{3^{-1}p^{-2}q^{-2/3}}\right)^{-2}$
$= \dfrac{p^{1/2}q^3}{3^2 p^4 q^{4/3}}$
$= \dfrac{q^{3-4/3}}{9p^{4-1/2}}$
$= \dfrac{q^{5/3}}{9p^{7/2}}$

Copyright © 2012 Pearson Education, Inc. Publishing as Addison-Wesley.

83. $p^{2/3}(p^{1/3} + 2p^{4/3})$
$= p^{2/3}p^{1/3} + p^{2/3}(2p^{4/3})$
$= p^{2/3+1/3} + 2p^{2/3+4/3}$
$= p^{3/3} + 2p^{6/3}$
$= p^1 + 2p^2$
$= p + 2p^2$

85. $k^{1/4}(k^{3/2} - k^{1/2})$
$= k^{1/4+3/2} - k^{1/4+1/2}$
$= k^{1/4+6/4} - k^{1/4+2/4}$
$= k^{7/4} - k^{3/4}$

87. $6a^{7/4}(a^{-7/4} + 3a^{-3/4})$
$= 6a^{7/4+(-7/4)} + 18a^{7/4+(-3/4)}$
$= 6a^0 + 18a^{4/4}$
$= 6(1) + 18a^1 = 6 + 18a$

89. $-5x^{7/6}(x^{5/6} - x^{-1/6})$
$= -5x^{7/6+5/6} + 5x^{7/6+(-1/6)}$
$= -5x^{12/6} + 5x^{6/6}$
$= -5x^2 + 5x$

91. $\sqrt[5]{x^3} \cdot \sqrt[4]{x} = x^{3/5} \cdot x^{1/4}$
$= x^{3/5+1/4}$
$= x^{12/20+5/20}$
$= x^{17/20}$

93. $\dfrac{\sqrt{x^5}}{\sqrt{x^8}} = \dfrac{(x^5)^{1/2}}{(x^8)^{1/2}}$
$= x^{5/2-(8/2)}$
$= x^{-3/2} = \dfrac{1}{x^{3/2}}$

95. $\sqrt{y} \cdot \sqrt[3]{yz} = y^{1/2} \cdot (yz)^{1/3}$
$= y^{1/2}y^{1/3}z^{1/3}$
$= y^{1/2+1/3}z^{1/3}$
$= y^{3/6+2/6}z^{1/3}$
$= y^{5/6}z^{1/3}$

97. $\sqrt[4]{\sqrt[3]{m}} = \sqrt[4]{m^{1/3}} = (m^{1/3})^{1/4} = m^{1/12}$

99. $\sqrt{\sqrt{\sqrt{x}}} = \sqrt{\sqrt{x^{1/2}}} = \sqrt{(x^{1/2})^{1/2}}$
$= \sqrt{x^{1/4}} = (x^{1/4})^{1/2} = x^{1/8}$

101. $\sqrt{\sqrt[3]{\sqrt[4]{x}}} = \sqrt{\sqrt[3]{x^{1/4}}}$
$= \sqrt{(x^{1/4})^{1/3}}$
$= (x^{1/12})^{1/2}$
$= x^{1/24}$

103. We are to show that, in general,
$$\sqrt{a^2 + b^2} \neq a + b.$$
When $a = 3$ and $b = 4$,
$$\sqrt{a^2 + b^2} = \sqrt{3^2 + 4^2}$$
$$= \sqrt{9 + 16}$$
$$= \sqrt{25} = 5,$$
but $a + b = 3 + 4 = 7$.
Since $5 \neq 7$, $\sqrt{a^2 + b^2} \neq a + b$.

105. Use $T(D) = 0.07D^{3/2}$ with $D = 16$.
$$T(16) = 0.07(16)^{3/2} = 0.07(16^{1/2})^3$$
$$= 0.07(4)^3 = 0.07(64) = 4.48$$
To the nearest tenth of an hour, time T is 4.5 hours.

107. Use a calculator and the formula
$$W = 35.74 + 0.6215T - 35.75V^{4/25} + 0.4275TV^{4/25}.$$
For $T = 30°F$ and $V = 15$ mph, $W \approx 19.0°F$. The table gives 19.0°F.

109. As in Exercise 107, for $T = 20°F$ and $V = 20$ mph, $W \approx 4.2°F$. The table gives 4°F.

111. $\sqrt{25} \cdot \sqrt{36} = 5 \cdot 6 = 30$
$\sqrt{25 \cdot 36} = \sqrt{900} = 30$
The results are the same.

8.3 Simplifying Radical Expressions

8.3 Now Try Exercises

N1. **(a)** $\sqrt{7} \cdot \sqrt{11} = \sqrt{7 \cdot 11} = \sqrt{77}$

  (b) $\sqrt{2mn} \cdot \sqrt{15} = \sqrt{2mn \cdot 15} = \sqrt{30mn}$

N2. **(a)** $\sqrt[3]{4} \cdot \sqrt[3]{5} = \sqrt[3]{4 \cdot 5} = \sqrt[3]{20}$

  (b) $\sqrt[4]{5t} \cdot \sqrt[4]{6r^3} = \sqrt[4]{5t \cdot 6r^3} = \sqrt[4]{30tr^3}$

  (c) $\sqrt[7]{20x} \cdot \sqrt[7]{3xy^3} = \sqrt[7]{20x \cdot 3xy^3} = \sqrt[7]{60x^2y^3}$

  (d) $\sqrt[3]{5} \cdot \sqrt[4]{9}$ cannot be simplified using the product rule for radicals, because the indexes (3 and 4) are different.

N3. **(a)** $\sqrt{\dfrac{49}{36}} = \dfrac{\sqrt{49}}{\sqrt{36}} = \dfrac{7}{6}$

  (b) $\sqrt{\dfrac{5}{144}} = \dfrac{\sqrt{5}}{\sqrt{144}} = \dfrac{\sqrt{5}}{12}$

  (c) $\sqrt[3]{-\dfrac{27}{1000}} = \sqrt[3]{\dfrac{-27}{1000}} = \dfrac{\sqrt[3]{-27}}{\sqrt[3]{1000}}$
$= \dfrac{-3}{10} = -\dfrac{3}{10}$

Copyright © 2012 Pearson Education, Inc. Publishing as Addison-Wesley.

(d) $\sqrt[4]{\dfrac{t}{16}} = \dfrac{\sqrt[4]{t}}{\sqrt[4]{16}} = \dfrac{\sqrt[4]{t}}{2}$

(e) $-\sqrt[5]{\dfrac{m^{15}}{243}} = -\dfrac{\sqrt[5]{m^{15}}}{\sqrt[5]{243}} = -\dfrac{m^3}{3}$

N4. **(a)** $\sqrt{50} = \sqrt{25 \cdot 2} = \sqrt{25} \cdot \sqrt{2} = 5\sqrt{2}$

(b) $\sqrt{192} = \sqrt{64 \cdot 3} = \sqrt{64} \cdot \sqrt{3} = 8\sqrt{3}$

(c) $\sqrt{42}$

No perfect square (other than 1) divides into 42, so $\sqrt{42}$ *cannot be simplified further.*

(d) $\sqrt[3]{108} = \sqrt[3]{27 \cdot 4} = \sqrt[3]{27} \cdot \sqrt[3]{4} = 3\sqrt[3]{4}$

(e) $-\sqrt[4]{80} = -\sqrt[4]{16 \cdot 5} = -\sqrt[4]{16} \cdot \sqrt[4]{5} = -2\sqrt[4]{5}$

N5. **(a)** $\sqrt{36x^5} = \sqrt{36x^4 \cdot x}$
$= \sqrt{36x^4} \cdot \sqrt{x} = 6x^2\sqrt{x}$

(b) $\sqrt{32m^5n^4} = \sqrt{16m^4n^4 \cdot 2m}$
$= \sqrt{16m^4n^4} \cdot \sqrt{2m}$
$= 4m^2n^2\sqrt{2m}$

(c) $\sqrt[3]{-125k^3p^7} = \sqrt[3]{-125k^3p^6 \cdot p}$
$= \sqrt[3]{-125k^3p^6} \cdot \sqrt[3]{p}$
$= -5kp^2\sqrt[3]{p}$

(d) $-\sqrt[4]{162x^7y^8} = -\sqrt[4]{81x^4y^8 \cdot 2x^3}$
$= -\sqrt[4]{81x^4y^8} \cdot \sqrt[4]{2x^3}$
$= -3xy^2\sqrt[4]{2x^3}$

N6. **(a)** $\sqrt[6]{7^2} = 7^{2/6} = 7^{1/3} = \sqrt[3]{7}$

(b) $\sqrt[6]{y^4} = y^{4/6} = y^{2/3} = \sqrt[3]{y^2}$

N7. $\sqrt[3]{3} \cdot \sqrt{6}$

The least common index of 3 and 2 is 6. Write each radical as a sixth root.

$\sqrt[3]{3} = 3^{1/3} = 3^{2/6} = \sqrt[6]{3^2} = \sqrt[6]{9}$

$\sqrt{6} = 6^{1/2} = 6^{3/6} = \sqrt[6]{6^3} = \sqrt[6]{216}$

Therefore,

$\sqrt[3]{3} \cdot \sqrt{6} = \sqrt[6]{9} \cdot \sqrt[6]{216} = \sqrt[6]{1944}.$

N8. **(a)** Substitute 5 for a and 8 for b in the Pythagorean theorem to find c.

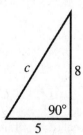

$c^2 = a^2 + b^2$
$c^2 = 5^2 + 8^2$
$c^2 = 89$
$c = \sqrt{89}$

The length of the hypotenuse is $\sqrt{89}$.

(b) Substitute 6 for a and 12 for c to find b.

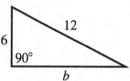

$a^2 + b^2 = c^2$
$b^2 = c^2 - a^2$
$b^2 = 12^2 - 6^2$
$b^2 = 108$
$b = \sqrt{108}$
$b = \sqrt{36 \cdot 3}$
$b = \sqrt{36} \cdot \sqrt{3}$
$b = 6\sqrt{3}$

The length of the unknown side is $6\sqrt{3}$.

N9. Use the distance formula,

$$d = \sqrt{(x_2 - x_1)^2 + (y_2 - y_1)^2}.$$

$(x_1, y_1) = (-4, -3)$ and $(x_2, y_2) = (-8, 6)$

$d = \sqrt{[-8 - (-4)]^2 + [6 - (-3)]^2}$
$= \sqrt{(-4)^2 + 9^2}$
$= \sqrt{16 + 81}$
$= \sqrt{97}$

8.3 Section Exercises

1. $\sqrt{3} \cdot \sqrt{3} = \sqrt{3 \cdot 3} = \sqrt{9}$, or 3

3. $\sqrt{18} \cdot \sqrt{2} = \sqrt{18 \cdot 2} = \sqrt{36}$, or 6

5. $\sqrt{5} \cdot \sqrt{6} = \sqrt{5 \cdot 6} = \sqrt{30}$

7. $\sqrt{14} \cdot \sqrt{x} = \sqrt{14 \cdot x} = \sqrt{14x}$

9. $\sqrt{14} \cdot \sqrt{3pqr} = \sqrt{14 \cdot 3pqr} = \sqrt{42pqr}$

11. $\sqrt[3]{2} \cdot \sqrt[3]{5} = \sqrt[3]{2 \cdot 5} = \sqrt[3]{10}$

13. $\sqrt[3]{7x} \cdot \sqrt[3]{2y} = \sqrt[3]{7x \cdot 2y} = \sqrt[3]{14xy}$

15. $\sqrt[4]{11} \cdot \sqrt[4]{3} = \sqrt[4]{11 \cdot 3} = \sqrt[4]{33}$

17. $\sqrt[4]{2x} \cdot \sqrt[4]{3x^2} = \sqrt[4]{2x \cdot 3x^2} = \sqrt[4]{6x^3}$

19. $\sqrt[3]{7} \cdot \sqrt[4]{3}$ cannot be multiplied using the product rule, because the indexes (3 and 4) are different.

21. $\sqrt{\dfrac{64}{121}} = \dfrac{\sqrt{64}}{\sqrt{121}} = \dfrac{8}{11}$

Copyright © 2012 Pearson Education, Inc. Publishing as Addison-Wesley.

23. $\sqrt{\dfrac{3}{25}} = \dfrac{\sqrt{3}}{\sqrt{25}} = \dfrac{\sqrt{3}}{5}$

25. $\sqrt{\dfrac{x}{25}} = \dfrac{\sqrt{x}}{\sqrt{25}} = \dfrac{\sqrt{x}}{5}$

27. $\sqrt{\dfrac{p^6}{81}} = \dfrac{\sqrt{p^6}}{\sqrt{81}} = \dfrac{\sqrt{(p^3)^2}}{9} = \dfrac{p^3}{9}$

29. $\sqrt[3]{-\dfrac{27}{64}} = \sqrt[3]{\dfrac{-27}{64}} = \dfrac{\sqrt[3]{-27}}{\sqrt[3]{64}} = \dfrac{-3}{4} = -\dfrac{3}{4}$

31. $\sqrt[3]{\dfrac{r^2}{8}} = \dfrac{\sqrt[3]{r^2}}{\sqrt[3]{8}} = \dfrac{\sqrt[3]{r^2}}{2}$

33. $-\sqrt[4]{\dfrac{81}{x^4}} = -\dfrac{\sqrt[4]{3^4}}{\sqrt[4]{x^4}} = -\dfrac{3}{x}$

35. $\sqrt[5]{\dfrac{1}{x^{15}}} = \dfrac{\sqrt[5]{1}}{\sqrt[5]{(x^3)^5}} = \dfrac{1}{x^3}$

37. $\sqrt{12} = \sqrt{4 \cdot 3} = \sqrt{4} \cdot \sqrt{3} = 2\sqrt{3}$

39. $\sqrt{288} = \sqrt{144 \cdot 2} = \sqrt{144} \cdot \sqrt{2} = 12\sqrt{2}$

41. $-\sqrt{32} = -\sqrt{16 \cdot 2} = -\sqrt{16} \cdot \sqrt{2} = -4\sqrt{2}$

43. $-\sqrt{28} = -\sqrt{4 \cdot 7} = -\sqrt{4} \cdot \sqrt{7} = -2\sqrt{7}$

45. $\sqrt{30}$ cannot be simplified further.

47. $\sqrt[3]{128} = \sqrt[3]{64 \cdot 2} = \sqrt[3]{64} \cdot \sqrt[3]{2} = 4\sqrt[3]{2}$

49. $\sqrt[3]{-16} = \sqrt[3]{-8 \cdot 2} = \sqrt[3]{-8} \cdot \sqrt[3]{2} = -2\sqrt[3]{2}$

51. $\sqrt[3]{40} = \sqrt[3]{8 \cdot 5} = \sqrt[3]{8} \cdot \sqrt[3]{5} = 2\sqrt[3]{5}$

53. $-\sqrt[4]{512} = -\sqrt[4]{256 \cdot 2} = -\sqrt[4]{4^4} \cdot \sqrt[4]{2} = -4\sqrt[4]{2}$

55. $\sqrt[5]{64} = \sqrt[5]{32 \cdot 2} = \sqrt[5]{2^5} \cdot \sqrt[5]{2} = 2\sqrt[5]{2}$

57. $-\sqrt[5]{486} = -\sqrt[5]{243 \cdot 2} = -\sqrt[5]{3^5} \cdot \sqrt[5]{2} = -3\sqrt[5]{2}$

59. $\sqrt[6]{128} = \sqrt[6]{64 \cdot 2} = \sqrt[6]{2^6} \cdot \sqrt[6]{2} = 2\sqrt[6]{2}$

61. His reasoning was incorrect. The radicand 14 must be written as a product of two factors (not a sum of two terms) where one of the two factors is a perfect cube.

63. $\sqrt{72k^2} = \sqrt{36k^2 \cdot 2} = \sqrt{36k^2} \cdot \sqrt{2} = 6k\sqrt{2}$

65. $\sqrt{144x^3y^9} = \sqrt{144x^2y^8 \cdot xy}$
$\qquad = \sqrt{(12xy^4)^2} \cdot \sqrt{xy}$
$\qquad = 12xy^4\sqrt{xy}$

67. $\sqrt{121x^6} = \sqrt{(11x^3)^2} = 11x^3$

69. $-\sqrt[3]{27t^{12}} = -\sqrt[3]{(3t^4)^3} = -3t^4$

71. $-\sqrt{100m^8z^4} = -\sqrt{(10m^4z^2)^2} = -10m^4z^2$

73. $-\sqrt[3]{-125a^6b^9c^{12}} = -\sqrt[3]{(-5a^2b^3c^4)^3}$
$\qquad = -(-5a^2b^3c^4)$
$\qquad = 5a^2b^3c^4$

75. $\sqrt[4]{\dfrac{1}{16}r^8t^{20}} = \sqrt[4]{(\tfrac{1}{2}r^2t^5)^4} = \tfrac{1}{2}r^2t^5$

77. $\sqrt{50x^3} = \sqrt{25x^2 \cdot 2x} = \sqrt{(5x)^2} \cdot \sqrt{2x}$
$\qquad = 5x\sqrt{2x}$

79. $-\sqrt{500r^{11}} = -\sqrt{100r^{10} \cdot 5r}$
$\qquad = -\sqrt{(10r^5)^2} \cdot \sqrt{5r} = -10r^5\sqrt{5r}$

81. $\sqrt{13x^7y^8} = \sqrt{(x^6y^8)(13x)}$
$\qquad = \sqrt{x^6y^8} \cdot \sqrt{13x}$
$\qquad = x^3y^4\sqrt{13x}$

83. $\sqrt[3]{8z^6w^9} = \sqrt[3]{(2z^2w^3)^3} = 2z^2w^3$

85. $\sqrt[3]{-16z^5t^7} = \sqrt[3]{(-2^3z^3t^6)(2z^2t)}$
$\qquad = -2zt^2\sqrt[3]{2z^2t}$

87. $\sqrt[4]{81x^{12}y^{16}} = \sqrt[4]{(3x^3y^4)^4} = 3x^3y^4$

89. $-\sqrt[4]{162r^{15}s^{10}} = -\sqrt[4]{81r^{12}s^8(2r^3s^2)}$
$\qquad = -\sqrt[4]{81r^{12}s^8} \cdot \sqrt[4]{2r^3s^2}$
$\qquad = -3r^3s^2\sqrt[4]{2r^3s^2}$

91. $\sqrt{\dfrac{y^{11}}{36}} = \dfrac{\sqrt{y^{11}}}{\sqrt{36}} = \dfrac{\sqrt{y^{10} \cdot y}}{6} = \dfrac{y^5\sqrt{y}}{6}$

93. $\sqrt[3]{\dfrac{x^{16}}{27}} = \dfrac{\sqrt[3]{x^{15} \cdot x}}{\sqrt[3]{27}} = \dfrac{x^5\sqrt[3]{x}}{3}$

95. $\sqrt[4]{48^2} = 48^{2/4} = 48^{1/2} = \sqrt{48} = \sqrt{16 \cdot 3}$
$\qquad = \sqrt{16} \cdot \sqrt{3} = 4\sqrt{3}$

97. $\sqrt[4]{25} = 25^{1/4} = (5^2)^{1/4} = 5^{2/4} = 5^{1/2} = \sqrt{5}$

99. $\sqrt[10]{x^{25}} = x^{25/10} = x^{5/2} = \sqrt{x^5}$
$\qquad = \sqrt{x^4 \cdot x} = x^2\sqrt{x}$

101. $\sqrt[3]{4} \cdot \sqrt{3}$

The least common index of 3 and 2 is 6. Write each radical as a sixth root.

$$\sqrt[3]{4} = 4^{1/3} = 4^{2/6} = \sqrt[6]{4^2} = \sqrt[6]{16}$$
$$\sqrt{3} = 3^{1/2} = 3^{3/6} = \sqrt[6]{3^3} = \sqrt[6]{27}$$

Therefore,

$$\sqrt[3]{4} \cdot \sqrt{3} = \sqrt[6]{16} \cdot \sqrt[6]{27}$$
$$= \sqrt[6]{16 \cdot 27} = \sqrt[6]{432}.$$

103. $\sqrt[4]{3} \cdot \sqrt[3]{4}$

The least common index of 4 and 3 is 12. Write each radical as a twelfth root.

Copyright © 2012 Pearson Education, Inc. Publishing as Addison-Wesley.

$$\sqrt[4]{3} = 3^{1/4} = 3^{3/12} = \sqrt[12]{3^3} = \sqrt[12]{27}$$
$$\sqrt[3]{4} = 4^{1/3} = 4^{4/12} = \sqrt[12]{4^4} = \sqrt[12]{256}$$

Therefore,

$$\sqrt[4]{3} \cdot \sqrt[3]{4} = \sqrt[12]{27} \cdot \sqrt[12]{256}$$
$$= \sqrt[12]{27 \cdot 256} = \sqrt[12]{6912}.$$

105. $\sqrt{x} \cdot \sqrt[3]{x}$

$$\sqrt{x} = x^{1/2} = x^{3/6} = \sqrt[6]{x^3}$$
$$\sqrt[3]{x} = x^{1/3} = x^{2/6} = \sqrt[6]{x^2}$$

So $\sqrt{x} \cdot \sqrt[3]{x} = \sqrt[6]{x^3} \cdot \sqrt[6]{x^2}$
$$= \sqrt[6]{x^3 \cdot x^2} = \sqrt[6]{x^5}.$$

107. Substitute 3 for a and 4 for b in the Pythagorean theorem to find c.

$$c^2 = a^2 + b^2$$
$$c = \sqrt{a^2 + b^2} = \sqrt{3^2 + 4^2}$$
$$= \sqrt{9 + 16} = \sqrt{25} = 5$$

The length of the hypotenuse is 5.

109. Substitute 4 for a and 12 for c in the Pythagorean theorem to find b.

$$a^2 + b^2 = c^2$$
$$b = \sqrt{c^2 - a^2} = \sqrt{12^2 - 4^2}$$
$$= \sqrt{144 - 16} = \sqrt{128}$$
$$= \sqrt{64}\sqrt{2} = 8\sqrt{2}$$

The length of the unknown leg is $8\sqrt{2}$.

111. Substitute 5 for b and 9 for c in the Pythagorean theorem to find a.

$$a^2 + b^2 = c^2$$
$$a = \sqrt{c^2 - b^2} = \sqrt{9^2 - 5^2}$$
$$= \sqrt{81 - 25} = \sqrt{56}$$
$$= \sqrt{4}\sqrt{14} = 2\sqrt{14}$$

The length of the unknown leg is $2\sqrt{14}$.

In Exercises 113–124, use the distance formula

$$d = \sqrt{(x_2 - x_1)^2 + (y_2 - y_1)^2}.$$

113. $(6, 13)$ and $(1, 1)$

$$d = \sqrt{(6 - 1)^2 + (13 - 1)^2}$$
$$= \sqrt{5^2 + 12^2} = \sqrt{25 + 144}$$
$$= \sqrt{169} = 13$$

115. $(-6, 5)$ and $(3, -4)$

$$d = \sqrt{(-6 - 3)^2 + [5 - (-4)]^2}$$
$$= \sqrt{(-9)^2 + (9)^2} = \sqrt{81 + 81}$$
$$= \sqrt{162} = \sqrt{81} \cdot \sqrt{2} = 9\sqrt{2}$$

117. $(-8, 2)$ and $(-4, 1)$

$$d = \sqrt{[-8 - (-4)]^2 + (2 - 1)^2}$$
$$= \sqrt{(-4)^2 + 1^2}$$
$$= \sqrt{16 + 1} = \sqrt{17}$$

119. $(4.7, 2.3)$ and $(1.7, -1.7)$

$$d = \sqrt{(4.7 - 1.7)^2 + [2.3 - (-1.7)]^2}$$
$$= \sqrt{3^2 + 4^2} = \sqrt{9 + 16}$$
$$= \sqrt{25} = 5$$

121. $(\sqrt{2}, \sqrt{6})$ and $(-2\sqrt{2}, 4\sqrt{6})$

$$d = \sqrt{\left[\sqrt{2} - \left(-2\sqrt{2}\right)\right]^2 + \left(\sqrt{6} - 4\sqrt{6}\right)^2}$$
$$= \sqrt{\left(3\sqrt{2}\right)^2 + \left(-3\sqrt{6}\right)^2}$$
$$= \sqrt{9 \cdot 2 + 9 \cdot 6} = \sqrt{18 + 54}$$
$$= \sqrt{72} = \sqrt{36} \cdot \sqrt{2} = 6\sqrt{2}$$

123. $(x + y, y)$ and $(x - y, x)$

$$d = \sqrt{[(x + y) - (x - y)]^2 + (y - x)^2}$$
$$= \sqrt{(2y)^2 + (y - x)^2}$$
$$= \sqrt{4y^2 + y^2 - 2xy + x^2}$$
$$= \sqrt{5y^2 - 2xy + x^2}$$

125. To find the lengths of the three sides of the triangle, use the distance formula to find the distance between each pair of points. Then add the distances to find the perimeter.

$$P = \sqrt{(-3 - 2)^2 + (-3 - 6)^2}$$
$$+ \sqrt{(2 - 6)^2 + (6 - 2)^2}$$
$$+ \sqrt{[6 - (-3)]^2 + [2 - (-3)]^2}$$
$$= \sqrt{(-5)^2 + (-9)^2} + \sqrt{(-4)^2 + 4^2}$$
$$+ \sqrt{9^2 + 5^2}$$
$$= \sqrt{25 + 81} + \sqrt{16 + 16} + \sqrt{81 + 25}$$
$$= \sqrt{106} + \sqrt{32} + \sqrt{106}$$
$$= 2\sqrt{106} + \sqrt{16} \cdot \sqrt{2}$$
$$= 2\sqrt{106} + 4\sqrt{2}$$

127. $d = 1.224\sqrt{h}$
$$= 1.224\sqrt{156} \approx 15.3 \text{ miles}$$

129. Substitute 21.7 for a and 16 for b in the Pythagorean theorem to find c.

$$c^2 = a^2 + b^2 = (21.7)^2 + 16^2$$
$$c = \sqrt{726.89} \approx 26.96$$

To the nearest tenth of an inch, the length of the diagonal of the screen is 27.0 inches.

Copyright © 2012 Pearson Education, Inc. Publishing as Addison-Wesley.

131. Use $f_1 = f_2 \sqrt{\dfrac{F_1}{F_2}}$ with $F_1 = 300$, $F_2 = 60$, and $f_2 = 260$ to find f_1.

$$f_1 = 260 \sqrt{\tfrac{300}{60}} = 260\sqrt{5} \approx 581$$

133. $\sqrt{9} + \sqrt{9} = 3 + 3 = 6$, and $\sqrt{4} = 2$; $6 \neq 2$, so the statement is false.

135. For the model MSX-77, we have a right triangle with legs x and x, and hypotenuse 26. Substitute x for a, x for b, and 26 for c in the Pythagorean theorem, then solve for x.

$$a^2 + b^2 = c^2$$
$$x^2 + x^2 = 26^2$$
$$2x^2 = 676$$
$$x^2 = 338$$
$$x = \sqrt{338} \approx 18.4$$

The length of the leg is about 18.4 inches. Similarly, for the model MSX-83 with $c = 24$, we get $\sqrt{288} \approx 17.0$ inches, and for the model MSX-60 with $c = 20$, we get $\sqrt{200} \approx 14.1$ inches.

137. $13x^4 - 12x^3 + 9x^4 + 2x^3$

$$= 13x^4 + 9x^4 - 12x^3 + 2x^3$$
$$= 22x^4 - 10x^3$$

139. $9q^2 + 2q - 5q - q^2$

$$= 9q^2 - q^2 + 2q - 5q$$
$$= 8q^2 - 3q$$

8.4 Adding and Subtracting Radical Expressions

8.4 Now Try Exercises

N1. (a) $\sqrt{12} + \sqrt{75}$

$$= \sqrt{4} \cdot \sqrt{3} + \sqrt{25} \cdot \sqrt{3}$$
$$= 2\sqrt{3} + 5\sqrt{3}$$
$$= (2 + 5)\sqrt{3}$$
$$= 7\sqrt{3}$$

(b) $-\sqrt{63t} + 3\sqrt{28t}$, $t \geq 0$

$$= -\sqrt{9} \cdot \sqrt{7t} + 3\sqrt{4} \cdot \sqrt{7t}$$
$$= -3\sqrt{7t} + 3 \cdot 2 \cdot \sqrt{7t}$$
$$= -3\sqrt{7t} + 6\sqrt{7t}$$
$$= (-3 + 6)\sqrt{7t}$$
$$= 3\sqrt{7t}$$

(c) $6\sqrt{7} - 2\sqrt{3}$

Here the radicals differ and are already simplified, so this expression *cannot be simplified further*.

N2. (a) $3\sqrt[3]{2000} - 4\sqrt[3]{128}$

$$= 3\sqrt[3]{1000 \cdot 2} - 4\sqrt[3]{64 \cdot 2}$$
$$= 3\sqrt[3]{1000}\sqrt[3]{2} - 4\sqrt[3]{64}\sqrt[3]{2}$$
$$= 3 \cdot 10 \sqrt[3]{2} - 4 \cdot 4 \sqrt[3]{2}$$
$$= 30\sqrt[3]{2} - 16\sqrt[3]{2}$$
$$= (30 - 16)\sqrt[3]{2}$$
$$= 14\sqrt[3]{2}$$

(b) $5\sqrt[4]{a^5 b^3} + \sqrt[4]{81ab^7}$

$$= 5\sqrt[4]{a^4 \cdot ab^3} + \sqrt[4]{81b^4 \cdot ab^3}$$
$$= 5\sqrt[4]{a^4} \cdot \sqrt[4]{ab^3} + \sqrt[4]{81b^4} \cdot \sqrt[4]{ab^3}$$
$$= 5a\sqrt[4]{ab^3} + 3b\sqrt[4]{ab^3}$$
$$= (5a + 3b)\sqrt[4]{ab^3}$$

(c) $\sqrt[3]{128t^4} - 2\sqrt{72t^3}$

$$= \sqrt[3]{64t^3 \cdot 2t} - 2\sqrt{36t^2 \cdot 2t}$$
$$= \sqrt[3]{64t^3} \cdot \sqrt[3]{2t} - 2\sqrt{36t^2} \cdot \sqrt{2t}$$
$$= 4t\sqrt[3]{2t} - 2 \cdot 6t\sqrt{2t}$$
$$= 4t\sqrt[3]{2t} - 12t\sqrt{2t}$$

N3. (a) $5\dfrac{\sqrt{5}}{\sqrt{45}} - 4\sqrt{\dfrac{28}{9}}$

$$= 5\dfrac{\sqrt{5}}{\sqrt{9 \cdot 5}} - 4\dfrac{\sqrt{4 \cdot 7}}{\sqrt{9}}$$
$$= 5\left(\dfrac{\sqrt{5}}{3\sqrt{5}}\right) - 4\left(\dfrac{2\sqrt{7}}{3}\right)$$
$$= \dfrac{5}{3} - \dfrac{8\sqrt{7}}{3}$$
$$= \dfrac{5 - 8\sqrt{7}}{3}$$

(b) $6\sqrt[3]{\dfrac{16}{x^{12}}} + 7\sqrt[3]{\dfrac{9}{x^9}}$

$$= 6\dfrac{\sqrt[3]{8 \cdot 2}}{\sqrt[3]{x^{12}}} + 7\dfrac{\sqrt[3]{9}}{\sqrt[3]{x^9}}$$
$$= \dfrac{6\sqrt[3]{8}\sqrt[3]{2}}{x^4} + \dfrac{7\sqrt[3]{9}}{x^3}$$
$$= \dfrac{6 \cdot 2\sqrt[3]{2}}{x^4} + \dfrac{7x\sqrt[3]{9}}{x \cdot x^3}$$
$$= \dfrac{12\sqrt[3]{2} + 7x\sqrt[3]{9}}{x^4}$$

8.4 Section Exercises

1. Simplify each radical and subtract.

$$\sqrt{36} - \sqrt{100} = 6 - 10 = -4$$

3. $-2\sqrt{48} + 3\sqrt{75}$

$$= -2\sqrt{16 \cdot 3} + 3\sqrt{25 \cdot 3}$$
$$= -2 \cdot 4\sqrt{3} + 3 \cdot 5\sqrt{3}$$
$$= -8\sqrt{3} + 15\sqrt{3} = 7\sqrt{3}$$

Copyright © 2012 Pearson Education, Inc. Publishing as Addison-Wesley.

5. $\sqrt[3]{16} + 4\sqrt[3]{54}$
$= \sqrt[3]{8 \cdot 2} + 4\sqrt[3]{27 \cdot 2}$
$= \sqrt[3]{8}\sqrt[3]{2} + 4\sqrt[3]{27}\sqrt[3]{2}$
$= 2\sqrt[3]{2} + 4 \cdot 3\sqrt[3]{2}$
$= 2\sqrt[3]{2} + 12\sqrt[3]{2} = 14\sqrt[3]{2}$

7. $\sqrt[4]{32} + 3\sqrt[4]{2}$
$= \sqrt[4]{16 \cdot 2} + 3\sqrt[4]{2}$
$= \sqrt[4]{16}\sqrt[4]{2} + 3\sqrt[4]{2}$
$= 2\sqrt[4]{2} + 3\sqrt[4]{2} = 5\sqrt[4]{2}$

9. $6\sqrt{18} - \sqrt{32} + 2\sqrt{50}$
$= 6\sqrt{9 \cdot 2} - \sqrt{16 \cdot 2} + 2\sqrt{25 \cdot 2}$
$= 6 \cdot 3\sqrt{2} - 4\sqrt{2} + 2 \cdot 5\sqrt{2}$
$= 18\sqrt{2} - 4\sqrt{2} + 10\sqrt{2}$
$= 24\sqrt{2}$

11. $5\sqrt{6} + 2\sqrt{10}$

The radicals differ and are already simplified, so the expression cannot be simplified further.

13. $2\sqrt{5} + 3\sqrt{20} + 4\sqrt{45}$
$= 2\sqrt{5} + 3\sqrt{4 \cdot 5} + 4\sqrt{9 \cdot 5}$
$= 2\sqrt{5} + 3 \cdot 2\sqrt{5} + 4 \cdot 3\sqrt{5}$
$= 2\sqrt{5} + 6\sqrt{5} + 12\sqrt{5}$
$= 20\sqrt{5}$

15. $\sqrt{72x} - \sqrt{8x}$
$= \sqrt{36 \cdot 2x} - \sqrt{4 \cdot 2x}$
$= 6\sqrt{2x} - 2\sqrt{2x}$
$= 4\sqrt{2x}$

17. $3\sqrt{72m^2} - 5\sqrt{32m^2} - 3\sqrt{18m^2}$
$= 3\sqrt{36m^2 \cdot 2} - 5\sqrt{16m^2 \cdot 2} - 3\sqrt{9m^2 \cdot 2}$
$= 3 \cdot 6m\sqrt{2} - 5 \cdot 4m\sqrt{2} - 3 \cdot 3m\sqrt{2}$
$= 18m\sqrt{2} - 20m\sqrt{2} - 9m\sqrt{2}$
$= (18m - 20m - 9m)\sqrt{2} = -11m\sqrt{2}$

19. $2\sqrt[3]{16} + \sqrt[3]{54} = 2\sqrt[3]{8 \cdot 2} + \sqrt[3]{27 \cdot 2}$
$= 2 \cdot 2\sqrt[3]{2} + 3\sqrt[3]{2}$
$= 4\sqrt[3]{2} + 3\sqrt[3]{2}$
$= 7\sqrt[3]{2}$

21. $2\sqrt[3]{27x} - 2\sqrt[3]{8x}$
$= 2\sqrt[3]{27 \cdot x} - 2\sqrt[3]{8 \cdot x}$
$= 2 \cdot 3\sqrt[3]{x} - 2 \cdot 2\sqrt[3]{x}$
$= 6\sqrt[3]{x} - 4\sqrt[3]{x}$
$= 2\sqrt[3]{x}$

23. $3\sqrt[3]{x^2y} - 5\sqrt[3]{8x^2y}$
$= 3\sqrt[3]{x^2y} - 5\sqrt[3]{8}\sqrt[3]{x^2y}$
$= 3\sqrt[3]{x^2y} - 5 \cdot 2\sqrt[3]{x^2y}$
$= (3 - 10)\sqrt[3]{x^2y}$
$= -7\sqrt[3]{x^2y}$

25. $3x\sqrt[3]{xy^2} - 2\sqrt[3]{8x^4y^2}$
$= 3x\sqrt[3]{xy^2} - 2\sqrt[3]{8x^3} \cdot \sqrt[3]{xy^2}$
$= 3x\sqrt[3]{xy^2} - 2 \cdot 2x \cdot \sqrt[3]{xy^2}$
$= (3x - 4x)\sqrt[3]{xy^2} = -x\sqrt[3]{xy^2}$

27. $5\sqrt[4]{32} + 3\sqrt[4]{162}$
$= 5\sqrt[4]{16 \cdot 2} + 3\sqrt[4]{81 \cdot 2}$
$= 5 \cdot 2\sqrt[4]{2} + 3 \cdot 3\sqrt[4]{2}$
$= 10\sqrt[4]{2} + 9\sqrt[4]{2}$
$= 19\sqrt[4]{2}$

29. $3\sqrt[4]{x^5y} - 2x\sqrt[4]{xy}$
$= 3\sqrt[4]{x^4 \cdot xy} - 2x\sqrt[4]{xy}$
$= 3x\sqrt[4]{xy} - 2x\sqrt[4]{xy}$
$= (3x - 2x)\sqrt[4]{xy} = x\sqrt[4]{xy}$

31. $2\sqrt[4]{32a^3} + 5\sqrt[4]{2a^3}$
$= 2\sqrt[4]{16} \cdot \sqrt[4]{2a^3} + 5\sqrt[4]{2a^3}$
$= 2 \cdot 2 \cdot \sqrt[4]{2a^3} + 5\sqrt[4]{2a^3}$
$= (4 + 5)\sqrt[4]{2a^3} = 9\sqrt[4]{2a^3}$

33. $\sqrt[3]{64xy^2} + \sqrt[3]{27x^4y^5}$
$= \sqrt[3]{64 \cdot xy^2} + \sqrt[3]{27x^3y^3 \cdot xy^2}$
$= 4\sqrt[3]{xy^2} + 3xy\sqrt[3]{xy^2}$
$= (4 + 3xy)\sqrt[3]{xy^2}$

35. $\sqrt[3]{192st^4} - \sqrt{27s^3t}$
$= \sqrt[3]{64t^3 \cdot 3st} - \sqrt{9s^2 \cdot 3st}$
$= 4t\sqrt[3]{3st} - 3s\sqrt{3st}$

37. $2\sqrt[3]{8x^4} + 3\sqrt[4]{16x^5}$
$= 2\sqrt[3]{8x^3 \cdot x} + 3\sqrt[4]{16x^4 \cdot x}$
$= 2 \cdot 2x\sqrt[3]{x} + 3 \cdot 2x\sqrt[4]{x}$
$= 4x\sqrt[3]{x} + 6x\sqrt[4]{x}$

39. $\sqrt{8} - \dfrac{\sqrt{64}}{\sqrt{16}} = \sqrt{4 \cdot 2} - \dfrac{8}{4}$
$= \sqrt{4} \cdot \sqrt{2} - 2$
$= 2\sqrt{2} - 2$

41. $\dfrac{2\sqrt{5}}{3} + \dfrac{\sqrt{5}}{6} = \dfrac{4\sqrt{5}}{6} + \dfrac{1\sqrt{5}}{6}$
$= \dfrac{4\sqrt{5} + 1\sqrt{5}}{6}$
$= \dfrac{5\sqrt{5}}{6}$

Copyright © 2012 Pearson Education, Inc. Publishing as Addison-Wesley.

43. $\sqrt{\dfrac{8}{9}} + \sqrt{\dfrac{18}{36}} = \dfrac{\sqrt{8}}{\sqrt{9}} + \dfrac{\sqrt{18}}{\sqrt{36}}$

$= \dfrac{\sqrt{4}\sqrt{2}}{3} + \dfrac{\sqrt{9}\sqrt{2}}{6}$

$= \dfrac{2\sqrt{2}}{3} + \dfrac{3\sqrt{2}}{6}$

$= \dfrac{4\sqrt{2}}{6} + \dfrac{3\sqrt{2}}{6}$

$= \dfrac{4\sqrt{2} + 3\sqrt{2}}{6} = \dfrac{7\sqrt{2}}{6}$

45. $\dfrac{\sqrt{32}}{3} + \dfrac{2\sqrt{2}}{3} - \dfrac{\sqrt{2}}{\sqrt{9}}$

$= \dfrac{\sqrt{16}\sqrt{2}}{3} + \dfrac{2\sqrt{2}}{3} - \dfrac{\sqrt{2}}{3}$

$= \dfrac{4\sqrt{2} + 2\sqrt{2} - \sqrt{2}}{3} = \dfrac{5\sqrt{2}}{3}$

47. $3\sqrt{\dfrac{50}{9}} + 8\dfrac{\sqrt{2}}{\sqrt{8}} = 3\dfrac{\sqrt{50}}{\sqrt{9}} + 8\dfrac{\sqrt{2}}{2\sqrt{2}}$

$= 3 \cdot \dfrac{5\sqrt{2}}{3} + 8 \cdot \dfrac{1}{2}$

$= 5\sqrt{2} + 4$

49. $\sqrt{\dfrac{25}{x^8}} + \sqrt{\dfrac{9}{x^6}} = \dfrac{\sqrt{25}}{\sqrt{x^8}} + \dfrac{\sqrt{9}}{\sqrt{x^6}}$

$= \dfrac{5}{x^4} + \dfrac{3}{x^3}$

$= \dfrac{5}{x^4} + \dfrac{3 \cdot x}{x^3 \cdot x} \qquad LCD = x^4$

$= \dfrac{5 + 3x}{x^4}$

51. $3\sqrt[3]{\dfrac{m^5}{27}} - 2m\sqrt[3]{\dfrac{m^2}{64}}$

$= \dfrac{3\sqrt[3]{m^5}}{\sqrt[3]{27}} - \dfrac{2m\sqrt[3]{m^2}}{\sqrt[3]{64}}$

$= \dfrac{3\sqrt[3]{m^3}\sqrt[3]{m^2}}{3} - \dfrac{2m\sqrt[3]{m^2}}{4}$

$= \dfrac{m\sqrt[3]{m^2}}{1} - \dfrac{m\sqrt[3]{m^2}}{2}$

$= \dfrac{2m\sqrt[3]{m^2} - m\sqrt[3]{m^2}}{2} = \dfrac{m\sqrt[3]{m^2}}{2}$

53. $3\sqrt[3]{\dfrac{2}{x^6}} - 4\sqrt[3]{\dfrac{5}{x^9}} = 3\dfrac{\sqrt[3]{2}}{\sqrt[3]{x^6}} - 4\dfrac{\sqrt[3]{5}}{\sqrt[3]{x^9}}$

$= 3\dfrac{\sqrt[3]{2}}{x^2} - 4\dfrac{\sqrt[3]{5}}{x^3}$

$= \dfrac{3 \cdot x \cdot \sqrt[3]{2}}{x^2 \cdot x} - \dfrac{4\sqrt[3]{5}}{x^3}$

$\qquad\qquad\qquad LCD = x^3$

$= \dfrac{3x\sqrt[3]{2} - 4\sqrt[3]{5}}{x^3}$

55. Only choice **B** has like radical terms, so it can be simplified without first simplifying the individual radical expressions.

$$3\sqrt{6} + 9\sqrt{6} = 12\sqrt{6}$$

57. $\sqrt{64} + \sqrt[3]{125} + \sqrt[4]{16} = \sqrt{8^2} + \sqrt[3]{5^3} + \sqrt[4]{2^4}$

$\qquad\qquad = 8 + 5 + 2 = 15$

This sum can be found easily since each radicand has a whole number power corresponding to the index of the radical; that is, each radical expression simplifies to a whole number.

59. Let $L = \sqrt{192} \approx \sqrt{196} = 14$ and $W = \sqrt{48} \approx \sqrt{49} = 7$. An estimate of the perimeter is $2L + 2W = 2(14) + 2(7) = 42$ meters. **(A)**

61. The perimeter, P, of a triangle is the sum of the measures of the sides.

$P = 3\sqrt{20} + 2\sqrt{45} + \sqrt{75}$

$= 3\sqrt{4 \cdot 5} + 2\sqrt{9 \cdot 5} + \sqrt{25 \cdot 3}$

$= 3 \cdot 2\sqrt{5} + 2 \cdot 3\sqrt{5} + 5\sqrt{3}$

$= 6\sqrt{5} + 6\sqrt{5} + 5\sqrt{3}$

$= 12\sqrt{5} + 5\sqrt{3}$

The perimeter is $\left(12\sqrt{5} + 5\sqrt{3}\right)$ inches.

63. To find the perimeter, add the lengths of the sides.

$4\sqrt{18} + \sqrt{108} + 2\sqrt{72} + 3\sqrt{12}$

$= 4\sqrt{9}\sqrt{2} + \sqrt{36}\sqrt{3} + 2\sqrt{36}\sqrt{2} + 3\sqrt{4}\sqrt{3}$

$= 4 \cdot 3\sqrt{2} + 6\sqrt{3} + 2 \cdot 6\sqrt{2} + 3 \cdot 2\sqrt{3}$

$= 12\sqrt{2} + 6\sqrt{3} + 12\sqrt{2} + 6\sqrt{3}$

$= 24\sqrt{2} + 12\sqrt{3}$

The perimeter is $\left(24\sqrt{2} + 12\sqrt{3}\right)$ inches.

65. $5xy(2x^2y^3 - 4x) = 5xy(2x^2y^3) - 5xy(4x)$

$\qquad\qquad\qquad = 10x^3y^4 - 20x^2y$

67. $(a^2 + b)(a^2 - b) = (a^2)^2 - b^2$

$\qquad\qquad\qquad = a^4 - b^2$

69. $(4x^3 + 3)^2 = (4x^3)^2 + 2 \cdot 4x^3 \cdot 3 + 3^2$

$\qquad\qquad\quad = 16x^6 + 24x^3 + 9$

Now multiply by $4x^3 + 3$.

$$
\begin{array}{r}
16x^6 + 24x^3 + 9 \\
4x^3 + 3 \\
\hline
48x^6 + 72x^3 + 27 \\
64x^9 + 96x^6 + 36x^3 \\
\hline
64x^9 + 144x^6 + 108x^3 + 27
\end{array}
$$

Thus, $(4x^3 + 3)^3 = 64x^9 + 144x^6 + 108x^3 + 27$.

Copyright © 2012 Pearson Education, Inc. Publishing as Addison-Wesley.

71. $\dfrac{8x^2 - 10x}{6x^2} = \dfrac{2x(4x - 5)}{2x \cdot 3x}$

$\qquad\qquad = \dfrac{4x - 5}{3x}$

8.5 Multiplying and Dividing Radical Expressions

8.5 Now Try Exercises

N1. Use the FOIL method and multiply as two binomials.

(a) $\left(8 - \sqrt{5}\right)\left(9 - \sqrt{2}\right)$

$\qquad\quad\ \mathbf{F}\qquad\ \mathbf{O}\qquad\ \ \mathbf{I}\qquad\ \ \mathbf{L}$

$= 8 \cdot 9 - 8\sqrt{2} - 9\sqrt{5} + \sqrt{5} \cdot \sqrt{2}$

$= 72 - 8\sqrt{2} - 9\sqrt{5} + \sqrt{10}$

(b) $\left(\sqrt{7} + \sqrt{5}\right)\left(\sqrt{7} - \sqrt{5}\right)$

$= \sqrt{7} \cdot \sqrt{7} - \sqrt{7} \cdot \sqrt{5}$

$\quad + \sqrt{5} \cdot \sqrt{7} - \sqrt{5} \cdot \sqrt{5}$

$= 7 - 5$

$= 2$

(c) $\left(\sqrt{15} - 4\right)^2$

$= \left(\sqrt{15} - 4\right)\left(\sqrt{15} - 4\right)$

$= \sqrt{15} \cdot \sqrt{15} - 4\sqrt{15} - 4\sqrt{15} + 4 \cdot 4$

$= 15 - 8\sqrt{15} + 16$

$= 31 - 8\sqrt{15}$

(d) $\left(8 + \sqrt[3]{5}\right)\left(8 - \sqrt[3]{5}\right)$

$= 8 \cdot 8 - 8\sqrt[3]{5} + 8\sqrt[3]{5} - \sqrt[3]{5} \cdot \sqrt[3]{5}$

$= 64 - \sqrt[3]{25}$

(e) For this one, we use the difference of squares.

$\left(\sqrt{m} - \sqrt{n}\right)\left(\sqrt{m} + \sqrt{n}\right)$

$= \left(\sqrt{m}\right)^2 - \left(\sqrt{n}\right)^2$

$= m - n, \quad m \geq 0 \text{ and } n \geq 0$

N2. (a) $\dfrac{8}{\sqrt{13}} = \dfrac{8 \cdot \sqrt{13}}{\sqrt{13} \cdot \sqrt{13}} = \dfrac{8\sqrt{13}}{13}$

(b) $\dfrac{9\sqrt{7}}{\sqrt{3}} = \dfrac{9\sqrt{7} \cdot \sqrt{3}}{\sqrt{3} \cdot \sqrt{3}} = \dfrac{9\sqrt{21}}{3} = 3\sqrt{21}$

(c) $\dfrac{-10}{\sqrt{20}} = \dfrac{-10}{\sqrt{4 \cdot 5}} = \dfrac{-10}{2\sqrt{5}} = \dfrac{-10 \cdot \sqrt{5}}{2\sqrt{5} \cdot \sqrt{5}}$

$= \dfrac{-10\sqrt{5}}{2 \cdot 5} = \dfrac{-10\sqrt{5}}{10} = -\sqrt{5}$

N3. (a) $-\sqrt{\dfrac{27}{80}} = -\dfrac{\sqrt{27}}{\sqrt{80}} = -\dfrac{\sqrt{9 \cdot 3}}{\sqrt{16 \cdot 5}}$

$= -\dfrac{3\sqrt{3}}{4\sqrt{5}} = -\dfrac{3\sqrt{3} \cdot \sqrt{5}}{4\sqrt{5} \cdot \sqrt{5}}$

$= -\dfrac{3\sqrt{15}}{4 \cdot 5} = -\dfrac{3\sqrt{15}}{20}$

(b) $\sqrt{\dfrac{48x^8}{y^3}} = \dfrac{\sqrt{16x^8 \cdot 3}}{\sqrt{y^2 \cdot y}}$

$= \dfrac{4x^4\sqrt{3}}{y\sqrt{y}}$

$= \dfrac{4x^4\sqrt{3} \cdot \sqrt{y}}{y\sqrt{y} \cdot \sqrt{y}}$

$= \dfrac{4x^4\sqrt{3y}}{y \cdot y}$

$= \dfrac{4x^4\sqrt{3y}}{y^2} \quad (y > 0)$

N4. (a) $\sqrt[3]{\dfrac{8}{81}} = \dfrac{\sqrt[3]{8}}{\sqrt[3]{81}} = \dfrac{2}{\sqrt[3]{27} \cdot \sqrt[3]{3}} = \dfrac{2}{3\sqrt[3]{3}}$

$= \dfrac{2 \cdot \sqrt[3]{9}}{3\sqrt[3]{3} \cdot \sqrt[3]{9}} = \dfrac{2\sqrt[3]{9}}{3\sqrt[3]{27}}$

$= \dfrac{2\sqrt[3]{9}}{3 \cdot 3} = \dfrac{2\sqrt[3]{9}}{9}$

(b) $\sqrt[4]{\dfrac{7x}{y}} = \dfrac{\sqrt[4]{7x}}{\sqrt[4]{y}} \cdot \dfrac{\sqrt[4]{y^3}}{\sqrt[4]{y^3}}$

$= \dfrac{\sqrt[4]{7xy^3}}{\sqrt[4]{y^4}}$

$= \dfrac{\sqrt[4]{7xy^3}}{y} \quad (x \geq 0, y > 0)$

N5. (a) $\dfrac{4}{1 + \sqrt{3}}$

Multiply both the numerator and denominator by the conjugate of the denominator, $1 - \sqrt{3}$.

$= \dfrac{4(1 - \sqrt{3})}{(1 + \sqrt{3})(1 - \sqrt{3})}$

$= \dfrac{4(1 - \sqrt{3})}{1 - 3} = \dfrac{4(1 - \sqrt{3})}{-2}$

$= -2(1 - \sqrt{3}), \text{ or } -2 + 2\sqrt{3}$

(b) $\dfrac{4}{5 + \sqrt{7}}$

Multiply both the numerator and denominator by the conjugate of the denominator, $5 - \sqrt{7}$.

$= \dfrac{4(5 - \sqrt{7})}{(5 + \sqrt{7})(5 - \sqrt{7})}$

$= \dfrac{4(5 - \sqrt{7})}{25 - 7}$

$= \dfrac{4(5 - \sqrt{7})}{18} = \dfrac{2(5 - \sqrt{7})}{9}$

Copyright © 2012 Pearson Education, Inc. Publishing as Addison-Wesley.

(c) $\dfrac{\sqrt{3}+\sqrt{7}}{\sqrt{5}-\sqrt{2}}$

$= \dfrac{\left(\sqrt{3}+\sqrt{7}\right)\left(\sqrt{5}+\sqrt{2}\right)}{\left(\sqrt{5}-\sqrt{2}\right)\left(\sqrt{5}+\sqrt{2}\right)}$

$= \dfrac{\sqrt{15}+\sqrt{6}+\sqrt{35}+\sqrt{14}}{5-2}$

$= \dfrac{\sqrt{15}+\sqrt{6}+\sqrt{35}+\sqrt{14}}{3}$

(d) $\dfrac{8}{\sqrt{3x}-\sqrt{y}}$

$= \dfrac{8\left(\sqrt{3x}+\sqrt{y}\right)}{\left(\sqrt{3x}-\sqrt{y}\right)\left(\sqrt{3x}+\sqrt{y}\right)}$

$= \dfrac{8\left(\sqrt{3x}+\sqrt{y}\right)}{3x-y}$

$$(3x \neq y,\ x > 0, y > 0)$$

N6. (a) $\dfrac{15-6\sqrt{2}}{18} = \dfrac{3(5-2\sqrt{2})}{18} = \dfrac{5-2\sqrt{2}}{6}$

(b) $\dfrac{15k+\sqrt{50k^2}}{20k} = \dfrac{15k+\sqrt{25k^2 \cdot 2}}{20k}$

$= \dfrac{15k+5k\sqrt{2}}{20k} \quad (k > 0)$

$= \dfrac{5k(3+\sqrt{2})}{5k \cdot 4}$

$= \dfrac{3+\sqrt{2}}{4}$

8.5 Section Exercises

1. $\left(A+\sqrt{B}\right)\left(A-\sqrt{B}\right)$

$= A^2 - \left(\sqrt{B}\right)^2$

$= A^2 - B \quad \textbf{(E)}$

3. $\left(\sqrt{A}+\sqrt{B}\right)\left(\sqrt{A}-\sqrt{B}\right)$

$= \left(\sqrt{A}\right)^2 - \left(\sqrt{B}\right)^2$

$= A - B \quad \textbf{(A)}$

5. $\left(\sqrt{A}-\sqrt{B}\right)^2$

$= \left(\sqrt{A}\right)^2 - 2\sqrt{A}\sqrt{B} + \left(\sqrt{B}\right)^2$

$= A - 2\sqrt{AB} + B \quad \textbf{(D)}$

7. $\sqrt{6}\left(3+\sqrt{2}\right) = 3\sqrt{6} + \sqrt{6}\cdot\sqrt{2}$

$= 3\sqrt{6} + \sqrt{12}$

$= 3\sqrt{6} + \sqrt{4}\cdot\sqrt{3}$

$= 3\sqrt{6} + 2\sqrt{3}$

9. $5\left(\sqrt{72}-\sqrt{8}\right) = 5\sqrt{72} - 5\sqrt{8}$

$= 5\sqrt{36}\cdot\sqrt{2} - 5\sqrt{4}\cdot\sqrt{2}$

$= 5\cdot6\cdot\sqrt{2} - 5\cdot2\cdot\sqrt{2}$

$= 30\sqrt{2} - 10\sqrt{2}$

$= 20\sqrt{2}$

11. $\left(\sqrt{7}+3\right)\left(\sqrt{7}-3\right) = \left(\sqrt{7}\right)^2 - 3^2$

$= 7 - 9 = -2$

13. $\left(\sqrt{2}-\sqrt{3}\right)\left(\sqrt{2}+\sqrt{3}\right)$

$= \left(\sqrt{2}\right)^2 - \left(\sqrt{3}\right)^2$

$= 2 - 3 = -1$

15. $\left(\sqrt{8}-\sqrt{2}\right)\left(\sqrt{8}+\sqrt{2}\right)$

$= \left(\sqrt{8}\right)^2 - \left(\sqrt{2}\right)^2$

$= 8 - 2 = 6$

17. $\left(\sqrt{2}+1\right)\left(\sqrt{3}-1\right)$

$\qquad \textbf{F} \qquad \textbf{O} \qquad \textbf{I} \qquad \textbf{L}$

$= \sqrt{2}\cdot\sqrt{3} - 1\sqrt{2} + 1\sqrt{3} - 1\cdot1$

$= \sqrt{6} - \sqrt{2} + \sqrt{3} - 1$

19. $\left(\sqrt{11}-\sqrt{7}\right)\left(\sqrt{2}+\sqrt{5}\right)$

$\qquad \textbf{F} \qquad\quad \textbf{O} \qquad\quad \textbf{I} \qquad\quad \textbf{L}$

$= \sqrt{11}\cdot\sqrt{2} + \sqrt{11}\cdot\sqrt{5} - \sqrt{7}\cdot\sqrt{2} - \sqrt{7}\cdot\sqrt{5}$

$= \sqrt{22} + \sqrt{55} - \sqrt{14} - \sqrt{35}$

21. $\left(2\sqrt{3}+\sqrt{5}\right)\left(3\sqrt{3}-2\sqrt{5}\right)$

$= \left(2\sqrt{3}\right)\left(3\sqrt{3}\right) + \left(2\sqrt{3}\right)\left(-2\sqrt{5}\right)$

$\quad + \left(\sqrt{5}\right)\left(3\sqrt{3}\right) + \left(\sqrt{5}\right)\left(-2\sqrt{5}\right)$

$= 2\cdot3\cdot3 - 2\cdot2\sqrt{3\cdot5} + 3\sqrt{5\cdot3} - 2\cdot5$

$= 18 - 4\sqrt{15} + 3\sqrt{15} - 10$

$= 8 - \sqrt{15}$

23. $\left(\sqrt{5}+2\right)^2 = \left(\sqrt{5}\right)^2 + 2\cdot\sqrt{5}\cdot2 + 2^2$

$= 5 + 4\sqrt{5} + 4$

$= 9 + 4\sqrt{5}$

25. $\left(\sqrt{21}-\sqrt{5}\right)^2$

$= \left(\sqrt{21}\right)^2 - 2\cdot\sqrt{21}\cdot\sqrt{5} + \left(\sqrt{5}\right)^2$

$= 21 - 2\sqrt{105} + 5$

$= 26 - 2\sqrt{105}$

27. $\left(2+\sqrt[3]{6}\right)\left(2-\sqrt[3]{6}\right) = 2^2 - \left(\sqrt[3]{6}\right)^2$

$= 4 - \sqrt[3]{6}\cdot\sqrt[3]{6}$

$= 4 - \sqrt[3]{36}$

29. $\left(2+\sqrt[3]{2}\right)\left(4-2\sqrt[3]{2}+\sqrt[3]{4}\right)$

$= 2\cdot4 - 2\cdot2\sqrt[3]{2} + 2\sqrt[3]{4}$

$\quad + 4\sqrt[3]{2} - 2\sqrt[3]{2}\cdot\sqrt[3]{2} + \sqrt[3]{2}\cdot\sqrt[3]{4}$

$= 8 - 4\sqrt[3]{2} + 2\sqrt[3]{4} + 4\sqrt[3]{2} - 2\sqrt[3]{4} + \sqrt[3]{8}$

$= 8 + 2$

$= 10$

Copyright © 2012 Pearson Education, Inc. Publishing as Addison-Wesley.

31. $\left(3\sqrt{x} - \sqrt{5}\right)\left(2\sqrt{x} + 1\right)$
$= \left(3\sqrt{x}\right)\left(2\sqrt{x}\right) + 3\sqrt{x} - 2\sqrt{5}\sqrt{x} - \sqrt{5}$
$= 6x + 3\sqrt{x} - 2\sqrt{5x} - \sqrt{5}$

33. $\left(3\sqrt{r} - \sqrt{s}\right)\left(3\sqrt{r} + \sqrt{s}\right)$
$= \left(3\sqrt{r}\right)^2 - \left(\sqrt{s}\right)^2$
$= 9r - s$

35. $\left(\sqrt[3]{2y} - 5\right)\left(4\sqrt[3]{2y} + 1\right)$
$= \sqrt[3]{2y} \cdot 4\sqrt[3]{2y} + \sqrt[3]{2y} - 5 \cdot 4\sqrt[3]{2y} - 5$
$= 4\sqrt[3]{4y^2} + \sqrt[3]{2y} - 20\sqrt[3]{2y} - 5$
$= 4\sqrt[3]{4y^2} - 19\sqrt[3]{2y} - 5$

37. $\left(\sqrt{3x} + 2\right)\left(\sqrt{3x} - 2\right) = \left(\sqrt{3x}\right)^2 - 2^2$
$= 3x - 4$

39. $\left(2\sqrt{x} + \sqrt{y}\right)\left(2\sqrt{x} - \sqrt{y}\right)$
$= \left(2\sqrt{x}\right)^2 - \left(\sqrt{y}\right)^2$
$= 2^2\left(\sqrt{x}\right)^2 - y$
$= 4x - y$

41. $\left[\left(\sqrt{2} + \sqrt{3}\right) - \sqrt{6}\right]\left[\left(\sqrt{2} + \sqrt{3}\right) + \sqrt{6}\right]$
$= \left(\sqrt{2} + \sqrt{3}\right)^2 - \left(\sqrt{6}\right)^2$
$= \left[\left(\sqrt{2}\right)^2 + 2\sqrt{2}\sqrt{3} + \left(\sqrt{3}\right)^2\right] - 6$
$= \left(2 + 2\sqrt{6} + 3\right) - 6$
$= 2\sqrt{6} - 1$

43. $\dfrac{7}{\sqrt{7}} = \dfrac{7 \cdot \sqrt{7}}{\sqrt{7} \cdot \sqrt{7}} = \dfrac{7\sqrt{7}}{7} = \sqrt{7}$

45. $\dfrac{15}{\sqrt{3}} = \dfrac{15 \cdot \sqrt{3}}{\sqrt{3} \cdot \sqrt{3}} = \dfrac{15\sqrt{3}}{3} = 5\sqrt{3}$

47. $\dfrac{\sqrt{3}}{\sqrt{2}} = \dfrac{\sqrt{3} \cdot \sqrt{2}}{\sqrt{2} \cdot \sqrt{2}} = \dfrac{\sqrt{6}}{2}$

49. $\dfrac{9\sqrt{3}}{\sqrt{5}} = \dfrac{9\sqrt{3} \cdot \sqrt{5}}{\sqrt{5} \cdot \sqrt{5}} = \dfrac{9\sqrt{15}}{5}$

51. $\dfrac{-7}{\sqrt{48}} = \dfrac{-7}{4\sqrt{3}} = \dfrac{-7 \cdot \sqrt{3}}{4\sqrt{3} \cdot \sqrt{3}}$
$= \dfrac{-7\sqrt{3}}{4 \cdot 3} = -\dfrac{7\sqrt{3}}{12}$

53. $\sqrt{\dfrac{7}{2}} = \dfrac{\sqrt{7}}{\sqrt{2}} = \dfrac{\sqrt{7} \cdot \sqrt{2}}{\sqrt{2} \cdot \sqrt{2}} = \dfrac{\sqrt{14}}{2}$

55. $-\sqrt{\dfrac{7}{50}} = -\dfrac{\sqrt{7}}{\sqrt{50}} = -\dfrac{\sqrt{7}}{\sqrt{25 \cdot 2}} = -\dfrac{\sqrt{7}}{5\sqrt{2}}$
$= -\dfrac{\sqrt{7} \cdot \sqrt{2}}{5\sqrt{2} \cdot \sqrt{2}} = -\dfrac{\sqrt{14}}{5 \cdot 2} = -\dfrac{\sqrt{14}}{10}$

57. $\sqrt{\dfrac{24}{x}} = \dfrac{\sqrt{24}}{\sqrt{x}} = \dfrac{\sqrt{4 \cdot 6}}{\sqrt{x}} = \dfrac{2\sqrt{6}}{\sqrt{x}}$
$= \dfrac{2\sqrt{6} \cdot \sqrt{x}}{\sqrt{x} \cdot \sqrt{x}} = \dfrac{2\sqrt{6x}}{x}$

59. $\dfrac{-8\sqrt{3}}{\sqrt{k}} = \dfrac{-8\sqrt{3} \cdot \sqrt{k}}{\sqrt{k} \cdot \sqrt{k}} = \dfrac{-8\sqrt{3k}}{k}$

61. $-\sqrt{\dfrac{150m^5}{n^3}}$
$= \dfrac{-\sqrt{150m^5}}{\sqrt{n^3}} = \dfrac{-\sqrt{25m^4 \cdot 6m}}{\sqrt{n^2 \cdot n}}$
$= \dfrac{-5m^2\sqrt{6m}}{n\sqrt{n}} = \dfrac{-5m^2\sqrt{6m} \cdot \sqrt{n}}{n\sqrt{n} \cdot \sqrt{n}}$
$= \dfrac{-5m^2\sqrt{6mn}}{n \cdot n} = \dfrac{-5m^2\sqrt{6mn}}{n^2}$

63. $\sqrt{\dfrac{288x^7}{y^9}} = \dfrac{\sqrt{288x^7}}{\sqrt{y^9}} = \dfrac{\sqrt{144x^6 \cdot 2x}}{\sqrt{y^8 \cdot y}}$
$= \dfrac{12x^3\sqrt{2x}}{y^4\sqrt{y}} = \dfrac{12x^3\sqrt{2x} \cdot \sqrt{y}}{y^4\sqrt{y} \cdot \sqrt{y}}$
$= \dfrac{12x^3\sqrt{2xy}}{y^4 \cdot y} = \dfrac{12x^3\sqrt{2xy}}{y^5}$

65. $\dfrac{5\sqrt{2m}}{\sqrt{y^3}} = \dfrac{5\sqrt{2m}}{\sqrt{y^3}} \cdot \dfrac{\sqrt{y}}{\sqrt{y}}$
$= \dfrac{5\sqrt{2my}}{\sqrt{y^4}}$
$= \dfrac{5\sqrt{2my}}{y^2}$

67. $-\sqrt{\dfrac{48k^2}{z}} = -\dfrac{\sqrt{16k^2 \cdot 3}}{\sqrt{z}} \cdot \dfrac{\sqrt{z}}{\sqrt{z}}$
$= -\dfrac{4k\sqrt{3z}}{z}$

69. $\sqrt[3]{\dfrac{2}{3}} = \dfrac{\sqrt[3]{2}}{\sqrt[3]{3}} = \dfrac{\sqrt[3]{2} \cdot \sqrt[3]{9}}{\sqrt[3]{3} \cdot \sqrt[3]{9}} = \dfrac{\sqrt[3]{18}}{\sqrt[3]{27}} = \dfrac{\sqrt[3]{18}}{3}$

71. $\sqrt[3]{\dfrac{4}{9}} = \dfrac{\sqrt[3]{4}}{\sqrt[3]{9}} = \dfrac{\sqrt[3]{4}}{\sqrt[3]{3^2}} = \dfrac{\sqrt[3]{4} \cdot \sqrt[3]{3}}{\sqrt[3]{3^2} \cdot \sqrt[3]{3}}$
$= \dfrac{\sqrt[3]{12}}{\sqrt[3]{3^3}} = \dfrac{\sqrt[3]{12}}{3}$

73. $\sqrt[3]{\dfrac{9}{32}} = \dfrac{\sqrt[3]{9}}{\sqrt[3]{32}} = \dfrac{\sqrt[3]{9}}{\sqrt[3]{8} \cdot \sqrt[3]{4}} \cdot \dfrac{\sqrt[3]{2}}{\sqrt[3]{2}}$
$= \dfrac{\sqrt[3]{9 \cdot 2}}{2 \cdot \sqrt[3]{8}}$
$= \dfrac{\sqrt[3]{18}}{4}$

Copyright © 2012 Pearson Education, Inc. Publishing as Addison-Wesley.

75. $-\sqrt[3]{\dfrac{2p}{r^2}} = -\dfrac{\sqrt[3]{2p}}{\sqrt[3]{r^2}} = -\dfrac{\sqrt[3]{2p}\cdot\sqrt[3]{r}}{\sqrt[3]{r^2}\cdot\sqrt[3]{r}}$

$\quad = -\dfrac{\sqrt[3]{2pr}}{\sqrt[3]{r^3}} = -\dfrac{\sqrt[3]{2pr}}{r}$

77. $\sqrt[3]{\dfrac{x^6}{y}} = \dfrac{\sqrt[3]{x^6}}{\sqrt[3]{y}} = \dfrac{x^2}{\sqrt[3]{y}}\cdot\dfrac{\sqrt[3]{y^2}}{\sqrt[3]{y^2}}$

$\quad = \dfrac{x^2\sqrt[3]{y^2}}{\sqrt[3]{y^3}}$

$\quad = \dfrac{x^2\sqrt[3]{y^2}}{y}$

79. $\sqrt[4]{\dfrac{16}{x}} = \dfrac{\sqrt[4]{16}}{\sqrt[4]{x}} = \dfrac{2}{\sqrt[4]{x}} = \dfrac{2\cdot\sqrt[4]{x^3}}{\sqrt[4]{x}\cdot\sqrt[4]{x^3}}$

$\quad = \dfrac{2\sqrt[4]{x^3}}{\sqrt[4]{x^4}} = \dfrac{2\sqrt[4]{x^3}}{x}$

81. $\sqrt[4]{\dfrac{2y}{z}} = \dfrac{\sqrt[4]{2y}}{\sqrt[4]{z}}\cdot\dfrac{\sqrt[4]{z^3}}{\sqrt[4]{z^3}}$

$\quad = \dfrac{\sqrt[4]{2yz^3}}{z}$

83. $\dfrac{3}{4+\sqrt{5}}$

Multiply both the numerator and denominator by the conjugate of the denominator, $4-\sqrt{5}$.

$\quad = \dfrac{3\left(4-\sqrt{5}\right)}{\left(4+\sqrt{5}\right)\left(4-\sqrt{5}\right)}$

$\quad = \dfrac{3\left(4-\sqrt{5}\right)}{16-5} = \dfrac{3\left(4-\sqrt{5}\right)}{11}$

85. $\dfrac{\sqrt{8}}{3-\sqrt{2}}$

Multiply both the numerator and denominator by the conjugate of the denominator, $3+\sqrt{2}$.

$\quad = \dfrac{\sqrt{4}\sqrt{2}\left(3+\sqrt{2}\right)}{\left(3-\sqrt{2}\right)\left(3+\sqrt{2}\right)}$

$\quad = \dfrac{2\sqrt{2}\left(3+\sqrt{2}\right)}{3^2-2}$

$\quad = \dfrac{2\cdot 3\sqrt{2}+2\cdot 2}{7}$

$\quad = \dfrac{6\sqrt{2}+4}{7}$

87. $\dfrac{2}{3\sqrt{5}+2\sqrt{3}}$

Multiply both the numerator and denominator by the conjugate of the denominator, $3\sqrt{5}-2\sqrt{3}$.

$\quad = \dfrac{2\left(3\sqrt{5}-2\sqrt{3}\right)}{\left(3\sqrt{5}+2\sqrt{3}\right)\left(3\sqrt{5}-2\sqrt{3}\right)}$

$\quad = \dfrac{2\left(3\sqrt{5}-2\sqrt{3}\right)}{3^2\cdot 5 - 2^2\cdot 3}$

$\quad = \dfrac{2\left(3\sqrt{5}-2\sqrt{3}\right)}{45-12} = \dfrac{2\left(3\sqrt{5}-2\sqrt{3}\right)}{33}$

89. $\dfrac{\sqrt{2}-\sqrt{3}}{\sqrt{6}-\sqrt{5}}$

Multiply both the numerator and denominator by the conjugate of the denominator, $\sqrt{6}+\sqrt{5}$.

$\quad = \dfrac{\left(\sqrt{2}-\sqrt{3}\right)\left(\sqrt{6}+\sqrt{5}\right)}{\left(\sqrt{6}-\sqrt{5}\right)\left(\sqrt{6}+\sqrt{5}\right)}$

$\quad = \dfrac{\sqrt{12}+\sqrt{10}-\sqrt{18}-\sqrt{15}}{\left(\sqrt{6}\right)^2-\left(\sqrt{5}\right)^2}$

$\quad = \dfrac{\sqrt{4}\cdot\sqrt{3}+\sqrt{10}-\sqrt{9}\cdot\sqrt{2}-\sqrt{15}}{6-5}$

$\quad = 2\sqrt{3}+\sqrt{10}-3\sqrt{2}-\sqrt{15}$

91. $\dfrac{m-4}{\sqrt{m}+2}$

Multiply both the numerator and denominator by the conjugate of the denominator, $\sqrt{m}-2$.

$\quad = \dfrac{(m-4)\left(\sqrt{m}-2\right)}{\left(\sqrt{m}+2\right)\left(\sqrt{m}-2\right)}$

$\quad = \dfrac{(m-4)\left(\sqrt{m}-2\right)}{m-4}$

$\quad = \sqrt{m}-2$

93. $\dfrac{4}{\sqrt{x}-2\sqrt{y}}$

Multiply both the numerator and denominator by the conjugate of the denominator, $\sqrt{x}+2\sqrt{y}$.

$\quad = \dfrac{4\left(\sqrt{x}+2\sqrt{y}\right)}{\left(\sqrt{x}-2\sqrt{y}\right)\left(\sqrt{x}+2\sqrt{y}\right)}$

$\quad = \dfrac{4\left(\sqrt{x}+2\sqrt{y}\right)}{x-4y}$

95. $\dfrac{\sqrt{x}-\sqrt{y}}{\sqrt{x}+\sqrt{y}}$

Multiply both the numerator and denominator by the conjugate of the denominator, $\sqrt{x}-\sqrt{y}$.

$\quad = \dfrac{\left(\sqrt{x}-\sqrt{y}\right)\left(\sqrt{x}-\sqrt{y}\right)}{\left(\sqrt{x}+\sqrt{y}\right)\left(\sqrt{x}-\sqrt{y}\right)}$

$\quad = \dfrac{\left(\sqrt{x}\right)^2-2\sqrt{x}\sqrt{y}+\left(\sqrt{y}\right)^2}{\left(\sqrt{x}\right)^2-\left(\sqrt{y}\right)^2}$

$\quad = \dfrac{x-2\sqrt{xy}+y}{x-y}$

Copyright © 2012 Pearson Education, Inc. Publishing as Addison-Wesley.

97. $\dfrac{5\sqrt{k}}{2\sqrt{k}+\sqrt{q}}$

Multiply both the numerator and denominator by the conjugate of the denominator, $2\sqrt{k}-\sqrt{q}$.

$$= \frac{5\sqrt{k}\left(2\sqrt{k}-\sqrt{q}\right)}{\left(2\sqrt{k}+\sqrt{q}\right)\left(2\sqrt{k}-\sqrt{q}\right)}$$

$$= \frac{5\sqrt{k}\left(2\sqrt{k}-\sqrt{q}\right)}{4k-q}$$

99. $\dfrac{30-20\sqrt{6}}{10}=\dfrac{10\left(3-2\sqrt{6}\right)}{10}=3-2\sqrt{6}$

101. $\dfrac{3-3\sqrt{5}}{3}=\dfrac{3\left(1-\sqrt{5}\right)}{3}=1-\sqrt{5}$

103. $\dfrac{16-4\sqrt{8}}{12}=\dfrac{16-4\left(2\sqrt{2}\right)}{12}=\dfrac{16-8\sqrt{2}}{12}$

$$= \frac{4\left(4-2\sqrt{2}\right)}{4\cdot 3}=\frac{4-2\sqrt{2}}{3}$$

105. $\dfrac{6p+\sqrt{24p^3}}{3p}$

$$= \frac{6p+\sqrt{4p^2\cdot 6p}}{3p}=\frac{6p+2p\sqrt{6p}}{3p}$$

$$= \frac{p\left(6+2\sqrt{6p}\right)}{3p}=\frac{6+2\sqrt{6p}}{3}$$

107. $\dfrac{3}{\sqrt{x+y}}=\dfrac{3}{\sqrt{x+y}}\cdot\dfrac{\sqrt{x+y}}{\sqrt{x+y}}$

$$= \frac{3\cdot\sqrt{x+y}}{\left(\sqrt{x+y}\right)^2}$$

$$= \frac{3\sqrt{x+y}}{x+y}$$

109. $\dfrac{p}{\sqrt{p+2}}=\dfrac{p}{\sqrt{p+2}}\cdot\dfrac{\sqrt{p+2}}{\sqrt{p+2}}$

$$= \frac{p\sqrt{p+2}}{p+2}$$

111. $\dfrac{1}{\sqrt{2}}\cdot\dfrac{\sqrt{3}}{2}-\dfrac{1}{\sqrt{2}}\cdot\dfrac{1}{2}=\dfrac{\sqrt{3}}{2\sqrt{2}}-\dfrac{1}{2\sqrt{2}}$

$$= \frac{\sqrt{3}-1}{2\sqrt{2}}=\frac{\left(\sqrt{3}-1\right)\sqrt{2}}{\left(2\sqrt{2}\right)\sqrt{2}}$$

$$= \frac{\sqrt{6}-\sqrt{2}}{2\cdot 2}=\frac{\sqrt{6}-\sqrt{2}}{4}$$

Using a calculator,

$$\frac{1}{\sqrt{2}}\cdot\frac{\sqrt{3}}{2}-\frac{1}{\sqrt{2}}\cdot\frac{1}{2}\approx 0.2588190451 \text{ and}$$

$$\frac{\sqrt{6}-\sqrt{2}}{4}\approx 0.2588190451.$$

113. $\dfrac{6-\sqrt{3}}{8}=\dfrac{6-\sqrt{3}}{8}\cdot\dfrac{6+\sqrt{3}}{6+\sqrt{3}}$

$$= \frac{6^2-\left(\sqrt{3}\right)^2}{8\left(6+\sqrt{3}\right)}$$

$$= \frac{36-3}{8\left(6+\sqrt{3}\right)}$$

$$= \frac{33}{8\left(6+\sqrt{3}\right)}$$

115. $\dfrac{2\sqrt{x}-\sqrt{y}}{3x}=\dfrac{2\sqrt{x}-\sqrt{y}}{3x}\cdot\dfrac{2\sqrt{x}+\sqrt{y}}{2\sqrt{x}+\sqrt{y}}$

$$= \frac{\left(2\sqrt{x}\right)^2-\left(\sqrt{y}\right)^2}{3x\left(2\sqrt{x}+\sqrt{y}\right)}$$

$$= \frac{4x-y}{3x\left(2\sqrt{x}+\sqrt{y}\right)}$$

117. $-8x+7=4$

$$-8x=-3$$

$$x=\tfrac{3}{8}$$

The solution set is $\left\{\tfrac{3}{8}\right\}$.

119. $6x^2-7x=3$

$$6x^2-7x-3=0$$

$$(3x+1)(2x-3)=0$$

$3x+1=0$ or $2x-3=0$

$\qquad 3x=-1$ or $2x=3$

$\qquad\quad x=-\tfrac{1}{3}$ or $x=\tfrac{3}{2}$

The solution set is $\left\{-\tfrac{1}{3},\tfrac{3}{2}\right\}$.

Summary Exercises on Operations with Radicals and Rational Exponents

1. $6\sqrt{10}-12\sqrt{10}=(6-12)\sqrt{10}$

$$= -6\sqrt{10}$$

3. $\left(1-\sqrt{3}\right)\left(2+\sqrt{6}\right)$

    F   **O**   **I**   **L**

$$= 2+\sqrt{6}-2\sqrt{3}-\sqrt{18}$$

$$= 2+\sqrt{6}-2\sqrt{3}-\sqrt{9}\cdot\sqrt{2}$$

$$= 2+\sqrt{6}-2\sqrt{3}-3\sqrt{2}$$

5. $\left(3\sqrt{5}+2\sqrt{7}\right)^2$

$$= \left(3\sqrt{5}\right)^2+2\left(3\sqrt{5}\right)\left(2\sqrt{7}\right)+\left(2\sqrt{7}\right)^2$$

$$= 9\cdot 5+12\sqrt{35}+4\cdot 7$$

$$= 45+12\sqrt{35}+28$$

$$= 73+12\sqrt{35}$$

7. $\dfrac{8}{\sqrt{7}+\sqrt{5}}=\dfrac{8}{\sqrt{7}+\sqrt{5}}\cdot\dfrac{\sqrt{7}-\sqrt{5}}{\sqrt{7}-\sqrt{5}}$

$$= \frac{8\left(\sqrt{7}-\sqrt{5}\right)}{7-5}$$

$$= \frac{8\left(\sqrt{7}-\sqrt{5}\right)}{2}$$

$$= 4\left(\sqrt{7}-\sqrt{5}\right)$$

Copyright © 2012 Pearson Education, Inc. Publishing as Addison-Wesley.

9. $\left(\sqrt{5}+7\right)\left(\sqrt{5}-7\right) = \left(\sqrt{5}\right)^2 - 7^2$
$$= 5 - 49$$
$$= -44$$

11. $\sqrt[3]{8a^3b^5c^9} = \sqrt[3]{8a^3b^3c^9} \cdot \sqrt[3]{b^2}$
$$= 2abc^3\sqrt[3]{b^2}$$

13. $\dfrac{3}{\sqrt{5}+2} = \dfrac{3}{\sqrt{5}+2} \cdot \dfrac{\sqrt{5}-2}{\sqrt{5}-2}$
$$= \dfrac{3\left(\sqrt{5}-2\right)}{5-4}$$
$$= 3\left(\sqrt{5}-2\right)$$

15. $\dfrac{16\sqrt{3}}{5\sqrt{12}} = \dfrac{16\sqrt{3}}{5 \cdot \sqrt{4} \cdot \sqrt{3}}$
$$= \dfrac{16}{5 \cdot 2}$$
$$= \tfrac{8}{5}$$

17. $\dfrac{-10}{\sqrt[3]{10}} = \dfrac{-10}{\sqrt[3]{10}} \cdot \dfrac{\sqrt[3]{100}}{\sqrt[3]{100}}$
$$= \dfrac{-10\sqrt[3]{100}}{\sqrt[3]{1000}}$$
$$= \dfrac{-10\sqrt[3]{100}}{10} = -\sqrt[3]{100}$$

19. $\sqrt{12x} - \sqrt{75x} = \sqrt{4} \cdot \sqrt{3x} - \sqrt{25} \cdot \sqrt{3x}$
$$= 2\sqrt{3x} - 5\sqrt{3x}$$
$$= -3\sqrt{3x}$$

21. $\sqrt[3]{\dfrac{13}{81}} = \dfrac{\sqrt[3]{13}}{\sqrt[3]{81}}$
$$= \dfrac{\sqrt[3]{13}}{\sqrt[3]{27} \cdot \sqrt[3]{3}} \cdot \dfrac{\sqrt[3]{9}}{\sqrt[3]{9}}$$
$$= \dfrac{\sqrt[3]{13 \cdot 9}}{3 \cdot \sqrt[3]{27}}$$
$$= \dfrac{\sqrt[3]{117}}{3 \cdot 3} = \dfrac{\sqrt[3]{117}}{9}$$

23. $\dfrac{6}{\sqrt[4]{3}} = \dfrac{6}{\sqrt[4]{3}} \cdot \dfrac{\sqrt[4]{3^3}}{\sqrt[4]{3^3}}$
$$= \dfrac{6\sqrt[4]{27}}{3} = 2\sqrt[4]{27}$$

25. $\sqrt[3]{\dfrac{x^2y}{x^{-3}y^4}} = \sqrt[3]{x^{2-(-3)}y^{1-4}}$
$$= \sqrt[3]{x^5y^{-3}}$$
$$= \dfrac{\sqrt[3]{x^5}}{\sqrt[3]{y^3}}$$
$$= \dfrac{\sqrt[3]{x^3 \cdot x^2}}{y} = \dfrac{x\sqrt[3]{x^2}}{y}$$

27. $\dfrac{x^{-2/3}y^{4/5}}{x^{-5/3}y^{-2/5}} = x^{-2/3-(-5/3)}y^{4/5-(-2/5)}$
$$= x^{3/3}y^{6/5}$$
$$= xy^{6/5}$$

29. $\left(125x^3\right)^{-2/3} = \dfrac{1}{\left(125x^3\right)^{2/3}}$
$$= \dfrac{1}{\left(\sqrt[3]{125x^3}\right)^2}$$
$$= \dfrac{1}{\left(5x\right)^2}$$
$$= \dfrac{1}{25x^2}$$

31. $\sqrt[3]{16x^2} - \sqrt[3]{54x^2} + \sqrt[3]{128x^2}$
$$= \sqrt[3]{8} \cdot \sqrt[3]{2x^2} - \sqrt[3]{27} \cdot \sqrt[3]{2x^2} + \sqrt[3]{64} \cdot \sqrt[3]{2x^2}$$
$$= 2\sqrt[3]{2x^2} - 3\sqrt[3]{2x^2} + 4\sqrt[3]{2x^2}$$
$$= (2 - 3 + 4)\sqrt[3]{2x^2}$$
$$= 3\sqrt[3]{2x^2}$$

33. **(a)** $\sqrt{64} = 8$

     (b) $x^2 = 64$
$$x = -\sqrt{64} \quad \text{or} \quad x = \sqrt{64}$$
$$x = -8 \quad \text{or} \quad x = 8$$

     The solution set is $\{-8, 8\}$.

35. **(a)** $x^2 = 16$
$$x = -\sqrt{16} \quad \text{or} \quad x = \sqrt{16}$$
$$x = -4 \quad \text{or} \quad x = 4$$

     The solution set is $\{-4, 4\}$.

     (b) $-\sqrt{16} = -\left(\sqrt{16}\right) = -4$

37. **(a)** Since $\left(\tfrac{9}{11}\right)^2 = \tfrac{81}{121}$, $-\sqrt{\tfrac{81}{121}} = -\tfrac{9}{11}$.

     (b) $x^2 = \tfrac{81}{121}$
$$x = -\sqrt{\tfrac{81}{121}} \quad \text{or} \quad x = \sqrt{\tfrac{81}{121}}$$
$$x = -\tfrac{9}{11} \quad \text{or} \quad x = \tfrac{9}{11}$$

     The solution set is $\left\{-\tfrac{9}{11}, \tfrac{9}{11}\right\}$.

39. **(a)** $x^2 = 0.04$
$$x = -\sqrt{0.04} \quad \text{or} \quad x = \sqrt{0.04}$$
$$x = -0.2 \quad \text{or} \quad x = 0.2$$

     The solution set is $\{-0.2, 0.2\}$.

     (b) Since $(0.2)^2 = 0.04$, $\sqrt{0.04} = 0.2$.

Copyright © 2012 Pearson Education, Inc. Publishing as Addison-Wesley.

8.6 Solving Equations with Radicals

8.6 Now Try Exercises

N1.
$$\sqrt{9x+7} = 5$$
$$\left(\sqrt{9x+7}\right)^2 = 5^2 \quad \textit{Square each side.}$$
$$9x + 7 = 25$$
$$9x = 18$$
$$x = 2$$

Check 2 in the original equation.

$$\sqrt{9x+7} = 5$$
$$\sqrt{9\cdot 2 + 7} \overset{?}{=} 5 \quad \textit{Let x = 2.}$$
$$\sqrt{25} \overset{?}{=} 5$$
$$5 = 5 \quad \textit{True}$$

The solution set is $\{2\}$.

N2. $\sqrt{3x+4} + 5 = 0$
Step 1
Isolate the radical on one side of the equation.
$$\sqrt{3x+4} = -5$$

The equation has no solution, because the square root of a real number must be nonnegative.

The solution set is $\emptyset$.

N3.
$$\sqrt{16-x} = x+4$$
$$\left(\sqrt{16-x}\right)^2 = (x+4)^2 \quad \textit{Square.}$$
$$16 - x = x^2 + 8x + 16$$
$$0 = x^2 + 9x$$
$$0 = x(x+9)$$

$$x = 0 \quad \text{or} \quad x + 9 = 0$$
$$\text{or} \quad x = -9$$

Check $x = 0$: $\sqrt{16} = 4$ *True*
Check $x = -9$: $\sqrt{25} = -5$ *False*

The solution set is $\{0\}$.

N4.
$$\sqrt{x^2 - 3x + 18} = x + 3$$
$$\left(\sqrt{x^2-3x+18}\right)^2 = (x+3)^2 \quad \textit{Square.}$$
$$x^2 - 3x + 18 = x^2 + 6x + 9$$
$$-9x = -9$$
$$x = 1$$

Check $x = 1$: $\sqrt{16} = 4$ *True*

The solution set is $\{1\}$.

N5. $\sqrt{3x+1} - \sqrt{x+4} = 1$
Get one radical on each
side of the equals sign.

$$\sqrt{3x+1} = 1 + \sqrt{x+4}$$
Square each side.
$$\left(\sqrt{3x+1}\right)^2 = \left(1 + \sqrt{x+4}\right)^2$$
$$3x + 1 = 1 + 2\sqrt{x+4} + (x+4)$$
Isolate the remaining radical.
$$3x = 2\sqrt{x+4} + x + 4$$
$$2x - 4 = 2\sqrt{x+4}$$
$$x - 2 = \sqrt{x+4}$$
Square each side again.
$$(x-2)^2 = \left(\sqrt{x+4}\right)^2$$
$$x^2 - 4x + 4 = x + 4$$
$$x^2 - 5x = 0$$
$$x(x-5) = 0$$

$$x = 0 \quad \text{or} \quad x - 5 = 0$$
$$\text{or} \quad x = 5$$

Check $x = 0$: $\sqrt{1} - \sqrt{4} = 1$ *False*
Check $x = 5$: $\sqrt{16} - \sqrt{9} = 1$ *True*

The solution set is $\{5\}$.

N6.
$$\sqrt[3]{4x-5} = \sqrt[3]{3x+2}$$
$$\left(\sqrt[3]{4x-5}\right)^3 = \left(\sqrt[3]{3x+2}\right)^3 \quad \textit{Cube each side.}$$
$$4x - 5 = 3x + 2$$
$$x = 7$$

Check $x = 7$: $\sqrt[3]{23} = \sqrt[3]{23}$ *True*
The solution set is $\{7\}$.

N7. Solve $x = \sqrt{\dfrac{y+2}{a}}$ for a.

$$x^2 = \frac{y+2}{a} \qquad \textit{Square each side.}$$
$$ax^2 = y + 2 \qquad \textit{Multiply by a.}$$
$$a = \frac{y+2}{x^2} \qquad \textit{Divide by } x^2.$$

8.6 Section Exercises

1. $\sqrt{3x+18} - x = 0$

(a) Check $x = 6$.

$$\sqrt{3(6)+18} - 6 \overset{?}{=} 0$$
$$\sqrt{18+18} - 6 \overset{?}{=} 0$$
$$\sqrt{36} - 6 \overset{?}{=} 0$$
$$6 - 6 = 0 \quad \textit{True}$$

The number 6 is a solution.

(b) Check $x = -3$.

$\sqrt{3(-3)+18} - (-3) = 0$ is a false statement since the principal square root of a number is nonnegative, and adding 3 to the radical makes the left side at least 3. The number -3 is not a solution.

Copyright © 2012 Pearson Education, Inc. Publishing as Addison-Wesley.

3. $\sqrt{x+2} - \sqrt{9x-2} = -2\sqrt{x-1}$

(a) Check $x = 2.$

$$\sqrt{2+2} - \sqrt{9(2)-2} \overset{?}{=} -2\sqrt{2-1}$$
$$\sqrt{4} - \sqrt{16} \overset{?}{=} -2\sqrt{1}$$
$$2 - 4 \overset{?}{=} -2$$
$$-2 = -2 \quad True$$

The number 2 is a solution.

(b) Check $x = 7.$

$$\sqrt{7+2} - \sqrt{9(7)-2} \overset{?}{=} -2\sqrt{7-1}$$
$$\sqrt{9} - \sqrt{61} \overset{?}{=} -2\sqrt{6}$$
$$3 - \sqrt{61} = -2\sqrt{6} \quad False$$

The number 7 is not a solution.

5. $\sqrt{9} = 3$, not -3. There is no solution of $\sqrt{x} = -3$ since the value of a principal square root cannot equal a negative number.

In Exercises 7–34, check each solution in the original equation.

7. $\sqrt{x-2} = 3$
$$\left(\sqrt{x-2}\right)^2 = 3^2 \quad Square\ each\ side.$$
$$x - 2 = 9$$
$$x = 11$$

Check the proposed solution, 11.
Check $x = 11$: $\sqrt{9} = 3$ *True*
The solution set is $\{11\}$.

9. $\sqrt{6k-1} = 1$
$$\left(\sqrt{6k-1}\right)^2 = 1^2 \qquad Square\ each\ side.$$
$$6k - 1 = 1$$
$$6k = 2$$
$$k = \tfrac{2}{6} = \tfrac{1}{3}$$

Check the proposed solution, $\tfrac{1}{3}$.
Check $k = \tfrac{1}{3}$: $\sqrt{1} = 1$ *True*
The solution set is $\left\{\tfrac{1}{3}\right\}$.

11. $\sqrt{4r+3} + 1 = 0$
$$\sqrt{4r+3} = -1 \quad Isolate\ the\ radical.$$

This equation has no solution, because $\sqrt{4r+3}$ cannot be negative.

The solution set is $\emptyset$.

13. $\sqrt{3x+1} - 4 = 0$
$$\sqrt{3x+1} = 4 \qquad Isolate\ the\ radical.$$
$$\left(\sqrt{3x+1}\right)^2 = 4^2 \quad Square\ each\ side.$$
$$3x + 1 = 16$$
$$3x = 15$$
$$x = 5$$

Check $x = 5$: $\sqrt{16} - 4 = 0$ *True*
The solution set is $\{5\}$.

15. $4 - \sqrt{x-2} = 0$
$$4 = \sqrt{x-2} \qquad Isolate.$$
$$4^2 = \left(\sqrt{x-2}\right)^2 \quad Square.$$
$$16 = x - 2$$
$$18 = x$$

Check $x = 18$: $4 - \sqrt{16} = 0$ *True*
The solution set is $\{18\}$.

17. $\sqrt{9x-4} = \sqrt{8x+1}$
$$\left(\sqrt{9x-4}\right)^2 = \left(\sqrt{8x+1}\right)^2 \quad Square.$$
$$9x - 4 = 8x + 1$$
$$x = 5$$

Check $x = 5$: $\sqrt{41} = \sqrt{41}$ *True*
The solution set is $\{5\}$.

19. $2\sqrt{x} = \sqrt{3x+4}$
$$\left(2\sqrt{x}\right)^2 = \left(\sqrt{3x+4}\right)^2 \quad Square.$$
$$4x = 3x + 4$$
$$x = 4$$

Check $x = 4$: $4 = \sqrt{16}$ *True*
The solution set is $\{4\}$.

21. $3\sqrt{x-1} = 2\sqrt{2x+2}$
$$\left(3\sqrt{x-1}\right)^2 = \left(2\sqrt{2x+2}\right)^2 \quad Square.$$
$$9(x-1) = 4(2x+2)$$
$$9x - 9 = 8x + 8$$
$$x = 17$$

Check $x = 17$: $3(4) = 2(6)$ *True*
The solution set is $\{17\}$.

23. $x = \sqrt{x^2+4x-20}$
$$x^2 = \left(\sqrt{x^2+4x-20}\right)^2 \quad Square.$$
$$x^2 = x^2 + 4x - 20$$
$$20 = 4x$$
$$5 = x$$

Check $x = 5$: $5 = \sqrt{25}$ *True*
The solution set is $\{5\}$.

25. $x = \sqrt{x^2+3x+9}$
$$x^2 = \left(\sqrt{x^2+3x+9}\right)^2 \quad Square.$$
$$x^2 = x^2 + 3x + 9$$
$$-3x = 9$$
$$x = -3$$

Substituting -3 for x makes the left side of the original equation negative, but the right side is nonnegative, so the solution set is $\emptyset$.

Copyright © 2012 Pearson Education, Inc. Publishing as Addison-Wesley.

27.
$$\sqrt{9-x} = x+3$$
$$(\sqrt{9-x})^2 = (x+3)^2 \qquad \text{Square.}$$
$$9-x = x^2 + 6x + 9$$
$$0 = x^2 + 7x$$
$$0 = x(x+7)$$
$$x = 0 \quad \text{or} \quad x+7 = 0$$
$$x = -7$$
Check $x = -7$: $\sqrt{16} = -4$ *False*
Check $x = 0$: $\sqrt{9} = 3$ *True*
The solution set is $\{0\}$.

29.
$$\sqrt{k^2+2k+9} = k+3$$
$$(\sqrt{k^2+2k+9})^2 = (k+3)^2 \qquad \text{Square.}$$
$$k^2 + 2k + 9 = k^2 + 6k + 9$$
$$0 = 4k$$
$$0 = k$$
Check $k = 0$: $\sqrt{9} = 3$ *True*
The solution set is $\{0\}$.

31.
$$\sqrt{x^2+12x-4} = x-4$$
$$(\sqrt{x^2+12x-4})^2 = (x-4)^2 \qquad \text{Square.}$$
$$x^2 + 12x - 4 = x^2 - 8x + 16$$
$$20x = 20$$
$$x = 1$$
Substituting 1 for x makes the left side of the original equation positive, but the right side is zero, so the solution set is $\emptyset$.

33.
$$\sqrt{r^2+9r+15} - r - 4 = 0$$
$$\sqrt{r^2+9r+15} = r+4 \qquad \text{Isolate.}$$
$$(\sqrt{r^2+9r+15})^2 = (r+4)^2 \qquad \text{Square.}$$
$$r^2 + 9r + 15 = r^2 + 8r + 16$$
$$r = 1$$
Check $r = 1$: $\sqrt{25} - 1 - 4 = 0$ *True*
The solution set is $\{1\}$.

35.
$$\sqrt{3x+4} = 8 - x$$
$(8-x)^2$ equals $64 - 16x + x^2$, not $64 + x^2$. The first step should be
$$3x + 4 = 64 - 16x + x^2.$$
Then we have
$$0 = x^2 - 19x + 60.$$
$$0 = (x-4)(x-15)$$
$$x - 4 = 0 \quad \text{or} \quad x - 15 = 0$$
$$x = 4 \quad \text{or} \qquad x = 15$$
Check $x = 4$: $\sqrt{16} = 8 - 4$ *True*
Check $x = 15$: $\sqrt{49} = 8 - 15$ *False*
The solution set is $\{4\}$.

37.
$$\sqrt[3]{2x+5} = \sqrt[3]{6x+1}$$
$$(\sqrt[3]{2x+5})^3 = (\sqrt[3]{6x+1})^3 \qquad \text{Cube each side.}$$
$$2x + 5 = 6x + 1$$
$$4 = 4x$$
$$1 = x$$
Check $x = 1$: $\sqrt[3]{7} = \sqrt[3]{7}$ *True*
The solution set is $\{1\}$.

39.
$$\sqrt[3]{x^2+5x+1} = \sqrt[3]{x^2+4x}$$
$$(\sqrt[3]{x^2+5x+1})^3 = (\sqrt[3]{x^2+4x})^3 \qquad \text{Cube.}$$
$$x^2 + 5x + 1 = x^2 + 4x$$
$$x = -1$$
Check $x = -1$: $\sqrt[3]{-3} = \sqrt[3]{-3}$ *True*
The solution set is $\{-1\}$.

41.
$$\sqrt[3]{2m-1} = \sqrt[3]{m+13}$$
$$(\sqrt[3]{2m-1})^3 = (\sqrt[3]{m+13})^3 \qquad \text{Cube.}$$
$$2m - 1 = m + 13$$
$$m = 14$$
Check $m = 14$: $\sqrt[3]{27} = \sqrt[3]{27}$ *True*
The solution set is $\{14\}$.

43.
$$\sqrt[4]{x+12} = \sqrt[4]{3x-4}$$
Raise each side to the fourth power.
$$(\sqrt[4]{x+12})^4 = (\sqrt[4]{3x-4})^4$$
$$x + 12 = 3x - 4$$
$$16 = 2x$$
$$8 = x$$
Check $x = 8$: $\sqrt[4]{20} = \sqrt[4]{20}$ *True*
The solution set is $\{8\}$.

45.
$$\sqrt[3]{x-8} + 2 = 0$$
$$\sqrt[3]{x-8} = -2 \qquad \text{Isolate.}$$
$$(\sqrt[3]{x-8})^3 = (-2)^3 \qquad \text{Cube.}$$
$$x - 8 = -8$$
$$x = 0$$
Check $x = 0$: $\sqrt[3]{-8} + 2 = 0$ *True*
The solution set is $\{0\}$.

47.
$$\sqrt[4]{2k-5} + 4 = 0$$
$$\sqrt[4]{2k-5} = -4 \quad \text{Isolate.}$$
This equation has no solution, because $\sqrt[4]{2k-5}$ cannot be negative.

The solution set is $\emptyset$.

Copyright © 2012 Pearson Education, Inc. Publishing as Addison-Wesley.

49. $\sqrt{k+2} - \sqrt{k-3} = 1$

Get one radical on each side of the equals sign.

$$\sqrt{k+2} = 1 + \sqrt{k-3}$$
$$\left(\sqrt{k+2}\right)^2 = \left(1 + \sqrt{k-3}\right)^2 \quad \text{Square.}$$
$$k+2 = 1 + 2\sqrt{k-3} + k - 3$$
$$4 = 2\sqrt{k-3} \quad \text{Isolate.}$$
$$2 = \sqrt{k-3} \quad \text{Divide by 2.}$$
$$2^2 = \left(\sqrt{k-3}\right)^2 \quad \text{Square again.}$$
$$4 = k - 3$$
$$7 = k$$

Check $k = 7$: $\sqrt{9} - \sqrt{4} = 1$ *True*

The solution set is $\{7\}$.

51. $\sqrt{2r+11} - \sqrt{5r+1} = -1$

Get one radical on each side of the equals sign.

$$\sqrt{2r+11} = -1 + \sqrt{5r+1}$$

Square each side.

$$\left(\sqrt{2r+11}\right)^2 = \left(-1 + \sqrt{5r+1}\right)^2$$
$$2r + 11 = 1 - 2\sqrt{5r+1} + 5r + 1$$

Isolate the remaining radical.

$$2\sqrt{5r+1} = 3r - 9$$

Square each side again.

$$\left(2\sqrt{5r+1}\right)^2 = (3r-9)^2$$
$$4(5r+1) = 9r^2 - 54r + 81$$
$$20r + 4 = 9r^2 - 54r + 81$$
$$0 = 9r^2 - 74r + 77$$
$$0 = (9r - 11)(r - 7)$$

$$9r - 11 = 0 \quad \text{or} \quad r - 7 = 0$$
$$r = \tfrac{11}{9} \qquad\qquad r = 7$$

Check $r = \tfrac{11}{9}$: $\tfrac{11}{3} - \tfrac{8}{3} = -1$ *False*

Check $r = 7$: $5 - 6 = -1$ *True*

The solution set is $\{7\}$.

53. $\sqrt{3p+4} - \sqrt{2p-4} = 2$

Get one radical on each side of the equals sign.

$$\sqrt{3p+4} = 2 + \sqrt{2p-4}$$

Square each side.

$$\left(\sqrt{3p+4}\right)^2 = \left(2 + \sqrt{2p-4}\right)^2$$
$$3p + 4 = 4 + 4\sqrt{2p-4} + 2p - 4$$

Isolate the remaining radical.

$$p + 4 = 4\sqrt{2p-4}$$

Square each side again.

$$(p+4)^2 = \left(4\sqrt{2p-4}\right)^2$$
$$p^2 + 8p + 16 = 16(2p - 4)$$
$$p^2 + 8p + 16 = 32p - 64$$
$$p^2 - 24p + 80 = 0$$
$$(p - 4)(p - 20) = 0$$

$$p - 4 = 0 \quad \text{or} \quad p - 20 = 0$$
$$p = 4 \qquad\qquad p = 20$$

Check $p = 4$: $\sqrt{16} - \sqrt{4} = 2$ *True*

Check $p = 20$: $\sqrt{64} - \sqrt{36} = 2$ *True*

The solution set is $\{4, 20\}$.

55. $\sqrt{3-3p} - 3 = \sqrt{3p+2}$

Square each side.

$$\left(\sqrt{3-3p} - 3\right)^2 = \left(\sqrt{3p+2}\right)^2$$
$$3 - 3p - 6\sqrt{3-3p} + 9 = 3p + 2$$

Isolate the remaining radical.

$$-6\sqrt{3-3p} = 6p - 10$$
$$-3\sqrt{3-3p} = 3p - 5$$

Square each side again.

$$\left(-3\sqrt{3-3p}\right)^2 = (3p-5)^2$$
$$9(3 - 3p) = 9p^2 - 30p + 25$$
$$27 - 27p = 9p^2 - 30p + 25$$
$$0 = 9p^2 - 3p - 2$$
$$0 = (3p + 1)(3p - 2)$$

$$3p + 1 = 0 \quad \text{or} \quad 3p - 2 = 0$$
$$p = -\tfrac{1}{3} \qquad\qquad p = \tfrac{2}{3}$$

Check $p = -\tfrac{1}{3}$: $\sqrt{4} - 3 = \sqrt{1}$ *False*

Check $p = \tfrac{2}{3}$: $\sqrt{1} - 3 = \sqrt{4}$ *False*

The solution set is $\emptyset$.

57. $\sqrt{2\sqrt{x+11}} = \sqrt{4x+2}$

$$2\sqrt{x+11} = 4x + 2 \qquad \text{Square.}$$
$$\left(2\sqrt{x+11}\right)^2 = (4x+2)^2 \qquad \text{Square again.}$$
$$4(x + 11) = 16x^2 + 16x + 4$$
$$4x + 44 = 16x^2 + 16x + 4$$
$$0 = 16x^2 + 12x - 40$$
$$0 = 4x^2 + 3x - 10$$
$$0 = (x + 2)(4x - 5)$$

$$x + 2 = 0 \quad \text{or} \quad 4x - 5 = 0$$
$$x = -2 \qquad\qquad x = \tfrac{5}{4}$$

Check $x = -2$: $\sqrt{6} = \sqrt{-6}$ *False*

Check $x = \tfrac{5}{4}$: $\sqrt{7} = \sqrt{7}$ *True*

The solution set is $\left\{\tfrac{5}{4}\right\}$.

59. $(2x - 9)^{1/2} = 2 + (x - 8)^{1/2}$

$$\sqrt{2x-9} = 2 + \sqrt{x-8}$$
$$\left(\sqrt{2x-9}\right)^2 = \left(2 + \sqrt{x-8}\right)^2$$
$$2x - 9 = 4 + 4\sqrt{x-8} + x - 8$$
$$x - 5 = 4\sqrt{x-8}$$
$$(x-5)^2 = \left(4\sqrt{x-8}\right)^2$$
$$x^2 - 10x + 25 = 16(x - 8)$$

Copyright © 2012 Pearson Education, Inc. Publishing as Addison-Wesley.

$$x^2 - 10x + 25 = 16x - 128$$
$$x^2 - 26x + 153 = 0$$
$$(x - 9)(x - 17) = 0$$
$$x = 9 \quad \text{or} \quad x = 17$$

Check $x = 9$: $\quad 9^{1/2} \overset{?}{=} 2 + 1^{1/2}$
$$3 = 2 + 1 \quad \textit{True}$$

Check $x = 17$: $\quad 25^{1/2} \overset{?}{=} 2 + 9^{1/2}$
$$5 = 2 + 3 \quad \textit{True}$$

The solution set is $\{9, 17\}$.

61. $\quad (2w - 1)^{2/3} - w^{1/3} = 0$
$$\sqrt[3]{(2w - 1)^2} = \sqrt[3]{w}$$
$$\left[\sqrt[3]{(2w - 1)^2}\right]^3 = \left(\sqrt[3]{w}\right)^3$$
$$(2w - 1)^2 = w$$
$$4w^2 - 4w + 1 = w$$
$$4w^2 - 5w + 1 = 0$$
$$(4w - 1)(w - 1) = 0$$
$$4w - 1 = 0 \quad \text{or} \quad w - 1 = 0$$
$$w = \tfrac{1}{4} \quad \text{or} \qquad w = 1$$

Check $w = \tfrac{1}{4}$: $\quad \left(-\tfrac{1}{2}\right)^{2/3} - \left(\tfrac{1}{4}\right)^{1/3} \overset{?}{=} 0$
$$\left(\tfrac{1}{4}\right)^{1/3} - \left(\tfrac{1}{4}\right)^{1/3} = 0 \quad \textit{True}$$

Check $w = 1$: $\qquad 1^{2/3} - 1^{1/3} = 0 \quad \textit{True}$

The solution set is $\left\{\tfrac{1}{4}, 1\right\}$.

63. Solve $Z = \sqrt{\dfrac{L}{C}}$ for L.
$$(Z)^2 = \left(\sqrt{\dfrac{L}{C}}\right)^2 \qquad \textit{Square.}$$
$$Z^2 = \dfrac{L}{C}$$
$$CZ^2 = L \qquad \textit{Multiply by C.}$$

65. Solve $V = \sqrt{\dfrac{2K}{m}}$ for K.
$$(V)^2 = \left(\sqrt{\dfrac{2K}{m}}\right)^2 \qquad \textit{Square.}$$
$$V^2 = \dfrac{2K}{m} \qquad \textit{Multiply by } \dfrac{m}{2}.$$
$$\dfrac{V^2 m}{2} = K$$

67. Solve $r = \sqrt{\dfrac{Mm}{F}}$ for M.
$$(r)^2 = \left(\sqrt{\dfrac{Mm}{F}}\right)^2 \qquad \textit{Square.}$$
$$r^2 = \dfrac{Mm}{F}$$
$$Fr^2 = Mm$$
$$\dfrac{Fr^2}{m} = M$$

69. Solve $N = \dfrac{1}{2\pi}\sqrt{\dfrac{a}{r}}$ for r.
$$2\pi N = \sqrt{\dfrac{a}{r}}$$
$$(2\pi N)^2 = \left(\sqrt{\dfrac{a}{r}}\right)^2$$
$$4\pi^2 N^2 = \dfrac{a}{r}$$
$$4\pi^2 N^2 r = a$$
$$r = \dfrac{a}{4\pi^2 N^2}$$

71. $\quad (5 + 9x) + (-4 - 8x)$
$$= 5 - 4 + 9x - 8x$$
$$= 1 + x$$

73. $\quad (x + 3)(2x - 5) = 2x^2 - 5x + 6x - 15$
$$= 2x^2 + x - 15$$

75. $\quad \dfrac{-7}{5 - \sqrt{2}} = \dfrac{-7}{5 - \sqrt{2}} \cdot \dfrac{5 + \sqrt{2}}{5 + \sqrt{2}}$
$$= \dfrac{-7(5 + \sqrt{2})}{5^2 - (\sqrt{2})^2}$$
$$= \dfrac{-7(5 + \sqrt{2})}{25 - 2} = \dfrac{-7(5 + \sqrt{2})}{23}$$

8.7 Complex Numbers

8.7 Now Try Exercises

N1. **(a)** $\sqrt{-49} = i\sqrt{49} = 7i$

(b) $-\sqrt{-121} = -i\sqrt{121} = -11i$

(c) $\sqrt{-3} = i\sqrt{3}$

(d) $\sqrt{-32} = i\sqrt{32} = i\sqrt{16 \cdot 2} = 4i\sqrt{2}$

N2. **(a)** $\sqrt{-4} \cdot \sqrt{-16} = i\sqrt{4} \cdot i\sqrt{16}$
$$= i \cdot 2 \cdot i \cdot 4$$
$$= 8i^2$$
$$= 8(-1)$$
$$= -8$$

(b) $\sqrt{-5} \cdot \sqrt{-11} = i\sqrt{5} \cdot i\sqrt{11}$
$$= i^2\sqrt{5 \cdot 11}$$
$$= (-1)\sqrt{55}$$
$$= -\sqrt{55}$$

(c) $\sqrt{-3} \cdot \sqrt{-12} = i\sqrt{3} \cdot i\sqrt{12}$
$$= i^2\sqrt{3 \cdot 12}$$
$$= (-1)\sqrt{36}$$
$$= -6$$

(d) $\sqrt{13} \cdot \sqrt{-2} = \sqrt{13} \cdot i\sqrt{2}$
$$= i\sqrt{13 \cdot 2}$$
$$= i\sqrt{26}$$

Copyright © 2012 Pearson Education, Inc. Publishing as Addison-Wesley.

N3. (a) $\dfrac{\sqrt{-72}}{\sqrt{-8}} = \dfrac{i\sqrt{72}}{i\sqrt{8}} = \sqrt{\dfrac{72}{8}} = \sqrt{9} = 3$

(b) $\dfrac{\sqrt{-48}}{\sqrt{3}} = \dfrac{i\sqrt{48}}{\sqrt{3}} = i\sqrt{\dfrac{48}{3}} = i\sqrt{16} = 4i$

N4. To add complex numbers, add the real parts and add the imaginary parts.

(a) $(-3 + 2i) + (4 + 7i)$
$= (-3 + 4) + (2 + 7)i$
$= 1 + 9i$

(b) $(5 - i) + (-3 + 3i) + (6 - 4i)$
$= [5 + (-3) + 6] + [(-1) + 3 + (-4)]i$
$= 8 - 2i$

N5. To subtract complex numbers, subtract the real parts and subtract the imaginary parts.

(a) $(7 + 10i) - (3 + 5i)$
$= (7 - 3) + (10 - 5)i$
$= 4 + 5i$

(b) $(5 - 2i) - (9 - 7i)$
$= (5 - 9) + [-2 - (-7)]i$
$= -4 + 5i$

(c) $(-1 + 12i) - (-1 - i)$
$= [-1 - (-1)] + [12 - (-1)]i$
$= 0 + 13i = 13i$

N6. (a) $8i(3 - 5i) = 8i(3) - 8i(5i)$
$= 24i - 40i^2$
$= 24i - 40(-1)$
$= 40 + 24i$

(b) $(7 - 2i)(4 + 3i)$

$\qquad$ F $\qquad$ O $\qquad$ I $\qquad$ L

$= 7(4) + 7(3i) + 4(-2i) + 3i(-2i)$
$= 28 + 21i - 8i - 6i^2$
$= 28 + 13i - 6(-1)$
$= 28 + 13i + 6$
$= 34 + 13i$

N7. In each case, multiply the numerator and denominator by the conjugate of the denominator.

(a) $\dfrac{4 + 2i}{1 + 3i} = \dfrac{(4 + 2i)(1 - 3i)}{(1 + 3i)(1 - 3i)}$

$= \dfrac{4 - 12i + 2i - 6i^2}{1^2 - (3i)^2}$

$= \dfrac{4 - 10i - 6(-1)}{1 - (-9)}$

$= \dfrac{4 - 10i + 6}{10}$

$= \dfrac{10}{10} - \dfrac{10}{10}i = 1 - i$

(b) $\dfrac{5 - 4i}{i}$ $\quad$ The conjugate of i is $-i$.

$= \dfrac{(5 - 4i)(-i)}{i(-i)}$

$= \dfrac{-5i + 4i^2}{-i^2}$

$= \dfrac{-5i + 4(-1)}{-(-1)}$

$= \dfrac{-5i - 4}{1}$

$= -4 - 5i$

N8. (a) $i^{16} = (i^4)^4 = 1^4 = 1$

(b) $i^{21} = i^{20} \cdot i^1 = (i^4)^5 \cdot i = 1^5 \cdot i = i$

(c) $i^{-6} = \dfrac{1}{i^6} = \dfrac{1}{i^4 \cdot i^2} = \dfrac{1}{1 \cdot (-1)} = \dfrac{1}{-1} = -1$

(d) $i^{-13} = \dfrac{1}{i^{13}} = \dfrac{1}{i^{12} \cdot i} = \dfrac{1}{(i^4)^3 \cdot i} = \dfrac{1}{1^3 \cdot i} = \dfrac{1}{i}$

Now multiply by $-i$, the conjugate of i.

$\dfrac{1}{i} = \dfrac{1(-i)}{i(-i)} = \dfrac{-i}{-i^2} = \dfrac{-i}{-(-1)} = \dfrac{-i}{1} = -i$

8.7 Section Exercises

1. $\sqrt{-1} = i$

3. $i^2 = -1$

5. $\dfrac{1}{i} = \dfrac{1}{i} \cdot \dfrac{-i}{-i}$ $\quad$ $-i$ *is the conjugate of* i

$= \dfrac{-i}{-i^2} = \dfrac{i}{i^2} = \dfrac{i}{-1} = -i$

7. $\sqrt{-169} = i\sqrt{169} = 13i$

9. $-\sqrt{-144} = -i\sqrt{144} = -12i$

11. $\sqrt{-5} = i\sqrt{5}$

13. $\sqrt{-48} = i\sqrt{48} = i\sqrt{16 \cdot 3} = 4i\sqrt{3}$

15. $\sqrt{-7} \cdot \sqrt{-15} = i\sqrt{7} \cdot i\sqrt{15} = i^2\sqrt{7 \cdot 15}$
$= -1\sqrt{105} = -\sqrt{105}$

17. $\sqrt{-4} \cdot \sqrt{-25} = i\sqrt{4} \cdot i\sqrt{25} = 2i \cdot 5i = 10i^2$
$= 10(-1) = -10$

19. $\sqrt{-3} \cdot \sqrt{11} = i\sqrt{3} \cdot \sqrt{11}$
$= i\sqrt{33}$

21. $\dfrac{\sqrt{-300}}{\sqrt{-100}} = \dfrac{i\sqrt{300}}{i\sqrt{100}} = \sqrt{\dfrac{300}{100}} = \sqrt{3}$

23. $\dfrac{\sqrt{-75}}{\sqrt{3}} = \dfrac{i\sqrt{75}}{\sqrt{3}} = i\sqrt{\dfrac{75}{3}} = i\sqrt{25} = 5i$

25. $\dfrac{-\sqrt{-64}}{\sqrt{-16}} = \dfrac{-i\sqrt{64}}{i\sqrt{16}} = -\dfrac{8}{4} = -2$

Copyright © 2012 Pearson Education, Inc. Publishing as Addison-Wesley.

27. Any real number a can be written as $a + 0i$, a complex number with imaginary part 0.

29. $(3 + 2i) + (-4 + 5i)$
$= [(3 + (-4)] + (2 + 5)i$
$= -1 + 7i$

31. $(5 - i) + (-5 + i)$
$= (5 - 5) + (-1 + 1)i$
$= 0$

33. $(4 + i) - (-3 - 2i)$
$= [(4 - (-3)] + [(1 - (-2)]i$
$= 7 + 3i$

35. $(-3 - 4i) - (-1 - 4i)$
$= [-3 - (-1)] + [-4 - (-4)]i$
$= -2$

37. $(-4 + 11i) + (-2 - 4i) + (7 + 6i)$
$= (-4 - 2 + 7) + (11 - 4 + 6)i$
$= 1 + 13i$

39. $[(7 + 3i) - (4 - 2i)] + (3 + i)$
Work inside the brackets first.
$= [(7 - 4) + (3 + 2)i] + (3 + i)$
$= (3 + 5i) + (3 + i)$
$= (3 + 3) + (5 + 1)i$
$= 6 + 6i$

41. If $a - c = b$, then $b + c = a$.
So, $(4 + 2i) - (3 + i) = 1 + i$ implies that
$$(1 + i) + (3 + i) = \underline{4 + 2i}.$$

43. $(3i)(27i) = 81i^2 = 81(-1) = -81$

45. $(-8i)(-2i) = 16i^2 = 16(-1) = -16$

47. $5i(-6 + 2i) = (5i)(-6) + (5i)(2i)$
$= -30i + 10i^2$
$= -30i + 10(-1)$
$= -10 - 30i$

49. $(4 + 3i)(1 - 2i)$
$\quad\quad\quad$ **F** $\quad$ **O** $\quad$ **I** $\quad$ **L**
$= (4)(1) + 4(-2i) + (3i)(1) + (3i)(-2i)$
$= 4 - 8i + 3i - 6i^2$
$= 4 - 5i - 6(-1)$
$= 4 - 5i + 6 = 10 - 5i$

51. $(4 + 5i)^2 = 4^2 + 2(4)(5i) + (5i)^2$
$= 16 + 40i + 25i^2$
$= 16 + 40i + 25(-1)$
$= 16 + 40i - 25$
$= -9 + 40i$

53. $2i(-4 - i)^2$
$= 2i[(-4)^2 - 2(-4)(i) + i^2]$
$= 2i(16 + 8i - 1)$
$= 2i(15 + 8i) = 30i + 16i^2$
$= 30i + 16(-1) = -16 + 30i$

55. $(12 + 3i)(12 - 3i)$
$= 12^2 - (3i)^2 = 144 - 9i^2$
$= 144 - 9(-1) = 144 + 9 = 153$

57. $(4 + 9i)(4 - 9i)$
$= 4^2 - (9i)^2 = 16 - 81i^2$
$= 16 - 81(-1) = 16 + 81 = 97$

59. $(1 + i)^2(1 - i)^2$
$= [(1 + i)(1 - i)]^2$
$= (1^2 - i^2)^2$
$= [1 - (-1)]^2$
$= 2^2 = 4$

61. The conjugate of $a + bi$ is $a - bi$.

63. $\dfrac{2}{1 - i}$
Multiply the numerator and the denominator by the conjugate of the denominator, $1 + i$.
$= \dfrac{2(1 + i)}{(1 - i)(1 + i)} = \dfrac{2(1 + i)}{1^2 - i^2}$
$= \dfrac{2(1 + i)}{1 - (-1)} = \dfrac{2(1 + i)}{2} = 1 + i$

65. $\dfrac{8i}{2 + 2i}$
Write in lowest terms.
$= \dfrac{2 \cdot 4i}{2(1 + i)} = \dfrac{4i}{1 + i}$
Multiply the numerator and the denominator by the conjugate of the denominator, $1 - i$.
$= \dfrac{4i(1 - i)}{(1 + i)(1 - i)}$
In the denominator, we make use of the fact that $(a + bi)(a - bi) = a^2 + b^2$.
$= \dfrac{4(i - i^2)}{1^2 + 1^2} = \dfrac{4(i + 1)}{2}$
$= 2(i + 1) = 2 + 2i$

67. $\dfrac{-7 + 4i}{3 + 2i}$
Multiply the numerator and the denominator by the conjugate of the denominator, $3 - 2i$.
$= \dfrac{(-7 + 4i)(3 - 2i)}{(3 + 2i)(3 - 2i)}$
In the denominator, we make use of the fact that $(a + bi)(a - bi) = a^2 + b^2$.
$= \dfrac{-21 + 14i + 12i + 8}{3^2 + 2^2}$
$= \dfrac{-13 + 26i}{13} = \dfrac{13(-1 + 2i)}{13} = -1 + 2i$

Copyright © 2012 Pearson Education, Inc. Publishing as Addison-Wesley.

69. $\dfrac{2-3i}{2+3i}$

Multiply the numerator and the denominator by the conjugate of the denominator, $2-3i$.

$= \dfrac{(2-3i)(2-3i)}{(2+3i)(2-3i)} = \dfrac{2^2 - 2(2)(3i) + (3i)^2}{2^2 + 3^2}$

$= \dfrac{4 - 12i + 9i^2}{4+9} = \dfrac{4 - 12i - 9}{13}$

$= \dfrac{-5 - 12i}{13} = -\dfrac{5}{13} - \dfrac{12}{13}i$

71. $\dfrac{3+i}{i}$

Multiply the numerator and the denominator by the conjugate of the denominator, $-i$.

$= \dfrac{(3+i)(-i)}{i(-i)} = \dfrac{-3i - i^2}{-i^2}$

$= \dfrac{-3i - (-1)}{-(-1)} = \dfrac{-3i + 1}{1}$

$= 1 - 3i$

73. $\dfrac{3-i}{-i}$

Multiply the numerator and the denominator by the conjugate of the denominator, i.

$= \dfrac{(3-i)(i)}{-i(i)} = \dfrac{3i - i^2}{-i^2} = \dfrac{3i - (-1)}{-(-1)}$

$= \dfrac{3i + 1}{1} = 1 + 3i$

75. $i^{18} = i^{16} \cdot i^2 = (i^4)^4 \cdot i^2$

$= 1^4 \cdot (-1) = 1 \cdot (-1) = -1$

77. $i^{89} = i^{88} \cdot i = (i^4)^{22} \cdot i = 1^{22} \cdot i$

$= 1 \cdot i = i$

79. $i^{38} = i^{36} \cdot i^2$

$= (i^4)^9 \cdot i^2 = 1^9(-1) = -1$

81. $i^{43} = i^{40} \cdot i^2 \cdot i$

$= (i^4)^{10} \cdot (-1) \cdot i$

$= 1^{10} \cdot (-i) = -i$

83. $i^{-5} = \dfrac{1}{i^5} = \dfrac{1}{i^4 \cdot i} = \dfrac{1}{1 \cdot i} = \dfrac{1}{i}$

From Exercise 5, $\dfrac{1}{i} = -i$.

85. Since $i^{20} = (i^4)^5 = 1^5 = 1$, the student multiplied by 1, which is justified by the identity property for multiplication.

87. $I = \dfrac{E}{R + (X_L - X_c)i}$

Substitute $2 + 3i$ for E, 5 for R, 4 for X_L, and 3 for X_c.

$I = \dfrac{2 + 3i}{5 + (4 - 3)i} = \dfrac{2 + 3i}{5 + i}$

$= \dfrac{(2+3i)(5-i)}{(5+i)(5-i)} = \dfrac{10 - 2i + 15i - 3i^2}{5^2 + 1^2}$

$= \dfrac{10 + 3 + 13i}{25 + 1} = \dfrac{13 + 13i}{26}$

$= \dfrac{13(1+i)}{13 \cdot 2} = \dfrac{1+i}{2} = \dfrac{1}{2} + \dfrac{1}{2}i$

89. To show that $1 + 5i$ is a solution of the equation $x^2 - 2x + 26 = 0$, substitute $1 + 5i$ for x.

$$x^2 - 2x + 26 = 0$$
$$(1+5i)^2 - 2(1+5i) + 26 \overset{?}{=} 0$$
$$(1 + 10i + 25i^2) - 2 - 10i + 26 \overset{?}{=} 0$$
$$1 + 10i - 25 - 2 - 10i + 26 \overset{?}{=} 0$$
$$(1 - 25 - 2 + 26) + (10 - 10)i \overset{?}{=} 0$$
$$0 = 0 \quad True$$

Thus, $1 + 5i$ is a solution of the given equation.

Now substitute $1 - 5i$ for x.

$$(1-5i)^2 - 2(1-5i) + 26 \overset{?}{=} 0$$
$$(1 - 10i + 25i^2) - 2 + 10i + 26 \overset{?}{=} 0$$
$$1 - 10i - 25 - 2 + 10i + 26 \overset{?}{=} 0$$
$$(1 - 25 - 2 + 26) + (-10 + 10)i \overset{?}{=} 0$$
$$0 = 0 \quad True$$

Thus, $1 - 5i$ is also a solution of the given equation.

91. $\dfrac{3}{2-i} + \dfrac{5}{1+i}$

$= \dfrac{3(2+i)}{(2-i)(2+i)} + \dfrac{5(1-i)}{(1+i)(1-i)}$

$= \dfrac{6 + 3i}{4+1} + \dfrac{5 - 5i}{1+1}$

$= \dfrac{2(6+3i)}{2 \cdot 5} + \dfrac{5(5-5i)}{5 \cdot 2}$

$= \dfrac{12 + 6i}{10} + \dfrac{25 - 25i}{10}$

$= \dfrac{37 - 19i}{10} = \dfrac{37}{10} - \dfrac{19}{10}i$

93. $\dfrac{2+i}{2-i} + \dfrac{i}{1+i}$

$= \dfrac{(2+i)(2+i)}{(2-i)(2+i)} + \dfrac{i(1-i)}{(1+i)(1-i)}$

$= \dfrac{4 + 4i + i^2}{4+1} + \dfrac{i - i^2}{1+1}$

$= \dfrac{3 + 4i}{5} + \dfrac{1+i}{2}$

$= \dfrac{2(3+4i)}{2 \cdot 5} + \dfrac{5(1+i)}{5 \cdot 2}$

$= \dfrac{6 + 8i}{10} + \dfrac{5 + 5i}{10}$

Copyright © 2012 Pearson Education, Inc. Publishing as Addison-Wesley.

$$= \frac{11 + 13i}{10} = \frac{11}{10} + \frac{13}{10}i$$

Multiplying by i gives us

$$\left(\frac{2+i}{2-i} + \frac{i}{1+i}\right)i = \left(\frac{11}{10} + \frac{13}{10}i\right)i$$

$$= \frac{11}{10}i + \frac{13}{10}i^2$$

$$= -\frac{13}{10} + \frac{11}{10}i.$$

95. $6x + 13 = 0$

$$6x = -13$$

$$x = -\tfrac{13}{6}$$

The solution set is $\left\{-\frac{13}{6}\right\}$.

97. $x(x + 3) = 40$

$$x^2 + 3x = 40$$

$$x^2 + 3x - 40 = 0$$

$$(x + 8)(x - 5) = 0$$

$$x + 8 = 0 \qquad \text{or} \qquad x - 5 = 0$$

$$x = -8 \qquad \text{or} \qquad x = 5$$

Check $x = -8$: $-8(-5) = 40$ *True*
Check $x = 5$: $5(8) = 40$ *True*

The solution set is $\{-8, 5\}$.

99. $5x^2 - 3x = 2$

$$5x^2 - 3x - 2 = 0$$

$$(5x + 2)(x - 1) = 0$$

$$5x + 2 = 0 \qquad \text{or} \qquad x - 1 = 0$$

$$x = -\tfrac{2}{5} \qquad \text{or} \qquad x = 1$$

Check $x = -\frac{2}{5}$: $\frac{4}{5} + \frac{6}{5} = 2$ *True*
Check $x = 1$: $5 - 3 = 2$ *True*

The solution set is $\left\{-\frac{2}{5}, 1\right\}$.

Chapter 8 Review Exercises

1. $\sqrt{1764} = 42$, because $42^2 = 1764$.

2. $-\sqrt{289} = -(17) = -17$, since $17^2 = 289$.

3. $\sqrt[3]{216} = 6$, because $6^3 = 216$.

4. $\sqrt[3]{-125} = -5$, because $(-5)^3 = -125$.

5. $-\sqrt[3]{27} = -(3) = -3$, since $3^3 = 27$.

6. $\sqrt[5]{-32} = -2$, because $(-2)^5 = -32$.

7. $\sqrt[n]{a}$ is not a real number if n is even and a is negative.

8.   **(a)** $\sqrt{x^2} = |x|$

   (b) $-\sqrt{x^2} = -|x|$

   (c) $\sqrt[3]{x^3} = x$

9. $-\sqrt{47} \approx -6.856$

10. $\sqrt[3]{-129} \approx -5.053$

11. $\sqrt[4]{605} \approx 4.960$

12. $\sqrt[4]{500^{-3}} \approx 0.009$

13. $-\sqrt[3]{500^4} \approx -3968.503$

14. $-\sqrt{28^{-1}} \approx -0.189$

15. $f(x) = \sqrt{x - 1}$
For the radicand to be nonnegative, we must have

$$x - 1 \geq 0 \quad \text{or} \quad x \geq 1.$$

Thus, the domain is $[1, \infty)$.
The function values are positive or zero (the result of the radical), so the range is $[0, \infty)$.

x	$f(x) = \sqrt{x - 1}$
1	$\sqrt{1 - 1} = 0$
2	$\sqrt{2 - 1} = 1$
5	$\sqrt{5 - 1} = 2$

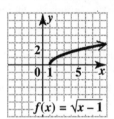

16. $f(x) = \sqrt[3]{x} + 4$
Since we can take the cube root of any real number, the domain is $(-\infty, \infty)$.
The result of a cube root can be any real number, so the range is $(-\infty, \infty)$. (The "+4" does not affect that range.)

x	$f(x) = \sqrt[3]{x} + 4$
-8	$\sqrt[3]{-8} + 4 = 2$
-1	$\sqrt[3]{-1} + 4 = 3$
0	$\sqrt[3]{0} + 4 = 4$
1	$\sqrt[3]{1} + 4 = 5$
8	$\sqrt[3]{8} + 4 = 6$

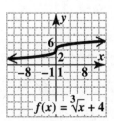

Copyright © 2012 Pearson Education, Inc. Publishing as Addison-Wesley.

17. The base $\sqrt{38}$ is closest to $\sqrt{36} = 6$. The height $\sqrt{99}$ is closest to $\sqrt{100} = 10$. Use the estimates $b = 6$ and $h = 10$ in $A = \frac{1}{2}bh$ to find an estimate of the area.

$$A \approx \frac{1}{2}(6)(10) = 30$$

Choice **B** is the best estimate.

18. One way to evaluate $8^{2/3}$ is to first find the *cube (or third)* root of *8*, which is *2*. Then raise that result to the *second* power, to get an answer of *4*. Therefore, $8^{2/3} = $ *4*.

19. **A.** $(-27)^{2/3} = \left[(-27)^{1/3}\right]^2$
This number is a square, so it is a positive number.
B. $(-64)^{5/3} = \left[(-64)^{1/3}\right]^5$
This number is the odd power of an odd root of a negative number, so it is a negative number.

C. $(-100)^{1/2}$
This number is the square root of a negative number, so it is not a real number.

D. $(-32)^{1/5}$
This number is an odd root of a negative number, so it is a negative number.
The only positive number is choice **A**.

20. $a^{m/n} = \sqrt[n]{a^m}$
Since n is odd, $\sqrt[n]{a^m}$ is positive if a^m is positive and negative if a^m is negative. Since a is negative, a^m is positive if m is even and negative if m is odd.

(a) If a is negative and n is odd, then $a^{m/n}$ is positive if m is even.

(b) If a is negative and n is odd, then $a^{m/n}$ is negative if m is odd.

21. If a is negative and n is even, then $a^{1/n}$ is *not* a real number. An example is $(-4)^{1/2}$, which is *not* a real number.

22. $49^{1/2} = \sqrt{49} = 7$

23. $-121^{1/2} = -\sqrt{121} = -11$

24. $16^{5/4} = \left(16^{1/4}\right)^5 = 2^5 = 32$

25. $-8^{2/3} = -\left(8^{1/3}\right)^2 = -2^2 = -4$

26. $-\left(\frac{36}{25}\right)^{3/2} = -\left(\sqrt{\frac{36}{25}}\right)^3$
$= -\left(\frac{6}{5}\right)^3 = -\frac{216}{125}$

27. $\left(-\frac{1}{8}\right)^{-5/3} = (-8)^{5/3} = \left[(-8)^{1/3}\right]^5$
$= (-2)^5 = -32$

28. $\left(\frac{81}{10{,}000}\right)^{-3/4} = \left(\frac{10{,}000}{81}\right)^{3/4}$
$= \left[\left(\frac{10{,}000}{81}\right)^{1/4}\right]^3$
$= \left(\frac{10}{3}\right)^3 = \frac{1000}{27}$

29. The base, -16, is negative, and $(-16)^3$ is still negative. The index, 4, is even, so $(-16)^{3/4}$ is not a real number.

30. First way: $8^{2/3} = \left(8^{1/3}\right)^2 = \left(\sqrt[3]{8}\right)^2$

Second way: $8^{2/3} = \left(8^2\right)^{1/3} = \sqrt[3]{8^2}$

31. The expression with fractional exponents, $a^{m/n}$, is equivalent to the radical expression, $\sqrt[n]{a^m}$. The denominator of the exponent is the index of the radical. For example, $\sqrt[3]{8^2} = \sqrt[3]{64} = 4$, and $8^{2/3} = (8^{1/3})^2 = 2^2 = 4$.

32. $(m + 3n)^{1/2} = \sqrt{m + 3n}$

33. $(3a + b)^{-5/3} = \dfrac{1}{(3a + b)^{5/3}}$
$= \dfrac{1}{((3a + b)^{1/3})^5}$
$= \dfrac{1}{\left(\sqrt[3]{3a + b}\right)^5}$
or $\dfrac{1}{\sqrt[3]{(3a + b)^5}}$

34. $\sqrt{7^9} = (7^9)^{1/2} = 7^{9/2}$

35. $\sqrt[5]{p^4} = \left(p^4\right)^{1/5} = p^{4/5}$

36. $5^{1/4} \cdot 5^{7/4} = 5^{1/4 + 7/4} = 5^{8/4}$
$= 5^2, \text{ or } 25$

37. $\dfrac{96^{2/3}}{96^{-1/3}} = 96^{2/3 - (-1/3)} = 96^1 = 96$

38. $\dfrac{(a^{1/3})^4}{a^{2/3}} = \dfrac{a^{4/3}}{a^{2/3}} = a^{4/3 - 2/3} = a^{2/3}$

39. $\dfrac{y^{-1/3} \cdot y^{5/6}}{y} = \dfrac{y^{-2/6} y^{5/6}}{y^{6/6}}$
$= y^{-2/6 + 5/6 - 6/6}$
$= y^{-3/6} = y^{-1/2} = \dfrac{1}{y^{1/2}}$

40. $\left(\dfrac{z^{-1} x^{-3/5}}{2^{-2} z^{-1/2} x}\right)^{-1} = \dfrac{z^1 x^{3/5}}{2^2 z^{1/2} x^{-1}}$
$= \frac{1}{4} z^{1 - 1/2} x^{3/5 - (-1)}$
$= \dfrac{z^{1/2} x^{8/5}}{4}$

Copyright © 2012 Pearson Education, Inc. Publishing as Addison-Wesley.

41. $r^{-1/2}(r + r^{3/2})$
$= r^{-1/2}(r) + r^{-1/2}(r^{3/2})$
$= r^{-1/2+1} + r^{-1/2+3/2}$
$= r^{1/2} + r^{2/2}$
$= r^{1/2} + r$

42. $\sqrt[8]{s^4} = (s^4)^{1/8} = s^{4/8} = s^{1/2}$

43. $\sqrt[6]{r^9} = (r^9)^{1/6} = r^{9/6} = r^{3/2}$

44. $\dfrac{\sqrt{p^5}}{p^2} = \dfrac{p^{5/2}}{p^2} = p^{5/2-2}$
$= p^{5/2-4/2} = p^{1/2}$

45. $\sqrt[4]{k^3} \cdot \sqrt{k^3} = (k^3)^{1/4}(k^3)^{1/2}$
$= k^{3/4}k^{3/2} = k^{3/4+3/2}$
$= k^{3/4+6/4} = k^{9/4}$

46. $\sqrt[3]{m^5} \cdot \sqrt[3]{m^8} = (m^5)^{1/3}(m^8)^{1/3}$
$= m^{5/3}m^{8/3} = m^{5/3+8/3} = m^{13/3}$

47. $\sqrt[4]{\sqrt[3]{z}} = \sqrt[4]{z^{1/3}} = (z^{1/3})^{1/4} = z^{1/12}$

48. $\sqrt{\sqrt{\sqrt{x}}} = \sqrt{\sqrt{x^{1/2}}} = \sqrt{(x^{1/2})^{1/2}}$
$= \sqrt{x^{1/4}} = (x^{1/4})^{1/2} = x^{1/8}$

49. $\sqrt[3]{\sqrt[5]{x}} = \sqrt[3]{x^{1/5}} = (x^{1/5})^{1/3} = x^{1/15}$

50. $\sqrt{\sqrt[6]{\sqrt[3]{x}}} = \sqrt{\sqrt[6]{x^{1/3}}} = \sqrt{(x^{1/3})^{1/6}}$
$= \sqrt{x^{1/18}} = (x^{1/18})^{1/2} = x^{1/36}$

51. The product rule for exponents applies only if the bases are the same. The bases in $3^{1/4} \cdot 2^{1/5}$ are 3 and 2, respectively.

52. $\sqrt{6} \cdot \sqrt{11} = \sqrt{6 \cdot 11} = \sqrt{66}$

53. $\sqrt{5} \cdot \sqrt{r} = \sqrt{5 \cdot r} = \sqrt{5r}$

54. $\sqrt[3]{6} \cdot \sqrt[3]{5} = \sqrt[3]{6 \cdot 5} = \sqrt[3]{30}$

55. $\sqrt[4]{7} \cdot \sqrt[4]{3} = \sqrt[4]{7 \cdot 3} = \sqrt[4]{21}$

56. $\sqrt{20} = \sqrt{4 \cdot 5} = \sqrt{4}\sqrt{5} = 2\sqrt{5}$

57. $\sqrt{75} = \sqrt{25 \cdot 3} = \sqrt{25}\sqrt{3} = 5\sqrt{3}$

58. $-\sqrt{125} = -\sqrt{25 \cdot 5} = -5\sqrt{5}$

59. $\sqrt[3]{-108} = \sqrt[3]{-27 \cdot 4} = -3\sqrt[3]{4}$

60. $\sqrt{100y^7} = \sqrt{100y^6 \cdot y} = 10y^3\sqrt{y}$

61. $\sqrt[3]{64p^4q^6} = \sqrt[3]{64p^3q^6 \cdot p} = 4pq^2\sqrt[3]{p}$

62. $\sqrt[3]{108a^8b^5} = \sqrt[3]{27a^6b^3 \cdot 4a^2b^2}$
$= 3a^2b\sqrt[3]{4a^2b^2}$

63. $\sqrt[3]{632r^8t^4} = \sqrt[3]{8r^6t^3 \cdot 79r^2t}$
$= 2r^2t\sqrt[3]{79r^2t}$

64. $\sqrt{\dfrac{y^3}{144}} = \dfrac{\sqrt{y^3}}{\sqrt{144}} = \dfrac{\sqrt{y^2 \cdot y}}{12} = \dfrac{y\sqrt{y}}{12}$

65. $\sqrt[3]{\dfrac{m^{15}}{27}} = \dfrac{\sqrt[3]{m^{15}}}{\sqrt[3]{27}} = \dfrac{\sqrt[3]{(m^5)^3}}{\sqrt[3]{3^3}} = \dfrac{m^5}{3}$

66. $\sqrt[3]{\dfrac{r^2}{8}} = \dfrac{\sqrt[3]{r^2}}{\sqrt[3]{8}} = \dfrac{\sqrt[3]{r^2}}{2}$

67. $\sqrt[4]{\dfrac{a^9}{81}} = \dfrac{\sqrt[4]{a^9}}{\sqrt[4]{81}} = \dfrac{\sqrt[4]{a^8 \cdot a}}{3} = \dfrac{a^2\sqrt[4]{a}}{3}$

68. $\sqrt[6]{15^3} = 15^{3/6} = 15^{1/2} = \sqrt{15}$

69. $\sqrt[4]{p^6} = (p^6)^{1/4} = p^{6/4} = p^{3/2}$
$= p^{2/2}p^{1/2} = p\sqrt{p}$

70. $\sqrt[3]{2} \cdot \sqrt[4]{5} = 2^{1/3} \cdot 5^{1/4}$
$= 2^{4/12} \cdot 5^{3/12}$
$= (2^4 \cdot 5^3)^{1/12}$
$= \sqrt[12]{16 \cdot 125} = \sqrt[12]{2000}$

71. $\sqrt{x} \cdot \sqrt[5]{x} = x^{1/2} \cdot x^{1/5}$
$= x^{5/10} \cdot x^{2/10}$
$= x^{7/10} = \sqrt[10]{x^7}$

72. Substitute 8 for a and 6 for b in the Pythagorean theorem to find the hypotenuse, x.
$$x^2 = a^2 + b^2$$
$$x = \sqrt{a^2 + b^2} = \sqrt{8^2 + 6^2}$$
$$= \sqrt{64 + 36} = \sqrt{100} = 10$$

The length of the hypotenuse is 10.

73. The given points are $(-4, 7)$ and $(10, 6)$.
$$d = \sqrt{(x_2 - x_1)^2 + (y_2 - y_1)^2}$$
$$= \sqrt{[10 - (-4)]^2 + (6 - 7)^2}$$
$$= \sqrt{14^2 + (-1)^2}$$
$$= \sqrt{196 + 1} = \sqrt{197}$$

74. $2\sqrt{8} - 3\sqrt{50} = 2\sqrt{4 \cdot 2} - 3\sqrt{25 \cdot 2}$
$= 2 \cdot 2\sqrt{2} - 3 \cdot 5\sqrt{2}$
$= 4\sqrt{2} - 15\sqrt{2} = -11\sqrt{2}$

75. $8\sqrt{80} - 3\sqrt{45} = 8\sqrt{16 \cdot 5} - 3\sqrt{9 \cdot 5}$
$= 8 \cdot 4\sqrt{5} - 3 \cdot 3\sqrt{5}$
$= 32\sqrt{5} - 9\sqrt{5} = 23\sqrt{5}$

Copyright © 2012 Pearson Education, Inc. Publishing as Addison-Wesley.

76. $-\sqrt{27y} + 2\sqrt{75y} = -\sqrt{9 \cdot 3y} + 2\sqrt{25 \cdot 3y}$

$\qquad = -3\sqrt{3y} + 2 \cdot 5\sqrt{3y}$

$\qquad = -3\sqrt{3y} + 10\sqrt{3y}$

$\qquad = 7\sqrt{3y}$

77. $2\sqrt{54m^3} + 5\sqrt{96m^3}$

$\qquad = 2\sqrt{9m^2 \cdot 6m} + 5\sqrt{16m^2 \cdot 6m}$

$\qquad = 2 \cdot 3m\sqrt{6m} + 5 \cdot 4m\sqrt{6m}$

$\qquad = 6m\sqrt{6m} + 20m\sqrt{6m} = 26m\sqrt{6m}$

78. $3\sqrt[3]{54} + 5\sqrt[3]{16} = 3\sqrt[3]{27 \cdot 2} + 5\sqrt[3]{8 \cdot 2}$

$\qquad = 3 \cdot 3\sqrt[3]{2} + 5 \cdot 2\sqrt[3]{2}$

$\qquad = 9\sqrt[3]{2} + 10\sqrt[3]{2} = 19\sqrt[3]{2}$

79. $-6\sqrt[4]{32} + \sqrt[4]{512} = -6\sqrt[4]{16 \cdot 2} + \sqrt[4]{256 \cdot 2}$

$\qquad = -6 \cdot 2\sqrt[4]{2} + 4\sqrt[4]{2}$

$\qquad = -12\sqrt[4]{2} + 4\sqrt[4]{2} = -8\sqrt[4]{2}$

80. Add the measures of the sides.

$P = a + b + c + d$

$P = 4\sqrt{8} + 6\sqrt{12} + 8\sqrt{2} + 3\sqrt{48}$

$\qquad = 4\sqrt{4 \cdot 2} + 6\sqrt{4 \cdot 3} + 8\sqrt{2} + 3\sqrt{16 \cdot 3}$

$\qquad = 4 \cdot 2\sqrt{2} + 6 \cdot 2\sqrt{3} + 8\sqrt{2} + 3 \cdot 4\sqrt{3}$

$\qquad = 8\sqrt{2} + 12\sqrt{3} + 8\sqrt{2} + 12\sqrt{3}$

$\qquad = 16\sqrt{2} + 24\sqrt{3}$

The perimeter is $\left(16\sqrt{2} + 24\sqrt{3}\right)$ feet.

81. Add the measures of the sides.

$P = a + b + c$

$P = 2\sqrt{27} + \sqrt{108} + \sqrt{50}$

$\qquad = 2\sqrt{9 \cdot 3} + \sqrt{36 \cdot 3} + \sqrt{25 \cdot 2}$

$\qquad = 2 \cdot 3\sqrt{3} + 6\sqrt{3} + 5\sqrt{2}$

$\qquad = 6\sqrt{3} + 6\sqrt{3} + 5\sqrt{2}$

$\qquad = 12\sqrt{3} + 5\sqrt{2}$

The perimeter is $\left(12\sqrt{3} + 5\sqrt{2}\right)$ feet.

82. $\left(\sqrt{3} + 1\right)\left(\sqrt{3} - 2\right) = 3 - 2\sqrt{3} + \sqrt{3} - 2$

$\qquad = 1 - \sqrt{3}$

83. $\left(\sqrt{7} + \sqrt{5}\right)\left(\sqrt{7} - \sqrt{5}\right) = \left(\sqrt{7}\right)^2 - \left(\sqrt{5}\right)^2$

$\qquad = 7 - 5 = 2$

84. $\left(3\sqrt{2} + 1\right)\left(2\sqrt{2} - 3\right)$

$\qquad = 6 \cdot 2 - 9\sqrt{2} + 2\sqrt{2} - 3$

$\qquad = 12 - 7\sqrt{2} - 3 = 9 - 7\sqrt{2}$

85. $\left(\sqrt{13} - \sqrt{2}\right)^2$

$\qquad = \left(\sqrt{13}\right)^2 - 2 \cdot \sqrt{13} \cdot \sqrt{2} + \left(\sqrt{2}\right)^2$

$\qquad = 13 - 2\sqrt{26} + 2 = 15 - 2\sqrt{26}$

86. $\left(\sqrt[3]{2} + 3\right)\left(\sqrt[3]{4} - 3\sqrt[3]{2} + 9\right)$

$\qquad = \sqrt[3]{2} \cdot \sqrt[3]{4} - \sqrt[3]{2} \cdot 3\sqrt[3]{2} + 9\sqrt[3]{2}$

$\qquad\quad + 3\sqrt[3]{4} - 3 \cdot 3\sqrt[3]{2} + 27$

$\qquad = \sqrt[3]{8} - 3\sqrt[3]{4} + 9\sqrt[3]{2} + 3\sqrt[3]{4} - 9\sqrt[3]{2} + 27$

$\qquad = 2 + 27 = 29$

87. $\left(\sqrt[3]{4y} - 1\right)\left(\sqrt[3]{4y} + 3\right)$

$\qquad = \sqrt[3]{16y^2} + 3\sqrt[3]{4y} - \sqrt[3]{4y} - 3$

$\qquad = \sqrt[3]{8 \cdot 2y^2} + 2\sqrt[3]{4y} - 3$

$\qquad = 2\sqrt[3]{2y^2} + 2\sqrt[3]{4y} - 3$

88. Show that $15 - 2\sqrt{26} \neq 13\sqrt{26}$.

Find a calculator approximation of each term.

$\qquad 4.801960973 \neq 66.28725368$

Therefore, $15 - 2\sqrt{26} \neq 13\sqrt{26}$.

89. Multiplying by $\sqrt[3]{6}$ still results in an expression with a radical in the denominator.

$$\frac{5\sqrt[3]{6}}{\sqrt[3]{6} \cdot \sqrt[3]{6}} = \frac{5\sqrt[3]{6}}{\sqrt[3]{36}}$$

To rationalize the denominator, multiply by $\sqrt[3]{6^2}$ or $\sqrt[3]{36}$.

$$\frac{5\sqrt[3]{6^2}}{\sqrt[3]{6} \cdot \sqrt[3]{6^2}} = \frac{5\sqrt[3]{36}}{\sqrt[3]{6^3}} = \frac{5\sqrt[3]{36}}{6}$$

90. $\dfrac{\sqrt{6}}{\sqrt{5}} = \dfrac{\sqrt{6} \cdot \sqrt{5}}{\sqrt{5} \cdot \sqrt{5}} = \dfrac{\sqrt{30}}{5}$

91. $\dfrac{-6\sqrt{3}}{\sqrt{2}} = \dfrac{-6\sqrt{3} \cdot \sqrt{2}}{\sqrt{2} \cdot \sqrt{2}} = \dfrac{-6\sqrt{6}}{2} = -3\sqrt{6}$

92. $\dfrac{3\sqrt{7p}}{\sqrt{y}} = \dfrac{3\sqrt{7p} \cdot \sqrt{y}}{\sqrt{y} \cdot \sqrt{y}} = \dfrac{3\sqrt{7py}}{y}$

93. $\sqrt{\dfrac{11}{8}} = \dfrac{\sqrt{11}}{\sqrt{8}} = \dfrac{\sqrt{11}}{\sqrt{4 \cdot 2}} = \dfrac{\sqrt{11}}{2\sqrt{2}} = \dfrac{\sqrt{11} \cdot \sqrt{2}}{2\sqrt{2} \cdot \sqrt{2}}$

$\qquad = \dfrac{\sqrt{22}}{2 \cdot 2} = \dfrac{\sqrt{22}}{4}$

94. $-\sqrt[3]{\dfrac{9}{25}} = -\dfrac{\sqrt[3]{9}}{\sqrt[3]{5^2}} = -\dfrac{\sqrt[3]{9} \cdot \sqrt[3]{5}}{\sqrt[3]{5^2} \cdot \sqrt[3]{5}}$

$\qquad = -\dfrac{\sqrt[3]{45}}{\sqrt[3]{5^3}} = -\dfrac{\sqrt[3]{45}}{5}$

95. $\sqrt[3]{\dfrac{108m^3}{n^5}} = \dfrac{\sqrt[3]{108m^3}}{\sqrt[3]{n^5}} = \dfrac{\sqrt[3]{27m^3 \cdot 4}}{\sqrt[3]{n^3 \cdot n^2}}$

$\qquad = \dfrac{3m\sqrt[3]{4}}{n\sqrt[3]{n^2}} = \dfrac{3m\sqrt[3]{4} \cdot \sqrt[3]{n}}{n\sqrt[3]{n^2} \cdot \sqrt[3]{n}}$

$\qquad = \dfrac{3m\sqrt[3]{4n}}{n \cdot n} = \dfrac{3m\sqrt[3]{4n}}{n^2}$

Copyright © 2012 Pearson Education, Inc. Publishing as Addison-Wesley.

96. $\dfrac{1}{\sqrt{2}+\sqrt{7}}$

Multiply the numerator and denominator by the conjugate of the denominator, $\sqrt{2}-\sqrt{7}$.

$$=\dfrac{1(\sqrt{2}-\sqrt{7})}{(\sqrt{2}+\sqrt{7})(\sqrt{2}-\sqrt{7})}$$
$$=\dfrac{\sqrt{2}-\sqrt{7}}{2-7}=\dfrac{\sqrt{2}-\sqrt{7}}{-5}$$

97. $\dfrac{-5}{\sqrt{6}-3}$

Multiply the numerator and denominator by the conjugate of the denominator, $\sqrt{6}+3$.

$$=\dfrac{-5(\sqrt{6}+3)}{(\sqrt{6}-3)(\sqrt{6}+3)}=\dfrac{-5(\sqrt{6}+3)}{6-9}$$
$$=\dfrac{-5(\sqrt{6}+3)}{-3}=\dfrac{5(\sqrt{6}+3)}{3}$$

98. $\dfrac{2-2\sqrt{5}}{8}=\dfrac{2(1-\sqrt{5})}{2\cdot4}=\dfrac{1-\sqrt{5}}{4}$

99. $\dfrac{4-8\sqrt{8}}{12}=\dfrac{4(1-2\sqrt{8})}{3\cdot4}=\dfrac{1-2\sqrt{8}}{3}$
$$=\dfrac{1-2\sqrt{4\cdot2}}{3}=\dfrac{1-4\sqrt{2}}{3}$$

100. $\dfrac{-18+\sqrt{27}}{6}=\dfrac{-18+\sqrt{9}\cdot\sqrt{3}}{6}=\dfrac{-18+3\sqrt{3}}{6}$
$$=\dfrac{3(-6+\sqrt{3})}{3\cdot2}=\dfrac{-6+\sqrt{3}}{2}$$

101. $\sqrt{8x+9}=5$
$(\sqrt{8x+9})^2=5^2$ *Square.*
$8x+9=25$
$8x=16$
$x=2$

Check $x=2$: $\sqrt{25}=5$ *True*
The solution set is $\{2\}$.

102. $\sqrt{2x-3}-3=0$
$\sqrt{2x-3}=3$ *Isolate.*
$(\sqrt{2x-3})^2=3^2$ *Square.*
$2x-3=9$
$2x=12$
$x=6$

Check $x=6$: $\sqrt{9}-3=0$ *True*
The solution set is $\{6\}$.

103. $\sqrt{3x+1}-2=-3$
$\sqrt{3x+1}=-1$ *Isolate.*

This equation has no solution, because $\sqrt{3x+1}$ cannot be negative.
The solution set is $\emptyset$.

104. $\sqrt{7x+1}=x+1$
$(\sqrt{7x+1})^2=(x+1)^2$ *Square.*
$7x+1=x^2+2x+1$
$0=x^2-5x$
$0=x(x-5)$
$x=0$ or $x=5$

Check $x=0$: $\sqrt{1}=1$ *True*
Check $x=5$: $\sqrt{36}=6$ *True*
The solution set is $\{0,5\}$.

105. $3\sqrt{x}=\sqrt{10x-9}$
$(3\sqrt{x})^2=(\sqrt{10x-9})^2$ *Square.*
$9x=10x-9$
$9=x$

Check $x=9$: $3\sqrt{9}=\sqrt{81}$ *True*
The solution set is $\{9\}$.

106. $\sqrt{x^2+3x+7}=x+2$
$(\sqrt{x^2+3x+7})^2=(x+2)^2$ *Square.*
$x^2+3x+7=x^2+4x+4$
$3=x$

Check $x=3$: $\sqrt{25}=5$ *True*
The solution set is $\{3\}$.

107. $\sqrt{x+2}-\sqrt{x-3}=1$
Get one radical on each side of the equals sign.
$\sqrt{x+2}=1+\sqrt{x-3}$
Square each side.
$(\sqrt{x+2})^2=(1+\sqrt{x-3})^2$
$x+2=1+2\sqrt{x-3}+x-3$
$4=2\sqrt{x-3}$
$2=\sqrt{x-3}$
Square each side again.
$2^2=(\sqrt{x-3})^2$
$4=x-3$
$7=x$

Check $x=7$: $\sqrt{9}-\sqrt{4}=1$ *True*
The solution set is $\{7\}$.

108. $\sqrt[3]{5x-1}=\sqrt[3]{3x-2}$
Cube each side.
$(\sqrt[3]{5x-1})^3=(\sqrt[3]{3x-2})^3$
$5x-1=3x-2$
$2x=-1$
$x=-\frac{1}{2}$

Check $x=-\frac{1}{2}$: $\sqrt[3]{-\frac{7}{2}}=\sqrt[3]{-\frac{7}{2}}$ *True*
The solution set is $\{-\frac{1}{2}\}$.

Copyright © 2012 Pearson Education, Inc. Publishing as Addison-Wesley.

109. $\sqrt[3]{2x^2 + 3x - 7} = \sqrt[3]{2x^2 + 4x + 6}$
Cube each side.
$$\left(\sqrt[3]{2x^2 + 3x - 7}\right)^3 = \left(\sqrt[3]{2x^2 + 4x + 6}\right)^3$$
$$2x^2 + 3x - 7 = 2x^2 + 4x + 6$$
$$-13 = x$$
Check $x = -13$: $\sqrt[3]{292} = \sqrt[3]{292}$ *True*
The solution set is $\{-13\}$.

110. $\sqrt[3]{3x^2 - 4x + 6} = \sqrt[3]{3x^2 - 2x + 8}$
$$\left(\sqrt[3]{3x^2 - 4x + 6}\right)^3 = \left(\sqrt[3]{3x^2 - 2x + 8}\right)^3$$
$$3x^2 - 4x + 6 = 3x^2 - 2x + 8$$
$$-2 = 2x$$
$$-1 = x$$
Check $x = -1$: $\sqrt[3]{13} = \sqrt[3]{13}$ *True*
The solution set is $\{-1\}$.

111. $\sqrt[3]{1 - 2x} - \sqrt[3]{-x - 13} = 0$
$$\sqrt[3]{1 - 2x} = \sqrt[3]{-x - 13}$$
Cube each side.
$$\left(\sqrt[3]{1 - 2x}\right)^3 = \left(\sqrt[3]{-x - 13}\right)^3$$
$$1 - 2x = -x - 13$$
$$14 = x$$
Check $x = 14$: $\sqrt[3]{-27} - \sqrt[3]{-27} = 0$ *True*
The solution set is $\{14\}$.

112. $\sqrt[3]{11 - 2x} - \sqrt[3]{-1 - 5x} = 0$
$$\sqrt[3]{11 - 2x} = \sqrt[3]{-1 - 5x}$$
$$\left(\sqrt[3]{11 - 2x}\right)^3 = \left(\sqrt[3]{-1 - 5x}\right)^3$$
$$11 - 2x = -1 - 5x$$
$$3x = -12$$
$$x = -4$$
Check $x = -4$: $\sqrt[3]{19} - \sqrt[3]{19} = 0$ *True*
The solution set is $\{-4\}$.

113. $\sqrt[4]{x - 1} + 2 = 0$
$$\sqrt[4]{x - 1} = -2$$
This equation has no solution, because $\sqrt[4]{x - 1}$
cannot be negative.
The solution set is $\emptyset$.

114. $\sqrt[4]{2x + 3} + 1 = 0$
$$\sqrt[4]{2x + 3} = -1$$
This equation has no solution, because $\sqrt[4]{2x + 3}$
cannot be negative.
The solution set is $\emptyset$.

115. $\sqrt[4]{x + 7} = \sqrt[4]{2x}$
Raise each side to the fourth power.
$$\left(\sqrt[4]{x + 7}\right)^4 = \left(\sqrt[4]{2x}\right)^4$$
$$x + 7 = 2x$$
$$7 = x$$
Check $x = 7$: $\sqrt[4]{14} = \sqrt[4]{14}$ *True*
The solution set is $\{7\}$.

116. $\sqrt[4]{x + 8} = \sqrt[4]{3x}$
Raise each side to the fourth power.
$$\left(\sqrt[4]{x + 8}\right)^4 = \left(\sqrt[4]{3x}\right)^4$$
$$x + 8 = 3x$$
$$8 = 2x$$
$$4 = x$$
Check $x = 4$: $\sqrt[4]{12} = \sqrt[4]{12}$ *True*
The solution set is $\{4\}$.

117. (a) Solve $L = \sqrt{H^2 + W^2}$ for H.
$$L^2 = H^2 + W^2$$
$$L^2 - W^2 = H^2$$
$$\sqrt{L^2 - W^2} = H$$

(b) Substitute 12 for L and 9 for W in
$H = \sqrt{L^2 - W^2}$.
$$H = \sqrt{12^2 - 9^2}$$
$$= \sqrt{144 - 81} = \sqrt{63}$$

To the nearest tenth of a foot, the height is
approximately 7.9 feet.

118. $\sqrt{-25} = i\sqrt{25} = 5i$

119. $\sqrt{-200} = i\sqrt{100 \cdot 2} = 10i\sqrt{2}$

120. If a is a positive real number, then $-a$ is negative.
So, $\sqrt{-a}$ is not a real number. Therefore, $-\sqrt{-a}$
is not a real number either.

121. $(-2 + 5i) + (-8 - 7i)$
$$= [-2 + (-8)] + [5 + (-7)]i$$
$$= -10 - 2i$$

122. $(5 + 4i) - (-9 - 3i)$
$$= [(5 - (-9)] + [(4 - (-3)]i$$
$$= 14 + 7i$$

123. $\sqrt{-5} \cdot \sqrt{-7} = i\sqrt{5} \cdot i\sqrt{7} = i^2\sqrt{35}$
$$= -1\left(\sqrt{35}\right) = -\sqrt{35}$$

124. $\sqrt{-25} \cdot \sqrt{-81} = 5i \cdot 9i = 45i^2$
$$= 45(-1) = -45$$

125. $\dfrac{\sqrt{-72}}{\sqrt{-8}} = \dfrac{i\sqrt{72}}{i\sqrt{8}} = \sqrt{\dfrac{72}{8}} = \sqrt{9} = 3$

126. $(2 + 3i)(1 - i) = 2 - 2i + 3i - 3i^2$
$$= 2 + i - 3(-1)$$
$$= 2 + i + 3$$
$$= 5 + i$$

127. $(6 - 2i)^2 = 6^2 - 2 \cdot 6 \cdot 2i + (2i)^2$
$$= 36 - 24i + 4i^2$$
$$= 36 - 24i + 4(-1)$$
$$= 36 - 24i - 4$$
$$= 32 - 24i$$

Copyright © 2012 Pearson Education, Inc. Publishing as Addison-Wesley.

128. $\dfrac{3-i}{2+i}$

Multiply by the conjugate of the denominator, $2-i$.

$$= \dfrac{(3-i)(2-i)}{(2+i)(2-i)} = \dfrac{6-3i-2i+i^2}{4-i^2}$$

$$= \dfrac{6-5i-1}{4-(-1)} = \dfrac{5-5i}{5} = \dfrac{5(1-i)}{5}$$

$$= 1-i$$

129. $\dfrac{5+14i}{2+3i}$

Multiply by the conjugate of the denominator, $2-3i$.

$$= \dfrac{(5+14i)(2-3i)}{(2+3i)(2-3i)} = \dfrac{10-15i+28i-42i^2}{4-9i^2}$$

$$= \dfrac{10+13i-42(-1)}{4-9(-1)} = \dfrac{52+13i}{13}$$

$$= \dfrac{13(4+i)}{13} = 4+i$$

130. $i^{11} = i^8 \cdot i^3 = (i^4)^2 \cdot i^2 \cdot i$

$$= 1^2 \cdot (-1) \cdot i = -i$$

131. $i^{36} = (i^4)^9 = 1^9 = 1$

132. $i^{-10} = \dfrac{1}{i^{10}} = \dfrac{1}{i^8 \cdot i^2} = \dfrac{1}{(i^4)^2 \cdot (-1)}$

$$= \dfrac{1}{1^2 \cdot (-1)} = \dfrac{1}{-1} = -1$$

Another method:

$i^{-10} = i^{-10} \cdot i^{12} = i^2 = -1$

Note that $i^{12} = (i^4)^3 = 1^3 = 1$.

133. $i^{-8} = \dfrac{1}{i^8} = \dfrac{1}{(i^4)^2} = \dfrac{1}{1^2} = \dfrac{1}{1} = 1$

134. [8.1] $-\sqrt[4]{256} = -\left(\sqrt[4]{4^4}\right) = -(4) = -4$

135. [8.2] $1000^{-2/3} = \dfrac{1}{1000^{2/3}} = \dfrac{1}{\left[(10^3)^{1/3}\right]^2}$

$$= \dfrac{1}{10^2} = \dfrac{1}{100}$$

136. [8.2] $\dfrac{z^{-1/5} \cdot z^{3/10}}{z^{7/10}} = z^{-2/10+3/10-7/10}$

$$= z^{-6/10} = \dfrac{1}{z^{6/10}} = \dfrac{1}{z^{3/5}}$$

137. [8.2] $\sqrt[4]{k^{24}} = k^{24/4} = k^6$

138. [8.3] $\sqrt[3]{54z^9t^8} = \sqrt[3]{27z^9t^6 \cdot 2t^2} = 3z^3t^2\sqrt[3]{2t^2}$

139. [8.4] $-5\sqrt{18} + 12\sqrt{72}$

$$= -5\sqrt{9\cdot2} + 12\sqrt{36\cdot2}$$

$$= -5\cdot3\sqrt{2} + 12\cdot6\sqrt{2}$$

$$= -15\sqrt{2} + 72\sqrt{2} = 57\sqrt{2}$$

140. [8.5] $\dfrac{-1}{\sqrt{12}} = \dfrac{-1}{\sqrt{4\cdot3}} = \dfrac{-1}{2\sqrt{3}} = \dfrac{-1\cdot\sqrt{3}}{2\sqrt{3}\cdot\sqrt{3}}$

$$= \dfrac{-\sqrt{3}}{2\cdot3} = \dfrac{-\sqrt{3}}{6}$$

141. [8.5] $\sqrt[3]{\dfrac{12}{25}} = \dfrac{\sqrt[3]{12}}{\sqrt[3]{25}} = \dfrac{\sqrt[3]{12}}{\sqrt[3]{5^2}}$

$$= \dfrac{\sqrt[3]{12}\cdot\sqrt[3]{5}}{\sqrt[3]{5^2}\cdot\sqrt[3]{5}} = \dfrac{\sqrt[3]{60}}{\sqrt[3]{5^3}} = \dfrac{\sqrt[3]{60}}{5}$$

142. [8.7] $i^{-1000} = \dfrac{1}{i^{1000}} = \dfrac{1}{(i^4)^{250}}$

$$= \dfrac{1}{1^{250}} = \dfrac{1}{1} = 1$$

143. [8.7] $\sqrt{-49} = i\sqrt{49} = 7i$

144. [8.7] $(4-9i) + (-1+2i)$

$$= (4-1) + (-9+2)i$$

$$= 3-7i$$

145. [8.7] $\dfrac{\sqrt{50}}{\sqrt{-2}} = \dfrac{\sqrt{25\cdot2}}{i\sqrt{2}} = \dfrac{5\sqrt{2}}{i\sqrt{2}} = \dfrac{5}{i}$

The conjugate of i is $-i$.

$$\dfrac{5(-i)}{i(-i)} = \dfrac{-5i}{-i^2} = \dfrac{-5i}{-(-1)} = -5i$$

146. [8.5] $\dfrac{3+\sqrt{54}}{6} = \dfrac{3+\sqrt{9\cdot6}}{6} = \dfrac{3+3\sqrt{6}}{6}$

$$= \dfrac{3(1+\sqrt{6})}{2\cdot3} = \dfrac{1+\sqrt{6}}{2}$$

147. [8.7] $(3+2i)^2 = 3^2 + 2\cdot3\cdot2i + (2i)^2$

$$= 9 + 12i + 4i^2$$

$$= 9 + 12i + 4(-1)$$

$$= 5 + 12i$$

148. [8.4] $8\sqrt[3]{x^3y^2} - 2x\sqrt[3]{y^2} = 8x\sqrt[3]{y^2} - 2x\sqrt[3]{y^2}$

$$= 6x\sqrt[3]{y^2}$$

149. [8.4] $9\sqrt{5} - 4\sqrt{15}$ cannot be simplified further.

150. [8.5] $\left(\sqrt{5} - \sqrt{3}\right)\left(\sqrt{7} + \sqrt{3}\right)$

$$= \sqrt{35} + \sqrt{15} - \sqrt{21} - 3$$

151. [8.6] $\sqrt{x+4} = x-2$

$$\left(\sqrt{x+4}\right)^2 = (x-2)^2 \qquad Square.$$

$$x+4 = x^2 - 4x + 4$$

$$0 = x^2 - 5x$$

$$0 = x(x-5)$$

$$x = 0 \quad \text{or} \quad x-5 = 0$$

$$x = 5$$

Check $x = 0$: $\sqrt{4} = -2$ *False*

Check $x = 5$: $\sqrt{9} = 3$ *True*

The solution set is $\{5\}$.

Copyright © 2012 Pearson Education, Inc. Publishing as Addison-Wesley.

152. [8.6] $\sqrt[3]{2x-9} = \sqrt[3]{5x+3}$
Cube each side.
$$\left(\sqrt[3]{2x-9}\right)^3 = \left(\sqrt[3]{5x+3}\right)^3$$
$$2x - 9 = 5x + 3$$
$$-3x = 12$$
$$x = -4$$

Check $x = -4$: $\sqrt[3]{-17} = \sqrt[3]{-17}$ *True*
The solution set is $\{-4\}$.

153. [8.6] $\sqrt{6+2x} - 1 = \sqrt{7-2x}$
Square each side.
$$\left(\sqrt{6+2x}-1\right)^2 = \left(\sqrt{7-2x}\right)^2$$
$$6 + 2x - 2\sqrt{6+2x} + 1 = 7 - 2x$$
$$4x = 2\sqrt{6+2x}$$
$$2x = \sqrt{6+2x}$$
Square each side again.
$$(2x)^2 = \left(\sqrt{6+2x}\right)^2$$
$$4x^2 = 6 + 2x$$
$$4x^2 - 2x - 6 = 0$$
$$2x^2 - x - 3 = 0$$
$$(2x-3)(x+1) = 0$$

$2x - 3 = 0$ or $x + 1 = 0$
$\quad x = \frac{3}{2}$ or $\qquad x = -1$

Check $x = \frac{3}{2}$: $\sqrt{9} - 1 = \sqrt{4}$ *True*
Check $x = -1$: $\sqrt{4} - 1 = \sqrt{9}$ *False*
The solution set is $\left\{\frac{3}{2}\right\}$.

154. [8.6] $\sqrt{7x+11} - 5 = 0$
$$\sqrt{7x+11} = 5$$
$$\left(\sqrt{7x+11}\right)^2 = 5^2$$
$$7x + 11 = 25$$
$$7x = 14$$
$$x = 2$$

Check $x = 2$: $\sqrt{25} - 5 = 0$ *True*
The solution set is $\{2\}$.

155. [8.6] $\sqrt{6x+2} - \sqrt{5x+3} = 0$
Get one radical on each side of the equals sign.
$$\sqrt{6x+2} = \sqrt{5x+3}$$
Square each side.
$$\left(\sqrt{6x+2}\right)^2 = \left(\sqrt{5x+3}\right)^2$$
$$6x + 2 = 5x + 3$$
$$x = 1$$

Check $x = 1$: $\sqrt{8} - \sqrt{8} = 0$ *True*
The solution set is $\{1\}$.

156. [8.6] $\sqrt{3+5x} - \sqrt{x+11} = 0$
$$\sqrt{3+5x} = \sqrt{x+11}$$
$$\left(\sqrt{3+5x}\right)^2 = \left(\sqrt{x+11}\right)^2$$
$$3 + 5x = x + 11$$
$$4x = 8$$
$$x = 2$$

Check $x = 2$: $\sqrt{13} - \sqrt{13} = 0$ *True*
The solution set is $\{2\}$.

157. [8.6] $3\sqrt{x} = \sqrt{8x+9}$
$$\left(3\sqrt{x}\right)^2 = \left(\sqrt{8x+9}\right)^2 \quad \text{Square.}$$
$$9x = 8x + 9$$
$$x = 9$$

Check $x = 9$: $3\sqrt{9} = \sqrt{81}$ *True*
The solution set is $\{9\}$.

158. [8.6] $6\sqrt{x} = \sqrt{30x+24}$
$$\left(6\sqrt{x}\right)^2 = \left(\sqrt{30x+24}\right)^2$$
$$36x = 30x + 24$$
$$6x = 24$$
$$x = 4$$

Check $x = 4$: $6\sqrt{4} = \sqrt{144}$ *True*
The solution set is $\{4\}$.

159. [8.6] $\sqrt{11+2x} + 1 = \sqrt{5x+1}$
Square each side.
$$\left(\sqrt{11+2x}+1\right)^2 = \left(\sqrt{5x+1}\right)^2$$
$$\left(\sqrt{11+2x}\right)^2 + 2 \cdot \sqrt{11+2x} \cdot 1 + 1$$
$$= 5x + 1$$
$$11 + 2x + 2\sqrt{11+2x} + 1 = 5x + 1$$
Isolate the remaining radical.
$$2\sqrt{11+2x} = 3x - 11$$
Square each side again.
$$\left(2\sqrt{11+2x}\right)^2 = (3x-11)^2$$
$$4(11+2x) = 9x^2 - 66x + 121$$
$$44 + 8x = 9x^2 - 66x + 121$$
$$0 = 9x^2 - 74x + 77$$
$$0 = (9x-11)(x-7)$$

$9x - 11 = 0$ or $x - 7 = 0$
$\quad x = \frac{11}{9}$ or $\qquad x = 7$

Check $x = \frac{11}{9}$: $\sqrt{\frac{121}{9}} + 1 = \sqrt{\frac{64}{9}}$ *False*
Check $x = 7$: $\sqrt{25} + 1 = \sqrt{36}$ *True*
The solution set is $\{7\}$.

Copyright © 2012 Pearson Education, Inc. Publishing as Addison-Wesley.

160. [8.6]

$$\sqrt{5x+6} - \sqrt{x+3} = 3$$

Get one radical on each side.

$$\sqrt{5x+6} = 3 + \sqrt{x+3}$$
$$\left(\sqrt{5x+6}\right)^2 = \left(3 + \sqrt{x+3}\right)^2$$
$$5x+6 = 9 + 6\sqrt{x+3} + x + 3$$

Isolate the remaining radical.

$$4x - 6 = 6\sqrt{x+3}$$
$$2x - 3 = 3\sqrt{x+3}$$
$$(2x-3)^2 = \left(3\sqrt{x+3}\right)^2$$
$$4x^2 - 12x + 9 = 9(x+3)$$
$$4x^2 - 12x + 9 = 9x + 27$$
$$4x^2 - 21x - 18 = 0$$
$$(4x+3)(x-6) = 0$$

$$4x + 3 = 0 \quad \text{or} \quad x - 6 = 0$$
$$x = -\tfrac{3}{4} \qquad\qquad x = 6$$

Check $x = -\tfrac{3}{4}$: $\quad \sqrt{\tfrac{9}{4}} - \sqrt{\tfrac{9}{4}} = 3 \quad$ *False*

Check $x = 6$: $\quad \sqrt{36} - \sqrt{9} = 3 \quad$ *True*

The solution set is $\{6\}$.

Chapter 8 Test

1. $\sqrt{841} = 29$, because $29^2 = 841$, so $-\sqrt{841} = -29$.

2. $\sqrt[3]{-512} = -8$, because $(-8)^3 = -512$.

3. $125^{1/3} = \sqrt[3]{125} = 5$, because $5^3 = 125$.

4. $\sqrt{146.25} \approx \sqrt{144} = 12$, because $12^2 = 144$, so choice **C** is the best estimate.

5. $\sqrt{478} \approx 21.863$

6. $\sqrt[3]{-832} \approx -9.405$

7. $f(x) = \sqrt{x+6}$
For the radicand to be nonnegative, we must have

$$x + 6 \ge 0 \quad \text{or} \quad x \ge -6.$$

Thus, the domain is $[-6, \infty)$.
The function values are positive or zero (the result of the radical), so the range is $[0, \infty)$.

x	$f(x) = \sqrt{x+6}$
-6	$\sqrt{-6+6} = 0$
-5	$\sqrt{-5+6} = 1$
-2	$\sqrt{-2+6} = 2$

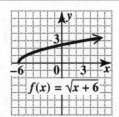

8. $\left(\dfrac{16}{25}\right)^{-3/2} = \left(\dfrac{25}{16}\right)^{3/2} = \left[\left(\dfrac{25}{16}\right)^{1/2}\right]^3$
$= \left(\dfrac{5}{4}\right)^3 = \dfrac{125}{64}$

9. $(-64)^{-4/3} = \dfrac{1}{(-64)^{4/3}} = \dfrac{1}{\left[(-64)^{1/3}\right]^4}$
$= \dfrac{1}{(-4)^4} = \dfrac{1}{256}$

10. $\dfrac{3^{2/5}x^{-1/4}y^{2/5}}{3^{-8/5}x^{7/4}y^{1/10}}$
$= 3^{2/5-(-8/5)}x^{-1/4-7/4}y^{2/5-1/10}$
$= 3^{10/5}x^{-8/4}y^{4/10-1/10}$
$= 3^2 x^{-2}y^{3/10} = \dfrac{9y^{3/10}}{x^2}$

11. $\left(\dfrac{x^{-4}y^{-6}}{x^{-2}y^3}\right)^{-2/3} = \left(\dfrac{x^2}{x^4y^6y^3}\right)^{-2/3}$
$= \left(\dfrac{1}{x^2y^9}\right)^{-2/3} = (x^2y^9)^{2/3}$
$= (x^2)^{2/3}(y^9)^{2/3} = x^{4/3}y^6$

12. $7^{3/4} \cdot 7^{-1/4} = 7^{3/4+(-1/4)}$
$= 7^{1/2}, \quad \text{or} \quad \sqrt{7}$

13. $\sqrt[3]{a^4} \cdot \sqrt[3]{a^7} = \sqrt[3]{a^{4+7}}$
$= \sqrt[3]{a^{11}} \quad (\text{or } a^{11/3})$
$= \sqrt[3]{a^9 \cdot a^2} = a^3\sqrt[3]{a^2}$

14. $a^2 + b^2 = c^2$
$12^2 + b^2 = 17^2 \qquad$ *Let $a = 12$, $c = 17$.*
$144 + b^2 = 289$
$b^2 = 145$
$b = \sqrt{145}$

15. The given points are $(-4, 2)$ and $(2, 10)$.
$d = \sqrt{(x_2 - x_1)^2 + (y_2 - y_1)^2}$
$= \sqrt{[2-(-4)]^2 + (10-2)^2}$
$= \sqrt{6^2 + 8^2} = \sqrt{36+64} = \sqrt{100} = 10$

16. $\sqrt{54x^5y^6} = \sqrt{9x^4y^6 \cdot 6x} = 3x^2y^3\sqrt{6x}$

17. $\sqrt[4]{32a^7b^{13}} = \sqrt[4]{16a^4b^{12} \cdot 2a^3b}$
$= 2ab^3\sqrt[4]{2a^3b}$

18. $\sqrt{2} \cdot \sqrt[3]{5} = 2^{1/2} \cdot 5^{1/3} = 2^{3/6} \cdot 5^{2/6}$
$= (2^3 \cdot 5^2)^{1/6} = \sqrt[6]{2^3 \cdot 5^2}$
$= \sqrt[6]{8 \cdot 25} = \sqrt[6]{200}$

Copyright © 2012 Pearson Education, Inc. Publishing as Addison-Wesley.

19. $3\sqrt{20} - 5\sqrt{80} + 4\sqrt{500}$

$\qquad = 3\sqrt{4 \cdot 5} - 5\sqrt{16 \cdot 5} + 4\sqrt{100 \cdot 5}$

$\qquad = 3 \cdot 2\sqrt{5} - 5 \cdot 4\sqrt{5} + 4 \cdot 10\sqrt{5}$

$\qquad = 6\sqrt{5} - 20\sqrt{5} + 40\sqrt{5} = 26\sqrt{5}$

20. $\sqrt[3]{16t^3s^5} - \sqrt[3]{54t^6s^2}$

$\qquad = \sqrt[3]{8t^3s^3} \cdot \sqrt[3]{2s^2} - \sqrt[3]{27t^6} \cdot \sqrt[3]{2s^2}$

$\qquad = 2ts\sqrt[3]{2s^2} - 3t^2\sqrt[3]{2s^2}$

$\qquad = (2ts - 3t^2)\sqrt[3]{2s^2}, \text{ or } t(2s - 3t)\sqrt[3]{2s^2}$

21. $(7\sqrt{5} + 4)(2\sqrt{5} - 1)$

$\qquad = 14 \cdot 5 - 7\sqrt{5} + 8\sqrt{5} - 4$

$\qquad = 70 + \sqrt{5} - 4 = 66 + \sqrt{5}$

22. $(\sqrt{3} - 2\sqrt{5})^2$

$\qquad = (\sqrt{3})^2 - 2 \cdot \sqrt{3} \cdot 2\sqrt{5} + (2\sqrt{5})^2$

$\qquad = 3 - 4\sqrt{15} + 4 \cdot 5$

$\qquad = 3 - 4\sqrt{15} + 20 = 23 - 4\sqrt{15}$

23. $\dfrac{-5}{\sqrt{40}} = \dfrac{-5}{\sqrt{4 \cdot 10}} = \dfrac{-5}{2\sqrt{10}} = \dfrac{-5 \cdot \sqrt{10}}{2\sqrt{10} \cdot \sqrt{10}}$

$\qquad = \dfrac{-5\sqrt{10}}{2 \cdot 10} = \dfrac{-5\sqrt{10}}{20} = -\dfrac{\sqrt{10}}{4}$

24. $\dfrac{2}{\sqrt[3]{5}} = \dfrac{2 \cdot \sqrt[3]{5^2}}{\sqrt[3]{5}\sqrt[3]{5^2}} = \dfrac{2\sqrt[3]{25}}{5}$

25. $\dfrac{-4}{\sqrt{7} + \sqrt{5}}$

Multiply the numerator and denominator by the conjugate of the denominator, $\sqrt{7} - \sqrt{5}$.

$\qquad = \dfrac{-4(\sqrt{7} - \sqrt{5})}{(\sqrt{7} + \sqrt{5})(\sqrt{7} - \sqrt{5})}$

$\qquad = \dfrac{-4(\sqrt{7} - \sqrt{5})}{7 - 5}$

$\qquad = \dfrac{-4(\sqrt{7} - \sqrt{5})}{2} = -2(\sqrt{7} - \sqrt{5})$

26. $\dfrac{6 + \sqrt{24}}{2} = \dfrac{6 + \sqrt{4 \cdot 6}}{2} = \dfrac{6 + 2\sqrt{6}}{2}$

$\qquad = \dfrac{2(3 + \sqrt{6})}{2} = 3 + \sqrt{6}$

27. **(a)** Substitute 50 for V_0, 0.01 for k, and 30 for T in the formula.

$\qquad V = \dfrac{V_0}{\sqrt{1 - kT}}$

$\qquad V = \dfrac{50}{\sqrt{1 - (0.01)(30)}}$

$\qquad = \dfrac{50}{\sqrt{1 - 0.3}} = \dfrac{50}{\sqrt{0.7}} \approx 59.8$

The velocity is about 59.8.

(b) $\qquad V = \dfrac{V_0}{\sqrt{1 - kT}}$

$\qquad V^2 = \dfrac{V_0^2}{1 - kT} \qquad$ *Square.*

$\qquad V^2(1 - kT) = V_0^2$

$\qquad V^2 - V^2kT = V_0^2$

$\qquad V^2 - V_0^2 = V^2kT$

$\qquad T = \dfrac{V^2 - V_0^2}{V^2k}$

or $\qquad T = \dfrac{V_0^2 - V^2}{-V^2k}$

28. $\sqrt[3]{5x} = \sqrt[3]{2x - 3}$

$\qquad (\sqrt[3]{5x})^3 = (\sqrt[3]{2x - 3})^3$

$\qquad 5x = 2x - 3$

$\qquad 3x = -3$

$\qquad x = -1$

Check $x = -1$: $\sqrt[3]{-5} = \sqrt[3]{-5}$ *True*
The solution set is $\{-1\}$.

29. $x + \sqrt{x + 6} = 9 - x$
Isolate the radical.

$\qquad \sqrt{x + 6} = 9 - 2x$

$\qquad (\sqrt{x + 6})^2 = (9 - 2x)^2 \qquad$ *Square.*

$\qquad x + 6 = 81 - 36x + 4x^2$

$\qquad 0 = 4x^2 - 37x + 75$

$\qquad 0 = (x - 3)(4x - 25)$

$\qquad x - 3 = 0 \quad \text{or} \quad 4x - 25 = 0$

$\qquad x = 3 \qquad\qquad x = \frac{25}{4}$

Check $x = 3$: $3 + \sqrt{9} = 6$ *True*
Check $x = \frac{25}{4}$: $\frac{25}{4} + \sqrt{\frac{49}{4}} = \frac{11}{4}$ *False*
The solution set is $\{3\}$.

30. $\sqrt{x + 4} - \sqrt{1 - x} = -1$

$\qquad \sqrt{x + 4} = -1 + \sqrt{1 - x}$

$\qquad (\sqrt{x + 4})^2 = (-1 + \sqrt{1 - x})^2$

$\qquad x + 4 = 1 - 2\sqrt{1 - x} + 1 - x$

$\qquad 2x + 2 = -2\sqrt{1 - x}$

$\qquad x + 1 = -\sqrt{1 - x}$

$\qquad (x + 1)^2 = (-\sqrt{1 - x})^2$

$\qquad x^2 + 2x + 1 = 1 - x$

$\qquad x^2 + 3x = 0$

$\qquad x(x + 3) = 0$

$\qquad x = 0 \quad \text{or} \quad x + 3 = 0$

$\qquad\qquad\qquad\qquad x = -3$

Check $x = -3$: $1 - 2 = -1$ *True*
Check $x = 0$: $2 - 1 = -1$ *False*
The solution set is $\{-3\}$.

Copyright © 2012 Pearson Education, Inc. Publishing as Addison-Wesley.

31. $(-2 + 5i) - (3 + 6i) - 7i$
$= (-2 - 3) + (5 - 6 - 7)i$
$= -5 - 8i$

32. $(1 + 5i)(3 + i)$
$= 3 + i + 15i + 5i^2$
$= 3 + 16i + 5(-1)$
$= -2 + 16i$

33. $\dfrac{7 + i}{1 - i}$
Multiply the numerator and denominator
by the conjugate of the denominator, $1 + i$.
$= \dfrac{(7 + i)(1 + i)}{(1 - i)(1 + i)}$
$= \dfrac{7 + 7i + i + i^2}{1 - i^2}$
$= \dfrac{7 + 8i - 1}{1 - (-1)}$
$= \dfrac{6 + 8i}{2} = \dfrac{2(3 + 4i)}{2} = 3 + 4i$

34. $i^{37} = i^{36} \cdot i = (i^4)^9 \cdot i = 1^9 \cdot i = 1 \cdot i = i$

35. **(a)** $i^2 = -1$ is *true*.

(b) $i = \sqrt{-1}$ is *true*.

(c) $i = -1$ is *false*; $i = \sqrt{-1}$.

(d) $\sqrt{-3} = i\sqrt{3}$ is *true*.

Cumulative Review Exercises (Chapters 1–8)

In Exercises 1 and 2, $a = -3$, $b = 5$, and $c = -4$.

1. $|2a^2 - 3b + c|$
$= |2(-3)^2 - 3(5) + (-4)|$
$= |2(9) - 15 - 4|$
$= |18 - 15 - 4| = |-1| = 1$

2. $\dfrac{(a + b)(a + c)}{3b - 6}$
$= \dfrac{(-3 + 5)[(-3 + (-4)]}{3(5) - 6}$
$= \dfrac{(2)(-7)}{15 - 6} = -\dfrac{14}{9}$

3. $3(x + 2) - 4(2x + 3) = -3x + 2$
$3x + 6 - 8x - 12 = -3x + 2$
$-5x - 6 = -3x + 2$
$-2x = 8$
$x = -4$

Check $x = -4$: $-6 + 20 = 14$ *True*
The solution set is $\{-4\}$.

4. $\frac{1}{3}x + \frac{1}{4}(x + 8) = x + 7$
Multiply by the LCD, 12.
$12[\frac{1}{3}x + \frac{1}{4}(x + 8)] = 12(x + 7)$
$4x + 3(x + 8) = 12x + 84$
$4x + 3x + 24 = 12x + 84$
$7x + 24 = 12x + 84$
$-5x = 60$
$x = -12$

Check $x = -12$: $-4 - 1 = -5$ *True*
The solution set is $\{-12\}$.

5. $0.04x + 0.06(100 - x) = 5.88$
Multiply each side by 100 to clear decimals.
$4x + 6(100 - x) = 588$
$4x + 600 - 6x = 588$
$-2x = -12$
$x = 6$

Check $x = 6$: $0.24 + 5.64 = 5.88$ *True*
The solution set is $\{6\}$.

6. $|6x + 7| = 13$

$6x + 7 = 13$ or $6x + 7 = -13$
$6x = 6$ $\qquad\qquad 6x = -20$
$x = 1$ or $\qquad x = -\frac{20}{6} = -\frac{10}{3}$

Check $x = 1$: $\qquad |13| = 13$ *True*
Check $x = -\frac{10}{3}$: $|-13| = 13$ *True*
The solution set is $\left\{-\frac{10}{3}, 1\right\}$.

7. $-5 - 3(x - 2) < 11 - 2(x + 2)$
$-5 - 3x + 6 < 11 - 2x - 4$
$1 - 3x < 7 - 2x$
$-x < 6$
Multiply by -1; reverse the inequality.
$x > -6$

The solution set is $(-6, \infty)$.

8. Let $x =$ the number of nickels and then
$100 - x =$ the number of quarters.

Value of nickels	+	value of quarters	=	total value.
$0.05x$	+	$0.25(100 - x)$	=	17.80

Multiply by 100 to clear the decimals.

$5x + 25(100 - x) = 1780$
$5x + 2500 - 25x = 1780$
$-20x = -720$
$x = 36$

Since $x = 36$, $100 - x = 100 - 36 = 64$.

There are 36 nickels and 64 quarters.

Copyright © 2012 Pearson Education, Inc. Publishing as Addison-Wesley.

9. Let $x =$ the amount of pure alcohol.

Make a table.

Number of Liters	Percent (as a decimal)	Liters of Pure Alcohol
x	$100\% = 1$	$1 \cdot x = x$
40	$18\% = 0.18$	$0.18(40) = 7.2$
$x + 40$	$22\% = 0.22$	$0.22(x + 40)$

From the last column:

$$x + 7.2 = 0.22x + 8.8$$
$$0.78x = 1.6$$
$$x = \frac{1.6}{0.78} = \frac{160}{78} = \frac{80}{39} \quad \text{or} \quad 2\frac{2}{39}$$

The required amount is $\frac{80}{39}$ or $2\frac{2}{39}$ liters of pure alcohol.

10. $4x - 3y = 12$

Let $x = 0$ to find the y-intercept, $(0, -4)$.
Let $y = 0$ to find the x-intercept, $(3, 0)$.
Draw a line through the intercepts.

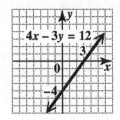

11. The given points are $(-4, 6)$ and $(2, -3)$.

$$m = \frac{-3 - 6}{2 - (-4)} = \frac{-9}{6} = -\frac{3}{2}$$

Use the point-slope form:

$$y - 6 = -\frac{3}{2}[x - (-4)]$$
$$y - 6 = -\frac{3}{2}(x + 4)$$
$$y - 6 = -\frac{3}{2}x - 6$$
$$y = -\frac{3}{2}x$$

12. $f(x) = 3x - 7$
$$f(-10) = 3(-10) - 7 = -30 - 7 = -37$$

13. $3x - y = 23$ (1)
$2x + 3y = 8$ (2)

To eliminate y, multiply equation (1) by 3 and add the result to equation (2).

$$\begin{array}{rcll} 9x - 3y &=& 69 & 3 \times (1) \\ 2x + 3y &=& 8 & (2) \\ \hline 11x &=& 77 & \\ x &=& 7 & \end{array}$$

Substitute 7 for x in (2),

$$2(7) + 3y = 8$$
$$14 + 3y = 8$$
$$3y = -6$$
$$y = -2$$

The solution set is $\{(7, -2)\}$.

14. $\begin{aligned} x + y + z &= 1 \\ x - y - z &= -3 \\ x + y - z &= -1 \end{aligned}$

Write the augmented matrix.

$$\begin{bmatrix} 1 & 1 & 1 & | & 1 \\ 1 & -1 & -1 & | & -3 \\ 1 & 1 & -1 & | & -1 \end{bmatrix}$$

$$\begin{bmatrix} 1 & 1 & 1 & | & 1 \\ 0 & -2 & -2 & | & -4 \\ 0 & 0 & -2 & | & -2 \end{bmatrix} \begin{array}{l} \\ -R_1 + R_2 \\ -R_1 + R_3 \end{array}$$

$$\begin{bmatrix} 1 & 1 & 1 & | & 1 \\ 0 & 1 & 1 & | & 2 \\ 0 & 0 & 1 & | & 1 \end{bmatrix} \begin{array}{l} \\ -\frac{1}{2}R_2 \\ -\frac{1}{2}R_3 \end{array}$$

This matrix gives the system

$$\begin{aligned} x + y + z &= 1 \quad (1) \\ y + z &= 2 \quad (2) \\ z &= 1. \end{aligned}$$

Substitute 1 for z in (2).

$$y + 1 = 2$$
$$y = 1$$

Substitute 1 for y and 1 for z in (1).

$$x + 1 + 1 = 1$$
$$x = -1$$

The solution set is $\{(-1, 1, 1)\}$.

15. Let $x =$ the number of 2-ounce letters and $y =$ the number of 3-ounce letters.

$$\begin{aligned} 5x + 3y &= 5.39 \quad (1) \\ 3x + 5y &= 5.73 \quad (2) \end{aligned}$$

To eliminate x, multiply (1) by -3 and (2) by 5 and add the results.

$$\begin{array}{rcll} -15x - 9y &=& -16.17 & -3 \times (1) \\ 15x + 25y &=& 28.65 & 5 \times (2) \\ \hline 16y &=& 12.48 & \\ y &=& 0.78 & \end{array}$$

Substitute $y = 0.78$ in (1).

$$\begin{aligned} 5x + 3y &= 5.39 \quad (1) \\ 5x + 3(0.78) &= 5.39 \\ 5x + 2.34 &= 5.39 \\ 5x &= 3.05 \\ x &= 0.61 \end{aligned}$$

The 2010 postage rate for a 2-ounce letter was $0.61 and for a 3-ounce letter, $0.78.

Copyright © 2012 Pearson Education, Inc. Publishing as Addison-Wesley.

16. $(3k^3 - 5k^2 + 8k - 2) - (4k^3 + 11k + 7)$
$\quad + (2k^2 - 5k)$
$\quad = 3k^3 - 4k^3 - 5k^2 + 2k^2$
$\qquad + 8k - 11k - 5k - 2 - 7$
$\quad = -k^3 - 3k^2 - 8k - 9$

17. $(8x - 7)(x + 3)$
$\quad = 8x^2 + 24x - 7x - 21$
$\quad = 8x^2 + 17x - 21$

18. $\dfrac{6y^4 - 3y^3 + 5y^2 + 6y - 9}{2y + 1}$

$$
\begin{array}{r}
3y^3 \;-\; 3y^2 \;+\; 4y \;+\; 1 \\
2y + 1\overline{\smash{\big)}\,6y^4 \;-\; 3y^3 \;+\; 5y^2 \;+\; 6y \;-\; 9} \\
\underline{6y^4 \;+\; 3y^3} \\
-\,6y^3 \;+\; 5y^2 \\
\underline{-\,6y^3 \;-\; 3y^2} \\
8y^2 \;+\; 6y \\
\underline{8y^2 \;+\; 4y} \\
2y \;-\; 9 \\
\underline{2y \;+\; 1} \\
-\,10
\end{array}
$$

The answer is

$$3y^3 - 3y^2 + 4y + 1 + \frac{-10}{2y + 1}.$$

19. $2p^2 - 5pq + 3q^2 = (2p - 3q)(p - q)$

20. $3k^4 + k^2 - 4 = (3k^2 + 4)(k^2 - 1)$
$\qquad\qquad\qquad\; = (3k^2 + 4)(k + 1)(k - 1)$

21. $x^3 + 512 = x^3 + 8^3$
$\qquad\qquad\; = (x + 8)(x^2 - 8x + 64)$

22. $2x^2 + 11x + 15 = 0$
$\quad (x + 3)(2x + 5) = 0$

$\qquad x + 3 = 0 \quad$ or $\quad 2x + 5 = 0$
$\qquad\quad x = -3 \quad$ or $\qquad\quad x = -\frac{5}{2}$

Check $x = -3$: $\quad 18 - 33 + 15 = 0 \quad$ *True*
Check $x = -\frac{5}{2}$: $\quad \frac{25}{2} - \frac{55}{2} + \frac{30}{2} = 0 \quad$ *True*
The solution set is $\left\{-3, -\frac{5}{2}\right\}$.

23. $\qquad 5x(x - 1) = 2(1 - x)$
$\qquad\quad 5x^2 - 5x = 2 - 2x$
$\quad 5x^2 - 3x - 2 = 0$
$\quad (5x + 2)(x - 1) = 0$

$\qquad 5x + 2 = 0 \quad$ or $\quad x - 1 = 0$
$\qquad\quad x = -\frac{2}{5} \quad$ or $\qquad x = 1$

Check $x = -\frac{2}{5}$: $\;\; -2\left(-\frac{7}{5}\right) = 2\left(\frac{7}{5}\right) \quad$ *True*
Check $x = 1$: $\qquad\quad 5(0) = 2(0) \quad$ *True*
The solution set is $\left\{-\frac{2}{5}, 1\right\}$.

24. $f(x) = \dfrac{4}{x^2 - 9} = \dfrac{4}{(x + 3)(x - 3)}$

The numbers -3 and 3 make the denominator 0 so they must be excluded from the set of all real numbers for the domain of f.
Domain: $\{x \mid x \neq \pm 3\}$

25. $\dfrac{y^2 + y - 12}{y^3 + 9y^2 + 20y} \div \dfrac{y^2 - 9}{y^3 + 3y^2}$

$\quad = \dfrac{y^2 + y - 12}{y(y^2 + 9y + 20)} \cdot \dfrac{y^3 + 3y^2}{y^2 - 9}$

$\quad = \dfrac{(y + 4)(y - 3)}{y(y + 4)(y + 5)} \cdot \dfrac{y^2(y + 3)}{(y + 3)(y - 3)}$

$\quad = \dfrac{y}{y + 5}$

26. $\dfrac{1}{x + y} + \dfrac{3}{x - y} \qquad$ *The LCD is*
$\qquad\qquad\qquad\qquad (x + y)(x - y).$

$\quad = \dfrac{1(x - y)}{(x + y)(x - y)} + \dfrac{3(x + y)}{(x - y)(x + y)}$

$\quad = \dfrac{(x - y) + 3(x + y)}{(x + y)(x - y)}$

$\quad = \dfrac{x - y + 3x + 3y}{(x + y)(x - y)}$

$\quad = \dfrac{4x + 2y}{(x + y)(x - y)}$

27. $\dfrac{\dfrac{-6}{x - 2}}{\dfrac{8}{3x - 6}} = \dfrac{-6}{x - 2} \div \dfrac{8}{3x - 6}$

$\qquad = \dfrac{-6}{x - 2} \cdot \dfrac{3x - 6}{8}$

$\qquad = \dfrac{-6}{x - 2} \cdot \dfrac{3(x - 2)}{8}$

$\qquad = \dfrac{-2 \cdot 3 \cdot 3}{2 \cdot 4} = -\dfrac{9}{4}$

28. $\dfrac{\dfrac{1}{a} - \dfrac{1}{b}}{\dfrac{a}{b} - \dfrac{b}{a}} \qquad$ *The LCD of both the numerator and the denominator is ab.*

$\quad = \dfrac{\dfrac{b - a}{ab}}{\dfrac{a^2 - b^2}{ab}}$

$\quad = \dfrac{b - a}{ab} \div \dfrac{a^2 - b^2}{ab}$

$\quad = \dfrac{b - a}{ab} \cdot \dfrac{ab}{a^2 - b^2}$

$\quad = \dfrac{b - a}{a^2 - b^2}$

$\quad = \dfrac{-(a - b)}{(a - b)(a + b)} = \dfrac{-1}{a + b}$

Copyright © 2012 Pearson Education, Inc. Publishing as Addison-Wesley.

29. $\dfrac{x^{-1}}{y - x^{-1}} = \dfrac{\frac{1}{x}}{y - \frac{1}{x}}$

Multiply the numerator and denominator by x.

$$= \dfrac{x\left(\dfrac{1}{x}\right)}{x\left(y - \dfrac{1}{x}\right)} = \dfrac{1}{xy - 1}$$

30. $\dfrac{x + 1}{x - 3} = \dfrac{4}{x - 3} + 6$

Multiply by the LCD, $x - 3$.

$$(x - 3)\left(\dfrac{x + 1}{x - 3}\right) = (x - 3)\left(\dfrac{4}{x - 3} + 6\right)$$
$$x + 1 = 4 + 6(x - 3)$$
$$x + 1 = 4 + 6x - 18$$
$$x + 1 = -14 + 6x$$
$$-5x = -15$$
$$x = 3$$

Substituting 3 in the original equation results in division by zero, so 3 cannot be a solution. The solution set is $\emptyset$.

31. Let $x =$ Richard's speed.
Then $x + 4 =$ Danielle's speed.

Use $d = rt$, or $t = \frac{d}{r}$, to make a table.

	Distance	Rate	Time
Richard	24	x	$\dfrac{24}{x}$
Danielle	48	$x + 4$	$\dfrac{48}{x + 4}$

Since the times are the same,

$$\dfrac{24}{x} = \dfrac{48}{x + 4}.$$

Multiply by the LCD, $x(x + 4)$.

$$x(x + 4)\left(\dfrac{24}{x}\right) = x(x + 4)\left(\dfrac{48}{x + 4}\right)$$
$$24(x + 4) = 48x$$
$$24x + 96 = 48x$$
$$-24x = -96$$
$$x = 4$$

Since $x = 4$, $x + 4 = 4 + 4 = 8$. Richard's speed is 4 mph; Danielle's speed is 8 mph.

32. $27^{-2/3} = \dfrac{1}{27^{2/3}} = \dfrac{1}{\left(27^{1/3}\right)^2} = \dfrac{1}{3^2} = \dfrac{1}{9}$

33. $\sqrt[3]{16x^2y} \cdot \sqrt[3]{3x^3y}$

$$= \sqrt[3]{8} \cdot \sqrt[3]{2x^2y} \cdot \sqrt[3]{x^3} \cdot \sqrt[3]{3y}$$
$$= 2\sqrt[3]{2x^2y} \cdot x\sqrt[3]{3y}$$
$$= 2x\sqrt[3]{6x^2y^2}$$

34. $\sqrt{50} + \sqrt{8} = \sqrt{25 \cdot 2} + \sqrt{4 \cdot 2}$
$$= 5\sqrt{2} + 2\sqrt{2} = 7\sqrt{2}$$

35. $\dfrac{1}{\sqrt{10} - \sqrt{8}}$

Multiply the numerator and denominator by the conjugate of the denominator, $\sqrt{10} + \sqrt{8}$.

$$= \dfrac{1\left(\sqrt{10} + \sqrt{8}\right)}{\left(\sqrt{10} - \sqrt{8}\right)\left(\sqrt{10} + \sqrt{8}\right)}$$
$$= \dfrac{\sqrt{10} + \sqrt{8}}{10 - 8} = \dfrac{\sqrt{10} + \sqrt{4 \cdot 2}}{2}$$
$$= \dfrac{\sqrt{10} + 2\sqrt{2}}{2}$$

36. The given points are $(-4, 4)$ and $(-2, 9)$.

$$d = \sqrt{[(-2) - (-4)]^2 + (9 - 4)^2}$$
$$= \sqrt{2^2 + 5^2}$$
$$= \sqrt{4 + 25} = \sqrt{29}$$

37. $\sqrt{3x - 8} = x - 2$

Square each side.

$$\left(\sqrt{3x - 8}\right)^2 = (x - 2)^2$$
$$3x - 8 = x^2 - 4x + 4$$
$$0 = x^2 - 7x + 12$$
$$0 = (x - 3)(x - 4)$$

$$x - 3 = 0 \quad \text{or} \quad x - 4 = 0$$
$$x = 3 \quad \text{or} \quad x = 4$$

Check $x = 3$: $\sqrt{1} = 1$ *True*
Check $x = 4$: $\sqrt{4} = 2$ *True*
The solution set is $\{3, 4\}$.

38. $\dfrac{6 - 2i}{1 - i}$

Multiply the numerator and denominator by the conjugate of the denominator, $1 + i$.

$$= \dfrac{(6 - 2i)(1 + i)}{(1 - i)(1 + i)}$$
$$= \dfrac{6 + 6i - 2i - 2i^2}{1^2 + 1^2}$$
$$= \dfrac{6 + 4i + 2}{1 + 1}$$
$$= \dfrac{8 + 4i}{2} = \dfrac{2(4 + 2i)}{2} = 4 + 2i$$

Copyright © 2012 Pearson Education, Inc. Publishing as Addison-Wesley.

CHAPTER 9   QUADRATIC EQUATIONS AND INEQUALITIES

9.1   The Square Root Property and Completing the Square

9.1   Now Try Exercises

N1. $2x^2 + 5x - 12 = 0$
$(2x - 3)(x + 4) = 0$

$$2x - 3 = 0 \quad \text{or} \quad x + 4 = 0$$
$$x = \tfrac{3}{2} \quad \text{or} \quad x = -4$$

Solution set: $\left\{-4, \tfrac{3}{2}\right\}$

N2.   **(a)** $t^2 = 10$

By the square root property,

$$t = \sqrt{10} \quad \text{or} \quad t = -\sqrt{10}.$$

Solution set: $\left\{-\sqrt{10}, \sqrt{10}\right\}$

(b) $2x^2 - 90 = 0$
$$2x^2 = 90$$
$$x^2 = 45$$

By the square root property,

$$x = \sqrt{45} \quad \text{or} \quad x = -\sqrt{45}$$
$$x = 3\sqrt{5} \quad \text{or} \quad x = -3\sqrt{5}$$

Check $x = \pm 3\sqrt{5}$: $2(45) - 90 = 0$ *True*

Solution set: $\left\{3\sqrt{5}, -3\sqrt{5}\right\}$

N3. $d = 16t^2$
$25 = 16t^2$ *Let d = 25.*
$\tfrac{25}{16} = t^2$

By the square root property,

$$t = \tfrac{5}{4} \quad \text{or} \quad t = -\tfrac{5}{4}.$$

Time cannot be negative, so we discard the negative solution. Therefore, 1.25 sec elapses between the dropping of the coin and the shot.

N4. $(x + 3)^2 = 25$

$$x + 3 = \sqrt{25} \quad \text{or} \quad x + 3 = -\sqrt{25}$$
$$x = -3 + 5 \quad \text{or} \quad x = -3 - 5$$
$$x = 2 \quad \text{or} \quad x = -8$$

Check $x = 2$: $(5)^2 = 25$ *True*
Check $x = -8$: $(-5)^2 = 25$ *True*

Solution set: $\{-8, 2\}$

N5. $(5x - 4)^2 = 27$

$$5x - 4 = \sqrt{27} \quad \text{or} \quad 5x - 4 = -\sqrt{27}$$
$$5x = 4 + 3\sqrt{3} \quad \text{or} \quad 5x = 4 - 3\sqrt{3}$$
$$x = \frac{4 + 3\sqrt{3}}{5} \quad \text{or} \quad x = \frac{4 - 3\sqrt{3}}{5}$$

Check $x = \dfrac{4 + 3\sqrt{3}}{5}$:

$(5x - 4)^2 = 27$ *Original equation*

$\left[5\left(\dfrac{4 + 3\sqrt{3}}{5}\right) - 4\right]^2 \overset{?}{=} 27$ *Let x = $\dfrac{4 + 3\sqrt{3}}{5}$.*

$\left(4 + 3\sqrt{3} - 4\right)^2 \overset{?}{=} 27$ *Multiply.*

$\left(3\sqrt{3}\right)^2 \overset{?}{=} 27$ *Simplify.*

$27 = 27$ *True*

The check for the other solution is similar.

Solution set: $\left\{\dfrac{4 \pm 3\sqrt{3}}{5}\right\}$

N6. $x^2 + 6x - 2 = 0$
$$x^2 + 6x = 2$$

Complete the square.

$$\left[\tfrac{1}{2}(6)\right]^2 = (3)^2 = 9$$

Add 9 to each side.

$$x^2 + 6x + 9 = 2 + 9$$
$$(x + 3)^2 = 11$$

Use the square root property.

$$x + 3 = \sqrt{11} \quad \text{or} \quad x + 3 = -\sqrt{11}$$
$$x = -3 + \sqrt{11} \quad \text{or} \quad x = -3 - \sqrt{11}$$

Check $x = -3 + \sqrt{11}$:
$$\left(-3 + \sqrt{11}\right)^2 + 6\left(-3 + \sqrt{11}\right) - 2 \overset{?}{=} 0$$
$$9 - 6\sqrt{11} + 11 - 18 + 6\sqrt{11} - 2 = 0 \quad \text{*True*}$$

Solution set: $\left\{-3 \pm \sqrt{11}\right\}$

N7. $x^2 + x - 3 = 0$
$$x^2 + x = 3$$

Complete the square.

$$\left[\tfrac{1}{2}(1)\right]^2 = \left(\tfrac{1}{2}\right)^2 = \tfrac{1}{4}$$

Add $\tfrac{1}{4}$ to each side.

$$x^2 + x + \tfrac{1}{4} = 3 + \tfrac{1}{4}$$
$$\left(x + \tfrac{1}{2}\right)^2 = \tfrac{13}{4}$$

Use the square root property.

Copyright © 2012 Pearson Education, Inc. Publishing as Addison-Wesley.

$$x + \tfrac{1}{2} = \sqrt{\tfrac{13}{4}} \qquad \text{or} \quad x + \tfrac{1}{2} = -\sqrt{\tfrac{13}{4}}$$

$$x + \tfrac{1}{2} = \tfrac{\sqrt{13}}{2} \qquad \text{or} \quad x + \tfrac{1}{2} = -\tfrac{\sqrt{13}}{2}$$

$$x = \frac{-1 + \sqrt{13}}{2} \quad \text{or} \qquad x = \frac{-1 - \sqrt{13}}{2}$$

Check that the solution set is $\left\{ \frac{-1 \pm \sqrt{13}}{2} \right\}$.

N8. $3x^2 + 12x - 5 = 0$

$$3x^2 + 12x = 5$$

$$x^2 + 4x = \tfrac{5}{3}$$

Complete the square.

$$\left[\tfrac{1}{2}(4) \right]^2 = (2)^2 = 4$$

Add 4 to each side.

$$x^2 + 4x + 4 = \tfrac{5}{3} + 4$$

$$(x + 2)^2 = \tfrac{17}{3}$$

Use the square root property.

$$x + 2 = \sqrt{\tfrac{17}{3}} \qquad \text{or} \quad x + 2 = -\sqrt{\tfrac{17}{3}}$$

$$x = -2 + \sqrt{\tfrac{17}{3}} \quad \text{or} \quad x = -2 - \sqrt{\tfrac{17}{3}}$$

$$x = -2 + \frac{\sqrt{51}}{3} \quad \text{or} \quad x = -2 - \frac{\sqrt{51}}{3}$$

$$x = \frac{-6 + \sqrt{51}}{3} \quad \text{or} \quad x = \frac{-6 - \sqrt{51}}{3}$$

Check that the solution set is $\left\{ \frac{-6 \pm \sqrt{51}}{3} \right\}$.

N9. **(a)** $t^2 = -24$

$$t = \sqrt{-24} \quad \text{or} \quad t = -\sqrt{-24}$$

$$t = 2i\sqrt{6} \quad \text{or} \quad t = -2i\sqrt{6}$$

Solution set: $\left\{ \pm 2i\sqrt{6} \right\}$

(b) $(x + 4)^2 = -36$

$$x + 4 = \sqrt{-36} \quad \text{or} \quad x + 4 = -\sqrt{-36}$$

$$x + 4 = 6i \qquad \text{or} \quad x + 4 = -6i$$

$$x = -4 + 6i \quad \text{or} \qquad x = -4 - 6i$$

Solution set: $\{-4 \pm 6i\}$

(c) $x^2 + 8x + 21 = 0$

$$x^2 + 8x = -21$$

Complete the square.

$$\left[\tfrac{1}{2}(8) \right]^2 = (4)^2 = 16$$

Add 16 to each side.

$$x^2 + 8x + 16 = -21 + 16$$

$$(x + 4)^2 = -5$$

$$x + 4 = \sqrt{-5} \qquad \text{or} \quad x + 4 = -\sqrt{-5}$$

$$x + 4 = i\sqrt{5} \qquad \text{or} \quad x + 4 = -i\sqrt{5}$$

$$x = -4 + i\sqrt{5} \quad \text{or} \qquad x = -4 - i\sqrt{5}$$

Solution set: $\left\{ -4 \pm i\sqrt{5} \right\}$

9.1 Section Exercises

1. **B** and **C** are quadratic equations because they are equations with a squared term and no terms of higher degree.

3. The zero-factor property requires a product equal to 0. The first step should have been to rewrite the equation with 0 on one side.

5. $x^2 + 3x + 2 = 0$

$$(x + 1)(x + 2) = 0$$

$$x + 1 = 0 \qquad \text{or} \qquad x + 2 = 0$$

$$x = -1 \qquad \text{or} \qquad x = -2$$

Solution set: $\{-2, -1\}$

7. $3x^2 + 8x - 3 = 0$

$$(x + 3)(3x - 1) = 0$$

$$x + 3 = 0 \qquad \text{or} \quad 3x - 1 = 0$$

$$x = -3 \qquad \text{or} \qquad x = \tfrac{1}{3}$$

Solution set: $\left\{ -3, \tfrac{1}{3} \right\}$

9. $2x^2 = 9x - 4$

$$2x^2 - 9x + 4 = 0$$

$$(2x - 1)(x - 4) = 0$$

$$2x - 1 = 0 \qquad \text{or} \qquad x - 4 = 0$$

$$x = \tfrac{1}{2} \qquad \text{or} \qquad x = 4$$

Solution set: $\left\{ \tfrac{1}{2}, 4 \right\}$

11. $x^2 = 81$

$$x = 9 \quad \text{or} \quad x = -9$$

Solution set: $\{\pm 9\}$

13. $x^2 = 17$

$$x = \sqrt{17} \quad \text{or} \quad x = -\sqrt{17}$$

Solution set: $\left\{ \pm\sqrt{17} \right\}$

15. $x^2 = 32$

$$x = \sqrt{32} \quad \text{or} \quad x = -\sqrt{32}$$

$$x = 4\sqrt{2} \quad \text{or} \quad x = -4\sqrt{2}$$

Solution set: $\left\{ \pm 4\sqrt{2} \right\}$

Copyright © 2012 Pearson Education, Inc. Publishing as Addison-Wesley.

17. $x^2 - 20 = 0$

$x^2 = 20$

$x = \sqrt{20}$ or $x = -\sqrt{20}$

$x = 2\sqrt{5}$ or $x = -2\sqrt{5}$

Solution set: $\left\{ \pm 2\sqrt{5} \right\}$

19. $3x^2 - 72 = 0$

$3x^2 = 72$

$x = 24$

$x = \sqrt{24}$ or $x = -\sqrt{24}$

$x = 2\sqrt{6}$ or $x = -2\sqrt{6}$

Solution set: $\left\{ \pm 2\sqrt{6} \right\}$

21. $(x + 2)^2 = 25$

$x + 2 = \sqrt{25}$ or $x + 2 = -\sqrt{25}$

$x + 2 = 5$ $\qquad x + 2 = -5$

$x = 3$ or $\qquad x = -7$

Solution set: $\{-7, 3\}$

23. $(x - 6)^2 = 49$

$x - 6 = \sqrt{49}$ or $x - 6 = -\sqrt{49}$

$x = 6 + 7$ or $\qquad x = 6 - 7$

$x = 13$ or $\qquad x = -1$

Solution set: $\{-1, 13\}$

25. $(x - 4)^2 = 3$

$x - 4 = \sqrt{3}$ or $x - 4 = -\sqrt{3}$

$x = 4 + \sqrt{3}$ or $\qquad x = 4 - \sqrt{3}$

Solution set: $\left\{ 4 \pm \sqrt{3} \right\}$

27. $(t + 5)^2 = 48$

$t + 5 = \sqrt{48}$ or $t + 5 = -\sqrt{48}$

$t + 5 = 4\sqrt{3}$ $\qquad t + 5 = -4\sqrt{3}$

$t = -5 + 4\sqrt{3}$ or $\qquad t = -5 - 4\sqrt{3}$

Solution set: $\left\{ -5 \pm 4\sqrt{3} \right\}$

29. $(3x - 1)^2 = 7$

$3x - 1 = \sqrt{7}$ or $3x - 1 = -\sqrt{7}$

$3x = 1 + \sqrt{7}$ $\qquad 3x = 1 - \sqrt{7}$

$x = \dfrac{1 + \sqrt{7}}{3}$ or $\qquad x = \dfrac{1 - \sqrt{7}}{3}$

Solution set: $\left\{ \frac{1 \pm \sqrt{7}}{3} \right\}$

31. $(4p + 1)^2 = 24$

$4p + 1 = \sqrt{24}$ or $4p + 1 = -\sqrt{24}$

$4p + 1 = 2\sqrt{6}$ $\qquad 4p + 1 = -2\sqrt{6}$

$4p = -1 + 2\sqrt{6}$ $\qquad 4p = -1 - 2\sqrt{6}$

$p = \dfrac{-1 + 2\sqrt{6}}{4}$ or $\qquad p = \dfrac{-1 - 2\sqrt{6}}{4}$

Solution set: $\left\{ \frac{-1 \pm 2\sqrt{6}}{4} \right\}$

33. $(2 - 5t)^2 = 12$

$2 - 5t = \sqrt{12}$ or $2 - 5t = -\sqrt{12}$

$-5t = -2 + \sqrt{12}$ or $\qquad -5t = -2 - \sqrt{12}$

$5t = 2 - \sqrt{12}$ or $\qquad 5t = 2 + \sqrt{12}$

$t = \dfrac{2 - 2\sqrt{3}}{5}$ or $\qquad t = \dfrac{2 + 2\sqrt{3}}{5}$

Solution set: $\left\{ \frac{2 \pm 2\sqrt{3}}{5} \right\}$

35. $d = 16t^2$

$500 = 16t^2$

$t^2 = \frac{500}{16} = 31.25$

$t = \sqrt{31.25}$ or $t = -\sqrt{31.25}$

$t \approx 5.6$ or $t \approx -5.6$

It would take the rock about 5.6 seconds (time must be positive) to fall to the ground.

37. Solve $(2x + 1)^2 = 5$ by the square root property. Solve $x^2 + 4x = 12$ by completing the square.

39. $x^2 + 6x + \underline{\quad}$

We need to add the square of half the coefficient of x to get a perfect square trinomial.

$\frac{1}{2}(6) = 3$ and $3^2 = 9$

Add 9 to $x^2 + 6x$ to get a perfect square trinomial.

$x^2 + 6x + 9 = (x + 3)^2$

41. $p^2 - 12p + \underline{\quad}$

$\frac{1}{2}(-12) = -6$ and $(-6)^2 = 36$

$p^2 - 12p + 36 = (p - 6)^2$

43. $q^2 + 9q + \underline{\quad}$

$\frac{1}{2}(9) = \frac{9}{2}$ and $\left(\frac{9}{2} \right)^2 = \frac{81}{4}$

$q^2 + 9q + \frac{81}{4} = \left(q + \frac{9}{2} \right)^2$

45. $x^2 + \frac{1}{4}x + \underline{\quad}$

$\frac{1}{2}\left(\frac{1}{4} \right) = \frac{1}{8}$ and $\left(\frac{1}{8} \right)^2 = \frac{1}{64}$

$x^2 + \frac{1}{4}x + \frac{1}{64} = \left(x + \frac{1}{8} \right)^2$

47. $x^2 - 0.8x + \underline{\quad}$

$\frac{1}{2}(-0.8) = -0.4$ and $(-0.4)^2 = 0.16$

$x^2 - 0.8x + 0.16 = (x - 0.4)^2$

Copyright © 2012 Pearson Education, Inc. Publishing as Addison-Wesley.

49. $x^2 + 4x - 2 = 0$

$\quad x^2 + 4x = 2$

$\left[\frac{1}{2}(4)\right]^2 = 2^2 = 4$

51. $x^2 + 10x + 18 = 0$

$\quad x^2 + 10x = -18$

$\left[\frac{1}{2}(10)\right]^2 = 5^2 = 25$

53. $3w^2 - w - 24 = 0$

$\quad w^2 - \frac{1}{3}w - 8 = 0 \quad$ *Divide by 3.*

$\quad\quad w^2 - \frac{1}{3}w = 8$

$\left[\frac{1}{2}\left(-\frac{1}{3}\right)\right]^2 = \left(-\frac{1}{6}\right)^2 = \frac{1}{36}$

55. $x^2 - 2x - 24 = 0$

Get the variable terms alone on the left side.

$$x^2 - 2x = 24$$

Complete the square by taking half of -2, the coefficient of x, and squaring the result.

$$\left[\frac{1}{2}(-2)\right]^2 = (-1)^2 = 1$$

Add 1 to each side.

$$x^2 - 2x + 1 = 24 + 1$$

Factor the left side.

$$(x - 1)^2 = 25$$

Use the square root property.

$\quad x - 1 = \sqrt{25} \quad$ or $\quad x - 1 = -\sqrt{25}$

$\quad x - 1 = 5 \quad\quad$ or $\quad x - 1 = -5$

$\quad\quad x = 6 \quad\quad$ or $\quad\quad x = -4$

Solution set: $\{-4, 6\}$

57. $x^2 + 4x - 2 = 0$

$\quad x^2 + 4x = 2$

$\quad x^2 + 4x + 4 = 2 + 4 \quad\quad \left[\frac{1}{2}(4)\right]^2 = 4$

$\quad\quad (x + 2)^2 = 6$

$\quad x + 2 = \sqrt{6} \quad$ or $\quad x + 2 = -\sqrt{6}$

$\quad\quad x = -2 + \sqrt{6} \quad$ or $\quad x = -2 - \sqrt{6}$

Solution set: $\left\{-2 \pm \sqrt{6}\right\}$

59. $x^2 + 10x + 18 = 0$

$\quad x^2 + 10x = -18$

$\quad x^2 + 10x + 25 = -18 + 25 \quad \left[\frac{1}{2}(10)\right]^2 = 25$

$\quad\quad (x + 5)^2 = 7$

$\quad x + 5 = \sqrt{7} \quad$ or $\quad x + 5 = -\sqrt{7}$

$\quad\quad x = -5 + \sqrt{7} \quad$ or $\quad x = -5 - \sqrt{7}$

Solution set: $\left\{-5 \pm \sqrt{7}\right\}$

61. $\quad 3w^2 - w = 24$

$\quad w^2 - \frac{1}{3}w = 8 \quad\quad$ *Divide by 3.*

Complete the square by taking half of $\frac{1}{3}$, the coefficient of w, and squaring the result.

$$\left[\frac{1}{2}\left(-\frac{1}{3}\right)\right]^2 = \left(-\frac{1}{6}\right)^2 = \frac{1}{36}$$

Add $\frac{1}{36}$ to each side.

$\quad w^2 - \frac{1}{3}w + \frac{1}{36} = 8 + \frac{1}{36}$

$\quad\quad \left(w - \frac{1}{6}\right)^2 = \frac{288}{36} + \frac{1}{36}$

$\quad\quad \left(w - \frac{1}{6}\right)^2 = \frac{289}{36}$

$\quad w - \frac{1}{6} = \sqrt{\frac{289}{36}} \quad$ or $\quad w - \frac{1}{6} = -\sqrt{\frac{289}{36}}$

$\quad\quad w = \frac{1}{6} + \frac{\sqrt{289}}{\sqrt{36}} \quad\quad w = \frac{1}{6} - \frac{\sqrt{289}}{\sqrt{36}}$

$\quad\quad w = \frac{1}{6} + \frac{17}{6} \quad\quad\quad w = \frac{1}{6} - \frac{17}{6}$

$\quad\quad w = \frac{18}{6} \quad\quad\quad\quad w = -\frac{16}{6}$

$\quad\quad w = 3 \quad\quad$ or $\quad\quad w = -\frac{8}{3}$

Solution set: $\left\{-\frac{8}{3}, 3\right\}$

63. $x^2 + 7x - 1 = 0$

$\quad x^2 + 7x = 1$

$\quad x^2 + 7x + \frac{49}{4} = 1 + \frac{49}{4} \quad \left[\frac{1}{2}(7)^2\right] = \frac{49}{4}$

$\quad\quad \left(x + \frac{7}{2}\right)^2 = \frac{53}{4}$

$\quad x + \frac{7}{2} = \sqrt{\frac{53}{4}} \quad$ or $\quad x + \frac{7}{2} = -\sqrt{\frac{53}{4}}$

$\quad\quad x = -\frac{7}{2} + \frac{\sqrt{53}}{2} \quad\quad x = -\frac{7}{2} - \frac{\sqrt{53}}{2}$

$\quad\quad x = \frac{-7+\sqrt{53}}{2} \quad$ or $\quad x = \frac{-7-\sqrt{53}}{2}$

Solution set: $\left\{\frac{-7\pm\sqrt{53}}{2}\right\}$

65. $2k^2 + 5k - 2 = 0$

$\quad 2k^2 + 5k = 2$

$\quad k^2 + \frac{5}{2}k = 1 \quad\quad$ *Divide by 2.*

Complete the square.

$$\left(\frac{1}{2} \cdot \frac{5}{2}\right)^2 = \left(\frac{5}{4}\right)^2 = \frac{25}{16}$$

Add $\frac{25}{16}$ to each side.

$\quad k^2 + \frac{5}{2}k + \frac{25}{16} = 1 + \frac{25}{16}$

$\quad\quad \left(k + \frac{5}{4}\right)^2 = \frac{41}{16}$

$\quad k + \frac{5}{4} = \sqrt{\frac{41}{16}} \quad$ or $\quad k + \frac{5}{4} = -\sqrt{\frac{41}{16}}$

$\quad\quad k = -\frac{5}{4} + \frac{\sqrt{41}}{4} \quad\quad k = -\frac{5}{4} - \frac{\sqrt{41}}{4}$

$\quad\quad k = \frac{-5+\sqrt{41}}{4} \quad$ or $\quad k = \frac{-5-\sqrt{41}}{4}$

Solution set: $\left\{\frac{-5\pm\sqrt{41}}{4}\right\}$

Copyright © 2012 Pearson Education, Inc. Publishing as Addison-Wesley.

67. $5x^2 - 10x + 2 = 0$

$$5x^2 - 10x = -2$$
$$x^2 - 2x = -\tfrac{2}{5} \qquad \textit{Divide by 5.}$$

Complete the square.

$$\left[\tfrac{1}{2}(-2)\right]^2 = (-1)^2 = 1$$

Add 1 to each side.

$$x^2 - 2x + 1 = -\tfrac{2}{5} + 1$$
$$(x-1)^2 = \tfrac{3}{5}$$

$$x - 1 = \sqrt{\tfrac{3}{5}} \qquad \text{or} \qquad x - 1 = -\sqrt{\tfrac{3}{5}}$$
$$x - 1 = \tfrac{\sqrt{3}}{\sqrt{5}} \cdot \tfrac{\sqrt{5}}{\sqrt{5}} \qquad\qquad x - 1 = -\tfrac{\sqrt{3}}{\sqrt{5}} \cdot \tfrac{\sqrt{5}}{\sqrt{5}}$$
$$x = \tfrac{5}{5} + \tfrac{\sqrt{15}}{5} \qquad\qquad x = \tfrac{5}{5} - \tfrac{\sqrt{15}}{5}$$
$$x = \tfrac{5+\sqrt{15}}{5} \qquad \text{or} \qquad x = \tfrac{5-\sqrt{15}}{5}$$

Solution set: $\left\{ \tfrac{5 \pm \sqrt{15}}{5} \right\}$

69. $9x^2 - 24x = -13$

$$x^2 - \tfrac{24}{9}x = \tfrac{-13}{9} \qquad \textit{Divide by 9.}$$
$$x^2 - \tfrac{8}{3}x = \tfrac{-13}{9}$$

Complete the square.

$$\left[\tfrac{1}{2}\left(-\tfrac{8}{3}\right)\right]^2 = \left(-\tfrac{4}{3}\right)^2 = \tfrac{16}{9}$$

Add $\tfrac{16}{9}$ to each side.

$$x^2 - \tfrac{8}{3}x + \tfrac{16}{9} = \tfrac{-13}{9} + \tfrac{16}{9}$$
$$\left(x - \tfrac{4}{3}\right)^2 = \tfrac{3}{9}$$

$$x - \tfrac{4}{3} = \sqrt{\tfrac{3}{9}} \qquad \text{or} \qquad x - \tfrac{4}{3} = -\sqrt{\tfrac{3}{9}}$$
$$x = \tfrac{4}{3} + \tfrac{\sqrt{3}}{3} \qquad\qquad x = \tfrac{4}{3} - \tfrac{\sqrt{3}}{3}$$
$$x = \tfrac{4+\sqrt{3}}{3} \qquad \text{or} \qquad x = \tfrac{4-\sqrt{3}}{3}$$

Solution set: $\left\{ \tfrac{4 \pm \sqrt{3}}{3} \right\}$

71. $z^2 - \tfrac{4}{3}z = -\tfrac{1}{9}$

Complete the square.

$$\left[\tfrac{1}{2}\left(-\tfrac{4}{3}\right)\right]^2 = \left(-\tfrac{2}{3}\right)^2 = \tfrac{4}{9}$$

Add $\tfrac{4}{9}$ to each side.

$$z^2 - \tfrac{4}{3}z + \tfrac{4}{9} = -\tfrac{1}{9} + \tfrac{4}{9}$$
$$\left(z - \tfrac{2}{3}\right)^2 = \tfrac{3}{9}$$

$$z - \tfrac{2}{3} = \sqrt{\tfrac{3}{9}} \qquad \text{or} \qquad z - \tfrac{2}{3} = -\sqrt{\tfrac{3}{9}}$$
$$z = \tfrac{2}{3} + \tfrac{\sqrt{3}}{3} \qquad\qquad z = \tfrac{2}{3} - \tfrac{\sqrt{3}}{3}$$
$$z = \tfrac{2+\sqrt{3}}{3} \qquad \text{or} \qquad z = \tfrac{2-\sqrt{3}}{3}$$

Solution set: $\left\{ \tfrac{2 \pm \sqrt{3}}{3} \right\}$

73. $0.1x^2 - 0.2x - 0.1 = 0$

Multiply each side by 10 to clear the decimals.

$$x^2 - 2x - 1 = 0$$
$$x^2 - 2x = 1$$

Complete the square.

$$\left[\tfrac{1}{2}(-2)\right]^2 = (-1)^2 = 1$$

Add 1 to each side.

$$x^2 - 2x + 1 = 1 + 1$$
$$(x-1)^2 = 2$$

$$x - 1 = \sqrt{2} \qquad \text{or} \qquad x - 1 = -\sqrt{2}$$
$$x = 1 + \sqrt{2} \quad \text{or} \qquad x = 1 - \sqrt{2}$$

Solution set: $\left\{ 1 \pm \sqrt{2} \right\}$

75. $x^2 = -12$

$$x = \sqrt{-12} \quad \text{or} \quad x = -\sqrt{-12}$$
$$x = i\sqrt{12} \qquad\qquad x = -i\sqrt{12}$$
$$x = 2i\sqrt{3} \quad \text{or} \quad x = -2i\sqrt{3}$$

Solution set: $\left\{ \pm 2i\sqrt{3} \right\}$

77. $(r - 5)^2 = -4$

$$r - 5 = \sqrt{-4} \quad \text{or} \quad r - 5 = -\sqrt{-4}$$
$$r = 5 + 2i \quad \text{or} \qquad r = 5 - 2i$$

Solution set: $\{5 \pm 2i\}$

79. $(6x - 1)^2 = -8$

$$6x - 1 = \sqrt{-8} \qquad \text{or} \quad 6x - 1 = -\sqrt{-8}$$
$$6x - 1 = i\sqrt{8} \qquad\qquad 6x - 1 = -i\sqrt{8}$$
$$6x - 1 = 2i\sqrt{2} \qquad\qquad 6x - 1 = -2i\sqrt{2}$$
$$6x = 1 + 2i\sqrt{2} \qquad\qquad 6x = 1 - 2i\sqrt{2}$$
$$x = \tfrac{1 + 2i\sqrt{2}}{6} \quad \text{or} \qquad x = \tfrac{1 - 2i\sqrt{2}}{6}$$

Solution set: $\left\{ \tfrac{1}{6} \pm \tfrac{\sqrt{2}}{3}i \right\}$

81. $m^2 + 4m + 13 = 0$

$$m^2 + 4m = -13$$

Complete the square.

$$\left(\tfrac{1}{2} \cdot 4\right)^2 = 2^2 = 4$$

Add 4 to each side.

$$m^2 + 4m + 4 = -13 + 4$$
$$(m + 2)^2 = -9$$

$$m + 2 = \sqrt{-9} \qquad \text{or} \qquad m + 2 = -\sqrt{-9}$$
$$m = -2 + 3i \quad \text{or} \qquad m = -2 - 3i$$

Solution set: $\{-2 \pm 3i\}$

Copyright © 2012 Pearson Education, Inc. Publishing as Addison-Wesley.

83. $3r^2 + 4r + 4 = 0$

$3r^2 + 4r = -4$

$r^2 + \frac{4}{3}r = \frac{-4}{3}$ *Divide by 3.*

Complete the square.

$$\left(\frac{1}{2} \cdot \frac{4}{3}\right)^2 = \left(\frac{2}{3}\right)^2 = \frac{4}{9}$$

Add $\frac{4}{9}$ to each side.

$r^2 + \frac{4}{3}r + \frac{4}{9} = \frac{-4}{3} + \frac{4}{9}$

$\left(r + \frac{2}{3}\right)^2 = \frac{-8}{9}$

$r + \dfrac{2}{3} = \dfrac{\sqrt{-8}}{\sqrt{9}}$ or $r + \dfrac{2}{3} = -\dfrac{\sqrt{-8}}{\sqrt{9}}$

$r = -\dfrac{2}{3} + \dfrac{2i\sqrt{2}}{3}$ $r = -\dfrac{2}{3} - \dfrac{2i\sqrt{2}}{3}$

Solution set: $\left\{-\frac{2}{3} \pm \frac{2\sqrt{2}}{3}i\right\}$

85. $-m^2 - 6m - 12 = 0$

Multiply each side by -1.

$m^2 + 6m + 12 = 0$

$m^2 + 6m = -12$

Complete the square.

$$\left(\frac{1}{2} \cdot 6\right)^2 = 3^2 = 9$$

Add 9 to each side.

$m^2 + 6m + 9 = -12 + 9$

$(m + 3)^2 = -3$

$m + 3 = \sqrt{-3}$ or $m + 3 = -\sqrt{-3}$

$m = -3 + i\sqrt{3}$ or $m = -3 - i\sqrt{3}$

Solution set: $\left\{-3 \pm i\sqrt{3}\right\}$

♦♦♦ Relating Concepts 87–92 ♦♦♦

87. The area of the original square is $x \cdot x$, or x^2.

88. Each rectangular strip has length x and width 1, so each strip has an area of $x \cdot 1$, or x.

89. From Exercise 88, the area of a rectangular strip is x. The area of 6 rectangular strips is $6x$.

90. These are 1 by 1 squares, so each has an area of $1 \cdot 1$, or 1.

91. There are 9 small squares, each with area 1 (from Exercise 90), so the total area is $9 \cdot 1$, or 9.

92. The area of the larger square is $(x + 3)^2$. Using the results from Exercises 87–91,

$$(x + 3)^2, \text{ or } x^2 + 6x + 9.$$

93. $x^2 - b = 0$

$x^2 = b$

$x = \sqrt{b}$ or $x = -\sqrt{b}$

Solution set: $\left\{\pm\sqrt{b}\right\}$

95. $4x^2 = b^2 + 16$

$x^2 = \dfrac{b^2 + 16}{4}$

$x = \sqrt{\dfrac{b^2 + 16}{4}}$ or $x = -\sqrt{\dfrac{b^2 + 16}{4}}$

$x = \dfrac{\sqrt{b^2 + 16}}{2}$ or $x = -\dfrac{\sqrt{b^2 + 16}}{2}$

Solution set: $\left\{\pm\frac{\sqrt{b^2+16}}{2}\right\}$

97. $(5x - 2b)^2 = 3a$

$5x - 2b = \sqrt{3a}$ or $5x - 2b = -\sqrt{3a}$

$5x = 2b + \sqrt{3a}$ $5x = 2b - \sqrt{3a}$

$x = \dfrac{2b + \sqrt{3a}}{5}$ or $x = \dfrac{2b - \sqrt{3a}}{5}$

Solution set: $\left\{\frac{2b \pm \sqrt{3a}}{5}\right\}$

99. $a = 3, b = 1, c = -1$

$\sqrt{b^2 - 4ac} = \sqrt{1^2 - 4(3)(-1)}$

$= \sqrt{1 + 12}$

$= \sqrt{13}$

101. $a = 6, b = 7, c = 2$

$\sqrt{b^2 - 4ac} = \sqrt{7^2 - 4(6)(2)}$

$= \sqrt{49 - 48}$

$= \sqrt{1} = 1$

9.2 The Quadratic Formula

9.2 Now Try Exercises

N1. $2x^2 + 3x - 20 = 0$

Here $a = 2$, $b = 3$, and $c = -20$.

$x = \dfrac{-b \pm \sqrt{b^2 - 4ac}}{2a}$

$x = \dfrac{-3 \pm \sqrt{3^2 - 4(2)(-20)}}{2(2)}$

$= \dfrac{-3 \pm \sqrt{9 + 160}}{4}$

$= \dfrac{-3 \pm \sqrt{169}}{4}$

$= \dfrac{-3 \pm 13}{4}$

$x = \dfrac{-3 + 13}{4} = \dfrac{10}{4} = \dfrac{5}{2}$

or $x = \dfrac{-3 - 13}{4} = \dfrac{-16}{4} = -4$

Solution set: $\left\{-4, \frac{5}{2}\right\}$

Copyright © 2012 Pearson Education, Inc. Publishing as Addison-Wesley.

N2.
$$3x^2 + 1 = -5x$$
$$3x^2 + 5x + 1 = 0$$
Here $a = 3$, $b = 5$, and $c = 1$.

$$x = \frac{-b \pm \sqrt{b^2 - 4ac}}{2a}$$
$$x = \frac{-5 \pm \sqrt{5^2 - 4(3)(1)}}{2(3)}$$
$$= \frac{-5 \pm \sqrt{25 - 12}}{6} = \frac{-5 \pm \sqrt{13}}{6}$$

Solution set: $\left\{ \dfrac{-5 \pm \sqrt{13}}{6} \right\}$

N3.
$$(x + 5)(x - 1) = -18$$
$$x^2 + 4x - 5 = -18$$
$$x^2 + 4x + 13 = 0$$
Here $a = 1$, $b = 4$, and $c = 13$.

$$x = \frac{-b \pm \sqrt{b^2 - 4ac}}{2a}$$
$$x = \frac{-4 \pm \sqrt{4^2 - 4(1)(13)}}{2(1)}$$
$$= \frac{-4 \pm \sqrt{16 - 52}}{2}$$
$$= \frac{-4 \pm \sqrt{-36}}{2} = \frac{-4 \pm 6i}{2}$$
$$= \frac{2(-2 \pm 3i)}{2} = -2 \pm 3i$$

Solution set: $\{-2 \pm 3i\}$

N4. **(a)** $8x^2 - 6x - 5 = 0$

Here $a = 8$, $b = -6$, and $c = -5$, so the discriminant is

$$b^2 - 4ac = (-6)^2 - 4(8)(-5)$$
$$= 36 + 160$$
$$= 196.$$

The discriminant is a perfect square and a, b, and c are integers, so there are two different rational solutions and the equation can be solved by factoring.

(b)
$$9x^2 = 24x - 16$$
$$9x^2 - 24x + 16 = 0$$

Here $a = 9$, $b = -24$, and $c = 16$, so the discriminant is

$$b^2 - 4ac = (-24)^2 - 4(9)(16)$$
$$= 576 - 576$$
$$= 0.$$

The discriminant is zero, so there is one rational solution and the equation can be solved by factoring.

(c)
$$3x^2 + 2x = -1$$
$$3x^2 + 2x + 1 = 0$$

Here $a = 3$, $b = 2$, and $c = 1$, so the discriminant is

$$b^2 - 4ac = 2^2 - 4(3)(1)$$
$$= 4 - 12$$
$$= -8.$$

The discriminant is negative but not a perfect square and a, b, and c are integers, so there are two nonreal complex solutions and the equation is best solved using the quadratic formula.

N5. $4x^2 + kx + 25 = 0$

The equation will have only one rational solution if the discriminant is 0. Since $a = 4$, $b = k$, and $c = 25$, the discriminant is

$$b^2 - 4ac = k^2 - 4(4)(25)$$
$$= k^2 - 400.$$

Set the discriminant equal to 0 and solve for k.

$$k^2 - 400 = 0$$
$$k^2 = 400 \quad \textit{Add 400.}$$
$$k = 20 \quad \text{or} \quad k = -20 \quad \textit{Square root property}$$

The equation will have only one rational solution if $k = 20$ or $k = -20$.

9.2 Section Exercises

1. The documentation was incorrect, since the fraction bar should extend under the term $-b$. The correct formula is
$$x = \frac{-b \pm \sqrt{b^2 - 4ac}}{2a}.$$

3. The last step is wrong. Because 5 is not a common factor in the numerator, the fraction cannot be reduced. The solution set is $\left\{ \frac{5 \pm \sqrt{5}}{10} \right\}$.

5.
$$x^2 - 8x + 15 = 0$$
Here $a = 1$, $b = -8$, and $c = 15$.

$$x = \frac{-b \pm \sqrt{b^2 - 4ac}}{2a}$$
$$x = \frac{-(-8) \pm \sqrt{(-8)^2 - 4(1)(15)}}{2(1)}$$
$$= \frac{8 \pm \sqrt{64 - 60}}{2}$$
$$= \frac{8 \pm \sqrt{4}}{2} = \frac{8 \pm 2}{2}$$
$$x = \frac{8 + 2}{2} = \frac{10}{2} = 5 \text{ or}$$
$$x = \frac{8 - 2}{2} = \frac{6}{2} = 3$$

Solution set: $\{3, 5\}$

Copyright © 2012 Pearson Education, Inc. Publishing as Addison-Wesley.

7. $2x^2 + 4x + 1 = 0$
Here $a = 2$, $b = 4$, and $c = 1$.

$$x = \frac{-b \pm \sqrt{b^2 - 4ac}}{2a}$$

$$x = \frac{-4 \pm \sqrt{4^2 - 4(2)(1)}}{2(2)}$$

$$= \frac{-4 \pm \sqrt{16 - 8}}{4}$$

$$= \frac{-4 \pm \sqrt{8}}{4} = \frac{-4 \pm 2\sqrt{2}}{4}$$

$$= \frac{2\left(-2 \pm \sqrt{2}\right)}{2 \cdot 2} = \frac{-2 \pm \sqrt{2}}{2}$$

Solution set: $\left\{\frac{-2 \pm \sqrt{2}}{2}\right\}$

9. $2x^2 - 2x = 1$
$2x^2 - 2x - 1 = 0$
Here $a = 2$, $b = -2$, and $c = -1$.

$$x = \frac{-b \pm \sqrt{b^2 - 4ac}}{2a}$$

$$x = \frac{-(-2) \pm \sqrt{(-2)^2 - 4(2)(-1)}}{2(2)}$$

$$= \frac{2 \pm \sqrt{4 + 8}}{4} = \frac{2 \pm \sqrt{12}}{4} = \frac{2 \pm 2\sqrt{3}}{4}$$

$$= \frac{2\left(1 \pm \sqrt{3}\right)}{2 \cdot 2} = \frac{1 \pm \sqrt{3}}{2}$$

Solution set: $\left\{\frac{1 \pm \sqrt{3}}{2}\right\}$

11. $x^2 + 18 = 10x$
$x^2 - 10x + 18 = 0$
Here $a = 1$, $b = -10$, and $c = 18$.

$$x = \frac{-b \pm \sqrt{b^2 - 4ac}}{2a}$$

$$x = \frac{-(-10) \pm \sqrt{(-10)^2 - 4(1)(18)}}{2(1)}$$

$$= \frac{10 \pm \sqrt{100 - 72}}{2} = \frac{10 \pm \sqrt{28}}{2}$$

$$= \frac{10 \pm 2\sqrt{7}}{2} = \frac{2\left(5 \pm \sqrt{7}\right)}{2} = 5 \pm \sqrt{7}$$

Solution set: $\left\{5 \pm \sqrt{7}\right\}$

13. $4x^2 + 4x - 1 = 0$
Here $a = 4$, $b = 4$, and $c = -1$.

$$x = \frac{-b \pm \sqrt{b^2 - 4ac}}{2a}$$

$$x = \frac{-4 \pm \sqrt{4^2 - 4(4)(-1)}}{2(4)}$$

$$= \frac{-4 \pm \sqrt{16 + 16}}{8} = \frac{-4 \pm \sqrt{32}}{8}$$

$$= \frac{-4 \pm 4\sqrt{2}}{8} = \frac{4\left(-1 \pm \sqrt{2}\right)}{2 \cdot 4}$$

$$= \frac{-1 \pm \sqrt{2}}{2}$$

Solution set: $\left\{\frac{-1 \pm \sqrt{2}}{2}\right\}$

15. $2 - 2x = 3x^2$
$0 = 3x^2 + 2x - 2$
Here $a = 3$, $b = 2$, and $c = -2$.

$$x = \frac{-b \pm \sqrt{b^2 - 4ac}}{2a}$$

$$x = \frac{-2 \pm \sqrt{2^2 - 4(3)(-2)}}{2(3)}$$

$$= \frac{-2 \pm \sqrt{4 + 24}}{6} = \frac{-2 \pm \sqrt{28}}{6}$$

$$= \frac{-2 \pm 2\sqrt{7}}{6} = \frac{2\left(-1 \pm \sqrt{7}\right)}{2 \cdot 3} = \frac{-1 \pm \sqrt{7}}{3}$$

Solution set: $\left\{\frac{-1 \pm \sqrt{7}}{3}\right\}$

17. $\dfrac{x^2}{4} - \dfrac{x}{2} = 1$

$\dfrac{x^2}{4} - \dfrac{x}{2} - 1 = 0$

$x^2 - 2x - 4 = 0$ *Multiply by 4.*
Here $a = 1$, $b = -2$, and $c = -4$.

$$x = \frac{-b \pm \sqrt{b^2 - 4ac}}{2a}$$

$$x = \frac{-(-2) \pm \sqrt{(-2)^2 - 4(1)(-4)}}{2(1)}$$

$$= \frac{2 \pm \sqrt{4 + 16}}{2} = \frac{2 \pm \sqrt{20}}{2}$$

$$= \frac{2 \pm 2\sqrt{5}}{2} = 1 \pm \sqrt{5}$$

Solution set: $\left\{1 \pm \sqrt{5}\right\}$

Copyright © 2012 Pearson Education, Inc. Publishing as Addison-Wesley.

19.
$$-2t(t+2) = -3$$
$$-2t^2 - 4t = -3$$
$$-2t^2 - 4t + 3 = 0$$
Here $a = -2$, $b = -4$, and $c = 3$.
$$t = \frac{-b \pm \sqrt{b^2 - 4ac}}{2a}$$
$$t = \frac{-(-4) \pm \sqrt{(-4)^2 - 4(-2)(3)}}{2(-2)}$$
$$= \frac{4 \pm \sqrt{16 + 24}}{-4} = \frac{4 \pm \sqrt{40}}{-4}$$
$$= \frac{4 \pm 2\sqrt{10}}{-4} = \frac{2\left(2 \pm \sqrt{10}\right)}{-2 \cdot 2}$$
$$= \frac{2 \pm \sqrt{10}}{-2} \cdot \frac{-1}{-1} = \frac{-2 \mp \sqrt{10}}{2}$$
$$= \frac{-2 \pm \sqrt{10}}{2}$$
Solution set: $\left\{ \frac{-2 \pm \sqrt{10}}{2} \right\}$

21.
$$(r-3)(r+5) = 2$$
$$r^2 + 2r - 15 = 2$$
$$r^2 + 2r - 17 = 0$$
Here $a = 1$, $b = 2$, and $c = -17$.
$$r = \frac{-b \pm \sqrt{b^2 - 4ac}}{2a}$$
$$r = \frac{-2 \pm \sqrt{2^2 - 4(1)(-17)}}{2(1)}$$
$$= \frac{-2 \pm \sqrt{4 + 68}}{2} = \frac{-2 \pm \sqrt{72}}{2}$$
$$= \frac{-2 \pm 6\sqrt{2}}{2} = \frac{2\left(-1 \pm 3\sqrt{2}\right)}{2}$$
$$= -1 \pm 3\sqrt{2}$$
Solution set: $\left\{ -1 \pm 3\sqrt{2} \right\}$

23.
$$(x+2)(x-3) = 1$$
$$x^2 - x - 6 = 1$$
$$x^2 - x - 7 = 0$$
Here $a = 1$, $b = -1$, and $c = -7$.
$$x = \frac{-b \pm \sqrt{b^2 - 4ac}}{2a}$$
$$x = \frac{-(-1) \pm \sqrt{(-1)^2 - 4(1)(-7)}}{2(1)}$$
$$= \frac{1 \pm \sqrt{1 + 28}}{2} = \frac{1 \pm \sqrt{29}}{2}$$
Solution set: $\left\{ \frac{1 \pm \sqrt{29}}{2} \right\}$

25.
$$p = \frac{5(5-p)}{3(p+1)}$$
$$3p(p+1) = 5(5-p)$$
$$3p^2 + 3p = 25 - 5p$$
$$3p^2 + 8p - 25 = 0$$
Here $a = 3$, $b = 8$, and $c = -25$.
$$p = \frac{-b \pm \sqrt{b^2 - 4ac}}{2a}$$
$$p = \frac{-8 \pm \sqrt{8^2 - 4(3)(-25)}}{2(3)}$$
$$= \frac{-8 \pm \sqrt{64 + 300}}{6} = \frac{-8 \pm \sqrt{364}}{6}$$
$$= \frac{-8 \pm 2\sqrt{91}}{6} = \frac{2\left(-4 \pm \sqrt{91}\right)}{2 \cdot 3} = \frac{-4 \pm \sqrt{91}}{3}$$
Solution set: $\left\{ \frac{-4 \pm \sqrt{91}}{3} \right\}$

27.
$$(2x+1)^2 = x+4$$
$$4x^2 + 4x + 1 = x + 4$$
$$4x^2 + 3x - 3 = 0$$
Here $a = 4$, $b = 3$, and $c = -3$.
$$x = \frac{-b \pm \sqrt{b^2 - 4ac}}{2a}$$
$$x = \frac{-3 \pm \sqrt{3^2 - 4(4)(-3)}}{2(4)}$$
$$= \frac{-3 \pm \sqrt{9 + 48}}{8} = \frac{-3 \pm \sqrt{57}}{8}$$
Solution set: $\left\{ \frac{-3 \pm \sqrt{57}}{8} \right\}$

29.
$$x^2 - 3x + 6 = 0$$
Here $a = 1$, $b = -3$, and $c = 6$.
$$x = \frac{-b \pm \sqrt{b^2 - 4ac}}{2a}$$
$$x = \frac{-(-3) \pm \sqrt{(-3)^2 - 4(1)(6)}}{2(1)}$$
$$= \frac{3 \pm \sqrt{9 - 24}}{2} = \frac{3 \pm \sqrt{-15}}{2}$$
$$= \frac{3 \pm i\sqrt{15}}{2} = \frac{3}{2} \pm \frac{\sqrt{15}}{2}i$$
Solution set: $\left\{ \frac{3}{2} \pm \frac{\sqrt{15}}{2}i \right\}$

Copyright © 2012 Pearson Education, Inc. Publishing as Addison-Wesley.

31. $r^2 - 6r + 14 = 0$

Here $a = 1$, $b = -6$, and $c = 14$.

$$r = \frac{-b \pm \sqrt{b^2 - 4ac}}{2a}$$

$$r = \frac{-(-6) \pm \sqrt{(-6)^2 - 4(1)(14)}}{2(1)}$$

$$= \frac{6 \pm \sqrt{36 - 56}}{2}$$

$$= \frac{6 \pm \sqrt{-20}}{2} = \frac{6 \pm 2i\sqrt{5}}{2}$$

$$= \frac{2\left(3 \pm i\sqrt{5}\right)}{2} = 3 \pm i\sqrt{5}$$

Solution set: $\left\{3 \pm i\sqrt{5}\right\}$

33. $4x^2 - 4x = -7$

$4x^2 - 4x + 7 = 0$

Here $a = 4$, $b = -4$, and $c = 7$.

$$x = \frac{-b \pm \sqrt{b^2 - 4ac}}{2a}$$

$$x = \frac{-(-4) \pm \sqrt{(-4)^2 - 4(4)(7)}}{2(4)}$$

$$= \frac{4 \pm \sqrt{16 - 112}}{8} = \frac{4 \pm \sqrt{-96}}{8}$$

$$= \frac{4 \pm 4i\sqrt{6}}{8} = \frac{4\left(1 \pm i\sqrt{6}\right)}{2 \cdot 4}$$

$$= \frac{1 \pm i\sqrt{6}}{2} = \frac{1}{2} \pm \frac{\sqrt{6}}{2}i$$

Solution set: $\left\{\frac{1}{2} \pm \frac{\sqrt{6}}{2}i\right\}$

35. $x(3x + 4) = -2$

$3x^2 + 4x = -2$

$3x^2 + 4x + 2 = 0$

Here $a = 3$, $b = 4$, and $c = 2$.

$$x = \frac{-b \pm \sqrt{b^2 - 4ac}}{2a}$$

$$x = \frac{-4 \pm \sqrt{4^2 - 4(3)(2)}}{2(3)}$$

$$= \frac{-4 \pm \sqrt{16 - 24}}{6} = \frac{-4 \pm \sqrt{-8}}{6}$$

$$= \frac{-4 \pm 2i\sqrt{2}}{6} = \frac{2\left(-2 \pm i\sqrt{2}\right)}{2 \cdot 3}$$

$$= \frac{-2 \pm i\sqrt{2}}{3} = -\frac{2}{3} \pm \frac{\sqrt{2}}{3}i$$

Solution set: $\left\{-\frac{2}{3} \pm \frac{\sqrt{2}}{3}i\right\}$

37. $(2x - 1)(8x - 4) = -1$

$16x^2 - 16x + 4 = -1$

$16x^2 - 16x + 5 = 0$

Here $a = 16$, $b = -16$, and $c = 5$.

$$x = \frac{-b \pm \sqrt{b^2 - 4ac}}{2a}$$

$$x = \frac{-(-16) \pm \sqrt{(-16)^2 - 4(16)(5)}}{2(16)}$$

$$= \frac{16 \pm \sqrt{256 - 320}}{32}$$

$$= \frac{16 \pm \sqrt{-64}}{32} = \frac{16 \pm 8i}{32}$$

$$= \frac{16}{32} \pm \frac{8}{32}i = \frac{1}{2} \pm \frac{1}{4}i$$

Solution set: $\left\{\frac{1}{2} \pm \frac{1}{4}i\right\}$

Note: We could also solve this equation without the quadratic formula as follows:

$$(2x - 1)(8x - 4) = -1$$
$$(2x - 1)4(2x - 1) = -1$$
$$(2x - 1)^2 = -\frac{1}{4}$$
$$2x - 1 = \pm\sqrt{-\frac{1}{4}}$$
$$2x = 1 \pm \frac{1}{2}i$$
$$x = \frac{1}{2} \pm \frac{1}{4}i$$

39. $25x^2 + 70x + 49 = 0$

Here $a = 25$, $b = 70$, and $c = 49$, so the discriminant is

$$b^2 - 4ac = 70^2 - 4(25)(49)$$
$$= 4900 - 4900$$
$$= 0.$$

Since the discriminant is 0, the quantity under the radical in the quadratic formula is 0, and there is only one rational solution. The answer is **B** and the equation can be solved by factoring.

41. $x^2 + 4x + 2 = 0$

Here $a = 1$, $b = 4$, and $c = 2$, so the discriminant is

$$b^2 - 4ac = 4^2 - 4(1)(2)$$
$$= 16 - 8$$
$$= 8.$$

Since the discriminant is positive, but not a perfect square, there are two distinct irrational number solutions. The answer is **C** and the equation is best solved using the quadratic formula.

43. $3x^2 = 5x + 2$

$3x^2 - 5x - 2 = 0$

Here $a = 3$, $b = -5$, and $c = -2$, so the discriminant is

$$b^2 - 4ac = (-5)^2 - 4(3)(-2)$$
$$= 25 + 24$$
$$= 49.$$

Since the discriminant is a perfect square, there are two distinct rational solutions. The answer is **A** and the equation can be solved by factoring.

45. $3m^2 - 10m + 15 = 0$

Here $a = 3$, $b = -10$, and $c = 15$, so the discriminant is

$$b^2 - 4ac = (-10)^2 - 4(3)(15)$$
$$= 100 - 180$$
$$= -80.$$

Since the discriminant is negative, there are two distinct nonreal complex number solutions. The answer is **D** and the equation is best solved using the quadratic formula.

47. $25x^2 + 70x + 49 = 0$
$$(5x + 7)^2 = 0$$
$$5x + 7 = 0$$
$$x = -\tfrac{7}{5}$$

Solution set: $\left\{-\tfrac{7}{5}\right\}$

49.
$$3x^2 = 5x + 2$$
$$3x^2 - 5x - 2 = 0$$
$$(3x + 1)(x - 2) = 0$$

$$3x + 1 = 0 \quad \text{or} \quad x - 2 = 0$$
$$x = -\tfrac{1}{3} \quad \text{or} \quad x = 2$$

Solution set: $\left\{-\tfrac{1}{3}, 2\right\}$

51. **(a)** $\qquad 3x^2 + 13x = -12$
$$3x^2 + 13x + 12 = 0$$

Here $a = 3$, $b = 13$, and $c = 12$, so the discriminant is

$$b^2 - 4ac = 13^2 - 4(3)(12)$$
$$= 169 - 144$$
$$= 25.$$

The discriminant is a perfect square, so the equation can be solved by factoring.

$$3x^2 + 13x + 12 = 0$$
$$(3x + 4)(x + 3) = 0$$

$$3x + 4 = 0 \quad \text{or} \quad x + 3 = 0$$
$$x = -\tfrac{4}{3} \qquad\qquad x = -3$$

Solution set: $\left\{-\tfrac{4}{3}, -3\right\}$

(b) $\qquad 2x^2 + 19 = 14x$
$$2x^2 - 14x + 19 = 0$$

Here $a = 2$, $b = -14$, and $c = 19$, so the discriminant is

$$b^2 - 4ac = (-14)^2 - 4(2)(19)$$
$$= 196 - 152$$
$$= 44.$$

The discriminant is not a perfect square, so use the quadratic formula.

$$x = \frac{-b \pm \sqrt{b^2 - 4ac}}{2a}$$
$$x = \frac{-(-14) \pm \sqrt{44}}{2 \cdot 2}$$
$$= \frac{14 \pm 2\sqrt{11}}{2 \cdot 2} = \frac{7 \pm \sqrt{11}}{2}$$

Solution set: $\left\{\frac{7 \pm \sqrt{11}}{2}\right\}$

53. $p^2 + bp + 25 = 0$

For there to be only one rational solution, $b^2 - 4ac$ must equal zero. Since $a = 1$ and $c = 25$,

$$b^2 - 4(1)(25) = 0$$
$$b^2 - 100 = 0$$
$$b^2 = 100$$
$$b = \pm \sqrt{100}$$
$$b = 10 \quad \text{or} \quad b = -10.$$

55. $am^2 + 8m + 1 = 0$

For there to be only one rational solution, $b^2 - 4ac$ must equal zero. Since $b = 8$ and $c = 1$,

$$8^2 - 4(a)(1) = 0$$
$$64 - 4a = 0$$
$$-4a = -64$$
$$a = 16.$$

57. $9x^2 - 30x + c = 0$

For there to be only one rational solution, $b^2 - 4ac$ must equal zero. Since $a = 9$ and $b = -30$,

$$(-30)^2 - 4(9)(c) = 0$$
$$900 - 36c = 0$$
$$-36c = -900$$
$$c = 25.$$

59. Substitute $-\tfrac{5}{2}$ for x and solve for b.

$$4x^2 + bx - 3 = 0$$
$$4\left(-\tfrac{5}{2}\right)^2 + b\left(-\tfrac{5}{2}\right) - 3 = 0$$
$$25 - \tfrac{5}{2}b - 3 = 0$$
$$22 = \tfrac{5}{2}b$$
$$\tfrac{44}{5} = b$$

So the equation is

$$4x^2 + \tfrac{44}{5}x - 3 = 0.$$

Now multiply by 5.

$$20x^2 + 44x - 15 = 0$$

Since $-\tfrac{5}{2}$ is a solution, $2x + 5$ must be one factor.

$$(2x + 5)(10x - 3) = 0$$

So the other solution is $\tfrac{3}{10}$.

Copyright © 2012 Pearson Education, Inc. Publishing as Addison-Wesley.

61. $\frac{3}{4}x + \frac{1}{2}x = -10$

$\quad 3x + 2x = -40 \quad$ *Multiply by 4.*

$\quad\quad\quad 5x = -40$

$\quad\quad\quad\quad x = -8$

Solution set: $\{-8\}$.

63. $\quad \sqrt{2x+6} = x - 1$

$\quad \left(\sqrt{2x+6}\right)^2 = (x-1)^2$

$\quad\quad 2x + 6 = x^2 - 2x + 1$

$\quad\quad\quad\quad 0 = x^2 - 4x - 5$

$\quad\quad\quad\quad 0 = (x-5)(x+1)$

$x - 5 = 0 \quad$ or $\quad x + 1 = 0$

$\quad x = 5 \quad$ or $\quad\quad x = -1$

Check $x = 5$: $\sqrt{16} = 4 \quad$ *True*

Check $x = -1$: $\sqrt{4} = -2 \quad$ *False*

Solution set: $\{5\}$

9.3 Equations Quadratic in Form

9.3 Now Try Exercises

N1. $\dfrac{2}{x} + \dfrac{3}{x+2} = 1$

Multiply each term by the LCD, $x(x+2)$.

$$x(x+2)\left(\frac{2}{x} + \frac{3}{x+2}\right) = x(x+2)(1)$$

$$2(x+2) + 3x = x^2 + 2x$$

$$2x + 4 + 3x = x^2 + 2x$$

$$x^2 - 3x - 4 = 0$$

$$(x+1)(x-4) = 0$$

$x + 1 = 0 \quad$ or $\quad x - 4 = 0$

$\quad x = -1 \quad$ or $\quad\quad x = 4$

Check $x = -1$: $-2 + 3 = 1 \quad$ *True*

Check $x = 4$: $\frac{1}{2} + \frac{1}{2} = 1 \quad$ *True*

Solution set: $\{-1, 4\}$

N2. Let x = the rate of the current. Make a table. Use $t = d/r$.

	d	r	t
Up	8	$18 - x$	$\dfrac{8}{18-x}$
Down	8	$18 + x$	$\dfrac{8}{18+x}$

The time going upriver added to the time going downriver is $\frac{9}{10}$ hr.

$$\frac{8}{18-x} + \frac{8}{18+x} = \frac{9}{10}$$

Multiply each term by the LCD, $10(18-x)(18+x)$.

$$10(18-x)(18+x)\frac{8}{18-x}$$

$$+ \; 10(18-x)(18+x)\frac{8}{18+x}$$

$$= 10(18-x)(18+x)\left(\tfrac{9}{10}\right)$$

$$80(18+x) + 80(18-x) = 9(18-x)(18+x)$$

$$1440 + 80x + 1440 - 80x = 9(324 - x^2)$$

$$2880 = 2916 - 9x^2$$

$$9x^2 = 36$$

$$x^2 = 4$$

$$x = -2 \quad \text{or} \quad x = 2$$

The rate can't be negative, so the rate of the current is 2 mph.

N3. Let x = the slower electrician's time alone. Then $x - 2$ = the faster electrician's time alone. Make a table.

	Rate	Time Working Together	Fractional Part of the Job Done
Slower Elec.	$\dfrac{1}{x}$	6	$\dfrac{6}{x}$
Faster Elec.	$\dfrac{1}{x-2}$	6	$\dfrac{6}{x-2}$

Since together they complete 1 whole job,

$$\frac{6}{x} + \frac{6}{x-2} = 1.$$

Multiply each term by the LCD, $x(x-2)$.

$$x(x-2)\left(\frac{6}{x}\right) + x(x-2)\left(\frac{6}{x-2}\right)$$

$$= x(x-2)(1)$$

$$6(x-2) + 6x = x(x-2)$$

$$6x - 12 + 6x = x^2 - 2x$$

$$0 = x^2 - 14x + 12$$

Here $a = 1$, $b = -14$, and $c = 12$.

$$x = \frac{-b \pm \sqrt{b^2 - 4ac}}{2a}$$

$$= \frac{-(-14) \pm \sqrt{(-14)^2 - 4(1)(12)}}{2(1)}$$

$$= \frac{14 \pm \sqrt{196 - 48}}{2}$$

$$= \frac{14 \pm \sqrt{148}}{2} = \frac{14 \pm 2\sqrt{37}}{2}$$

$$= \frac{2\left(7 \pm \sqrt{37}\right)}{2} = 7 \pm \sqrt{37}$$

$$x = 7 + \sqrt{37} \approx 13.1 \quad \text{or} \quad x = 7 - \sqrt{37} \approx 0.9$$

Copyright © 2012 Pearson Education, Inc. Publishing as Addison-Wesley.

The slower electrician's time cannot be 0.9 since the faster electrician's time would then be $0.9 - 2$ or -1.1.

So the slower electrician's time working alone is 13.1 hr and the faster electrician's time working alone is $13.1 - 2 = 11.1$ hr.

N4. (a)
$$x = \sqrt{9x - 20}$$
$$x^2 = \left(\sqrt{9x - 20}\right)^2 \quad \textit{Square.}$$
$$x^2 = 9x - 20$$
$$x^2 - 9x + 20 = 0$$
$$(x - 4)(x - 5) = 0$$
$$x - 4 = 0 \quad \text{or} \quad x - 5 = 0$$
$$x = 4 \quad \text{or} \quad x = 5$$

Check $x = 4$: $4 = \sqrt{16}$ *True*
Check $x = 5$: $5 = \sqrt{25}$ *True*

Solution set: $\{4, 5\}$

(b) $x + \sqrt{x} = 20$
$$\sqrt{x} = 20 - x \qquad \textit{Isolate.}$$
$$\left(\sqrt{x}\right)^2 = (20 - x)^2 \qquad \textit{Square.}$$
$$x = 400 - 40x + x^2$$
$$0 = x^2 - 41x + 400$$
$$0 = (x - 16)(x - 25)$$
$$x - 16 = 0 \quad \text{or} \quad x - 25 = 0$$
$$x = 16 \quad \text{or} \quad x = 25$$

Check $x = 16$: $16 + 4 = 20$ *True*
Check $x = 25$: $25 + 5 = 20$ *False*

Solution set: $\{16\}$

N5. (a) $x^4 - 10x^2 + 9 = 0$

Since $x^4 = (x^2)^2$, define $u = x^2$, and the original equation becomes $u^2 - 10u + 9 = 0$.

(b) $6(x + 2)^2 - 11(x + 2) + 4 = 0$

Because this equation involves both $(x + 2)^2$ and $(x + 2)$, define $u = x + 2$, and the original equation becomes $6u^2 - 11u + 4 = 0$.

N6. (a) $x^4 - 17x^2 + 16 = 0$
$$\text{Let } y = x^2, \text{ so } y^2 = (x^2)^2 = x^4.$$
$$y^2 - 17y + 16 = 0$$
$$(y - 1)(y - 16) = 0$$
$$y - 1 = 0 \quad \text{or} \quad y - 16 = 0$$
$$y = 1 \quad \text{or} \quad y = 16$$
To find x, substitute x^2 for y.
$$x^2 = 1 \quad \text{or} \quad x^2 = 16$$
$$x = \pm 1 \quad \text{or} \quad x = \pm 4$$

Check $x = \pm 1$: $1 - 17 + 16 = 0$ *True*
Check $x = \pm 4$: $256 - 272 + 16 = 0$ *True*

Solution set: $\{\pm 1, \pm 4\}$

(b)
$$x^4 + 4 = 8x^2$$
$$x^4 - 8x^2 + 4 = 0$$
$$\text{Let } y = x^2, \text{ so } (x^2)^2 = x^4.$$
$$y^2 - 8y + 4 = 0$$

Use the quadratic formula with $a = 1$, $b = -8$, and $c = 4$.

$$y = \frac{-b \pm \sqrt{b^2 - 4ac}}{2a}$$
$$y = \frac{-(-8) \pm \sqrt{(-8)^2 - 4(1)(4)}}{2(1)}$$
$$= \frac{8 \pm \sqrt{64 - 16}}{2} = \frac{8 \pm \sqrt{48}}{2}$$
$$= \frac{8 \pm 4\sqrt{3}}{2} = 4 \pm 2\sqrt{3}$$

To find x, substitute x^2 for y.

$$x^2 = 4 \pm 2\sqrt{3}$$
$$x = \pm\sqrt{4 \pm 2\sqrt{3}}$$

Check $x = \pm\sqrt{4 - 2\sqrt{3}}$:
$$\left(\pm\sqrt{4 - 2\sqrt{3}}\right)^4 + 4 \overset{?}{=} 8\left(\pm\sqrt{4 - 2\sqrt{3}}\right)^2$$
$$\left(4 - 2\sqrt{3}\right)^2 + 4 \overset{?}{=} 8\left(4 - 2\sqrt{3}\right)$$
$$\left(16 - 16\sqrt{3} + 12\right) + 4 \overset{?}{=} 32 - 16\sqrt{3}$$
$$32 - 16\sqrt{3} = 32 - 16\sqrt{3} \quad \textit{True}$$

The check for $x = \pm\sqrt{4 + 2\sqrt{3}}$ is similar.

The solution set contains four numbers:

$$\left\{ \pm\sqrt{4 + 2\sqrt{3}},\ \pm\sqrt{4 - 2\sqrt{3}} \right\}$$

N7. (a) $6(x - 4)^2 + 11(x - 4) - 10 = 0$
Let $y = x - 4$. The equation becomes:
$$6y^2 + 11y - 10 = 0$$
$$(2y + 5)(3y - 2) = 0$$
$$2y + 5 = 0 \quad \text{or} \quad 3y - 2 = 0$$
$$y = -\frac{5}{2} \quad \text{or} \quad y = \frac{2}{3}$$
To find x, substitute $x - 4$ for y.
$$x - 4 = -\frac{5}{2} \quad \text{or} \quad x - 4 = \frac{2}{3}$$
$$x = \frac{3}{2} \quad \text{or} \quad x = \frac{14}{3}$$

Check $x = \frac{3}{2}$: $\frac{75}{2} - \frac{55}{2} - 10 = 0$ *True*
Check $x = \frac{14}{3}$: $\frac{8}{3} + \frac{22}{3} - 10 = 0$ *True*

Solution set: $\left\{\frac{3}{2}, \frac{14}{3}\right\}$

Copyright © 2012 Pearson Education, Inc. Publishing as Addison-Wesley.

(b) $2x^{2/3} - 7x^{1/3} + 3 = 0$

Let $y = x^{1/3}$, so $y^2 = (x^{1/3})^2 = x^{2/3}$.

$$2y^2 - 7y + 3 = 0$$
$$(2y - 1)(y - 3) = 0$$

$2y - 1 = 0$ $\quad$ or $\quad$ $y - 3 = 0$
$y = \frac{1}{2}$ $\quad$ or $\quad$ $y = 3$

To find x, substitute $x^{1/3}$ for y.

$x^{1/3} = \frac{1}{2}$ $\quad$ or $\quad$ $x^{1/3} = 3$

Cube each side of each equation.

$(x^{1/3})^3 = (\frac{1}{2})^3$ $\quad$ or $\quad$ $(x^{1/3})^3 = 3^3$
$x = \frac{1}{8}$ $\quad$ or $\quad$ $x = 27$

Check $x = \frac{1}{8}$: $\quad$ $\frac{1}{2} - \frac{7}{2} + 3 = 0$ $\quad$ *True*
Check $x = 27$: $18 - 21 + 3 = 0$ $\quad$ *True*

Solution set: $\left\{\frac{1}{8}, 27\right\}$

9.3 Section Exercises

1. $\dfrac{14}{x} = x - 5$

This is a rational equation, so multiply each side by the LCD, x.

3. $(x^2 + x)^2 - 8(x^2 + x) + 12 = 0$

This is quadratic in form, so substitute a variable for $x^2 + x$.

5. The proposed solution -1 does not check.
Solution set: $\{4\}$

7. $\dfrac{14}{x} = x - 5$

Multiply each term by the LCD, x.

$$x\left(\frac{14}{x}\right) = x(x) - 5x$$
$$14 = x^2 - 5x$$
$$-x^2 + 5x + 14 = 0$$
$$x^2 - 5x - 14 = 0$$
$$(x + 2)(x - 7) = 0$$

$x + 2 = 0$ $\quad$ or $\quad$ $x - 7 = 0$
$x = -2$ $\quad$ or $\quad$ $x = 7$

Check $x = -2$: $\quad -7 = -7$ $\quad$ *True*
Check $x = 7$: $\quad 2 = 2$ $\quad$ *True*

Solution set: $\{-2, 7\}$

9. $1 - \dfrac{3}{x} - \dfrac{28}{x^2} = 0$

Multiply each term by the LCD, x^2.

$$x^2(1) - x^2\left(\frac{3}{x}\right) - x^2\left(\frac{28}{x^2}\right) = x^2 \cdot 0$$
$$x^2 - 3x - 28 = 0$$
$$(x + 4)(x - 7) = 0$$

$x + 4 = 0$ $\quad$ or $\quad$ $x - 7 = 0$
$x = -4$ $\quad$ or $\quad$ $x = 7$

Check $x = -4$: $\quad 1 + \frac{3}{4} - \frac{7}{4} = 0$ $\quad$ *True*
Check $x = 7$: $\quad 1 - \frac{3}{7} - \frac{4}{7} = 0$ $\quad$ *True*

Solution set: $\{-4, 7\}$

11. $3 - \dfrac{1}{t} = \dfrac{2}{t^2}$

Multiply each term by the LCD, t^2.

$$t^2(3) - t^2\left(\frac{1}{t}\right) = t^2\left(\frac{2}{t^2}\right)$$
$$3t^2 - t = 2$$
$$3t^2 - t - 2 = 0$$
$$(3t + 2)(t - 1) = 0$$

$3t + 2 = 0$ $\quad$ or $\quad$ $t - 1 = 0$
$t = -\frac{2}{3}$ $\quad$ or $\quad$ $t = 1$

Check $t = -\frac{2}{3}$: $\quad 3 + \frac{3}{2} = \frac{9}{2}$ $\quad$ *True*
Check $t = 1$: $\quad 3 - 1 = 2$ $\quad$ *True*

Solution set: $\left\{-\frac{2}{3}, 1\right\}$

13. $\dfrac{1}{x} + \dfrac{2}{x + 2} = \dfrac{17}{35}$

Multiply each term by the LCD, $35x(x + 2)$.

$$35x(x+2)\left(\frac{1}{x}\right) + 35x(x+2)\left(\frac{2}{x+2}\right)$$
$$= 35x(x+2)\left(\tfrac{17}{35}\right)$$
$$35(x + 2) + 35x(2) = 17x(x + 2)$$
$$35x + 70 + 70x = 17x^2 + 34x$$
$$70 + 105x = 17x^2 + 34x$$
$$0 = 17x^2 - 71x - 70$$
$$0 = (17x + 14)(x - 5)$$

$17x + 14 = 0$ $\quad$ or $\quad$ $x - 5 = 0$
$x = -\frac{14}{17}$ $\quad$ or $\quad$ $x = 5$

Check $x = -\frac{14}{17}$: $\quad -\frac{17}{14} + \frac{17}{10} = \frac{17}{35}$ $\quad$ *True*
Check $x = 5$: $\qquad \frac{1}{5} + \frac{2}{7} = \frac{17}{35}$ $\quad$ *True*

Solution set: $\left\{-\frac{14}{17}, 5\right\}$

15. $\dfrac{2}{x + 1} + \dfrac{3}{x + 2} = \dfrac{7}{2}$

Multiply each term by the LCD, $2(x + 1)(x + 2)$.

$$2(x+1)(x+2)\left(\frac{2}{x+1} + \frac{3}{x+2}\right)$$
$$= 2(x+1)(x+2)\left(\tfrac{7}{2}\right)$$
$$2(x + 2)(2) + 2(x + 1)(3)$$
$$= (x + 1)(x + 2)(7)$$
$$4x + 8 + 6x + 6 = (x^2 + 3x + 2)(7)$$
$$10x + 14 = 7x^2 + 21x + 14$$
$$0 = 7x^2 + 11x$$
$$0 = x(7x + 11)$$

$x = 0$ $\quad$ or $\quad$ $7x + 11 = 0$
$\qquad\qquad\qquad x = -\frac{11}{7}$

Check $x = -\frac{11}{7}$: $\quad -\frac{7}{2} + 7 = \frac{7}{2}$ $\quad$ *True*

Check $x = 0$: $\qquad 2 + \frac{3}{2} = \frac{7}{2}$ $\quad$ *True*

Solution set: $\quad \left\{ -\frac{11}{7}, 0 \right\}$

17. $\dfrac{3}{2x} - \dfrac{1}{2(x+2)} = 1$

Multiply each term by the LCD, $2x(x+2)$.

$$2x(x+2)\left(\frac{3}{2x} - \frac{1}{2(x+2)} \right)$$
$$= 2x(x+2) \cdot 1$$
$$3(x+2) - x(1) = 2x(x+2)$$
$$3x + 6 - x = 2x^2 + 4x$$
$$0 = 2x^2 + 2x - 6$$
$$0 = x^2 + x - 3$$

Use $a = 1$, $b = 1$, $c = -3$ in the quadratic formula.

$$x = \frac{-b \pm \sqrt{b^2 - 4ac}}{2a}$$
$$x = \frac{-1 \pm \sqrt{1^2 - 4(1)(-3)}}{2(1)}$$
$$= \frac{-1 \pm \sqrt{1 + 12}}{2} = \frac{-1 \pm \sqrt{13}}{2}$$

Use a calculator to check both proposed solutions. Both solutions check.

Solution set: $\quad \left\{ \frac{-1 \pm \sqrt{13}}{2} \right\}$

19. $3 = \dfrac{1}{t+2} + \dfrac{2}{(t+2)^2}$

Multiply each term by the LCD, $(t+2)^2$.

$$3(t+2)^2 = 1(t+2) + 2$$
$$3(t^2 + 4t + 4) = t + 2 + 2$$
$$3t^2 + 12t + 12 = t + 4$$
$$3t^2 + 11t + 8 = 0$$
$$(3t + 8)(t + 1) = 0$$

$3t + 8 = 0 \quad$ or $\quad t + 1 = 0$

$\qquad t = -\frac{8}{3} \qquad\qquad\quad t = -1$

Check $t = -\frac{8}{3}$: $\quad 3 = -\frac{3}{2} + \frac{9}{2}$ $\quad$ *True*

Check $t = -1$: $\quad 3 = 1 + 2$ $\quad$ *True*

Solution set: $\quad \left\{ -\frac{8}{3}, -1 \right\}$

21. $\dfrac{6}{p} = 2 + \dfrac{p}{p+1}$

Multiply each term by the LCD, $p(p+1)$.

$$6(p+1) = 2p(p+1) + p \cdot p$$
$$6p + 6 = 2p^2 + 2p + p^2$$
$$0 = 3p^2 - 4p - 6$$

Use $a = 3$, $b = -4$, and $c = -6$ in the quadratic formula.

$$p = \frac{-b \pm \sqrt{b^2 - 4ac}}{2a}$$
$$p = \frac{-(-4) \pm \sqrt{(-4)^2 - 4(3)(-6)}}{2(3)}$$
$$= \frac{4 \pm \sqrt{16 + 72}}{2(3)} = \frac{4 \pm \sqrt{88}}{2(3)}$$
$$= \frac{4 \pm 2\sqrt{22}}{2(3)} = \frac{2 \pm \sqrt{22}}{3}$$

Use a calculator to check both proposed solutions. Both solutions check.

Solution set: $\quad \left\{ \frac{2 \pm \sqrt{22}}{3} \right\}$

23. $1 - \dfrac{1}{2x+1} - \dfrac{1}{(2x+1)^2} = 0$

Multiply each term by the LCD, $(2x+1)^2$.

$$(2x+1)^2 - 1(2x+1) - 1 = 0$$
$$4x^2 + 4x + 1 - 2x - 1 - 1 = 0$$
$$4x^2 + 2x - 1 = 0$$

Use $a = 4$, $b = 2$, $c = -1$ in the quadratic formula.

$$x = \frac{-b \pm \sqrt{b^2 - 4ac}}{2a}$$
$$x = \frac{-2 \pm \sqrt{2^2 - 4(4)(-1)}}{2(4)}$$
$$= \frac{-2 \pm \sqrt{4 + 16}}{2(4)} = \frac{-2 \pm \sqrt{20}}{2(4)}$$
$$= \frac{-2 \pm 2\sqrt{5}}{2(4)} = \frac{-1 \pm \sqrt{5}}{4}$$

Use a calculator to check both proposed solutions. Both solutions check.

Solution set: $\quad \left\{ \frac{-1 \pm \sqrt{5}}{4} \right\}$

25. Rate in still water: 20 mph
Rate of current: t mph

(a) When the boat travels upstream, the current works against the rate of the boat in still water, so the rate is $(20 - t)$ mph.

(b) When the boat travels downstream, the current works with the rate of the boat in still water, so the rate is $(20 + t)$ mph.

27. Let x = rate of the boat in still water.
With the rate of the current at 15 mph, then
$x - 15$ = rate going upstream and
$x + 15$ = rate going downstream.
Complete a table using the information in the problem, the rates given above, and the formula $d = rt$ or $t = \frac{d}{r}$.

Copyright © 2012 Pearson Education, Inc. Publishing as Addison-Wesley.

	d	r	t
Upstream	4	$x - 15$	$\dfrac{4}{x - 15}$
Downstream	16	$x + 15$	$\dfrac{16}{x + 15}$

The time, 48 min, is written as $\frac{48}{60} = \frac{4}{5}$ hr. The time upstream plus the time downstream equals $\frac{4}{5}$. So, from the table, the equation is written as

$$\frac{4}{x - 15} + \frac{16}{x + 15} = \frac{4}{5}.$$

Multiply by the LCD, $5(x - 15)(x + 15)$.

$$5(x - 15)(x + 15)\left(\frac{4}{x - 15} + \frac{16}{x + 15}\right)$$
$$= 5(x - 15)(x + 15) \cdot \frac{4}{5}$$
$$20(x + 15) + 80(x - 15)$$
$$= 4(x - 15)(x + 15)$$
$$20x + 300 + 80x - 1200$$
$$= 4(x^2 - 225)$$
$$100x - 900 = 4x^2 - 900$$
$$0 = 4x^2 - 100x$$
$$0 = 4x(x - 25)$$

$$4x = 0 \quad \text{or} \quad x - 25 = 0$$
$$x = 0 \quad \text{or} \quad x = 25$$

Reject $x = 0$ mph as a possible boat speed. William's boat had a top speed of 25 mph.

29. Let $x =$ Adrian's rate from Jackson to Lodi. Then $x + 10 =$ his rate from Lodi to Manteca.

Make a table. Use $t = \frac{d}{r}$.

	d	r	t
Jackson to Lodi	40	x	$\dfrac{40}{x}$
Lodi to Manteca	40	$x + 10$	$\dfrac{40}{x + 10}$

Driving time for the entire trip was 88 minutes, or $\frac{88}{60} = \frac{22}{15}$ hours.

$$\frac{40}{x} + \frac{40}{x + 10} = \frac{22}{15}$$

Multiply each term by the LCD, $15x(x + 10)$.

$$15x(x + 10)\left(\frac{40}{x} + \frac{40}{x + 10}\right)$$
$$= 15x(x + 10)\left(\tfrac{22}{15}\right)$$
$$600(x + 10) + 600x = 22x(x + 10)$$
$$600x + 6000 + 600x = 22x^2 + 220x$$
$$0 = 22x^2 - 980x - 6000$$
$$0 = 11x^2 - 490x - 3000$$
$$0 = (11x + 60)(x - 50)$$

$$11x + 60 = 0 \quad \text{or} \quad x - 50 = 0$$
$$x = -\tfrac{60}{11} \quad \text{or} \quad x = 50$$

Reject $-\frac{60}{11}$ since speed cannot be negative. Adrian's rate from Jackson to Lodi is 50 mph.

31. Let x be the time in hours required for the faster worker to cut the lawn. Then the slower worker requires $x + 1$ hours. Complete the chart.

	Rate	Time Working Together	Fractional Part of the Job Done
Faster Worker	$\dfrac{1}{x}$	2	$\dfrac{2}{x}$
Slower Worker	$\dfrac{1}{x + 1}$	2	$\dfrac{2}{x + 1}$

Part done by faster worker plus Part done by slower worker equals one whole job.

$$\frac{2}{x} + \frac{2}{x + 1} = 1$$

Multiply each side by the LCD, $x(x + 1)$.

$$x(x + 1)\left(\frac{2}{x} + \frac{2}{x + 1}\right) = x(x + 1) \cdot 1$$
$$2(x + 1) + 2x = x^2 + x$$
$$2x + 2 + 2x = x^2 + x$$
$$0 = x^2 - 3x - 2$$

Solve for x using the quadratic formula with $a = 1$, $b = -3$, and $c = -2$.

$$x = \frac{-(-3) \pm \sqrt{(-3)^2 - 4(1)(-2)}}{2(1)}$$
$$= \frac{3 \pm \sqrt{9 + 8}}{2} = \frac{3 \pm \sqrt{17}}{2}$$
$$x = \frac{3 + \sqrt{17}}{2} \quad \text{or} \quad x = \frac{3 - \sqrt{17}}{2}$$
$$x \approx 3.6 \quad \text{or} \quad x \approx -0.6$$

Discard -0.6 as a solution since time cannot be negative. It would take the faster worker approximately 3.6 hours.

33. Let x represent the time in hours it takes Nancy to plant the flowers. Then $x + 2$ is the time it takes Rusty. Organize the information in a chart.

Worker	Rate	Time Working Together	Fractional Part of the Job Done
Nancy	$\dfrac{1}{x}$	12	$\dfrac{12}{x}$
Rusty	$\dfrac{1}{x + 2}$	12	$\dfrac{12}{x + 2}$

Part done by Nancy plus part done by Rusty equals one whole job.

$$\frac{12}{x} + \frac{12}{x + 2} = 1$$

Copyright © 2012 Pearson Education, Inc. Publishing as Addison-Wesley.

Multiply each side by the LCD, $x(x + 2)$.

$$x(x+2)\left(\frac{12}{x} + \frac{12}{x+2}\right) = x(x+2) \cdot 1$$
$$12(x+2) + 12x = x^2 + 2x$$
$$12x + 24 + 12x = x^2 + 2x$$
$$0 = x^2 - 22x - 24$$

Solve for x using the quadratic formula with $a = 1$, $b = -22$, and $c = -24$.

$$x = \frac{-(-22) \pm \sqrt{(-22)^2 - 4(1)(-24)}}{2(1)}$$
$$= \frac{22 \pm \sqrt{580}}{2}$$
$$x = \frac{22 + \sqrt{580}}{2} \approx 23.0 \quad \text{or}$$
$$x = \frac{22 - \sqrt{580}}{2} \approx -1.0$$

Since x represents time, discard the negative solution.
Nancy takes about 23.0 hours planting flowers alone while Rusty takes about 25.0 hours planting alone.

35. Let $x =$ the number of hours it takes for the faster pipe alone to fill the tank.
$x + 3 =$ the number of hours it takes for the slower pipe alone to fill the tank.

Working together, both pipes can fill the tank in 2 hours. Make a chart.

Pipe	Rate	Time	Fractional Part of Tank Filled
Faster	$\dfrac{1}{x}$	2	$\dfrac{2}{x}$
Slower	$\dfrac{1}{x+3}$	2	$\dfrac{2}{x+3}$

Since together the faster and slower pipes fill one tank, the sum of their fractional parts is 1; that is,

$$\frac{2}{x} + \frac{2}{x+3} = 1.$$

Multiply each term by the LCD, $x(x + 3)$.
$$2(x+3) + 2(x) = x(x+3)$$
$$2x + 6 + 2x = x^2 + 3x$$
$$0 = x^2 - x - 6$$
$$0 = (x-3)(x+2)$$

$$x - 3 = 0 \quad \text{or} \quad x + 2 = 0$$
$$x = 3 \quad \text{or} \quad x = -2$$

Reject -2. The faster pipe takes 3 hours to fill the tank alone and the slower pipe takes 6 hours to fill the tank alone.

37.
$$x = \sqrt{7x - 10}$$
$$(x)^2 = \left(\sqrt{7x - 10}\right)^2$$
$$x^2 = 7x - 10$$
$$x^2 - 7x + 10 = 0$$
$$(x-2)(x-5) = 0$$

$$x - 2 = 0 \quad \text{or} \quad x - 5 = 0$$
$$x = 2 \quad \text{or} \quad x = 5$$

Check $x = 2$: $\quad 2 = \sqrt{4} \quad$ *True*
Check $x = 5$: $\quad 5 = \sqrt{25} \quad$ *True*

Solution set: $\{2, 5\}$

39.
$$2x = \sqrt{11x + 3}$$
$$(2x)^2 = \left(\sqrt{11x + 3}\right)^2$$
$$4x^2 = 11x + 3$$
$$4x^2 - 11x - 3 = 0$$
$$(4x+1)(x-3) = 0$$

$$4x + 1 = 0 \quad \text{or} \quad x - 3 = 0$$
$$x = -\tfrac{1}{4} \quad \text{or} \quad x = 3$$

Check $x = -\tfrac{1}{4}$: $\quad -\tfrac{1}{2} = \sqrt{\tfrac{1}{4}} \quad$ *False*
Check $x = 3$: $\quad 6 = \sqrt{36} \quad$ *True*

Solution set: $\{3\}$

41.
$$3x = \sqrt{16 - 10x}$$
$$(3x)^2 = \left(\sqrt{16 - 10x}\right)^2$$
$$9x^2 = 16 - 10x$$
$$9x^2 + 10x - 16 = 0$$
$$(9x - 8)(x + 2) = 0$$

$$9x - 8 = 0 \quad \text{or} \quad x + 2 = 0$$
$$x = \tfrac{8}{9} \quad \text{or} \quad x = -2$$

Check $x = \tfrac{8}{9}$: $\quad \tfrac{8}{3} = \sqrt{\tfrac{64}{9}} \quad$ *True*
Check $x = -2$: $-6 = \sqrt{36} \quad$ *False*

Solution set: $\left\{\tfrac{8}{9}\right\}$

43. $t + \sqrt{t} = 12$
$$\sqrt{t} = 12 - t$$
$$\left(\sqrt{t}\right)^2 = (12 - t)^2$$
$$t = 144 - 24t + t^2$$
$$0 = t^2 - 25t + 144$$
$$0 = (t-9)(t-16)$$

$$t - 9 = 0 \quad \text{or} \quad t - 16 = 0$$
$$t = 9 \quad\quad\quad t = 16$$

Check $t = 9$: $\quad 9 + 3 = 12 \quad$ *True*
Check $t = 16$: $16 + 4 = 12 \quad$ *False*

Solution set: $\{9\}$

Copyright © 2012 Pearson Education, Inc. Publishing as Addison-Wesley.

45.
$$x = \sqrt{\frac{6 - 13x}{5}}$$
$$x^2 = \frac{6 - 13x}{5}$$
$$5x^2 = 6 - 13x$$
$$5x^2 + 13x - 6 = 0$$
$$(5x - 2)(x + 3) = 0$$

$$5x - 2 = 0 \quad \text{or} \quad x + 3 = 0$$
$$x = \tfrac{2}{5} \quad \text{or} \quad x = -3$$

Check $x = \tfrac{2}{5}$: $\tfrac{2}{5} = \sqrt{\tfrac{4}{25}}$ *True*

Check $x = -3$: $-3 = \sqrt{9}$ *False*

Solution set: $\left\{\tfrac{2}{5}\right\}$

47.
$$-x = \sqrt{\frac{8 - 2x}{3}}$$
$$(-x)^2 = \left(\sqrt{\frac{8 - 2x}{3}}\right)^2$$
$$x^2 = \frac{8 - 2x}{3}$$
$$3x^2 = 8 - 2x$$
$$3x^2 + 2x - 8 = 0$$
$$(3x - 4)(x + 2) = 0$$

$$3x - 4 = 0 \quad \text{or} \quad x + 2 = 0$$
$$x = \tfrac{4}{3} \quad \text{or} \quad x = -2$$

Check $x = \tfrac{4}{3}$: $-\tfrac{4}{3} = \sqrt{\tfrac{16}{9}}$ *False*

Check $x = -2$: $2 = \sqrt{4}$ *True*

Solution set: $\{-2\}$

49. $x^4 - 29x^2 + 100 = 0$

Let $u = x^2$, so $u^2 = x^4$.
$$u^2 - 29u + 100 = 0$$
$$(u - 4)(u - 25) = 0$$

$$u - 4 = 0 \quad \text{or} \quad u - 25 = 0$$
$$u = 4 \quad \text{or} \quad u = 25$$

To find x, substitute x^2 for u.
$$x^2 = 4 \quad \text{or} \quad x^2 = 25$$
$$x = \pm 2 \quad \text{or} \quad x = \pm 5$$

Check $x = \pm 2$: $16 - 116 + 100 = 0$ *True*

Check $x = \pm 5$: $625 - 725 + 100 = 0$ *True*

Solution set: $\{\pm 2, \pm 5\}$

51. $4q^4 - 13q^2 + 9 = 0$

Let $u = q^2$, so $u^2 = q^4$.
$$4u^2 - 13u + 9 = 0$$
$$(4u - 9)(u - 1) = 0$$

$$4u - 9 = 0 \quad \text{or} \quad u - 1 = 0$$
$$u = \tfrac{9}{4} \quad \text{or} \quad u = 1.$$

To find q, substitute q^2 for u.
$$q^2 = \tfrac{9}{4} \quad \text{or} \quad q^2 = 1$$
$$q = \pm \tfrac{3}{2} \quad \text{or} \quad q = \pm 1$$

Check $q = \pm \tfrac{3}{2}$: $\tfrac{81}{4} - \tfrac{117}{4} + 9 = 0$ *True*

Check $q = \pm 1$: $4 - 13 + 9 = 0$ *True*

Solution set: $\left\{\pm 1, \pm \tfrac{3}{2}\right\}$

53.
$$x^4 + 48 = 16x^2$$
$$x^4 - 16x^2 + 48 = 0$$

Let $u = x^2$, so $u^2 = x^4$.
$$u^2 - 16u + 48 = 0$$
$$(u - 4)(u - 12) = 0$$

$$u - 4 = 0 \quad \text{or} \quad u - 12 = 0$$
$$u = 4 \quad \text{or} \quad u = 12$$

To find x substitute x^2 for u.
$$x^2 = 4 \qquad\qquad x^2 = 12$$
$$x = \pm \sqrt{4} \qquad\qquad x = \pm \sqrt{12}$$
$$x = \pm 2 \quad \text{or} \quad x = \pm 2\sqrt{3}$$

Check $x = \pm 2$: $16 + 48 = 64$ *True*

Check $x = \pm 2\sqrt{3}$: $144 + 48 = 192$ *True*

Solution set: $\left\{\pm 2, \pm 2\sqrt{3}\right\}$

55. $(x + 3)^2 + 5(x + 3) + 6 = 0$

Let $u = x + 3$, so $u^2 = (x + 3)^2$.
$$u^2 + 5u + 6 = 0$$
$$(u + 3)(u + 2) = 0$$

$$u + 3 = 0 \quad \text{or} \quad u + 2 = 0$$
$$u = -3 \quad \text{or} \quad u = -2$$

To find x, substitute $x + 3$ for u.
$$x + 3 = -3 \quad \text{or} \quad x + 3 = -2$$
$$x = -6 \quad \text{or} \quad x = -5$$

Check $x = -6$: $9 - 15 + 6 = 0$ *True*

Check $x = -5$: $4 - 10 + 6 = 0$ *True*

Solution set: $\{-6, -5\}$

57. $3(m + 4)^2 - 8 = 2(m + 4)$

Let $u = m + 4$, so $u^2 = (m + 4)^2$.
$$3u^2 - 8 = 2u$$
$$3u^2 - 2u - 8 = 0$$
$$(3u + 4)(u - 2) = 0$$

$$3u + 4 = 0 \quad \text{or} \quad u - 2 = 0$$
$$u = -\tfrac{4}{3} \qquad\qquad u = 2$$

$$m + 4 = -\tfrac{4}{3} \quad \text{or} \quad m + 4 = 2$$
$$m = -\tfrac{16}{3} \qquad\qquad m = -2$$

Copyright © 2012 Pearson Education, Inc. Publishing as Addison-Wesley.

Check $m = -\frac{16}{3}$: $\quad \frac{16}{3} - 8 = -\frac{8}{3}$ *True*
Check $m = -2$: $\quad 12 - 8 = 4$ *True*

Solution set: $\left\{ -\frac{16}{3}, -2 \right\}$

59. $x^{2/3} + x^{1/3} - 2 = 0$

Let $u = x^{1/3}$, so $u^2 = x^{2/3}$.

$$u^2 + u - 2 = 0$$
$$(u + 2)(u - 1) = 0$$

$$u + 2 = 0 \quad \text{or} \quad u - 1 = 0$$
$$u = -2 \quad \text{or} \quad u = 1$$

To find x, substitute $x^{1/3}$ for u.

$$x^{1/3} = -2 \quad \text{or} \quad x^{1/3} = 1$$

Cube each side of each equation.

$$\left(x^{1/3}\right)^3 = (-2)^3 \qquad \left(x^{1/3}\right)^3 = 1^3$$
$$x = -8 \quad \text{or} \quad x = 1$$

Check $x = -8$: $\quad 4 - 2 - 2 = 0$ *True*
Check $x = 1$: $\quad 1 + 1 - 2 = 0$ *True*

Solution set: $\{-8, 1\}$

61. $r^{2/3} + r^{1/3} - 12 = 0$

Let $u = r^{1/3}$, so $u^2 = r^{2/3}$.

$$u^2 + u - 12 = 0$$
$$(u + 4)(u - 3) = 0$$

$$u + 4 = 0 \quad \text{or} \quad u - 3 = 0$$
$$u = -4 \quad \text{or} \quad u = 3$$

To find r, substitute $r^{1/3}$ for u.

$$r^{1/3} = -4 \qquad \text{or} \qquad r^{1/3} = 3$$
$$\left(r^{1/3}\right)^3 = (-4)^3 \qquad \left(r^{1/3}\right)^3 = 3^3$$
$$r = -64 \quad \text{or} \quad r = 27$$

Check $r = -64$: $\quad 16 - 4 - 12 = 0$ *True*
Check $r = 27$: $\quad 9 + 3 - 12 = 0$ *True*

Solution set: $\{-64, 27\}$

63. $4x^{4/3} - 13x^{2/3} + 9 = 0$

Let $u = x^{2/3}$, so $u^2 = x^{4/3}$.

$$4u^2 - 13u + 9 = 0$$
$$(4u - 9)(u - 1) = 0$$

$$4u - 9 = 0 \qquad \text{or} \qquad u - 1 = 0$$
$$u = \frac{9}{4} \qquad \text{or} \qquad u = 1$$
$$x^{2/3} = \frac{9}{4} \qquad \text{or} \qquad x^{2/3} = 1$$
$$\left(x^{2/3}\right)^{1/2} = \left(\frac{9}{4}\right)^{1/2} \qquad \text{or} \qquad \left(x^{2/3}\right)^{1/2} = 1^{1/2}$$
$$x^{1/3} = \pm\frac{3}{2} \qquad \text{or} \qquad x^{1/3} = \pm 1$$
$$\left(x^{1/3}\right)^3 = \left(\pm\frac{3}{2}\right)^3 \qquad \text{or} \qquad \left(x^{1/3}\right)^3 = (\pm 1)^3$$
$$x = \pm\frac{27}{8} \qquad \text{or} \qquad x = \pm 1$$

Check $x = \pm\frac{27}{8}$:

$$4\left(\frac{81}{16}\right) - 13\left(\frac{9}{4}\right) + 9 \stackrel{?}{=} 0$$
$$\frac{81}{4} - \frac{117}{4} + \frac{36}{4} = 0 \quad \text{*True*}$$

Check $x = \pm 1$:

$$4 - 13 + 9 = 0 \quad \text{*True*}$$

Solution set: $\left\{ \pm 1, \pm\frac{27}{8} \right\}$

65. $2 + \dfrac{5}{3x - 1} = \dfrac{-2}{(3x - 1)^2}$

Let $u = 3x - 1$, so $u^2 = (3x - 1)^2$.

$$2 + \frac{5}{u} = -\frac{2}{u^2}$$

Multiply each term by the LCD, u^2.

$$u^2\left(2 + \frac{5}{u}\right) = u^2\left(-\frac{2}{u^2}\right)$$
$$2u^2 + 5u = -2$$
$$2u^2 + 5u + 2 = 0$$
$$(2u + 1)(u + 2) = 0$$

$$2u + 1 = 0 \quad \text{or} \quad u + 2 = 0$$
$$u = -\frac{1}{2} \qquad u = -2$$

To find x, substitute $3x - 1$ for u.

$$3x - 1 = -\frac{1}{2} \quad \text{or} \quad 3x - 1 = -2$$
$$3x = \frac{1}{2} \qquad 3x = -1$$
$$x = \frac{1}{6} \quad \text{or} \quad x = -\frac{1}{3}$$

Check $x = \frac{1}{6}$: $\quad 2 - 10 = -8$ *True*
Check $x = -\frac{1}{3}$: $\quad 2 - \frac{5}{2} = -\frac{1}{2}$ *True*

Solution set: $\left\{ -\frac{1}{3}, \frac{1}{6} \right\}$

67. $2 - 6(z - 1)^{-2} = (z - 1)^{-1}$

Let $u = (z - 1)^{-1}$, so $u^2 = (z - 1)^{-2}$.

$$2 - 6u^{-2} = u^{-1}$$
$$2 - \frac{6}{u^2} = \frac{1}{u}.$$

Multiply each term by the LCD, u^2.

$$2u^2 - 6 = u$$
$$2u^2 - u - 6 = 0$$
$$(2u + 3)(u - 2) = 0$$

$$2u + 3 = 0 \quad \text{or} \quad u - 2 = 0$$
$$u = -\frac{3}{2} \quad \text{or} \quad u = 2$$

To find z, substitute $z - 1$ for u.

$$z - 1 = -\frac{3}{2} \quad \text{or} \quad z - 1 = 2$$
$$z = -\frac{1}{2} \quad \text{or} \quad z = 3$$

Check $z = -\frac{1}{2}$: $\quad 2 - \frac{8}{3} = -\frac{2}{3}$ *True*
Check $z = 3$: $\quad 2 - \frac{3}{2} = \frac{1}{2}$ *True*

Solution set: $\left\{ -\frac{1}{2}, 3 \right\}$

Copyright © 2012 Pearson Education, Inc. Publishing as Addison-Wesley.

69. $12x^4 - 11x^2 + 2 = 0$

Let $u = x^2$, so $u^2 = x^4$.

$12u^2 - 11u + 2 = 0$

$(4u - 1)(3u - 2) = 0$

$$4u - 1 = 0 \quad \text{or} \quad 3u - 2 = 0$$
$$u = \tfrac{1}{4} \quad \text{or} \quad u = \tfrac{2}{3}$$
$$x^2 = \tfrac{1}{4} \qquad\qquad x^2 = \tfrac{2}{3}$$
$$x = \pm \tfrac{1}{2} \qquad\qquad x = \pm \sqrt{\tfrac{2}{3}}$$

Note: $x = \pm \sqrt{\tfrac{2}{3}} = \pm \tfrac{\sqrt{2}}{\sqrt{3}} \cdot \tfrac{\sqrt{3}}{\sqrt{3}} = \pm \tfrac{\sqrt{6}}{3}$

Check $x = \pm \tfrac{1}{2}$: $\tfrac{3}{4} - \tfrac{11}{4} + 2 = 0$ *True*

Check $x = \pm \tfrac{\sqrt{6}}{3}$: $\tfrac{16}{3} - \tfrac{22}{3} + 2 = 0$ *True*

Solution set: $\left\{ \pm \tfrac{\sqrt{6}}{3}, \pm \tfrac{1}{2} \right\}$

71. $\sqrt{2x + 3} = 2 + \sqrt{x - 2}$

Square each side.

$$\left(\sqrt{2x + 3} \right)^2 = \left(2 + \sqrt{x - 2} \right)^2$$
$$2x + 3 = 4 + 4\sqrt{x - 2} + (x - 2)$$
$$2x + 3 = x + 2 + 4\sqrt{x - 2}$$

Isolate the radical term on one side.

$$x + 1 = 4\sqrt{x - 2}$$

Square each side again.

$$(x + 1)^2 = \left(4\sqrt{x - 2} \right)^2$$
$$x^2 + 2x + 1 = 16(x - 2)$$
$$x^2 + 2x + 1 = 16x - 32$$
$$x^2 - 14x + 33 = 0$$
$$(x - 11)(x - 3) = 0$$

$$x - 11 = 0 \quad \text{or} \quad x - 3 = 0$$
$$x = 11 \quad \text{or} \quad x = 3$$

Check $x = 11$: $\sqrt{25} = 2 + \sqrt{9}$ *True*

Check $x = 3$: $\sqrt{9} = 2 + \sqrt{1}$ *True*

Solution set: $\{3, 11\}$

73. $2\left(1 + \sqrt{r} \right)^2 = 13\left(1 + \sqrt{r} \right) - 6$

Let $u = 1 + \sqrt{r}$, so $u^2 = \left(1 + \sqrt{r} \right)^2$.

$$2u^2 = 13u - 6$$
$$2u^2 - 13u + 6 = 0$$
$$(2u - 1)(u - 6) = 0$$

$$2u - 1 = 0 \quad \text{or} \quad u - 6 = 0$$
$$u = \tfrac{1}{2} \quad \text{or} \quad u = 6$$

Replace u with $1 + \sqrt{r}$.

$$1 + \sqrt{r} = \tfrac{1}{2} \quad \text{or} \quad 1 + \sqrt{r} = 6$$
$$\sqrt{r} = -\tfrac{1}{2} \qquad\qquad \sqrt{r} = 5$$
$$\text{\emph{Not possible,}} \qquad\qquad r = 25$$
$$\text{\emph{since }} \sqrt{r} \geq 0.$$

Check $r = 25$: $72 = 78 - 6$ *True*

Solution set: $\{25\}$

75. $2m^6 + 11m^3 + 5 = 0$

Let $y = m^3$, so $y^2 = m^6$.

$$2y^2 + 11y + 5 = 0$$
$$(2y + 1)(y + 5) = 0$$

$$2y + 1 = 0 \quad \text{or} \quad y + 5 = 0$$
$$y = -\tfrac{1}{2} \quad \text{or} \quad y = -5$$

To find m, substitute m^3 for y.

$$m^3 = -\tfrac{1}{2} \quad \text{or} \quad m^3 = -5$$

Take the cube root of each side of each equation.

$$m = \sqrt[3]{-\tfrac{1}{2}} \qquad\qquad \text{or} \quad m = \sqrt[3]{-5}$$
$$m = -\sqrt[3]{\tfrac{1}{2}} \qquad\qquad m = -\sqrt[3]{5}$$
$$= -\frac{\sqrt[3]{1}}{\sqrt[3]{2}} \cdot \frac{\sqrt[3]{2^2}}{\sqrt[3]{2^2}}$$
$$= -\frac{\sqrt[3]{4}}{2}$$

Check $m = -\dfrac{\sqrt[3]{4}}{2}$: $\tfrac{1}{2} - \tfrac{11}{2} + 5 = 0$ *True*

Check $m = -\sqrt[3]{5}$: $50 - 55 + 5 = 0$ *True*

Solution set: $\left\{ -\sqrt[3]{5}, -\tfrac{\sqrt[3]{4}}{2} \right\}$

77. $6 = 7(2w - 3)^{-1} + 3(2w - 3)^{-2}$

Let $u = (2w - 3)^{-1}$, so $u^2 = (2w - 3)^{-2}$.

$$6 = 7u + 3u^2$$
$$0 = 3u^2 + 7u - 6$$
$$0 = (3u - 2)(u + 3)$$

$$3u - 2 = 0 \qquad \text{or} \qquad u + 3 = 0$$
$$u = \tfrac{2}{3} \qquad \text{or} \qquad u = -3$$
$$(2w - 3)^{-1} = \tfrac{2}{3} \qquad\qquad (2w - 3)^{-1} = -3$$
$$\frac{1}{2w - 3} = \frac{2}{3} \qquad\qquad \frac{1}{2w - 3} = -3$$
$$3 = 2(2w - 3) \qquad\qquad 1 = -3(2w - 3)$$
$$3 = 4w - 6 \qquad\qquad 1 = -6w + 9$$
$$9 = 4w \qquad\qquad -8 = -6w$$
$$\tfrac{9}{4} = w \qquad\qquad \tfrac{4}{3} = w$$

Check $w = \tfrac{4}{3}$: $6 = -21 + 27$ *True*

Check $w = \tfrac{9}{4}$: $6 = \tfrac{14}{3} + \tfrac{4}{3}$ *True*

Solution set: $\left\{ \tfrac{4}{3}, \tfrac{9}{4} \right\}$

Copyright © 2012 Pearson Education, Inc. Publishing as Addison-Wesley.

79.
$$2x^4 - 9x^2 = -2$$
$$2x^4 - 9x^2 + 2 = 0$$
Let $u = x^2$, so $u^2 = x^4$.
$$2u^2 - 9u + 2 = 0$$

Use $a = 2$, $b = -9$, and $c = 2$ in the quadratic formula.

$$u = \frac{-b \pm \sqrt{b^2 - 4ac}}{2a}$$

$$u = \frac{-(-9) \pm \sqrt{(-9)^2 - 4(2)(2)}}{2(2)}$$

$$= \frac{9 \pm \sqrt{81 - 16}}{4} = \frac{9 \pm \sqrt{65}}{4}$$

To find x, substitute x^2 for u.

$$x^2 = \frac{9 \pm \sqrt{65}}{4}$$

$$x = \pm \sqrt{\frac{9 \pm \sqrt{65}}{4}} = \pm \frac{\sqrt{9 \pm \sqrt{65}}}{2}$$

Note: The last expression represents four numbers. All four proposed solutions check.

Solution set: $\left\{ \pm \frac{\sqrt{9 + \sqrt{65}}}{2}, \pm \frac{\sqrt{9 - \sqrt{65}}}{2} \right\}$

81.
$$2x^4 + x^2 - 3 = 0$$
Let $m = x^2$, so $m^2 = x^4$.
$$2m^2 + m - 3 = 0$$
$$(2m + 3)(m - 1) = 0$$

$$2m + 3 = 0 \quad \text{or} \quad m - 1 = 0$$
$$m = -\tfrac{3}{2} \quad \text{or} \quad m = 1$$

To find x, substitute x^2 for m.
$$x^2 = -\tfrac{3}{2} \quad \text{or} \quad x^2 = 1$$

$$x^2 = -\tfrac{3}{2} \qquad \text{or} \quad x^2 = 1$$
$$x = \pm \sqrt{-\tfrac{3}{2}} \qquad x = \pm \sqrt{1}$$
$$x = \pm \frac{\sqrt{3}}{\sqrt{2}} \cdot \frac{\sqrt{2}}{\sqrt{2}} i \qquad x = \pm 1$$
$$x = \pm \frac{\sqrt{6}}{2} i$$

Check $x = \pm \frac{\sqrt{6}}{2} i$: $\frac{9}{2} - \frac{3}{2} - 3 = 0$ *True*
Check $x = \pm 1$: $2 + 1 - 3 = 0$ *True*

Solution set: $\left\{ \pm 1, \pm \frac{\sqrt{6}}{2} i \right\}$

83. Solve $P = 2L + 2W$ for W.
$$P - 2L = 2W$$
$$\frac{P - 2L}{2} = W, \quad \text{or} \quad W = \frac{P}{2} - L$$

85. Solve $F = \frac{9}{5}C + 32$ for C.
$$F - 32 = \frac{9}{5}C$$
$$\tfrac{5}{9}(F - 32) = C$$

Summary Exercises on Solving Quadratic Equations

1. $(2x + 3)^2 = 4$

Since the equation has the form $(ax + b)^2 = c$, use the *square root property*.

3. $x^2 + 5x - 8 = 0$

The discriminant is
$$b^2 - 4ac = 5^2 - 4(1)(-8)$$
$$= 25 + 32 = 57.$$

Since the discriminant is not a perfect square, use the *quadratic formula*.

5.
$$3x^2 = 2 - 5x$$
$$3x^2 + 5x - 2 = 0$$

The discriminant is
$$b^2 - 4ac = 5^2 - 4(3)(-2)$$
$$= 25 + 24 = 49.$$

Since the discriminant is a perfect square, use *factoring*.

7. $p^2 = 7$

$$p = \sqrt{7} \quad \text{or} \quad p = -\sqrt{7}$$

Solution set: $\left\{ \pm \sqrt{7} \right\}$

9. $n^2 + 6n + 4 = 0$
$$n^2 + 6n = -4$$
$$n^2 + 6n + 9 = -4 + 9 \quad \left[\tfrac{1}{2}(6)\right]^2 = 9$$
$$(n + 3)^2 = 5$$

$$n + 3 = \sqrt{5} \qquad \text{or} \quad n + 3 = -\sqrt{5}$$
$$n = -3 + \sqrt{5} \quad \text{or} \qquad n = -3 - \sqrt{5}$$

Solution set: $\left\{ -3 \pm \sqrt{5} \right\}$

11. $\dfrac{5}{x} + \dfrac{12}{x^2} = 2$

Multiply each term by the LCD, x^2.
$$5x + 12 = 2x^2$$
$$0 = 2x^2 - 5x - 12$$
$$0 = (2x + 3)(x - 4)$$

$$2x + 3 = 0 \quad \text{or} \quad x - 4 = 0$$
$$x = -\tfrac{3}{2} \quad \text{or} \qquad x = 4$$

Solution set: $\left\{ -\tfrac{3}{2}, 4 \right\}$

Copyright © 2012 Pearson Education, Inc. Publishing as Addison-Wesley.

13. $2r^2 - 4r + 1 = 0$

Use $a = 2$, $b = -4$, and $c = 1$ in the quadratic formula.

$$r = \frac{-b \pm \sqrt{b^2 - 4ac}}{2a}$$

$$r = \frac{-(-4) \pm \sqrt{(-4)^2 - 4(2)(1)}}{2(2)}$$

$$= \frac{4 \pm \sqrt{16 - 8}}{2(2)} = \frac{4 \pm \sqrt{8}}{2(2)}$$

$$= \frac{4 \pm 2\sqrt{2}}{2(2)} = \frac{2 \pm \sqrt{2}}{2}$$

Solution set: $\left\{ \frac{2 \pm \sqrt{2}}{2} \right\}$

15. $\qquad x\sqrt{2} = \sqrt{5x - 2}$

$$\left(x\sqrt{2} \right)^2 = \left(\sqrt{5x - 2} \right)^2$$

$$x^2 \cdot 2 = 5x - 2$$

$$2x^2 - 5x + 2 = 0$$

$$(2x - 1)(x - 2) = 0$$

$2x - 1 = 0 \quad$ or $\quad x - 2 = 0$

$x = \frac{1}{2} \qquad\qquad x = 2$

Check $x = \frac{1}{2}$: $\quad \frac{1}{2}\sqrt{2} = \sqrt{\frac{1}{2}} \qquad$ *True*

Check $x = 2$: $\quad 2\sqrt{2} = \sqrt{8} \qquad$ *True*

Solution set: $\left\{ \frac{1}{2}, 2 \right\}$

17. $(2x + 3)^2 = 8$

$2x + 3 = \sqrt{8} \qquad$ or $\quad 2x + 3 = -\sqrt{8}$

$2x = -3 + 2\sqrt{2} \qquad\qquad 2x = -3 - 2\sqrt{2}$

$x = \dfrac{-3 + 2\sqrt{2}}{2} \quad$ or $\quad x = \dfrac{-3 - 2\sqrt{2}}{2}$

Solution set: $\left\{ \frac{-3 \pm 2\sqrt{2}}{2} \right\}$

19. $\qquad t^4 + 14 = 9t^2$

$$t^4 - 9t^2 + 14 = 0$$

$$\left(t^2 - 2 \right)\left(t^2 - 7 \right) = 0$$

$t^2 - 2 = 0 \qquad$ or $\quad t^2 - 7 = 0$

$t^2 = 2 \qquad\qquad\qquad t^2 = 7$

$t = \pm\sqrt{2} \quad$ or $\qquad t = \pm\sqrt{7}$

Solution set: $\left\{ \pm\sqrt{2}, \pm\sqrt{7} \right\}$

21. $z^2 + z + 1 = 0$

Use $a = 1$, $b = 1$, and $c = 1$ in the quadratic formula.

$$z = \frac{-b \pm \sqrt{b^2 - 4ac}}{2a}$$

$$z = \frac{-1 \pm \sqrt{1^2 - 4(1)(1)}}{2(1)}$$

$$= \frac{-1 \pm \sqrt{1 - 4}}{2}$$

$$= \frac{-1 \pm \sqrt{-3}}{2}$$

$$= \frac{-1 \pm i\sqrt{3}}{2}$$

$$= -\frac{1}{2} \pm \frac{\sqrt{3}}{2}i$$

Solution set: $\left\{ -\frac{1}{2} \pm \frac{\sqrt{3}}{2}i \right\}$

23. $\qquad 4t^2 - 12t + 9 = 0$

$$(2t - 3)(2t - 3) = 0$$

$$(2t - 3)^2 = 0$$

$$2t - 3 = 0$$

$$t = \frac{3}{2}$$

Solution set: $\left\{ \frac{3}{2} \right\}$

25. $r^2 - 72 = 0$

$$r^2 = 72$$

$$r = \pm\sqrt{72} = \pm 6\sqrt{2}$$

Solution set: $\left\{ \pm 6\sqrt{2} \right\}$

27. $\qquad x^2 - 5x - 36 = 0$

$$(x + 4)(x - 9) = 0$$

$x + 4 = 0 \quad$ or $\quad x - 9 = 0$

$x = -4 \quad$ or $\qquad x = 9$

Solution set: $\{-4, 9\}$

29. $\qquad 3p^2 = 6p - 4$

$$3p^2 - 6p + 4 = 0$$

Use $a = 3$, $b = -6$, and $c = 4$ in the quadratic formula.

$$p = \frac{-b \pm \sqrt{b^2 - 4ac}}{2a}$$

$$p = \frac{-(-6) \pm \sqrt{(-6)^2 - 4(3)(4)}}{2(3)}$$

$$= \frac{6 \pm \sqrt{36 - 48}}{2(3)} = \frac{6 \pm \sqrt{-12}}{2(3)}$$

$$= \frac{6 \pm 2i\sqrt{3}}{2(3)} = \frac{3 \pm i\sqrt{3}}{3} = 1 \pm \frac{\sqrt{3}}{3}i$$

Solution set: $\left\{ 1 \pm \frac{\sqrt{3}}{3}i \right\}$

31. $\qquad \dfrac{4}{r^2} + 3 = \dfrac{1}{r}$

Multiply each term by the LCD, r^2.

$$4 + 3r^2 = r$$

$$3r^2 - r + 4 = 0$$

Use $a = 3$, $b = -1$, and $c = 4$ in the quadratic formula.

$$r = \frac{-b \pm \sqrt{b^2 - 4ac}}{2a}$$

$$r = \frac{-(-1) \pm \sqrt{(-1)^2 - 4(3)(4)}}{2(3)}$$

$$= \frac{1 \pm \sqrt{1 - 48}}{6} = \frac{1 \pm \sqrt{-47}}{6}$$

$$= \frac{1 \pm i\sqrt{47}}{6} = \frac{1}{6} \pm \frac{\sqrt{47}}{6}i$$

Solution set: $\left\{ \frac{1}{6} \pm \frac{\sqrt{47}}{6}i \right\}$

9.4 Formulas and Further Applications

9.4 Now Try Exercises

N1. **(a)** Solve $n = \dfrac{ab}{E^2}$ for E.

$$E^2 n = ab \qquad \textit{Multiply by } E^2.$$

$$E^2 = \frac{ab}{n} \qquad \textit{Divide by n.}$$

$$E = \pm\sqrt{\frac{ab}{n}} \qquad \begin{array}{l}\textit{Square root}\\\textit{property}\end{array}$$

$$E = \frac{\pm\sqrt{ab}}{\sqrt{n}} \cdot \frac{\sqrt{n}}{\sqrt{n}} \qquad \begin{array}{l}\textit{Rationalize}\\\textit{the}\\\textit{denominator.}\end{array}$$

$$E = \frac{\pm\sqrt{abn}}{n}$$

(b) Solve $S = \sqrt{\dfrac{pq}{n}}$ for p.

$$S^2 = \frac{pq}{n} \qquad \begin{array}{l}\textit{Square both}\\\textit{sides.}\end{array}$$

$$nS^2 = pq \qquad \textit{Multiply by n.}$$

$$\frac{nS^2}{q} = p \qquad \textit{Divide by q.}$$

N2. Solve $r^2 + 9r = -c$ for r.

$$r^2 + 9r + c = 0$$

Use $a = 1$, $b = 9$, and $c = c$ in the quadratic formula.

$$r = \frac{-b \pm \sqrt{b^2 - 4ac}}{2a}$$

$$r = \frac{-9 \pm \sqrt{9^2 - 4(1)c}}{2(1)}$$

$$r = \frac{-9 \pm \sqrt{81 - 4c}}{2}$$

The solutions are

$$r = \frac{-9 + \sqrt{81 - 4c}}{2} \text{ and } r = \frac{-9 - \sqrt{81 - 4c}}{2}.$$

N3. *Step 2*

Let $x = $ the width of the barn. Then $x + 10 = $ the length of the barn.

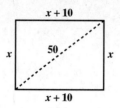

Step 3

The diagonal of the footprint is 50 feet, so a right triangle is formed. Use the Pythagorean theorem.

$$a^2 + b^2 = c^2$$
$$x^2 + (x + 10)^2 = 50^2$$

Step 4

$$x^2 + x^2 + 20x + 100 = 2500$$
$$2x^2 + 20x - 2400 = 0$$
$$x^2 + 10x - 1200 = 0$$
$$(x + 40)(x - 30) = 0$$

$$x + 40 = 0 \quad \text{or} \quad x - 30 = 0$$
$$x = -40 \quad \text{or} \qquad\quad x = 30$$

Step 5

The width cannot be -40, so the width of the barn is 30 feet and the length of the barn is $30 + 10 = 40$ feet.

Step 6

$30^2 + 40^2 = 50^2$ and 40 is 10 more than 30, as required.

N4. *Step 2*

Let $x = $ the width of the grass strip.

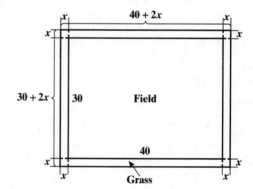

Step 3

The width of the large rectangle is $30 + 2x$, and the length is $40 + 2x$.

$$\begin{array}{ccc}\text{Area of} & \text{area of} & \text{area of}\\\text{rectangle} & - \quad \text{field} & = \quad \text{grass.}\end{array}$$

$$(40 + 2x)(30 + 2x) - 40(30) = 296$$

Step 4

$$1200 + 140x + 4x^2 - 1200 = 296$$
$$4x^2 + 140x - 296 = 0$$
$$x^2 + 35x - 74 = 0$$
$$(x + 37)(x - 2) = 0$$

Copyright © 2012 Pearson Education, Inc. Publishing as Addison-Wesley.

$$x + 37 = 0 \quad \text{or} \quad x - 2 = 0$$
$$x = -37 \quad \text{or} \quad x = 2$$

Step 5
The width cannot be -37, so the grass strip should be 2 yards wide.

Step 6
If $x = 2$, then the area of the large rectangle is

$$(40 + 2 \cdot 2)(30 + 2 \cdot 2) = 44 \cdot 34 = 1496 \text{ yd}^2.$$

The area of the field is $40 \cdot 30 = 1200 \text{ yd}^2$. So, the area of the grass strip is $1496 - 1200 = 296 \text{ yd}^2$, as required. The answer is correct.

N5. $-16t^2 + 60t + 120 = s(t)$

$$-16t^2 + 60t + 120 = 0 \qquad \textit{Let s(t) = 0.}$$
$$4t^2 - 15t - 30 = 0 \qquad \textit{Divide by −4.}$$

Use $a = 4$, $b = -15$, and $c = -30$ in the quadratic formula.

$$t = \frac{-b \pm \sqrt{b^2 - 4ac}}{2a}$$

$$t = \frac{-(-15) \pm \sqrt{(-15)^2 - 4(4)(-30)}}{2(4)}$$

$$= \frac{15 \pm \sqrt{225 + 480}}{8} = \frac{15 \pm \sqrt{705}}{8}$$

$$t = \frac{15 + \sqrt{705}}{8} \approx 5.2 \quad \text{or} \quad t = \frac{15 - \sqrt{705}}{8} \approx -1.4$$

The object will hit the ground about 5.2 seconds after it is projected.

N6. **(a)** For 2005, $x = 2005 - 1980 = 25$.

$$f(x) = -0.065x^2 + 14.8x + 249$$
$$f(25) = -0.065(25)^2 + 14.8(25) + 249$$
$$= 578.375$$

The CPI for 2005 was about 578.

(b) Find the value of x that makes $f(x) = 500$.

$$f(x) = -0.065x^2 + 14.8x + 249$$
$$500 = -0.065x^2 + 14.8x + 249$$
$$0 = -0.065x^2 + 14.8x - 251$$

Now use $a = -0.065$, $b = 14.8$, and $c = -251$ in the quadratic formula.

$$x = \frac{-14.8 \pm \sqrt{14.8^2 - 4(-0.065)(-251)}}{2(-0.065)}$$

$$= \frac{-14.8 \pm \sqrt{153.78}}{-0.13}$$

$$x \approx 18.5 \quad \text{or} \quad x \approx 209.2$$

The solution $x \approx 209.2$ is rejected since it corresponds to a year far beyond the period covered by the model. The solution $x \approx 18.5$ is valid for this problem. The corresponding year is 1998.

9.4 Section Exercises

1. The first step in solving a formula that has the specified variable in the denominator is to multiply each side by the LCD to clear the equation of fractions.

3. We must recognize that a formula like

$$gw^2 = kw + 24$$

is quadratic in w. So the first step is to write the formula in standard form (with 0 on one side, in decreasing powers of w). This allows us to apply the quadratic formula to solve for w.

5. Since the triangle is a right triangle, use the Pythagorean theorem with legs m and n and hypotenuse p.

$$m^2 + n^2 = p^2$$
$$m^2 = p^2 - n^2$$
$$m = \sqrt{p^2 - n^2}$$

Only the positive square root is given since m represents the side of a triangle.

7. Solve $d = kt^2$ for t.

$$kt^2 = d$$
$$t^2 = \frac{d}{k} \qquad \textit{Divide by k.}$$
$$t = \pm\sqrt{\frac{d}{k}} \qquad \textit{Use square root property.}$$
$$= \frac{\pm\sqrt{d}}{\sqrt{k}} \cdot \frac{\sqrt{k}}{\sqrt{k}} \qquad \textit{Rationalize denominator.}$$
$$t = \frac{\pm\sqrt{dk}}{k} \qquad \textit{Simplify.}$$

9. Solve $I = \frac{ks}{d^2}$ for d.

$$Id^2 = ks \qquad \textit{Multiply by } d^2.$$
$$d^2 = \frac{ks}{I} \qquad \textit{Divide by I.}$$
$$d = \pm\sqrt{\frac{ks}{I}} \qquad \textit{Use square root property.}$$
$$= \pm\frac{\sqrt{ks}}{\sqrt{I}} \cdot \frac{\sqrt{I}}{\sqrt{I}} \qquad \textit{Rationalize denominator.}$$
$$d = \frac{\pm\sqrt{ksI}}{I} \qquad \textit{Simplify.}$$

11. Solve $F = \frac{kA}{v^2}$ for v.

$$v^2 F = kA \qquad \textit{Multiply by } v^2.$$
$$v^2 = \frac{kA}{F} \qquad \textit{Divide by F.}$$
$$v = \pm\sqrt{\frac{kA}{F}} \qquad \textit{Use square root property.}$$
$$= \frac{\pm\sqrt{kA}}{\sqrt{F}} \cdot \frac{\sqrt{F}}{\sqrt{F}} \qquad \textit{Rationalize denominator.}$$
$$v = \frac{\pm\sqrt{kAF}}{F} \qquad \textit{Simplify.}$$

Copyright © 2012 Pearson Education, Inc. Publishing as Addison-Wesley.

13. Solve $V = \frac{1}{3}\pi r^2 h$ for r.

$3V = \pi r^2 h$ $\qquad$ *Multiply by 3.*

$\dfrac{3V}{\pi h} = r^2$ $\qquad$ *Divide by πh.*

$r = \pm\sqrt{\dfrac{3V}{\pi h}}$ $\qquad$ *Use square root property.*

$= \dfrac{\pm\sqrt{3V}\cdot\sqrt{\pi h}}{\sqrt{\pi h}\cdot\sqrt{\pi h}}$ $\qquad$ *Rationalize denominator.*

$r = \dfrac{\pm\sqrt{3\pi V h}}{\pi h}$ $\qquad$ *Simplify.*

15. Solve $At^2 + Bt = -C$ for t.

$At^2 + Bt + C = 0$

Use the quadratic formula.

$$t = \frac{-B \pm \sqrt{B^2 - 4AC}}{2A}$$

17. Solve $D = \sqrt{kh}$ for h.

$D^2 = kh$ $\qquad$ *Square each side.*

$\dfrac{D^2}{k} = h$ $\qquad$ *Divide by k.*

19. Solve $p = \sqrt{\dfrac{k\ell}{g}}$ for ℓ.

$p^2 = \dfrac{k\ell}{g}$ $\qquad$ *Square each side.*

$p^2 g = k\ell$ $\qquad$ *Multiply by g.*

$\dfrac{p^2 g}{k} = \ell$ $\qquad$ *Divide by k.*

21. Solve $S = 4\pi r^2$ for r.

$\dfrac{S}{4\pi} = r^2$ $\qquad$ *Divide by 4π.*

$r = \pm\sqrt{\dfrac{S}{4\pi}}$ $\qquad$ *Use square root property.*

$= \dfrac{\pm\sqrt{S}\cdot\sqrt{\pi}}{\sqrt{4\pi}\cdot\sqrt{\pi}}$ $\qquad$ *Rationalize denominator.*

$r = \dfrac{\pm\sqrt{S\pi}}{2\pi}$ $\qquad$ *Simplify.*

23. Solve $p = \dfrac{E^2 R}{(r+R)^2}$ for R $(E > 0)$.

$p(r+R)^2 = E^2 R$

$p(r^2 + 2rR + R^2) = E^2 R$

$pr^2 + 2prR + pR^2 = E^2 R$

$pR^2 + 2prR - E^2 R + pr^2 = 0$

$pR^2 + (2pr - E^2)R + pr^2 = 0$

Here $a = p$, $b = 2pr - E^2$, and $c = pr^2$.

$R = \dfrac{-(2pr - E^2) \pm \sqrt{(2pr - E^2)^2 - 4p\cdot pr^2}}{2p}$

$= \dfrac{E^2 - 2pr \pm \sqrt{4p^2 r^2 - 4prE^2 + E^4 - 4p^2 r^2}}{2p}$

$= \dfrac{E^2 - 2pr \pm \sqrt{E^4 - 4prE^2}}{2p}$

$= \dfrac{E^2 - 2pr \pm \sqrt{E^2(E^2 - 4pr)}}{2p}$

$R = \dfrac{E^2 - 2pr \pm E\sqrt{E^2 - 4pr}}{2p}$

25. Solve $10p^2 c^2 + 7pcr = 12r^2$ for r.

$0 = 12r^2 - 7pcr - 10p^2 c^2$

Here $a = 12$, $b = -7pc$, and $c = -10p^2 c^2$.

$r = \dfrac{-(-7pc) \pm \sqrt{(-7pc)^2 - 4(12)(-10p^2 c^2)}}{2(12)}$

$= \dfrac{7pc \pm \sqrt{49p^2 c^2 + 480p^2 c^2}}{24}$

$= \dfrac{7pc \pm \sqrt{529p^2 c^2}}{24} = \dfrac{7pc \pm 23pc}{24}$

$r = \dfrac{7pc + 23pc}{24} = \dfrac{30pc}{24} = \dfrac{5pc}{4}$ or

$r = \dfrac{7pc - 23pc}{24} = \dfrac{-16pc}{24} = -\dfrac{2pc}{3}$

27. Solve $LI^2 + RI + \dfrac{1}{c} = 0$ for I.

$cLI^2 + cRI + 1 = 0$ $\quad$ *Multiply by c.*

Here $a = cL$, $b = cR$, and $c = 1$.

$I = \dfrac{-cR \pm \sqrt{(cR)^2 - 4(cL)(1)}}{2(cL)}$

$= \dfrac{-cR \pm \sqrt{c^2 R^2 - 4cL}}{2cL}$

29. Apply the Pythagorean theorem.

$(x + 4)^2 = x^2 + (x + 1)^2$

$x^2 + 8x + 16 = x^2 + x^2 + 2x + 1$

$0 = x^2 - 6x - 15$

Here $a = 1$, $b = -6$, and $c = -15$.

$x = \dfrac{-(-6) \pm \sqrt{(-6)^2 - 4(1)(-15)}}{2(1)}$

$= \dfrac{6 \pm \sqrt{36 + 60}}{2} = \dfrac{6 \pm \sqrt{96}}{2}$

$x = \dfrac{6 + \sqrt{96}}{2} \approx 7.9$ or

$x = \dfrac{6 - \sqrt{96}}{2} \approx -1.9$

Reject the negative solution.
If $x = 7.9$, then

$x + 4 = 11.9$ and $x + 1 = 8.9$.

The lengths of the sides of the triangle are approximately 7.9, 8.9, and 11.9.

31. Let $x =$ the distance traveled by the eastbound ship. Then $x + 70 =$ the distance traveled by the southbound ship.

Copyright © 2012 Pearson Education, Inc. Publishing as Addison-Wesley.

Since the ships are traveling at right angles to one another, the distance d between them can be found using the Pythagorean theorem.

$$c^2 = a^2 + b^2$$
$$d^2 = x^2 + (x + 70)^2$$

Let $d = 170$, and solve for x.

$$170^2 = x^2 + (x + 70)^2$$
$$28{,}900 = x^2 + x^2 + 140x + 4900$$
$$0 = 2x^2 + 140x - 24{,}000$$
$$0 = x^2 + 70x - 12{,}000$$
$$0 = (x + 150)(x - 80)$$

$$x + 150 = 0 \quad \text{or} \quad x - 80 = 0$$
$$x = -150 \quad \text{or} \quad x = 80$$

Distance cannot be negative, so reject -150. If $x = 80$, then $x + 70 = 150$. The eastbound ship traveled 80 miles, and the southbound ship traveled 150 miles.

33. Let $\quad x = $ length of the shorter leg;
$\quad 2x - 1 = $ length of the longer leg;
$\quad 2x - 1 + 2 = $ length of the hypotenuse.

Use the Pythagorean theorem.

$$x^2 + (2x - 1)^2 = (2x + 1)^2$$
$$x^2 + 4x^2 - 4x + 1 = 4x^2 + 4x + 1$$
$$5x^2 - 4x + 1 = 4x^2 + 4x + 1$$
$$x^2 - 8x = 0$$
$$x(x - 8) = 0$$

$$x = 0 \quad \text{or} \quad x - 8 = 0$$
$$x = 8$$

Since x represents length, discard 0 as a solution. If $x = 8$, then

$$2x - 1 = 2(8) - 1 = 15 \quad \text{and}$$
$$2x - 1 + 2 = 2(8) + 1 = 17.$$

The lengths are 8 inches, 15 inches, and 17 inches.

35. Let $x = $ the width of the rug.
Then $2x + 4 = $ the length of the rug.

Use the Pythagorean theorem.

$$x^2 + (2x + 4)^2 = 26^2$$
$$x^2 + 4x^2 + 16x + 16 = 676$$
$$5x^2 + 16x - 660 = 0$$
$$(5x + 66)(x - 10) = 0$$

$$5x + 66 = 0 \quad \text{or} \quad x - 10 = 0$$
$$x = -\tfrac{66}{5} \quad \text{or} \quad x = 10$$

Discard the negative solution. If $x = 10$, then $2x + 4 = 24$. The width of the rug is 10 feet, and the length is 24 feet.

37. Let $x = $ the width of the border. Then the width of the pool and the border is $30 + 2x$ ft and the length of the pool and border is $40 + 2x$ ft.

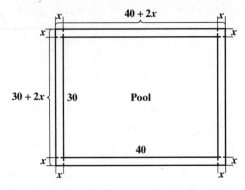

Since the area of the pool is $30 \cdot 40 = 1200$ ft^2, we can write an equation using the total area of the pool and the border.

The area of the pool and border	is	the area of the pool	plus	the area of the border.
$(30 + 2x)(40 + 2x)$	$=$	1200	$+$	296

$$1200 + 140x + 4x^2 = 1496$$
$$4x^2 + 140x - 296 = 0$$
$$x^2 + 35x - 74 = 0$$
$$(x - 2)(x + 37) = 0$$

$$x - 2 = 0 \quad \text{or} \quad x + 37 = 0$$
$$x = 2 \quad \text{or} \quad x = -37$$

Discard the negative solution. The strip can be 2 feet wide.

39. Let $x = $ original width of the rectangle. Then $2x - 2$ represents the original length of the rectangle.

Now $x + 5$ is the new width that makes the rectangle a square. Thus, the new width must equal the original length since the sides of a square are equal.

$$x + 5 = 2x - 2$$
$$7 = x$$

The dimensions of the original rectangle are 7 meters and $2(7) - 2 = 12$ meters. Note that the area of the square is $(12)(12) = 144$ m^2.

41. Let x be the width of the sheet metal. Then the length is $2x - 4$.

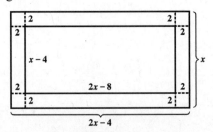

Copyright © 2012 Pearson Education, Inc. Publishing as Addison-Wesley.

By cutting out 2-inch squares from each corner we get a rectangle with width $x - 4$ and length $(2x - 4) - 4 = 2x - 8$. The uncovered box then has height 2 inches, length $2x - 8$ inches, and width $x - 4$ inches.

Use the formula $V = LWH$ or $V = HLW$.

$$256 = 2(2x - 8)(x - 4)$$
$$256 = 4(x - 4)(x - 4) \qquad \text{Factor out 2.}$$
$$64 = (x - 4)^2 \qquad \text{Divide by 4.}$$

Use the square root property.

$$\pm 8 = x - 4$$

$$x - 4 = 8 \quad \text{or} \quad x - 4 = -8$$
$$x = 12 \quad \text{or} \quad x = -4$$

Since x represents width, discard the negative solution. The width is 12 inches, and the length is $2(12) - 4 = 20$ inches.

43.
$$s = 144t - 16t^2$$
$$128 = 144t - 16t^2 \qquad \text{Let } s = 128.$$
$$0 = -16t^2 + 144t - 128$$
$$0 = t^2 - 9t + 8 \qquad \text{Divide by } -16.$$
$$0 = (t - 8)(t - 1)$$

$$t - 8 = 0 \quad \text{or} \quad t - 1 = 0$$
$$t = 8 \quad \text{or} \quad t = 1$$

The object will be 128 feet above the ground at two times, going up and coming down, or at 1 second and at 8 seconds.

45.
$$s = -16t^2 + 128t$$
$$213 = -16t^2 + 128t \qquad \text{Let } s = 213.$$
$$0 = -16t^2 + 128t - 213$$

Here $a = -16$, $b = 128$, and $c = -213$.

$$t = \frac{-b \pm \sqrt{b^2 - 4ac}}{2a}$$
$$t = \frac{-128 \pm \sqrt{128^2 - 4(-16)(-213)}}{2(-16)}$$
$$= \frac{-128 \pm \sqrt{16{,}384 - 13{,}632}}{-32}$$
$$= \frac{-128 \pm \sqrt{2752}}{-32}$$
$$t = \frac{-128 + \sqrt{2752}}{-32} \approx 2.4 \quad \text{or}$$
$$t = \frac{-128 - \sqrt{2752}}{-32} \approx 5.6$$

The ball will be 213 feet from the ground after 2.4 seconds and again after 5.6 seconds.

47.
$$D(t) = 13t^2 - 100t$$
$$180 = 13t^2 - 100t \qquad \text{Let } D(t) = 180.$$
$$0 = 13t^2 - 100t - 180$$

Here $a = 13$, $b = -100$, and $c = -180$.

$$t = \frac{-b \pm \sqrt{b^2 - 4ac}}{2a}$$
$$t = \frac{-(-100) \pm \sqrt{(-100)^2 - 4(13)(-180)}}{2(13)}$$
$$= \frac{100 \pm \sqrt{10{,}000 + 9360}}{2(13)}$$
$$= \frac{100 \pm \sqrt{19{,}360}}{2(13)}$$
$$= \frac{100 \pm 44\sqrt{10}}{2(13)} = \frac{50 \pm 22\sqrt{10}}{13}$$
$$t = \frac{50 + 22\sqrt{10}}{13} \approx 9.2 \quad \text{or}$$
$$t = \frac{50 - 22\sqrt{10}}{13} \approx -1.5$$

Discard the negative solution. The car will skid 180 feet in approximately 9.2 seconds.

49.
$$s(t) = -16t^2 + 160t$$
$$400 = -16t^2 + 160t \qquad \text{Let } s(t) = 400.$$
$$0 = -16t^2 + 160t - 400$$
$$0 = t^2 - 10t + 25 \qquad \text{Divide by } -16.$$
$$0 = (t - 5)(t - 5)$$
$$0 = (t - 5)^2$$
$$0 = t - 5$$
$$5 = t$$

The ball reaches a height of 400 feet after 5 seconds. This is its maximum height since this is the only time it reaches 400 feet.

51. Supply and demand are equal when

$$3p - 200 = \frac{3200}{p}.$$

Solve for p.

$$3p^2 - 200p = 3200$$
$$3p^2 - 200p - 3200 = 0$$

Use the quadratic formula with $a = 3$, $b = -200$, and $c = -3200$.

$$p = \frac{-(-200) \pm \sqrt{(-200)^2 - 4(3)(-3200)}}{2(3)}$$
$$= \frac{200 \pm \sqrt{40{,}000 + 38{,}400}}{6}$$
$$= \frac{200 \pm \sqrt{78{,}400}}{6} = \frac{200 \pm 280}{6}$$
$$p = \frac{480}{6} = 80 \quad \text{or} \quad p = \frac{-80}{6} = -\frac{40}{3}$$

Discard the negative solution. The supply and demand are equal when the price is 80 cents or $0.80.

Copyright © 2012 Pearson Education, Inc. Publishing as Addison-Wesley.

53. Let $r =$ the interest rate. Let $A = 2142.45$, $P = 2000$, and solve for r.

$$A = P(1 + r)^2$$
$$2142.45 = 2000(1 + r)^2$$
$$1.071225 = (1 + r)^2$$

Use the square root property.

$$1 + r = 1.035 \quad \text{or} \quad 1 + r = -1.035$$
$$r = 0.035 \quad \text{or} \quad r = -2.035$$

Since the interest rate cannot be negative, reject -2.035. The interest rate is 3.5%.

55. Let F denote the Froude number.

Solve $F = \dfrac{v^2}{g\ell}$ for v.

$$v^2 = Fg\ell$$
$$v = \pm\sqrt{Fg\ell}$$

v is positive, so

$$v = \sqrt{Fg\ell}.$$

For the rhinoceros, $\ell = 1.2$ and $F = 2.57$.

$$v = \sqrt{(2.57)(9.8)(1.2)} \approx 5.5$$

or 5.5 meters per second.

57. Write a proportion.

$$\frac{x - 4}{3x - 19} = \frac{4}{x - 3}$$

Multiply by the LCD, $(3x - 19)(x - 3)$.

$$(3x - 19)(x - 3)\left(\frac{x - 4}{3x - 19}\right)$$
$$= (3x - 19)(x - 3)\left(\frac{4}{x - 3}\right)$$

$$(x - 3)(x - 4) = (3x - 19)4$$
$$x^2 - 7x + 12 = 12x - 76$$
$$x^2 - 19x + 88 = 0$$
$$(x - 8)(x - 11) = 0$$

$$x - 8 = 0 \quad \text{or} \quad x - 11 = 0$$
$$x = 8 \quad \text{or} \quad x = 11$$

If $x = 8$, then $3x - 19 = 3(8) - 19 = 5$.

If $x = 11$, then $3x - 19 = 3(11) - 19 = 14$.

Thus, $AC = 5$ or $AC = 14$.

59. **(a)** From the graph, the spending on physician and clinical services in 2005 appears to be $420 billion (to the nearest $10 billion).

(b) For 2005, $x = 2005 - 2000 = 5$.

$$f(x) = 0.3214x^2 + 25.06x + 288.2$$
$$f(5) = 0.3214(5)^2 + 25.06(5) + 288.2$$
$$= 421.535$$

To the nearest $10 billion, the model gives $420 billion, the same as the estimate in part (a).

61. Use $f(x) = 0.3214x^2 + 25.06x + 288.2$ with $f(x) = 400$.

$$400 = 0.3214x^2 + 25.06x + 288.2$$
$$0 = 0.3214x^2 + 25.06x - 111.8$$

Here $a = 0.3214$, $b = 25.06$, and $c = -111.8$.

$$x = \frac{-b \pm \sqrt{b^2 - 4ac}}{2a}$$

$$x = \frac{-25.06 \pm \sqrt{(25.06)^2 - 4(0.3214)(-111.8)}}{2(0.3214)}$$

$$= \frac{-25.06 \pm \sqrt{771.73368}}{0.6428}$$

$$\approx 4.23 \text{ or } -82.20$$

Reject the negative value. The model indicates that the spending on physician and clinical services reached, and then exceeded, $400 billion in the year $2000 + 4 = 2004$. The graph indicates that spending first exceeded $400 billion in 2005.

63. $\quad 3 - x \le 5$

$\quad\quad -x \le 2 \quad$ *Subtract 3.*

$\quad\quad x \ge -2 \quad$ *Multiply by -1; reverse inequality.*

Solution set: $[-2, \infty)$

65. $\quad -\frac{1}{2}x - 3 > 5$

$\quad\quad -\frac{1}{2}x > 8 \quad$ *Add 3.*

$\quad\quad x < -16 \quad$ *Multiply by -2; reverse inequality.*

Solution set: $(-\infty, -16)$

9.5 Polynomial and Rational Inequalities

9.5 Now Try Exercises

N1. $\quad x^2 + 2x - 8 > 0$

Use factoring to solve the quadratic equation

$$x^2 + 2x - 8 = 0.$$
$$(x + 4)(x - 2) = 0$$

$$x + 4 = 0 \quad \text{or} \quad x - 2 = 0$$
$$x = -4 \quad \text{or} \quad x = 2$$

Locate the numbers -4 and 2 that divide the number line into three intervals A, B, and C.

Choose a number from each interval to substitute in the inequality

$$x^2 + 2x - 8 > 0.$$

Interval A: Let $x = -5$.

$$(-5)^2 + 2(-5) - 8 \overset{?}{>} 0$$
$$25 - 10 - 8 \overset{?}{>} 0$$
$$7 > 0 \qquad \textit{True}$$

Interval B: Let $x = 0$.

$$(0)^2 + 2(0) - 8 \overset{?}{>} 0$$
$$-8 > 0 \qquad \textit{False}$$

Interval C: Let $x = 3$.

$$(3)^2 + 2(3) - 8 \overset{?}{>} 0$$
$$9 + 6 - 8 \overset{?}{>} 0$$
$$7 > 0 \qquad \textit{True}$$

The numbers in Intervals A and C are solutions. The numbers -4 and 2 are excluded because of $>$.

Solution set: $(-\infty, -4) \cup (2, \infty)$

N2. **(a)** $(4x - 1)^2 > -3$

The square of any real number is always greater than or equal to 0, so any real number satisfies this inequality.

Solution set: $(-\infty, \infty)$

(b) $(4x - 1)^2 < -3$

The square of a real number is never negative.

Solution set: $\varnothing$

N3. $(x + 4)(x - 3)(2x + 1) \le 0$
Solve the equation
$(x + 4)(x - 3)(2x + 1) = 0$.
$x + 4 = 0$ or $x - 3 = 0$ or $2x + 1 = 0$
$x = -4$ or $x = 3$ or $x = -\frac{1}{2}$
Locate these numbers and the intervals A, B, C, and D on a number line.

Test a number from each interval in the inequality

$$(x + 4)(x - 3)(2x + 1) \le 0.$$

Interval A: Let $x = -5$.

$$(-1)(-8)(-9) \overset{?}{\le} 0$$
$$-72 \le 0 \qquad \textit{True}$$

Interval B: Let $x = -2$.

$$(2)(-5)(-3) \overset{?}{\le} 0$$
$$30 \le 0 \qquad \textit{False}$$

Interval C: Let $x = 0$.

$$(4)(-3)(1) \overset{?}{\le} 0$$
$$-12 \le 0 \qquad \textit{True}$$

Interval D: Let $x = 4$.

$$(8)(1)(9) \overset{?}{\le} 0$$
$$72 \le 0 \qquad \textit{False}$$

The numbers in Intervals A and C, including $-4, -\frac{1}{2}$, and 3 because of $\le$, are solutions.

Solution set: $(-\infty, -4] \cup [-\frac{1}{2}, 3]$

N4. $\dfrac{3}{x + 1} > 4$

Write the inequality so that 0 is on one side.

$$\frac{3}{x + 1} - 4 > 0$$
$$\frac{3}{x + 1} - \frac{4(x + 1)}{x + 1} > 0$$
$$\frac{3 - 4x - 4}{x + 1} > 0$$
$$\frac{-4x - 1}{x + 1} > 0$$

The number $-\frac{1}{4}$ makes the numerator 0, and -1 makes the denominator 0. These two numbers determine three intervals.

Test a number from each interval in the inequality

$$\frac{3}{x + 1} > 4.$$

Interval A: Let $x = -2$.

$$\frac{3}{-2 + 1} \overset{?}{>} 4$$
$$-3 > 4 \qquad \textit{False}$$

Interval B: Let $x = -\frac{1}{2}$.

$$\frac{3}{-\frac{1}{2} + 1} \overset{?}{>} 4$$
$$6 > 4 \qquad \textit{True}$$

Interval C: Let $x = 0$.

$$\frac{3}{0 + 1} \overset{?}{>} 4$$
$$3 > 4 \qquad \textit{False}$$

The solution set includes numbers in Interval B, excluding endpoints.

Solution set: $(-1, -\frac{1}{4})$

Copyright © 2012 Pearson Education, Inc. Publishing as Addison-Wesley.

N5.
$$\frac{x-3}{x+3} \le 2$$

Write the inequality so that 0 is on one side.

$$\frac{x-3}{x+3} - 2 \le 0$$

$$\frac{x-3}{x+3} - \frac{2(x+3)}{x+3} \le 0$$

$$\frac{x-3-2x-6}{x+3} \le 0$$

$$\frac{-x-9}{x+3} \le 0$$

The number -9 makes the numerator 0, and -3 makes the denominator 0. These two numbers determine three intervals.

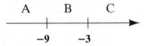

Test a number from each interval in the inequality

$$\frac{x-3}{x+3} \le 2.$$

Interval A: Let $x = -10$.
$$\frac{-10-3}{-10+3} \overset{?}{\le} 2$$
$$\frac{13}{7} \le 2 \qquad \textit{True}$$

Interval B: Let $x = -5$.
$$\frac{-5-3}{-5+3} \overset{?}{\le} 2$$
$$4 \le 2 \qquad \textit{False}$$

Interval C: Let $x = 0$.
$$\frac{0-3}{0+3} \overset{?}{\le} 2$$
$$-1 \le 2 \qquad \textit{True}$$

The numbers in Intervals A and C are solutions. -3 is not in the solution set (since it makes the denominator 0), but -9 is.

Solution set: $(-\infty, -9] \cup (-3, \infty)$

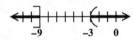

9.5 Section Exercises

1.
$$x^2 - 2x - 3 > 0 \quad \textit{Original inequality}$$
$$0^2 - 2(0) - 3 \overset{?}{>} 0 \quad \textit{Choose 0 in } (-1, 3).$$
$$-3 > 0 \quad \textit{False}$$

3.
$$2x^2 - 7x - 15 > 0 \quad \textit{Original inequality}$$
$$2(6)^2 - 7(6) - 15 \overset{?}{>} 0 \quad \textit{Choose 6 in } (5, \infty).$$
$$72 - 42 - 15 \overset{?}{>} 0$$
$$15 > 0 \quad \textit{True}$$

5.
$$(x+1)(x-5) > 0$$
Solve the equation
$$(x+1)(x-5) = 0.$$

$$x + 1 = 0 \quad \text{or} \quad x - 5 = 0$$
$$x = -1 \quad \text{or} \quad x = 5$$

The numbers -1 and 5 divide a number line into three intervals: A, B, and C.

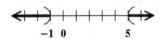

Test a number from each interval in the original inequality, $(x+1)(x-5) > 0$.

Interval A: Let $x = -2$.
$$(-2+1)(-2-5) \overset{?}{>} 0$$
$$-1(-7) \overset{?}{>} 0$$
$$7 > 0 \qquad \textit{True}$$

Interval B: Let $x = 0$.
$$(0+1)(0-5) \overset{?}{>} 0$$
$$-5 > 0 \qquad \textit{False}$$

Interval C: Let $x = 6$.
$$(6+1)(6-5) \overset{?}{>} 0$$
$$7 > 0 \qquad \textit{True}$$

The solution set includes the numbers in Intervals A and C, excluding -1 and 5 because of $>$.

Solution set: $(-\infty, -1) \cup (5, \infty)$

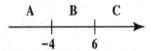

7.
$$(x+4)(x-6) < 0$$
Solve the equation
$$(x+4)(x-6) = 0.$$

$$x + 4 = 0 \quad \text{or} \quad x - 6 = 0$$
$$x = -4 \quad \text{or} \quad x = 6$$

The numbers -4 and 6 divide a number line into three intervals: A, B, and C.

Test a number from each interval in the original inequality, $(x+4)(x-6) < 0$.

Interval A: Let $x = -5$.
$$(-5+4)(-5-6) \overset{?}{<} 0$$
$$-1(-11) \overset{?}{<} 0$$
$$11 < 0 \qquad \textit{False}$$

Copyright © 2012 Pearson Education, Inc. Publishing as Addison-Wesley.

Interval B: Let $x = 0$.
$$4(-6) \overset{?}{<} 0$$
$$-24 < 0 \qquad \textit{True}$$
Interval C: Let $x = 7$.
$$(7+4)(7-6) \overset{?}{<} 0$$
$$11(1) \overset{?}{<} 0$$
$$11 < 0 \qquad \textit{False}$$

The solution set includes Interval B, where the expression is negative.

Solution set: $(-4, 6)$

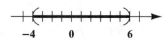

9. $x^2 - 4x + 3 \geq 0$
Solve the equation
$$x^2 - 4x + 3 = 0.$$
$$(x-1)(x-3) = 0$$

$$x - 1 = 0 \quad \text{or} \quad x - 3 = 0$$
$$x = 1 \quad \text{or} \quad x = 3$$

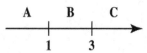

Test a number from each interval in the original inequality, $x^2 - 4x + 3 \geq 0$.

Interval A: Let $x = 0$.
$$3 \geq 0 \qquad \textit{True}$$
Interval B: Let $x = 2$.
$$2^2 - 4(2) + 3 \overset{?}{\geq} 0$$
$$-1 \geq 0 \qquad \textit{False}$$
Interval C: Let $x = 4$.
$$4^2 - 4(4) + 3 \overset{?}{\geq} 0$$
$$3 \geq 0 \qquad \textit{True}$$

The solution set includes the numbers in Intervals A and C, including 1 and 3 because of $\geq$.

Solution set: $(-\infty, 1] \cup [3, \infty)$

11. $10x^2 + 9x \geq 9$
$$10x^2 + 9x - 9 \geq 0$$
Solve the equation
$$10x^2 + 9x - 9 = 0.$$
$$(2x+3)(5x-3) = 0$$

$$2x + 3 = 0 \quad \text{or} \quad 5x - 3 = 0$$
$$x = -\tfrac{3}{2} \quad \text{or} \quad x = \tfrac{3}{5}$$

Test a number from each interval in the original inequality, $10x^2 + 9x \geq 9$.

Interval A: Let $x = -2$.
$$10(-2)^2 + 9(-2) \overset{?}{\geq} 9$$
$$40 - 18 \overset{?}{\geq} 9$$
$$22 \geq 9 \qquad \textit{True}$$
Interval B: Let $x = 0$.
$$0 \geq 9 \qquad \textit{False}$$
Interval C: Let $x = 1$.
$$10(1)^2 + 9(1) \overset{?}{\geq} 9$$
$$10 + 9 \overset{?}{\geq} 9$$
$$19 \geq 9 \qquad \textit{True}$$

The solution set includes the numbers in Intervals A and C, including $-\frac{3}{2}$ and $\frac{3}{5}$ because of $\geq$.

Solution set: $\left(-\infty, -\frac{3}{2}\right] \cup \left[\frac{3}{5}, \infty\right)$

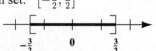

13. $4x^2 - 9 \leq 0$
Solve the equation
$$4x^2 - 9 = 0.$$
$$(2x+3)(2x-3) = 0$$

$$2x + 3 = 0 \quad \text{or} \quad 2x - 3 = 0$$
$$x = -\tfrac{3}{2} \quad \text{or} \quad x = \tfrac{3}{2}$$

Test a number from each interval in the original inequality, $4x^2 - 9 \leq 0$.

Interval A: Let $x = -2$.
$$4(-2)^2 - 9 \overset{?}{\leq} 0$$
$$7 \leq 0 \qquad \textit{False}$$
Interval B: Let $x = 0$.
$$-9 \leq 0 \qquad \textit{True}$$
Interval C: Let $x = 2$.
$$4(2)^2 - 9 \overset{?}{\leq} 0$$
$$7 \leq 0 \qquad \textit{False}$$

The solution set includes Interval B, including the endpoints.

Solution set: $\left[-\frac{3}{2}, \frac{3}{2}\right]$

Copyright © 2012 Pearson Education, Inc. Publishing as Addison-Wesley.

15.
$$6x^2 + x \geq 1$$
$$6x^2 + x - 1 \geq 0$$

Solve the equation

$$6x^2 + x - 1 = 0.$$
$$(2x + 1)(3x - 1) = 0$$

$$2x + 1 = 0 \quad \text{or} \quad 3x - 1 = 0$$
$$x = -\tfrac{1}{2} \quad \text{or} \quad x = \tfrac{1}{3}$$

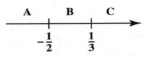

Test a number from each interval in the original inequality, $6x^2 + x \geq 1$.

Interval A: Let $x = -1$.
$$6(-1)^2 + (-1) \overset{?}{\geq} 1$$
$$5 \geq 1 \qquad \textit{True}$$

Interval B: Let $x = 0$.
$$0 \geq 1 \qquad \textit{False}$$

Interval C: Let $x = 1$.
$$6(1)^2 + 1 \overset{?}{\geq} 1$$
$$7 \geq 1 \qquad \textit{True}$$

The solution set includes the numbers in Intervals A and C, including $-\tfrac{1}{2}$ and $\tfrac{1}{3}$ because of $\geq$.

Solution set: $\left(-\infty, -\tfrac{1}{2}\right] \cup \left[\tfrac{1}{3}, \infty\right)$

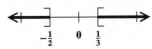

17. $z^2 - 4z \geq 0$

Solve the equation

$$z^2 - 4z = 0.$$
$$z(z - 4) = 0$$

$$z = 0 \quad \text{or} \quad z - 4 = 0$$
$$z = 4$$

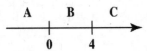

Test a number from each interval in the original inequality, $z^2 - 4z \geq 0$.

Interval A: Let $z = -1$.
$$(-1)^2 - 4(-1) \overset{?}{\geq} 0$$
$$5 \geq 0 \qquad \textit{True}$$

Interval B: Let $z = 2$.
$$2^2 - 4(2) \overset{?}{\geq} 0$$
$$-4 \geq 0 \qquad \textit{False}$$

Interval C: Let $z = 5$.
$$5^2 - 4(5) \overset{?}{\geq} 0$$
$$5 \geq 0 \qquad \textit{True}$$

The solution set includes the numbers in Intervals A and C, including 0 and 4 because of $\geq$.

Solution set: $(-\infty, 0] \cup [4, \infty)$

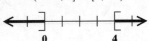

19. $3x^2 - 5x \leq 0$

Solve the equation

$$3x^2 - 5x = 0.$$
$$x(3x - 5) = 0$$

$$x = 0 \quad \text{or} \quad 3x - 5 = 0$$
$$x = \tfrac{5}{3}$$

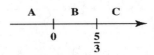

Test a number from each interval in the original inequality, $3x^2 - 5x \leq 0$.

Interval A: Let $x = -1$.
$$3(-1)^2 - 5(-1) \overset{?}{\leq} 0$$
$$8 \leq 0 \qquad \textit{False}$$

Interval B: Let $x = 1$.
$$3(1)^2 - 5(1) \overset{?}{\leq} 0$$
$$-2 \leq 0 \qquad \textit{True}$$

Interval C: Let $x = 2$.
$$3(2)^2 - 5(2) \overset{?}{\leq} 0$$
$$2 \leq 0 \qquad \textit{False}$$

The solution set includes the numbers in Interval B, including the endpoints.

Solution set: $\left[0, \tfrac{5}{3}\right]$

21. $x^2 - 6x + 6 \geq 0$

Solve the equation

$$x^2 - 6x + 6 = 0.$$

Since $x^2 - 6x + 6$ does not factor, let $a = 1$, $b = -6$, and $c = 6$ in the quadratic formula.

$$x = \frac{-(-6) \pm \sqrt{(-6)^2 - 4(1)(6)}}{2(1)}$$

$$= \frac{6 \pm \sqrt{12}}{2} = \frac{6 \pm 2\sqrt{3}}{2}$$

$$= \frac{2\left(3 \pm \sqrt{3}\right)}{2} = 3 \pm \sqrt{3}$$

$$x = 3 + \sqrt{3} \approx 4.7 \text{ or}$$
$$x = 3 - \sqrt{3} \approx 1.3$$

Copyright © 2012 Pearson Education, Inc. Publishing as Addison-Wesley.

Test a number from each interval in the original inequality, $x^2 - 6x + 6 \geq 0$.

Interval A: Let $x = 0$.
$$6 \geq 0 \qquad \textit{True}$$
Interval B: Let $x = 3$.
$$3^2 - 6(3) + 6 \overset{?}{\geq} 0$$
$$-3 \geq 0 \qquad \textit{False}$$
Interval C: Let $x = 5$.
$$5^2 - 6(5) + 6 \overset{?}{\geq} 0$$
$$1 \geq 0 \qquad \textit{True}$$

The solution set includes the numbers in Intervals A and C, including $3 - \sqrt{3}$ and $3 + \sqrt{3}$ because of $\geq$.

Solution set: $\left(-\infty, 3 - \sqrt{3}\right] \cup \left[3 + \sqrt{3}, \infty\right)$

23. $(4 - 3x)^2 \geq -2$

Since $(4 - 3x)^2$ is either 0 or positive, $(4 - 3x)^2$ will always be greater than -2. Therefore, the solution set is $(-\infty, \infty)$.

25. $(3x + 5)^2 \leq -4$

Since $(3x + 5)^2$ is never negative, $(3x + 5)^2$ will never be less than or equal to a negative number. Therefore, the solution set is $\emptyset$.

27. $(x - 1)(x - 2)(x - 4) < 0$
The numbers 1, 2, and 4 are solutions of the cubic equation
$$(x - 1)(x - 2)(x - 4) = 0.$$

These numbers divide a number line into four intervals.

Test a number from each interval in the inequality
$$(x - 1)(x - 2)(x - 4) < 0.$$

Interval A: Let $x = 0$.
$$-1(-2)(-4) \overset{?}{<} 0$$
$$-8 < 0 \qquad \textit{True}$$

Interval B: Let $x = 1.5$.
$$(1.5 - 1)(1.5 - 2)(1.5 - 4) \overset{?}{<} 0$$
$$0.5(-0.5)(-2.5) \overset{?}{<} 0$$
$$0.625 < 0 \qquad \textit{False}$$
Interval C: Let $x = 3$.
$$(3 - 1)(3 - 2)(3 - 4) \overset{?}{<} 0$$
$$2(1)(-1) \overset{?}{<} 0$$
$$-2 < 0 \qquad \textit{True}$$
Interval D: Let $x = 5$.
$$(5 - 1)(5 - 2)(5 - 4) \overset{?}{<} 0$$
$$4(3)(1) \overset{?}{<} 0$$
$$12 < 0 \qquad \textit{False}$$

The numbers in Intervals A and C, not including 1, 2, or 4, are solutions.

Solution set: $(-\infty, 1) \cup (2, 4)$

29. $(x - 4)(2x + 3)(3x - 1) \geq 0$
The numbers 4, $-\frac{3}{2}$, and $\frac{1}{3}$ are solutions of the cubic equation
$$(x - 4)(2x + 3)(3x - 1) = 0.$$

These numbers divide a number line into 4 intervals.

Interval A: Let $x = -2$.
$$-6(-1)(-7) \overset{?}{\geq} 0$$
$$-42 \geq 0 \qquad \textit{False}$$
Interval B: Let $x = 0$.
$$-4(3)(-1) \overset{?}{\geq} 0$$
$$12 \geq 0 \qquad \textit{True}$$
Interval C: Let $x = 1$.
$$-3(5)(2) \overset{?}{\geq} 0$$
$$-30 \geq 0 \qquad \textit{False}$$
Interval D: Let $x = 5$.
$$1(13)(14) \overset{?}{\geq} 0$$
$$182 \geq 0 \qquad \textit{True}$$

The solution set includes numbers in Intervals B and D, including the endpoints.

Solution set: $\left[-\frac{3}{2}, \frac{1}{3}\right] \cup [4, \infty)$

Copyright © 2012 Pearson Education, Inc. Publishing as Addison-Wesley.

31. $\dfrac{x-1}{x-4} > 0$

The number 1 makes the numerator 0, and 4 makes the denominator 0. These two numbers determine three intervals.

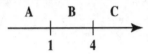

Test a number from each interval in the inequality

$$\dfrac{x-1}{x-4} > 0.$$

Interval A: Let $x = 0$.

$$\dfrac{0-1}{0-4} \overset{?}{>} 0$$

$$\dfrac{1}{4} > 0 \qquad \textit{True}$$

Interval B: Let $x = 2$.

$$\dfrac{2-1}{2-4} \overset{?}{>} 0$$

$$\dfrac{1}{-2} > 0 \qquad \textit{False}$$

Interval C: Let $x = 5$.

$$\dfrac{5-1}{5-4} \overset{?}{>} 0$$

$$4 > 0 \qquad \textit{True}$$

The solution set includes numbers in Intervals A and C, excluding endpoints.

Solution set: $(-\infty, 1) \cup (4, \infty)$

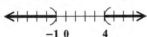

33. $\dfrac{2x+3}{x-5} \le 0$

The number $-\frac{3}{2}$ makes the numerator 0, and 5 makes the denominator 0. These two numbers determine three intervals.

Test a number from each interval in the inequality

$$\dfrac{2x+3}{x-5} \le 0.$$

Interval A: Let $x = -2$.

$$\dfrac{2(-2)+3}{(-2)-5} \overset{?}{\le} 0$$

$$\tfrac{1}{7} \le 0 \qquad \textit{False}$$

Interval B: Let $x = 0$.

$$\dfrac{2(0)+3}{0-5} \overset{?}{\le} 0$$

$$-\tfrac{3}{5} \le 0 \qquad \textit{True}$$

Interval C: Let $x = 6$.

$$\dfrac{2(6)+3}{6-5} \overset{?}{\le} 0$$

$$15 \le 0 \qquad \textit{False}$$

The solution set includes the points in Interval B. The endpoint 5 is not included since it makes the left side undefined. The endpoint $-\frac{3}{2}$ is included because it makes the left side equal to 0.

Solution set: $\left[-\frac{3}{2}, 5\right)$

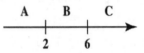

35. $\dfrac{8}{x-2} \ge 2$

Write the inequality so that 0 is on one side.

$$\dfrac{8}{x-2} - 2 \ge 0$$

$$\dfrac{8}{x-2} - \dfrac{2(x-2)}{x-2} \ge 0$$

$$\dfrac{8-2x+4}{x-2} \ge 0$$

$$\dfrac{-2x+12}{x-2} \ge 0$$

The number 6 makes the numerator 0, and 2 makes the denominator 0. These two numbers determine three intervals.

Test a number from each interval in the inequality

$$\dfrac{8}{x-2} \ge 2.$$

Interval A: Let $x = 0$.

$$\dfrac{8}{0-2} \overset{?}{\ge} 2$$

$$-4 \ge 2 \qquad \textit{False}$$

Interval B: Let $x = 3$.

$$\dfrac{8}{3-2} \overset{?}{\ge} 2$$

$$8 \ge 2 \qquad \textit{True}$$

Interval C: Let $x = 7$.

$$\dfrac{8}{7-2} \overset{?}{\ge} 2$$

$$\tfrac{8}{5} \ge 2 \qquad \textit{False}$$

The solution set includes numbers in Interval B, including 6 but excluding 2, which makes the fraction undefined.

Solution set: $(2, 6]$

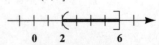

Copyright © 2012 Pearson Education, Inc. Publishing as Addison-Wesley.

37.
$$\frac{3}{2x-1} < 2$$

Write the inequality so that 0 is on one side.

$$\frac{3}{2x-1} - 2 < 0$$

$$\frac{3}{2x-1} - \frac{2(2x-1)}{2x-1} < 0$$

$$\frac{3 - 4x + 2}{2x-1} < 0$$

$$\frac{-4x+5}{2x-1} < 0$$

The number $\frac{5}{4}$ makes the numerator 0, and $\frac{1}{2}$ makes the denominator 0. These two numbers determine three intervals.

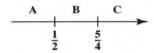

Test a number from each interval in the inequality

$$\frac{3}{2x-1} < 2.$$

Interval A: Let $x = 0$.

$$\frac{3}{2(0)-1} \overset{?}{<} 2$$

$$-3 < 2 \qquad \textit{True}$$

Interval B: Let $x = 1$.

$$\frac{3}{2(1)-1} \overset{?}{<} 2$$

$$3 < 2 \qquad \textit{False}$$

Interval C: Let $x = 2$.

$$\frac{3}{2(2)-1} \overset{?}{<} 2$$

$$1 < 2 \qquad \textit{True}$$

The solution set includes numbers in Intervals A and C, excluding endpoints.

Solution set: $\left(-\infty, \frac{1}{2}\right) \cup \left(\frac{5}{4}, \infty\right)$

39.
$$\frac{x-3}{x+2} \geq 2$$

Write the inequality so that 0 is on one side.

$$\frac{x-3}{x+2} - 2 \geq 0$$

$$\frac{x-3}{x+2} - \frac{2(x+2)}{x+2} \geq 0$$

$$\frac{x-3-2x-4}{x+2} \geq 0$$

$$\frac{-x-7}{x+2} \geq 0$$

The number -7 makes the numerator 0, and -2 makes the denominator 0. These two numbers determine three intervals.

Test a number from each interval in the inequality

$$\frac{x-3}{x+2} \geq 2.$$

Interval A: Let $x = -8$.

$$\frac{-11}{-6} \overset{?}{\geq} 2$$

$$\frac{11}{6} \geq 2 \qquad \textit{False}$$

Interval B: Let $x = -4$.

$$\frac{-7}{-2} \overset{?}{\geq} 2$$

$$\frac{7}{2} \geq 2 \qquad \textit{True}$$

Interval C: Let $x = 0$.

$$\frac{-3}{2} \geq 2 \qquad \textit{False}$$

The solution set includes numbers in Interval B, including -7 but excluding -2, which makes the fraction undefined.

Solution set: $[-7, -2)$

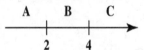

41.
$$\frac{x-8}{x-4} < 3$$

Write the inequality so that 0 is on one side.

$$\frac{x-8}{x-4} - 3 < 0$$

$$\frac{x-8}{x-4} - \frac{3(x-4)}{x-4} < 0$$

$$\frac{x-8-3x+12}{x-4} < 0$$

$$\frac{-2x+4}{x-4} < 0$$

The number 2 makes the numerator 0, and 4 makes the denominator 0. These two numbers determine three intervals.

Test a number from each interval in the inequality

$$\frac{x-8}{x-4} < 3.$$

Interval A: Let $x = 0$.

$$\frac{-8}{-4} \overset{?}{<} 3$$

$$2 < 3 \qquad \textit{True}$$

Interval B: Let $x = 3$.

$$\frac{-5}{-1} \overset{?}{<} 3$$

$$5 < 3 \qquad \textit{False}$$

Interval C: Let $x = 5$.

$$\frac{-3}{1} < 3 \qquad \textit{True}$$

Copyright © 2012 Pearson Education, Inc. Publishing as Addison-Wesley.

The solution set includes numbers in Intervals A and C, excluding endpoints.

Solution set: $(-\infty, 2) \cup (4, \infty)$

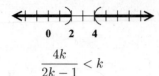

43.
$$\frac{4k}{2k-1} < k$$

Write the inequality so that 0 is on one side.

$$\frac{4k}{2k-1} - k < 0$$

$$\frac{4k}{2k-1} - \frac{k(2k-1)}{2k-1} < 0$$

$$\frac{4k - 2k^2 + k}{2k-1} < 0$$

$$\frac{-2k^2 + 5k}{2k-1} < 0$$

$$\frac{k(-2k + 5)}{2k-1} < 0$$

The numbers 0 and $\frac{5}{2}$ make the numerator 0, and $\frac{1}{2}$ makes the denominator 0. These three numbers determine four intervals.

Test a number from each interval in the inequality

$$\frac{4k}{2k-1} < k.$$

Interval A: Let $k = -1$.

$$\frac{4(-1)}{2(-1)-1} \overset{?}{<} -1$$

$$\frac{4}{3} < -1 \qquad \text{False}$$

Interval B: Let $k = \frac{1}{4}$.

$$\frac{4\left(\frac{1}{4}\right)}{2\left(\frac{1}{4}\right)-1} \overset{?}{<} \frac{1}{4}$$

$$-2 < \frac{1}{4} \qquad \text{True}$$

Interval C: Let $k = 1$.

$$\frac{4(1)}{2(1)-1} \overset{?}{<} 1$$

$$4 < 1 \qquad \text{False}$$

Interval D: Let $k = 3$.

$$\frac{4(3)}{2(3)-1} \overset{?}{<} 3$$

$$\frac{12}{5} < 3 \qquad \text{True}$$

The solution set includes numbers in Intervals B and D. None of the endpoints are included.

Solution set: $\left(0, \frac{1}{2}\right) \cup \left(\frac{5}{2}, \infty\right)$

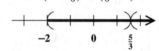

45. $\dfrac{2x-3}{x^2+1} \geq 0$

The denominator is positive for all real numbers x, so it has no effect on the solution set for the inequality.

$$2x - 3 \geq 0$$
$$2x \geq 3$$
$$x \geq \tfrac{3}{2}$$

Solution set: $\left[\frac{3}{2}, \infty\right)$

47. $\dfrac{(3x-5)^2}{x+2} > 0$

The numerator is positive for all real numbers x except $x = \frac{5}{3}$, which makes it equal to 0. If we solve the inequality $x + 2 > 0$, then we only have to be sure to exclude $\frac{5}{3}$ from that solution set to determine the solution set of the original inequality.

$$x + 2 > 0$$
$$x > -2$$

Solution set: $\left(-2, \frac{5}{3}\right) \cup \left(\frac{5}{3}, \infty\right)$

49. $f(x) = x^2 + 4x - 3$
$$f(2) = 2^2 + 4(2) - 3$$
$$= 4 + 8 - 3$$
$$= 9$$

51. $f(x) = 8x^2 - 2x - 3$
$$0 = 8x^2 - 2x - 3$$
$$0 = (2x + 1)(4x - 3)$$

$$2x + 1 = 0 \qquad \text{or} \qquad 4x - 3 = 0$$
$$x = -\tfrac{1}{2} \qquad \text{or} \qquad x = \tfrac{3}{4}$$

Solution set: $\left\{-\frac{1}{2}, \frac{3}{4}\right\}$

Chapter 9 Review Exercises

1. $t^2 = 121$

$$t = 11 \quad \text{or} \quad t = -11$$

Solution set: $\{\pm 11\}$

2. $p^2 = 3$

$$p = \sqrt{3} \quad \text{or} \quad p = -\sqrt{3}$$

Solution set: $\left\{\pm\sqrt{3}\right\}$

Copyright © 2012 Pearson Education, Inc. Publishing as Addison-Wesley.

3. $(2x+5)^2 = 100$

$2x+5 = 10 \quad$ or $\quad 2x+5 = -10$

$2x = 5 \qquad\qquad 2x = -15$

$x = \frac{5}{2} \quad$ or $\qquad x = -\frac{15}{2}$

Solution set: $\left\{-\frac{15}{2}, \frac{5}{2}\right\}$

4. $(3x-2)^2 = -25$

$3x-2 = \sqrt{-25} \quad$ or $\quad 3x-2 = -\sqrt{-25}$

$3x-2 = 5i \qquad\qquad 3x-2 = -5i$

$3x = 2+5i \qquad\qquad 3x = 2-5i$

$x = \dfrac{2+5i}{3} \qquad\qquad x = \dfrac{2-5i}{3}$

$x = \frac{2}{3} + \frac{5}{3}i \quad$ or $\qquad x = \frac{2}{3} - \frac{5}{3}i$

Solution set: $\left\{\frac{2}{3} \pm \frac{5}{3}i\right\}$

5. $x^2 + 4x = 15$

Complete the square.

$\left(\frac{1}{2} \cdot 4\right)^2 = 2^2 = 4$

Add 4 to each side.

$x^2 + 4x + 4 = 15 + 4$

$(x+2)^2 = 19$

$x+2 = \sqrt{19} \qquad$ or $\quad x+2 = -\sqrt{19}$

$x = -2+\sqrt{19} \quad$ or $\qquad x = -2-\sqrt{19}$

Solution set: $\left\{-2 \pm \sqrt{19}\right\}$

6. $2x^2 - 3x = -1$

$x^2 - \frac{3}{2}x = -\frac{1}{2} \qquad\qquad$ *Divide by 2.*

Complete the square.

$\left[\frac{1}{2}\left(-\frac{3}{2}\right)\right]^2 = \left(-\frac{3}{4}\right)^2 = \frac{9}{16}$

Add $\frac{9}{16}$ to each side.

$x^2 - \frac{3}{2}x + \frac{9}{16} = -\frac{1}{2} + \frac{9}{16}$

$\left(x-\frac{3}{4}\right)^2 = -\frac{8}{16} + \frac{9}{16}$

$\left(x-\frac{3}{4}\right)^2 = \frac{1}{16}$

$x-\frac{3}{4} = \sqrt{\frac{1}{16}} \quad$ or $\quad x-\frac{3}{4} = -\sqrt{\frac{1}{16}}$

$x-\frac{3}{4} = \frac{1}{4} \qquad\qquad x-\frac{3}{4} = -\frac{1}{4}$

$x = \frac{3}{4} + \frac{1}{4} \qquad\qquad x = \frac{3}{4} - \frac{1}{4}$

$x = 1 \qquad$ or $\qquad x = \frac{1}{2}$

Solution set: $\left\{\frac{1}{2}, 1\right\}$

7. By the square root property, the first step should be $x = \sqrt{12}$ or $x = -\sqrt{12}$. The solution set is $\left\{\pm 2\sqrt{3}\right\}$.

8. $4.9t^2 = d$

$4.9t^2 = 165 \qquad$ *Let d = 165.*

$t^2 = \dfrac{165}{4.9}$

$t = \sqrt{\dfrac{165}{4.9}} \qquad t \geq 0$

$t \approx 5.8$ seconds

It would take about 5.8 seconds for the wallet to fall 165 meters.

9. $2x^2 + x - 21 = 0$

Here $a = 2$, $b = 1$, and $c = -21$.

$x = \dfrac{-b \pm \sqrt{b^2 - 4ac}}{2a}$

$x = \dfrac{-1 \pm \sqrt{1^2 - 4(2)(-21)}}{2(2)}$

$= \dfrac{-1 \pm \sqrt{1 + 168}}{4}$

$= \dfrac{-1 \pm \sqrt{169}}{4} = \dfrac{-1 \pm 13}{4}$

$x = \dfrac{-1 + 13}{4} = \dfrac{12}{4} = 3$ or

$x = \dfrac{-1 - 13}{4} = -\dfrac{14}{4} = -\dfrac{7}{2}$

Solution set: $\left\{-\frac{7}{2}, 3\right\}$

10. $x^2 + 5x = 7$

$x^2 + 5x - 7 = 0$

Here $a = 1$, $b = 5$, and $c = -7$.

$x = \dfrac{-b \pm \sqrt{b^2 - 4ac}}{2a}$

$x = \dfrac{-5 \pm \sqrt{5^2 - 4(1)(-7)}}{2(1)}$

$= \dfrac{-5 \pm \sqrt{25 + 28}}{2} = \dfrac{-5 \pm \sqrt{53}}{2}$

Solution set: $\left\{\frac{-5 \pm \sqrt{53}}{2}\right\}$

11. $(t+3)(t-4) = -2$

$t^2 - t - 12 = -2$

$t^2 - t - 10 = 0$

Here $a = 1$, $b = -1$, and $c = -10$.

$t = \dfrac{-b \pm \sqrt{b^2 - 4ac}}{2a}$

$t = \dfrac{-(-1) \pm \sqrt{(-1)^2 - 4(1)(-10)}}{2(1)}$

$= \dfrac{1 \pm \sqrt{1 + 40}}{2} = \dfrac{1 \pm \sqrt{41}}{2}$

Solution set: $\left\{\frac{1 \pm \sqrt{41}}{2}\right\}$

Copyright © 2012 Pearson Education, Inc. Publishing as Addison-Wesley.

12. $2x^2 + 3x + 4 = 0$
Here $a = 2$, $b = 3$, and $c = 4$.

$$x = \frac{-b \pm \sqrt{b^2 - 4ac}}{2a}$$

$$x = \frac{-3 \pm \sqrt{3^2 - 4(2)(4)}}{2(2)}$$

$$= \frac{-3 \pm \sqrt{9 - 32}}{4} = \frac{-3 \pm \sqrt{-23}}{4}$$

$$= \frac{-3 \pm i\sqrt{23}}{4} = -\frac{3}{4} \pm \frac{\sqrt{23}}{4} i$$

Solution set: $\left\{ -\frac{3}{4} \pm \frac{\sqrt{23}}{4} i \right\}$

13.
$$3p^2 = 2(2p - 1)$$
$$3p^2 = 4p - 2$$
$$3p^2 - 4p + 2 = 0$$
Here $a = 3$, $b = -4$, and $c = 2$.

$$p = \frac{-b \pm \sqrt{b^2 - 4ac}}{2a}$$

$$p = \frac{-(-4) \pm \sqrt{(-4)^2 - 4(3)(2)}}{2(3)}$$

$$= \frac{4 \pm \sqrt{16 - 24}}{6} = \frac{4 \pm \sqrt{-8}}{6}$$

$$= \frac{4 \pm 2i\sqrt{2}}{6} = \frac{2\left(2 \pm i\sqrt{2}\right)}{6}$$

$$= \frac{2 \pm i\sqrt{2}}{3} = \frac{2}{3} \pm \frac{\sqrt{2}}{3} i$$

Solution set: $\left\{ \frac{2}{3} \pm \frac{\sqrt{2}}{3} i \right\}$

14. $x(2x - 7) = 3x^2 + 3$
$$2x^2 - 7x = 3x^2 + 3$$
$$0 = x^2 + 7x + 3$$
Here $a = 1$, $b = 7$, and $c = 3$.

$$x = \frac{-b \pm \sqrt{b^2 - 4ac}}{2a}$$

$$x = \frac{-7 \pm \sqrt{7^2 - 4(1)(3)}}{2(1)}$$

$$= \frac{-7 \pm \sqrt{49 - 12}}{2} = \frac{-7 \pm \sqrt{37}}{2}$$

Solution set: $\left\{ \frac{-7 \pm \sqrt{37}}{2} \right\}$

15. $x^2 + 5x + 2 = 0$
Here $a = 1$, $b = 5$, and $c = 2$.
$$b^2 - 4ac = 5^2 - 4(1)(2)$$
$$= 25 - 8 = 17$$
Since the discriminant is positive, but not a perfect square, there are two distinct irrational number solutions. The answer is **C**.

16.
$$4t^2 = 3 - 4t$$
$$4t^2 + 4t - 3 = 0$$
Here $a = 4$, $b = 4$, and $c = -3$.
$$b^2 - 4ac = 4^2 - 4(4)(-3)$$
$$= 16 + 48$$
$$= 64 \text{ or } 8^2$$
Since the discriminant is positive, and a perfect square, there are two distinct rational number solutions. The answer is **A**.

17.
$$4x^2 = 6x - 8$$
$$4x^2 - 6x + 8 = 0$$
Here $a = 4$, $b = -6$, and $c = 8$.
$$b^2 - 4ac = (-6)^2 - 4(4)(8)$$
$$= 36 - 128 = -92$$

Since the discriminant is negative, there are two distinct nonreal complex number solutions. The answer is **D**.

18. $9z^2 + 30z + 25 = 0$
Here $a = 9$, $b = 30$, and $c = 25$.
$$b^2 - 4ac = 30^2 - 4(9)(25)$$
$$= 900 - 900 = 0$$
Since the discriminant is zero, there is exactly one rational number solution. The answer is **B**.

19.
$$\frac{15}{x} = 2x - 1$$
$$x\left(\frac{15}{x}\right) = x(2x - 1) \qquad \textit{Multiply by the LCD, x.}$$
$$15 = 2x^2 - x$$
$$0 = 2x^2 - x - 15$$
$$0 = (2x + 5)(x - 3)$$

$$2x + 5 = 0 \qquad \text{or} \quad x - 3 = 0$$
$$x = -\tfrac{5}{2} \quad \text{or} \qquad x = 3$$

Check $x = -\frac{5}{2}$: $-6 = -5 - 1$ *True*
Check $x = 3$: $5 = 6 - 1$ *True*
Solution set: $\left\{ -\frac{5}{2}, 3 \right\}$

20.
$$\frac{1}{n} + \frac{2}{n + 1} = 2$$
$$n(n + 1)\left(\frac{1}{n} + \frac{2}{n + 1}\right) \qquad \textit{Multiply by the LCD, n(n+1).}$$
$$= n(n + 1) \cdot 2$$
$$(n + 1) + 2n = 2n^2 + 2n$$
$$0 = 2n^2 - n - 1$$
$$0 = (2n + 1)(n - 1)$$

$$2n + 1 = 0 \qquad \text{or} \quad n - 1 = 0$$
$$n = -\tfrac{1}{2} \quad \text{or} \qquad n = 1$$

Check $n = -\frac{1}{2}$: $-2 + 4 = 2$ *True*
Check $n = 1$: $1 + 1 = 2$ *True*
Solution set: $\left\{ -\frac{1}{2}, 1 \right\}$

Copyright © 2012 Pearson Education, Inc. Publishing as Addison-Wesley.

21.
$$-2r = \sqrt{\frac{48 - 20r}{2}}$$

Square each side.

$$(-2r)^2 = \left(\sqrt{\frac{48 - 20r}{2}}\right)^2$$

$$4r^2 = \frac{48 - 20r}{2}$$

$$4r^2 = 24 - 10r$$

$$4r^2 + 10r - 24 = 0$$

$$2r^2 + 5r - 12 = 0$$

$$(r + 4)(2r - 3) = 0$$

$$r + 4 = 0 \quad \text{or} \quad 2r - 3 = 0$$
$$r = -4 \quad \text{or} \quad r = \tfrac{3}{2}$$

Check $r = -4$: $\quad 8 = \sqrt{64} \quad$ *True*
Check $r = \tfrac{3}{2}$: $\quad -3 = \sqrt{9} \quad$ *False*
Solution set: $\{-4\}$

22. $8(3x + 5)^2 + 2(3x + 5) - 1 = 0$
Let $u = 3x + 5$, so $u^2 = (3x + 5)^2$.
$$8u^2 + 2u - 1 = 0$$
$$(2u + 1)(4u - 1) = 0$$

$$2u + 1 = 0 \quad \text{or} \quad 4u - 1 = 0$$
$$u = -\tfrac{1}{2} \quad \text{or} \quad u = \tfrac{1}{4}$$

To find x, substitute $3x + 5$ for u.

$$3x + 5 = -\tfrac{1}{2} \quad \text{or} \quad 3x + 5 = \tfrac{1}{4}$$
$$3x = -\tfrac{11}{2} \qquad\qquad 3x = -\tfrac{19}{4}$$
$$x = -\tfrac{11}{6} \quad \text{or} \qquad x = -\tfrac{19}{12}$$

Check $x = -\tfrac{11}{6}$: $\quad 2 - 1 - 1 = 0 \quad$ *True*
Check $x = -\tfrac{19}{12}$: $0.5 + 0.5 - 1 = 0 \quad$ *True*
Solution set: $\left\{-\tfrac{11}{6}, -\tfrac{19}{12}\right\}$

23. $2x^{2/3} - x^{1/3} - 28 = 0$
Let $u = x^{1/3}$, so $u^2 = (x^{1/3})^2 = x^{2/3}$.
$$2u^2 - u - 28 = 0$$
$$(2u + 7)(u - 4) = 0$$

$$2u + 7 = 0 \quad \text{or} \quad u - 4 = 0$$
$$u = -\tfrac{7}{2} \quad \text{or} \quad u = 4$$

To find x, substitute $x^{1/3}$ for u.

$$x^{1/3} = -\tfrac{7}{2} \quad \text{or} \quad x^{1/3} = 4$$
$$(x^{1/3})^3 = \left(-\tfrac{7}{2}\right)^3 \quad (x^{1/3})^3 = 4^3$$
$$x = -\tfrac{343}{8} \quad \text{or} \qquad x = 64$$

Check $x = -\tfrac{343}{8}$: $\quad 24.5 + 3.5 - 28 = 0 \quad$ *True*
Check $x = 64$: $\qquad 32 - 4 - 28 = 0 \quad$ *True*
Solution set: $\left\{-\tfrac{343}{8}, 64\right\}$

24. $p^4 - 10p^2 + 9 = 0$
Let $x = p^2$, so $x^2 = p^4$.
$$x^2 - 10x + 9 = 0$$
$$(x - 1)(x - 9) = 0$$

$$x - 1 = 0 \quad \text{or} \quad x - 9 = 0$$
$$x = 1 \quad \text{or} \qquad x = 9$$

To find p, substitute p^2 for x.

$$p^2 = 1 \qquad \text{or} \quad p^2 = 9$$
$$p = \pm\sqrt{1} \qquad\qquad p = \pm\sqrt{9}$$
$$p = \pm 1 \quad \text{or} \qquad p = \pm 3$$

Check $p = \pm 1$: $\quad 1 - 10 + 9 = 0 \qquad$ *True*
Check $p = \pm 3$: $\quad 81 - 90 + 9 = 0 \qquad$ *True*
Solution set: $\{-3, -1, 1, 3\}$

25. Let $x =$ Bahaa's rate.
Make a table. Use $d = rt$, or $t = \frac{d}{r}$.

	Distance	Rate	Time
Upstream	20	$x - 3$	$\dfrac{20}{x - 3}$
Downstream	20	$x + 3$	$\dfrac{20}{x + 3}$

Time | | time
upstream plus downstream equals 7 hr.

$$\frac{20}{x - 3} + \frac{20}{x + 3} = 7$$

Multiply each side by the LCD, $(x + 3)(x - 3)$.

$$(x + 3)(x - 3)\left(\frac{20}{x - 3} + \frac{20}{x + 3}\right) = (x + 3)(x - 3)(7)$$
$$20(x + 3) + 20(x - 3) = 7(x^2 - 9)$$
$$20x + 60 + 20x - 60 = 7x^2 - 63$$
$$0 = 7x^2 - 40x - 63$$
$$0 = (x - 7)(7x + 9)$$

$$x - 7 = 0 \quad \text{or} \quad 7x + 9 = 0$$
$$x = 7 \quad \text{or} \qquad x = -\tfrac{9}{7}$$

Reject $-\tfrac{9}{7}$. Bahaa's rate was 7 mph.

26. Let $x =$ Carol-Ann's rate on the trip to pick up her friend. Make a chart. Use $d = rt$, or $t = \frac{d}{r}$.

	Distance	Rate	Time
To Her Friend	8	x	$\dfrac{8}{x}$
To the Mall	11	$x + 15$	$\dfrac{11}{x + 15}$

Time to pick plus time to equals 24 min
up her friend mall (or 0.4 hr).

$$\frac{8}{x} + \frac{11}{x + 15} = 0.4$$

Copyright © 2012 Pearson Education, Inc. Publishing as Addison-Wesley.

Multiply each side by the LCD, $x(x+15)$.

$$x(x+15)\left(\frac{8}{x} + \frac{11}{x+15}\right) = x(x+15)(0.4)$$
$$8(x+15) + 11x = 0.4x(x+15)$$
$$8x + 120 + 11x = 0.4x^2 + 6x$$
$$0 = 0.4x^2 - 13x - 120$$

Multiply by 5 to clear the decimal.

$$0 = 2x^2 - 65x - 600$$
$$0 = (x-40)(2x+15)$$

$$x - 40 = 0 \quad \text{or} \quad 2x + 15 = 0$$
$$x = 40 \quad \text{or} \quad x = -\frac{15}{2}$$

Reject $-\frac{15}{2}$. Carol-Ann's rate on the trip to pick up her friend was 40 mph.

27. Let $x =$ the amount of time for the old machine alone and
$x - 1 =$ the amount of time for the new machine alone.

Make a chart.

Machine	Rate	Time Together	Fractional Part of the Job Done
Old	$\dfrac{1}{x}$	2	$\dfrac{2}{x}$
New	$\dfrac{1}{x-1}$	2	$\dfrac{2}{x-1}$

Part done by old machine plus Part done by new machine equals 1 whole job.

$$\frac{2}{x} + \frac{2}{x-1} = 1$$

Multiply each term by the LCD, $x(x-1)$.

$$x(x-1)\left(\frac{2}{x} + \frac{2}{x-1}\right) = x(x-1) \cdot 1$$
$$2(x-1) + 2x = x^2 - x$$
$$2x - 2 + 2x = x^2 - x$$
$$0 = x^2 - 5x + 2$$

Use the quadratic formula.

$$x = \frac{-b \pm \sqrt{b^2 - 4ac}}{2a}$$
$$x = \frac{-(-5) \pm \sqrt{(-5)^2 - 4(1)(2)}}{2(1)}$$
$$= \frac{5 \pm \sqrt{25 - 8}}{2} = \frac{5 \pm \sqrt{17}}{2}$$
$$x = \frac{5 + \sqrt{17}}{2} \approx 4.6 \text{ or}$$
$$x = \frac{5 - \sqrt{17}}{2} \approx 0.4$$

Reject 0.4 as the time for the old machine, because that would yield a negative time for the new machine. Thus, the old machine takes about 4.6 hours.

28. Let $x =$ the time for Claude alone and
$x - 1 =$ the time for Zoran alone.

Worker	Rate	Time Together	Fractional Part of the Job Done
Claude	$\dfrac{1}{x}$	1.5	$\dfrac{1.5}{x}$
Zoran	$\dfrac{1}{x-1}$	1.5	$\dfrac{1.5}{x-1}$

Part by Claude plus part by Zoran equals 1 whole job.

$$\frac{1.5}{x} + \frac{1.5}{x-1} = 1$$

Multiply each term by the LCD, $x(x-1)$.

$$x(x-1)\left(\frac{1.5}{x} + \frac{1.5}{x-1}\right) = x(x-1) \cdot 1$$
$$1.5(x-1) + 1.5x = x^2 - x$$
$$1.5x - 1.5 + 1.5x = x^2 - x$$
$$0 = x^2 - 4x + 1.5$$

Use the quadratic formula.

$$x = \frac{-b \pm \sqrt{b^2 - 4ac}}{2a}$$
$$x = \frac{-(-4) \pm \sqrt{(-4)^2 - 4(1)(1.5)}}{2(1)}$$
$$= \frac{4 \pm \sqrt{16 - 6}}{2} = \frac{4 \pm \sqrt{10}}{2}$$
$$x = \frac{4 + \sqrt{10}}{2} \approx 3.6 \text{ or}$$
$$x = \frac{4 - \sqrt{10}}{2} \approx 0.4$$

Reject 0.4 as Claude's time, because that would make Zoran's time negative. Thus, Claude's time alone is about 3.6 hours and Zoran's time alone is about $x - 1 = 2.6$ hours.

29. Solve $k = \dfrac{rF}{wv^2}$ for v.

Multiply each side by v^2, then divide by k.

$$v^2 = \frac{rF}{kw}$$
$$v = \pm\sqrt{\frac{rF}{kw}} = \frac{\pm\sqrt{rF}}{\sqrt{kw}}$$
$$= \frac{\pm\sqrt{rF}}{\sqrt{kw}} \cdot \frac{\sqrt{kw}}{\sqrt{kw}} = \frac{\pm\sqrt{rFkw}}{kw}$$

Copyright © 2012 Pearson Education, Inc. Publishing as Addison-Wesley.

30. Solve $p = \sqrt{\dfrac{yz}{6}}$ for y.

$$p^2 = \left(\sqrt{\dfrac{yz}{6}}\right)^2 \qquad \textit{Square each side.}$$

$$p^2 = \dfrac{yz}{6}$$

$$\dfrac{6p^2}{z} = y, \quad \text{or} \quad y = \dfrac{6p^2}{z}$$

31. Solve $mt^2 = 3mt + 6$ for t.

$$mt^2 - 3mt - 6 = 0$$

Use the quadratic formula with $a = m$, $b = -3m$, and $c = -6$.

$$t = \dfrac{-b \pm \sqrt{b^2 - 4ac}}{2a}$$

$$t = \dfrac{3m \pm \sqrt{(-3m)^2 - 4(m)(-6)}}{2m}$$

$$= \dfrac{3m \pm \sqrt{9m^2 + 24m}}{2m}$$

32. Let $x =$ the length of the longer leg;

$\dfrac{3}{4}x =$ the length of the shorter leg;

$2x - 9 =$ the length of the hypotenuse.

Use the Pythagorean theorem.

$$c^2 = a^2 + b^2$$

$$(2x - 9)^2 = x^2 + \left(\tfrac{3}{4}x\right)^2$$

$$4x^2 - 36x + 81 = x^2 + \tfrac{9}{16}x^2$$

$$16\left(4x^2 - 36x + 81\right) = 16\left(x^2 + \tfrac{9}{16}x^2\right)$$

$$64x^2 - 576x + 1296 = 16x^2 + 9x^2$$

$$39x^2 - 576x + 1296 = 0$$

$$13x^2 - 192x + 432 = 0 \qquad \textit{Divide by 3.}$$

$$(13x - 36)(x - 12) = 0$$

$$13x - 36 = 0 \quad \text{or} \quad x - 12 = 0$$

$$x = \tfrac{36}{13} \quad \text{or} \quad x = 12$$

Reject $x = \tfrac{36}{13}$ since $2\left(\tfrac{36}{13}\right) - 9$ is negative. If $x = 12$, then

$$\tfrac{3}{4}x = \tfrac{3}{4}(12) = 9$$

and

$$2x - 9 = 2(12) - 9 = 15.$$

The lengths of the three sides are 9 feet, 12 feet, and 15 feet.

33. Let $x =$ the amount removed from one dimension.

The area of the square is 256 cm^2, so the length of one side is $\sqrt{256}$ or 16 cm. The dimensions of the new rectangle are $16 + x$ and $16 - x$ cm. The area of the new rectangle is 16 cm^2 less than the area of the square.

$$(16 + x)(16 - x) = 256 - 16$$

$$256 - x^2 = 240$$

$$-x^2 = -16$$

$$x^2 = 16$$

$$x = \pm\sqrt{16} = \pm 4$$

Length cannot be negative, so reject -4. If $x = 4$, then $16 + x = 20$, and $16 - x = 12$. The dimensions are 20 cm by 12 cm.

34. Let $x =$ the width of the border.

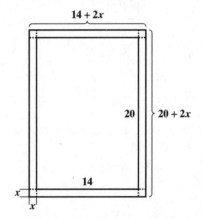

Area of mat $=$ length $\bullet$ width

$$352 = (2x + 20)(2x + 14)$$

$$352 = 4x^2 + 68x + 280$$

$$0 = 4x^2 + 68x - 72$$

$$0 = x^2 + 17x - 18$$

$$0 = (x + 18)(x - 1)$$

$$x + 18 = 0 \qquad \text{or} \quad x - 1 = 0$$

$$x = -18 \quad \text{or} \qquad x = 1$$

Reject the negative answer for length. The mat is 1 inch wide.

35.
$$V = 3(x - 6)^2$$

$$432 = 3(x - 6)^2 \qquad \textit{Let V = 432.}$$

$$144 = (x - 6)^2 \qquad \textit{Divide by 3.}$$

Use the square root property.

$$\pm\sqrt{144} = x - 6$$

$$6 \pm 12 = x$$

$$x = 6 + 12 = 18$$

$$\text{or} \qquad x = 6 - 12 = -6$$

Discard the negative solution since x represents the length. The original length is 18 inches.

36. $f(t) = -16t^2 + 45t + 400$

$$200 = -16t^2 + 45t + 400 \qquad \textit{Let f(t) = 200.}$$

$$0 = -16t^2 + 45t + 200$$

$$0 = 16t^2 - 45t - 200$$

Here $a = 16$, $b = -45$, and $c = -200$.

Copyright © 2012 Pearson Education, Inc. Publishing as Addison-Wesley.

$$t = \frac{-b \pm \sqrt{b^2 - 4ac}}{2a}$$

$$t = \frac{-(-45) \pm \sqrt{(-45)^2 - 4(16)(-200)}}{2(16)}$$

$$= \frac{45 \pm \sqrt{2025 + 12{,}800}}{32}$$

$$= \frac{45 \pm \sqrt{14{,}825}}{32}$$

$$t = \frac{45 + \sqrt{14{,}825}}{32} \approx 5.2 \quad \text{or}$$

$$t = \frac{45 - \sqrt{14{,}825}}{32} \approx -2.4$$

Reject the negative solution since time cannot be negative. The ball will reach a height of 200 ft above the ground after about 5.2 seconds.

37. $f(t) = 100t^2 - 300t$ is the distance of the light from the starting point at t minutes. When the light returns to the starting point, the value of $f(t)$ will be 0.

$$0 = 100t^2 - 300t$$
$$0 = t^2 - 3t \qquad \textit{Divide by 100.}$$
$$0 = t(t - 3)$$
$$t = 0 \quad \text{or} \quad t - 3 = 0$$
$$t = 3$$

Since $t = 0$ represents the starting time, the light will return to the starting point in 3 minutes.

38. **(a)** Use $f(x) = 230.5x^2 - 252.9x + 5987$ with $x = 7$.

$$f(7) = 230.5(7)^2 - 252.9(7) + 5987$$
$$= 15{,}511.2 \approx 15{,}511$$

The value of \$15,511 million is close to the number suggested by the graph.

(b) Let $f(x) = 14{,}000$.

$$14{,}000 = 230.5x^2 - 252.9x + 5987$$
$$0 = 230.5x^2 - 252.9x - 8013$$

Here $a = 230.5$, $b = -252.9$, and $c = -8013$.

$$x = \frac{-b \pm \sqrt{b^2 - 4ac}}{2a}$$

$$x = \frac{-(-252.9) \pm \sqrt{(-252.9)^2 - 4(230.5)(-8013)}}{2(230.5)}$$

$$= \frac{252.9 \pm \sqrt{7{,}451{,}944.41}}{461}$$

$$x = \frac{252.9 + \sqrt{7{,}451{,}944.41}}{461} \approx 6.47 \quad \text{or}$$

$$x = \frac{252.9 - \sqrt{7{,}451{,}944.41}}{461} \approx -5.37$$

Since time cannot be negative, the correct answer is $x \approx 6$, which represents 2006. Based on the graph, the revenue in 2006 was closer to \$13,000 million than \$14,000 million.

39. $(x - 4)(2x + 3) > 0$

Solve the equation
$(x - 4)(2x + 3) = 0.$

$$x - 4 = 0 \quad \text{or} \quad 2x + 3 = 0$$
$$x = 4 \quad \text{or} \qquad x = -\tfrac{3}{2}$$

The numbers $-\frac{3}{2}$ and 4 divide a number line into three intervals.

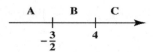

Test a number from each interval in the inequality

$$(x - 4)(2x + 3) > 0.$$

Interval A: Let $x = -2$.

$$-6(-1) \overset{?}{>} 0$$
$$6 > 0 \qquad \textit{True}$$

Interval B: Let $x = 0$.

$$-4(3) \overset{?}{>} 0$$
$$-12 > 0 \qquad \textit{False}$$

Interval C: Let $x = 5$.

$$1(13) \overset{?}{>} 0$$
$$13 > 0 \qquad \textit{True}$$

The solution set includes numbers in Intervals A and C, excluding endpoints.

Solution set: $\left(-\infty, -\frac{3}{2}\right) \cup (4, \infty)$

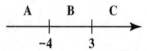

40. $x^2 + x \le 12$

Solve the equation
$$x^2 + x = 12.$$
$$x^2 + x - 12 = 0$$
$$(x + 4)(x - 3) = 0$$

$$x + 4 = 0 \quad \text{or} \quad x - 3 = 0$$
$$x = -4 \quad \text{or} \qquad x = 3$$

The numbers -4 and 3 divide a number line into three intervals.

Test a number from each interval in the inequality

$$x^2 + x \le 12.$$

Interval A: Let $x = -5$.

$$25 - 5 \overset{?}{\le} 12$$
$$20 \le 12 \qquad \textit{False}$$

Interval B: Let $x = 0$.

$$0 \le 12 \qquad \textit{True}$$

Copyright © 2012 Pearson Education, Inc. Publishing as Addison-Wesley.

Interval C: Let $x = 4$.

$$16 + 4 \overset{?}{\leq} 12$$
$$20 \leq 12 \quad \textit{False}$$

The numbers in Interval B, including -4 and 3, are solutions.

Solution set: $[-4, 3]$

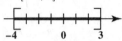

41. $(x + 2)(x - 3)(x + 5) \leq 0$

The numbers -2, 3, and -5 are solutions of the cubic equation

$$(x + 2)(x - 3)(x + 5) = 0.$$

These numbers divide a number line into four intervals.

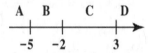

Test a number from each interval in the inequality

$$(x + 2)(x - 3)(x + 5) \leq 0.$$

Interval A: Let $x = -6$.
$$(-4)(-9)(-1) \leq 0 \qquad \textit{True}$$
Interval B: Let $x = -3$.
$$(-5)(-6)(2) \leq 0 \qquad \textit{False}$$
Interval C: Let $x = 0$.
$$(2)(-3)(5) \leq 0 \qquad \textit{True}$$
Interval D: Let $x = 4$.
$$(6)(1)(9) \leq 0 \qquad \textit{False}$$

The numbers in Intervals A and C, including -5, -2, and 3, are solutions.

Solution set: $(-\infty, -5] \cup [-2, 3]$

42. $(4x + 3)^2 \leq -4$

The square of a real number is never negative. So, the solution set of this inequality is $\emptyset$.

43. $\dfrac{6}{2z - 1} < 2$

Write the inequality so that 0 is on one side.

$$\frac{6}{2z - 1} - 2 < 0$$
$$\frac{6}{2z - 1} - \frac{2(2z - 1)}{2z - 1} < 0$$
$$\frac{6 - 4z + 2}{2z - 1} < 0$$
$$\frac{-4z + 8}{2z - 1} < 0$$

The number 2 makes the numerator 0, and $\frac{1}{2}$ makes the denominator 0. These two numbers determine three intervals.

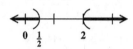

Test a number from each interval in the inequality

$$\frac{6}{2z - 1} < 2.$$

Interval A: Let $z = 0$.
$$-6 < 2 \qquad \textit{True}$$
Interval B: Let $z = 1$.
$$6 < 2 \qquad \textit{False}$$
Interval C: Let $z = 3$.
$$\frac{6}{5} < 2 \qquad \textit{True}$$

The solution set includes numbers in Intervals A and C, excluding endpoints.

Solution set: $\left(-\infty, \frac{1}{2}\right) \cup (2, \infty)$

44. $\dfrac{3t + 4}{t - 2} \leq 1$

Write the inequality so that 0 is on one side.

$$\frac{3t + 4}{t - 2} - 1 \leq 0$$
$$\frac{3t + 4}{t - 2} - \frac{1(t - 2)}{t - 2} \leq 0$$
$$\frac{3t + 4 - t + 2}{t - 2} \leq 0$$
$$\frac{2t + 6}{t - 2} \leq 0$$

The number -3 makes the numerator 0, and 2 makes the denominator 0. These two numbers determine three intervals.

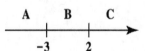

Test a number from each interval in the inequality

$$\frac{3t + 4}{t - 2} \leq 1.$$

Interval A: Let $t = -4$.
$$\frac{-8}{-6} \overset{?}{\leq} 1$$
$$\frac{4}{3} \leq 1 \qquad \textit{False}$$
Interval B: Let $t = 0$.
$$\frac{4}{-2} \overset{?}{\leq} 1$$
$$-2 \leq 1 \qquad \textit{True}$$
Interval C: Let $t = 3$.
$$\frac{13}{1} \overset{?}{\leq} 1$$
$$13 \leq 1 \qquad \textit{False}$$

Copyright © 2012 Pearson Education, Inc. Publishing as Addison-Wesley.

The numbers in Interval B, including -3 but not 2, are solutions.

Solution set: $[-3, 2)$

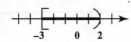

45. **[9.4]** Solve $V = r^2 + R^2 h$ for R.
$$V - r^2 = R^2 h$$
$$R^2 h = V - r^2$$
$$R^2 = \frac{V - r^2}{h}$$
$$R = \pm \sqrt{\frac{V - r^2}{h}} = \frac{\pm \sqrt{V - r^2}}{\sqrt{h}}$$
$$= \frac{\pm \sqrt{V - r^2}}{\sqrt{h}} \cdot \frac{\sqrt{h}}{\sqrt{h}}$$
$$= \frac{\pm \sqrt{Vh - r^2 h}}{h}$$

46. **[9.2]** $3t^2 - 6t = -4$
$$3t^2 - 6t + 4 = 0$$
Use the quadratic formula.

$$t = \frac{-b \pm \sqrt{b^2 - 4ac}}{2a}$$
$$t = \frac{-(-6) \pm \sqrt{(-6)^2 - 4(3)(4)}}{2(3)}$$
$$= \frac{6 \pm \sqrt{-12}}{6} = \frac{6 \pm 2i\sqrt{3}}{6}$$
$$= \frac{2\left(3 \pm i\sqrt{3}\right)}{6} = \frac{3 \pm i\sqrt{3}}{3} = 1 \pm \frac{\sqrt{3}}{3} i$$

Solution set: $\left\{ 1 \pm \frac{\sqrt{3}}{3} i \right\}$

47. **[9.1]** $(3x + 11)^2 = 7$

$$3x + 11 = \sqrt{7} \qquad \text{or } 3x + 11 = -\sqrt{7}$$
$$3x = -11 + \sqrt{7} \qquad\qquad 3x = -11 - \sqrt{7}$$
$$x = \frac{-11 + \sqrt{7}}{3} \quad \text{or} \qquad x = \frac{-11 - \sqrt{7}}{3}$$

Solution set: $\left\{ \frac{-11 \pm \sqrt{7}}{3} \right\}$

48. **[9.4]** Solve $S = \dfrac{I d^2}{k}$ for d.

Multiply each side by k, then divide by I.
$$\frac{Sk}{I} = d^2$$
$$d = \pm \sqrt{\frac{Sk}{I}} = \frac{\pm \sqrt{Sk}}{\sqrt{I}}$$
$$= \frac{\pm \sqrt{Sk}}{\sqrt{I}} \cdot \frac{\sqrt{I}}{\sqrt{I}} = \frac{\pm \sqrt{SkI}}{I}$$

49. **[9.5]** $(8x - 7)^2 \geq -1$

The square of any real number is always greater than or equal to 0, so any real number satisfies this inequality.

Solution set: $(-\infty, \infty)$

50. **[9.3]** $2x - \sqrt{x} = 6$
$$2x - \sqrt{x} - 6 = 0$$
Let $u = \sqrt{x}$, so $u^2 = x$.
$$2u^2 - u - 6 = 0$$
$$(2u + 3)(u - 2) = 0$$
$$2u + 3 = 0 \qquad \text{or} \quad u - 2 = 0$$
$$u = -\tfrac{3}{2} \quad \text{or} \qquad u = 2$$
$$\sqrt{x} = -\tfrac{3}{2} \quad \text{or} \qquad \sqrt{x} = 2$$

Since $\sqrt{x} \geq 0$, we must have $\sqrt{x} = 2$, or $x = 4$.

Check $x = 4$: $8 - 2 = 6$ *True*

Solution set: $\{4\}$

51. **[9.3]** $x^4 - 8x^2 = -1$
$$x^4 - 8x^2 + 1 = 0$$

Use the quadratic formula with $a = 1$, $b = -8$, and $c = 1$ to solve for x^2, not x.

$$x^2 = \frac{-b \pm \sqrt{b^2 - 4ac}}{2a}$$
$$x^2 = \frac{-(-8) \pm \sqrt{(-8)^2 - 4(1)(1)}}{2(1)}$$
$$= \frac{8 \pm \sqrt{64 - 4}}{2} = \frac{8 \pm \sqrt{60}}{2}$$
$$= \frac{8 \pm 2\sqrt{15}}{2} = 4 \pm \sqrt{15}$$

Both $4 + \sqrt{15}$ and $4 - \sqrt{15}$ are positive and can be set equal to x^2.

$$x^2 = 4 + \sqrt{15} \qquad \text{or} \quad x^2 = 4 - \sqrt{15}$$
$$x = \pm \sqrt{4 + \sqrt{15}} \qquad\quad x = \pm \sqrt{4 - \sqrt{15}}$$

The proposed solutions all check.

Solution set: $\left\{ \pm \sqrt{4 + \sqrt{15}}, \ \pm \sqrt{4 - \sqrt{15}} \right\}$

52. **[9.5]** $\dfrac{-2}{x + 5} \leq -5$

Write the inequality so that 0 is on one side.

$$\frac{-2}{x + 5} + 5 \leq 0$$
$$\frac{-2}{x + 5} + \frac{5(x + 5)}{x + 5} \leq 0$$
$$\frac{-2 + 5x + 25}{x + 5} \leq 0$$
$$\frac{5x + 23}{x + 5} \leq 0$$

Copyright © 2012 Pearson Education, Inc. Publishing as Addison-Wesley.

The number $-\frac{23}{5}$ makes the numerator 0, and -5 makes the denominator 0. These two numbers determine three intervals.

Test a number from each interval in the inequality

$$\frac{-2}{x+5} \le -5.$$

Interval A: Let $x = -6$.
$$\frac{-2}{-1} \stackrel{?}{\le} -5$$
$$2 \le -5 \qquad \textit{False}$$

Interval B: Let $x = -\frac{24}{5}$.
$$\frac{-2}{\frac{1}{5}} \stackrel{?}{\le} -5$$
$$-10 \le -5 \qquad \textit{True}$$

Interval C: Let $x = 0$.
$$\frac{-2}{5} \le -5 \qquad \textit{False}$$

The numbers in Interval B, including $-\frac{23}{5}$ but not -5, are solutions.

Solution set: $\left(-5, -\frac{23}{5}\right]$

53.   **[9.3]** $6 + \dfrac{15}{s^2} = -\dfrac{19}{s}$

Multiply each term by the LCD, s^2.

$$s^2\left(6 + \frac{15}{s^2}\right) = s^2\left(-\frac{19}{s}\right)$$
$$6s^2 + 15 = -19s$$
$$6s^2 + 19s + 15 = 0$$
$$(3s + 5)(2s + 3) = 0$$

$$3s + 5 = 0 \quad \text{or} \quad 2s + 3 = 0$$
$$s = -\tfrac{5}{3} \quad \text{or} \quad\quad s = -\tfrac{3}{2}$$

Check $s = -\frac{5}{3}$: $6 + \frac{27}{5} = \frac{57}{5}$ *True*
Check $s = -\frac{3}{2}$: $6 + \frac{20}{3} = \frac{38}{3}$ *True*

Solution set: $\left\{-\frac{5}{3}, -\frac{3}{2}\right\}$

54.   **[9.3]** $\left(x^2 - 2x\right)^2 = 11\left(x^2 - 2x\right) - 24$

Let $u = x^2 - 2x$, so $u^2 = \left(x^2 - 2x\right)^2$.
$$u^2 = 11u - 24$$
$$u^2 - 11u + 24 = 0$$
$$(u - 8)(u - 3) = 0$$

$$u - 8 = 0 \quad \text{or} \quad u - 3 = 0$$
$$u = 8 \quad \text{or} \quad\quad u = 3$$

To find x, substitute $x^2 - 2x$ for u.

$$x^2 - 2x = 8 \qquad \text{or} \qquad x^2 - 2x = 3$$
$$x^2 - 2x - 8 = 0 \qquad\qquad x^2 - 2x - 3 = 0$$
$$(x - 4)(x + 2) = 0 \qquad (x - 3)(x + 1) = 0$$
$$x - 4 = 0 \text{ or } x + 2 = 0 \quad x - 3 = 0 \text{ or } x + 1 = 0$$
$$x = 4 \quad \text{or} \quad x = -2 \quad \text{or} \quad x = 3 \quad \text{or} \quad x = -1$$

The proposed solutions all check.

Solution set: $\{-2, -1, 3, 4\}$

55.   **[9.5]** $(r - 1)(2r + 3)(r + 6) < 0$
Solve the equation
$$(r - 1)(2r + 3)(r + 6) = 0.$$

$$r - 1 = 0 \quad \text{or} \quad 2r + 3 = 0 \quad \text{or} \quad r + 6 = 0$$
$$r = 1 \quad \text{or} \quad\quad r = -\tfrac{3}{2} \quad \text{or} \quad\quad r = -6$$

The numbers -6, $-\frac{3}{2}$, and 1 divide a number line into four intervals.

Test a number from each interval in the inequality

$$(r - 1)(2r + 3)(r + 6) < 0.$$

Interval A: Let $r = -7$.
$$-8(-11)(-1) \stackrel{?}{<} 0$$
$$-88 < 0 \qquad \textit{True}$$

Interval B: Let $r = -2$.
$$-3(-1)(4) \stackrel{?}{<} 0$$
$$12 < 0 \qquad \textit{False}$$

Interval C: Let $r = 0$.
$$-1(3)(6) \stackrel{?}{<} 0$$
$$-18 < 0 \qquad \textit{True}$$

Interval D: Let $r = 2$.
$$1(7)(8) \stackrel{?}{<} 0$$
$$56 < 0 \qquad \textit{False}$$

The numbers in Intervals A and C, not including -6, $-\frac{3}{2}$, or 1, are solutions.

Solution set: $(-\infty, -6) \cup \left(-\frac{3}{2}, 1\right)$

56.   **[9.4]** Let $x =$ the rate of the boat in still water. Use $t = d/r$ to make a table.

	Distance	Rate	Time
Upstream	15	$x - 5$	$\dfrac{15}{x - 5}$
Downstream	15	$x + 5$	$\dfrac{15}{x + 5}$

The total time is 4 hours.

Copyright © 2012 Pearson Education, Inc. Publishing as Addison-Wesley.

$$\frac{15}{x-5} + \frac{15}{x+5} = 4$$

Multiply each term by the LCD, $(x+5)(x-5)$.

$$15(x+5) + 15(x-5) = 4(x+5)(x-5)$$
$$15x + 75 + 15x - 75 = 4x^2 - 100$$
$$0 = 4x^2 - 30x - 100$$
$$0 = 2x^2 - 15x - 50$$
$$0 = (2x+5)(x-10)$$

$$2x+5 = 0 \quad \text{or} \quad x-10 = 0$$
$$x = -\frac{5}{2} \quad \text{or} \quad x = 10$$

The rate must be positive, so check $x = 10$. The 15-mile trip upstream would take 3 hours at 5 mph. The trip downstream would take 1 hour at 15 mph. Thus, the total time would be 4 hours and we have the rate of the boat in still water equal to 10 mph.

57. **[9.4]** Let $x =$ the width of the rectangle.
Then $2x - 1 =$ the length of the rectangle.

Use the Pythagorean theorem.

$$x^2 + (2x-1)^2 = (2.5)^2$$
$$x^2 + 4x^2 - 4x + 1 = 6.25$$
$$5x^2 - 4x - 5.25 = 0$$

Multiply by 4.

$$20x^2 - 16x - 21 = 0$$
$$(2x-3)(10x+7) = 0$$

$$2x - 3 = 0 \quad \text{or} \quad 10x + 7 = 0$$
$$x = \frac{3}{2} \quad \text{or} \quad x = -\frac{7}{10}$$

Discard the negative solution.
If $x = 1.5$, then

$$2x - 1 = 2(1.5) - 1 = 2.$$

The width of the rectangle is 1.5 cm, and the length is 2 cm.

58. **[9.4]** Let $x =$ the length of the wire.
Use the Pythagorean formula.

$$x^2 = 100^2 + 400^2 \quad \textit{Height=400.}$$
$$= 10,000 + 160,000$$
$$= 170,000$$
$$x = \pm\sqrt{170,000} \approx \pm 412.3$$

Reject the negative solution. The length of the wire is about 412.3 feet.

Chapter 9 Test

1. $t^2 = 54$

$$t = \sqrt{54} \quad \text{or} \quad t = -\sqrt{54}$$
$$t = 3\sqrt{6} \quad \text{or} \quad t = -3\sqrt{6}$$

Solution set: $\left\{ \pm 3\sqrt{6} \right\}$

2. $(7x+3)^2 = 25$

$$7x + 3 = 5 \quad \text{or} \quad 7x + 3 = -5$$
$$7x = 2 \qquad\qquad 7x = -8$$
$$x = \tfrac{2}{7} \quad \text{or} \qquad x = -\tfrac{8}{7}$$

Solution set: $\left\{ -\frac{8}{7}, \frac{2}{7} \right\}$

3. $x^2 + 2x = 4$
$$x^2 + 2x + 1 = 4 + 1$$
$$(x+1)^2 = 5$$

$$x + 1 = \sqrt{5} \quad \text{or} \quad x + 1 = -\sqrt{5}$$
$$x = -1 + \sqrt{5} \quad \text{or} \quad x = -1 - \sqrt{5}$$

Solution set: $\left\{ -1 \pm \sqrt{5} \right\}$

4. $2x^2 - 3x - 1 = 0$
Here $a = 2, b = -3$, and $c = -1$.

$$x = \frac{-b \pm \sqrt{b^2 - 4ac}}{2a}$$
$$x = \frac{-(-3) \pm \sqrt{(-3)^2 - 4(2)(-1)}}{2(2)}$$
$$= \frac{3 \pm \sqrt{17}}{4}$$

Solution set: $\left\{ \frac{3 \pm \sqrt{17}}{4} \right\}$

5. $3t^2 - 4t = -5$
$$3t^2 - 4t + 5 = 0$$
Here $a = 3, b = -4$, and $c = 5$.

$$t = \frac{-b \pm \sqrt{b^2 - 4ac}}{2a}$$
$$t = \frac{-(-4) \pm \sqrt{(-4)^2 - 4(3)(5)}}{2(3)}$$
$$= \frac{4 \pm \sqrt{-44}}{6} = \frac{4 \pm 2i\sqrt{11}}{6}$$
$$= \frac{2\left(2 \pm i\sqrt{11}\right)}{6} = \frac{2 \pm i\sqrt{11}}{3}$$
$$= \frac{2}{3} \pm \frac{\sqrt{11}}{3}i$$

Solution set: $\left\{ \frac{2}{3} \pm \frac{\sqrt{11}}{3}i \right\}$

6. $$3x = \sqrt{\frac{9x+2}{2}}$$
$$9x^2 = \frac{9x+2}{2} \quad \textit{Square each side.}$$
$$18x^2 = 9x + 2$$
$$18x^2 - 9x - 2 = 0$$
Here $a = 18, b = -9$, and $c = -2$.

Copyright © 2012 Pearson Education, Inc. Publishing as Addison-Wesley.

$$x = \frac{-b \pm \sqrt{b^2 - 4ac}}{2a}$$

$$x = \frac{-(-9) \pm \sqrt{(-9)^2 - 4(18)(-2)}}{2(18)}$$

$$= \frac{9 \pm \sqrt{225}}{36} = \frac{9 \pm 15}{36}$$

$$x = \frac{9 + 15}{36} = \frac{24}{36} = \frac{2}{3} \text{ or}$$

$$x = \frac{9 - 15}{36} = \frac{-6}{36} = -\frac{1}{6}$$

Check $x = \frac{2}{3}$: $\quad 2 = \sqrt{4} \quad$ *True*

Check $x = -\frac{1}{6}$: $-\frac{1}{2} = \sqrt{\frac{1}{4}} \quad$ *False*

Solution set: $\left\{ \frac{2}{3} \right\}$

7. If k is a negative number, then $4k$ is also negative, so the equation $x^2 = 4k$ will have two nonreal complex solutions. The answer is **A**.

8. $2x^2 - 8x - 3 = 0$

$$b^2 - 4ac = (-8)^2 - 4(2)(-3)$$
$$= 64 + 24 = 88$$

The discriminant, 88, is positive but not a perfect square, so there will be two distinct irrational number solutions.

9. $\qquad 3 - \dfrac{16}{x} - \dfrac{12}{x^2} = 0$

Multiply each term by the LCD, x^2.

$$x^2 \left(3 - \frac{16}{x} - \frac{12}{x^2} \right) = x^2 \cdot 0$$

$$3x^2 - 16x - 12 = 0$$
$$(3x + 2)(x - 6) = 0$$

$$3x + 2 = 0 \quad \text{or} \quad x - 6 = 0$$
$$x = -\frac{2}{3} \quad \text{or} \quad x = 6$$

Check $x = -\frac{2}{3}$: $\quad 3 + 24 - 27 = 0 \quad$ *True*

Check $x = 6$: $\qquad 3 - \frac{8}{3} - \frac{1}{3} = 0 \quad$ *True*

Solution set: $\left\{ -\frac{2}{3}, 6 \right\}$

10. $4x^2 + 7x - 3 = 0$

Use the quadratic formula with $a = 4$, $b = 7$, and $c = -3$.

$$x = \frac{-b \pm \sqrt{b^2 - 4ac}}{2a}$$

$$x = \frac{-7 \pm \sqrt{7^2 - 4(4)(-3)}}{2(4)}$$

$$= \frac{-7 \pm \sqrt{97}}{8}$$

Solution set: $\left\{ \frac{-7 \pm \sqrt{97}}{8} \right\}$

11. $\qquad 9x^4 + 4 = 37x^2$
$$9x^4 - 37x^2 + 4 = 0$$
Let $u = x^2$, so $u^2 = (x^2)^2 = x^4$.
$$9u^2 - 37u + 4 = 0$$
$$(9u - 1)(u - 4) = 0$$

$$9u - 1 = 0 \quad \text{or} \quad u - 4 = 0$$
$$u = \frac{1}{9} \quad \text{or} \qquad u = 4$$

To find x, substitute x^2 for u.

$$x^2 = \frac{1}{9} \qquad \text{or} \qquad x^2 = 4$$
$$x = \pm \sqrt{\frac{1}{9}} \qquad\qquad x = \pm \sqrt{4}$$
$$x = \pm \frac{1}{3} \qquad \text{or} \qquad x = \pm 2$$

Check $x = \pm \frac{1}{3}$: $\quad \frac{1}{9} + 4 = \frac{37}{9} \qquad$ *True*

Check $x = \pm 2$: $144 + 4 = 37(4) \qquad$ *True*

Solution set: $\left\{ \pm \frac{1}{3}, \pm 2 \right\}$

12. $12 = (2n + 1)^2 + (2n + 1)$
Let $u = 2n + 1$, so $u^2 = (2n + 1)^2$.
$$12 = u^2 + u$$
$$0 = u^2 + u - 12$$
$$0 = (u + 4)(u - 3)$$

$$u + 4 = 0 \quad \text{or} \quad u - 3 = 0$$
$$u = -4 \quad \text{or} \qquad u = 3$$

To find n, substitute $2n + 1$ for u.

$$2n + 1 = -4 \quad \text{or} \quad 2n + 1 = 3$$
$$2n = -5 \qquad\qquad 2n = 2$$
$$n = -\frac{5}{2} \quad \text{or} \qquad n = 1$$

Check $n = -\frac{5}{2}$: $\quad 12 = 16 - 4 \quad$ *True*

Check $n = 1$: $\qquad 12 = 9 + 3 \qquad$ *True*

Solution set: $\left\{ -\frac{5}{2}, 1 \right\}$

13. $(2x + 1)(3x + 1) = 4$
$$6x^2 + 5x + 1 = 4$$
$$6x^2 + 5x - 3 = 0$$
Here $a = 6$, $b = 5$, and $c = -3$.

$$x = \frac{-b \pm \sqrt{b^2 - 4ac}}{2a}$$

$$x = \frac{-5 \pm \sqrt{5^2 - 4(6)(-3)}}{2(6)}$$

$$= \frac{-5 \pm \sqrt{97}}{12}$$

Solution set: $\left\{ \frac{-5 \pm \sqrt{97}}{12} \right\}$

Copyright © 2012 Pearson Education, Inc. Publishing as Addison-Wesley.

14. Solve $S = 4\pi r^2$ for r.

$$\frac{S}{4\pi} = r^2$$

$$r = \pm\sqrt{\frac{S}{4\pi}} = \frac{\pm\sqrt{S}}{2\sqrt{\pi}}$$

$$= \frac{\pm\sqrt{S}}{2\sqrt{\pi}} \cdot \frac{\sqrt{\pi}}{\sqrt{\pi}}$$

$$r = \frac{\pm\sqrt{\pi S}}{2\pi}$$

15. Let $x =$ Terry's time alone.
Then $x - 2 =$ Callie's time alone. Make a table.

	Rate	Time Together	Fractional Part of the Job Done
Terry	$\frac{1}{x}$	5	$\frac{5}{x}$
Callie	$\frac{1}{x-2}$	5	$\frac{5}{x-2}$

Part done by Terry	plus	part done by Callie	equals	1 whole job.
↓	↓	↓	↓	↓
$\frac{5}{x}$	$+$	$\frac{5}{x-2}$	$=$	1

Multiply each side by the LCD, $x(x-2)$.

$$x(x-2)\left(\frac{5}{x} + \frac{5}{x-2}\right) = x(x-2) \cdot 1$$

$$5x - 10 + 5x = x^2 - 2x$$

$$0 = x^2 - 12x + 10$$

Use the quadratic formula with $a = 1$, $b = -12$, and $c = 10$.

$$x = \frac{-b \pm \sqrt{b^2 - 4ac}}{2a}$$

$$x = \frac{-(-12) \pm \sqrt{(-12)^2 - 4(1)(10)}}{2(1)}$$

$$= \frac{12 \pm \sqrt{104}}{2} = \frac{12 \pm 2\sqrt{26}}{2}$$

$$= \frac{2\left(6 \pm \sqrt{26}\right)}{2} = 6 \pm \sqrt{26}$$

$$x = 6 + \sqrt{26} \approx 11.1 \quad \text{or}$$

$$x = 6 - \sqrt{26} \approx 0.9$$

Reject 0.9 for Terry's time, because that would yield a negative time for Callie. Thus, Terry's time is about 11.1 hours and Callie's time is $x - 2 \approx 9.1$ hours.

16. Let $x =$ Qihong's rate.
Make a table. Use $d = rt$, or $t = \frac{d}{r}$.

	Distance	Rate	Time
Upstream	10	$x - 3$	$\frac{10}{x-3}$
Downstream	10	$x + 3$	$\frac{10}{x+3}$

Time upstream	plus	Time downstream	equals 3.5 hr.
↓	↓	↓	↓	↓
$\frac{10}{x-3}$	$+$	$\frac{10}{x+3}$	$=$	$\frac{7}{2}$

Multiply each side by the LCD, $2(x+3)(x-3)$.

$$2(x+3)(x-3)\left(\frac{10}{x-3} + \frac{10}{x+3}\right) = 2(x+3)(x-3)\left(\frac{7}{2}\right)$$

$$20(x+3) + 20(x-3) = 7\left(x^2 - 9\right)$$

$$20x + 60 + 20x - 60 = 7x^2 - 63$$

$$0 = 7x^2 - 40x - 63$$

$$0 = (x - 7)(7x + 9)$$

$$x - 7 = 0 \quad \text{or} \quad 7x + 9 = 0$$

$$x = 7 \quad \text{or} \quad x = -\frac{9}{7}$$

Reject $-\frac{9}{7}$ since the rate can't be negative. Qihong's rate was 7 mph.

17. Let $x =$ the width of the walk.
The area of the walk is equal to the area of the outer figure minus the area of the pool.

$$152 = (10 + 2x)(24 + 2x) - (24)(10)$$

$$152 = 240 + 68x + 4x^2 - 240$$

$$0 = 4x^2 + 68x - 152$$

$$0 = x^2 + 17x - 38$$

$$0 = (x + 19)(x - 2)$$

$$x + 19 = 0 \quad \text{or} \quad x - 2 = 0$$

$$x = -19 \quad \text{or} \quad x = 2$$

Reject -19 since width can't be negative. The walk is 2 feet wide.

18. Let $x =$ the height of the tower. Then $2x + 2 =$ the distance from the point to the top.

The distance from the base to the point is 30 m. These three segments form a right triangle, so the Pythagorean theorem applies.

$$a^2 + b^2 = c^2$$

$$x^2 + 30^2 = (2x + 2)^2$$

$$x^2 + 900 = 4x^2 + 8x + 4$$

$$0 = 3x^2 + 8x - 896$$

$$0 = (x - 16)(3x + 56)$$

$$x - 16 = 0 \quad \text{or} \quad 3x + 56 = 0$$

$$x = 16 \quad \text{or} \quad x = -\frac{56}{3}$$

Reject $-\frac{56}{3}$ since height can't be negative. The tower is 16 m high.

Copyright © 2012 Pearson Education, Inc. Publishing as Addison-Wesley.

19.
$$2x^2 + 7x > 15$$
$$2x^2 + 7x - 15 > 0$$
Solve the equation
$$2x^2 + 7x - 15 = 0.$$
$$(2x - 3)(x + 5) = 0$$

$$2x - 3 = 0 \quad \text{or} \quad x + 5 = 0$$
$$x = \tfrac{3}{2} \quad \text{or} \qquad x = -5$$

The numbers -5 and $\frac{3}{2}$ divide a number line into three intervals.

```
        A    B    C
    ————+————+————————▶
       -5    3
             ─
             2
```

Test a number from each interval in the original inequality, $2x^2 + 7x > 15$.

Interval A: Let $x = -6$.
$$72 - 42 \overset{?}{>} 15$$
$$30 > 15 \qquad \textit{True}$$
Interval B: Let $x = 0$.
$$0 > 15 \qquad \textit{False}$$
Interval C: Let $x = 2$.
$$8 + 14 \overset{?}{>} 15$$
$$22 > 15 \qquad \textit{True}$$

The numbers in Intervals A and C, not including -5 and $\frac{3}{2}$, are solutions.

Solution set: $(-\infty, -5) \cup \left(\frac{3}{2}, \infty\right)$

```
    ◀———)——————————(———▶
       -5      0    3
                    ─
                    2
```

20. $\dfrac{5}{t - 4} \leq 1$

Write the inequality so that 0 is on one side.

$$\frac{5}{t - 4} - 1 \leq 0$$
$$\frac{5}{t - 4} - \frac{1(t - 4)}{t - 4} \leq 0$$
$$\frac{5 - t + 4}{t - 4} \leq 0$$
$$\frac{-t + 9}{t - 4} \leq 0$$

The number 9 makes the numerator 0, and 4 makes the denominator 0. These two numbers determine three intervals.

```
        A    B    C
    ————+————+————————▶
        4    9
```

Test a number from each interval in the inequality

$$\frac{5}{t - 4} \leq 1.$$

Interval A: Let $t = 0$.
$$\frac{5}{-4} \leq 1 \qquad \textit{True}$$
Interval B: Let $t = 7$.
$$\frac{5}{3} \leq 1 \qquad \textit{False}$$
Interval C: Let $t = 10$.
$$\frac{5}{6} \leq 1 \qquad \textit{True}$$

The numbers in Intervals A and C, including 9 but not 4, are solutions.

Solution set: $(-\infty, 4) \cup [9, \infty)$

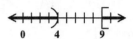

Cumulative Review Exercises (Chapters 1–9)

1. $S = \left\{ -\frac{7}{3}, -2, -\sqrt{3}, 0, 0.7, \sqrt{12}, \sqrt{-8}, 7, \frac{32}{3} \right\}$

(a) The elements of S that are integers are $-2, 0$, and 7.

(b) The elements of S that are rational numbers are $-\frac{7}{3}, -2, 0, 0.7, 7$, and $\frac{32}{3}$.

(c) All the elements of S except $\sqrt{-8}$ are real numbers.

(d) All the elements of S are complex numbers.

2.
$$7 - (4 + 3t) + 2t = -6(t - 2) - 5$$
$$7 - 4 - 3t + 2t = -6t + 12 - 5$$
$$3 - t = -6t + 7$$
$$5t = 4$$
$$t = \frac{4}{5}$$

Check $t = \frac{4}{5}$: $\frac{11}{5} = \frac{11}{5}$ *True*

Solution set: $\left\{\frac{4}{5}\right\}$

3. $|6x - 9| = |-4x + 2|$

$$6x - 9 = -4x + 2 \quad \text{or} \quad 6x - 9 = -(-4x + 2)$$
$$10x = 11 \qquad\qquad\qquad 6x - 9 = 4x - 2$$
$$\qquad\qquad\qquad\qquad\qquad 2x = 7$$
$$x = \frac{11}{10} \qquad \text{or} \qquad\qquad x = \frac{7}{2}$$

Check $x = \frac{11}{10}$: $\left|-\frac{24}{10}\right| = \left|-\frac{24}{10}\right|$ *True*
Check $x = \frac{7}{2}$: $|12| = |-12|$ *True*
Solution set: $\left\{\frac{11}{10}, \frac{7}{2}\right\}$

Copyright © 2012 Pearson Education, Inc. Publishing as Addison-Wesley.

4.
$$2x = \sqrt{\frac{5x+2}{3}}$$
$$(2x)^2 = \left(\sqrt{\frac{5x+2}{3}}\right)^2 \quad \text{Square.}$$
$$4x^2 = \frac{5x+2}{3}$$
$$12x^2 = 5x+2$$
$$12x^2 - 5x - 2 = 0$$
$$(3x-2)(4x+1) = 0$$

$$3x - 2 = 0 \quad \text{or} \quad 4x+1 = 0$$
$$x = \tfrac{2}{3} \quad \text{or} \qquad x = -\tfrac{1}{4}$$

Check $x = \tfrac{2}{3}$: $\quad \tfrac{4}{3} = \sqrt{\tfrac{16}{9}} \quad$ *True*

Check $x = -\tfrac{1}{4}$: $\quad -\tfrac{1}{2} = \sqrt{\tfrac{1}{4}} \quad$ *False*

Solution set: $\left\{\tfrac{2}{3}\right\}$

5.
$$\frac{3}{x-3} - \frac{2}{x-2} = \frac{3}{x^2 - 5x + 6}$$
$$\frac{3}{x-3} - \frac{2}{x-2} = \frac{3}{(x-3)(x-2)}$$
Multiply each term by the LCD, $(x-3)(x-2)$.
$$(x-3)(x-2)\left(\frac{3}{x-3} - \frac{2}{x-2}\right)$$
$$= (x-3)(x-2)\left[\frac{3}{(x-3)(x-2)}\right]$$
$$3(x-2) - 2(x-3) = 3$$
$$3x - 6 - 2x + 6 = 3$$
$$x = 3$$

The number 3 is not allowed as a solution since it makes the denominator 0. Solution set: $\emptyset$

6.
$$(r-5)(2r+3) = 1$$
$$2r^2 - 7r - 15 = 1$$
$$2r^2 - 7r - 16 = 0$$
Here $a = 2$, $b = -7$, and $c = -16$.

$$r = \frac{-b \pm \sqrt{b^2 - 4ac}}{2a}$$
$$r = \frac{-(-7) \pm \sqrt{(-7)^2 - 4(2)(-16)}}{2(2)}$$
$$= \frac{7 \pm \sqrt{49 + 128}}{4} = \frac{7 \pm \sqrt{177}}{4}$$

Solution set: $\left\{\frac{7 \pm \sqrt{177}}{4}\right\}$

7.
$$x^4 - 5x^2 + 4 = 0$$
Let $u = x^2$, so $u^2 = (x^2)^2 = x^4$.
$$u^2 - 5u + 4 = 0$$
$$(u-4)(u-1) = 0$$

$$u - 4 = 0 \quad \text{or} \quad u - 1 = 0$$
$$u = 4 \quad \text{or} \qquad u = 1$$

To find x, substitute x^2 for u.
$$x^2 = 4 \qquad \text{or} \qquad x^2 = 1$$
$$x = 2 \text{ or } x = -2 \quad \text{or} \quad x = 1 \text{ or } x = -1$$
The proposed solutions check.

Solution set: $\{\pm 1, \pm 2\}$

8.
$$-2x + 4 \le -x + 3$$
$$-x \le -1$$
Multiply by -1, and reverse the direction of the inequality.
$$x \ge 1$$

Solution set: $[1, \infty)$

9.
$$|3x - 7| \le 1$$

$$-1 \le 3x - 7 \le 1$$
$$6 \le 3x \qquad \le 8 \quad \text{Add 7.}$$
$$2 \le x \qquad \le \tfrac{8}{3} \quad \text{Divide by 3.}$$

Solution set: $\left[2, \tfrac{8}{3}\right]$

10.
$$x^2 - 4x + 3 < 0$$
Solve the equation
$$x^2 - 4x + 3 = 0.$$
$$(x-3)(x-1) = 0$$

$$x - 3 = 0 \quad \text{or} \quad x - 1 = 0$$
$$x = 3 \quad \text{or} \qquad x = 1$$

The numbers 1 and 3 divide a number line into three intervals.

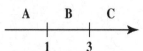

Test a number from each interval in the original inequality, $x^2 - 4x + 3 < 0$.

Interval A: Let $x = 0$.
$$3 < 0 \quad \text{\textit{False}}$$
Interval B: Let $x = 2$.
$$4 - 8 + 3 \overset{?}{<} 0$$
$$-1 < 0 \quad \text{\textit{True}}$$
Interval C: Let $x = 4$.
$$16 - 16 + 3 \overset{?}{<} 0$$
$$3 < 0 \quad \text{\textit{False}}$$

The numbers from Interval B, not including 1 or 3, are solutions.

Solution set: $(1, 3)$

11.
$$\frac{3}{p+2} > 1$$
Write the inequality so that 0 is on one side.
$$\frac{3}{p+2} - 1 > 0$$

Copyright © 2012 Pearson Education, Inc. Publishing as Addison-Wesley.

$$\frac{3}{p+2} - \frac{1(p+2)}{p+2} > 0$$

$$\frac{3-p-2}{p+2} > 0$$

$$\frac{-p+1}{p+2} > 0$$

The number 1 makes the numerator 0, and -2 makes the denominator 0. These two numbers determine three intervals.

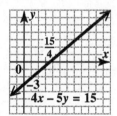

Test a number from each interval in the inequality

$$\frac{3}{p+2} > 1.$$

Interval A: Let $p = -4$.
$$\frac{3}{-2} > 1 \qquad \textit{False}$$

Interval B: Let $p = 0$.
$$\frac{3}{2} > 1 \qquad \textit{True}$$

Interval C: Let $p = 2$.
$$\frac{3}{4} > 1 \qquad \textit{False}$$

The numbers from Interval B, not including -2 or 1, are solutions.

Solution set: $(-2, 1)$

12. $4x - 5y = 15$

Draw the line through its intercepts, $\left(\frac{15}{4}, 0\right)$ and $(0, -3)$. The graph passes the vertical line test, so the relation is a function. As with any line that is not horizontal or vertical, the domain and range are both $(-\infty, \infty)$.

Solve the equation for y.

$$4x - 5y = 15$$
$$-5y = 15 - 4x$$
$$y = \frac{15 - 4x}{-5}, \quad \text{or} \quad \frac{4x - 15}{5}$$

Thus, $f(x) = \frac{4}{5}x - 3$.

13. $4x - 5y < 15$

Draw a dashed line through the points $\left(\frac{15}{4}, 0\right)$ and $(0, -3)$. Check the origin:

$$4(0) - 5(0) \overset{?}{<} 15$$
$$0 < 15 \qquad \textit{True}$$

Shade the region that contains the origin.

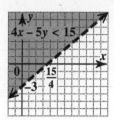

The relation is not a function since for any value of x, there is more than one value of y.

14.   The graph of $x = 5$ is a vertical line. A horizontal line has equation $y = k$.

15. $-2x + 7y = 16$

Solve the equation for y.

$$7y = 2x + 16$$
$$y = \frac{2}{7}x + \frac{16}{7}$$

So the slope is $\frac{2}{7}$ and the y-intercept is $\left(0, \frac{16}{7}\right)$. Let $y = 0$ in $-2x + 7y = 16$ to find the x-intercept.

$$-2x + 7(0) = 16$$
$$-2x = 16$$
$$x = -8$$

The x-intercept is $(-8, 0)$.

16.   **(a)**   Solve $5x + 2y = 6$ for y.
$$2y = -5x + 6$$
$$y = -\frac{5}{2}x + 3$$

So the slope of the given line and the desired line is $-\frac{5}{2}$. The required form is

$$y = -\frac{5}{2}x + b.$$
$$-3 = -\frac{5}{2}(2) + b \quad \textit{Let x=2, y=−3.}$$
$$-3 = -5 + b$$
$$2 = b$$

The equation is $y = -\frac{5}{2}x + 2$.

(b)   The negative reciprocal of the slope in part (a) is

$$-\frac{1}{-\frac{5}{2}} = \frac{2}{5},$$

which is the slope of the line perpendicular to the given line.

So $y = \frac{2}{5}x + b$.
$$1 = \frac{2}{5}(-4) + b \quad \textit{Let x = −4, y = 1.}$$
$$1 = -\frac{8}{5} + b$$
$$\frac{13}{5} = b$$

The equation is $y = \frac{2}{5}x + \frac{13}{5}$.

Copyright © 2012 Pearson Education, Inc. Publishing as Addison-Wesley.

17. $2x - 4y = 10$ (1)
$9x + 3y = 3$ (2)

Simplify the equations.

$x - 2y = 5$ (3) $\frac{1}{2} \times$ (1)
$3x + y = 1$ (4) $\frac{1}{3} \times$ (2)

To eliminate y, multiply (4) by 2 and add the result to (3).

$$\begin{array}{rl} x - 2y = 5 & (3) \\ 6x + 2y = 2 & 2 \times (4) \\ \hline 7x = 7 & \\ x = 1 & \end{array}$$

Substitute $x = 1$ into (4).

$$\begin{array}{rl} 3x + y = 1 & (4) \\ 3(1) + y = 1 & \\ y = -2 & \end{array}$$

Solution set: $\{(1, -2)\}$

18. $x + y + 2z = 3$ (1)
$-x + y + z = -5$ (2)
$2x + 3y - z = -8$ (3)

Eliminate z by adding (2) and (3).

$$\begin{array}{rl} -x + y + z = -5 & (2) \\ 2x + 3y - z = -8 & (3) \\ \hline x + 4y = -13 & (4) \end{array}$$

To get another equation without z, multiply equation (3) by 2 and add the result to equation (1).

$$\begin{array}{rl} x + y + 2z = 3 & (1) \\ 4x + 6y - 2z = -16 & 2 \times (3) \\ \hline 5x + 7y = -13 & (5) \end{array}$$

To eliminate x, multiply (4) by -5 and add the result to (5).

$$\begin{array}{rl} -5x - 20y = 65 & -5 \times (4) \\ 5x + 7y = -13 & (5) \\ \hline -13y = 52 & \\ y = -4 & \end{array}$$

Use (4) to find x.

$$\begin{array}{rl} x + 4y = -13 & (4) \\ x + 4(-4) = -13 & \\ x - 16 = -13 & \\ x = 3 & \end{array}$$

Use (2) to find z.

$$\begin{array}{rl} -x + y + z = -5 & (2) \\ -3 - 4 + z = -5 & \\ -7 + z = -5 & \\ z = 2 & \end{array}$$

Solution set: $\{(3, -4, 2)\}$

19. Let $x =$ the amount of revenue for Microsoft and $y =$ the amount of revenue for Oracle.

The combined revenues were \$82.8 billion, so

$$x + y = 82.8. \quad (1)$$

Revenues for Microsoft were \$6.8 billion less than three times those of Oracle, so

$$x = 3y - 6.8. \quad (2)$$

Substitute $3y - 6.8$ for x in (1), then solve for y.

$$\begin{array}{l} (3y - 6.8) + y = 82.8 \\ 4y = 89.6 \\ y = 22.4 \end{array}$$

From (2), $x = 3(22.4) - 6.8 = 60.4$. Thus, in 2009, Microsoft had \$60.4 billion in revenue and Oracle had \$22.4 billion in revenue.

20. $\left(\dfrac{x^{-3}y^2}{x^5 y^{-2}}\right)^{-1} = \left(x^{-3-5}y^{2-(-2)}\right)^{-1}$

$= \left(x^{-8}y^4\right)^{-1}$

$= x^8 y^{-4}$

$= \dfrac{x^8}{y^4}$

21. $\dfrac{(4x^{-2})^2(2y^3)}{8x^{-3}y^5} = \dfrac{16x^{-4}(2y^3)}{8x^{-3}y^5}$

$= \dfrac{4x^{-4}y^3}{x^{-3}y^5}$

$= 4x^{-4-(-3)}y^{3-5}$

$= 4x^{-1}y^{-2}$

$= \dfrac{4}{xy^2}$

22. $\left(\frac{2}{3}t + 9\right)^2 = \left(\frac{2}{3}t\right)^2 + 2\left(\frac{2}{3}t\right)(9) + 9^2$

$= \frac{4}{9}t^2 + 12t + 81$

23. Divide $4x^3 + 2x^2 - x + 26$ by $x + 2$.

$$\begin{array}{r} 4x^2 - 6x + 11 \\ x + 2 \overline{\smash{\big)}\, 4x^3 + 2x^2 - x + 26} \\ \underline{4x^3 + 8x^2} \\ -6x^2 - x \\ \underline{-6x^2 - 12x} \\ 11x + 26 \\ \underline{11x + 22} \\ 4 \end{array}$$

$(4x^3 + 2x^2 - x + 26)$ divided by $(x + 2)$ is

$$4x^2 - 6x + 11 + \frac{4}{x + 2}.$$

Copyright © 2012 Pearson Education, Inc. Publishing as Addison-Wesley.

24. $24m^2 + 2m - 15$

The two integers whose product is $24(-15) = -360$ and whose sum is 2 are 20 and -18.

$$24m^2 + 2m - 15$$
$$= 24m^2 + 20m - 18m - 15$$
$$= 4m(6m + 5) - 3(6m + 5)$$
$$= (6m + 5)(4m - 3)$$

25. $8x^3 + 27y^3$

Use the sum of cubes formula,

$$a^3 + b^3 = (a + b)(a^2 - ab + b^2),$$

with $a = 2x$ and $b = 3y$.

$$8x^3 + 27y^3$$
$$= (2x + 3y)\left[(2x)^2 - (2x)(3y) + (3y)^2\right]$$
$$= (2x + 3y)\left(4x^2 - 6xy + 9y^2\right)$$

26. $9x^2 - 30xy + 25y^2$

Use the perfect square formula,

$$a^2 - 2ab + b^2 = (a - b)^2,$$

with $a = 3x$ and $b = 5y$.

$$9x^2 - 30xy + 25y^2$$
$$= \left[(3x)^2 - 2(3x)(5y) + (5y)^2\right]$$
$$= (3x - 5y)^2$$

27. $\dfrac{5x + 2}{-6} \div \dfrac{15x + 6}{5}$

Multiply by the reciprocal.

$$= \frac{5x + 2}{-6} \cdot \frac{5}{15x + 6}$$
$$= \frac{5x + 2}{-6} \cdot \frac{5}{3(5x + 2)}$$
$$= \frac{5}{-18}, \text{ or } -\frac{5}{18}$$

28. $\dfrac{3}{2 - x} - \dfrac{5}{x} + \dfrac{6}{x^2 - 2x}$

$$= \frac{3}{2 - x} - \frac{5}{x} + \frac{6}{x(x - 2)}$$
$$= \frac{-3}{x - 2} - \frac{5}{x} + \frac{6}{x(x - 2)}$$

The LCD is $x(x - 2)$.

$$= \frac{-3x}{(x - 2)x} - \frac{5(x - 2)}{x(x - 2)} + \frac{6}{x(x - 2)}$$
$$= \frac{-3x - 5(x - 2) + 6}{x(x - 2)}$$
$$= \frac{-3x - 5x + 10 + 6}{x(x - 2)}$$
$$= \frac{-8x + 16}{x(x - 2)}$$
$$= \frac{-8(x - 2)}{x(x - 2)} = -\frac{8}{x}$$

29. $\dfrac{\dfrac{r}{s} - \dfrac{s}{r}}{\dfrac{r}{s} + 1}$

Multiply the numerator and denominator by the LCD of all the fractions, rs.

$$= \frac{\left(\dfrac{r}{s} - \dfrac{s}{r}\right)rs}{\left(\dfrac{r}{s} + 1\right)rs} = \frac{r^2 - s^2}{r^2 + rs}$$
$$= \frac{(r - s)(r + s)}{r(r + s)} = \frac{r - s}{r}$$

30. $\sqrt[3]{\dfrac{27}{16}} = \dfrac{\sqrt[3]{27}}{\sqrt[3]{16}} = \dfrac{\sqrt[3]{3^3}}{\sqrt[3]{8 \cdot 2}} = \dfrac{3}{2\sqrt[3]{2}}$

$$= \frac{3 \cdot \sqrt[3]{4}}{2\sqrt[3]{2} \cdot \sqrt[3]{4}} = \frac{3\sqrt[3]{4}}{2\sqrt[3]{8}}$$
$$= \frac{3\sqrt[3]{4}}{2 \cdot 2} = \frac{3\sqrt[3]{4}}{4}$$

31. $\dfrac{2}{\sqrt{7} - \sqrt{5}} = \dfrac{2(\sqrt{7} + \sqrt{5})}{(\sqrt{7} - \sqrt{5})(\sqrt{7} + \sqrt{5})}$

$$= \frac{2(\sqrt{7} + \sqrt{5})}{7 - 5}$$
$$= \frac{2(\sqrt{7} + \sqrt{5})}{2} = \sqrt{7} + \sqrt{5}$$

32. Let $x =$ the distance traveled by the southbound car and $2x - 38 =$ the distance traveled by the eastbound car.

Since the cars are traveling at right angles with one another, the Pythagorean theorem can be applied.

$$a^2 + b^2 = c^2$$
$$x^2 + (2x - 38)^2 = 95^2$$
$$x^2 + 4x^2 - 152x + 1444 = 9025$$
$$5x^2 - 152x - 7581 = 0$$

$$x = \frac{-b \pm \sqrt{b^2 - 4ac}}{2a}$$
$$x = \frac{-(-152) \pm \sqrt{(-152)^2 - 4(5)(-7581)}}{2(5)}$$
$$= \frac{152 \pm \sqrt{174{,}724}}{10} = \frac{152 \pm 418}{10}$$

Thus, $x = \dfrac{152 + 418}{10} = 57$ (the other value is negative). The southbound car traveled 57 miles, and the eastbound car traveled $2x - 38 = 2(57) - 38 = 76$ miles.

Copyright © 2012 Pearson Education, Inc. Publishing as Addison-Wesley.

Copyright © 2012 Pearson Education, Inc. Publishing as Addison-Wesley.

CHAPTER 10 ADDITIONAL GRAPHS OF FUNCTIONS AND RELATIONS

10.1 Review of Operations and Composition

10.1 Now Try Exercises

N1. $f(x) = 3x - 4$, $g(x) = 2x^2 + 5$

(a) $(f + g)(2) = f(2) + g(2)$
$$= [3(2) - 4] + [2(2)^2 + 5]$$
$$= 2 + 13$$
$$= 15$$

(b) $(f - g)(-1) = f(-1) - g(-1)$
$$= [3(-1) - 4] - [2(-1)^2 + 5]$$
$$= -7 - (7)$$
$$= -14$$

(c) $(fg)(0) = f(0) \cdot g(0)$
$$= [3(0) - 4] \cdot [2(0)^2 + 5]$$
$$= -4 \cdot 5$$
$$= -20$$

(d) $\left(\dfrac{f}{g}\right)(3) = \dfrac{f(3)}{g(3)} = \dfrac{3(3) - 4}{2(3)^2 + 5} = \dfrac{5}{23}$

N2. $f(x) = 4x - 2$, $g(x) = \sqrt{3x + 12}$

(a) $(f + g)(x) = f(x) + g(x)$
$$= 4x - 2 + \sqrt{3x + 12}$$

(b) $(f - g)(x) = f(x) - g(x)$
$$= 4x - 2 - \sqrt{3x + 12}$$

(c) $(fg)(x) = f(x) \cdot g(x)$
$$= (4x - 2)\sqrt{3x + 12}$$

(d) $\left(\dfrac{f}{g}\right)(x) = \dfrac{f(x)}{g(x)} = \dfrac{4x - 2}{\sqrt{3x + 12}}$

The domain of f is $(-\infty, \infty)$ and the domain of g is $[-4, \infty)$ [since $3x + 12 \geq 0$ if $x \geq -4$], so the domain in parts (a), (b), and (c) is the intersection of $(-\infty, \infty)$ and $[-4, \infty)$, that is, $[-4, \infty)$.

In part (d) we must exclude the values that make the denominator 0. Thus, we exclude -4 and the domain in part (d) is $(-4, \infty)$.

N3.
$$f(x) = 2x^2 + 3$$
$$f(x + h) = 2(x + h)^2 + 3$$
$$= 2(x^2 + 2xh + h^2) + 3$$
$$= 2x^2 + 4xh + 2h^2 + 3$$

Now subtract $f(x)$ from $f(x + h)$.

$$f(x + h) - f(x)$$
$$= 2x^2 + 4xh + 2h^2 + 3 - (2x^2 + 3)$$
$$= 4xh + 2h^2$$

Then divide the last result by h.

$$\frac{4xh + 2h^2}{h} = \frac{h(4x + 2h)}{h} = 4x + 2h$$

Thus,

$$\frac{f(x + h) - f(x)}{h} = 4x + 2h.$$

N4. $f(x) = \dfrac{5}{x + 2}$, $g(x) = x - 4$

(a) $(f \circ g)(-3) = f(g(-3))$
$$= f(-7) = \frac{5}{-5} = -1$$

(b) $(g \circ f)(-3) = g(f(-3))$
$$= g(-5) = -9$$

N5. $f(x) = 2x^2 - x$, $g(x) = x - 4$

(a) $(g \circ f)(x) = g(f(x))$
$$= g(2x^2 - x)$$
$$= (2x^2 - x) - 4$$
$$= 2x^2 - x - 4$$

The domain is $(-\infty, \infty)$.

(b) $(f \circ g)(x) = f(g(x))$
$$= f(x - 4)$$
$$= 2(x - 4)^2 - (x - 4)$$
$$= 2(x^2 - 8x + 16) - x + 4$$
$$= 2x^2 - 16x + 32 - x + 4$$
$$= 2x^2 - 17x + 36$$

The domain is $(-\infty, \infty)$.

N6. $f(x) = \sqrt{x + 5}$, $g(x) = 2x - 1$

$(f \circ g)(x) = f(g(x))$
$$= f(2x - 1)$$
$$= \sqrt{(2x - 1) + 5}$$
$$= \sqrt{2x + 4}$$

The radical expression $\sqrt{2x + 4}$ is a nonnegative real number only when $2x + 4 \geq 0$, or $x \geq -2$. Thus, the domain of $f \circ g$ is the interval $[-2, \infty)$.

$(g \circ f)(x) = g(f(x))$
$$= g\left(\sqrt{x + 5}\right)$$
$$= 2\sqrt{x + 5} - 1$$

The radical expression $\sqrt{x + 5}$ is a nonnegative real number only when $x + 5 \geq 0$, or $x \geq -5$. Thus, the domain of $g \circ f$ is the interval $[-5, \infty)$.

Copyright © 2012 Pearson Education, Inc. Publishing as Addison-Wesley.

N7. $(f \circ g)(x) = 3\sqrt{x+4} - 1$ ▪ We could substitute $x + 4$ for x in the radical. In this case, $f(x) = 3\sqrt{x} - 1$ and $g(x) = x + 4$.

Another choice would be to substitute $\sqrt{x+4}$ for x in the polynomial. In this case, $f(x) = 3x - 1$ and $g(x) = \sqrt{x+4}$.

10.1 Section Exercises

In Exercises 1–18, $f(x) = 2x^2 - 4$ and $g(x) = 3x + 1$.

1. $(f+g)(3) = f(3) + g(3)$
$$= \left[2(3)^2 - 4\right] + \left[3(3) + 1\right]$$
$$= (18 - 4) + (9 + 1)$$
$$= 14 + 10 = 24$$

3. $(f-g)(1) = f(1) - g(1)$
$$= \left[2(1)^2 - 4\right] - \left[3(1) + 1\right]$$
$$= (2 - 4) - (3 + 1)$$
$$= -2 - 4 = -6$$

5. $(fg)(4) = f(4) \cdot g(4)$
$$= \left[2(4)^2 - 4\right] \cdot \left[3(4) + 1\right]$$
$$= (32 - 4) \cdot (12 + 1)$$
$$= 28 \cdot 13 = 364$$

7. $(gf)(-3) = g(-3) \cdot f(-3)$
$$= \left[3(-3) + 1\right] \cdot \left[2(-3)^2 - 4\right]$$
$$= (-9 + 1) \cdot (18 - 4)$$
$$= -8 \cdot 14 = -112$$

9. $\left(\dfrac{g}{f}\right)(-1) = \dfrac{g(-1)}{f(-1)}$
$$= \dfrac{3(-1) + 1}{2(-1)^2 - 4}$$
$$= \dfrac{-3 + 1}{2 - 4} = \dfrac{-2}{-2} = 1$$

11. $\left(\dfrac{f}{g}\right)(4) = \dfrac{f(4)}{g(4)}$
$$= \dfrac{2(4)^2 - 4}{3(4) + 1}$$
$$= \dfrac{32 - 4}{12 + 1} = \dfrac{28}{13}$$

13. $(f \circ g)(2) = f(g(2))$
$$= f(3(2) + 1)$$
$$= f(7)$$
$$= 2(7)^2 - 4$$
$$= 98 - 4 = 94$$

15. $(g \circ f)(2) = g(f(2))$
$$= g\left(2(2)^2 - 4\right)$$
$$= g(8 - 4)$$
$$= g(4)$$
$$= 3(4) + 1 = 13$$

17. $(f \circ g)(-2) = f(g(-2))$
$$= f(3(-2) + 1)$$
$$= f(-5)$$
$$= 2(-5)^2 - 4$$
$$= 50 - 4 = 46$$

19. $f(x) = 4x - 1$, $g(x) = 6x + 3$

(a) $(f+g)(x) = f(x) + g(x)$
$$= (4x - 1) + (6x + 3)$$
$$= 10x + 2$$

(b) $(f-g)(x) = f(x) - g(x)$
$$= (4x - 1) - (6x + 3)$$
$$= -2x - 4$$

(c) $(fg)(x) = f(x)g(x)$
$$= (4x - 1)(6x + 3)$$
$$= 24x^2 + 6x - 3$$

(d) $\left(\dfrac{f}{g}\right)(x) = \dfrac{f(x)}{g(x)}$
$$= \dfrac{4x - 1}{6x + 3}$$

The domains of $f + g$, $f - g$, and fg are the set of all real numbers, or $(-\infty, \infty)$.
The domain of $\frac{f}{g}$ is the set of all real numbers except $-\frac{1}{2}$, since if $x = -\frac{1}{2}$, the denominator is 0. This set is written in interval notation as $\left(-\infty, -\frac{1}{2}\right) \cup \left(-\frac{1}{2}, \infty\right)$.

21. $f(x) = 3x^2 - 2x$, $g(x) = x^2 - 2x + 1$

(a) $(f+g)(x)$
$$= f(x) + g(x)$$
$$= (3x^2 - 2x) + \left(x^2 - 2x + 1\right)$$
$$= 4x^2 - 4x + 1$$

(b) $(f-g)(x)$
$$= f(x) - g(x)$$
$$= (3x^2 - 2x) - \left(x^2 - 2x + 1\right)$$
$$= 2x^2 - 1$$

(c) $(fg)(x) = f(x)g(x)$
$$= \left(3x^2 - 2x\right)\left(x^2 - 2x + 1\right)$$

(d) $\left(\dfrac{f}{g}\right)(x) = \dfrac{f(x)}{g(x)}$
$$= \dfrac{3x^2 - 2x}{x^2 - 2x + 1}$$

The domains of $f + g$, $f - g$, and fg are the set of all real numbers, or $(-\infty, \infty)$.
The domain of $\frac{f}{g}$ is the set of all real numbers x such that $x^2 - 2x + 1 \neq 0$. Since $x^2 - 2x + 1 = (x - 1)^2$, the only number which gives this denominator a value of 0 is $x = 1$. Therefore, the domain is the set of all real numbers except 1, or $(-\infty, 1) \cup (1, \infty)$.

Copyright © 2012 Pearson Education, Inc. Publishing as Addison-Wesley.

23. $f(x) = \sqrt{2x+5}$, $g(x) = \sqrt{4x+9}$

  (a) $(f+g)(x) = \sqrt{2x+5} + \sqrt{4x+9}$

  (b) $(f-g)(x) = \sqrt{2x+5} - \sqrt{4x+9}$

  (c) $(fg)(x) = \sqrt{2x+5} \cdot \sqrt{4x+9}$

$$= \sqrt{(2x+5)(4x+9)}$$

  (d) $\left(\dfrac{f}{g}\right)(x) = \dfrac{\sqrt{2x+5}}{\sqrt{4x+9}} = \sqrt{\dfrac{2x+5}{4x+9}}$

25. $f(x) = 2x^2 - 1$

  (a) $f(x+h)$

$$= 2(x+h)^2 - 1$$
$$= 2(x^2 + 2xh + h^2) - 1$$
$$= 2x^2 + 4xh + 2h^2 - 1$$

$f(x+h) - f(x)$
$$= (2x^2 + 4xh + 2h^2 - 1) - (2x^2 - 1)$$
$$= 2x^2 + 4xh + 2h^2 - 1 - 2x^2 + 1$$
$$= 4xh + 2h^2$$

  (b) $\dfrac{f(x+h) - f(x)}{h} = \dfrac{4xh + 2h^2}{h}$

$$= \dfrac{h(4x+2h)}{h}$$
$$= 4x + 2h$$

27. $f(x) = x^2 + 4x$

  (a) $f(x+h)$

$$= (x+h)^2 + 4(x+h)$$
$$= x^2 + 2xh + h^2 + 4x + 4h$$

$f(x+h) - f(x)$
$$= (x^2 + 2xh + h^2 + 4x + 4h) - (x^2 + 4x)$$
$$= x^2 + 2xh + h^2 + 4x + 4h - x^2 - 4x$$
$$= 2xh + h^2 + 4h$$

  (b) $\dfrac{f(x+h) - f(x)}{h} = \dfrac{2xh + h^2 + 4h}{h}$

$$= \dfrac{h(2x+h+4)}{h}$$
$$= 2x + h + 4$$

29. $f(x) = 5x + 3$, $g(x) = -x^2 + 4x + 3$

$(f \circ g)(x) = f(g(x))$
$$= f(-x^2 + 4x + 3)$$
$$= 5(-x^2 + 4x + 3) + 3$$
$$= -5x^2 + 20x + 18$$

$(g \circ f)(x)$
$$= g(f(x))$$
$$= g(5x + 3)$$
$$= -(5x+3)^2 + 4(5x+3) + 3$$
$$= -(25x^2 + 30x + 9) + 20x + 12 + 3$$
$$= -25x^2 - 10x + 6$$

The domain of $f \circ g$ and $g \circ f$ is $(-\infty, \infty)$.

31. $f(x) = \dfrac{1}{x}$, $g(x) = x^2$

The domain of f is $(-\infty, 0) \cup (0, \infty)$.
The domain of g is $(-\infty, \infty)$.

$$(f \circ g)(x) = f(g(x))$$
$$= f(x^2)$$
$$= \dfrac{1}{x^2}$$

The value 0 must be excluded from the domain of $f \circ g$, so the domain is $(-\infty, 0) \cup (0, \infty)$.

$$(g \circ f)(x) = g(f(x))$$
$$= g\left(\dfrac{1}{x}\right)$$
$$= \left(\dfrac{1}{x}\right)^2$$
$$= \dfrac{1}{x^2}$$

The value 0 must be excluded from the domain of $g \circ f$, so the domain is $(-\infty, 0) \cup (0, \infty)$.

33. $f(x) = \sqrt{x+2}$, $g(x) = 8x - 6$

The domain of f is the set of all x such that $x + 2 \geq 0$, or $x \geq -2$, or $[-2, \infty)$.
The domain of g is $(-\infty, \infty)$.

$$(f \circ g)(x) = f(g(x))$$
$$= \sqrt{(8x-6) + 2} = \sqrt{8x - 4}$$
$$= \sqrt{4(2x-1)} = 2\sqrt{2x-1}$$

The domain of $f \circ g$ is the set of all real numbers for which $2x - 1 \geq 0$, or $x \geq \frac{1}{2}$, which may be written $\left[\frac{1}{2}, \infty\right)$.

$$(g \circ f)(x) = g(f(x))$$
$$= g(\sqrt{x+2})$$
$$= 8\sqrt{x+2} - 6$$

The domain of $g \circ f$ is the set of all real numbers for which $x + 2 \geq 0$, or $x \geq -2$, which may be written $[-2, \infty)$.

35. $f(x) = \dfrac{1}{x-5}$, $g(x) = \dfrac{2}{x}$

The domain of f is $(-\infty, 5) \cup (5, \infty)$.
The domain of g is $(-\infty, 0) \cup (0, \infty)$.

$$(f \circ g)(x) = f(g(x))$$
$$= f\left(\dfrac{2}{x}\right)$$
$$= \dfrac{1}{\left(\frac{2}{x} - 5\right)} \cdot \dfrac{x}{x}$$
$$= \dfrac{x}{2 - 5x}$$

Copyright © 2012 Pearson Education, Inc. Publishing as Addison-Wesley.

To find the domain of $f \circ g$, focus on the denominators of the fractions that are involved. For $\frac{2}{x}$ to be defined, we must have $x \neq 0$. For $\frac{x}{2-5x}$ to be defined, we must have $2 - 5x \neq 0$, or $x \neq \frac{2}{5}$. In interval notation, the domain of $f \circ g$ is $(-\infty, 0) \cup \left(0, \frac{2}{5}\right) \cup \left(\frac{2}{5}, \infty\right)$.

$$(g \circ f)(x) = g(f(x))$$
$$= g\left(\frac{1}{x-5}\right)$$
$$= \frac{2}{\frac{1}{x-5}}$$
$$= 2(x-5)$$

The domain of $g \circ f$ is the set of all x such that $x \neq 5$ [because f is not defined], or $(-\infty, 5) \cup (5, \infty)$.

37. $(f+g)(1) = f(1) + g(1)$

The point $(1, 1)$ is on the graph of $y = f(x)$, so $f(1) = 1$.

The point $(1, 3)$ is on the graph of $y = g(x)$, so $g(1) = 3$.

$$f(1) + g(1) = 1 + 3 = 4$$

39. $(f-g)(0) = f(0) - g(0)$

The point $(0, 1)$ is on the graph of $y = f(x)$, so $f(0) = 1$.

The point $(0, 4)$ is on the graph of $y = g(x)$, so $g(0) = 4$.

$$f(0) - g(0) = 1 - 4 = -3$$

41. $f(-2) \cdot g(4)$
$= -1 \cdot 0$ *(-2, -1) is on f, (4, 0) is on g*
$= 0$

43. $(f \circ g)(2) = f(g(2))$
           $= f(2)$ *(2, 2) is on g*
           $= 1$ *(2, 1) is on f*

45. $(g \circ f)(2) = g(f(2))$
           $= g(1)$ *(2, 1) is on f*
           $= 3$ *(1, 3) is on g*

47. $[f(0)]^6 = (1)^6$ *(0, 1) is on f*
          $= 1$

49. $(f \circ g)(2) = f(g(2))$
           $= f(3) = 1$

51. $(g \circ f)(3) = g(f(3))$
           $= g(1) = 9$

53. $(f \circ f)(4) = f(f(4))$
           $= f(3) = 1$

55. $(f \circ g)(1) = f(g(1))$
           $= f(9)$

However, $f(9)$ cannot be determined from the given table.

57. $h(x) = (6x - 2)^2$ ▪ One choice is $f(x) = x^2$ and $g(x) = 6x - 2$. Then

$$(f \circ g)(x) = f(g(x))$$
$$= f(6x - 2)$$
$$= (6x - 2)^2 = h(x).$$

Other correct answers are possible.

59. $h(x) = \frac{1}{x^2 + 2}$ ▪ One choice is $f(x) = \frac{1}{x+2}$ and $g(x) = x^2$. Then

$$(f \circ g)(x) = f(g(x))$$
$$= f(x^2)$$
$$= \frac{1}{x^2 + 2} = h(x).$$

Other correct answers are possible.

♦♦♦ Relating Concepts 61–68 ♦♦♦

61. Because $\underline{0}$ is the identity element for addition, $a + \underline{0} = \underline{0} + a$ for all real numbers.

62. Because $\underline{1}$ is the identity element for multiplication, $a \cdot \underline{1} = \underline{1} \cdot a$ for all real numbers a.

63. If we choose the function

$$g(x) = 2x^2 + 3x - 1,$$

we find that
$$(f \circ g)(x) = f(g(x)) = f(2x^2 + 3x - 1)$$
$$= 2x^2 + 3x - 1$$

and
$$(g \circ f)(x) = g(f(x)) = g(x)$$
$$= 2x^2 + 3x - 1.$$

Thus,
$$(f \circ g)(x) = g(x)$$

and
$$(g \circ f)(x) = g(x).$$

In each case, we get $g(x)$. This will happen no matter what function is chosen as g.

64. Because composition of the function $f(x) = x$ with any function leaves that function unchanged, $f(x) = x$ is called the *identity function*.

65. The inverse property of addition says that for every real number a, there exists a unique real number $\underline{-a}$ such that $a + \underline{-a} = \underline{-a} + a = 0$.

66. The inverse property of multiplication says that for every nonzero real number a, there exists a unique real number $\frac{1}{a}$ such that $a \cdot \frac{1}{a} = \frac{1}{a} \cdot a = 1$.

Copyright © 2012 Pearson Education, Inc. Publishing as Addison-Wesley.

67. $f(x) = x^3 + 2, g(x) = \sqrt[3]{x-2}$

$$(f \circ g)(x) = f(g(x))$$
$$= f\left(\sqrt[3]{x-2}\right)$$
$$= \left(\sqrt[3]{x-2}\right)^3 + 2$$
$$= x - 2 + 2$$
$$= x$$

$$(g \circ f)(x) = g(f(x))$$
$$= g\left(x^3 + 2\right)$$
$$= \sqrt[3]{(x^3 + 2) - 2}$$
$$= \sqrt[3]{x^3}$$
$$= x$$

Thus, $(f \circ g)(x) = x$ and $(g \circ f)(x) = x$. In each case, we get x.

68. f and g are called inverses.

69. $f(x) = 3x, g(x) = 1760x$

$$(f \circ g)(x) = f(g(x))$$
$$= f(1760x)$$
$$= 3(1760x)$$
$$= 5280x$$

$(f \circ g)(x)$ computes the number of feet in x miles.

71. $D(p) = \dfrac{-p^2}{100} + 500$ and $p(c) = 2c - 10$

$$D(p(c)) = D(2c - 10)$$
$$= \frac{-(2c-10)^2}{100} + 500$$
$$= \frac{-(4c^2 - 40c + 100)}{100} + 500$$
$$= \frac{-4(c^2 - 10c + 25)}{100} + 500$$
$$= \frac{-c^2 + 10c - 25}{25} + 500$$

Since this expresses the demand in terms of the cost, it may be labeled as $D(c)$.

73. $f(x) = x^2 + 4x - 3$
$$f(2) = 2^2 + 4(2) - 3$$
$$= 4 + 8 - 3$$
$$= 9$$

75. $f(x) = 8x^2 - 2x - 3$
$$0 = 8x^2 - 2x - 3$$
$$0 = (2x + 1)(4x - 3)$$

$$2x + 1 = 0 \quad\quad \text{or} \quad\quad 4x - 3 = 0$$
$$x = -\tfrac{1}{2} \quad\quad \text{or} \quad\quad x = \tfrac{3}{4}$$

The solution set is $\left\{-\tfrac{1}{2}, \tfrac{3}{4}\right\}$.

10.2 Graphs of Quadratic Functions

10.2 Now Try Exercises

N1. $f(x) = x^2 - 3$

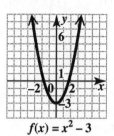

$$f(x) = x^2 - 3$$

The graph of $f(x) = x^2 - 3$ has the same shape as the graph of $f(x) = x^2$, but the graph is shifted down 3 units and has vertex $(0, -3)$. The graph of this parabola is symmetric about its axis $x = 0$.

Since x can be any real number, the domain is $(-\infty, \infty)$. The value of y is always greater than or equal to -3, so the range is $[-3, \infty)$.

N2. $f(x) = (x + 1)^2$

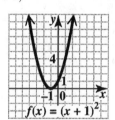

$$f(x) = (x + 1)^2$$

The parabola $f(x) = (x + 1)^2$ has the same shape as $f(x) = x^2$ but is shifted 1 unit to the left, since -1 would cause $x + 1$ to equal 0. The vertex is $(-1, 0)$ and the axis is $x = -1$. The domain is $(-\infty, \infty)$ and the range is $[0, \infty)$.

N3. $f(x) = (x + 1)^2 - 2$

$$f(x) = (x + 1)^2 - 2$$

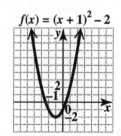

$f(x) = (x + 1)^2 - 2$ has the same shape as $f(x) = x^2$ but is shifted 1 unit to the left (since $x + 1 = 0$ if $x = -1$) and 2 units down (because of the -2). The vertex is $(-1, -2)$ and the axis is $x = -1$. The domain is $(-\infty, \infty)$ and the range is $[-2, \infty)$.

Copyright © 2012 Pearson Education, Inc. Publishing as Addison-Wesley.

N4. $f(x) = -3x^2$

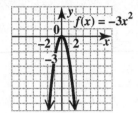

The coefficient -3 indicates that the parabola opens down (because of the negative sign) and is narrower than the graph of $y = x^2$. The vertex is $(0, 0)$ and the axis is $x = 0$. The domain is $(-\infty, \infty)$ and the range is $(-\infty, 0]$.

N5. $f(x) = 2(x - 1)^2 + 2$

This equation has a graph like $f(x) = x^2$ but is shifted 1 unit to the right and 2 units up. It has vertex $(1, 2)$. Since $a = 2 > 0$, the parabola opens up. Since $|a| = |2| = 2 > 1$, the parabola is narrower than the graph of $f(x) = x^2$. If $x = 0$, then $f(x) = 4$, so $(0, 4)$ is on the graph. By symmetry about the axis $x = 1$, the point $(2, 4)$ is also on the graph.

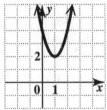

$f(x) = 2(x - 1)^2 + 2$

N6. Use the points with

$$y = ax^2 + bx + c.$$

Use $(0, 4973)$: $a(0)^2 + b(0) + c = 4973$
Use $(4, 7321)$: $a(4)^2 + b(4) + c = 7321$
Use $(8, 7663)$: $a(8)^2 + b(8) + c = 7663$

Simplifying the system gives us:

$$
\begin{aligned}
c &= 4973 \quad (1) \\
16a + 4b + c &= 7321 \quad (2) \\
64a + 8b + c &= 7663 \quad (3)
\end{aligned}
$$

Substituting 4973 for c and simplifying gives us:

$$
\begin{aligned}
16a + 4b &= 2348 \quad (4) \\
64a + 8b &= 2690 \quad (5)
\end{aligned}
$$

Dividing (4) by 4 and (5) by 2 gives us:

$$
\begin{aligned}
4a + b &= 587 \quad (6) \\
32a + 4b &= 1345 \quad (7)
\end{aligned}
$$

From (6) we see that

$$b = 587 - 4a \quad (8),$$

so we substitute $587 - 4a$ for b in (7), and then solve for a.

$$
\begin{aligned}
32a + 4(587 - 4a) &= 1345 \\
32a + 2348 - 16a &= 1345 \\
16a &= -1003 \\
a &= -62.6875
\end{aligned}
$$

To find b, use (8).

$$
\begin{aligned}
b &= 587 - 4a \\
&= 587 - 4(-62.6875) = 837.75
\end{aligned}
$$

So using these three points gives us the quadratic model

$$y = -62.6875x^2 + 837.75x + 4973.$$

10.2 Section Exercises

1. A parabola with equation $f(x) = a(x - h)^2 + k$ has vertex $V(h, k)$. We'll identify the vertex for each quadratic function.

(a) $f(x) = (x + 2)^2 - 1$

$V(-2, -1)$, choice **B**

(b) $f(x) = (x + 2)^2 + 1$

$V(-2, 1)$, choice **C**

(c) $f(x) = (x - 2)^2 - 1$

$V(2, -1)$, choice **A**

(d) $f(x) = (x - 2)^2 + 1$

$V(2, 1)$, choice **D**

For Exercises 3–12, we write $f(x)$ in the form $f(x) = a(x - h)^2 + k$ and then list the vertex (h, k).

3. $f(x) = -3x^2 = -3(x - 0)^2 + 0$
The vertex (h, k) is $(0, 0)$.

5. $f(x) = x^2 + 4 = 1(x - 0)^2 + 4$
The vertex (h, k) is $(0, 4)$.

7. $f(x) = (x - 1)^2 = 1(x - 1)^2 + 0$
The vertex (h, k) is $(1, 0)$.

9. $f(x) = (x + 3)^2 - 4 = 1[x - (-3)]^2 - 4$
The vertex (h, k) is $(-3, -4)$.

11. $f(x) = -(x - 5)^2 + 6 = -1(x - 5)^2 + 6$
The vertex (h, k) is $(5, 6)$.

13. $f(x) = -\frac{2}{5}x^2$

Since $a = -\frac{2}{5} < 0$, the graph opens down. Since $|a| = \left|-\frac{2}{5}\right| = \frac{2}{5} < 1$, the graph is wider than the graph of $f(x) = x^2$.

15. $f(x) = 3x^2 + 1$

Since $a = 3 > 0$, the graph opens up. Since $|a| = |3| = 3 > 1$, the graph is narrower than the graph of $f(x) = x^2$.

Copyright © 2012 Pearson Education, Inc. Publishing as Addison-Wesley.

17. $f(x) = -4(x + 2)^2 + 5$

Since $a = -4 < 0$, the graph opens down.
Since $|a| = |-4| = 4 > 1$, the graph is narrower than the graph of $f(x) = x^2$.

19. **(a)** $f(x) = (x - 4)^2 - 2 = 1(x - 4)^2 - 2$ has vertex $(4, -2)$. Because $a = 1 > 0$, the graph opens up. The correct answer is **D**.

(b) $f(x) = (x - 2)^2 - 4 = 1(x - 2)^2 - 4$ has vertex $(2, -4)$. Because $a = 1 > 0$, the graph opens up. The correct answer is **B**.

(c) $f(x) = -(x - 4)^2 - 2 = -1(x - 4)^2 - 2$ has vertex $(4, -2)$. Because $a = -1 < 0$, the graph opens down. The correct answer is **C**.

(d) $f(x) = -(x - 2)^2 - 4 = -1(x - 2)^2 - 4$ has vertex $(2, -4)$. Because $a = -1 < 0$, the graph opens down. The correct answer is **A**.

21. $f(x) = -2x^2$ written in the form
$f(x) = a(x - h)^2 + k$ is
$f(x) = -2(x - 0)^2 + 0$.

Here, $h = 0$ and $k = 0$, so the vertex (h, k) is $(0, 0)$. Since $a = -2 < 0$, the graph opens down. Since $|a| = |-2| = 2 > 1$, the graph is narrower than the graph of $f(x) = x^2$. By evaluating the function with $x = 2$ and $x = -2$, we see that the points $(2, -8)$ and $(-2, -8)$ are on the graph.

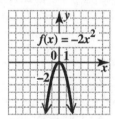

23. $f(x) = x^2 - 1$ written in the form
$f(x) = a(x - h)^2 + k$ is
$f(x) = 1(x - 0)^2 + (-1)$.

Here, $h = 0$ and $k = -1$, so the vertex is $(0, -1)$. The graph opens up and has the same shape as $f(x) = x^2$ because $a = 1$. Two other points on the graph are $(-2, 3)$ and $(2, 3)$.

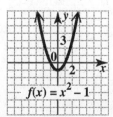

25. $f(x) = -x^2 + 2$ written in the form
$f(x) = a(x - h)^2 + k$ is
$f(x) = -1(x - 0)^2 + 2$.

Here, $h = 0$ and $k = 2$, so the vertex (h, k) is $(0, 2)$. Since $a = -1 < 0$, the graph opens down. Since $|a| = |-1| = 1$, the graph has the same shape as $f(x) = x^2$. The points $(2, -2)$ and $(-2, -2)$ are on the graph.

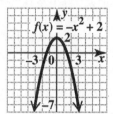

27. $f(x) = (x - 4)^2$ written in the form
$f(x) = a(x - h)^2 + k$ is
$f(x) = 1(x - 4)^2 + 0$.

Here, $h = 4$ and $k = 0$, so the vertex (h, k) is $(4, 0)$ and the axis is $x = 4$. The graph opens up since a is positive and has the same shape as $f(x) = x^2$ because $|a| = 1$. Two other points on the graph are $(2, 4)$ and $(6, 4)$. We can substitute any value for x, so the domain is $(-\infty, \infty)$. The range is $[0, \infty)$ since the smallest y-value is 0.

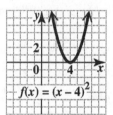

29. $f(x) = (x + 2)^2 - 1$ written in the form
$f(x) = a(x - h)^2 + k$ is
$f(x) = 1[x - (-2)]^2 + (-1)$.

Since $h = -2$ and $k = -1$, the vertex (h, k) is $(-2, -1)$ and the axis is $x = -2$. Here, $a = 1$, so the graph opens up and has the same shape as $f(x) = x^2$. The points $(-1, 0)$ and $(-3, 0)$ are on the graph. We can substitute any value for x, so the domain is $(-\infty, \infty)$. The range is $[-1, \infty)$ since the smallest y-value is -1.

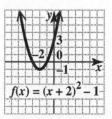

Copyright © 2012 Pearson Education, Inc. Publishing as Addison-Wesley.

31. $f(x) = 2(x-2)^2 - 4$ written in the form
$f(x) = a(x-h)^2 + k$ is
$f(x) = 2(x-2)^2 + (-4)$.

Here, $h = 2$ and $k = -4$, so the vertex (h, k) is
$(2, -4)$ and the axis is $x = 2$. The graph opens up
and is narrower than $f(x) = x^2$ because
$|a| = |2| > 1$. Two other points on the graph are
$(0, 4)$ and $(4, 4)$. We can substitute any value for
x, so the domain is $(-\infty, \infty)$. The value of y is
greater than or equal to -4, so the range is
$[-4, \infty)$.

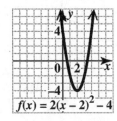

$f(x) = 2(x-2)^2 - 4$

33. $f(x) = -\frac{1}{2}(x+1)^2 + 2$ written in the form
$f(x) = a(x-h)^2 + k$ is
$f(x) = -\frac{1}{2}[x - (-1)]^2 + 2$.

Since $h = -1$ and $k = 2$, the vertex (h, k) is
$(-1, 2)$ and the axis is $x = -1$. Here,
$a = -0.5 < 0$, so the graph opens down. Also,
$|a| = |-0.5| = 0.5 < 1$, so the graph is wider than
the graph of $f(x) = x^2$. The points $(1, 0)$ and
$(-3, 0)$ are on the graph. We can substitute any
value for x, so the domain is $(-\infty, \infty)$. The value
of y is less than or equal to 2, so the range is
$(-\infty, 2]$.

$f(x) = -\frac{1}{2}(x+1)^2 + 2$

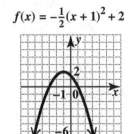

35. $f(x) = 2(x-2)^2 - 3$ written in the form
$f(x) = a(x-h)^2 + k$ is
$f(x) = 2(x-2)^2 + (-3)$.

Here, $h = 2$ and $k = -3$, so the vertex (h, k) is
$(2, -3)$ and the axis is $x = 2$. The graph opens up
and is narrower than $f(x) = x^2$ because
$|a| = |2| > 1$. Two other points on the graph are
$(3, -1)$ and $(1, -1)$. We can substitute any value
for x, so the domain is $(-\infty, \infty)$. The value of y
is greater than or equal to -3, so the range is
$[-3, \infty)$.

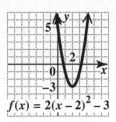

$f(x) = 2(x-2)^2 - 3$

37. The points appear to lie on a line, so a *linear*
function would be a more appropriate model. The
line would rise, so it would have a *positive* slope.

39. The points appear to lie on a parabola, so a
quadratic function would be a more appropriate
model. The parabola would open up, so a would
be *positive*.

41. The points appear to lie on a parabola, so a
quadratic function would be a more appropriate
model. The parabola would open down, so a
would be *negative*.

43. (a)

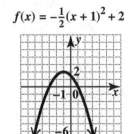

(b) Since the arrangement of the data points is
approximately parabolic, a quadratic function
would be the more appropriate model for the data
set. The coefficient of x^2 should be positive, since
the roughly parabolic shape of the graphed data set
opens upward.

(c) Use $ax^2 + bx + c = y$ with $(0, 1825)$,
$(3, 3921)$, and $(6, 7805)$.

$$0a + 0b + c = 1825 \quad (1)$$
$$9a + 3b + c = 3921 \quad (2)$$
$$36a + 6b + c = 7805 \quad (3)$$

From (1), $c = 1825$, so the system becomes

$$9a + 3b = 2096 \quad (4)$$
$$36a + 6b = 5980 \quad (5)$$

Now eliminate b.

$$
\begin{array}{rcll}
-18a - 6b &=& -4192 & -2 \times (4) \\
36a + 6b &=& 5980 & (5) \\
\hline
18a &=& 1788 & \\
a &=& \frac{1788}{18} = 99\frac{1}{3} \approx 99.3 &
\end{array}
$$

From (5) with $a = 99\frac{1}{3} = \frac{298}{3}$,

Copyright © 2012 Pearson Education, Inc. Publishing as Addison-Wesley.

$$36\left(\tfrac{298}{3}\right) + 6b = 5980$$
$$3576 + 6b = 5980$$
$$6b = 2404$$
$$b = \tfrac{2404}{6} \approx 400.7$$

The quadratic function is approximately

$$y = f(x) = 99.3x^2 + 400.7x + 1825.$$

(d) $x = 2007 - 2000 = 7$ and $f(7) \approx 9496$. The sales of digital cameras in the United States in 2007 were about $9496 million.

(e) No. The number of digital cameras sold in 2007 is far below the number approximated by the model. Rather than continuing to increase, sales of digital cameras fell in 2007.

45. (a) $y = f(x) = -69.15x^2 + 863.6x + 4973$
$x = 2006 - 1995 = 11$ and
$f(11) = 6105.45 \approx 6105$

(b) The approximation using the model is low.

47. $x^2 - x - 20 = 0$
From the screens, we see that the x-values of the x-intercepts are -4 and 5, so the solution set is $\{-4, 5\}$.

49. $-2x^2 + 6x = -2(x^2 - 3x)$

51.
$$x^2 + 3x - 4 = 0$$
$$(x + 4)(x - 1) = 0$$
$$x + 4 = 0 \quad \text{or} \quad x - 1 = 0$$
$$x = -4 \quad \text{or} \quad x = 1$$

The solution set is $\{-4, 1\}$.

53.
$$x^2 + 6x - 3 = 0$$
$$x^2 + 6x = 3$$
$$x^2 + 6x + 9 = 3 + 9$$
$$(x + 3)^2 = 12$$
$$x + 3 = \pm\sqrt{12}$$
$$x = -3 \pm 2\sqrt{3}$$

The solution set is $\left\{-3 \pm 2\sqrt{3}\right\}$.

10.3 More About Parabolas and Their Applications

10.3 Now Try Exercises

N1. $f(x) = x^2 + 2x - 8$

To find the vertex, first complete the square on $x^2 + 2x$.

$$\left[\tfrac{1}{2}(2)\right]^2 = 1^2 = 1$$

Now add and subtract 1 on the right.

$$f(x) = (x^2 + 2x + 1 - 1) - 8$$
$$= (x^2 + 2x + 1) - 1 - 8$$
$$f(x) = (x + 1)^2 - 9$$

The vertex of this parabola is $(-1, -9)$.

N2. $f(x) = -4x^2 + 16x - 10$

Factor out -4 from the first two terms.

$$f(x) = -4(x^2 - 4x) - 10$$

Complete the square on $x^2 - 4x$.

$$\left[\tfrac{1}{2}(-4)\right]^2 = (-2)^2 = 4$$

Add and subtract 4.

$$f(x) = -4(x^2 - 4x + 4 - 4) - 10$$
$$= -4(x^2 - 4x + 4) - 4(-4) - 10$$
$$= -4(x^2 - 4x + 4) + 16 - 10$$
$$f(x) = -4(x - 2)^2 + 6$$

The vertex of this parabola is $(2, 6)$.

N3. For $f(x) = 3x^2 - 2x + 8$, $a = 3$, $b = -2$, and $c = 8$.

The vertex is $\left(\dfrac{-b}{2a}, f\left(\dfrac{-b}{2a}\right)\right)$, so the x-coordinate is

$$\frac{-b}{2a} = \frac{-(-2)}{2(3)} = \frac{2}{6} = \frac{1}{3},$$

and the y-coordinate is

$$f\left(\tfrac{1}{3}\right) = 3\left(\tfrac{1}{3}\right)^2 - 2\left(\tfrac{1}{3}\right) + 8$$
$$= 3\left(\tfrac{1}{9}\right) - \tfrac{2}{3} + 8$$
$$= \tfrac{1}{3} - \tfrac{2}{3} + \tfrac{24}{3}$$
$$= \tfrac{23}{3}.$$

The vertex is $\left(\tfrac{1}{3}, \tfrac{23}{3}\right)$.

N4. $f(x) = x^2 + 2x - 3$

Step 1
The graph opens up because $a = 1 > 0$.

Step 2
To find the vertex, complete the square on $x^2 + 2x$.

$$\left[\tfrac{1}{2}(2)\right]^2 = (1)^2 = 1$$

Add and subtract 1.

$$f(x) = x^2 + 2x + 1 - 1 - 3$$
$$f(x) = (x + 1)^2 - 4$$

The vertex is at $(-1, -4)$.

Copyright © 2012 Pearson Education, Inc. Publishing as Addison-Wesley.

Step 3

Find any x-intercepts. Let $f(x) = 0$.

$$f(x) = x^2 + 2x - 3$$
$$0 = x^2 + 2x - 3$$
$$0 = (x + 3)(x - 1)$$

$$x + 3 = 0 \quad \text{or} \quad x - 1 = 0$$
$$x = -3 \quad \text{or} \quad x = 1$$

The x-intercepts are $(-3, 0)$ and $(1, 0)$.

Find the y-intercept. Let $x = 0$.

$$f(0) = 0^2 + 2(0) - 3 = -3$$

The y-intercept is $(0, -3)$.

Step 4

The axis is $x = -1$. By symmetry, another point on the graph is $(-2, -3)$. Since x can be any real number, the domain is $(-\infty, \infty)$. Since the lowest point of the parabola is $(-1, -4)$, the range is $[-4, \infty)$.

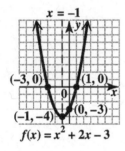

N5. **(a)** $f(x) = -2x^2 + 3x - 2$

Here $a = -2$, $b = 3$, and $c = -2$. The discriminant is

$$b^2 - 4ac = 3^2 - 4(-2)(-2)$$
$$= 9 - 16$$
$$= -7.$$

Since the discriminant is negative, the graph has no x-intercepts.

(b) $f(x) = 3x^2 + 2x - 1$

Here $a = 3$, $b = 2$, and $c = -1$. The discriminant is

$$b^2 - 4ac = 2^2 - 4(3)(-1)$$
$$= 4 + 12$$
$$= 16.$$

Since the discriminant is positive, the graph has two x-intercepts.

(c) $f(x) = 4x^2 - 12x + 9$

Here $a = 4$, $b = -12$, and $c = 9$. The discriminant is

$$b^2 - 4ac = (-12)^2 - 4(4)(9)$$
$$= 144 - 144$$
$$= 0.$$

Since the discriminant is 0, the graph has only one x-intercept.

N6. In Example 6, replace 120 with 80, so

$$A(x) = -2x^2 + 80x.$$

Here $a = -2$, $b = 80$, and $c = 0$, so the x-coordinate of the vertex is

$$x = \frac{-b}{2a} = \frac{-80}{2(-2)} = 20,$$

and the y-coordinate of the vertex, $A(x)$, is

$$A(20) = -2(20)^2 + 80(20)$$
$$= -2(400) + 1600$$
$$= -800 + 1600$$
$$= 800.$$

The graph is a parabola that opens down with vertex at $(20, 800)$. The vertex of the graph shows that the maximum area will be 800 ft^2. This area will occur if the width, x, is 20 feet and the length is $80 - 2x$, or 40 feet.

N7. For $s(t) = -16t^2 + 48t$, $a = -16$, $b = 48$, and $c = 0$. Use the vertex formula.

$$t = \frac{-b}{2a} = \frac{-48}{2(-16)} = \frac{3}{2} = 1.5$$

This indicates that the maximum height is attained at 1.5 seconds. To find this maximum height, calculate $s(1.5)$.

$$s(t) = -16t^2 + 48t$$
$$s(1.5) = -16(1.5)^2 + 48(1.5)$$
$$= -16(2.25) + 72$$
$$= -36 + 72 = 36$$

Therefore, the stomp rocket reaches a maximum height of 36 feet in 1.5 seconds.

N8. $x = (y + 2)^2 - 1$

Since the roles of x and y are reversed, this graph has its vertex at $(-1, -2)$. It opens to the right, the positive x-direction, and has the same shape as $y = x^2$. The axis is $y = -2$, the domain is $[-1, \infty)$, and the range is $(-\infty, \infty)$. The points $(0, -1)$ and $(0, -3)$ are on the graph.

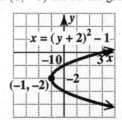

Copyright © 2012 Pearson Education, Inc. Publishing as Addison-Wesley.

N9. $x = -3y^2 - 6y - 5$
$x = -3(y^2 + 2y) - 5$

Complete the square on $y^2 + 2y$.

$$\left[\tfrac{1}{2}(2)\right]^2 = (1)^2 = 1$$

Add and subtract 1.

$$x = -3(y^2 + 2y + 1 - 1) - 5$$
$$= -3(y^2 + 2y + 1) - 3(-1) - 5$$
$$x = -3(y + 1)^2 - 2$$

The vertex is $(-2, -1)$ and the axis is $y = -1$.
Since $a = -3 < 0$, the graph opens to the left.
The domain is $(-\infty, -2]$, and the range is
$(-\infty, \infty)$.

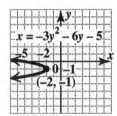

10.3 Section Exercises

1. If x is squared, the parabola has a vertical axis.
If y is squared, the parabola has a horizontal axis.

3. Use the discriminant, $b^2 - 4ac$, of the function. If
it is positive, there are two x-intercepts. If it is
zero, there is one x-intercept (at the vertex), and if
it is negative, there is no x-intercept.

5. As in Example 1, we'll complete the square to find
the vertex.

$$f(x) = x^2 + 8x + 10$$
$$= x^2 + 8x + \underline{16} + 10 - \underline{16} \quad \left[\tfrac{1}{2}(8)\right]^2 = 16$$
$$= (x + 4)^2 - 6$$

The vertex is $(-4, -6)$.

7. As in Example 2, we'll complete the square to find
the vertex.

$$f(x) = -2x^2 + 4x - 5$$
$$= -2(x^2 - 2x) - 5$$
$$= -2(x^2 - 2x + 1 - 1) - 5$$
$$= -2(x^2 - 2x + 1) + (-2)(-1) - 5$$
$$= -2(x - 1)^2 - 3$$

The vertex is $(1, -3)$.

9. As in Example 3, we'll use the vertex formula to
find the vertex.

$$f(x) = x^2 + x - 7$$

The x-coordinate of the vertex is

$$\frac{-b}{2a} = \frac{-1}{2(1)} = -\frac{1}{2}.$$

The y-coordinate of the vertex is

$$f\left(-\tfrac{1}{2}\right) = \tfrac{1}{4} - \tfrac{1}{2} - 7 = -\tfrac{29}{4}.$$

The vertex is $\left(-\tfrac{1}{2}, -\tfrac{29}{4}\right)$.

11. As in Example 2, we'll complete the square to find
the vertex.

$$f(x) = 2x^2 + 4x + 5$$
$$= 2(x^2 + 2x) + 5$$
$$= 2(x^2 + 2x + 1 - 1) + 5$$
$$= 2(x^2 + 2x + 1) + 2(-1) + 5$$
$$= 2(x + 1)^2 - 2 + 5$$
$$f(x) = 2(x + 1)^2 + 3$$

The vertex is $(-1, 3)$.
Because $a = 2 > 1$, the graph opens up and is
narrower than the graph of $y = x^2$.

For $y = f(x) = 2x^2 + 4x + 5$, $a = 2$, $b = 4$, and
$c = 5$. The discriminant is

$$b^2 - 4ac = 4^2 - 4(2)(5)$$
$$= 16 - 40 = -24.$$

The discriminant is negative, so the parabola has
no x-intercepts.

13. $f(x) = -x^2 + 5x + 3$

Use the vertex formula with $a = -1$ and $b = 5$.

The x-coordinate of the vertex is

$$\frac{-b}{2a} = \frac{-5}{2(-1)} = \frac{5}{2}.$$

The y-coordinate of the vertex is

$$f\left(\frac{-b}{2a}\right) = f\left(\frac{5}{2}\right)$$
$$= -\left(\tfrac{5}{2}\right)^2 + 5\left(\tfrac{5}{2}\right) + 3$$
$$= -\tfrac{25}{4} + \tfrac{25}{2} + 3$$
$$= \frac{-25 + 50 + 12}{4} = \frac{37}{4}.$$

The vertex is

$$\left(\frac{-b}{2a}, f\left(\frac{-b}{2a}\right)\right) = \left(\frac{5}{2}, \frac{37}{4}\right).$$

Because $a = -1$, the parabola opens down and
has the same shape as the graph of $y = x^2$.

$$b^2 - 4ac = 5^2 - 4(-1)(3)$$
$$= 25 + 12 = 37$$

The discriminant is positive, so the parabola has
two x-intercepts.

Copyright © 2012 Pearson Education, Inc. Publishing as Addison-Wesley.

15. Complete the square on the y-terms to find the vertex.

$$x = \tfrac{1}{3}y^2 + 6y + 24$$
$$= \tfrac{1}{3}(y^2 + 18y) + 24$$
$$= \tfrac{1}{3}(y^2 + 18y + 81 - 81) + 24$$
$$= \tfrac{1}{3}(y^2 + 18y + 81) + \tfrac{1}{3}(-81) + 24$$
$$= \tfrac{1}{3}(y + 9)^2 - 27 + 24$$
$$x = \tfrac{1}{3}(y + 9)^2 - 3$$

The vertex is $(-3, -9)$.
The graph is a horizontal parabola. The graph opens to the right since $a = 0.\overline{3} > 0$ and is wider than the graph of $y = x^2$ since $|a| = |0.\overline{3}| < 1$.

17. The graph of $y = 2x^2 + 4x - 3$ is a vertical parabola opening up, so choice F is correct. **(F)**

19. The graph of $y = -\tfrac{1}{2}x^2 - x + 1$ is a vertical parabola opening down, so choices A and C are possibilities. The graph in C is wider than the graph in A, so it must correspond to $a = -\tfrac{1}{2}$ while the graph in A must correspond to $a = -1$. **(C)**

21. The graph of $x = -y^2 - 2y + 4$ is a horizontal parabola opening to the left, so choice D is correct. **(D)**

23. $y = f(x) = x^2 + 8x + 10$

Step 1
Since $a = 1 > 0$, the graph opens up and is the same shape as the graph of $y = x^2$.

Step 2
From Exercise 5, the vertex is $(-4, -6)$. Since the graph opens up, the axis goes through the x-coordinate of the vertex—its equation is $x = -4$.

Step 3
To find the y-intercept, let $x = 0$.
$f(0) = 10$, so the y-intercept is $(0, 10)$.
To find the x-intercepts, let $y = 0$.

$$0 = x^2 + 8x + 10$$
$$x = \frac{-8 \pm \sqrt{64 - 40}}{2} = \frac{-8 \pm \sqrt{24}}{2}$$
$$= \frac{-8 \pm 2\sqrt{6}}{2} = -4 \pm \sqrt{6}$$

The x-intercepts are approximately $(-6.45, 0)$ and $(-1.55, 0)$.

Step 4
For an additional point on the graph, let $x = -2$ (two units to the right of the axis) to get $f(-2) = -2$. So the point $(-2, -2)$ is on the graph. By symmetry, the point $(-6, -2)$ (two units to the left of the axis) is on the graph.

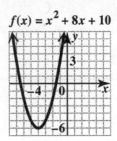
$f(x) = x^2 + 8x + 10$

From the graph, we see that the domain is $(-\infty, \infty)$ and the range is $[-6, \infty)$.

25. $y = f(x) = -2x^2 + 4x - 5$

Step 1
Since $a = -2$, the graph opens down and is narrower than the graph of $y = x^2$.

Step 2
From Exercise 7, the vertex is $(1, -3)$. Since the graph opens down, the axis goes through the x-coordinate of the vertex—its equation is $x = 1$.

Step 3
If $x = 0$, $y = -5$, so the y-intercept is $(0, -5)$. To find the x-intercepts, let $y = 0$.

$$0 = -2x^2 + 4x - 5$$
$$x = \frac{-4 \pm \sqrt{16 - 40}}{2(-2)}$$

The discriminant is negative, so there are no x-intercepts.

Step 4
By symmetry, $(2, -5)$ is also on the graph.

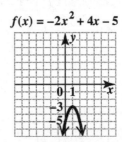
$f(x) = -2x^2 + 4x - 5$

From the graph, we see that the domain is $(-\infty, \infty)$ and the range is $(-\infty, -3]$.

27. $x = (y + 2)^2 + 1 = y^2 + 4y + 5$

The roles of x and y are reversed, so this is a horizontal parabola.

Step 1
The coefficient of y^2 is $1 > 0$, so the graph opens to the right.

Step 2
The vertex can be identified from the given form of the equation. When $y = -2$, $x = 1$, so the vertex is $(1, -2)$. Since the graph opens right, the axis goes through the y-coordinate of the vertex— its equation is $y = -2$.

Copyright © 2012 Pearson Education, Inc. Publishing as Addison-Wesley.

Step 3
To find the x-intercept, let $y = 0$.
$x = 0^2 + 4(0) + 5 = 5$, so the x-intercept is $(5, 0)$.

To find the y-intercepts, let $x = 0$.

$$0 = (y + 2)^2 + 1$$
$$-1 = (y + 2)^2$$

Since $(y + 2)^2$ cannot be negative, there are no y-intercepts.

Step 4
For an additional point on the graph, let $y = -4$ (two units below the axis) to get
$x = (-4 + 2)^2 + 1 = 5$.

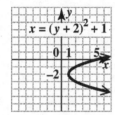

From the graph, we see the domain is $[1, \infty)$ and the range is $(-\infty, \infty)$.

29. $x = -\frac{1}{5}y^2 + 2y - 4$

The roles of x and y are reversed, so this is a horizontal parabola.

Step 1
Since $a = -\frac{1}{5} < 0$, the graph opens to the left and is wider than the graph of $y = x^2$.

Step 2
The y-coordinate of the vertex is

$$\frac{-b}{2a} = \frac{-2}{2\left(-\frac{1}{5}\right)} = \frac{-2}{-\frac{2}{5}} = 5.$$

The x-coordinate of the vertex is

$$-\frac{1}{5}(5)^2 + 2(5) - 4 = -5 + 10 - 4 = 1.$$

Thus, the vertex is $(1, 5)$. Since the graph opens left, the axis goes through the y-coordinate of the vertex—its equation is $y = 5$.

Step 3
To find the x-intercept, let $y = 0$.
If $y = 0$, $x = -4$, so the x-intercept is $(-4, 0)$.
To find the y-intercepts, let $x = 0$.

$$0 = -\frac{1}{5}y^2 + 2y - 4$$
$$0 = y^2 - 10y + 20 \quad \textit{Multiply by } -5.$$
$$y = \frac{10 \pm \sqrt{100 - 80}}{2} = \frac{10 \pm \sqrt{20}}{2}$$
$$= \frac{10 \pm 2\sqrt{5}}{2} = 5 \pm \sqrt{5}$$

The y-intercepts are approximately $(0, 7.2)$ and $(0, 2.8)$.

Step 4
For an additional point on the graph, let $y = 7$ (two units above the axis) to get $x = \frac{1}{5}$. So the point $\left(\frac{1}{5}, 7\right)$ is on the graph. By symmetry, the point $\left(\frac{1}{5}, 3\right)$ (two units below the axis) is on the graph.

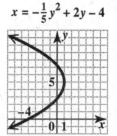

From the graph, we see that the domain is $(-\infty, 1]$ and the range is $(-\infty, \infty)$.

31. $x = 3y^2 + 12y + 5$

The roles of x and y are reversed, so this is a horizontal parabola.

Step 1
Since $a = 3 > 0$, the graph opens to the right and is narrower than the graph of $y = x^2$.

Step 2
Use the formula to find the y-value of the vertex.

$$\frac{-b}{2a} = \frac{-12}{2(3)} = -2$$

If $y = -2$, $x = -7$, so the vertex is $(-7, -2)$. Since the graph opens right, the axis goes through the y-coordinate of the vertex—its equation is $y = -2$.

Step 3
If $y = 0$, $x = 5$, so the x-intercept is $(5, 0)$.
To find the y-intercepts, let $x = 0$.

$$0 = 3y^2 + 12y + 5$$
$$y = \frac{-12 \pm \sqrt{144 - 60}}{6} = \frac{-12 \pm \sqrt{84}}{6}$$
$$= \frac{-12 \pm 2\sqrt{21}}{6} = \frac{-6 \pm \sqrt{21}}{3}$$

The y-intercepts are approximately $(0, -0.5)$ and $(0, -3.5)$.

Step 4
By symmetry, $(5, -4)$ is also on the graph.

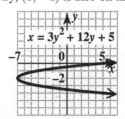

From the graph, we see that the domain is $[-7, \infty)$ and the range is $(-\infty, \infty)$.

Copyright © 2012 Pearson Education, Inc. Publishing as Addison-Wesley.

33. Let $x =$ one number, $40 - x =$ the other number, and $P =$ the product.

$$P = x(40 - x)$$
$$= 40x - x^2 \text{ or } -x^2 + 40x$$

This parabola opens down so the maximum occurs at the vertex.

Here $a = -1$, $b = 40$, and $c = 0$.

$$\frac{-b}{2a} = \frac{-40}{2(-1)} = 20$$

$x = 20$ when the product is a maximum.

Since $x = 20$, $40 - x = 20$, and the two numbers are 20 and 20.

35. Let x represent the length of the two equal sides, and let $280 - 2x$ represent the length of the remaining side (the side parallel to the highway). Then substitute x for L and $280 - 2x$ for W in the formula for the area of a rectangle, $A = LW$.

$$A = x(280 - 2x)$$
$$= 280x - 2x^2 \text{ or } -2x^2 + 280x$$

The maximum area will occur at the vertex.

$$x = \frac{-b}{2a} = \frac{-280}{2(-2)} = 70$$

When $x = 70$, $280 - 2x = 140$, and $A = 9800$. Thus, the dimensions of the lot with maximum area are 140 feet by 70 feet and the maximum area is 9800 square feet.

37. $s(t) = -16t^2 + 32t$

Here, $a = -16 < 0$, so the parabola opens down. The time it takes to reach the maximum height and the maximum height are given by the vertex of the parabola. Use the vertex formula to find that

$$t = \frac{-b}{2a} = \frac{-32}{2(-16)} = \frac{-32}{-32} = 1,$$
and $s(t) = -16(1)^2 + 32(1)$
$$= -16 + 32 = 16.$$

The vertex is $(1, 16)$, so the maximum height is 16 feet which occurs when the time is 1 second. The object hits the ground when $s = 0$.

$$0 = -16t^2 + 32t$$
$$0 = -16t(t - 2)$$
$$-16t = 0 \quad \text{or} \quad t - 2 = 0$$
$$t = 0 \quad \text{or} \quad t = 2$$

It takes 2 seconds for the object to hit the ground.

39. The graph of the height of the projectile,

$$s(t) = -16t^2 + 64t + 1,$$

is a parabola that opens down since $a = -16 < 0$.

The time at which the cork reaches its maximum height and the maximum height are the t and s coordinates of the vertex.

$$t = \frac{-b}{2a} = \frac{-64}{2(-16)} = 2$$
$$s(2) = -16(2)^2 + 64(2) + 1$$
$$= -64 + 128 + 1 = 65$$

The cork reaches a maximum height of 65 feet after 2 seconds.

41. $C(x) = x^2 - 40x + 610$

Find the vertex of the graph of C.

$$\frac{-b}{2a} = \frac{-(-40)}{2(1)} = 20$$
$$C(20) = 20^2 - 40(20) + 610 = 210$$

He should sell 20 tacos per day to minimize his costs of $210.

43. $f(x) = 22.88x^2 - 141.3x + 1044$

(a) Since the graph opens up, the y-value of the vertex is a minimum.

(b) The x-value of the vertex is given by

$$x = \frac{-b}{2a} = \frac{-(-141.3)}{2(22.88)} \approx 3.088$$

The year was $2000 + 3 = 2003$.

$f(3.088) \approx \$825.8$ billion, which is the minimum amount of total receipts from individual taxes.

45. $f(x) = -20.57x^2 + 758.9x - 3140$

(a) The coefficient of x^2 is negative because a parabola that models the data must open down.

(b) Use the vertex formula.

$$x = \frac{-b}{2a} = \frac{-758.9}{2(-20.57)} \approx 18.45$$
$$f(18.45) \approx 3860$$

The vertex is approximately $(18.45, 3860)$.

(c) 18 corresponds to 2018, so in 2018 social security assets will reach their maximum value of $3860 billion.

47. The number of people on the plane is $100 - x$ since x is the number of unsold seats. The price per seat is $200 + 4x$.

(a) The total revenue received for the flight is found by multiplying the number of seats by the price per seat. Thus, the revenue is

$$R(x) = (100 - x)(200 + 4x)$$
$$= 20,000 + 200x - 4x^2.$$

Copyright © 2012 Pearson Education, Inc. Publishing as Addison-Wesley.

(b) Use the formula for the vertex.

$$x = \frac{-b}{2a} = \frac{-200}{2(-4)} = 25$$

$R(25) = 22{,}500$, so the vertex is $(25,\ 22{,}500)$. $R(0) = 20{,}000$, so the R-intercept is $(0,\ 20{,}000)$. From the factored form for R, we see that the positive x-intercept is $(100, 0)$. (The factor $200 + 4x$ leads to a negative x-intercept, meaningless in this problem.)

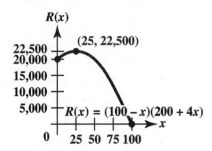

$R(x)$

(25, 22,500)

$R(x) = (100 - x)(200 + 4x)$

(c) The number of unsold seats x that produce the maximum revenue is 25, the x-value of the vertex.

(d) The maximum revenue is $22{,}500$, the y-value of the vertex.

49. $f(x) = x^2 + 4$

$$f(-x) = (-x)^2 + 4$$
$$= x^2 + 4$$

51. $f(x) = 3x^2 - 3$

$$f(0) = 3(0)^2 - 3 = -3$$
$$f(1) = 3(1)^2 - 3 = 0$$

Since $0 > -3$, $f(1)$ is greater.

10.4 Symmetry; Increasing and Decreasing Functions

10.4 Now Try Exercises

N1. $g(x) = 3\sqrt{x}$

Some values for x, $\sqrt{x}$, and $3\sqrt{x}$ are given in the table.

x	$\sqrt{x}$	$3\sqrt{x}$
0	0	0
1	1	3
4	2	6
9	3	9

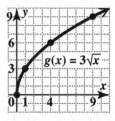

$g(x) = 3\sqrt{x}$

The graph of $g(x) = 3\sqrt{x}$ is stretched vertically compared with the graph of $f(x) = \sqrt{x}$. Every y-value of $g(x)$ is three times the y-value of $f(x)$.

N2. $x = y^2 + 1$

(1) To test for symmetry with respect to the x-axis, replace y with $-y$.

$$x = (-y)^2 + 1$$
$$x = y^2 + 1$$

(2) To test for symmetry with respect to the y-axis, replace x with $-x$.

$$-x = y^2 + 1$$
$$x = -y^2 - 1$$

These tests show that the graph of $x = y^2 + 1$ is symmetric with respect to the x-axis, but not symmetric with respect to the y-axis.

N3. (a) To test for symmetry with respect to the origin, replace x with $-x$ and y with $-y$.

$$x = y^2 - 2$$
$$(-x) = (-y)^2 - 2$$
$$-x = y^2 - 2$$
$$x = -y^2 + 2$$

The last equation is not equivalent to the original equation, so the graph is not symmetric with respect to the origin.

(b) To test for symmetry with respect to the origin, replace x with $-x$ and y with $-y$.

$$y = 2x$$
$$(-y) = 2(-x)$$
$$y = 2x$$

The last equation is equivalent to the original equation, so the graph is symmetric with respect to the origin.

N4. The function is increasing if $x \leq -1$, decreasing if $-1 \leq x \leq 1$, and increasing if $x \geq 1$. In terms of intervals, the function is increasing on $(-\infty, -1]$ and $[1, \infty)$, and is decreasing on $[-1, 1]$. Note that the y-values of the function are *not* listed. Also note that it is permissible for a number to be listed in two intervals.

10.4 Section Exercises

1. (a) $y = -f(x)$

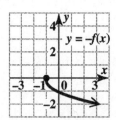

$y = -f(x)$

The graph of $y = f(x)$ is reflected about the x-axis to obtain the graph of $y = -f(x)$.

Copyright © 2012 Pearson Education, Inc. Publishing as Addison-Wesley.

(b) $y = 2f(x)$

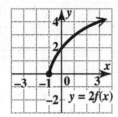

The graph of $y = 2f(x)$ has the same shape as that of $y = f(x)$, but is stretched vertically by a factor of 2.

3. **(a)** The graph of an equation is symmetric with respect to the x-axis if we get an equivalent equation when we replace every y in the equation with $-y$.

    (b) The graph of an equation is symmetric with respect to the y-axis if we get an equivalent equation when we replace every x in the equation with $-x$.

    (c) The graph of an equation is symmetric with respect to the origin if we get an equivalent equation when we replace every x with $-x$ and every y with $-y$ at the same time.

5. **(a)** The point $(-4, 2)$ is symmetric to $(-4, -2)$ with respect to the x-axis since it has the opposite y-value.

    (b) The point $(4, -2)$ is symmetric to $(-4, -2)$ with respect to the y-axis since it has the opposite x-value.

    (c) The point $(4, 2)$ is symmetric to $(-4, -2)$ with respect to the origin since it has opposite x- and y-values.

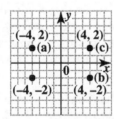

7. **(a)** The point that is symmetric to $(-8, 0)$ with respect to the x-axis is $(-8, 0)$ itself because this point lies on the x-axis.

    (b) The point $(8, 0)$ is symmetric to $(-8, 0)$ with respect to the y-axis since it has the opposite x-value.

    (c) The point $(8, 0)$ is symmetric to $(-8, 0)$ with respect to the origin since it has the opposite x-value and 0 is its own opposite.

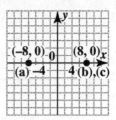

9. $x^2 + y^2 = 5$

    (1) To test for symmetry with respect to the x-axis, replace y with $-y$.

    $$x^2 + (-y)^2 = 5$$
    $$x^2 + y^2 = 5$$

    (2) To test for symmetry with respect to the y-axis, replace x with $-x$.

    $$(-x)^2 + y^2 = 5$$
    $$x^2 + y^2 = 5$$

    (3) To test for symmetry with respect to the origin, replace x with $-x$ and y with $-y$.

    $$(-x)^2 + (-y)^2 = 5$$
    $$x^2 + y^2 = 5$$

    Since each time an equation is obtained which is equivalent to the original one, the graph of the relation $x^2 + y^2 = 5$ is symmetric with respect to the x-axis, the y-axis, and the origin.

11. $y = x^2 - 8x$

    (1) Replace y with $-y$.

    $$-y = x^2 - 8x$$
    $$y = -x^2 + 8x$$

    (2) Replace x with $-x$.

    $$y = (-x)^2 - 8(-x)$$
    $$y = x^2 + 8x$$

    (3) Replace x with $-x$ and y with $-y$.

    $$-y = (-x)^2 - 8(-x)$$
    $$-y = x^2 + 8x$$
    $$y = -x^2 - 8x$$

    None of the equations obtained by these replacements is equivalent to the original equation. Therefore, for the graph of $y = x^2 - 8x$, none of the symmetries apply.

13. $y = |x|$

    (1) Replace y with $-y$.

    $$-y = |x|$$
    $$y = -|x|$$

    (2) Replace x with $-x$.

    $$y = |-x|$$
    $$y = |x|$$

Copyright © 2012 Pearson Education, Inc. Publishing as Addison-Wesley.

(3) Replace x with $-x$ and y with $-y$.

$$-y = |-x|$$
$$-y = |x|$$
$$y = -|x|$$

These tests show that the graph of $y = |x|$ is symmetric with respect to the y-axis but not to either the x-axis or the origin.

15. $y = x^3$

(1) Replace y with $-y$.

$$-y = x^3$$
$$y = -x^3$$

(2) Replace x with $-x$.

$$y = (-x)^3$$
$$y = -x^3$$

(3) Replace x with $-x$ and y with $-y$.

$$-y = (-x)^3$$
$$-y = -x^3$$
$$y = x^3$$

These tests show that the graph of $y = x^3$ is symmetric with respect to the origin, but not to either the x-axis or the y-axis.

17. $f(x) = y = \dfrac{1}{1 + x^2}$

(1) Replace y with $-y$.

$$-y = \frac{1}{1 + x^2}$$
$$y = -\frac{1}{1 + x^2}$$

(2) Replace x with $-x$.

$$y = \frac{1}{1 + (-x)^2}$$
$$y = \frac{1}{1 + x^2}$$

(3) Replace x with $-x$ and y with $-y$.

$$-y = \frac{1}{1 + (-x)^2}$$
$$-y = \frac{1}{1 + x^2}$$
$$y = -\frac{1}{1 + x^2}$$

The graph of $y = \dfrac{1}{1 + x^2}$ is symmetric with respect to the y-axis, but not with respect to the x-axis or the origin.

19. $xy = 2$

(1) Replace y with $-y$.

$$x(-y) = 2$$
$$-xy = 2$$

(2) Replace x with $-x$.

$$(-x)y = 2$$
$$-xy = 2$$

(3) Replace x with $-x$ and y with $-y$.

$$(-x)(-y) = 2$$
$$xy = 2$$

The graph of $xy = 2$ is symmetric with respect to the origin but not with respect to either axis.

♦♦♦ Relating Concepts 21–24 ♦♦♦

21. If $n = 2$, $f(x) = x^2$.
$$f(-x) = (-x)^2 = x^2 = f(x),$$
so f is an even function.

If $n = 4$, $f(x) = x^4$.
$$f(-x) = (-x)^4 = x^4 = f(x),$$
so f is an even function.

If $n = 6$, $f(x) = x^6$.
$$f(-x) = (-x)^6 = x^6 = f(x),$$
so f is an even function.

22. If $n = 1$, $f(x) = x$.
$$f(-x) = -x = -f(x),$$
so f is an odd function.

If $n = 3$, $f(x) = x^3$.
$$f(-x) = (-x)^3 = -x^3 = -f(x),$$
so f is an odd function.

If $n = 5$, $f(x) = x^5$.
$$f(-x) = (-x)^5 = -x^5 = -f(x),$$
so f is an odd function.

23. The definition of an even function states that $f(-x) = f(x)$ for all x in the domain of f. This is equivalent to saying that for every point (x, y) on the graph of f, $(-x, y)$ is also on the graph. Thus, if a function is even, its graph is symmetric with respect to the y-axis.

24. The definition of an odd function states that $f(-x) = -f(x)$ for all x in the domain of f. This is equivalent to saying that for every point (x, y) on the graph of f, $(-x, -y)$ is also on the graph. Thus, if a function is odd, its graph is symmetric with respect to the origin.

25. From the graph, we see that f is increasing on the interval $(-\infty, -3]$, and decreasing on the interval $[0, \infty)$.

Copyright © 2012 Pearson Education, Inc. Publishing as Addison-Wesley.

27. From the graph, we see that f is increasing on the intervals $(-\infty, -2]$ and $[1, \infty)$, and decreasing on the interval $[-2, 1]$.

29. $f(x) = 2x^2 + 1$

The graph of this function is a parabola symmetric with respect to the y-axis, opening upward. The vertex is $(0, 1)$.

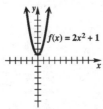

From the graph, we see that f is increasing on $[0, \infty)$, and decreasing on $(-\infty, 0]$.

31. From the graph, we see that the function is increasing over the interval $[2002, 2008]$.

33. **(a)** If we fold the figure about the line, the sides would match. Thus, the rectangle is symmetric with respect to the line.

(b) If we rotate the figure $180°$, we get the same figure. Thus, it is also symmetric with respect to the point.

35. **(a)** If we fold the figure about the line, the sides would not match. Thus, the figure is not symmetric with respect to the line.

(b) If we rotate the figure $180°$, we get the same figure. Thus, it is symmetric with respect to the point.

37. Since $f(2) = 3$, the point $(2, 3)$ is on the graph. Since the graph is symmetric with respect to the origin, $(-2, -3)$ is also on the graph. Therefore, $f(-2) = -3$.

39. $f(x)$ is symmetric with respect to the line $x = 3$; $f(2) = 3$

$f(2) = 3$ means that $(2, 3)$ is an ordered pair of the function, and this point is 1 unit to the left of the line $x = 3$. The reflection of $(2, 3)$ across $x = 3$ is the point $(4, 3)$, since it is 1 unit to the right of $x = 3$. The ordered pair $(4, 3)$ indicates that another value of the function is $f(4) = 3$.

41. The condition $f(-x) = f(x)$ indicates that the graph is symmetric with respect to the y-axis, so draw a mirror image of the given figure on the left side of the y-axis. Doing this causes the original "V" shape to blossom into a "W" shape.

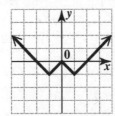

43. $|2x|$

(a) $|2(3)| = |6| = 6$

(b) $|2(-5)| = |-10| = 10$

45. $|5 - x|$

(a) $|5 - 3| = |2| = 2$

(b) $|5 - (-5)| = |10| = 10$

47. $|3x + 1|$

(a) $|3(3) + 1| = |10| = 10$

(b) $|3(-5) + 1| = |-14| = 14$

10.5 Piecewise Linear Functions

10.5 Now Try Exercises

N1. $f(x) = 2|x + 1| - 3$

The graph of $y = |x|$ is shown in the text. Replacing x with $x + 1$ shifts that graph to the left one unit. Multiplying by 2 makes the graph narrower than the graph of $y = |x|$ (the slopes of the partial lines are 2 and -2). The " $- 3$" shifts the graph three units down. The vertex is at the point $(-1, -3)$.

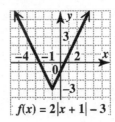

N2. $f(x) = \begin{cases} \frac{1}{2}x + 2 & \text{if } x \geq -2 \\ -x - 3 & \text{if } x < -2 \end{cases}$

If $x \geq -2$, then $f(x) = \frac{1}{2}x + 2$. This is the graph of a line with slope $\frac{1}{2}$ and left endpoint $(-2, 1)$ (the solid circle indicates that the graph includes that point).

If $x < -2$, then $f(x) = -x - 3$. This is the graph of a line with slope -1 and right endpoint $(-2, -1)$ (the open circle indicates that the graph does not include that point).

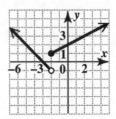

$f(x) = \begin{cases} \frac{1}{2}x + 2 & \text{if } x \geq -2 \\ -x - 3 & \text{if } x < -2 \end{cases}$

Copyright © 2012 Pearson Education, Inc. Publishing as Addison-Wesley.

N3. $f(x) = \begin{cases} 2.833x + 799 & \text{if } 0 \le x \le 6 \\ -26.25x + 973.5 & \text{if } 6 < x \le 10 \end{cases}$

The year 2000 corresponds to
$x = 2000 - 1998 = 2$.

$$f(2) = 2.833(2) + 799 = 804.666 \approx 805$$

The year 2006 corresponds to
$x = 2006 - 1998 = 8$.

$$f(8) = -26.25(8) + 973.5 = 763.5 \approx 764$$

N4. **(a)** $[\![5]\!] = 5 \quad (5 \le 5)$

(b) $[\![-6]\!] = -6 \quad (-6 \le -6)$

(c) $[\![3.5]\!] = 3 \quad (3 \le 3.5)$

(d) $[\![-4.1]\!] = -5 \quad (-5 \le -4.1)$

N5. $f(x) = [\![x - 1]\!]$

The graph of f is the same as the graph of $y = [\![x]\!]$ shifted one unit to the right. Shifting the graph of $y = [\![x]\!]$ (right, left, up, down) does not change the domain and range. The domain of f is $(-\infty, \infty)$ and the range of f is the set of integers, that is, $\{\dots, -2, -1, 0, 1, 2, \dots\}$.

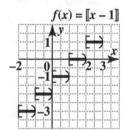

$f(x) = [\![x - 1]\!]$

N6. $f(x) = [\![\frac{1}{3}x - 1]\!]$

To get $[\![\frac{1}{3}x - 1]\!] = 0$, we need
$$\begin{aligned} 0 &\le \tfrac{1}{3}x - 1 < 1 \\ 1 &\le \tfrac{1}{3}x \qquad < 2 \\ 3 &\le x \qquad\quad < 6. \end{aligned}$$

To get $[\![\frac{1}{3}x - 1]\!] = 1$, we need
$$\begin{aligned} 1 &\le \tfrac{1}{3}x - 1 < 2 \\ 2 &\le \tfrac{1}{3}x \qquad < 3 \\ 6 &\le x \qquad\quad < 9. \end{aligned}$$

Notice that the length of each step is 3. Follow this pattern to graph the step function.

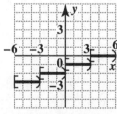

$f(x) = \left[\!\left[\tfrac{1}{3}x - 1\right]\!\right]$

N7. This function is similar to the greatest integer function, but in this case, we use the integer that is *greater than* or equal to the number. For example, for $\frac{1}{2}$ hour of parking, you would be charged for 1 hour.

| | Hours | |
Interval	Charged for	Cost
$(0, 1]$	1	$4.00
$(1, 2]$	2	$6.00
$(2, 3]$	3	$8.00
$(3, 4]$	4	$10.00

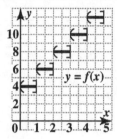

10.5 Section Exercises

1. $f(x) = |x - 2| + 2$

The graph of this function has its "vertex" at $(2, 2)$, so the correct graph is **B**.

3. $f(x) = |x - 2| - 2$

The graph of this function has its "vertex" at $(2, -2)$, so the correct graph is **A**.

5. $f(x) = |x + 1|$

Since x can be any real number, the domain is $(-\infty, \infty)$.

The value of y is always greater than or equal to 0, so the range is $[0, \infty)$.

The graph of $y = |x + 1|$ looks like the graph of the absolute value function $y = |x|$, but the graph is translated 1 unit to the left. The x-value of its "vertex" is obtained by setting $x + 1 = 0$ and solving for x:

$$\begin{aligned} x + 1 &= 0 \\ x &= -1. \end{aligned}$$

Since the corresponding y-value is 0, the "vertex" is $(-1, 0)$. Some additional points are $(-3, 2)$, $(-2, 1)$, $(0, 1)$, and $(1, 2)$.

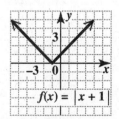

$f(x) = |x + 1|$

Copyright © 2012 Pearson Education, Inc. Publishing as Addison-Wesley.

7. $f(x) = |2 - x|$

The x-value of the "vertex" is obtained by setting $2 - x = 0$ and solving for x:

$$2 - x = 0$$
$$2 = x$$

When $x = 2$, $y = |2 - 2| = |0| = 0$, so the "vertex" is $(2, 0)$. Some additional points are $(0, 2)$, $(1, 1)$, $(3, 1)$, and $(4, 2)$.

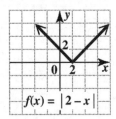

9. $y = |x| + 4$

The graph is the same as $y = |x|$ except translated 4 units up. The "vertex" is $(0, 4)$. The domain is $(-\infty, \infty)$ and the range is $[4, \infty)$.

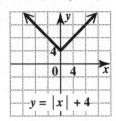

11. $y = 3|x - 2| - 1$

The graph is the same as $y = 3|x|$ except translated 2 units to the right and 1 unit down. The coefficient 3 indicates that the rays that form this graph are steeper than the rays that form the graph of $y = |x|$. The "vertex" is $(2, -1)$. Two other points on the graph are $(0, 5)$ and $(4, 5)$. The domain is $(-\infty, \infty)$ and the range is $[-1, \infty)$.

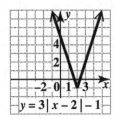

13. $f(x) = \begin{cases} 2x & \text{if } x \leq -1 \\ x - 1 & \text{if } x > -1 \end{cases}$

(a) Since $-5 \leq -1$, use $f(x) = 2x$.
$f(-5) = 2(-5) = -10$

(b) Since $-1 \leq -1$, use $f(x) = 2x$.
$f(-1) = 2(-1) = -2$

(c) Since $0 > -1$, use $f(x) = x - 1$.
$f(0) = 0 - 1 = -1$

(d) Since $3 > -1$, use $f(x) = x - 1$.
$f(3) = 3 - 1 = 2$

(e) Since $5 > -1$, use $f(x) = x - 1$.
$f(5) = 5 - 1 = 4$

15. $f(x) = \begin{cases} 2 & \text{if } x \leq 0 \\ -6 & \text{if } x > 0 \end{cases}$

(a) Since $-5 \leq 0$, use $f(x) = 2$. $f(-5) = 2$

(b) Since $-1 \leq 0$, use $f(x) = 2$. $f(-1) = 2$

(c) Since $0 \leq 0$, use $f(x) = 2$. $f(0) = 2$

(d) Since $3 > 0$, use $f(x) = -6$. $f(3) = -6$

(e) Since $5 > 0$, use $f(x) = -6$. $f(5) = -6$

17. $f(x) = \begin{cases} 4 - x & \text{if } x < 2 \\ 1 + 2x & \text{if } x \geq 2 \end{cases}$

Graph the line $y = 4 - x$ to the left of $x = 2$, using two points such as $(-2, 6)$ and $(0, 4)$. Use an open endpoint (open circle) at $(2, 2)$ to show that this point is not part of the graph. Graph the line $y = 1 + 2x$ to the right of $x = 2$, including an endpoint at $(2, 5)$. Notice that the two portions of this graph do not meet.

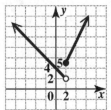

$$f(x) = \begin{cases} 4 - x & \text{if } x < 2 \\ 1 + 2x & \text{if } x \geq 2 \end{cases}$$

19. $f(x) = \begin{cases} x - 1 & \text{if } x \leq 3 \\ 2 & \text{if } x > 3 \end{cases}$

Graph the line $y = x - 1$ to the left of $x = 3$ as a ray, including the endpoint at $(3, 2)$. Graph the horizontal line $y = 2$ to the right of $x = 3$. Notice that the two portions of this graph meet at the point $(3, 2)$.

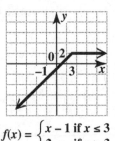

$$f(x) = \begin{cases} x - 1 & \text{if } x \leq 3 \\ 2 & \text{if } x > 3 \end{cases}$$

21. $f(x) = \begin{cases} 2x + 1 & \text{if } x \geq 0 \\ x & \text{if } x < 0 \end{cases}$

Graph the line $y = 2x + 1$ to the right of $x = 0$, including the endpoint $(0, 1)$. Graph the line $y = x$ to the left of $x = 0$. Use an open endpoint at $(0, 0)$ to show that this point is not part of the graph. Notice that the two portions do not meet.

Copyright © 2012 Pearson Education, Inc. Publishing as Addison-Wesley.

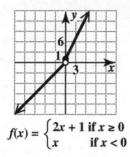

$$f(x) = \begin{cases} 2x+1 & \text{if } x \geq 0 \\ x & \text{if } x < 0 \end{cases}$$

23. $f(x) = \begin{cases} 2+x & \text{if } x < -4 \\ -x^2 & \text{if } x \geq -4 \end{cases}$

Graph the line $y = 2 + x$ to the left of $x = -4$. Use an open endpoint at $(-4, -2)$ to show that this point is not part of the graph. Draw the portion of the parabola $y = -x^2$ that is to the right of $x = -4$, including the endpoint at $(-4, -16)$. Notice that the two portions do not meet.

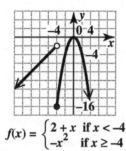

$$f(x) = \begin{cases} 2+x & \text{if } x < -4 \\ -x^2 & \text{if } x \geq -4 \end{cases}$$

25. $f(x) = \begin{cases} |x| & \text{if } x > -2 \\ x^2 - 2 & \text{if } x \leq -2 \end{cases}$

Draw the portion of the absolute value graph $y = |x|$ that is to the right of $x = -2$, but do not include the endpoint. Draw the portion of the parabola $y = x^2 - 2$ that is to the left of $x = -2$, including the endpoint at $(-2, 2)$. Notice that the two portions of this graph meet at the point $(-2, 2)$.

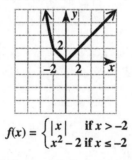

$$f(x) = \begin{cases} |x| & \text{if } x > -2 \\ x^2 - 2 & \text{if } x \leq -2 \end{cases}$$

In Exercises 27–36, the answer is the greatest integer that is less than or equal to the number in the greatest integer symbol.

27. $[\![3]\!] = 3 \quad (3 \leq 3)$

29. $[\![4.5]\!] = 4 \quad (4 \leq 4.5)$

31. $[\![\frac{1}{2}]\!] = 0 \quad (0 \leq \frac{1}{2})$

33. $[\![-14]\!] = -14 \quad (-14 \leq -14)$

35. $[\![-10.1]\!] = -11 \quad (-11 \leq -10.1)$

37. $f(x) = [\![x]\!] - 1$

This is the graph of the greatest integer function, $g(x) = [\![x]\!]$, shifted down 1 unit.

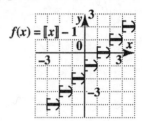

39. $f(x) = [\![x - 3]\!]$

This is the graph of the greatest integer function, $g(x) = [\![x]\!]$, shifted 3 units to the right.

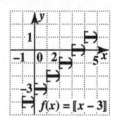

$$f(x) = [\![x - 3]\!]$$

41. $f(x) = [\![-x]\!]$

Evaluate the function for several values of x to help determine a pattern. Remember, the greatest integer of a number is the integer to the *left* of the number on the number line. If the number is an integer, then the greatest integer of the number is just itself.

x	$-x$	$y = f(x) = [\![-x]\!]$
-2	2	2
-1.5	1.5	1
-1	1	1
-0.5	0.5	0
0	0	0
0.5	-0.5	-1
1	-1	-1
1.5	-1.5	-2
2	-2	-2

Observe that for any x in the interval $(-2, -1]$, $y = 1$.
For x in $(-1, 0]$, $y = 0$.
For x in $(0, 1]$, $y = -1$.
The domain is $(-\infty, \infty)$ and the range is the set of integers. Sketch the graph of this step function.

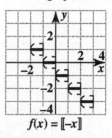

$$f(x) = [\![-x]\!]$$

Copyright © 2012 Pearson Education, Inc. Publishing as Addison-Wesley.

43. $f(x) = [\![2x - 1]\!]$

To get $[\![2x - 1]\!] = 0$, we need
$$0 \le 2x - 1 < 1$$
$$1 \le 2x \quad < 2$$
$$\tfrac{1}{2} \le x \quad < 1.$$

So for x in $\left[\tfrac{1}{2}, 1\right)$, $[\![2x - 1]\!] = 0$.

To get $[\![2x - 1]\!] = 1$, we need
$$1 \le 2x - 1 < 2$$
$$2 \le 2x \quad < 3$$
$$1 \le x \quad < \tfrac{3}{2}.$$

So for x in $\left[1, \tfrac{3}{2}\right)$, $[\![2x - 1]\!] = 1$.

Notice that the length of each step is $\tfrac{1}{2}$. Follow this pattern to graph the step function.

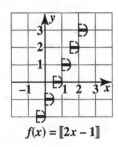

$$f(x) = [\![2x - 1]\!]$$

45. $f(x) = [\![3x]\!]$

To get $[\![3x]\!] = 0$, we need
$$0 \le 3x < 1$$
$$0 \le x \quad < \tfrac{1}{3}.$$

So for x in $\left[0, \tfrac{1}{3}\right)$, $[\![3x + 1]\!] = 0$.

To get $[\![3x]\!] = 1$, we need
$$1 \le 3x < 2$$
$$\tfrac{1}{3} \le x \quad < \tfrac{2}{3}.$$

So for x in $\left[\tfrac{1}{3}, \tfrac{2}{3}\right)$, $[\![3x + 1]\!] = 1$.

Notice that the length of each step is $\tfrac{1}{3}$. Follow this pattern to graph the step function.

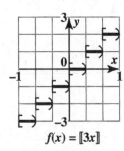

$$f(x) = [\![3x]\!]$$

47. Both of the graphing calculator screens show the graphs of
$$y_1 = |x + 4| \quad \text{and} \quad y_2 = 6.$$

(a) $|x + 4| = 6$

From the two screens, we see that the graphs intersect at the points $(-10, 6)$ and $(2, 6)$. The x-coordinates of these points of intersection are the solutions of the equation, so the solution set is $\{-10, 2\}$.

(b) $|x + 4| < 6$

The graph of y_1 is *below* the graph of y_2 for values of x between -10 and 2. Thus, the solution set for this inequality is $(-10, 2)$.

(c) $|x + 4| > 6$

The graph of y_1 is *above* the graph of y_2 for values of x less than -10 or greater than 2. Thus, the solution set of this inequality is $(-\infty, -10) \cup (2, \infty)$.

49. $|x - 5| = 4$
$$x - 5 = 4 \quad \text{or} \quad x - 5 = -4$$
$$x = 9 \quad \text{or} \quad x = 1$$

The solution set is $\{1, 9\}$.

Now graph $y_1 = |x - 5|$ and $y_2 = 4$ and find the intersection points of the two graphs.

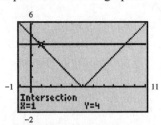

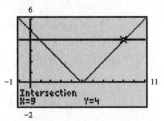

From these two screens, we see that the graphs intersect at two points $(1, 4)$ and $(9, 4)$. The x-coordinates of these points of intersection are the solutions of the equation, so the solution set is $\{1, 9\}$.

The answers obtained by the analytic and graphical methods are the same.

51. $|7 - 4x| > 1$
$$7 - 4x > 1 \quad \text{or} \quad 7 - 4x < -1$$
$$-4x > -6 \quad \text{or} \quad -4x < -8$$
$$x < \tfrac{3}{2} \quad \text{or} \quad x > 2$$

The solution set is $\left(-\infty, \tfrac{3}{2}\right) \cup (2, \infty)$.

Now graph $y_1 = |7 - 4x|$ and $y_2 = 1$ and find their intersection points.

Copyright © 2012 Pearson Education, Inc. Publishing as Addison-Wesley.

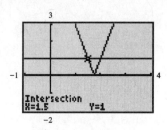

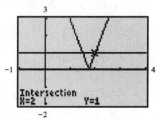

From these two screens, we see that the graphs intersect at $(1.5, 1)$ and $(2, 1)$. The graph of y_1 lies *above* the graph of y_2 for x-values less than 1.5 or greater than 2. Thus, the solution set is $(-\infty, 1.5) \cup (2, \infty)$.

53. $|0.5x - 2| < 1$

$$-1 < 0.5x - 2 < 1$$
$$1 < 0.5x \quad\; < 3$$
$$2 < x \quad\;\;\; < 6 \quad \textit{Multiply by 2.}$$

The solution set is $(2, 6)$.

Now graph $y_1 = |0.5x - 2|$ and $y_2 = 1$ and find their intersection points.

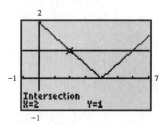

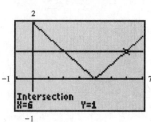

The graph of y_1 lies *below* the graph of y_2 for x-values between 2 and 6. Thus, the solution set is $(2, 6)$.

55. For $0 \le x \le 10$, use the points $(0, 12.252)$ and $(10, 15.188)$. The slope is

$$m = \frac{15.188 - 12.252}{10 - 0} = 0.2936.$$

Use the slope-intercept form to get an equation of the line.

$$y = 0.2936x + 12.252.$$

For $10 < x \le 18$, use the points $(10, 15.188)$ and $(18, 12.495)$. The slope is

$$m = \frac{12.495 - 15.188}{18 - 10} = -0.336625.$$

Use the point-slope form to get an equation of the line.

$$y - 15.188 = -0.336625(x - 10)$$
$$y - 15.188 = -0.336625x + 3.36625$$
$$y = -0.336625x + 18.55425$$

The piecewise linear function (rounded) is

$$f(x) = \begin{cases} 0.294x + 12.252 & \text{if } 0 \le x \le 10 \\ -0.337x + 18.554 & \text{if } 10 < x \le 18 \end{cases}$$

57. (a) For 1 day, the cost is

$$S(1) = 4 + 7 \cdot 1 = \$11.$$

(b) For 1.25 days, the cost is the same as for 2 days.

$$S(1.25) = 4 + 7 \cdot 2$$
$$= 4 + 14$$
$$= \$18$$

(c) For 3.5 days, the cost is the same as for 4 days.

$$S(3.5) = 4 + 7 \cdot 4$$
$$= 4 + 28$$
$$= \$32$$

(d)

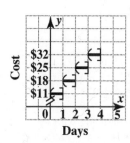

59. The following table summarizes the number of hours, x, and the cost of parking, $f(x)$, on the interval $(0, 2]$.

x	$(0, 0.5]$	$(0.5, 1]$	$(1, 1.5]$	$(1.5, 2]$
$f(x)$	\$3	\$5	\$7	\$9

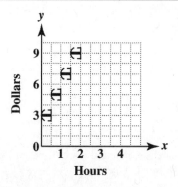

Copyright © 2012 Pearson Education, Inc. Publishing as Addison-Wesley.

61. $\{(0,1),(1,2),(2,4),(3,8)\}$

The domain is the set of x-values: $\{0,1,2,3\}$.
The range is the set of y-values: $\{1,2,4,8\}$.

63. Using the vertical line test, we find any vertical line will intersect the graph at most once. This indicates that the graph represents a function.

65. $f(x) = 3x + 8, g(x) = \frac{1}{3}(x - 8)$

$$(f \circ g)(x) = f(g(x))$$
$$= f\left(\tfrac{1}{3}(x - 8)\right)$$
$$= 3\left(\tfrac{1}{3}(x - 8)\right) + 8$$
$$= (x - 8) + 8 = x$$

$$(g \circ f)(x) = g(f(x))$$
$$= g(3x + 8)$$
$$= \tfrac{1}{3}[(3x + 8) - 8]$$
$$= \tfrac{1}{3}(3x) = x$$

Chapter 10 Review Exercises

In Exercises 1–6, $f(x) = x^2 - 2x$ and $g(x) = 5x + 3$.

1. $(f + g)(x) = f(x) + g(x)$
$$= (x^2 - 2x) + (5x + 3)$$
$$= x^2 + 3x + 3$$

The domains of f and g are $(-\infty, \infty)$, so the domain of $f + g$ is also $(-\infty, \infty)$.

2. $(f - g)(x) = f(x) - g(x)$
$$= (x^2 - 2x) - (5x + 3)$$
$$= x^2 - 7x - 3$$

The domains of f and g are $(-\infty, \infty)$, so the domain of $f - g$ is also $(-\infty, \infty)$.

3. $(fg)(x) = f(x) \cdot g(x)$
$$= (x^2 - 2x)(5x + 3)$$

The domains of f and g are $(-\infty, \infty)$, so the domain of fg is also $(-\infty, \infty)$.

4. $\left(\dfrac{f}{g}\right)(x) = \dfrac{f(x)}{g(x)}$
$$= \dfrac{x^2 - 2x}{5x + 3}$$

The domains of f and g are $(-\infty, \infty)$, but since the denominator g equals 0 when $x = -\frac{3}{5}$, we must exclude $-\frac{3}{5}$, so the domain of $\frac{f}{g}$ is $\left(-\infty, -\frac{3}{5}\right) \cup \left(-\frac{3}{5}, \infty\right)$.

5. $(g \circ f)(x) = g(f(x))$
$$= g\left(x^2 - 2x\right)$$
$$= 5\left(x^2 - 2x\right) + 3$$
$$= 5x^2 - 10x + 3$$

The domains of f and g are $(-\infty, \infty)$, so the domain of $g \circ f$ is also $(-\infty, \infty)$.

6. $(f \circ g)(x) = f(g(x))$
$$= f(5x + 3)$$
$$= (5x + 3)^2 - 2(5x + 3)$$
$$= 25x^2 + 30x + 9 - 10x - 6$$
$$= 25x^2 + 20x + 3$$

The domains of f and g are $(-\infty, \infty)$, so the domain of $f \circ g$ is also $(-\infty, \infty)$.

In Exercises 7–12, $f(x) = 2x - 3$ and $g(x) = \sqrt{x}$.

7. $(f - g)(4) = f(4) - g(4)$
$$= (2 \cdot 4 - 3) - \sqrt{4}$$
$$= 5 - 2 = 3$$

8. $\left(\dfrac{f}{g}\right)(9) = \dfrac{f(9)}{g(9)}$
$$= \dfrac{2(9) - 3}{\sqrt{9}}$$
$$= \dfrac{15}{3} = 5$$

9. $(fg)(5) = f(5) \cdot g(5)$
$$= (2 \cdot 5 - 3)\left(\sqrt{5}\right)$$
$$= 7\sqrt{5}$$

10. $(f + g)(5) = f(5) + g(5)$
$$= (2 \cdot 5 - 3) + \sqrt{5}$$
$$= 7 + \sqrt{5}$$

11. $(fg)(2b) = f(2b) \cdot g(2b)$
$$= (2 \cdot 2b - 3)\left(\sqrt{2b}\right)$$
$$= (4b - 3)\left(\sqrt{2b}\right), b \geq 0$$

12. $(g \circ f)(2) = g(f(2))$
$$= g(1)$$
$$= \sqrt{1} = 1$$

13. $(f \circ g)(2) = f(g(2))$
$$= f\left(\sqrt{2}\right)$$
$$= 2\sqrt{2} - 3$$

The answers are not equal, so composition of functions is not commutative.

14. $f(x) = \sqrt{9 - 2x}$
If we try to find $f(5)$, we obtain

$$f(5) = \sqrt{9 - 2 \cdot 5} = \sqrt{9 - 10} = \sqrt{-1},$$

which is not a real number. Since the x-values and y-values in a function must be real numbers so that (x, y) corresponds to a point in the plane, 5 cannot be in the domain of f.

15. $y = 6 - 2x^2$
Write in $y = a(x - h)^2 + k$ form as
$y = -2(x - 0)^2 + 6$. The vertex (h, k) is $(0, 6)$.

Copyright © 2012 Pearson Education, Inc. Publishing as Addison-Wesley.

16. $f(x) = -(x - 1)^2$
Write in $y = a(x - h)^2 + k$ form as
$y = -1(x - 1)^2 + 0$. The vertex (h, k) is $(1, 0)$.

17. $y = (x - 3)^2 + 7$
The equation is in the form $y = a(x - h)^2 + k$, so
the vertex (h, k) is $(3, 7)$.

18. $y = f(x) = -3x^2 + 4x - 2$

Use the vertex formula with $a = -3$ and $b = 4$.

The x-coordinate of the vertex is

$$\frac{-b}{2a} = \frac{-4}{2(-3)} = \frac{2}{3}.$$

The y-coordinate of the vertex is

$$f\left(\frac{-b}{2a}\right) = f\left(\frac{2}{3}\right)$$
$$= -3\left(\tfrac{2}{3}\right)^2 + 4\left(\tfrac{2}{3}\right) - 2$$
$$= -\tfrac{4}{3} + \tfrac{8}{3} - 2$$
$$= -\tfrac{2}{3}.$$

The vertex is $\left(\tfrac{2}{3}, -\tfrac{2}{3}\right)$.

19. $x = (y - 3)^2 - 4$

When $y = 3$, $x = -4$, so the vertex is $(-4, 3)$.

20. $f(x) = -5x^2$

The vertex is $(0, 0)$ and the axis is $x = 0$.
Since $a = -5 < 0$, the parabola opens down, and
since $|a| = 5 > 1$, the parabola is narrower than
the graph of $f(x) = x^2$. Two other points on the
graph are $(-1, -5)$ and $(1, -5)$.

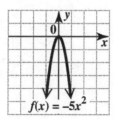

The domain is $(-\infty, \infty)$. The largest y-value is 0,
so the range is $(-\infty, 0]$.

21. $f(x) = 3x^2 - 2$
$\quad = 3(x - 0)^2 - 2$

The vertex is $(0, -2)$ and the axis is $x = 0$.
Since $a = 3 > 0$, the parabola opens up, and since
$|a| = 3 > 1$, the parabola is narrower than the
graph of $f(x) = x^2$. Two other points on the
graph are $(-1, 1)$ and $(1, 1)$.

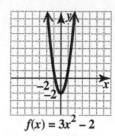

$f(x) = 3x^2 - 2$

The domain is $(-\infty, \infty)$. The smallest y-value is
-2, so the range is $[-2, \infty)$.

22. $y = (x + 2)^2$
$\quad = (x + 2)^2 + 0$

The vertex is $(-2, 0)$ and the axis is $x = -2$.
The parabola opens up and has the same shape as
the graph of $y = x^2$. This graph is simply the
graph of $y = x^2$ translated 2 units to the left.

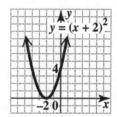

$y = (x + 2)^2$

The domain is $(-\infty, \infty)$. The smallest y-value is
0, so the range is $[0, \infty)$.

23. $y = 2(x - 2)^2 - 3$

The vertex is $(2, -3)$ and the axis is $x = 2$.
Since $a = 2 > 0$, the parabola opens up, and since
$|a| = 2 > 1$, the parabola is narrower than the
graph of $f(x) = x^2$. Two other points on the
graph are $(0, 5)$ and $(4, 5)$.

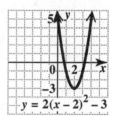

$y = 2(x - 2)^2 - 3$

The domain is $(-\infty, \infty)$. The smallest y-value is
-3, so the range is $[-3, \infty)$.

24. $f(x) = -2x^2 + 8x - 5$
Complete the square to find the vertex.

$$f(x) = -2\left(x^2 - 4x\right) - 5$$
$$= -2\left(x^2 - 4x + 4 - 4\right) - 5$$
$$= -2\left(x^2 - 4x + 4\right) - 2(-4) - 5$$
$$= -2(x - 2)^2 + 8 - 5$$
$$= -2(x - 2)^2 + 3$$

The equation is in the form $y = a(x - h)^2 + k$, so
the vertex (h, k) is $(2, 3)$ and the axis is $x = 2$.
Here, $a = -2 < 0$, so the parabola opens down.

Copyright © 2012 Pearson Education, Inc. Publishing as Addison-Wesley.

Also, $|a| = |-2| = 2 > 1$, so the graph is narrower than the graph of $y = x^2$. The points $(0, -5)$, $(1, 1)$, and $(3, 1)$ are on the graph.

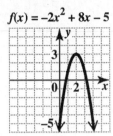

$$f(x) = -2x^2 + 8x - 5$$

The domain is $(-\infty, \infty)$. The largest y-value is 3, so the range is $(-\infty, 3]$.

25. $y = x^2 + 3x + 2$

Complete the square to find the vertex. (We could also use the vertex formula.)

$$y - 2 = x^2 + 3x$$
Move constant to the left.
$$y - 2 + \tfrac{9}{4} = x^2 + 3x + \tfrac{9}{4}$$
$$\left(\tfrac{1}{2}\right)(3) = \tfrac{3}{2}, \text{ and } \left(\tfrac{3}{2}\right)^2 = \tfrac{9}{4}$$
$$y + \tfrac{1}{4} = \left(x + \tfrac{3}{2}\right)^2$$
Combine constants on the left.
$$y = \left(x + \tfrac{3}{2}\right)^2 - \tfrac{1}{4}$$

The vertex is $\left(-\tfrac{3}{2}, -\tfrac{1}{4}\right)$ and the axis is $x = -\tfrac{3}{2}$. The graph opens up and has the same shape as the graph of $y = x^2$. The graph also contains $(-3, 2)$ and $(0, 2)$.

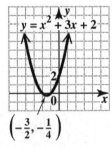

$$\left(-\tfrac{3}{2}, -\tfrac{1}{4}\right)$$

The domain is $(-\infty, \infty)$. The smallest y-value is $-\tfrac{1}{4}$, so the range is $\left[-\tfrac{1}{4}, \infty\right)$.

26. $x = (y - 1)^2 + 2$

The graph is a horizontal parabola with vertex $(2, 1)$ and axis $y = 1$.
Since $a = 1$, the parabola opens to the right and has the same shape as the graph of $x = y^2$. Two other points on the graph are $(6, -1)$ and $(6, 3)$.

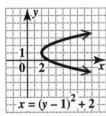

$$x = (y - 1)^2 + 2$$

The smallest x-value is 2, so the domain is $[2, \infty)$. The range is $(-\infty, \infty)$.

27. $x = 2(y + 3)^2 - 4 = 2[y - (-3)]^2 + (-4)$

Since the roles of x and y are reversed, this is a horizontal parabola. The equation is in the form

$$x = a(y - k)^2 + h,$$

so the vertex (h, k) is $(-4, -3)$ and the axis is $y = -3$. Here, $a = 2 > 0$, so the parabola opens to the right and is narrower than the graph of $y = x^2$. Two other points on the graph are $(4, -1)$ and $(4, -5)$.

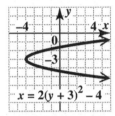

$$x = 2(y + 3)^2 - 4$$

The smallest x-value is -4, so the domain is $[-4, \infty)$. The range is $(-\infty, \infty)$.

28. $x = -\tfrac{1}{2}y^2 + 6y - 14$

Since the roles of x and y are reversed, this is a horizontal parabola. Complete the square to find the vertex.

$$x = -\tfrac{1}{2}y^2 + 6y - 14$$
$$= -\tfrac{1}{2}(y^2 - 12y) - 14$$
$$= -\tfrac{1}{2}(y^2 - 12y + 36 - 36) - 14$$
$$= -\tfrac{1}{2}(y^2 - 12y + 36) - \tfrac{1}{2}(-36) - 14$$
$$= -\tfrac{1}{2}(y - 6)^2 + 18 - 14$$
$$x = -\tfrac{1}{2}(y - 6)^2 + 4$$

The equation is in the form $x = a(y - k)^2 + h$, so the vertex (h, k) is $(4, 6)$ and the axis is $y = 6$. Here, $a = -\tfrac{1}{2} < 0$, so the parabola opens to the left.
Also, $|a| = \left|-\tfrac{1}{2}\right| = \tfrac{1}{2} < 1$, so the graph is wider than the graph of $y = x^2$. The points $(-14, 0)$, $(2, 4)$, and $(2, 8)$ are on the graph.

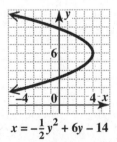

$$x = -\tfrac{1}{2}y^2 + 6y - 14$$

The largest x-value is 4, so the domain is $(-\infty, 4]$. The range is $(-\infty, \infty)$.

Copyright © 2012 Pearson Education, Inc. Publishing as Addison-Wesley.

29. (a) Use $ax^2 + bx + c = y$ with $(0, 2.9)$, $(10, 24.3)$, and $(20, 56.5)$.

$$
\begin{array}{rl}
c = & 2.9 \quad (1) \\
100a + 10b + c = & 24.3 \quad (2) \\
400a + 20b + c = & 56.5 \quad (3)
\end{array}
$$

(b) Rewrite equations (2) and (3) with $c = 2.9$.

$$100a + 10b + 2.9 = 24.3$$
$$400a + 20b + 2.9 = 56.5$$

$$100a + 10b = 21.4 \quad (4)$$
$$400a + 20b = 53.6 \quad (5)$$

Now eliminate b.

$$
\begin{array}{rl}
-200a - 20b = & -42.8 \quad -2 \times (4) \\
\underline{400a + 20b = 53.6} & (5) \\
200a = & 10.8 \\
a = & \frac{10.8}{200} = 0.054
\end{array}
$$

Use (4) to find b.

$$
\begin{aligned}
100a + 10b &= 21.4 \quad (4) \\
100(0.054) + 10b &= 21.4 \\
5.4 + 10b &= 21.4 \\
10b &= 16 \\
b &= \tfrac{16}{10} = 1.6
\end{aligned}
$$

Thus, we get the quadratic function

$$f(x) = 0.054x^2 + 1.6x + 2.9.$$

(c) Use $f(x) = 0.054x^2 + 1.6x + 2.9$ with $x = 21$ for the year 2006.

$$
\begin{aligned}
f(21) &= 0.054(21)^2 + 1.6(21) + 2.9 \\
&= 60.314 \approx \$60.3 \text{ billion}
\end{aligned}
$$

The result using the model is close to the table value of $61.4 billion, but slightly low.

30. $s(t) = -16t^2 + 160t$

The equation represents a parabola. Since $a = -16 < 0$, the parabola opens down. The time and maximum height occur at the vertex of the parabola, given by

$$\left(\frac{-b}{2a}, s\left(\frac{-b}{2a} \right) \right).$$

Using the standard form of the equation, $a = -16$ and $b = 160$, so

$$t = \frac{-b}{2a} = \frac{-160}{2(-16)} = 5,$$

and $\quad s(t) = -16(5)^2 + 160(5)$
$$= -400 + 800 = 400.$$

The vertex is $(5, 400)$. The time at which the maximum height is reached is 5 seconds. The maximum height is 400 feet.

31. Let $L =$ the length of the rectangle and $W =$ the width.

The perimeter of the rectangle is 200 m, so

$$
\begin{aligned}
2L + 2W &= 200 \\
2W &= 200 - 2L \\
W &= 100 - L.
\end{aligned}
$$

Since the area is length times width, substitute $100 - L$ for W.

$$
\begin{aligned}
A &= LW \\
&= L(100 - L) \\
&= 100L - L^2 \quad \text{or} \quad -L^2 + 100L
\end{aligned}
$$

Use the vertex formula.

$$L = \frac{-b}{2a} = \frac{-100}{2(-1)} = 50$$

So $L = 50$ meters and
$W = 100 - L = 100 - 50 = 50$ meters.
The maximum area is

$$50 \cdot 50 = 2500 \text{ m}^2.$$

32. Let $x =$ one number.
Then $10 - x =$ the other number.

Let $P(x)$ be their product.

$$
\begin{aligned}
P(x) &= x(10 - x) \\
&= -x^2 + 10x \\
&= -(x^2 - 10x) \\
&= -(x^2 - 10x + 25) + 25 \\
&= -(x - 5)^2 + 25
\end{aligned}
$$

The product is maximum when $x = 5$, the x-coordinate of the vertex. The two numbers are 5 and $10 - 5$ or 5.

33. $2x^2 - y^2 = 4$

(1) Replace y with $-y$.

$$
\begin{aligned}
2x^2 - (-y)^2 &= 4 \\
2x^2 - y^2 &= 4
\end{aligned}
$$

The graph is symmetric with respect to the x-axis.

(2) Replace x with $-x$.

$$
\begin{aligned}
2(-x)^2 - y^2 &= 4 \\
2x^2 - y^2 &= 4
\end{aligned}
$$

The graph is symmetric with respect to the y-axis.

(3) Replace x with $-x$ and y with $-y$.

$$
\begin{aligned}
2(-x)^2 - (-y)^2 &= 4 \\
2x^2 - y^2 &= 4
\end{aligned}
$$

The graph is symmetric with respect to the origin. Thus, the graph has all three symmetries.

Copyright © 2012 Pearson Education, Inc. Publishing as Addison-Wesley.

34. $3x^2 + 4y^2 = 12$

(1) Replace y with $-y$.

$$3x^2 + 4(-y)^2 = 12$$
$$3x^2 + 4y^2 = 12$$

The graph is symmetric with respect to the x-axis.

(2) Replace x with $-x$.

$$3(-x)^2 + 4y^2 = 12$$
$$3x^2 + 4y^2 = 12$$

The graph is symmetric with respect to the y-axis.

(3) Replace x with $-x$ and y with $-y$.

$$3(-x)^2 + 4(-y)^2 = 12$$
$$3x^2 + 4y^2 = 12$$

The graph is symmetric with respect to the origin.
Thus, the graph has all three symmetries.

35. $2x - y^2 = 8$

(1) Replace y with $-y$.

$$2x - (-y)^2 = 8$$
$$2x - y^2 = 8$$

The graph is symmetric with respect to the x-axis.

(2) Replace x with $-x$.

$$2(-x) - y^2 = 8$$
$$-2x - y^2 = 8$$

The graph is not symmetric with respect to the y-axis.

(3) Replace x with $-x$ and y with $-y$.

$$2(-x) - (-y)^2 = 8$$
$$-2x - y^2 = 8$$

The graph is not symmetric with respect to the origin.
Thus, the graph is symmetric with respect to the x-axis only.

36. $y = 2x^2 + 3$

(1) Replace y with $-y$.

$$-y = 2x^2 + 3$$
$$y = -2x^2 - 3$$

The graph is not symmetric with respect to the x-axis.

(2) Replace x with $-x$.

$$y = 2(-x)^2 + 3$$
$$y = 2x^2 + 3$$

The graph is symmetric with respect to the y-axis.

(3) Replace x with $-x$ and y with $-y$.

$$-y = 2(-x)^2 + 3$$
$$-y = 2x^2 + 3$$
$$y = -2x^2 - 3$$

The graph is not symmetric with respect to the origin.
Thus, the graph is symmetric with respect to the y-axis only.

37. $y = 2\sqrt{x} - 4$

(1) Replace y with $-y$.

$$-y = 2\sqrt{x} - 4$$
$$y = -2\sqrt{x} + 4$$

The graph is not symmetric with respect to the x-axis.

(2) Replace x with $-x$.

$$y = 2\sqrt{-x} - 4$$

The graph is not symmetric with respect to the y-axis.

(3) Replace x with $-x$ and y with $-y$.

$$-y = 2\sqrt{-x} - 4$$
$$y = -2\sqrt{-x} + 4$$

The graph is not symmetric with respect to the origin.
Thus, the graph has none of the symmetries.

38. $y = \dfrac{1}{x^2}$

(1) Replace y with $-y$.

$$-y = \frac{1}{x^2}$$
$$y = -\frac{1}{x^2}$$

The graph is not symmetric with respect to the x-axis.

(2) Replace x with $-x$.

$$y = \frac{1}{(-x)^2}$$
$$y = \frac{1}{x^2}$$

The graph is symmetric with respect to the y-axis.

(3) Replace x with $-x$ and y with $-y$.

$$-y = \frac{1}{(-x)^2}$$
$$-y = \frac{1}{x^2}$$
$$y = -\frac{1}{x^2}$$

The graph is not symmetric with respect to the origin.
Thus, the graph is symmetric with respect to the y-axis only.

39. It is true that a circle centered at the origin has symmetry with respect to both axes and the origin. However, a circle is not the graph of a function because it fails the vertical line test.

40. **(a)** For any point (x, y) on a circle with center at the origin, the point $(x, -y)$ is also on the circle, so the circle is symmetric with respect to the x-axis.

(b) For any point (x, y) on a circle with center at the origin, the point $(-x, y)$ is also on the circle, so the circle is symmetric with respect to the y-axis.

(c) For any point (x, y) on a circle with center at the origin, the point $(-x, -y)$ is also on the circle, so the circle is symmetric with respect to the origin.

41. $f(x) = mx + b, m < 0$

Recall that m represents the slope of the line. A line with negative slope goes downward from left to right; that is, as x increases, y decreases. Such a line is the graph of a function that is *decreasing* over all real numbers.

42. The function is increasing on $(-\infty, 2]$.
The function is decreasing on $[2, \infty)$.

43. The function is increasing on $[0, \infty)$.

44. The function is increasing on $(-\infty, -3]$ and $[-1, \infty)$.
The function is decreasing on $[-3, -1]$.

45. The function is decreasing on $(-\infty, 1]$ and increasing on $[1, \infty)$.

46. $f(x) = 2|x| + 3$

The domain is $(-\infty, \infty)$.
The range is $[3, \infty)$.
The "vertex" occurs at $(0, 3)$.
The graph also passes through the points $(-2, 7)$, $(-1, 5)$, $(1, 5)$, and $(2, 7)$. Notice that this graph is symmetric with respect to the y-axis.

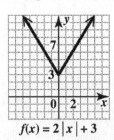

$f(x) = 2|x| + 3$

47. $f(x) = |x - 2|$

The "vertex" of this absolute value graph is at $(2, 0)$, and the graph also passes through $(0, 2)$, $(1, 1)$, $(3, 1)$, and $(4, 2)$. This is the graph of $y = |x|$ translated 2 units to the right.

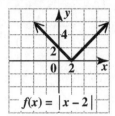

$f(x) = |x - 2|$

48. $f(x) = -|x - 1|$

The "vertex" occurs at $(1, 0)$.
Some other points on the graph are $(-2, -3)$, $(0, -1)$, $(2, -1)$, and $(4, -3)$.
This graph may also be obtained directly from the graph of $y = |x|$ by translating to the right 1 unit and then reflecting about the x-axis.

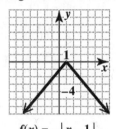

$f(x) = -|x - 1|$

49. $f(x) = \begin{cases} 2x + 1 & \text{if } x \le -1 \\ x + 3 & \text{if } x > -1 \end{cases}$

Graph the portion of the line $y = 2x + 1$ that is to the left of $x = -1$, with a solid endpoint at $(-1, -1)$. Graph the portion of the line $y = x + 3$ that is to the right of $x = -1$ with an open endpoint (open circle) to show that this point is not part of the graph. Notice that two portions of this graph do not meet.

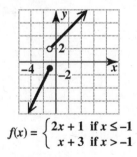

$f(x) = \begin{cases} 2x + 1 & \text{if } x \le -1 \\ x + 3 & \text{if } x > -1 \end{cases}$

50. $f(x) = \begin{cases} -x & \text{if } x \le 0 \\ x^2 & \text{if } x > 0 \end{cases}$

This is a function defined piecewise, so the graph will be made up of two parts.
Graph the portion of the line $y = -x$ that is to the left of $x = 0$, including the endpoint at $(0, 0)$.
Draw the portion of the parabola $y = x^2$ that is to the right of $x = 0$, but do not include the endpoint.

Copyright © 2012 Pearson Education, Inc. Publishing as Addison-Wesley.

Note that the two portions of this graph meet at $(0, 0)$.

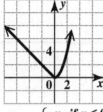

$$f(x) = \begin{cases} -x & \text{if } x \le 0 \\ x^2 & \text{if } x > 0 \end{cases}$$

51. $f(x) = -[\![x]\!]$

If x is in the interval $[0, 1)$, $f(x) = 0$. If x is in the interval $[1, 2)$, $f(x) = -1$. Following this pattern, graph $f(x)$.

Alternatively, since we know what the graph of $g(x) = [\![x]\!]$ looks like, and $f = -g$, we can think of f as a reflection of g through the x-axis.

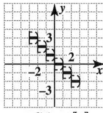

$f(x) = -[\![x]\!]$

52. $f(x) = [\![x + 1]\!]$

For any x in $[-2, -1)$, $y = -1$.

For any x in $[-1, 0)$, $y = 0$.

For any x in $[0, 1)$, $y = 1$.

For any x in $[1, 2)$, $y = 2$.

Follow this pattern to graph the step function.

Alternatively, we can shift the graph of $g(x) = [\![x]\!]$ one unit to the left to obtain the graph of f.

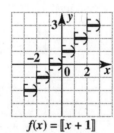

$f(x) = [\![x + 1]\!]$

53. The "+4" shifts the graph of $y = |x|$ four units to the left. The "2" makes the graph narrower. The "−3" shifts the graph of $y = 2|x + 4|$ down 3 units.

54. **(a)** $C(1) = \$0.90$

(b) $C(2.3) = C(3) = \$0.90 + 2(\$0.10)$
$$= \$1.10$$

(c) $C(8) = \$0.90 + 7(\$0.10)$
$$= \$1.60$$

(d)

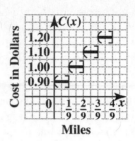

Miles

(e) Since x can be any positive number, the domain is $(0, \infty)$. The range is a set of possible taxi charges, $\{0.90, 1.00, 1.10, 1.20, \dots\}$.

55. **[10.2]** $g(x) = x^2 - 5$ ▪ The graph is shifted 5 units down from the graph of $f(x) = x^2$. The figure that most closely resembles the graph of $g(x)$ is Choice **F**.

56. **[10.2]** $h(x) = -x^2 + 4$ ▪ Because the coefficient of x^2 is negative, the parabola opens down. The graph is shifted 4 units up from the graph of $f(x) = -x^2$. The figure that most closely resembles the graph of $h(x)$ is Choice **B**.

57. **[10.2]** $F(x) = (x - 1)^2$ ▪ The graph is shifted 1 unit to the right of the graph of $f(x) = x^2$. The figure that most closely resembles the graph of $F(x)$ is Choice **C**.

58. **[10.2]** $G(x) = (x + 1)^2$ ▪ The graph is shifted 1 unit to the left of the graph of $f(x) = x^2$. The figure that most closely resembles the graph of $G(x)$ is Choice **A**.

59. **[10.2]** $H(x) = (x - 1)^2 + 1$ ▪ The graph is shifted 1 unit to the right and 1 unit up from the graph of $f(x) = x^2$. The figure that most closely resembles the graph of $H(x)$ is Choice **E**.

60. **[10.2]** $K(x) = (x + 1)^2 + 1$ ▪ The graph is shifted 1 unit to the left and 1 unit up from the graph of $f(x) = x^2$. The figure that most closely resembles the graph of $K(x)$ is choice **D**.

61. **[10.3]** $y = f(x) = 4x^2 + 4x - 2$

Complete the square to find the vertex.

$$y = 4(x^2 + x) - 2$$
$$= 4\left(x^2 + x + \tfrac{1}{4} - \tfrac{1}{4}\right) - 2$$
$$= 4\left(x^2 + x + \tfrac{1}{4}\right) + 4\left(-\tfrac{1}{4}\right) - 2$$
$$= 4\left(x + \tfrac{1}{2}\right)^2 - 1 - 2$$
$$= 4\left(x + \tfrac{1}{2}\right)^2 - 3$$
$$y = 4\left(x + \tfrac{1}{2}\right)^2 - 3$$

The equation is now in the form $y = a(x - h)^2 + k$, so the vertex (h, k) is $\left(-\tfrac{1}{2}, -3\right)$ and the axis is $x = -\tfrac{1}{2}$. Since $a = 4 > 0$, the parabola opens up. Also, $|a| = |4| = 4 > 1$, so the graph is narrower than the graph of $y = x^2$. The points $(-2, 6)$, $(0, -2)$, and $(1, 6)$ are on the graph.

Copyright © 2012 Pearson Education, Inc. Publishing as Addison-Wesley.

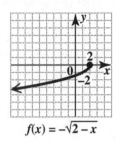

$f(x) = 4x^2 + 4x - 2$

The domain is $(-\infty, \infty)$. The smallest y-value is -3, so the range is $[-3, \infty)$.

62. **[10.4]** $f(x) = y = -\sqrt{2-x}$

Notice that y is nonpositive. Squaring both sides, we obtain $y^2 = 2 - x$ or $x = -y^2 + 2$. The graph is a parabola that opens to the left. Since y is nonpositive, use the lower half of the parabola only, including the endpoint at $(2, 0)$.

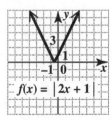

$f(x) = -\sqrt{2-x}$

63. **[10.5]** $f(x) = |2x + 1|$

The domain is $(-\infty, \infty)$ and the range is $[0, \infty)$. The "vertex" occurs where $y = 0$. Find the corresponding x-value.

$$2x + 1 = 0$$
$$2x = -1$$
$$x = -\tfrac{1}{2}$$

The "vertex" is $\left(-\tfrac{1}{2}, 0\right)$. Two other points on the graph are $(-2, 3)$ and $(1, 3)$.

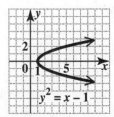

$f(x) = |2x + 1|$

64. **[10.3]** $y^2 = x - 1 \Leftrightarrow x = y^2 + 1$

This is a horizontal parabola. It has vertex $(1, 0)$ and opens to the right. Some other points on the graph are $(2, 1)$, $(2, -1)$, $(5, 2)$ and $(5, -2)$.

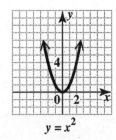

$y^2 = x - 1$

65. **[10.5]** $f(x) = [\![x]\!] - 2$

This graph may be obtained by translating the graph of the greatest integer function, $f(x) = [\![x]\!]$, down 2 units.

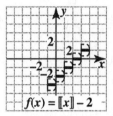

$f(x) = [\![x]\!] - 2$

66. **[10.5]** For $0 < x \le 50$, $F(x) = \$37$.

For $50 < x \le 75$, $F(x) = \$37 + \$10 = \$47$.

For $75 < x \le 100$, $F(x) = \$37 + 2 \cdot \$10 = \$57$.

The domain of F is $(0, \infty)$ (theoretically). The range of F is $\{37, 47, 57, 67, \dots\}$.

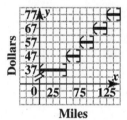

♦♦♦ Relating Concepts 67–70 ♦♦♦

67. Graph $y = x^2$ and $y = x^2 - 4$. The second is obtained by translating the first _4_ units in the _downward_ direction.

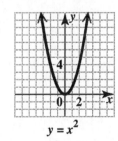

$y = x^2$ $y = x^2 - 4$

68. Graph $y = x^2$ and $y = (x + 3)^2$. The second is obtained by translating the first _3_ units to the _left_.

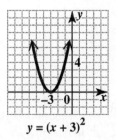

$y = x^2$ $y = (x + 3)^2$

Copyright © 2012 Pearson Education, Inc. Publishing as Addison-Wesley.

69. Graph $y = x^2$ and $y = (x - 4)^2$. The second is obtained by translating the first <u>4</u> units to the <u>right</u>.

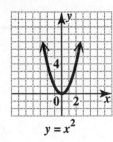

$$y = x^2$$

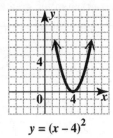

$$y = (x - 4)^2$$

70. The graph of $y = f(x - h)^2 + k$ is obtained by translating the graph of $y = f(x)$ h units to the right if $h > 0$, $|h|$ units to the left if $h < 0$, k units upward if $k > 0$, $|k|$ units downward if $k < 0$.

Chapter 10 Test

For Exercises 1–5, $f(x) = 4x + 2$ and $g(x) = -x^2 + 3$.

1. $g(1) = -1^2 + 3 = -1 + 3 = 2$

2. $(f + g)(-2)$
$= f(-2) + g(-2)$
$= [4(-2) + 2] + [-(-2)^2 + 3]$
$= (-8 + 2) + (-4 + 3)$
$= -6 + (-1) = -7$

3. $\left(\dfrac{f}{g}\right)(3) = \dfrac{f(3)}{g(3)}$
$= \dfrac{4(3) + 2}{-3^2 + 3}$
$= \dfrac{12 + 2}{-9 + 3}$
$= \dfrac{14}{-6} = -\dfrac{7}{3}$

4. $(f \circ g)(2) = f(g(2))$
$= f(-2^2 + 3)$
$= f(-1)$
$= 4(-1) + 2$
$= -4 + 2 = -2$

5. $(f - g)(x) = f(x) - g(x)$
$= (4x + 2) - (-x^2 + 3)$
$= 4x + 2 + x^2 - 3$
$= x^2 + 4x - 1$

The domains of f and g are $(-\infty, \infty)$, so the domain of $f - g$ will also be $(-\infty, \infty)$.

6. $f(x) = a(x - h)^2 + k$
Since $a < 0$, the parabola opens down. Since $h > 0$ and $k < 0$, the x-coordinate is positive and the y-coordinate is negative. Therefore, the vertex is in quadrant IV. The correct graph is **A**.

7. $f(x) = \frac{1}{2}x^2 - 2$
$f(x) = \frac{1}{2}(x - 0)^2 - 2$
The graph is a parabola in $f(x) = a(x - h)^2 + k$ form with vertex (h, k) at $(0, -2)$. The axis is $x = 0$. Since $a = \frac{1}{2} > 0$, the parabola opens up. Also, $|a| = \left|\frac{1}{2}\right| = \frac{1}{2} < 1$, so the graph of the parabola is wider than the graph of $f(x) = x^2$. The points $(2, 0)$ and $(-2, 0)$ are on the graph.

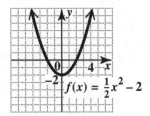

$$f(x) = \tfrac{1}{2}x^2 - 2$$

From the graph, we see that the x-values can be any real number, so the domain is $(-\infty, \infty)$. The y-values are greater than or equal to -2, so the range is $[-2, \infty)$.

8. $f(x) = -x^2 + 4x - 1$

The x-coordinate of the vertex is

$$x = \frac{-b}{2a} = \frac{-4}{2(-1)} = 2.$$

The y-coordinate of the vertex is

$$f(2) = -4 + 8 - 1 = 3.$$

The graph is a parabola with vertex (h, k) at $(2, 3)$ and axis $x = 2$. Since $a = -1 < 0$, the parabola opens down.
Also, $|a| = |-1| = 1$, so the graph has the same shape as the graph of $f(x) = x^2$. The points $(0, -1)$ and $(4, -1)$ are on the graph.

$$f(x) = -x^2 + 4x - 1$$

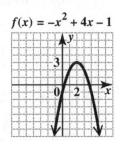

From the graph, we see that the x-values can be any real number, so the domain is $(-\infty, \infty)$.

The y-values are less than or equal to 3, so the range is $(-\infty, 3]$.

9. $x = -(y - 2)^2 + 2$
The equation is in $x = a(y - k)^2 + h$ form. The graph is a horizontal parabola with vertex (h, k) at $(2, 2)$ and axis $y = 2$. Since $a = -1 < 0$, the graph opens to the left. Also, $|a| = |-1| = 1$, so the graph has the same shape as the graph of $y = x^2$. The points $(-2, 0)$ and $(-2, 4)$ are on the same graph.

Copyright © 2012 Pearson Education, Inc. Publishing as Addison-Wesley.

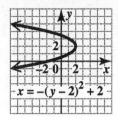

The largest value of x is 2, so the domain is $(-\infty, 2]$.

The y-values can be any real number, so the range is $(-\infty, \infty)$.

10. **(a)** Use $f(x) = -0.529x^2 + 8.00x + 115$ with $x = 4$.

$$f(4) = -0.529(4)^2 + 8.00(4) + 115$$
$$= 138.536 \approx 139 \text{ million}$$

In 2004, there were about 139 million civilians employed in the United States.

(b) Find the vertex.

$$\frac{-b}{2a} = \frac{-8.00}{2(-0.529)} \approx 7.6 \text{ (rounding)}$$

$$f(7.6) = -0.529(7.6)^2 + 8.00(7.6) + 115$$
$$\approx 145$$

In 2007, there were about 145 million civilians employed in the United States.

11. Let $x =$ the width of the lot.
Then $640 - 2x =$ the length of the lot.

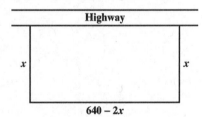

Area $\mathcal{A}$ is length times width.

$$\mathcal{A} = x(640 - 2x)$$
$$\mathcal{A}(x) = 640x - 2x^2 = -2x^2 + 640x$$

Use the vertex formula.

$$x = \frac{-b}{2a} = \frac{-640}{2(-2)} = 160$$

$$\mathcal{A}(160) = -2(160)^2 + 640(160)$$
$$= -51{,}200 + 102{,}400$$
$$= 51{,}200$$

The graph is a parabola that opens down, so the maximum occurs at the vertex $(160, 51{,}200)$. The maximum area is $51{,}200 \text{ ft}^2$ if the width x is 160 feet and the length is $640 - 2x = 640 - 2(160) = 320$ feet.

12. $f(x) = -x^2 + 1$

Replace $f(x)$ with y.

$$y = -x^2 + 1$$

(1) To test for symmetry with respect to the x-axis, replace y with $-y$.

$$-y = -x^2 + 1$$
$$y = x^2 - 1$$

The graph is not symmetric with respect to the x-axis.

(2) To test for symmetry with respect to the y-axis, replace x with $-x$.

$$y = -(-x)^2 + 1$$
$$y = -x^2 + 1$$

Therefore, the graph is symmetric with respect to the y-axis.

(3) To test for symmetry with respect to the origin, replace x with $-x$ and y with $-y$.

$$-y = -(-x)^2 + 1$$
$$-y = -x^2 + 1$$
$$y = x^2 - 1$$

The graph is not symmetric with respect to the origin.

Thus, the graph of $y = -x^2 + 1$ is symmetric with respect to the y-axis, but not with respect to the x-axis or the origin.

13. $x = y^2 + 7$

(1) Replace y with $-y$.

$$x = (-y)^2 + 7$$
$$x = y^2 + 7$$

The graph is symmetric with respect to the x-axis.

(2) Replace x with $-x$.

$$-x = y^2 + 7$$
$$x = -y^2 - 7$$

The graph is not symmetric with respect to the y-axis.

(3) Replace x with $-x$ and y with $-y$.

$$-x = (-y)^2 + 7$$
$$-x = y^2 + 7$$
$$x = -y^2 - 7$$

The graph is not symmetric with respect to the origin.

Thus, the graph of $x = y^2 + 7$ is symmetric with respect to the x-axis, but not with respect to the y-axis or the origin.

Copyright © 2012 Pearson Education, Inc. Publishing as Addison-Wesley.

14. $x^2 + y^2 = 4$

(1) Replace x with $-x$.

$$(-x)^2 + y^2 = 4$$
$$x^2 + y^2 = 4$$

The graph is symmetric with respect to the y-axis.

(2) Replace y with $-y$.

$$x^2 + (-y)^2 = 4$$
$$x^2 + y^2 = 4$$

The graph is symmetric with respect to the x-axis.

(3) Replace x with $-x$ and y with $-y$.

$$(-x)^2 + (-y)^2 = 4$$
$$x^2 + y^2 = 4$$

The graph is symmetric with respect to the origin. Thus, the graph of $x^2 + y^2 = 4$ is symmetric with respect to the x-axis, the y-axis, and the origin.

15. **(a)** $f(x) = |x - 2|$ is the graph of $g(x) = |x|$ shifted 2 units right. The graph is **C**.

(b) $f(x) = |x + 2|$ is the graph of $g(x) = |x|$ shifted 2 units left. The graph is **A**.

(c) $f(x) = |x| + 2$ is the graph of $g(x) = |x|$ shifted 2 units up. The graph is **D**.

(d) $f(x) = |x| - 2$ is the graph of $g(x) = |x|$ shifted 2 units down. The graph is **B**.

16. From the graph, we see that the function is decreasing on the interval $(-\infty, -2]$, increasing on the interval $[-2, 1]$, and constant on the interval $[1, \infty)$.

17. $f(x) = |x - 3| + 4$

This is the graph of the absolute value function, $g(x) = |x|$, shifted 3 units to the right (since $x - 3 = 0$ if $x = 3$) and 4 units upward (because of the $+4$). Its vertex is at $(3, 4)$.

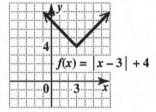

18. $f(x) = [\![2x]\!]$

To get $[\![2x]\!] = 0$, we need

$$0 \le 2x < 1$$
$$0 \le x < \tfrac{1}{2}.$$

So for x in $\left[0, \tfrac{1}{2}\right)$, $[\![2x]\!] = 0$.

To get $[\![2x]\!] = 1$, we need

$$1 \le 2x < 2$$
$$\tfrac{1}{2} \le x < 1.$$

So for x in $\left[\tfrac{1}{2}, 1\right)$, $[\![2x]\!] = 1$.

Notice that the length of each step is $\tfrac{1}{2}$. Follow this pattern to graph the step function.

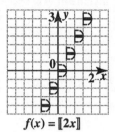

$$f(x) = [\![2x]\!]$$

19. $f(x) = \begin{cases} -x & \text{if } x \le 2 \\ x - 4 & \text{if } x > 2 \end{cases}$

This function is defined piecewise. The graph will be made up of two rays.

For $x \le 2$, graph the line $y = -x$ through $(0, 0)$ and $(1, -1)$, with endpoint at $(2, -2)$.

For $x > 2$, graph the line $y = x - 4$ through $(3, -1)$ and $(4, 0)$.

Notice that the two portions of the graph meet at the point $(2, -2)$.

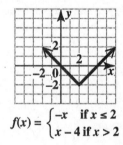

$$f(x) = \begin{cases} -x & \text{if } x \le 2 \\ x - 4 & \text{if } x > 2 \end{cases}$$

20. For any portion of the first ounce, the cost will be one 44¢ stamp. If the weight exceeds one ounce (up to two ounces), an additional 27¢ stamp is required. The following table summarizes the weight of a letter, x, and the number of stamps required, $p(x)$, on the interval $(0, 5]$.

x	$(0, 1]$	$(1, 2]$	$(2, 3]$	$(3, 4]$	$(4, 5]$
$p(x)$	1	2	3	4	5

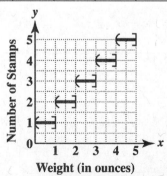

Copyright © 2012 Pearson Education, Inc. Publishing as Addison-Wesley.

Cumulative Review Exercises (Chapters 1–10)

1. $2(3x - 1) = -(4 - x) - 28$
$$6x - 2 = -4 + x - 28$$
$$6x - 2 = x - 32$$
$$5x = -30$$
$$x = -6$$

The solution set is $\{-6\}$.

2.   Solve $3p = q(p + q)$ for p.
$$3p = qp + q^2$$
$$3p - qp = q^2$$
$$p(3 - q) = q^2$$
$$p = \frac{q^2}{3 - q}$$

3. $2x + 3 \leq 5 - (x - 4)$
$$2x + 3 \leq 5 - x + 4$$
$$2x + 3 \leq -x + 9$$
$$3x \leq 6$$
$$x \leq 2$$

The solution set is $(-\infty, 2]$.

4. $2x + 1 > 5$ or $2 - x \geq 2$
$$\quad 2x > 4 \qquad\qquad -x \geq 0$$
$$\quad x > 2 \quad \text{or} \qquad x \leq 0$$

The solution set is $(-\infty, 0] \cup (2, \infty)$.

5. $|5 - 3x| = 12$

$5 - 3x = 12$ or $5 - 3x = -12$
$$-3x = 7 \qquad\qquad -3x = -17$$
$$x = -\frac{7}{3} \quad \text{or} \qquad x = \frac{17}{3}$$

The solution set is $\left\{-\frac{7}{3}, \frac{17}{3}\right\}$.

6. $|12x + 7| \geq 0$

Since the absolute value of an expression is always nonnegative (positive or zero), the inequality is true for any real number x.

The solution set is $(-\infty, \infty)$.

7.   Find the slope of the line through $(-2, 1)$ and $(5, -4)$.
$$m = \frac{y_2 - y_1}{x_2 - x_1} = \frac{-4 - 1}{5 - (-2)} = \frac{-5}{7} = -\frac{5}{7}$$

8.   Find the slope of the line with equation $2x - 3y = 6$. Solve for y.
$$-3y = -2x + 6$$
$$y = \tfrac{2}{3}x - 2$$

The slope is $\frac{2}{3}$.

9.   **(a)**   The graph of $y = -\frac{1}{2}x + 3$ is a line with slope $-\frac{1}{2}$. The slope of any line perpendicular to it has slope 2. Use the point-slope form of a line with $m = 2$ and $(x_1, y_1) = (1, 6)$.
$$y - y_1 = m(x - x_1)$$
$$y - 6 = 2(x - 1)$$
$$y - 6 = 2x - 2$$
$$y = 2x + 4$$

(b)   From part (a), $y = 2x + 4$.
$$-4 = 2x - y \qquad \textit{Standard form}$$

10. $f(x) = \sqrt{1 - 4x}$

The radicand, $1 - 4x$, must be nonnegative.
$$1 - 4x \geq 0$$
$$-4x \geq -1$$
$$x \leq \tfrac{1}{4}$$

The domain of f is $\left(-\infty, \frac{1}{4}\right]$.

11. $f(x) = x^2 - 1$
$$f(-3) = (-3)^2 - 1$$
$$= 9 - 1 = 8$$

12. $\begin{aligned} x + y + z &= 2 \quad (1) \\ 2x + y - z &= 5 \quad (2) \\ x - y + z &= -2 \quad (3) \end{aligned}$

To eliminate z, add (1) and (2)

$$\begin{array}{rl} x + y + z = 2 & (1) \\ \underline{2x + y - z = 5} & (2) \\ 3x + 2y \quad\;\; = 7 & (4) \end{array}$$

To eliminate z again, add (2) and (3).

$$\begin{array}{rl} 2x + y - z = 5 & (2) \\ \underline{x - y + z = -2} & (3) \\ 3x \qquad\qquad = 3 & (5) \end{array}$$

From (5), $x = 1$. Substitute 1 for x in (4).

$$3(1) + 2y = 7$$
$$2y = 4$$
$$y = 2$$

Substitute 1 for x and 2 for y in (1).

$$1 + 2 + z = 2$$
$$z = -1$$

The solution set is $\{(1, 2, -1)\}$.

Copyright © 2012 Pearson Education, Inc. Publishing as Addison-Wesley.

13.
$$x + 2y = 10 \quad (1)$$
$$3x - y = 9 \quad (2)$$

$$\begin{bmatrix} 1 & 2 & | & 10 \\ 3 & -1 & | & 9 \end{bmatrix}$$

$$\begin{bmatrix} 1 & 2 & | & 10 \\ 0 & -7 & | & -21 \end{bmatrix} \quad -3R_1 + R_2$$

$$\begin{bmatrix} 1 & 2 & | & 10 \\ 0 & 1 & | & 3 \end{bmatrix} \quad -\frac{1}{7}R_2$$

From the second row, $y = 3$. Substitute 3 for y in (1).

$$x + 2(3) = 10$$
$$x = 4$$

The solution set is $\{(4, 3)\}$.

14. $(4x^3 - 3x^2 + 2x + 3) \div (x - 1)$

$$\begin{array}{r}
4x^2 + x + 3 \\
x - 1 \overline{\smash{\big)}\ 4x^3 - 3x^2 + 2x - 3} \\
\underline{4x^3 - 4x^2} \\
x^2 + 2x \\
\underline{x^2 - x} \\
3x - 3 \\
\underline{3x - 3} \\
0
\end{array}$$

Answer: $4x^2 + x + 3$

15. $\dfrac{3p}{p+1} + \dfrac{4}{p-1} - \dfrac{6}{p^2-1}$

$$= \frac{3p(p-1)}{(p+1)(p-1)} + \frac{4(p+1)}{(p-1)(p+1)} - \frac{6}{p^2-1}$$

$$LCD = (p+1)(p-1)$$

$$= \frac{(3p^2 - 3p) + (4p + 4) - 6}{(p+1)(p-1)}$$

$$= \frac{3p^2 + p - 2}{(p+1)(p-1)}$$

$$= \frac{(3p - 2)(p + 1)}{(p+1)(p-1)}$$

$$= \frac{3p - 2}{p - 1}$$

16. $\dfrac{6x - 10}{3x} \cdot \dfrac{6}{18x - 30}$

$$= \frac{2(3x - 5)}{3x} \cdot \frac{6}{6(3x - 5)}$$

$$= \frac{2 \cdot 6(3x - 5)}{3x \cdot 6(3x - 5)}$$

$$= \frac{2}{3x}$$

17. $x^6 - y^6 = (x^3)^2 - (y^3)^2$

$$= (x^3 + y^3)(x^3 - y^3)$$

$$= (x + y)(x^2 - xy + y^2)$$

$$\quad \cdot (x - y)(x^2 + xy + y^2)$$

18. $2k^4 + k^2 - 3 = (2k^2 + 3)(k^2 - 1)$

$$= (2k^2 + 3)(k + 1)(k - 1)$$

19. $\dfrac{2(m^{-1})^{-4}}{(3m^{-3})^2} = \dfrac{2m^4}{3^2m^{-6}}$

$$= \frac{2m^4 m^6}{3^2}$$

$$= \frac{2}{9}m^{10}$$

20. $\sqrt[3]{k^2} \cdot \sqrt{k} = k^{2/3} \cdot k^{1/2}$

$$= k^{(2/3) + (1/2)}$$

$$= k^{(4/6) + (3/6)}$$

$$= k^{7/6}$$

In radical form, this is

$$\sqrt[6]{k^7} \quad \text{or} \quad k\sqrt[6]{k}.$$

21.
$$2x^3 - x^2 - 28x = 0$$
$$x(2x^2 - x - 28) = 0$$
$$x(2x + 7)(x - 4) = 0$$

$$x = 0 \quad \text{or} \quad 2x + 7 = 0 \quad \text{or} \quad x - 4 = 0$$
$$x = -\frac{7}{2} \qquad x = 4$$

The solution set is $\left\{-\frac{7}{2}, 0, 4\right\}$.

22.
$$\frac{4}{x+2} - \frac{11}{9} = \frac{5}{3x+6}$$

Multiply by the LCD, $9(x + 2)$.

$$9(x+2)\left(\frac{4}{x+2} - \frac{11}{9}\right) = 9(x+2)\left[\frac{5}{3(x+2)}\right]$$
$$9(4) - 11(x+2) = 3(5)$$
$$36 - 11x - 22 = 15$$
$$-1 = 11x$$
$$-\frac{1}{11} = x$$

Check $x = -\frac{1}{11}$: $\quad \frac{44}{21} - \frac{11}{9} = \frac{55}{63}$ *True*

The solution set is $\left\{-\frac{1}{11}\right\}$.

23. $\sqrt{x + 1} - \sqrt{x - 2} = 1$

$$\sqrt{x + 1} = 1 + \sqrt{x - 2}$$

Square both sides.

$$\left(\sqrt{x + 1}\right)^2 = \left(1 + \sqrt{x - 2}\right)^2$$
$$x + 1 = 1 + 2\sqrt{x - 2} + x - 2$$
$$2 = 2\sqrt{x - 2}$$
$$1 = \sqrt{x - 2}$$

Square both sides.

$$1 = x - 2$$
$$3 = x$$

Check $x = 3$: $\quad \sqrt{4} - \sqrt{1} = 1$ *True*

The solution set is $\{3\}$.

Copyright © 2012 Pearson Education, Inc. Publishing as Addison-Wesley.

24. $5x^2 + 3x = 3$

$5x^2 + 3x - 3 = 0$

Use the quadratic formula with $a = 5$, $b = 3$, and $c = -3$.

$$x = \frac{-b \pm \sqrt{b^2 - 4ac}}{2a}$$

$$= \frac{-3 \pm \sqrt{3^2 - 4(5)(-3)}}{2(5)}$$

$$= \frac{-3 \pm \sqrt{69}}{10}$$

The solution set is $\left\{ \frac{-3 \pm \sqrt{69}}{10} \right\}$.

25. Solve $S = 2\pi rh + \pi r^2$ for r.

We must recognize this equation as a quadratic equation in r.

$$\pi r^2 + (2\pi h)r - S = 0$$

Use the quadratic formula with $a = \pi$, $b = 2\pi h$, and $c = -S$.

$$r = \frac{-b \pm \sqrt{b^2 - 4ac}}{2a}$$

$$= \frac{-2\pi h \pm \sqrt{(2\pi h)^2 - 4(\pi)(-S)}}{2(\pi)}$$

$$= \frac{-2\pi h \pm \sqrt{4\pi^2 h^2 + 4\pi S}}{2\pi}$$

$$= \frac{-2\pi h \pm \sqrt{4(\pi^2 h^2 + \pi S)}}{2\pi}$$

$$= \frac{-2\pi h \pm 2\sqrt{\pi^2 h^2 + \pi S}}{2\pi}$$

$$= \frac{-\pi h \pm \sqrt{\pi^2 h^2 + \pi S}}{\pi}$$

26. $3x^2 - 8 > 10x$

Solve the equation

$3x^2 - 10x - 8 = 0$.

$(3x + 2)(x - 4) = 0$

$3x + 2 = 0 \quad$ or $\quad x - 4 = 0$

$x = -\frac{2}{3} \quad$ or $\quad x = 4$

The numbers $-\frac{2}{3}$ and 4 divide a number line into 3 intervals:

$$
\begin{array}{ccc}
A & B & C \\
\hline
& | & | \\
& -\frac{2}{3} & 4
\end{array}
$$

Test a number from each interval in the original inequality, $3x^2 - 8 > 10x$.

Interval A: Let $x = -1$.

$3 - 8 \overset{?}{>} -10$

$-5 > -10 \quad$ *True*

Interval B: Let $x = 0$.

$-8 > 0 \quad$ *False*

Interval C: Let $x = 5$.

$3(25) - 8 \overset{?}{>} 10(5)$

$67 > 50 \quad$ *True*

The solution set includes the numbers in Intervals A and C, excluding $-\frac{2}{3}$ and 4 because of $>$.

The solution set is $\left(-\infty, -\frac{2}{3}\right) \cup (4, \infty)$.

27. $h = -16t^2 + 256t$

(a) The rocket returns to the ground when $h = 0$.

$0 = -16t^2 + 256t$

$0 = -16t(t - 16)$

$t = 0 \quad$ or $\quad t = 16$

The rocket returns to the ground in 16 seconds.

(b) $768 = -16t^2 + 256t \qquad$ *Let $h = 768$.*

$0 = -16t^2 + 256t - 768$

$0 = -16(t^2 - 16t + 48)$

$0 = -16(t - 4)(t - 12)$

$t = 4 \quad$ or $\quad t = 12$

The rocket will be 768 feet above the ground after 4 seconds and after 12 seconds.

28. $2x - 3y = 6$

This is a linear equation, so its graph is a line. The x-intercept is $(3, 0)$ and the y-intercept is $(0, -2)$.

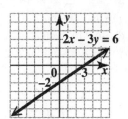

29. $x + 2y \le 4$

Graph the solid line $x + 2y = 4$ through its intercepts, $(4, 0)$ and $(0, 2)$. Testing $(0, 0)$ gives us $0 \le 4$, a true statement, so shade the region containing $(0, 0)$, that is, the region below the line.

The solution set is the shaded region below the line and the line.

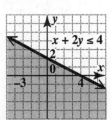

Copyright © 2012 Pearson Education, Inc. Publishing as Addison-Wesley.

30. $f(x) = -2x^2 + 5x + 3$

This is a quadratic function, so its graph is a parabola. The parabola opens down since $a = -2 < 0$. The y-intercept is $(0, 3)$. Find the x-intercepts by solving $f(x) = 0$.

$$-2x^2 + 5x + 3 = 0$$
$$2x^2 - 5x - 3 = 0$$
$$(2x + 1)(x - 3) = 0$$

The x-intercepts are $\left(-\frac{1}{2}, 0\right)$ and $(3, 0)$.
The x-value of the vertex is

$$x = -\frac{b}{2a} = \frac{-5}{2(-2)} = \frac{5}{4}.$$

The y-value of the vertex is

$$f\left(\tfrac{5}{4}\right) = \tfrac{49}{8} = 6.125.$$

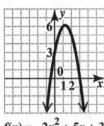

$$f(x) = -2x^2 + 5x + 3$$

31. $f(x) = |x + 1|$

Since x can be any real number, the domain is $(-\infty, \infty)$.

The value of y is always greater than or equal to 0, so the range is $[0, \infty)$.

The graph of $y = |x + 1|$ looks like the graph of the absolute value function $y = |x|$, but the graph is translated 1 unit to the left. The x-value of its "vertex" is obtained by setting $x + 1 = 0$ and solving for x:

$$x + 1 = 0$$
$$x = -1.$$

Since the corresponding y-value is 0, the "vertex" is $(-1, 0)$. Some additional points are $(-3, 2)$, $(-2, 1)$, $(0, 1)$, and $(1, 2)$.

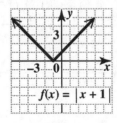

$$f(x) = |x + 1|$$

32. $y^2 = x + 3$

(1) Replace y with $-y$.

$$(-y)^2 = x + 3$$
$$y^2 = x + 3$$

The graph is symmetric with respect to the x-axis.

(2) Replace x with $-x$.

$$y^2 = -x + 3$$

The graph is not symmetric with respect to the y-axis.

(3) Replace x with $-x$ and y with $-y$.

$$(-y)^2 = -x + 3$$
$$y^2 = -x + 3$$

The graph is not symmetric with respect to the origin.

Thus, the graph is symmetric with respect to the x-axis, but not with respect to the y-axis or the origin.

33. $x^2 - 6y^2 = 18$

(1) Replace y with $-y$.

$$x^2 - 6(-y)^2 = 18$$
$$x^2 - 6y^2 = 18$$

The graph is symmetric with respect to the x-axis.

(2) Replace x with $-x$.

$$(-x)^2 - 6y^2 = 18$$
$$x^2 - 6y^2 = 18$$

The graph is symmetric with respect to the y-axis.

(3) Replace x with $-x$ and y with $-y$.

$$(-x)^2 - 6(-y)^2 = 18$$
$$x^2 - 6y^2 = 18$$

The graph is symmetric with respect to the origin. Thus, the graph is symmetric with respect to the x-axis, the y-axis, and the origin.

34. $f(x) = [\![x + 2]\!]$

(a) $f(3) = [\![3 + 2]\!] = [\![5]\!] = 5$

(b) $f(-3.1) = [\![-3.1 + 2]\!] = [\![-1.1]\!] = -2$

Copyright © 2012 Pearson Education, Inc. Publishing as Addison-Wesley.

CHAPTER 11 INVERSE, EXPONENTIAL, AND LOGARITHMIC FUNCTIONS

11.1 Inverse Functions

11.1 Now Try Exercises

N1. **(a)** $F = \{(-1,-2), (0,0), (1,-2), (2,-8)\}$

Since the y-value -2 corresponds to x-values -1 and 1, the function is not one-to-one. Therefore, it does not have an inverse.

(b) $G = \{(0,0), (1,1), (4,2), (9,3)\}$

Each x-value in the relation corresponds to only one y-value, so the relation is indeed a function. Furthermore, each y-value corresponds to only one x-value, so this is a one-to-one function.

The inverse function is found by interchanging the x- and y-values in each ordered pair.

The inverse is

$$G^{-1} = \{(0,0), (1,1), (2,4), (3,9)\}.$$

(c) Since the height 639 corresponds to the number of stories 31 and 41, the function is not one-to-one. Therefore, it does not have an inverse.

N2. **(a)** Since a horizontal line will intersect the graph in no more than one point, the function is one-to-one.

(b) Since a horizontal line could intersect the graph in two points, the function is not one-to-one.

N3. To find the equation of the inverse of a function,

  (1) determine if the function is one-to-one;
  (2) replace $f(x)$ with y;
  (3) interchange x and y;   *Step 1*
  (4) solve for y;   *Step 2*
  (5) replace y with $f^{-1}(x)$. *Step 3*

(a) $f(x) = 5x - 7$

The graph of f is a line. By the horizontal line test, it is a one-to-one function. To find the inverse, first replace $f(x)$ with y.

Step 1 Interchange x and y.
$$y = 5x - 7$$
$$x = 5y - 7$$

Step 2 Solve for y.
$$x + 7 = 5y$$
$$\frac{x+7}{5} = y$$

Step 3 Replace y with $f^{-1}(x)$.
$$f^{-1}(x) = \frac{x+7}{5}, \text{ or } f^{-1}(x) = \frac{1}{5}x + \frac{7}{5}$$

(b) $f(x) = (x+1)^2$

This equation has a vertical parabola as its graph, so some horizontal lines will intersect the graph at two points. For example, both $x = -2$ and $x = 0$ correspond to $y = 1$. Thus, the function is not a one-to-one function and does not have an inverse.

(c) $f(x) = x^3 - 4$

Refer to Section 5.3 to see from its graph that a cubing function like this is a one-to-one function. To find the inverse, first replace $f(x)$ with y.

Step 1 Interchange x and y.
$$y = x^3 - 4$$
$$x = y^3 - 4$$

Step 2 Solve for y.
$$x + 4 = y^3$$
Take the cube root on each side.
$$\sqrt[3]{x+4} = y$$

Step 3 Replace y with $f^{-1}(x)$.
$$f^{-1}(x) = \sqrt[3]{x+4}$$

N4. The given graph goes through $(-3,-2)$, $(-2,-1)$, $(-1,2)$, and $(0,7)$. So, the inverse will go through $(-2,-3)$, $(-1,-2)$, $(2,-1)$, and $(7,0)$. Use these points and symmetry about $y = x$ to complete the graph of the inverse.

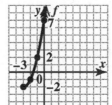 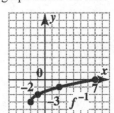

11.1 Section Exercises

1. This function is not one-to-one because both France and the United States are paired with the same trans fat percentage, 11.

3. The function in the table that pairs a city with a distance is a one-to-one function because for each city there is one distance and each distance has only one city to which it is paired.

If the distance from Indianapolis to Denver had 1 mile added to it, it would be $1058 + 1 = 1059$ mi, the same as the distance from Los Angeles to Denver. In this case, one distance would have two cities to which it is paired, and the function would not be one-to-one.

5. If a function is made up of ordered pairs in such a way that the same y-value appears in a correspondence with two different x-values, then the function is not one-to-one. Choice **B**

Copyright © 2012 Pearson Education, Inc. Publishing as Addison-Wesley.

7. All of the graphs pass the vertical line test, so they all represent functions. The graph in choice **A** is the only one that passes the horizontal line test, so it is the one-to-one function.

9. $\{(3,6), (2,10), (5,12)\}$ is a one-to-one function, since each x-value corresponds to only one y-value and each y-value corresponds to only one x-value. To find the inverse, interchange x and y in each ordered pair. The inverse is

$$\{(6,3), (10,2), (12,5)\}.$$

11. $\{(-1,3), (2,7), (4,3), (5,8)\}$ is not a one-to-one function. The ordered pairs $(-1,3)$ and $(4,3)$ have the same y-value for two different x-values.

13. The graph of $f(x) = 2x + 4$ is a nonvertical, nonhorizontal line. By the horizontal line test, $f(x)$ is a one-to-one function. To find the inverse, replace $f(x)$ with y.

$$y = 2x + 4$$

Interchange x and y.

$$x = 2y + 4$$

Solve for y.

$$2y = x - 4$$

$$y = \frac{x - 4}{2}$$

Replace y with $f^{-1}(x)$.

$$f^{-1}(x) = \frac{x-4}{2}, \quad \text{or} \quad f^{-1}(x) = \frac{1}{2}x - 2$$

15. Write $g(x) = \sqrt{x-3}$ as $y = \sqrt{x-3}$. Since $x \geq 3$, $y \geq 0$. The graph of g is half of a horizontal parabola that opens to the right. The graph passes the horizontal line test, so g is one-to-one. To find the inverse, interchange x and y to get

$$x = \sqrt{y-3}.$$

Note that now $y \geq 3$, so $x \geq 0$. Solve for y by squaring each side.

$$x^2 = y - 3$$
$$x^2 + 3 = y$$

Replace y with $g^{-1}(x)$.

$$g^{-1}(x) = x^2 + 3, \, x \geq 0$$

17. $f(x) = 3x^2 + 2$ is not a one-to-one function because two x-values, such as 1 and -1, both have the same y-value, in this case 5. The graph of this function is a vertical parabola which does not pass the horizontal line test.

19. The graph of $f(x) = x^3 - 4$ is the graph of $g(x) = x^3$ shifted down 4 units. (Recall that $g(x) = x^3$ is the elongated S-shaped curve.) The graph of f passes the horizontal line test, so f is one-to-one.

Replace $f(x)$ with y.

$$y = x^3 - 4$$

Interchange x and y.

$$x = y^3 - 4$$

Solve for y.

$$x + 4 = y^3$$

Take the cube root of each side.

$$\sqrt[3]{x+4} = y$$

Replace y with $f^{-1}(x)$.

$$f^{-1}(x) = \sqrt[3]{x+4}$$

In Exercises 21–24, $f(x) = 2^x$ is a one-to-one function.

21. **(a)** To find $f(3)$, substitute 3 for x.

$$f(x) = 2^x, \text{ so } f(3) = 2^3 = 8.$$

(b) Since f is one-to-one and $f(3) = 8$, it follows that $f^{-1}(8) = 3$.

23. **(a)** To find $f(0)$, substitute 0 for x.

$$f(x) = 2^x, \text{ so } f(0) = 2^0 = 1.$$

(b) Since f is one-to-one and $f(0) = 1$, it follows that $f^{-1}(1) = 0$.

25. **(a)** The function is one-to-one since any horizontal line intersects the graph at most once.

(b) In the graph, the two points marked on the line are $(-1, 5)$ and $(2, -1)$. Interchange x and y in each ordered pair to get $(5, -1)$ and $(-1, 2)$. Plot these points, then draw a dashed line through them to obtain the graph of the inverse function.

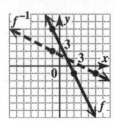

27. **(a)** The function is not one-to-one since there are horizontal lines that intersect the graph more than once. For example, the line $y = 1$ intersects the graph twice.

Copyright © 2012 Pearson Education, Inc. Publishing as Addison-Wesley.

29. **(a)** The function is one-to-one since any horizontal line intersects the graph at most once.

(b) In the graph, the four points marked on the curve are $(-4, 2)$, $(-1, 1)$, $(1, -1)$, and $(4, -2)$. Interchange x and y in each ordered pair to get $(2, -4)$, $(1, -1)$, $(-1, 1)$, and $(-2, 4)$. Plot these points, then draw a dashed curve (symmetric to the original graph about the line $y = x$) through them to obtain the graph of the inverse.

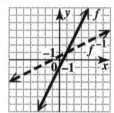

31. $f(x) = 2x - 1$ or $y = 2x - 1$
The graph is a line through $(-2, -5)$, $(0, -1)$, and $(3, 5)$. Plot these points and draw the solid line through them. Then the inverse will be a line through $(-5, -2)$, $(-1, 0)$, and $(5, 3)$. Plot these points and draw the dashed line through them.

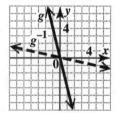

33. $g(x) = -4x$ or $y = -4x$
The graph is a line through $(0, 0)$ and $(1, -4)$. For the inverse, interchange x and y in each ordered pair to get the points $(0, 0)$ and $(-4, 1)$. Draw a dashed line through these points to obtain the graph of the inverse function.

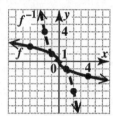

35. $f(x) = \sqrt{x}$, $x \geq 0$
Complete the table of values.

x	$f(x)$
0	0
1	1
4	2

Plot these points and connect them with a solid smooth curve.
Since $f(x)$ is one-to-one, make a table of values for $f^{-1}(x)$ by interchanging x and y.

x	$f^{-1}(x)$
0	0
1	1
2	4

Plot these points and connect them with a dashed smooth curve.

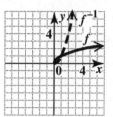

37. $f(x) = x^3 - 2$
Complete the table of values.

x	$f(x)$
-1	-3
0	-2
1	-1
2	6

Plot these points and connect them with a solid smooth curve.
Make a table of values for f^{-1}.

x	$f^{-1}(x)$
-3	-1
-2	0
-1	1
6	2

Plot these points and connect them with a dashed smooth curve.

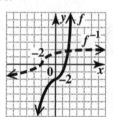

♦♦♦ Relating Concepts 39–42 ♦♦♦

39. $f(x) = 4x - 5$
Replace $f(x)$ with y.
         $y = 4x - 5$
Interchange x and y.
         $x = 4y - 5$
Solve for y.
         $x + 5 = 4y$
         $\dfrac{x + 5}{4} = y$
Replace y with $f^{-1}(x)$.
         $\dfrac{x + 5}{4} = f^{-1}(x)$,
or $f^{-1}(x) = \dfrac{1}{4}x + \dfrac{5}{4}$

Copyright © 2012 Pearson Education, Inc. Publishing as Addison-Wesley.

40. Insert each number in the inverse function found in Exercise 39,

$$f^{-1}(x) = \frac{x+5}{4}.$$

$$f^{-1}(47) = \frac{47+5}{4} = \frac{52}{4} = 13 = M,$$

$$f^{-1}(95) = \frac{95+5}{4} = \frac{100}{4} = 25 = Y,$$

and so on.

The decoded message is as follows:
My graphing calculator is the greatest thing since sliced bread.

41. A one-to-one code is essential to this process because if the code is not one-to-one, an encoded number would refer to two different letters.

42. Answers will vary according to the student's name. For example, Jane Doe is encoded as follows:

1004 5 2748 129 68 3379 129.

43. $Y_1 = f(x) = 2x - 7$
Replace $f(x)$ with y.
$$y = 2x - 7$$
Interchange x and y.
$$x = 2y - 7$$
Solve for y.
$$x + 7 = 2y$$
$$\frac{x+7}{2} = y$$
Replace y with $f^{-1}(x)$.
$$\frac{x+7}{2} = f^{-1}(x) = Y_2 \left(\text{or } \frac{1}{2}x + \frac{7}{2} \right)$$

Now graph Y_1 and Y_2.

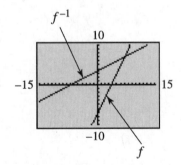

45. $Y_1 = f(x) = x^3 + 5$
Replace $f(x)$ with y.
$$y = x^3 + 5$$
Interchange x and y.
$$x = y^3 + 5$$
Solve for y.
$$x - 5 = y^3$$
Take the cube root of each side.
$$\sqrt[3]{x-5} = y$$

Replace y with $f^{-1}(x)$.
$$\sqrt[3]{x-5} = f^{-1}(x) = Y_2$$

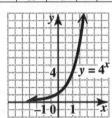

47. $f(x) = 4^x$, so $f(3) = 4^3 = 64$.

49. $f(x) = 4^x$, so $f\left(-\frac{1}{2}\right) = 4^{-1/2} = \frac{1}{4^{1/2}} = \frac{1}{2}$.

11.2 Exponential Functions

11.2 Now Try Exercises

In Now Try Exercises 1–3, choose values of x and find the corresponding values of y. Then plot the points, and draw a smooth curve through them.

N1. $f(x) = 4^x$

x	-2	-1	0	1	2
$f(x) = 4^x$	$\frac{1}{16}$	$\frac{1}{4}$	1	4	16

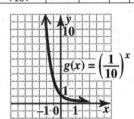

N2. $g(x) = \left(\frac{1}{10}\right)^x$

x	-2	-1	0	1	2
$g(x) = \left(\frac{1}{10}\right)^x$	100	10	1	$\frac{1}{10}$	$\frac{1}{100}$

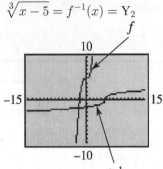

N3. $y = f(x) = 4^{2x-1}$

Make a table of values. It will help to find values for $2x - 1$ before you find y.

x	$2x-1$	$y = 4^{2x-1}$
0	-1	$\frac{1}{4}$
$\frac{1}{2}$	0	1
1	1	4
-1	-3	$\frac{1}{64}$

Copyright © 2012 Pearson Education, Inc. Publishing as Addison-Wesley.

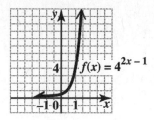

N4. $8^x = 16$

Step 1 Write each side with the base 2.
$$(2^3)^x = 2^4$$

Step 2 Simplify exponents.
$$2^{3x} = 2^4$$

Step 3 Set the exponents equal.
$$3x = 4$$

Step 4 Solve.
$$x = \frac{4}{3}$$

Check $x = \frac{4}{3}$ in the original equation.

$$8^{4/3} \overset{?}{=} 16$$
$$(8^{1/3})^4 \overset{?}{=} 16$$
$$2^4 \overset{?}{=} 16$$
$$16 = 16 \qquad \textit{True}$$

The solution set is $\left\{\frac{4}{3}\right\}$.

N5. **(a)** $3^{2x-1} = 27^{x+4}$

$$(3)^{2x-1} = (3^3)^{x+4} \quad \textit{Same base}$$
$$3^{2x-1} = 3^{3x+12} \quad \textit{Simplify exponents.}$$
$$2x - 1 = 3x + 12 \quad \textit{Set exponents equal.}$$
$$-13 = x$$

Check $x = -13$:

$$3^{-26-1} \overset{?}{=} 27^{-13+4}$$
$$3^{-27} \overset{?}{=} (3^3)^{-9}$$
$$3^{-27} = 3^{-27} \qquad \textit{True}$$

The solution set is $\{-13\}$.

(b) $5^x = \frac{1}{625}$

$$5^x = \frac{1}{5^4}$$
$$5^x = 5^{-4} \quad \textit{Same base}$$
$$x = -4 \quad \textit{Set exponents equal.}$$

Check $x = -4$: $\quad 5^{-4} = \frac{1}{5^4} = \frac{1}{625}$

The solution set is $\{-4\}$.

(c) $\left(\frac{2}{7}\right)^x = \frac{343}{8}$

$$\left(\frac{2}{7}\right)^x = \left(\frac{7}{2}\right)^3$$
$$\left(\frac{2}{7}\right)^x = \left(\frac{2}{7}\right)^{-3}$$
$$x = -3 \quad \textit{Set exponents equal.}$$

Check $x = -3$: $\quad \left(\frac{2}{7}\right)^{-3} = \left(\frac{7}{2}\right)^3 = \frac{343}{8}$

The solution set is $\{-3\}$.

N6. $x = 2000 - 1750 = 250$

$$f(x) = 266(1.001)^x$$
$$f(250) = 266(1.001)^{250}$$
$$f(250) \approx 342 \text{ parts per million}$$

N7. $f(x) = 1038(1.000134)^{-x}$
$$f(6000) = 1038(1.000134)^{-6000}$$
$$f(6000) \approx 465$$

The pressure is approximately 465 millibars.

11.2 Section Exercises

1. Since the graph of $F(x) = a^x$ always contains the point $(0, 1)$, the correct response is **C**.

3. Since the graph of $F(x) = a^x$ always approaches the x-axis, the correct response is **A**.

5. $f(x) = 3^x$
Make a table of values.
$$f(-2) = 3^{-2} = \frac{1}{3^2} = \frac{1}{9},$$
$$f(-1) = 3^{-1} = \frac{1}{3^1} = \frac{1}{3}, \text{ and so on.}$$

x	-2	-1	0	1	2
$f(x)$	$\frac{1}{9}$	$\frac{1}{3}$	1	3	9

Plot the points from the table and draw a smooth curve through them.

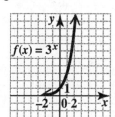

7. $g(x) = \left(\frac{1}{3}\right)^x$
Make a table of values.
$$g(-2) = \left(\frac{1}{3}\right)^{-2} = \left(\frac{3}{1}\right)^2 = 9,$$
$$g(-1) = \left(\frac{1}{3}\right)^{-1} = \left(\frac{3}{1}\right)^1 = 3, \text{ and so on.}$$

x	-2	-1	0	1	2
$g(x)$	9	3	1	$\frac{1}{3}$	$\frac{1}{9}$

Plot the points from the table and draw a smooth curve through them.

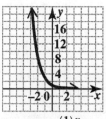

$$g(x) = \left(\frac{1}{3}\right)^x$$

Copyright © 2012 Pearson Education, Inc. Publishing as Addison-Wesley.

9. $y = 4^{-x}$

This equation can be rewritten as

$$y = (4^{-1})^x = \left(\tfrac{1}{4}\right)^x,$$

which shows that it is *falling* from left to right.
Make a table of values.

x	-2	-1	0	1	2
y	16	4	1	$\frac{1}{4}$	$\frac{1}{16}$

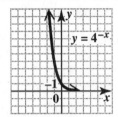

11. $y = 2^{2x-2}$

Make a table of values. It will help to find values
for $2x - 2$ before you find y.

x	-2	-1	0	1	2	3
$2x - 2$	-6	-4	-2	0	2	4
y	$\frac{1}{64}$	$\frac{1}{16}$	$\frac{1}{4}$	1	4	16

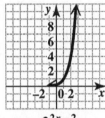

$y = 2^{2x-2}$

13.   For an exponential function defined by $f(x) = a^x$,
if $a > 1$, the graph <u>rises</u> from left to right. (See
Example 1, $f(x) = 2^x$, in your text.) If
$0 < a < 1$, the graph <u>falls</u> from left to right. (See
Example 2, $g(x) = \left(\tfrac{1}{2}\right)^x = 2^{-x}$, in your text.)

15. $6^x = 36$

Write each side as a power of 6.

$$6^x = 6^2$$

For $a > 0$ and $a \neq 1$, if $a^x = a^y$, then $x = y$.
Set the exponents equal to each other.

$$x = 2$$

Check $x = 2$: $6^2 = 36$ *True*
The solution set is $\{2\}$.

17. $100^x = 1000$

Write each side as a power of 10.

$$\left(10^2\right)^x = 10^3$$
$$10^{2x} = 10^3$$

For $a > 0$ and $a \neq 1$, if $a^x = a^y$, then $x = y$.
Set the exponents equal to each other.

$$2x = 3$$
$$x = \tfrac{3}{2}$$

Check $x = \tfrac{3}{2}$: $100^{3/2} = 1000$ *True*
The solution set is $\left\{\tfrac{3}{2}\right\}$.

19. $16^{2x+1} = 64^{x+3}$

Write each side as a power of 4.

$$\left(4^2\right)^{2x+1} = \left(4^3\right)^{x+3}$$
$$4^{4x+2} = 4^{3x+9}$$

Set the exponents equal.

$$4x + 2 = 3x + 9$$
$$x = 7$$

Check $x = 7$: $16^{15} = 64^{10}$ *True*
The solution set is $\{7\}$.

21. $5^x = \frac{1}{125}$

$$5^x = \left(\tfrac{1}{5}\right)^3$$

Write each side as a power of 5.

$$5^x = 5^{-3}$$

Set the exponents equal.

$$x = -3$$

Check $x = -3$: $5^{-3} = \frac{1}{125}$ *True*
The solution set is $\{-3\}$.

23. $5^x = 0.2$

$$5^x = \tfrac{2}{10} = \tfrac{1}{5}$$

Write each side as a power of 5.

$$5^x = 5^{-1}$$

Set the exponents equal.

$$x = -1$$

Check $x = -1$: $5^{-1} = 0.2$ *True*
The solution set is $\{-1\}$.

25. $\left(\tfrac{3}{2}\right)^x = \tfrac{8}{27}$

$$\left(\tfrac{3}{2}\right)^x = \left(\tfrac{2}{3}\right)^3$$

Write each side as a power of $\tfrac{3}{2}$.

$$\left(\tfrac{3}{2}\right)^x = \left(\tfrac{3}{2}\right)^{-3}$$

Set the exponents equal.

$$x = -3$$

Check $x = -3$: $\left(\tfrac{3}{2}\right)^{-3} = \tfrac{8}{27}$ *True*
The solution set is $\{-3\}$.

27. $12^{2.6} \approx 639.545$

29. $0.5^{3.921} \approx 0.066$

31. $2.718^{2.5} \approx 12.179$

33.   **(a)** $y = (1.046 \times 10^{-38})(1.0444^x)$
$\qquad = (1.046 \times 10^{-38})(1.0444^{2000})$
$\qquad \approx 0.57 \approx 0.6$

The increase for the exponential function in the
year 2000 is about 0.6°C.

Copyright © 2012 Pearson Education, Inc. Publishing as Addison-Wesley.

(b) $y = 0.009x - 17.67$
$= 0.009(2000) - 17.67$
$= 0.33 \approx 0.3$

The increase for the linear function in the year 2000 is about 0.3°C.

35. **(a)** $y = (1.046 \times 10^{-38})(1.0444^x)$
$= (1.046 \times 10^{-38})(1.0444^{2020})$
$\approx 1.35 \approx 1.4$

The increase for the exponential function in the year 2020 is about 1.4°C.

(b) $y = 0.009x - 17.67$
$= 0.009(2020) - 17.67$
$= 0.51 \approx 0.5$

The increase for the linear function in the year 2020 is about 0.5°C.

37. $f(x) = 4231(1.0174)^x$

(a) 1980 corresponds to $x = 10$.

$$f(10) = 4231(1.0174)^{10}$$
$$\approx 5028$$

The answer has units in millions of metric tons.

(b) 1995 corresponds to $x = 25$.

$$f(25) = 4231(1.0174)^{25} \approx 6512$$

(c) 2000 corresponds to $x = 30$.

$$f(30) = 4231(1.0174)^{30} \approx 7099$$

The actual amount, 6735 millions of metric tons, is less than the 7099 millions of metric tons that the model provides.

39. $V(t) = 5000(2)^{-0.15t}$
(a) The original value is found when $t = 0$.

$$V(0) = 5000(2)^{-0.15(0)}$$
$$= 5000(2)^0$$
$$= 5000(1) = 5000$$

The original value is $5000.

(b) The value after 5 years is found when $t = 5$.

$$V(5) = 5000(2)^{-0.15(5)}$$
$$= 5000(2)^{-0.75} \approx 2973.02$$

The value after 5 years is about $2973.

(c) The value after 10 years is found when $t = 10$.

$$V(10) = 5000(2)^{-0.15(10)}$$
$$= 5000(2)^{-1.5} \approx 1767.77$$

The value after 10 years is about $1768.

(d) Use the results of parts (a) – (c) to make a table of values.

t	0	5	10
$V(t)$	5000	2973	1768

Plot the points from the table and draw a smooth curve through them.

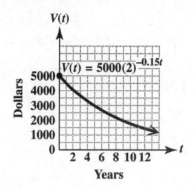

41. $V(t) = 5000(2)^{-0.15t}$
$2500 = 5000(2)^{-0.15t}$ *Let V(t) = 2500.*
$\frac{1}{2} = (2)^{-0.15t}$ *Divide by 5000.*
$2^{-1} = 2^{-0.15t}$
$-1 = -0.15t$ *Equate exponents.*
$t = \dfrac{-1}{-0.15} \approx 6.67$

The value of the machine will be $2500 in approximately 6.67 years after it was purchased.

43. $16 = 2 \cdot 2 \cdot 2 \cdot 2 = 2^4$, so $\square = 4$.

45. $2^0 = 1$, so $\square = 0$.

11.3 Logarithmic Functions

11.3 Now Try Exercises

N1. **(a)** For the exponential form $6^3 = 216$, the logarithmic form is $\log_6 216 = 3$.

(b) For the logarithmic form $\log_{64} 4 = \frac{1}{3}$, the exponential form is $64^{1/3} = 4$.

N2. **(a)** $\log_2 x = -5$
Write in exponential form.
$x = 2^{-5}$
$x = \dfrac{1}{2^5} = \dfrac{1}{32}$

The argument (the input of the logarithm) must be a positive number, so $x = \frac{1}{32}$ is acceptable.

The solution set is $\left\{ \frac{1}{32} \right\}$.

(b) $\log_{3/2}(2x - 1) = 3$
$2x - 1 = \left(\frac{3}{2}\right)^3$ *Exponential form*
$2x - 1 = \frac{27}{8}$ *Apply the exponent.*
$2x = \frac{35}{8}$ *Add 1.*
$x = \frac{35}{16}$ *Divide by 2.*

Copyright © 2012 Pearson Education, Inc. Publishing as Addison-Wesley.

Substitute $\frac{35}{16}$ for x in the argument, $2x - 1$,

$2\left(\frac{35}{16}\right) - 1 = \frac{35}{8} - 1 = \frac{27}{8} > 0$, so $\frac{35}{16}$ is acceptable.

The solution set is $\left\{\frac{35}{16}\right\}$.

(c) $\log_x 10 = 2$

$\quad\quad x^2 = 10 \quad\quad$ *Exponential form*

$\quad\quad x = \pm\sqrt{10} \quad$ *Take square roots.*

Reject $x = -\sqrt{10}$ since the base of a logarithm must be positive and not equal to 1.

The solution set is $\left\{\sqrt{10}\right\}$.

(d) $\log_{125} \sqrt[3]{5} = x$

$\quad\quad 125^x = \sqrt[3]{5} \quad$ *Exponential form*

$\quad\quad \left(5^3\right)^x = 5^{1/3} \quad$ *Same base*

$\quad\quad 5^{3x} = 5^{1/3} \quad$ *Power rule*

$\quad\quad 3x = \frac{1}{3} \quad\quad$ *Equate exponents.*

$\quad\quad x = \frac{1}{9} \quad\quad$ *Divide by 3.*

The solution set is $\left\{\frac{1}{9}\right\}$.

N3. For each expression, use the properties of logarithms,

$$\log_b b = 1 \quad \text{and} \quad \log_b 1 = 0,$$

with $b > 0$ and $b \neq 1$.

(a) $\log_{10} 10 = 1$ **(b)** $\log_8 1 = 0$

(c) $\log_{0.1} 1 = 0$

N4. $y = f(x) = \log_6 x$ is equivalent to $x = 6^y$.

Choose values for y and find x.

x	$\frac{1}{36}$	$\frac{1}{6}$	1	6	36
y	-2	-1	0	1	2

Plot the points, and draw a smooth curve through them.

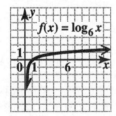

N5. $y = g(x) = \log_{1/4} x$ is equivalent to $x = \left(\frac{1}{4}\right)^y$.

Choose values for y and find x.

x	4	1	$\frac{1}{4}$	$\frac{1}{16}$
y	-1	0	1	2

Plot the points, and draw a smooth curve through them.

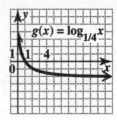

N6. $G(t) = 15.0 + 2.00 \log_{10} t$

(a) $G(1) = 15.0 + 2.00 \log_{10} 1 \quad$ *Let t = 1.*

$\quad\quad = 15.0 + 2.00(0) \quad\quad log_{10}\ 1 = 0$

$\quad\quad = 15.0$

The GNP in 2004 was \$15.0 million.

(b) $G(10) = 15.0 + 2.00 \log_{10} 10 \quad$ *Let t = 10.*

$\quad\quad = 15.0 + 2.00(1) \quad\quad log_{10}\ 10 = 1$

$\quad\quad = 17.0$

The GNP in 2013 is predicted to be \$17.0 million.

11.3 Section Exercises

1. **(a)** $\log_{1/3} 3 = -1$ is equivalent to $\left(\frac{1}{3}\right)^{-1} = 3$.
    (B)

    (b) $\log_5 1 = 0$ is equivalent to $5^0 = 1$.   **(E)**

    (c) $\log_2 \sqrt{2} = \frac{1}{2}$ is equivalent to $2^{1/2} = \sqrt{2}$.
    (D)

    (d) $\log_{10} 1000 = 3$ is equivalent to $10^3 = 1000$.
    (F)

    (e) $\log_8 \sqrt[3]{8} = \frac{1}{3}$ is equivalent to $8^{1/3} = \sqrt[3]{8}$.
    (A)

    (f) $\log_4 4 = 1$ is equivalent to $4^1 = 4$.   **(C)**

3. The base is 4, the exponent (logarithm) is 5, and the number is 1024, so $4^5 = 1024$ becomes $\log_4 1024 = 5$ in logarithmic form.

5. $\frac{1}{2}$ is the base and -3 is the exponent, so $\left(\frac{1}{2}\right)^{-3} = 8$ becomes $\log_{1/2} 8 = -3$ in logarithmic form.

7. The base is 10, the exponent (logarithm) is -3, and the number is 0.001, so $10^{-3} = 0.001$ becomes $\log_{10} 0.001 = -3$ in logarithmic form.

9. $\sqrt[4]{625} = 625^{1/4} = 5$

The base is 625, the exponent (logarithm) is $\frac{1}{4}$, and the number is 5, so $\sqrt[4]{625} = 5$ becomes $\log_{625} 5 = \frac{1}{4}$ in logarithmic form.

11. $8^{-2/3} = \frac{1}{4}$ becomes $\log_8 \frac{1}{4} = -\frac{2}{3}$ in logarithmic form.

13. $5^0 = 1$ becomes $\log_5 1 = 0$ in logarithmic form.

Copyright © 2012 Pearson Education, Inc. Publishing as Addison-Wesley.

15. In $\log_4 64 = 3$, 4 is the base and 3 is the logarithm (exponent), so $\log_4 64 = 3$ becomes $4^3 = 64$ in exponential form.

17. In $\log_{10} \frac{1}{10,000} = -4$, the base is 10, the logarithm (exponent) is -4, and the number is $\frac{1}{10,000}$, so $\log_{10} \frac{1}{10,000} = -4$ becomes $10^{-4} = \frac{1}{10,000}$ in exponential form.

19. In $\log_6 1 = 0$, 6 is the base and 0 is the logarithm (exponent), so $\log_6 1 = 0$ becomes $6^0 = 1$ in exponential form.

21. In $\log_9 3 = \frac{1}{2}$, the base is 9, the logarithm (exponent) is $\frac{1}{2}$, and the number is 3, so $\log_9 3 = \frac{1}{2}$ becomes $9^{1/2} = 3$ in exponential form.

23. $\log_{1/4} \frac{1}{2} = \frac{1}{2}$ becomes $\left(\frac{1}{4}\right)^{1/2} = \frac{1}{2}$ in exponential form.

25. $\log_5 5^{-1} = -1$ becomes $5^{-1} = 5^{-1}$ in exponential form.

27. Use the properties of logarithms,

$$\log_b b = 1 \quad \text{and} \quad \log_b 1 = 0,$$

for $b > 0$, $b \neq 1$.

(a) $\log_8 8 = 1$ **(C)**

(b) $\log_{16} 1 = 0$ **(B)**

(c) $\log_{0.3} 1 = 0$ **(B)**

(d) $\log_{\sqrt{7}} \sqrt{7} = 1$ **(C)**

29.
$$x = \log_{27} 3$$
Write in exponential form.
$$27^x = 3$$
Write each side as a power of 3.
$$(3^3)^x = 3$$
$$3^{3x} = 3^1$$
Set the exponents equal.
$$3x = 1$$
$$x = \frac{1}{3}$$
Check $x = \frac{1}{3}$: $\frac{1}{3} = \log_{27} 3$ since $27^{1/3} = 3$.

The solution set is $\left\{\frac{1}{3}\right\}$.

31.
$$\log_x 9 = \frac{1}{2}$$
Change to exponential form.
$$x^{1/2} = 9$$
$$(x^{1/2})^2 = 9^2 \quad \textit{Square.}$$
$$x^1 = 81$$
$$x = 81$$

$x = 81$ is an acceptable base since it is a positive number (not equal to 1).

Check $x = 81$: $\log_{81} 9 = \frac{1}{2}$ since $81^{1/2} = 9$

The solution set is $\{81\}$.

33.
$$\log_x 125 = -3$$
Write in exponential form.
$$x^{-3} = 125$$
$$\frac{1}{x^3} = 125$$
$$1 = 125\left(x^3\right)$$
$$\frac{1}{125} = x^3$$
Take the cube root of each side.
$$\sqrt[3]{\frac{1}{125}} = \sqrt[3]{x^3}$$
$$x = \sqrt[3]{\frac{1}{5^3}} = \frac{1}{5}$$

$x = \frac{1}{5}$ is an acceptable base since it is a positive number (not equal to 1).

The solution set is $\left\{\frac{1}{5}\right\}$.

35. $\log_{12} x = 0$
$$12^0 = x \quad \textit{Write in exponential form.}$$
$$1 = x$$

The argument (the input of the logarithm) must be a positive number, so $x = 1$ is acceptable.

The solution set is $\{1\}$.

37. $\log_x x = 1$
$$x^1 = x \quad \textit{Write in exponential form.}$$

This equation is true for all the numbers x that are allowed as the base of a logarithm; that is, all positive numbers x, $x \neq 1$.

The solution set is $\{x \mid x > 0, \ x \neq 1\}$.

39. $\log_x \frac{1}{25} = -2$
$$x^{-2} = \frac{1}{25} \quad \textit{Write in exponential form.}$$
$$\frac{1}{x^2} = \frac{1}{25}$$
$$x^2 = 25 \quad \textit{Denominators must be equal.}$$
$$x = \pm 5$$

Reject $x = -5$ since the base of a logarithm must be positive and not equal to 1.

The solution set is $\{5\}$.

41. $\log_8 32 = x$
$$8^x = 32 \quad \textit{Exponential form}$$
Write each side as a power of 2.
$$\left(2^3\right)^x = 2^5$$
$$2^{3x} = 2^5$$
$$3x = 5 \quad \textit{Equate exponents.}$$
$$x = \frac{5}{3}$$

Check $x = \frac{5}{3}$: $\log_8 32 = \frac{5}{3}$ since $8^{5/3} = 2^5 = 32$.

The solution set is $\left\{\frac{5}{3}\right\}$.

Copyright © 2012 Pearson Education, Inc. Publishing as Addison-Wesley.

43. $\log_\pi \pi^4 = x$

$\qquad \pi^x = \pi^4$ *Exponential form*

$\qquad x = 4$ *Equate exponents.*

Check $x = 4$: $\log_\pi \pi^4 = 4$ since $\pi^4 = \pi^4$.

The solution set is $\{4\}$.

45. $\log_6 \sqrt{216} = x$

$\log_6 216^{1/2} = x$ *Equivalent form*

$\qquad 6^x = 216^{1/2}$ *Exponential form*

$\qquad 6^x = \left(6^3\right)^{1/2}$ *Same base*

$\qquad 6^x = 6^{3/2}$

$\qquad x = \tfrac{3}{2}$ *Equate exponents.*

Check $x = \tfrac{3}{2}$:

$\qquad \log_6 \sqrt{216} = \tfrac{3}{2}$ since $6^{3/2} = \sqrt{6^3} = \sqrt{216}$.

The solution set is $\left\{\tfrac{3}{2}\right\}$.

47. $\log_4 (2x + 4) = 3$

$\qquad 2x + 4 = 4^3$ *Exponential form*

$\qquad 2x = 64 - 4$

$\qquad 2x = 60$

$\qquad x = 30$

Check $x = 30$: $\log_4 (2 \cdot 30 + 4) = \log_4 64 = 3$.

The solution set is $\{30\}$.

49. $\qquad y = \log_3 x$

$\qquad 3^y = x$ *Exponential form*

Refer to Section 11.2, Exercise 5, for the graph of $f(x) = 3^x$. Since $y = \log_3 x$ (or $3^y = x$) is the inverse of $f(x) = y = 3^x$, its graph is symmetric about the line $y = x$ to the graph of $f(x) = 3^x$. The graph can be plotted by reversing the ordered pairs in the table of values belonging to $f(x) = 3^x$.

x	$\tfrac{1}{9}$	$\tfrac{1}{3}$	1	3	9
y	-2	-1	0	1	2

Plot the points, and draw a smooth curve through them.

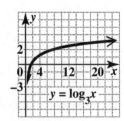

51. $\qquad y = \log_{1/3} x$

$\qquad \left(\tfrac{1}{3}\right)^y = x$ *Exponential form*

Refer to Section 11.2, Exercise 7, for the graph of $g(x) = \left(\tfrac{1}{3}\right)^x$. Since $y = \log_{1/3} x$ (or $\left(\tfrac{1}{3}\right)^y = x$) is the inverse of $y = \left(\tfrac{1}{3}\right)^x$, its graph is symmetric about the line $y = x$ to the graph of $y = \left(\tfrac{1}{3}\right)^x$.

The graph can be plotted by reversing the ordered pairs in the table of values belonging to $g(x) = \left(\tfrac{1}{3}\right)^x$.

x	9	3	1	$\tfrac{1}{3}$	$\tfrac{1}{9}$
y	-2	-1	0	1	2

Plot the points, and draw a smooth curve through them.

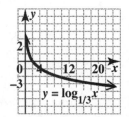

53.   The number 1 is not used as a base for a logarithmic function since the function would look like $x = 1^y$ in exponential form. Then, for any real value of y, the statement $1 = 1$ would always be the result since every power of 1 is equal to 1.

55.   The range of $f(x) = a^x$ is the domain of $g(x) = \log_a x$, that is, $\underline{(0, \infty)}$.

The domain of $f(x) = a^x$ is the range of $g(x) = \log_a x$, that is, $\underline{(-\infty, \infty)}$.

57.   The values of t are on the horizontal axis, and the values of $f(t)$ are on the vertical axis. Read the value of $f(t)$ from the graph for the given value of t. At $t = 0$, $f(0) = 8$.

59.   To find $f(60)$, find 60 on the t-axis, then go up to the graph and across to the $f(t)$ axis to read the value of $f(60)$. At $t = 60$, $f(60) = 24$.

61. $f(x) = 3800 + 585 \log_2 x$

(a) $x = 1982 - 1980 = 2$

$\qquad f(2) = 3800 + 585 \log_2 2$

$\qquad\quad\; = 3800 + 585(1)$

$\qquad\quad\; = 4385$

The model gives an approximate withdrawal of 4385 billion ft^3 of natural gas from crude oil wells in the United States for 1982.

(b) $x = 1988 - 1980 = 8$

$\qquad f(8) = 3800 + 585 \log_2 8$

$\qquad\quad\; = 3800 + 585(3)$

$\qquad\quad\; = 5555$

The model gives an approximate withdrawal of 5555 billion ft^3 of natural gas from crude oil wells in the United States for 1988.

Copyright © 2012 Pearson Education, Inc. Publishing as Addison-Wesley.

(c) $x = 1996 - 1980 = 16$

$$f(16) = 3800 + 585 \log_2 16$$
$$= 3800 + 585(4)$$
$$= 6140$$

The model gives an approximate withdrawal of 6140 billion ft^3 of natural gas from crude oil wells in the United States for 1996.

63. $S(t) = 100 + 30 \log_3 (2t + 1)$

(a) $S(1) = 100 + 30 \log_3 (2 \cdot 1 + 1)$
$$= 100 + 30 \log_3 (3)$$
$$= 100 + 30(1) = 130$$

After 1 year, the sales were 130 thousand units.

(b) $S(13) = 100 + 30 \log_3 (2 \cdot 13 + 1)$
$$= 100 + 30 \log_3 (27)$$
$$= 100 + 30(3) = 190$$

After 13 years, the sales were 190 thousand units.

(c) Make a table of values, plot the points they represent, and draw a smooth curve through them.

To make the table, find values of t such that $2t + 1 = 3^k$, where $k = 0, 1, 2, 3, 4$.

k	0	1	2	3	4
3^k	1	3	9	27	81
$2t+1$	1	3	9	27	81
$2t$	0	2	8	26	80
t	0	1	4	13	40
$S(t)$	100	130	160	190	220

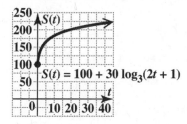

65. $R = \log_{10} \dfrac{x}{x_0}$

Change to exponential form.

$$10^R = \frac{x}{x_0}, \text{ so } x = x_0 \, 10^R.$$

Let $R = 6.7$ for the Northridge earthquake, with intensity x_1.

$$x_1 = x_0 10^{6.7}$$

Let $R = 7.3$ for the Landers earthquake, with intensity x_2.

$$x_2 = x_0 10^{7.3}$$

The ratio of x_2 to x_1 is

$$\frac{x_2}{x_1} = \frac{x_0 10^{7.3}}{x_0 10^{6.7}} = 10^{0.6} \approx 3.98.$$

The Landers earthquake was about 4 times more powerful than the Northridge earthquake.

67. $g(x) = \log_3 x$
On a TI-83/4, assign 3^x to Y_1. Then enter

$$\text{DrawInv } Y_1$$

on the home screen to obtain the figure that follows. DrawInv is choice 8 under the DRAW menu. Y_1 is choice 1 under VARS, Y-VARS, Function.

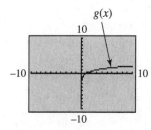

69. $g(x) = \log_{1/3} x$
Assign (1/3)^x to Y_1 and enter DrawInv Y_1.

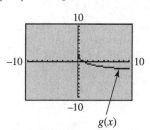

71. $4^7 \cdot 4^2 = 4^{7+2} = 4^9$

73. $\dfrac{7^8}{7^{-4}} = 7^{8-(-4)} = 7^{12}$

11.4 Properties of Logarithms

11.4 Now Try Exercises

N1. Use the product rule for logarithms.

$$\log_b xy = \log_b x + \log_b y$$

(a) $\log_{10} (7 \cdot 9) = \log_{10} 7 + \log_{10} 9$

(b) $\log_5 11 + \log_5 8 = \log_5 (11 \cdot 8)$
$$= \log_5 88$$

(c) $\log_5 (5x) = \log_5 5 + \log_5 x$
$$= 1 + \log_5 x \quad (x > 0)$$

(d) $\log_2 t^3 = \log_2 (t \cdot t \cdot t)$
$$= \log_2 t + \log_2 t + \log_2 t$$
$$= 3 \log_2 t \quad (t > 0)$$

Copyright © 2012 Pearson Education, Inc. Publishing as Addison-Wesley.

N2. Use the quotient rule for logarithms.

$$\log_b \frac{x}{y} = \log_b x - \log_b y$$

(a) $\log_{10} \frac{7}{9} = \log_{10} 7 - \log_{10} 9$

(b) $\log_4 x - \log_4 12 = \log_4 \frac{x}{12} \quad (x > 0)$

(c) $\log_5 \frac{25}{27} = \log_5 25 - \log_5 27$
$$= 2 - \log_5 27$$

N3. Use the power rule for logarithms.

$$\log_b x^r = r \log_b x$$

(a) $\log_7 5^3 = 3 \log_7 5$

(b) $\log_a \sqrt{10} = \log_a 10^{1/2}$
$$= \tfrac{1}{2} \log_a 10 \quad (a > 0)$$

(c) $\log_3 \sqrt[4]{x^3} = \log_3 x^{3/4}$
$$= \tfrac{3}{4} \log_3 x \quad (x > 0)$$

N4. Use the Special Properties,

$$b^{\log_b x} = x \quad \text{or} \quad \log_b b^x = x.$$

(a) By the second property,

$$\log_4 4^7 = 7.$$

(b) Using the second property,

$$\log_{10} 10,000 = \log_{10} 10^4 = 4.$$

(c) By the first property,

$$8^{\log_8 5} = 5.$$

N5. Use the properties of logarithms.

(a) $\log_3 9z^4$

$$= \log_3 9 + \log_3 z^4 \quad \textit{Product rule}$$
$$= 2 + 4 \log_3 z \quad \begin{array}{l}\log_3 3^2 = 2; \\ \textit{Power rule}\end{array}$$

(b) $\log_6 \sqrt{\frac{n}{3m}}$

$$= \log_6 \left(\tfrac{n}{3m}\right)^{1/2}$$
$$= \tfrac{1}{2} \log_6 \left(\tfrac{n}{3m}\right) \qquad \textit{Po. rule}$$
$$= \tfrac{1}{2}(\log_6 n - \log_6 3m) \qquad \textit{Qu. rule}$$
$$= \tfrac{1}{2}[\log_6 n - (\log_6 3 + \log_6 m)]$$
$$= \tfrac{1}{2}(\log_6 n - \log_6 3 - \log_6 m)$$

(c) $\log_2 x + 3 \log_2 y - \log_2 z$

$$= \log_2 x + \log_2 y^3 - \log_2 z \quad \textit{Power rule}$$
$$= \log_2 (xy^3) - \log_2 z \quad \textit{Product rule}$$
$$= \log_2 \frac{xy^3}{z} \qquad \textit{Quotient rule}$$

(d) $\log_5 (x + 10) + \log_5 (x - 10)$
$$\quad - \tfrac{3}{5} \log_5 x \quad (x > 10)$$
$$= \log_5 [(x + 10)(x - 10)] - \tfrac{3}{5} \log_5 x$$
$$\qquad\qquad \textit{Product rule}$$
$$= \log_5 (x^2 - 100) - \log_5 x^{3/5}$$
$$\qquad\qquad \textit{Power rule}$$
$$= \log_5 \frac{x^2 - 100}{x^{3/5}} \qquad \textit{Quotient rule}$$

(e) $\log_7 (49 + 2x)$ cannot be written as a sum of logarithms, since

$$\log_b (M + N) \neq \log_b M + \log_b N.$$

There is no property of logarithms to rewrite the logarithm of a sum.

N6. (a) $\log_2 70 = \log_2 (7 \cdot 10)$
$$= \log_2 7 + \log_2 10$$
$$= 2.8074 + 3.3219$$
$$= 6.1293$$

(b) $\log_2 0.7 = \log_2 \frac{7}{10}$
$$= \log_2 7 - \log_2 10$$
$$= 2.8074 - 3.3219$$
$$= -0.5145$$

(c) $\log_2 49 = \log_2 7^2$
$$= 2 \log_2 7$$
$$= 2(2.8074)$$
$$= 5.6148$$

N7. (a) $\log_2 16 + \log_2 16 = \log_2 32$

Evaluate each side.

$$LS = \log_2 16 + \log_2 16$$
$$= \log_2 2^4 + \log_2 2^4 = 4 + 4 = 8$$
$$RS = \log_2 32 = \log_2 2^5 = 5$$

The statement is *false* because $8 \neq 5$.

(b) $(\log_2 4)(\log_3 9) = \log_6 36$

Evaluate each side.

$$LS = (\log_2 4)(\log_3 9)$$
$$= \left(\log_2 2^2\right)\left(\log_3 3^2\right) = (2)(2) = 4$$
$$RS = \log_6 36 = \log_6 6^2 = 2$$

The statement is *false* because $4 \neq 2$.

11.4 Section Exercises

1.   By the product rule,

$$\log_{10} (7 \cdot 8) = \log_{10} 7 + \log_{10} 8.$$

3.   By a special property (see page 632 in the text),

$$3^{\log_3 4} = 4.$$

5.   By a special property (see page 632 in the text),

$$\log_3 3^9 = 9.$$

Copyright © 2012 Pearson Education, Inc. Publishing as Addison-Wesley.

7. Use the product rule for logarithms.

$$\log_7(4 \cdot 5) = \log_7 4 + \log_7 5$$

9. Use the quotient rule for logarithms.

$$\log_5 \tfrac{8}{3} = \log_5 8 - \log_5 3$$

11. Use the power rule for logarithms.

$$\log_4 6^2 = 2\log_4 6$$

13. $\log_3 \dfrac{\sqrt[3]{4}}{x^2 y} = \log_3 \dfrac{4^{1/3}}{x^2 y}$

Use the quotient rule for logarithms.

$$= \log_3 4^{1/3} - \log_3(x^2 y)$$

Use the product rule for logarithms.

$$= \log_3 4^{1/3} - \left(\log_3 x^2 + \log_3 y\right)$$

$$= \log_3 4^{1/3} - \log_3 x^2 - \log_3 y$$

Use the power rule for logarithms.

$$= \tfrac{1}{3}\log_3 4 - 2\log_3 x - \log_3 y$$

15. $\log_3 \sqrt{\dfrac{xy}{5}}$

$= \log_3 \left(\dfrac{xy}{5}\right)^{1/2}$

$= \tfrac{1}{2}\log_3\left(\dfrac{xy}{5}\right)$ *Power rule*

$= \tfrac{1}{2}\left[\log_3(xy) - \log_3 5\right]$ *Quotient rule*

$= \tfrac{1}{2}\left(\log_3 x + \log_3 y - \log_3 5\right)$ *Product rule*

$= \tfrac{1}{2}\log_3 x + \tfrac{1}{2}\log_3 y - \tfrac{1}{2}\log_3 5$

17. $\log_2 \dfrac{\sqrt[3]{x} \cdot \sqrt[5]{y}}{r^2}$

$= \log_2 \dfrac{x^{1/3} y^{1/5}}{r^2}$

$= \log_2\left(x^{1/3}y^{1/5}\right) - \log_2 r^2$ *Quotient rule*

$= \log_2 x^{1/3} + \log_2 y^{1/5} - \log_2 r^2$ *Product rule*

$= \tfrac{1}{3}\log_2 x + \tfrac{1}{5}\log_2 y - 2\log_2 r$ *Power rule*

19. In the notation $\log_a(x+y)$, the parentheses do not indicate multiplication. They indicate that $x+y$ is the result of raising a to some power.

21. By the product rule for logarithms,

$$\log_b x + \log_b y = \log_b xy.$$

23. By the quotient rule for logarithms,

$$\log_a m - \log_a n = \log_a \tfrac{m}{n}.$$

25. $(\log_a r - \log_a s) + 3\log_a t$

$= \log_a \dfrac{r}{s} + \log_a t^3$ *Quotient rule; Power rule*

$= \log_a \dfrac{rt^3}{s}$ *Product rule*

27. $3\log_a 5 - 4\log_a 3$

$= \log_a 5^3 - \log_a 3^4$ *Power rule*

$= \log_a \dfrac{5^3}{3^4}$ *Quotient rule*

$= \log_a \dfrac{125}{81}$

29. $\log_{10}(x+3) + \log_{10}(x-3)$

$= \log_{10}\left[(x+3)(x-3)\right]$ *Product rule*

$= \log_{10}(x^2 - 9)$

31. By the power rule for logarithms,

$3\log_p x + \tfrac{1}{2}\log_p y - \tfrac{3}{2}\log_p z - 3\log_p a$

$= \log_p x^3 + \log_p y^{1/2} - \log_p z^{3/2} - \log_p a^3.$

Group the terms into sums.

$= \left(\log_p x^3 + \log_p y^{1/2}\right) - \left(\log_p z^{3/2} + \log_p a^3\right)$

$= \log_p\left(x^3 y^{1/2}\right) - \log_p\left(z^{3/2}a^3\right)$ *Product rule*

$= \log_p \dfrac{x^3 y^{1/2}}{z^{3/2}a^3}$ *Quotient rule*

In Exercises 33–44,
$\log_{10} 2 \approx 0.3010$ and $\log_{10} 9 \approx 0.9542.$

33. $\log_{10} 18 = \log_{10}(2 \cdot 9)$

$= \log_{10} 2 + \log_{10} 9$ *Product rule*

$\approx 0.3010 + 0.9542$

$= 1.2552$

35. $\log_{10} \tfrac{2}{9} = \log_{10} 2 - \log_{10} 9$ *Quotient rule*

$\approx 0.3010 - 0.9542$

$= -0.6532$

37. Use the product and power rules for logarithms.

$\log_{10} 36 = \log_{10}(2^2 \cdot 9)$

$= 2\log_{10} 2 + \log_{10} 9$

$\approx 2(0.3010) + 0.9542$

$= 1.5562.$

39. $\log_{10} \sqrt[4]{9} = \log_{10} 9^{1/4}$

$= \tfrac{1}{4}\log_{10} 9$ *Power rule*

$\approx \tfrac{1}{4}(0.9542)$

$= 0.23855 \approx 0.2386$

41. $\log_{10} 3 = \log_{10} 9^{1/2}$ *Rename 3*

$= \tfrac{1}{2}\log_{10} 9$ *Power rule*

$\approx \tfrac{1}{2}(0.9542)$

$= 0.4771$

43. $\log_{10} 9^5 = 5\log_{10} 9$ *Power rule*

$\approx 5(0.9542)$

$= 4.7710$

45. $\text{LS} = \log_2(8 + 32) = \log_2 40$

$\text{RS} = \log_2 8 + \log_2 32 = \log_2(8 \cdot 32)$

$= \log_2 256$

$\text{LS} \neq \text{RS}$, so the given statement is *false*.

Copyright © 2012 Pearson Education, Inc. Publishing as Addison-Wesley.

47. $\log_3 7 + \log_3 7^{-1} = \log_3 7 + (-1)\log_3 7$
$\qquad\qquad\qquad\qquad\qquad = 0$

The given statement is *true*.

49. $\log_6 60 - \log_6 10 = \log_6 \frac{60}{10}$
$\qquad\qquad\qquad\qquad = \log_6 6 = 1$

The given statement is *true*.

51. $\dfrac{\log_{10} 7}{\log_{10} 14} \overset{?}{=} \dfrac{1}{2}$

$2\log_{10} 7 \overset{?}{=} 1\log_{10}(7 \cdot 2)$
$\qquad\qquad$ *Cross products are equal.*

$2\log_{10} 7 \overset{?}{=} \log_{10} 7 + \log_{10} 2$
$\quad\log_{10} 7 \neq \log_{10} 2$ *Subtract $\log_{10}$ 7.*

The given statement is *false*.

53. The exponent of a quotient is the difference between the exponent of the numerator and the exponent of the denominator.

55. We can't determine the logarithm of 0 because no number allowed as a logarithmic base can be raised to a power with a result of 0.

57. $10^4 = 10,000$ becomes $\log_{10} 10,000 = 4$.

59. $10^{-2} = 0.01$ becomes $\log_{10} 0.01 = -2$.

61. $\log_{10} 1 = 0$ becomes $10^0 = 1$.

11.5   Common and Natural Logarithms

11.5   Now Try Exercises

N1. Evaluate each logarithm using a calculator.

(a) $\log 115 \approx 2.0607$

(b) $\log 0.25 \approx -0.6021$

N2. $\mathrm{pH} = -\log [\mathrm{H_3O^+}]$
$\qquad = -\log(3.4 \times 10^{-5})$
$\qquad = -(\log 3.4 + \log 10^{-5})$
$\qquad = -(\log 3.4 - 5\log 10)$
$\qquad \approx -[0.5315 - 5(1)]$
$\qquad = -0.5315 + 5$
$\qquad = 4.4685 \approx 4.5$

Since the pH is between 4.0 and 6.0, the wetland is a poor fen.

N3. $\qquad \mathrm{pH} = -\log [\mathrm{H_3O^+}]$
$\qquad\quad 2.6 = -\log [\mathrm{H_3O^+}]$ *Let pH = 2.6.*
$\qquad -2.6 = \log [\mathrm{H_3O^+}]$
$\qquad -2.6 = \log_{10} [\mathrm{H_3O^+}]$
$\quad 10^{-2.6} = [\mathrm{H_3O^+}]$ *Exponential form*

$[\mathrm{H_3O^+}] \approx 2.5 \times 10^{-3}$

N4. $D = 10\log\left(\dfrac{I}{I_0}\right)$
$\qquad = 10\log\left(\dfrac{6.312 \times 10^{13} I_0}{I_0}\right)$
$\qquad = 10\log(6.312 \times 10^{13})$
$\qquad \approx 10(13.8) \approx 138$

The level is about 138 dB.

N5. Evaluate each logarithm using a calculator.

(a) $\ln 0.26 \approx -1.3471$

(b) $\ln 12 \approx 2.4849$

(c) $\ln 150 \approx 5.0106$

N6. $f(x) = 51,600 - 7457 \ln x$
$\quad f(600) = 51,600 - 7457 \ln 600$
$\qquad\qquad \approx 3898.1 \approx 3900$

Atmospheric pressure is 600 millibars at approximately 3900 m.

N7. The change-of-base rule is

$$\log_a x = \frac{\log_b x}{\log_b a}.$$

Use common logarithms ($b = 10$).

$$\log_8 60 = \frac{\log_{10} 60}{\log_{10} 8} = \frac{\log 60}{\log 8} \approx 1.9690$$

N8. $f(x) = 2014 + 384.7 \log_2 x$
2002 corresponds to $x = 2002 - 1989 = 13$.
$f(13) = 2014 + 384.7 \log_2 13$
$\qquad = 2014 + 384.7\left(\dfrac{\log 13}{\log 2}\right)$
$\qquad \approx 2014 + 384.7(3.7004)$
$\qquad \approx 3438$

The model gives about 3438 million barrels for 2002, which is greater than the actual amount of 3336 million barrels.

11.5   Section Exercises

1. Since $\log x = \log_{10} x$, the base is 10. The correct response is **C**.

3. $10^0 = 1$ and $10^1 = 10$, so $\log 1 = 0$ and $\log 10 = 1$. Thus, the value of $\log 6.3$ must lie between 0 and 1. The correct response is **C**.

5. By the special property, $\log_b b^x = x$,
$$\log 10^{31.6} = \log_{10} 10^{31.6} = 31.6.$$

7. To four decimal places,
$$\log 43 \approx 1.6335.$$

9. $\log 328.4 \approx 2.5164$

11. $\log 0.0326 \approx -1.4868$

Copyright © 2012 Pearson Education, Inc. Publishing as Addison-Wesley.

13. $\log (4.76 \times 10^9) \approx 9.6776$
On a TI-83/4, enter

$$\boxed{\text{LOG}}\ 4.76\ \boxed{\text{2nd}}\ \boxed{\text{EE}}\ 9\ ).$$

15. $\ln 7.84 \approx 2.0592$

17. $\ln 0.0556 \approx -2.8896$

19. $\ln 388.1 \approx 5.9613$

21. $\ln (8.59 \times e^2) \approx 4.1506$
On a TI-83/4, enter

$$\boxed{\text{LN}}\ 8.59\ \boxed{\text{X}}\ \boxed{\text{2nd}}\ \boxed{e^x}\ 2\ )\ ).$$

23. $\ln 10 \approx 2.3026$

25. **(a)** $\log 356.8 \approx 2.552\,424\,846$

  (b) $\log 35.68 \approx 1.552\,424\,846$

  (c) $\log 3.568 \approx 0.552\,424\,846$

  (d) The whole number part of the answers (2, 1, or 0) varies, whereas the decimal part (0.552 424 846) remains the same, indicating that the whole number part corresponds to the placement of the decimal point and the decimal part corresponds to the digits 3, 5, 6, and 8.

27. $\text{pH} = -\log [\text{H}_3\text{O}^+]$
$= -\log (3.1 \times 10^{-5}) \approx 4.5$
Since the pH is between 3.0 and 6.0, the wetland is classified as a *poor fen*.

29. $\text{pH} = -\log [\text{H}_3\text{O}^+]$
$= -\log (2.5 \times 10^{-2}) \approx 1.6$
Since the pH is less than 3.0, the wetland is classified as a *bog*.

31. $\text{pH} = -\log [\text{H}_3\text{O}^+]$
$= -\log (2.7 \times 10^{-7}) \approx 6.6$
Since the pH is between 6.0 and 7.5, the wetland is classified as a *rich fen*.

33. Ammonia has a hydronium ion concentration of 2.5×10^{-12}.
$\text{pH} = -\log [\text{H}_3\text{O}^+]$
$\text{pH} = -\log (2.5 \times 10^{-12}) \approx 11.6$

35. Grapes have a hydronium ion concentration of 5.0×10^{-5}.
$\text{pH} = -\log [\text{H}_3\text{O}^+]$
$\text{pH} = -\log (5.0 \times 10^{-5}) \approx 4.3$

37. Human blood plasma has a pH of 7.4.
$$\text{pH} = -\log [\text{H}_3\text{O}^+]$$
$$7.4 = -\log [\text{H}_3\text{O}^+]$$
$$\log [\text{H}_3\text{O}^+] = -7.4$$
$$[\text{H}_3\text{O}^+] = 10^{-7.4} \approx 4.0 \times 10^{-8}$$

39. Spinach has a pH value of 5.4.
$$\text{pH} = -\log [\text{H}_3\text{O}^+]$$
$$5.4 = -\log [\text{H}_3\text{O}^+]$$
$$\log [\text{H}_3\text{O}^+] = -5.4$$
$$[\text{H}_3\text{O}^+] = 10^{-5.4} \approx 4.0 \times 10^{-6}$$

41. $D = 10 \log \left(\dfrac{I}{I_0} \right)$

  (a) $D = 10 \log \left(\dfrac{5.012 \times 10^{10} I_0}{I_0} \right)$
  $= 10 \log (5.012 \times 10^{10}) \approx 107$

The average decibel level for *Avatar* is about 107 dB.

  (b) $D = 10 \log \left(\dfrac{10^{10} I_0}{I_0} \right)$
  $= 10 \log 10^{10} = 10 \cdot 10 = 100$

The average decibel level for *Iron Man 2* is 100 dB.

  (c) $D = 10 \log \left(\dfrac{6{,}310{,}000{,}000\, I_0}{I_0} \right)$
  $= 10 \log (6{,}310{,}000{,}000) \approx 98$

The average decibel level for *Clash of the Titans* is about 98 dB.

43. $N(r) = -5000 \ln r$

  (a) 85% (or 0.85)

  $N(0.85) = -5000 \ln 0.85 \approx 813 \approx 800$

  The number of years since the split for 85% of common words is about 800 years.

  (b) 35% (or 0.35)

  $N(0.35) = -5000 \ln 0.35 \approx 5249 \approx 5200$

  The number of years since the split for 35% of common words is about 5200 years.

  (c) 10% (or 0.10)

  $N(0.10) = -5000 \ln 0.10 \approx 11{,}513 \approx 11{,}500$

  The number of years since the split for 10% of common words is about 11,500 years.

45. $f(x) = -1317 + 304 \ln x$

  (a) $x = 1998 - 1900 = 98$

  $f(98) = -1317 + 304 \ln (98) \approx 77$

  The prediction for 1998 is 77%.

Copyright © 2012 Pearson Education, Inc. Publishing as Addison-Wesley.

(b) $f(x) = 50$

$50 = -1317 + 304 \ln x$

$1367 = 304 \ln x$

$\ln x = \frac{1367}{304}$

$e^{\ln x} = e^{1367/304}$

$x \approx 89.72$

The outpatient surgeries reached 50% in 1989.

47. $T = -0.642 - 189 \ln(1 - p)$

(a) $T = -0.642 - 189 \ln(1 - 0.25)$

$= -0.642 - 189 \ln(0.75) \approx 53.73$

About $54 per ton will reduce emissions by 25%.

(b) If $p = 0$, then $\ln(1 - p) = \ln 1 = 0$, so T would be negative. If $p = 1$, then $\ln(1 - p) = \ln 0$, but the domain of $\ln x$ is $(0, \infty)$.

49. The change-of-base rule is

$$\log_a x = \frac{\log_b x}{\log_b a}.$$

Use common logarithms ($b = 10$).

$\log_3 12 = \frac{\log_{10} 12}{\log_{10} 3} = \frac{\log 12}{\log 3} \approx 2.2619$

51. Use natural logarithms ($b = e$).

$\log_5 3 = \frac{\log_e 3}{\log_e 5} = \frac{\ln 3}{\ln 5} \approx 0.6826$

53. $\log_3 \sqrt{2} = \frac{\ln \sqrt{2}}{\ln 3} \approx 0.3155$

55. $\log_\pi e = \frac{\ln e}{\ln \pi} = \frac{1}{\ln \pi} \approx 0.8736$

57. $\log_e 12 = \frac{\ln 12}{\ln e} = \frac{\ln 12}{1} \approx 2.4849$

59. Let $m =$ the number of letters in your first name and $n =$ the number of letters in your last name.

Answers will vary, but suppose the name is Paul Bunyan, with $m = 4$ and $n = 6$.

(a) $\log_m n = \log_4 6$ is the exponent to which 4 must be raised in order to obtain 6.

(b) Use the change-of-base rule.

$$\log_4 6 = \frac{\log 6}{\log 4}$$

$$\approx 1.29248125$$

(c) Here, $m = 4$. Use the power key ($y^x, x^y, \wedge$) on your calculator.

$$4^{1.29248125} \approx 6$$

The result is 6, the value of n.

61. $f(x) = 3800 + 585 \log_2 x$

$x = 2003 - 1980 = 23$

$f(23) = 3800 + 585 \log_2 23$

$$= 3800 + 585 \left(\frac{\log 23}{\log 2}\right)$$

$$\approx 6446$$

The model gives an approximate withdrawal of 6446 billion ft³ of natural gas from crude oil wells in the United States for 2003.

63. $4^{2x} = 8^{3x+1}$

$\left(2^2\right)^{2x} = \left(2^3\right)^{3x+1}$

$2^{4x} = 2^{9x+3}$

$4x = 9x + 3$

$-3 = 5x$

$x = -\frac{3}{5}$

The solution set is $\left\{-\frac{3}{5}\right\}$.

65. $\log_3(x + 4) = 2$

$x + 4 = 3^2$

$x + 4 = 9$

$x = 5$

Substitute 5 for x in the argument, $x + 4$, $5 + 4 = 9 > 0$, so 5 is acceptable.

The solution set is $\{5\}$.

67. $\log_{1/2} 8 = x$

$\left(\frac{1}{2}\right)^x = 8$

$\left(2^{-1}\right)^x = 2^3$

$2^{-x} = 2^3$

$-x = 3$

$x = -3$

The solution set is $\{-3\}$.

69. $\log(x + 2) + \log(x + 3)$

$= \log[(x + 2)(x + 3)], \quad \text{or} \quad \log\left(x^2 + 5x + 6\right)$

11.6 Exponential and Logarithmic Equations; Further Applications

11.6 Now Try Exercises

N1. $5^x = 20$

$\log 5^x = \log 20$ *Property 3*

$x \log 5 = \log 20$ *Power rule*

$x = \frac{\log 20}{\log 5}$ *Divide.*

$x \approx 1.861$

Check $x = 1.861$: $5^{1.861} \approx 20$

The solution set is $\{1.861\}$.

Copyright © 2012 Pearson Education, Inc. Publishing as Addison-Wesley.

N2. $e^{0.12x} = 10$

Take base e logarithms on each side.

$\ln e^{0.12x} = \ln 10$

$0.12x \ln e = \ln 10$ *Power rule*

$0.12x = \ln 10$ *ln e = 1*

$x = \dfrac{\ln 10}{0.12} \approx 19.188$

Check $x = 19.188$: $e^{0.12(19.188)} \approx 10$

The solution set is $\{19.188\}$.

N3. $\log_5 (x-1)^3 = 2$

$(x-1)^3 = 5^2$ *Exponential form*

$(x-1)^3 = 25$

$x - 1 = \sqrt[3]{25}$ *Cube root*

$x = 1 + \sqrt[3]{25}$ *Add 1.*

Check $x = 1 + \sqrt[3]{25}$:

$\log_5 \left(1 + \sqrt[3]{25} - 1\right)^3 = \log_5 \left(\sqrt[3]{25}\right)^3$

$= \log_5 25 = \log_5 5^2 = 2$

The solution set is $\left\{1 + \sqrt[3]{25}\right\}$.

N4. $\log_4 (2x + 13)$

$- \log_4 (x + 1) = \log_4 10$ *Given equation*

$\log_4 \dfrac{2x+13}{x+1} = \log_4 10$ *Quotient rule*

$\dfrac{2x+13}{x+1} = 10$ *Property 4*

$2x + 13 = 10(x + 1)$ *Multiply by x + 1.*

$2x + 13 = 10x + 10$ *Distributive prop.*

$3 = 8x$ *Subtract 2x; 10.*

$\frac{3}{8} = x$ *Divide by 8.*

Check $x = \frac{3}{8}$:

$\log_4 \frac{55}{4} - \log_4 \frac{11}{8} = \log_4 \frac{\frac{55}{4}}{\frac{11}{8}} = \log_4 10$

The solution set is $\left\{\frac{3}{8}\right\}$.

N5. $\log_4 (x+2) + \log_4 2x = 2$

$\log_4 [2x(x+2)] = 2$ *Product rule*

$2x(x+2) = 4^2$ *Exponential form*

$2x^2 + 4x = 16$

$2x^2 + 4x - 16 = 0$

$x^2 + 2x - 8 = 0$

$(x+4)(x-2) = 0$

$x + 4 = 0$ or $x - 2 = 0$

$x = -4$ or $x = 2$

Reject $x = -4$ since it yields an equation in which the logarithm of a negative number must be found.

Check $x = 2$: $\log_4 4 + \log_4 4 = 1 + 1 = 2$

The solution set is $\{2\}$.

N6. Use $A = P\left(1 + \dfrac{r}{n}\right)^{nt}$ with $P = 10{,}000$, $r = 2.5\% = 0.025$, $n = 12$ (compounded monthly), and $t = 10$.

$A = 10{,}000\left(1 + \dfrac{0.025}{12}\right)^{12(10)}$

$= 10{,}000(1.002083)^{120} \approx 12{,}836.92$

There will be \$12,836.92 in the account.

N7. $A = P\left(1 + \dfrac{r}{n}\right)^{nt}$

$2P = P\left(1 + \dfrac{0.04}{4}\right)^{4 \cdot t}$ *Double, so A = 2P*
                                          Quarterly, n = 4

$2 = (1.01)^{4t}$ *Divide by P.*

$\log 2 = \log (1.01)^{4t}$ *Property 3*

$\log 2 = 4t \log 1.01$ *Power rule*

$4t = \dfrac{\log 2}{\log 1.01}$ *Divide by log 1.01.*

$t = \dfrac{\log 2}{4 \log 1.01}$ *Divide by 4.*

$t \approx 17.42$

It will take about 17.42 years for any amount of money in an account paying 4% interest compounded quarterly to double.

N8. **(a)** $A = Pe^{rt}$

$A = 4000e^{0.03(2)}$

$A \approx 4247.35$

It will grow to \$4247.35.

(b) $A = Pe^{rt}$

$2(4000) = 4000e^{0.03t}$

$2 = e^{0.03t}$ *Divide by 4000.*

$\ln 2 = \ln e^{0.03t}$ *Take natural logs.*

$\ln 2 = 0.03t$ *ln e^k = k*

$t = \dfrac{\ln 2}{0.03}$ *Divide by 0.03.*

$t \approx 23.1$

It would take about 23.1 years for the initial investment to double.

N9. $y = y_0 e^{-0.00043t}$

(a) $y = 4.5e^{-0.00043(150)}$ *Let t = 150.*

$y \approx 4.22$

After 150 years, there will be about 4.2 grams.

(b) $\frac{1}{2}(4.5) = 4.5e^{-0.00043t}$

$\frac{1}{2} = e^{-0.00043t}$

$\ln \frac{1}{2} = -0.00043t$

$t = \dfrac{\ln \frac{1}{2}}{-0.00043} \approx 1612$

The half-life is about 1612 years.

Copyright © 2012 Pearson Education, Inc. Publishing as Addison-Wesley.

11.6 Section Exercises

1.
$$7^x = 5$$
Take the logarithm of each side.
$$\log 7^x = \log 5$$
Use the power rule for logarithms.
$$x \log 7 = \log 5$$
$$x = \frac{\log 5}{\log 7} \approx 0.827$$

The solution set is $\{0.827\}$.

3.
$$9^{-x+2} = 13$$
$$\log 9^{-x+2} = \log 13$$
$$(-x + 2) \log 9 = \log 13 \, (*)$$
$$-x \log 9 + 2 \log 9 = \log 13$$
$$-x \log 9 = \log 13 - 2 \log 9$$
$$x \log 9 = 2 \log 9 - \log 13$$
$$x = \frac{2 \log 9 - \log 13}{\log 9}$$
$$\approx 0.833$$

$(*)$ Alternative solution steps:

$$(-x + 2) \log 9 = \log 13$$
$$-x + 2 = \frac{\log 13}{\log 9}$$
$$2 - \frac{\log 13}{\log 9} = x$$

The solution set is $\{0.833\}$.

5.
$$3^{2x} = 14$$
$$\log 3^{2x} = \log 14$$
$$2x \log 3 = \log 14$$
$$x = \frac{\log 14}{2 \log 3} \approx 1.201$$

The solution set is $\{1.201\}$.

7.
$$2^{x+3} = 5^x$$
$$\log 2^{x+3} = \log 5^x$$
$$(x + 3) \log 2 = x \log 5$$
$$x \log 2 + 3 \log 2 = x \log 5$$
Get x-terms on one side.
$$x \log 2 - x \log 5 = -3 \log 2$$
$$x (\log 2 - \log 5) = -3 \log 2 \quad \textit{Factor out x.}$$
$$x = \frac{-3 \log 2}{\log 2 - \log 5}$$
$$\approx 2.269$$

The solution set is $\{2.269\}$.

9.
$$2^{x+3} = 3^{x-4}$$
$$\log 2^{x+3} = \log 3^{x-4}$$
$$(x + 3) \log 2 = (x - 4) \log 3$$
$$x \log 2 + 3 \log 2 = x \log 3 - 4 \log 3$$
$$\textit{Distributive property}$$

Get x-terms on one side.
$$x \log 2 - x \log 3 = -3 \log 2 - 4 \log 3$$
Factor out x.
$$x(\log 2 - \log 3) = -3 \log 2 - 4 \log 3$$
$$x = \frac{-3 \log 2 - 4 \log 3}{\log 2 - \log 3}$$
$$\approx 15.967$$

The solution set is $\{15.967\}$.

11.
$$4^{2x+3} = 6^{x-1}$$
$$\log 4^{2x+3} = \log 6^{x-1}$$
$$(2x + 3) \log 4 = (x - 1) \log 6$$
$$2x \log 4 + 3 \log 4 = x \log 6 - 1 \log 6$$
$$\textit{Distributive property}$$
Get x-terms on one side.
$$2x \log 4 - x \log 6 = -3 \log 4 - \log 6$$
Factor out x.
$$x(2 \log 4 - \log 6) = -3 \log 4 - \log 6$$
$$x = \frac{-3 \log 4 - \log 6}{2 \log 4 - \log 6}$$
$$\approx -6.067$$

The solution set is $\{-6.067\}$.

13.
$$e^{0.012x} = 23$$
Take base e logarithms on each side.
$$\ln e^{0.012x} = \ln 23$$
$$0.012x (\ln e) = \ln 23$$
$$0.012x = \ln 23 \quad \textit{ln e = 1}$$
$$x = \frac{\ln 23}{0.012} \approx 261.291$$

The solution set is $\{261.291\}$.

15.
$$e^{-0.205x} = 9$$
Take base e logarithms on each side.
$$\ln e^{-0.205x} = \ln 9$$
$$-0.205x (\ln e) = \ln 9$$
$$-0.205x = \ln 9 \quad \textit{ln e = 1}$$
$$x = \frac{\ln 9}{-0.205} \approx -10.718$$

The solution set is $\{-10.718\}$.

17.
$$\ln e^{3x} = 9$$
$$3x = 9 \quad \textit{Special property}$$
$$x = 3$$

The solution set is $\{3\}$.

19.
$$\ln e^{0.45x} = \sqrt{7}$$
$$0.45x = \sqrt{7}$$
$$x = \frac{\sqrt{7}}{0.45} \approx 5.879$$

The solution set is $\{5.879\}$.

Copyright © 2012 Pearson Education, Inc. Publishing as Addison-Wesley.

21. $\ln e^{-x} = \pi$

$\quad -x = \pi$

$\quad x = -\pi \approx -3.142$

The solution set is $\{-\pi\}$, or $\{-3.142\}$.

23. $e^{\ln 2x} = e^{\ln (x+1)}$

$\quad 2x = x + 1 \qquad$ *Special property*

$\quad x = 1$

The solution set is $\{1\}$.

25. Let's try Exercise 14.

$$e^{0.006x} = 30$$

$$\log e^{0.006x} = \log 30$$

$$0.006x \, (\log e) = \log 30$$

$$x = \frac{\log 30}{0.006 \log e} \approx 566.866$$

The natural logarithm is a better choice because $\ln e = 1$, whereas $\log e$ needs to be calculated.

27. $\log_3 (6x + 5) = 2$

$\quad 6x + 5 = 3^2 \qquad$ *Exponential form*

$\quad 6x + 5 = 9$

$\quad 6x = 4$

$\quad x = \frac{4}{6} = \frac{2}{3}$

Check $x = \frac{2}{3}$: $\log_3 9 = \log_3 3^2 = 2$

The solution set is $\left\{\frac{2}{3}\right\}$.

29. $\log_2 (2x - 1) = 5$

$\quad 2x - 1 = 2^5 \quad$ *Exponential form*

$\quad 2x - 1 = 32$

$\quad 2x = 33$

$\quad x = \frac{33}{2}$

Check $x = \frac{33}{2}$: $\log_2 32 = \log_2 2^5 = 5$

The solution set is $\left\{\frac{33}{2}\right\}$.

31. $\log_7 (x + 1)^3 = 2$

$\quad (x + 1)^3 = 7^2 \qquad$ *Exponential form*

$\quad x + 1 = \sqrt[3]{49} \qquad$ *Cube root*

$\quad x = -1 + \sqrt[3]{49}$

Check $x = -1 + \sqrt[3]{49}$: $\log_7 49 = \log_7 7^2 = 2$

The solution set is $\left\{-1 + \sqrt[3]{49}\right\}$.

33. 2 cannot be a solution because $\log (2 - 3) = \log (-1)$, and -1 is not in the domain of $\log x$.

35. $\log (6x + 1) = \log 3$

$\quad 6x + 1 = 3 \qquad$ *Property 4*

$\quad 6x = 2$

$\quad x = \frac{2}{6} = \frac{1}{3}$

Check $x = \frac{1}{3}$: $\log (2 + 1) = \log 3 \quad$ *True*

The solution set is $\left\{\frac{1}{3}\right\}$.

37. $\log_5 (3t + 2) - \log_5 t = \log_5 4$

$$\log_5 \frac{3t + 2}{t} = \log_5 4$$

$$\frac{3t + 2}{t} = 4$$

$$3t + 2 = 4t$$

$$2 = t$$

Check $t = 2$: $\log_5 8 - \log_5 2 = \log_5 \frac{8}{2} = \log_5 4$

The solution set is $\{2\}$.

39. $\log 4x - \log (x - 3) = \log 2$

$$\log \frac{4x}{x - 3} = \log 2$$

$$\frac{4x}{x - 3} = 2$$

$$4x = 2(x - 3)$$

$$4x = 2x - 6$$

$$2x = -6$$

$$x = -3$$

Reject $x = -3$, because $4x = -12$, which yields an equation in which the logarithm of a negative number must be found.

The solution set is $\emptyset$.

41. $\log_2 x + \log_2 (x - 7) = 3$

$\quad \log_2 [x(x - 7)] = 3$

$\quad x(x - 7) = 2^3 \quad$ *Exponential form*

$\quad x^2 - 7x = 8$

$\quad x^2 - 7x - 8 = 0$

$\quad (x - 8)(x + 1) = 0$

$\quad x - 8 = 0 \quad$ or $\quad x + 1 = 0$

$\quad x = 8 \quad$ or $\quad x = -1$

Reject $x = -1$, because it yields an equation in which the logarithm of a negative number must be found.

Check $x = 8$: $\log_2 8 + \log_2 1 = \log_2 2^3 + 0 = 3$

The solution set is $\{8\}$.

43. $\log 5x - \log (2x - 1) = \log 4$

$$\log \frac{5x}{2x - 1} = \log 4$$

$$\frac{5x}{2x - 1} = 4$$

$$5x = 8x - 4$$

$$4 = 3x$$

$$\frac{4}{3} = x$$

Check $x = \frac{4}{3}$: $\log \frac{20}{3} - \log \frac{5}{3} = \log \frac{20/3}{5/3} = \log 4$

The solution set is $\left\{\frac{4}{3}\right\}$.

Copyright © 2012 Pearson Education, Inc. Publishing as Addison-Wesley.

45.
$$\log_2 x + \log_2 (x - 6) = 4$$
$$\log_2 [x(x - 6)] = 4$$
$$x(x - 6) = 2^4 \quad \textit{Exponential form}$$
$$x^2 - 6x = 16$$
$$x^2 - 6x - 16 = 0$$
$$(x - 8)(x + 2) = 0$$
$$x - 8 = 0 \quad \text{or} \quad x + 2 = 0$$
$$x = 8 \quad \text{or} \quad x = -2$$

Reject $x = -2$, because it yields an equation in which the logarithm of a negative number must be found.

Check $x = 8$:
$$\log_2 8 + \log_2 2 = \log_2 16 = \log_2 2^4 = 4$$

The solution set is $\{8\}$.

47. **(a)** Use the formula $A = P\left(1 + \frac{r}{n}\right)^{nt}$ with $P = 2000$, $r = 0.04$, $n = 4$, and $t = 6$.

$$A = 2000\left(1 + \frac{0.04}{4}\right)^{4 \cdot 6}$$
$$= 2000(1.01)^{24} \approx 2539.47$$

The account will contain $2539.47.

(b)
$$3000 = 2000\left(1 + \frac{0.04}{4}\right)^{4t}$$
$$\frac{3000}{2000} = (1.01)^{4t}$$
$$\log\left(\tfrac{3}{2}\right) = \log (1.01)^{4t}$$
$$\log\left(\tfrac{3}{2}\right) = 4t \log (1.01)$$
$$t = \frac{\log\left(\tfrac{3}{2}\right)}{4 \log (1.01)} \approx 10.2$$

It will take about 10.2 years for the account to grow to $3000.

49. **(a)** Use the formula $A = Pe^{rt}$ with $P = 4000$, $r = 0.035$, and $t = 6$.

$$A = 4000e^{(0.035)(6)}$$
$$= 4000e^{0.21} \approx 4934.71$$

There will be $4934.71 in the account.

(b) If the initial amount doubles, then $A = 2P$, or $8000.

$$8000 = 4000e^{0.035t}$$
$$2 = e^{0.035t} \qquad \textit{Divide by 4000.}$$
$$\ln 2 = \ln e^{0.035t}$$
$$\ln 2 = 0.035t$$
$$t = \frac{\ln 2}{0.035} \approx 19.8$$

The initial amount will double in about 19.8 years.

51. Use $A = P\left(1 + \frac{r}{n}\right)^{nt}$ with $P = 5000$, $r = 0.07$, and $t = 12$.

(a) If the interest is compounded annually, $n = 1$.

$$A = 5000\left(1 + \frac{0.07}{1}\right)^{1 \cdot 12}$$
$$= 5000(1.07)^{12} \approx 11,260.96$$

There will be $11,260.96 in the account.

(b) If the interest is compounded semiannually, $n = 2$.

$$A = 5000\left(1 + \frac{0.07}{2}\right)^{2 \cdot 12}$$
$$= 5000(1.035)^{24} \approx 11,416.64$$

There will be $11,416.64 in the account.

(c) If the interest is compounded quarterly, $n = 4$.

$$A = 5000\left(1 + \frac{0.07}{4}\right)^{4 \cdot 12}$$
$$= 5000(1.0175)^{48} \approx 11,497.99$$

There will be $11,497.99 in the account.

(d) If the interest is compounded daily, $n = 365$.

$$A = 5000\left(1 + \frac{0.07}{365}\right)^{365 \cdot 12}$$
$$\approx 11,580.90$$

There will be $11,580.90 in the account.

(e) Use the continuous compound interest formula.

$$A = Pe^{rt}$$
$$A = 5000e^{0.07(12)}$$
$$= 5000e^{0.84} \approx 11,581.83$$

There will be $11,581.83 in the account.

53. In the continuous compound interest formula, let $A = 1850$, $r = 0.065$, and $t = 40$.

$$A = Pe^{rt}$$
$$1850 = Pe^{0.065(40)}$$
$$1850 = Pe^{2.6}$$
$$P = \frac{1850}{e^{2.6}} \approx 137.41$$

Deposit $137.41 today.

55. $f(x) = 15.94e^{0.0656x}$

(a) 1980 corresponds to $x = 0$.

$$f(0) = 15.94e^{0.0656(0)}$$
$$= 15.94(1) = 15.94$$

The approximate volume of materials recovered from municipal solid waste collections in the United States in 1980 was 15.9 million tons.

(b) 1990 corresponds to $x = 10$.

$$f(10) \approx 30.7 \text{ million tons}$$

(c) 2000 corresponds to $x = 20$.

$$f(20) \approx 59.2 \text{ million tons}$$

(d) 2007 corresponds to $x = 27$.

$$f(27) \approx 93.7 \text{ million tons}$$

Copyright © 2012 Pearson Education, Inc. Publishing as Addison-Wesley.

57. $S(x) = 112{,}047e^{0.0827x}$

2007 corresponds to $x = 3$.

$S(3) = 112{,}047e^{0.0827(3)}$

$\quad\ \approx 143{,}598$

The approximate value of software-publisher revenues for 2007 was 143,598 million dollars.

59. $A(t) = 2.00e^{-0.053t}$

(a) Let $t = 4$.

$$A(4) = 2.00e^{-0.053(4)}$$
$$= 2.00e^{-0.212}$$
$$\approx 1.62$$

About 1.62 grams would be present.

(b) $A(10) = 2.00e^{-0.053(10)}$

$\qquad\quad \approx 1.18$

About 1.18 grams would be present.

(c) $A(20) = 2.00e^{-0.053(20)}$

$\qquad\quad \approx 0.69$

About 0.69 grams would be present.

(d) The initial amount is the amount $A(t)$ present at time $t = 0$.

$$A(0) = 2.00e^{-0.053(0)}$$
$$= 2.00e^{0} = 2.00(1) = 2.00$$

Initially, 2.00 grams were present.

61. **(a)** Find $A(t) = 400e^{-0.032t}$ when $t = 25$.

$$A(25) = 400e^{-0.032(25)}$$
$$= 400e^{-0.8} \approx 179.73$$

About 179.73 grams of lead will be left.

(b) Use $A(t) = 400e^{-0.032t}$, with $A(t) = \frac{1}{2}(400) = 200$.

$$200 = 400e^{-0.032t}$$
$$\tfrac{200}{400} = e^{-0.032t}$$
$$0.5 = e^{-0.032t}$$
$$\ln 0.5 = \ln e^{-0.032t}$$
$$\ln 0.5 = -0.032t(\ln e)$$
$$t = \frac{\ln 0.5}{-0.032} \approx 21.66$$

It would take about 21.66 years for the sample to decay to half its original amount.

63. $\qquad\quad f(x) = 15.94e^{0.0656x}$

$\qquad\qquad 130 = 15.94e^{0.0656x}$

$\qquad e^{0.0656x} = \tfrac{130}{15.94}$

$\qquad \ln e^{0.0656x} = \ln \tfrac{130}{15.94}$

$0.0656x\,(\ln e) = \ln \tfrac{130}{15.94}$

$\qquad\qquad 0.0656x = \ln \tfrac{130}{15.94}$

$$x = \frac{\ln \tfrac{130}{15.94}}{0.0656} \approx 32$$

Since $x = 0$ corresponds to 1980, $x = 32$ corresponds to 2012.

65. The display means that after 250 years, approximately 2.918 748 7 grams of the original sample of radium 226 remain.

67. $f(x) = 2x^2$

The graph of f is narrower than the graph of $y = x^2$.

x	0	± 1	± 2
y	0	2	8

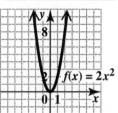

69. $f(x) = (x + 1)^2$

The graph of f is the graph of $y = x^2$ shifted *left* 1 unit.

x	-3	-2	-1	0	1
y	4	1	0	1	4

$f(x) = (x + 1)^2$

Chapter 11 Review Exercises

1. Since a horizontal line intersects the graph in two points, the function is not one-to-one.

2. Since every horizontal line intersects the graph in no more than one point, the function is one-to-one.

3. The function $f(x) = -3x + 7$ is a linear function. By the horizontal line test, it is a one-to-one function. To find the inverse, replace $f(x)$ with y.

$$y = -3x + 7$$

Interchange x and y.

$$x = -3y + 7$$

Solve for y.

$$x - 7 = -3y$$
$$\frac{x - 7}{-3} = y, \ \text{ or } \ \frac{7 - x}{3} = y$$

Replace y with $f^{-1}(x)$.

$$f^{-1}(x) = \frac{x - 7}{-3}, \ \text{ or } \ f^{-1}(x) = -\frac{1}{3}x + \frac{7}{3}$$

Copyright © 2012 Pearson Education, Inc. Publishing as Addison-Wesley.

4. $f(x) = \sqrt[3]{6x - 4}$

The function is one-to-one since each $f(x)$-value corresponds to exactly one x-value.

To find the inverse, replace $f(x)$ with y.

$$y = \sqrt[3]{6x - 4}$$

Interchange x and y.

$$x = \sqrt[3]{6y - 4}$$

Solve for y.

$$x^3 = 6y - 4 \quad \textit{Cube each side.}$$
$$x^3 + 4 = 6y$$
$$\frac{x^3 + 4}{6} = y$$

Replace y with $f^{-1}(x)$

$$\frac{x^3 + 4}{6} = f^{-1}(x)$$

5. $f(x) = -x^2 + 3$

This is an equation of a vertical parabola which opens down.

Since a horizontal line will intersect the graph in two points, the function is not one-to-one.

6. This function is not one-to-one because two sodas in the list have 41 mg of caffeine.

7. The graph is a linear function through $(0, 1)$ and $(3, 0)$. The graph of $f^{-1}(x)$ will include the points $(1, 0)$ and $(0, 3)$, found by interchanging x and y. Plot these points, and draw a straight line through them.

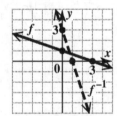

8. The graph is a curve through $(1, 2)$, $(0, 1)$, and $\left(-1, \frac{1}{2}\right)$. Interchange x and y to get $(2, 1)$, $(1, 0)$, and $\left(\frac{1}{2}, -1\right)$, which are on the graph of $f^{-1}(x)$. Plot these points, and draw a smooth curve through them.

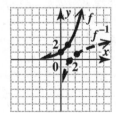

9. $f(x) = 3^x$

Make a table of values.

x	-2	-1	0	1	2
$f(x)$	$\frac{1}{9}$	$\frac{1}{3}$	1	3	9

Plot the points from the table and draw a smooth curve through them.

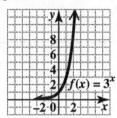

10. $f(x) = \left(\frac{1}{3}\right)^x$

Make a table of values.

x	-2	-1	0	1	2
$f(x)$	9	3	1	$\frac{1}{3}$	$\frac{1}{9}$

Plot the points from the table and draw a smooth curve through them.

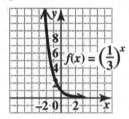

11. $y = 2^{2x+3}$

Make a table of values.

x	-2	$-\frac{3}{2}$	-1	0	$\frac{1}{2}$
y	$\frac{1}{2}$	1	2	8	16

Plot the points from the table and draw a smooth curve through them.

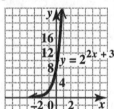

12. $5^{2x+1} = 25$

Write each side as a power of 5.

$$(5)^{2x+1} = (5)^2$$
$$2x + 1 = 2 \quad \textit{Equate exponents.}$$
$$2x = 1$$
$$x = \frac{1}{2}$$

Check $x = \frac{1}{2}$: $\quad 5^2 = 25 \quad$ *True*

The solution set is $\left\{\frac{1}{2}\right\}$.

13. $4^{3x} = 8^{x+4}$

Write each side as a power of 2.

$$\left(2^2\right)^{3x} = \left(2^3\right)^{x+4}$$
$$2^{6x} = 2^{3x+12}$$
$$6x = 3x + 12 \quad \textit{Equate exponents.}$$
$$3x = 12$$
$$x = 4$$

Copyright © 2012 Pearson Education, Inc. Publishing as Addison-Wesley.

Check $x = 4$: $4^{12} = 8^8$ *True*

The solution set is $\{4\}$.

14. $\left(\frac{1}{27}\right)^{x-1} = 9^{2x}$

$\left[\left(\frac{1}{3}\right)^3\right]^{x-1} = \left(3^2\right)^{2x}$

Write each side as a power of 3.

$\left(3^{-3}\right)^{x-1} = \left(3^2\right)^{2x}$

$3^{-3x+3} = 3^{4x}$

$-3x + 3 = 4x$ *Equate exponents.*

$3 = 7x$

$\frac{3}{7} = x$

Check $x = \frac{3}{7}$: $\left(\frac{1}{27}\right)^{-4/7} = 9^{6/7}$ *True*

The solution set is $\left\{\frac{3}{7}\right\}$.

15. $S(x) = 33.07(1.0241)^{-x}$

(a) $x = 1975 - 1970 = 5$

$S(5) = 33.07(1.0241)^{-5}$

≈ 29.4

The approximate amount of sulfur dioxide emissions in the United States in 1975 was 29.4 million tons.

(b) $x = 1995 - 1970 = 25$

$S(25) \approx 18.2$ million tons

(c) $x = 2005 - 1970 = 35$

$S(35) \approx 14.4$ million tons

16. $g(x) = \log_3 x$

Replace $g(x)$ with y, and write in exponential form.

$y = \log_3 x$

$3^y = x$

Make a table of values. Since $x = 3^y$ is the inverse of $f(x) = y = 3^x$ in Exercise 9, simply reverse the ordered pairs in the table of values belonging to $f(x) = 3^x$.

x	$\frac{1}{9}$	$\frac{1}{3}$	1	3	9
y	-2	-1	0	1	2

Plot the points from the table and draw a smooth curve through them.

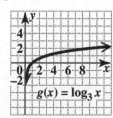

17. $g(x) = \log_{1/3} x$

Replace $g(x)$ with y, and write in exponential form.

$y = \log_{1/3} x$

$\left(\frac{1}{3}\right)^y = x$

Make a table of values. Since $x = \left(\frac{1}{3}\right)^y$ is the inverse of $f(x) = y = \left(\frac{1}{3}\right)^x$ in Exercise 10, simply reverse the ordered pairs in the table of values belonging to $f(x) = \left(\frac{1}{3}\right)^x$.

x	9	3	1	$\frac{1}{3}$	$\frac{1}{9}$
y	-2	-1	0	1	2

Plot the points from the table and draw a smooth curve through them.

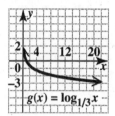

18. $\log_8 64 = x$

$8^x = 64$ *Exponential form*

Write each side as a power of 8.

$8^x = 8^2$

$x = 2$ *Equate exponents.*

The solution set is $\{2\}$.

19. $\log_2 \sqrt{8} = x$

$2^x = \sqrt{8}$ *Exponential form*

$2^x = 8^{1/2}$

Write each side as a power of 2.

$2^x = \left(2^3\right)^{1/2}$

$2^x = 2^{3/2}$

$x = \frac{3}{2}$ *Equate exponents.*

The solution set is $\left\{\frac{3}{2}\right\}$.

20. $\log_x \left(\frac{1}{49}\right) = -2$

$x^{-2} = \frac{1}{49}$ *Exponential form*

$\frac{1}{x^2} = \frac{1}{49}$

$x^2 = 49$

$x = \pm 7$

Since x is the base, we cannot have a negative number.

The solution set is $\{7\}$.

Copyright © 2012 Pearson Education, Inc. Publishing as Addison-Wesley.

21. $\log_4 x = \frac{3}{2}$

$\quad x = 4^{3/2}$ *Exponential form*

$\quad x = \left(\sqrt{4}\right)^3 = 2^3 = 8$

The argument (the input of the logarithm) must be a positive number, so $x = 8$ is acceptable.

The solution set is $\{8\}$.

22. $\log_k 4 = 1$

$\quad k^1 = 4$ *Exponential form*

$\quad k = 4$

$k = 4$ is an acceptable base since it is a positive number (not equal to 1).

The solution set is $\{4\}$.

23. $\log_b b^2 = 2$

$\quad b^2 = b^2$ *Exponential form*

This is an identity. Thus, b can be any real number, $b > 0$ and $b \neq 1$.

The solution set is $\{b \mid b > 0, \, b \neq 1\}$.

24. $\log_b a$ is the exponent to which b must be raised to obtain a.

25. From Exercise 24,

$$b^{\log_b a} = a.$$

26. $S(x) = 100 \log_2 (x + 2)$

(a) When $x = 6$,

$$S(6) = 100 \log_2 (6 + 2)$$
$$= 100(3) = 300.$$

After 6 weeks the sales were 300 thousand dollars or \$300,000.

(b) To graph the function, make a table of values that includes the ordered pair from above.

x	0	2	6
$S(x)$	100	200	300

Plot the ordered pairs and draw the graph through them.

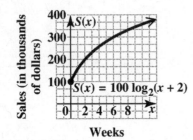

27. $\log_2 3xy^2$

$\quad = \log_2 3 + \log_2 x + \log_2 y^2$ *Product rule*

$\quad = \log_2 3 + \log_2 x + 2 \log_2 y$ *Power rule*

28. $\log_4 \dfrac{\sqrt{x} \cdot w^2}{z}$

$\quad = \log_4 \left(\sqrt{x} \cdot w^2\right) - \log_4 z$ *Quotient rule*

$\quad = \log_4 x^{1/2} + \log_4 w^2 - \log_4 z$ *Product rule*

$\quad = \frac{1}{2} \log_4 x + 2 \log_4 w - \log_4 z$ *Power rule*

29. $\log_b 3 + \log_b x - 2 \log_b y$

Use the product and power rules for logarithms.

$\quad = \log_b (3 \cdot x) - \log_b y^2$

$\quad = \log_b \dfrac{3x}{y^2}$ *Quotient rule*

30. $\log_3 (x + 7) - \log_3 (4x + 6)$

$\quad = \log_3 \left(\dfrac{x + 7}{4x + 6}\right)$ *Quotient rule*

31. $\log 28.9 \approx 1.4609$

32. $\log 0.257 \approx -0.5901$

33. $\ln 28.9 \approx 3.3638$

34. $\ln 0.257 \approx -1.3587$

35. $\log_{16} 13 = \dfrac{\log 13}{\log 16} \approx 0.9251$

36. $\log_4 12 = \dfrac{\log 12}{\log 4} \approx 1.7925$

37. Milk has a hydronium ion concentration of 4.0×10^{-7}.

$\quad \text{pH} = -\log \left[H_3O^+\right]$

$\quad \text{pH} = -\log \left(4.0 \times 10^{-7}\right) \approx 6.4$

38. Crackers have a hydronium ion concentration of 3.8×10^{-9}.

$\quad \text{pH} = -\log \left[H_3O^+\right]$

$\quad \text{pH} = -\log \left(3.8 \times 10^{-9}\right) \approx 8.4$

39. Orange juice has a pH of 4.6.

$\quad \text{pH} = -\log \left[H_3O^+\right]$

$\quad 4.6 = -\log \left[H_3O^+\right]$

$\quad \log_{10} \left[H_3O^+\right] = -4.6$

$\quad \left[H_3O^+\right] = 10^{-4.6} \approx 2.5 \times 10^{-5}$

40. Since we want to determine a ratio of intensities, we'll solve the given formula for the intensity I.

$$M = 6 - 2.5 \log \frac{I}{I_0}$$

$$2.5 \log \frac{I}{I_0} = 6 - M$$

$$\log_{10} \frac{I}{I_0} = \frac{6 - M}{2.5} \qquad \textit{Common log}$$

$$\frac{I}{I_0} = 10^{(6 - M)/2.5} \qquad \textit{Exponential form}$$

$$I = 10^{0.4(6 - M)} I_0 \qquad 1/2.5 = 0.4$$

Copyright © 2012 Pearson Education, Inc. Publishing as Addison-Wesley.

We now have I as a function of M. The ratio of intensities between stars of magnitude 1 and 3 is

$$\frac{I(1)}{I(3)} = \frac{10^{0.4(6-1)}I_0}{10^{0.4(6-3)}I_0} = \frac{10^2}{10^{1.2}} = 10^{0.8} \approx 6.3.$$

Thus, a star of magnitude 1 is about 6.3 times as intense as a star of magnitude 3.

41. $t(r) = \dfrac{\ln 2}{\ln (1 + r)}$

(a) $4\% = 0.04$; $t(0.04) = \dfrac{\ln 2}{\ln (1 + 0.04)} \approx 18$

At 4%, it would take about 18 years.

(b) $6\% = 0.06$; $t(0.06) = \dfrac{\ln 2}{\ln (1 + 0.06)} \approx 12$

At 6%, it would take about 12 years.

(c) $10\% = 0.10$; $t(0.10) = \dfrac{\ln 2}{\ln (1 + 0.10)} \approx 7$

At 10%, it would take about 7 years.

(d) $12\% = 0.12$; $t(0.12) = \dfrac{\ln 2}{\ln (1 + 0.12)} \approx 6$

At 12%, it would take about 6 years.

(e) Each comparison shows approximately the same number. For example, in part (a) the doubling time is 18 yr (rounded) and $\frac{72}{4} = 18$.

Thus, the formula $t = \dfrac{72}{100r}$ (called the *rule of 72*) is an excellent approximation of the doubling time formula.

42.
$$3^x = 9.42$$
$$\log 3^x = \log 9.42$$
$$x \log 3 = \log 9.42$$
$$x = \frac{\log 9.42}{\log 3} \approx 2.042$$

Check $x = 2.042$: $3^{2.042} \approx 9.425$

The solution set is $\{2.042\}$.

43.
$$2^{x-1} = 15$$
$$\log 2^{x-1} = \log 15$$
$$(x - 1) \log 2 = \log 15$$
$$x - 1 = \frac{\log 15}{\log 2}$$
$$x = \frac{\log 15}{\log 2} + 1 \approx 4.907$$

Check $x = 4.907$: $2^{3.907} \approx 15.0$

The solution set is $\{4.907\}$.

44. $e^{0.06x} = 3$

Take base e logarithms on each side.
$$\ln e^{0.06x} = \ln 3$$
$$0.06x \ln e = \ln 3$$
$$0.06x = \ln 3 \qquad \textit{ln e = 1}$$
$$x = \frac{\ln 3}{0.06} \approx 18.310$$

Check $x = 18.310$: $e^{1.0986} \approx 3.0$

The solution set is $\{18.310\}$.

45. $\log_3 (9x + 8) = 2$
$$9x + 8 = 3^2 \quad \textit{Exponential form}$$
$$9x + 8 = 9$$
$$9x = 1$$
$$x = \tfrac{1}{9}$$

Check $x = \tfrac{1}{9}$: $\log_3 9 = \log_3 3^2 = 2$

The solution set is $\left\{\tfrac{1}{9}\right\}$.

46. $\log_5 (x + 6)^3 = 2$
Change to exponential form.
$$(x + 6)^3 = 5^2$$
$$(x + 6)^3 = 25$$
Take the cube root of each side.
$$x + 6 = \sqrt[3]{25}$$
$$x = \sqrt[3]{25} - 6$$

Check $x = -6 + \sqrt[3]{25}$: $\log_5 25 = \log_5 5^2 = 2$

The solution set is $\left\{-6 + \sqrt[3]{25}\right\}$.

47. $\log_3 (x + 2) - \log_3 x = \log_3 2$
$$\log_3 \frac{x + 2}{x} = \log_3 2 \quad \textit{Quotient rule}$$
$$\frac{x + 2}{x} = 2 \qquad\quad \textit{Property 4}$$
$$x + 2 = 2x$$
$$2 = x$$

Check $x = 2$: $\log_3 4 - \log_3 2 = \log_3 \tfrac{4}{2} = \log_3 2$

The solution set is $\{2\}$.

48.
$$\log (2x + 3) = 1 + \log x$$
$$\log (2x + 3) - \log x = 1$$
$$\log_{10} \frac{2x + 3}{x} = 1 \qquad \textit{Quotient rule}$$
$$10^1 = \frac{2x + 3}{x} \quad \begin{array}{l}\textit{Exponential}\\ \textit{form}\end{array}$$
$$10x = 2x + 3$$
$$8x = 3$$
$$x = \tfrac{3}{8}$$

Check $x = \tfrac{3}{8}$:
$$LS = \log \left(\tfrac{3}{4} + 3\right) = \log \tfrac{15}{4}$$
$$RS = \log 10 + \log \tfrac{3}{8} = \log \tfrac{30}{8} = \log \tfrac{15}{4}$$

The solution set is $\left\{\tfrac{3}{8}\right\}$.

Copyright © 2012 Pearson Education, Inc. Publishing as Addison-Wesley.

49. $\log_4 x + \log_4 (8 - x) = 2$

$\qquad \log_4 [x(8 - x)] = 2$ *Product rule*

$\qquad\qquad x(8 - x) = 4^2$ *Exponential form*

$\qquad\qquad\quad 8x - x^2 = 16$

$\qquad\qquad x^2 - 8x + 16 = 0$

$\qquad\qquad (x - 4)(x - 4) = 0$

$\qquad\qquad\qquad\quad x - 4 = 0$

$\qquad\qquad\qquad\qquad\quad x = 4$

Check $x = 4$: $\log_4 4 + \log_4 4 = 1 + 1 = 2$

The solution set is $\{4\}$.

50. $\log_2 x + \log_2 (x + 15) = \log_2 16$

$\qquad \log_2 [x(x + 15)] = \log_2 16$ *Product rule*

$\qquad\qquad\quad x^2 + 15x = 16$ *Property 4*

$\qquad\qquad x^2 + 15x - 16 = 0$

$\qquad\qquad (x + 16)(x - 1) = 0$

$\qquad x + 16 = 0 \qquad \text{or} \quad x - 1 = 0$

$\qquad\quad x = -16 \quad \text{or} \qquad\quad x = 1$

Reject $x = -16$, because it yields an equation in which the logarithm of a negative number must be found.

Check $x = 1$:

$\quad \log_2 1 + \log_2 16 = 0 + \log_2 16 = \log_2 16$

The solution set is $\{1\}$.

51. When the power rule was applied in the second step, the domain was changed from $\{x \mid x \neq 0\}$ to $\{x \mid x > 0\}$. Instead of using the power rule for logarithms, we can change the original equation to the exponential form $x^2 = 10^2$ and get $x = \pm 10$. As you can see in the erroneous solution, the valid solution -10 was "lost." The solution set is $\{\pm 10\}$.

52. $A = P\left(1 + \dfrac{r}{n}\right)^{nt}$

Let $P = 20{,}000, r = 0.04$, and $t = 5$. For $n = 4$ (quarterly compounding),

$$A = 20{,}000\left(1 + \frac{0.04}{4}\right)^{4 \cdot 5} \approx 24{,}403.80.$$

There will be $\$24{,}403.80$ in the account after 5 years.

53. In the continuous compounding formula, let $P = 10{,}000, r = 0.0375$, and $t = 3$.

$\quad A = Pe^{rt}$

$\quad A = 10{,}000e^{0.0375(3)} \approx 11{,}190.72$

There will be $\$11{,}190.72$ in the account after 3 years.

54. Use $A = P\left(1 + \dfrac{r}{n}\right)^{nt}$.

Plan A:

Let $P = 1000, r = 0.04, n = 4$, and $t = 3$.

$$A = 1000\left(1 + \frac{0.04}{4}\right)^{4 \cdot 3} \approx 1126.83$$

Plan B:

Let $P = 1000, r = 0.039, n = 12$, and $t = 3$.

$$A = 1000\left(1 + \frac{0.039}{12}\right)^{12 \cdot 3} \approx 1123.91$$

Plan A is the better plan by $\$2.92$.

55. Let $Q(t) = \frac{1}{2}A_0$ to find the half-life of the radioactive substance.

$$Q(t) = A_0 e^{-0.05t}$$

$$\tfrac{1}{2}A_0 = A_0 e^{-0.05t}$$

$$0.5 = e^{-0.05t}$$

$$\ln 0.5 = \ln e^{-0.05t}$$

$$\ln 0.5 = -0.05t \,(\ln e)$$

$$-0.05t = \ln 0.5$$

$$t = \frac{\ln 0.5}{-0.05} \approx 13.9$$

The half-life is about 13.9 days.

56. $S = C(1 - r)^n$

(a) Let $C = 30{,}000, r = 0.15$, and $n = 12$.

$$S = 30{,}000(1 - 0.15)^{12}$$

$$= 30{,}000(0.85)^{12} \approx 4267$$

The scrap value is about $\$4267$.

(b) Let $S = \frac{1}{2}C$ and $n = 6$.

$$S = C(1 - r)^n$$

$$\tfrac{1}{2}C = C(1 - r)^6$$

$$0.5 = (1 - r)^6 \quad (*)$$

$$\ln 0.5 = \ln (1 - r)^6$$

$$\ln 0.5 = 6\ln (1 - r)$$

$$\ln (1 - r) = \frac{\ln 0.5}{6}$$

$$\ln (1 - r) \approx -0.1155$$

$$1 - r = e^{-0.1155}$$

$$1 - r \approx 0.89$$

$$r = 0.11$$

The rate is approximately 11%.

$(*)$ Alternative solution steps without logarithms:

$$0.5 = (1 - r)^6$$

$$\sqrt[6]{0.5} = 1 - r$$

$$r = 1 - \sqrt[6]{0.5} \approx 0.11$$

Note that $1 - r$ must be positive, so $\pm \sqrt[6]{0.5}$ is not needed.

Copyright © 2012 Pearson Education, Inc. Publishing as Addison-Wesley.

57. $N(r) = -5000 \ln r$

Replace $N(r)$ with 2000, and solve for r.

$$2000 = -5000 \ln r$$
$$\ln r = -\frac{2000}{5000}$$
$$\log_e r = -0.4$$
$$r = e^{-0.4} \qquad \textit{Exponential form}$$
$$r \approx 0.67 \qquad \textit{Approximate}$$

About 67% of the words are common to both of the evolving languages.

58. **A.** Solve $7^x = 23$ by using the power rule with common logarithms.

$$7^x = 23$$
$$\log 7^x = \log 23$$
$$x \log 7 = \log 23$$
$$x = \frac{\log 23}{\log 7}$$

B. Solve $7^x = 23$ by using the power rule with natural logarithms.

$$7^x = 23$$
$$\ln 7^x = \ln 23$$
$$x \ln 7 = \ln 23$$
$$x = \frac{\ln 23}{\ln 7}$$

C. Use the change-of-base rule with the solution from **A.**

$$x = \frac{\log 23}{\log 7} = \log_7 23$$

D. $x = \dfrac{\log 23}{\log 7} \neq \log_{23} 7$

The answer is **D.**

59. **[11.4]** $\log_2 128 = \log_2 2^7 = 7$, by a special property of logarithms.

60. **[11.4]** By a special property of logarithms,
$$5^{\log_5 36} = 36.$$

61. **[11.5]** $e^{\ln 4} = 4$ since $e^{\ln x} = x$.

62. **[11.5]** $10^{\log e} = e$ since $10^{\log x} = x$.

63. **[11.4]** $\log_3 3^{-5} = -5$ since $\log_a a^x = x$.

64. **[11.5]** $\ln e^{5.4} = 5.4$ since $\ln e^x = x$.

65. **[11.6]** $\log_3 (x + 9) = 4$
$$x + 9 = 3^4 \qquad \textit{Exponential form}$$
$$x + 9 = 81$$
$$x = 72$$

Check $x = 72$: $\log_3 81 = \log_3 3^4 = 4$

The solution set is $\{72\}$.

66. **[11.6]** $\ln e^x = 3$
$$x = 3$$

The solution set is $\{3\}$.

67. **[11.3]** $\log_x \frac{1}{81} = 2$
$$x^2 = \frac{1}{81} \qquad \textit{Exponential form}$$
$$x^2 = \left(\frac{1}{9}\right)^2$$
$$x = \pm \frac{1}{9} \qquad \textit{Square root property}$$

The base x cannot be negative.

Check $x = \frac{1}{9}$: $\log_{1/9} \frac{1}{81} = 2$ since $\left(\frac{1}{9}\right)^2 = \frac{1}{81}$

The solution set is $\left\{\frac{1}{9}\right\}$.

68. **[11.2]** $27^x = 81$

Write each side as a power of 3.
$$\left(3^3\right)^x = 3^4$$
$$3^{3x} = 3^4$$
$$3x = 4 \qquad \textit{Equate exponents.}$$
$$x = \frac{4}{3}$$

Check $x = \frac{4}{3}$: $27^{4/3} = 3^4 = 81$

The solution set is $\left\{\frac{4}{3}\right\}$.

69. **[11.2]** $2^{2x - 3} = 8$

Write each side as a power of 2.
$$2^{2x - 3} = 2^3$$
$$2x - 3 = 3 \qquad \textit{Equate exponents.}$$
$$2x = 6$$
$$x = 3$$

Check $x = 3$: $2^3 = 8$ *True*

The solution set is $\{3\}$.

70. **[11.2]** $5^{x + 2} = 25^{2x + 1}$
$$5^{x + 2} = \left(5^2\right)^{2x + 1} \qquad \textit{Equal bases}$$
$$5^{x + 2} = 5^{4x + 2}$$
$$x + 2 = 4x + 2 \qquad \textit{Equate exponents.}$$
$$0 = 3x$$
$$0 = x$$

Check $x = 0$: $5^2 = 25^1$ *True*

The solution set is $\{0\}$.

71. **[11.6]**
$$\log_3 (x + 1) - \log_3 x = 2$$
$$\log_3 \frac{x + 1}{x} = 2 \qquad \textit{Quotient rule}$$
$$\frac{x + 1}{x} = 3^2 \qquad \textit{Exponential form}$$
$$9x = x + 1$$
$$8x = 1$$
$$x = \frac{1}{8}$$

Check $x = \frac{1}{8}$:
$$\log_3 \frac{9}{8} - \log_3 \frac{1}{8} = \log_3 9 = \log_3 3^2 = 2$$

The solution set is $\left\{\frac{1}{8}\right\}$.

Copyright © 2012 Pearson Education, Inc. Publishing as Addison-Wesley.

72. **[11.6]** $\log(3x - 1) = \log 10$
$$3x - 1 = 10$$
$$3x = 11$$
$$x = \tfrac{11}{3}$$

Check $x = \tfrac{11}{3}$: $\log(11 - 1) = \log 10$ *True*

The solution set is $\left\{ \tfrac{11}{3} \right\}$.

73. **[11.6]** $\ln\left(x^2 + 3x + 4\right) = \ln 2$
$$x^2 + 3x + 4 = 2$$
$$x^2 + 3x + 2 = 0$$
$$(x + 2)(x + 1) = 0$$

$x + 2 = 0 \quad$ or $\quad x + 1 = 0$
$x = -2 \quad$ or $\qquad x = -1$

Check $x = -2$: $\ln(4 - 6 + 4) = \ln 2$ *True*
Check $x = -1$: $\ln(1 - 3 + 4) = \ln 2$ *True*

The solution set is $\{-2, -1\}$.

74. **[11.6] (a)** $\log(2x + 3) = \log x + 1$
$\log(2x + 3) - \log x = 1$
$$\log_{10} \frac{2x + 3}{x} = 1 \quad \textit{Quotient rule}$$
$$\frac{2x + 3}{x} = 10^1 \quad \begin{array}{l}\textit{Exponential}\\ \textit{form}\end{array}$$
$$2x + 3 = 10x$$
$$3 = 8x$$
$$\tfrac{3}{8} = x$$

Check $x = \tfrac{3}{8}$:
LS $= \log\left(\tfrac{3}{4} + 3\right) = \log \tfrac{15}{4}$
RS $= \log \tfrac{3}{8} + \log 10 = \log \tfrac{30}{8} = \log \tfrac{15}{4}$

The solution set is $\left\{ \tfrac{3}{8} \right\}$.

(b) From the graph, the x-value of the x-intercept is 0.375, the decimal equivalent of $\tfrac{3}{8}$. Note that the solutions of $Y_1 - Y_2 = 0$ are the same as the solutions of $Y_1 = Y_2$.

75. **[11.6]** For 2005, $x = 2005 - 1980 = 25$

$$R(x) = 0.0997e^{0.0470x}$$
$$R(25) = 0.0997e^{0.0470(25)}$$
$$\approx 0.3228$$

In 2005, about 32.28% of municipal solid waste was recovered.

76. **[11.5] (a)** There are 90 of one species and 10 of another, so

$$p_1 = \tfrac{90}{100} = 0.9 \quad \text{and} \quad p_2 = \tfrac{10}{100} = 0.1.$$

Thus, the index of diversity is

$$-(p_1 \ln p_1 + p_2 \ln p_2)$$
$$= -(0.9 \ln 0.9 + 0.1 \ln 0.1)$$
$$\approx 0.325.$$

(b) There are 60 of one species and 40 of another, so

$$p_1 = \tfrac{60}{100} = 0.6 \quad \text{and} \quad p_2 = \tfrac{40}{100} = 0.4.$$

Thus, the index of diversity is

$$-(p_1 \ln p_1 + p_2 \ln p_2)$$
$$= -(0.6 \ln 0.6 + 0.4 \ln 0.4)$$
$$\approx 0.673.$$

Chapter 11 Test

1.   **(a)** $f(x) = x^2 + 9$

This function is not one-to-one. The graph of $f(x)$ is a vertical parabola. A horizontal line will intersect the graph more than once.

(b) This function is one-to-one. A horizontal line will not intersect the graph in more than one point.

2.
$$f(x) = \sqrt[3]{x + 7}$$
Replace $f(x)$ with y.
$$y = \sqrt[3]{x + 7}$$
Interchange x and y.
$$x = \sqrt[3]{y + 7}$$
Solve for y.
$$x^3 = y + 7$$
$$x^3 - 7 = y$$
Replace y with $f^{-1}(x)$.
$$f^{-1}(x) = x^3 - 7$$

3. By the horizontal line test, $f(x)$ is a one-to-one function and has an inverse. Choose some points on the graph of $f(x)$, such as $(4, 0)$, $(3, -1)$, and $(0, -2)$. To graph the inverse, interchange the x- and y-values to get $(0, 4)$, $(-1, 3)$, and $(-2, 0)$. Plot these points and draw a smooth curve through them.

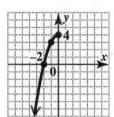

4. $f(x) = 6^x$
Make a table of values.

x	-2	-1	0	1
$f(x)$	$\frac{1}{36}$	$\frac{1}{6}$	1	6

Plot these points and draw a smooth exponential curve through them.

Copyright © 2012 Pearson Education, Inc. Publishing as Addison-Wesley.

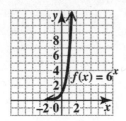

5. $g(x) = \log_6 x$
Make a table of values.

Powers of 6	6^{-2}	6^{-1}	6^0	6^1
x	$\frac{1}{36}$	$\frac{1}{6}$	1	6
$g(x)$	-2	-1	0	1

Plot these points and draw a smooth logarithmic curve through them.

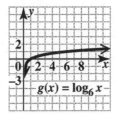

6. $y = 6^x$ and $y = \log_6 x$ are inverse functions. To use the graph from Exercise 4 to obtain the graph of the function in Exercise 5, interchange the x- and y-coordinates of the ordered pairs $\left(-2, \frac{1}{36}\right)$, $\left(-1, \frac{1}{6}\right)$, $(0, 1)$, and $(1, 6)$ to get $\left(\frac{1}{36}, -2\right)$, $\left(\frac{1}{6}, -1\right)$, $(1, 0)$, and $(6, 1)$.
Plot these points and draw a smooth logarithmic curve through them.

7.
$$5^x = \frac{1}{625}$$
$$5^x = \left(\frac{1}{5}\right)^4$$
Write each side as a power of 5.
$$5^x = 5^{-4}$$
$$x = -4 \qquad \textit{Equate exponents.}$$

The solution set is $\{-4\}$.

8.
$$2^{3x-7} = 8^{2x+2}$$
Write each side as a power of 2.
$$2^{3x-7} = \left(2^3\right)^{2x+2}$$
$$3x - 7 = 3(2x+2) \quad \textit{Equate exponents.}$$
$$3x - 7 = 6x + 6$$
$$-13 = 3x$$
$$-\frac{13}{3} = x$$

Check $x = -\frac{13}{3}$:
$$\text{LS} = 2^{-13-7} = 2^{-20}$$
$$\text{RS} = 8^{-20/3} = \left(2^3\right)^{-20/3} = 2^{-20}$$

The solution set is $\left\{-\frac{13}{3}\right\}$.

9. $f(x) = 46.9e^{0.0247x}$

(a) 2015: $x = 2015 - 2008 = 7$
$$f(7) = 46.9e^{0.0247(7)} \approx 55.8$$

The U.S. Hispanic population estimate for 2015 is 55.8 million.

(b) 2030: $x = 2030 - 2008 = 22$
$$f(22) = 46.9e^{0.0247(22)} \approx 80.8$$

The U.S. Hispanic population estimate for 2030 is 80.8 million.

10. The base is 4, the exponent (logarithm) is -2, and the number is 0.0625, so $4^{-2} = 0.0625$ becomes $\log_4 0.0625 = -2$ in logarithmic form.

11. The base is 7, the logarithm (exponent) is 2, and the number is 49, so $\log_7 49 = 2$ becomes $7^2 = 49$ in exponential form.

12. $\log_{1/2} x = -5$
$$x = \left(\frac{1}{2}\right)^{-5} \qquad \textit{Exponential form}$$
$$x = \left(\frac{2}{1}\right)^5 = 32$$

The argument (the input of the logarithm) must be a positive number, so $x = 32$ is acceptable.

Check $x = 32$:
$$\log_{1/2} 32 = -5 \text{ since } \left(\frac{1}{2}\right)^{-5} = 2^5 = 32$$

The solution set is $\{32\}$.

13.
$$x = \log_9 3$$
$$9^x = 3 \qquad \textit{Exponential form}$$
Write each side as a power of 3.
$$\left(3^2\right)^x = 3$$
$$3^{2x} = 3^1$$
$$2x = 1 \qquad \textit{Equate exponents.}$$
$$x = \frac{1}{2}$$

Check $x = \frac{1}{2}$:
$$\frac{1}{2} = \log_9 3 \text{ since } 9^{1/2} = \sqrt{9} = 3$$

The solution set is $\left\{\frac{1}{2}\right\}$.

14. $\log_x 16 = 4$
$$x^4 = 16 \qquad \textit{Exponential form}$$
$$x^2 = \pm 4 \qquad \textit{Square root property}$$
Reject -4 since $x^2 \geq 0$.
$$x^2 = 4$$
$$x = \pm 2 \qquad \textit{Square root property}$$

Reject -2 since the base cannot be negative.

Check $x = 2$: $\log_2 16 = 4$ since $2^4 = 16$

The solution set is $\{2\}$.

Copyright © 2012 Pearson Education, Inc. Publishing as Addison-Wesley.

15. The value of $\log_2 32$ is $\underline{5}$. This means that if we raise $\underline{2}$ to the $\underline{5\text{th}}$ power, the result is $\underline{32}$.

16. $\log_3 x^2 y$

$\quad = \log_3 x^2 + \log_3 y \quad \textit{Product rule}$

$\quad = 2\log_3 x + \log_3 y \quad \textit{Power rule}$

17. $\log_5 \left(\dfrac{\sqrt{x}}{yz} \right)$

$\quad = \log_5 \sqrt{x} - \log_5 yz \qquad\qquad \textit{Quotient rule}$

$\quad = \log_5 x^{1/2} - (\log_5 y + \log_5 z) \quad \textit{Product rule}$

$\quad = \frac{1}{2}\log_5 x - \log_5 y - \log_5 z \quad\ \ \textit{Power rule}$

18. $3\log_b s - \log_b t$

$\quad = \log_b s^3 - \log_b t \quad \textit{Power rule}$

$\quad = \log_b \dfrac{s^3}{t} \qquad\qquad \textit{Quotient rule}$

19. $\frac{1}{4}\log_b r + 2\log_b s - \frac{2}{3}\log_b t$

Use the power rule for logarithms.

$\quad = \log_b r^{1/4} + \log_b s^2 - \log_b t^{2/3}$

Use the product and quotient rules for logarithms.

$\quad = \log_b \dfrac{r^{1/4} s^2}{t^{2/3}}$

20. **(a)** $\log 23.1 \approx 1.3636$

$\quad$ **(b)** $\ln 0.82 \approx -0.1985$

21. **(a)** $\log_3 19 = \dfrac{\log_{10} 19}{\log_{10} 3} = \dfrac{\log 19}{\log 3}$

$\quad$ **(b)** $\log_3 19 = \dfrac{\log_e 19}{\log_e 3} = \dfrac{\ln 19}{\ln 3}$

$\quad$ **(c)** The four-decimal-place approximation of either fraction is 2.6801.

22. $\quad 3^x = 78$

$\quad \ln 3^x = \ln 78$

$\quad x\ln 3 = \ln 78 \qquad\qquad \textit{Power rule}$

$\quad x = \dfrac{\ln 78}{\ln 3} \approx 3.966$

Check $x = 3.966$: $\ 3^{3.966} \approx 78.0$

The solution set is $\{3.966\}$.

23. $\log_8 (x+5) + \log_8 (x-2) = 1$

Use the product rule for logarithms.

$\quad \log_8 [(x+5)(x-2)] = 1$

$\quad\quad (x+5)(x-2) = 8^1 \quad \textit{Exp. form}$

$\quad\quad\quad x^2 + 3x - 10 = 8$

$\quad\quad\quad x^2 + 3x - 18 = 0$

$\quad\quad\quad (x+6)(x-3) = 0$

$\quad x+6 = 0 \quad$ or $\quad x-3 = 0$

$\quad\quad x = -6 \quad$ or $\quad\quad x = 3$

Reject $x = -6$, because $x + 5 = -1$, which yields an equation in which the logarithm of a negative number must be found.

Check $x = 3$: $\ \log_8 8 + \log_8 1 = 1 + 0 = 1$

The solution set is $\{3\}$.

24. $A = P\left(1 + \dfrac{r}{n}\right)^{nt}$

$\quad A = 10{,}000\left(1 + \dfrac{0.045}{4}\right)^{4\cdot 5} \approx 12{,}507.51$

$10{,}000$ invested at 4.5% annual interest, compounded quarterly, will increase to $12{,}507.51 in 5 years.

25. $A = Pe^{rt}$

$\quad$ **(a)** $A = 15{,}000e^{0.05(5)} \approx 19{,}260.38$

There will be $19{,}260.38 in the account.

$\quad$ **(b)** Let $A = 2(15{,}000)$ and solve for t.

$\quad\quad 2(15{,}000) = 15{,}000e^{0.05t}$

$\quad\quad\quad 2 = e^{0.05t}$

$\quad\quad\quad \ln 2 = \ln e^{0.05t}$

$\quad\quad\quad \ln 2 = 0.05t\,(\ln e)$

$\quad\quad\quad 0.05t = \ln 2$

$\quad\quad\quad t = \dfrac{\ln 2}{0.05} \approx 13.9$

The principal will double in about 13.9 years.

Cumulative Review Exercises (Chapters 1–11)

For Exercises 1–4,

$$S = \left\{ -\tfrac{9}{4}, -2, -\sqrt{2}, 0, 0.6, \sqrt{11}, \sqrt{-8}, 6, \tfrac{30}{3} \right\}.$$

1. The integers are -2, 0, 6, and $\frac{30}{3}$ (or 10).

2. The rational numbers are $-\frac{9}{4}$, -2, 0, 0.6, 6, and $\frac{30}{3}$ (or 10). Each can be expressed as a quotient of two integers.

3. The irrational numbers are $-\sqrt{2}$ and $\sqrt{11}$.

4. $\quad |-8| + 6 - |-2| - (-6+2)$

$\quad = 8 + 6 - 2 - (-4)$

$\quad = 14 - 2 + 4 = 16$

5. $\quad 2(-5) + (-8)(4) - (-3)$

$\quad = -10 - 32 + 3 = -39$

6. $\quad 7 - (3 + 4x) + 2x = -5(x-1) - 3$

$\quad\quad 7 - 3 - 4x + 2x = -5x + 5 - 3$

$\quad\quad\quad 4 - 2x = -5x + 2$

$\quad\quad\quad\quad 3x = -2$

$\quad\quad\quad\quad x = -\frac{2}{3}$

The solution set is $\left\{ -\frac{2}{3} \right\}$.

Copyright © 2012 Pearson Education, Inc. Publishing as Addison-Wesley.

7. $2x + 2 \le 5x - 1$
$-3x \le -3$
Divide by -3; reverse the inequality.
$x \ge 1$

The solution set is $[1, \infty)$.

8. $|2x - 5| = 9$

$2x - 5 = 9$ or $2x - 5 = -9$
$2x = 14$ $\qquad$ $2x = -4$
$x = 7$ or $\qquad$ $x = -2$

The solution set is $\{-2, 7\}$.

9. $|4x + 2| > 10$

$4x + 2 > 10$ or $4x + 2 < -10$
$4x > 8$ $\qquad$ $4x < -12$
$x > 2$ or $\qquad$ $x < -3$

The solution set is $(-\infty, -3) \cup (2, \infty)$.

10. $5x + 2y = 10$

Find the x- and y-intercepts. To find the x-intercept, let $y = 0$.

$5x + 2(0) = 10$
$5x = 10$
$x = 2$

The x-intercept is $(2, 0)$.
To find the y-intercept, let $x = 0$.

$5(0) + 2y = 10$
$2y = 10$
$y = 5$

The y-intercept is $(0, 5)$.
Plot the intercepts and draw the line through them.

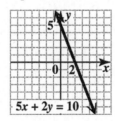

11. $-4x + y \le 5$

Graph the line $-4x + y = 5$, which has intercepts $(0, 5)$ and $\left(-\frac{5}{4}, 0\right)$, as a solid line because the inequality involves $\le$. Test $(0, 0)$, which yields $0 \le 5$, a true statement. Shade the region that includes $(0, 0)$.

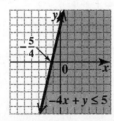

12. **(a)** Yes, this graph is the graph of a function because it passes the vertical line test.

(b) $(x_1, y_1) = (2003, 41{,}218)$ and $(x_2, y_2) = (2008, 57{,}949)$.

$$m = \frac{y_2 - y_1}{x_2 - x_1} = \frac{57{,}949 - 41{,}218}{2008 - 2003}$$
$$= \frac{16{,}731}{5} = 3346.2$$

The slope of the line in the graph is 3346.2 and can be interpreted as follows: The number of travelers increased by an average of 3346.2 thousand per year during the period 2003-2008.

13. Through $(5, -1)$; parallel to $3x - 4y = 12$
Find the slope of

$3x - 4y = 12$
$-4y = -3x + 12$
$y = \frac{3}{4}x - 3.$

The slope is $\frac{3}{4}$, so a line parallel to it also has slope $\frac{3}{4}$. Let $m = \frac{3}{4}$ and $(x_1, y_1) = (5, -1)$ in the point-slope form.

$y - y_1 = m(x - x_1)$
$y - (-1) = \frac{3}{4}(x - 5)$
$y + 1 = \frac{3}{4}x - \frac{15}{4}$
$y = \frac{3}{4}x - \frac{19}{4}$

14. $5x - 3y = 14$ (1)
$2x + 5y = 18$ (2)

Multiply equation (1) by 5 and equation (2) by 3. Then add the results.

$25x - 15y = 70 \qquad 5 \times (1)$
$\underline{6x + 15y = 54} \qquad 3 \times (2)$
$31x \qquad\quad = 124 \; Add$
$x = 4$

Substitute 4 for x in equation (1) to find y.

$5x - 3y = 14 \qquad (1)$
$5(4) - 3y = 14$
$20 - 3y = 14$
$-3y = -6$
$y = 2$

The solution set is $\{(4, 2)\}$.

15. $x + 2y + 3z = 11$ (1)
$3x - y + z = 8$ (2)
$2x + 2y - 3z = -12$ (3)

To eliminate z, add equations (1) and (3).

$x + 2y + 3z = 11 \qquad (1)$
$\underline{2x + 2y - 3z = -12} \qquad (3)$
$3x + 4y = -1 \qquad (4)$

Copyright © 2012 Pearson Education, Inc. Publishing as Addison-Wesley.

To eliminate z again, multiply equation (2) by 3 and add the result to equation (3).

$$
\begin{array}{rlr}
9x - 3y + 3z &= 24 & 3 \times (2) \\
2x + 2y - 3z &= -12 & (3) \\
\hline
11x - y &= 12 & (5)
\end{array}
$$

Multiply equation (5) by 4 and add the result to equation (4).

$$
\begin{array}{rlr}
44x - 4y &= 48 & 4 \times (5) \\
3x + 4y &= -1 & (4) \\
\hline
47x &= 47 & \\
x &= 1 &
\end{array}
$$

Substitute 1 for x in equation (5) to find y.

$$
\begin{aligned}
11x - y &= 12 \qquad (5) \\
11(1) - y &= 12 \\
11 - y &= 12 \\
-y &= 1 \\
y &= -1
\end{aligned}
$$

Substitute 1 for x and -1 for y in equation (2) to find z.

$$
\begin{aligned}
3x - y + z &= 8 \qquad (2) \\
3(1) - (-1) + z &= 8 \\
3 + 1 + z &= 8 \\
4 + z &= 8 \\
z &= 4
\end{aligned}
$$

The solution set is $\{(1, -1, 4)\}$.

16. Let x = the amount of candy at \$1.00 per pound.

	Number of Pounds	Price per Pound	Value
First Candy	x	\$1.00	$1.00x$
Second Candy	10	\$1.96	$1.96(10)$
Mixture	$x + 10$	\$1.60	$1.60(x + 10)$

The sum of the values of each candy must equal the value of the mixture.

$$1.00x + 1.96(10) = 1.60(x + 10)$$

Multiply by 10 to clear the decimals.

$$
\begin{aligned}
10x + 196 &= 16(x + 10) \\
10x + 196 &= 16x + 160 \\
36 &= 6x \\
6 &= x
\end{aligned}
$$

Use 6 pounds of the \$1.00 candy.

17. $(2p + 3)(3p - 1) = 6p^2 - 2p + 9p - 3$
$$= 6p^2 + 7p - 3$$

18. $(4k - 3)^2 = (4k)^2 - 2(4k)(3) + 3^2$
$$= 16k^2 - 24k + 9$$

19. $(3m^3 + 2m^2 - 5m) - (8m^3 + 2m - 4)$
$$
\begin{aligned}
&= 3m^3 + 2m^2 - 5m - 8m^3 - 2m + 4 \\
&= 3m^3 - 8m^3 + 2m^2 - 5m - 2m + 4 \\
&= -5m^3 + 2m^2 - 7m + 4
\end{aligned}
$$

20.

$$
\begin{array}{r}
2t^3 + 5t^2 - 3t + 4 \\
3t + 1 \overline{\smash{\big)}\ 6t^4 + 17t^3 - 4t^2 + 9t + 4} \\
\underline{6t^4 + 2t^3} \\
15t^3 - 4t^2 \\
\underline{15t^3 + 5t^2} \\
-9t^2 + 9t \\
\underline{-9t^2 - 3t} \\
12t + 4 \\
\underline{12t + 4} \\
0
\end{array}
$$

The quotient is $2t^3 + 5t^2 - 3t + 4$.

21. $8x + x^3$
Factor out the GCF, x.

$$8x + x^3 = x(8 + x^2)$$

22. $24y^2 - 7y - 6$

Two integer factors whose product is $(24)(-6) = -144$ and whose sum is -7 are -16 and 9.
Rewrite the trinomial in a form that can be factored by grouping.

$$
\begin{aligned}
24y^2 - 7y - 6 &= 24y^2 - 16y + 9y - 6 \\
&= 8y(3y - 2) + 3(3y - 2) \\
&= (3y - 2)(8y + 3)
\end{aligned}
$$

Thus, $24y^2 - 7y - 6 = (3y - 2)(8y + 3)$.

23. $5z^3 - 19z^2 - 4z$
Factor out the GCF, z, and then factor by trial and error.

$$
\begin{aligned}
5z^3 - 19z^2 - 4z &= z(5z^2 - 19z - 4) \\
&= z(5z + 1)(z - 4)
\end{aligned}
$$

24. $16a^2 - 25b^4$

Use the difference of squares formula,

$$\boldsymbol{x^2 - y^2 = (x + y)(x - y),}$$

where $x = 4a$ and $y = 5b^2$.

$$16a^2 - 25b^4 = (4a + 5b^2)(4a - 5b^2)$$

25. $8c^3 + d^3$

Use the sum of cubes formula,

$$\boldsymbol{x^3 + y^3 = (x + y)(x^2 - xy + y^2),}$$

where $x = 2c$ and $y = d$.

$$8c^3 + d^3 = (2c + d)(4c^2 - 2cd + d^2)$$

26. $16r^2 + 56rq + 49q^2$
$$= (4r)^2 + 2(4r)(7q) + (7q)^2$$

Use the perfect square formula,

$$x^2 + 2xy + y^2 = (x+y)^2,$$

where $x = 4r$ and $y = 7q$.

$$16r^2 + 56rq + 49q^2 = (4r + 7q)^2$$

27. $\dfrac{(5p^3)^4(-3p^7)}{2p^2(4p^4)} = \dfrac{(5^4 p^{12})(-3p^7)}{8p^6}$

$$= \dfrac{(625)(-3)p^{19}}{8p^6}$$

$$= -\dfrac{1875p^{13}}{8}$$

28. $\dfrac{x^2 - 9}{x^2 + 7x + 12} \div \dfrac{x - 3}{x + 5}$

Multiply by the reciprocal.

$$= \dfrac{x^2 - 9}{x^2 + 7x + 12} \cdot \dfrac{x + 5}{x - 3}$$

$$= \dfrac{(x+3)(x-3)}{(x+3)(x+4)} \cdot \dfrac{(x+5)}{(x-3)} \quad \textit{Factor}$$

$$= \dfrac{x + 5}{x + 4}$$

29. $\dfrac{2}{k+3} - \dfrac{5}{k-2}$

The LCD is $(k+3)(k-2)$.

$$= \dfrac{2(k-2)}{(k+3)(k-2)} - \dfrac{5(k+3)}{(k-2)(k+3)}$$

$$= \dfrac{2k - 4 - 5k - 15}{(k+3)(k-2)}$$

$$= \dfrac{-3k - 19}{(k+3)(k-2)}$$

30. $\sqrt{288} = \sqrt{144 \cdot 2} = \sqrt{144}\sqrt{2} = 12\sqrt{2}$

31. $2\sqrt{32} - 5\sqrt{98} = 2\sqrt{16 \cdot 2} - 5\sqrt{49 \cdot 2}$

$$= 2 \cdot 4\sqrt{2} - 5 \cdot 7\sqrt{2}$$

$$= 8\sqrt{2} - 35\sqrt{2}$$

$$= -27\sqrt{2}$$

32. $\sqrt{2x+1} - \sqrt{x} = 1$

$$\sqrt{2x+1} = 1 + \sqrt{x}$$

$$\left(\sqrt{2x+1}\right)^2 = \left(1 + \sqrt{x}\right)^2$$

$$2x + 1 = 1 + 2\sqrt{x} + x$$

$$x = 2\sqrt{x}$$

$$(x)^2 = \left(2\sqrt{x}\right)^2$$

$$x^2 = 4x$$

$$x^2 - 4x = 0$$

$$x(x - 4) = 0$$

$$x = 0 \quad \text{or} \quad x = 4$$

Check $x = 0$: $\sqrt{1} - \sqrt{0} = 1$ *True*
Check $x = 4$: $\sqrt{9} - \sqrt{4} = 1$ *True*

The solution set is $\{0, 4\}$.

33. $(5 + 4i)(5 - 4i) = 5^2 - (4i)^2$

$$= 25 - 16i^2$$

$$= 25 - 16(-1)$$

$$= 25 + 16 = 41$$

34. $3x^2 - x - 1 = 0$
Here $a = 3$, $b = -1$, and $c = -1$.
Use the quadratic formula.

$$x = \dfrac{-b \pm \sqrt{b^2 - 4ac}}{2a}$$

$$x = \dfrac{-(-1) \pm \sqrt{(-1)^2 - 4(3)(-1)}}{2(3)}$$

$$= \dfrac{1 \pm \sqrt{1 + 12}}{6} = \dfrac{1 \pm \sqrt{13}}{6}$$

The solution set is $\left\{ \dfrac{1 \pm \sqrt{13}}{6} \right\}$.

35. $x^2 + 2x - 8 > 0$
Solve the equation

$$x^2 + 2x - 8 = 0.$$
$$(x + 4)(x - 2) = 0$$

$$x + 4 = 0 \quad \text{or} \quad x - 2 = 0$$
$$x = -4 \quad \text{or} \quad x = 2$$

The numbers -4 and 2 divide a number line into three intervals.

Test a number from each interval in the inequality

$$x^2 + 2x - 8 > 0.$$

Interval A: Let $x = -5$.
$$25 - 10 - 8 \overset{?}{>} 0$$
$$7 > 0 \qquad \textit{True}$$

Interval B: Let $x = 0$.
$$-8 > 0 \qquad \textit{False}$$

Interval C: Let $x = 3$.
$$9 + 6 - 8 \overset{?}{>} 0$$
$$7 > 0 \qquad \textit{True}$$

The numbers in Intervals A and C, not including -4 or 2 because of $>$, are solutions. The solution set is $(-\infty, -4) \cup (2, \infty)$.

36. $x^4 - 5x^2 + 4 = 0$

Let $u = x^2$, so $u^2 = (x^2)^2 = x^4$.

$$u^2 - 5u + 4 = 0$$
$$(u - 1)(u - 4) = 0$$

$$u - 1 = 0 \quad \text{or} \quad u - 4 = 0$$
$$u = 1 \quad \text{or} \qquad u = 4$$

To find x, substitute x^2 for u.

$$x^2 = 1 \quad \text{or} \quad x^2 = 4$$
$$x = \pm 1 \quad \text{or} \quad x = \pm 2$$

The solution set is $\{\pm 1,\ \pm 2\}$.

37. $f(x) = \frac{1}{3}(x - 1)^2 + 2$ is in
$f(x) = a(x - h)^2 + k$ form. The graph is a
vertical parabola with vertex (h, k) at $(1, 2)$.
Since $a = \frac{1}{3} > 0$, the graph opens up.
Also, $|a| = \left|\frac{1}{3}\right| = \frac{1}{3} < 1$, so the graph is wider
than the graph of $f(x) = x^2$. The points $\left(0, 2\frac{1}{3}\right)$,
$(-2, 5)$, and $(4, 5)$ are also on the graph.

$$f(x) = \tfrac{1}{3}(x - 1)^2 + 2$$

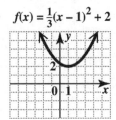

38. $f(x) = 2^x$

Make a table of values.

x	-2	-1	0	1	2
$f(x)$	$\frac{1}{4}$	$\frac{1}{2}$	1	2	4

Plot the ordered pairs from the table, and draw a
smooth exponential curve through the points.

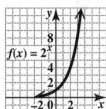

39.
$$5^{x+3} = \left(\tfrac{1}{25}\right)^{3x+2}$$
$$5^{x+3} = \left[\left(\tfrac{1}{5}\right)^2\right]^{3x+2}$$

Write each side to the power of 5.

$$5^{x+3} = \left(5^{-2}\right)^{(3x+2)}$$
$$5^{x+3} = 5^{-2(3x+2)}$$
$$x + 3 = -2(3x + 2)$$
$$x + 3 = -6x - 4$$
$$7x = -7$$
$$x = -1$$

Check $x = -1$: $5^2 = \left(\tfrac{1}{25}\right)^{-1}$ *True*
The solution set is $\{-1\}$.

40. $f(x) = \log_3 x$
Make a table of values.

Powers of 3	3^{-2}	3^{-1}	3^0	3^1	3^2
x	$\frac{1}{9}$	$\frac{1}{3}$	1	3	9
y	-2	-1	0	1	2

Plot the ordered pairs and draw a smooth
logarithmic curve through the points.

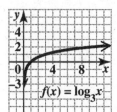

$$f(x) = \log_3 x$$

41. $\log \dfrac{x^3 \sqrt{y}}{z}$

$$= \log \frac{x^3 y^{1/2}}{z}$$
$$= \log \left(x^3 y^{1/2}\right) - \log z \qquad \text{\textit{Quotient rule}}$$
$$= \log x^3 + \log y^{1/2} - \log z \quad \text{\textit{Product rule}}$$
$$= 3 \log x + \tfrac{1}{2} \log y - \log z \quad \text{\textit{Power rule}}$$

42. $B(t) = 25{,}000e^{0.2t}$

(a) At noon, $t = 0$.

$$B(0) = 25{,}000e^{0.2(0)}$$
$$= 25{,}000e^0 = 25{,}000(1) = 25{,}000$$

25,000 bacteria are present at noon.

(b) At 1 P.M., $t = 1$.

$$B(1) = 25{,}000e^{0.2(1)} \approx 30{,}535$$

About 30,500 bacteria are present at 1 P.M.

(c) At 2 P.M., $t = 2$.

$$B(2) = 25{,}000e^{0.2(2)} \approx 37{,}296$$

About 37,300 bacteria are present at 2 P.M.

(d) The population doubles when $B(t) = 50{,}000$.

$$50{,}000 = 25{,}000e^{0.2t}$$
$$2 = e^{0.2t}$$
$$\ln 2 = \ln e^{0.2t}$$
$$\ln 2 = 0.2t \ln e$$
$$t = \frac{\ln 2}{0.2} \approx 3.5$$

The population will double in about 3.5 hours, or
at about 3:30 P.M.

Copyright © 2012 Pearson Education, Inc. Publishing as Addison-Wesley.

CHAPTER 12 POLYNOMIAL AND RATIONAL FUNCTIONS

12.1 Zeros of Polynomial Functions (I)

12.1 Now Try Exercises

N1. $\dfrac{3x^2 - 8x - 10}{x - 4}$

$$
\begin{array}{r|rrr}
4 & 3 & -8 & -10 \\
  &   & 12 & 16 \\
\hline
  & 3 & 4 & 6
\end{array}
\quad\leftarrow \textit{Coefficients of numerator}
$$

$\leftarrow$ *Remainder*

$$3x \quad +4 \qquad \textit{Write the answer from the bottom row.}$$

The quotient polynomial is $3x + 4$ and the remainder is 6.

Answer: $3x + 4 + \dfrac{6}{x - 4}$

N2. $\dfrac{2x^4 + 5x^3 + x - 10}{x + 3}$

Insert a 0 for the missing x^2-term.

$$
\begin{array}{r|rrrrr}
-3 & 2 & 5 & 0 & 1 & -10 \\
   &   & -6 & 3 & -9 & 24 \\
\hline
   & 2 & -1 & 3 & -8 & 14
\end{array}
\quad\leftarrow \textit{Remainder}
$$

Answer: $2x^3 - x^2 + 3x - 8 + \dfrac{14}{x + 3}$

N3. Find $f(-3)$ for $f(x) = 2x^3 - x^2 - 2x + 40$.

Divide $f(x)$ by $x - (-3) = x + 3$.

$$
\begin{array}{r|rrrr}
-3 & 2 & -1 & -2 & 40 \\
   &   & -6 & 21 & -57 \\
\hline
   & 2 & -7 & 19 & -17
\end{array}
\quad\leftarrow \textit{Remainder}
$$

By the remainder theorem, $f(-3) = -17$.

N4. To decide whether 1 is a zero of the polynomial function

$$f(x) = x^4 - 3x^3 + 2x^2 + 3x - 3,$$

use synthetic division to divide $f(x)$ by $x - 1$. If the remainder is 0, then 1 is a zero of $f(x)$. Otherwise, it is not.

$$
\begin{array}{r|rrrrr}
1 & 1 & -3 & 2 & 3 & -3 \\
  &   & 1 & -2 & 0 & 3 \\
\hline
  & 1 & -2 & 0 & 3 & 0
\end{array}
\quad\leftarrow \textit{Remainder}
$$

Since the remainder is 0, 1 is a zero of $f(x)$.

12.1 Section Exercises

1. $\dfrac{x^2 - 6x + 5}{x - 1}$

$$
\begin{array}{r|rrr}
1 & 1 & -6 & 5 \\
  &   & 1 & -5 \\
\hline
  & 1 & -5 & 0
\end{array}
\quad\leftarrow \textit{Coefficients of numerator}
$$

$$x \quad - 5 \qquad \textit{Write the answer from the bottom row.}$$

Answer: $x - 5$

3. $\dfrac{4m^2 + 19m - 5}{m + 5}$

$m + 5 = m - (-5)$, so use -5.

$$
\begin{array}{r|rrr}
-5 & 4 & 19 & -5 \\
   &   & -20 & 5 \\
\hline
   & 4 & -1 & 0
\end{array}
$$

Answer: $4m - 1$

5. $\dfrac{2a^2 + 8a + 13}{a + 2}$

$a + 2 = a - (-2)$, so use -2.

$$
\begin{array}{r|rrr}
-2 & 2 & 8 & 13 \\
   &   & -4 & -8 \\
\hline
   & 2 & 4 & 5
\end{array}
\quad\leftarrow \textit{Remainder}
$$

The quotient polynomial is $2a + 4$ and the remainder is 5.

Answer: $2a + 4 + \dfrac{5}{a + 2}$

7. $(p^2 - 3p + 5) \div (p + 1)$

$$
\begin{array}{r|rrr}
-1 & 1 & -3 & 5 \\
   &   & -1 & 4 \\
\hline
   & 1 & -4 & 9
\end{array}
$$

Answer: $p - 4 + \dfrac{9}{p + 1}$

9. $\dfrac{4a^3 - 3a^2 + 2a - 3}{a - 1}$

$$
\begin{array}{r|rrrr}
1 & 4 & -3 & 2 & -3 \\
  &   & 4 & 1 & 3 \\
\hline
  & 4 & 1 & 3 & 0
\end{array}
$$

Answer: $4a^2 + a + 3$

11. $(x^5 - 2x^3 + 3x^2 - 4x - 2) \div (x - 2)$

Insert 0 for the missing x^4-term.

$$
\begin{array}{r|rrrrrr}
2 & 1 & 0 & -2 & 3 & -4 & -2 \\
  &   & 2 & 4 & 4 & 14 & 20 \\
\hline
  & 1 & 2 & 2 & 7 & 10 & 18
\end{array}
\quad\leftarrow \textit{Remainder}
$$

Answer: $x^4 + 2x^3 + 2x^2 + 7x + 10 + \dfrac{18}{x - 2}$

Copyright © 2012 Pearson Education, Inc. Publishing as Addison-Wesley.

13. $\left(-4r^6 - 3r^5 - 3r^4 + 5r^3 - 6r^2 + 3r + 3\right) \div$
$(r - 1)$

$$1\,\overline{\,|\,-4\;-3\;-3\;\;\;\;5\;\;-6\;\;\;\;3\;\;\;\;3}$$
$$\;\;\;\;-4\;-7\;-10\;-5\;-11\;-8$$
$$\overline{\;-4\;-7\;-10\;-5\;-11\;\;-8\;-5\;\leftarrow\;\textit{Remainder}}$$

Answer:

$-4r^5 - 7r^4 - 10r^3 - 5r^2 - 11r - 8 + \dfrac{-5}{r-1}$

15. $\left(-3y^5 + 2y^4 - 5y^3 - 6y^2 - 1\right) \div (y + 2)$

Insert 0 for the missing y-term.

$$-2\,\overline{\,|\,-3\;\;2\;\;-5\;\;-6\;\;\;\;0\;\;-1}$$
$$\;\;\;\;\;\;6\;-16\;\;42\;-72\;\;144$$
$$\overline{\;-3\;\;8\;-21\;\;36\;-72\;\;143\;\leftarrow\;\textit{Remainder}}$$

Answer:

$-3y^4 + 8y^3 - 21y^2 + 36y - 72 + \dfrac{143}{y+2}$

17. $f(x) = 2x^3 + x^2 + x - 8;\ k = -1$

Use synthetic division to write the polynomial in
the form $f(x) = (x - k)q(x) + r$.

$$-1\,\overline{\,|\,2\;\;\;\;1\;\;\;\;1\;\;\;-8}$$
$$\;\;\;\;\;\;-2\;\;\;1\;\;\;-2$$
$$\overline{\;\;2\;\;-1\;\;\;2\;\;-10}$$

Note that $x - k = x - (-1) = x + 1$.
$f(x) = (x + 1)\left(2x^2 - x + 2\right) + (-10)$
or $\ (x + 1)\left(2x^2 - x + 2\right) - 10$

19. $f(x) = -x^3 + 2x^2 + 4;\ k = -2$

$$-2\,\overline{\,|\,-1\;\;\;\;2\;\;\;\;0\;\;\;\;4}$$
$$\;\;\;\;\;\;\;\;2\;\;-8\;\;16$$
$$\overline{\;-1\;\;\;\;4\;\;-8\;\;20}$$

$f(x) = (x + 2)(-x^2 + 4x - 8) + 20$

21. $f(x) = 4x^4 - 3x^3 - 20x^2 - x;\ k = 3$

$$3\,\overline{\,|\,4\;\;-3\;\;-20\;\;-1\;\;\;\;0}$$
$$\;\;\;\;\;\;12\;\;\;\;27\;\;\;\;21\;\;\;60$$
$$\overline{\;4\;\;\;\;9\;\;\;\;\;7\;\;\;\;20\;\;\;60}$$

$f(x) = (x - 3)(4x^3 + 9x^2 + 7x + 20) + 60$

23. $k = 3;\ f(x) = x^2 - 4x + 5$

$$3\,\overline{\,|\,1\;\;-4\;\;\;\;5}$$
$$\;\;\;\;\;\;3\;\;-3$$
$$\overline{\;1\;\;-1\;\;\;\;2}$$

From the remainder theorem, $f(k) = r$, so
$f(3) = 2$.

25. $k = 2;\ f(x) = 2x^2 - 3x - 3$

$$2\,\overline{\,|\,2\;\;-3\;\;-3}$$
$$\;\;\;\;\;\;4\;\;\;\;2$$
$$\overline{\;2\;\;\;\;1\;\;-1}$$

By the remainder theorem, $f(2) = -1$.

27. $k = -1;\ f(x) = x^3 - 4x^2 + 2x + 1$

$$-1\,\overline{\,|\,1\;\;-4\;\;\;\;2\;\;\;\;1}$$
$$\;\;\;\;\;\;-1\;\;\;\;5\;\;-7$$
$$\overline{\;1\;\;-5\;\;\;\;7\;\;-6}$$

By the remainder theorem, $f(-1) = -6$.

29. $k = 3;\ f(x) = 2x^5 - 10x^3 - 19x^2 - 45$

$$3\,\overline{\,|\,2\;\;\;\;0\;\;-10\;\;-19\;\;\;\;0\;\;-45}$$
$$\;\;\;\;\;\;6\;\;\;\;18\;\;\;\;24\;\;\;15\;\;\;45$$
$$\overline{\;2\;\;\;\;6\;\;\;\;8\;\;\;\;\;5\;\;\;15\;\;\;\;0}$$

By the remainder theorem, $f(3) = 0$.

31. $k = 2 + i;\ f(x) = x^2 - 5x + 1$

$$2 + i\,\overline{\,|\,1\;\;\;\;-5\;\;\;\;\;\;\;\;1}$$
$$\;\;\;\;\;\;\;\;\;\;2 + i\;\;\;-7 - i$$
$$\overline{\;1\;\;\;-3 + i\;\;\;-6 - i}$$

By the remainder theorem, $f(2 + i) = -6 - i$.
See Section 8.7 for examples of adding and
multiplying complex numbers.

33. $k = \frac{1}{3};\ f(x) = 9x^3 - 6x^2 + x$

$$\tfrac{1}{3}\,\overline{\,|\,9\;\;-6\;\;\;\;1\;\;\;\;0}$$
$$\phantom{\tfrac{1}{3}\,|}\;\;\;\;\;\;\;\;3\;\;-1\;\;\;\;0$$
$$\overline{\phantom{\tfrac{1}{3}\,|}\;9\;\;-3\;\;\;\;0\;\;\;\;0}$$

By the remainder theorem, $f\left(\tfrac{1}{3}\right) = 0$.

35. By the remainder theorem, a zero remainder
means that $f(k) = 0$; that is, k is a number that
makes $f(x) = 0$.

37. To determine if 3 is a zero of
$f(x) = 2x^3 - 6x^2 - 9x + 27$, divide
synthetically.

$$3\,\overline{\,|\,2\;\;-6\;\;-9\;\;\;\;27}$$
$$\;\;\;\;\;\;6\;\;\;\;0\;\;-27$$
$$\overline{\;2\;\;\;\;0\;\;-9\;\;\;\;0}$$

Yes, 3 is a zero of $f(x)$ because $f(3) = 0$.

39. To determine if -5 is a zero of
$f(x) = x^3 + 7x^2 + 10x$, divide synthetically.

$$-5\,\overline{\,|\,1\;\;\;\;7\;\;\;\;10\;\;\;\;0}$$
$$\;\;\;\;\;\;-5\;\;-10\;\;\;\;0$$
$$\overline{\;1\;\;\;\;2\;\;\;\;0\;\;\;\;0}$$

Yes, -5 is a zero of $f(x)$ because $f(-5) = 0$.

41. To determine if $\frac{2}{5}$ is a zero of
$f(x) = 5x^4 + 2x^3 - x + 15$, divide synthetically.

$$\tfrac{2}{5}\,\overline{\,|\,5\;\;2\;\;0\;\;\;\;-1\;\;\;\;15}$$
$$\phantom{\tfrac{2}{5}\,|}\;\;\;\;\;\;2\;\;\tfrac{8}{5}\;\;\tfrac{16}{25}\;\;-\tfrac{18}{125}$$
$$\overline{\phantom{\tfrac{2}{5}\,|}\;5\;\;4\;\;\tfrac{8}{5}\;\;-\tfrac{9}{25}\;\;\tfrac{1857}{125}}$$

No, $\frac{2}{5}$ is not a zero of $f(x)$ because $f\left(\tfrac{2}{5}\right) = \dfrac{1857}{125}$.

Copyright © 2012 Pearson Education, Inc. Publishing as Addison-Wesley.

43. To determine if $2 - i$ is a zero of $f(x) = x^2 + 3x + 4$, divide synthetically.

$$2 - i \overline{\smash{\big)}\ \begin{array}{ccc} 1 & 3 & 4 \\ & 2 - i & 9 - 7i \\ \hline 1 & 5 - i & 13 - 7i \end{array}}$$

No, $2 - i$ is not a zero of $f(x)$ because $f(2 - i) = 13 - 7i$.

♦ ♦ ♦ Relating Concepts 45–50 ♦ ♦ ♦

45. Factor $f(x) = 2x^2 + 5x - 12$. The two integers whose product is $2(-12) = -24$ and whose sum is 5 are 8 and -3.

$$\begin{aligned} 2x^2 + 5x - 12 \\ = 2x^2 - 3x + 8x - 12 \\ = x(2x - 3) + 4(2x - 3) \\ = (2x - 3)(x + 4) \end{aligned}$$

46. Solve $f(x) = 2x^2 + 5x - 12 = 0$.

$$\begin{aligned} 2x^2 + 5x - 12 &= 0 \\ (2x - 3)(x + 4) &= 0 \end{aligned}$$

$$\begin{array}{ccc} 2x - 3 = 0 & \text{or} & x + 4 = 0 \\ 2x = 3 & & x = -4 \\ x = \tfrac{3}{2} & & \end{array}$$

The solution set is $\left\{-4, \tfrac{3}{2}\right\}$.

47. $f(x) = 2x^2 + 5x - 12$

$$\begin{aligned} f(-4) &= 2(-4)^2 + 5(-4) - 12 \\ &= 2(16) - 20 - 12 \\ &= 32 - 20 - 12 = 0 \end{aligned}$$

Another method: We can divide $f(x)$ by $x + 4$. The remainder is equal to $f(-4)$.

$$-4 \overline{\smash{\big)}\ \begin{array}{ccc} 2 & 5 & -12 \\ & -8 & 12 \\ \hline 2 & -3 & 0 \end{array}} \leftarrow \quad \textit{Remainder}$$

By the remainder theorem, $f(-4) = 0$.

48. $f(x) = 2x^2 + 5x - 12$

$$\begin{aligned} f\left(\tfrac{3}{2}\right) &= 2\left(\tfrac{3}{2}\right)^2 + 5\left(\tfrac{3}{2}\right) - 12 \\ &= 2\left(\tfrac{9}{4}\right) + \tfrac{15}{2} - 12 \\ &= \tfrac{9}{2} + \tfrac{15}{2} - \tfrac{24}{2} = 0 \end{aligned}$$

49. If $f(a) = 0$, then $x - \underline{\ a\ }$ is a factor of $f(x)$.

50. $g(x) = 3x^3 - 4x^2 - 17x + 6$

$$\begin{aligned} g(3) &= 3(3)^3 - 4(3)^2 - 17(3) + 6 \\ &= 81 - 36 - 51 + 6 = 0 \end{aligned}$$

Since $g(3) = 0$, $x - 3$ is a factor of $g(x)$. To check, use synthetic division to see if 3 is a solution of the equation $g(x) = 0$.

$$3 \overline{\smash{\big)}\ \begin{array}{cccc} 3 & -4 & -17 & 6 \\ & 9 & 15 & -6 \\ \hline 3 & 5 & -2 & 0 \end{array}}$$

Therefore, $x - 3$ is a factor of $g(x)$ and

$$g(x) = (x - 3)(3x^2 + 5x - 2).$$

Now factor $3x^2 + 5x - 2$. The two integers whose product is $3(-2) = -6$ and whose sum is 5 are 6 and -1.

$$\begin{aligned} 3x^2 + 5x - 2 \\ = 3x^2 - x + 6x - 2 \\ = x(3x - 1) + 2(3x - 1) \\ = (3x - 1)(x + 2) \end{aligned}$$

Thus, $g(x)$ completely factored is

$$(x - 3)(3x - 1)(x + 2).$$

51. $f(x) = 2x^3 - x^2 + x - 6$

$$\begin{aligned} f(-1) &= 2(-1)^3 - (-1)^2 + (-1) - 6 \\ &= 2(-1) - (1) - 1 - 6 \\ &= -2 - 1 - 1 - 6 \\ &= -10 \end{aligned}$$

53. $x^3 + x^2 - x - 1$

$$\begin{aligned} &= \left(x^3 + x^2\right) + (-x - 1) \\ &= x^2(x + 1) - 1(x + 1) \\ &= (x + 1)\left(x^2 - 1\right) \\ &= (x + 1)(x + 1)(x - 1) \\ &= (x + 1)^2(x - 1) \end{aligned}$$

12.2   Zeros of Polynomial Functions (II)

12.2   Now Try Exercises

N1. $f(x) = x^3 - x^2 + x + 14;\ x + 2$

By the factor theorem, $x + 2$ will be a factor of $f(x)$ only if $f(-2) = 0$. Use synthetic division and the remainder theorem.

$$-2 \overline{\smash{\big)}\ \begin{array}{cccc} 1 & -1 & 1 & 14 \\ & -2 & 6 & -14 \\ \hline 1 & -3 & 7 & 0 \end{array}}$$

Since $f(-2) = 0$, $x + 2$ is a factor of $f(x)$.

N2. $f(x) = 6x^3 + 17x^2 + 6x - 8$

Since -2 is a zero of $f(x)$, $x + 2$ is a factor. Divide $f(x)$ by $x + 2$.

$$-2 \overline{\smash{\big)}\ \begin{array}{cccc} 6 & 17 & 6 & -8 \\ & -12 & -10 & 8 \\ \hline 6 & 5 & -4 & 0 \end{array}}$$

Thus,

$$\begin{aligned} f(x) &= (x + 2)\left(6x^2 + 5x - 4\right) \\ &= (x + 2)(2x - 1)(3x + 4). \end{aligned}$$

Copyright © 2012 Pearson Education, Inc. Publishing as Addison-Wesley.

N3. $f(x) = 8x^4 + 10x^3 - 11x^2 - 10x + 3$

(a) By the rational zeros theorem, if $\frac{p}{q}$ is a rational zero of $f(x)$, p must be a factor of the constant term $a_0 = 3$ and q must be a factor of the leading coefficient $a_4 = 8$. Thus, p can be ± 1 or ± 3, and q can be ± 1, ± 2, ± 4, or ± 8. The possible rational zeros are all possible quotients of the form $\frac{p}{q}$:

$$\pm 1, \; \pm \tfrac{1}{2}, \; \pm \tfrac{1}{4}, \; \pm \tfrac{1}{8}, \; \pm 3, \; \pm \tfrac{3}{2}, \; \pm \tfrac{3}{4}, \; \pm \tfrac{3}{8}.$$

(b) Use the remainder theorem to see if 1 is a zero.

$$\begin{array}{r|rrrrr}
1 & 8 & 10 & -11 & -10 & 3 \\
  &   & 8 & 18 & 7   & -3 \\
\hline
  & 8 & 18 & 7   & -3 & 0
\end{array}$$

The 0 remainder shows that 1 is a zero. Now, use the quotient polynomial $8x^3 + 18x^2 + 7x - 3$ and synthetic division to see if -1 is also a zero.

$$\begin{array}{r|rrrr}
-1 & 8 & 18 & 7 & -3 \\
   &   & -8 & -10 & 3 \\
\hline
   & 8 & 10 & -3 & 0
\end{array}$$

The new quotient polynomial is $8x^2 + 10x - 3$, which factors as $(2x + 3)(4x - 1)$, and is equal to 0 for $x = -\frac{3}{2}$ and $x = \frac{1}{4}$. Thus, the rational zeros are

$$1, -1, -\tfrac{3}{2}, \text{ and } \tfrac{1}{4}.$$

The factored form of f is

$$f(x) = (x - 1)(x + 1)(2x + 3)(4x - 1).$$

N4. Zeros of -1, 2, and 5; $f(8) = 54$

The polynomial function has the form

$$f(x) = a(x + 1)(x - 2)(x - 5)$$

for some real number a. To find a, use the fact that $f(8) = 54$.

$$\begin{aligned}
f(8) = a(8 + 1)(8 - 2)(8 - 5) &= 54 \\
a(9)(6)(3) &= 54 \\
3a &= 1 \\
a &= \tfrac{1}{3}
\end{aligned}$$

Thus,

$$\begin{aligned}
f(x) &= \tfrac{1}{3}(x + 1)(x - 2)(x - 5) \\
&= \tfrac{1}{3}(x^2 - x - 2)(x - 5) \\
&= \tfrac{1}{3}(x^3 - 6x^2 + 3x + 10) \\
&= \tfrac{1}{3}x^3 - 2x^2 + x + \tfrac{10}{3}
\end{aligned}$$

N5. The complex number $3 - 2i$ also must be a zero, so the polynomial function has at least three zeros, 4, $3 + 2i$, and $3 - 2i$. For the polynomial to be of least possible degree, these must be the only zeros. By the factor theorem there must be three factors, $x - 4$, $x - (3 + 2i)$, and $x - (3 - 2i)$. A polynomial function of least possible degree is

$$\begin{aligned}
f(x) &= (x - 4)[x - (3 + 2i)][x - (3 - 2i)] \\
&= (x - 4)[(x - 3) - 2i][(x - 3) + 2i] \\
&= (x - 4)[(x - 3)^2 - (2i)^2] \\
&= (x - 4)(x^2 - 6x + 9 + 4) \\
&= (x - 4)(x^2 - 6x + 13) \\
&= x^3 - 10x^2 + 37x - 52
\end{aligned}$$

N6. $f(x) = x^4 - x^3 + 6x^2 + 14x - 20$

Since $f(x)$ has only real coefficients and since $1 + 3i$ is a zero, by the conjugate zeros theorem $1 - 3i$ is also a zero. To find the remaining zeros, we first divide $f(x)$ by $x - (1 + 3i)$.

$$\begin{array}{r|rrrrr}
1 + 3i & 1 & -1 & 6 & 14 & -20 \\
       &   & 1 + 3i & -9 + 3i & -12 - 6i & 20 \\
\hline
       & 1 & 3i & -3 + 3i & 2 - 6i & 0
\end{array}$$

Now divide the quotient polynomial by $x - (1 - 3i)$.

$$\begin{array}{r|rrrr}
1 - 3i & 1 & 3i & -3 + 3i & 2 - 6i \\
       &   & 1 - 3i & 1 - 3i & -2 + 6i \\
\hline
       & 1 & 1 & -2 & 0
\end{array}$$

The last quotient polynomial is

$$x^2 + x - 2 = (x + 2)(x - 1).$$

Thus, the zeros of f are

$$1 + 3i, \; 1 - 3i, \; -2, \; \text{and} \; 1.$$

12.2 Section Exercises

1. *Statement:* Given that $x - 1$ is a factor of $f(x) = x^6 - x^4 + 2x^2 - 2$, we are assured that $f(1) = 0$.

This statement is justified by the factor theorem; therefore, it is *true*.

3. *Statement:* For the function defined by $f(x) = (x + 2)^4(x - 3)$, 2 is a zero of multiplicity 4.

The zeros of the function are -2 and 3. Since 2 is not a zero of the function, the statement is *false*. (It would be true to say that -2 is a zero of multiplicity 4.)

5. $4x^2 + 2x + 54$; $x - 4$

Let $f(x) = 4x^2 + 2x + 54$. By the factor theorem, $x - 4$ will be a factor of $f(x)$ only if $f(4) = 0$. Use synthetic division and the remainder theorem.

$$\begin{array}{r|rrr}
4 & 4 & 2 & 54 \\
  &   & 16 & 72 \\
\hline
  & 4 & 18 & 126
\end{array}$$

Since the remainder is 126, $f(4) = 126$, so $x - 4$ is not a factor of $f(x)$.

Copyright © 2012 Pearson Education, Inc. Publishing as Addison-Wesley.

7. $x^3 + 2x^2 - 3;\ x - 1$

Let $f(x) = x^3 + 2x^2 - 3$. By the factor theorem, $x - 1$ will be a factor of $f(x)$ only if $f(1) = 0$.

$$
\begin{array}{r|rrrr}
1 & 1 & 2 & 0 & -3 \\
  &   & 1 & 3 & 3 \\
\hline
  & 1 & 3 & 3 & 0
\end{array}
$$

Since $f(1) = 0$, $x - 1$ is a factor of $f(x)$.

9. $2x^4 + 5x^3 - 2x^2 + 5x + 6;\ x + 3$

Let $f(x) = 2x^4 + 5x^3 - 2x^2 + 5x + 6$. By the factor theorem, $x + 3$ will be a factor of $f(x)$ only if $f(-3) = 0$.

$$
\begin{array}{r|rrrrr}
-3 & 2 & 5 & -2 & 5 & 6 \\
   &   & -6 & 3 & -3 & -6 \\
\hline
   & 2 & -1 & 1 & 2 & 0
\end{array}
$$

Since $f(-3) = 0$, $x + 3$ is a factor of $f(x)$.

11. $f(x) = 2x^3 - 3x^2 - 17x + 30;\ k = 2$

Since 2 is a zero of $f(x)$, $x - 2$ is a factor. Divide $f(x)$ by $x - 2$.

$$
\begin{array}{r|rrrr}
2 & 2 & -3 & -17 & 30 \\
  &   & 4 & 2 & -30 \\
\hline
  & 2 & 1 & -15 & 0
\end{array}
$$

Thus,

$$
\begin{aligned}
f(x) &= (x - 2)\left(2x^2 + x - 15\right) \\
     &= (x - 2)(2x - 5)(x + 3).
\end{aligned}
$$

13. $f(x) = 6x^3 + 13x^2 - 14x + 3;\ k = -3$

Since -3 is a zero of $f(x)$, $x + 3$ is a factor. Divide $f(x)$ by $x + 3$.

$$
\begin{array}{r|rrrr}
-3 & 6 & 13 & -14 & 3 \\
   &   & -18 & 15 & -3 \\
\hline
   & 6 & -5 & 1 & 0
\end{array}
$$

Thus,

$$
\begin{aligned}
f(x) &= (x + 3)\left(6x^2 - 5x + 1\right) \\
     &= (x + 3)(3x - 1)(2x - 1).
\end{aligned}
$$

15. $f(x) = x^3 - x^2 - 4x - 6;\ 3$

Since 3 is a zero, first divide $f(x)$ by $x - 3$.

$$
\begin{array}{r|rrrr}
3 & 1 & -1 & -4 & -6 \\
  &   & 3 & 6 & 6 \\
\hline
  & 1 & 2 & 2 & 0
\end{array}
$$

This gives

$$
f(x) = (x - 3)(x^2 + 2x + 2).
$$

Use the quadratic formula with $a = 1$, $b = 2$, and $c = 2$ to find the remaining two zeros.

$$
\begin{aligned}
x &= \frac{-2 \pm \sqrt{2^2 - 4(1)(2)}}{2(1)} \\
  &= \frac{-2 \pm \sqrt{-4}}{2} = \frac{-2 \pm 2i}{2} \\
  &= \frac{2(-1 \pm i)}{2} = -1 \pm i
\end{aligned}
$$

The remaining zeros are $-1 \pm i$.

17. $f(x) = x^3 - 7x^2 + 17x - 15;\ 2 - i$

Since $2 - i$ is a zero, first divide $f(x)$ by $x - (2 - i)$.

$$
\begin{array}{r|rrrr}
2 - i & 1 & -7 & 17 & -15 \\
      &   & 2 - i & -11 + 3i & 15 \\
\hline
      & 1 & -5 - i & 6 + 3i & 0
\end{array}
$$

By the conjugate zeros theorem, $2 + i$ is also a zero, so divide the quotient polynomial from the first synthetic division by $x - (2 + i)$.

$$
\begin{array}{r|rrr}
2 + i & 1 & -5 - i & 6 + 3i \\
      &   & 2 + i & -6 - 3i \\
\hline
      & 1 & -3 & 0
\end{array}
$$

This gives

$$
f(x) = [x - (2 - i)][x - (2 + i)](x - 3).
$$

The remaining zeros are $2 + i$ and 3.

19. $f(x) = x^4 - 14x^3 + 51x^2 - 14x + 50;\ i$

Since i is a zero, first divide $f(x)$ by $x - i$.

$$
\begin{array}{r|rrrrr}
i & 1 & -14 & 51 & -14 & 50 \\
  &   & i & -1 - 14i & 14 + 50i & -50 \\
\hline
  & 1 & -14 + i & 50 - 14i & 50i & 0
\end{array}
$$

By the conjugate zeros theorem, $-i$ is also a zero, so divide the quotient polynomial from the first synthetic division by $-i$.

$$
\begin{array}{r|rrrr}
-i & 1 & -14 + i & 50 - 14i & 50i \\
   &   & -i & 14i & -50i \\
\hline
   & 1 & -14 & 50 & 0
\end{array}
$$

The remaining zeros will be the zeros of the new quotient polynomial, $x^2 - 14x + 50$. Use the quadratic formula to find these zeros.

$$
\begin{aligned}
x &= \frac{-(-14) \pm \sqrt{(-14)^2 - 4(1)(50)}}{2(1)} \\
  &= \frac{14 \pm \sqrt{-4}}{2} = \frac{14 \pm 2i}{2} \\
  &= \frac{2(7 \pm i)}{2} = 7 \pm i
\end{aligned}
$$

The other zeros are $-i$ and $7 \pm i$.

Copyright © 2012 Pearson Education, Inc. Publishing as Addison-Wesley.

21. $f(x) = x^3 - 2x^2 - 13x - 10$

(a) By the rational zeros theorem, if $\frac{p}{q}$ is a rational zero of $f(x)$, p must be a factor of the constant term $a_0 = -10$ and q must be a factor of the leading coefficient $a_3 = 1$. The possible values of p are ± 1, ± 2, ± 5, ± 10. The possible values of q are ± 1. The possible rational zeros are all possible quotients of the form $\frac{p}{q}$:

$$\pm 1, \pm 2, \pm 5, \pm 10$$

(b) We first show that -1 is a zero.

$$
\begin{array}{r|rrr}
-1 & 1 & -2 & -13 & -10 \\
   &   & -1 &   3 & 10 \\
\hline
   & 1 & -3 & -10 &   0 \\
\end{array}
$$

-1 is a zero, so $f(x)$ may be factored as

$$f(x) = (x+1)(x^2 - 3x - 10).$$

To find the remaining zeros, solve

$$
\begin{aligned}
x^2 - 3x - 10 &= 0. \\
(x+2)(x-5) &= 0 \\
x = -2 \ \text{ or } \ x &= 5
\end{aligned}
$$

The rational zeros are -1, -2, and 5.

(c) $f(x) = (x+1)(x+2)(x-5)$

23. $f(x) = x^3 + 6x^2 - x - 30$

(a) The possible rational zeros are

$$\pm 1, \pm 2, \pm 3, \pm 5, \pm 6, \pm 10, \pm 15, \pm 30.$$

(b) Check 2 using synthetic division.

$$
\begin{array}{r|rrr}
2 & 1 & 6 & -1 & -30 \\
  &   & 2 & 16 & 30 \\
\hline
  & 1 & 8 & 15 &   0 \\
\end{array}
$$

2 is a zero. The remaining zeros are the zeros of the quotient, $x^2 + 8x + 15$. These zeros may be found by factoring.

$$
\begin{aligned}
x^2 + 8x + 15 &= 0 \\
(x+3)(x+5) &= 0 \\
x = -3 \ \text{ or } \ x &= -5
\end{aligned}
$$

The rational zeros are 2, -3, and -5.

(c) $f(x) = (x+5)(x+3)(x-2)$

25. $f(x) = 6x^3 + 17x^2 - 31x - 12$

(a) The possible values of p are ± 1, ± 2, ± 3, ± 4, ± 6, ± 12. The possible values of q are ± 1, ± 2, ± 3, ± 6. The possible values of $\frac{p}{q}$ when $q = \pm 1$ are just the possible values of p. With $q = \pm 2$, we obtain the possible $\frac{p}{q}$ values

$$\pm \tfrac{1}{2}, \pm \tfrac{3}{2}.$$

With $q = \pm 3$, we get

$$\pm \tfrac{1}{3}, \pm \tfrac{2}{3}, \pm \tfrac{4}{3}.$$

With $q = \pm 6$, we get

$$\pm \tfrac{1}{6}.$$

(b)

$$
\begin{array}{r|rrr}
-4 & 6 & 17 & -31 & -12 \\
   &   & -24 & 28 & 12 \\
\hline
   & 6 & -7 & -3 &   0 \\
\end{array}
$$

$6x^2 - 7x - 3 = (3x+1)(2x-3)$
The rational zeros are -4, $-\frac{1}{3}$, and $\frac{3}{2}$.

(c) $f(x) = (x+4)(3x+1)(2x-3)$

27. $f(x) = 12x^3 + 20x^2 - x - 6$

(a) p: ± 1, ± 2, ± 3, ± 6
   q: ± 1, ± 2, ± 3, ± 4, ± 6, ± 12
   $\frac{p}{q}$: ± 1, ± 2, ± 3, ± 6,

$$\pm \tfrac{1}{2}, \pm \tfrac{3}{2}, \pm \tfrac{1}{3}, \pm \tfrac{2}{3},$$
$$\pm \tfrac{1}{4}, \pm \tfrac{3}{4}, \pm \tfrac{1}{6}, \pm \tfrac{1}{12}$$

(b)

$$
\begin{array}{r|rrr}
\frac{1}{2} & 12 & 20 & -1 & -6 \\
            &   & 6 & 13 & 6 \\
\hline
            & 12 & 26 & 12 & 0 \\
\end{array}
$$

$$
\begin{aligned}
12x^2 + 26x + 12 &= 2\left(6x^2 + 13x + 6\right) \\
&= 2(3x+2)(2x+3)
\end{aligned}
$$
The rational zeros are $\frac{1}{2}$, $-\frac{2}{3}$, and $-\frac{3}{2}$.

(c) $f(x) = \left(x - \tfrac{1}{2}\right) \cdot 2(3x+2)(2x+3)$
   $= (2x-1)(3x+2)(2x+3)$

29. $f(x) = (x+4)^2(x^2 - 7)(x+1)^4$

To find the zeros, let $f(x) = 0$. Set each factor equal to zero and solve for x. Clearly, $x = -4$ and $x = -1$ are zeros.

$$
\begin{aligned}
x^2 - 7 &= 0 \\
x^2 &= 7 \\
x &= \pm \sqrt{7}
\end{aligned}
$$

The zeros are -4 (multiplicity 2), $\pm \sqrt{7}$, and -1 (multiplicity 4).

31. $f(x) = 3x^3(x-2)(x+3)(x^2 - 1)$

To find the zeros, let $f(x) = 0$. Set each factor equal to zero and solve for x. We can ignore the factor 3 since $3 \neq 0$.

$$
\begin{aligned}
x^2 - 1 &= 0 \\
x^2 &= 1 \\
x &= \pm 1
\end{aligned}
$$

The zeros are 0 (multiplicity 3), 2, -3, 1, and -1.

Copyright © 2012 Pearson Education, Inc. Publishing as Addison-Wesley.

33. $f(x) = (9x + 7)^2(x^2 + 16)^2 = 0$

$$(9x + 7)^3 = 0$$
$$x = -\tfrac{7}{9} \text{ (multiplicity 2)}$$
$$(x^2 + 16)^2 = 0$$
$$x^2 + 16 = 0$$
$$x^2 = -16$$
$$x = \pm 4i \text{ (multiplicity 2)}$$

The zeros are $-\tfrac{7}{9}$ (multiplicity 2), $4i$ (multiplicity 2), and $-4i$ (multiplicity 2).

In Exercises 35–44, find a polynomial function of least possible degree with real coefficients having the given zeros. For each of these exercises, other answers are possible.

35. $3 + i$ and $3 - i$

$$\begin{aligned}
f(x) &= [x - (3 + i)][x - (3 - i)] \\
&= (x - 3 - i)(x - 3 + i) \\
&= [(x - 3) - i][(x - 3) + i] \\
&= (x - 3)^2 - (i)^2 \quad \text{\textit{Product of sum and}} \\
&\qquad\qquad\qquad\qquad \text{\textit{difference of terms}} \\
&= x^2 - 6x + 9 - i^2 \\
&= x^2 - 6x + 9 + 1 \\
&= x^2 - 6x + 10
\end{aligned}$$

37. $1 + \sqrt{2}, 1 - \sqrt{2}$, and 3

$$\begin{aligned}
f(x) &= \left[x - \left(1 + \sqrt{2}\right)\right]\left[x - \left(1 - \sqrt{2}\right)\right] \\
&\quad \bullet (x - 3) \\
&= \left(x - 1 - \sqrt{2}\right)\left(x - 1 + \sqrt{2}\right) \bullet (x - 3) \\
&= \left[(x - 1) - \sqrt{2}\right]\left[(x - 1) + \sqrt{2}\right] \\
&\quad \bullet (x - 3) \\
&= \left[(x - 1)^2 - \left(\sqrt{2}\right)^2\right](x - 3) \\
&\qquad \text{\textit{Product of sum and difference of terms}} \\
&= (x^2 - 2x + 1 - 2)(x - 3) \\
&= (x^2 - 2x - 1)(x - 3) \\
&= x^3 - 5x^2 + 5x + 3
\end{aligned}$$

39. $-2 + i, 3$, and -3 ($-2 - i$ is also a zero)

$$\begin{aligned}
f(x) &= [x - (-2 + i)][x - (-2 - i)] \\
&\quad \bullet (x - 3)(x + 3) \\
&= [(x + 2) - i][(x + 2) + i] \\
&\quad \bullet (x - 3)(x + 3) \\
&= \left[(x + 2)^2 - i^2\right]\left(x^2 - 3^2\right) \\
&\qquad \text{\textit{Product of sum and difference of terms}} \\
&= \left(x^2 + 4x + 4 + 1\right)\left(x^2 - 9\right) \\
&= \left(x^2 + 4x + 5\right)\left(x^2 - 9\right) \\
&= x^4 + 4x^3 - 4x^2 - 36x - 45
\end{aligned}$$

41. 2 and $3i$

By the conjugate zeros theorem, $-3i$ must also be a zero.

$$\begin{aligned}
f(x) &= (x - 2)(x - 3i)[x - (-3i)] \\
&= (x - 2)(x - 3i)(x + 3i) \\
&= (x - 2)\left(x^2 - 9i^2\right) \\
&= (x - 2)\left(x^2 + 9\right) \\
&= x^3 - 2x^2 + 9x - 18
\end{aligned}$$

43. $1 + 2i, 2$ (multiplicity 2)

By the conjugate zeros theorem, $1 - 2i$ must also be a zero.

$$\begin{aligned}
f(x) &= (x - 2)^2[x - (1 + 2i)] \\
&\quad \bullet [x - (1 - 2i)] \\
&= \left(x^2 - 4x + 4\right)[(x - 1) - 2i] \\
&\quad \bullet [(x - 1) + 2i] \\
&= \left(x^2 - 4x + 4\right)\left[(x - 1)^2 - (2i)^2\right] \\
&= \left(x^2 - 4x + 4\right)\left(x^2 - 2x + 1 + 4\right) \\
&= \left(x^2 - 4x + 4\right)\left(x^2 - 2x + 5\right) \\
&= x^4 - 6x^3 + 17x^2 - 28x + 20
\end{aligned}$$

45.   Zeros of $-3, 1$, and 4; $f(2) = 30$

The polynomial function has the form

$$f(x) = a(x + 3)(x - 1)(x - 4)$$

for some real number a. To find a, use the fact that $f(2) = 30$.

$$\begin{aligned}
f(2) &= a(2 + 3)(2 - 1)(2 - 4) = 30 \\
&\quad a(5)(1)(-2) = 30 \\
&\quad -10a = 30 \\
&\quad a = -3
\end{aligned}$$

Thus, $f(x) = -3(x + 3)(x - 1)(x - 4)$
$$\begin{aligned}
&= -3\left(x^2 + 2x - 3\right)(x - 4) \\
&= -3\left(x^3 - 2x^2 - 11x + 12\right) \\
&= -3x^3 + 6x^2 + 33x - 36.
\end{aligned}$$

47.   Zeros of $-2, 1$, and 0; $f(-1) = -1$

The polynomial function has the form

$$f(x) = a(x + 2)(x - 1)(x - 0)$$

for some real number a. To find a, use the fact that $f(-1) = -1$.

$$\begin{aligned}
f(-1) &= a(-1 + 2)(-1 - 1)(-1 - 0) = -1. \\
&\quad a(1)(-2)(-1) = -1 \\
&\quad 2a = -1 \\
&\quad a = -\tfrac{1}{2}
\end{aligned}$$

Thus, $f(x) = -\tfrac{1}{2}(x + 2)(x - 1)x$
$$\begin{aligned}
&= -\tfrac{1}{2}(x^2 + x - 2)x \\
&= -\tfrac{1}{2}x^3 - \tfrac{1}{2}x^2 + x.
\end{aligned}$$

Copyright © 2012 Pearson Education, Inc. Publishing as Addison-Wesley.

49. Zeros of 5, i, and $-i$; $f(-1) = 48$

The polynomial function has the form

$$f(x) = a(x - 5)(x - i)(x + i)$$

for some real number a. To find a, use the fact that $f(-1) = 48$.

$$f(-1) = a(-1 - 5)(-1 - i)(-1 + i) = 48$$
$$a(-6)(2) = 48$$
$$a = -4$$

Thus, $f(x) = -4(x - 5)(x - i)(x + i)$
$$= -4(x - 5)(x^2 + 1)$$
$$= -4(x^3 - 5x^2 + x - 5)$$
$$= -4x^3 + 20x^2 - 4x + 20.$$

♦ ♦ ♦ Relating Concepts 51–54 ♦ ♦ ♦

51. $f(x) = x^3 - 21x - 20$

Find the quotient $\dfrac{x^3 - 21x - 20}{x + 4}$.

$$-4 \,\overline{)\begin{array}{rrrr} 1 & 0 & -21 & -20 \\ & -4 & 16 & 20 \\ \hline 1 & -4 & -5 & 0 \end{array}}$$

$g(x) = x^2 - 4x - 5$

52.

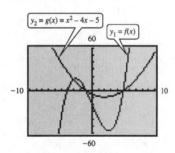

The function g is a quadratic function. The x-intercepts of g are also x-intercepts of f.

53. Find the quotient $\dfrac{x^2 - 4x - 5}{x - 5}$.

$$5 \,\overline{)\begin{array}{rrr} 1 & -4 & -5 \\ & 5 & 5 \\ \hline 1 & 1 & 0 \end{array}}$$

$h(x) = x + 1$

54.

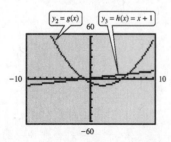

The function h is a linear function. The x-intercept of h is also an x-intercept of g.

55. If -2 is a zero of multiplicity 2, then

$$f(x) = x^4 + 2x^3 - 7x^2 - 20x - 12$$

can be divided by $x + 2$, and the resulting quotient polynomial can be divided by $x + 2$ again. Each time the remainder must be 0. This is demonstrated in the following.

$$-2 \,\overline{)\begin{array}{rrrrr} 1 & 2 & -7 & -20 & -12 \\ & -2 & 0 & 14 & 12 \\ \hline 1 & 0 & -7 & -6 & 0 \end{array}}$$

Divide the quotient polynomial by $x + 2$.

$$-2 \,\overline{)\begin{array}{rrrr} 1 & 0 & -7 & -6 \\ & -2 & 4 & 6 \\ \hline 1 & -2 & -3 & 0 \end{array}}$$

Factor the quotient polynomial.

$$x^2 - 2x - 3 = (x + 1)(x - 3)$$

The remaining zeros are -1 and 3, and

$$f(x) = (x + 2)^2(x + 1)(x - 3).$$

57. $f(x) = 0.86x^3 - 5.24x^2 + 3.55x + 7.84$

Enter this function into the calculator as

$$Y_1 = 0.86*x^3 - 5.24*x^2 + 3.55*x + 7.84.$$

From a graph of the function, we estimate the zeros as -1, 2, and 5, and use these values as our guesses.

solve(Y_1, X, -1) -0.88
solve(Y_1, X, 2)   2.12
solve(Y_1, X, 5)   4.86

The real zeros are approximately -0.88, 2.12, and 4.86.

59. $f(x) = 2.45x^4 - 3.22x^3 + 0.47x^2 - 6.54x + 3$

Enter this function into the calculator as

$$Y_1 = 2.45*x^4 - 3.22*x^3 + 0.47*x^2 - 6.54*x + 3.$$

solve(Y_1, X, 1)   0.44
solve(Y_1, X, 2)   1.81

The real zeros are approximately 0.44 and 1.81.

61. $f(x) = -\sqrt{7}x^3 + \sqrt{5}x + \sqrt{17}$

Enter this function into the calculator as

$$Y_1 = -\sqrt{(7)}*x^3 + \sqrt{(5)}*x + \sqrt{(17)}.$$

solve(Y_1, X, 1)   1.40

The only real zero is approximately 1.40.

Copyright © 2012 Pearson Education, Inc. Publishing as Addison-Wesley.

63. $f(x) = 2x^3 - 4x^2 + 2x + 7$

$f(x)$ has 2 variations in sign:

$$f(x) = +2x^3 - 4x^2 + 2x + 7$$

$$\underbrace{\qquad}_{1} \underbrace{\qquad}_{2}$$

Thus, f has either 2 or $2 - 2 = 0$ positive real zeros.

$f(-x)$ has 1 variation in sign:

$$f(-x) = 2(-x)^3 - 4(-x)^2 + 2(-x) + 7$$
$$= -2x^3 - 4x^2 - 2x + 7$$

$$\underbrace{\qquad}_{1}$$

Thus, f has 1 negative real zero.

65. $f(x) = 5x^4 + 3x^2 + 2x - 9$

$f(x)$ has 1 variation in sign:

$$f(x) = +5x^4 + 3x^2 + 2x - 9$$

$$\underbrace{\qquad}_{1}$$

Thus, f has 1 positive real zero.

$f(-x)$ has 1 variation in sign:

$$f(-x) = 5(-x)^4 + 3(-x)^2 + 2(-x) - 9$$
$$= 5x^4 + 3x^2 - 2x - 9$$

$$\underbrace{\qquad}_{1}$$

Thus, f has 1 negative real zero.

67. $f(x) = x^5 + 3x^4 - x^3 + 2x + 3$

$f(x)$ has 2 variations in sign:

$$f(x) = x^5 + 3x^4 - x^3 + 2x + 3$$

$$\underbrace{\qquad}_{1} \underbrace{\qquad}_{2}$$

Thus, f has 2 or $2 - 2 = 0$ positive real zeros.

$f(-x)$ has 3 variations in sign:

$$f(-x) = (-x)^5 + 3(-x)^4 - (-x)^3 + 2(-x) + 3$$
$$= -x^5 + 3x^4 + x^3 - 2x + 3$$

$$\underbrace{\qquad}_{1} \underbrace{\qquad}_{2} \underbrace{\qquad}_{3}$$

Thus, f has 3 or $3 - 2 = 1$ negative real zeros.

69. $f(x) = 2x^2 - x - 28$

$$f\left(-\tfrac{7}{2}\right) = 2\left(-\tfrac{7}{2}\right)^2 - \left(-\tfrac{7}{2}\right) - 28$$
$$= \tfrac{49}{2} + \tfrac{7}{2} - 28$$
$$= 28 - 28$$
$$= 0$$

Thus, $-\tfrac{7}{2}$ is a zero of $f(x)$.

12.3 Graphs and Applications of Polynomial Functions

12.3 Now Try Exercises

N1. The graph of $f(x) = -x^3$ is symmetric with respect to the origin. It is reflected across the x-axis compared to the graph of $g(x) = x^3$ (see Figure 1 in the text).

x	y
0	0
± 1	∓ 1
± 2	∓ 8

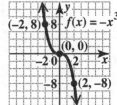

N2. $f(x) = \tfrac{1}{2}(x + 1)^4 - 2$

This function f has a graph like that of $g(x) = x^4$, but it is translated 1 unit to the left. The $\tfrac{1}{2}$ causes the graph to be broader than the graph of $g(x) = x^4$. The -2 translates the graph of $y = \tfrac{1}{2}(x + 1)^4$ two units down.

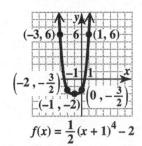

$$f(x) = \tfrac{1}{2}(x + 1)^4 - 2$$

N3. $f(x) = x^3 + 3x^2 - 6x - 8$

Step 1 The possible rational zeros are

$$\pm 1, \pm 2, \pm 4, \pm 8.$$

Check 2 using synthetic division.

$$\begin{array}{r|rrrr} 2 & 1 & 3 & -6 & -8 \\ & & 2 & 10 & 8 \\ \hline & 1 & 5 & 4 & 0 \end{array}$$

2 is a zero. The remaining zeros are the zeros of the quotient, $x^2 + 5x + 4$. These zeros may be found by factoring.

$$x^2 + 5x + 4 = 0$$
$$(x + 4)(x + 1) = 0$$
$$x = -4 \quad \text{or} \quad x = -1$$

The rational zeros are -4, -1, and 2.

Step 2 $f(0) = -8$, so plot $(0, -8)$.

Copyright © 2012 Pearson Education, Inc. Publishing as Addison-Wesley.

Step 3

Interval	Test point	Value of $f(x)$	Sign of $f(x)$
$(-\infty, -4)$	-5	-28	Negative
$(-4, -1)$	-2	8	Positive
$(-1, 2)$	0	-8	Negative
$(2, \infty)$	3	28	Positive

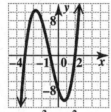

$$f(x) = x^3 + 3x^2 - 6x - 8$$

N4. $f(x) = x^3 + x^2 - 6x$

$$
\begin{array}{r|rrrr}
-4 & 1 & 1 & -6 & 0 \\
   &   & -4 & 12 & -24 \\
\hline
   & 1 & -3 & 6 & -24
\end{array}
$$

$$
\begin{array}{r|rrrr}
-2 & 1 & 1 & -6 & 0 \\
   &   & -2 & 2 & 8 \\
\hline
   & 1 & -1 & -4 & 8
\end{array}
$$

Since $f(-4) = -24 < 0$ (*negative*) and $f(-2) = 8 > 0$ (*positive*), the intermediate value theorem assures us that there must be a real zero between -4 and -2.

N5. Show that $f(x) = 2x^4 + 21x^3 - 36x^2 + x + 12$ has no real zero greater than 2.

$$
\begin{array}{r|rrrrr}
2 & 2 & 21 & -36 & 1 & 12 \\
  &   & 4 & 50 & 28 & 58 \\
\hline
  & 2 & 25 & 14 & 29 & 70
\end{array}
$$

By the boundedness theorem, since $2 > 0$ and the numbers in the bottom row of the synthetic division are nonnegative, f has no real zero greater than 2.

N6. $f(x) = x^3 - 3x^2 - 3x + 2$

Enter the function in your calculator and find the *zero* or *root* feature.

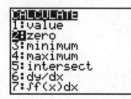

From the graph of f, we see that there is a zero between 3 and 4, so enter 3 as a guess for the left bound and 4 for the right bound.

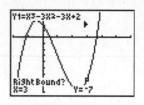

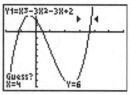

Guess any number between 3 and 4 and the calculator will approximate the zero.

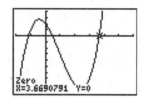

In this case, we see that the approximation is 3.67. Similarly, the other two zeros are -1.15 and 0.48.

N7.
$$f(x) = a(x - h)^2 + k \qquad \textit{Given form}$$
$$f(x) = a(x - 2)^2 + 1563 \quad \textit{(h, k) = (2, 1563)}$$
$$105{,}489 = a(8 - 2)^2 + 1563 \quad \textit{x = 8 for 1988}$$
$$103{,}926 = a(6)^2$$
$$a = \tfrac{103{,}926}{36} \approx 2886.8 \approx 2887$$

To approximate the number of cases in 1985, let $x = 5$ in $f(x) = 2887(x - 2)^2 + 1563$ to get 27,546, which is more than the actual number of 22,620.

12.3 Section Exercises

1. $f(x) = \tfrac{1}{4}x^6$ is in the form $f(x) = ax^n$.

$|a| = \tfrac{1}{4} < 1$, so the graph is broader than that of $f(x) = x^6$, but has the same general shape.

The graph includes the points $(\pm 2, 16)$, $\left(\pm 1, \tfrac{1}{4}\right)$, and $(0, 0)$. Connect these points with a smooth curve.

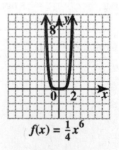

$$f(x) = \tfrac{1}{4}x^6$$

Copyright © 2012 Pearson Education, Inc. Publishing as Addison-Wesley.

3. $f(x) = -\frac{5}{4}x^5$ is in the form $f(x) = ax^n$.

Since $|a| = \frac{5}{4} > 1$, the graph is narrower than that of $f(x) = x^5$. Since $a = -\frac{5}{4}$ is negative, the graph is the reflection of $f(x) = \frac{5}{4}x^5$ about the x-axis.

The graph includes the points $(-2, 40)$, $\left(-1, \frac{5}{4}\right)$, $(0, 0)$, $\left(1, -\frac{5}{4}\right)$, and $(2, -40)$.

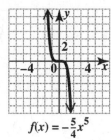

$$f(x) = -\frac{5}{4}x^5$$

5. The graph of $f(x) = \frac{1}{2}x^3 + 1$ looks like $y = x^3$ but is broader $\left(\text{since } |a| = \frac{1}{2} < 1\right)$ and is translated 1 unit up.

The graph includes the points $(-2, -3)$, $\left(-1, \frac{1}{2}\right)$, $(0, 1)$, $\left(1, \frac{3}{2}\right)$, and $(2, 5)$.

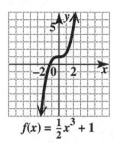

$$f(x) = \frac{1}{2}x^3 + 1$$

7. The graph of $f(x) = -(x + 1)^3$ can be obtained by reflecting the graph of $f(x) = x^3$ about the x-axis and then translating it 1 unit to the left.

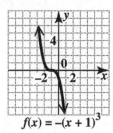

$$f(x) = -(x + 1)^3$$

9. The graph of $f(x) = (x - 1)^4 + 2$ has the same shape as $y = x^4$, but is translated 1 unit to the right and 2 units up.

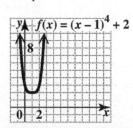

$$f(x) = (x - 1)^4 + 2$$

11. $f(x) = x^3 - 3x^2 - 6x + 8$

The maximum number of turning points in the graph of a polynomial function of degree n is $n - 1$.

This polynomial function is of degree 3, so it has a maximum of $3 - 1 = 2$ turning points.

13. $f(x) = -x^4 - 4x^3 + 3x^2 + 18x$

This polynomial function is of degree 4, so the maximum number of turning points in the graph is $4 - 1 = 3$.

15. $f(x) = 2x^4 - 9x^3 + 5x^2 + 57x - 45$

This polynomial function is of degree 4, so the maximum number of turning points in the graph is $4 - 1 = 3$.

17. $y = x^3 - 3x^2 - 6x + 8$

The range of an odd-degree polynomial is $(-\infty, \infty)$. The y-intercept of the graph is 8. The graph fitting these criteria is **A**.

19. Since graph **C** crosses the x-axis at one point, the graph has one real zero.

21. A polynomial of degree 3 can have at most 2 turning points. Graphs **B** and **D** have more than 2 turning points, so they cannot be graphs of cubic polynomial functions.

23. On the left side of the y-axis, graph **A** crosses the x-axis once, at -2. Therefore, graph **A** has one negative real zero.

25. The dominating term of f, $5x^5$, is of odd degree and its coefficient is positive, so its end behavior is symbolized as $\curvearrowright$.

27. The dominating term of f, $-4x^3$, is of odd degree and its coefficient is negative, so its end behavior is symbolized as $\curvearrowleft$.

29. The dominating term of f, $9x^6$, is of even degree and its coefficient is positive, so its end behavior is symbolized as $\smile$.

31. The dominating term of f, $-5x^{10}$, is of even degree and its coefficient is negative, so its end behavior is symbolized as $\frown$.

33. $f(x) = x^3 + 5x^2 + 2x - 8$

The possible rational roots are

$$\pm 1, \ \pm 2, \ \pm 4, \ \text{and} \ \pm 8.$$

Check 1 using synthetic division.

$$\begin{array}{r|rrrr} 1 & 1 & 5 & 2 & -8 \\ & & 1 & 6 & 8 \\ \hline & 1 & 6 & 8 & 0 \end{array}$$

The quotient polynomial is $x^2 + 6x + 8$, so

$$f(x) = (x - 1)(x + 2)(x + 4).$$

Copyright © 2012 Pearson Education, Inc. Publishing as Addison-Wesley.

The zeros are -4, -2, and 1, which divide the x-axis into four intervals (or regions).

Interval	Test point	Value of $f(x)$	Sign of $f(x)$
$(-\infty, -4)$	-5	-18	Negative
$(-4, -2)$	-3	4	Positive
$(-2, 1)$	0	-8	Negative
$(1, \infty)$	2	24	Positive

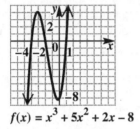

$$f(x) = x^3 + 5x^2 + 2x - 8$$

35. $f(x) = 2x(x - 3)(x + 2)$

The three zeros, 0, 3, and -2, divide the x-axis into four intervals.

Interval	Test point	Value of $f(x)$	Sign of $f(x)$
$(-\infty, -2)$	-3	-36	Negative
$(-2, 0)$	-1	8	Positive
$(0, 3)$	1	-12	Negative
$(3, \infty)$	4	48	Positive

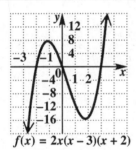

$$f(x) = 2x(x - 3)(x + 2)$$

37. $f(x) = x^2(x - 2)(x + 3)^2$

The three zeros, 0, 2, and -3, divide the x-axis into four intervals.

Interval	Test point	Value of $f(x)$	Sign of $f(x)$
$(-\infty, -3)$	-4	-96	Negative
$(-3, 0)$	-1	-12	Negative
$(0, 2)$	1	-16	Negative
$(2, \infty)$	3	324	Positive

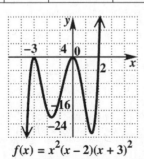

$$f(x) = x^2(x - 2)(x + 3)^2$$

39. $f(x) = (3x - 1)(x + 2)^2$

The two zeros, $\frac{1}{3}$ and -2, divide the x-axis into three intervals.

Interval	Test point	Value of $f(x)$	Sign of $f(x)$
$(-\infty, -2)$	-3	-10	Negative
$(-2, \frac{1}{3})$	0	-4	Negative
$(\frac{1}{3}, \infty)$	1	18	Positive

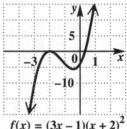

$$f(x) = (3x - 1)(x + 2)^2$$

41. $f(x) = x^3 + 5x^2 - x - 5$
$$= x^2(x + 5) - 1(x + 5)$$
$$= (x + 5)(x^2 - 1)$$
$$= (x + 5)(x + 1)(x - 1)$$

The three zeros, -5, -1, and 1, divide the x-axis into four intervals.

Interval	Test point	Value of $f(x)$	Sign of $f(x)$
$(-\infty, -5)$	-6	-35	Negative
$(-5, -1)$	-2	9	Positive
$(-1, 1)$	0	-5	Negative
$(1, \infty)$	2	21	Positive

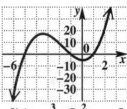

$$f(x) = x^3 + 5x^2 - x - 5$$

43. $f(x) = x^3 - x^2 - 2x$
$$= x(x^2 - x - 2)$$
$$= x(x + 1)(x - 2)$$

The three zeros, 0, -1, and 2, divide the x-axis into four intervals.

Interval	Test point	Value of $f(x)$	Sign of $f(x)$
$(-\infty, -1)$	-2	-8	Negative
$(-1, 0)$	$-\frac{1}{2}$	$\frac{5}{8}$	Positive
$(0, 2)$	1	-2	Negative
$(2, \infty)$	3	12	Positive

Copyright © 2012 Pearson Education, Inc. Publishing as Addison-Wesley.

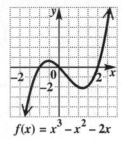

$$f(x) = x^3 - x^2 - 2x$$

45. $f(x) = 2x^3(x^2 - 4)(x - 1)$

$\qquad = 2x^3(x + 2)(x - 2)(x - 1)$

The four zeros, $0, -2, 2$, and 1, divide the x-axis into five intervals.

Interval	Test point	Value of $f(x)$	Sign of $f(x)$
$(-\infty, -2)$	-3	1080	Positive
$(-2, 0)$	-1	-12	Negative
$(0, 1)$	$\frac{1}{2}$	$\frac{15}{32}$	Positive
$(1, 2)$	$\frac{3}{2}$	$-\frac{189}{32}$	Negative
$(2, \infty)$	3	540	Positive

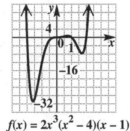

$$f(x) = 2x^3(x^2 - 4)(x - 1)$$

47. $f(x) = 2x^3 - 5x^2 - x + 6$

The possible rational zeros are

$\qquad \pm 1, \ \pm 2, \ \pm 3, \ \pm 6, \ \pm \frac{1}{2}, \ \pm \frac{3}{2}.$

Check -1 using synthetic division.

$$-1 \ \underline{\left|\ \begin{array}{rrrr} 2 & -5 & -1 & 6 \\ & -2 & 7 & -6 \end{array}\right.}$$
$$\begin{array}{rrrr} 2 & -7 & 6 & 0 \end{array}$$

The quotient polynomial can be factored as

$\qquad 2x^2 - 7x + 6 = (2x - 3)(x - 2).$

$f(x)$ may be written in factored form as

$\qquad f(x) = (x + 1)(2x - 3)(x - 2).$

The three zeros, $-1, \frac{3}{2}$, and 2, divide the x-axis into four intervals.

Interval	Test point	Value of $f(x)$	Sign of $f(x)$
$(-\infty, -1)$	-2	-28	Negative
$(-1, \frac{3}{2})$	0	6	Positive
$(\frac{3}{2}, 2)$	$\frac{7}{4}$	$-\frac{11}{32}$	Negative
$(2, \infty)$	3	12	Positive

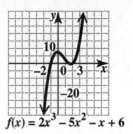

$$f(x) = 2x^3 - 5x^2 - x + 6$$

49. $f(x) = x^3 + x^2 - 8x - 12$

The possible rational zeros are

$\qquad \pm 1, \ \pm 2, \ \pm 3, \ \pm 4, \ \pm 6, \ \pm 12.$

Check 3 using synthetic division.

$$3 \ \underline{\left|\ \begin{array}{rrrr} 1 & 1 & -8 & -12 \\ & 3 & 12 & 12 \end{array}\right.}$$
$$\begin{array}{rrrr} 1 & 4 & 4 & 0 \end{array}$$

The quotient polynomial can be factored as

$\qquad x^2 + 4x + 4 = (x + 2)(x + 2).$

$f(x)$ may be written in factored form as

$\qquad f(x) = (x - 3)(x + 2)^2.$

The two zeros, 3 and -2 (multiplicity 2), divide the x-axis into three intervals.

Interval	Test point	Value of $f(x)$	Sign of $f(x)$
$(-\infty, -2)$	-3	-6	Negative
$(-2, 3)$	0	-12	Negative
$(3, \infty)$	4	36	Positive

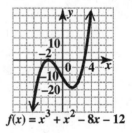

$$f(x) = x^3 + x^2 - 8x - 12$$

51. $f(x) = -x^3 - x^2 + 8x + 12$

The possible rational zeros are

$\qquad \pm 1, \ \pm 2, \ \pm 3, \ \pm 4, \ \pm 6, \ \pm 12.$

Check -2 using synthetic division.

$$-2 \ \underline{\left|\ \begin{array}{rrrr} -1 & -1 & 8 & 12 \\ & 2 & -2 & -12 \end{array}\right.}$$
$$\begin{array}{rrrr} -1 & 1 & 6 & 0 \end{array}$$

The quotient polynomial can be factored as

$\qquad -x^2 + x + 6 = -1(x^2 - x - 6)$
$$\qquad\qquad\qquad = -1(x - 3)(x + 2).$$

$f(x)$ may be written in factored form as

Copyright © 2012 Pearson Education, Inc. Publishing as Addison-Wesley.

$$f(x) = -1(x-3)(x+2)^2$$
$$= (-x+3)(x+2)^2.$$

The two zeros, 3 and -2 (multiplicity 2), divide the x-axis into three intervals.

Interval	Test point	Value of $f(x)$	Sign of $f(x)$
$(-\infty, -2)$	-3	6	Positive
$(-2, 3)$	0	12	Positive
$(3, \infty)$	4	-36	Negative

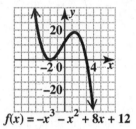

$f(x) = -x^3 - x^2 + 8x + 12$

53. $f(x) = x^4 - 18x^2 + 81$
$$= (x^2 - 9)(x^2 - 9)$$
$$= (x+3)(x-3)(x+3)(x-3)$$
$$= (x+3)^2(x-3)^2$$

f is easily factorable, so no synthetic division is required. From the factored form of $f(x)$, we see that there are two zeros: -3 (multiplicity 2) and 3 (multiplicity 2).

Interval	Test point	Value of $f(x)$	Sign of $f(x)$
$(-\infty, -3)$	-4	49	Positive
$(-3, 3)$	0	81	Positive
$(3, \infty)$	4	49	Positive

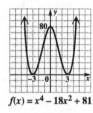

$f(x) = x^4 - 18x^2 + 81$

55. $f(x) = x^4 + x^3 - 6x^2 - 20x - 16$

(a) We need to find $f(-2)$ and $f(-1)$. We can use synthetic division or direct substitution.

$$
\begin{array}{r|rrrrr}
-2 & 1 & 1 & -6 & -20 & -16 \\
   &   & -2 & 2 & 8 & 24 \\
\hline
   & 1 & -1 & -4 & -12 & 8
\end{array}
$$

So $f(-2) = 8$.

$f(-1) = (-1)^4 + (-1)^3 - 6(-1)^2 - 20(-1) - 16$
$$= 1 - 1 - 6 + 20 - 16$$
$$= -2$$

Since $f(-2)$ is *positive* and $f(-1)$ is *negative*, the intermediate value theorem assures us that there must be a real zero between -2 and -1.

(b)

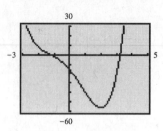

There are two zeros: -1.236 and 3.236.

57. $f(x) = x^4 - 4x^3 - 20x^2 + 32x + 12$

(a)
$$
\begin{array}{r|rrrrr}
-4 & 1 & -4 & -20 & 32 & 12 \\
   &   & -4 & 32 & -48 & 64 \\
\hline
   & 1 & -8 & 12 & -16 & 76
\end{array}
$$

$$
\begin{array}{r|rrrrr}
-3 & 1 & -4 & -20 & 32 & 12 \\
   &   & -3 & 21 & -3 & -87 \\
\hline
   & 1 & -7 & 1 & 29 & -75
\end{array}
$$

Since $f(-4) = 76 > 0$ (*positive*) and $f(-3) = -75 < 0$ (*negative*), the intermediate value theorem assures us that there must be a real zero between -4 and -3.

(b)

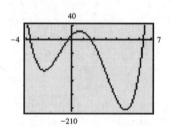

There are four zeros:

$$-3.646, -0.317, 1.646, \text{ and } 6.317.$$

59. $f(x) = x^4 - x^3 + 3x^2 - 8x + 8$;
no real zero greater than 2

$$
\begin{array}{r|rrrrr}
2 & 1 & -1 & 3 & -8 & 8 \\
  &   & 2 & 2 & 10 & 4 \\
\hline
  & 1 & 1 & 5 & 2 & 12
\end{array}
$$

By the boundedness theorem, since $2 > 0$ and all numbers in the bottom row of the synthetic division are nonnegative, f has no real zero greater than 2.

61. $f(x) = x^4 + x^3 - x^2 + 3$;
no real zero less than -2

$$
\begin{array}{r|rrrrr}
-2 & 1 & 1 & -1 & 0 & 3 \\
   &   & -2 & 2 & -2 & 4 \\
\hline
   & 1 & -1 & 1 & -2 & 7
\end{array}
$$

By the boundedness theorem, since $-2 < 0$ and the numbers in the bottom row of the synthetic division alternate sign, f has no real zero less than -2.

63. $f(x) = 3x^4 + 2x^3 - 4x^2 + x - 1$;
no real zero greater than 1

Copyright © 2012 Pearson Education, Inc. Publishing as Addison-Wesley.

$$\begin{array}{r|rrrr} 1 & 3 & 2 & -4 & 1 & -1 \\ & & 3 & 5 & 1 & 2 \\ \hline & 3 & 5 & 1 & 2 & 1 \end{array}$$

By the boundedness theorem, since $1 > 0$ and all numbers in the bottom row of the synthetic division are nonnegative, f has no real zero greater than 1.

65. The graph shows that the zeros are $-6, 2,$ and $5.$ The polynomial function has the form

$$f(x) = a(x+6)(x-2)(x-5).$$

Since $(0, 30)$ is on the graph, $f(0) = 30.$

$$f(0) = a(0+6)(0-2)(0-5)$$
$$30 = 60a$$
$$a = \tfrac{30}{60} = 0.5$$

A cubic polynomial that has the graph shown is

$$f(x) = 0.5(x+6)(x-2)(x-5)$$
$$= 0.5x^3 - 0.5x^2 - 16x + 30.$$

67. $f(x) = 0.86x^3 - 5.24x^2 + 3.55x + 7.84$

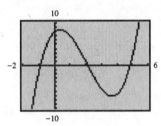

From the graph, the x-intercepts (the real zeros of f), to the nearest hundredth, are $-0.88, 2.12,$ and $4.86.$

69. $f(x) = \sqrt{7}x^3 + \sqrt{5}x^2 + \sqrt{17}$

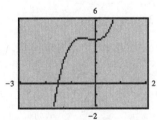

From the graph, the x-intercept (the real zero of f), to the nearest hundredth, is $-1.52.$

71. $f(x) = -\sqrt{15}x^4 - \sqrt{3}x^2 + 7$

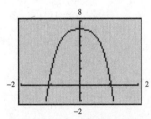

From the graph, the x-intercepts (the real zeros of f), to the nearest hundredth, are -1.07 and $1.07.$

73. $f(x) = x^3 + 4x^2 - 8x - 8, [-3.8, -3]$

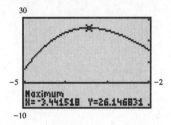

To the nearest hundredth, the coordinates of the turning point in the interval $[-3.8, -3]$ are $(-3.44, 26.15).$

75. $f(x) = 2x^3 - 5x^2 - x + 1, [-1, 0]$

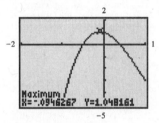

To the nearest hundredth, the coordinates of the turning point in the interval $[-1, 0]$ are $(-0.09, 1.05).$

77. $f(x) = -x^4 + x + 3, [0, 1]$

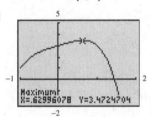

To the nearest hundredth, the coordinates of the turning point in the interval $[0, 1]$ are $(0.63, 3.47).$

♦♦♦ Relating Concepts 79–88 ♦♦♦

79.
$$f(x) = 2x^3$$
$$f(-x) = 2(-x)^3$$
$$= -2x^3$$
$$= -f(x)$$

Since $f(-x) = -f(x)$, this is an odd function.

80.
$$f(x) = -4x^5$$
$$f(-x) = -4(-x)^5$$
$$= 4x^5$$
$$= -f(x)$$

Since $f(-x) = -f(x)$, this is an odd function.

81.
$$f(x) = 0.2x^4$$
$$f(-x) = 0.2(-x)^4$$
$$= 0.2x^4$$
$$= f(x)$$

Since $f(-x) = f(x)$, this is an even function.

Copyright © 2012 Pearson Education, Inc. Publishing as Addison-Wesley.

82. $f(x) = -x^6$

   $f(-x) = -(-x)^6$

   $\qquad = -x^6$

   $\qquad = f(x)$

Since $f(-x) = f(x)$, this is an even function.

83. $f(x) = -x^5$

   $f(-x) = -(-x)^5$

   $\qquad = -(-x^5)$

   $\qquad = -f(x)$

Since $f(-x) = -f(x)$, this is an odd function.

84. $f(x) = -5x^7$

   $f(-x) = -5(-x)^7$

   $\qquad = 5x^7$

   $\qquad = -f(x)$

Since $f(-x) = -f(x)$, this is an odd function.

85. $f(x) = 2x^3 + 3x^2$

   $f(-x) = 2(-x)^3 + 3(-x)^2$

   $\qquad = -2x^3 + 3x^2$

Since $f(-x)$ is not equal to either $f(x)$ or $-f(x)$, this function is neither even nor odd.

86. $f(x) = 4x^5 - x^4$

   $f(-x) = 4(-x)^5 - (-x)^4$

   $\qquad = -4x^5 - x^4$

Since $f(-x)$ is not equal to either $f(x)$ or $-f(x)$, this function is neither even nor odd.

87. $f(x) = x^4 + 3x^2 + 5$

   $f(-x) = (-x)^4 + 3(-x)^2 + 5$

   $\qquad = x^4 + 3x^2 + 5$

   $\qquad = f(x)$

Since $f(-x) = f(x)$, this is an even function.

88. By the definition of an even function, if (a, b) lies on the graph of an even function, then so does $(-a, b)$. Therefore, the graph of an even function is symmetric with respect to the _y-axis_ .

If (a, b) lies on the graph of an odd function, then by definition, so does $(-a, -b)$. Therefore, the graph of an odd function is symmetric with respect to the _origin_ .

89. (a) Plot the ordered pairs with x varying from 2 (for 1982) to 13 (for 1993). See part (b) for the graph.

(b) If the vertex is $(2, 620)$, the equation will have the form

$$g(x) = a(x - h)^2 + k$$

$$g(x) = a(x - 2)^2 + 620.$$

We use the point $(13, 220{,}592)$ to find the value of a.

$$g(13) = a(13 - 2)^2 + 620 = 220{,}592$$

$$121a = 219{,}972$$

So $a = \dfrac{219{,}972}{121} \approx 1817.95 \approx 1818.$

Thus,

$$g(x) = 1818(x - 2)^2 + 620.$$

$g(x) = 1818(x - 2)^2 + 620$

(c) 1987 corresponds to $x = 7$.

$$g(7) = 1818(7 - 2)^2 + 620$$

$$= 46{,}070.$$

This figure is a bit higher than the figure $40{,}820$ given in the table.

91. (a) Let $x =$ the length and $20 - 2x =$ the width.

Both length and width must be positive, so

$$x > 0 \quad \text{and} \quad 20 - 2x > 0$$

$$-2x > -20$$

$$x < 10.$$

The restrictions on x are given by the inequality $0 < x < 10$.

(b) Use the formula $\mathcal{A} = WL$.

$$\mathcal{A}(x) = x(20 - 2x)$$

$$\text{or} \quad \mathcal{A}(x) = -2x^2 + 20x$$

(c) The x-value of the vertex of the graph of this quadratic function is the desired value of x. Use the vertex formula with $a = -2$ and $b = 20$.

$$x = \frac{-b}{2a} = \frac{-20}{2(-2)} = 5$$

The y-value of the vertex gives the maximum area.

$$\mathcal{A}(x) = -2x^2 + 20x$$

$$\mathcal{A}(5) = -2(5)^2 + 20(5)$$

$$= -50 + 100 = 50$$

The maximum cross-sectional area is 50 square inches.

(d) We must solve the quadratic inequality

$$-2x^2 + 20x < 40$$

$$\text{or} \quad -2x^2 + 20x - 40 < 0.$$

Solve the corresponding quadratic equation

Copyright © 2012 Pearson Education, Inc. Publishing as Addison-Wesley.

$$-2x^2 + 20x - 40 = 0$$
$$\text{or } x^2 - 10x + 20 = 0.$$

Use the quadratic formula with $a = 1$, $b = -10$, and $c = 20$.

$$x = \frac{-(-10) \pm \sqrt{(-10)^2 - 4(1)(20)}}{2(1)}$$
$$\approx \frac{10 \pm 4.472}{2}$$
$$x \approx 2.76 \quad \text{or} \quad x \approx 7.24$$

The values 2.76 and 7.24 divide the number line into three intervals: $(-\infty, 2.76)$, $(2.76, 7.24)$, and $(7.24, \infty)$. However, in (a), we saw that in this problem x is restricted to $0 < x < 10$; that is, the open interval $(0, 10)$. Therefore, we need to consider the intervals $(0, 2.76)$, $(2.76, 7.24)$, and $(7.24, 10)$.

Use a test point in each interval to determine where the expression $-2x^2 + 20x - 40$ is negative. We find that it is negative in the intervals $(0, 2.76)$ and $(7.24, 10)$. Therefore, the area of a cross section will be less than 40 square inches when x is between 0 and 2.76 or between 7.24 and 10.

93. Refer to the diagram in the textbook. Since the point $C(x, y)$ lies on the parabola $y = 9 - x^2$, we may write the coordinates of this point as $C(x, 9 - x^2)$. Examine the dimensions of the rectangle. The width is $2x$ and the length is $9 - x^2$. The area of this rectangle is

$$\mathcal{A}(x) = 2x(9 - x^2)$$
$$= 18x - 2x^3$$
$$\mathcal{A}(x) = -2x^3 + 18x.$$

We need to find the value of x for which $\mathcal{A}(x)$ has its maximum value. Use a graphing calculator to find the maximum, which is one of the turning points on the graph of this cubic function.

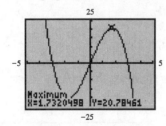

The x-value of the maximum point (the value of x that maximizes the area of the rectangle) is approximately 1.732.

95. $g(x) = -0.006x^4 + 0.140x^3 - 0.053x^2 + 1.79x$

Let $x = 20$.

$$g(20) = -0.006(20)^4 + 0.140(20)^3$$
$$- 0.053(20)^2 + 1.79(20)$$
$$= 174.6 \approx 175$$

The concentration after 20 seconds is approximately 175.

97. $f(x) = 0.05\overline{3}x - 0.04\overline{3}$

(a) Let $x = 10$.

$$f(10) = 0.05\overline{3}(10) - 0.04\overline{3}$$
$$= 0.49$$

Assuming the function continues to model the situation, about 49% of the patients will have AIDS after 10 years.

(b) Let $f(x) = 0.5$.

$$0.5 = 0.05\overline{3}x - 0.04\overline{3}$$
$$0.54\overline{3} = 0.05\overline{3}x$$
$$x = \frac{0.54\overline{3}}{0.05\overline{3}} \approx 10.2$$

After approximately 10.2 years, half the patients would have AIDS.

99. $\dfrac{x^2 - 9}{2x + 5}$

(a) The fraction will equal zero when the numerator equals zero.

$$x^2 - 9 = 0$$
$$x^2 = 9$$
$$x = \pm 3$$

(b) The fraction will be undefined when the denominator equals zero.

$$2x + 5 = 0$$
$$2x = -5$$
$$x = -\frac{5}{2}$$

101. $\dfrac{2x^2 + x - 21}{3x^2 - 17x - 6}$

(a) $\quad 2x^2 + x - 21 = 0$
$$\quad (2x + 7)(x - 3) = 0$$
$$\quad x = -\frac{7}{2} \quad \text{or} \quad x = 3$$

(b) $\quad 3x^2 - 17x - 6 = 0$
$$\quad (3x + 1)(x - 6) = 0$$
$$\quad x = -\frac{1}{3} \quad \text{or} \quad x = 6$$

Summary Exercises on Polynomial Functions and Graphs

1. $f(x) = x^4 + 3x^3 - 3x^2 - 11x - 6$

(a) $f(x)$ has 1 variation in sign, so f has exactly 1 positive real zero.

$$f(-x) = x^4 - 3x^3 - 3x^2 + 11x - 6$$

has 3 variations in sign, so f has either 3 or $3 - 2 = 1$ negative real zeros.

Copyright © 2012 Pearson Education, Inc. Publishing as Addison-Wesley.

(b) The possible rational zeros are

$$\pm 1, \ \pm 2, \ \pm 3, \ \pm 6.$$

(c)
$$-3 \ \overline{\big)\ \begin{array}{rrrrr} 1 & 3 & -3 & -11 & -6 \\ & -3 & 0 & 9 & 6 \\ \hline 1 & 0 & -3 & -2 & 0 \end{array}}$$

$$-1 \ \overline{\big)\ \begin{array}{rrrr} 1 & 0 & -3 & -2 \\ & -1 & 1 & 2 \\ \hline 1 & -1 & -2 & 0 \end{array}}$$

The quotient polynomial can be factored as

$$x^2 - x - 2 = (x - 2)(x + 1)$$

The rational zeros of f are -3, -1 (multiplicity 2), and 2.

(d) f has no other real zeros.

(e) f has no other complex zeros.

(f) Since the zeros of f are -3, -1, and 2, the x-intercepts of the graph are $(-3, 0)$, $(-1, 0)$, and $(2, 0)$.

(g) $f(0) = -6$, so the y-intercept of the graph is $(0, -6)$.

(h) Find $f(4)$.

$$4 \ \overline{\big)\ \begin{array}{rrrrr} 1 & 3 & -3 & -11 & -6 \\ & 4 & 28 & 100 & 356 \\ \hline 1 & 7 & 25 & 89 & 350 \end{array}}$$

$f(4) = 350$, so the point $(4, 350)$ is on the graph.

(i)

$f(x) = x^4 + 3x^3 - 3x^2 - 11x - 6$

3. $f(x) = 2x^5 - 10x^4 + x^3 - 5x^2 - x + 5$

(a) $f(x)$ has 4 variations in sign, so f has either 4 or 2 or 0 positive zeros.

$$f(-x) = -2x^5 - 10x^4 - x^3 - 5x^2 + x + 5$$

has 1 variation in sign, so f has 1 negative zero.

(b) The possible rational zeros are

$$\pm 1, \ \pm 5, \pm \tfrac{1}{2}, \ \pm \tfrac{5}{2}.$$

(c)
$$5 \ \overline{\big)\ \begin{array}{rrrrrr} 2 & -10 & 1 & -5 & -1 & 5 \\ & 10 & 0 & 5 & 0 & -5 \\ \hline 2 & 0 & 1 & 0 & -1 & 0 \end{array}}$$

The remaining zeros are the zeros of the quotient, $2x^4 + x^2 - 1$. This polynomial can be factored as the product of two binomials.

$$2x^4 + x^2 - 1 = 0$$
$$(2x^2 - 1)(x^2 + 1) = 0$$

$$\begin{array}{lll} 2x^2 - 1 = 0 & \text{or} & x^2 + 1 = 0 \\ x^2 = \tfrac{1}{2} & \text{or} & x^2 = -1 \\ x = \pm \tfrac{\sqrt{2}}{2} & \text{or} & x = \pm i \end{array}$$

The only rational zero of f is 5.

(d) The other real zeros (the irrational zeros) of f are $-\frac{\sqrt{2}}{2}$ and $\frac{\sqrt{2}}{2}$.

(e) The other complex zeros (the nonreal zeros) of f are $-i$ and i.

(f) Corresponding to the three real zeros, the graph has x-intercepts at $\left(-\frac{\sqrt{2}}{2}, 0\right)$, $\left(\frac{\sqrt{2}}{2}, 0\right)$, and $(5, 0)$.

(g) $f(0) = 5$, so the y-intercept of the graph is $(0, 5)$.

(h) Find $f(4)$.

$$4 \ \overline{\big)\ \begin{array}{rrrrrr} 2 & -10 & 1 & -5 & -1 & 5 \\ & 8 & -8 & -28 & -132 & -532 \\ \hline 2 & -2 & -7 & -33 & -133 & -527 \end{array}}$$

$f(4) = -527$, so the point $(4, -527)$ is on the graph.

(i)

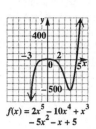

$$f(x) = 2x^5 - 10x^4 + x^3 - 5x^2 - x + 5$$

5. $f(x) = -2x^4 - x^3 + x + 2$

(a) $f(x)$ has 1 variation in sign, so f has 1 positive zero.

$$f(-x) = -2x^4 + x^3 - x + 2$$

has 3 variations in sign, so f has 3 or 1 negative zeros.

(b) The possible rational zeros are

$$\pm 1, \ \pm 2, \ \pm \tfrac{1}{2}.$$

(c)
$$1 \ \overline{\big)\ \begin{array}{rrrrr} -2 & -1 & 0 & 1 & 2 \\ & -2 & -3 & -3 & -2 \\ \hline -2 & -3 & -3 & -2 & 0 \end{array}}$$

$$-1 \ \overline{\big)\ \begin{array}{rrrr} -2 & -3 & -3 & -2 \\ & 2 & 1 & 2 \\ \hline -2 & -1 & -2 & 0 \end{array}}$$

$$-2x^2 - x - 2 = 0$$
$$2x^2 + x + 2 = 0$$

Copyright © 2012 Pearson Education, Inc. Publishing as Addison-Wesley.

$$x = \frac{-1 \pm \sqrt{1^2 - 4(2)(2)}}{2(2)}$$

$$= \frac{-1 \pm \sqrt{-15}}{4}$$

$$= -\frac{1}{4} \pm i\frac{\sqrt{15}}{4}$$

The rational zeros of f are -1 and 1.

(d) f has no other real zeros.

(e) The other complex zeros of f are $-\frac{1}{4} + i\frac{\sqrt{15}}{4}$ and $-\frac{1}{4} - i\frac{\sqrt{15}}{4}$.

(f) Corresponding to the two real zeros of f, the graph has x-intercepts at $(-1, 0)$ and $(1, 0)$.

(g) $f(0) = 2$, so the y-intercept of the graph is $(0, 2)$.

(h) Find $f(4)$.

$$4 \overline{\smash{\big)}\ \begin{array}{rrrrr} -2 & -1 & 0 & 1 & 2 \\ & -8 & -36 & -144 & -572 \\ \hline -2 & -9 & -36 & -143 & -570 \end{array}}$$

$f(4) = -570$, so the point $(4, -570)$ is on the graph.

(i)

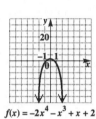

$f(x) = -2x^4 - x^3 + x + 2$

7. $f(x) = 3x^4 - 14x^2 - 5$

(a) $f(x)$ has 1 variation in sign, so f has 1 positive zero.

$$f(-x) = 3x^4 - 14x^2 - 5$$

has 1 variation in sign, so f has 1 negative zero.

(b) The possible rational zeros of f are

$$\pm 1, \ \pm 5, \pm \tfrac{1}{3}, \ \pm \tfrac{5}{3}.$$

(c) None of the 8 possible rational zeros is a zero of the polynomial, so f has no rational zeros.

(d) Factor the polynomial as the product of the binomials.

$$3x^4 - 14x^2 - 5 = 0$$
$$(3x^2 + 1)(x^2 - 5) = 0$$
$$3x^2 + 1 = 0 \qquad \text{or} \quad x^2 - 5 = 0$$
$$x^2 = -\tfrac{1}{3} \qquad \text{or} \qquad x^2 = 5$$
$$x = \pm i\frac{\sqrt{3}}{3} \quad \text{or} \qquad x = \pm\sqrt{5}$$

The real zeros of f are $-\sqrt{5}$ and $\sqrt{5}$.

(e) The other complex zeros of f are $-i\frac{\sqrt{3}}{3}$ and $i\frac{\sqrt{3}}{3}$.

(f) Corresponding to the 2 real zeros of f, the x-intercepts of the graph are $\left(-\sqrt{5}, 0\right)$ and $\left(\sqrt{5}, 0\right)$.

(g) $f(0) = -5$, so the y-intercept of the graph is $(0, -5)$.

(h) Find $f(4)$.

$$4 \overline{\smash{\big)}\ \begin{array}{rrrrr} 3 & 0 & -14 & 0 & -5 \\ & 12 & 48 & 136 & 544 \\ \hline 3 & 12 & 34 & 136 & 539 \end{array}}$$

$f(4) = 539$, so the point $(4, 539)$ is on the graph.

(i)

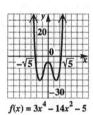

$f(x) = 3x^4 - 14x^2 - 5$

9. $f(x) = -3x^4 + 22x^3 - 55x^2 + 52x - 12$

(a) $f(x)$ has 4 variations in sign, so f has 4, 2, or 0 positive zeros.

$$f(-x) = -3x^4 - 22x^3 - 55x^2 - 52x - 12$$

has 0 variations in sign, so f has 0 negative zeros.

(b) The possible rational zeros are

$$\pm 1, \ \pm 2, \ \pm 3, \ \pm 4, \ \pm 6,$$
$$\pm 12, \ \pm \tfrac{1}{3}, \ \pm \tfrac{2}{3}, \ \pm \tfrac{4}{3}.$$

(c) $3 \overline{\smash{\big)}\ \begin{array}{rrrrr} -3 & 22 & -55 & 52 & -12 \\ & -9 & 39 & -48 & 12 \\ \hline -3 & 13 & -16 & 4 & 0 \end{array}}$

$2 \overline{\smash{\big)}\ \begin{array}{rrrr} -3 & 13 & -16 & 4 \\ & -6 & 14 & -4 \\ \hline -3 & 7 & -2 & 0 \end{array}}$

$$-3x^2 + 7x - 2 = 0$$
$$3x^2 - 7x + 2 = 0$$
$$(3x - 1)(x - 2) = 0$$

$$3x - 1 = 0 \quad \text{or} \quad x - 2 = 0$$
$$x = \tfrac{1}{3} \quad \text{or} \qquad x = 2$$

The rational zeros of f are $\tfrac{1}{3}$, 2 (multiplicity 2), and 3.

(d) All of the zeros are rational, so there are no other real zeros of f.

(e) f has no other complex zeros.

Copyright © 2012 Pearson Education, Inc. Publishing as Addison-Wesley.

(f) Corresponding to the real zeros of f, the x-intercepts of the graph are $\left(\frac{1}{3}, 0\right)$, $(2, 0)$, and $(3, 0)$.

(g) $f(0) = -12$, so the y-intercept of the graph is $(0, -12)$.

(h) Find $f(4)$.

$$\begin{array}{r|rrrrr} 4 & -3 & 22 & -55 & 52 & -12 \\ & & -12 & 40 & -60 & -32 \\ \hline & -3 & 10 & -15 & -8 & -44 \end{array}$$

$f(4) = -44$, so the point $(4, -44)$ is on the graph.

(i)

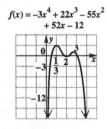

$$f(x) = -3x^4 + 22x^3 - 55x^2 + 52x - 12$$

12.4 Graphs and Applications of Rational Functions

12.4 Now Try Exercises

N1. $f(x) = -\dfrac{5}{x} = -5 \cdot \dfrac{1}{x}$

Compared to $g(x) = \frac{1}{x}$, the graph will be reflected about the x-axis (because of the negative sign), and each point will be five times as far from the x-axis. The x- and y-axes are the horizontal and vertical asymptotes.

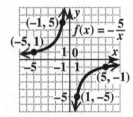

N2. $f(x) = \dfrac{1}{x + 3} = \dfrac{1}{x - (-3)}$

The domain of this function is the set of all real numbers except -3. The graph is that of $g(x) = \frac{1}{x}$, translated 3 units to the left.

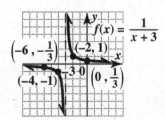

N3. **(a)** $f(x) = \dfrac{x - 2}{2x^2 - 3x - 9} = \dfrac{x - 2}{(2x + 3)(x - 3)}$

f is in lowest terms. The denominator is 0 if $x = -\frac{3}{2}$ or $x = 3$, so $x = -\frac{3}{2}$ and $x = 3$ are the equations of the vertical asymptotes. The numerator has lesser degree than the denominator ($1 < 2$), so $y = 0$ is a horizontal asymptote.

(b) $f(x) = \dfrac{x - 2}{x + 1}$

f is in lowest terms. The denominator is 0 if $x = -1$, so $x = -1$ is the equation of the vertical asymptote. Since the degree of the numerator equals the degree of the denominator, the graph has a horizontal asymptote at $y = \frac{1}{1} = 1$.

(c) $f(x) = \dfrac{x^2 + 3x + 3}{x + 2}$

f is in lowest terms. The denominator is 0 if $x = -2$, so $x = -2$ is the equation of the vertical asymptote. The degree of the numerator, 2, is greater than the degree of the denominator, 1. We can find an equation of the oblique asymptote by dividing the numerator by the denominator (in this case we can use synthetic division).

$$\begin{array}{r|rrr} -2 & 1 & 3 & 3 \\ & & -2 & -2 \\ \hline & 1 & 1 & 1 \end{array}$$

Thus, we can write f as $f(x) = x + 1 + \dfrac{1}{x + 2}$ and an equation of the oblique (slant) asymptote is $y = x + 1$ (so $f(x)$ will get close to that line as $|x| \to \infty$).

N4. $f(x) = \dfrac{x - 2}{2x^2 - 3x - 9} = \dfrac{x - 2}{(2x + 3)(x - 3)}$

Step 1
f is in lowest terms. The denominator is 0 if $x = -\frac{3}{2}$ or $x = 3$, so $x = -\frac{3}{2}$ and $x = 3$ are the equations of the vertical asymptotes.

Step 2
The numerator has lesser degree than the denominator ($1 < 2$), so $y = 0$ is a horizontal asymptote.

Step 3
$f(0) = \frac{2}{9}$, so the y-intercept is $\left(0, \frac{2}{9}\right)$.

Step 4
The only zero of the numerator is 2, so the only x-intercept is $(2, 0)$.

Copyright © 2012 Pearson Education, Inc. Publishing as Addison-Wesley.

Step 5
Since the horizontal asymptote is the x-axis, we know from Step 4 that the graph intersects the horizontal asymptote at $(2, 0)$.

Step 6

x	-3	-1	1	4
y	$-\frac{5}{18}$	$\frac{3}{4}$	$\frac{1}{10}$	$\frac{2}{11}$

Step 7

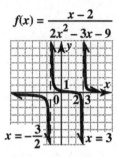

$$f(x) = \frac{x - 2}{2x^2 - 3x - 9}$$

N5. $f(x) = \dfrac{x - 2}{x + 1}$

The vertical asymptote is $x = -1$. The degree of the numerator and the degree of the denominator are equal, so as $|x| \to \infty$, $f(x) \to$ ratio of leading coefficients, that is, the horizontal asymptote is $y = \frac{1}{1} = 1$. Since $f(0) = -2$, the y-intercept is $(0, -2)$. The x-intercept is $(2, 0)$. Now solve $f(x) = 1$.

$$1 = \frac{x - 2}{x + 1}$$
$$1(x + 1) = x - 2$$
$$x + 1 = x - 2$$
$$1 = -2$$

The last equation indicates that $f(x)$ is *never* equal to 1, that is, the graph *never* crosses the horizontal asymptote. An additional point on the graph is $(-4, 2)$, and we can now graph the function.

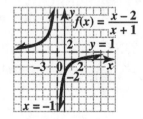

N6. $f(x) = \dfrac{2x(x + 3)}{(x + 2)^2} = \dfrac{2x^2 + 6x}{x^2 + 4x + 4}$

The vertical asymptote is $x = -2$. The degree of the numerator and the degree of the denominator are equal, so as $|x| \to \infty$, $f(x) \to$ ratio of leading coefficients, that is, the horizontal asymptote is $y = \frac{2}{1} = 2$. Since $f(0) = 0$, the y-intercept is $(0, 0)$. The x-intercepts are $(0, 0)$ and $(-3, 0)$. Now solve $f(x) = 2$.

$$2 = \frac{2x^2 + 6x}{x^2 + 4x + 4}$$
$$2x^2 + 8x + 8 = 2x^2 + 6x$$
$$2x = -8$$
$$x = -4$$

Thus, the graph crosses the horizontal asymptote *only* at the point $(-4, 2)$. An additional point on the graph is $(-1, -4)$, and we can now graph the function.

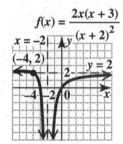

$$f(x) = \frac{2x(x + 3)}{(x + 2)^2}$$

N7. $f(x) = \dfrac{x^2 + 3x + 3}{x + 2}$

The vertical asymptote is $x = -2$. The degree of the numerator is greater than the degree of the denominator, so we divide to find the oblique asymptote.

$$-2 \, \overline{\left)\begin{array}{ccc} 1 & 3 & 3 \\ & -2 & -2 \end{array}\right.}$$
$$ \overline{\begin{array}{ccc} 1 & 1 & 1 \end{array}}$$

The result indicates that $f(x) = x + 1 + \frac{1}{x + 2}$, so the oblique asymptote is $y = x + 1$. The y-intercept is $\left(0, \frac{3}{2}\right)$ and there are no x-intercepts since $x^2 + 3x + 3 \neq 0$ for any real number x. Solving $f(x) = x + 1$ gives us $2 = 3$, indicating that the graph never crosses the oblique asymptote. An additional point on the graph is $(-4, -3.5)$, and we can now graph the function.

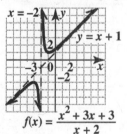

$$f(x) = \frac{x^2 + 3x + 3}{x + 2}$$

N8. $f(x) = \dfrac{2x^2 - 5x - 3}{x - 3}$
$$= \frac{(2x + 1)(x - 3)}{x - 3}$$
$$= 2x + 1 \quad (x \neq 3)$$

The graph of f is the same as the graph of $y = 2x + 1$, except there is a hole at $x = 3$ since the original function is undefined for $x = 3$. Substituting 3 for x in $y = 2x + 1$ gives $y = 7$, so the hole has coordinates $(3, 7)$.

graph on next page

Copyright © 2012 Pearson Education, Inc. Publishing as Addison-Wesley.

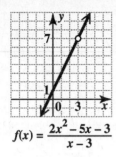

$$f(x) = \frac{2x^2 - 5x - 3}{x - 3}$$

N9. (a) $x = \frac{3}{3.2} = 0.9375$ $\left[\text{or } \frac{15}{16}\right]$

(b) $f(x) = \dfrac{x^2}{2(1 - x)}$

$f(0.9375) = \dfrac{0.9375^2}{2(1 - 0.9375)}$

$= 7.03125 \approx 7.03$

$\left[\text{or } f\left(\frac{15}{16}\right) = \frac{225}{32}\right]$

12.4 Section Exercises

1. $f(x) = -\frac{3}{x}$

The expression on the right side of the equation can be rewritten so that

$$f(x) = -3 \bullet \frac{1}{x}.$$

Compared to $f(x) = \frac{1}{x}$, each point will be 3 times as far from the x-axis. Because of the negative sign, the graph will be reflected about the x-axis. Just as with the graph of $f(x) = \frac{1}{x}$, $y = 0$ is the horizontal asymptote and $x = 0$ is the vertical asymptote.

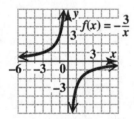

3. $f(x) = \dfrac{1}{x + 2}$

Since $\frac{1}{x+2} = \frac{1}{x-(-2)}$, the graph of $f(x) = \frac{1}{x+2}$ will be similar to the graph of $f(x) = \frac{1}{x}$, except that each point will be translated 2 units to the left. Just as with $f(x) = \frac{1}{x}$, $y = 0$ is the horizontal asymptote, but this graph has $x = -2$ as its vertical asymptote.

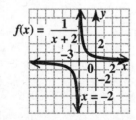

5. $f(x) = \frac{1}{x} + 1$

The graph of this function will be similar to the graph of $f(x) = \frac{1}{x}$, except that each point will be translated 1 unit upward. Just as with $f(x) = \frac{1}{x}$, $x = 0$ is the vertical asymptote, but this graph has $y = 1$ as its horizontal asymptote.

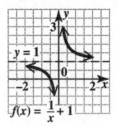

7. Graphs **A**, **B**, and **C** have a domain of $(-\infty, 3) \cup (3, \infty)$.

9. Graph **A** has a range of $(-\infty, 0) \cup (0, \infty)$.

11. Graphs **A**, **C**, and **D** have the x-axis as a horizontal asymptote.

13. (a) The graph of $f(x) = \dfrac{1}{(x - 3)^2}$ is similar to the graph of $f(x) = \frac{1}{x^2}$, but it is translated 3 units to the right.

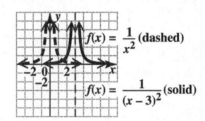

(b) The graph of $f(x) = -\frac{2}{x^2}$ is the reflection of $f(x) = \frac{1}{x^2}$ about the x-axis and with each point twice as far from the x-axis.

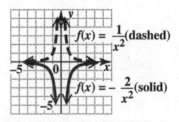

(c) The graph of $f(x) = \dfrac{-2}{(x - 3)^2}$ is the graph in part (b) translated 3 units to the right.

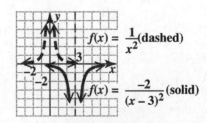

Copyright © 2012 Pearson Education, Inc. Publishing as Addison-Wesley.

15. $f(x) = \dfrac{-8}{3x - 7}$

To find any vertical asymptotes, solve $3x - 7 = 0$. Thus, $x = \frac{7}{3}$ is a vertical asymptote. Since the numerator has a lower degree than the denominator, $y = 0$ is the horizontal asymptote.

17. $f(x) = \dfrac{x + 3}{3x^2 + x - 10}$

To find any vertical asymptotes, set the denominator equal to 0 and solve for x.

$$3x^2 + x - 10 = 0$$
$$(3x - 5)(x + 2) = 0$$

$$3x - 5 = 0 \quad \text{or} \quad x + 2 = 0$$
$$x = \tfrac{5}{3} \quad \text{or} \quad x = -2$$

The vertical asymptotes are $x = \frac{5}{3}$ and $x = -2$. Since the numerator has a lower degree than the denominator, $y = 0$ is the horizontal asymptote.

19. $f(x) = \dfrac{2 - x}{x + 2}$

Setting the denominator equal to 0 shows that the vertical asymptote is $x = -2$.
Since the numerator and the denominator have the same degree, 1, to find the horizontal asymptote, divide the numerator and denominator by x^1 to get

$$f(x) = \dfrac{\frac{2}{x} - \frac{x}{x}}{\frac{x}{x} + \frac{2}{x}} = \dfrac{\frac{2}{x} - 1}{1 + \frac{2}{x}}.$$

As $|x|$ gets larger, $\frac{2}{x}$ approaches 0, and the value of $f(x)$ approaches

$$\dfrac{0 - 1}{1 + 0} = -1.$$

The line $y = -1$ is therefore the horizontal asymptote.

21. $f(x) = \dfrac{3x - 5}{2x + 9}$

$f(x)$ has vertical asymptote $x = -\frac{9}{2}$ since the denominator is 0 if $x = -\frac{9}{2}$. Dividing the numerator and denominator by x, we find that the horizontal asymptote is $y = \frac{3}{2}$.

23. $f(x) = \dfrac{2}{x^2 - 4x + 3}$

To find any vertical asymptotes, set the denominator equal to 0 and solve for x.

$$x^2 - 4x + 3 = 0$$
$$(x - 3)(x - 1) = 0$$

$$x - 3 = 0 \quad \text{or} \quad x - 1 = 0$$
$$x = 3 \quad \text{or} \quad x = 1$$

The vertical asymptotes are $x = 3$ and $x = 1$. Since the numerator has a lower degree than the denominator, $y = 0$ is the horizontal asymptote.

25. $f(x) = \dfrac{x^2 - 1}{x + 3}$

The vertical asymptote is $x = -3$, found by solving $x + 3 = 0$.
Since the degree of the numerator is one more than the degree of the denominator, the graph has an oblique asymptote. To find it, divide $x^2 - 1$ by $x + 3$ and disregard any remainder.

$$-3 \,\overline{\big)\, 1 \quad 0 \quad -1 \,}$$
$$\underline{\quad\quad -3 \quad 9\,}$$
$$1 \quad -3 \quad 8$$

Thus,

$$f(x) = \dfrac{x^2 - 1}{x + 3} = x - 3 + \dfrac{8}{x + 3}.$$

As $|x| \to \infty$, the graph approaches the oblique asymptote $y = x - 3$.

27. $f(x) = \dfrac{(x - 3)(x + 1)}{(x + 2)(2x - 5)}$

The vertical asymptotes are $x = -2$ and $x = \frac{5}{2}$, since these values make the denominator equal to 0.
Multiply the factors in the numerator and denominator to get

$$f(x) = \dfrac{x^2 - 2x - 3}{2x^2 - x - 10}.$$

Thus, the horizontal asymptote, found by dividing the numerator and denominator by x^2, is $y = \frac{1}{2}$.

29. A. $f(x) = \dfrac{1}{x^2 + 2}$ has a graph that does not have a vertical asymptote. This is because there is no real number that can make the denominator $x^2 + 2$ become 0.

31. $f(x) = \dfrac{x + 1}{x - 4}$

The vertical asymptote is $x = 4$.
Since the degree of the numerator equals the degree of the denominator, the graph has a horizontal asymptote at $y = \frac{1}{1} = 1$.
($\frac{1}{1}$ is the ratio of leading coefficients.)
$f(0) = -\frac{1}{4}$, so the y-intercept is $\left(0, -\frac{1}{4}\right)$.
The only zero of the numerator is -1, so the only x-intercept is $(-1, 0)$.
To the right of $x = 4$, we have the point $(5, 6)$.

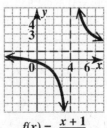

$$f(x) = \dfrac{x + 1}{x - 4}$$

Copyright © 2012 Pearson Education, Inc. Publishing as Addison-Wesley.

33. $f(x) = \dfrac{-x}{x^2 - 4} = \dfrac{-x}{(x+2)(x-2)}$

The vertical asymptotes are $x = 2$ and $x = -2$.
Since the numerator has lesser degree than the
denominator, $y = 0$ is the horizontal asymptote.
$f(0) = 0$, so the y-intercept is $(0, 0)$, and this is
also the only x-intercept.

x	-3	-1	1	3
y	$\frac{3}{5}$	$-\frac{1}{3}$	$\frac{1}{3}$	$-\frac{3}{5}$

Plot these points and sketch the graph.

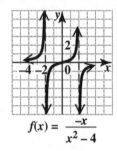

$$f(x) = \frac{-x}{x^2 - 4}$$

35. $f(x) = \dfrac{3x}{(x+1)(x-2)}$

The vertical asymptotes are $x = -1$ and $x = 2$.
Since the numerator has lesser degree than the
denominator, $y = 0$ is the horizontal asymptote.
$f(0) = 0$, so the y-intercept is $(0, 0)$, and this is
also the only x-intercept.

x	-2	$-\frac{1}{2}$	1	3
y	$-\frac{3}{2}$	$\frac{6}{5}$	$-\frac{3}{2}$	$\frac{9}{4}$

Plot these points and sketch the graph.

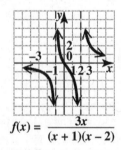

$$f(x) = \frac{3x}{(x+1)(x-2)}$$

37. $f(x) = \dfrac{2x+1}{(x+2)(x+4)} = \dfrac{2x+1}{x^2+6x+8}$

The vertical asymptotes are $x = -2$ and $x = -4$.
Since the numerator has lesser degree than the
denominator, $y = 0$ is the horizontal asymptote.
$f(0) = \frac{1}{8}$, so the y-intercept is $\left(0, \frac{1}{8}\right)$.
The only zero of the numerator is $-\frac{1}{2}$, so the only
x-intercept is $\left(-\frac{1}{2}, 0\right)$.

x	-5	-3	-1	1
y	-3	5	$-\frac{1}{3}$	$\frac{1}{5}$

Plot these points and sketch the graph.

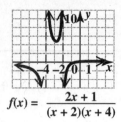

$$f(x) = \frac{2x+1}{(x+2)(x+4)}$$

39. $f(x) = \dfrac{3x}{x-1}$

The vertical asymptote is $x = 1$.
Since the degree of the numerator equals the
degree of the denominator, the graph has a
horizontal asymptote at $y = \frac{3}{1} = 3$.
$f(0) = 0$, so the y-intercept is $(0, 0)$, and this is
also the only x-intercept.
To the right of $x = 1$, we have the point $(2, 6)$.

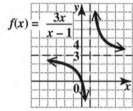

$$f(x) = \frac{3x}{x-1}$$

41. $f(x) = \dfrac{1}{(x+5)(x-2)} = \dfrac{1}{x^2+3x-10}$

The vertical asymptotes are $x = 2$ and $x = -5$.
Since the numerator has lesser degree than the
denominator, $y = 0$ is the horizontal asymptote.
$f(0) = -\frac{1}{10}$, so the y-intercept is $\left(0, -\frac{1}{10}\right)$.
There are no x-intercepts because the numerator is
never equal to 0.

x	-6	3
y	$\frac{1}{8}$	$\frac{1}{8}$

Plot these points and sketch the graph.

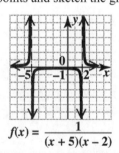

$$f(x) = \frac{1}{(x+5)(x-2)}$$

43. $f(x) = \dfrac{3}{(x+4)^2} = \dfrac{3}{x^2+8x+16}$

The vertical asymptote is $x = -4$.
Since the numerator has lesser degree than the
denominator, $y = 0$ is the horizontal asymptote.
$f(0) = \frac{3}{16}$, so the y-intercept is $\left(0, \frac{3}{16}\right)$.
There are no x-intercepts because the numerator is
never equal to 0. Notice that the value of
$y = f(x)$ is always positive.
To the left of $x = -4$, we have the point $(-5, 3)$.

Copyright © 2012 Pearson Education, Inc. Publishing as Addison-Wesley.

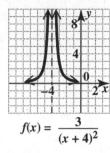

$$f(x) = \frac{3}{(x+4)^2}$$

45. $f(x) = \frac{(x-3)(x+1)}{(x-1)^2} = \frac{x^2 - 2x - 3}{x^2 - 2x + 1}$

The vertical asymptote is $x = 1$.
Since the degree of the numerator equals the
degree of the denominator, the graph has a
horizontal asymptote at $y = \frac{1}{1} = 1$.
$f(0) = -3$, so the y-intercept is $(0, -3)$.
The x-intercepts are $(3, 0)$ and $(-1, 0)$.

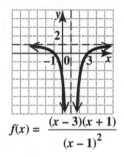

$$f(x) = \frac{(x-3)(x+1)}{(x-1)^2}$$

47. $f(x) = \frac{x^2 + 1}{x + 3}$

The vertical asymptote is $x = -3$.
Since the degree of the numerator is one more than
the degree of the denominator, the graph has an
oblique asymptote, which can be found by
division. Use synthetic division.

$$-3 \begin{array}{|rrr} 1 & 0 & 1 \\ & -3 & 9 \\ \hline 1 & -3 & 10 \end{array}$$

Thus, $f(x) = \frac{x^2 + 1}{x + 3} = x - 3 + \frac{10}{x + 3}$, and the
oblique asymptote is $y = x - 3$.
$f(0) = \frac{1}{3}$, so the y-intercept is $\left(0, \frac{1}{3}\right)$.
Note that $f(x)$ can never be zero, so the graph has
no x-intercepts.
To the left of $x = -3$, we have the point
$(-4, -17)$.

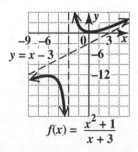

$$f(x) = \frac{x^2 + 1}{x + 3}$$

49. $f(x) = \frac{2x^2 + 3}{x - 4}$

The vertical asymptote is $x = 4$.
Since the degree of the numerator is one more than
the degree of the denominator, the graph has an
oblique asymptote, which can be found by
division. Divide $2x^2 + 3$ by $x - 4$.

$$4 \begin{array}{|rrr} 2 & 0 & 3 \\ & 8 & 32 \\ \hline 2 & 8 & 35 \end{array}$$

Thus, $f(x) = \frac{2x^2 + 3}{x - 4} = 2x + 8 + \frac{35}{x - 4}$, and
the oblique asymptote is the line $y = 2x + 8$.
$f(0) = -\frac{3}{4}$, so the y-intercept is $\left(0, -\frac{3}{4}\right)$.
Note that $f(x)$ can never be zero, so the graph has
no x-intercepts.
To the right of $x = 4$, we have the point $(5, 53)$.

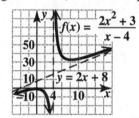

51. $f(x) = \frac{(x-5)(x-2)}{x^2 + 9} = \frac{x^2 - 7x + 10}{x^2 + 9}$

The graph has no vertical asymptotes since
$x^2 + 9 = 0$ has no real solutions.
Since the degree of the numerator equals the
degree of the denominator, the graph has a
horizontal asymptote at $y = \frac{1}{1} = 1$.
$f(0) = \frac{10}{9}$, so the y-intercept is $\left(0, \frac{10}{9}\right)$.
The x-intercepts are $(5, 0)$ and $(2, 0)$.
Solve the equation $f(x) = 1$ to determine if the
graph intersects its horizontal asymptote.

$$f(x) = \frac{x^2 - 7x + 10}{x^2 + 9} = 1$$
$$x^2 - 7x + 10 = x^2 + 9$$
$$-7x = -1$$
$$x = \frac{1}{7}$$

Thus, the graph intersects the horizontal asymptote
$y = 1$ at the point $\left(\frac{1}{7}, 1\right)$.

x	-10	-3	-2
y	≈ 1.65	≈ 2.22	≈ 2.15

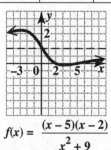

$$f(x) = \frac{(x-5)(x-2)}{x^2 + 9}$$

Copyright © 2012 Pearson Education, Inc. Publishing as Addison-Wesley.

53. $f(x) = \dfrac{x^2 - 9}{x + 3} = \dfrac{(x + 3)(x - 3)}{x + 3}$
$$= x - 3 \ (x \neq -3).$$

The graph of this function will be the same as the graph of $y = g(x) = x - 3$ (a line), with the exception of the point with x-value -3. A "hole" appears in the graph at $(-3, g(-3)) = (-3, -6)$.

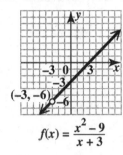

$$f(x) = \dfrac{x^2 - 9}{x + 3}$$

55. $f(x) = \dfrac{25 - x^2}{x - 5} = \dfrac{(5 + x)(5 - x)}{x - 5}$
$$= -1(5 + x) = -x - 5 \ (x \neq 5)$$

The graph of this function will be the same as the graph of $y = g(x) = -x - 5$ (a line), with the exception of the point with x-value 5. A "hole" appears in the graph at $(5, g(5)) = (5, -10)$.

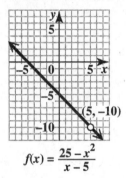

$$f(x) = \dfrac{25 - x^2}{x - 5}$$

♦ ♦ ♦ Relating Concepts 57–62 ♦ ♦ ♦

In Exercises 57 – 62, consider the rational function

$$f(x) = \dfrac{x^2 - 16}{x + 4}.$$

57. Graph this function as y_1 on a graphing calculator. On a TI-83/4, press $\boxed{\text{ZOOM}}$ $\boxed{8}$ $\boxed{\text{ENTER}}$ to produce the following graph.

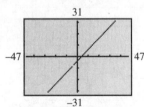

58. The graph looks like a line except for a "gap" at the point where $x = -4$. (This unlit portion of the screen is called a pixel.)

59. Trying to evaluate the function for $x = -4$ leads to an error message. Since -4 is not in the domain of f, the graphing calculator cannot assign a value to $f(-4)$.

60. $f(x) = \dfrac{x^2 - 16}{x + 4} = \dfrac{(x + 4)(x - 4)}{x + 4}$
$$= x - 4 \ (x \neq -4).$$

Let $g(x) = x - 4$. The domain of g is $(-\infty, \infty)$, whereas the domain of f is $(-\infty, -4) \cup (-4, \infty)$.

61. $g(x) = x - 4$

Graph this function as y_2 on a graphing calculator.

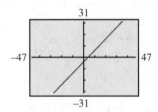

This graph is a line with slope 1 and y-intercept $(0, -4)$. Observe that there is no "gap" in the graph at the point where $x = -4$. This graph is the same as that of y_1 except that the pixel at $(-4, -8)$ is lit.

62. $g(x) = x - 4$
$g(-4) = -4 - 4 = -8$

There is no difficulty obtaining the value of $g(-4)$ since -4 is in the domain of g.

63. **(a)** $f(x) = \dfrac{1}{(x - 2)^2}$

Notice that no matter what value x takes on $(x \neq 2)$, $(x - 2)^2$ will be greater than zero. That means $f(x) > 0$, which is graph **C**.

(b) $f(x) = \dfrac{1}{x - 2}$

Notice if $x > 2$, $x - 2 > 0$, and $f(x) > 0$. If $x < 2$, $x - 2 < 0$, and $f(x) < 0$. This is graph **A**.

(c) $f(x) = \dfrac{-1}{x - 2}$

Notice if $x > 2$, $f(x) < 0$. If $x < 2$, $f(x) > 0$. This is graph **B**.

(d) $f(x) = \dfrac{-1}{(x - 2)^2}$

Notice that no matter what value x takes on $(x \neq 2)$, $(x - 2)^2$ will be greater than zero. Thus $f(x) < 0$. This is graph **D**.

Copyright © 2012 Pearson Education, Inc. Publishing as Addison-Wesley.

65. (a) The ratios are listed in the table below.

Year	Deaths/Cases
1982	0.397
1983	0.457
1984	0.516
1985	0.554
1986	0.589
1987	0.581
1988	0.585
1989	0.606
1990	0.620
1991	0.623
1992	0.607
1993	0.610

(b) After 1985, the ratio becomes fairly constant and is equal to 0.6, rounded to the nearest tenth.

(c) Let $h(x) = \dfrac{g(x)}{f(x)} = \dfrac{1818(x-2)^2 + 620}{2975(x-2)^2 + 1563}$.
Graph $h(x)$ on the interval $[2, 20]$.

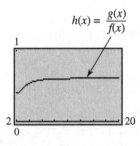

The graph of h quickly becomes horizontal with a value of approximately 0.61. The model predicts the ratios in the table quite well. Both are 0.6, rounded to the nearest tenth.

(d) Since $h(x) = \dfrac{g(x)}{f(x)} \approx 0.6$, it follows that

$$g(x) \approx 0.6f(x).$$

(e) Using this model, we predict that in 1994 the total number of deaths caused by AIDS since the disease began was

$$0.6(4{,}000{,}000) = 2{,}400{,}000.$$

67. (a) $T(r) = \dfrac{2r - k}{2r^2 - 2kr}$. If $k = 25$, then

$$T(r) = \dfrac{2r - 25}{2r^2 - 2(25)r} = \dfrac{2r - 25}{2r^2 - 50r}.$$

Graph $Y_1 = \dfrac{2x - 25}{2x^2 - 50x}$ and $Y_2 = 0.5$ (since 30 sec $= 0.5$ min) on the same screen.

For $r = x$,

$$y = \frac{2x - 25}{2x^2 - 50x} \qquad y = 0.5$$

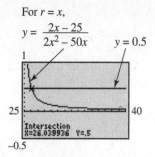

Using the "intersect" option in the CALC menu, we find that the graphs intersect at $x \approx 26$, which represents $r \approx 26$ in the given function. Therefore, there must be an average admittance rate of 26 vehicles per minute.

(b) $\dfrac{26}{5.3} \approx 4.9$ or 5 parking attendants must be on duty to keep the wait less than 30 seconds.

In Exercises 69–72, use the distance formula

$$d = \sqrt{(x_2 - x_1)^2 + (y_2 - y_1)^2}.$$

69. $(2, -1)$ and $(4, 3)$

$$d = \sqrt{(4 - 2)^2 + [3 - (-1)]^2}$$
$$= \sqrt{2^2 + 4^2} = \sqrt{4 + 16}$$
$$= \sqrt{20} = \sqrt{4} \cdot \sqrt{5} = 2\sqrt{5}$$

71. (x, y) and $(-2, 5)$

$$d = \sqrt{[x - (-2)]^2 + (y - 5)^2}$$
$$= \sqrt{(x + 2)^2 + (y - 5)^2}$$

Chapter 12 Review Exercises

1. $\dfrac{3x^2 - x - 2}{x - 1}$

$$\begin{array}{r|rrr}
1 & 3 & -1 & -2 \\
  &   & 3 & 2 \\
\hline
  & 3 & 2 & 0
\end{array}$$

The answer is $3x + 2$.

2. $\dfrac{10x^2 - 3x - 15}{x + 2}$

$$\begin{array}{r|rrr}
-2 & 10 & -3 & -15 \\
   &   & -20 & 46 \\
\hline
   & 10 & -23 & 31
\end{array}$$

The answer is $10x - 23 + \dfrac{31}{x + 2}$.

3. $(2x^3 - 5x^2 + 12) \div (x - 3)$

$$\begin{array}{r|rrrr}
3 & 2 & -5 & 0 & 12 \\
  &   & 6 & 3 & 9 \\
\hline
  & 2 & 1 & 3 & 21
\end{array}$$

The answer is $2x^2 + x + 3 + \dfrac{21}{x - 3}$.

Copyright © 2012 Pearson Education, Inc. Publishing as Addison-Wesley.

4. $\left(-x^4 + 19x^2 + 18x + 15\right) \div (x + 4)$

$$
\begin{array}{r|rrrrr}
-4 & -1 & 0 & 19 & 18 & 15 \\
   &   & 4 & -16 & -12 & -24 \\
\hline
   & -1 & 4 & 3 & 6 & -9
\end{array}
$$

The answer is $-x^3 + 4x^2 + 3x + 6 + \dfrac{-9}{x+4}$.

5. $2x^3 + 8x^2 - 14x - 20 = 0$

$$
\begin{array}{r|rrrr}
-5 & 2 & 8 & -14 & -20 \\
   &   & -10 & 10 & 20 \\
\hline
   & 2 & -2 & -4 & 0
\end{array}
$$

From the remainder theorem, $f(-5) = 0$, so -5 is a solution of the equation.

6. $-3x^4 + 2x^3 + 5x^2 - 9x + 1 = 0$

$$
\begin{array}{r|rrrrr}
-5 & -3 & 2 & 5 & -9 & 1 \\
   &   & 15 & -85 & 400 & -1955 \\
\hline
   & -3 & 17 & -80 & 391 & -1954
\end{array}
$$

$f(-5) = -1954 \neq 0$, so -5 is not a solution of the equation.

7. $f(x) = 3x^3 - 5x^2 + 4x - 1; \ k = -1$

$$
\begin{array}{r|rrrr}
-1 & 3 & -5 & 4 & -1 \\
   &   & -3 & 8 & -12 \\
\hline
   & 3 & -8 & 12 & -13
\end{array}
$$

Thus, $f(k) = f(-1) = -13$.

8. $f(x) = x^4 - 2x^3 - 9x - 5; \ k = 3$

$$
\begin{array}{r|rrrrr}
3 & 1 & -2 & 0 & -9 & -5 \\
  &   & 3 & 3 & 9 & 0 \\
\hline
  & 1 & 1 & 3 & 0 & -5
\end{array}
$$

Thus, $f(k) = f(3) = -5$.

9. $f(x) = 2x^3 - 9x^2 - 6x + 5$

The possible rational zeros are

$$\pm 1, \ \pm 5, \ \pm\tfrac{1}{2}, \ \pm\tfrac{5}{2}.$$

$$
\begin{array}{r|rrrr}
-1 & 2 & -9 & -6 & 5 \\
   &   & -2 & 11 & -5 \\
\hline
   & 2 & -11 & 5 & 0
\end{array}
$$

So -1 is a zero. The remaining zeros can be found by factoring the quotient.

$$2x^2 - 11x + 5 = 0$$
$$(2x - 1)(x - 5) = 0$$
$$x = \tfrac{1}{2} \ \text{ or } \ x = 5$$

By using synthetic division and factoring, we find that the rational zeros of $f(x)$ are -1, $\tfrac{1}{2}$, and 5.

10. $f(x) = 3x^3 - 10x^2 - 27x + 10$

The possible rational zeros are

$$\pm 1, \ \pm 2, \ \pm 5, \ \pm 10, \ \pm\tfrac{1}{3}, \ \pm\tfrac{2}{3}, \ \pm\tfrac{5}{3}, \ \pm\tfrac{10}{3}.$$

$$
\begin{array}{r|rrrr}
-2 & 3 & -10 & -27 & 10 \\
   &   & -6 & 32 & -10 \\
\hline
   & 3 & -16 & 5 & 0
\end{array}
$$

So -2 is a zero. Factoring gives us $3x^2 - 16x + 5 = (3x - 1)(x - 5)$.   By using synthetic division and factoring, we find that the rational zeros of $f(x)$ are -2, $\tfrac{1}{3}$, and 5.

11. $f(x) = x^3 - \tfrac{17}{6}x^2 - \tfrac{13}{3}x - \tfrac{4}{3}$

Multiply by 6 to clear fractions.

$$6f(x) = 6x^3 - 17x^2 - 26x - 8$$

has the same zeros as $f(x)$.

Possible rational zeros are

$$\pm 1, \ \pm 2, \ \pm 4, \ \pm 8, \ \pm\tfrac{1}{2},$$
$$\pm\tfrac{1}{3}, \ \pm\tfrac{2}{3}, \ \pm\tfrac{4}{3}, \ \pm\tfrac{8}{3}, \ \pm\tfrac{1}{6}.$$

$$
\begin{array}{r|rrrr}
4 & 6 & -17 & -26 & -8 \\
  &   & 24 & 28 & 8 \\
\hline
  & 6 & 7 & 2 & 0
\end{array}
$$

So 4 is a zero. Factoring gives us $6x^2 + 7x + 2 = (2x + 1)(3x + 2)$.   By using synthetic division and factoring, we find that the rational zeros of $f(x)$ are 4, $-\tfrac{1}{2}$, and $-\tfrac{2}{3}$.

12. $f(x) = 8x^4 - 14x^3 - 29x^2 - 4x + 3$

The possible rational zeros are

$$\pm 1, \ \pm 3, \ \pm\tfrac{1}{2}, \ \pm\tfrac{3}{2}, \ \pm\tfrac{1}{4}, \ \pm\tfrac{3}{4}, \ \pm\tfrac{1}{8}, \ \pm\tfrac{3}{8}.$$

$$
\begin{array}{r|rrrrr}
-1 & 8 & -14 & -29 & -4 & 3 \\
   &   & -8 & 22 & 7 & -3 \\
\hline
   & 8 & -22 & -7 & 3 & 0
\end{array}
$$

$$
\begin{array}{r|rrrr}
3 & 8 & -22 & -7 & 3 \\
  &   & 24 & 6 & -3 \\
\hline
  & 8 & 2 & -1 & 0
\end{array}
$$

So -1 and 3 are zeros. Factoring gives us $8x^2 + 2x - 1 = (4x - 1)(2x + 1)$.   By using synthetic division and factoring, we find that the rational zeros of $f(x)$ are -1, 3, $-\tfrac{1}{2}$, and $\tfrac{1}{4}$.

13. $f(x) = 2x^4 + x^3 - 4x^2 + 3x - 2$

Use synthetic division to determine whether -1 is a zero of f.

$$
\begin{array}{r|rrrrr}
-1 & 2 & 1 & -4 & 3 & -2 \\
   &   & -2 & 1 & 3 & -6 \\
\hline
   & 2 & -1 & -3 & 6 & -8
\end{array}
$$

Since the remainder is -8 and it is not 0, -1 is not a zero of f.

Copyright © 2012 Pearson Education, Inc. Publishing as Addison-Wesley.

14. $f(x) = 2x^4 + x^3 - 4x^2 + 3x - 2$

Use synthetic division to determine whether -2 is a zero of f.

$$
\begin{array}{r|rrrrr}
-2 & 2 & 1 & -4 & 3 & -2 \\
   &   & -4 & 6 & -4 & 2 \\
\hline
   & 2 & -3 & 2 & -1 & 0
\end{array}
$$

Since the remainder is 0, -2 is a zero of f.

15. $f(x) = x^3 + 2x^2 + 3x - 1$

$x + 1$ is a factor of $f(x)$ if $f(-1) = 0$.

$$
\begin{array}{r|rrrr}
-1 & 1 & 2 & 3 & -1 \\
   &   & -1 & -1 & -2 \\
\hline
   & 1 & 1 & 2 & -3
\end{array}
$$

Since $f(-1) = -3 \neq 0$, $x + 1$ is not a factor of f.

16. $f(x) = 2x^3 - x^2 + x + 4$

$x + 1$ is a factor of $f(x)$ if $f(-1) = 0$.

$$
\begin{array}{r|rrrr}
-1 & 2 & -1 & 1 & 4 \\
   &   & -2 & 3 & -4 \\
\hline
   & 2 & -3 & 4 & 0
\end{array}
$$

Since $f(-1) = 0$, $x + 1$ is a factor of $f(x)$.

17. Zeros are -2, 1, and 4; $f(2) = 16$.

$f(x) = a(x + 2)(x - 1)(x - 4)$
Use $f(2) = 16$ to determine the value of a.

$$f(2) = a(2 + 2)(2 - 1)(2 - 4) = 16$$
$$\textit{Let } x = 2; f(2) = 16.$$
$$a(4)(1)(-2) = 16$$
$$-8a = 16$$
$$a = -2$$

$$f(x) = -2(x + 2)(x - 1)(x - 4)$$
$$= -2(x + 2)(x^2 - 5x + 4)$$
$$= -2(x^3 - 3x^2 - 6x + 8)$$
$$= -2x^3 + 6x^2 + 12x - 16$$

In Exercises 18–20, we give only one such polynomial. Others are possible.

18. Zeros of 2, -2, and $-i$

If the polynomial has zeros of 2, -2, and $-i$ and is required to have real coefficients, then the conjugate of $-i$, which is i, must also be a zero.

$$f(x) = (x - 2)(x + 2)(x + i)(x - i)$$
$$= (x^2 - 4)(x^2 + 1)$$
$$= x^4 - 3x^2 - 4$$

19. Zeros of 2, -3, and $5i$

If 2, -3, and $5i$ are zeros of a polynomial with real coefficients, then the conjugate of $5i$, which is $-5i$, must also be a zero.

$$f(x) = (x - 2)(x + 3)(x - 5i)(x + 5i)$$
$$= (x^2 + x - 6)(x^2 + 25)$$
$$= x^4 + x^3 + 19x^2 + 25x - 150$$

20. Zeros of -3 and $1 - i$

If -3 and $1 - i$ are zeros of a polynomial with real coefficients, then the conjugate of $1 - i$, which is $1 + i$, must also be a zero.

$$f(x) = (x + 3)$$
$$\bullet\, [x - (1 - i)][x - (1 + i)]$$
$$= (x + 3)$$
$$\bullet\, [(x - 1) + i][(x - 1) - i]$$
$$= (x + 3)\big[(x - 1)^2 + 1\big]$$
$$= (x + 3)(x^2 - 2x + 1 + 1)$$
$$= (x + 3)(x^2 - 2x + 2)$$
$$= x^3 + x^2 - 4x + 6$$

21. $f(x) = x^4 - 3x^3 - 8x^2 + 22x - 24$; $1 - i$

Divide by $x - (1 - i)$; then by $x - (1 + i)$.

$$
\begin{array}{r|rrrrr}
1 - i & 1 & -3 & -8 & 22 & -24 \\
      &   & 1 - i & -3 + i & -10 + 12i & 24 \\
\hline
      & 1 & -2 - i & -11 + i & 12 + 12i & 0
\end{array}
$$

$$
\begin{array}{r|rrrr}
1 + i & 1 & -2 - i & -11 + i & 12 + 12i \\
      &   & 1 + i & -1 - i & -12 - 12i \\
\hline
      & 1 & -1 & -12 & 0
\end{array}
$$

The remaining quotient is

$$x^2 - x - 12 = (x - 4)(x + 3).$$

Thus, the zeros are $1 - i$, $1 + i$, 4, and -3. In factored form,

$$f(x) = (x - 1 + i)(x - 1 - i)$$
$$\bullet\, (x - 4)(x + 3).$$

22. If a polynomial function has six real zeros, then the degree of the polynomial must be at least 6. Each zero corresponds to a factor of the form $x - k$ for the polynomial, and when six binomials of that form are multiplied together, the degree of the resulting polynomial will be 6. (If any of the zeros has multiplicity greater than one, the degree of the polynomial will be greater than 6.) Thus, it is not possible for the polynomial to be of degree 5.

23. $f(x) = x^3 - 9x$ ■   The degree of the polynomial is 3, so the maximum number of turning points of the graph is $3 - 1 = 2$.

24. $f(x) = 4x^4 - 6x^2 + 2$ ■   The degree of the polynomial is 4, so the maximum number of turning points of the graph is $4 - 1 = 3$.

Copyright © 2012 Pearson Education, Inc. Publishing as Addison-Wesley.

25. $f(x) = x^3 + 5$

The graph will be the same as that of $f(x) = x^3$, but translated 5 units up.

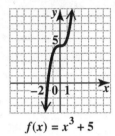

$f(x) = x^3 + 5$

26. $f(x) = 1 - x^4$

Since $f(x) = (1 - x^2)(1 + x^2)$
$= (1 + x)(1 - x)(1 + x^2)$,
the zeros are 1 and -1.
Since $f(x) = f(-x)$, the graph is symmetric with respect to the y-axis. The graph will be the same as that of $f(x) = x^4$ but reflected about the x-axis and translated up 1 unit.

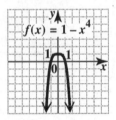

$f(x) = 1 - x^4$

27. $f(x) = x^2(2x + 1)(x - 2)$

The zeros are 0, $-\frac{1}{2}$, and 2.
The zeros divide the x-axis into four intervals:

$(-\infty, -\frac{1}{2})$, $(-\frac{1}{2}, 0)$, $(0, 2)$, and $(2, \infty)$.

Test a point in each interval to find the sign of $f(x)$ in that interval.

Interval	Test point	Value of $f(x)$	Sign of $f(x)$
$(-\infty, -\frac{1}{2})$	-1	3	Positive
$(-\frac{1}{2}, 0)$	$-\frac{1}{4}$	$-\frac{9}{128}$	Negative
$(0, 2)$	1	-3	Negative
$(2, \infty)$	3	63	Positive

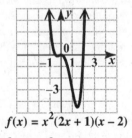

$f(x) = x^2(2x + 1)(x - 2)$

28. $f(x) = 2x^3 + 13x^2 + 15x$
$= x(2x^2 + 13x + 15)$
$= x(2x + 3)(x + 5)$

The zeros are 0, $-\frac{3}{2}$, and -5.

Interval	Test point	Value of $f(x)$	Sign of $f(x)$
$(-\infty, -5)$	-6	-54	Negative
$(-5, -\frac{3}{2})$	-2	6	Positive
$(-\frac{3}{2}, 0)$	-1	-4	Negative
$(0, \infty)$	1	30	Positive

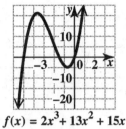

$f(x) = 2x^3 + 13x^2 + 15x$

29. $f(x) = 12x^3 - 13x^2 - 5x + 6$

The possible rational zeros are

$$\pm 1, \pm 2, \pm 3, \pm 6, \pm \tfrac{1}{2}, \pm \tfrac{3}{2},$$

$$\pm \tfrac{1}{3}, \pm \tfrac{2}{3}, \pm \tfrac{1}{4}, \pm \tfrac{3}{4}, \pm \tfrac{1}{6}, \pm \tfrac{1}{12}.$$

$$
\begin{array}{r|rrrr}
1 & 12 & -13 & -5 & 6 \\
  &   & 12 & -1 & -6 \\
\hline
  & 12 & -1 & -6 & 0
\end{array}
$$

So 1 is a zero and since
$12x^2 - x - 6 = (4x - 3)(3x + 2)$,

$$f(x) = (4x - 3)(3x + 2)(x - 1).$$

The zeros are $\frac{3}{4}$, $-\frac{2}{3}$, and 1.

Interval	Test point	Value of $f(x)$	Sign of $f(x)$
$(-\infty, -\frac{2}{3})$	-1	-14	Negative
$(-\frac{2}{3}, \frac{3}{4})$	0	6	Positive
$(\frac{3}{4}, 1)$	$\frac{7}{8}$	$-\frac{37}{128}$	Negative
$(1, \infty)$	2	40	Positive

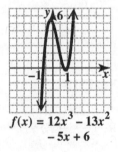

$f(x) = 12x^3 - 13x^2 - 5x + 6$

30. $f(x) = x^4 - 2x^3 - 5x^2 + 6x$
$= x(x^3 - 2x^2 - 5x + 6)$

The possible rational zeros are

$$\pm 1, \pm 2, \pm 3, \pm 6.$$

$$
\begin{array}{r|rrrr}
1 & 1 & -2 & -5 & 6 \\
  &   & 1 & -1 & -6 \\
\hline
  & 1 & -1 & -6 & 0
\end{array}
$$

Copyright © 2012 Pearson Education, Inc. Publishing as Addison-Wesley.

So 1 is a zero and since
$x^2 - x - 6 = (x + 2)(x - 3)$,

$$f(x) = x(x - 1)(x + 2)(x - 3).$$

The zeros are 0, 1, −2, and 3.

Interval	Test point	Value of $f(x)$	Sign of $f(x)$
$(-\infty, -2)$	−3	72	Positive
$(-2, 0)$	−1	−8	Negative
$(0, 1)$	$\frac{1}{2}$	$\frac{25}{16}$	Positive
$(1, 3)$	2	−8	Negative
$(3, \infty)$	4	72	Positive

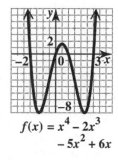

$$f(x) = x^4 - 2x^3 - 5x^2 + 6x$$

31. Show $f(x) = 3x^3 - 8x^2 + x + 2$ has real zeros in $[-1, 0]$ and $[2, 3]$.

$f(-1) = -10$ and $f(0) = 2$. By the intermediate value theorem, since $f(-1) < 0$ and $f(0) > 0$, there is a zero in $[-1, 0]$.
$f(2) = -4$ and $f(3) = 14$. Since $f(2) < 0$ and $f(3) > 0$, there is a zero in $[2, 3]$.

32. Show $f(x) = 4x^3 - 37x^2 + 50x + 60$ has real zeros in $[2, 3]$ and $[7, 8]$.

$f(2) = 44 > 0$ and $f(3) = -15 < 0$, so there is a zero in $[2, 3]$.
$f(7) = -31 < 0$ and $f(8) = 140 > 0$, so there is a zero in $[7, 8]$.

33. Show $f(x) = x^3 + 2x^2 - 22x - 8$ has real zeros in $[-1, 0]$ and $[-6, -5]$.

$f(-1) = 15 > 0$ and $f(0) = -8 < 0$, so there is a zero in $[-1, 0]$.
$f(-6) = -20 < 0$ and $f(-5) = 27 > 0$, so there is a zero in $[-6, -5]$.

34. Show that $f(x) = 2x^4 - x^3 - 21x^2 + 51x - 36$ has no real zeros greater than 4.

$$
\begin{array}{r|rrrrr}
4 & 2 & -1 & -21 & 51 & -36 \\
  &   & 8 & 28 & 28 & 316 \\
\hline
  & 2 & 7 & 7 & 79 & 280 \\
\end{array}
$$

Since $4 > 0$ and all the numbers in the last row are nonnegative, f has no real zero greater than 4.

35. Show that $f(x) = 6x^4 + 13x^3 - 11x^2 - 3x + 5$ has no real zero greater than 1 or less than −3.

$$
\begin{array}{r|rrrrr}
1 & 6 & 13 & -11 & -3 & 5 \\
  &   & 6 & 19 & 8 & 5 \\
\hline
  & 6 & 19 & 8 & 5 & 10 \\
\end{array}
$$

All numbers in the bottom row are nonnegative and $1 > 0$, so by the boundedness theorem, f has no real zero greater than 1.

$$
\begin{array}{r|rrrrr}
-3 & 6 & 13 & -11 & -3 & 5 \\
   &   & -18 & 15 & -12 & 45 \\
\hline
   & 6 & -5 & 4 & -15 & 50 \\
\end{array}
$$

The signs alternate and $-3 < 0$, so, by the boundedness theorem, f has no real zero less than −3.

36. (a) $f(x) = 2x^3 - 11x^2 - 2x + 2$

The graph (see Exercise 37) has three x-intercepts, so all three of the zeros are real. Rounding to the nearest thousandth where necessary, we see from the x-intercepts that the zeros are −0.5, 0.354, and 5.646.

(b) $f(x) = x^4 - 4x^3 - 5x^2 + 14x - 15$

The graph (see Exercise 38) has two x-intercepts, so f has two real zeros. (The other two zeros are nonreal complex numbers.) Rounding to the nearest thousandth, we see from the x-intercepts that the zeros are −2.259 and 4.580.

(c) $f(x) = x^3 + 3x^2 - 4x - 2$

The graph (see Exercise 39) has three x-intercepts, so f has three real zeros. To the nearest thousandth, the zeros of f are −3.895, −0.397, and 1.292.

37. $f(x) = 2x^3 - 11x^2 - 2x + 2$

From Exercise 36(a), we know that the zeros are −0.5 (exact) and 0.354 and 5.646 (approximate). Find a few additional points.

x	−1	0	1	4	6
y	−9	2	−9	−54	26

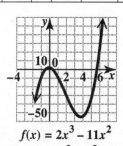

$$f(x) = 2x^3 - 11x^2 - 2x + 2$$

Copyright © 2012 Pearson Education, Inc. Publishing as Addison-Wesley.

38. $f(x) = x^4 - 4x^3 - 5x^2 + 14x - 15$

From Exercise 36(b), we know that f has two real zeros which are approximately -2.259 and 4.580.

x	-3	-2	0	4	5
y	87	-15	-15	-39	55

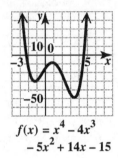

$$f(x) = x^4 - 4x^3$$
$$- 5x^2 + 14x - 15$$

39. $f(x) = x^3 + 3x^2 - 4x - 2$

From Exercise 36(c), we know that there are three real zeros, and that to the nearest thousandth, the zeros are -3.895, -0.397, and 1.292.

x	-4	-3	-1	0	1	2
y	-2	10	4	-2	-2	10

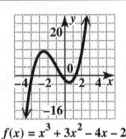

$$f(x) = x^3 + 3x^2 - 4x - 2$$

40. $f(x) = 2x^4 - 3x^3 + 4x^2 + 5x - 1$

$f(x)$ has 3 sign variations, so f has 3 or 1 positive zeros.
$f(-x) = 2x^4 + 3x^3 + 4x^2 - 5x - 1$
$f(-x)$ has 1 sign variation, so f has 1 negative zero.

x	-2	-1	0	1	2
y	61	3	-1	7	33

The sign changes indicate that there is a zero in $[-1, 0]$ and $[0, 1]$.

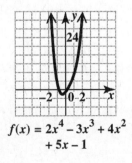

$$f(x) = 2x^4 - 3x^3 + 4x^2$$
$$+ 5x - 1$$

41. $f(x) = \dfrac{8}{x} = 8\left(\dfrac{1}{x}\right)$

The graph is similar to that of $f(x) = \frac{1}{x}$, but each value is multiplied by 8. The x- and y-axes are the asymptotes.

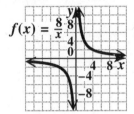

$$f(x) = \frac{8}{x}$$

42. $f(x) = \dfrac{2}{3x - 1}$

The graph has a vertical asymptote, $x = \frac{1}{3}$, and a horizontal asymptote, $y = 0$.
Since $f(0) = -2$, the y-intercept is $(0, -2)$. The numerator can never be equal to 0, so there are no x-intercepts. To the right of $x = \frac{1}{3}$, we have the point $(1, 1)$.

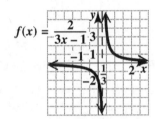

$$f(x) = \frac{2}{3x - 1}$$

43. $f(x) = \dfrac{4x - 2}{3x + 1} = \dfrac{2(2x - 1)}{3x + 1}$

The graph has a vertical asymptote, $x = -\frac{1}{3}$.
The degree of the numerator equals the degree of the denominator, so the horizontal asymptote is $y = \frac{4}{3}$ (the ratio of the leading coefficients).
$f(0) = -2$, so the y-intercept is $(0, -2)$.
The x-intercept is $\left(\frac{1}{2}, 0\right)$.

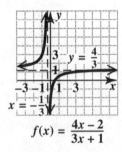

$$f(x) = \frac{4x - 2}{3x + 1}$$

44. $f(x) = \dfrac{6x}{(x - 1)(x + 2)}$

The vertical asymptotes are $x = 1$ and $x = -2$.
The horizontal asymptote is $y = 0$.
$f(0) = 0$, so the y-intercept is $(0, 0)$. The only zero of the numerator is 0, so the only x-intercept is also $(0, 0)$.

Copyright © 2012 Pearson Education, Inc. Publishing as Addison-Wesley.

x	-3	-1	$\frac{1}{2}$	2
y	$-\frac{9}{2}$	3	$-\frac{12}{5}$	3

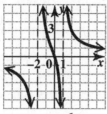

$$f(x) = \frac{6x}{(x-1)(x+2)}$$

45. $f(x) = \dfrac{2x}{x^2-1} = \dfrac{2x}{(x+1)(x-1)}$

The vertical asymptotes are $x = 1$ and $x = -1$. Since the numerator has lower degree than the denominator, the horizontal asymptote is $y = 0$. Notice that

$$f(-x) = \frac{2(-x)}{(-x)^2 - 1}$$
$$= \frac{-2x}{x^2 - 1}$$
$$= -f(x),$$

so the graph is symmetric with respect to the origin. The x- and y-intercepts are both $(0,0)$.

x	$\frac{1}{2}$	2	5
y	$-\frac{4}{3}$	$\frac{4}{3}$	$\frac{5}{12}$

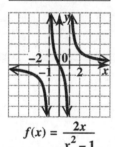

$$f(x) = \frac{2x}{x^2-1}$$

46. $f(x) = \dfrac{x^2+4}{x+2}$

The vertical asymptote is $x = -2$. Since the degree of the numerator is exactly one more than the degree of the denominator, use synthetic division to find the oblique asymptote.

$$-2 \begin{array}{|rrr} 1 & 0 & 4 \\ & -2 & 4 \\ \hline 1 & -2 & 8 \end{array}$$

Thus, $f(x) = \dfrac{x^2+4}{x+2} = x - 2 + \dfrac{8}{x+2}$, and the oblique asymptote is $y = x - 2$.

$f(0) = 2$, so the y-intercept is $(0, 2)$.

Because the numerator has no real zeros, there are no x-intercepts.

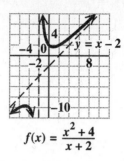

$$f(x) = \frac{x^2+4}{x+2}$$

47. $f(x) = \dfrac{x^2-1}{x} = \dfrac{(x+1)(x-1)}{x}$

There is one vertical asymptote, $x = 0$. Since the numerator is of degree exactly one more than the denominator, there is no horizontal asymptote, but there is an oblique asymptote.

$$f(x) = \frac{x^2-1}{x} = x - \frac{1}{x},$$

so as $|x| \to \infty$, $f(x) \to x$.

Thus, the line $y = x$ is an oblique asymptote.

There is no y-intercept because $f(0)$ is undefined. The zeros of the numerator are -1 and 1, so the graph has two x-intercepts, $(-1, 0)$ and $(1, 0)$.

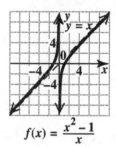

$$f(x) = \frac{x^2-1}{x}$$

48. $f(x) = \dfrac{x^2+6x+5}{x-3} = \dfrac{(x+1)(x+5)}{x-3}$

There is one vertical asymptote, $x = 3$. Since the numerator is of degree exactly one more than the denominator, there is no horizontal asymptote, but there is an oblique asymptote. Divide $x^2 + 6x + 5$ by $x - 3$ synthetically.

$$3 \begin{array}{|rrr} 1 & 6 & 5 \\ & 3 & 27 \\ \hline 1 & 9 & 32 \end{array}$$

The division shows that

$$\frac{x^2+6x+5}{x-3} = x + 9 + \frac{32}{x-3}.$$

As $|x| \to \infty$, $f(x) \to x + 9$.

Thus, the line $y = x + 9$ is an oblique asymptote.

$f(0) = -\frac{5}{3}$, so the y-intercept is $\left(0, -\frac{5}{3}\right)$.

The zeros of the numerator are -5 and -1, so the x-intercepts are $(-5, 0)$ and $(-1, 0)$.

graph on next page

Copyright © 2012 Pearson Education, Inc. Publishing as Addison-Wesley.

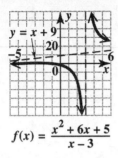

$$f(x) = \frac{x^2 + 6x + 5}{x - 3}$$

49. $f(x) = \dfrac{4x^2 - 9}{2x + 3} = \dfrac{(2x + 3)(2x - 3)}{2x + 3}$

$\qquad = 2x - 3 \; \left(x \neq -\frac{3}{2}\right)$

The graph is the same as that of $f(x) = 2x - 3$, the line with slope 2 and y-intercept $(0, -3)$, except that the point with x-value $-\frac{3}{2}$ is missing. This is shown by graphing the line $y = 2x - 3$ with an open circle at $\left(-\frac{3}{2}, -6\right)$ to indicate the missing point.

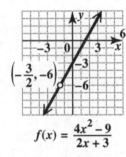

$$f(x) = \frac{4x^2 - 9}{2x + 3}$$

50. $f(x) = \dfrac{(x + 4)(2x + 5)}{x - 1} = \dfrac{2x^2 + 13x + 20}{x - 1}$

The graph has one vertical asymptote, $x = 1$. We see that the degree of the numerator is exactly one more than the degree of the denominator, so there is an oblique asymptote. Divide synthetically.

$$\begin{array}{r|rrr} 1 & 2 & 13 & 20 \\ & & 2 & 15 \\ \hline & 2 & 15 & 35 \end{array}$$

Thus, $f(x) = 2x + 15 + \dfrac{35}{x - 1}$, and the oblique asymptote is $y = 2x + 15$.

$f(0) = -20$, so the y-intercept is $(0, -20)$. The zeros of the numerator are -4 and $-\frac{5}{2}$, so the x-intercepts are $(-4, 0)$ and $\left(-\frac{5}{2}, 0\right)$.

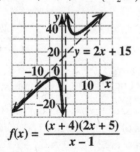

$$f(x) = \frac{(x + 4)(2x + 5)}{x - 1}$$

51. $R(x) = \dfrac{80x - 8000}{x - 110}$

(a) $R(55) = \dfrac{80(55) - 8000}{55 - 110}$

$\qquad \approx \$65.5$ tens of millions

(b) $R(60) = \dfrac{80(60) - 8000}{60 - 110}$

$\qquad = \$64$ tens of millions

(c) $R(70) = \dfrac{80(70) - 8000}{70 - 110}$

$\qquad = \$60$ tens of millions

(d) $R(90) = \dfrac{80(90) - 8000}{90 - 110}$

$\qquad = \$40$ tens of millions

(e) $R(100) = \dfrac{80(100) - 8000}{100 - 110}$

$\qquad = \$0$

(f)

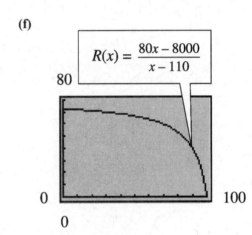

52. $R(x) = \dfrac{60x - 6000}{x - 120}$

(a) $R(50) = \dfrac{60(50) - 6000}{50 - 120}$

$\qquad \approx \$42.9$ million

(b) $R(60) = \dfrac{60(60) - 6000}{60 - 120}$

$\qquad = \$40$ million

(c) $R(80) = \dfrac{60(80) - 6000}{80 - 120}$

$\qquad = \$30$ million

(d) $R(100) = \dfrac{60(100) - 6000}{100 - 120}$

$\qquad = \$0$

Copyright © 2012 Pearson Education, Inc. Publishing as Addison-Wesley.

(e)

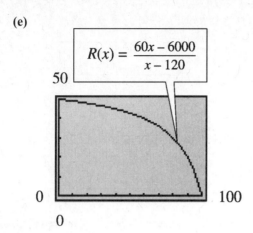

$$R(x) = \frac{60x - 6000}{x - 120}$$

53. [12.4] $f(x) = -\dfrac{1}{x^3}$

The vertical asymptote is $x = 0$.
The horizontal asymptote is $y = 0$.

$f(-x) = -\dfrac{1}{(-x)^3} = \dfrac{1}{x^3} = -f(x)$, so the graph

is symmetric with respect to the origin. Use the asymptotes and some points in quadrants II and IV to sketch the graph.

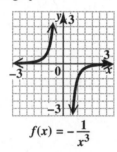

$$f(x) = -\frac{1}{x^3}$$

54. [12.4] $f(x) = \dfrac{-4x + 3}{2x + 1}$

The vertical asymptote is $x = -\frac{1}{2}$.
The degree of the numerator equals the degree of the denominator, so the horizontal asymptote is $y = \frac{-4}{2} = -2$ (the ratio of the leading coefficients).
$f(0) = 3$, so the y-intercept is $(0, 3)$.
The x-intercept is $\left(\frac{3}{4}, 0\right)$.

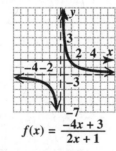

$$f(x) = \frac{-4x + 3}{2x + 1}$$

55. [12.4] $f(x) = \dfrac{x^3 + 1}{x + 1}$

This rational function is not written in lowest terms. Factor the numerator as the sum of cubes

and write the resulting rational expression in lowest terms.

$$\frac{x^3 + 1}{x + 1} = \frac{(x + 1)(x^2 - x + 1)}{x + 1}$$
$$= x^2 - x + 1 \quad (x \neq -1)$$

The graph of this function looks like the parabola $y = x^2 - x + 1$, which has vertex $\left(\frac{1}{2}, \frac{3}{4}\right)$ and opens upward, except there is a "hole" at the point where $x = -1$, which is the point $(-1, 3)$.

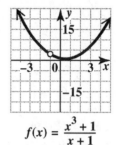

$$f(x) = \frac{x^3 + 1}{x + 1}$$

56. [12.3] $f(x) = 3x^3 + 2x^2 - 27x - 18$

To find the zeros, factor the polynomial by grouping.

$$\begin{aligned} f(x) &= 3x^3 + 2x^2 - 27x - 18 \\ &= x^2(3x + 2) - 9(3x + 2) \\ &= (3x + 2)(x^2 - 9) \\ &= (3x + 2)(x + 3)(x - 3) \end{aligned}$$

The zeros are zeros at $-\frac{2}{3}$, -3 and 3.

Interval	Test point	Value of $f(x)$	Sign of $f(x)$
$(-\infty, -3)$	-4	-70	Negative
$\left(-3, -\frac{2}{3}\right)$	-1	8	Positive
$\left(-\frac{2}{3}, 3\right)$	0	-18	Negative
$(3, \infty)$	4	98	Positive

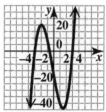

$$f(x) = 3x^3 + 2x^2 - 27x - 18$$

57. [12.1] $\dfrac{2x^3 + x - 6}{x + 2}$

$$-2 \begin{array}{|rrrr} 2 & 0 & 1 & -6 \\ & -4 & 8 & -18 \\ \hline 2 & -4 & 9 & -24 \end{array}$$

So $\dfrac{2x^3 + x - 6}{x + 2} = 2x^2 - 4x + 9 - \dfrac{24}{x + 2}$.

Multiplying each side by $x + 2$ gives the form:

$$2x^3 + x - 6 = (x + 2)(2x^2 - 4x + 9) + (-24)$$

Copyright © 2012 Pearson Education, Inc. Publishing as Addison-Wesley.

58. **[12.2]** $1, -1,$ and $3i$ are zeros; $f(2) = 39$.

Since the polynomial is required to have real coefficients, $-3i$, which is the complex conjugate of $3i$, must also be a zero.

$$f(x) = a(x-1)(x+1)(x-3i)(x+3i)$$
$$= a(x^2-1)(x^2+9)$$
$$= a(x^4+8x^2-9)$$

Use $f(2) = 39$ to determine the value of a.

$$39 = f(2) = a(16+32-9)$$
$$39 = a(39)$$
$$a = 1$$

Thus, $f(x) = x^4 + 8x^2 - 9$.

59. **[12.3]** For the polynomial function defined by $f(x) = x^3 - 3x^2 - 7x + 12$, $f(4) = 0$. Therefore, we can say that 4 is a <u>zero</u> of the function, 4 is a <u>solution</u> of the equation $x^3 - 3x^2 - 7x + 12 = 0$, and that $(4, 0)$ is an <u>x-intercept</u> of the graph of the function.

60. **[12.3]** **C.** $f(x) = \frac{1}{x}$ is not a polynomial function because it has a variable in the denominator. All other choices are polynomial functions.

Chapter 12 Test

1. $(2x^3 - 3x - 10) \div (x - 2)$

$$2 \begin{array}{|rrrr} 2 & 0 & -3 & -10 \\ & 4 & 8 & 10 \\ \hline 2 & 4 & 5 & 0 \end{array}$$

The quotient is $2x^2 + 4x + 5$.

2. $\dfrac{x^4 + 2x^3 - x^2 + 3x - 5}{x + 1}$

$$-1 \begin{array}{|rrrrr} 1 & 2 & -1 & 3 & -5 \\ & -1 & -1 & 2 & -5 \\ \hline 1 & 1 & -2 & 5 & -10 \end{array}$$

So $\dfrac{x^4 + 2x^3 - x^2 + 3x - 5}{x + 1} =$

$x^3 + x^2 - 2x + 5 - \dfrac{10}{x + 1}.$

Multiplying each side by $x + 1$ gives the form:

$$x^4 + 2x^3 - x^2 + 3x - 5 =$$
$$(x + 1)(x^3 + x^2 - 2x + 5) + (-10)$$

3. $f(x) = x^4 - 2x^3 - 15x^2 - 4x - 12$

$$-3 \begin{array}{|rrrrr} 1 & -2 & -15 & -4 & -12 \\ & -3 & 15 & 0 & 12 \\ \hline 1 & -5 & 0 & -4 & 0 \end{array}$$

When $f(x)$ is divided by $x + 3$, the remainder is 0, so by the remainder theorem, $f(-3) = 0$. Therefore, by the factor theorem, $x + 3$ is a factor of $f(x)$.

4. $f(x) = 3x^3 - 4x^2 - 5x + 9$

$$-4 \begin{array}{|rrrr} 3 & -4 & -5 & 9 \\ & -12 & 64 & -236 \\ \hline 3 & -16 & 59 & -227 \end{array}$$

$f(-4) = -227$

5. $f(x) = 6x^4 - 11x^3 - 35x^2 + 34x + 24$

$$3 \begin{array}{|rrrrr} 6 & -11 & -35 & 34 & 24 \\ & 18 & 21 & -42 & -24 \\ \hline 6 & 7 & -14 & -8 & 0 \end{array}$$

When $f(x)$ is divided by $x - 3$, the remainder is 0. Therefore, by the remainder theorem, 3 is a zero of f.

6. A polynomial having zeros $2, -1,$ and $-i$ must also have i (the conjugate of $-i$) as a zero if its coefficients are real. The polynomial has the form

$$f(x) = a(x - 2)(x + 1)(x + i)(x - i).$$

Use $f(3) = 80$ to find the value of a.

$$f(3) = a(3 - 2)(3 + 1)(3 + i)(3 - i)$$
$$80 = a(1)(4)(10)$$
$$80 = 40a$$
$$a = 2$$

$$f(x) = 2(x - 2)(x + 1)(x + i)(x - i)$$
$$= 2(x^2 - x - 2)(x^2 + 1)$$
$$= 2(x^4 - x^3 - x^2 - x - 2)$$
$$= 2x^4 - 2x^3 - 2x^2 - 2x - 4$$

7. $f(x) = 6x^3 - 25x^2 + 12x + 7$

(a) The possible rational zeros are

$$\pm 1, \ \pm \tfrac{1}{2}, \ \pm \tfrac{1}{3}, \ \pm \tfrac{1}{6}, \ \pm 7, \ \pm \tfrac{7}{2}, \ \pm \tfrac{7}{3}, \ \pm \tfrac{7}{6}.$$

(b) $1 \begin{array}{|rrrr} 6 & -25 & 12 & 7 \\ & 6 & -19 & -7 \\ \hline 6 & -19 & -7 & 0 \end{array}$

1 is a zero. The other zeros may be found by factoring the quotient.

$$6x^2 - 19x - 7 = (2x - 7)(3x + 1)$$

The rational zeros are $-\tfrac{1}{3}, 1,$ and $\tfrac{7}{2}$.

8. $f(x) = 2x^4 - 3x^3 + 4x^2 - 5x - 1$

$$2 \begin{array}{|rrrrr} 2 & -3 & 4 & -5 & -1 \\ & 4 & 2 & 12 & 14 \\ \hline 2 & 1 & 6 & 7 & 13 \end{array}$$

Since $2 > 0$ and all numbers in the bottom row are nonnegative, the boundedness theorem guarantees that f has no real zero greater than 2.

Copyright © 2012 Pearson Education, Inc. Publishing as Addison-Wesley.

$$-1 \overline{)\begin{array}{rrrrr} 2 & -3 & 4 & -5 & -1 \\ & -2 & 5 & -9 & 14 \\ \hline 2 & -5 & 9 & -14 & 13 \end{array}}$$

Since $-1 < 0$ and the numbers in the bottom row alternate in sign, the boundedness theorem guarantees that f has no real zero less than -1.

9. $f(x) = 2x^3 - x + 3$

(a) $f(-2) = -11 < 0$ and $f(-1) = 2 > 0$, so by the intermediate value theorem, there is a zero between -2 and -1.

(b)

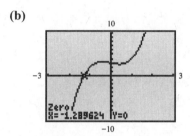

There is one x-intercept at $(-1.290, 0)$. To the nearest thousandth, the only real zero of f is -1.290.

10. $f(x) = 2x^3 - 9x^2 + 4x + 8$

(a) The graph of a polynomial function of degree 3 can have a maximum of 3 x-intercepts. This is because a polynomial of degree 3 has at most 3 zeros.

(b) The degree of this function is 3, so the maximum number of turning points of the graph is $3 - 1 = 2$.

11. $f(x) = (x - 1)^4$

Translate the graph of $y = x^4$ one unit to the right.

$f(x) = (x - 1)^4$

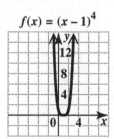

12. $f(x) = x(x + 1)(x - 2)$

The zeros of f are 0, -1, and 2. These three zeros divide the x-axis into four intervals.

Interval	Test point	Value of $f(x)$	Sign of $f(x)$
$(-\infty, -1)$	-2	-8	Negative
$(-1, 0)$	$-\frac{1}{2}$	$\frac{5}{8}$	Positive
$(0, 2)$	1	-2	Negative
$(2, \infty)$	3	12	Positive

$f(x) = x(x + 1)(x - 2)$

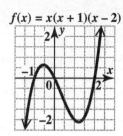

13. $f(x) = 2x^3 - 7x^2 + 2x + 3$

The possible rational zeros are

$$\pm 1, \ \pm 3, \ \pm \tfrac{1}{2}, \ \pm \tfrac{3}{2}.$$

$$3 \overline{)\begin{array}{rrrr} 2 & -7 & 2 & 3 \\ & 6 & -3 & -3 \\ \hline 2 & -1 & -1 & 0 \end{array}}$$

So 3 is a zero and since

$$2x^2 - x - 1 = (2x + 1)(x - 1),$$

the factored form of f is

$$f(x) = (2x + 1)(x - 1)(x - 3).$$

The x-intercepts are $\left(-\frac{1}{2}, 0\right)$, $(1, 0)$, and $(3, 0)$. $f(0) = 3$, so the y-intercept is $(0, 3)$.

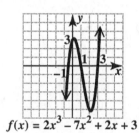

$f(x) = 2x^3 - 7x^2 + 2x + 3$

14. $f(x) = x^4 - 5x^2 + 6$
$ = (x^2 - 2)(x^2 - 3)$
$ = (x + \sqrt{2})(x - \sqrt{2})(x + \sqrt{3})(x - \sqrt{3})$

The x-intercepts are $(\pm \sqrt{2}, 0)$ and $(\pm \sqrt{3}, 0)$. The y-intercept is $(0, 6)$.

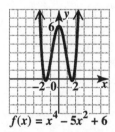

$f(x) = x^4 - 5x^2 + 6$

15. $f(x) = -0.184x^3 + 1.45x^2 + 10.7x - 27.9$

June is the sixth month, so find $f(6)$.

$$f(6) = 48.756 \approx 49$$

The average temperature in June is about $49°$ F.

Copyright © 2012 Pearson Education, Inc. Publishing as Addison-Wesley.

16. $f(x) = \dfrac{-2}{x+3} = (-2) \cdot \dfrac{1}{x-(-3)}$

Compared to $y = \frac{1}{x}$, the graph will be reflected about the x-axis (because of the negative sign) and each point will be twice as far from the axis and translated 3 units to the left.
The vertical asymptote is $x = -3$ and the horizontal asymptote is $y = 0$.

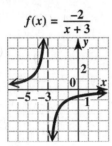

17. $f(x) = \dfrac{3x-1}{x-2}$

There is one vertical asymptote, $x = 2$ (denominator equals 0).
The degree of the numerator equals the degree of the denominator, so the horizontal asymptote is $y = \frac{3}{1} = 3$ (the ratio of the leading coefficients).
$f(0) = \frac{1}{2}$, so the y-intercept is $\left(0, \frac{1}{2}\right)$.
The only zero of the numerator is $\frac{1}{3}$, so $\left(\frac{1}{3}, 0\right)$ is the only x-intercept.

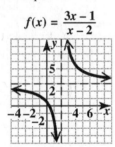

18. $f(x) = \dfrac{x^2-1}{x^2-9} = \dfrac{(x+1)(x-1)}{(x+3)(x-3)}$

There are vertical asymptotes at $x = -3$ and $x = 3$.
The degree of the numerator equals the degree of the denominator, so the horizontal asymptote is $y = \frac{1}{1} = 1$ (the ratio of the leading coefficients).
The y-intercept is $\left(0, \frac{1}{9}\right)$.
The x-intercepts are $(-1, 0)$ and $(1, 0)$.

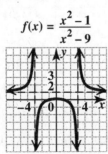

19. $f(x) = \dfrac{2x^2 + x - 6}{x-1}$

To find the oblique asymptote, divide synthetically.

$$1 \begin{array}{|rrr} 2 & 1 & -6 \\ & 2 & 3 \\ \hline 2 & 3 & -3 \end{array}$$

Thus, $f(x) = 2x + 3 + \dfrac{-3}{x-1}$, and the oblique asymptote is $y = 2x + 3$.

20. $x^2 + 4$ is positive, so $f(x) = \dfrac{1}{x^2+4}$ can never equal 0. Thus, choice **D** has a graph with no x-intercepts.

Cumulative Review Exercises (Chapters 1–12)

1. $\begin{aligned} 4x + 8x &= 17x - 10 \\ 12x &= 17x - 10 \\ -5x &= -10 \\ x &= 2 \end{aligned}$

The solution set is $\{2\}$.

2. $0.10(x-6) + 0.05x = 0.06(50)$
Multiply by 100 to clear decimals.
$\begin{aligned} 10(x-6) + 5x &= 6(50) \\ 10x - 60 + 5x &= 300 \\ 15x &= 360 \\ x &= 24 \end{aligned}$

The solution set is $\{24\}$.

3. $\dfrac{x+1}{3} + \dfrac{2x}{3} = x + \dfrac{1}{3}$
Multiply by the LCD, 3, to clear fractions.
$3\left(\dfrac{x+1}{3} + \dfrac{2x}{3}\right) = 3\left(x + \dfrac{1}{3}\right)$
$\begin{aligned} x + 1 + 2x &= 3x + 1 \\ 3x + 1 &= 3x + 1 \end{aligned}$

This equation is satisfied by all real numbers. The solution set is {all real numbers}.

4. $\begin{aligned} 4(x+2) &\geq 6x - 8 \\ 4x + 8 &\geq 6x - 8 \\ -2x &\geq -16 \\ x &\leq 8 \end{aligned}$

The solution set is $(-\infty, 8]$.

5. $\begin{aligned} 5 < 3x - 4 &< 9 \\ 9 < 3x &< 13 \\ 3 < x &< \tfrac{13}{3} \end{aligned}$

The solution set is $\left(3, \frac{13}{3}\right)$.

Copyright © 2012 Pearson Education, Inc. Publishing as Addison-Wesley.

6. $|x + 2| - 3 > 2$

$\qquad |x + 2| > 5$

$\qquad x + 2 < -5 \quad$ or $\quad x + 2 > 5$

$\qquad\quad x < -7 \quad$ or $\qquad\quad x > 3$

The solution set is $(-\infty, -7) \cup (3, \infty)$.

7.   Let $x =$ the number of nickels;

$\qquad y =$ the number of dimes.

The jar contains 38 nickels and dimes, so

$$x + y = 38. \quad (1)$$

The value of the money is \$2.50, so

$$0.05x + 0.10y = 2.50. \quad (2)$$

Solve the system.

$$
\begin{array}{rrcll}
-5x & - & 5y & = & -190 & \quad -5 \times (1) \\
5x & + & 10y & = & 250 & \quad 100 \times (2) \\
\hline
& & 5y & = & 60 & \quad \textit{Add.} \\
& & y & = & 12 &
\end{array}
$$

From (1), $x = 26$.

There are 26 nickels and 12 dimes in the jar.

8.   The slope of the line $y = \frac{1}{3}x - 6$ is $\frac{1}{3}$. Any line perpendicular to that line has slope -3, which is the negative reciprocal of $\frac{1}{3}$. Use the point-slope form to find the equation of the line through $(5, -3)$ with slope -3.

$$
\begin{aligned}
y - y_1 &= m(x - x_1) \\
y - (-3) &= -3(x - 5) \\
y + 3 &= -3x + 15 \\
y &= -3x + 12
\end{aligned}
$$

9. $-3x + 5y = -15$

If $y = 0$, then $x = 5$ and the x-intercept is $(5, 0)$.
If $x = 0$, then $y = -3$ and the y-intercept is $(0, -3)$.
Draw the line through the intercepts.

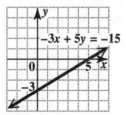

10. $y \le -2x + 7$

Draw a solid line through the intercepts $\left(\frac{7}{2}, 0\right)$ and $(0, 7)$.
Testing $(0, 0)$ gives $0 \le 7$, a true statement.
Shade the region containing $(0, 0)$.

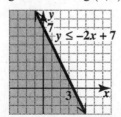

11. $3x + 2y = -4 \qquad (1)$

$\qquad\quad y = 2x + 5 \qquad (2)$

Since equation (2) is already solved for y, substitute $2x + 5$ for y in equation (1).

$$
\begin{aligned}
3x + 2y &= -4 \qquad (1) \\
3x + 2(2x + 5) &= -4 \qquad \textit{Let } y = 2x + 5. \\
3x + 4x + 10 &= -4 \\
7x &= -14 \\
x &= -2
\end{aligned}
$$

Substitute -2 for x in (2).

$$y = 2(-2) + 5 = 1$$

The solution $(-2, 1)$ checks.
The solution set is $\{(-2, 1)\}$.

12.
$$
\begin{array}{rrrrcr}
-2x & + & y & - & 4z & = & 2 \\
3x & + & 2y & - & z & = & -3 \\
-x & - & 4y & - & 2z & = & -17
\end{array}
$$

Write the augmented matrix.

$$
\left[\begin{array}{rrr|r}
-2 & 1 & -4 & 2 \\
3 & 2 & -1 & -3 \\
-1 & -4 & -2 & -17
\end{array}\right]
$$

$$
\left[\begin{array}{rrr|r}
0 & 9 & 0 & 36 \\
0 & -10 & -7 & -54 \\
-1 & -4 & -2 & -17
\end{array}\right]
\begin{array}{l}
-2R_3 + R_1 \\
3R_3 + R_2 \\
\\
\end{array}
$$

This matrix gives the system

$$
\begin{aligned}
9y &= 36 \\
-10y - 7z &= -54 \\
-x - 4y - 2z &= -17.
\end{aligned}
$$

From the first equation,

$$
\begin{aligned}
9y &= 36 \\
y &= 4.
\end{aligned}
$$

Substitute $y = 4$ in the second equation.

$$
\begin{aligned}
-10y - 7z &= -54 \\
-10(4) - 7z &= -54 \\
-7z &= -14 \\
z &= 2
\end{aligned}
$$

Substitute $y = 4$ and $z = 2$ in the third equation.

$$
\begin{aligned}
-x - 4y - 2z &= -17 \\
-x - 4(4) - 2(2) &= -17 \\
-x - 16 - 4 &= -17 \\
-x &= 3 \\
x &= -3
\end{aligned}
$$

The solution set is $\{(-3, 4, 2)\}$.

Copyright © 2012 Pearson Education, Inc. Publishing as Addison-Wesley.

13.
$$2x + y + 3z = 1$$
$$x - 2y + z = -3$$
$$-3x + y - 2z = -4$$

Write the augmented matrix.

$$\begin{bmatrix} 2 & 1 & 3 & | & 1 \\ 1 & -2 & 1 & | & -3 \\ -3 & 1 & -2 & | & -4 \end{bmatrix}$$

$$\begin{bmatrix} 1 & -2 & 1 & | & -3 \\ 2 & 1 & 3 & | & 1 \\ -3 & 1 & -2 & | & -4 \end{bmatrix} \quad R_1 \leftrightarrow R_2$$

$$\begin{bmatrix} 1 & -2 & 1 & | & -3 \\ 0 & 5 & 1 & | & 7 \\ 0 & -5 & 1 & | & -13 \end{bmatrix} \quad \begin{matrix} -2R_1 + R_2 \\ 3R_1 + R_3 \end{matrix}$$

$$\begin{bmatrix} 1 & -2 & 1 & | & -3 \\ 0 & 5 & 1 & | & 7 \\ 0 & 0 & 2 & | & -6 \end{bmatrix} \quad R_2 + R_3$$

This matrix gives the system

$$x - 2y + z = -3$$
$$5y + z = 7$$
$$2z = -6.$$

From the third equation,

$$2z = -6$$
$$z = -3.$$

Substitute $z = -3$ in the second equation.

$$5y + z = 7$$
$$5y + (-3) = 7$$
$$5y = 10$$
$$y = 2$$

Substitute $y = 2$ and $z = -3$ in the first equation.

$$x - 2y + z = -3$$
$$x - 2(2) + (-3) = -3$$
$$x - 4 - 3 = -3$$
$$x = 4$$

The solution set is $\{(4, 2, -3)\}$.

14. $f(x) = x^2$ is a parabola opening up with vertex $(0, 0)$.

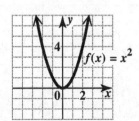

15. $(r^4 - 2r^3 + 6)(3r - 1)$

Multiply vertically.

$$\begin{array}{r} r^4 - 2r^3 + 0r^2 + 0r + 6 \\ 3r - 1 \\ \hline -r^4 + 2r^3 \qquad\qquad - 6 \\ 3r^5 - 6r^4 \qquad\qquad + 18r \\ \hline 3r^5 - 7r^4 + 2r^3 \qquad + 18r - 6 \end{array}$$

16. $[(k - 5h) + 2]^2$
$$= (k - 5h)^2 + 2(2)(k - 5h) + 2^2$$
Square of a binomial
$$= k^2 - 10kh + 25h^2 + 4(k - 5h) + 4$$
$$= k^2 - 10kh + 25h^2 + 4k - 20h + 4$$

17. $6x^2 - 15x - 9 = 3(2x^2 - 5x - 3)$
$$= 3(x - 3)(2x + 1)$$

18. $729 + 8y^6$
$$= 9^3 + (2y^2)^3$$
$$= (9 + 2y^2)[9^2 - 9(2y^2) + (2y^2)^2]$$
$$= (9 + 2y^2)(81 - 18y^2 + 4y^4)$$

19.
$$x^3 + 3x^2 - x - 3 = 0$$
$$(x^3 + 3x^2) - (x + 3) = 0$$
$$x^2(x + 3) - 1(x + 3) = 0$$
$$(x + 3)(x^2 - 1) = 0$$
$$(x + 3)(x + 1)(x - 1) = 0$$

$$x + 3 = 0 \quad \text{or} \quad x + 1 = 0 \quad \text{or} \quad x - 1 = 0$$
$$x = -3 \quad \text{or} \quad x = -1 \quad \text{or} \quad x = 1$$

The solution set is $\{-3, -1, 1\}$.

20. $\dfrac{x^2 - 36}{4x^2 - 21x - 18}$

(a) The expression is undefined when the denominator is 0.

$$4x^2 - 21x - 18 = 0$$
$$(4x + 3)(x - 6) = 0$$
$$x = -\tfrac{3}{4}, 6$$

(b) $\dfrac{x^2 - 36}{4x^2 - 21x - 18} = \dfrac{(x + 6)(x - 6)}{(4x + 3)(x - 6)}$
$$= \dfrac{x + 6}{4x + 3}$$

21. $\dfrac{2r + 4}{5r} \cdot \dfrac{3r}{5r + 10}$
$$= \dfrac{2(r + 2)}{5r} \cdot \dfrac{3r}{5(r + 2)}$$
$$= \tfrac{2}{5} \cdot \tfrac{3}{5} = \tfrac{6}{25}$$

Copyright © 2012 Pearson Education, Inc. Publishing as Addison-Wesley.

22. $\dfrac{y^2 - 2y - 3}{y^2 + 4y + 4} \div \dfrac{y^2 - 1}{y^2 + y - 2}$

$= \dfrac{y^2 - 2y - 3}{y^2 + 4y + 4} \cdot \dfrac{y^2 + y - 2}{y^2 - 1}$

$= \dfrac{(y-3)(y+1)}{(y+2)(y+2)} \cdot \dfrac{(y+2)(y-1)}{(y+1)(y-1)}$

$= \dfrac{y-3}{y+2}$

23. $\dfrac{3x + 12}{2x + 7} + \dfrac{-7x - 26}{2x + 7}$

$= \dfrac{3x + 12 - 7x - 26}{2x + 7}$

$= \dfrac{-4x - 14}{2x + 7}$

$= \dfrac{-2(2x + 7)}{2x + 7} = -2$

24. $\dfrac{2}{r-2} - \dfrac{r+3}{r-1}$

$= \dfrac{2(r-1)}{(r-2)(r-1)} - \dfrac{(r+3)(r-2)}{(r-1)(r-2)}$

$\qquad\qquad LCD = (r-2)(r-1)$

$= \dfrac{2r - 2}{(r-2)(r-1)} - \dfrac{r^2 + r - 6}{(r-1)(r-2)}$

$= \dfrac{(2r - 2) - (r^2 + r - 6)}{(r-2)(r-1)}$

$= \dfrac{2r - 2 - r^2 - r + 6}{(r-2)(r-1)}$

$= \dfrac{-r^2 + r + 4}{(r-2)(r-1)}$

25. $\dfrac{\dfrac{1}{y} + \dfrac{1}{y-1}}{\dfrac{1}{y} - \dfrac{2}{y-1}}$

$= \dfrac{\left(\dfrac{1}{y} + \dfrac{1}{y-1}\right) \cdot y(y-1)}{\left(\dfrac{1}{y} - \dfrac{2}{y-1}\right) \cdot y(y-1)}$

$\qquad\qquad LCD = y(y-1)$

$= \dfrac{\dfrac{1}{y} \cdot y(y-1) + \dfrac{1}{y-1} \cdot y(y-1)}{\dfrac{1}{y} \cdot y(y-1) - \dfrac{2}{y-1} \cdot y(y-1)}$

$= \dfrac{(y-1) + y}{(y-1) - 2(y)}$

$= \dfrac{2y - 1}{-y - 1},\ \text{ or } \ \dfrac{1 - 2y}{y + 1}$

26. $\dfrac{10}{x^2} - \dfrac{3}{x} = 1$

Multiply by the LCD, x^2, to clear fractions.

$x^2 \left(\dfrac{10}{x^2} - \dfrac{3}{x}\right) = x^2 \cdot 1$

$10 - 3x = x^2$

$0 = x^2 + 3x - 10$

$0 = (x + 5)(x - 2)$

$x + 5 = 0 \quad \text{or} \quad x - 2 = 0$

$x = -5 \quad \text{or} \quad x = 2$

The solution set is $\{-5, 2\}$.

27. $\dfrac{3}{x+1} = \dfrac{1}{x-1} - \dfrac{2}{x^2 - 1}$

$\dfrac{3}{x+1} = \dfrac{1}{x-1} - \dfrac{2}{(x+1)(x-1)}$

Multiply by the LCD, $(x+1)(x-1)$.

$(x+1)(x-1)\left(\dfrac{3}{x+1}\right)$

$\qquad = (x+1)(x-1)$

$\qquad \cdot \left(\dfrac{1}{x-1} - \dfrac{2}{(x+1)(x-1)}\right)$

$3(x-1) = 1(x+1) - 2$

$3x - 3 = x + 1 - 2$

$3x - 3 = x - 1$

$2x = 2$

$x = 1$

However, $x = 1$ makes two of the original denominators equal zero, so it is not acceptable as a solution.

The solution set is $\emptyset$.

28. $\sqrt{32} - \sqrt{128} + \sqrt{162}$

$= \sqrt{16 \cdot 2} - \sqrt{64 \cdot 2} + \sqrt{81 \cdot 2}$

$= 4\sqrt{2} - 8\sqrt{2} + 9\sqrt{2}$

$= (4 - 8 + 9)\sqrt{2} = 5\sqrt{2}$

29. $\dfrac{3 - 4\sqrt{2}}{1 - \sqrt{2}}$

Multiply numerator and denominator by the conjugate of the denominator.

$= \dfrac{3 - 4\sqrt{2}}{1 - \sqrt{2}} \cdot \dfrac{1 + \sqrt{2}}{1 + \sqrt{2}}$

$= \dfrac{3 + 3\sqrt{2} - 4\sqrt{2} - 4 \cdot 2}{1^2 - \left(\sqrt{2}\right)^2}$

$= \dfrac{3 + (3 - 4)\sqrt{2} - 8}{1 - 2}$

$= \dfrac{-5 - \sqrt{2}}{-1} = 5 + \sqrt{2}$

Copyright © 2012 Pearson Education, Inc. Publishing as Addison-Wesley.

30. $x^2 + 4x + 2 = 0$

Here $a = 1$, $b = 4$, and $c = 2$. The discriminant is

$$b^2 - 4ac = 4^2 - 4(1)(2)$$
$$= 16 - 8 = 8.$$

31. Since the discriminant, 8, is positive but not a perfect square, the solutions of the quadratic equation $x^2 + 4x + 2 = 0$ are irrational.

32. Use the quadratic formula to solve $x^2 + 4x + 2 = 0$.

$$x = \frac{-b \pm \sqrt{b^2 - 4ac}}{2a}$$
$$= \frac{-4 \pm \sqrt{8}}{2(1)} = \frac{-4 \pm 2\sqrt{2}}{2}$$
$$= \frac{2\left(-2 \pm \sqrt{2}\right)}{2} = -2 \pm \sqrt{2}$$

The solution set is $\left\{-2 \pm \sqrt{2}\right\}$.

33. $3x^2 - 13x - 10 \le 0$

Use factoring to solve the quadratic equation
$3x^2 - 13x - 10 = 0$.
$(3x + 2)(x - 5) = 0$

$$3x + 2 = 0 \quad \text{or} \quad x - 5 = 0$$
$$x = -\tfrac{2}{3} \quad \text{or} \quad x = 5$$

Locate the numbers $-\tfrac{2}{3}$ and 5 that divide the number line into three intervals A, B, and C.

$$\begin{array}{ccc} \text{A} & \text{B} & \text{C} \end{array}$$

Choose a number from each interval to substitute in the inequality

$$3x^2 - 13x - 10 \le 0.$$

Interval A: Let $x = -1$.

$$3(-1)^2 - 13(-1) - 10 \overset{?}{\le} 0$$
$$3 + 13 - 10 \overset{?}{\le} 0$$
$$6 \le 0 \qquad \textit{False}$$

Interval B: Let $x = 0$.
$$-10 \le 0 \qquad \textit{True}$$

Interval C: Let $x = 6$.

$$3 \cdot 6^2 - 13 \cdot 6 - 10 \overset{?}{\le} 0$$
$$108 - 78 - 10 \overset{?}{\le} 0$$
$$20 \le 0 \qquad \textit{False}$$

Every number in Interval B is a solution of the original inequality, as are the endpoints $-\tfrac{2}{3}$ and 5. The solution set is $\left[-\tfrac{2}{3}, 5\right]$.

34. $x = y^2 + 3$

(1) To test for symmetry with respect to the x-axis, replace y with $-y$.

$$x = (-y)^2 + 3$$
$$x = y^2 + 3$$

(2) To test for symmetry with respect to the y-axis, replace x with $-x$.

$$-x = y^2 + 3$$
$$x = -y^2 - 3$$

(3) To test for symmetry with respect to the origin, replace x with $-x$ and y with $-y$.

$$-x = (-y)^2 + 3$$
$$-x = y^2 + 3$$
$$x = -y^2 - 3$$

These tests show that the graph of the relation $x = y^2 + 3$ is symmetric with respect to the x-axis, but not with respect to the y-axis or the origin.

35. $y = x^2 + 3$

(1) To test for symmetry with respect to the x-axis, replace y with $-y$.

$$-y = x^2 + 3$$
$$y = -x^2 - 3$$

(2) To test for symmetry with respect to the y-axis, replace x with $-x$.

$$y = (-x)^2 + 3$$
$$y = x^2 + 3$$

(3) To test for symmetry with respect to the origin, replace x with $-x$ and y with $-y$.

$$-y = (-x)^2 + 3$$
$$-y = x^2 + 3$$
$$y = -x^2 - 3$$

These tests show that the graph of the relation $y = x^2 + 3$ is symmetric with respect to the y-axis, but not with respect to the x-axis or the origin.

36. $y = 5x^3$

(1) To test for symmetry with respect to the x-axis, replace y with $-y$.

$$-y = 5x^3$$
$$y = -5x^3$$

(2) To test for symmetry with respect to the y-axis, replace x with $-x$.

$$y = 5(-x)^3$$
$$y = -5x^3$$

Copyright © 2012 Pearson Education, Inc. Publishing as Addison-Wesley.

(3) To test for symmetry with respect to the origin, replace x with $-x$ and y with $-y$.

$$-y = 5(-x)^3$$
$$-y = -5x^3$$
$$y = 5x^3$$

These tests show that the graph of the relation $y = 5x^3$ is symmetric with respect to the origin, but not with respect to the x-axis or the y-axis.

37. $f(x) = -x^2 + 3$

The graph is a parabola with vertex at $(0, 3)$ and opening down. The left half of the parabola is where the graph is increasing, and that is the interval $(-\infty, 0]$.

38. If $x \le 3$, then $f(x) = x^2 - 6$. Thus,

$$f(3) = 3^2 - 6 = 9 - 6 = 3.$$

39.
$$f(x) = \sqrt[3]{3x + 5}$$
Replace $f(x)$ with y.
$$y = \sqrt[3]{3x + 5}$$
Interchange x and y.
$$x = \sqrt[3]{3y + 5}$$
Solve for y; cube both sides.
$$x^3 = 3y + 5$$
$$x^3 - 5 = 3y$$
$$\frac{x^3 - 5}{3} = y$$
Replace y with $f^{-1}(x)$.
$$f^{-1}(x) = \frac{x^3 - 5}{3}$$

40. Since 8 and 16 are multiples of 2, we'll use the change-of-base rule with base 2 logarithms.

$$\log_{16} \frac{1}{8} = \frac{\log_2 \frac{1}{8}}{\log_2 16}$$
$$= \frac{\log_2 2^{-3}}{\log_2 2^4}$$
$$= \frac{-3}{4} = -\frac{3}{4}$$

Another solution:

$$\log_{16} \tfrac{1}{8} = x$$
$$\tfrac{1}{8} = 16^x \qquad \textit{Exponential form}$$
$$2^{-3} = \left(2^4\right)^x$$
$$2^{-3} = 2^{4x}$$
$$-3 = 4x \qquad \textit{Equate exponents.}$$
$$x = -\tfrac{3}{4}$$

41.
$$\log_2 x + \log_2 (x + 2) - 3 = 0$$
$$\log_2 [x(x + 2)] = 3$$
$$x^2 + 2x = 2^3$$
$$x^2 + 2x - 8 = 0$$
$$(x + 4)(x - 2) = 0$$

$$x + 4 = 0 \quad \text{or} \quad x - 2 = 0$$
$$x = -4 \quad \text{or} \qquad x = 2$$

Reject $x = -4$, because it yields an equation in which the logarithm of a negative number must be found.
The solution set is $\{2\}$.

42. $\left(x^4 + 7x^3 - 5x^2 + 2x + 13\right) \div (x + 1)$

$$
\begin{array}{r|rrrrr}
-1 & 1 & 7 & -5 & 2 & 13 \\
   &   & -1 & -6 & 11 & -13 \\
\hline
   & 1 & 6 & -11 & 13 & 0
\end{array}
$$

The quotient is

$$x^3 + 6x^2 - 11x + 13.$$

43. $f(x) = x^3 + 7x^2 + 7x - 15$

(a) -3 is a zero of f.

$$
\begin{array}{r|rrrr}
-3 & 1 & 7 & 7 & -15 \\
   &   & -3 & -12 & 15 \\
\hline
   & 1 & 4 & -5 & 0
\end{array}
$$

$$x^2 + 4x - 5 = (x + 5)(x - 1)$$

The other zeros of f are -5 and 1.

(b) The zeros of f are -3, -5, and 1, so

$$f(x) = (x + 3)(x + 5)(x - 1).$$

(c) The zeros of f are -3, -5, and 1, so the x-intercepts of the graph are $(-3, 0)$, $(-5, 0)$, and $(1, 0)$.
Also, $f(0) = -15$, so the y-intercept of the graph is $(0, -15)$.

(d)

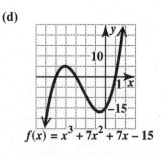

$f(x) = x^3 + 7x^2 + 7x - 15$

Copyright © 2012 Pearson Education, Inc. Publishing as Addison-Wesley.

44. $f(x) = \dfrac{x^2 - 4}{x^2 - 9} = \dfrac{(x+2)(x-2)}{(x+3)(x-3)}$

(a) To find the vertical asymptotes, set the denominator equal to 0 and solve for x. We get $x = -3$ and $x = 3$, so the equations of the vertical asymptotes of the graph are $x = -3$ and $x = 3$.

(b) To find the x-intercepts, find the zeros of the numerator. We get $x = -2$ and $x = 2$, so the x-intercepts of the graph are $(-2, 0)$ and $(2, 0)$.

(c) $f(x) = \dfrac{x^2 - 4}{x^2 - 9} = \dfrac{1x^2 - 4}{1x^2 - 9}$

The numerator and denominator have the same degree, so dividing by x^2 in the numerator and denominator produces the horizontal asymptote. The equation of the horizontal asymptote of the graph is $y = \frac{1}{1}$, or $y = 1$.

(d)

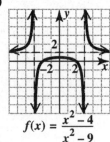

$f(x) = \dfrac{x^2 - 4}{x^2 - 9}$

$f(0) = \frac{4}{9}$, so the y-intercept is $\left(0, \frac{4}{9}\right)$.

Copyright © 2012 Pearson Education, Inc. Publishing as Addison-Wesley.

CHAPTER 13 CONIC SECTIONS AND NONLINEAR SYSTEMS

13.1 The Circle and the Ellipse

13.1 Now Try Exercises

N1. If the point (x, y) is on the circle, the distance from (x, y) to the center $(0, 0)$ is 6. Use the distance formula.

$$\sqrt{(x_2 - x_1)^2 + (y_2 - y_1)^2} = d$$
$$\sqrt{(x - 0)^2 + (y - 0)^2} = 6$$
$$\sqrt{x^2 + y^2} = 6$$
$$x^2 + y^2 = 36 \quad \textit{Square.}$$

An equation of this circle is $x^2 + y^2 = 36$.

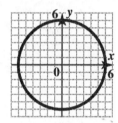

N2. Center at $(-2, 2)$; radius 3

$$\sqrt{(x_2 - x_1)^2 + (y_2 - y_1)^2} = d$$
$$\sqrt{[x - (-2)]^2 + (y - 2)^2} = 3$$
$$\sqrt{(x + 2)^2 + (y - 2)^2} = 3$$
$$(x + 2)^2 + (y - 2)^2 = 9 \quad \textit{Square.}$$

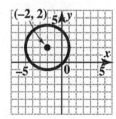

N3. Center at $(-5, 4)$; radius $\sqrt{6}$

Substitute $h = -5$, $k = 4$, and $r = \sqrt{6}$ in the center-radius form of the equation of a circle.

$$(x - h)^2 + (y - k)^2 = r^2$$
$$[x - (-5)]^2 + (y - 4)^2 = \left(\sqrt{6}\right)^2$$
$$(x + 5)^2 + (y - 4)^2 = 6$$

N4. To find the center and radius, add the appropriate constants to complete the squares on x and y.

$$x^2 + y^2 - 8x + 10y - 8 = 0$$
$$(x^2 - 8x) + (y^2 + 10y) = 8$$
$$(x^2 - 8x + 16) + (y^2 + 10y + 25) = 8 + 16 + 25$$
$$(x - 4)^2 + (y + 5)^2 = 49$$
$$(x - 4)^2 + [y - (-5)]^2 = 7^2$$

The circle has center at $(4, -5)$ and radius 7.

N5. $\dfrac{x^2}{16} + \dfrac{y^2}{25} = 1$

This ellipse has an equation of the form

$$\frac{x^2}{a^2} + \frac{y^2}{b^2} = 1.$$

Here, $a^2 = 16$, so $a = 4$ and the x-intercepts are $(4, 0)$ and $(-4, 0)$. Similarly, $b^2 = 25$, so $b = 5$ and the y-intercepts are $(0, 5)$ and $(0, -5)$. Plot the intercepts, and draw the ellipse through them.

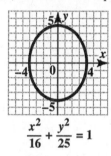

$$\frac{x^2}{16} + \frac{y^2}{25} = 1$$

N6. $\dfrac{(x - 3)^2}{36} + \dfrac{(y - 4)^2}{4} = 1$

The center is $(3, 4)$.

The ellipse passes through the four points:

$$(3 - 6, 4) \qquad = (-3, 4)$$
$$(3 + 6, 4) \qquad = (9, 4)$$
$$(3, 4 + 2) = (3, 6)$$
$$(3, 4 - 2) = (3, 2)$$

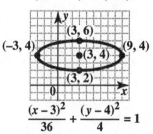

$$\frac{(x - 3)^2}{36} + \frac{(y - 4)^2}{4} = 1$$

13.1 Section Exercises

1. **(a)** $x^2 + y^2 = 25$ can be written in the center-radius form as

$$(x - 0)^2 + (y - 0)^2 = 5^2.$$

The center is the point $(0, 0)$.

(b) The radius is 5.

Copyright © 2012 Pearson Education, Inc. Publishing as Addison-Wesley.

(c) The x-intercepts are $(5, 0)$ and $(-5, 0)$. The y-intercepts are $(0, 5)$ and $(0, -5)$.

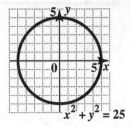

$x^2 + y^2 = 25$

3. $(x - 3)^2 + (y - 2)^2 = 25$ is an equation of a circle with center $(3, 2)$ and radius 5, choice **B**.

5. $(x + 3)^2 + (y - 2)^2 = 25$ is an equation of a circle with center $(-3, 2)$ and radius 5, choice **D**.

7. Center: $(-4, 3)$; radius: 2

Substitute $h = -4$, $k = 3$, and $r = 2$ in the center-radius form of the equation of a circle.

$$(x - h)^2 + (y - k)^2 = r^2$$
$$[x - (-4)]^2 + (y - 3)^2 = 2^2$$
$$(x + 4)^2 + (y - 3)^2 = 4$$

9. Center: $(-8, -5)$; radius: $\sqrt{5}$

Substitute $h = -8$, $k = -5$, and $r = \sqrt{5}$ in the center-radius form of the equation of a circle.

$$(x - h)^2 + (y - k)^2 = r^2$$
$$[x - (-8)]^2 + [y - (-5)]^2 = \left(\sqrt{5}\right)^2$$
$$(x + 8)^2 + (y + 5)^2 = 5$$

11. $x^2 + y^2 + 4x + 6y + 9 = 0$

Rewrite the equation keeping only the variable terms on the left and grouping the x-terms and y-terms.

$$x^2 + 4x + y^2 + 6y = -9$$

Complete both squares on the left, and add the same constants to the right.

$$\left(x^2 + 4x + \underline{4}\right) + \left(y^2 + 6y + \underline{9}\right) = -9 + \underline{4} + \underline{9}$$
$$(x + 2)^2 + (y + 3)^2 = 4$$

From the form $(x - h)^2 + (y - k)^2 = r^2$, we have $h = -2$, $k = -3$, and $r = 2$. The center is $(-2, -3)$, and the radius r is 2.

13.
$$x^2 + y^2 + 10x - 14y - 7 = 0$$
$$\left(x^2 + 10x \right) + \left(y^2 - 14y \right) = 7$$
$$\left(x^2 + 10x + \underline{25}\right) + \left(y^2 - 14y + \underline{49}\right)$$
$$= 7 + \underline{25} + \underline{49}$$
$$(x + 5)^2 + (y - 7)^2 = 81$$

The center is $(-5, 7)$, and the radius is $\sqrt{81} = 9$.

15. $3x^2 + 3y^2 - 12x - 24y + 12 = 0$
$$x^2 + y^2 - 4x - 8y + 4 = 0 \quad \textit{Divide by 3.}$$

Complete the square on x and the square on y.

$$\left(x^2 - 4x \quad\right) + \left(y^2 - 8y \quad\right) = -4$$
$$\left(x^2 - 4x + \underline{4}\right) + \left(y^2 - 8y + \underline{16}\right)$$
$$= -4 + \underline{4} + \underline{16}$$
$$(x - 2)^2 + (y - 4)^2 = 16$$

The center is $(2, 4)$, and the radius is $\sqrt{16} = 4$.

17.
$$x^2 + y^2 = 9$$
$$(x - 0)^2 + (y - 0)^2 = 3^2$$

Here, $h = 0$, $k = 0$, and $r = 3$, so the graph is a circle with center $(0, 0)$ and radius 3.

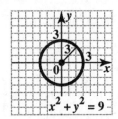

$x^2 + y^2 = 9$

19.
$$2y^2 = 10 - 2x^2$$
$$2x^2 + 2y^2 = 10$$
$$x^2 + y^2 = 5 \quad \textit{Divide by 2.}$$
$$(x - 0)^2 + (y - 0)^2 = \left(\sqrt{5}\right)^2$$

Here, $h = 0$, $k = 0$, and $r = \sqrt{5} \approx 2.2$, so the graph is a circle with center $(0, 0)$ and radius $\sqrt{5}$.

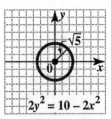

$2y^2 = 10 - 2x^2$

21. $(x + 3)^2 + (y - 2)^2 = 9$

Here, $h = -3$, $k = 2$, and $r = \sqrt{9} = 3$. The graph is a circle with center $(-3, 2)$ and radius 3.

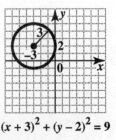

$(x + 3)^2 + (y - 2)^2 = 9$

Copyright © 2012 Pearson Education, Inc. Publishing as Addison-Wesley.

23.
$$x^2 + y^2 - 4x - 6y + 9 = 0$$
$$\left(x^2 - 4x \quad \right) + \left(y^2 - 6y \quad \right) = -9$$
Complete the square for each variable.
$$\left(x^2 - 4x + \underline{4}\right) + \left(y^2 - 6y + \underline{9}\right)$$
$$= -9 + \underline{4} + \underline{9}$$
$$(x-2)^2 + (y-3)^2 = 4$$
Here, $h = 2$, $k = 3$, and $r = \sqrt{4} = 2$. The graph is a circle with center $(2, 3)$ and radius 2.

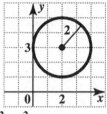

$$x^2 + y^2 - 4x - 6y + 9 = 0$$

25.
$$x^2 + y^2 + 6x - 6y + 9 = 0$$
$$\left(x^2 + 6x \quad \right) + \left(y^2 - 6y \quad \right) = -9$$
Complete the square for each variable.
$$\left(x^2 + 6x + \underline{9}\right) + \left(y^2 - 6y + \underline{9}\right)$$
$$= -9 + \underline{9} + \underline{9}$$
$$(x+3)^2 + (y-3)^2 = 9$$

The center is $(-3, 3)$, and the radius is $\sqrt{9} = 3$.

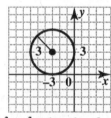

$$x^2 + y^2 + 6x - 6y + 9 = 0$$

27. This method works because the pencil is always the same distance from the fastened end. The fastened end works as the center, and the length of the string from the fastened end to the pencil is the radius.

29. $\dfrac{x^2}{9} + \dfrac{y^2}{25} = 1$ is in the form $\dfrac{x^2}{a^2} + \dfrac{y^2}{b^2} = 1$. The graph is an ellipse with $a^2 = 9$ and $b^2 = 25$, so $a = 3$ and $b = 5$. The x-intercepts are $(3, 0)$ and $(-3, 0)$. The y-intercepts are $(0, 5)$ and $(0, -5)$. Plot the intercepts, and draw the ellipse through them.

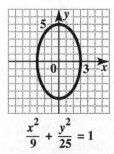

$$\frac{x^2}{9} + \frac{y^2}{25} = 1$$

31. $\dfrac{x^2}{36} + \dfrac{y^2}{16} = 1$ is in the form $\dfrac{x^2}{a^2} + \dfrac{y^2}{b^2} = 1$. The graph is an ellipse with $a^2 = 36$ and $b^2 = 16$, so $a = 6$ and $b = 4$. The x-intercepts are $(6, 0)$ and $(-6, 0)$. The y-intercepts are $(0, 4)$ and $(0, -4)$. Plot the intercepts, and draw the ellipse through them.

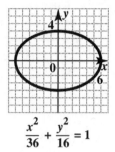

$$\frac{x^2}{36} + \frac{y^2}{16} = 1$$

33. $\dfrac{x^2}{16} + \dfrac{y^2}{4} = 1$ is in the form $\dfrac{x^2}{a^2} + \dfrac{y^2}{b^2} = 1$. The graph is an ellipse with $a^2 = 16$ and $b^2 = 4$, so $a = 4$ and $b = 2$. The x-intercepts are $(4, 0)$ and $(-4, 0)$. The y-intercepts are $(0, 2)$ and $(0, -2)$. Plot the intercepts, and draw the ellipse through them.

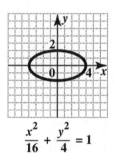

$$\frac{x^2}{16} + \frac{y^2}{4} = 1$$

35. $\dfrac{y^2}{25} = 1 - \dfrac{x^2}{49} \Leftrightarrow \dfrac{x^2}{49} + \dfrac{y^2}{25} = 1$ is in the form $\dfrac{x^2}{a^2} + \dfrac{y^2}{b^2} = 1$. The graph is an ellipse with $a^2 = 49$ and $b^2 = 25$, so $a = 7$ and $b = 5$. The x-intercepts are $(7, 0)$ and $(-7, 0)$. The y-intercepts are $(0, 5)$ and $(0, -5)$. Plot the intercepts, and draw the ellipse through them.

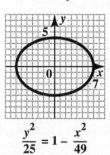

$$\frac{y^2}{25} = 1 - \frac{x^2}{49}$$

Copyright © 2012 Pearson Education, Inc. Publishing as Addison-Wesley.

37. $\dfrac{(x+1)^2}{64} + \dfrac{(y-2)^2}{49} = 1$ is in the form

$\dfrac{(x-h)^2}{a^2} + \dfrac{(y-k)^2}{b^2} = 1$, so the center of the

ellipse is at $(-1, 2)$. Since $a^2 = 64$, $a = 8$. Since $b^2 = 49$, $b = 7$. The graph includes the points $(-1 \pm 8, 2)$ and $(-1, 2 \pm 7)$, or equivalently, $(7, 2)$, $(-9, 2)$, $(-1, 9)$, and $(-1, -5)$. Plot the points, and draw the ellipse through them.

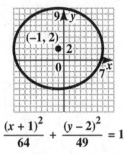

$$\dfrac{(x+1)^2}{64} + \dfrac{(y-2)^2}{49} = 1$$

39. $\dfrac{(x-2)^2}{16} + \dfrac{(y-1)^2}{9} = 1$

The center of the ellipse is at $(2, 1)$. Since $a^2 = 16$, $a = 4$. Since $b^2 = 9$, $b = 3$. Some points on the ellipse are $(2 \pm 4, 1)$ and $(2, 1 \pm 3)$, or equivalently, $(6, 1)$, $(-2, 1)$, $(2, 4)$, and $(2, -2)$. Plot the points, and draw the ellipse through them.

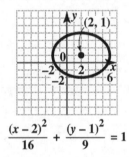

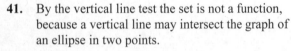
$$\dfrac{(x-2)^2}{16} + \dfrac{(y-1)^2}{9} = 1$$

41. By the vertical line test the set is not a function, because a vertical line may intersect the graph of an ellipse in two points.

43. $(x+2)^2 + (y-4)^2 = 16$

$(y-4)^2 = 16 - (x+2)^2$

Take the square root of each side.

$$y - 4 = \pm\sqrt{16 - (x+2)^2}$$
$$y = 4 \pm \sqrt{16 - (x+2)^2}$$

Therefore, the two functions used to obtain the graph were

$$y_1 = 4 + \sqrt{16 - (x+2)^2} \quad \text{and}$$
$$y_2 = 4 - \sqrt{16 - (x+2)^2}.$$

45. $x^2 + y^2 = 36$

$y^2 = 36 - x^2$

Take the square root of both sides.

$$y = \pm\sqrt{36 - x^2}$$

Therefore,
$$y_1 = \sqrt{36 - x^2} \quad \text{and} \quad y_2 = -\sqrt{36 - x^2}.$$

Use these two functions to obtain the graph.

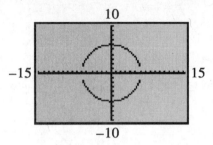

47. $\dfrac{x^2}{16} + \dfrac{y^2}{4} = 1$

$\dfrac{y^2}{4} = 1 - \dfrac{x^2}{16}$

$y^2 = 4\left(1 - \dfrac{x^2}{16}\right)$

Take the square root of both sides.

$$y = \pm 2\sqrt{1 - \dfrac{x^2}{16}}$$

Therefore,

$$y_1 = 2\sqrt{1 - \dfrac{x^2}{16}} \quad \text{and} \quad y_2 = -2\sqrt{1 - \dfrac{x^2}{16}}.$$

Use these two functions to obtain the graph.

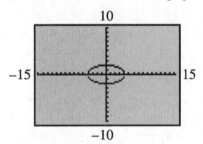

49. $\dfrac{x^2}{36} + \dfrac{y^2}{9} = 1$ is in the form $\dfrac{x^2}{a^2} + \dfrac{y^2}{b^2} = 1$.

$c^2 = a^2 - b^2 = 36 - 9 = 27$, so

$c = \sqrt{27} = 3\sqrt{3}$.

The kidney stone and the source of the beam must be placed $3\sqrt{3}$ units from the center of the ellipse.

51. (a) $100x^2 + 324y^2 = 32{,}400$

$\dfrac{x^2}{324} + \dfrac{y^2}{100} = 1$ *Divide by 32,400.*

$\dfrac{x^2}{18^2} + \dfrac{y^2}{10^2} = 1$

The height in the center is the y-coordinate of the positive y-intercept. The height is 10 meters.

(b) The width of the ellipse is the distance between the x-intercepts, $(-18, 0)$ and $(18, 0)$. The width across the bottom of the arch is $18 + 18 = 36$ meters.

Copyright © 2012 Pearson Education, Inc. Publishing as Addison-Wesley.

53. $\dfrac{x^2}{141.7^2} + \dfrac{y^2}{141.1^2} = 1$

   (a) $c^2 = a^2 - b^2$, so

   $$c = \sqrt{a^2 - b^2} = \sqrt{141.7^2 - 141.1^2}$$
   $$= \sqrt{169.68} \approx 13.0$$

   From the figure, the greatest distance (the **apogee**) is $a + c = 141.7 + 13.0 = 154.7$ million miles.

   (b) The least distance (the **perigee**) is $a - c = 141.7 - 13.0 = 128.7$ million miles.

55. Plot the points $(3, 4)$, $(-3, 4)$, $(3, -4)$, and $(-3, -4)$.

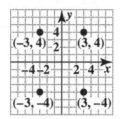

57. $4x + 3y = 12$

   If $y = 0$, then $4x = 12$, so $x = 3$ and $(3, 0)$ is the x-intercept.
   If $x = 0$, then $3y = 12$, so $y = 4$ and $(0, 4)$ is the y-intercept.

13.2 The Hyperbola and Functions Defined by Radicals

13.2 Now Try Exercises

N1. $\dfrac{x^2}{25} - \dfrac{y^2}{9} = 1$ or $\dfrac{x^2}{5^2} - \dfrac{y^2}{3^2} = 1$

   Step 1
   The graph is a hyperbola with $a = 5$ and $b = 3$. It has left and right branches with x-intercepts $(5, 0)$ and $(-5, 0)$.

   Step 2
   The four points $(5, 3)$, $(-5, 3)$, $(-5, -3)$, and $(5, -3)$ are the vertices of the fundamental rectangle.

   Steps 3 and 4
   The equations of the asymptotes are $y = \pm \frac{3}{5}x$, and the hyperbola approaches these lines as x and y get larger and larger in absolute value.

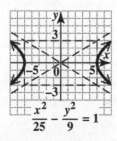

N2. $\dfrac{y^2}{9} - \dfrac{x^2}{16} = 1$ or $\dfrac{y^2}{3^2} - \dfrac{x^2}{4^2} = 1$

   The graph is a hyperbola with $b = 3$ and $a = 4$. It has upper and lower branches with y-intercepts $(0, 3)$ and $(0, -3)$. The four points $(4, 3)$, $(-4, 3)$, $(-4, -3)$, and $(4, -3)$ are the vertices of the fundamental rectangle. Draw the asymptotes, $y = \pm \frac{3}{4}x$, through the corners of the fundamental rectangle. Then draw the branches of the hyperbola so that they approach the asymptotes.

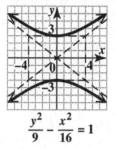

N3. **(a)** $y^2 - 10 = -x^2$
   $x^2 + y^2 = 10$

   Since the x^2- and y^2-terms have equal positive coefficients, the graph of the equation is a **circle**.

   (b) $y - 2x^2 = 8$
   $y = 2x^2 + 8$

   This is an equation of a vertical **parabola** since only one variable, x, is squared.

   (c) $3x^2 + y^2 = 4$
   $\dfrac{3x^2}{4} + \dfrac{y^2}{4} = 1$ *Divide by 4.*

   Since the x^2- and y^2-terms have different positive coefficients, the graph of the equation is an **ellipse**.

N4. $f(x) = \sqrt{64 - x^2}$
   $y = \sqrt{64 - x^2}$ $f(x) = y$
   $y^2 = 64 - x^2$ *Square both sides.*
   $x^2 + y^2 = 64$ *Add x^2.*

   The last equation is the graph of a circle with center $(0, 0)$ and radius 8. Since $f(x)$ represents a principal square root in the original equation, $f(x)$ is nonnegative and its graph is the upper half of the circle.

   The domain of the function is $[-8, 8]$ and its range is $[0, 8]$.

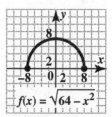

Copyright © 2012 Pearson Education, Inc. Publishing as Addison-Wesley.

N5.
$$\frac{y}{4} = -\sqrt{1 - \frac{x^2}{9}}$$

$$\frac{y^2}{16} = 1 - \frac{x^2}{9} \qquad \textit{Square both sides.}$$

$$\frac{x^2}{9} + \frac{y^2}{16} = 1 \qquad \textit{Add } \frac{x^2}{9}.$$

The last equation is the graph of an ellipse with intercepts $(3, 0)$, $(-3, 0)$, $(0, 4)$, and $(0, -4)$. Since $\frac{y}{4}$ equals a negative square root in the original equation, y must be nonpositive, restricting the graph to the lower half of the ellipse.

The domain of the function is $[-3, 3]$ and its range is $[-4, 0]$.

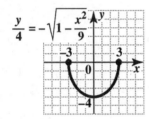

13.2 Section Exercises

1. $\dfrac{x^2}{25} + \dfrac{y^2}{9} = 1$

This is the standard form for the equation of an ellipse with x-intercepts $(5, 0)$ and $(-5, 0)$ and y-intercepts $(0, 3)$ and $(0, -3)$. This is graph **C**.

3. $\dfrac{x^2}{9} - \dfrac{y^2}{25} = 1$

This is the standard form for the equation of a hyperbola that opens left and right. Its x-intercepts are $(3, 0)$ and $(-3, 0)$. This is graph **D**.

5. $\dfrac{x^2}{16} - \dfrac{y^2}{9} = 1$ is in the form $\dfrac{x^2}{a^2} - \dfrac{y^2}{b^2} = 1$. The graph is a hyperbola with $a = 4$ and $b = 3$. The x-intercepts are $(4, 0)$ and $(-4, 0)$. There are no y-intercepts. The vertices of the fundamental rectangle are $(4, 3)$, $(4, -3)$, $(-4, -3)$, and $(-4, 3)$. Extend the diagonals of the rectangle through these points to get the asymptotes. Graph a branch of the hyperbola through each intercept and approaching the asymptotes.

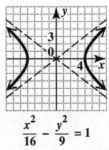

$$\frac{x^2}{16} - \frac{y^2}{9} = 1$$

7. $\dfrac{y^2}{4} - \dfrac{x^2}{25} = 1$ is in the form $\dfrac{y^2}{b^2} - \dfrac{x^2}{a^2} = 1$. The graph is a hyperbola with $a = 5$ and $b = 2$. The y-intercepts are $(0, 2)$ and $(0, -2)$. There are no x-intercepts. The vertices of the fundamental rectangle are $(5, 2)$, $(5, -2)$, $(-5, -2)$, and $(-5, 2)$. Extend the diagonals of the rectangle through these points to get the asymptotes. Graph a branch of the hyperbola through each intercept and approaching the asymptotes.

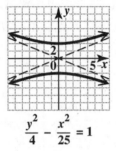

$$\frac{y^2}{4} - \frac{x^2}{25} = 1$$

9. $\dfrac{x^2}{25} - \dfrac{y^2}{36} = 1$ is in the form $\dfrac{x^2}{a^2} - \dfrac{y^2}{b^2} = 1$. The graph is a hyperbola with $a = 5$ and $b = 6$. The x-intercepts are $(5, 0)$ and $(-5, 0)$. There are no y-intercepts. The vertices of the fundamental rectangle are $(5, 6)$, $(5, -6)$, $(-5, -6)$ and $(-5, 6)$.

Extend the diagonals of the rectangle through these points to get the asymptotes. Graph a branch of the hyperbola through each intercept and approaching the asymptotes.

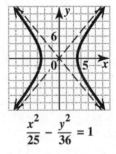

$$\frac{x^2}{25} - \frac{y^2}{36} = 1$$

11. $\dfrac{y^2}{16} - \dfrac{x^2}{16} = 1$ is in the form $\dfrac{y^2}{b^2} - \dfrac{x^2}{a^2} = 1$. The graph is a hyperbola with $a = 4$ and $b = 4$. The y-intercepts are $(0, 4)$ and $(0, -4)$. There are no x-intercepts. The vertices of the fundamental rectangle are $(4, 4)$, $(4, -4)$, $(-4, -4)$, and $(-4, 4)$. Extend the diagonals of the rectangle through these points to get the asymptotes. Graph a branch of the hyperbola through each intercept and approaching the asymptotes.

Copyright © 2012 Pearson Education, Inc. Publishing as Addison-Wesley.

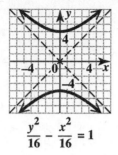

$$\frac{y^2}{16} - \frac{x^2}{16} = 1$$

13. $x^2 - y^2 = 16$

$$\frac{x^2}{16} - \frac{y^2}{16} = 1 \quad \textit{Divide by 16.}$$

This equation is in the form

$$\frac{x^2}{a^2} - \frac{y^2}{b^2} = 1$$

with $a = 4$ and $b = 4$. The graph is a **hyperbola** with x-intercepts $(4, 0)$ and $(-4, 0)$ and no y-intercepts. One asymptote passes through $(4, 4)$ and $(-4, -4)$. The other asymptote passes through $(-4, 4)$ and $(4, -4)$. Sketch the graph through the intercepts and approaching the asymptotes.

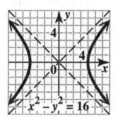

$$x^2 - y^2 = 16$$

15. $4x^2 + y^2 = 16$

$$\frac{x^2}{4} + \frac{y^2}{16} = 1 \quad \textit{Divide by 16.}$$

This equation is in the form

$$\frac{x^2}{a^2} + \frac{y^2}{b^2} = 1$$

with $a = 2$ and $b = 4$. The graph is an **ellipse**. The x-intercepts $(2, 0)$ and $(-2, 0)$. The y-intercepts are $(0, 4)$ and $(0, -4)$. Plot the intercepts and draw the ellipse through them.

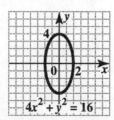

$$4x^2 + y^2 = 16$$

17.
$$y^2 = 36 - x^2$$
$$x^2 + y^2 = 36$$
$$(x - 0)^2 + (y - 0)^2 = 36$$

The graph is a **circle** with center at $(0, 0)$ and radius $\sqrt{36} = 6$.

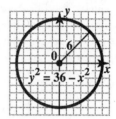

$$y^2 = 36 - x^2$$

19. $x^2 - 2y = 0$
$$x^2 = 2y$$
$$\tfrac{1}{2}x^2 = y$$

The equation is in the form

$$f(x) = a(x - h)^2 + k,$$

with $a = \frac{1}{2}$, $h = 0$, and $k = 0$. The graph is a **parabola** that opens up and is wider than the graph of $f(x) = x^2$ because $|a| = \frac{1}{2} < 1$. The vertex is at $(0, 0)$.

x	1	2	3
y	$\frac{1}{2}$	2	4.5

Use symmetry about $x = 0$ to draw the graph.

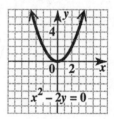

$$x^2 - 2y = 0$$

21.
$$y^2 = 4 + x^2$$
$$y^2 - x^2 = 4$$
$$\frac{y^2}{4} - \frac{x^2}{4} = 1 \quad \textit{Divide by 4.}$$

The graph is a **hyperbola** with y-intercepts $(0, 2)$ and $(0, -2)$. One asymptote passes through $(2, 2)$ and $(-2, -2)$. The other asymptote passes through $(-2, 2)$ and $(2, -2)$.

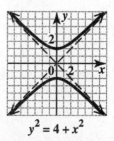

$$y^2 = 4 + x^2$$

Copyright © 2012 Pearson Education, Inc. Publishing as Addison-Wesley.

23. $f(x) = \sqrt{16 - x^2}$

Replace $f(x)$ with y and square both sides to get the equation

$$y^2 = 16 - x^2 \quad \text{or} \quad x^2 + y^2 = 16.$$

This is the graph of a circle with center $(0, 0)$ and radius 4. Since $f(x)$, or y, represents a principal square root in the original equation, $f(x)$ must be nonnegative. This restricts the graph to the upper half of the circle.

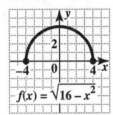

The domain is $[-4, 4]$, and the range is $[0, 4]$.

25. $f(x) = -\sqrt{36 - x^2}$

$y = -\sqrt{36 - x^2}$ *Replace f(x) with y.*

$y^2 = 36 - x^2$ *Square each side.*

$x^2 + y^2 = 36$

This is a circle centered at the origin with radius $\sqrt{36} = 6$. Since $f(x)$, or y, represents a nonpositive square root in the original equation, $f(x)$ must be nonpositive. This restricts the graph to the bottom half of the circle.

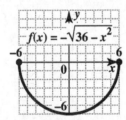

The domain is $[-6, 6]$, and the range is $[-6, 0]$.

27. $y = -2\sqrt{1 - \dfrac{x^2}{9}}$

$y^2 = 4\left(1 - \dfrac{x^2}{9}\right)$ *Square each side.*

$\dfrac{y^2}{4} = 1 - \dfrac{x^2}{9}$ *Divide by 4.*

$\dfrac{x^2}{9} + \dfrac{y^2}{4} = 1$

This is the equation of an ellipse with intercepts $(3, 0)$, $(-3, 0)$, $(0, 2)$, and $(0, -2)$. Since y equals a negative square root in the original equation, y must be nonpositive, restricting the graph to the lower half of the ellipse.

The domain of the function is $[-3, 3]$ and its range is $[-2, 0]$.

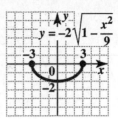

29. $\dfrac{y}{3} = \sqrt{1 + \dfrac{x^2}{9}}$

$\dfrac{y^2}{9} = 1 + \dfrac{x^2}{9}$ *Square each side.*

$\dfrac{y^2}{9} - \dfrac{x^2}{9} = 1$

This is a hyperbola opening up and down with y-intercepts $(0, 3)$ and $(0, -3)$. The four points $(3, 3)$, $(3, -3)$, $(-3, 3)$, and $(-3, -3)$ are the vertices of the rectangle that determine the asymptotes. Since $f(x)$, or y, represents a square root in the original equation, $f(x)$ must be nonnegative. This restricts the graph to the upper half of the hyperbola.

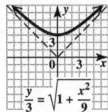

The domain is $(-\infty, \infty)$, and the range is $[3, \infty)$.

31. $\dfrac{(x - 2)^2}{4} - \dfrac{(y + 1)^2}{9} = 1$ is a hyperbola centered at $(2, -1)$, with $a = 2$ and $b = 3$. The hyperbola goes through the points $(2 \pm 2, -1)$, or $(4, -1)$ and $(0, -1)$. The asymptotes are the extended diagonals of the rectangle with vertices $(2, 3)$, $(2, -3)$, $(-2, -3)$ and $(-2, 3)$ shifted 2 units right and 1 unit down, or $(4, 2)$, $(4, -4)$, $(0, -4)$ and $(0, 2)$. Draw the hyperbola.

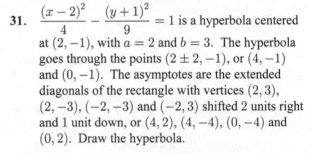

33. $\dfrac{y^2}{36} - \dfrac{(x - 2)^2}{49} = 1$ is a hyperbola centered at $(2, 0)$ with $a = 7$, and $b = 6$. The lowest point on the upper branch is $(2, 6)$ and the highest point on the lower branch is $(2, -6)$. The asymptotes are the extended diagonals of the rectangle with vertices $(7, 6)$, $(7, -6)$, $(-7, -6)$, and $(-7, 6)$

Copyright © 2012 Pearson Education, Inc. Publishing as Addison-Wesley.

shifted right 2 units, or $(9, 6)$, $(9, -6)$, $(-5, -6)$, and $(-5, 6)$. Draw the hyperbola.

$$\frac{y^2}{36} - \frac{(x-2)^2}{49} = 1$$

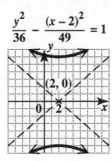

35. **(a)** $400x^2 - 625y^2 = 250{,}000$

$$\frac{x^2}{625} - \frac{y^2}{400} = 1 \quad \textit{Divide by 250,000.}$$

$$\frac{x^2}{25^2} - \frac{y^2}{20^2} = 1$$

The x-intercepts are $(25, 0)$ and $(-25, 0)$. The distance between the buildings is the distance between the x-intercepts. The buildings are $25 + 25 = 50$ meters apart at their closest point.

(b) At $x = 50$, $y = \dfrac{d}{2}$, so $d = 2y$.

$$400(50)^2 - 625y^2 = 250{,}000$$

$$1{,}000{,}000 - 625y^2 = 250{,}000$$

$$-625y^2 = -750{,}000$$

$$y^2 = 1200$$

$$y = \sqrt{1200}$$

The distance d is $2\sqrt{1200} \approx 69.3$ meters.

37. $$\frac{x^2}{9} - y^2 = 1$$

$$\frac{x^2}{9} - 1 = y^2 \quad \textit{Isolate } y^2.$$

$$\pm\sqrt{\frac{x^2}{9} - 1} = y \quad \textit{Take square roots.}$$

The two functions used to obtain the graph were

$$y_1 = \sqrt{\frac{x^2}{9} - 1} \quad \text{and} \quad y_2 = -\sqrt{\frac{x^2}{9} - 1}.$$

39. $$\frac{x^2}{25} - \frac{y^2}{49} = 1$$

$$\frac{x^2}{25} - 1 = \frac{y^2}{49}$$

$$49\left(\frac{x^2}{25} - 1\right) = y^2 \quad \textit{Multiply by 49.}$$

$$\pm 7\sqrt{\frac{x^2}{25} - 1} = y \quad \textit{Take square roots.}$$

To obtain the graph, use the two functions

$$y_1 = 7\sqrt{\frac{x^2}{25} - 1} \quad \text{and} \quad y_2 = -7\sqrt{\frac{x^2}{25} - 1}.$$

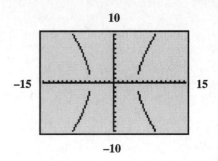

41. $y^2 - 9x^2 = 9$

$$y^2 = 9 + 9x^2$$

$$y^2 = 9(1 + x^2) \quad \textit{Factor out 9.}$$

$$y = \pm 3\sqrt{1 + x^2} \quad \textit{Take square roots.}$$

To obtain the graphs, use the two functions

$$y_1 = 3\sqrt{1 + x^2} \quad \text{and} \quad y_2 = -3\sqrt{1 + x^2}.$$

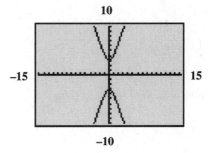

43. $2x + y = 13 \quad (1)$
$y = 3x + 3 \quad (2)$

Substitute $3x + 3$ for y in (1).

$$2x + (3x + 3) = 13$$

$$5x + 3 = 13$$

$$5x = 10$$

$$x = 2$$

Substitute 2 for x in (2).

$$y = 3(2) + 3 = 9$$

The solution set is $\{(2, 9)\}$.

45. $4x - 3y = -10 \quad (1)$
$4x + 6y = 8 \quad (2)$

To eliminate x, multiply equation (1) by -1 and add the result to equation (2).

$$\begin{array}{rcll} -4x + 3y &=& 10 & -1 \times (1) \\ 4x + 6y &=& 8 & (2) \\ \hline 9y &=& 18 & \textit{Add.} \\ y &=& 2 & \textit{Solve for } y. \end{array}$$

To find x, substitute 2 for y in equation (1) or (2).

$$4x + 6y = 8 \quad (2)$$

$$4x + 6(2) = 8$$

$$4x + 12 = 8$$

$$4x = -4$$

$$x = -1$$

The solution set is $\{(-1, 2)\}$.

Copyright © 2012 Pearson Education, Inc. Publishing as Addison-Wesley.

47. $2x^4 - 5x^2 - 3 = 0$

Let $u = x^2$, so that $u^2 = x^4$.

$2u^2 - 5u - 3 = 0$

$(2u + 1)(u - 3) = 0$

$2u + 1 = 0$ or $u - 3 = 0$

     $u = -\frac{1}{2}$ or $u = 3$

     $x^2 = -\frac{1}{2}$ or $x^2 = 3$

     $x = \pm\sqrt{-\frac{2}{4}}$ or $x = \pm\sqrt{3}$

     $x = \pm\frac{\sqrt{2}}{2}i$

The solution set is $\left\{-\sqrt{3}, \sqrt{3}, -\frac{\sqrt{2}}{2}i, \frac{\sqrt{2}}{2}i\right\}$.

13.3 Nonlinear Systems of Equations

13.3 Now Try Exercises

N1. $4x^2 + y^2 = 36$ (1)

      $x - y = 3$ (2)

Solve equation (2) for y.

$$y = x - 3 \quad (3)$$

Substitute $x - 3$ for y in equation (1).

$$4x^2 + y^2 = 36 \quad (1)$$
$$4x^2 + (x - 3)^2 = 36$$
$$4x^2 + (x^2 - 6x + 9) = 36$$
$$5x^2 - 6x - 27 = 0$$
$$(x - 3)(5x + 9) = 0$$

$x - 3 = 0$ or $5x + 9 = 0$

    $x = 3$ or $x = -\frac{9}{5}$

From (3) with $x = 3, y = 3 - 3 = 0$.

From (3) with $x = -\frac{9}{5}, y = -\frac{9}{5} - 3 = -\frac{24}{5}$.

Check $(3, 0)$: $36 + 0 = 36$; $3 - 0 = 3$

Check $\left(-\frac{9}{5}, -\frac{24}{5}\right)$: $\frac{324}{25} + \frac{576}{25} = 36$; $-\frac{9}{5} + \frac{24}{5} = 3$

The solution set is $\left\{(3, 0), \left(-\frac{9}{5}, -\frac{24}{5}\right)\right\}$.

N2. $xy = 2$ (1)

   $x - 3y = 1$ (2)

Solve equation (1) for y.

$$y = \frac{2}{x} \quad (3)$$

Substitute $\dfrac{2}{x}$ for y in equation (2).

$$x - 3y = 1 \quad (2)$$
$$x - 3\left(\frac{2}{x}\right) = 1$$
$$x - \frac{6}{x} = 1$$
$$x^2 - 6 = x \quad \textit{Multiply by x.}$$
$$x^2 - x - 6 = 0$$
$$(x - 3)(x + 2) = 0$$

$x - 3 = 0$ or $x + 2 = 0$

    $x = 3$ or $x = -2$

From (3) with $x = 3, y = \frac{2}{3}$.

From (3) with $x = -2, y = \frac{2}{-2} = -1$.

Check $\left(3, \frac{2}{3}\right)$: $3\left(\frac{2}{3}\right) = 2$; $3 - 2 = 1$

Check $(-2, -1)$: $(-2)(-1) = 2$; $-2 + 3 = 1$

The solution set is $\left\{(-2, -1), \left(3, \frac{2}{3}\right)\right\}$.

N3. $x^2 + y^2 = 16$ (1)

    $4x^2 + 13y^2 = 100$ (2)

Multiply equation (1) by -4 and add the result to (2).

$$\begin{array}{rll} -4x^2 - 4y^2 &= -64 & -4 \times (1) \\ 4x^2 + 13y^2 &= 100 & \\ \hline 9y^2 &= 36 & \textit{Add.} \\ y^2 &= 4 & \textit{Divide by 9.} \end{array}$$

$y = 2$ or $y = -2$ *Square root property*

Substitute 4 for y^2 in equation (1).

$$x^2 + 4 = 16$$
$$x^2 = 12 \qquad \textit{Add 5.}$$
$$x = 2\sqrt{3} \text{ or } x = -2\sqrt{3} \quad \textit{Square rt. prop.}$$

Check: If $x = \pm 2\sqrt{3}$, $x^2 = 12$, and if $y = \pm 2$, $y^2 = 4$. Thus, in any case, we get $12 + 4 = 16$ in (1) and $48 + 52 = 100$ in (2). All *four* ordered pairs check.

The solution set is $\left\{\left(2\sqrt{3}, 2\right), \left(2\sqrt{3}, -2\right), \left(-2\sqrt{3}, 2\right), \left(-2\sqrt{3}, -2\right)\right\}$.

N4. $x^2 + 3xy - y^2 = 23$ (1)

    $x^2 - y^2 = 5$ (2)

Multiply equation (2) by -1 and add the result to equation (1).

$$\begin{array}{rll} x^2 + 3xy - y^2 &= 23 & (1) \\ -x^2 + y^2 &= -5 & -1 \times (2) \\ \hline 3xy &= 18 & \\ xy &= 6 & \end{array}$$

Solving $xy = 6$ for y gives us $y = \frac{6}{x}$ (3).

Substitute $\frac{6}{x}$ for y in equation (2).

$$x^2 - y^2 = 5 \qquad (2)$$
$$x^2 - \frac{36}{x^2} = 5$$
$$x^4 - 36 = 5x^2$$
$$x^4 - 5x^2 - 36 = 0$$
$$(x^2 - 9)(x^2 + 4) = 0$$

$$x^2 - 9 = 0 \quad \text{or} \quad x^2 + 4 = 0$$
$$x^2 = 9 \quad \text{or} \quad x^2 = -4$$
$$x = \pm 3 \quad \text{or} \quad x = \pm 2i$$

If $x = \pm 3$, then from (3), $y = \pm \frac{6}{3} = \pm 2$.

If $x = \pm 2i$, then from (3),

$$y = \pm \frac{6}{2i} \cdot \frac{i}{i} = \pm \frac{3i}{-1} = \mp 3i.$$

The solution set is

$$\{(3, 2), (-3, -2), (2i, -3i), (-2i, 3i)\}.$$

13.3 Section Exercises

1. The line intersects the ellipse in exactly one point, so there is *one* point in the solution set of the system.

3. The line does not intersect the hyperbola, so there are no points in the solution set of the system.

5. A line and a circle; no points ▪ Draw any circle, and then draw a line that does not cross the circle.

7. A line and a hyperbola; one point ▪ The line is tangent to the hyperbola.

9. A circle and an ellipse; four points ▪ Draw any ellipse, and then draw a circle with the same center whose radius is just large enough so that there are four points of intersection. (If the radius of the circle is too large or too small, there may be fewer points of intersection.)

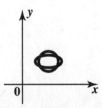

11. A parabola and an ellipse; four points ▪ Draw any parabola, and then draw an ellipse large enough so that there are four points of intersection. (If the ellipse is too large or too small, there may be fewer points of intersection.)

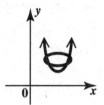

13. $y = 4x^2 - x$ (1)
     $y = x$ (2)

Substitute x for y in equation (1).

$$y = 4x^2 - x \qquad (1)$$
$$x = 4x^2 - x$$
$$0 = 4x^2 - 2x$$
$$0 = 2x(2x - 1)$$

$$2x = 0 \quad \text{or} \quad 2x - 1 = 0$$
$$x = 0 \quad \text{or} \quad x = \tfrac{1}{2}$$

Use equation (2) to find y for each x-value.

If $x = 0$, then $y = 0$.
If $x = \frac{1}{2}$, then $y = \frac{1}{2}$.

The solution set is $\left\{(0, 0), \left(\frac{1}{2}, \frac{1}{2}\right)\right\}$.

15. $y = x^2 + 6x + 9$ (1)
   $x + y = 3$ (2)

Substitute $x^2 + 6x + 9$ for y in equation (2).

$$x + y = 3 \quad (2)$$
$$x + (x^2 + 6x + 9) = 3$$
$$x^2 + 7x + 9 = 3$$
$$x^2 + 7x + 6 = 0$$
$$(x + 6)(x + 1) = 0$$

$$x + 6 = 0 \quad \text{or} \quad x + 1 = 0$$
$$x = -6 \quad \text{or} \quad x = -1$$

Substitute these values for x in equation (2) and solve for y.

If $x = -6$, then $x + y = 3$ (2)
           $-6 + y = 3$
               $y = 9.$

If $x = -1$, then $x + y = 3$ (2)
           $-1 + y = 3$
               $y = 4.$

The solution set is $\{(-6, 9), (-1, 4)\}$.

Copyright © 2012 Pearson Education, Inc. Publishing as Addison-Wesley.

17. $x^2 + y^2 = 2$ (1)
   $2x + y = 1$ (2)

Solve equation (2) for y.

$$y = 1 - 2x \quad (3)$$

Substitute $1 - 2x$ for y in equation (1).

$$x^2 + y^2 = 2 \quad (1)$$
$$x^2 + (1 - 2x)^2 = 2$$
$$x^2 + 1 - 4x + 4x^2 = 2$$
$$5x^2 - 4x - 1 = 0$$
$$(5x + 1)(x - 1) = 0$$

$$5x + 1 = 0 \quad \text{or} \quad x - 1 = 0$$
$$x = -\tfrac{1}{5} \quad \text{or} \quad x = 1$$

Use equation (3) to find y for each x-value.

If $x = -\tfrac{1}{5}$, then

$$y = 1 - 2\left(-\tfrac{1}{5}\right) = 1 + \tfrac{2}{5} = \tfrac{7}{5}.$$

If $x = 1$, then

$$y = 1 - 2(1) = -1.$$

The solution set is $\left\{\left(-\tfrac{1}{5}, \tfrac{7}{5}\right), (1, -1)\right\}$.

19. $xy = 4$ (1)
   $3x + 2y = -10$ (2)

Solve equation (1) for y to get $y = \tfrac{4}{x}$.

Substitute $\tfrac{4}{x}$ for y in equation (2) to find x.

$$3x + 2y = -10 \quad (2)$$
$$3x + 2\left(\frac{4}{x}\right) = -10$$

Multiply by the LCD, x.

$$3x^2 + 8 = -10x$$
$$3x^2 + 10x + 8 = 0$$
$$(3x + 4)(x + 2) = 0$$

$$3x + 4 = 0 \quad \text{or} \quad x + 2 = 0$$
$$x = -\tfrac{4}{3} \quad \text{or} \quad x = -2$$

Since $y = \tfrac{4}{x}$, if $x = -\tfrac{4}{3}$, then $y = \tfrac{4}{-\frac{4}{3}} = -3$.

If $x = -2$, then $y = \tfrac{4}{-2} = -2$.

The solution set is $\left\{(-2, -2), \left(-\tfrac{4}{3}, -3\right)\right\}$.

21. $xy = -3$ (1)
   $x + y = -2$ (2)

Solve equation (2) for y.

$$y = -x - 2 \quad (3)$$

Substitute $-x - 2$ for y in equation (1).

$$xy = -3 \quad (1)$$
$$x(-x - 2) = -3$$
$$-x^2 - 2x = -3$$
$$-x^2 - 2x + 3 = 0$$
$$x^2 + 2x - 3 = 0 \qquad \textit{Multiply by } -1.$$
$$(x + 3)(x - 1) = 0$$

$$x + 3 = 0 \quad \text{or} \quad x - 1 = 0$$
$$x = -3 \quad \text{or} \quad x = 1$$

Use equation (3) to find y for each x-value.

If $x = -3$, then

$$y = -(-3) - 2 = 1.$$

If $x = 1$, then

$$y = -(1) - 2 = -3.$$

The solution set is $\{(-3, 1), (1, -3)\}$.

23. $y = 3x^2 + 6x$ (1)
   $y = x^2 - x - 6$ (2)

Substitute $x^2 - x - 6$ for y in equation (1) to find x.

$$y = 3x^2 + 6x \qquad (1)$$
$$x^2 - x - 6 = 3x^2 + 6x$$
$$0 = 2x^2 + 7x + 6$$
$$0 = (2x + 3)(x + 2)$$

$$2x + 3 = 0 \quad \text{or} \quad x + 2 = 0$$
$$x = -\tfrac{3}{2} \quad \text{or} \quad x = -2$$

Substitute $-\tfrac{3}{2}$ for x in equation (1) to find y.

$$y = 3x^2 + 6x \qquad (1)$$
$$y = 3\left(-\tfrac{3}{2}\right)^2 + 6\left(-\tfrac{3}{2}\right)$$
$$= 3\left(\tfrac{9}{4}\right) + 6\left(-\tfrac{6}{4}\right)$$
$$= \tfrac{27}{4} - \tfrac{36}{4} = -\tfrac{9}{4}$$

Substitute -2 for x in equation (1) to find y.

$$y = 3x^2 + 6x \qquad (1)$$
$$y = 3(-2)^2 + 6(-2)$$
$$= 12 - 12 = 0$$

The solution set is $\left\{\left(-\tfrac{3}{2}, -\tfrac{9}{4}\right), (-2, 0)\right\}$.

25. $2x^2 - y^2 = 6$ (1)
   $y = x^2 - 3$ (2)

Substitute $x^2 - 3$ for y in equation (1).

$$2x^2 - y^2 = 6 \quad (1)$$
$$2x^2 - \left(x^2 - 3\right)^2 = 6$$
$$2x^2 - \left(x^4 - 6x^2 + 9\right) = 6$$
$$-x^4 + 8x^2 - 9 = 6$$
$$-x^4 + 8x^2 - 15 = 0$$
$$x^4 - 8x^2 + 15 = 0 \qquad \textit{Multiply by } -1.$$

Copyright © 2012 Pearson Education, Inc. Publishing as Addison-Wesley.

Let $z = x^2$, so $z^2 = x^4$.

$$z^2 - 8z + 15 = 0$$
$$(z - 3)(z - 5) = 0$$

$$z - 3 = 0 \quad \text{or} \quad z - 5 = 0$$
$$z = 3 \quad \text{or} \qquad z = 5$$

Since $z = x^2$,

$$x^2 = 3 \quad \text{or} \quad x^2 = 5.$$

Use the square root property.

$$x = \pm\sqrt{3} \quad \text{or} \quad x = \pm\sqrt{5}$$

Use equation (2) to find y for each x-value.

If $x = \sqrt{3}$ or $-\sqrt{3}$, then

$$y = \left(\pm\sqrt{3}\right)^2 - 3 = 0.$$

(Note: We could substitute 3 for x^2 in (2) to obtain the same result.)

If $x = \sqrt{5}$ or $-\sqrt{5}$, then

$$y = \left(\pm\sqrt{5}\right)^2 - 3 = 2.$$

The solution set is
$$\left\{ \left(-\sqrt{3}, 0\right), \left(\sqrt{3}, 0\right), \left(-\sqrt{5}, 2\right), \left(\sqrt{5}, 2\right) \right\}.$$

27. $\quad x^2 - xy + y^2 = 0 \qquad$ (1)
$\quad x - 2y = 1 \qquad\qquad$ (2)

Solve equation (2) for x.

$$x = 2y + 1 \quad (3)$$

Substitute $2y + 1$ for x in equation (1).

$$(2y + 1)^2 - (2y + 1)y + y^2 = 0$$
$$4y^2 + 4y + 1 - 2y^2 - y + y^2 = 0$$
$$3y^2 + 3y + 1 = 0$$

Use the quadratic formula with $a = 3$, $b = 3$, and $c = 1$.

$$y = \frac{-b \pm \sqrt{b^2 - 4ac}}{2a}$$
$$y = \frac{-3 \pm \sqrt{9 - 12}}{6}$$
$$= \frac{-3 \pm \sqrt{-3}}{6} = \frac{-3 \pm i\sqrt{3}}{6}$$

Use equation (3) to solve for x.

$$x = 2y + 1 \qquad\qquad (3)$$
$$= 2\left(\frac{-3 \pm i\sqrt{3}}{6}\right) + 1$$
$$= \frac{-3 \pm i\sqrt{3}}{3} + \frac{3}{3}$$
$$= \pm\frac{i\sqrt{3}}{3}$$

The solution set is
$$\left\{ \left(\tfrac{\sqrt{3}}{3}i, -\tfrac{1}{2} + \tfrac{\sqrt{3}}{6}i\right), \left(-\tfrac{\sqrt{3}}{3}i, -\tfrac{1}{2} - \tfrac{\sqrt{3}}{6}i\right) \right\}.$$

29. $\quad 3x^2 + 2y^2 = 12 \quad$ (1)
$\qquad x^2 + 2y^2 = 4 \quad\;$ (2)

Multiply equation (2) by -1 and add the result to equation (1).

$$\begin{array}{rcl} 3x^2 + 2y^2 &=& 12 \quad (1) \\ -x^2 - 2y^2 &=& -4 \qquad -1 \times (2) \\ \hline 2x^2 \quad\;\; &=& 8 \\ x^2 &=& 4 \\ x &=& \pm 2 \end{array}$$

Substitute ± 2 for x in equation (2) to find y.

$$x^2 + 2y^2 = 4 \qquad (2)$$
$$(\pm 2)^2 + 2y^2 = 4$$
$$4 + 2y^2 = 4$$
$$2y^2 = 0$$
$$y^2 = 0$$
$$y = 0$$

The solution set is $\{(-2, 0), (2, 0)\}$.

31. $\quad 2x^2 + 3y^2 = 6 \quad$ (1)
$\qquad x^2 + 3y^2 = 3 \quad$ (2)

Multiply equation (2) by -1 and add the result to equation (1).

$$\begin{array}{rcl} 2x^2 + 3y^2 &=& 6 \quad (1) \\ -x^2 - 3y^2 &=& -3 \qquad -1 \times (2) \\ \hline x^2 \qquad\;\; &=& 3 \\ x &=& \pm\sqrt{3} \end{array}$$

Substitute $\pm\sqrt{3}$ for x in equation (2).

$$x^2 + 3y^2 = 3 \qquad (2)$$
$$\left(\pm\sqrt{3}\right)^2 + 3y^2 = 3$$
$$3 + 3y^2 = 3$$
$$y^2 = 0$$
$$y = 0$$

The solution set is $\left\{ \left(\sqrt{3}, 0\right), \left(-\sqrt{3}, 0\right) \right\}$.

33. $\quad 2x^2 + y^2 = 28 \quad$ (1)
$\qquad 4x^2 - 5y^2 = 28 \quad$ (2)

Multiply (1) by 5 and add the result to equation (2).

$$\begin{array}{rcl} 10x^2 + 5y^2 &=& 140 \qquad 5 \times (1) \\ 4x^2 - 5y^2 &=& 28 \quad (2) \\ \hline 14x^2 \qquad\;\; &=& 168 \\ x^2 &=& 12 \\ x &=& \pm\sqrt{12} = \pm 2\sqrt{3} \end{array}$$

Copyright © 2012 Pearson Education, Inc. Publishing as Addison-Wesley.

Let $x = \pm 2\sqrt{3}$ in (1).

$$2\left(\pm 2\sqrt{3}\right)^2 + y^2 = 28$$
$$2(12) + y^2 = 28$$
$$y^2 = 4$$
$$y = \pm 2$$

The solution set is $\left\{ \left(2\sqrt{3}, 2\right), \left(2\sqrt{3}, -2\right), \right.$

$$\left. \left(-2\sqrt{3}, 2\right), \left(-2\sqrt{3}, -2\right) \right\}.$$

35. $2x^2 = 8 - 2y^2$ (1)
      $3x^2 = 24 - 4y^2$ (2)

Multiply equation (1) by -2 and add the result to equation (2).

$$\begin{array}{rcl} -4x^2 &=& -16 + 4y^2 \qquad -2 \times (1) \\ \underline{3x^2 &=& 24 - 4y^2 \qquad (2)} \\ -x^2 &=& 8 \\ x^2 &=& -8 \\ x &=& \pm\sqrt{-8} = \pm 2i\sqrt{2} \end{array}$$

Substitute $\pm 2i\sqrt{2}$ for x in equation (1).

$$2x^2 = 8 - 2y^2 \quad (1)$$
$$2\left(\pm 2i\sqrt{2}\right)^2 = 8 - 2y^2$$
$$2(-8) = 8 - 2y^2$$
$$-16 = 8 - 2y^2$$
$$2y^2 = 24$$
$$y^2 = 12$$
$$y = \pm\sqrt{12} = \pm 2\sqrt{3}$$

Since $2i\sqrt{2}$ can be paired with either $2\sqrt{3}$ or $-2\sqrt{3}$ and $-2i\sqrt{2}$ can be paired with either $2\sqrt{3}$ or $-2\sqrt{3}$, there are four possible solutions.

The solution set is

$$\left\{ \left(-2i\sqrt{2}, -2\sqrt{3}\right), \left(-2i\sqrt{2}, 2\sqrt{3}\right), \right.$$

$$\left. \left(2i\sqrt{2}, -2\sqrt{3}\right), \left(2i\sqrt{2}, 2\sqrt{3}\right) \right\}.$$

37. $x^2 + xy + y^2 = 15$ (1)
           $x^2 + y^2 = 10$ (2)

Multiply equation (2) by -1 and add the result to equation (1).

$$\begin{array}{rcl} x^2 + xy + y^2 &=& 15 \quad (1) \\ \underline{-x^2 \qquad - y^2 &=& -10 \qquad -1 \times (2)} \\ xy &=& 5 \\ y &=& \dfrac{5}{x} \end{array}$$

Substitute $\frac{5}{x}$ for y in equation (2).

$$x^2 + y^2 = 10 \quad (2)$$
$$x^2 + \left(\frac{5}{x}\right)^2 = 10$$
$$x^2 + \frac{25}{x^2} = 10$$
$$x^4 + 25 = 10x^2 \quad \textit{Multiply by } x^2.$$
$$x^4 - 10x^2 + 25 = 0$$

Let $z = x^2$, so $z^2 = x^4$.

$$z^2 - 10z + 25 = 0$$
$$(z - 5)^2 = 0$$
$$z - 5 = 0$$
$$z = 5$$

Since $z = x^2$,

$$x^2 = 5, \text{ and } x = \pm\sqrt{5}.$$

Using the equation $y = \frac{5}{x}$, we get the following.

If $x = -\sqrt{5}$, then

$$y = \frac{5}{-\sqrt{5}} = \frac{5 \cdot \sqrt{5}}{-\sqrt{5} \cdot \sqrt{5}} = \frac{5\sqrt{5}}{-5} = -\sqrt{5}.$$

Similarly, if $x = \sqrt{5}$, then $y = \sqrt{5}$.

The solution set is $\left\{ \left(-\sqrt{5}, -\sqrt{5}\right), \left(\sqrt{5}, \sqrt{5}\right) \right\}$.

39. $3x^2 + 2xy - 3y^2 = 5$ (1)
        $-x^2 - 3xy + y^2 = 3$ (2)

Multiply equation (2) by 3 and add the result to equation (1).

$$\begin{array}{rcl} 3x^2 + 2xy - 3y^2 &=& 5 \quad (1) \\ \underline{-3x^2 - 9xy + 3y^2 &=& 9 \qquad 3 \times (2)} \\ -7xy &=& 14 \\ x &=& \dfrac{14}{-7y} = -\dfrac{2}{y} \end{array}$$

Substitute $-\frac{2}{y}$ for x in equation (2).

$$-x^2 - 3xy + y^2 = 3 \quad (2)$$
$$-\left(-\frac{2}{y}\right)^2 - 3\left(-\frac{2}{y}\right)y + y^2 = 3$$
$$-\left(\frac{4}{y^2}\right) + 6 + y^2 = 3$$
$$y^2 + 3 - \frac{4}{y^2} = 0$$
$$y^4 + 3y^2 - 4 = 0 \quad \textit{Multiply by } y^2.$$
$$(y^2 + 4)(y^2 - 1) = 0$$

$$\begin{array}{ccc} y^2 + 4 = 0 & \text{or} & y^2 - 1 = 0 \\ y^2 = -4 & & y^2 = 1 \\ y = \pm 2i & \text{or} & y = \pm 1 \end{array}$$

Since $x = -\frac{2}{y}$, substitute these values for y to find the values of x.

Copyright © 2012 Pearson Education, Inc. Publishing as Addison-Wesley.

If $y = 2i$, then

$$x = -\frac{2}{2i} = -\frac{1}{i} = -\frac{1}{i} \cdot \frac{i}{i} = \frac{-i}{-1} = i.$$

If $y = -2i$, then

$$x = -\frac{2}{-2i} = \frac{1}{i} = \frac{1}{i} \cdot \frac{i}{i} = \frac{i}{-1} = -i.$$

If $y = 1$, then $x = -\frac{2}{1} = -2$.

If $y = -1$, then $x = -\frac{2}{-1} = 2$.

The solution set is
$\{(i, 2i), (-i, -2i), (2, -1), (-2, 1)\}$.

41. $\quad xy = -6 \quad (1)$
$\quad x + y = -1 \quad (2)$

Solve both equations for y.

$$y = -\frac{6}{x} \quad \text{and} \quad y = -x - 1$$

Graph

$$Y_1 = -\frac{6}{X} \quad \text{and} \quad Y_2 = -X - 1$$

on a graphing calculator to obtain the solution set
$\{(2, -3), (-3, 2)\}$.

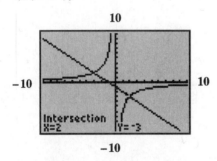

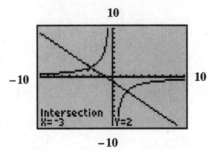

Now solve the system algebraically using substitution.

$$xy = -6 \quad (1)$$
$$x + y = -1 \quad (2)$$

Solve equation (2) for y.

$$y = -1 - x \quad (3)$$

Substitute $-1 - x$ for y in equation (1).

$$x(-1 - x) = -6$$
$$-x - x^2 = -6$$
$$0 = x^2 + x - 6$$
$$0 = (x + 3)(x - 2)$$

$$x + 3 = 0 \quad \text{or} \quad x - 2 = 0$$
$$x = -3 \quad \text{or} \quad x = 2$$

Substitute these values for x in (3) to find y.

If $x = -3$, then $y = -1 - (-3) = 2$.

If $x = 2$, then $y = -1 - 2 = -3$.

The solution set, $\{(2, -3), (-3, 2)\}$, is the same as
that obtained using a graphing calculator.

43. Let $W = $ the width, and $L = $ the length.

The formula for the area of a rectangle is
$LW = \mathcal{A}$, so

$$LW = 84. \quad (1)$$

The perimeter of a rectangle is given by
$2L + 2W = P$, so

$$2L + 2W = 38. \quad (2)$$

Solve equation (2) for L to get

$$L = 19 - W. \quad (3)$$

Substitute $19 - W$ for L in equation (1).

$$LW = 84 \quad (1)$$
$$(19 - W)W = 84$$
$$19W - W^2 = 84$$
$$-W^2 + 19W - 84 = 0$$
$$W^2 - 19W + 84 = 0 \qquad \textit{Multiply by } -1.$$
$$(W - 7)(W - 12) = 0$$

$$W - 7 = 0 \quad \text{or} \quad W - 12 = 0$$
$$W = 7 \quad \text{or} \quad W = 12$$

Using equation (3), with $W = 7$,

$$L = 19 - 7 = 12.$$

If $W = 12$, then $L = 7$, which are the same two
numbers. Length must be greater than width, so
the length is 12 feet and the width is 7 feet.

45. $\quad px = 16 \qquad (1)$
$\quad p = 10x + 12 \qquad (2)$

Substitute $10x + 12$ for p in equation (1).

$$px = 16 \qquad (1)$$
$$(10x + 12)x = 16$$
$$10x^2 + 12x = 16$$
$$10x^2 + 12x - 16 = 0$$
$$5x^2 + 6x - 8 = 0$$
$$(5x - 4)(x + 2) = 0$$

$$5x - 4 = 0 \quad \text{or} \quad x + 2 = 0$$
$$x = \tfrac{4}{5} \quad \text{or} \quad x = -2$$

Since x cannot be negative, eliminate -2 as a value of x. Substitute $\tfrac{4}{5}$ for x in equation (2) to find p.

$$p = 10x + 12 \qquad (2)$$
$$p = 10\left(\tfrac{4}{5}\right) + 12$$
$$= 8 + 12 = 20$$

Since the demand x is in thousands,

$$\text{demand} = \tfrac{4}{5}(1000) = 800.$$

The equilibrium price is \$20. The supply/demand at that price is 800 calculators.

47. $2x - y \leq 4$

Draw a solid line through the intercepts $(2, 0)$ and $(0, -4)$. Checking $(0, 0)$ gives us $0 \leq 4$, a true statement, so shade the side containing $(0, 0)$.

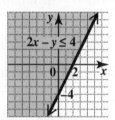

13.4 Second-Degree Inequalities, Systems of Inequalities, and Linear Programming

13.4 Now Try Exercises

N1. $x^2 + y^2 \geq 9$

The boundary, $x^2 + y^2 = 9$, is a circle with radius 3. It is drawn as a solid curve because of the $\geq$ symbol. Use the point $(0, 0)$ as a test point.

$$x^2 + y^2 \geq 9$$
$$0^2 + 0^2 \overset{?}{\geq} 9$$
$$0 \geq 9 \quad \textit{False}$$

Shade the region that does not contain $(0, 0)$. This is the region outside the circle.

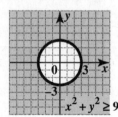

N2. $y \geq -(x + 2)^2 + 1$

The boundary, $y = -(x + 2)^2 + 1$ is a parabola opening downward with its vertex at $(-2, 1)$. It is

drawn as a solid curve because of the $\geq$ symbol. Use the point $(0, 0)$ as a test point.

$$0 \overset{?}{\geq} -(0 + 2)^2 + 1$$
$$0 \overset{?}{\geq} -4 + 1$$
$$0 \geq -3 \qquad \textit{True}$$

Shade the region that contains $(0, 0)$. This is the region outside the parabola.

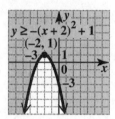

N3. $25x^2 - 16y^2 > 400$

$$\frac{x^2}{16} - \frac{y^2}{25} > 1 \qquad \textit{Divide by 400.}$$

The boundary is the hyperbola

$$\frac{x^2}{16} - \frac{y^2}{25} = 1,$$

which has intercepts $(4, 0)$ and $(-4, 0)$. It is drawn as a dashed curve because of the $>$ symbol. Test $(0, 0)$.

$$25(0)^2 - 16(0)^2 \overset{?}{>} 400$$
$$0 > 400 \quad \textit{False}$$

Shade the region that does not contain $(0, 0)$. This is the region inside the branches of the hyperbola.

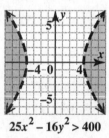

N4. $x^2 + y^2 > 9$
$$y > x^2 - 1$$

The boundary, $x^2 + y^2 = 9$, is a circle with radius 3. Draw this boundary as a dashed curve. Test $(0, 0)$.

$$0^2 + 0^2 \overset{?}{>} 9$$
$$0 > 9 \quad \textit{False}$$

Shade the region outside the circle.
The boundary $y = x^2 - 1$ is an upward opening parabola with vertical axis $x = 0$ and vertex $(0, -1)$. Draw this boundary as a dashed curve. Test $(0, 0)$.

Copyright © 2012 Pearson Education, Inc. Publishing as Addison-Wesley.

$$0 \overset{?}{\geq} 0^2 - 1$$
$$0 \geq -1 \qquad True$$

Shade the region inside the parabola.
The graph of the solution set of the system includes all points common to the two graphs, that is, the intersection of the two shaded regions.

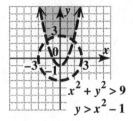

N5. $3x + 2y > 6$
$\qquad y \geq \frac{1}{2}x - 2$
$\qquad x \geq 0$

The boundary $3x + 2y = 6$ is a dashed line with intercepts $(2, 0)$ and $(0, 3)$. Test $(0, 0)$.

$$3(0) + 2(0) \overset{?}{>} 6$$
$$0 > 6 \qquad False$$

Shade the side of the line that does not contain $(0, 0)$.

The boundary $y = \frac{1}{2}x - 2$ is a solid line with intercepts $(4, 0)$ and $(0, -2)$. Test $(0, 0)$.

$$0 \overset{?}{\geq} \frac{1}{2}(0) - 2$$
$$0 \geq -2 \qquad True$$

Shade the side of the line that contains $(0, 0)$.

The boundary $x = 0$ is a solid vertical line through $(0, 0)$. Shade the region to the right of the line since $x \geq 0$.

The intersection of the three shaded regions is the graph of the solution set of the system.

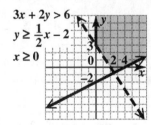

N6. $\dfrac{x^2}{4} + \dfrac{y^2}{16} \leq 1$
$\qquad y \leq x^2 - 2$
$\qquad y + 3 > 0$

The boundary $\dfrac{x^2}{4} + \dfrac{y^2}{16} = 1$ is an ellipse with intercepts $(2, 0)$, $(-2, 0)$, $(0, 4)$, and $(0, -4)$. It is drawn as a solid curve because of the $\leq$ symbol. Test $(0, 0)$.

$$\frac{0^2}{4} + \frac{0^2}{16} \overset{?}{\leq} 1$$
$$0 \leq 1 \qquad True$$

Shade the region that contains $(0, 0)$. This is the region inside the ellipse.

The boundary $y = x^2 - 2$ is a parabola opening upward with vertex at $(0, -2)$. It is drawn as a solid curve because of the $\leq$ symbol. Test $(0, 0)$.

$$0 \overset{?}{\leq} 0^2 - 2$$
$$0 \leq -2 \qquad False$$

Shade the region that does not contain $(0, 0)$. This is the region outside the parabola.

The boundary $y + 3 = 0$, or $y = -3$, is a dashed horizontal line through $(0, -3)$. Shade the region above the line since $y + 3 > 0$.

The graph of the solution set of the system includes all points common to the three graphs, that is, the intersection of the three shaded regions.

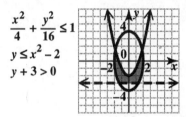

N7. Let x = the amount of money invested in Treasury bills, and y = the amount in corporate bonds.

The total amount cannot exceed $21,000, so one constraint is

$$x + y \leq 21{,}000.$$

Hazel invests no more than twice as much money in bonds as she does in Treasury bills, so another constraint is

$$y \leq 2x.$$

Since x and y cannot be negative, two further constraints are

$$x \geq 0 \quad \text{and} \quad y \geq 0.$$

Thus, the inequalities that we consider for this system are

$$x + y \leq 21{,}000$$
$$y \leq 2x$$
$$x \geq 0$$
$$y \geq 0.$$

Below is a graph of the system. Solve systems of equations to find the vertices. For example, solving the system $x + y = 21{,}000$ and $y = 2x$ gives us the solution $x = 7000$ and $y = 14{,}000$.

Copyright © 2012 Pearson Education, Inc. Publishing as Addison-Wesley.

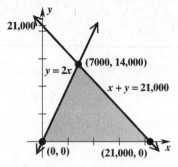

The vertices of the feasible region are $(0, 0)$, $(21,000, 0)$, and $(7000, 14,000)$.

Since the Treasury bills yield 3% simple interest and the corporate bonds yield 4%, the objective function is $0.03x + 0.04y$.

The following table shows that $(7000, 14,000)$ corresponds to the maximum value of the objective function.

Point	Income $= 0.03x + 0.04y$
$(0, 0)$	$0.03(0) + 0.04(0) = 0$
$(21,000, 0)$	$0.03(21,000) + 0.04(0) = 630$
$(7000, 14,000)$	$0.03(7000) + 0.04(14,000) = 770 \leftarrow$

Therefore, the maximum yearly income from these two investments will be $770 if $7000 is invested in Treasury bills and $14,000 is invested in corporate bonds.

N8. *Step 1*
Let $x =$ the number of medical kits, and $y =$ the number of containers of water.

The total weight cannot exceed 80,000 pounds. Each medical kit weighs 10 lbs and each container of water weighs 20 lbs, so one constraint is

$$10x + 20y \leq 80,000, \quad \text{or} \quad x + 2y \leq 8000.$$

The total volume cannot exceed 6000 cubic feet. Each medical kit takes up 1 ft^3 and each container of water takes up 1 ft^3, so another constraint is

$$x + y \leq 6000.$$

Since x and y cannot be negative, two further constraints are

$$x \geq 0 \quad \text{and} \quad y \geq 0.$$

Step 2
Thus, the inequalities that we consider for this system are

$$x + 2y \leq 8000$$
$$x + y \leq 6000$$
$$x \geq 0$$
$$y \geq 0.$$

Below is a graph of the system. Solve systems of equations to find the vertices. For example,

solving the system $x + 2y = 8000$ and $x + y = 6000$ gives us the solution $x = 4000$ and $y = 2000$.

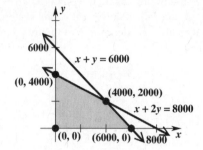

Step 3
The vertices of the feasible region are $(0, 0)$, $(0, 4000)$, $(6000, 0)$, and $(4000, 2000)$.

Step 4
Since each medical kit will aid 6 people and each container of water will serve 10 people, the objective function is $6x + 10y$.

The following table shows that $(4000, 2000)$ corresponds to the maximum value of the objective function.

Point	Value $= 6x + 10y$
$(0, 0)$	$6(0) + 10(0) = 0$
$(0, 4000)$	$6(0) + 10(4000) = 40,000$
$(6000, 0)$	$6(6000) + 10(0) = 36,000$
$(4000, 2000)$	$6(4000) + 10(2000) = 44,000 \leftarrow$

Step 5
Therefore, 44,000 people will be assisted if 4000 medical kits and 2000 containers of water are sent.

13.4 Section Exercises

1. $x^2 + y^2 < 25$
     $y > -2$

The boundary, $x^2 + y^2 = 25$, is a circle (dashed because of the $<$ sign) with center $(0, 0)$ and radius 5. When $(0, 0)$ is tested, a true statement, $0 < 25$, results, so the inside of the circle is shaded. The graph of $y = -2$ is a horizontal line through $(0, -2)$ with shading above the dashed line, since $y > -2$. The correct answer is **C**.

3. $y \geq x^2 + 4$

This is an inequality whose boundary is a solid parabola, opening up, with vertex $(0, 4)$. The inside of the parabola is shaded since $(0, 0)$ makes the inequality false. This is graph **B**.

5. $y < x^2 + 4$

This is an inequality whose boundary is a dashed parabola, opening up, with vertex $(0, 4)$. The outside of the parabola is shaded since $(0, 0)$ makes the inequality true. This is graph **A**.

Copyright © 2012 Pearson Education, Inc. Publishing as Addison-Wesley.

7.
$$y^2 > 4 + x^2$$
$$y^2 - x^2 > 4$$
$$\frac{y^2}{4} - \frac{x^2}{4} > 1$$

The boundary, $\frac{y^2}{4} - \frac{x^2}{4} = 1$, is a hyperbola with y-intercepts $(0,2)$ and $(0,-2)$ and asymptotes formed by the extended diagonals of the fundamental rectangle with vertices at $(2,2)$, $(2,-2)$, $(-2,-2)$, and $(-2,2)$. The hyperbola has dashed branches because of $>$. Test $(0,0)$.

$$0^2 \overset{?}{>} 4 + 0^2$$
$$0 > 4 \qquad \textit{False}$$

Shade the sides of the hyperbola that do not contain $(0,0)$. These are the regions inside the branches of the hyperbola.

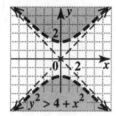

9.
$$y \geq x^2 - 2$$
$$y \geq (x - 0)^2 - 2$$

Graph the solid vertical parabola $y = x^2 - 2$ with vertex $(0,-2)$. Two other points on the parabola are $(2,2)$ and $(-2,2)$. Test a point not on the parabola, say $(0,0)$, in $y \geq x^2 - 2$ to get $0 \geq -2$, a true statement. Shade that portion of the graph that contains the point $(0,0)$. This is the region inside the parabola.

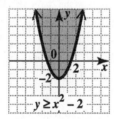

11.
$$2y^2 \geq 8 - x^2$$
$$x^2 + 2y^2 \geq 8$$
$$\frac{x^2}{8} + \frac{y^2}{4} \geq 1$$

The boundary, $\frac{x^2}{8} + \frac{y^2}{4} = 1$, is the ellipse with intercepts $\left(2\sqrt{2},0\right)$, $\left(-2\sqrt{2},0\right)$, $(0,2)$, and $(0,-2)$, drawn as a solid curve because of $\geq$. Test $(0,0)$.

$$2(0)^2 \overset{?}{\geq} 8 - 0^2$$
$$0 \geq 8 \qquad \textit{False}$$

Shade the region of the ellipse that does not contain $(0,0)$. This is the region outside the ellipse.

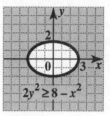

13. $y \leq x^2 + 4x + 2$

Graph the solid vertical parabola $y = x^2 + 4x + 2$. Use the vertex formula $x = \frac{-b}{2a}$ to obtain the vertex $(-2,-2)$. Two other points on the parabola are $(0,2)$ and $(1,7)$. Test a point not on the parabola, say $(0,0)$, in $y \leq x^2 + 4x + 2$ to get $0 \leq 2$, a true statement. Shade outside the parabola, since this region contains $(0,0)$.

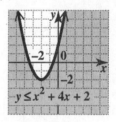

15.
$$9x^2 > 16y^2 + 144$$
$$9x^2 - 16y^2 > 144$$
$$\frac{x^2}{16} - \frac{y^2}{9} > 1$$

The boundary, $\frac{x^2}{16} - \frac{y^2}{9} = 1$, is a hyperbola with x-intercepts $(4,0)$ and $(-4,0)$ and asymptotes formed by the extended diagonals of the fundamental rectangle with vertices at $(4,3)$, $(4,-3)$, $(-4,-3)$, and $(-4,3)$. The hyperbola has dashed branches because of $>$. Test $(0,0)$.

$$9(0)^2 \overset{?}{>} 16(0)^2 + 144$$
$$0 > 144 \qquad \textit{False}$$

Shade the sides of the hyperbola that do not contain $(0,0)$. These are the regions inside the branches of the hyperbola.

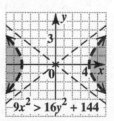

Copyright © 2012 Pearson Education, Inc. Publishing as Addison-Wesley.

17. $x^2 - 4 \geq -4y^2$

$x^2 + 4y^2 \geq 4$

$\dfrac{x^2}{4} + \dfrac{y^2}{1} \geq 1$

Graph the solid ellipse $\frac{x^2}{4} + \frac{y^2}{1} = 1$ through the x-intercepts $(2, 0)$ and $(-2, 0)$ and y-intercepts $(0, 1)$ and $(0, -1)$. Test a point not on the ellipse, say $(0, 0)$, in $x^2 - 4 \geq -4y^2$ to get $-4 \geq 0$, a false statement. Shade outside the ellipse, since this region does *not* include $(0, 0)$.

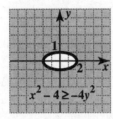

19. $x \leq -y^2 + 6y - 7$

Complete the square to find the vertex.

$$x = -y^2 + 6y - 7$$
$$= -\left(y^2 - 6y\right) - 7$$
$$= -\left(y^2 - 6y + 9\right) + 9 - 7$$
$$x = -(y - 3)^2 + 2$$

The boundary, $x = -(y - 3)^2 + 2$, is a solid horizontal parabola with vertex at $(2, 3)$ that opens to the left. Test $(0, 0)$.

$$0 \overset{?}{\leq} -0^2 + 6(0) - 7$$
$$0 \leq -7 \qquad \textit{False}$$

Shade the region of the parabola that does not contain $(0, 0)$. This is the region inside the parabola.

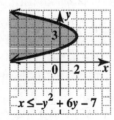

21. $2x + 5y < 10$

$\quad\;\; x - 2y < 4$

Graph $2x + 5y = 10$ as a dashed line through $(5, 0)$ and $(0, 2)$. Test $(0, 0)$.

$$2x + 5y < 10$$
$$2(0) + 5(0) \overset{?}{<} 10$$
$$0 < 10 \quad \textit{True}$$

Shade the region containing $(0, 0)$.
Graph $x - 2y = 4$ as a dashed line through $(4, 0)$ and $(0, -2)$. Test $(0, 0)$.

$$x - 2y < 4$$
$$0 - 2(0) \overset{?}{<} 4$$
$$0 < 4 \quad \textit{True}$$

Shade the region containing $(0, 0)$. The graph of the system is the intersection of the two shaded regions.

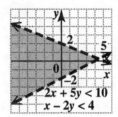

23. $5x - 3y \leq 15$

$\quad\; 4x + y \geq 4$

The boundary, $5x - 3y = 15$, is a solid line with intercepts $(3, 0)$ and $(0, -5)$. Test $(0, 0)$.

$$5(0) - 3(0) \overset{?}{\leq} 15$$
$$0 \leq 15 \quad \textit{True}$$

Shade the side of the line that contains $(0, 0)$. The boundary, $4x + y = 4$, is a solid line with intercepts $(1, 0)$ and $(0, 4)$. Test $(0, 0)$.

$$4(0) + 0 \overset{?}{\geq} 4$$
$$0 \geq 4 \quad \textit{False}$$

Shade the side of the line that does not contain $(0, 0)$.
The graph of the system is the intersection of the two shaded regions.

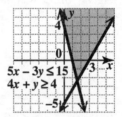

25. $x \leq 5$

$\quad y \leq 4$

Graph $x = 5$ as a solid vertical line through $(5, 0)$. Since $x \leq 5$, shade the left side of the line.
Graph $y = 4$ as a solid horizontal line through $(0, 4)$. Since $y \leq 4$, shade below the line.
The graph of the system is the intersection of the two shaded regions.

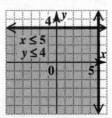

27. $y > x^2 - 4$
$\quad\ y < -x^2 + 3$

The boundary, $y = x^2 - 4$, is a dashed parabola with vertex $(0, -4)$ that opens up. Test $(0, 0)$.

$$0 \overset{?}{>} 0^2 - 4$$
$$0 > -4 \qquad \textit{True}$$

Shade the side of the parabola that contains $(0, 0)$. This is the region inside the parabola.
The boundary, $y = -x^2 + 3$, is a dashed parabola with vertex $(0, 3)$ that opens down. Test $(0, 0)$.

$$0 \overset{?}{<} -0^2 + 3$$
$$0 < 3 \qquad \textit{True}$$

Shade the side of the parabola that contains $(0, 0)$. This is the region inside the parabola.
The graph of the system is the intersection of the two shaded regions.

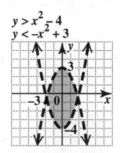

29. $x^2 + y^2 \geq 4 \quad (1)$
$\qquad x + y \leq 5 \quad (2)$
$\qquad\quad x \geq 0 \quad (3)$
$\qquad\quad y \geq 0 \quad (4)$

The graph of (1) is the region outside the solid circle boundary $x^2 + y^2 = 4$ with center $(0, 0)$ and radius 2.
The graph of (2), equivalent to $y \leq -x + 5$, is the region below the solid line $x + y = 5$ with intercepts $(5, 0)$ and $(0, 5)$.
The intersection of the graphs of (3) and (4) is quadrant I where both x and y are positive. The positive x- and y-axes are included.
The graph of the system is the intersection of the graphs of (1) and (2) in quadrant I.

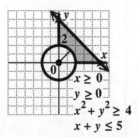

31. $y \leq -x^2$
$\quad\ y \geq x - 3$
$\quad\ y \leq -1$
$\quad\ x < 1$

The boundary, $y = -x^2$, is a solid parabola with vertex at $(0, 0)$ that opens down. Test $(0, -1)$.

$$-1 \overset{?}{\leq} -(0)^2$$
$$-1 \leq 0 \qquad \textit{True}$$

Shade the side of the parabola that contains $(0, -1)$. This is the region inside the parabola.
The boundary, $y = x - 3$, is a solid line with intercepts $(3, 0)$ and $(0, -3)$. Test $(0, 0)$.

$$0 \overset{?}{\geq} 0 - 3$$
$$0 \geq -3 \qquad \textit{True}$$

Shade the side of the line that contains $(0, 0)$.
For $y \leq -1$, shade below the solid horizontal line $y = -1$.
For $x < 1$, shade to the left of the dashed vertical line $x = 1$.
The intersection of the four shaded regions is the graph of the system.

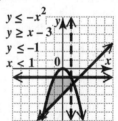

33. $x^2 + y^2 > 36,\ x \geq 0$

This is a circle of radius 6 centered at the origin. The graph is a dashed curve. Since $(0, 0)$ does not satisfy the inequality, the shading will be outside the circle. By including the restriction $x \geq 0$, we consider only the shading to the right of, and including, the y-axis.

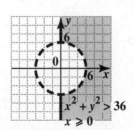

Copyright © 2012 Pearson Education, Inc. Publishing as Addison-Wesley.

35. $x < y^2 - 3$, $x < 0$

Consider the equation.

$$x = y^2 - 3, \text{ or}$$
$$x = (y - 0)^2 - 3.$$

This is a parabola with vertex $(-3, 0)$, opening to the right having the same shape as $x = y^2$. The graph is a dashed curve. The shading will be outside the parabola, since $(0, 0)$ does not satisfy the inequality. By including the restriction $x < 0$, we consider only the shading to the left of, but not including, the y-axis. The y-axis is a dashed line.

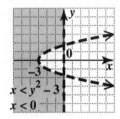

37. $4x^2 - y^2 > 16$, $x < 0$

Consider the equation

$$4x^2 - y^2 = 16, \text{ or}$$
$$\frac{x^2}{4} - \frac{y^2}{16} = 1.$$

This is a hyperbola with x-intercepts $(2, 0)$ and $(-2, 0)$. The graph is a dashed curve. The shading will be to the left of the left branch and to the right of the right branch of the hyperbola, since $(0, 0)$ does not satisfy the inequality. By including the restriction $x < 0$, we consider only the shading to the left of the y-axis.

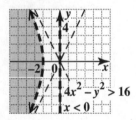

39. $x^2 + 4y^2 \geq 1$, $x \geq 0$, $y \geq 0$

Consider the equation

$$x^2 + 4y^2 = 1, \text{ or}$$
$$\frac{x^2}{1} + \frac{y^2}{\frac{1}{4}} = 1.$$

The graph is a solid ellipse with x-intercepts $(1, 0)$ and $(-1, 0)$ and y-intercepts $\left(0, \frac{1}{2}\right)$ and $\left(0, -\frac{1}{2}\right)$. The shading will be outside the ellipse, since $(0, 0)$ does not satisfy the inequality. By including the restrictions $x \geq 0$ and $y \geq 0$, we consider only the shading in quadrant I and portions of the x- and y-axis.

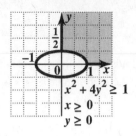

41. $y \geq x - 3$
$y \leq -x + 4$

The graphs of both inequalities include the points on the lines as part of the solution because of $\geq$ and $\leq$.

To produce the graph of the system on the TI-83/4, make the following Y-assignments:

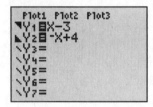

To get the upper and lower darkened triangles to the left of Y_1 and Y_2, simply place the cursor in that spot and press $\boxed{\text{ENTER}}$ to cycle through the graphing choices.

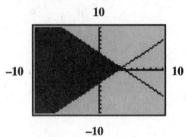

43. $y < x^2 + 4x + 4$
$y > -3$

The graphs do *not* include the points on the parabola or the line as part of the solution because of $<$ and $>$.

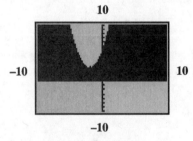

45.

Point	Value $= 3x + 5y$
$(1, 1)$	$3(1) + 5(1) = 8 \leftarrow$
$(2, 7)$	$3(2) + 5(7) = 41$
$(5, 10)$	$3(5) + 5(10) = 65 \leftarrow$
$(6, 3)$	$3(6) + 5(3) = 33$

The maximum value is 65 and the minimum value is 8.

Copyright © 2012 Pearson Education, Inc. Publishing as Addison-Wesley.

47.

Point	Value $= 40x + 75y$
$(0,0)$	$40(0) + 75(0) = 0$ ←
$(0,12)$	$40(0) + 75(12) = 900$ ←
$(4,8)$	$40(4) + 75(8) = 760$
$(8,3)$	$40(8) + 75(3) = 545$
$(9,0)$	$40(9) + 75(0) = 360$

The maximum value is 900 and the minimum value is 0.

49. Maximize $5x + 2y$ subject to

$$2x + 3y \leq 6 \quad \big| \quad x \geq 0$$
$$4x + y \leq 6 \quad \big| \quad y \geq 0$$

Graph the solid lines $x = 0$, $y = 0$, $2x + 3y = 6$ and $4x + y = 6$. Notice that three vertices are at $(0,0)$, $(0,2)$, and $\left(\frac{3}{2}, 0\right)$.

To find the point of intersection of

$$2x + 3y = 6 \textbf{ (1)} \quad \text{and} \quad 4x + y = 6 \textbf{ (2)},$$

solve this system by elimination.

$$\begin{array}{rll} 2x + 3y = & 6 & (1) \\ -12x - 3y = & -18 & -3 \times (2) \\ \hline -10x = & -12 & \textit{Add.} \\ x = & \frac{6}{5} & \end{array}$$

To find y, substitute $\frac{6}{5}$ for x in (2).

$$4\left(\tfrac{6}{5}\right) + y = 6, \text{ so } y = 6 - \tfrac{24}{5} = \tfrac{6}{5}.$$

Thus, $\left(\frac{6}{5}, \frac{6}{5}\right)$ is the fourth vertex.
A graph of the solution is shown on the next page.

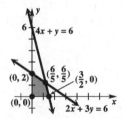

Substitute the coordinates of each vertex into the expression $5x + 2y$ to see where this value is maximized.

Point	Value $= 5x + 2y$
$(0,0)$	$5(0) + 2(0) = 0$
$(0,2)$	$5(0) + 2(2) = 4$
$\left(\frac{6}{5}, \frac{6}{5}\right)$	$5\left(\frac{6}{5}\right) + 2\left(\frac{6}{5}\right) = \frac{42}{5}$ ←
$\left(\frac{3}{2}, 0\right)$	$5\left(\frac{3}{2}\right) + 2(0) = \frac{15}{2}$

The maximum value is $\frac{42}{5}$ at $\left(\frac{6}{5}, \frac{6}{5}\right)$.

51. Minimize $2x + y$ subject to

$$3x - y \geq 12 \quad \big| \quad x \geq 2$$
$$x + y \leq 15 \quad \big| \quad y \geq 5$$

The intersection of the above constraints is the triangular region bounded by the lines $3x - y = 12$, $y = 5$, and $x + y = 15$. The

vertices of the triangle can be obtained as follows. Substitute $y = 5$ into $3x - y = 12$.

$$3x - 5 = 12 \quad \text{or} \quad x = \tfrac{17}{3}$$

$\left(\frac{17}{3}, 5\right)$ is one vertex.

Substitute $y = 5$ into $x + y = 15$.

$$x + y = 15 \quad \text{or} \quad x = 10$$

$(10, 5)$ is another vertex.
To find the point of intersection of the lines $3x - y = 12$ and $x + y = 15$, solve this system by elimination.

$$\begin{array}{rcll} 3x - y & = & 12 & (1) \\ x + y & = & 15 & (2) \\ \hline 4x & = & 27 & \textit{Add.} \\ x & = & \frac{27}{4} & \end{array}$$

To find y, substitute $\frac{27}{4}$ for x in (2).

$$\tfrac{27}{4} + y = 15, \text{ so } y = 15 - \tfrac{27}{4} = \tfrac{33}{4}.$$

Thus, $\left(\frac{27}{4}, \frac{33}{4}\right)$ is the third vertex.
A graph of the solution is shown.

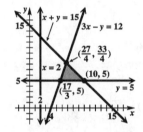

Substitute the coordinates of each vertex into the expression $2x + y$ to see where this value is minimized.

Point	Value $= 2x + y$
$\left(\frac{17}{3}, 5\right)$	$2\left(\frac{17}{3}\right) + 5 = \frac{49}{3} = 16\frac{1}{3}$ ←
$\left(\frac{27}{4}, \frac{33}{4}\right)$	$2\left(\frac{27}{4}\right) + \frac{33}{4} = \frac{87}{4} = 21\frac{3}{4}$
$(10, 5)$	$2(10) + 5 = 25$

The minimum value is $\frac{49}{3}$ at $\left(\frac{17}{3}, 5\right)$.

53. *Step 1*
Let $P =$ the number of pigs and $G =$ the number of geese. The objective function is $80G + 40P$. The constraints are:

$$\begin{array}{ll} P \geq 0 & \textit{Nonnegative no. of pigs} \\ G \geq 0 & \textit{Nonnegative no. of geese} \\ P + G \leq 16 & \textit{Total number of animals} \\ G \leq 12 & \textit{No more than 12 geese} \\ 50P + 20G \leq 500 & \textit{\$500 available to spend} \end{array}$$

Step 2
Graph the region of feasible solutions by using a horizontal P-axis and a vertical G-axis.

Copyright © 2012 Pearson Education, Inc. Publishing as Addison-Wesley.

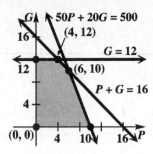

Step 3
The vertices are at $(0,0)$, $(0,12)$, $(4,12)$, $(10,0)$, and $(6,10)$, which is the intersection of $P + G = 16$ and $50P + 20G = 500$.

Step 4
Check the value of $80G + 40P$ at each vertex to find the maximum profit.

Point (P,G)	$80G + 40P = $ **Profit**
$(0,0)$	$80(0) + 40(0) = 0$
$(0,12)$	$80(12) + 40(0) = 960$
$(4,12)$	$80(12) + 40(4) = 1120 \leftarrow$
$(6,10)$	$80(10) + 40(6) = 1040$
$(10,0)$	$80(0) + 40(10) = 400$

Step 5
The maximum profit is $1120 with 4 pigs and 12 geese.

55. *Step 1*
Let $x = $ the number of cabinet #1 and $y = $ the number of cabinet #2. The objective function is $8x + 12y$. The constraints are:

$$x, y \geq 0 \quad \text{Nonnegative no. of cabinets}$$
$$20x + 40y \leq 280 \quad \text{Cost constraint}$$
$$6x + 8y \leq 72 \quad \text{Space constraint}$$

Step 2
Graph the region of feasible solutions.

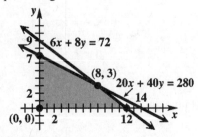

Step 3
The vertices are at $(0,0)$, $(0,7)$, $(12,0)$, and $(8,3)$, which is the intersection of $20x + 40y = 280$ and $6x + 8y = 72$.

Step 4
Check the value of $8x + 12y$ at each vertex to find the maximum storage capacity.

Point	$8x + 12y = $ **Storage capacity**
$(0,0)$	$8(0) + 12(0) = 0$
$(0,7)$	$8(0) + 12(7) = 84$
$(8,3)$	$8(8) + 12(3) = 100 \leftarrow$
$(12,0)$	$8(12) + 12(0) = 96$

Step 5
$8x + 12y$ is maximized at $(8,3)$. She should get 8 #1 cabinets and 3 #2 cabinets. This corresponds to maximum storage capacity of 100 cubic feet.

57. *Step 1*
Let $x = $ the number of gallons of gasoline, and $y = $ the number of gallons of fuel oil. The objective function is $1.9x + 1.5y$. The constraints are:

$x, y \geq 0$	*Nonnegative no. of gallons*
$x \geq 2y$	*Requirement*
$y \geq 3,000,000$	*Fuel oil demand*
$x \leq 6,400,000$	*Gasoline demand*

Step 2
Graph the region of feasible solutions.

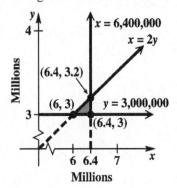

Step 3
The vertices are at $(6.4, 3)$, $(6, 3)$, and $(6.4, 3.2)$.

Step 4
Check the value of $1.9x + 1.5y$ at each vertex to find the maximum revenue.

Point	$1.9x + 1.5y = $ **Revenue**
$(6.4, 3)$	$1.9(6.4) + 1.5(3) = 16.66$
$(6, 3)$	$1.9(6) + 1.5(3) = 15.9$
$(6.4, 3.2)$	$1.9(6.4) + 1.5(3.2) = 16.96 \leftarrow$

Step 5
6.4 million gallons of gasoline and 3.2 million gallons of fuel oil should be produced for maximum revenue of $16.96 million, or $16,960,000.

59. $\dfrac{n+5}{n}$

(a) $[n = 1]$ $\dfrac{1+5}{1} = \dfrac{6}{1} = 6$

(b) $[n = 2]$ $\dfrac{2+5}{2} = \dfrac{7}{2}$

(c) $[n = 3]$ $\dfrac{3+5}{3} = \dfrac{8}{3}$

(d) $[n = 4]$ $\dfrac{4+5}{4} = \dfrac{9}{4}$

61. $n^2 - n$

(a) $[n = 1]$ $1^2 - 1 = 1 - 1 = 0$

(b) $[n = 2]$ $2^2 - 2 = 4 - 2 = 2$

Copyright © 2012 Pearson Education, Inc. Publishing as Addison-Wesley.

(c) $[n = 3]$ $3^2 - 3 = 9 - 3 = 6$

(d) $[n = 4]$ $4^2 - 4 = 16 - 4 = 12$

Chapter 13 Review Exercises

1. $x^2 + y^2 = 121$

This is an equation of a *circle* with center $(0, 0)$ and radius $\sqrt{121} = 11$.

2. $x^2 + 4y^2 = 4$

$\dfrac{x^2}{4} + y^2 = 1$ *Divide by 4.*

This is an equation of an *ellipse* with center $(0, 0)$, x-intercepts $(\pm 2, 0)$, and y-intercepts $(0, \pm 1)$.

3. $\dfrac{x^2}{16} + \dfrac{y^2}{25} = 1$

This is an equation of an *ellipse* with center $(0, 0)$, x-intercepts $(\pm 4, 0)$, and y-intercepts $(0, \pm 5)$.

4. $3x^2 + 3y^2 = 300$

$x^2 + y^2 = 100$ *Divide by 3.*

This is an equation of a *circle* with center $(0, 0)$ and radius $\sqrt{100} = 10$.

5. Center $(-2, 4)$, $r = 3$
Here $h = -2$, $k = 4$, and $r = 3$, so an equation of the circle is

$$(x - h)^2 + (y - k)^2 = r^2$$
$$[x - (-2)]^2 + (y - 4)^2 = 3^2$$
$$(x + 2)^2 + (y - 4)^2 = 9.$$

6. Center $(-1, -3)$, $r = 5$
Here $h = -1$, $k = -3$, and $r = 5$, so an equation of the circle is

$$(x - h)^2 + (y - k)^2 = r^2$$
$$[x - (-1)]^2 + [y - (-3)]^2 = 5^2$$
$$(x + 1)^2 + (y + 3)^2 = 25.$$

7. Center $(4, 2)$, $r = 6$
Here $h = 4$, $k = 2$, and $r = 6$, so an equation of the circle is

$$(x - h)^2 + (y - k)^2 = r^2$$
$$(x - 4)^2 + (y - 2)^2 = 6^2$$
$$(x - 4)^2 + (y - 2)^2 = 36.$$

8. $x^2 + y^2 + 6x - 4y - 3 = 0$

Write the equation in center-radius form,

$$(x - h)^2 + (y - k)^2 = r^2,$$

by completing the squares on x and y.

$$(x^2 + 6x \quad) + (y^2 - 4y \quad) = 3$$
$$(x^2 + 6x + \underline{9}) + (y^2 - 4y + \underline{4})$$
$$= 3 + \underline{9} + \underline{4}$$
$$(x + 3)^2 + (y - 2)^2 = 16$$
$$[x - (-3)]^2 + (y - 2)^2 = 16$$

The circle has center (h, k) at $(-3, 2)$ and radius $\sqrt{16} = 4$.

9. $x^2 + y^2 - 8x - 2y + 13 = 0$

Write the equation in center-radius form,

$$(x - h)^2 + (y - k)^2 = r^2,$$

by completing the squares on x and y.

$$(x^2 - 8x \quad) + (y^2 - 2y \quad) = -13$$
$$(x^2 - 8x + \underline{16}) + (y^2 - 2y + \underline{1})$$
$$= -13 + \underline{16} + \underline{1}$$
$$(x - 4)^2 + (y - 1)^2 = 4$$

The circle has center (h, k) at $(4, 1)$ and radius $\sqrt{4} = 2$.

10. $2x^2 + 2y^2 + 4x + 20y = -34$
$x^2 + y^2 + 2x + 10y = -17$

Write the equation in center-radius form,

$$(x - h)^2 + (y - k)^2 = r^2,$$

by completing the squares on x and y.

$$(x^2 + 2x \quad) + (y^2 + 10y \quad) = -17$$
$$(x^2 + 2x + \underline{1}) + (y^2 + 10y + \underline{25})$$
$$= -17 + \underline{1} + \underline{25}$$
$$(x + 1)^2 + (y + 5)^2 = 9$$
$$[x - (-1)]^2 + [y - (-5)]^2 = 9$$

The circle has center (h, k) at $(-1, -5)$ and radius $\sqrt{9} = 3$.

11. $4x^2 + 4y^2 - 24x + 16y = 48$
$x^2 + y^2 - 6x + 4y = 12$

Write the equation in center-radius form,

$$(x - h)^2 + (y - k)^2 = r^2,$$

by completing the squares on x and y.

$$(x^2 - 6x \quad) + (y^2 + 4y \quad) = 12$$
$$(x^2 - 6x + \underline{9}) + (y^2 + 4y + \underline{4})$$
$$= 12 + \underline{9} + \underline{4}$$
$$(x - 3)^2 + (y + 2)^2 = 25$$
$$(x - 3)^2 + [y - (-2)]^2 = 25$$

The circle has center (h, k) at $(3, -2)$ and radius $\sqrt{25} = 5$.

Copyright © 2012 Pearson Education, Inc. Publishing as Addison-Wesley.

12.
$$x^2 + y^2 = 16$$
$$(x-0)^2 + (y-0)^2 = 4^2$$
This is a circle with center $(0,0)$ and radius 4.

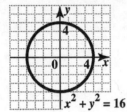

13. $\dfrac{x^2}{16} + \dfrac{y^2}{9} = 1$ is in $\dfrac{x^2}{a^2} + \dfrac{y^2}{b^2} = 1$ form with $a = 4$ and $b = 3$. The graph is an ellipse with x-intercepts $(4,0)$ and $(-4,0)$ and y-intercepts $(0,3)$ and $(0,-3)$. Plot the intercepts, and draw the ellipse through them.

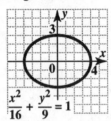

14. $\dfrac{x^2}{49} + \dfrac{y^2}{25} = 1$ is in $\dfrac{x^2}{a^2} + \dfrac{y^2}{b^2} = 1$ form with $a = 7$ and $b = 5$. The graph is an ellipse with x-intercepts $(7,0)$ and $(-7,0)$ and y-intercepts $(0,5)$ and $(0,-5)$. Plot the intercepts, and draw the ellipse through them.

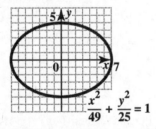

15. The total distance on the horizontal axis is $160 + 16{,}000 = 16{,}160$ km. This represents $2a$, so $a = \frac{1}{2}(16{,}160) = 8080$. The distance from Earth to the center of the ellipse is
$$8080 - 160 = 7920,$$
which is the value of c. From the hint, we know that $c^2 = a^2 - b^2$, so
$$b^2 = a^2 - c^2.$$
$$b^2 = 8080^2 - 7920^2$$
$$= 2{,}560{,}000$$
Thus, $b = \sqrt{2{,}560{,}000} = 1600$ and the equation is
$$\frac{x^2}{8080^2} + \frac{y^2}{1600^2} = 1$$
or $\dfrac{x^2}{65{,}286{,}400} + \dfrac{y^2}{2{,}560{,}000} = 1.$

16. **(a)** The distance between the foci is $2c$, where c can be found using the relationship
$$c = \sqrt{a^2 - b^2}.$$
$$= \sqrt{310^2 - (513/2)^2}$$
$$\approx 174.1 \text{ feet}$$

So the distance is about 348.2 feet.

(b) The approximate perimeter of the Roman Colosseum is
$$P \approx 2\pi\sqrt{\frac{a^2 + b^2}{2}}.$$
$$= 2\pi\sqrt{\frac{310^2 + (513/2)^2}{2}}$$
$$\approx 1787.6 \text{ feet}$$

17. $\dfrac{x^2}{16} - \dfrac{y^2}{25} = 1$ is in $\dfrac{x^2}{a^2} - \dfrac{y^2}{b^2} = 1$ form with $a = 4$ and $b = 5$. The graph is a hyperbola with x-intercepts $(4,0)$ and $(-4,0)$ and asymptotes that are the extended diagonals of the rectangle with vertices $(4,5)$, $(4,-5)$, $(-4,-5)$, and $(-4,5)$. Graph a branch of the hyperbola through each intercept approaching the asymptotes.

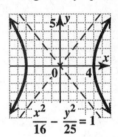

18. $\dfrac{y^2}{25} - \dfrac{x^2}{4} = 1$ is in $\dfrac{y^2}{b^2} - \dfrac{x^2}{a^2} = 1$ form with $a = 2$ and $b = 5$. The graph is a hyperbola with y-intercepts $(0,5)$ and $(0,-5)$ and asymptotes that are the extended diagonals of the rectangle with vertices $(2,5)$, $(2,-5)$, $(-2,-5)$, and $(-2,5)$. Graph a branch of the hyperbola through each intercept approaching the asymptotes.

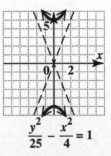

Copyright © 2012 Pearson Education, Inc. Publishing as Addison-Wesley.

19.
$$f(x) = -\sqrt{16 - x^2}$$
Replace $f(x)$ with y.
$$y = -\sqrt{16 - x^2}$$
Square both sides.
$$y^2 = 16 - x^2$$
$$x^2 + y^2 = 16$$

This equation is the graph of a circle with center $(0, 0)$ and radius 4. Since $f(x)$ represents a nonpositive square root, $f(x)$ is nonpositive and its graph is the lower half of the circle.

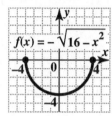

20.
$$x^2 + y^2 = 64$$
$$(x - 0)^2 + (y - 0)^2 = 8^2$$

The last equation is in $(x - h)^2 + (y - k)^2 = r^2$ form. The graph is a *circle*.

21. $y = 2x^2 - 3$
$$y = 2(x - 0)^2 - 3$$

The last equation is in $y = a(x - h)^2 + k$ form. The graph is a *parabola*.

22.
$$y^2 = 2x^2 - 8$$
$$2x^2 - y^2 = 8$$
$$\frac{x^2}{4} - \frac{y^2}{8} = 1$$

The last equation is in $\frac{x^2}{a^2} - \frac{y^2}{b^2} = 1$ form, so the graph is a *hyperbola*.

23.
$$y^2 = 8 - 2x^2$$
$$2x^2 + y^2 = 8$$
$$\frac{x^2}{4} + \frac{y^2}{8} = 1$$

The last equation is in $\frac{x^2}{a^2} + \frac{y^2}{b^2} = 1$ form, so the graph is an *ellipse*.

24. $x = y^2 + 4$
$$x = (y - 0)^2 + 4$$

The last equation is in $x = a(y - k)^2 + h$ form, so the graph is a *parabola*.

25.
$$x^2 - y^2 = 64$$
$$\frac{x^2}{64} - \frac{y^2}{64} = 1$$

The last equation is in $\frac{x^2}{a^2} - \frac{y^2}{b^2} = 1$ form, so the graph is a *hyperbola*.

26. A hyperbola is defined as the set of all points in a plane such that the absolute value of the *difference* of the distances from two fixed points (called *foci*) is constant. The hyperbola shown in the text will have an equation of the form
$$\frac{x^2}{a^2} - \frac{y^2}{b^2} = 1. \quad (1)$$

The constant difference for this hyperbola is
$$|d_1 - d_2| = |80 - 30| = 50.$$

Let Q be the point on the right branch of the hyperbola that is on $\overline{MS}$. Let $w = \overline{MQ}$ and $z = \overline{QS}$. Since $\overline{MS} = 100$ and $\overline{MQ} - \overline{QS}$ must be 50 (since 50 is the constant difference for this hyperbola), we have the following system.

$$
\begin{array}{rll}
w + z & = & 100 \quad (2) \\
w - z & = & 50 \\
\hline
2w & = & 150 \quad Add. \\
w & = & 75
\end{array}
$$

From (2), we see that $z = 25$. The distance from the center of the hyperbola, C, to S, is 50, so $a = \overline{CQ} = 50 - 25 = 25$.

To find b^2, we'll find a point on the hyperbola, substitute for a, x, and y, and then solve for b^2. Let P have coordinates (d, e). If we draw a perpendicular line from P to $\overline{MS}$, and call the point of intersection R, we see two right triangles, PRM and PRS. Using the Pythagorean formula, we get the following system of equations.

$$
\begin{array}{rll}
(50 + d)^2 + e^2 & = & 80^2 \\
(50 - d)^2 + e^2 & = & 30^2 \quad (3) \\
\hline
(2500 + 100d + d^2) - (2500 - 100d + d^2) & & \\
& = & 6400 - 900 \quad Subtract. \\
200d & = & 5500 \\
d & = & 27.5
\end{array}
$$

From (3) with $d = 27.5$, we get $e^2 = 30^2 - 22.5^2$, so $e = \sqrt{393.75}$. Now substitute 25 for a, 27.5 for x, and $\sqrt{393.75}$ for y in (1) to solve for b^2.

$$\frac{27.5^2}{25^2} - \frac{393.75}{b^2} = 1$$
$$\frac{756.25}{625} - \frac{625}{625} = \frac{393.75}{b^2}$$
$$\frac{131.25}{625} = \frac{393.75}{b^2}$$
$$\frac{625}{131.25} = \frac{b^2}{393.75}$$
$$b^2 = \frac{625}{131.25}(393.75)$$
$$b^2 = 1875$$

Thus, equation (1) becomes
$$\frac{x^2}{625} - \frac{y^2}{1875} = 1.$$

Copyright © 2012 Pearson Education, Inc. Publishing as Addison-Wesley.

27.
$$2y = 3x - x^2 \quad (1)$$
$$x + 2y = -12 \quad (2)$$

Substitute $3x - x^2$ for $2y$ in equation (2).

$$x + 2y = -12 \quad (2)$$
$$x + (3x - x^2) = -12$$
$$-x^2 + 4x + 12 = 0$$
$$x^2 - 4x - 12 = 0$$
$$(x - 6)(x + 2) = 0$$

$$x - 6 = 0 \quad \text{or} \quad x + 2 = 0$$
$$x = 6 \quad \text{or} \quad x = -2$$

Substitute these values for x in equation (2) to find y.

If $x = 6$, then
$$x + 2y = -12 \quad (2)$$
$$6 + 2y = -12$$
$$2y = -18$$
$$y = -9.$$

If $x = -2$, then
$$x + 2y = -12 \quad (2)$$
$$-2 + 2y = -12$$
$$2y = -10$$
$$y = -5.$$

The solution set is $\{(6, -9), (-2, -5)\}$.

28.
$$y + 1 = x^2 + 2x \quad (1)$$
$$y + 2x = 4 \quad (2)$$

Solve equation (2) for y.

$$y = 4 - 2x \quad (3)$$

Substitute $4 - 2x$ for y in (1).

$$(4 - 2x) + 1 = x^2 + 2x$$
$$0 = x^2 + 4x - 5$$
$$0 = (x + 5)(x - 1)$$

$$x + 5 = 0 \quad \text{or} \quad x - 1 = 0$$
$$x = -5 \quad \text{or} \quad x = 1$$

Substitute these values for x in equation (3) to find y.

If $x = -5$, then $y = 4 - 2(-5) = 14$.
If $x = 1$, then $y = 4 - 2(1) = 2$.

The solution set is $\{(1, 2), (-5, 14)\}$.

29.
$$x^2 + 3y^2 = 28 \quad (1)$$
$$y - x = -2 \quad (2)$$

Solve equation (2) for y.

$$y = x - 2$$

Substitute $x - 2$ for y in equation (1).

$$x^2 + 3y^2 = 28 \quad (1)$$
$$x^2 + 3(x - 2)^2 = 28$$
$$x^2 + 3(x^2 - 4x + 4) - 28 = 0$$
$$x^2 + 3x^2 - 12x + 12 - 28 = 0$$
$$4x^2 - 12x - 16 = 0$$
$$4(x^2 - 3x - 4) = 0$$
$$4(x - 4)(x + 1) = 0$$

$$x - 4 = 0 \quad \text{or} \quad x + 1 = 0$$
$$x = 4 \quad \text{or} \quad x = -1$$

Since $y = x - 2$, if $x = 4$, then $y = 4 - 2 = 2$.

If $x = -1$, then $y = -1 - 2 = -3$.

The solution set is $\{(4, 2), (-1, -3)\}$.

30.
$$xy = 8 \quad (1)$$
$$x - 2y = 6 \quad (2)$$

Solve equation (2) for x.

$$x = 2y + 6 \quad (3)$$

Substitute $2y + 6$ for x in equation (1) to find y.

$$xy = 8 \quad (1)$$
$$(2y + 6)y = 8$$
$$2y^2 + 6y - 8 = 0$$
$$2(y^2 + 3y - 4) = 0$$
$$2(y + 4)(y - 1) = 0$$

$$y + 4 = 0 \quad \text{or} \quad y - 1 = 0$$
$$y = -4 \quad \text{or} \quad y = 1$$

Substitute these values for y in equation (3) to find x.
If $y = -4$, then $x = 2(-4) + 6 = -2$.
If $y = 1$, then $x = 2(1) + 6 = 8$.

The solution set is $\{(-2, -4), (8, 1)\}$.

31.
$$x^2 + y^2 = 6 \quad (1)$$
$$x^2 - 2y^2 = -6 \quad (2)$$

Multiply equation (2) by -1 and add the result to equation (1).

$$
\begin{array}{rcll}
x^2 + y^2 &=& 6 & (1) \\
-x^2 + 2y^2 &=& 6 & -1 \times (2) \\
\hline
3y^2 &=& 12 & \\
y^2 &=& 4 &
\end{array}
$$

$$y = 2 \quad \text{or} \quad y = -2$$

Substitute these values for y in equation (1) to find x.
If $y = \pm 2$, then

Copyright © 2012 Pearson Education, Inc. Publishing as Addison-Wesley.

$$x^2 + y^2 = 6 \qquad (1)$$
$$x^2 + (\pm 2)^2 = 6$$
$$x^2 + 4 = 6$$
$$x^2 = 2.$$

$$x = \sqrt{2} \quad \text{or} \quad x = -\sqrt{2}$$

Since each value of x can be paired with each value of y, there are four points and the solution set is
$$\left\{ (\sqrt{2}, 2), (-\sqrt{2}, 2), (\sqrt{2}, -2), (-\sqrt{2}, -2) \right\}.$$

32. $3x^2 - 2y^2 = 12 \quad (1)$
$\qquad x^2 + 4y^2 = 18 \quad (2)$

Multiply equation (1) by 2 and add the result to equation (2).

$$
\begin{array}{rl}
6x^2 - 4y^2 = 24 & 2 \times (1) \\
\underline{x^2 + 4y^2 = 18} & (2) \\
7x^2 = 42 & \\
x^2 = 6 &
\end{array}
$$

$$x = \sqrt{6} \quad \text{or} \quad x = -\sqrt{6}$$

Substitute these values for x in equation (2) to find y.
If $x = \pm\sqrt{6}$, then

$$x^2 + 4y^2 = 18. \qquad (2)$$
$$\left(\pm\sqrt{6}\right)^2 + 4y^2 = 18$$
$$6 + 4y^2 = 18$$
$$4y^2 = 12$$
$$y^2 = 3$$

$$y = \sqrt{3} \quad \text{or} \quad y = -\sqrt{3}$$

The solution set is
$$\left\{ (\sqrt{6}, \sqrt{3}), (\sqrt{6}, -\sqrt{3}), \right.$$
$$\left. (-\sqrt{6}, \sqrt{3}), (-\sqrt{6}, -\sqrt{3}) \right\}.$$

33. A circle and a line can intersect in zero, one, or two points, so zero, one, or two solutions are possible.

34. A parabola and a hyperbola can intersect in zero, one, two, three, or four points, so zero, one, two, three, or four solutions are possible.

35. $\qquad 9x^2 \geq 16y^2 + 144$
$$9x^2 - 16y^2 \geq 144$$
$$\frac{x^2}{16} - \frac{y^2}{9} \geq 1$$

The boundary, $\frac{x^2}{16} - \frac{y^2}{9} = 1$, is a solid hyperbola with x-intercepts $(4, 0)$ and $(-4, 0)$. The asymptotes are the extended diagonals of the rectangle with vertices $(4, 3), (4, -3), (-4, -3),$ and $(-4, 3)$. Test $(0, 0)$.

$$9(0)^2 \overset{?}{\geq} 16(0)^2 + 144$$
$$0 \geq 144 \qquad \textit{False}$$

Shade the sides of the hyperbola that do not contain $(0, 0)$. These are the regions inside the branches of the hyperbola.

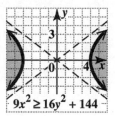

36. $4x^2 + y^2 \geq 16$
$$\frac{x^2}{4} + \frac{y^2}{16} \geq 1$$

The boundary, $\frac{x^2}{4} + \frac{y^2}{16} = 1$, is a solid ellipse with intercepts $(2, 0), (-2, 0), (0, 4),$ and $(0, -4)$. Test $(0, 0)$.

$$4(0)^2 + 0^2 \overset{?}{\geq} 16$$
$$0 \geq 16 \qquad \textit{False}$$

Shade the side of ellipse that does not contain $(0, 0)$. This is the region outside the ellipse.

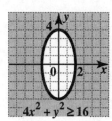

37. $y < -(x + 2)^2 + 1$

The boundary, $y = -(x + 2)^2 + 1$, is a dashed vertical parabola with vertex $(-2, 1)$. Since $a = -1 < 0$, the parabola opens down. Also, $|a| = |-1| = 1$, so the graph has the same shape as the graph of $y = x^2$. Test $(0, 0)$.

$$0 \overset{?}{<} -(0 + 2)^2 + 1$$
$$0 \overset{?}{<} -(4) + 1$$
$$0 < -3 \qquad \textit{False}$$

Shade the side of the parabola that does not contain $(0, 0)$. This is the region inside the parabola.

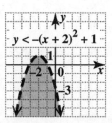

Copyright © 2012 Pearson Education, Inc. Publishing as Addison-Wesley.

38. $2x + 5y \leq 10$
$3x - y \leq 6$

The boundary, $2x + 5y = 10$ is a solid line with intercepts $(5, 0)$ and $(0, 2)$. Test $(0, 0)$.

$$2(0) + 5(0) \overset{?}{\leq} 10$$
$$0 \leq 10 \quad True$$

Shade the side of the line that contains $(0, 0)$.
The boundary, $3x - y = 6$, is a solid line with intercepts $(2, 0)$ and $(0, -6)$. Test $(0, 0)$.

$$3(0) - 0 \overset{?}{\leq} 6$$
$$0 \leq 6 \quad True$$

Shade the side of the line that contains $(0, 0)$.
The graph of the system is the intersection of the two shaded regions.

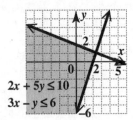

$2x + 5y \leq 10$
$3x - y \leq 6$

39. $|x| \leq 2$
$|y| > 1$
$4x^2 + 9y^2 \leq 36$

The equation of the boundary, $|x| = 2$, can be written as

$$x = -2 \text{ or } x = 2.$$

The graph is these two solid vertical lines. Since $|0| \leq 2$ is true, the region between the lines, containing $(0, 0)$, is shaded.
The boundary, $|y| = 1$, consists of the two dashed horizontal lines $y = 1$ and $y = -1$. Since $|0| > 1$ is false, the regions above $y = 1$ and below $y = -1$, not containing $(0, 0)$, are shaded.
The boundary given by

$$4x^2 + 9y^2 = 36$$
$$\text{or } \frac{x^2}{9} + \frac{y^2}{4} = 1$$

is graphed as a solid ellipse with intercepts $(3, 0)$, $(-3, 0)$, $(0, 2)$, and $(0, -2)$. Test $(0, 0)$.

$$4(0)^2 + 9(0)^2 \overset{?}{\leq} 36$$
$$0 \leq 36 \quad True$$

The region inside the ellipse, containing $(0, 0)$, is shaded.
The graph of the system consists of the regions that include the common points of the three shaded regions.

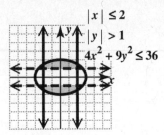

$|x| \leq 2$
$|y| > 1$
$4x^2 + 9y^2 \leq 36$

40. $9x^2 \leq 4y^2 + 36$
$x^2 + y^2 \leq 16$

The equation of the first boundary is

$$9x^2 = 4y^2 + 36$$
$$9x^2 - 4y^2 = 36$$
$$\frac{x^2}{4} - \frac{y^2}{9} = 1.$$

The graph is a solid hyperbola with x-intercepts $(2, 0)$ and $(-2, 0)$. The asymptotes are the extended diagonals of the rectangle with vertices $(2, 3)$, $(2, -3)$, $(-2, -3)$, and $(-2, 3)$. Test $(0, 0)$.

$$9(0)^2 \overset{?}{\leq} 4(0)^2 + 36$$
$$0 \leq 36 \quad True$$

Shade the region between the branches of the hyperbola that contains $(0, 0)$.
The equation of the second boundary is $x^2 + y^2 = 16$. This is a solid circle with center $(0, 0)$ and radius 4. Test $(0, 0)$.

$$0^2 + 0^2 \overset{?}{\leq} 16$$
$$0 \leq 16 \quad True$$

Shade the region inside the circle.
The graph of the system is the intersection of the shaded regions which is between the two branches of the hyperbola and inside the circle.

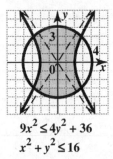

$9x^2 \leq 4y^2 + 36$
$x^2 + y^2 \leq 16$

41. Let x = the number of batches of cakes, and y = the number of batches of cookies.

The time required in the ovens is $2x + \frac{3}{2}y \leq 15$ (no more than 15 hours).
The time required in the decorating room is $3x + \frac{2}{3}y \leq 13$ (no more than 13 hours).
The number of batches cannot be negative, so we also have $x \geq 0$ and $y \geq 0$.

Copyright © 2012 Pearson Education, Inc. Publishing as Addison-Wesley.

Graph $2x + \frac{3}{2}y = 15$ as a solid line and shade the region below it.

Graph $3x + \frac{2}{3}y = 13$ as a solid line and shade the region below it.

Shade above the x-axis and to the right of the y-axis. The solution of the system is the intersection of the shaded regions.

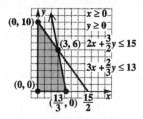

42. Let x = the number of units of basic pizza, and y = the number of units of plain pizza.

The company sells at least 3 units per day of basic $(x \geq 3)$.

The company sells at least 2 units per day of plain $(y \geq 2)$.

Beef costs $5 per unit for basic, and $4 per unit for plain. They can spend no more than $50 per day on beef $(5x + 4y \leq 50)$.

Dough for basic is $2 per unit, while dough for plain is $1 per unit. The company can spend no more than $16 per day on dough $(2x + y \leq 16)$.

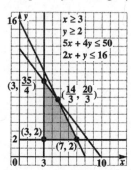

43. The feasible region in the graph for Exercise 41 has vertices at $(0,0)$, $\left(\frac{13}{3}, 0\right)$, $(3,6)$, and $(0,10)$. (The lines $2x + \frac{3}{2}y = 15$ and $3x + \frac{2}{3}y = 13$ intersect at $(3,6)$, which may be verified by solving the system consisting of those two equations.) The profit is $30x + 20y$, where, as before, x is the number of batches of cakes and y is the number of batches of cookies.

Point	$30x + 20y$ = Profit
$(0,0)$	$30(0) + 20(0) = 0$
$\left(\frac{13}{3}, 0\right)$	$30\left(\frac{13}{3}\right) + 20(0) = 130$
$(3,6)$	$30(3) + 20(6) = 210$ ←
$(0,10)$	$30(0) + 20(10) = 200$

A maximum profit of $210 is achieved when the bakery makes 3 batches of cakes and 6 batches of cookies.

44. The feasible region in the graph for Exercise 42 has vertices at $(3,2)$, $(7,2)$, $\left(\frac{14}{3}, \frac{20}{3}\right)$, and $\left(3, \frac{35}{4}\right)$. Since basic sells for $20 per unit and plain for $15 per unit, we want to maximize $20x + 15y$. Evaluate $20x + 15y$ at the vertices.

Point	$20x + 15y$ = Revenue
$(3,2)$	$20(3) + 15(2) = 90$
$(7,2)$	$20(7) + 15(2) = 170$
$\left(3, \frac{35}{4}\right)$	$20(3) + 15\left(\frac{35}{4}\right) = 191\frac{1}{4}$
$\left(\frac{14}{3}, \frac{20}{3}\right)$	$20\left(\frac{14}{3}\right) + 15\left(\frac{20}{3}\right) = 193\frac{1}{3}$ ←

The maximum revenue is $193.33. This occurs when $\frac{14}{3}$ units of basic pizza and $\frac{20}{3}$ units of plain pizza are produced.

45. **[13.2]** $\dfrac{y^2}{4} - 1 = \dfrac{x^2}{9}$

$$\frac{y^2}{4} - \frac{x^2}{9} = 1$$

The equation is in $\frac{y^2}{b^2} - \frac{x^2}{a^2} = 1$ form with $a = 3$ and $b = 2$. The graph is a hyperbola with y-intercepts $(0,2)$ and $(0,-2)$ and asymptotes that are the extended diagonals of the rectangle with vertices $(3,2)$, $(3,-2)$, $(-3,-2)$, and $(-3,2)$. Draw a branch of the hyperbola through each intercept and approaching the asymptotes.

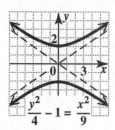

46. **[13.1]** $x^2 + y^2 = 25$ is in $(x-h)^2 + (y-k)^2 = r^2$ form. The graph is a circle with center at $(0,0)$ and radius 5.

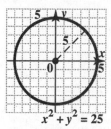

47. **[13.1]** $x^2 + 9y^2 = 9$ in $\frac{x^2}{a^2} + \frac{y^2}{b^2} = 1$ form is $\frac{x^2}{9} + \frac{y^2}{1} = 1$ with $a = 3$ and $b = 1$. The graph is an ellipse with x-intercepts $(3,0)$ and $(-3,0)$ and y-intercepts $(0,1)$ and $(0,-1)$. Plot the intercepts, and draw the ellipse through them.

graph on next page

Copyright © 2012 Pearson Education, Inc. Publishing as Addison-Wesley.

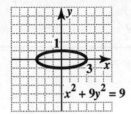

48. **[13.2]** $x^2 - 9y^2 = 9$

$$\frac{x^2}{9} - \frac{y^2}{1} = 1$$

The equation is in the $\frac{x^2}{a^2} - \frac{y^2}{b^2} = 1$ form with $a = 3$ and $b = 1$. The graph is a hyperbola with x-intercepts $(3, 0)$ and $(-3, 0)$ and asymptotes that are the extended diagonals of the rectangle with vertices $(3, 1)$, $(3, -1)$, and $(-3, -1)$, and $(-3, 1)$. Graph a branch of the hyperbola through each intercept and approaching the asymptotes.

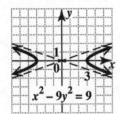

49. **[13.2]** $f(x) = \sqrt{4 - x}$
Replace $f(x)$ with y.

$$y = \sqrt{4 - x}$$

Square both sides.

$$y^2 = 4 - x$$
$$x = -y^2 + 4$$
$$x = -1(y - 0)^2 + 4$$

This equation is the graph of a horizontal parabola with vertex $(4, 0)$. Since $a = -1 < 0$, the graph opens to the left. Also, $|a| = |-1| = 1$, so the graph has the same shape as the graph of $x = y^2$. The points $(0, 2)$ and $(3, 1)$ are on the graph. Since $f(x)$ represents a square root, $f(x)$ is nonnegative and its graph is the upper half of the parabola.

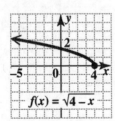

50. **[13.4]** $4y > 3x - 12$
$x^2 < 16 - y^2$

The boundary $4y = 3x - 12$ is a dashed line with intercepts $(0, -3)$ and $(4, 0)$. Test $(0, 0)$.

$$4(0) \overset{?}{>} 3(0) - 12$$
$$0 > -12 \qquad \textit{True}$$

Shade the side of the line that contains $(0, 0)$. The boundary $x^2 = 16 - y^2$, or $x^2 + y^2 = 16$, is a dashed circle with center at $(0, 0)$ and radius 4. Test $(0, 0)$.

$$0^2 \overset{?}{<} 16 - 0^2$$
$$0 < 16 \qquad \textit{True}$$

Shade the region inside the circle. The graph of the system is the intersection of the two shaded regions.

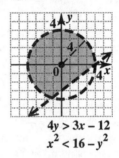

$$4y > 3x - 12$$
$$x^2 < 16 - y^2$$

51. **[13.1]** **(a)** $\dfrac{x^2}{3352} + \dfrac{y^2}{3211} = 1$

$c^2 = a^2 - b^2 = 3352 - 3211$, so $c^2 = 141$ and $c = \sqrt{141}$.

From Exercise 53 in Section 13.1, the apogee is $a + c = \sqrt{3352} + \sqrt{141} \approx 69.8$ million kilometers.

(b) The perigee is
$a - c = \sqrt{3352} - \sqrt{141} \approx 46.0$ million kilometers.

52. **[13.4]** Maximize $2x + 5y$ subject to

$$3x + 2y \le 6 \quad \Big| \quad x \ge 0$$
$$-2x + 4y \le 8 \quad \Big| \quad y \ge 0$$

First sketch a graph of the solution of the system.

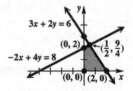

Now evaluate $2x + 5y$ at the vertices.

Point	Value = $2x + 5y$
$(0, 0)$	$2(0) + 5(0) = 0$
$(0, 2)$	$2(0) + 5(2) = 10$
$\left(\frac{1}{2}, \frac{9}{4}\right)$	$2\left(\frac{1}{2}\right) + 5\left(\frac{9}{4}\right) = 12\frac{1}{4}$ ←
$(2, 0)$	$2(2) + 5(0) = 4$

At $\left(\frac{1}{2}, \frac{9}{4}\right)$, the maximum value $12\frac{1}{4}$ is found.

Copyright © 2012 Pearson Education, Inc. Publishing as Addison-Wesley.

Chapter 13 Test

1. The equation $x^2 + y^2 = 1$ is of the form $x^2 + y^2 = r^2$, which has a circle as its graph, so choice **D** is correct.

2. The circle $x^2 + y^2 = 1$ can be written as

$$(x - 0)^2 + (y - 0)^2 = 1^2,$$

so its center is $(0, 0)$ and its radius is 1.

3.
$$(x - 2)^2 + (y + 3)^2 = 16$$
$$(x - 2)^2 + [y - (-3)]^2 = 4^2$$

The graph is a circle with center $(2, -3)$ and radius 4.

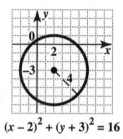

$$(x-2)^2 + (y+3)^2 = 16$$

4. $x^2 + y^2 + 8x - 2y = 8$

To find the center and radius, complete the squares on x and y.

$$\left(x^2 + 8x \quad\right) + \left(y^2 - 2y \quad\right) = 8$$
$$\left(x^2 + 8x + \underline{16}\right) + \left(y^2 - 2y + \underline{1}\right) = 8 + \underline{16} + \underline{1}$$
$$(x + 4)^2 + (y - 1)^2 = 25$$

The graph is a circle with center $(-4, 1)$ and radius $\sqrt{25} = 5$.

5.
$$f(x) = \sqrt{9 - x^2}$$
Replace $f(x)$ with y.
$$y = \sqrt{9 - x^2}$$
Square both sides.
$$y^2 = 9 - x^2$$
$$x^2 + y^2 = 9$$

The graph of $x^2 + y^2 = 9$ is a circle of radius $\sqrt{9} = 3$ centered at the origin. Since $f(x)$ is nonnegative, only the top half of the circle is graphed.

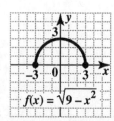

6. $4x^2 + 9y^2 = 36$
$$\frac{x^2}{9} + \frac{y^2}{4} = 1$$

The equation is in $\dfrac{x^2}{a^2} + \dfrac{y^2}{b^2} = 1$ form with $a = 3$ and $b = 2$. The graph is an ellipse with intercepts $(3, 0)$, $(-3, 0)$, $(0, 2)$, and $(0, -2)$. Plot these intercepts, and draw the ellipse through them.

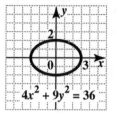

7. $16y^2 - 4x^2 = 64$
$$\frac{y^2}{4} - \frac{x^2}{16} = 1$$

The equation is in $\dfrac{y^2}{b^2} - \dfrac{x^2}{a^2} = 1$ form with $a = 4$ and $b = 2$. The graph is a hyperbola with y-intercepts $(0, 2)$ and $(0, -2)$ and asymptotes that are the extended diagonals of the rectangle with vertices $(4, 2)$, $(4, -2)$, $(-4, -2)$, and $(-4, 2)$.

Draw a branch of the hyperbola through each intercept and approaching the asymptotes.

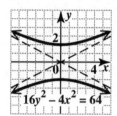

8.
$$\frac{y}{2} = -\sqrt{1 - \frac{x^2}{9}}$$

Square both sides.
$$\frac{y^2}{4} = 1 - \frac{x^2}{9}$$
$$\frac{x^2}{9} + \frac{y^2}{4} = 1$$

This is an ellipse with x-intercepts $(3, 0)$ and $(-3, 0)$ and y-intercepts $(0, 2)$ and $(0, -2)$. Since y represents a negative square root in the original equation, y must be nonpositive. This restricts the graph to the lower half of the ellipse.

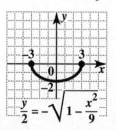

Copyright © 2012 Pearson Education, Inc. Publishing as Addison-Wesley.

9. $6x^2 + 4y^2 = 12$

We have the *sum* of squares with different coefficients equal to a positive number, so this is an equation of an *ellipse*.

10. $16x^2 = 144 + 9y^2$
$16x^2 - 9y^2 = 144$

We have the *difference* of squares equal to a positive number, so this is an equation of a *hyperbola*.

11. $y^2 = 20 - x^2$
$x^2 + y^2 = 20$

We have the *sum* of squares with the same coefficients equal to a positive number, so this is an equation of an *circle*.

12. $4y^2 + 4x = 9$
$4x = -4y^2 + 9$

We have an x-term and a y^2-term, so this is an equation of a horizontal *parabola*.

13. $2x - y = 9$ (1)
$xy = 5$ (2)

Solve equation (1) for y.

$$y = 2x - 9 \quad (3)$$

Substitute $2x - 9$ for y in equation (2).

$$\begin{aligned} xy &= 5 \quad (2) \\ x(2x - 9) &= 5 \\ 2x^2 - 9x &= 5 \\ 2x^2 - 9x - 5 &= 0 \\ (2x + 1)(x - 5) &= 0 \end{aligned}$$

$$\begin{aligned} 2x + 1 &= 0 \quad \text{or} \quad x - 5 = 0 \\ x &= -\tfrac{1}{2} \quad \text{or} \quad x = 5 \end{aligned}$$

Substitute these values for x in equation (3) to find y.

If $x = -\tfrac{1}{2}$, then $y = 2\left(-\tfrac{1}{2}\right) - 9 = -10$.

If $x = 5$, then $y = 2(5) - 9 = 1$.

The solution set is $\left\{\left(-\tfrac{1}{2}, -10\right), (5, 1)\right\}$.

14. $x - 4 = 3y$ (1)
$x^2 + y^2 = 8$ (2)

Solve equation (1) for x.

$$x = 3y + 4$$

Substitute $3y + 4$ for x in equation (2).

$$\begin{aligned} x^2 + y^2 &= 8 \quad (2) \\ (3y + 4)^2 + y^2 &= 8 \\ 9y^2 + 24y + 16 + y^2 &= 8 \\ 10y^2 + 24y + 8 &= 0 \\ 2\left(5y^2 + 12y + 4\right) &= 0 \\ 2(5y + 2)(y + 2) &= 0 \end{aligned}$$

$$\begin{aligned} 5y + 2 &= 0 \quad \text{or} \quad y + 2 = 0 \\ y &= -\tfrac{2}{5} \quad \text{or} \quad y = -2 \end{aligned}$$

Since $x = 3y + 4$, substitute these values for y to find x.

If $y = -\tfrac{2}{5}$, then

$$x = 3\left(-\tfrac{2}{5}\right) + 4 = -\tfrac{6}{5} + 4 = \tfrac{14}{5}.$$

If $y = -2$, then $x = 3(-2) + 4 = -2$.

The solution set is $\left\{(-2, -2), \left(\tfrac{14}{5}, -\tfrac{2}{5}\right)\right\}$.

15. $x^2 + y^2 = 25$ (1)
$x^2 - 2y^2 = 16$ (2)

Multiply equation (1) by 2 and add the result to equation (2).

$$\begin{array}{rll} 2x^2 + 2y^2 &= 50 & 2 \times (1) \\ x^2 - 2y^2 &= 16 & (2) \\ \hline 3x^2 &= 66 & \\ x^2 &= 22 & \end{array}$$

$$x = \sqrt{22} \quad \text{or} \quad x = -\sqrt{22}$$

Substitute 22 for x^2 in equation (1).

$$\begin{aligned} x^2 + y^2 &= 25 \quad (1) \\ 22 + y^2 &= 25 \\ y^2 &= 3 \end{aligned}$$

$$y = \sqrt{3} \quad \text{or} \quad y = -\sqrt{3}$$

The solution set is

$$\left\{\left(\sqrt{22}, \sqrt{3}\right), \left(\sqrt{22}, -\sqrt{3}\right),\right.$$
$$\left.\left(-\sqrt{22}, \sqrt{3}\right), \left(-\sqrt{22}, -\sqrt{3}\right)\right\}.$$

Copyright © 2012 Pearson Education, Inc. Publishing as Addison-Wesley.

16. $y < x^2 - 2$

The boundary, $y = x^2 - 2$, is a dashed parabola in $y = a(x - h)^2 + k$ form with vertex (h, k) at $(0, -2)$. Since $a = 1 > 0$, the parabola opens up. It also has the same shape as $y = x^2$. The points $(2, 2)$ and $(-2, 2)$ are on the graph. Test $(0, 0)$.

$$0 \overset{?}{\le} (0)^2 - 2$$
$$0 \le -2 \qquad \textit{False}$$

Shade the side of the parabola that does not contain $(0, 0)$. This is the region outside the parabola.

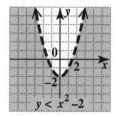

17. $x^2 + 25y^2 \le 25$
$x^2 + y^2 \le 9$

The first boundary, $\frac{x^2}{25} + \frac{y^2}{1} = 1$, is a solid ellipse with intercepts $(5, 0)$, $(-5, 0)$, $(0, 1)$, and $(0, -1)$. Test $(0, 0)$.

$$0^2 + 25 \cdot 0^2 \overset{?}{\le} 25$$
$$0 \le 25 \qquad \textit{True}$$

Shade the region inside the ellipse.
The second boundary, $x^2 + y^2 = 9$, is a solid circle with center $(0, 0)$ and radius 3. Test $(0, 0)$.

$$0^2 + 0^2 \overset{?}{\le} 9$$
$$0 \le 9 \qquad \textit{True}$$

Shade the region inside the circle. The solution of the system is the intersection of the two shaded regions.

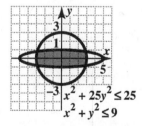

18. Find the value of $2x + 4y$ at each vertex.

Point	$2x + 4y =$ Value
$(2, 0)$	$2(2) + 4(0) = 4$
$(0, 8)$	$2(0) + 4(8) = 32$
$(8, 8)$	$2(8) + 4(8) = 48 \leftarrow$
$(5, 2)$	$2(5) + 4(2) = 18$

The maximum value is at $(x, y) = (8, 8)$.

19. The maximum value of $2x + 4y$ is 48.

20. Let $x =$ the number of radios manufactured, and $y =$ the number of DVD players manufactured. The objective function is $15x + 35y$. The constraints are:

$5 \le x \le 25$ *Radios constraint*

$x \le y$ *Number of radios cannot exceed the number of DVD players*

$0 \le y \le 30$ *DVD players constraint*

Graph the region of feasible solutions.

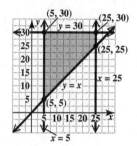

The vertices are at $(5, 5)$, $(5, 30)$, $(25, 30)$, and $(25, 25)$.

Check the value of $15x + 35y$ at each vertex to find the maximum profit.

Point	$15x + 35y =$ Profit
$(5, 5)$	$15(5) + 35(5) = 250$
$(5, 30)$	$15(5) + 35(30) = 1125$
$(25, 30)$	$15(25) + 35(30) = 1425 \leftarrow$
$(25, 25)$	$15(25) + 35(25) = 1250$

The company should manufacture 25 radios and 30 DVD players to obtain a maximum daily profit of \$1425.

Cumulative Review Exercises (Chapters 1–13)

1.
$$4 - (2x + 3) + x = 5x - 3$$
$$4 - 2x - 3 + x = 5x - 3$$
$$-x + 1 = 5x - 3$$
$$-6x = -4$$
$$x = \tfrac{2}{3}$$

The solution set is $\left\{\tfrac{2}{3}\right\}$.

2.
$$-4x + 7 \ge 6x + 1$$
$$-10x \ge -6$$
$$x \le \frac{-6}{-10} \qquad \textit{Divide by } -10; \textit{ reverse}$$
$$\phantom{x \le \frac{-6}{-10} \qquad} \textit{the inequality direction.}$$
$$x \le \tfrac{3}{5}$$

The solution set is $\left(-\infty, \tfrac{3}{5}\right]$.

Copyright © 2012 Pearson Education, Inc. Publishing as Addison-Wesley.

3. $|5x| - 6 = 14$

$|5x| = 20$

$5x = 20$ or $5x = -20$

$x = 4$ or $x = -4$

The solution set is $\{-4, 4\}$.

4. $|2p - 5| > 15$

$2p - 5 > 15$ or $2p - 5 < -15$

$2p > 20$ $\qquad$ $2p < -10$

$p > 10$ or $\qquad$ $p < -5$

The solution set is $(-\infty, -5) \cup (10, \infty)$.

5. Let $(x_1, y_1) = (2, 5)$ and $(x_2, y_2) = (-4, 1)$.

$$m = \frac{y_2 - y_1}{x_2 - x_1} = \frac{1 - 5}{-4 - 2} = \frac{-4}{-6} = \frac{2}{3}$$

6. Through $(-3, -2)$; perpendicular to $2x - 3y = 7$

Write $2x - 3y = 7$ in slope-intercept form.

$$-3y = -2x + 7$$
$$y = \tfrac{2}{3}x - \tfrac{7}{3}$$

The slope is $\frac{2}{3}$. Perpendicular lines have slopes that are negative reciprocals of each other, so a line perpendicular to the given line will have slope $-\frac{3}{2}$. Let $m = -\frac{3}{2}$ and $(x_1, y_1) = (-3, -2)$ in the point-slope form.

$$y - y_1 = m(x - x_1)$$
$$y - (-2) = -\tfrac{3}{2}[x - (-3)]$$
$$y + 2 = -\tfrac{3}{2}(x + 3)$$

Multiply by 2 to clear the fraction.

$$2y + 4 = -3(x + 3)$$
$$2y + 4 = -3x - 9$$
$$3x + 2y = -13$$

7. $3x - y = 12$ $\quad$ (1)

$2x + 3y = -3$ $\quad$ (2)

Multiply equation (1) by 3 and add the result to equation (2).

$$
\begin{array}{rcll}
9x - 3y &=& 36 & \quad 3 \times (1) \\
2x + 3y &=& -3 & \quad (2) \\
\hline
11x &=& 33 & \\
x &=& 3 &
\end{array}
$$

Substitute 3 for x in equation (1) to find y.

$$3x - y = 12 \quad (1)$$
$$3(3) - y = 12$$
$$9 - y = 12$$
$$-y = 3$$
$$y = -3$$

The solution set is $\{(3, -3)\}$.

8. $x + y - 2z = 9$ $\quad$ (1)

$2x + y + z = 7$ $\quad$ (2)

$3x - y - z = 13$ $\quad$ (3)

Add equation (2) and equation (3).

$$
\begin{array}{rcll}
2x + y + z &=& 7 & \quad (2) \\
3x - y - z &=& 13 & \quad (3) \\
\hline
5x &=& 20 & \\
x &=& 4 &
\end{array}
$$

Multiply equation (1) by -1 and add the result to equation (2).

$$
\begin{array}{rcll}
-x - y + 2z &=& -9 & \\
2x + y + z &=& 7 & \quad (2) \\
\hline
x + 3z &=& -2 & \quad (4)
\end{array}
$$

Substitute 4 for x in equation (4) to find z.

$$x + 3z = -2 \quad (4)$$
$$4 + 3z = -2$$
$$3z = -6$$
$$z = -2$$

Substitute 4 for x and -2 for z in equation (2) to find y.

$$2x + y + z = 7 \quad (2)$$
$$2(4) + y - 2 = 7$$
$$y + 6 = 7$$
$$y = 1$$

The solution set is $\{(4, 1, -2)\}$.

9. $xy = -5$ $\quad$ (1)

$2x + y = 3$ $\quad$ (2)

Solve equation (2) for y.

$$y = -2x + 3 \quad (3)$$

Substitute $-2x + 3$ for y in equation (1).

$$xy = -5 \quad (1)$$
$$x(-2x + 3) = -5$$
$$-2x^2 + 3x = -5$$
$$-2x^2 + 3x + 5 = 0$$
$$2x^2 - 3x - 5 = 0$$
$$(2x - 5)(x + 1) = 0$$

$2x - 5 = 0$ or $x + 1 = 0$

$x = \tfrac{5}{2}$ or $\qquad$ $x = -1$

Substitute these values for x in equation (3) to find y.

If $x = \frac{5}{2}$, then $y = -2\left(\frac{5}{2}\right) + 3 = -2$.

If $x = -1$, then $y = -2(-1) + 3 = 5$.

The solution set is $\left\{(-1, 5), \left(\frac{5}{2}, -2\right)\right\}$.

Copyright © 2012 Pearson Education, Inc. Publishing as Addison-Wesley.

10. Let s = Al's rate, $2s$ = Bev's rate, t = Bev's time, and $t + \frac{1}{2}$ = Al's time.

	Distance	Rate	Time
Al	20	s	$t + \frac{1}{2}$
Bev	20	$2s$	t

Since $d = rt$, the system of equations is

$$s\left(t + \tfrac{1}{2}\right) = 20 \qquad (1)$$
$$2st = 20. \qquad (2)$$

Solve equation (2) for s.

$$2st = 20 \qquad (2)$$
$$s = \frac{20}{2t} = \frac{10}{t}$$

Substitute $\frac{10}{t}$ for s in equation (1) to find t.

$$s\left(t + \tfrac{1}{2}\right) = 20 \qquad (1)$$
$$\frac{10}{t}\left(t + \frac{1}{2}\right) = 20$$
$$10 + \frac{5}{t} = 20$$

Multiply each term by the LCD, t.

$$10t + 5 = 20t$$
$$5 = 10t$$
$$t = \tfrac{5}{10} = \tfrac{1}{2}$$

Since $s = \frac{10}{t}$ and $t = \frac{1}{2}$,

$$s = \frac{10}{\frac{1}{2}} = 20.$$

So Al's rate was 20 mph and Bev's rate was $2 \cdot 20 = 40$ mph.

Another solution:

$$st + \tfrac{1}{2}s = 20 \qquad (1)$$
$$2st = 20 \qquad (2)$$

Multiply equation (1) by -2 and add the result to equation (2).

$$
\begin{array}{rcll}
-2st - s &=& -40 & -2 \times (1) \\
2st &=& 20 & (2) \\
\hline
-s &=& -20 & \\
s &=& 20 &
\end{array}
$$

Substitute 20 for s in equation (2) to find $t = \frac{1}{2}$.

11. $(5y - 3)^2 = (5y)^2 - 2(5y)3 + 3^2$
$$= 25y^2 - 30y + 9$$

12. $\dfrac{8x^4 - 4x^3 + 2x^2 + 13x + 8}{2x + 1}$

$$
\begin{array}{r}
4x^3 - 4x^2 + 3x + 5 \\
2x + 1 \overline{)\,8x^4 - 4x^3 + 2x^2 + 13x + 8} \\
\underline{8x^4 + 4x^3\phantom{{}+2x^2+13x+8}} \\
-8x^3 + 2x^2\phantom{{}+13x+8} \\
\underline{-8x^3 - 4x^2\phantom{{}+13x+8}} \\
6x^2 + 13x\phantom{{}+8} \\
\underline{6x^2 + 3x\phantom{{}+8}} \\
10x + 8 \\
\underline{10x + 5} \\
3
\end{array}
$$

The answer is

$$4x^3 - 4x^2 + 3x + 5 + \frac{3}{2x + 1}.$$

13. $12x^2 - 7x - 10$

Two integer factors whose product is $(12)(-10) = -120$ and whose sum is -7 are 8 and -15. Rewrite the trinomial in a form that can be factored by grouping.

$$12x^2 - 7x - 10$$
$$= 12x^2 + 8x - 15x - 10$$
$$= 4x(3x + 2) - 5(3x + 2)$$
$$= (3x + 2)(4x - 5)$$

14. $z^4 - 1 = \left(z^2 + 1\right)\left(z^2 - 1\right)$
$$= \left(z^2 + 1\right)(z + 1)(z - 1)$$

15. $a^3 - 27b^3 = a^3 - (3b)^3$
$$= (a - 3b)\left(a^2 + 3ab + 9b^2\right)$$

16. $\dfrac{y^2 - 4}{y^2 - y - 6} \div \dfrac{y^2 - 2y}{y - 1}$

Multiply by the reciprocal.

$$= \frac{y^2 - 4}{y^2 - y - 6} \cdot \frac{y - 1}{y^2 - 2y}$$

Factor and simplify.

$$= \frac{(y + 2)(y - 2)}{(y - 3)(y + 2)} \cdot \frac{(y - 1)}{y(y - 2)}$$
$$= \frac{y - 1}{y(y - 3)}$$

17. $\dfrac{5}{c + 5} - \dfrac{2}{c + 3}$

The LCD is $(c + 5)(c + 3)$.

$$= \frac{5(c + 3)}{(c + 5)(c + 3)} - \frac{2(c + 5)}{(c + 3)(c + 5)}$$
$$= \frac{5c + 15 - 2c - 10}{(c + 5)(c + 3)}$$
$$= \frac{3c + 5}{(c + 5)(c + 3)}$$

Copyright © 2012 Pearson Education, Inc. Publishing as Addison-Wesley.

18.
$$\frac{p}{p^2+p} + \frac{1}{p^2+p} = \frac{p+1}{p^2+p}$$
$$= \frac{p+1}{p(p+1)} = \frac{1}{p}$$

19. Let $x =$ the time to do the job working together. Make a chart.

Worker	Rate	Time Together	Fractional Part of the Job Done
Henry	$\frac{1}{3}$	x	$\frac{x}{3}$
Lawrence	$\frac{1}{2}$	x	$\frac{x}{2}$

Part done by Henry plus part done by Lawrence equals 1 whole job.
$$\frac{x}{3} \qquad + \qquad \frac{x}{2} \qquad = \qquad 1$$

Multiply by the LCD, 6.
$$6\left(\frac{x}{3} + \frac{x}{2}\right) = 6 \cdot 1$$
$$2x + 3x = 6$$
$$5x = 6$$
$$x = \frac{6}{5} \ \text{ or } \ 1\frac{1}{5}$$

It takes $\frac{6}{5}$ or $1\frac{1}{5}$ hours to do the job together.

20.
$$\frac{(2a)^{-2}a^4}{a^{-3}} = \frac{2^{-2}a^{-2}a^4}{a^{-3}} = \frac{2^{-2}a^2}{a^{-3}}$$
$$= \frac{a^2a^3}{2^2} = \frac{a^5}{4}$$

21.
$$4\sqrt[3]{16} - 2\sqrt[3]{54} = 4\sqrt[3]{8 \cdot 2} - 2\sqrt[3]{27 \cdot 2}$$
$$= 4 \cdot 2\sqrt[3]{2} - 2 \cdot 3\sqrt[3]{2}$$
$$= 8\sqrt[3]{2} - 6\sqrt[3]{2} = 2\sqrt[3]{2}$$

22.
$$\frac{3\sqrt{5x}}{\sqrt{2x}} = \frac{3\sqrt{5x} \cdot \sqrt{2x}}{\sqrt{2x} \cdot \sqrt{2x}} = \frac{3\sqrt{10x^2}}{2x}$$
$$= \frac{3x\sqrt{10}}{2x} = \frac{3\sqrt{10}}{2}$$

23.
$$\frac{5+3i}{2-i}$$

Multiply the numerator and denominator by the conjugate of the denominator.
$$= \frac{(5+3i)(2+i)}{(2-i)(2+i)}$$
$$= \frac{10 + 5i + 6i + 3i^2}{4 - i^2}$$
$$= \frac{10 + 11i + 3(-1)}{4 - (-1)}$$
$$= \frac{7 + 11i}{5} = \frac{7}{5} + \frac{11}{5}i$$

24. $2\sqrt{x} = \sqrt{5x+3}$
$$4x = 5x + 3 \qquad \textit{Square each side.}$$
$$-x = 3$$
$$x = -3$$

Since x must be nonnegative so that $\sqrt{x}$ is a real number, -3 cannot be a solution. The solution set is $\emptyset$.

25.
$$10q^2 + 13q = 3$$
$$10q^2 + 13q - 3 = 0$$
$$(5q - 1)(2q + 3) = 0$$
$$5q - 1 = 0 \quad \text{or} \quad 2q + 3 = 0$$
$$q = \tfrac{1}{5} \quad \text{or} \qquad q = -\tfrac{3}{2}$$

The solution set is $\left\{\frac{1}{5}, -\frac{3}{2}\right\}$.

26. $3x^2 - 3x - 2 = 0$

Use the quadratic formula with $a = 3$, $b = -3$, and $c = -2$.
$$x = \frac{-b \pm \sqrt{b^2 - 4ac}}{2a}$$
$$x = \frac{-(-3) \pm \sqrt{(-3)^2 - 4(3)(-2)}}{2(3)}$$
$$= \frac{3 \pm \sqrt{9 + 24}}{6} = \frac{3 \pm \sqrt{33}}{6}$$

The solution set is $\left\{\frac{3 \pm \sqrt{33}}{6}\right\}$.

27. $2(x^2 - 3)^2 - 5(x^2 - 3) = 12$
Let $u = (x^2 - 3)$, so $u^2 = (x^2 - 3)^2$.
$$2u^2 - 5u = 12$$
$$2u^2 - 5u - 12 = 0$$
$$(2u + 3)(u - 4) = 0$$
$$2u + 3 = 0 \qquad \text{or} \quad u - 4 = 0$$
$$u = -\tfrac{3}{2} \quad \text{or} \qquad u = 4$$

Substitute $x^2 - 3$ for u to find x.

If $u = -\frac{3}{2}$, then
$$x^2 - 3 = -\tfrac{3}{2}$$
$$x^2 = \tfrac{3}{2}$$
$$x = \pm\sqrt{\frac{3}{2}} = \pm\frac{\sqrt{3}}{\sqrt{2}} \cdot \frac{\sqrt{2}}{\sqrt{2}} = \pm\frac{\sqrt{6}}{2}.$$

If $u = 4$, then
$$x^2 - 3 = 4$$
$$x^2 = 7$$
$$x = \pm\sqrt{7}.$$

The solution set is $\left\{\pm\frac{\sqrt{6}}{2}, \pm\sqrt{7}\right\}$.

Copyright © 2012 Pearson Education, Inc. Publishing as Addison-Wesley.

28. $\log(x+2)+\log(x-1)=1$
$$\log_{10}[(x+2)(x-1)]=1$$
$$(x+2)(x-1)=10^1$$
$$x^2+x-2=10$$
$$x^2+x-12=0$$
$$(x+4)(x-3)=0$$

$x+4=0$　　or　　$x-3=0$
$x=-4$　or　　　$x=3$

The original equation is undefined when $x=-4$.

The solution set is $\{3\}$.

29. Solve $F=\dfrac{kwv^2}{r}$ for v.
$$Fr=kwv^2$$
$$v^2=\frac{Fr}{kw}$$

Take the square root of each side.

$$v=\pm\sqrt{\frac{Fr}{kw}}=\frac{\pm\sqrt{Fr}}{\sqrt{kw}}\cdot\frac{\sqrt{kw}}{\sqrt{kw}}=\frac{\pm\sqrt{Frkw}}{kw}$$

30. $f(x)=y=x^3+4$

Interchange x and y and solve for y.

$$x=y^3+4$$
$$y^3=x-4$$
$$y=\sqrt[3]{x-4}$$
$$f^{-1}(x)=\sqrt[3]{x-4}$$

31. **(a)** $a^{\log_a x}=x$, so $3^{\log_3 4}=4$.

　　(b) $e^{\ln x}=x$, so $e^{\ln 7}=7$.

32. $2\log(3x+7)-\log 4=\log(3x+7)^2-\log 4$
$$=\log\frac{(3x+7)^2}{4}$$

33. **(a)** $y=28.43(1.25)^x$
　　　$y=28.43(1.25)^5$　　　*Let $x=5$.*
　　　≈ 86.8 billion dollars

The sales are estimated to be \$86.8 billion in 2005.

　　(b) $y=28.43(1.25)^8$　　　*Let $x=8$.*
　　　≈ 169.5 billion dollars

The sales are estimated to be \$169.5 billion in 2008.

34. Divide $f(x)=2x^3-4x^2+5x-10$ by $x-3$ using synthetic division to find $f(3)$.

$$3\,\begin{array}{|rrrr}2 & -4 & 5 & -10\\ & 6 & 6 & 33\\\hline 2 & 2 & 11 & 23\end{array}$$

Thus, by the remainder theorem, $f(3)=23$.

35. Use synthetic division to divide
$$5x^4+10x^3+6x^2+8x-8 \text{ by } x+2.$$

$$-2\,\begin{array}{|rrrrr}5 & 10 & 6 & 8 & -8\\ & -10 & 0 & -12 & 8\\\hline 5 & 0 & 6 & -4 & 0\end{array}$$

$f(-2)=0$ so $x-(-2)=x+2$ is a factor by the factor theorem. The other factor is $5x^3+6x-4$.

36. $f(x)=3x^3+x^2-22x-24$; -2 is a zero

$$-2\,\begin{array}{|rrrr}3 & 1 & -22 & -24\\ & -6 & 10 & 24\\\hline 3 & -5 & -12 & 0\end{array}$$

$3x^2-5x-12=(3x+4)(x-3)$
The zeros of $f(x)$ are -2, $-\frac{4}{3}$, and 3.

37. $f(x)=-3x+5$

The equation is in slope-intercept form, so the y-intercept is $(0,5)$ and $m=-3$ or $\frac{-3}{1}$.

Plot $(0,5)$. From $(0,5)$, move down 3 units and right 1 unit to the point $(1,2)$. Draw the line through these two points.

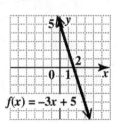

38. $f(x)=-2(x-1)^2+3$

The graph is a parabola that has been shifted 1 unit to the right and 3 units upward from $(0,0)$, so its vertex is at $(1,3)$. Since $a=-2<0$, the parabola opens down. Also $|a|=|-2|=2>1$, so the graph is narrower than the graph of $f(x)=x^2$. The points $(0,1)$ and $(2,1)$ are on the graph.

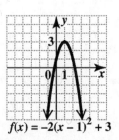

Copyright © 2012 Pearson Education, Inc. Publishing as Addison-Wesley.

39. $\dfrac{x^2}{25} + \dfrac{y^2}{16} \le 1$

The boundary, $\dfrac{x^2}{25} + \dfrac{y^2}{16} = 1$, is a solid ellipse in

$\dfrac{x^2}{a^2} + \dfrac{y^2}{b^2} = 1$ form with intercepts $(5, 0)$, $(-5, 0)$,

$(0, 4)$, and $(0, -4)$. Test $(0, 0)$.

$$\dfrac{0^2}{25} + \dfrac{0^2}{16} \overset{?}{\le} 1$$
$$0 \le 1 \quad \textit{True}$$

Shade the region inside the ellipse.

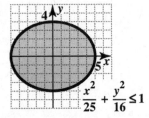

40. $f(x) = \sqrt{x - 2}$

This is the graph of $g(x) = \sqrt{x}$ shifted 2 units
right.

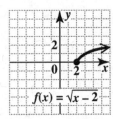

41. $\dfrac{x^2}{4} - \dfrac{y^2}{16} = 1$ is in $\dfrac{x^2}{a^2} - \dfrac{y^2}{b^2} = 1$ form.

The graph is a hyperbola with x-intercepts $(2, 0)$
and $(-2, 0)$ and asymptotes that are the extended
diagonals of the rectangle with vertices $(2, 4)$,
$(2, -4)$, $(-2, -4)$, and $(-2, 4)$. Draw a branch of
the hyperbola through each intercept approaching
the asymptotes.

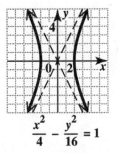

42. $f(x) = 3^x$

The graph of f is an increasing exponential.

x	-1	0	1	2
$f(x)$	$\frac{1}{3}$	1	3	9

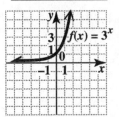

Copyright © 2012 Pearson Education, Inc. Publishing as Addison-Wesley.

CHAPTER 14 FURTHER TOPICS IN ALGEBRA

14.1 Sequences and Series

14.1 Now Try Exercises

N1. $a_n = 5 - 3n$

To get a_3, the third term, replace n with 3.

$$a_3 = 5 - 3(3) = -4$$

N2. $-3, 9, -27, 81, \ldots$ can be written as

$(-3)^1, (-3)^2, (-3)^3, (-3)^4, \ldots$, so $a_n = (-3)^n$.

N3. Make a table as follows:

Month	Interest	Payment	Unpaid balance
0			8,000
1	$8000(0.02) = 160$	$400 + 160 = 560$	$8000 - 400 = 7600$
2	$7600(0.02) = 152$	$400 + 152 = 552$	$7600 - 400 = 7200$
3	$7200(0.02) = 144$	$400 + 144 = 544$	$7200 - 400 = 6800$
4	$6800(0.02) = 136$	$400 + 136 = 536$	$6800 - 400 = 6400$

The payments are $560, $552, $544, and $536; the unpaid balance is $6400.

N4. $\displaystyle\sum_{i=1}^{5}(i^2 - 4)$

$$= (1^2 - 4) + (2^2 - 4) + (3^2 - 4)$$
$$+ (4^2 - 4) + (5^2 - 4)$$
$$= (1 - 4) + (4 - 4) + (9 - 4)$$
$$+ (16 - 4) + (25 - 4)$$
$$= -3 + 0 + 5 + 12 + 21$$
$$= 35$$

N5. (a) $3 + 5 + 7 + 9 + 11$

The terms increase by 2, so $2i$ is part of the general term. If $i = 1$, $2i = 2$, but since the first term is 3, we must add 1 to $2i$ to get 3. There are five terms, so

$$3 + 5 + 7 + 9 + 11 = \sum_{i=1}^{5}(2i + 1)$$

(b) $-1 - 4 - 9 - 16 - 25$
$$= -1^2 - 2^2 - 3^2 - 4^2 - 5^2$$
$$= \sum_{i=1}^{5} -i^2$$

N6. $\displaystyle \overline{x} = \frac{\displaystyle\sum_{i=1}^{n} x_i}{n} = \frac{\displaystyle\sum_{i=1}^{5} x_i}{5}$

$$= \frac{\left(\begin{array}{c}461{,}054 + 188{,}381 + 136{,}583 \\ + 107{,}630 + 93{,}958\end{array}\right)}{5}$$

$$= \frac{987{,}606}{5} = 197{,}521.2$$

To the nearest whole number, the average number of Quarter Horses registered per state in the top five states is 197,521.

14.1 Section Exercises

1. $a_n = n + 1$
$a_1 = 1 + 1 = 2$ *Replace n with 1.*
$a_2 = 2 + 1 = 3$ *Replace n with 2.*
$a_3 = 3 + 1 = 4$ *Replace n with 3.*
$a_4 = 4 + 1 = 5$ *Replace n with 4.*
$a_5 = 5 + 1 = 6$ *Replace n with 5.*

Answer: $2, 3, 4, 5, 6$

3. $a_n = \dfrac{n + 3}{n}$

To get a_i, replace n with i, where $i = 1, 2, 3, 4, 5$.

$a_1 = \dfrac{1 + 3}{1} = \dfrac{4}{1} = 4$ $a_2 = \dfrac{2 + 3}{2} = \dfrac{5}{2}$

$a_3 = \dfrac{3 + 3}{3} = \dfrac{6}{3} = 2$ $a_4 = \dfrac{4 + 3}{4} = \dfrac{7}{4}$

$a_5 = \dfrac{5 + 3}{5} = \dfrac{8}{5}$ **Answer:** $4, \dfrac{5}{2}, 2, \dfrac{7}{4}, \dfrac{8}{5}$

5. $a_n = 3^n$
To get a_i, replace n with i, where $i = 1, 2, 3, 4, 5$.

$a_1 = 3^1 = 3$ $a_2 = 3^2 = 9$

$a_3 = 3^3 = 27$ $a_4 = 3^4 = 81$

$a_5 = 3^5 = 243$ **Answer:** $3, 9, 27, 81, 243$

7. $a_n = \dfrac{1}{n^2}$
To get a_i, replace n with i, where $i = 1, 2, 3, 4, 5$.

$a_1 = \dfrac{1}{1^2} = 1$ $a_2 = \dfrac{1}{2^2} = \dfrac{1}{4}$

$a_3 = \dfrac{1}{3^2} = \dfrac{1}{9}$ $a_4 = \dfrac{1}{4^2} = \dfrac{1}{16}$

$a_5 = \dfrac{1}{5^2} = \dfrac{1}{25}$ **Answer:** $1, \dfrac{1}{4}, \dfrac{1}{9}, \dfrac{1}{16}, \dfrac{1}{25}$

9. $a_n = 5(-1)^{n-1}$

$a_1 = 5(-1)^{1-1} = 5(-1)^0 = 5(1) = 5$

$a_2 = 5(-1)^{2-1} = 5(-1)^1 = 5(-1) = -5$

$a_3 = 5(-1)^{3-1} = 5(-1)^2 = 5(1) = 5$

$a_4 = 5(-1)^{4-1} = 5(-1)^3 = 5(-1) = -5$

$a_5 = 5(-1)^{5-1} = 5(-1)^4 = 5(1) = 5$

Answer: $5, -5, 5, -5, 5$

Copyright © 2012 Pearson Education, Inc. Publishing as Addison-Wesley.

11. $a_n = n - \dfrac{1}{n}$

$a_1 = 1 - \dfrac{1}{1} = 0 \qquad a_2 = 2 - \dfrac{1}{2} = \dfrac{3}{2}$

$a_3 = 3 - \dfrac{1}{3} = \dfrac{8}{3} \qquad a_4 = 4 - \dfrac{1}{4} = \dfrac{15}{4}$

$a_5 = 5 - \dfrac{1}{5} = \dfrac{24}{5}$ **Answer:** $0, \dfrac{3}{2}, \dfrac{8}{3}, \dfrac{15}{4}, \dfrac{24}{5}$

13. $a_n = -9n + 2$
To find a_8, replace n with 8.

$a_8 = -9(8) + 2 = -72 + 2 = -70$

15. $a_n = \dfrac{3n + 7}{2n - 5}$
To find a_{14}, replace n with 14.

$a_{14} = \dfrac{3(14) + 7}{2(14) - 5} = \dfrac{42 + 7}{28 - 5} = \dfrac{49}{23}$

17. $a_n = (n + 1)(2n + 3)$
To find a_8, replace n with 8.

$\begin{aligned} a_8 &= (8 + 1)[2(8) + 3] \\ &= 9(19) \\ &= 171 \end{aligned}$

19. $4, 8, 12, 16, \ldots$ can be written as
$4 \cdot 1, 4 \cdot 2, 4 \cdot 3, 4 \cdot 4, \ldots$, so $a_n = 4n$.

21. $-8, -16, -24, -32, \ldots$ can be written as
$-8 \cdot 1, -8 \cdot 2, -8 \cdot 3, -8 \cdot 4, \ldots$, so $a_n = -8n$.

23. $\dfrac{1}{3}, \dfrac{1}{9}, \dfrac{1}{27}, \dfrac{1}{81}, \ldots$ can be written as
$\dfrac{1}{3^1}, \dfrac{1}{3^2}, \dfrac{1}{3^3}, \dfrac{1}{3^4}, \ldots$, so $a_n = \dfrac{1}{3^n}$.

25. $\dfrac{2}{5}, \dfrac{3}{6}, \dfrac{4}{7}, \dfrac{5}{8}, \ldots$ can be written as
$\dfrac{1 + 1}{1 + 4}, \dfrac{2 + 1}{2 + 4}, \dfrac{3 + 1}{3 + 4}, \dfrac{4 + 1}{4 + 4}, \ldots$, so $a_n = \dfrac{n + 1}{n + 4}$.

27. Make a table as follows:

Month	Interest	Payment	Unpaid balance
0			1000
1	$1000(0.01) = 10$	$100 + 10 = 110$	$1000 - 100 = 900$
2	$900(0.01) = 9$	$100 + 9 = 109$	$900 - 100 = 800$
3	$800(0.01) = 8$	$100 + 8 = 108$	$800 - 100 = 700$
4	$700(0.01) = 7$	$100 + 7 = 107$	$700 - 100 = 600$
5	$600(0.01) = 6$	$100 + 6 = 106$	$600 - 100 = 500$
6	$500(0.01) = 5$	$100 + 5 = 105$	$500 - 100 = 400$

The payments are \$110, \$109, \$108, \$107, \$106, and \$105; the unpaid balance is \$400.

29. When new, the car is worth \$20,000.
Let a_n = the value of the car after the nth year.
The car retains $\frac{4}{5}$ of its value each year.

$a_1 = \frac{4}{5}(20{,}000) = \$16{,}000$
(value after the first year)

$a_2 = \frac{4}{5}(16{,}000) = \$12{,}800$

$a_3 = \frac{4}{5}(12{,}800) = \$10{,}240$

$a_4 = \frac{4}{5}(10{,}240) = \$8192$

$a_5 = \frac{4}{5}(8192) = \$6553.60$

The value of the car after 5 years is about \$6554.

31. $\displaystyle\sum_{i=1}^{5} (i + 3)$

$\begin{aligned} &= (1 + 3) + (2 + 3) + (3 + 3) + (4 + 3) + (5 + 3) \\ &\qquad \qquad \qquad \textit{Let } i = 1 \textit{ to 5 and add terms.} \\ &= 4 + 5 + 6 + 7 + 8 \\ &= 30 \end{aligned}$

33. $\displaystyle\sum_{i=1}^{3} (i^2 + 2)$

$\begin{aligned} &= (1^2 + 2) + (2^2 + 2) + (3^2 + 2) \\ &\qquad \qquad \textit{Let } i = 1 \textit{ to 3 and add terms.} \\ &= 3 + 6 + 11 \\ &= 20 \end{aligned}$

35. $\displaystyle\sum_{i=1}^{6} (-1)^i$

$\begin{aligned} &= (-1)^1 + (-1)^2 + (-1)^3 + (-1)^4 \\ &\quad + (-1)^5 + (-1)^6 \\ &= -1 + 1 - 1 + 1 - 1 + 1 \\ &= 0 \end{aligned}$

37. $\displaystyle\sum_{i=3}^{7} (i - 3)(i + 2)$

$\begin{aligned} &= (3 - 3)(3 + 2) + (4 - 3)(4 + 2) \\ &\quad + (5 - 3)(5 + 2) + (6 - 3)(6 + 2) \\ &\quad + (7 - 3)(7 + 2) \\ &= 0(5) + 1(6) + 2(7) + 3(8) + 4(9) \\ &= 0 + 6 + 14 + 24 + 36 \\ &= 80 \end{aligned}$

39. $3 + 4 + 5 + 6 + 7$

$\begin{aligned} &= (1 + 2) + (2 + 2) + (3 + 2) \\ &\quad + (4 + 2) + (5 + 2) \\ &= \sum_{i=1}^{5} (i + 2) \end{aligned}$

41. $-2 + 4 - 8 + 16 - 32$

$\begin{aligned} &= (-1)^1 \cdot 2^1 + (-1)^2 \cdot 2^2 + (-1)^3 \cdot 2^3 \\ &\quad + (-1)^4 \cdot 2^4 + (-1)^5 \cdot 2^5 \\ &= \sum_{i=1}^{5} (-1)^i \cdot 2^i, \text{ or } \sum_{i=1}^{5} (-2)^i \end{aligned}$

43. $1 + 4 + 9 + 16$

$\begin{aligned} &= 1^2 + 2^2 + 3^2 + 4^2 \\ &= \sum_{i=1}^{4} i^2 \end{aligned}$

Copyright © 2012 Pearson Education, Inc. Publishing as Addison-Wesley.

45. A sequence is a list of terms in a specific order, while a series is the indicated sum of the terms of a sequence.

47. $\bar{x} = \dfrac{\displaystyle\sum_{i=1}^{n} x_i}{n} = \dfrac{\displaystyle\sum_{i=1}^{7} x_i}{7}$

$= \dfrac{8 + 11 + 14 + 9 + 7 + 6 + 8}{7} = \dfrac{63}{7} = 9$

49. $\bar{x} = \dfrac{5 + 9 + 8 + 2 + 4 + 7 + 3 + 2 + 0}{9} = \dfrac{40}{9}$

51. $\bar{x} = \dfrac{8041 + 7975 + 8117 + 8024 + 8022}{5}$

$= \dfrac{40{,}179}{5} = 8035.8$

The average number of funds available for this five-year period was about 8036.

53. $\begin{array}{ll} a + 3d = 12 & (1) \\ a + 8d = 22 & (2) \end{array}$

Multiply (1) by -1 and add the result to (2).

$\begin{array}{lll} -a - 3d = -12 & & -1 \times (1) \\ \underline{a + 8d = 22} & (2) & \\ 5d = 10 & & Add. \\ d = 2 & & \end{array}$

Substitute 2 for d in (1).

$$a + 3d = 12 \qquad (1)$$
$$a + 3(2) = 12$$
$$a + 6 = 12$$
$$a = 6$$

Thus, $a = 6$ and $d = 2$.

55. Given $a = -2$, $n = 5$, and $d = 3$,

$$a + (n - 1)d = -2 + (5 - 1)3$$
$$= -2 + (4)3$$
$$= -2 + 12$$
$$= 10.$$

14.2 Arithmetic Sequences

14.2 Now Try Exercises

N1. $-4, -13, -22, -31, -40, \ldots$

You should find the difference for all pairs of adjacent terms to determine if the sequence is arithmetic. In this case, we are *given* that the sequence is arithmetic, so d is the difference between any two adjacent terms. Choose the terms -13 and -4.

$$d = -13 - (-4) = -13 + 4 = -9$$

N2. Given $a_1 = 10$ and $d = -8$,

$$a_2 = a_1 + d = 10 - 8 = 2$$
$$a_3 = a_2 + d = 2 - 8 = -6$$
$$a_4 = a_3 + d = -6 - 8 = -14$$
$$a_5 = a_4 + d = -14 - 8 = -22$$

The first five terms of the sequence are 10, 2, -6, -14, -22.

N3. $-5, 0, 5, 10, 15, \ldots$

To find d, subtract any two adjacent terms.

$$d = 5 - 0 = 5$$

The first term is $a_1 = -5$. Use it to find a_n.

$$a_n = a_1 + (n - 1)d$$
$$= -5 + (n - 1)(5)$$
$$= -5 + 5n - 5$$
$$= 5n - 10$$

Now $a_{20} = 5(20) - 10 = 100 - 10 = 90$.

N4. After 1 month, the account will have

$$\$1000 + 1 \cdot \$120 = \$1120.$$

After 2 months, the account will have

$$\$1000 + 2 \cdot \$120 = \$1240.$$

In general, after n months, the account will have

$$\$1000 + n \cdot \$120.$$

Thus, after 96 months, the account will have

$$\$1000 + 96 \cdot \$120 = \$12{,}520.$$

N5. **(a)** Given $a_1 = 21$ and $d = -3$, find a_{22}.

$$a_n = a_1 + (n - 1)d$$
$$a_{22} = a_1 + (22 - 1)d$$
$$= 21 + 21(-3)$$
$$= -42$$

(b) Given $a_7 = 25$ and $a_{12} = 40$, find a_{19}. Use $a_n = a_1 + (n - 1)d$ to write a system of equations.

$$\begin{array}{ll} a_7 = a_1 + (7 - 1)d & \\ 25 = a_1 + 6d & (1) \\ a_{12} = a_1 + (12 - 1)d & \\ 40 = a_1 + 11d & (2) \end{array}$$

To eliminate a_1, multiply (1) by -1 and add the result to (2).

$$\begin{array}{lll} -25 = -a_1 - 6d & & -1 \times (1) \\ \underline{40 = a_1 + 11d} & (2) & \\ 15 = 5d & & Add. \\ 3 = d & & \end{array}$$

From (1), $25 = a_1 + 6(3)$, so $a_1 = 7$. Now find a_{19}.

Copyright © 2012 Pearson Education, Inc. Publishing as Addison-Wesley.

$$a_{19} = a_1 + (19 - 1)d$$
$$= 7 + 18(3)$$
$$= 61$$

An alternative approach: There are 5 differences from a_7 to a_{12}, so

$$a_{12} - a_7 = 5d.$$
$$40 - 25 = 5d$$
$$15 = 5d$$
$$3 = d$$

There are 7 differences from a_{12} to a_{19}, so

$$a_{19} = a_{12} + 7d.$$
$$= 40 + 7(3)$$
$$= 61$$

N6. $1, \frac{4}{3}, \frac{5}{3}, 2, \ldots, 11$

$a_n = a_1 + (n-1)d$	*Formula for a_n*
$11 = 1 + (n-1)(\frac{1}{3})$	$d = \frac{4}{3} - 1 = \frac{1}{3}$
$33 = 3 + (n-1)$	*Multiply by 3.*
$33 = n + 2$	*Combine terms.*
$31 = n$	*Subtract 2.*

The sequence has 31 terms.

N7. Since we want the sum of the first seven terms, we'll find a_1 and a_7 using $a_n = 5n - 7$.

$$a_1 = 5(1) - 7 = -2$$
$$a_7 = 5(7) - 7 = 28$$

Now use the formula for the sum of the first n terms of an arithmetic sequence.

$S_n = \frac{n}{2}(a_1 + a_n)$	*Formula for S_n.*
$S_7 = \frac{7}{2}(a_1 + a_7)$	*Let $n = 7$.*
$S_7 = \frac{7}{2}(-2 + 28)$	*Substitute.*
$= \frac{7}{2}(26)$	*Add.*
$= 91$	*Multiply.*

N8. We are given $a_1 = -8$, $d = -5$, and $n = 9$. Use the second formula for the sum of an arithmetic sequence.

$$S_n = \frac{n}{2}[2a_1 + (n-1)d]$$
$$S_9 = \frac{9}{2}[2(-8) + (9-1)(-5)]$$
$$= \frac{9}{2}[-16 + 8(-5)]$$
$$= \frac{9}{2}[-16 - 40]$$
$$= \frac{9}{2}(-56) = -252$$

N9. Evaluate $\displaystyle\sum_{i=1}^{11}(5i - 7)$.

To find the first and last (11th) terms, let $n = 1$ and $n = 11$ in $a_n = 5n - 7$.

$$a_1 = 5(1) - 7 = -2$$
$$a_{11} = 5(11) - 7 = 48$$

Now find S_{11}.

$$S_n = \frac{n}{2}(a_1 + a_n)$$
$$S_{11} = \frac{11}{2}(-2 + 48)$$
$$= \frac{11}{2}(46) = 253$$

14.2 Section Exercises

1. $1, 2, 3, 4, 5, \ldots$

d is the difference between any two adjacent terms. Choose the terms 3 and 2.

$$d = 3 - 2 = 1$$

The terms 2 and 1 would give

$$d = 2 - 1 = 1,$$

the same result. Therefore, the common difference is $d = 1$.

Note: You should find the difference for all pairs of adjacent terms to determine if the sequence is arithmetic.

3. $2, -4, 6, -8, 10, -12, \ldots$

The difference between the first two terms is $-4 - 2 = -6$, but the difference between the second and third terms is $6 - (-4) = 10$. The differences are not the same so the sequence is *not arithmetic.*

5. $10, 5, 0, -5, -10, \ldots$

Choose the terms 5 and 10, and find the difference.

$$d = 5 - 10 = -5$$

The terms -10 and -5 would give

$$d = -10 - (-5) = -10 + 5 = -5,$$

the same result. Therefore, the common difference is $d = -5$.

7. $a_1 = 5, d = 4$
$a_2 = a_1 + d = 5 + 4 = 9$
$a_3 = a_2 + d = 9 + 4 = 13$
$a_4 = a_3 + d = 13 + 4 = 17$
$a_5 = a_4 + d = 17 + 4 = 21$

Answer: $5, 9, 13, 17, 21$

9. $a_1 = -2, d = -4$
$a_2 = a_1 + d = -2 + (-4) = -6$
$a_3 = a_2 + d = -6 + (-4) = -10$
$a_4 = a_3 + d = -10 + (-4) = -14$
$a_5 = a_4 + d = -14 + (-4) = -18$

Answer: $-2, -6, -10, -14, -18$

Copyright © 2012 Pearson Education, Inc. Publishing as Addison-Wesley.

11. $a_1 = 2$, $d = 5$

$$\begin{aligned} a_n &= a_1 + (n-1)d \\ &= 2 + (n-1)5 \\ &= 2 + 5n - 5 \\ &= 5n - 3 \end{aligned}$$

13. $3, \frac{15}{4}, \frac{9}{2}, \frac{21}{4}, \ldots$

To find d, subtract any two adjacent terms.

$$d = \frac{15}{4} - 3 = \frac{15}{4} - \frac{12}{4} = \frac{3}{4}$$

The first term is $a_1 = 3$. Now find a_n.

$$\begin{aligned} a_n &= a_1 + (n-1)d \\ &= 3 + (n-1)\left(\frac{3}{4}\right) \\ &= 3 + \frac{3}{4}n - \frac{3}{4} \\ &= \frac{3}{4}n + \frac{9}{4} \end{aligned}$$

15. $-3, 0, 3, \ldots$

To find d, subtract any two adjacent terms.

$$d = 0 - (-3) = 3$$

The first term is $a_1 = -3$. Now find a_n.

$$\begin{aligned} a_n &= a_1 + (n-1)d \\ &= -3 + (n-1)3 \\ &= -3 + 3n - 3 \\ &= 3n - 6 \end{aligned}$$

17. Given $a_1 = 4$ and $d = 3$, find a_{25}.

$$\begin{aligned} a_n &= a_1 + (n-1)d \\ a_{25} &= 4 + (25-1)3 \\ &= 4 + 72 \\ &= 76 \end{aligned}$$

19. Given $2, 4, 6, \ldots$, find a_{24}.

Here, $a_1 = 2$ and $d = 4 - 2 = 2$.

$$\begin{aligned} a_n &= a_1 + (n-1)d \\ a_{24} &= 2 + (24-1)2 \\ &= 2 + 46 \\ &= 48 \end{aligned}$$

21. Given $a_{12} = -45$ and $a_{10} = -37$, find a_1.

Use $a_n = a_1 + (n-1)d$ to write a system of equations.

$$\begin{aligned} a_{12} &= a_1 + (12-1)d \\ -45 &= a_1 + 11d \qquad (1) \\ a_{10} &= a_1 + (10-1)d \\ -37 &= a_1 + 9d \qquad (2) \end{aligned}$$

To eliminate d, multiply equation (1) by -9 and equation (2) by 11. Then add the results.

$$\begin{array}{rll} 405 = & -9a_1 - 99d & -9 \times (1) \\ -407 = & 11a_1 + 99d & 11 \times (2) \\ \hline -2 = & 2a_1 & \\ -1 = & a_1 & \end{array}$$

23. $3, 5, 7, \ldots, 33$

Let n represent the number of terms in the sequence. So, $a_n = 33$, $a_1 = 3$, and $d = 5 - 3 = 2$.

$$\begin{aligned} a_n &= a_1 + (n-1)d \\ 33 &= 3 + (n-1)2 \\ 33 &= 3 + 2n - 2 \\ 33 &= 2n + 1 \\ 32 &= 2n \\ n &= 16 \end{aligned}$$

The sequence has 16 terms.

25. $\frac{3}{4}, 3, \frac{21}{4}, \ldots, 12$

Let n represent the number of terms in the sequence. So, $a_n = 12$, $a_1 = \frac{3}{4}$, and $d = 3 - \frac{3}{4} = \frac{9}{4}$.

$$\begin{aligned} a_n &= a_1 + (n-1)d \\ 12 &= \frac{3}{4} + (n-1)\left(\frac{9}{4}\right) \\ \frac{45}{4} &= (n-1)\left(\frac{9}{4}\right) \\ 5 &= n - 1 \qquad \textit{Multiply by } \frac{4}{9}. \\ 6 &= n \end{aligned}$$

The sequence has 6 terms.

27. n represents the number of terms.

29. Find S_6 given $a_1 = 6$, $d = 3$, and $n = 6$.

$$\begin{aligned} S_n &= \frac{n}{2}\left[2a_1 + (n-1)d\right] \\ S_6 &= \frac{6}{2}\left[2 \cdot 6 + (6-1)3\right] \\ &= 3(12 + 5 \cdot 3) \\ &= 3(27) \\ &= 81 \end{aligned}$$

31. Find S_6 given $a_1 = 7$, $d = -3$, and $n = 6$.

$$\begin{aligned} S_n &= \frac{n}{2}\left[2a_1 + (n-1)d\right] \\ S_6 &= \frac{6}{2}\left[2 \cdot 7 + (6-1)(-3)\right] \\ &= 3\left[14 + 5(-3)\right] \\ &= 3(14 - 15) \\ &= 3(-1) \\ &= -3 \end{aligned}$$

33. Find S_6 given $a_n = 4 + 3n$.

Find the first and last terms.

$$\begin{aligned} a_1 &= 4 + 3(1) = 7 \\ a_6 &= 4 + 3(6) = 22 \end{aligned}$$

Now find the sum.

$$\begin{aligned} S_n &= \frac{n}{2}(a_1 + a_n) \\ S_6 &= \frac{6}{2}(7 + 22) \\ &= 3(29) \\ &= 87 \end{aligned}$$

Copyright © 2012 Pearson Education, Inc. Publishing as Addison-Wesley.

35. $\sum\limits_{i=1}^{10} (8i - 5)$

$$a_n = 8n - 5$$
$$a_1 = 8(1) - 5 = 3$$
$$a_{10} = 8(10) - 5 = 75$$

Use $S_n = \frac{n}{2}(a_1 + a_n)$ with $n = 10$, $a_1 = 3$, and $a_{10} = 75$.

$$S_{10} = \frac{10}{2}(3 + 75)$$
$$= 5(78)$$
$$= 390$$

37. $\sum\limits_{i=1}^{20} \left(\frac{3}{2}i + 4\right)$

$$a_n = \frac{3}{2}n + 4$$
$$a_1 = \frac{3}{2}(1) + 4 = \frac{11}{2}$$
$$a_{20} = \frac{3}{2}(20) + 4 = 34$$

Use $S_n = \frac{n}{2}(a_1 + a_n)$ with $n = 20$, $a_1 = \frac{11}{2}$, and $a_{20} = 34$.

$$S_{20} = \frac{20}{2}\left(\frac{11}{2} + 34\right)$$
$$= \frac{20}{2}\left(\frac{79}{2}\right)$$
$$= 395$$

39. $\sum\limits_{i=1}^{250} i$ ■ Here, $a_n = n$, $a_1 = 1$, and $a_{250} = 250$.

Use $S_n = \frac{n}{2}(a_1 + a_n)$ with $n = 250$, $a_1 = 1$, and $a_{250} = 250$.

$$S_{250} = \frac{250}{2}(1 + 250)$$
$$= 125(251)$$
$$= 31,375$$

41. The sequence is $1, 2, 3, \ldots, 30$.

$$S_n = \frac{n}{2}(a_1 + a_n)$$
$$S_{30} = \frac{30}{2}(1 + 30)$$
$$= 15(31)$$
$$= 465$$

The account will have \$465 deposited in it over the entire month.

43. Cherian's salary at six-month intervals form an arithmetic sequence with $a_1 = 1600$ and $d = 50$. Since Cherian's salary is increased every 6 months, or $2(5) = 10$ times, after 5 years it will equal the term a_{11}.

$$a_n = a_1 + (n - 1)d$$
$$a_{11} = 1600 + (11 - 1)50$$
$$= 1600 + 500$$
$$= 2100$$

Cherian's salary will be \$2100/month.

45. Given the sequence $20, 22, 24, \ldots$, for 25 terms, find a_{25}. Here, $a_1 = 20$, $d = 22 - 20 = 2$, and $n = 25$.

$$a_n = a_1 + (n - 1)d$$
$$a_{25} = 20 + (25 - 1)2$$
$$= 20 + 48$$
$$= 68$$

There are 68 seats in the last row. Now find S_{25}.

$$S_n = \frac{n}{2}(a_1 + a_n)$$
$$S_{25} = \frac{25}{2}(20 + 68)$$
$$= \frac{25}{2}(88)$$
$$= 25(44)$$
$$= 1100$$

There are 1100 seats in the section.

47. Given the sequence $35, 31, 27, \ldots$, can the sequence end in 1? If not, find the last positive value. If the sequence ends in 1, we can find n, a whole number.

$$d = 31 - 35 = -4$$
$$a_n = a_1 + (n - 1)d$$
$$1 = 35 + (n - 1)(-4)$$
$$1 = 35 - 4n + 4$$
$$-38 = -4n$$
$$9.5 = n$$

Since n is not a whole number, the sequence cannot end in 1. The largest n possible is $n = 9$.

$$a_n = a_1 + (n - 1)d$$
$$a_9 = 35 + (9 - 1)(-4)$$
$$= 35 - 32 = 3$$

She can build 9 rows. There are 3 blocks in the last row.

49. $a = 2, r = 3, n = 2$

$$ar^n = 2(3)^2 = 2(9) = 18$$

51. $a = 4, r = \frac{1}{2}, n = 3$

$$ar^n = 4\left(\frac{1}{2}\right)^3 = 4\left(\frac{1}{8}\right) = \frac{1}{2}$$

14.3 Geometric Sequences

14.3 Now Try Exercises

N1. $\frac{1}{4}, -1, 4, -16, 64, \ldots$

You should find the ratio for all pairs of adjacent terms to determine if the sequence is geometric. In this case, we are *given* that the sequence is geometric, so to find r, choose any two adjacent terms and divide the second one by the first one. Choose the terms 4 and -16.

$$r = \frac{-16}{4} = -4$$

Copyright © 2012 Pearson Education, Inc. Publishing as Addison-Wesley.

Notice that any other two adjacent terms could have been used with the same result. The common ratio is $r = -4$.

N2. $\frac{1}{4}, -1, 4, -16, 64, \ldots$

The common ratio is

$$r = \frac{-16}{4} = -4.$$

Substitute into the formula for a_n.

$$a_n = a_1 r^{n-1} = \frac{1}{4}(-4)^{n-1}$$

N3. (a) Given $a_1 = 3$ and $r = 2$, find a_8.

$$a_n = a_1 r^{n-1}$$
$$a_8 = 3(2)^{8-1}$$
$$= 3(2)^7 = 3(128) = 384$$

(b) Given the geometric sequence $10, 2, \frac{2}{5}, \frac{2}{25},$ $\ldots$, find a_7.

The common ratio is

$$r = \frac{2}{10} = \frac{1}{5}.$$

Substitute into the formula for a_n.

$$a_n = a_1 r^{n-1}$$
$$a_7 = 10(\tfrac{1}{5})^{7-1} = \frac{10}{15,625} = \frac{2}{3125}$$

N4. $a_1 = 25$

$$a_2 = a_1 r = 25(-\tfrac{1}{5}) = -5$$
$$a_3 = a_2 r = -5(-\tfrac{1}{5}) = 1$$
$$a_4 = a_3 r = 1(-\tfrac{1}{5}) = -\tfrac{1}{5}$$
$$a_5 = a_4 r = -\tfrac{1}{5}(-\tfrac{1}{5}) = \tfrac{1}{25}$$

N5. Given $a_1 = 4$ and $r = 2$, find S_6.

$$S_n = \frac{a_1(1 - r^n)}{1 - r}$$
$$S_6 = \frac{4(1 - 2^6)}{1 - 2}$$
$$= \frac{4(1 - 64)}{-1}$$
$$= 4(63) = 252$$

N6. For $\displaystyle\sum_{i=1}^{5} 8\left(\tfrac{1}{2}\right)^i$, $a_1 = 4$ and $r = \tfrac{1}{2}$. Find S_5.

$$S_n = \frac{a_1(1 - r^n)}{1 - r}$$
$$S_5 = \frac{4\left[1 - \left(\tfrac{1}{2}\right)^5\right]}{1 - \tfrac{1}{2}}$$
$$= \frac{4\left(1 - \tfrac{1}{32}\right)}{\tfrac{1}{2}}$$
$$= 8\left(\tfrac{32}{32} - \tfrac{1}{32}\right) = 8\left(\tfrac{31}{32}\right) = \tfrac{31}{4}$$

N7. (a) Use the formula for the future value of an ordinary annuity with $R = 600$, $i = 0.025$, and $n = 18$.

$$S = R\left[\frac{(1 + i)^n - 1}{i}\right]$$
$$S = 600\left[\frac{(1 + 0.025)^{18} - 1}{0.025}\right]$$
$$= 13,431.81$$

The future value of the annuity is \$13,431.81.

(b) Now use $r = \$100$, $i = \frac{0.03}{12} = 0.0025$, and $n = 12(18) = 216$.

$$S = \$100\left[\frac{(1 + 0.0025)^{216} - 1}{0.0025}\right]$$
$$= \$28,594.03$$

The future value of the annuity is \$28,594.03.

N8. Use the formula for the sum of the terms of an infinite geometric sequence with $a_1 = -4$ and $r = \frac{2}{3}$.

$$S = \frac{a_1}{1 - r} = \frac{-4}{1 - \frac{2}{3}} = \frac{-4}{\frac{1}{3}} = -12$$

N9. For $\displaystyle\sum_{i=1}^{\infty} \left(\frac{5}{8}\right)\left(\frac{3}{4}\right)^i$, $a_1 = \frac{15}{32}$ and $r = \frac{3}{4}$.

$$S = \frac{a_1}{1 - r} = \frac{\frac{15}{32}}{1 - \frac{3}{4}} = \frac{\frac{15}{32}}{\frac{1}{4}} = \frac{15}{8}$$

14.3 Section Exercises

1. $4, 8, 16, 32, \ldots$

To find r, choose any two adjacent terms and divide the second one by the first one.

$$r = \tfrac{8}{4} = 2$$

Notice that any two other adjacent terms could have been used with the same result. The common ratio is $r = 2$.

Note: You should find the ratio for all pairs of adjacent terms to determine if the sequence is geometric.

3. $\frac{1}{3}, \frac{2}{3}, \frac{3}{3}, \frac{4}{3}, \ldots$

Choose any two adjacent terms and divide the second by the first.

$$r = \frac{\frac{2}{3}}{\frac{1}{3}} = \frac{2}{3} \cdot 3 = 2$$

Confirm this result with any two other adjacent terms.

$$r = \frac{\frac{3}{3}}{\frac{2}{3}} = 1 \cdot \frac{3}{2} = \frac{3}{2}$$

Since $2 \neq \frac{3}{2}$, the ratios are not the same. The sequence is *not geometric*.

Copyright © 2012 Pearson Education, Inc. Publishing as Addison-Wesley.

5. $1, -3, 9, -27, 81, \ldots$

[1st and 2nd terms] $\quad r = \frac{-3}{1} = -3$

[2nd and 3rd terms] $\quad r = \frac{9}{-3} = -3$

The common ratio is $r = -3$.

7. $1, -\frac{1}{2}, \frac{1}{4}, -\frac{1}{8}, \ldots$

$\left[\dfrac{\text{2nd term}}{\text{1st term}}\right] \quad r = \dfrac{-\frac{1}{2}}{1} = -\dfrac{1}{2}$

$\left[\dfrac{\text{3rd term}}{\text{2nd term}}\right] \quad r = \dfrac{\frac{1}{4}}{-\frac{1}{2}} = \dfrac{1}{4}\left(-\dfrac{2}{1}\right) = -\dfrac{1}{2}$

The common ratio is $r = -\frac{1}{2}$.

9. Find a general term for $-5, -10, -20, \ldots$.

First, find r. $\quad r = \frac{-10}{-5} = 2$

Use $a_1 = -5$ and $r = 2$ to find a_n.

$$a_n = a_1 r^{n-1}$$
$$a_n = -5(2)^{n-1}$$

11. Find a general term for $-2, \frac{2}{3}, -\frac{2}{9}, \ldots$.

First, find r. $\quad r = \dfrac{\frac{2}{3}}{-2} = \dfrac{2}{3}\left(-\dfrac{1}{2}\right) = -\dfrac{1}{3}$

Use $a_1 = -2$ and $r = -\frac{1}{3}$ to find a_n.

$$a_n = a_1 r^{n-1}$$
$$a_n = -2\left(-\tfrac{1}{3}\right)^{n-1}$$

13. Find a general term for $10, -2, \frac{2}{5}, \ldots$.

First, find r. $\quad r = \frac{-2}{10} = -\frac{1}{5}$

Use $a_1 = 10$ and $r = -\frac{1}{5}$ to find a_n.

$$a_n = a_1 r^{n-1}$$
$$a_n = 10\left(-\tfrac{1}{5}\right)^{n-1}$$

15. Substitute $a_1 = 2$, $r = 5$, and $n = 10$ in the nth-term formula to find a_{10}.

$$a_n = a_1 r^{n-1}$$
$$a_{10} = a_1(r)^{10-1}$$
$$= 2(5)^9 = 3{,}906{,}250$$

17. Given $\frac{1}{2}, \frac{1}{6}, \frac{1}{18}, \ldots$, find a_{12}.

First, find r. $\quad r = \dfrac{\frac{1}{6}}{\frac{1}{2}} = \dfrac{1}{6}\left(\dfrac{2}{1}\right) = \dfrac{1}{3}$

Substitute $a_1 = \frac{1}{2}$, $r = \frac{1}{3}$, and $n = 12$ in the nth-term formula.

$$a_n = a_1 r^{n-1}$$
$$a_{12} = a_1(r)^{12-1}$$
$$= \tfrac{1}{2}\left(\tfrac{1}{3}\right)^{11} \text{ or } \tfrac{1}{354{,}294}$$

19. Given $a_3 = \frac{1}{2}$ and $a_7 = \frac{1}{32}$, find a_{25}.

Find a_1 and r using the general term $a_n = a_1 r^{n-1}$.

$$a_3 = a_1 r^{3-1}$$
$$\tfrac{1}{2} = a_1 r^2 \qquad (1)$$
$$a_7 = a_1 r^{7-1}$$
$$\tfrac{1}{32} = a_1 r^6 \qquad (2)$$

Solve (1) for a_1.

$$a_1 = \dfrac{1}{2r^2}$$

Substitute $\frac{1}{2r^2}$ for a_1 in (2).

$$\dfrac{1}{32} = \dfrac{1}{2r^2}r^6$$
$$\tfrac{1}{16} = r^4$$
$$r^2 = \pm\tfrac{1}{4}$$

Since r^2 is positive,

$$r^2 = \tfrac{1}{4} \quad \text{and} \quad r = \pm\tfrac{1}{2}.$$

Substitute $\frac{1}{4}$ for r^2 in (1).

$$\tfrac{1}{2} = a_1\left(\tfrac{1}{4}\right)$$
$$2 = a_1$$

Use $a_1 = 2$ and $r = \frac{1}{2}$ $\left(\text{or } -\frac{1}{2}\right)$ to find a_{25}.

$$a_{25} = a_1(r)^{25-1}$$
$$= 2\left(\tfrac{1}{2}\right)^{24} = \dfrac{1}{2^{23}}$$

21. $a_1 = 2$, $r = 3$; use $a_n = a_1 r^{n-1}$.

$$a_2 = a_1 r^1 = 2(3) = 6$$
$$a_3 = a_1 r^2 = 2(3)^2 = 18$$
$$a_4 = a_1 r^3 = 2(3)^3 = 54$$
$$a_5 = a_1 r^4 = 2(3)^4 = 162$$

Answer: $2, 6, 18, 54, 162$

23. $a_1 = 5$, $r = -\frac{1}{5}$; use $a_n = a_1 r^{n-1}$.

$$a_2 = a_1 r^1 = 5\left(-\tfrac{1}{5}\right) = -1$$
$$a_3 = a_1 r^2 = 5\left(-\tfrac{1}{5}\right)^2 = \tfrac{1}{5}$$
$$a_4 = a_1 r^3 = 5\left(-\tfrac{1}{5}\right)^3 = -\tfrac{1}{25}$$
$$a_5 = a_1 r^4 = 5\left(-\tfrac{1}{5}\right)^4 = \tfrac{1}{125}$$

Answer: $5, -1, \frac{1}{5}, -\frac{1}{25}, \frac{1}{125}$

25. $\frac{1}{3}, \frac{1}{9}, \frac{1}{27}, \frac{1}{81}, \frac{1}{243}$

$a_1 = \frac{1}{3}$, $n = 5$, and $r = \dfrac{\frac{1}{9}}{\frac{1}{3}} = \dfrac{1}{9} \cdot 3 = \dfrac{1}{3}$.

$$S_n = \dfrac{a_1(1 - r^n)}{1 - r}$$
$$S_5 = \dfrac{\frac{1}{3}\left[1 - \left(\frac{1}{3}\right)^5\right]}{1 - \frac{1}{3}}$$

Copyright © 2012 Pearson Education, Inc. Publishing as Addison-Wesley.

$$= \frac{\frac{1}{3}\left(1 - \frac{1}{243}\right)}{\frac{2}{3}}$$

$$= \frac{\frac{1}{3}\left(\frac{242}{243}\right)}{\frac{2}{3}} = \frac{121}{243}$$

27. $-\frac{4}{3}, -\frac{4}{9}, -\frac{4}{27}, -\frac{4}{81}, -\frac{4}{243}, -\frac{4}{729}$

$a_1 = -\frac{4}{3}, n = 6$, and $r = \frac{-\frac{4}{9}}{-\frac{4}{3}} = -\frac{4}{9}\left(-\frac{3}{4}\right) = \frac{1}{3}$.

$$S_n = \frac{a_1(1 - r^n)}{1 - r}$$

$$S_6 = \frac{-\frac{4}{3}\left[1 - \left(\frac{1}{3}\right)^6\right]}{1 - \frac{1}{3}}$$

$$= \frac{-\frac{4}{3}\left(1 - \frac{1}{729}\right)}{\frac{2}{3}}$$

$$= \frac{-\frac{4}{3}\left(\frac{728}{729}\right)}{\frac{2}{3}} = -\frac{1456}{729} \approx -1.997$$

29. $\sum_{i=1}^{7} 4\left(\frac{2}{5}\right)^i$

Use $a_1 = 4\left(\frac{2}{5}\right) = \frac{8}{5}, n = 7$, and $r = \frac{2}{5}$.

$$S_n = \frac{a_1(1 - r^n)}{1 - r}$$

$$S_7 = \frac{\frac{8}{5}\left[1 - \left(\frac{2}{5}\right)^7\right]}{1 - \frac{2}{5}}$$

$$= \frac{\frac{8}{5}\left[1 - \left(\frac{2}{5}\right)^7\right]}{\frac{3}{5}}$$

$$= \frac{8}{3}\left[1 - \left(\frac{2}{5}\right)^7\right] \approx 2.662$$

31. $\sum_{i=1}^{10} (-2)\left(\frac{3}{5}\right)^i$

Use $a_1 = (-2)\left(\frac{3}{5}\right) = -\frac{6}{5}, n = 10$, and $r = \frac{3}{5}$.

$$S_n = \frac{a_1(1 - r^n)}{1 - r}$$

$$S_{10} = \frac{-\frac{6}{5}\left[1 - \left(\frac{3}{5}\right)^{10}\right]}{1 - \frac{3}{5}}$$

$$= \frac{-\frac{6}{5}\left[1 - \left(\frac{3}{5}\right)^{10}\right]}{\frac{2}{5}}$$

$$= -3\left[1 - \left(\frac{3}{5}\right)^{10}\right] \approx -2.982$$

33. There are 21 deposits, so $n = 21$.

$$S = R\left[\frac{(1 + i)^n - 1}{i}\right]$$

$$= 1000\left[\frac{(1 + 0.044)^{21} - 1}{0.044}\right]$$

$$= 33,410.84$$

There will be $33,410.84 in the account.

35. Quarterly deposits for 10 years give us $n = 4 \cdot 10 = 40$. The interest rate per period is $i = \frac{0.05}{4} = 0.0125$.

$$S = R\left[\frac{(1 + i)^n - 1}{i}\right]$$

$$= 1200\left[\frac{(1 + 0.0125)^{40} - 1}{0.0125}\right]$$

$$= 61,787.47$$

We now use the compound interest formula to determine the value of this money after 5 more years.

$$A = P\left(1 + \frac{r}{n}\right)^{nt}$$

$$= 61,787.47\left(1 + \frac{0.06}{12}\right)^{12(5)}$$

$$= 83,342.04$$

The woman is also saving $300 per month, so we use the annuity formula to determine that value.

$$S = R\left[\frac{(1 + i)^n - 1}{i}\right]$$

$$= 300\left[\frac{\left(1 + \frac{0.06}{12}\right)^{12(5)} - 1}{\frac{0.06}{12}}\right]$$

$$= 20,931.01$$

Adding $20,931.01 to $83,342.04 gives a total of $104,273.05 in the account.

37. Find the sum if $a_1 = 6$ and $r = \frac{1}{3}$.
Since $|r| < 1$, the sum exists.

$$S = \frac{a_1}{1 - r} = \frac{6}{1 - \frac{1}{3}} = \frac{6}{\frac{2}{3}} = 6 \cdot \frac{3}{2} = 9$$

39. Find the sum if $a_1 = 1000$ and $r = -\frac{1}{10}$.
Since $|r| < 1$, the sum exists.

$$S = \frac{a_1}{1 - r} = \frac{1000}{1 - \left(-\frac{1}{10}\right)} = \frac{1000}{\frac{11}{10}}$$

$$= 1000 \cdot \frac{10}{11} = \frac{10,000}{11}$$

41. $\sum_{i=1}^{\infty} \frac{9}{8}\left(-\frac{2}{3}\right)^i$

$a_1 = \frac{9}{8}\left(-\frac{2}{3}\right)^1 = -\frac{3}{4}$ and $r = -\frac{2}{3}$.
Since $|r| < 1$, the sum exists.

$$S = \frac{a_1}{1 - r} = \frac{-\frac{3}{4}}{1 - \left(-\frac{2}{3}\right)} = \frac{-\frac{3}{4}}{\frac{5}{3}}$$

$$= -\frac{3}{4} \cdot \frac{3}{5} = -\frac{9}{20}$$

43. $\sum_{i=1}^{\infty} \frac{12}{5}\left(\frac{5}{4}\right)^i$

Since $|r| = \frac{5}{4} > 1$, the sum *does not exist*.

Copyright © 2012 Pearson Education, Inc. Publishing as Addison-Wesley.

45. The ball is dropped from a height of 10 feet and will rebound $\frac{3}{5}$ of its original height.

Let a_n = the ball's height on the nth rebound.

$$a_1 = 10 \quad \text{and} \quad r = \frac{3}{5}.$$

Since we must find the height after the fourth bounce, $n = 5$ (since a_1 is the starting point). Use $a_n = a_1 r^{n-1}$.

$$a_5 = 10\left(\tfrac{3}{5}\right)^{5-1} = 10\left(\tfrac{3}{5}\right)^4 \approx 1.3$$

The ball will rebound approximately 1.3 feet after the fourth bounce.

47. This exercise can be modeled by a geometric sequence with $a_1 = 256$ and $r = \frac{1}{2}$. First we need to find n so that $a_n = 32$.

$$32 = a_1 r^{n-1}$$
$$32 = 256\left(\tfrac{1}{2}\right)^{n-1}$$
$$\tfrac{1}{8} = \left(\tfrac{1}{2}\right)^{n-1}$$
$$\left(\tfrac{1}{2}\right)^3 = \left(\tfrac{1}{2}\right)^{n-1}$$
$$3 = n - 1$$
$$4 = n$$

Since n is 4, this means that 32 grams will be present on the day which corresponds to the 4th term of the sequence. That would be on day 3. To find what is left after the tenth day, we need to find a_{11} since we started with a_1.

$$a_{11} = a_1 r^{11-1} = 256\left(\tfrac{1}{2}\right)^{10} = \tfrac{256}{1024} = \tfrac{1}{4}$$

There will be $\frac{1}{4}$ gram of the substance after 10 days.

49. **(a)** Here, $a_1 = 1.1$ billion and $r = 106\% = 1.06$. Since we must find the consumption after 5 years, $n = 6$ (since a_1 is the starting point).

$$a_6 = a_1 r^{n-1}$$
$$a_6 = 1.1(1.06)^{6-1}$$
$$= 1.1(1.06)^5 \approx 1.5$$

The community will use about 1.5 billion units 5 years from now.

(b) If consumption doubles, then the consumption would be $2a_1$.

$$2a_1 = a_1(1.06)^{n-1}$$
$$2 = (1.06)^{n-1}$$
$$\ln 2 = \ln(1.06)^{n-1}$$
$$\ln 2 = (n-1)\ln(1.06)$$
$$n - 1 = \frac{\ln 2}{\ln 1.06}$$
$$n - 1 \approx 12$$
$$n \approx 13$$

Since n is about 13, that would represent the 13th term of the sequence, which represents about 12 years after the start.

51. Since the machine depreciates by $\frac{1}{4}$ of its value, it retains $1 - \frac{1}{4} = \frac{3}{4}$ of its value. Since the cost of the machine new is \$50,000, $a_1 = 50,000$. We want the value after 8 years so since the original cost is a_1, we need to find a_9.

$$a_9 = a_1 r^{9-1} = 50{,}000\left(\tfrac{3}{4}\right)^8 \approx 5005.65$$

The machine's value after 8 years is about \$5005.65.

♦♦♦ Relating Concepts 53–58 ♦♦♦

53. $\frac{1}{3} = 0.33333\ldots$

54. $\frac{2}{3} = 0.66666\ldots$

55. $0.33333\ldots$
$$\underline{+\ 0.66666\ldots}$$
$$0.99999\ldots$$

56. $S = \dfrac{a_1}{1-r} = \dfrac{0.9}{1-0.1} = \dfrac{0.9}{0.9} = 1$

Therefore, $0.99999\ldots = 1$.

57. From Exercise 56, $0.99999\ldots = 1$, so choice **B** is correct.

58. $0.49999\ldots = 0.4 + 0.09999\ldots$
$$= \tfrac{4}{10} + \tfrac{1}{10}(0.9999\ldots)$$
$$= \tfrac{4}{10} + \tfrac{1}{10}(1)$$
$$= \tfrac{5}{10} = \tfrac{1}{2}$$

59. $(3x + 2y)^2 = (3x)^2 + 2(3x)(2y) + (2y)^2$
$$= 9x^2 + 12xy + 4y^2$$

61. $(a-b)^3$
$$= (a-b)^2(a-b)^1$$
$$= \left(a^2 - 2ab + b^2\right)(a-b)$$
$$= a^3 - 2a^2b + ab^2 - a^2b + 2ab^2 - b^3$$
$$= a^3 - 3a^2b + 3ab^2 - b^3$$

14.4 The Binomial Theorem

14.4 Now Try Exercises

N1. $7! = 7 \cdot 6 \cdot 5 \cdot 4 \cdot 3 \cdot 2 \cdot 1 = 5040$

N2. **(a)** $\dfrac{8!}{6!\,2!} = \dfrac{8 \cdot 7 \cdot 6 \cdot 5 \cdot 4 \cdot 3 \cdot 2 \cdot 1}{(6 \cdot 5 \cdot 4 \cdot 3 \cdot 2 \cdot 1)(2 \cdot 1)}$
$$= \dfrac{8 \cdot 7}{2 \cdot 1} = \dfrac{4 \cdot 7}{1} = 28$$

(b) $\dfrac{8!}{5!\,3!} = \dfrac{8 \cdot 7 \cdot 6 \cdot 5 \cdot 4 \cdot 3 \cdot 2 \cdot 1}{(5 \cdot 4 \cdot 3 \cdot 2 \cdot 1)(3 \cdot 2 \cdot 1)}$
$$= \dfrac{8 \cdot 7 \cdot 6}{3 \cdot 2 \cdot 1} = \dfrac{8 \cdot 7}{1} = 56$$

Copyright © 2012 Pearson Education, Inc. Publishing as Addison-Wesley.

(c) $\dfrac{6!}{6!\,0!} = \dfrac{1}{0!}$ [cancel $6!$] $= \dfrac{1}{1} = 1$

(d) $\dfrac{6!}{5!\,1!} = \dfrac{6\cdot5\cdot4\cdot3\cdot2\cdot1}{(5\cdot4\cdot3\cdot2\cdot1)(1)} = \dfrac{6}{1} = 6$

N3. $_7C_2 = \dfrac{7!}{2!\,(7-2)!} = \dfrac{7!}{2!\,5!}$

$= \dfrac{7\cdot6\cdot5\cdot4\cdot3\cdot2\cdot1}{(2\cdot1)(5\cdot4\cdot3\cdot2\cdot1)} = 21$

N4. $(a+3b)^5$

$= (a)^5 + \dfrac{5!}{1!\,4!}(a)^4(3b)^1 + \dfrac{5!}{2!\,3!}(a)^3(3b)^2$

$+ \dfrac{5!}{3!\,2!}(a)^2(3b)^3 + \dfrac{5!}{4!\,1!}(a)^1(3b)^4 + (3b)^5$

$= a^5 + 5a^4(3b) + 10a^3(9b^2)$

$\quad + 10a^2(27b^3) + 5a(81b^4) + 243b^5$

$= a^5 + 15a^4b + 90a^3b^2$

$\quad + 270a^2b^3 + 405ab^4 + 243b^5$

N5. $\left(\dfrac{x}{3} - 2y\right)^4$

$= \left(\dfrac{x}{3}\right)^4 + \dfrac{4!}{1!\,3!}\left(\dfrac{x}{3}\right)^3(-2y)^1 + \dfrac{4!}{2!\,2!}\left(\dfrac{x}{3}\right)^2(-2y)^2$

$+ \dfrac{4!}{3!\,1!}\left(\dfrac{x}{3}\right)^1(-2y)^3 + (-2y)^4$

$= \dfrac{x^4}{3^4} + 4\cdot\dfrac{x^3}{3^3}(-2y) + 6\cdot\dfrac{x^2}{3^2}(4y^2)$

$+ 4\cdot\dfrac{x}{3}(-8y^3) + 16y^4$

$= \tfrac{1}{81}x^4 - \tfrac{8}{27}x^3y + \tfrac{8}{3}x^2y^2 - \tfrac{32}{3}xy^3 + 16y^4$

N6. In the sixth term of the expansion of $\left(2m - n^2\right)^8$, the exponent on $(-n^2)$ is $6 - 1 = 5$ and the exponent on $2m$ is $8 - 5 = 3$. The sixth term is

$\dfrac{8!}{5!\,3!}(2m)^3(-n^2)^5 = \dfrac{8\cdot7\cdot6}{3\cdot2\cdot1}\cdot2^3m^3\cdot(-n^{10})$

$= -56\cdot8m^3n^{10}$

$= -448m^3n^{10}$

14.4 Section Exercises

1. $6! = 6\cdot5\cdot4\cdot3\cdot2\cdot1 = 720$

3. $8! = 8\cdot7\cdot6\cdot5\cdot4\cdot3\cdot2\cdot1 = 40{,}320$

5. $\dfrac{6!}{4!\,2!} = \dfrac{6\cdot5\cdot4\cdot3\cdot2\cdot1}{(4\cdot3\cdot2\cdot1)(2\cdot1)} = \dfrac{6\cdot5}{2\cdot1} = 15$

7. $\dfrac{4!}{0!\,4!} = \dfrac{1}{0!}$ [cancel $4!$] $= \dfrac{1}{1} = 1$

9. $4!\cdot5 = (4\cdot3\cdot2\cdot1)\cdot5$

$= 24\cdot5 = 120$

11. $_6C_2 = \dfrac{6!}{2!\,(6-2)!} = \dfrac{6!}{2!\,4!} = 15$,

by Exercise 5.

13. $_{13}C_{11} = \dfrac{13!}{11!\,(13-11)!} = \dfrac{13!}{11!\,2!}$

$= \dfrac{13\cdot12}{2\cdot1}$ *11! cancels*

$= 13\cdot6 = 78$

15. $(m+n)^4$

$= m^4 + \dfrac{4!}{1!\,3!}m^3n^1 + \dfrac{4!}{2!\,2!}m^2n^2$

$+ \dfrac{4!}{3!\,1!}m^1n^3 + n^4$

$= m^4 + 4m^3n + 6m^2n^2 + 4mn^3 + n^4$

17. $(a-b)^5 = [a+(-b)]^5$

$= a^5 + \dfrac{5!}{1!\,4!}a^4(-b)^1 + \dfrac{5!}{2!\,3!}a^3(-b)^2$

$+ \dfrac{5!}{3!\,2!}a^2(-b)^3 + \dfrac{5!}{4!\,1!}a^1(-b)^4 + (-b)^5$

$= a^5 - 5a^4b + 10a^3b^2 - 10a^2b^3 + 5ab^4 - b^5$

19. $(2x+3)^3$

$= (2x)^3 + \dfrac{3!}{1!\,2!}(2x)^2(3)^1 + \dfrac{3!}{2!\,1!}(2x)(3)^2$

$+ (3)^3$

$= 8x^3 + 3(4x^2)(3) + 3(2x)(9) + 27$

$= 8x^3 + 36x^2 + 54x + 27$

21. $\left(\dfrac{x}{2} - y\right)^4 = \left[\left(\dfrac{x}{2}\right) + (-y)\right]^4$

$= \left(\dfrac{x}{2}\right)^4 + \dfrac{4!}{1!\,3!}\left(\dfrac{x}{2}\right)^3(-y)^1 + \dfrac{4!}{2!\,2!}\left(\dfrac{x}{2}\right)^2(-y)^2$

$+ \dfrac{4!}{3!\,1!}\left(\dfrac{x}{2}\right)^1(-y)^3 + (-y)^4$

$= \dfrac{x^4}{16} - \dfrac{x^3y}{2} + \dfrac{3x^2y^2}{2} - 2xy^3 + y^4$

23. $(x^2+1)^4$

$= \left(x^2\right)^4 + \dfrac{4!}{1!\,3!}\left(x^2\right)^3(1)^1 + \dfrac{4!}{2!\,2!}\left(x^2\right)^2(1)^2$

$+ \dfrac{4!}{3!\,1!}\left(x^2\right)^1(1)^3 + (1)^4$

$= x^8 + 4x^6 + 6x^4 + 4x^2 + 1$

25. $(3x^2 - y^2)^3 = \left[3x^2 + (-y^2)\right]^3$

$= \left(3x^2\right)^3 + \dfrac{3!}{1!\,2!}\left(3x^2\right)^2\left(-y^2\right)^1$

$+ \dfrac{3!}{2!\,1!}\left(3x^2\right)^1\left(-y^2\right)^2 + \left(-y^2\right)^3$

$= 27x^6 - 3(9x^4)y^2 + 3(3x^2)y^4 - y^6$

$= 27x^6 - 27x^4y^2 + 9x^2y^4 - y^6$

27. $(r+2s)^{12} = r^{12} + \dfrac{12!}{1!\,11!}r^{11}(2s)^1$

$+ \dfrac{12!}{2!\,10!}r^{10}(2s)^2 + \dfrac{12!}{3!\,9!}r^9(2s)^3 + \cdots$

The first four terms of the expansion are

$r^{12} + 24r^{11}s + 264r^{10}s^2 + 1760r^9s^3.$

Copyright © 2012 Pearson Education, Inc. Publishing as Addison-Wesley.

29. $(3x - y)^{14} = [3x + (-y)]^{14}$

$= (3x)^{14} + \dfrac{14!}{1!\,13!}(3x)^{13}(-y)^1$

$+ \dfrac{14!}{2!\,12!}(3x)^{12}(-y)^2$

$+ \dfrac{14!}{3!\,11!}(3x)^{11}(-y)^3 + \cdots$

The first four terms of the expansion are

$3^{14}x^{14} - 14(3^{13})x^{13}y$

$+ 91(3^{12})x^{12}y^2 - 364(3^{11})x^{11}y^3.$

31. $(t^2 + u^2)^{10}$

$= (t^2)^{10} + \dfrac{10!}{1!\,9!}(t^2)^9(u^2)^1$

$+ \dfrac{10!}{2!\,8!}(t^2)^8(u^2)^2$

$+ \dfrac{10!}{3!\,7!}(t^2)^7(u^2)^3 + \cdots$

The first four terms of the expansion are

$t^{20} + 10t^{18}u^2 + 45t^{16}u^4 + 120t^{14}u^6.$

33. The rth term of the expansion of $(x + y)^n$ is

$$\dfrac{n!}{(r-1)!\,[n-(r-1)]!}(x)^{n-(r-1)}(y)^{r-1}.$$

Start with the exponent on y, which is 1 less than the term number r. In this case, we are looking for the fourth term, so $r = 4$ and $r - 1 = 3$.
Thus, the fourth term of $(2m + n)^{10}$ is

$$\dfrac{10!}{3!\,(10-3)!}(2m)^{10-3}(n)^3$$

$$= \dfrac{10!}{3!\,7!}2^7m^7n^3$$

$$= 120(2^7)m^7n^3.$$

35. The seventh term of $\left(x + \dfrac{y}{2}\right)^8$ is

$$\dfrac{8!}{6!\,(8-6)!}(x)^{8-6}\left(\dfrac{y}{2}\right)^6$$

$$= \dfrac{8!}{6!\,2!}x^2\dfrac{y^6}{2^6}$$

$$= 28\,\dfrac{x^2y^6}{2^6} = \dfrac{7x^2y^6}{16}.$$

37. The third term of $(k - 1)^9$ is

$$\dfrac{9!}{2!\,(9-2)!}k^{9-2}(-1)^2 = \dfrac{9!}{2!\,7!}k^7 = 36k^7.$$

39. The expansion of $(x^2 + 2y)^6$ has seven terms, so the middle term is the fourth.
The fourth term of $(x^2 + 2y)^6$ is

$$\dfrac{6!}{3!\,(6-3)!}(x^2)^{6-3}(2y)^3$$

$$= \dfrac{6!}{3!\,3!}(x^2)^3(8y^3)$$

$$= 20x^6(8y^3) = 160x^6y^3.$$

41. The term of the expansion of $(3x^3 - 4y^2)^5$ with x^9y^4 in it is the term with $(3x^3)^3(-4y^2)^2$, since $(x^3)^3(y^2)^2 = x^9y^4$. The term is

$$\dfrac{5!}{2!\,3!}(3x^3)^3(-4y^2)^2$$

$$= 10(27x^9)(16y^4) = 4320x^9y^4.$$

43. $\dfrac{5n(n+1)}{2}$

(a) $\dfrac{5(1)(1+1)}{2}$ Let $n = 1$.

$= \dfrac{5(2)}{2} = 5$

(b) $\dfrac{5(2)(2+1)}{2}$ Let $n = 2$.

$= \dfrac{5(2)(3)}{2} = 15$

45. $\dfrac{n^2(n+1)^2}{4}$

(a) $\dfrac{1^2(1+1)^2}{4}$ Let $n = 1$.

$= \dfrac{1(4)}{4} = 1$

(b) $\dfrac{2^2(2+1)^2}{4}$ Let $n = 2$.

$= \dfrac{4(9)}{4} = 9$

14.5 Mathematical Induction

14.5 Now Try Exercises

N1. Let S_n represent the statement

$$6 + 12 + 18 + \cdots + 6n = 3n(n + 1).$$

Prove that S_n is true for every positive integer n.

Step 1
S_1 is the statement

$$6(1) = 3(1)(1 + 1)$$
$$6 = 3(2),$$

which is true.

Step 2
Show that if S_k is true, then S_{k+1} is also true.
S_k is the statement

$$6 + 12 + 18 + \cdots + 6k = 3k(k + 1).$$

Add the $(k + 1)$st term, $6(k + 1)$, to each side.

Copyright © 2012 Pearson Education, Inc. Publishing as Addison-Wesley.

$$6 + 12 + 18 + \cdots + 6k + 6(k+1)$$
$$= 3k(k+1) + 6(k+1)$$
$$= (k+1)(3k+6)$$
$$= 3(k+1)(k+2)$$

So $6 + 12 + 18 + \cdots + 6k + 6(k+1)$
$$= 3(k+1)(k+2).$$

The final result is the statement S_{k+1}. Therefore, if S_k is true, then S_{k+1} is true. The two steps required for a proof by mathematical induction are completed, so the general statement S_n is true for every positive integer value of n.

N2. Prove that

$$8 + 11 + 14 + \cdots + (3n+5) = \tfrac{1}{2}n(3n+13).$$

for all positive integers n.

Step 1
S_1 is the statement

$$3(1) + 5 = \tfrac{1}{2}(1)[3(1)+13]$$
$$3 + 5 = \tfrac{1}{2}(16),$$

which is true.

Step 2
Show that if S_k is true, then S_{k+1} is also true.
S_k is the statement

$$8 + 11 + 14 + \cdots + (3k+5) = \tfrac{1}{2}k(3k+13).$$

Add the $(k+1)$st term, $3(k+1)+5$, to each side.

$$8 + 11 + 14 + \cdots + 3k + 5 + 3(k+1) + 5$$
$$= \tfrac{1}{2}k(3k+13) + 3(k+1) + 5$$
$$= \tfrac{3}{2}k^2 + \tfrac{13}{2}k + 3k + 3 + 5$$
$$= \tfrac{3}{2}k^2 + \tfrac{19}{2}k + 8$$
$$= \tfrac{1}{2}(3k^2 + 19k + 16)$$
$$= \tfrac{1}{2}(k+1)(3k+16)$$
$$= \tfrac{1}{2}(k+1)[3(k+1)+13]$$

So $8 + 11 + 14 + \cdots + 3k + 5 + 3(k+1) + 5$
$$= \tfrac{1}{2}(k+1)[3(k+1)+13].$$

The final result is the statement S_{k+1}. Therefore, if S_k is true, then S_{k+1} is true. The two steps required for a proof by mathematical induction are completed, so the general statement S_n is true for every positive integer value of n.

N3. Let S_n represent the statement

$$2^n > n.$$

Prove that S_n is true for every positive integer n.

Step 1
S_1 is the statement

$$2^1 > 1,$$

which is true.

Step 2
Show that if S_k is true, then S_{k+1} is also true.

$$2^k > k \qquad S_k \text{ statement}$$
$$2 \cdot 2^k > 2 \cdot k \qquad \text{Multiply by 2.}$$
$$2^1 \cdot 2^k > 2k$$
$$2^{k+1} > k + k$$
$$2^{k+1} > k + 1 \qquad \text{Since } k \geq 1.$$

The final result is the statement S_{k+1}. Therefore, if S_k is true, then S_{k+1} is true. The two steps required for a proof by mathematical induction are completed, so the general statement S_n is true for every positive integer value of n.

14.5 Section Exercises

1. A proof by mathematical induction allows us to prove that a statement is true for all <u>positive integers</u>.

3. Let S_n represent the statement

$$3 + 6 + 9 + \cdots + 3n = \frac{3n(n+1)}{2}.$$

Prove that S_n is true for every positive integer n.

Step 1
S_1 is the statement

$$3 = \frac{3 \cdot 1(1+1)}{2}$$
$$3 = \tfrac{6}{2},$$

which is true.

Step 2
Show that if S_k is true, then S_{k+1} is also true.
S_k is the statement

$$3 + 6 + 9 + \cdots + 3k = \frac{3k(k+1)}{2}.$$

Add the $(k+1)$st term, $3(k+1)$, to each side.

$$3 + 6 + 9 + \cdots + 3k + 3(k+1)$$
$$= \frac{3k(k+1)}{2} + 3(k+1)$$
$$= \frac{3k(k+1)}{2} + \frac{6(k+1)}{2}$$
$$= \frac{(k+1)(3k+6)}{2}$$

So $3 + 6 + 9 + \cdots + 3k + 3(k+1)$
$$= \frac{3(k+1)(k+2)}{2}.$$

The final result is the statement S_{k+1}. Therefore, if S_k is true, then S_{k+1} is true. The two steps required for a proof by mathematical induction are completed, so the general statement S_n is true for every positive integer value of n.

Copyright © 2012 Pearson Education, Inc. Publishing as Addison-Wesley.

5.   Let S_n represent the statement

$$1 + 3 + 5 + \cdots + (2n - 1) = n^2.$$

Prove that S_n is true for every positive integer n.

Step 1

S_1 is the statement

$$1 = 1^2,$$

which is true.

Step 2

Show that if S_k is true, then S_{k+1} is also true.
S_k is the statement

$$1 + 3 + 5 + \cdots + (2k - 1) = k^2.$$

Add the $(k + 1)$st term, $2(k + 1) - 1 = 2k + 1$,
to each side.

$$1 + 3 + 5 + \cdots + (2k - 1) + (2k + 1) = k^2 + (2k + 1)$$
$$= k^2 + 2k + 1$$
$$= (k + 1)^2$$

So $1 + 3 + 5 + \cdots + (2k - 1) + (2k + 1) = (k + 1)^2$.

The final result is the statement S_{k+1}. Therefore,
if S_k is true, then S_{k+1} is true. The two steps
required for a proof by mathematical induction are
completed, so the general statement S_n is true for
every positive integer value of n.

7.   Let S_n represent the statement

$$1^3 + 2^3 + 3^3 + \cdots + n^3 = \frac{n^2(n + 1)^2}{4}.$$

Prove that S_n is true for every positive integer n.

Step 1

S_1 is the statement

$$1^3 = \frac{1^2(1 + 1)^2}{4}$$
$$1 = \frac{4}{4},$$

which is true.

Step 2

Show that if S_k is true, then S_{k+1} is also true.
S_k is the statement

$$1^3 + 2^3 + 3^3 + \cdots + k^3 = \frac{k^2(k + 1)^2}{4}.$$

Add the $(k + 1)$st term, $(k + 1)^3$, to each side.

$$1^3 + 2^3 + 3^3 + \cdots + k^3 + (k + 1)^3$$
$$= \frac{k^2(k + 1)^2}{4} + (k + 1)^3$$
$$= \frac{k^2(k + 1)^2 + 4(k + 1)^3}{4}$$
$$= \frac{(k + 1)^2[k^2 + 4(k + 1)]}{4}$$
$$= \frac{(k + 1)^2(k^2 + 4k + 4)}{4}$$

So $1^3 + 2^3 + 3^3 + \cdots + k^3 + (k + 1)^3$
$$= \frac{(k + 1)^2(k + 2)^2}{4}.$$

The final result is the statement S_{k+1}. Therefore,
if S_k is true, then S_{k+1} is true. The two steps
required for a proof by mathematical induction are
completed, so the general statement S_n is true for
every positive integer value of n.

9.   Let S_n represent the statement

$$7 \cdot 8 + 7 \cdot 8^2 + 7 \cdot 8^3 + \cdots + 7 \cdot 8^n = 8(8^n - 1).$$

Prove that S_n is true for every positive integer n.

Step 1

S_1 is the statement

$$7 \cdot 8 = 8(8^1 - 1)$$
$$7 \cdot 8 = 8 \cdot 7,$$

which is true.

Step 2

Show that if S_k is true, then S_{k+1} is also true.
S_k is the statement

$$7 \cdot 8 + 7 \cdot 8^2 + 7 \cdot 8^3 + \cdots + 7 \cdot 8^k = 8(8^k - 1).$$

Add the $(k + 1)$st term, $7 \cdot 8^{k+1}$, to each side.

$$7 \cdot 8 + 7 \cdot 8^2 + 7 \cdot 8^3 + \cdots + 7 \cdot 8^k + 7 \cdot 8^{k+1}$$
$$= 8(8^k - 1) + 7 \cdot 8^{k+1}$$
$$= 8[(8^k - 1) + 7 \cdot 8^k]$$
$$= 8(1 \cdot 8^k - 1 + 7 \cdot 8^k)$$
$$= 8(8 \cdot 8^k - 1)$$

So $7 \cdot 8 + 7 \cdot 8^2 + 7 \cdot 8^3 + \cdots + 7 \cdot 8^k + 7 \cdot 8^{k+1}$
$$= 8(8^{k+1} - 1)$$

The final result is the statement S_{k+1}. Therefore,
if S_k is true, then S_{k+1} is true. The two steps
required for a proof by mathematical induction are
completed, so the general statement S_n is true for
every positive integer value of n.

Copyright © 2012 Pearson Education, Inc. Publishing as Addison-Wesley.

11. Let S_n represent the statement

$$\frac{1}{1\cdot 2} + \frac{1}{2\cdot 3} + \frac{1}{3\cdot 4} + \cdots + \frac{1}{n(n+1)} = \frac{n}{n+1}.$$

Prove that S_n is true for every positive integer n.

Step 1
S_1 is the statement

$$\frac{1}{1\cdot 2} = \frac{1}{1+1},$$

which is true.

Step 2
Show that if S_k is true, then S_{k+1} is also true.
S_k is the statement

$$\frac{1}{1\cdot 2} + \frac{1}{2\cdot 3} + \frac{1}{3\cdot 4} + \cdots + \frac{1}{k(k+1)} = \frac{k}{k+1}.$$

Add the $(k+1)$st term, $\frac{1}{(k+1)(k+2)}$, to each side.

$$\frac{1}{1\cdot 2} + \frac{1}{2\cdot 3} + \cdots + \frac{1}{k(k+1)} + \frac{1}{(k+1)(k+2)}$$
$$= \frac{k}{k+1} + \frac{1}{(k+1)(k+2)}$$
$$= \frac{k(k+2)+1}{(k+1)(k+2)}$$
$$= \frac{k^2+2k+1}{(k+1)(k+2)}$$
$$= \frac{(k+1)^2}{(k+1)(k+2)}$$
$$= \frac{k+1}{k+2}$$

The final result is the statement S_{k+1}. Therefore, if S_k is true, then S_{k+1} is true. The two steps required for a proof by mathematical induction are completed, so the general statement S_n is true for every positive integer value of n.

13. Let S_n represent the statement

$$\frac{4}{5} + \frac{4}{5^2} + \frac{4}{5^3} + \cdots + \frac{4}{5^n} = 1 - \frac{1}{5^n}.$$

Prove that S_n is true for every positive integer n.

Step 1
S_1 is the statement

$$\frac{4}{5} = 1 - \frac{1}{5^1},$$

which is true.

Step 2
Show that if S_k is true, then S_{k+1} is also true.
S_k is the statement

$$\frac{4}{5} + \frac{4}{5^2} + \frac{4}{5^3} + \cdots + \frac{4}{5^k} = 1 - \frac{1}{5^k}.$$

Add the $(k+1)$st term, $\frac{4}{5^{k+1}}$, to each side.

$$\frac{4}{5} + \frac{4}{5^2} + \frac{4}{5^3} + \cdots + \frac{4}{5^k} + \frac{4}{5^{k+1}}$$
$$= 1 - \frac{1}{5^k} + \frac{4}{5^{k+1}}$$
$$= 1 - \frac{5\cdot 1}{5\cdot 5^k} + \frac{4}{5^{k+1}}$$
$$= 1 - \frac{5}{5^{k+1}} + \frac{4}{5^{k+1}}$$
$$= 1 + \frac{-5}{5^{k+1}} + \frac{4}{5^{k+1}}$$
$$= 1 - \frac{1}{5^{k+1}}$$

The final result is the statement S_{k+1}. Therefore, if S_k is true, then S_{k+1} is true. The two steps required for a proof by mathematical induction are completed, so the general statement S_n is true for every positive integer value of n.

15. Let S_n represent the statement

$$x^{2n} + x^{2n-1}y + \cdots + xy^{2n-1} + y^{2n}$$
$$= \frac{x^{2n+1} - y^{2n+1}}{x - y}.$$

Prove that S_n is true for every positive integer n.

Step 1
S_1 is the statement

$$x^2 + xy + y^2 = \frac{x^3 - y^3}{x - y}.$$

To verify that S_1 is true, factor $x^3 - y^3$ as the difference of cubes.

$$x^2 + xy + y^2 \overset{?}{=} \frac{(x-y)(x^2+xy+y^2)}{x - y}$$
$$x^2 + xy + y^2 = x^2 + xy + y^2$$

Thus, S_1 is true.

Step 2
Show that if S_k is true, then S_{k+1} is also true.
S_k is the statement

$$x^{2k} + x^{2k-1}y + \cdots + xy^{2k-1} + y^{2k}$$
$$= \frac{x^{2k+1} - y^{2k+1}}{x - y}.$$

S_{k+1} is the statement

$$x^{2(k+1)} + x^{2(k+1)-1}y + x^{2(k+1)-2}y^2$$
$$+ \cdots + x^2 y^{2(k+1)-2} + xy^{2(k+1)-1} + y^{2(k+1)}$$
$$= \frac{x^{2(k+1)+1} - y^{2(k+1)+1}}{x - y}.$$

Assume S_k is true. We will work with the left side of the S_{k+1} statement and show that it is equal to the right side.

Copyright © 2012 Pearson Education, Inc. Publishing as Addison-Wesley.

$$x^{2(k+1)} + x^{2(k+1)-1}y + x^{2(k+1)-2}y^2$$
$$+ \cdots + x^2 y^{2(k+1)-2} + xy^{2(k+1)-1} + y^{2(k+1)}$$
$$= x^{2(k+1)} + xy\left[x^{2k} + x^{2k-1}y + \cdots + xy^{2k-1} + y^{2k}\right]$$
$$+ y^{2(k+1)}$$
$$= x^{2(k+1)} + xy\left(\frac{x^{2k+1} - y^{2k+1}}{x - y}\right) + y^{2(k+1)}$$

$$\textit{assuming } S_k \textit{ is true}$$

$$= \frac{x^{2k+2}(x - y) + x^{2k+2}y - xy^{2k+2} + y^{2k+2}(x - y)}{x - y}$$
$$= \frac{x^{2k+3} - yx^{2k+2} + yx^{2k+2} - xy^{2k+2} + xy^{2k+2} - y^{2k+3}}{x - y}$$
$$= \frac{x^{2k+3} - y^{2k+3}}{x - y}$$
$$= \frac{x^{2(k+1)+1} - y^{2(k+1)+1}}{x - y}$$

The final result is the statement S_{k+1}. Therefore, if S_k is true, then S_{k+1} is true. The two steps required for a proof by mathematical induction are completed, so the general statement S_n is true for every positive integer value of n.

17. Let S_n represent the statement $(a^m)^n = a^{mn}$. (Assume that a and m are constant.)

Prove that S_n is true for every positive integer n.

Step 1
S_1 is the statement

$$(a^m)^1 = a^{m \cdot 1},$$

which is true.

Step 2
Show that if S_k is true, then S_{k+1} is also true. S_k is the statement $(a^m)^k = a^{mk}$, and S_{k+1} is the statement $(a^m)^{k+1} = a^{m(k+1)}$.

$$(a^m)^{k+1} = (a^m)^k (a^m)^1$$
$$= a^{mk} \cdot a^m \qquad \textit{Using the assumption that } S_k \textit{ is true}$$
$$= a^{mk+m}$$
$$= a^{m(k+1)}$$

The final result is the statement S_{k+1}. Therefore, if S_k is true, then S_{k+1} is true. The two steps required for a proof by mathematical induction are completed, so the general statement S_n is true for every positive integer value of n.

19. Let S_n represent the statement:

If $a > 1$, then $a^n > 1$.

Prove that S_n is true for every positive integer n.

Step 1
S_1 is the statement:

If $a > 1$, then $a^1 > 1$,

which is true.

Step 2
Show that if S_k is true, then S_{k+1} is also true. Assume $a > 1$.

$$a^k > 1 \qquad S_k \textit{ statement}$$
$$a \cdot a^k > a \cdot 1 \qquad \textit{Multiply by } a.$$
$$a^1 \cdot a^k > a$$
$$a^{k+1} > a$$
$$a^{k+1} > 1 \qquad \textit{Since } a > 1.$$

The final result is the statement S_{k+1}. Therefore, if S_k is true, then S_{k+1} is true. The two steps required for a proof by mathematical induction are completed, so the general statement S_n is true for every positive integer value of n.

21. Let S_n represent the statement:

If $0 < a < 1$, then $a^n < a^{n-1}$.

Prove that S_n is true for every positive integer n.

Step 1
S_1 is the statement:

If $0 < a < 1$, then $a^1 < a^0$,

which is true since $0 < a < 1$ and $a^0 = 1$.

Step 2
Show that if S_k is true, then S_{k+1} is also true. Assume $0 < a < 1$.

$$a^k < a^{k-1} \qquad S_k \textit{ statement}$$
$$a \cdot a^k < a \cdot a^{k-1} \qquad \textit{Multiply by } a.$$
$$a^{1+k} < a^{1+(k-1)}$$
$$a^{k+1} < a^k$$

The final result is the statement S_{k+1}. Therefore, if S_k is true, then S_{k+1} is true. The two steps required for a proof by mathematical induction are completed, so the general statement S_n is true for every positive integer value of n.

23. Let S_n represent the statement that the number of points of intersection of n lines is $\dfrac{n^2 - n}{2}$.

Since 2 is the smallest number of lines that can intersect, we need to prove this statement for every positive integer $n \geq 2$.

Step 1
S_2 is the statement that 2 nonparallel lines intersect in one point, since

$$\frac{2^2 - 2}{2} = \frac{2}{2} = 1.$$

It is obvious that S_2 is true.

Step 2
Show that if S_k is true, then S_{k+1} is also true. S_k is the statement:

k lines in a plane (no lines parallel, no three lines passing through the same point) intersect in $\dfrac{k^2 - k}{2}$ points.

Copyright © 2012 Pearson Education, Inc. Publishing as Addison-Wesley.

If one more line is added to the set of k lines, this new $(k + 1)$st line will intersect each of the previous k lines in one point. Thus, there will be k additional intersection points. So the number of intersection points for $k + 1$ lines is

$$\frac{k^2 - k}{2} + k$$
$$= \frac{k^2 - k}{2} + \frac{2k}{2}$$
$$= \frac{k^2 - k + 2k}{2}$$
$$= \frac{k^2 + 2k + 1 - 1 - k}{2}$$
$$= \frac{(k^2 + 2k + 1) - (k + 1)}{2}$$
$$= \frac{(k + 1)^2 - (k + 1)}{2}.$$

The final result is the statement S_{k+1}. Therefore, if S_k is true, then S_{k+1} is true. The two steps required for a proof by mathematical induction are completed, so the general statement S_n is true for every positive integer value of n, $n \geq 2$.

25. Let $a_n =$ the number of sides of the nth figure.

$a_1 = 3$ the triangle has 3 sides

$a_2 = 3 \cdot 4$ since each side of the first figure will develop into 4 sides.

$a_3 = 3 \cdot 4^2$, and so on.

This gives a geometric sequence with $a_1 = 3$ and $r = 4$, so

$$a_n = 3 \cdot 4^{n-1}.$$

To find the perimeter of each figure, multiply the number of sides by the length of each side. In each figure, the lengths of the sides are $\frac{1}{3}$ the lengths of the sides in the preceding figure. Thus, if $P_n =$ perimeter of the nth figure,

$$P_1 = 3(1) = 3$$
$$P_2 = 3 \cdot 4\left(\tfrac{1}{3}\right) = 4$$
$$P_3 = 3 \cdot 4^2\left(\tfrac{1}{9}\right) = \tfrac{16}{3}, \text{ and so on.}$$

This gives a geometric sequence with

$$P_1 = 3 \quad \text{and} \quad r = \tfrac{4}{3}.$$

Thus, $P_n = a_1 r^{n-1} = 3\left(\tfrac{4}{3}\right)^{n-1}.$

The result may also be written as

$$P_n = \frac{3^1 \cdot 4^{n-1}}{3^{n-1}} = \frac{4^{n-1}}{3^{-1} \cdot 3^{n-1}} = \frac{4^{n-1}}{3^{n-2}}.$$

27. $5! = 5 \cdot 4 \cdot 3 \cdot 2 \cdot 1 = 120$

29. $_8C_3 = \dfrac{8!}{3! \, (8 - 3)!} = \dfrac{8!}{3! \, 5!}$

$\quad\quad = \dfrac{8 \cdot 7 \cdot 6}{3 \cdot 2 \cdot 1} = 8 \cdot 7 = 56$

14.6 Counting Theory

14.6 Now Try Exercises

N1. Three independent events are involved: selecting a cell phone, selecting a messaging package, and selecting a data service. We can apply the fundamental principle of counting to find the total number of options available.

$$21 \cdot 5 \cdot 4 = 420$$

N2. There are 7 ways to pick the first brand, then 6 ways to pick the second brand, and so on. Thus, using all seven brands, the number of ways to display the brands in a row is

$$7 \cdot 6 \cdot 5 \cdot 4 \cdot 3 \cdot 2 \cdot 1 = 5040.$$

N3. There are 7 ways to pick the first exercise, then 6 ways to pick the second exercise, and so on. Thus, the number of ways to arrange the exercises is

$$7 \cdot 6 \cdot 5 \cdot 4 = 840.$$

N4. There are 7 ways to choose the first place winner, 6 ways to choose the second place winner, and 5 ways to choose the third place winner. Using the fundamental principle of counting, the number of ways to choose the winners is

$$7 \cdot 6 \cdot 5 = 210.$$

Alternatively, we can think of this problem as 7 things taken 3 at a time and use the permutation formula.

$$_7P_3 = \frac{7!}{(7 - 3)!} = \frac{7!}{4!} = \frac{7 \cdot 6 \cdot 5 \cdot 4!}{4!} = 210$$

N5. We can think of the number of ways for this problem as 10! or $_{10}P_{10}$. Evaluating gives us

$$_{10}P_{10} = \frac{10!}{(10 - 10)!} = \frac{10!}{0!} = 10! = 3{,}628{,}800.$$

N6. There are two choices for the first letter. Since the other 2 letters can be repeated, there are 26 choices for each letter. There are 10 choices for each digit. Using the fundamental principle of counting, the number of possible employee identification numbers are

$$2 \cdot 26 \cdot 26 \cdot 10 \cdot 10 = 135{,}200.$$

N7. Think of this problem as 10 elements taken 4 at a time. Since the order of the elements is *not* important, use combinations.

$$_{10}C_4 = \frac{10!}{4! \, (10 - 4)!} = \frac{10!}{4! \, 6!}$$
$$= \frac{10 \cdot 9 \cdot 8 \cdot 7 \cdot 6!}{4 \cdot 3 \cdot 2 \cdot 1 \cdot 6!} = 210$$

Copyright © 2012 Pearson Education, Inc. Publishing as Addison-Wesley.

N8. (a) Order is not important, so we get

$$_{12}C_4 = \frac{12!}{4!\,8!} = \frac{12 \cdot 11 \cdot 10 \cdot 9 \cdot 8!}{4 \cdot 3 \cdot 2 \cdot 1 \cdot 8!} = 495.$$

(b) Since the department chair must be included, we are now selecting 3 from a group of 11. Order not being important, we get

$$_{11}C_3 = \frac{11!}{3!\,8!} = \frac{11 \cdot 10 \cdot 9 \cdot 8!}{3 \cdot 2 \cdot 1 \cdot 8!} = 165.$$

N9. The number of ways to select 1 of the 5 freshmen is $_5C_1$. The number of ways to select 3 of the 8 sophomores is $_8C_3$. Now apply the fundamental principle of counting.

$$_5C_1 \cdot {}_8C_3 = 5 \cdot 56 = 280$$

N10. (a) The ordering of the visits is important, so use a *permutation*.

(b) The ordering of the jurors is not important, so use a *combination*.

(c) The ordering of the offices is important, so use a *permutation*.

14.6 Section Exercises

1.
$$_6P_4 = \frac{6!}{(6-4)!} = \frac{6!}{2!} = \frac{6 \cdot 5 \cdot 4 \cdot 3 \cdot 2!}{2!}$$
$$= 6 \cdot 5 \cdot 4 \cdot 3 = 360$$

3.
$$_9P_2 = \frac{9!}{(9-2)!} = \frac{9!}{7!} = \frac{9 \cdot 8 \cdot 7!}{7!}$$
$$= 9 \cdot 8 = 72$$

5.
$$_5P_0 = \frac{5!}{(5-0)!} = \frac{5!}{5!} = 1$$

7.
$$_4C_2 = \frac{4!}{2!\,(4-2)!} = \frac{4!}{2!\,2!}$$
$$= \frac{4 \cdot 3 \cdot 2!}{2 \cdot 2!} = 6$$

9.
$$_6C_1 = \frac{6!}{1!\,(6-1)!} = \frac{6!}{1!\,5!} = \frac{6 \cdot 5!}{1 \cdot 5!} = 6$$

11.
$$_3C_0 = \frac{3!}{0!\,(3-0)!} = \frac{3!}{0!\,3!} = \frac{3!}{1 \cdot 3!} = 1$$

13. By the fundamental principle of counting, if two events are independent and one can occur in m ways and the other can occur in n ways, then both events can occur in mn ways. We can extend this principle to three events, so the number of different types of homes available is

$$6 \cdot 4 \cdot 2 = 48.$$

15. The number of ways in which 8 people can be seated in 8 seats in a row is given by $_8P_8$.

$$_8P_8 = \frac{8!}{(8-8)!} = \frac{8!}{0!} = \frac{8!}{1} = 8! = 40{,}320$$

17. The number of ways in which 7 of 10 mice can be arranged in a row is given by $_{10}P_7$.

$$_{10}P_7 = \frac{10!}{(10-7)!} = \frac{10!}{3!}$$
$$= \frac{10 \cdot 9 \cdot 8 \cdot 7 \cdot 6 \cdot 5 \cdot 4 \cdot 3!}{3!} = 604{,}800$$

19. $400 - 20 = 380$ of the courses are not mathematics courses; 4 of these are to be arranged in an order to form a schedule.

$$_{380}P_4 = \frac{380!}{(380-4)!} = \frac{380!}{376!}$$
$$= \frac{380 \cdot 379 \cdot 378 \cdot 377 \cdot 376!}{376!}$$
$$= 380 \cdot 379 \cdot 378 \cdot 377$$
$$= 20{,}523{,}714{,}120$$

There are $20{,}523{,}714{,}120$ or $2.052371412 \times 10^{10}$ possible schedules that do not include a math course.

21. The number of ways in which 3 people can be chosen from 35 people is given by $_{35}P_3$.

$$_{35}P_3 = \frac{35!}{(35-3)!} = \frac{35!}{32!}$$
$$= \frac{35 \cdot 34 \cdot 33 \cdot 32!}{32!} = 39{,}270$$

23. The number of different 6-member committees that can be selected from 50 members is $_{50}C_6$.

$$_{50}C_6 = \frac{50!}{6!\,(50-6)!} = \frac{50!}{6!\,44!}$$
$$= \frac{50 \cdot 49 \cdot 48 \cdot 47 \cdot 46 \cdot 45 \cdot 44!}{6 \cdot 5 \cdot 4 \cdot 3 \cdot 2 \cdot 1 \cdot 44!}$$
$$= 15{,}890{,}700$$

25. The number of ways to select 4 of the 7 extras is $_7C_4$.

$$_7C_4 = \frac{7!}{4!\,(7-4)!} = \frac{7!}{4!\,3!} = \frac{7 \cdot 6 \cdot 5 \cdot 4!}{3 \cdot 2 \cdot 1 \cdot 4!} = 35$$

There are 35 different hamburger combinations.

27. This problem involves choosing 3 elements from a set of 7 elements. There are $_7C_3$ such subsets.

$$_7C_3 = \frac{7!}{3!\,(7-3)!} = \frac{7!}{3!\,4!} = \frac{7 \cdot 6 \cdot 5 \cdot 4!}{3 \cdot 2 \cdot 1 \cdot 4!} = 35$$

There are 35 different 3-card combinations.

29. There are 9 black marbles in the bag, so the number of different samples containing 3 black marbles is $_9C_3$.

$$_9C_3 = \frac{9!}{3!\,(9-3)!} = \frac{9!}{3!\,6!} = \frac{9 \cdot 8 \cdot 7 \cdot 6!}{3 \cdot 2 \cdot 1 \cdot 6!} = 84$$

Consider drawing a sample of 3 marbles which contains 2 black marbles as a two-stage task:

drawing the 2 black marbles and then drawing the 1 non-black marble.

The number of samples containing 2 black marbles drawn from the 9 black marbles in the bag is $_9C_2$.

$$_9C_2 = \frac{9!}{2!\,(9-2)!} = \frac{9!}{2!\,7!} = \frac{9 \cdot 8 \cdot 7!}{2 \cdot 1 \cdot 7!} = 36$$

The number of samples containing 1 non-black marble drawn from the 9 non-black marbles in the bag is $_9C_1$.

$$_9C_1 = \frac{9!}{1!\,(9-1)!} = \frac{9!}{1!\,8!} = \frac{9 \cdot 8!}{1 \cdot 8!} = 9$$

Applying the fundamental principle of counting, we find that the number of samples containing exactly 2 black marbles is

$$36 \cdot 9 = 324.$$

31. The number of samples with 4 defective bulbs that can be drawn from the 5 defective bulbs is $_5C_4$.

$$_5C_4 = \frac{5!}{(5-4)!\,4!} = \frac{5!}{1!\,4!} = \frac{5 \cdot 4!}{1 \cdot 4!} = 5$$

Selecting a sample of 2 good bulbs and 2 defective bulbs is a two-stage task. To find the number of such samples, proceed as follows: Find the number of ways to draw 2 good bulbs from the $24 - 5 = 19$ good bulbs, then find the number of ways to draw 2 defective bulbs from the 5 defective bulbs, and, finally, use the fundamental principle of counting to multiply the results. The number of samples with 2 good bulbs and 2 defective bulbs is $_{19}C_2 \cdot {_5C_2}$.

$$_{19}C_2 \cdot {_5C_2} = \frac{19!}{2!\,17!} \cdot \frac{5!}{2!\,3!} = 171 \cdot 10 = 1710$$

33. **(a)** A five-digit postal zip code is an example of a permutation since the order of numbers makes a difference.

(b) A hand of cards is an example of a combination since the order in which the cards are held does not make a difference.

(c) A committee of school board members is an example of a combination since no matter in what order these people are chosen, it still is the same committee. The word "committee" suggests a combination problem.

35. Order is important since the secretaries will be assigned to 3 different managers, so this is a permutation problem.

$$_7P_3 = \frac{7!}{(7-3)!} = \frac{7!}{4!} = \frac{7 \cdot 6 \cdot 5 \cdot 4!}{4!}$$
$$= 7 \cdot 6 \cdot 5 = 210$$

The assignments can be made in 210 ways.

37. An office has 8 men and 11 women workers. There are a total of $8 + 11 = 19$ workers, and 5 of them are to be chosen. Order is not important, so this is a combination problem.

(a) Choose all men.
This means choose 5 of the 8 men.

$$_8C_5 = \frac{8!}{5!\,(8-5)!} = \frac{8!}{5!\,3!}$$
$$= \frac{8 \cdot 7 \cdot 6 \cdot 5!}{3 \cdot 2 \cdot 1 \cdot 5!} = 56$$

56 committees having 5 men can be chosen.

(b) Choose all women.
This means choose 5 of the 11 women.

$$_{11}C_5 = \frac{11!}{5!\,(11-5)!} = \frac{11!}{5!\,6!}$$
$$= \frac{11 \cdot 10 \cdot 9 \cdot 8 \cdot 7 \cdot 6!}{5 \cdot 4 \cdot 3 \cdot 2 \cdot 1 \cdot 6!} = 462$$

462 committees having 5 women can be chosen.

(c) Choose 3 men and 2 women.
Since choosing the men and choosing the women are independent events, we can use the fundamental principle of counting to find the number of committees with 3 men and 2 women.

$$_8C_3 \cdot {_{11}C_2} = \frac{8!}{3!\,(8-3)!} \cdot \frac{11!}{2!\,(11-2)!}$$
$$= \frac{8!}{3!\,5!} \cdot \frac{11!}{2!\,9!}$$
$$= \frac{8 \cdot 7 \cdot 6 \cdot 5!}{3 \cdot 2 \cdot 1 \cdot 5!} \cdot \frac{11 \cdot 10 \cdot 9!}{2 \cdot 1 \cdot 9!}$$
$$= 56 \cdot 55 = 3080$$

3080 committees having 3 men and 2 women can be chosen.

(d) Choose no more than 3 women.
This means choose 0 women (and 5 men) or choose 1 woman (and 4 men) or choose 2 women (and 3 men) or choose 3 women (and 2 men). Thus, the number of possible committees with no more than 3 women is

$$_{11}C_0 \cdot {_8C_5} + {_{11}C_1} \cdot {_8C_4} + {_{11}C_2} \cdot {_8C_3} + {_{11}C_3} \cdot {_8C_2}$$
$$= 1 \cdot 56 + 11 \cdot 70 + 55 \cdot 56 + 165 \cdot 28$$
$$= 56 + 770 + 3080 + 4620$$
$$= 8526.$$

8526 committees having no more than 3 women can be chosen.

39. In electing people to a committee, order is not important, so we use combinations. Since 4 people are to be elected from 10 candidates, the number of possible different committees is

$$_{10}C_4 = \frac{10!}{4!\,(10-4)!} = \frac{10!}{4!\,6!}$$
$$= \frac{10 \cdot 9 \cdot 8 \cdot 7 \cdot 6!}{4 \cdot 3 \cdot 2 \cdot 1 \cdot 6!} = 210.$$

On the other hand, if each person has a different responsibility, order is important, so we use permutations.

In this case, the number of different ways that the committee can be formed is

$$_{10}P_4 = \frac{10!}{(10-4)!} = \frac{10!}{6!}$$
$$= \frac{10 \cdot 9 \cdot 8 \cdot 7 \cdot 6!}{6!} = 5040.$$

41. The order of the ingredients is not important, so we use combinations.

The number of ways to select 4 of the 7 available vegetables is

$$_7C_4 = \frac{7!}{4!\,(7-4)!} = \frac{7!}{4!\,3!}$$
$$= \frac{7 \cdot 6 \cdot 5 \cdot 4!}{3 \cdot 2 \cdot 1 \cdot 4!} = 35.$$

There are 35 different soups that she can make.

43. Order is not important in selecting members of a delegation, so we can use combinations. (A delegation is a subset of the group of workers.)

(a) The number of ways to select a delegation of 3 from the group of 12 workers is

$$_{12}C_3 = \frac{12!}{3!\,9!} = 220.$$

(b) There is only one way to select the first delegate because this must be the foreman. The other two delegates must be chosen from the remaining 11 workers, which can be done in $_{11}C_2$ ways. Thus, by the fundamental principle of counting, the number of possible delegations that include the foreman is

$$1 \cdot {}_{11}C_2 = \frac{11!}{2!\,9!} = 55.$$

(c) If the delegation includes exactly one woman, it must include exactly two men. Since there are 5 women and 7 men, the number of possible delegations is

$$_5C_1 \cdot {}_7C_2 = 5 \cdot 21 = 105.$$

45. $$_nP_{n-1} = \frac{n!}{[n-(n-1)]!} = \frac{n!}{1!} = \frac{n!}{1} = n!$$

and $$_nP_n = \frac{n!}{(n-n)!} = \frac{n!}{0!} = n!$$

Therefore, $_nP_{n-1} = {}_nP_n$.

47. $$_nP_0 = \frac{n!}{(n-0)!} = \frac{n!}{n!} = 1$$

49. $$_nC_0 = \frac{n!}{0!\,(n-0)!} = \frac{n!}{0!\,n!} = \frac{n!}{1 \cdot n!} = \frac{n!}{n!} = 1$$

51. $$\frac{5}{36} + \frac{1}{36} - \frac{2}{36} = \frac{5+1-2}{36} = \frac{4}{36} = \frac{1}{9}$$

53. $$\frac{12}{52} + \frac{1}{26} - \frac{1}{52} = \frac{12+2-1}{52} = \frac{13}{52} = \frac{1}{4}$$

14.7 Basics of Probability

14.7 Now Try Exercises

N1. **(a)** $1, 2$, and 3 are less than 4, so $E_7 = \{1, 2, 3\}$.

$$P(E_7) = \frac{n(E_7)}{n(S)} = \frac{3}{6} = \frac{1}{2}$$

(b) All numbers on a die are greater than 0, so $E_8 = \{1, 2, 3, 4, 5, 6\}$.

$$P(E_8) = \frac{n(E_8)}{n(S)} = \frac{6}{6} = 1$$

N2. There are 26 red cards in a deck of 52 cards, so $n(E) = 26$, $n(S) = 52$, and

$$P(E) = \frac{n(E)}{n(S)} = \frac{26}{52} = \frac{1}{2}.$$

There are $52 - 26 = 26$ cards that are not red cards, so

$$P(E') = \frac{n(E')}{n(S)} = \frac{26}{52} = \frac{1}{2}.$$

N3. **(a)** The odds in favor of rolling a die and getting a 4 are

$$P(E) \text{ to } P(E') = \frac{1}{6} \text{ to } \frac{5}{6} = \frac{\frac{1}{6}}{\frac{5}{6}}$$
$$= \frac{1}{5}, \text{ or } 1 \text{ to } 5.$$

(b) The odds against the event are

$$\frac{P(E')}{P(E)} = \frac{\frac{8}{10}}{\frac{2}{10}} = \frac{8}{2} = \frac{4}{1}, \text{ or } 4 \text{ to } 1.$$

N4. **(a)** There are 26 red cards and 13 spades. Let R be the event that the card is a red card and S the event that the card is a spade. R and S are mutually exclusive, so

$$P(R \text{ or } S) = P(R) + P(S)$$
$$= \tfrac{26}{52} + \tfrac{13}{52} = \tfrac{39}{52} \text{ or } \tfrac{3}{4}$$

(b) There are 4 kings. Let K be the event that the card is a king. R and K are not mutually exclusive, so

$$P(R \text{ or } K) = P(R) + P(K) - P(R \text{ and } K)$$
$$= \tfrac{26}{52} + \tfrac{4}{52} - \tfrac{2}{52}$$
$$= \tfrac{28}{52} \text{ or } \tfrac{7}{13}$$

Copyright © 2012 Pearson Education, Inc. Publishing as Addison-Wesley.

N5. The sample space includes 36 possible pairs. There are 6 ways to roll a 7 with two dice:

$(1,6), (2,5), (3,4), (4,3), (5,2),$ and $(6,1)$.

There are 2 ways to roll an 11: $(5,6)$ and $(6,5)$.

$$
\begin{aligned}
&P(7 \text{ or } 11) \\
&= P(7) + P(11) \\
&= \tfrac{6}{36} + \tfrac{2}{36} \\
&= \tfrac{8}{36}, \text{ or } \tfrac{2}{9}
\end{aligned}
$$

14.7 Section Exercises

1. Two ordinary coins are tossed. For each coin, either heads, H, or tails, T, lands up. The sample space is

$$S = \{HH, HT, TH, TT\}.$$

3. Two slips are drawn. For each slip, a number, 1, 2, 3, 4, or 5, shows.

The sample space may be written

$$
\begin{aligned}
S = \{&(1,2), (1,3), (1,4), (1,5), (2,3), \\
&(2,4), (2,5), (3,4), (3,5), (4,5)\}.
\end{aligned}
$$

5. The sample space is $S = \{HH, HT, TH, TT\}$, so $n(S) = 4$.

(a) The event E_1 of both coins showing the same face is

$$E_1 = \{HH, TT\}, \text{ so } n(E_1) = 2.$$

The probability is

$$P(E_1) = \frac{n(E_1)}{n(S)} = \frac{2}{4} = \frac{1}{2}.$$

(b) The event E_2 where at least one head appears is

$$E_2 = \{HH, HT, TH\},$$

so $n(E_2) = 3$. Then,

$$P(E_2) = \frac{n(E_2)}{n(S)} = \frac{3}{4}.$$

7. See the sample space listed for Exercise 3. There are 10 possible outcomes, so $n(S) = 10$.

(a) Event E_1 (both numbers even) $= \{(2,4)\}$, so $n(E_1) = 1$.

$$P(E_1) = \frac{n(E_1)}{n(S)} = \frac{1}{10}$$

(b) Event E_2 (both numbers odd) $= \{(1,3), (1,5), (3,5)\}$, so $n(E_2) = 3$.

$$P(E_2) = \frac{n(E_2)}{n(S)} = \frac{3}{10}$$

(c) Event E_3 (both numbers the same) $= \emptyset$, so $n(E_3) = 0$.

$$P(E_3) = \frac{n(E_3)}{n(S)} = \frac{0}{10} = 0$$

(d) Event E_4 (one odd number, one even number) $= \{(1,2), (1,4), (2,3), (2,5), (3,4), (4,5)\}$, so $n(E_4) = 6$.

$$P(E_4) = \frac{n(E_4)}{n(S)} = \frac{6}{10} = \frac{3}{5}$$

Note that the answers for parts (a), (b), and (d) add up to one. This must be true because we have exhausted all possibilities: both even, both odd, one of each.

9. There are $3 + 4 + 8 = 15$ marbles, so $n(S) = 15$.

(a) E_1: yellow marble is drawn
There are 3 yellow marbles, so $n(E_1) = 3$.

$$P(E_1) = \tfrac{3}{15} = \tfrac{1}{5}$$

(b) E_2: blue marble is drawn
There are 8 blue marbles, so $n(E_2) = 8$.

$$P(E_2) = \tfrac{8}{15}$$

(c) E_3: black marble is drawn
There are no black marbles, so $n(E_3) = 0$.

$$P(E_3) = \tfrac{0}{15} = 0$$

11. The probability of any event is a number between 0 and 1, inclusive. Since $\tfrac{6}{5}$ is greater than 1, it is impossible for this to be the answer to a probability problem.

♦♦♦ Relating Concepts 13–16 ♦♦♦

13. The number of ways to choose the sample of 3 from the 12 engines is

$$_{12}C_3 = \frac{12!}{3!\,9!} = 220.$$

14. This is a two-step task. Using the fundamental principle of counting, we multiply the number of ways of choosing 1 defective engine and the number of ways of choosing 2 good engines. Since 1 defective engine and 11 good engines are in the container, the number of ways to choose a sample of 3 with 1 defective engine and 2 good engines is

$$_{1}C_1 \cdot {}_{11}C_2 = 1 \cdot \frac{11!}{2!\,9!} = 55.$$

15. From Exercise 14, there are 55 samples that contain the defective engine, so $n(E) = 55$.

From Exercise 13, there are 220 possible samples of 3 engines. These samples make up the sample space, so $n(S) = 220$.

16. Using the definition of probability, we see that the probability that the defective engine is in the sample is

$$P(E) = \frac{n(E)}{n(S)} = \frac{55}{220} = 0.25.$$

Copyright © 2012 Pearson Education, Inc. Publishing as Addison-Wesley.

17. Using the definition of probability, we see that the probability of drawing a face card is

$$P(E) = \frac{n(E)}{n(S)} = \frac{4 \cdot 3}{52} = \frac{3}{13}.$$

Also, $P(E') = 1 - P(E) = 1 - \frac{3}{13} = \frac{10}{13}.$

19. If the event E is "such a bank making a small business loan," then

$$P(E) = 0.002.$$

Thus $P(E') = 1 - 0.002 = 0.998.$
The odds against the event are

$$\frac{P(E')}{P(E)} = \frac{0.998}{0.002} = \frac{499}{1}, \quad \text{or} \quad 499 \text{ to } 1.$$

21. There are 10 relatives, all equally likely to arrive first, so $n(S) = 10$. Let M be the event "first guest is mother," U be the event "first guest is an uncle," B be the event "first guest is a brother," and C be the event "first guest is a cousin." In each part of this exercise, the compound event involves an alternative of two mutually exclusive events, so we can find the probability of the compound event by adding the two probabilities.

(a) $P(U \text{ or } C) = P(U) + P(C)$
$$= \frac{2}{10} + \frac{4}{10} = \frac{6}{10} = \frac{3}{5}$$

(b) $P(B \text{ or } C) = P(B) + P(C)$
$$= \frac{3}{10} + \frac{4}{10} = \frac{7}{10}$$

(c) $P(U \text{ or } M) = P(U) + P(M)$
$$= \frac{2}{10} + \frac{1}{10} = \frac{3}{10}$$

23. There are 52 cards in the deck, $n(S) = 52$.

(a) Since there are 3 face cards (K, Q, J) in each of the 4 suits, there are 12 face cards.

$$P(\text{face card}) = \frac{12}{52} = \frac{3}{13}$$

(b) This is the probability of a compound event.

$P(\text{red or } 3)$
$$= P(\text{red}) + P(3) - P(\text{red and } 3)$$
$$= \frac{n(\text{red})}{n(S)} + \frac{n(3)}{n(S)} - \frac{n(\text{red and } 3)}{n(S)}$$
$$= \frac{26}{52} + \frac{4}{52} - \frac{2}{52} = \frac{28}{52} = \frac{7}{13}$$

(c) The events represented by "ace," "2," and "3" are mutually exclusive.

$P(\text{less than } 4)$
$$= P(\text{ace or 2 or 3})$$
$$= P(\text{ace}) + P(2) + P(3)$$
$$= \frac{n(\text{ace})}{n(S)} + \frac{n(2)}{n(S)} + \frac{n(3)}{n(S)}$$
$$= \frac{4}{52} + \frac{4}{52} + \frac{4}{52}$$
$$= \frac{12}{52} = \frac{3}{13}$$

25. There are $2 + 5 + 3 = 10$ marbles in the bag, so $n(S) = 10$. In (a) and (b), the compound events involve alternative events that are mutually exclusive, so we may simply add the individual probabilities.

(a) There are 2 yellow marbles and 3 blue marbles, so

$$P(Y \text{ or } B) = P(Y) + P(B)$$
$$= \frac{2}{10} + \frac{3}{10}$$
$$= \frac{5}{10} = \frac{1}{2}$$

(b) There are 2 yellow marbles and 5 red marbles, so

$$P(Y \text{ or } R) = P(Y) + P(R)$$
$$= \frac{2}{10} + \frac{5}{10} = \frac{7}{10}$$

(c) There are no green marbles, so

$$P(G) = \frac{0}{10} = 0$$

27. $P(\text{more than 5 years}) = 0.30$
$P(\text{female}) = 0.28$
$P(\text{retirement plan contributor}) = 0.65$
$P(\text{retirement plan contributor and female})$
$= \frac{1}{2}(0.28) = 0.14$

(a) $P(\text{male}) = 1 - P(\text{female})$
$$= 1 - 0.28 = 0.72$$

(b) $P(\text{5 years or less})$
$$= 1 - P(\text{more than 5 years})$$
$$= 1 - 0.30 = 0.70$$

(c) $P(\text{retirement plan contributor or female})$
$$= P(\text{contributor}) + P(\text{female})$$
$$- P(\text{contributor and female})$$
$$= 0.65 + 0.28 - 0.14 = 0.79$$

29. $P(\text{no more than four good toes})$
$$= 0.77 + 0.13 = 0.90$$

31. $E = \{s_1, s_2, s_5\}$, so
$$P(E) = 0.17 + 0.03 + 0.21 = 0.41.$$

33. $E \cap F = \{s_5\}$, so
$$P(E \cap F) = 0.21.$$

35. $E' = \{s_3, s_4, s_6\}$ and $F' = \{s_1, s_2, s_3, s_6\}$, so $E' \cup F' = \{s_1, s_2, s_3, s_4, s_6\}$. Thus,

$P(E' \cup F')$
$$= 0.17 + 0.03 + 0.09 + 0.46 + 0.04$$
$$= 0.79.$$

Alternatively, since s_5 is the only outcome not in $E' \cup F'$,

$$P(E' \cup F') = 1 - P(\{s_5\})$$
$$= 1 - 0.21 = 0.79.$$

Copyright © 2012 Pearson Education, Inc. Publishing as Addison-Wesley.

Chapter 14 Review Exercises

1. $a_n = 2n - 3$
$a_1 = 2(1) - 3 = -1 \qquad a_2 = 2(2) - 3 = 1$
$a_3 = 2(3) - 3 = 3 \qquad a_4 = 2(4) - 3 = 5$

Answer: $-1, 1, 3, 5$

2. $a_n = \dfrac{n-1}{n}$
$a_1 = \dfrac{1-1}{1} = 0 \qquad a_2 = \dfrac{2-1}{2} = \dfrac{1}{2}$
$a_3 = \dfrac{3-1}{3} = \dfrac{2}{3} \qquad a_4 = \dfrac{4-1}{4} = \dfrac{3}{4}$

Answer: $0, \frac{1}{2}, \frac{2}{3}, \frac{3}{4}$

3. $a_n = n^2$
$a_1 = (1)^2 = 1 \qquad a_2 = (2)^2 = 4$
$a_3 = (3)^2 = 9 \qquad a_4 = (4)^2 = 16$

Answer: $1, 4, 9, 16$

4. $a_n = \left(\frac{1}{2}\right)^n$
$a_1 = \left(\frac{1}{2}\right)^1 = \frac{1}{2} \qquad a_2 = \left(\frac{1}{2}\right)^2 = \frac{1}{4}$
$a_3 = \left(\frac{1}{2}\right)^3 = \frac{1}{8} \qquad a_4 = \left(\frac{1}{2}\right)^4 = \frac{1}{16}$

Answer: $\frac{1}{2}, \frac{1}{4}, \frac{1}{8}, \frac{1}{16}$

5. $a_n = (n+1)(n-1)$
$a_1 = (1+1)(1-1) = 2(0) = 0$
$a_2 = (2+1)(2-1) = 3(1) = 3$
$a_3 = (3+1)(3-1) = 4(2) = 8$
$a_4 = (4+1)(4-1) = 5(3) = 15$

Answer: $0, 3, 8, 15$

6. $a_n = n(-1)^{n-1}$
$a_1 = 1(-1)^{1-1} = 1(1) = 1$
$a_2 = 2(-1)^{2-1} = 2(-1) = -2$
$a_3 = 3(-1)^{3-1} = 3(1) = 3$
$a_4 = 4(-1)^{4-1} = 4(-1) = -4$

Answer: $1, -2, 3, -4$

7. $\displaystyle\sum_{i=1}^{5} i^2 = 1^2 + 2^2 + 3^2 + 4^2 + 5^2$
$= 1 + 4 + 9 + 16 + 25$

8. $\displaystyle\sum_{i=1}^{6} (i+1) = (1+1) + (2+1) + (3+1)$
$+ (4+1) + (5+1) + (6+1)$
$= 2 + 3 + 4 + 5 + 6 + 7$

9. $\displaystyle\sum_{i=3}^{6} (5i - 4)$
$= (15 - 4) + (20 - 4) + (25 - 4) + (30 - 4)$
Let i = 3 to 6.
$= 11 + 16 + 21 + 26$

10. $\displaystyle\sum_{i=1}^{4} (i + 2)$
$= (1+2) + (2+2) + (3+2) + (4+2)$
$= 3 + 4 + 5 + 6$
$= 18$

11. $\displaystyle\sum_{i=1}^{6} 2^i$
$= 2^1 + 2^2 + 2^3 + 2^4 + 2^5 + 2^6$
$= 2 + 4 + 8 + 16 + 32 + 64$
$= 126$

12. $\displaystyle\sum_{i=4}^{7} \frac{i}{i+1}$
$= \frac{4}{4+1} + \frac{5}{5+1} + \frac{6}{6+1} + \frac{7}{7+1}$
$= \frac{4}{5} + \frac{5}{6} + \frac{6}{7} + \frac{7}{8} \qquad LCD = 2^3 \cdot 3 \cdot 5 \cdot 7 = 840$
$= \frac{672}{840} + \frac{700}{840} + \frac{720}{840} + \frac{735}{840} = \frac{2827}{840}$

13. $\bar{x} = \dfrac{13{,}778 + 14{,}862 + 16{,}680 + 17{,}916 + 13{,}985}{5}$
$\bar{x} = \dfrac{77{,}221}{5} = 15{,}444.2$

The average total of retirement assets of Americans for the years 2004–2008 was \$15,444 billion.

14. $2, 5, 8, 11, \ldots$ is an *arithmetic* sequence with
$$d = 5 - 2 = 3.$$

Note: You should find the difference for all pairs of adjacent terms to determine if the sequence is arithmetic.

15. $-6, -2, 2, 6, 10, \ldots$ is an *arithmetic* sequence with
$$d = -2 - (-6) = 4.$$

16. $\frac{2}{3}, -\frac{1}{3}, \frac{1}{6}, -\frac{1}{12}, \ldots$ is a *geometric* sequence with
$$r = \frac{-\frac{1}{3}}{\frac{2}{3}} = -\frac{1}{3} \cdot \frac{3}{2} = -\frac{1}{2}.$$

Note: You should find the ratio for all pairs of adjacent terms to determine if the sequence is geometric.

17. $-1, 1, -1, 1, -1, \ldots$ is a *geometric* sequence with
$$r = \frac{1}{-1} = -1.$$

18. $64, 32, 8, \frac{1}{2}, \ldots$

Find two differences:
$32 - 64 = -32$ and $8 - 32 = -24$, so the sequence is not arithmetic.

Find two ratios:
$\frac{32}{64} \neq \frac{8}{32}$, so the sequence is not geometric.

Therefore, the sequence is *neither*.

Copyright © 2012 Pearson Education, Inc. Publishing as Addison-Wesley.

19. 64, 32, 16, 8, ... is a *geometric* sequence with

$$r = \frac{32}{64} = \frac{1}{2}.$$

20. We are given the first six terms: 1, 1, 2, 3, 5, 8

7th term: $5 + 8 = 13$ 8th term: $8 + 13 = 21$
9th term: $13 + 21 = 34$ 10th term: $21 + 34 = 55$
11th term: $34 + 55 = 89$

21. Given $a_1 = -2$ and $d = 5$, find a_{16}.

$$a_n = a_1 + (n-1)d$$
$$a_{16} = -2 + (16-1)5$$
$$= -2 + 15(5)$$
$$= -2 + 75 = 73$$

22. Given $a_6 = 12$ and $a_8 = 18$, find a_{25}.

$$a_n = a_1 + (n-1)d$$
$$a_6 = a_1 + 5d, \quad \text{so} \quad 12 = a_1 + 5d \quad (1)$$
$$a_8 = a_1 + 7d, \quad \text{so} \quad 18 = a_1 + 7d \quad (2)$$

Multiply equation (1) by -1 and add the result to equation (2).

$$-12 = -a_1 - 5d \qquad -1 \times (1)$$
$$\underline{18 = \quad a_1 + 7d \quad (2)}$$
$$6 = \qquad 2d$$
$$3 = d$$

To find a_1, substitute $d = 3$ in equation (1).

$$12 = a_1 + 5d \qquad (1)$$
$$12 = a_1 + 5(3)$$
$$12 = a_1 + 15$$
$$-3 = a_1$$

Use $a_1 = -3$ and $d = 3$ to find a_{25}.

$$a_n = a_1 + (n-1)d$$
$$a_{25} = -3 + (25-1)3$$
$$= -3 + 24(3)$$
$$= -3 + 72 = 69$$

23. $a_1 = -4, \ d = -5$

$$a_n = a_1 + (n-1)d$$
$$a_n = -4 + (n-1)(-5)$$
$$= -4 - 5n + 5$$
$$= -5n + 1$$

24. 6, 3, 0, −3, ...

To get the general term, a_n, first find d.
$d = 3 - 6 = -3$

$$a_n = a_1 + (n-1)d$$
$$a_n = 6 + (n-1)(-3)$$
$$= 6 - 3n + 3$$
$$= -3n + 9$$

25. 7, 10, 13, ..., 49

Here, $a_1 = 7$ and $d = 10 - 7 = 3$.
Now find n, the number of terms.

$$a_n = a_1 + (n-1)d$$
$$49 = 7 + (n-1)(3)$$
$$42 = 3(n-1)$$
$$14 = n - 1 \qquad \textit{Divide by 3.}$$
$$15 = n$$

There are 15 terms in this sequence.

26. 5, 1, −3, ... , −79

Here, $a_1 = 5$ and $d = 1 - 5 = -4$.
Now find n, the number of terms.

$$a_n = a_1 + (n-1)d$$
$$-79 = 5 + (n-1)(-4)$$
$$-79 = 9 - 4n$$
$$-88 = -4n$$
$$n = 22$$

There are 22 terms in this sequence.

27. Find S_8 if $a_1 = -2$ and $d = 6$. Find a_8 first.

$$a_8 = a_1 + (8-1)d$$
$$= -2 + 7(6)$$
$$= -2 + 42 = 40$$

Now find the sum.

$$S_n = \frac{n}{2}(a_1 + a_n)$$
$$S_8 = \frac{8}{2}(-2 + 40)$$
$$= 4(38) = 152$$

28. Find S_8 if $a_n = -2 + 5n$.
Find the first and last terms.

$$a_1 = -2 + 5(1) = 3$$
$$a_8 = -2 + 5(8) = 38$$

Now find the sum.

$$S_n = \frac{n}{2}(a_1 + a_n)$$
$$S_8 = \frac{8}{2}(3 + 38)$$
$$= 4(41)$$
$$= 164$$

29. Find the general term for the geometric sequence
$-1, -4, -16, \ldots$.

We determine that $a_1 = -1$ and $r = \frac{-4}{-1} = 4$.

Using the formula $a_n = a_1 r^{n-1}$, we get

$$a_n = -1(4)^{n-1}.$$

Copyright © 2012 Pearson Education, Inc. Publishing as Addison-Wesley.

30. Find the general term for the geometric sequence $\frac{2}{3}, \frac{2}{15}, \frac{2}{75}, \ldots$.

We determine that $a_1 = \frac{2}{3}$ and

$$r = \frac{\frac{2}{15}}{\frac{2}{3}} = \frac{2}{15} \cdot \frac{3}{2} = \frac{1}{5}.$$

Using the formula $a_n = a_1 r^{n-1}$, we get

$$a_n = \frac{2}{3}\left(\frac{1}{5}\right)^{n-1}.$$

31. Find a_{11} for $2, -6, 18, \ldots$.

We determine that $a_1 = 2$ and $r = \frac{-6}{2} = -3$.

$$a_n = a_1 r^{n-1}$$
$$a_{11} = 2(-3)^{11-1}$$
$$= 2(-3)^{10} = 118{,}098$$

32. Given $a_3 = 20$ and $a_5 = 80$, find a_{10}. Using the formula $a_n = a_1 r^{n-1}$, we get the following.

For a_3: $\quad a_3 = a_1 r^{3-1}$
$$20 = a_1 r^2$$
For a_5: $\quad a_5 = a_1 r^{5-1}$
$$80 = a_1 r^4$$

The ratio of a_5 to a_3 is

$$\frac{80}{20} = \frac{a_1 r^4}{a_1 r^2}$$
$$4 = r^2$$
$$r = \pm 2.$$

Since $20 = a_1 r^2$ and $r^2 = 4$, we get $20 = a_1(4)$, and hence, $5 = a_1$. Now find a_{10}.

$$a_n = a_1 r^{n-1}$$
$$a_{10} = 5(\pm 2)^{10-1}$$
$$= 5(\pm 2)^9$$

Two answers are possible for a_{10}:

$$5(2)^9 = 2560 \quad \textbf{or} \quad 5(-2)^9 = -2560.$$

33. $\displaystyle\sum_{i=1}^{5} \left(\frac{1}{4}\right)^i$ ■ $a_1 = \frac{1}{4}, r = \frac{1}{4}$, and $n = 5$.

$$S_n = \frac{a_1(1-r^n)}{1-r}$$
$$S_5 = \frac{\frac{1}{4}\left[1 - \left(\frac{1}{4}\right)^5\right]}{1 - \frac{1}{4}}$$
$$= \frac{\frac{1}{4}\left(1 - \frac{1}{1024}\right)}{\frac{3}{4}}$$
$$= \frac{1}{3}\left(\frac{1023}{1024}\right) = \frac{341}{1024}$$

34. $\displaystyle\sum_{i=1}^{8} \frac{3}{4}(-1)^i$ ■ $a_1 = -\frac{3}{4}, r = -1$, and $n = 8$.

$$S_n = \frac{a_1(1-r^n)}{1-r}$$
$$S_8 = \frac{-\frac{3}{4}\left[1 - (-1)^8\right]}{1 - (-1)}$$
$$= \frac{-\frac{3}{4}(1 - 1)}{2}$$
$$= \frac{-\frac{3}{4}(0)}{2} = 0$$

35. $\displaystyle\sum_{i=1}^{\infty} 4\left(\frac{1}{5}\right)^i$

The terms are the terms of an infinite geometric sequence with $a_1 = \frac{4}{5}$ and $r = \frac{1}{5}$.

$$S = \frac{a_1}{1-r} = \frac{\frac{4}{5}}{1 - \frac{1}{5}} = \frac{\frac{4}{5}}{\frac{4}{5}} = 1$$

36. $\displaystyle\sum_{i=1}^{\infty} 2(3)^i$

The terms are the terms of an infinite geometric sequence with $a_1 = 6$ and $r = 3$.

$$S = \frac{a_1}{1-r} \quad \text{if} \quad |r| < 1,$$

but $r = 3$, so S does not exist and hence the sum does not exist.

37. $(2p - q)^5 = [2p + (-q)]^5$
$$= (2p)^5 + \frac{5!}{1!\,4!}(2p)^4(-q)^1$$
$$+ \frac{5!}{2!\,3!}(2p)^3(-q)^2 + \frac{5!}{3!\,2!}(2p)^2(-q)^3$$
$$+ \frac{5!}{4!\,1!}(2p)^1(-q)^4 + (-q)^5$$
$$= 32p^5 + 5\left(16p^4\right)(-q) + 10\left(8p^3\right)q^2$$
$$+ 10\left(4p^2\right)\left(-q^3\right) + 5(2p)q^4 - q^5$$
$$= 32p^5 - 80p^4q + 80p^3q^2 - 40p^2q^3$$
$$+ 10pq^4 - q^5$$

38. $(x^2 + 3y)^4$
$$= \left(x^2\right)^4 + \frac{4!}{1!\,3!}\left(x^2\right)^3(3y)^1$$
$$+ \frac{4!}{2!\,2!}\left(x^2\right)^2(3y)^2$$
$$+ \frac{4!}{3!\,1!}\left(x^2\right)^1(3y)^3 + (3y)^4$$
$$= x^8 + 4\left(x^6\right)(3y) + 6\left(x^4\right)\left(9y^2\right)$$
$$+ 4\left(x^2\right)\left(27y^3\right) + 81y^4$$
$$= x^8 + 12x^6y + 54x^4y^2 + 108x^2y^3$$
$$+ 81y^4$$

Copyright © 2012 Pearson Education, Inc. Publishing as Addison-Wesley.

39. $(3t^3 - s^2)^4 = [3t^3 + (-s^2)]^4$

$\quad = (3t^3)^4 + \dfrac{4!}{1!\,3!}(3t^3)^3(-s^2)^1$

$\quad\quad + \dfrac{4!}{2!\,2!}(3t^3)^2(-s^2)^2$

$\quad\quad + \dfrac{4!}{3!\,1!}(3t^3)^1(-s^2)^3 + (-s^2)^4$

$\quad = 3^4 t^{12} - 4(3^3 t^9)(s^2) + 6(9t^6)(s^4)$

$\quad\quad - 4(3t^3)(s^6) + s^8$

$\quad = 81t^{12} - 108t^9 s^2 + 54t^6 s^4 - 12t^3 s^6 + s^8$

40. The fourth term ($r = 4$, so $r - 1 = 3$) of $(3a + 2b)^{19}$ is

$$\dfrac{19!}{3!\,16!}(3a)^{16}(2b)^3$$

$$= 969(3)^{16}(a^{16})(8)b^3$$

$$= 7752(3)^{16}a^{16}b^3.$$

41. Let S_n represent the statement

$$2 + 6 + 10 + 14 + \cdots + (4n - 2) = 2n^2.$$

Prove that S_n is true for every positive integer n.

Step 1
S_1 is the statement

$$4 \cdot 1 - 2 = 2(1)^2,$$

which is true.

Step 2
Show that if S_k is true, then S_{k+1} is also true.
S_k is the statement

$$2 + 6 + 10 + 14 + \cdots + (4k - 2) = 2k^2.$$

Add $4(k + 1) - 2 = 4k + 2$ to each side.

$$2 + 6 + 10 + 14 + \cdots + (4k - 2)$$
$$+ 4k + 2$$
$$= (2k^2) + (4k + 2)$$
$$= 2(k^2 + 2k + 1)$$
$$= 2(k + 1)^2$$

The final result is the statement S_{k+1}. Therefore, if S_k is true, then S_{k+1} is true. The two steps required for a proof by mathematical induction are completed, so the general statement S_n is true for every positive integer value of n.

42. Let S_n represent the statement

$$2^2 + 4^2 + 6^2 + \cdots + (2n)^2$$
$$= \dfrac{2n(n + 1)(2n + 1)}{3}.$$

Prove that S_n is true for every positive integer n.

Step 1
S_1 is the statement

$$2^2 = \dfrac{2 \cdot 1(1 + 1)(2 \cdot 1 + 1)}{3}$$

$$4 = \dfrac{2 \cdot 2 \cdot 3}{3},$$

which is true.

Step 2
Show that if S_k is true, then S_{k+1} is also true.
S_k is the statement

$$2^2 + 4^2 + 6^2 + \cdots + (2k)^2$$
$$= \dfrac{2k(k + 1)(2k + 1)}{3}.$$

Add the $(k + 1)$st term, $[2(k + 1)]^2 = (2k + 2)^2$, to each side.

$$2^2 + 4^2 + 6^2 + \cdots + (2k)^2 + (2k + 2)^2$$
$$= \dfrac{2k(k + 1)(2k + 1)}{3} + \dfrac{3[2(k + 1)]^2}{3}$$
$$= \dfrac{2(k + 1)[k(2k + 1) + 3 \cdot 2(k + 1)]}{3}$$

$$\qquad\qquad\qquad\qquad \textit{Factor out 2(k + 1)}$$

$$= \dfrac{2(k + 1)(2k^2 + k + 6k + 6)}{3}$$
$$= \dfrac{2(k + 1)(2k^2 + 7k + 6)}{3}$$
$$= \dfrac{2(k + 1)(k + 2)(2k + 3)}{3}$$

The final result is the statement S_{k+1}. Therefore, if S_k is true, then S_{k+1} is true. The two steps required for a proof by mathematical induction are completed, so the general statement S_n is true for every positive integer value of n.

43. Let S_n represent the statement

$$2 + 2^2 + 2^3 + \cdots + 2^n = 2(2^n - 1).$$

Prove that S_n is true for every positive integer n.

Step 1
S_1 is the statement

$$2 = 2(2^1 - 1)$$
$$2 = 2 \cdot 1,$$

which is true.

Step 2
Show that if S_k is true, then S_{k+1} is also true.
S_k is the statement

$$2 + 2^2 + 2^3 + \cdots + 2^k = 2(2^k - 1).$$

Add the $(k + 1)$st term, 2^{k+1}, to each side.

$$2 + 2^2 + 2^3 + \cdots + 2^k + 2^{k+1}$$
$$= 2(2^k - 1) + 2^{k+1}$$
$$= 2^{k+1} - 2 + 2^{k+1}$$
$$= 2 \cdot 2^{k+1} - 2 \cdot 1$$
$$= 2(2^{k+1} - 1)$$

Copyright © 2012 Pearson Education, Inc. Publishing as Addison-Wesley.

The final result is the statement S_{k+1}. Therefore, if S_k is true, then S_{k+1} is true. The two steps required for a proof by mathematical induction are completed, so the general statement S_n is true for every positive integer value of n.

44. Let S_n represent the statement

$$1^3 + 3^3 + 5^3 + \cdots + (2n-1)^3$$
$$= n^2(2n^2 - 1).$$

Prove that S_n is true for every positive integer n.

Step 1
S_1 is the statement

$$1^3 = 1^2(2 \cdot 1^2 - 1)$$
$$1 = 1(2-1),$$

which is true.

Step 2
Show that if S_k is true, then S_{k+1} is also true. S_k is the statement

$$1^3 + 3^3 + 5^3 + \cdots + (2k-1)^3$$
$$= k^2(2k^2 - 1).$$

Add the $(k+1)$st term, $[2(k+1)-1]^3 = (2k+1)^3$, to each side.

$$1^3 + 3^3 + 5^3 + \cdots + (2k-1)^3 + (2k+1)^3$$
$$= k^2(2k^2 - 1) + (2k+1)^3$$
$$= 2k^4 - k^2 + 8k^3 + 12k^2 + 6k + 1$$
$$= 2k^4 + 8k^3 + 11k^2 + 6k + 1$$

The final result is the statement S_{k+1}. Therefore, if S_k is true, then S_{k+1} is true. The two steps required for a proof by mathematical induction are completed, so the general statement S_n is true for every positive integer value of n.

45. $_5P_5 = \dfrac{5!}{(5-5)!} = \dfrac{5!}{0!} = \dfrac{5!}{1} = 120$

46. $_9P_2 = \dfrac{9!}{(9-2)!} = \dfrac{9!}{7!} = \dfrac{9 \cdot 8 \cdot 7!}{7!} = 72$

47. $_7C_3 = \dfrac{7!}{3!(7-3)!} = \dfrac{7!}{3!\,4!} = \dfrac{7 \cdot 6 \cdot 5 \cdot 4!}{3 \cdot 2 \cdot 1 \cdot 4!} = 35$

48. $_8C_5 = \dfrac{8!}{5!(8-5)!} = \dfrac{8!}{5!\,3!} = \dfrac{8 \cdot 7 \cdot 6 \cdot 5!}{3 \cdot 2 \cdot 1 \cdot 5!} = 56$

49. Use the fundamental principle of counting.

$$2 \cdot 4 \cdot 3 \cdot 2 = 48$$

48 different wedding arrangements are possible.

50. Three independent events are involved. There are 5 choices of style, 3 choices of fabric, and 6 choices of color. Therefore, using the fundamental principle of counting, there are

$$5 \cdot 3 \cdot 6 = 90 \text{ different couches.}$$

51. There are 4 choices for the first job, 3 choices for the second job, and so on.

$$4 \cdot 3 \cdot 2 \cdot 1 = 24$$

There are 24 ways in which the jobs can be assigned.

52. There are 4 spots to be filled by 26 letters and 3 spots by 10 digits.
If repeats are allowed,

$$26 \cdot 10 \cdot 10 \cdot 10 \cdot 26 \cdot 26 \cdot 26 = 456{,}976{,}000$$

different license plates can be formed.

If no repeats are allowed,

$$26 \cdot 10 \cdot 9 \cdot 8 \cdot 25 \cdot 24 \cdot 23 = 258{,}336{,}000$$

different license plates can be formed.

53. (a) Let E be the event "drawing a green marble."

$$n(E) = 4$$
$$n(S) = 4 + 5 + 6 = 15$$
$$\text{(total number of marbles)}$$
$$P(E) = \frac{n(E)}{n(S)} = \frac{4}{15}$$

(b) Let E be the event "drawing a black marble."

$$P(E) = \frac{n(E)}{n(S)} = \frac{5}{15} = \frac{1}{3}$$

The probability that the marble is not black is given by

$$P(E') = 1 - P(E)$$
$$= 1 - \frac{1}{3} = \frac{2}{3}.$$

(c) Let E be the event "drawing a blue marble." (There are none.)

$$P(E) = \frac{n(E)}{n(S)} = \frac{0}{15} = 0$$

54. (a) Let E be the event "green marble is drawn."

$$P(E) = \frac{4}{15}$$
$$P(E') = 1 - P(E)$$
$$= 1 - \frac{4}{15} = \frac{11}{15}$$

The odds in favor of drawing a green marble are

$$\frac{P(E)}{P(E')} = \frac{\frac{4}{15}}{\frac{11}{15}} = \frac{4}{11} \quad \text{or} \quad 4 \text{ to } 11.$$

(b) Let F be the event "white marble is drawn."

$$P(F) = \frac{6}{15} = \frac{2}{5}$$
$$P(F') = 1 - \frac{2}{5} = \frac{3}{5}$$

The odds against drawing a white marble are

$$\frac{P(F')}{P(F)} = \frac{\frac{3}{5}}{\frac{2}{5}} = \frac{3}{2} \quad \text{or} \quad 3 \text{ to } 2.$$

Copyright © 2012 Pearson Education, Inc. Publishing as Addison-Wesley.

For Exercises 55–60, the 52 cards in a standard deck form the sample space, so $n(S) = 52$.

55. $P(\text{black king}) = \dfrac{n(\text{black kings})}{n(S)}$

$= \frac{2}{52} = \frac{1}{26}$

56. Let F be the event that a face card is drawn and A be the event that an ace is drawn. The events are mutually exclusive, so

$P(F \text{ or } A) = P(F) + P(A)$

$= \frac{12}{52} + \frac{4}{52} = \frac{16}{52}$ or $\frac{4}{13}$.

57. Let A be the event that an ace is drawn and D be the event that a diamond is drawn. The events are *not* mutually exclusive, so

$P(A \text{ or } D)$

$= P(A) + P(D) - P(A \text{ and } D)$

$= \frac{4}{52} + \frac{13}{52} - \frac{1}{52} = \frac{16}{52}$ or $\frac{4}{13}$.

58. Let D be the event that a diamond is drawn, so $P(D) = \frac{13}{52} = \frac{1}{4}$. The probability that the card drawn is *not* a diamond is

$P(D') = 1 - P(D)$

$= 1 - \frac{1}{4} = \frac{3}{4}$.

59. Let D be the event that a diamond is drawn, so $P(D) = \frac{13}{52}$ and $P(D') = \frac{39}{52}$.

Let B be the event that a black card is drawn, so $P(B) = \frac{26}{52}$ and $P(B') = \frac{26}{52}$. The events are *not* mutually exclusive (hearts are not diamonds and not black), so the probability that the card drawn is *not* a diamond or *not* black is

$P(D' \text{ or } B')$

$= P(D') + P(B') - P(D' \text{ and } B')$

$= \frac{39}{52} + \frac{26}{52} - P(\text{a heart})$

$= \frac{39}{52} + \frac{26}{52} - \frac{13}{52} = \frac{52}{52}$ or 1.

Thus, $P(D' \text{ or } B')$ is a certainty since every card drawn satisfy one of the two given events.

60. $P(8, 9, \text{ or } 10) = P(8) + P(9) + P(10)$

$= \frac{4}{52} + \frac{4}{52} + \frac{4}{52}$

$= \frac{12}{52}$ or $\frac{3}{13}$

61. **[14.3]** The geometric sequence $-3, 6, -12, \ldots$ has $a_1 = -3$ and $r = \frac{6}{-3} = -2$. First find a_{10}.

$a_n = a_1 r^{n-1}$

$a_{10} = -3(-2)^9$

$= -3(-512) = 1536$

Now find the sum of the first ten terms.

$S_n = \dfrac{a_1(1 - r^n)}{1 - r}$

$S_{10} = \dfrac{-3\left[1 - (-2)^{10}\right]}{1 - (-2)}$

$= \dfrac{-3(1 - 1024)}{3}$

$= -(-1023) = 1023$

62. **[14.2]** The arithmetic sequence $1, 7, 13, \ldots$ has $a_1 = 1$ and $d = 7 - 1 = 6$. First find a_{40}.

$a_n = a_1 + (n - 1)d$

$a_{40} = 1 + (40 - 1)6$

$= 1 + (39)6$

$= 1 + 234 = 235$

Now find the tenth term so that we can find the sum of the first ten terms.

$a_{10} = 1 + (10 - 1)6$

$= 1 + (9)6 = 55$

Now find the sum of the first ten terms.

$S_n = \dfrac{n}{2}(a_1 + a_n)$

$S_{10} = \dfrac{10}{2}(1 + 55)$

$= 5(56) = 280$

63. **[14.2]** The arithmetic sequence with $a_1 = -4$ and $d = 3$ is $-4, -1, 2, 5, \ldots$. First find a_{15}.

$a_n = a_1 + (n - 1)d$

$a_{15} = -4 + (15 - 1)3$

$= -4 + 42 = 38$

Now find the sum of the first ten terms.

$S_n = \dfrac{n}{2}\left[2a_1 + (n - 1)d\right]$

$S_{10} = \dfrac{10}{2}\left[2(-4) + (10 - 1)3\right]$

$= 5(-8 + 27)$

$= 5(19) = 95$

64. **[14.3]** The geometric sequence has $a_1 = 1$ and $r = -3$. First find a_9.

$a_n = a_1 r^{n-1}$

$a_9 = 1(-3)^{9-1}$

$= (-3)^8 = 6561$

Now find the sum of the first ten terms.

$S_n = \dfrac{a_1(r^n - 1)}{r - 1}$

$S_{10} = \dfrac{1\left[(-3)^{10} - 1\right]}{-3 - 1}$

$= -\frac{1}{4}(3^{10} - 1)$

$= -\frac{1}{4}(59{,}049 - 1)$

$= -14{,}762$

65. **[14.3]** $2, 8, 32, \ldots$ is a geometric sequence with $a_1 = 2$ and $r = \frac{8}{2} = 4$.

$a_n = a_1 r^{n-1}$

$a_n = 2(4)^{n-1}$

66. **[14.2]** $2, 7, 12, \ldots$ is an arithmetic sequence with $a_1 = 2$ and $d = 7 - 2 = 5$.

$a_n = a_1 + (n - 1)d$

$a_n = 2 + (n - 1)5$

$= 2 + 5n - 5$

$= 5n - 3$

Copyright © 2012 Pearson Education, Inc. Publishing as Addison-Wesley.

67. **[14.2]** $12, 9, 6, \ldots$ is an arithmetic sequence with $a_1 = 12$ and $d = -3$.

$$a_n = a_1 + (n-1)d$$
$$a_n = 12 + (n-1)(-3)$$
$$= 12 - 3n + 3$$
$$= -3n + 15$$

68. **[14.3]** $27, 9, 3, \ldots$ is a geometric sequence with $a_1 = 27$ and $r = \frac{9}{27} = \frac{1}{3}$.

$$a_n = a_1 r^{n-1}$$
$$a_n = 27\left(\frac{1}{3}\right)^{n-1}$$

69. **[14.2]** The distances traveled in successive seconds are

$$3, 7, 11, 15, 19, \ldots.$$

This is an arithmetic sequence with $a_1 = 3$ and $d = 4$. Since we know a_1 and d, we'll use the second formula for the sum of an arithmetic sequence with $S_n = 210$.

$$S_n = \frac{n}{2}\left[2a_1 + (n-1)d\right]$$
$$210 = \frac{n}{2}\left[2(3) + (n-1)4\right]$$
$$420 = n(6 + 4n - 4)$$
$$420 = 6n + 4n^2 - 4n$$
$$0 = 4n^2 + 2n - 420$$
$$0 = 2n^2 + n - 210$$
$$0 = (2n + 21)(n - 10)$$

$$2n + 21 = 0 \quad \text{or} \quad n - 10 = 0$$
$$n = -\tfrac{21}{2} \quad \text{or} \quad n = 10$$

Discard $-\frac{21}{2}$ since time cannot be negative. It takes her 10 seconds.

70. **[14.3]** Use the formula for the future value of an ordinary annuity with $R = 672$, $i = \frac{0.045}{4} = 0.01125$, and $n = 7(4) = 28$.

$$S = R\left[\frac{(1+i)^n - 1}{i}\right]$$
$$S = 672\left[\frac{(1 + 0.01125)^{28} - 1}{0.01125}\right]$$
$$= 21{,}973.00$$

The future value of the annuity is \$21,973.00.

71. **[14.1]** Since $100\% - 3\% = 97\% = 0.97$, the population after 1 year is $0.97(50{,}000)$, after 2 years is $0.97\left[0.97(50{,}000)\right]$ or $(0.97)^2(50{,}000)$, and after n years is $(0.97)^n(50{,}000)$. After 6 years, the population is

$$(0.97)^6(50{,}000) \approx 41{,}649 \approx 42{,}000.$$

72. **[14.1]** $\left(\frac{1}{2}\right)^n$ is left after n strokes. So $\left(\frac{1}{2}\right)^7 = \frac{1}{128} = 0.0078125$ is left after 7 strokes.

73. **[14.6]** There are 6 people on the student body council.

(a) In selecting a 3-member delegation, $_6C_3$ choices are possible.

$$_6C_3 = \frac{6!}{3!\,3!} = \frac{6 \cdot 5 \cdot 4 \cdot 3!}{3 \cdot 2 \cdot 1 \cdot 3!} = 20$$

20 delegations are possible.

(b) If the president must attend, then there are two slots available for the remaining 5 members, which means there are $_5C_2$ choices.

$$_5C_2 = \frac{5!}{2!\,3!} = \frac{5 \cdot 4 \cdot 3!}{2 \cdot 1 \cdot 3!} = 10$$

10 delegations are possible if the president must attend.

74. **[14.6]** Order is important, so this is a permutation problem.

$$_9P_3 = \frac{9!}{6!} = \frac{9 \cdot 8 \cdot 7 \cdot 6!}{6!} = 9 \cdot 8 \cdot 7 = 504$$

The winners can be determined 504 ways.

75. **[14.7]** **(a)** $P(\text{no more than 3})$
$$= P(0 \text{ or } 1 \text{ or } 2 \text{ or } 3)$$
$$= P(0) + P(1) + P(2) + P(3)$$
$$= 0.31 + 0.25 + 0.18 + 0.12$$
$$= 0.86$$

(b) $P(\text{at least 2})$
$$= P(2 \text{ or } 3 \text{ or } 4 \text{ or } 5)$$
$$= P(2) + P(3) + P(4) + P(5)$$
$$= 0.18 + 0.12 + 0.08 + 0.06$$
$$= 0.44$$

Chapter 14 Test

1. $a_n = (-1)^n + 1$
$a_1 = (-1)^1 + 1 = 0$
$a_2 = (-1)^2 + 1 = 1 + 1 = 2$
$a_3 = (-1)^3 + 1 = -1 + 1 = 0$
$a_4 = (-1)^4 + 1 = 1 + 1 = 2$
$a_5 = (-1)^5 + 1 = -1 + 1 = 0$

Answer: $0, 2, 0, 2, 0$

2. $a_1 = 4, \ d = 2$
$a_2 = a_1 + d = 4 + 2 = 6$
$a_3 = a_2 + d = 6 + 2 = 8$
$a_4 = a_3 + d = 8 + 2 = 10$
$a_5 = a_4 + d = 10 + 2 = 12$

Answer: $4, 6, 8, 10, 12$

Copyright © 2012 Pearson Education, Inc. Publishing as Addison-Wesley.

3. $a_4 = 6$, $r = \frac{1}{2}$

First find a_1.

$$a_n = a_1 r^{n-1}$$
$$a_4 = a_1 \left(\frac{1}{2}\right)^{4-1}$$
$$6 = a_1 \left(\frac{1}{8}\right)$$
$$a_1 = 48$$

Now find the remaining terms.

$$a_2 = \frac{1}{2}a_1 = \frac{1}{2}(48) = 24$$
$$a_3 = \frac{1}{2}a_2 = \frac{1}{2}(24) = 12$$
$$a_4 = \frac{1}{2}a_3 = \frac{1}{2}(12) = 6$$
$$a_5 = \frac{1}{2}a_4 = \frac{1}{2}(6) = 3$$

Answer: $48, 24, 12, 6, 3$

4. Given $a_1 = 6$ and $d = -2$, find a_4.

$$a_n = a_1 + (n-1)d$$
$$a_4 = a_1 + (4-1)d$$
$$= 6 + (3)(-2)$$
$$= 6 - 6 = 0$$

5. Given $a_5 = 16$ and $a_7 = 9$, find a_4.
This is a geometric sequence, so

$$a_6 = a_5 r \quad \text{and} \quad a_7 = a_6 r.$$

From the last equation, $a_6 = \dfrac{a_7}{r}$, so by substitution,

$$a_5 r = \frac{a_7}{r}.$$

Multiply by r and divide by a_5 to get

$$r^2 = \frac{a_7}{a_5}$$
$$r^2 = \frac{9}{16}$$
$$r = \pm \sqrt{\frac{9}{16}} = \pm \frac{3}{4}$$

Since $a_5 = a_4 r$, we know that $a_4 = \dfrac{16}{r}$.
Substituting each value of r in the last equation gives us two values of a_4.

$$a_4 = \frac{16}{\frac{3}{4}} = \frac{64}{3} \quad \text{or} \quad a_4 = \frac{16}{-\frac{3}{4}} = -\frac{64}{3}$$

6. Given the arithmetic sequence with $a_2 = 12$ and $a_3 = 15$, find S_5. First find d.

$$d = a_3 - a_2 = 15 - 12 = 3$$

Now find the first and fifth terms.

$$a_1 = a_2 - 3 = 12 - 3 = 9$$
$$a_5 = a_4 + 3 = a_3 + 6 = 15 + 6 = 21$$

Now find the sum of the first five terms.

$$S_n = \frac{n}{2}(a_1 + a_n)$$
$$S_5 = \frac{5}{2}(9 + 21)$$
$$= \frac{5}{2}(30) = 75$$

7. Given the geometric sequence with $a_5 = 4$ and $a_7 = 1$, find S_5.

$$r^2 = \frac{a_7}{a_5} = \frac{1}{4}, \text{ so } r = \frac{1}{2} \text{ or } r = -\frac{1}{2}.$$

Use $a_7 = a_1 r^6$ to get $1 = a_1 \left(\pm \frac{1}{2}\right)^6$, and so $a_1 = 64$. Use $S_n = \dfrac{a_1(1 - r^n)}{1 - r}$.

$$S_5 = \frac{64\left[1 - \left(\frac{1}{2}\right)^5\right]}{1 - \frac{1}{2}} \quad \text{or} \quad S_5 = \frac{64\left[1 - \left(-\frac{1}{2}\right)^5\right]}{1 - \left(-\frac{1}{2}\right)}$$

$$= \frac{64}{\frac{1}{2}}\left(1 - \frac{1}{32}\right) \qquad = \frac{64}{\frac{3}{2}}\left(1 + \frac{1}{32}\right)$$

$$= 128\left(\frac{31}{32}\right) \qquad\qquad = \frac{128}{3}\left(\frac{33}{32}\right)$$

$$= 124 \qquad\qquad\qquad = 44$$

8. $\overline{x} = \dfrac{\text{total}}{5}$, where total $=$
$78{,}473 + 80{,}967 + 83{,}860 + 86{,}150 + 97{,}103.$

Thus, $\overline{x} = \dfrac{426{,}553}{5} = 85{,}310.6 \approx 85{,}311.$

The average number of commercial bank offices for this five-year period was $85{,}311$.

9. $S = R\left[\dfrac{(1+i)^n - 1}{i}\right]$

$$= 4000\left[\frac{\left(1 + \frac{0.06}{4}\right)^{4(7)} - 1}{\frac{0.06}{4}}\right]$$

$$= 4000\left[\frac{(1.015)^{28} - 1}{0.015}\right]$$

$$\approx 137{,}925.91$$

The account will have $\$137{,}925.91$ at the end of this term.

10. An infinite geometric series has a sum if $|r| < 1$, where r is the common ratio.

11. $\displaystyle\sum_{i=1}^{5}(2i + 8)$

$$= [2(1) + 8] + [2(2) + 8] + [2(3) + 8]$$
$$+ [2(4) + 8] + [2(5) + 8]$$
$$= 10 + 12 + 14 + 16 + 18$$
$$= 70$$

12. $\displaystyle\sum_{i=1}^{6}(3i - 5)$

Find the first and sixth terms.

$$a_1 = 3(1) - 5 = -2$$
$$a_6 = 3(6) - 5 = 13$$

Now find the sum of the first six terms.

$$S_6 = \frac{n}{2}(a_1 + a_6)$$
$$S_6 = \frac{6}{2}(-2 + 13)$$
$$= 3(11) = 33$$

Copyright © 2012 Pearson Education, Inc. Publishing as Addison-Wesley.

13. $\displaystyle\sum_{i=1}^{500} i$ ▪ Use the formula $S_{500} = \frac{n}{2}(a_1 + a_{500})$ with $a_1 = 1$ and $a_{500} = 500$.

$$S_{500} = \frac{500}{2}(1 + 500)$$
$$= 250(501) = 125{,}250$$

14. $\displaystyle\sum_{i=1}^{3} \frac{1}{2}(4^i) = \frac{1}{2}\sum_{i=1}^{3}(4^i)$
$$= \frac{1}{2}(4^1 + 4^2 + 4^3)$$
$$= \frac{1}{2}(4 + 16 + 64)$$
$$= \frac{1}{2}(84) = 42$$

15. $\displaystyle\sum_{i=1}^{\infty} \left(\frac{1}{4}\right)^i$
This is an infinite geometric series with $a_1 = \frac{1}{4}$ and $r = \frac{1}{4}$. Since $|r| = \frac{1}{4} < 1$, the sum exists.

$$S = \frac{a_1}{1-r} = \frac{\frac{1}{4}}{1 - \frac{1}{4}} = \frac{\frac{1}{4}}{\frac{3}{4}} = \frac{1}{4} \cdot \frac{4}{3} = \frac{1}{3}$$

16. $\displaystyle\sum_{i=1}^{\infty} 6\left(\frac{3}{2}\right)^i$
This is an infinite geometric series with $a_1 = 6\left(\frac{3}{2}\right)^1 = 9$ and $r = \frac{3}{2}$.
Since $|r| = \frac{3}{2} > 1$, the sum does not exist.

17. $8! = 8 \cdot 7 \cdot 6 \cdot 5 \cdot 4 \cdot 3 \cdot 2 \cdot 1$
$$= 40{,}320$$

18. By definition, $0! = 1$.

19. $\dfrac{6!}{4!\,2!} = \dfrac{6 \cdot 5 \cdot 4 \cdot 3 \cdot 2 \cdot 1}{(4 \cdot 3 \cdot 2 \cdot 1)(2 \cdot 1)} = \dfrac{6 \cdot 5}{2 \cdot 1} = 15$

20. $_{12}C_{10} = \dfrac{12!}{10!\,(12-10)!}$
$$= \dfrac{12 \cdot 11 \cdot 10!}{10!\,(2!)} = \dfrac{12 \cdot 11}{2} = 66$$

21. $(3k - 5)^4 = [3k + (-5)]^4$
$$= (3k)^4 + \frac{4!}{1!\,3!}(3k)^3(-5)^1$$
$$+ \frac{4!}{2!\,2!}(3k)^2(-5)^2 + \frac{4!}{3!\,1!}(3k)^1(-5)^3$$
$$+ (-5)^4$$
$$= 81k^4 - 4(27k^3)(5) + 6(9k^2)(25)$$
$$- 4(3k)(125) + 625$$
$$= 81k^4 - 540k^3 + 1350k^2 - 1500k + 625$$

22. The fifth term ($r = 5$, so $r - 1 = 4$) of $\left(2x - \dfrac{y}{3}\right)^{12}$ is
$$\frac{12!}{4!\,(12-4)!}(2x)^{12-4}\left(-\frac{y}{3}\right)^4$$
$$= \frac{12!}{4!\,8!}(2x)^8\left(-\frac{y}{3}\right)^4$$
$$= \frac{14{,}080x^8 y^4}{9}.$$

23. The amounts of unpaid balance during 15 months form an arithmetic sequence
$$300, 280, 260, \ldots, 40, 20,$$
which is the sequence with $n = 15$, $a_1 = 300$, and $a_{15} = 20$. Find the sum of these balances.
$$S_n = \frac{n}{2}(a_1 + a_n)$$
$$S_{15} = \frac{15}{2}(300 + 20)$$
$$= \frac{15}{2}(320) = 2400$$

Since 1% interest is paid on this total, the interest paid is 1% of \$2400 or \$24. The sewing machine cost \$300 (paid monthly at \$20), so the total cost is \$300 + \$24 = \$324.

24. The weekly populations form a geometric sequence with $a_1 = 20$ and $r = 3$ since the colony begins with 20 insects and triples each week. Find the general term of this geometric sequence.
$$a_n = a_1 r^{n-1}$$
$$a_n = 20(3)^{n-1}$$

We're assuming that from the beginning of July to the end of September is 12 weeks, so find a_{12}.
$$a_n = 20(3)^{n-1}$$
$$a_{12} = 20(3)^{11}$$

At the end of September, $20(3)^{11} = 3{,}542{,}940$ insects will be present in the colony.

25. Let S_n represent the statement
$$8 + 14 + 20 + \cdots + (6n + 2) = 3n^2 + 5n.$$
Prove that S_n is true for every positive integer n.

Step 1
S_1 is the statement
$$6(1) + 2 = 3(1)^2 + 5(1)$$
$$6 + 2 = 3 + 5,$$
which is true.

Step 2
Show that if S_k is true, then S_{k+1} is also true. S_k is the statement
$$8 + 14 + 20 + \cdots + (6k + 2) = 3k^2 + 5k.$$
Add the $(k+1)$st term, $[6(k+1)+2]$, to each side.
$$8 + 14 + 20 + \cdots + (6k + 2) + [6(k+1) + 2]$$
$$= 3k^2 + 5k + [6(k+1) + 2]$$
$$= 3k^2 + 5k + 6k + 6 + 2$$
$$= (3k^2 + 6k + 3) + (5k + 5)$$
$$= 3(k^2 + 2k + 1) + 5(k + 1)$$
$$= 3(k+1)^2 + 5(k+1)$$

Copyright © 2012 Pearson Education, Inc. Publishing as Addison-Wesley.

So $8 + 14 + 20 + \cdots + (6k + 2) + [6(k + 1) + 2]$
$\qquad = 3(k + 1)^2 + 5(k + 1)$

The final result is the statement S_{k+1}. Therefore, if S_k is true, then S_{k+1} is true. The two steps required for a proof by mathematical induction are completed, so the general statement S_n is true for every positive integer value of n.

26. $_{11}P_3 = \dfrac{11!}{(11 - 3)!} = \dfrac{11!}{8!} = \dfrac{11 \cdot 10 \cdot 9 \cdot 8!}{8!} = 990$

27. $_{45}C_1 = \dfrac{45!}{1!(45 - 1)!} = \dfrac{45!}{1! \, 44!} = \dfrac{45 \cdot 44!}{1 \cdot 44!} = 45$

28. Use the fundamental principle of counting. The first event, selecting a style, can occur in 4 ways. The second event, selecting a fabric, can occur in 3 ways. The third event, selecting a color, can occur in 5 ways. Thus, there are

$$4 \cdot 3 \cdot 5 = 60 \text{ different coats.}$$

29. Use permutations since order is important.

$$_{30}P_3 = \dfrac{30!}{(30 - 3)!} = \dfrac{30!}{27!} = 30 \cdot 29 \cdot 28$$
$$= 24{,}360$$

The offices can be filled in 24,360 ways.

30. This is a combination problem since order is not important. $_9C_3$ represents the number of ways 3 elements can be chosen from a group of 9 elements.

$$_9C_3 = \dfrac{9!}{3!(9 - 3)!} = \dfrac{9!}{3! \, 6!} = \dfrac{9 \cdot 8 \cdot 7 \cdot 6!}{3 \cdot 2 \cdot 1 \cdot 6!} = 84$$

The committee can be chosen in 84 ways.

For Exercises 31–34, the sample space is the set of all cards in a standard deck, so $n(S) = 52$.

31. Let E be the event drawing a red three. There are 2 red threes in a deck, so $n(E) = 2$.

$$P(E) = \dfrac{n(E)}{n(S)} = \dfrac{2}{52} = \dfrac{1}{26}$$

The probability of drawing a red three is $\frac{1}{26}$.

32. Let E be the event drawing a face card. Each suit contains 3 face cards (jack, queen, and king), so the deck contains 12 face cards. Thus,

$$n(E) = 12 \quad \text{and} \quad P(E) = \tfrac{12}{52}.$$

The probability of drawing a card that is *not* a face card is

$$P(E') = 1 - P(E)$$
$$= \tfrac{52}{52} - \tfrac{12}{52} = \tfrac{40}{52} \text{ or } \tfrac{10}{13}.$$

33. Let K be the event that a king is drawn and S be the event that a spade is drawn. The events are *not* mutually exclusive, so

$$P(K \text{ or } S)$$
$$= P(K) + P(S) - P(K \text{ and } S)$$
$$= \tfrac{4}{52} + \tfrac{13}{52} - \tfrac{1}{52} = \tfrac{16}{52} \text{ or } \tfrac{4}{13}.$$

The probability of drawing a king or a spade is $\frac{4}{13}$.

34. Let E be the event drawing a face card. As shown in the solution to Exercise 32,

$$P(E) = \tfrac{3}{13} \quad \text{and} \quad P(E') = \tfrac{10}{13}.$$

The odds in favor of drawing a face card are

$$\dfrac{P(E)}{P(E')} = \dfrac{\frac{3}{13}}{\frac{10}{13}} = \dfrac{3}{10} \quad \text{or} \quad 3 \text{ to } 10.$$

Cumulative Review Exercises (Chapters 1–14)

1. $|-7| + 6 - |-10| - (-8 + 3)$
$\quad = 7 + 6 - 10 - (-5)$
$\quad = 13 - 10 + 5 = 8$

2. $4(-6) + (-8)(5) - (-9)$
$\quad = -24 - 40 + 9 = -55$

In Exercises 3–4, use the set

$$P = \left\{ -\tfrac{8}{3}, 10, 0, \sqrt{13}, -\sqrt{3}, \tfrac{45}{15}, \sqrt{-7}, 0.82, -3 \right\}.$$

3. The rational numbers are

$$-\tfrac{8}{3}, 10, 0, \tfrac{45}{15} \text{ (or 3), } 0.82, \text{ and } -3.$$

4. The irrational numbers are $\sqrt{13}$ and $-\sqrt{3}$.

5. $9 - (5 + 3x) + 5x = -4(x - 3) - 7$
$\quad 9 - 5 - 3x + 5x = -4x + 12 - 7$
$\qquad\qquad 4 + 2x = -4x + 5$
$\qquad\qquad\qquad 6x = 1$
$\qquad\qquad\qquad\ x = \tfrac{1}{6}$

The solution set is $\left\{ \tfrac{1}{6} \right\}$.

6. $7x + 18 \le 9x - 2$
$\quad\ 7x \le 9x - 20 \quad$ *Subtract 18.*
$\ -2x \le -20 \qquad$ *Subtract 9x.*
$\qquad x \ge 10 \qquad\quad$ *Divide by -2;*
$\qquad\qquad\qquad\qquad$ *reverse symbol.*

The solution set is $[10, \infty)$.

7. $|4x - 3| = 21$

$4x - 3 = 21 \quad$ or $\quad 4x - 3 = -21$
$\quad 4x = 24 \qquad\qquad\qquad 4x = -18$
$\quad\ x = 6 \qquad$ or $\quad\ x = -\tfrac{18}{4} = -\tfrac{9}{2}$

The solution set is $\left\{ -\tfrac{9}{2}, 6 \right\}$.

Copyright © 2012 Pearson Education, Inc. Publishing as Addison-Wesley.

8.
$$\frac{x+3}{12} - \frac{x-3}{6} = 0$$
Multiply by the LCD, 12.
$$12\left(\frac{x+3}{12} - \frac{x-3}{6}\right) = 12(0)$$
$$x + 3 - 2(x - 3) = 0$$
$$x + 3 - 2x + 6 = 0$$
$$9 - x = 0$$
$$9 = x$$

Check $x = 9$: $1 - 1 = 0$ *True*
The solution set is $\{9\}$.

9.
$$2x > 8 \quad \text{or} \quad -3x > 9$$
$$x > 4 \quad \text{or} \quad x < -3$$

The solution set is $(-\infty, -3) \cup (4, \infty)$.

10. $|2x - 5| \geq 11$
$$2x - 5 \geq 11 \quad \text{or} \quad 2x - 5 \leq -11$$
$$2m \geq 16 \qquad \qquad 2m \leq -6$$
$$m \geq 8 \quad \text{or} \qquad m \leq -3$$

The solution set is $(-\infty, -3] \cup [8, \infty)$.

11. Let $(x_1, y_1) = (4, -5)$
and $(x_2, y_2) = (-12, -17)$. Then
$$m = \frac{y_2 - y_1}{x_2 - x_1} = \frac{-17 - (-5)}{-12 - 4} = \frac{-12}{-16} = \frac{3}{4}.$$
The slope is $\frac{3}{4}$.

12. To find the equation of the line through $(-2, 10)$ and parallel to $3x + y = 7$, find the slope of
$$3x + y = 7$$
$$y = -3x + 7.$$

The slope is -3, so a line parallel to it also has slope -3. Use $m = -3$ and $(x_1, y_1) = (-2, 10)$ in the point-slope form.
$$y - y_1 = m(x - x_1)$$
$$y - 10 = -3[x - (-2)]$$
$$y - 10 = -3(x + 2)$$

Write in standard form.
$$y - 10 = -3x - 6$$
$$3x + y = 4$$

Alternative solution: The line must be of the form $3x + y = k$ since it is parallel to $3x + y = 7$. Substitute -2 for x and 10 for y to find k.
$$3(-2) + 10 = k$$
$$4 = k$$

The equation is $3x + y = 4$.

13. $x - 3y = 6$

Find the x- and y-intercepts. To find the x-intercept, let $y = 0$.
$$x - 3(0) = 6$$
$$x = 6$$

The x-intercept is $(6, 0)$.

To find the y-intercept, let $x = 0$.
$$0 - 3y = 6$$
$$y = -2$$

The y-intercept is $(0, -2)$.

Plot the intercepts and draw the line through them.

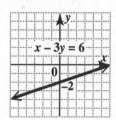

14. $4x - y < 4$

Graph the line $4x - y = 4$, which has intercepts $(0, -4)$ and $(1, 0)$, as a dashed line because the inequality involves $<$. Test $(0, 0)$, which yields $0 < 4$, a true statement. Shade the region on the side of the line that includes $(0, 0)$.

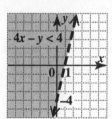

15.
$$y = 5x + 3 \qquad (1)$$
$$2x + 3y = -8 \qquad (2)$$

From equation (1), substitute $5x + 3$ for y in equation (2). Then solve for x.
$$2x + 3y = -8 \qquad (2)$$
$$2x + 3(5x + 3) = -8$$
$$2x + 15x + 9 = -8$$
$$17x = -17$$
$$x = -1$$

From (1), $y = 5(-1) + 3 = -2$.

The solution $(-1, -2)$ checks.

The solution set is $\{(-1, -2)\}$.

Copyright © 2012 Pearson Education, Inc. Publishing as Addison-Wesley.

16.
$$x + 2y + z = 8$$
$$2x - y + 3z = 15$$
$$-x + 3y - 3z = -11$$

Write the augmented matrix.

$$\begin{bmatrix} 1 & 2 & 1 & | & 8 \\ 2 & -1 & 3 & | & 15 \\ -1 & 3 & -3 & | & -11 \end{bmatrix}$$

$$\begin{bmatrix} 1 & 2 & 1 & | & 8 \\ 0 & -5 & 1 & | & -1 \\ 0 & 5 & -2 & | & -3 \end{bmatrix} \quad \begin{matrix} -2R_1 + R_2 \\ R_1 + R_3 \end{matrix}$$

$$\begin{bmatrix} 1 & 2 & 1 & | & 8 \\ 0 & 1 & -\frac{1}{5} & | & \frac{1}{5} \\ 0 & 5 & -2 & | & -3 \end{bmatrix} \quad -\frac{1}{5}R_2$$

$$\begin{bmatrix} 1 & 2 & 1 & | & 8 \\ 0 & 1 & -\frac{1}{5} & | & \frac{1}{5} \\ 0 & 0 & -1 & | & -4 \end{bmatrix} \quad -5R_2 + R_3$$

$$\begin{bmatrix} 1 & 2 & 1 & | & 8 \\ 0 & 1 & -\frac{1}{5} & | & \frac{1}{5} \\ 0 & 0 & 1 & | & 4 \end{bmatrix} \quad -R_3$$

This matrix gives the system

$$x + 2y + z = 8$$
$$y - \tfrac{1}{5}z = \tfrac{1}{5}$$
$$z = 4.$$

Substitute $z = 4$ in the second equation.

$$y - \tfrac{1}{5}(4) = \tfrac{1}{5}$$
$$y = \tfrac{1}{5} + \tfrac{4}{5} = 1$$

Substitute $y = 1$ and $z = 4$ in the first equation.

$$x + 2(1) + 4 = 8$$
$$x + 6 = 8$$
$$x = 2$$

The solution set is $\{(2, 1, 4)\}$.

17. Let x = the number of pounds of $3 per lb nuts.

	Number of Pounds	Price per Pound	Value
$3/lb nuts	x	3	$3x$
$4.25/lb nuts	8	4.25	4.25(8)
Mixture	$x + 8$	4	$4(x + 8)$

The last column gives the equation.

$$3x + 4.25(8) = 4(x + 8)$$
$$3x + 34 = 4x + 32$$
$$2 = x$$

Use 2 pounds of the $3 per pound nuts.

18. $(4p + 2)(5p - 3)$
$$\quad \text{\textbf{F} \quad \textbf{O} \quad \textbf{I} \quad \textbf{L}}$$
$$= 20p^2 - 12p + 10p - 6$$
$$= 20p^2 - 2p - 6$$

19. $(2m^3 - 3m^2 + 8m) - (7m^3 + 5m - 8)$
$$= 2m^3 - 7m^3 - 3m^2 + 8m - 5m + 8$$
$$= -5m^3 - 3m^2 + 3m + 8$$

20. Divide $6t^4 + 5t^3 - 18t^2 + 14t - 1$ by $3t - 2$.

$$
\begin{array}{r}
2t^3 + 3t^2 - 4t + 2 \\
3t - 2 \overline{\smash{)}\, 6t^4 + 5t^3 - 18t^2 + 14t - 1} \\
\underline{6t^4 - 4t^3} \\
9t^3 - 18t^2 \\
\underline{9t^3 - 6t^2} \\
-12t^2 + 14t \\
\underline{-12t^2 + 8t} \\
6t - 1 \\
\underline{6t - 4} \\
3
\end{array}
$$

The quotient is $2t^3 + 3t^2 - 4t + 2$.

The remainder is 3.

Answer: $2t^3 + 3t^2 - 4t + 2 + \dfrac{3}{3t - 2}$

21. $6z^3 + 5z^2 - 4z = z(6z^2 + 5z - 4)$

To factor $6z^2 + 5z - 4$, look for two integers whose product is $(6)(-4) = -24$ and whose sum is 5. The required numbers are 8 and -3.

$$6z^2 + 5z - 4$$
$$= 6z^2 + 8z - 3z - 4$$
$$= 2z(3z + 4) - 1(3z + 4)$$
$$= (3z + 4)(2z - 1)$$

Thus, $6z^3 + 5z^2 - 4z = z(3z + 4)(2z - 1)$.

22. $49a^4 - 9b^2 = (7a^2)^2 - (3b)^2$
$$= (7a^2 + 3b)(7a^2 - 3b)$$

23. $c^3 + 27d^3 = c^3 + (3d)^3$
$$= (c + 3d)(c^2 - 3cd + 9d^2)$$

24.
$$2x^2 + x = 10$$
$$2x^2 + x - 10 = 0$$
$$(2x + 5)(x - 2) = 0$$

$$2x + 5 = 0 \quad \text{or} \quad x - 2 = 0$$
$$x = -\tfrac{5}{2} \quad \text{or} \quad x = 2$$

Check $x = -\frac{5}{2}$: $\frac{25}{2} - \frac{5}{2} = 10$ *True*
Check $x = 2$: $8 + 2 = 10$ *True*

The solution set is $\left\{-\frac{5}{2}, 2\right\}$.

Copyright © 2012 Pearson Education, Inc. Publishing as Addison-Wesley.

25. $\dfrac{(3p^2)^3(-2p^6)}{4p^3(5p^7)}$

$= \dfrac{3^3p^6(-2)p^6}{20p^{10}}$

$= \dfrac{-54p^{12}}{20p^{10}}$

$= -\dfrac{27}{10}p^{12-10}$

$= -\dfrac{27p^2}{10}$

26. $\dfrac{x^2-16}{x^2+2x-8} \div \dfrac{x-4}{x+7}$

$= \dfrac{x^2-16}{x^2+2x-8} \cdot \dfrac{x+7}{x-4}$

$= \dfrac{(x+4)(x-4)(x+7)}{(x+4)(x-2)(x-4)}$

$= \dfrac{x+7}{x-2}$

27. $\dfrac{5}{p^2+3p} - \dfrac{2}{p^2-4p}$

$= \dfrac{5}{p(p+3)} - \dfrac{2}{p(p-4)}$

The LCD is $p(p+3)(p-4)$.

$= \dfrac{5(p-4)}{p(p+3)(p-4)} - \dfrac{2(p+3)}{p(p-4)(p+3)}$

$= \dfrac{5p-20-2p-6}{p(p+3)(p-4)}$

$= \dfrac{3p-26}{p(p+3)(p-4)}$

28. $\dfrac{4}{x-3} - \dfrac{6}{x+3} = \dfrac{24}{x^2-9}$

$\dfrac{4}{x-3} - \dfrac{6}{x+3} = \dfrac{24}{(x+3)(x-3)}$

Multiply by the LCD, $(x+3)(x-3)$. $(x \neq \pm 3)$

$4(x+3) - 6(x-3) = 24$

$4x+12-6x+18 = 24$

$-2x+30 = 24$

$-2x = -6$

$x = 3$

$x \neq 3$ because that would make a denominator 0.
The solution set is $\emptyset$.

29. $6x^2 + 5x = 8$

$6x^2 + 5x - 8 = 0$

Use the quadratic formula with $a = 6$, $b = 5$, and $c = -8$.

$x = \dfrac{-b \pm \sqrt{b^2-4ac}}{2a}$

$x = \dfrac{-5 \pm \sqrt{5^2 - 4(6)(-8)}}{2(6)}$

$= \dfrac{-5 \pm \sqrt{25+192}}{12}$

$= \dfrac{-5 \pm \sqrt{217}}{12}$

The solution set is $\left\{ \dfrac{-5 \pm \sqrt{217}}{12} \right\}$.

30. $3^{2x-1} = 81$

$3^{2x-1} = 3^4$

$2x - 1 = 4$ *Equate exponents.*

$2x = 5$

$x = \frac{5}{2}$

Check $x = \frac{5}{2}$: $3^{5-1} = 3^4 = 81$

The solution set is $\left\{ \frac{5}{2} \right\}$.

31. $\log_8 x + \log_8(x+2) = 1$

$\log_8 x(x+2) = 1$ *Product rule*

$x(x+2) = 8^1$ *Exponential form*

$x^2 + 2x - 8 = 0$

$(x+4)(x-2) = 0$

$x+4 = 0$ or $x-2 = 0$

$x = -4$ or $x = 2$

$x \neq -4$ because $\log_8(-4)$ does not exist.

Check $x = 2$: $\log_8 2 + \log_8 4 = \log_8(2 \cdot 4) = 1$

The solution set is $\{2\}$.

32. $5\sqrt{72} - 4\sqrt{50}$

$= 5\sqrt{36 \cdot 2} - 4\sqrt{25 \cdot 2}$

$= 5 \cdot 6\sqrt{2} - 4 \cdot 5\sqrt{2}$

$= 30\sqrt{2} - 20\sqrt{2}$

$= 10\sqrt{2}$

Copyright © 2012 Pearson Education, Inc. Publishing as Addison-Wesley.

33. $f(x) = 2(x-2)^2 - 3$ is in
$f(x) = a(x-h)^2 + k$ form.

The graph is a vertical parabola with vertex (h, k) at $(2, -3)$. Since $a = 2 > 0$, the graph opens up. Also, $|a| = |2| = 2 > 1$, so the graph is narrower than the graph of $f(x) = x^2$. The points $(0, 5)$ and $(4, 5)$ are on the graph.

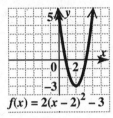

$f(x) = 2(x-2)^2 - 3$

34. Graph $g(x) = \left(\frac{1}{3}\right)^x$.

Make a table of values.

x	-2	-1	0	1	2
$g(x)$	9	3	1	$\frac{1}{3}$	$\frac{1}{9}$

Plot these points, and draw a smooth decreasing exponential curve through them.

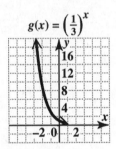

$g(x) = \left(\frac{1}{3}\right)^x$

35. Graph $y = \log_{1/3} x$.
Change to exponential form.

$$y = \log_{1/3} x \quad \Leftrightarrow \quad \left(\frac{1}{3}\right)^y = x$$

This is the inverse of the graph of $g(x) = y = \left(\frac{1}{3}\right)^x$ in Exercise 34. To find points on the graph, interchange the x- and y-values in the table.

x	9	3	1	$\frac{1}{3}$	$\frac{1}{9}$
y	-2	-1	0	1	2

Plot these points, and draw a smooth decreasing logarithmic curve through them.

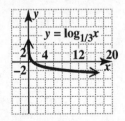

$y = \log_{1/3} x$

36. $f(x) = \dfrac{2}{x-3}$

Setting the denominator equal to zero shows that the vertical asymptote is $x = 3$. Since the numerator has lower degree than the denominator, $y = 0$ is the horizontal asymptote.

$f(0) = -\frac{2}{3}$, so the y-intercept is $\left(0, -\frac{2}{3}\right)$.

The numerator can never be equal to the zero, so there are no x-intercepts.

To the right of $x = 3$, we have the point $(4, 2)$.

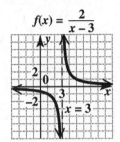

$f(x) = \dfrac{2}{x-3}$

37. $\dfrac{x^2}{9} + \dfrac{y^2}{25} = 1$ is in $\dfrac{x^2}{a^2} + \dfrac{y^2}{b^2} = 1$ form with $a = 3$ and $b = 5$. The graph is an ellipse centered at $(0, 0)$ with x-intercepts $(3, 0)$ and $(-3, 0)$ and y-intercepts $(0, 5)$ and $(0, -5)$. Plot the intercepts and draw the ellipse through them.

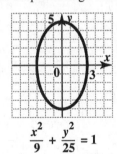

$\dfrac{x^2}{9} + \dfrac{y^2}{25} = 1$

38. $x^2 - y^2 = 9$
$$\dfrac{x^2}{9} - \dfrac{y^2}{9} = 1$$

The graph is a hyperbola centered at $(0, 0)$ with x-intercepts $(3, 0)$ and $(-3, 0)$. The asymptotes are $y = \pm x$. Draw the right and left branches through the intercepts and approaching the asymptotes.

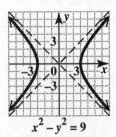

$x^2 - y^2 = 9$

Copyright © 2012 Pearson Education, Inc. Publishing as Addison-Wesley.

39. $f(x) = 2x^3 + 9x^2 + 3x - 4; \; f(-4) = 0$

Divide $f(x)$ by $x - (-4) = x + 4$.

$$
\begin{array}{r|rrrr}
-4 & 2 & 9 & 3 & -4 \\
   &   & -8 & -4 & 4 \\
\hline
   & 2 & 1 & -1 & 0
\end{array}
$$

The quotient, $2x^2 + x - 1$, has factors $(2x - 1)(x + 1)$. Thus,

$$f(x) = (2x - 1)(x + 4)(x + 1).$$

40.
$$
\begin{array}{ll}
xy = -5 & (1) \\
2x + y = 3 & (2)
\end{array}
$$

Solve equation (2) for y.

$$y = -2x + 3 \quad (3)$$

Substitute $-2x + 3$ for y in equation (1).

$$
\begin{aligned}
xy &= -5 \quad (1) \\
x(-2x + 3) &= -5 \\
-2x^2 + 3x &= -5 \\
-2x^2 + 3x + 5 &= 0 \\
2x^2 - 3x - 5 &= 0 \\
(2x - 5)(x + 1) &= 0
\end{aligned}
$$

$$
\begin{array}{lll}
2x - 5 = 0 & \text{or} & x + 1 = 0 \\
x = \frac{5}{2} & \text{or} & x = -1
\end{array}
$$

Substitute these values for x in equation (3) to find y.
If $x = \frac{5}{2}$, then $y = -2\left(\frac{5}{2}\right) + 3 = -2$.

If $x = -1$, then $y = -2(-1) + 3 = 5$.

The solution set is $\left\{(-1, 5), \left(\frac{5}{2}, -2\right)\right\}$.

41. Center at $(-5, 12)$; radius 9
Use the equation of a circle with $h = -5$, $k = 12$, and $r = 9$.

$$
\begin{aligned}
(x - h)^2 + (y - k)^2 &= r^2 \\
[x - (-5)]^2 + (y - 12)^2 &= 9^2 \\
(x + 5)^2 + (y - 12)^2 &= 81
\end{aligned}
$$

42. $a_n = 5n - 12$
$a_1 = 5(1) - 12 = 5 - 12 = -7$
$a_2 = 5(2) - 12 = 10 - 12 = -2$
$a_3 = 5(3) - 12 = 15 - 12 = 3$
$a_4 = 5(4) - 12 = 20 - 12 = 8$
$a_5 = 5(5) - 12 = 25 - 12 = 13$

Answer: $-7, -2, 3, 8, 13$

43. (a) $a_1 = 8, \; d = 2$

$$
\begin{aligned}
S_n &= \frac{n}{2}[2a_1 + (n - 1)d] \\
S_6 &= \frac{6}{2}[2(8) + (6 - 1)2] \\
&= 3(16 + 10) \\
&= 3(26) \\
&= 78
\end{aligned}
$$

(b) $15 - 6 + \frac{12}{5} - \frac{24}{25} + \cdots$

This is an infinite geometric series with $a_1 = 15$ and $r = \frac{-6}{15} = -\frac{2}{5}$. The sum is

$$S = \frac{a_1}{1 - r} = \frac{15}{1 - \left(-\frac{2}{5}\right)} = \frac{15}{\frac{7}{5}} = 15 \cdot \frac{5}{7} = \frac{75}{7}.$$

44. $\displaystyle\sum_{i=1}^{4} 3i = 3\sum_{i=1}^{4} i$

$$
\begin{aligned}
&= 3(1 + 2 + 3 + 4) \\
&= 3(10) \\
&= 30
\end{aligned}
$$

45. $(2a - 1)^5 = [2a + (-1)]^5$

$$
\begin{aligned}
&= (2a)^5 + \frac{5!}{1! \, 4!}(2a)^4(-1)^1 \\
&\quad + \frac{5!}{2! \, 3!}(2a)^3(-1)^2 + \frac{5!}{3! \, 2!}(2a)^2(-1)^3 \\
&\quad + \frac{5!}{4! \, 1!}(2a)^1(-1)^4 + (-1)^5 \\
&= 32a^5 + 5(16a^4)(-1) + 10(8a^3)(1) \\
&\quad + 10(4a^2)(-1) + 5(2a)(1) + (-1) \\
&= 32a^5 - 80a^4 + 80a^3 - 40a^2 + 10a - 1
\end{aligned}
$$

46. The fourth term ($r = 4$, so $r - 1 = 3$) of $\left(3x^4 - \frac{1}{2}y^2\right)^5$ is

$$
\begin{aligned}
&\frac{5!}{3! \, (5 - 3)!}(3x^4)^{5-3}\left(-\frac{1}{2}y^2\right)^3 \\
&= \frac{5!}{3! \, 2!}(3x^4)^2\left(-\frac{1}{2}\right)^3(y^2)^3 \\
&= 10(9x^8)\left(-\frac{1}{8}\right)y^6 \\
&= -\frac{45x^8y^6}{4}.
\end{aligned}
$$

Copyright © 2012 Pearson Education, Inc. Publishing as Addison-Wesley.

47. Let S_n represent the statement

$$4 + 8 + 12 + 16 + \cdots + 4n = 2n(n+1).$$

Prove that S_n is true for every positive integer n.

Step 1
S_1 is the statement

$$4(1) = 2(1)(1+1)$$
$$4 = 2(2),$$

which is true.

Step 2
Show that if S_k is true, then S_{k+1} is also true.
S_k is the statement

$$4 + 8 + 12 + 16 + \cdots + 4k = 2k(k+1).$$

Add the $(k+1)$st term, $4(k+1)$, to each side.

$$4 + 8 + 12 + 16 + \cdots + 4k + 4(k+1)$$
$$= 2k(k+1) + 4(k+1)$$
$$= (k+1)(2k+4)$$
$$= 2(k+1)(k+2)$$

So $4 + 8 + 12 + 16 + \cdots + 4k + 4(k+1)$
$$= 2(k+1)(k+2).$$

The final result is the statement S_{k+1}. Therefore, if S_k is true, then S_{k+1} is true. The two steps required for a proof by mathematical induction are completed, so the general statement S_n is true for every positive integer value of n.

48. (a) $9! = 9 \cdot 8 \cdot 7 \cdot 6 \cdot 5 \cdot 4 \cdot 3 \cdot 2 \cdot 1$
$$= 362{,}880$$

(b) $_7P_3 = \dfrac{7!}{(7-3)!} = \dfrac{7!}{4!} = \dfrac{7 \cdot 6 \cdot 5 \cdot 4!}{4!} = 210$

(c) $_{10}C_4 = \dfrac{10!}{4!\,(10-4)!} = \dfrac{10!}{4!\,6!}$
$$= \dfrac{10 \cdot 9 \cdot 8 \cdot 7 \cdot 6!}{4 \cdot 3 \cdot 2 \cdot 1 \cdot 6!} = 10 \cdot 3 \cdot 7 = 210$$

49. There are only two ways to roll a sum of 11 (call this E) with two dice, 6 and 5 or 5 and 6. There are $6 \cdot 6 = 36$ possible rolls with two dice.

$$P(E) = \frac{n(E)}{n(S)} = \frac{2}{36} = \frac{1}{18}$$

50. The odds that it will rain (call this E) are 3 to 7. Thus, it will rain 3 times for every 7 times it will not rain. The sample space is the total number of times considered or $3 + 7 = 10$. Therefore,

$$P(E) = \frac{3}{3+7} = \frac{3}{10}.$$

Copyright © 2012 Pearson Education, Inc. Publishing as Addison-Wesley.

APPENDIX A PROPERTIES OF MATRICES

Appendix A   Now Try Exercises

N1. (a) $\begin{bmatrix} 9 & a \\ b & 13 \end{bmatrix} = \begin{bmatrix} c & 4 \\ -2 & d \end{bmatrix}$

Corresponding elements are equal, so $a = 4$, $b = -2$, $c = 9$, and $d = 13$.

(b) This statement can never be true since the matrices have different dimensions.

N2. (a) $\begin{bmatrix} 6 & -3 & 5 \\ 8 & 8 & 10 \end{bmatrix} + \begin{bmatrix} -9 & 12 & -14 \\ 3 & 5 & 4 \end{bmatrix}$

$= \begin{bmatrix} 6 + (-9) & -3 + 12 & 5 + (-14) \\ 8 + 3 & 8 + 5 & 10 + 4 \end{bmatrix}$

$= \begin{bmatrix} -3 & 9 & -9 \\ 11 & 13 & 14 \end{bmatrix}$

(b) The first matrix is 1×2 and the second matrix is 2×1. Since the matrices have different dimensions, they cannot be added.

N3. (a) $\begin{bmatrix} -3 & 7 \\ -4 & 12 \end{bmatrix} - \begin{bmatrix} -5 & 8 \\ 12 & 0 \end{bmatrix}$

$= \begin{bmatrix} -3 - (-5) & 7 - 8 \\ -4 - 12 & 12 - 0 \end{bmatrix}$

$= \begin{bmatrix} 2 & -1 \\ -16 & 12 \end{bmatrix}$

(b) The first matrix is 2×1 and the second matrix is 1×2. Since the matrices have different dimensions, they cannot be subtracted.

N4. $-\frac{2}{3} \begin{bmatrix} 6 & -18 \\ 12 & 9 \end{bmatrix} = \begin{bmatrix} -\frac{2}{3}(6) & -\frac{2}{3}(-18) \\ -\frac{2}{3}(12) & -\frac{2}{3}(9) \end{bmatrix}$

$= \begin{bmatrix} -4 & 12 \\ -8 & -6 \end{bmatrix}$

N5. (a) Let A denote the first matrix and B the second. The matrix A is 2×2 and the matrix B is 2×1. Since the number of columns of A, 2, equals the number of rows of B, 2, the product AB can be found, and it is 2×1.

(b) Let A denote the first matrix and B the second. The matrix A is 2×1 and the matrix B is 2×2. Since the number of columns of A, 1, does not equal the number of rows of B, 2, the product AB can not be found.

N6. $AB = \begin{bmatrix} 1 & -2 & 3 \\ 5 & 0 & 4 \\ -8 & 7 & -7 \end{bmatrix} \begin{bmatrix} 1 \\ -2 \\ 3 \end{bmatrix}$

$= \begin{bmatrix} 1(1) + (-2)(-2) + 3(3) \\ 5(1) + 0(-2) + 4(3) \\ -8(1) + 7(-2) + (-7)(3) \end{bmatrix}$

$= \begin{bmatrix} 14 \\ 17 \\ -43 \end{bmatrix}$

BA is not defined.

N7. $AB = \begin{bmatrix} -1 & 0 \\ 2 & 5 \end{bmatrix} \begin{bmatrix} -3 & 5 \\ 2 & 1 \end{bmatrix}$

$= \begin{bmatrix} -1(-3) + 0(2) & -1(5) + 0(1) \\ 2(-3) + 5(2) & 2(5) + 5(1) \end{bmatrix}$

$= \begin{bmatrix} 3 + 0 & -5 + 0 \\ -6 + 10 & 10 + 5 \end{bmatrix} = \begin{bmatrix} 3 & -5 \\ 4 & 15 \end{bmatrix}$

$BA = \begin{bmatrix} -3 & 5 \\ 2 & 1 \end{bmatrix} \begin{bmatrix} -1 & 0 \\ 2 & 5 \end{bmatrix}$

$= \begin{bmatrix} -3(-1) + 5(2) & -3(0) + 5(5) \\ 2(-1) + 1(2) & 2(0) + 1(5) \end{bmatrix}$

$= \begin{bmatrix} 3 + 10 & 0 + 25 \\ -2 + 2 & 0 + 5 \end{bmatrix} = \begin{bmatrix} 13 & 25 \\ 0 & 5 \end{bmatrix}$

N8. Below is the solution for Exercise 48.

(a) $\begin{bmatrix} 1 & 4 & \frac{1}{4} & \frac{1}{4} & 1 \\ 0 & 3 & 0 & \frac{1}{4} & 0 \\ 4 & 3 & 2 & 1 & 1 \\ 0 & 1 & 0 & \frac{1}{3} & 0 \end{bmatrix} \begin{bmatrix} 5 & 5 \\ 8 & 10 \\ 10 & 12 \\ 12 & 15 \\ 5 & 6 \end{bmatrix}$

$= \begin{bmatrix} 47.5 & 57.75 \\ 27 & 33.75 \\ 81 & 95 \\ 12 & 15 \end{bmatrix}$

(b) $\begin{bmatrix} 20 & 200 & 50 & 60 \end{bmatrix}$

$\begin{bmatrix} 20 & 200 & 50 & 60 \end{bmatrix} \begin{bmatrix} 1 & 4 & \frac{1}{4} & \frac{1}{4} & 1 \\ 0 & 3 & 0 & \frac{1}{4} & 0 \\ 4 & 3 & 2 & 1 & 1 \\ 0 & 1 & 0 & \frac{1}{3} & 0 \end{bmatrix}$

$= \begin{bmatrix} 220 & 890 & 105 & 125 & 70 \end{bmatrix}$

(c) $\begin{bmatrix} 20 & 200 & 50 & 60 \end{bmatrix} \begin{bmatrix} 47.5 & 57.75 \\ 27 & 33.75 \\ 81 & 95 \\ 12 & 15 \end{bmatrix}$

$= \begin{bmatrix} 11{,}120 & 13{,}555 \end{bmatrix}$

Appendix A   Section Exercises

1. $\begin{bmatrix} w & x \\ y & z \end{bmatrix} = \begin{bmatrix} 3 & 2 \\ -1 & 4 \end{bmatrix}$

Corresponding elements are equal, so $w = 3$, $x = 2$, $y = -1$, and $z = 4$.

Copyright © 2012 Pearson Education, Inc. Publishing as Addison-Wesley.

3. $\begin{bmatrix} 0 & 5 & x \\ -1 & 3 & y+2 \\ 4 & 1 & z \end{bmatrix} = \begin{bmatrix} 0 & w+3 & 6 \\ -1 & 3 & 0 \\ 4 & 1 & 8 \end{bmatrix}$

Corresponding elements are equal, so

$$x = 6;$$
$$w + 3 = 5 \Rightarrow w = 2;$$
$$y + 2 = 0 \Rightarrow y = -2;$$
$$z = 8.$$

5. $\begin{bmatrix} -7+z & 4r & 8s \\ 6p & 2 & 5 \end{bmatrix} + \begin{bmatrix} -9 & 8r & 3 \\ 2 & 5 & 19 \end{bmatrix}$

$$= \begin{bmatrix} 2 & 36 & 27 \\ 20 & 7 & 12a \end{bmatrix}$$

$\begin{bmatrix} -16+z & 12r & 8s+3 \\ 6p+2 & 7 & 24 \end{bmatrix} = \begin{bmatrix} 2 & 36 & 27 \\ 20 & 7 & 12a \end{bmatrix}$

Corresponding elements must be equal:

$$-16 + z = 2 \quad \Rightarrow \quad z = 18$$
$$12r = 36 \quad \Rightarrow \quad r = 3$$
$$8s + 3 = 27 \quad \Rightarrow \quad 8s = 24 \quad \Rightarrow \quad s = 3$$
$$6p + 2 = 20 \quad \Rightarrow \quad 6p = 18 \quad \Rightarrow \quad p = 3$$
$$24 = 12a \quad \Rightarrow \quad a = 2$$

7. $\begin{bmatrix} -4 & 8 \\ 2 & 3 \end{bmatrix}$

The matrix has 2 rows and 2 columns, so it is a 2×2 *square* matrix.

9. $\begin{bmatrix} 2 \\ 4 \end{bmatrix}$

This matrix has 2 rows and 1 column, so it is a 2×1 *column* matrix.

11. $\begin{bmatrix} -6 & 8 & 0 & 0 \\ 4 & 1 & 9 & 2 \\ 3 & -5 & 7 & 1 \end{bmatrix}$

This matrix has 3 rows and 4 columns, so it is a 3×4 matrix.

13. $\begin{bmatrix} 6 & -9 & 2 \\ 4 & 1 & 3 \end{bmatrix} + \begin{bmatrix} -8 & 2 & 5 \\ 6 & -3 & 4 \end{bmatrix}$

$$= \begin{bmatrix} 6+(-8) & -9+2 & 2+5 \\ 4+6 & 1+(-3) & 3+4 \end{bmatrix}$$

$$= \begin{bmatrix} -2 & -7 & 7 \\ 10 & -2 & 7 \end{bmatrix}$$

15. $\begin{bmatrix} -6 & 8 \\ 0 & 0 \end{bmatrix} - \begin{bmatrix} 0 & 0 \\ -4 & -2 \end{bmatrix}$

$$= \begin{bmatrix} -6 & 8 \\ 0 & 0 \end{bmatrix} + \begin{bmatrix} 0 & 0 \\ 4 & 2 \end{bmatrix}$$

$$= \begin{bmatrix} -6 & 8 \\ 4 & 2 \end{bmatrix}$$

17. $\begin{bmatrix} 3x+y & x-2y & 2x \\ 5x & 3y & x+y \end{bmatrix}$

$$+ \begin{bmatrix} 2x & 3y & 5x+y \\ 3x+2y & x & 2x \end{bmatrix}$$

$$= \begin{bmatrix} 5x+y & x+y & 7x+y \\ 8x+2y & x+3y & 3x+y \end{bmatrix}$$

19. $\begin{bmatrix} 3 \\ 2 \end{bmatrix} + [2 \quad 3]$

These two matrices do not have the same dimensions, so they cannot be added.

In Exercises 21–26,

$$A = \begin{bmatrix} -2 & 4 \\ 0 & 3 \end{bmatrix} \quad \text{and} \quad B = \begin{bmatrix} -6 & 2 \\ 4 & 0 \end{bmatrix}.$$

21. $2A = 2\begin{bmatrix} -2 & 4 \\ 0 & 3 \end{bmatrix} = \begin{bmatrix} -4 & 8 \\ 0 & 6 \end{bmatrix}$

23. $2A - B = 2\begin{bmatrix} -2 & 4 \\ 0 & 3 \end{bmatrix} - \begin{bmatrix} -6 & 2 \\ 4 & 0 \end{bmatrix}$

$$= \begin{bmatrix} -4 & 8 \\ 0 & 6 \end{bmatrix} + \begin{bmatrix} 6 & -2 \\ -4 & 0 \end{bmatrix}$$

$$= \begin{bmatrix} 2 & 6 \\ -4 & 6 \end{bmatrix}$$

25. $-A + \frac{1}{2}B = -\begin{bmatrix} -2 & 4 \\ 0 & 3 \end{bmatrix} + \frac{1}{2}\begin{bmatrix} -6 & 2 \\ 4 & 0 \end{bmatrix}$

$$= \begin{bmatrix} 2 & -4 \\ 0 & -3 \end{bmatrix} + \begin{bmatrix} -3 & 1 \\ 2 & 0 \end{bmatrix}$$

$$= \begin{bmatrix} -1 & -3 \\ 2 & -3 \end{bmatrix}$$

27. The matrix A is 2×3 and the matrix B is 3×5. Since the number of columns of A, 3, equals the number of rows of B, 3, the product AB can be found, and it is 2×5.

29. The matrix B is 3×5 and the matrix A is 2×3. Since the number of columns of B, 5, does not equal the number of rows of A, 2, the product BA can *not* be found.

31. $\begin{bmatrix} 1 & 2 \\ 3 & 4 \end{bmatrix}\begin{bmatrix} -1 \\ 7 \end{bmatrix}$

$$= \begin{bmatrix} 1(-1) + 2(7) \\ 3(-1) + 4(7) \end{bmatrix}$$

$$= \begin{bmatrix} -1 + 14 \\ -3 + 28 \end{bmatrix} = \begin{bmatrix} 13 \\ 25 \end{bmatrix}$$

Copyright © 2012 Pearson Education, Inc. Publishing as Addison-Wesley.

33. $\begin{bmatrix} 3 & -4 & 1 \\ 5 & 0 & 2 \end{bmatrix} \begin{bmatrix} -1 \\ 4 \\ 2 \end{bmatrix}$

$= \begin{bmatrix} 3(-1) + (-4)(4) + 1(2) \\ 5(-1) + 0(4) + 2(2) \end{bmatrix}$

$= \begin{bmatrix} -3 - 16 + 2 \\ -5 + 0 + 4 \end{bmatrix} = \begin{bmatrix} -17 \\ -1 \end{bmatrix}$

35. $\begin{bmatrix} 5 & 2 \\ -1 & 4 \end{bmatrix} \begin{bmatrix} 3 & -2 \\ 1 & 0 \end{bmatrix}$

$= \begin{bmatrix} 5(3) + 2(1) & 5(-2) + 2(0) \\ -1(3) + 4(1) & -1(-2) + 4(0) \end{bmatrix}$

$= \begin{bmatrix} 15 + 2 & -10 + 0 \\ -3 + 4 & 2 + 0 \end{bmatrix} = \begin{bmatrix} 17 & -10 \\ 1 & 2 \end{bmatrix}$

37. $\begin{bmatrix} 2 & 2 & -1 \\ 3 & 0 & 1 \end{bmatrix} \begin{bmatrix} 0 & 2 \\ -1 & 4 \\ 0 & 2 \end{bmatrix}$

$= \begin{bmatrix} 2(0) + 2(-1) + (-1)(0) & 2(2) + 2(4) + (-1)(2) \\ 3(0) + 0(-1) + 1(0) & 3(2) + 0(4) + 1(2) \end{bmatrix}$

$= \begin{bmatrix} -2 & 10 \\ 0 & 8 \end{bmatrix}$

39. $\begin{bmatrix} -1 & 2 & 0 \\ 0 & 3 & 2 \\ 0 & 1 & 4 \end{bmatrix} \begin{bmatrix} 2 & -1 & 2 \\ 0 & 2 & 1 \\ 3 & 0 & -1 \end{bmatrix}$

$= \begin{bmatrix} -1(2) + 2(0) + 0(3) \\ 0(2) + 3(0) + 2(3) \\ 0(2) + 1(0) + 4(3) \end{bmatrix}$

$\begin{matrix} (-1)(-1) + 2(2) + 0(0) \\ 0(-1) + 3(2) + 2(0) \\ 0(-1) + 1(2) + 4(0) \end{matrix}$

$\begin{matrix} (-1)(2) + 2(1) + 0(-1) \\ 0(2) + 3(1) + 2(-1) \\ 0(2) + 1(1) + 4(-1) \end{matrix}$

$= \begin{bmatrix} -2 & 5 & 0 \\ 6 & 6 & 1 \\ 12 & 2 & -3 \end{bmatrix}$

41. $\begin{bmatrix} -2 & 4 & 1 \end{bmatrix} \begin{bmatrix} 3 & -2 & 4 \\ 2 & 1 & 0 \\ 0 & -1 & 4 \end{bmatrix}$

$= [-2(3) + 4(2) + 1(0)$
$\quad (-2)(-2) + 4(1) + 1(-1)$
$\quad (-2)(4) + 4(0) + 1(4)]$

$= \begin{bmatrix} 2 & 7 & -4 \end{bmatrix}$

43. $\begin{bmatrix} -3 & 0 & 2 & 1 \\ 4 & 0 & 2 & 6 \end{bmatrix} \begin{bmatrix} -4 & 2 \\ 0 & 1 \end{bmatrix}$

It is not possible to find this matrix product because the number of columns of the first matrix (four) is not equal to the number of rows of the second matrix (two).

45. The given information may be written as the 3×2 matrix

	Flat-head	Round-head
Adelphi	100	150
Beltsville	125	50
College Park	175	200

or as the 2×3 matrix

	Adelphi	Beltsville	College Park
Flat-head	100	125	175
Round-head	150	50	200

47. **(a)** The sales figure information may be written as the 3×3 matrix

	Nonfat	Regular	Super Creamy
I	50	100	30
II	10	90	50
III	60	120	40

(b) The income per gallon information may be written as the 3×1 matrix

Nonfat	12
Regular	10
Super Creamy	15

(If the matrix in part (a) had been written with its rows and columns interchanged, then this income per gallon information would be written instead as a 1×3 matrix.)

(c) $\begin{bmatrix} 50 & 100 & 30 \\ 10 & 90 & 50 \\ 60 & 120 & 40 \end{bmatrix} \begin{bmatrix} 12 \\ 10 \\ 15 \end{bmatrix} = \begin{bmatrix} 2050 \\ 1770 \\ 2520 \end{bmatrix}$

(This result may be written as a 1×3 matrix instead.)

(d) $2050 + 1770 + 2520 = 6340$

The total daily income from the three locations is $6340.

49. $A + B = \begin{bmatrix} a_{11} & a_{12} \\ a_{21} & a_{22} \end{bmatrix} + \begin{bmatrix} b_{11} & b_{12} \\ b_{21} & b_{22} \end{bmatrix}$

$= \begin{bmatrix} a_{11} + b_{11} & a_{12} + b_{12} \\ a_{21} + b_{21} & a_{22} + b_{22} \end{bmatrix}$

Since addition of real numbers is commutative, we can rewrite the last matrix.

$= \begin{bmatrix} b_{11} + a_{11} & b_{12} + a_{12} \\ b_{21} + a_{21} & b_{22} + a_{22} \end{bmatrix}$

$= \begin{bmatrix} b_{11} & b_{12} \\ b_{21} & b_{22} \end{bmatrix} + \begin{bmatrix} a_{11} & a_{12} \\ a_{21} & a_{22} \end{bmatrix}$

$= B + A$

Copyright © 2012 Pearson Education, Inc. Publishing as Addison-Wesley.

51. $(AB)C$

$$= \left(\begin{bmatrix} a_{11} & a_{12} \\ a_{21} & a_{22} \end{bmatrix} \begin{bmatrix} b_{11} & b_{12} \\ b_{21} & b_{22} \end{bmatrix} \right) \begin{bmatrix} c_{11} & c_{12} \\ c_{21} & c_{22} \end{bmatrix}$$

$$= \begin{bmatrix} a_{11}b_{11} + a_{12}b_{21} & a_{11}b_{12} + a_{12}b_{22} \\ a_{21}b_{11} + a_{22}b_{21} & a_{21}b_{12} + a_{22}b_{22} \end{bmatrix} \begin{bmatrix} c_{11} & c_{12} \\ c_{21} & c_{22} \end{bmatrix}$$

$$= \begin{bmatrix} (a_{11}b_{11} + a_{12}b_{21})c_{11} + (a_{11}b_{12} + a_{12}b_{22})c_{21} & (a_{11}b_{11} + a_{12}b_{21})c_{12} + (a_{11}b_{12} + a_{12}b_{22})c_{22} \\ (a_{21}b_{11} + a_{22}b_{21})c_{11} + (a_{21}b_{12} + a_{22}b_{22})c_{21} & (a_{21}b_{11} + a_{22}b_{21})c_{12} + (a_{21}b_{12} + a_{22}b_{22})c_{22} \end{bmatrix}$$

$$= \begin{bmatrix} a_{11}b_{11}c_{11} + a_{12}b_{21}c_{11} + a_{11}b_{12}c_{21} + a_{12}b_{22}c_{21} & a_{11}b_{11}c_{12} + a_{12}b_{21}c_{12} + a_{11}b_{12}c_{22} + a_{12}b_{22}c_{22} \\ a_{21}b_{11}c_{11} + a_{22}b_{21}c_{11} + a_{21}b_{12}c_{21} + a_{22}b_{22}c_{21} & a_{21}b_{11}c_{12} + a_{22}b_{21}c_{12} + a_{21}b_{12}c_{22} + a_{22}b_{22}c_{22} \end{bmatrix}$$

$$= \begin{bmatrix} (a_{11}b_{11}c_{11} + a_{11}b_{12}c_{21}) + (a_{12}b_{21}c_{11} + a_{12}b_{22}c_{21}) & (a_{11}b_{11}c_{12} + a_{11}b_{12}c_{22}) + (a_{12}b_{21}c_{12} + a_{12}b_{22}c_{22}) \\ (a_{21}b_{11}c_{11} + a_{21}b_{12}c_{21}) + (a_{22}b_{21}c_{11} + a_{22}b_{22}c_{21}) & (a_{21}b_{11}c_{12} + a_{21}b_{12}c_{22}) + (a_{22}b_{21}c_{12} + a_{22}b_{22}c_{22}) \end{bmatrix}$$

$$= \begin{bmatrix} a_{11}(b_{11}c_{11} + b_{12}c_{21}) + a_{12}(b_{21}c_{11} + b_{22}c_{21}) & a_{11}(b_{11}c_{12} + b_{12}c_{22}) + a_{12}(b_{21}c_{12} + b_{22}c_{22}) \\ a_{21}(b_{11}c_{11} + b_{12}c_{21}) + a_{22}(b_{21}c_{11} + b_{22}c_{21}) & a_{21}(b_{11}c_{12} + b_{12}c_{22}) + a_{22}(b_{21}c_{12} + b_{22}c_{22}) \end{bmatrix}$$

$$= \begin{bmatrix} a_{11} & a_{12} \\ a_{21} & a_{22} \end{bmatrix} \begin{bmatrix} b_{11}c_{11} + b_{12}c_{21} & b_{11}c_{12} + b_{12}c_{22} \\ b_{21}c_{11} + b_{22}c_{21} & b_{21}c_{12} + b_{22}c_{22} \end{bmatrix}$$

$$= \begin{bmatrix} a_{11} & a_{12} \\ a_{21} & a_{22} \end{bmatrix} \left(\begin{bmatrix} b_{11} & b_{12} \\ b_{21} & b_{22} \end{bmatrix} \begin{bmatrix} c_{11} & c_{12} \\ c_{21} & c_{22} \end{bmatrix} \right)$$

$$= A(BC)$$

53. $c(A + B)$

$$= c \left(\begin{bmatrix} a_{11} & a_{12} \\ a_{21} & a_{22} \end{bmatrix} + \begin{bmatrix} b_{11} & b_{12} \\ b_{21} & b_{22} \end{bmatrix} \right)$$

$$= c \begin{bmatrix} a_{11} + b_{11} & a_{12} + b_{12} \\ a_{21} + b_{21} & a_{22} + b_{22} \end{bmatrix}$$

$$= \begin{bmatrix} c(a_{11} + b_{11}) & c(a_{12} + b_{12}) \\ c(a_{21} + b_{21}) & c(a_{22} + b_{22}) \end{bmatrix}$$

$$= \begin{bmatrix} ca_{11} + cb_{11} & ca_{12} + cb_{12} \\ ca_{21} + cb_{21} & ca_{22} + cb_{22} \end{bmatrix}$$

$$= \begin{bmatrix} ca_{11} & ca_{12} \\ ca_{21} & ca_{22} \end{bmatrix} + \begin{bmatrix} cb_{11} & cb_{12} \\ cb_{21} & cb_{22} \end{bmatrix}$$

$$= c \begin{bmatrix} a_{11} & a_{12} \\ a_{21} & a_{22} \end{bmatrix} + c \begin{bmatrix} b_{11} & b_{12} \\ b_{21} & b_{22} \end{bmatrix}$$

$$= cA + cB$$

55. Let c and d be any real numbers.

$$c(A)d = c \begin{bmatrix} a_{11} & a_{12} \\ a_{21} & a_{22} \end{bmatrix} d$$

$$= \begin{bmatrix} ca_{11} & ca_{12} \\ ca_{21} & ca_{22} \end{bmatrix} d$$

$$= \begin{bmatrix} (ca_{11})d & (ca_{12})d \\ (ca_{21})d & (ca_{22})d \end{bmatrix}$$

$$= \begin{bmatrix} cd(a_{11}) & cd(a_{12}) \\ cd(a_{21}) & cd(a_{22}) \end{bmatrix}$$

$$= cd \begin{bmatrix} a_{11} & a_{12} \\ a_{21} & a_{22} \end{bmatrix}$$

$$= cd(A)$$

Copyright © 2012 Pearson Education, Inc. Publishing as Addison-Wesley.

APPENDIX B   MATRIX INVERSES

Appendix B   Now Try Exercises

N1. $AI_4 = \begin{bmatrix} 2 & 1 & 3 & 9 \\ 5 & -2 & 8 & 6 \\ 6 & 2 & 1 & 0 \\ 5 & 3 & 7 & 7 \end{bmatrix} \begin{bmatrix} 1 & 0 & 0 & 0 \\ 0 & 1 & 0 & 0 \\ 0 & 0 & 1 & 0 \\ 0 & 0 & 0 & 1 \end{bmatrix}$

$= \begin{bmatrix} 2(1) + 1(0) + 3(0) + 9(0) \\ 5(1) - 2(0) + 8(0) + 6(0) \\ 6(1) + 2(0) + 1(0) + 0(0) \\ 5(1) + 3(0) + 7(0) + 7(0) \end{bmatrix}$

$\begin{matrix} 2(0) + 1(1) + 3(0) + 9(0) \\ 5(0) - 2(1) + 8(0) + 6(0) \\ 6(0) + 2(1) + 1(0) + 0(0) \\ 5(0) + 3(1) + 7(0) + 7(0) \end{matrix}$

$\begin{matrix} 2(0) + 1(0) + 3(1) + 9(0) \\ 5(0) - 2(0) + 8(1) + 6(0) \\ 6(0) + 2(0) + 1(1) + 0(0) \\ 5(0) + 3(0) + 7(1) + 7(0) \end{matrix}$

$\begin{matrix} 2(0) + 1(0) + 3(0) + 9(1) \\ 5(0) - 2(0) + 8(0) + 6(1) \\ 6(0) + 2(0) + 1(0) + 0(1) \\ 5(0) + 3(0) + 7(0) + 7(1) \end{matrix}$

$= \begin{bmatrix} 2 & 1 & 3 & 9 \\ 5 & -2 & 8 & 6 \\ 6 & 2 & 1 & 0 \\ 5 & 3 & 7 & 7 \end{bmatrix} = A$

N2. Find the inverse of $A = \begin{bmatrix} -6 & 5 \\ -5 & 4 \end{bmatrix}$.

We use the symbol $\leftrightarrow$ for "interchange."

$[A \mid I_2] = \begin{bmatrix} -6 & 5 & | & 1 & 0 \\ -5 & 4 & | & 0 & 1 \end{bmatrix}$

$\begin{bmatrix} 6 & -5 & | & -1 & 0 \\ -5 & 4 & | & 0 & 1 \end{bmatrix} \quad -1R_2$

$\begin{bmatrix} 1 & -1 & | & -1 & 1 \\ -5 & 4 & | & 0 & 1 \end{bmatrix} \quad R_2 + R_1$

$\begin{bmatrix} 1 & -1 & | & -1 & 1 \\ 0 & -1 & | & -5 & 6 \end{bmatrix} \quad 5R_1 + R_2$

$\begin{bmatrix} 1 & -1 & | & -1 & 1 \\ 0 & 1 & | & 5 & -6 \end{bmatrix} \quad -1R_2$

$\begin{bmatrix} 1 & 0 & | & 4 & -5 \\ 0 & 1 & | & 5 & -6 \end{bmatrix} \quad R_2 + R_1$

$A^{-1} = \begin{bmatrix} 4 & -5 \\ 5 & -6 \end{bmatrix}$

N3. Find A^{-1} if $A = \begin{bmatrix} -4 & 2 & 0 \\ 1 & -1 & 2 \\ 0 & 1 & 4 \end{bmatrix}$.

We use the symbol $\leftrightarrow$ for "interchange."

$\begin{bmatrix} -4 & 2 & 0 & | & 1 & 0 & 0 \\ 1 & -1 & 2 & | & 0 & 1 & 0 \\ 0 & 1 & 4 & | & 0 & 0 & 1 \end{bmatrix}$

$\begin{bmatrix} 1 & -1 & 2 & | & 0 & 1 & 0 \\ -4 & 2 & 0 & | & 1 & 0 & 0 \\ 0 & 1 & 4 & | & 0 & 0 & 1 \end{bmatrix} \quad R_1 \leftrightarrow R_2$

$\begin{bmatrix} 1 & -1 & 2 & | & 0 & 1 & 0 \\ 0 & -2 & 8 & | & 1 & 4 & 0 \\ 0 & 1 & 4 & | & 0 & 0 & 1 \end{bmatrix} \quad 4R_1 + R_2$

$\begin{bmatrix} 1 & -1 & 2 & | & 0 & 1 & 0 \\ 0 & 1 & 4 & | & 0 & 0 & 1 \\ 0 & -2 & 8 & | & 1 & 4 & 0 \end{bmatrix} \quad R_2 \leftrightarrow R_3$

$\begin{bmatrix} 1 & 0 & 6 & | & 0 & 1 & 1 \\ 0 & 1 & 4 & | & 0 & 0 & 1 \\ 0 & 0 & 16 & | & 1 & 4 & 2 \end{bmatrix} \quad \begin{matrix} R_2 + R_1 \\ \\ 2R_2 + R_3 \end{matrix}$

$\begin{bmatrix} 1 & 0 & 6 & | & 0 & 1 & 1 \\ 0 & 1 & 4 & | & 0 & 0 & 1 \\ 0 & 0 & 1 & | & \frac{1}{16} & \frac{1}{4} & \frac{1}{8} \end{bmatrix} \quad \frac{1}{16}R_3$

$\begin{bmatrix} 1 & 0 & 0 & | & -\frac{3}{8} & -\frac{1}{2} & \frac{1}{4} \\ 0 & 1 & 0 & | & -\frac{1}{4} & -1 & \frac{1}{2} \\ 0 & 0 & 1 & | & \frac{1}{16} & \frac{1}{4} & \frac{1}{8} \end{bmatrix} \quad \begin{matrix} -6R_3 + R_1 \\ -4R_3 + R_2 \end{matrix}$

$A^{-1} = \begin{bmatrix} -\frac{3}{8} & -\frac{1}{2} & \frac{1}{4} \\ -\frac{1}{4} & -1 & \frac{1}{2} \\ \frac{1}{16} & \frac{1}{4} & \frac{1}{8} \end{bmatrix}$

Copyright © 2012 Pearson Education, Inc. Publishing as Addison-Wesley.

N4. Find A^{-1}, if possible, given

$$A = \begin{bmatrix} 4 & -2 & 5 \\ 0 & 1 & 0 \\ -8 & 4 & -10 \end{bmatrix}.$$

$$\left[\begin{array}{ccc|ccc} 4 & -2 & 5 & 1 & 0 & 0 \\ 0 & 1 & 0 & 0 & 1 & 0 \\ -8 & 4 & -10 & 0 & 0 & 1 \end{array}\right]$$

There are many possibilities for selecting row operations. Here, since the second row is in the form "0-1-0", we'll obtain zeros in C_2.

$$\left[\begin{array}{ccc|ccc} 4 & 0 & 5 & 1 & 2 & 0 \\ 0 & 1 & 0 & 0 & 1 & 0 \\ -8 & 0 & -10 & 0 & -4 & 1 \end{array}\right] \begin{array}{l} 2R_2 + R_1 \\ \\ -4R_2 + R_3 \end{array}$$

Now, we could multiply R_1 by $\frac{1}{4}$, but that would introduce fractions. Instead, we'll obtain zeros in R_3 by noting the relationship between R_1 and R_3.

$$\left[\begin{array}{ccc|ccc} 4 & 0 & 5 & 1 & 2 & 0 \\ 0 & 1 & 0 & 0 & 1 & 0 \\ 0 & 0 & 0 & 2 & 0 & 1 \end{array}\right] \begin{array}{l} \\ \\ 2R_1 + R_3 \end{array}$$

At this point, the matrix should be changed so that the third-row, third-column element will be 1. Since that element is now 0, the desired transformation cannot be completed. Therefore, the inverse of the given matrix *does not exist*.

Note: You may have immediately noticed that each element of the third row of A is -2 times the corresponding element in the first row. Using the row operator $2R_1 + R_3$ would lead to a row of 0's, which leads us to the same conclusion in fewer steps.

N5. $5x + 2y = -1$
$2x + 3y = 15$

Let $A = \begin{bmatrix} 5 & 2 \\ 2 & 3 \end{bmatrix}$, $X = \begin{bmatrix} x \\ y \end{bmatrix}$, and $B = \begin{bmatrix} -1 \\ 15 \end{bmatrix}$.

Find A^{-1} as in Now Try Exercise 2.

$$A^{-1} = \begin{bmatrix} \frac{3}{11} & -\frac{2}{11} \\ -\frac{2}{11} & \frac{5}{11} \end{bmatrix}$$

Now find $A^{-1}B$.

$$A^{-1}B = \begin{bmatrix} \frac{3}{11} & -\frac{2}{11} \\ -\frac{2}{11} & \frac{5}{11} \end{bmatrix} \begin{bmatrix} -1 \\ 15 \end{bmatrix} = \begin{bmatrix} -3 \\ 7 \end{bmatrix}$$

Since $X = A^{-1}B$,

$$X = \begin{bmatrix} x \\ y \end{bmatrix} = \begin{bmatrix} -3 \\ 7 \end{bmatrix}.$$

The solution set of the system is $\{(-3, 7)\}$.

Appendix B Section Exercises

1. $AI_2 = \begin{bmatrix} 4 & -2 \\ 3 & 1 \end{bmatrix} \begin{bmatrix} 1 & 0 \\ 0 & 1 \end{bmatrix}$

$= \begin{bmatrix} 4(1) + (-2)(0) & 4(0) + (-2)(1) \\ 3(1) + 1(0) & 3(0) + 1(1) \end{bmatrix}$

$= \begin{bmatrix} 4 & -2 \\ 3 & 1 \end{bmatrix} = A$

3. $\begin{bmatrix} 5 & 7 \\ 2 & 3 \end{bmatrix} \begin{bmatrix} 3 & -7 \\ -2 & 5 \end{bmatrix}$

$= \begin{bmatrix} 5(3) + 7(-2) & 5(-7) + 7(5) \\ 2(3) + 3(-2) & 2(-7) + 3(5) \end{bmatrix}$

$= \begin{bmatrix} 1 & 0 \\ 0 & 1 \end{bmatrix}$

Since the product obtained by multiplying the matrices is the 2×2 identity matrix I_2, the given matrices are inverses of each other.

5. $\begin{bmatrix} -1 & 2 \\ 3 & -5 \end{bmatrix} \begin{bmatrix} -5 & -2 \\ -3 & -1 \end{bmatrix} = \begin{bmatrix} -1 & 0 \\ 0 & -1 \end{bmatrix}$

Since this product is not the 2×2 identity matrix I_2, the given matrices are not inverses of each other.

7. $\begin{bmatrix} 0 & 1 & 0 \\ 0 & 0 & -2 \\ 1 & -1 & 0 \end{bmatrix} \begin{bmatrix} 1 & 0 & 1 \\ 1 & 0 & 0 \\ 0 & -1 & 0 \end{bmatrix} = \begin{bmatrix} 1 & 0 & 0 \\ 0 & 2 & 0 \\ 0 & 0 & 1 \end{bmatrix}$

Since this product is not the 3×3 identity matrix I_3, the given matrices are not inverses of each other.

9. $\begin{bmatrix} -1 & -1 & -1 \\ 4 & 5 & 0 \\ 0 & 1 & -3 \end{bmatrix} \begin{bmatrix} 15 & 4 & -5 \\ -12 & -3 & 4 \\ -4 & -1 & 1 \end{bmatrix}$

$= \begin{bmatrix} 1 & 0 & 0 \\ 0 & 1 & 0 \\ 0 & 0 & 1 \end{bmatrix} = I_3$

The given matrices are inverses of each other.

11. Find the inverse of $A = \begin{bmatrix} -1 & -2 \\ 3 & 5 \end{bmatrix}$, if it exists.

$$[A \mid I_2] = \left[\begin{array}{cc|cc} -1 & -2 & 1 & 0 \\ 3 & 5 & 0 & 1 \end{array}\right]$$

$$\left[\begin{array}{cc|cc} -1 & -2 & 1 & 0 \\ 0 & -1 & 3 & 1 \end{array}\right] \begin{array}{l} \\ 3R_1 + R_2 \end{array}$$

$$\left[\begin{array}{cc|cc} 1 & 2 & -1 & 0 \\ 0 & -1 & 3 & 1 \end{array}\right] \begin{array}{l} -1R_1 \\ \end{array}$$

$$\left[\begin{array}{cc|cc} 1 & 0 & 5 & 2 \\ 0 & -1 & 3 & 1 \end{array}\right] \begin{array}{l} 2R_2 + R_1 \\ \end{array}$$

$$\left[\begin{array}{cc|cc} 1 & 0 & 5 & 2 \\ 0 & 1 & -3 & -1 \end{array}\right] \begin{array}{l} \\ -1R_2 \end{array}$$

$$A^{-1} = \begin{bmatrix} 5 & 2 \\ -3 & -1 \end{bmatrix}$$

Copyright © 2012 Pearson Education, Inc. Publishing as Addison-Wesley.

13. Find the inverse of $\begin{bmatrix} 5 & 10 \\ -3 & -6 \end{bmatrix}$, if it exists.

$$[A \,|\, I_2] = \begin{bmatrix} 5 & 10 & | & 1 & 0 \\ -3 & -6 & | & 0 & 1 \end{bmatrix}$$

$$\begin{bmatrix} 1 & 2 & | & \frac{1}{5} & 0 \\ -3 & -6 & | & 0 & 1 \end{bmatrix} \quad \frac{1}{5}R_1$$

$$\begin{bmatrix} 1 & 2 & | & \frac{1}{5} & 0 \\ 0 & 0 & | & \frac{3}{5} & 1 \end{bmatrix} \quad 3R_1 + R_2$$

At this point, the matrix should be changed so that the second-row, second-column element will be 1. Since that element is now 0, the desired transformation cannot be completed. Therefore, the inverse of the given matrix does not exist.

15. $[A \,|\, I_3] =$

$$\begin{bmatrix} 1 & 0 & 1 & | & 1 & 0 & 0 \\ 0 & -1 & 0 & | & 0 & 1 & 0 \\ 2 & 1 & 1 & | & 0 & 0 & 1 \end{bmatrix}$$

$$\begin{bmatrix} 1 & 0 & 1 & | & 1 & 0 & 0 \\ 0 & -1 & 0 & | & 0 & 1 & 0 \\ 0 & 1 & -1 & | & -2 & 0 & 1 \end{bmatrix} \quad -2R_1 + R_3$$

$$\begin{bmatrix} 1 & 0 & 1 & | & 1 & 0 & 0 \\ 0 & 1 & 0 & | & 0 & -1 & 0 \\ 0 & 1 & -1 & | & -2 & 0 & 1 \end{bmatrix} \quad -1R_2$$

$$\begin{bmatrix} 1 & 0 & 1 & | & 1 & 0 & 0 \\ 0 & 1 & 0 & | & 0 & -1 & 0 \\ 0 & 0 & -1 & | & -2 & 1 & 1 \end{bmatrix} \quad -1R_2 + R_3$$

$$\begin{bmatrix} 1 & 0 & 1 & | & 1 & 0 & 0 \\ 0 & 1 & 0 & | & 0 & -1 & 0 \\ 0 & 0 & 1 & | & 2 & -1 & -1 \end{bmatrix} \quad -1R_3$$

$$\begin{bmatrix} 1 & 0 & 0 & | & -1 & 1 & 1 \\ 0 & 1 & 0 & | & 0 & -1 & 0 \\ 0 & 0 & 1 & | & 2 & -1 & -1 \end{bmatrix} \quad -1R_3 + R_1$$

$$A^{-1} = \begin{bmatrix} -1 & 1 & 1 \\ 0 & -1 & 0 \\ 2 & -1 & -1 \end{bmatrix}$$

17. $[A \,|\, I_3] = \begin{bmatrix} 3 & 3 & -1 & | & 1 & 0 & 0 \\ 2 & 6 & 0 & | & 0 & 1 & 0 \\ -6 & -6 & 2 & | & 0 & 0 & 1 \end{bmatrix}$

Notice the relationship between R_1 and R_3: each element in R_3 is -2 times the element in R_1.

$$\begin{bmatrix} 3 & 3 & -1 & | & 1 & 0 & 0 \\ 2 & 6 & 0 & | & 0 & 1 & 0 \\ 0 & 0 & 0 & | & 2 & 0 & 1 \end{bmatrix} \quad 2R_1 + R_3$$

Just as in Now Try Exercise 4, we have a row of 0's. None of these 0's can be changed to a 1, so the desired transformation cannot be completed. Therefore, the inverse of the given matrix *does not exist*.

19. $[A \,|\, I_3] =$

$$\begin{bmatrix} 2 & 2 & -4 & | & 1 & 0 & 0 \\ 2 & 6 & 0 & | & 0 & 1 & 0 \\ -3 & -3 & 5 & | & 0 & 0 & 1 \end{bmatrix}$$

$$\begin{bmatrix} 1 & 1 & -2 & | & \frac{1}{2} & 0 & 0 \\ 2 & 6 & 0 & | & 0 & 1 & 0 \\ -3 & -3 & 5 & | & 0 & 0 & 1 \end{bmatrix} \quad \frac{1}{2}R_1$$

$$\begin{bmatrix} 1 & 1 & -2 & | & \frac{1}{2} & 0 & 0 \\ 0 & 4 & 4 & | & -1 & 1 & 0 \\ 0 & 0 & -1 & | & \frac{3}{2} & 0 & 1 \end{bmatrix} \quad \begin{matrix} -2R_1 + R_2 \\ 3R_1 + R_3 \end{matrix}$$

$$\begin{bmatrix} 1 & 1 & -2 & | & \frac{1}{2} & 0 & 0 \\ 0 & 1 & 1 & | & -\frac{1}{4} & \frac{1}{4} & 0 \\ 0 & 0 & -1 & | & \frac{3}{2} & 0 & 1 \end{bmatrix} \quad \frac{1}{4}R_2$$

$$\begin{bmatrix} 1 & 0 & -3 & | & \frac{3}{4} & -\frac{1}{4} & 0 \\ 0 & 1 & 1 & | & -\frac{1}{4} & \frac{1}{4} & 0 \\ 0 & 0 & -1 & | & \frac{3}{2} & 0 & 1 \end{bmatrix} \quad -1R_2 + R_1$$

$$\begin{bmatrix} 1 & 0 & -3 & | & \frac{3}{4} & -\frac{1}{4} & 0 \\ 0 & 1 & 1 & | & -\frac{1}{4} & \frac{1}{4} & 0 \\ 0 & 0 & 1 & | & -\frac{3}{2} & 0 & -1 \end{bmatrix} \quad -1R_3$$

$$\begin{bmatrix} 1 & 0 & 0 & | & -\frac{15}{4} & -\frac{1}{4} & -3 \\ 0 & 1 & 0 & | & \frac{5}{4} & \frac{1}{4} & 1 \\ 0 & 0 & 1 & | & -\frac{3}{2} & 0 & -1 \end{bmatrix} \quad \begin{matrix} 3R_3 + R_1 \\ -1R_3 + R_2 \end{matrix}$$

$$A^{-1} = \begin{bmatrix} -\frac{15}{4} & -\frac{1}{4} & -3 \\ \frac{5}{4} & \frac{1}{4} & 1 \\ -\frac{3}{2} & 0 & -1 \end{bmatrix}$$

Copyright © 2012 Pearson Education, Inc. Publishing as Addison-Wesley.

21. $[A \mid I_4] =$

$$\begin{bmatrix} 1 & 1 & 0 & 2 & | & 1 & 0 & 0 & 0 \\ 2 & -1 & 1 & -1 & | & 0 & 1 & 0 & 0 \\ 3 & 3 & 2 & -2 & | & 0 & 0 & 1 & 0 \\ 1 & 2 & 1 & 0 & | & 0 & 0 & 0 & 1 \end{bmatrix}$$

$$\begin{bmatrix} 1 & 1 & 0 & 2 & | & 1 & 0 & 0 & 0 \\ 0 & -3 & 1 & -5 & | & -2 & 1 & 0 & 0 \\ 0 & 0 & 2 & -8 & | & -3 & 0 & 1 & 0 \\ 0 & 1 & 1 & -2 & | & -1 & 0 & 0 & 1 \end{bmatrix} \quad \begin{matrix} -2R_1 + R_2 \\ -3R_1 + R_3 \\ -1R_1 + R_4 \end{matrix}$$

$$\begin{bmatrix} 1 & 1 & 0 & 2 & | & 1 & 0 & 0 & 0 \\ 0 & 1 & -\frac{1}{3} & \frac{5}{3} & | & \frac{2}{3} & -\frac{1}{3} & 0 & 0 \\ 0 & 0 & 2 & -8 & | & -3 & 0 & 1 & 0 \\ 0 & 1 & 1 & -2 & | & -1 & 0 & 0 & 1 \end{bmatrix} \quad -\frac{1}{3}R_2$$

$$\begin{bmatrix} 1 & 0 & \frac{1}{3} & \frac{1}{3} & | & \frac{1}{3} & \frac{1}{3} & 0 & 0 \\ 0 & 1 & -\frac{1}{3} & \frac{5}{3} & | & \frac{2}{3} & -\frac{1}{3} & 0 & 0 \\ 0 & 0 & 2 & -8 & | & -3 & 0 & 1 & 0 \\ 0 & 0 & \frac{4}{3} & -\frac{11}{3} & | & -\frac{5}{3} & \frac{1}{3} & 0 & 1 \end{bmatrix} \quad \begin{matrix} -1R_2 + R_1 \\ \\ \\ -1R_2 + R_4 \end{matrix}$$

$$\begin{bmatrix} 1 & 0 & \frac{1}{3} & \frac{1}{3} & | & \frac{1}{3} & \frac{1}{3} & 0 & 0 \\ 0 & 1 & -\frac{1}{3} & \frac{5}{3} & | & \frac{2}{3} & -\frac{1}{3} & 0 & 0 \\ 0 & 0 & 1 & -4 & | & -\frac{3}{2} & 0 & \frac{1}{2} & 0 \\ 0 & 0 & \frac{4}{3} & -\frac{11}{3} & | & -\frac{5}{3} & \frac{1}{3} & 0 & 1 \end{bmatrix} \quad \frac{1}{2}R_3$$

$$\begin{bmatrix} 1 & 0 & 0 & \frac{5}{3} & | & \frac{5}{6} & \frac{1}{3} & -\frac{1}{6} & 0 \\ 0 & 1 & 0 & \frac{1}{3} & | & \frac{1}{6} & -\frac{1}{3} & \frac{1}{6} & 0 \\ 0 & 0 & 1 & -4 & | & -\frac{3}{2} & 0 & \frac{1}{2} & 0 \\ 0 & 0 & 0 & \frac{5}{3} & | & \frac{1}{3} & \frac{1}{3} & -\frac{2}{3} & 1 \end{bmatrix} \quad \begin{matrix} -\frac{1}{3}R_3 + R_1 \\ \frac{1}{3}R_3 + R_2 \\ \\ -\frac{4}{3}R_3 + R_4 \end{matrix}$$

$$\begin{bmatrix} 1 & 0 & 0 & \frac{5}{3} & | & \frac{5}{6} & \frac{1}{3} & -\frac{1}{6} & 0 \\ 0 & 1 & 0 & \frac{1}{3} & | & \frac{1}{6} & -\frac{1}{3} & \frac{1}{6} & 0 \\ 0 & 0 & 1 & -4 & | & -\frac{3}{2} & 0 & \frac{1}{2} & 0 \\ 0 & 0 & 0 & 1 & | & \frac{1}{5} & \frac{1}{5} & -\frac{2}{5} & \frac{3}{5} \end{bmatrix} \quad \frac{3}{5}R_4$$

$$\begin{bmatrix} 1 & 0 & 0 & 0 & | & \frac{1}{2} & 0 & \frac{1}{2} & -1 \\ 0 & 1 & 0 & 0 & | & \frac{1}{10} & -\frac{2}{5} & \frac{3}{10} & -\frac{1}{5} \\ 0 & 0 & 1 & 0 & | & -\frac{7}{10} & \frac{4}{5} & -\frac{11}{10} & \frac{12}{5} \\ 0 & 0 & 0 & 1 & | & \frac{1}{5} & \frac{1}{5} & -\frac{2}{5} & \frac{3}{5} \end{bmatrix} \quad \begin{matrix} -\frac{5}{3}R_4 + R_1 \\ -\frac{1}{3}R_4 + R_2 \\ 4R_4 + R_3 \end{matrix}$$

$$A^{-1} = \begin{bmatrix} \frac{1}{2} & 0 & \frac{1}{2} & -1 \\ \frac{1}{10} & -\frac{2}{5} & \frac{3}{10} & -\frac{1}{5} \\ -\frac{7}{10} & \frac{4}{5} & -\frac{11}{10} & \frac{12}{5} \\ \frac{1}{5} & \frac{1}{5} & -\frac{2}{5} & \frac{3}{5} \end{bmatrix}$$

23. $\quad -x + y = 1$
$\qquad\;\; 2x - y = 1$

Let $A = \begin{bmatrix} -1 & 1 \\ 2 & -1 \end{bmatrix}$, $X = \begin{bmatrix} x \\ y \end{bmatrix}$, and $B = \begin{bmatrix} 1 \\ 1 \end{bmatrix}$.

Find A^{-1}.

$$[A \mid I_2] = \begin{bmatrix} -1 & 1 & | & 1 & 0 \\ 2 & -1 & | & 0 & 1 \end{bmatrix}$$

$$\begin{bmatrix} 1 & -1 & | & -1 & 0 \\ 2 & -1 & | & 0 & 1 \end{bmatrix} \quad -1R_1$$

$$\begin{bmatrix} 1 & -1 & | & -1 & 0 \\ 0 & 1 & | & 2 & 1 \end{bmatrix} \quad -2R_1 + R_2$$

$$\begin{bmatrix} 1 & 0 & | & 1 & 1 \\ 0 & 1 & | & 2 & 1 \end{bmatrix} \quad R_2 + R_1$$

Thus, $A^{-1} = \begin{bmatrix} 1 & 1 \\ 2 & 1 \end{bmatrix}$.

$$X = A^{-1}B = \begin{bmatrix} 1 & 1 \\ 2 & 1 \end{bmatrix} \begin{bmatrix} 1 \\ 1 \end{bmatrix} = \begin{bmatrix} 2 \\ 3 \end{bmatrix}$$

The solution set is $\{(2,3)\}$.

25. $\quad 2x - y = -8$
$\qquad\;\; 3x + y = -2$

Let $A = \begin{bmatrix} 2 & -1 \\ 3 & 1 \end{bmatrix}$, $X = \begin{bmatrix} x \\ y \end{bmatrix}$, and $B = \begin{bmatrix} -8 \\ -2 \end{bmatrix}$.

Find A^{-1}.

$$[A \mid I_2] = \begin{bmatrix} 2 & -1 & | & 1 & 0 \\ 3 & 1 & | & 0 & 1 \end{bmatrix}$$

$$\begin{bmatrix} 1 & -\frac{1}{2} & | & \frac{1}{2} & 0 \\ 3 & 1 & | & 0 & 1 \end{bmatrix} \quad \frac{1}{2}R_1$$

$$\begin{bmatrix} 1 & -\frac{1}{2} & | & \frac{1}{2} & 0 \\ 0 & \frac{5}{2} & | & -\frac{3}{2} & 1 \end{bmatrix} \quad -3R_1 + R_2$$

$$\begin{bmatrix} 1 & -\frac{1}{2} & | & \frac{1}{2} & 0 \\ 0 & 1 & | & -\frac{3}{5} & \frac{2}{5} \end{bmatrix} \quad \frac{2}{5}R_2$$

$$\begin{bmatrix} 1 & 0 & | & \frac{1}{5} & \frac{1}{5} \\ 0 & 1 & | & -\frac{3}{5} & \frac{2}{5} \end{bmatrix} \quad \frac{1}{2}R_2 + R_1$$

Thus, $A^{-1} = \begin{bmatrix} \frac{1}{5} & \frac{1}{5} \\ -\frac{3}{5} & \frac{2}{5} \end{bmatrix}$.

$$X = A^{-1}B = \begin{bmatrix} \frac{1}{5} & \frac{1}{5} \\ -\frac{3}{5} & \frac{2}{5} \end{bmatrix} \begin{bmatrix} -8 \\ -2 \end{bmatrix} = \begin{bmatrix} -2 \\ 4 \end{bmatrix}$$

The solution set is $\{(-2,4)\}$.

Copyright © 2012 Pearson Education, Inc. Publishing as Addison-Wesley.

27. $2x + 3y = -10$
$3x + 4y = -12$

Let $A = \begin{bmatrix} 2 & 3 \\ 3 & 4 \end{bmatrix}$, $X = \begin{bmatrix} x \\ y \end{bmatrix}$, and $B = \begin{bmatrix} -10 \\ -12 \end{bmatrix}$.

Find A^{-1}.

$$[A \mid I_2] = \begin{bmatrix} 2 & 3 & | & 1 & 0 \\ 3 & 4 & | & 0 & 1 \end{bmatrix}$$

$$\begin{bmatrix} 1 & \frac{3}{2} & | & \frac{1}{2} & 0 \\ 3 & 4 & | & 0 & 1 \end{bmatrix} \qquad \frac{1}{2}R_1$$

$$\begin{bmatrix} 1 & \frac{3}{2} & | & \frac{1}{2} & 0 \\ 0 & -\frac{1}{2} & | & -\frac{3}{2} & 1 \end{bmatrix} \qquad -3R_1 + R_2$$

$$\begin{bmatrix} 1 & \frac{3}{2} & | & \frac{1}{2} & 0 \\ 0 & 1 & | & 3 & -2 \end{bmatrix} \qquad -2R_2$$

$$\begin{bmatrix} 1 & 0 & | & -4 & 3 \\ 0 & 1 & | & 3 & -2 \end{bmatrix} \qquad -\frac{3}{2}R_2 + R_1$$

Thus, $A^{-1} = \begin{bmatrix} -4 & 3 \\ 3 & -2 \end{bmatrix}$.

$$X = A^{-1}B = \begin{bmatrix} -4 & 3 \\ 3 & -2 \end{bmatrix} \begin{bmatrix} -10 \\ -12 \end{bmatrix} = \begin{bmatrix} 4 \\ -6 \end{bmatrix}$$

The solution set is $\{(4, -6)\}$.

29. $x + 3y + 3z = 1$
$x + 4y + 3z = 0$
$x + 3y + 4z = -1$

$$A = \begin{bmatrix} 1 & 3 & 3 \\ 1 & 4 & 3 \\ 1 & 3 & 4 \end{bmatrix}, X = \begin{bmatrix} x \\ y \\ z \end{bmatrix}, B = \begin{bmatrix} 1 \\ 0 \\ -1 \end{bmatrix}$$

Find A^{-1}.

$$[A \mid I_3] =$$

$$\begin{bmatrix} 1 & 3 & 3 & | & 1 & 0 & 0 \\ 1 & 4 & 3 & | & 0 & 1 & 0 \\ 1 & 3 & 4 & | & 0 & 0 & 1 \end{bmatrix}$$

$$\begin{bmatrix} 1 & 3 & 3 & | & 1 & 0 & 0 \\ 0 & 1 & 0 & | & -1 & 1 & 0 \\ 0 & 0 & 1 & | & -1 & 0 & 1 \end{bmatrix} \qquad \begin{array}{l} -1R_1 + R_2 \\ -1R_1 + R_3 \end{array}$$

$$\begin{bmatrix} 1 & 0 & 3 & | & 4 & -3 & 0 \\ 0 & 1 & 0 & | & -1 & 1 & 0 \\ 0 & 0 & 1 & | & -1 & 0 & 1 \end{bmatrix} \qquad -3R_2 + R_1$$

$$\begin{bmatrix} 1 & 0 & 0 & | & 7 & -3 & -3 \\ 0 & 1 & 0 & | & -1 & 1 & 0 \\ 0 & 0 & 1 & | & -1 & 0 & 1 \end{bmatrix} \qquad -3R_3 + R_1$$

Thus, $A^{-1} = \begin{bmatrix} 7 & -3 & -3 \\ -1 & 1 & 0 \\ -1 & 0 & 1 \end{bmatrix}$.

$$X = A^{-1}B = \begin{bmatrix} 7 & -3 & -3 \\ -1 & 1 & 0 \\ -1 & 0 & 1 \end{bmatrix} \begin{bmatrix} 1 \\ 0 \\ -1 \end{bmatrix} = \begin{bmatrix} 10 \\ -1 \\ -2 \end{bmatrix}$$

The solution set is $\{(10, -1, -2)\}$.

31. $2x + 2y - 4z = 12$
$2x + 6y = 16$
$-3x - 3y + 5z = -20$

$$A = \begin{bmatrix} 2 & 2 & -4 \\ 2 & 6 & 0 \\ -3 & -3 & 5 \end{bmatrix}, X = \begin{bmatrix} x \\ y \\ z \end{bmatrix}, B = \begin{bmatrix} 12 \\ 16 \\ -20 \end{bmatrix}$$

Using Exercise 19,

$$X = A^{-1}B$$

$$= \begin{bmatrix} -\frac{15}{4} & -\frac{1}{4} & -3 \\ \frac{5}{4} & \frac{1}{4} & 1 \\ -\frac{3}{2} & 0 & -1 \end{bmatrix} \begin{bmatrix} 12 \\ 16 \\ -20 \end{bmatrix} = \begin{bmatrix} 11 \\ -1 \\ 2 \end{bmatrix}.$$

The solution set is $\{(11, -1, 2)\}$.

Copyright © 2012 Pearson Education, Inc. Publishing as Addison-Wesley.

33.
$$\begin{aligned} x + y \qquad + 2w &= 3 \\ 2x - y + z - w &= 3 \\ 3x + 3y + 2z - 2w &= 5 \\ x + 2y + z \qquad &= 3 \end{aligned}$$

$$A = \begin{bmatrix} 1 & 1 & 0 & 2 \\ 2 & -1 & 1 & -1 \\ 3 & 3 & 2 & -2 \\ 1 & 2 & 1 & 0 \end{bmatrix}, \; X = \begin{bmatrix} x \\ y \\ z \\ w \end{bmatrix}, \; B = \begin{bmatrix} 3 \\ 3 \\ 5 \\ 3 \end{bmatrix}$$

Using Exercise 21,

$$X = A^{-1}B$$

$$= \begin{bmatrix} \frac{1}{2} & 0 & \frac{1}{2} & -1 \\ \frac{1}{10} & -\frac{2}{5} & \frac{3}{10} & -\frac{1}{5} \\ -\frac{7}{10} & \frac{4}{5} & -\frac{11}{10} & \frac{12}{5} \\ \frac{1}{5} & \frac{1}{5} & -\frac{2}{5} & \frac{3}{5} \end{bmatrix} \begin{bmatrix} 3 \\ 3 \\ 5 \\ 3 \end{bmatrix} = \begin{bmatrix} 1 \\ 0 \\ 2 \\ 1 \end{bmatrix}.$$

The solution set is $\{(1, 0, 2, 1)\}$.

35. (a) $S = a + bA + cB$

$$\begin{aligned} 602.7 &= a + 5.543b + 37.14c \\ 656.7 &= a + 6.933b + 41.30c \\ 778.5 &= a + 7.638b + 45.62c \end{aligned}$$

(b) $C = \begin{bmatrix} 1 & 5.543 & 37.14 \\ 1 & 6.933 & 41.30 \\ 1 & 7.638 & 45.62 \end{bmatrix}, \; X = \begin{bmatrix} a \\ b \\ c \end{bmatrix}$

$$D = \begin{bmatrix} 602.7 \\ 656.7 \\ 778.5 \end{bmatrix}, \text{ and } X = C^{-1}D.$$

Using a graphing calculator with matrix capabilities, we obtain

$$\begin{bmatrix} a \\ b \\ c \end{bmatrix} = \begin{bmatrix} -490.547375 \\ -89 \\ 42.71875 \end{bmatrix}.$$

Thus, $a \approx -490.547$, $b = -89$, and $c = 42.71875$.

(c) $S = -490.547 - 89A + 42.71875B$

(d) If $A = 7.752$ and $B = 47.38$, the predicted value of S is given by

$$\begin{aligned} S &= -490.547 - 89(7.752) \\ &\quad + 42.71875(47.38) \\ &= 843.539375 \approx 843.5. \end{aligned}$$

The predicted value is approximately 843.5.

(e) If $A = 8.9$ and $B = 66.25$, the predicted value of S is given by

$$\begin{aligned} S &= -490.547 - 89(8.9) \\ &\quad + 42.71875(66.25) \\ &= 1547.4701875 \approx 1547.5. \end{aligned}$$

The predicted value is approximately 1547.5. Using only three consecutive years to forecast six years into the future is probably not wise.

37. Two ways are using row operations on the augmented matrix of the system, and using the inverse of the coefficient matrix. Inverse matrices cannot be used for inconsistent systems or systems with dependent equations. Using row operations is usually an efficient method. The inverse matrix method is useful if there are several systems to be solved with the same coefficient matrix, but different constants.

Copyright © 2012 Pearson Education, Inc. Publishing as Addison-Wesley.

APPENDIX C DETERMINANTS AND CRAMER'S RULE

Appendix C   Now Try Exercises

N1. The value of a 2×2 determinant is given by

$$\begin{vmatrix} a & b \\ c & d \end{vmatrix} = ad - bc.$$

$$\begin{vmatrix} 6 & 3 \\ 4 & -5 \end{vmatrix} = 6(-5) - 3(4)$$
$$= -30 - 12 = -42$$

N2. $\begin{vmatrix} 0 & -2 & 3 \\ 4 & 1 & -5 \\ 6 & -1 & 5 \end{vmatrix}$

Expand by minors about column 1. Note that the sign of the number in the second row, first column is changed.

$$= 0\begin{vmatrix} 1 & -5 \\ -1 & 5 \end{vmatrix} - 4\begin{vmatrix} -2 & 3 \\ -1 & 5 \end{vmatrix} + 6\begin{vmatrix} -2 & 3 \\ 1 & -5 \end{vmatrix}$$
$$= 0[1(5) - (-5)(-1)] - 4[-2(5) - 3(-1)]$$
$$+ 6[-2(-5) - 3(1)]$$
$$= 0 - 4(-7) + 6(7)$$
$$= 28 + 42 = 70$$

N3. $\begin{vmatrix} 0 & -2 & 3 \\ 4 & 1 & -5 \\ 6 & -1 & 5 \end{vmatrix}$

$$= -(-2)\begin{vmatrix} 4 & -5 \\ 6 & 5 \end{vmatrix} + 1\begin{vmatrix} 0 & 3 \\ 6 & 5 \end{vmatrix} - (-1)\begin{vmatrix} 0 & 3 \\ 4 & -5 \end{vmatrix}$$
$$= 2[4(5) - (-5)(6)] + 1[0(5) - 3(6)]$$
$$+ 1[0(-5) - 3(4)]$$
$$= 2(50) + 1(-18) + 1(-12)$$
$$= 70$$

N4. $3x - 2y = -33$
$2x + 3y = -9$

$$D = \begin{vmatrix} 3 & -2 \\ 2 & 3 \end{vmatrix} = 3(3) - (-2)(2) = 13$$

$$D_x = \begin{vmatrix} -33 & -2 \\ -9 & 3 \end{vmatrix} = -33(3) - (-2)(-9) = -117$$

$$D_y = \begin{vmatrix} 3 & -33 \\ 2 & -9 \end{vmatrix} = 3(-9) - (-33)(2) = 39$$

$$x = \frac{D_x}{D} = \frac{-117}{13} = -9; \quad y = \frac{D_y}{D} = \frac{39}{13} = 3$$

The solution set is $\{(-9, 3)\}$.

N5. $4x + 2y + z = 15$
$-2x + 5y - 2z = -6$
$x - 3y + 4z = 0$

$$D = \begin{vmatrix} 4 & 2 & 1 \\ -2 & 5 & -2 \\ 1 & -3 & 4 \end{vmatrix}$$

Expand by minors about row 3.

$$= 1\begin{vmatrix} 2 & 1 \\ 5 & -2 \end{vmatrix} - (-3)\begin{vmatrix} 4 & 1 \\ -2 & -2 \end{vmatrix} + 4\begin{vmatrix} 4 & 2 \\ -2 & 5 \end{vmatrix}$$
$$= -9 - 18 + 96 = 69$$

$$D_x = \begin{vmatrix} 15 & 2 & 1 \\ -6 & 5 & -2 \\ 0 & -3 & 4 \end{vmatrix}$$

Expand by minors about row 3.

$$= 0\begin{vmatrix} 2 & 1 \\ 5 & -2 \end{vmatrix} - (-3)\begin{vmatrix} 15 & 1 \\ -6 & -2 \end{vmatrix} + 4\begin{vmatrix} 15 & 2 \\ -6 & 5 \end{vmatrix}$$
$$= 0 - 72 + 348 = 276$$

$$D_y = \begin{vmatrix} 4 & 15 & 1 \\ -2 & -6 & -2 \\ 1 & 0 & 4 \end{vmatrix}$$

Expand by minors about row 3.

$$= 1\begin{vmatrix} 15 & 1 \\ -6 & -2 \end{vmatrix} - 0\begin{vmatrix} 4 & 1 \\ -2 & -2 \end{vmatrix} + 4\begin{vmatrix} 4 & 15 \\ -2 & -6 \end{vmatrix}$$
$$= -24 - 0 + 24 = 0$$

$$D_z = \begin{vmatrix} 4 & 2 & 15 \\ -2 & 5 & -6 \\ 1 & -3 & 0 \end{vmatrix}$$

Expand by minors about row 3.

$$= 1\begin{vmatrix} 2 & 15 \\ 5 & -6 \end{vmatrix} - (-3)\begin{vmatrix} 4 & 15 \\ -2 & -6 \end{vmatrix} + 0\begin{vmatrix} 4 & 2 \\ -2 & 5 \end{vmatrix}$$
$$= -87 + 18 + 0 = -69$$

$$x = \frac{D_x}{D} = \frac{276}{69} = 4; \quad y = \frac{D_y}{D} = \frac{0}{69} = 0$$
$$z = \frac{D_z}{D} = \frac{-69}{69} = -1$$

The solution set is $\{(4, 0, -1)\}$.

N6. $5x + 3y + z = 1$
$x - 2y + 3z = 6$
$10x + 6y + 2z = 3$

$$D = \begin{vmatrix} 5 & 3 & 1 \\ 1 & -2 & 3 \\ 10 & 6 & 2 \end{vmatrix}$$

Expand by minors about row 3.

$$= 10\begin{vmatrix} 3 & 1 \\ -2 & 3 \end{vmatrix} - 6\begin{vmatrix} 5 & 1 \\ 1 & 3 \end{vmatrix} + 2\begin{vmatrix} 5 & 3 \\ 1 & -2 \end{vmatrix}$$
$$= 10(11) - 6(14) + 2(-13)$$
$$= 110 - 84 - 26 = 0$$

Because $D = 0$, Cramer's rule does not apply.

Copyright © 2012 Pearson Education, Inc. Publishing as Addison-Wesley.

Appendix C Section Exercises

1. (a) *True*, a matrix is an array of numbers, while a determinant is a single number.

    (b) *True*, a square matrix has the same number of rows as columns.

    (c) *False*, the determinant $\begin{vmatrix} a & b \\ c & d \end{vmatrix}$ is equal to $ad - bc$.

    (d) The value $\begin{vmatrix} 0 & 0 \\ x & y \end{vmatrix}$ is zero for any replacements of x and y.
    $$\begin{vmatrix} 0 & 0 \\ x & y \end{vmatrix} = 0(y) - 0(x) = 0$$

    No matter what replacements are used for x and y, the value of the determinant is zero since both x and y are being multiplied by zero. The statement is *true*.

3. $\begin{vmatrix} -2 & 5 \\ -1 & 4 \end{vmatrix} = -2(4) - 5(-1)$
    $$= -8 + 5 = -3$$

5. $\begin{vmatrix} 1 & -2 \\ 7 & 0 \end{vmatrix} = 1(0) - (-2)7$
    $$= 0 + 14 = 14$$

7. $\begin{vmatrix} 0 & 4 \\ 0 & 4 \end{vmatrix} = 0(4) - 4(0)$
    $$= 0 - 0 = 0$$

9. $\begin{vmatrix} -1 & 2 & 4 \\ -3 & -2 & -3 \\ 2 & -1 & 5 \end{vmatrix}$ Expand by minors about the first column.

    $= -1\begin{vmatrix} -2 & -3 \\ -1 & 5 \end{vmatrix} - (-3)\begin{vmatrix} 2 & 4 \\ -1 & 5 \end{vmatrix}$
    $\quad + 2\begin{vmatrix} 2 & 4 \\ -2 & -3 \end{vmatrix}$
    $= -1[-2(5) - (-3)(-1)]$
    $\quad + 3[2(5) - 4(-1)] + 2[2(-3) - 4(-2)]$
    $= -1(-13) + 3(14) + 2(2)$
    $= 13 + 42 + 4 = 59$

11. $\begin{vmatrix} 1 & 0 & -2 \\ 0 & 2 & 3 \\ 1 & 0 & 5 \end{vmatrix}$ There are two 0's in column 2. We'll expand about that column since there is only 1 minor to evaluate.

    $= -0\begin{vmatrix} 0 & 3 \\ 1 & 5 \end{vmatrix} + 2\begin{vmatrix} 1 & -2 \\ 1 & 5 \end{vmatrix} - 0\begin{vmatrix} 1 & -2 \\ 0 & 3 \end{vmatrix}$
    $= 0 + 2[1(5) - (-2)(1)] - 0$
    $= 2[5 + 2] = 2(7) = 14$

13. $\begin{vmatrix} 3 & -1 & 2 \\ 1 & 5 & -2 \\ 0 & 2 & 0 \end{vmatrix}$ Expand about row 3.

    $= 0 - 2\begin{vmatrix} 3 & 2 \\ 1 & -2 \end{vmatrix} + 0$
    $= -2[3(-2) - 2(1)]$
    $= -2[-6 - 2] = -2[-8] = 16$

15. $\begin{vmatrix} 0 & 0 & 3 \\ 4 & 0 & -2 \\ 2 & -1 & 3 \end{vmatrix}$ Expand about row 1.

    $= 0 - 0 + 3\begin{vmatrix} 4 & 0 \\ 2 & -1 \end{vmatrix}$
    $= 3[4(-1) - 0(2)]$
    $= 3(-4) = -12$

17. $\begin{vmatrix} 1 & 1 & 2 \\ 5 & 5 & 7 \\ 3 & 3 & 1 \end{vmatrix}$ Expand about row 1.

    $= 1\begin{vmatrix} 5 & 7 \\ 3 & 1 \end{vmatrix} - 1\begin{vmatrix} 5 & 7 \\ 3 & 1 \end{vmatrix} + 2\begin{vmatrix} 5 & 5 \\ 3 & 3 \end{vmatrix}$
    $= 1[5(1) - 7(3)] - 1[5(1) - 7(3)]$
    $\quad + 2[5(3) - 5(3)]$
    $= 1(-16) - 1(-16) + 2(0)$
    $= -16 + 16 + 0 = 0$

19. $x = \dfrac{D_x}{D} = \dfrac{-43}{-43} = 1$
    $y = \dfrac{D_y}{D} = \dfrac{0}{-43} = 0$
    $z = \dfrac{D_z}{D} = \dfrac{43}{-43} = -1$

    The solution set is $\{(1, 0, -1)\}$.

21. $5x + 2y = -3$
    $4x - 3y = -30$

    $D = \begin{vmatrix} 5 & 2 \\ 4 & -3 \end{vmatrix} = -15 - 8 = -23$

    $D_x = \begin{vmatrix} -3 & 2 \\ -30 & -3 \end{vmatrix} = 9 + 60 = 69$

    $D_y = \begin{vmatrix} 5 & -3 \\ 4 & -30 \end{vmatrix} = -150 + 12 = -138$

    $x = \dfrac{D_x}{D} = \dfrac{69}{-23} = -3; y = \dfrac{D_y}{D} = \dfrac{-138}{-23} = 6$

    The solution set is $\{(-3, 6)\}$.

Copyright © 2012 Pearson Education, Inc. Publishing as Addison-Wesley.

23. $3x - y = 9$
   $2x + 5y = 8$

$$D = \begin{vmatrix} 3 & -1 \\ 2 & 5 \end{vmatrix} = 15 + 2 = 17$$

$$D_x = \begin{vmatrix} 9 & -1 \\ 8 & 5 \end{vmatrix} = 45 + 8 = 53$$

$$D_y = \begin{vmatrix} 3 & 9 \\ 2 & 8 \end{vmatrix} = 24 - 18 = 6$$

$$x = \frac{D_x}{D} = \frac{53}{17}; y = \frac{D_y}{D} = \frac{6}{17}$$

The solution set is $\left\{ \left(\frac{53}{17}, \frac{6}{17} \right) \right\}$.

25. $4x + 5y = 6$
   $7x + 8y = 9$

$$D = \begin{vmatrix} 4 & 5 \\ 7 & 8 \end{vmatrix} = 32 - 35 = -3$$

$$D_x = \begin{vmatrix} 6 & 5 \\ 9 & 8 \end{vmatrix} = 48 - 45 = 3$$

$$D_y = \begin{vmatrix} 4 & 6 \\ 7 & 9 \end{vmatrix} = 36 - 42 = -6$$

$$x = \frac{D_x}{D} = \frac{3}{-3} = -1; y = \frac{D_y}{D} = \frac{-6}{-3} = 2$$

The solution set is $\{(-1, 2)\}$.

27. $x - y + 6z = 19$
   $3x + 3y - z = 1$
   $x + 9y + 2z = -19$

$$D = \begin{vmatrix} 1 & -1 & 6 \\ 3 & 3 & -1 \\ 1 & 9 & 2 \end{vmatrix} \text{ Expand about column 1.}$$

$$= 1 \begin{vmatrix} 3 & -1 \\ 9 & 2 \end{vmatrix} - 3 \begin{vmatrix} -1 & 6 \\ 9 & 2 \end{vmatrix} + 1 \begin{vmatrix} -1 & 6 \\ 3 & -1 \end{vmatrix}$$

$$= 1(6 + 9) - 3(-2 - 54) + 1(1 - 18)$$

$$= 15 + 168 - 17 = 166$$

$$D_x = \begin{vmatrix} 19 & -1 & 6 \\ 1 & 3 & -1 \\ -19 & 9 & 2 \end{vmatrix} \text{ Expand about column 3.}$$

$$= 6 \begin{vmatrix} 1 & 3 \\ -19 & 9 \end{vmatrix} - (-1) \begin{vmatrix} 19 & -1 \\ -19 & 9 \end{vmatrix}$$

$$+ 2 \begin{vmatrix} 19 & -1 \\ 1 & 3 \end{vmatrix}$$

$$= 6(9 + 57) + 1(171 - 19) + 2(57 + 1)$$

$$= 396 + 152 + 116 = 664$$

$$D_y = \begin{vmatrix} 1 & 19 & 6 \\ 3 & 1 & -1 \\ 1 & -19 & 2 \end{vmatrix} \text{ Expand about column 1.}$$

$$= 1 \begin{vmatrix} 1 & -1 \\ -19 & 2 \end{vmatrix} - 3 \begin{vmatrix} 19 & 6 \\ -19 & 2 \end{vmatrix} + 1 \begin{vmatrix} 19 & 6 \\ 1 & -1 \end{vmatrix}$$

$$= 1(2 - 19) - 3(38 + 114) + 1(-19 - 6)$$

$$= -17 - 456 - 25 = -498$$

$$D_z = \begin{vmatrix} 1 & -1 & 19 \\ 3 & 3 & 1 \\ 1 & 9 & -19 \end{vmatrix} \text{ Expand about column 1.}$$

$$= 1 \begin{vmatrix} 3 & 1 \\ 9 & -19 \end{vmatrix} - 3 \begin{vmatrix} -1 & 19 \\ 9 & -19 \end{vmatrix} + 1 \begin{vmatrix} -1 & 19 \\ 3 & 1 \end{vmatrix}$$

$$= 1(-57 - 9) - 3(19 - 171) + 1(-1 - 57)$$

$$= -66 + 456 - 58 = 332$$

$$x = \frac{D_x}{D} = \frac{664}{166} = 4; y = \frac{D_y}{D} = \frac{-498}{166} = -3$$

$$z = \frac{D_z}{D} = \frac{332}{166} = 2$$

The solution set is $\{(4, -3, 2)\}$.

29. $7x + y - z = 4$
   $2x - 3y + z = 2$
   $-6x + 9y - 3z = -6$

$$D = \begin{vmatrix} 7 & 1 & -1 \\ 2 & -3 & 1 \\ -6 & 9 & -3 \end{vmatrix} \text{ Expand about column 3.}$$

$$= -1 \begin{vmatrix} 2 & -3 \\ -6 & 9 \end{vmatrix} - 1 \begin{vmatrix} 7 & 1 \\ -6 & 9 \end{vmatrix} - 3 \begin{vmatrix} 7 & 1 \\ 2 & -3 \end{vmatrix}$$

$$= -1(18 - 18) - 1(63 + 6) - 3(-21 - 2)$$

$$= 0 - 69 + 69 = 0$$

Because $D = 0$, Cramer's rule does not apply.

Copyright © 2012 Pearson Education, Inc. Publishing as Addison-Wesley.

31. $-x + 2y \qquad = 4$
$3x + y \qquad = -5$
$2x \qquad + z = -1$

$D = \begin{vmatrix} -1 & 2 & 0 \\ 3 & 1 & 0 \\ 2 & 0 & 1 \end{vmatrix}$ Expand about column 3.

$= 0 - 0 + 1 \begin{vmatrix} -1 & 2 \\ 3 & 1 \end{vmatrix}$

$= 1(-1 - 6) = -7$

$D_x = \begin{vmatrix} 4 & 2 & 0 \\ -5 & 1 & 0 \\ -1 & 0 & 1 \end{vmatrix}$ Expand about column 3.

$= 0 - 0 + 1 \begin{vmatrix} 4 & 2 \\ -5 & 1 \end{vmatrix}$

$= 1(4 + 10) = 14$

$D_y = \begin{vmatrix} -1 & 4 & 0 \\ 3 & -5 & 0 \\ 2 & -1 & 1 \end{vmatrix}$ Expand about column 3.

$= 0 - 0 + 1 \begin{vmatrix} -1 & 4 \\ 3 & -5 \end{vmatrix}$

$= 1(5 - 12) = -7$

$D_z = \begin{vmatrix} -1 & 2 & 4 \\ 3 & 1 & -5 \\ 2 & 0 & -1 \end{vmatrix}$ Expand about row 3.

$= 2 \begin{vmatrix} 2 & 4 \\ 1 & -5 \end{vmatrix} - 0 - 1 \begin{vmatrix} -1 & 2 \\ 3 & 1 \end{vmatrix}$

$= 2(-10 - 4) - 1(-1 - 6)$

$= -28 + 7 = -21$

$x = \dfrac{D_x}{D} = \dfrac{14}{-7} = -2; \, y = \dfrac{D_y}{D} = \dfrac{-7}{-7} = 1$

$z = \dfrac{D_z}{D} = \dfrac{-21}{-7} = 3$

The solution set is $\{(-2, 1, 3)\}$.

33. $-5x - y \qquad = -10$
$3x + 2y + z = -3$
$ - y - 2z = -13$

$D = \begin{vmatrix} -5 & -1 & 0 \\ 3 & 2 & 1 \\ 0 & -1 & -2 \end{vmatrix}$ Expand about column 3.

$= 0 - 1 \begin{vmatrix} -5 & -1 \\ 0 & -1 \end{vmatrix} - 2 \begin{vmatrix} -5 & -1 \\ 3 & 2 \end{vmatrix}$

$= -1(5 - 0) - 2(-10 + 3)$

$= -5 + 14 = 9$

$D_x = \begin{vmatrix} -10 & -1 & 0 \\ -3 & 2 & 1 \\ -13 & -1 & -2 \end{vmatrix}$ Expand about column 3.

$= 0 - 1 \begin{vmatrix} -10 & -1 \\ -13 & -1 \end{vmatrix} - 2 \begin{vmatrix} -10 & -1 \\ -3 & 2 \end{vmatrix}$

$= -1(10 - 13) - 2(-20 - 3)$

$= 3 + 46 = 49$

$D_y = \begin{vmatrix} -5 & -10 & 0 \\ 3 & -3 & 1 \\ 0 & -13 & -2 \end{vmatrix}$ Expand about column 3.

$= 0 - 1 \begin{vmatrix} -5 & -10 \\ 0 & -13 \end{vmatrix} - 2 \begin{vmatrix} -5 & -10 \\ 3 & -3 \end{vmatrix}$

$= -1(65) - 2(15 + 30)$

$= -65 - 90 = -155$

$D_z = \begin{vmatrix} -5 & -1 & -10 \\ 3 & 2 & -3 \\ 0 & -1 & -13 \end{vmatrix}$ Expand about row 3.

$= 0 - (-1) \begin{vmatrix} -5 & -10 \\ 3 & -3 \end{vmatrix} - 13 \begin{vmatrix} -5 & -1 \\ 3 & 2 \end{vmatrix}$

$= 1(15 + 30) - 13(-10 + 3)$

$= 45 + 91 = 136$

$x = \dfrac{D_x}{D} = \dfrac{49}{9}; \, y = \dfrac{D_y}{D} = \dfrac{-155}{9} = -\dfrac{155}{9}$

$z = \dfrac{D_z}{D} = \dfrac{136}{9}$

The solution set is $\left\{ \left(\dfrac{49}{9}, -\dfrac{155}{9}, \dfrac{136}{9} \right) \right\}$.

35. $\begin{vmatrix} 4 & x \\ 2 & 3 \end{vmatrix} = 8$

Evaluate the determinant.

$\begin{vmatrix} 4 & x \\ 2 & 3 \end{vmatrix} = 4(3) - x(2)$

$= 12 - 2x$

Now solve the original equation.

$12 - 2x = 8$
$-2x = -4$
$x = 2$

The solution set is $\{2\}$.

37. $\begin{vmatrix} x & 4 \\ x & -3 \end{vmatrix} = 0$

Evaluate the determinant.

$\begin{vmatrix} x & 4 \\ x & -3 \end{vmatrix} = x(-3) - 4x$

$= -3x - 4x = -7x$

Now solve the original equation.

$-7x = 0$
$x = 0$

The solution set is $\{0\}$.